INTERNATIONAL SERIES OF MONOGRAPHS ON PHYSICS

SERIES EDITORS

J. BIRMAN   CITY UNIVERSITY OF NEW YORK
S. F. EDWARDS   UNIVERSITY OF CAMBRIDGE
R. FRIEND   UNIVERSITY OF CAMBRIDGE
M. REES   UNIVERSITY OF CAMBRIDGE
D. SHERRINGTON   UNIVERSITY OF OXFORD
G. VENEZIANO   CERN, GENEVA

INTERNATIONAL SERIES OF MONOGRAPHS ON PHYSICS

145. M. Bordag, G. L. Klimchitskaya, U. Mohideen, V. M. Mostepanenko: *Advances in the Casimir effect*
144. T. R. Field: *Electromagnetic scattering from random media*
143. W. Götze: *Complex dynamics of glass-forming liquids – a mode-coupling theory*
142. V.M. Agranovich: *Excitations in organic solids*
141. W.T. Grandy: *Entropy and the time evolution of macroscopic systems*
140. M. Alcubierre: *Introduction to 3 + 1 numerical relativity*
139. A. L. Ivanov, S. G. Tikhodeev: *Problems of condensed matter physics – quantum coherence phenomena in electron-hole and coupled matter-light systems*
138. I. M. Vardavas, F. W. Taylor: *Radiation and climate*
137. A. F. Borghesani: *Ions and electrons in liquid helium*
136. C. Kiefer: *Quantum gravity, Second edition*
135. V. Fortov, I. Iakubov, A. Khrapak: *Physics of strongly coupled plasma*
134. G. Fredrickson: *The equilibrium theory of inhomogeneous polymers*
133. H. Suhl: *Relaxation processes in micromagnetics*
132. J. Terning: *Modern supersymmetry*
131. M. Mariño: *Chern-Simons theory, matrix models, and topological strings*
130. V. Gantmakher: *Electrons and disorder in solids*
129. W. Barford: *Electronic and optical properties of conjugated polymers*
128. R. E. Raab, O. L. de Lange: *Multipole theory in electromagnetism*
127. A. Larkin, A. Varlamov: *Theory of fluctuations in superconductors*
126. P. Goldbart, N. Goldenfeld, D. Sherrington: *Stealing the gold*
125. S. Atzeni, J. Meyer-ter-Vehn: *The physics of inertial fusion*
123. T. Fujimoto: *Plasma spectroscopy*
122. K. Fujikawa, H. Suzuki: *Path integrals and quantum anomalies*
121. T. Giamarchi: *Quantum physics in one dimension*
120. M. Warner, E. Terentjev: *Liquid crystal elastomers*
119. L. Jacak, P. Sitko, K. Wieczorek, A. Wojs: *Quantum Hall systems*
118. J. Wesson: *Tokamaks, Third edition*
117. G. Volovik: *The Universe in a helium droplet*
116. L. Pitaevskii, S. Stringari: *Bose–Einstein condensation*
115. G. Dissertori, I.G. Knowles, M. Schmelling: *Quantum chromodynamics*
114. B. DeWitt: *The global approach to quantum field theory*
113. J. Zinn-Justin: *Quantum field theory and critical phenomena, Fourth edition*
112. R.M. Mazo: *Brownian motion – fluctuations, dynamics, and applications*
111. H. Nishimori: *Statistical physics of spin glasses and information processing – an introduction*
110. N.B. Kopnin: *Theory of nonequilibrium superconductivity*
109. A. Aharoni: *Introduction to the theory of ferromagnetism, Second edition*
108. R. Dobbs: *Helium three*
107. R. Wigmans: *Calorimetry*
106. J. Kübler: *Theory of itinerant electron magnetism*
105. Y. Kuramoto, Y. Kitaoka: *Dynamics of heavy electrons*
104. D. Bardin, G. Passarino: *The Standard Model in the making*
103. G.C. Branco, L. Lavoura, J.P. Silva: *CP Violation*
102. T.C. Choy: *Effective medium theory*
101. H. Araki: *Mathematical theory of quantum fields*
100. L. M. Pismen: *Vortices in nonlinear fields*
99. L. Mestel: *Stellar magnetism*
98. K. H. Bennemann: *Nonlinear optics in metals*
94. S. Chikazumi: *Physics of ferromagnetism*
91. R. A. Bertlmann: *Anomalies in quantum field theory*
90. P. K. Gosh: *Ion traps*
87. P. S. Joshi: *Global aspects in gravitation and cosmology*
86. E. R. Pike, S. Sarkar: *The quantum theory of radiation*
83. P. G. de Gennes, J. Prost: *The physics of liquid crystals*
73. M. Doi, S. F. Edwards: *The theory of polymer dynamics*
69. S. Chandrasekhar: *The mathematical theory of black holes*
51. C. Møller: *The theory of relativity*
46. H. E. Stanley: *Introduction to phase transitions and critical phenomena*
32. A. Abragam: *Principles of nuclear magnetism*
27. P. A. M. Dirac: *Principles of quantum mechanics*
23. R. E. Peierls: *Quantum theory of solids*

Advances in the Casimir Effect

M. Bordag
Center for Theoretical Studies and Institute for Theoretical Physics,
Universität Leipzig, Germany

G.L. Klimchitskaya
Department of Physics, North-West Technical University, St. Petersburg, Russia
and
Center for Theoretical Studies and Institute for Theoretical Physics,
Universität Leipzig, Germany

U. Mohideen
Department of Physics and Astronomy, University of California, Riverside, USA

V.M. Mostepanenko
Noncommercial Partnership 'Scientific Instruments', Moscow, Russia
and
Center for Theoretical Studies and Institute for Theoretical Physics,
Universität Leipzig, Germany

OXFORD
UNIVERSITY PRESS

Great Clarendon Street, Oxford OX2 6DP

Oxford University Press is a department of the University of Oxford.
It furthers the University's objective of excellence in research, scholarship,
and education by publishing worldwide in

Oxford New York

Auckland Cape Town Dar es Salaam Hong Kong Karachi
Kuala Lumpur Madrid Melbourne Mexico City Nairobi
New Delhi Shanghai Taipei Toronto

With offices in

Argentina Austria Brazil Chile Czech Republic France Greece
Guatemala Hungary Italy Japan Poland Portugal Singapore
South Korea Switzerland Thailand Turkey Ukraine Vietnam

Oxford is a registered trade mark of Oxford University Press
in the UK and in certain other countries

Published in the United States
by Oxford University Press Inc., New York

© M. Bordag, G.L. Klimchitskaya, U. Mohideen, and V.M. Mostepanenko, 2009

The moral rights of the authors have been asserted
Database right Oxford University Press (maker)

First Published 2009
Reprinted 2010

All rights reserved. No part of this publication may be reproduced,
stored in a retrieval system, or transmitted, in any form or by any means,
without the prior permission in writing of Oxford University Press,
or as expressly permitted by law, or under terms agreed with the appropriate
reprographics rights organization. Enquiries concerning reproduction
outside the scope of the above should be sent to the Rights Department,
Oxford University Press, at the address above

You must not circulate this book in any other binding or cover
and you must impose the same condition on any acquirer

British Library Cataloguing in Publication Data
Data available

Library of Congress Cataloging in Publication Data
Data available

Printed in Great Britain
on acid-free paper by
CPI Antony Rowe, Chippenham, Wiltshire

ISBN 978–0–19–923874–3 (Hbk)

3 5 7 9 10 8 6 4 2

PREFACE

The general subject of this book is the physical phenomenon named after the Dutch physicist Hendrik Brugt Gerhard Casimir, who predicted it in 1948. In the last fifteen years the Casimir effect has received widespread attention in both fundamental and applied physics. This is due to the Casimir effect being a direct manifestation of the most intriguing yet basic type of physical reality, i.e. the quantum vacuum. The Casimir effect in its simplest form is the attraction of a pair of neutral, parallel conducting plates resulting from the modification of the electromagnetic vacuum by the boundaries. It is a purely quantum effect. There is no force acting between neutral plates in classical electrodynamics. The roots of the Casimir effect date back to the introduction by Planck in 1911 of half-quanta in the context of black-body radiation. In quantum field theory, this results in an infinite energy of the vacuum state (the so-called zero-point energy). Casimir was the first to provide a method to subtract the infinite vacuum energy in free Minkowski space from the infinite vacuum energy in the presence of plates. Both infinite quantities were made finite through the use of a procedure called regularization. After subtraction, the regularization was removed, leaving a finite result, which leads to the Casimir force.

The Casimir force is closely connected with the well-known phenomenon of the van der Waals force. It provides an extension of the van der Waals interactions to larger separation distances between the interacting bodies, where relativistic effects come into play. The Casimir effect has become an interdisciplinary subject. It plays an important role in various fields of physics such as condensed matter physics, quantum field theory, atomic and molecular physics, gravitation and cosmology, and also in mathematical physics. Most recently, the Casimir effect has been applied to nanotechnology and for obtaining constraints on the predictions of unification theories beyond the Standard Model. In many physical phenomena the Casimir effect plays a primary role, while in many others it must be taken into account to provide a complete quantitative description. Some examples are the formation of hadron masses, the interaction of thin films, surface tension, bulk and surface critical phenomena, Bose–Einstein condensation, atom–surface interactions, the problem of the cosmological constant, the interaction of cosmic strings, compactification of extra dimensions, stiction in microdevices, and absorption phenomena in carbon nanotubes.

This book attempts to presents a comprehensive picture of the extensive studies in the field of the Casimir effect and its applications. Equal emphasis is placed on experiment and theory, and on fundamental and applied aspects. The book is not a monograph in the literal sense of the word, because it covers a wider range of diverse topics and even many subdisciplines of physics, for example quantum field theory, condensed matter physics, atomic and molecular physics, gravitation and cosmology, mathematical physics, and nanotechnology. This is due to the extraordinary role played in modern physics by the concepts of the quantum vacuum and zero-point oscillations. Another unique feature of this book has already been mentioned. It presents both experiment and theory, including their mutual influence. To comply with the requirements of experiments, special attention is paid to the Casimir force acting between real media, including effects of nonzero skin depth, surface roughness, nonzero temperature, etc.

Although the book is not intended as a textbook (because it requires some prior basic knowledge of both theory and experiment), it will serve as an introduction to the subject. On problems where there is no consensus, the authors provide a critical analysis of all the pros and cons. It is intended for all physicists, both experimentalists and theorists, who are working on the various manifestations of vacuum oscillations. This includes not only experts in van der Waals and Casimir forces but also those in elementary particle physics, condensed matter physics, atomic physics, and gravitation and cosmology. Applied physicists working, for instance, on single-electron transistors or carbon nanostructures, or on the design of new generations of microchips, nanotweezers, and nanoscale actuators will find in the book material that is both interesting and illuminating, and useful for their research.

Most of the text is written with sufficient detail, with explanations and links to publications, that it can be used by advanced students, both undergraduate and graduate, who are beginning work in the field of van der Waals or Casimir forces and related subjects. It is expected that students will have a prior acquaintance with basic courses in electrodynamics and quantum mechanics. Some initial knowledge of the elements of quantum field theory is also desirable. Postgraduate scientists working on all the above-mentioned subjects can use this book as a basic source of information and reference in their research.

Leipzig	M. B.
St. Petersburg and Leipzig	G. L. K.
Riverside	U. M.
St. Petersburg and Leipzig	V. M. M.
December, 2008	

ACKNOWLEDGEMENTS

The authors are grateful to the co-authors of their joint papers and to other colleagues, specifically J. F. Babb, G. Barton, V. B. Bezerra, G. Bimonte, E. V. Blagov, F. Chen, R. S. Decca, E. Elizalde, E. Fischbach, B. Geyer, N. R. Khusnutdinov, K. Kirsten, D. E. Krause, D. López, A. A. Maradudin, P. W. Milonni, K. A. Milton, D. Robaschik, C. Romero, A. A. Saharian, and D. V. Vassilevich, for helpful discussions of various aspects of the Casimir effect. Special gratitude is due to B. Geyer, who read the whole manuscript and suggested numerous useful corrections and improvements. However, the responsibility for any shortcomings rests with the authors.

We acknowledge the Americal Physical Society, Institute of Physics Publishing, Springer, Elsevier, and World Scientific for permission to reproduce some of the figures included in the book, which were originally published in Journal papers co-authored by one of us. All such figures contain a reference to the original source.

A large amount of the research reflected in the book has been financially supported by the National Science Foundation and the Department of Energy of the USA, by the European Community's Sixth Framework Programme within the STRP project PARNASS, and by the Deutsche Forschungsgemeinschaft, Germany. G. L. K. and V. M. M. gratefully acknowledge the hospitality of the Institute for Theoretical Physics, Leipzig University, and the Department of Physics and Astronomy, University of California, Riverside.

CONTENTS

1 Introduction   1
    1.1 Zero-point oscillations and their manifestations   1
    1.2 Connection between van der Waals and Casimir forces   5
    1.3 The Casimir effect as a multidisciplinary subject   7
    1.4 A guide to this book   8

I PHYSICAL AND MATHEMATICAL FOUNDATIONS OF THE CASIMIR EFFECT FOR IDEAL BOUNDARIES

2 Simple models of the Casimir effect   17
    2.1 The scalar Casimir effect on an interval   17
    2.2 The Abel–Plana formula and regularization   21
    2.3 The scalar Casimir effect on a circle   24
    2.4 Local and global descriptions of the Casimir effect   27
    2.5 Elementary approach to the Casimir force between two parallel planes   29

3 Field quantization and vacuum energy in the presence of boundaries   33
    3.1 Field equations for fields of various spins   33
    3.2 Various boundaries and boundary conditions   38
    3.3 Canonical quantization and the vacuum energy as a mode expansion   40
    3.4 Vacuum energy in terms of Green's functions   46
    3.5 Path-integral quantization   48
    3.6 Propagators with boundary conditions   51

4 Regularization and renormalization of the vacuum energy   55
    4.1 Regularization schemes   56
    4.2 The divergent part of the vacuum energy   57
       4.2.1 The divergent part in the cutoff regularization   57
       4.2.2 The divergent part in the zeta function regularization and the heat kernel expansion   59
    4.3 Renormalization of the vacuum energy   65
       4.3.1 Smooth background fields   66
       4.3.2 Singular background fields and boundary conditions   69
       4.3.3 Finiteness of the Casimir force between separate bodies   71

5 The Casimir effect at nonzero temperature   73
    5.1 The Matsubara formulation   73
    5.2 The Casimir effect at low and high temperature   79

6	**Approximate and numerical approaches**	84
	6.1 The multiple-reflection expansion	85
	6.2 Semiclassical approaches	88
	6.3 World line numerical methods	91
	6.4 Pairwise summation	93
	6.5 The proximity force approximation	97
7	**The Casimir effect for two ideal-metal planes**	103
	7.1 The scalar Casimir effect for parallel planes	103
	7.1.1 Dirichlet boundary conditions	103
	7.1.2 Mixed boundary conditions	106
	7.2 The electromagnetic Casimir effect between parallel planes	107
	7.2.1 Ideal-metal planes	107
	7.2.2 An ideal-metal plane and an infinitely permeable plane	110
	7.3 The radiative corrections to the Casimir force	112
	7.4 Two parallel planes at nonzero temperature	117
	7.4.1 General case	117
	7.4.2 The limit of low temperature	122
	7.4.3 The limit of high temperature	124
	7.5 The spinor Casimir effect between parallel planes	125
	7.6 The Casimir effect for a wedge	128
	7.7 The dynamical Casimir effect	131
	7.7.1 Uniformly moving plane	131
	7.7.2 Particle creation from an accelerated plane	131
8	**The Casimir effect in rectangular boxes**	136
	8.1 The scalar Casimir effect in a rectangle	136
	8.1.1 Regularization using the Abel–Plana formula	137
	8.1.2 Regularization using the Epstein zeta function	139
	8.1.3 A Casimir piston in a rectangle	142
	8.2 The scalar Casimir effect in a three-dimensional box	143
	8.3 The electromagnetic Casimir effect in a three-dimensional box	148
	8.4 Rectangular boxes with different boundary conditions	152
	8.5 Rectangular boxes at nonzero temperature	155
	8.5.1 The scalar Casimir effect	156
	8.5.2 The electromagnetic Casimir effect	160
9	**Single spherical and cylindrical boundaries**	166
	9.1 Separation of variables and mode summation	168
	9.1.1 Spherical symmetry	168
	9.1.2 Mode summation for the interior problem	170
	9.1.3 Mode summation for the exterior problem	172
	9.1.4 Cylindrical symmetry	175
	9.2 The scalar Casimir effect for a spherical shell	178
	9.2.1 Boundary conditions and mode-generating functions	178

	9.2.2 Analytic continuation for regularized vacuum energy and divergent contributions	180
	9.2.3 The renormalized vacuum energy for a massive scalar field	185
	9.2.4 The vacuum energy for a massless scalar field	188
9.3	The electromagnetic Casimir effect for a spherical shell and for a dielectric ball	193
	9.3.1 Boundary conditions and separation of polarizations	193
	9.3.2 The mode-generating functions	195
	9.3.3 The electromagnetic Casimir effect for a conducting spherical shell	196
	9.3.4 The Casimir effect for a dielectric ball	200
9.4	The spinor Casimir effect for a sphere	207
9.5	Spherical shell at nonzero temperature	212
	9.5.1 Low-temperature expansion	212
	9.5.2 High-temperature expansion	214
9.6	The Casimir effect for a cylinder	215
	9.6.1 Conducting cylindrical shell	216
	9.6.2 Dielectric cylinder	221

10 The Casimir force between objects of arbitrary shape — 227

10.1	Various approaches to the calculation of the Casimir energy	228
	10.1.1 Functional-determinant representation for the case of boundary conditions on separate bodies	228
	10.1.2 T-matrix approach for potentials with disjoint support	232
10.2	Casimir attraction between two bodies	236
10.3	Application to cylindrical geometry	239
	10.3.1 Two parallel cylinders and a cylinder parallel to a plane	240
	10.3.2 Cylinder parallel to a plane at large separation	245
	10.3.3 The limit of short separations, and corrections beyond the proximity force approximation	246
10.4	Applications to spherical geometry	249
	10.4.1 General formulas for two spheres and for a sphere in front of a plane	250
	10.4.2 A sphere and a plane at large separation	254
	10.4.3 Corrections beyond the proximity force approximation at small separations	255
10.5	Corrugated planes	258

11 Spaces with non-Euclidean topology — 262

11.1	Topologically nontrivial flat spaces	262
	11.1.1 Three-dimensional space–time	262
	11.1.2 Four-dimensional space–time	264
11.2	Topologically nontrivial curved spaces	265
	11.2.1 Three-dimensional space–time	266

	11.2.2 Four-dimensional space–time	268
11.3	Nontrivial topologies in cosmology	270
11.4	Compactification of extra dimensions	274
11.5	Topological defects	276

II THE CASIMIR FORCE BETWEEN REAL BODIES

12 The Lifshitz theory of the van der Waals and Casimir forces between plane dielectrics — 281

- 12.1 The Lifshitz formula for two semispaces at zero temperature — 282
  - 12.1.1 Representation in terms of imaginary frequencies — 284
  - 12.1.2 Representation in terms of real frequencies — 288
  - 12.1.3 The limiting cases of small and large separations — 289
- 12.2 The Lifshitz formula for stratified and magnetic media — 290
- 12.3 Two semispaces at nonzero temperature — 294
  - 12.3.1 Representation in terms of Matsubara frequencies — 294
  - 12.3.2 Representation in terms of real frequencies — 298
- 12.4 Correlation of energy and free energy — 299
- 12.5 Asymptotic properties of the Lifshitz formula at low and high temperature — 301
  - 12.5.1 Finite static dielectric permittivity — 302
  - 12.5.2 Static conductivity of the dielectric material and the third law of thermodynamics — 307
- 12.6 Computational results for typical dielectrics — 310
  - 12.6.1 Dielectric permittivity along the imaginary frequency axis — 310
  - 12.6.2 Free energy and pressure as functions of separation and temperature — 312
  - 12.6.3 The inclusion of dc conductivity — 316
- 12.7 Problems with polar dielectrics — 319
- 12.8 The Lifshitz formula for anisotropic plates — 321
  - 12.8.1 Uniaxial crystals — 321
  - 12.8.2 Casimir torque — 322
- 12.9 Lifshitz-type formula for radiative heat transfer — 323
- 12.10 Application region of the Lifshitz formula — 324

13 The Casimir interaction between real-metal plates at zero temperature — 328

- 13.1 Perturbation theory in the relative skin depth, and the plasma model — 328
- 13.2 Drude model and the Lifshitz formula at zero temperature — 331
  - 13.2.1 The Drude dielectric permittivity — 331
  - 13.2.2 Computations using the plasma and Drude models — 334
- 13.3 Computations using tabulated optical data — 335
- 13.4 Surface impedance approach — 339

	13.4.1 The concept of the Leontovich impedance	339
	13.4.2 The Lifshitz formula with the Leontovich impedance	341
13.5	The generalized plasma-like dielectric permittivity	343
	13.5.1 Generalized plasma-like permittivity and optical data	344
	13.5.2 Generalized Kramers–Kronig relations for the plasma and plasma-like permittivities	346
	13.5.3 Computations using the generalized plasma-like model	349

14 The Casimir interaction between real metals at nonzero temperature   351

14.1	The problem associated with the zero-frequency term in the Lifshitz formula	352
14.2	Perturbation theory for metals described by the plasma model	355
	14.2.1 Casimir free energy per unit area and Casimir pressure	356
	14.2.2 Agreement with the Nernst heat theorem	360
14.3	Metals described by the Drude model	361
	14.3.1 Prediction of large thermal corrections below $1\,\mu$m	362
	14.3.2 Violation of Nernst's theorem for Drude metals with perfect crystal lattices	365
	14.3.3 The role of impurities	371
	14.3.4 Why the Drude model is not applicable in the Lifshitz theory	374
	14.3.5 Attempts at modifying the reflection coefficients	376
14.4	Leontovich impedance approach at nonzero temperature	380
	14.4.1 Impedance in the frequency region of the normal skin effect	381
	14.4.2 Impedance in the region of the anomalous skin effect	382
	14.4.3 Impedance in the region of infrared optics	383
	14.4.4 Impedance using the Drude model	385
14.5	The role of evanescent and propagating waves	387
14.6	Metals described by the generalized plasma-like model	392
	14.6.1 Computational results	393
	14.6.2 Perturbation theory for the generalized plasma model	395
	14.6.3 Agreement with the Nernst heat theorem	399

15 The Casimir interaction between a metal and a dielectric   401

15.1	An ideal-metal plate and a plate with constant permittivity	401
	15.1.1 The asymptotic behavior at low and high temperature	402
	15.1.2 The Casimir energy and pressure at zero temperature	405
15.2	Metal and dielectric plates with permittivities depending on frequency	406
	15.2.1 The low- and high-temperature limits	407
	15.2.2 Analytical results at zero temperature	411
15.3	Computational results	413
15.4	Conductivity of a dielectric plate and the Nernst heat theorem	415

16 The Lifshitz theory of atom–wall interactions — 419
- 16.1 The van der Waals and Casimir–Polder interatomic potentials — 419
- 16.2 The Lifshitz formula for an atom above a plate — 422
- 16.3 Interaction of atoms with a metal wall — 425
  - 16.3.1 Atom near an ideal-metal plane — 425
  - 16.3.2 A real-metal plate and an atom — 429
  - 16.3.3 Asymptotic behavior at low temperature — 433
  - 16.3.4 The case of short separations — 437
- 16.4 Interaction of atoms with a dielectric wall — 439
  - 16.4.1 Asymptotic properties at low and high temperature for a finite static permittivity of the wall material — 439
  - 16.4.2 Computations of the free energy — 441
  - 16.4.3 Various approaches to including the dc conductivity, and the Nernst theorem — 444
- 16.5 The impact of magnetic properties on atom–wall interaction — 449
- 16.6 Atom–wall interactions in the nonequilibrium case — 451
- 16.7 Anisotropic materials: interaction of hydrogen atoms with graphite — 453
  - 16.7.1 Dielectric permittivity of graphite along the imaginary frequency axis — 453
  - 16.7.2 Computational results for plates of different thickness — 456

17 The Casimir force between rough surfaces and corrugated surfaces — 460
- 17.1 Method of pairwise summation for real bodies with rough surfaces — 461
  - 17.1.1 Formulation of the method — 461
  - 17.1.2 Perturbation theory in the roughness amplitudes for two parallel plates — 466
  - 17.1.3 Applications to large-scale roughness — 471
  - 17.1.4 Perturbation theory for a sphere above a plate — 477
  - 17.1.5 Stochastic roughness with large correlation length — 480
- 17.2 The proximity force approximation for real rough bodies — 482
  - 17.2.1 Geometrical averaging for regular roughness — 482
  - 17.2.2 Geometrical averaging for stochastic roughness — 485
- 17.3 Nonparallel plates as large-scale roughness — 486
- 17.4 Various approaches for short-scale roughness — 490
- 17.5 Sinusoidally corrugated surfaces — 494
  - 17.5.1 The Casimir energy and pressure — 494
  - 17.5.2 The lateral Casimir force — 500
  - 17.5.3 Application regions of approximate methods — 502
  - 17.5.4 The role of roughness and corrugations in atom–plate interactions — 506

Contents

III MEASUREMENTS OF THE CASIMIR FORCE AND THEIR APPLICATIONS IN BOTH FUNDAMENTAL PHYSICS AND NANOTECHNOLOGY

18 General requirements for Casimir force measurements — 513
  18.1 Primary achievements of older measurements — 513
    18.1.1 Experiment with parallel plates by Sparnaay — 513
    18.1.2 Experiments by Derjaguin et al. — 515
    18.1.3 Experiments by Tabor, Winterton, and Israelachvili — 515
    18.1.4 Experiments by van Blockland and Overbeek — 516
    18.1.5 Dynamical measurements by Hunklinger and Arnold et al. — 518
    18.1.6 Measurements of the Casimir–Polder force by Sukenik and Hinds et al. — 518
  18.2 General requirements following from the older measurements — 519
  18.3 Rigorous procedures for comparison of experiment and theory — 520
    18.3.1 Experimental errors and precision — 521
    18.3.2 Theoretical uncertainties for real materials — 525
    18.3.3 Statistical framework for the comparison of theory with experiment — 527

19 Measurements of the Casimir force between metals — 530
  19.1 Experiment with torsion pendulum — 530
  19.2 Experiments with an atomic force microscope — 533
    19.2.1 First AFM experiment with aluminum surfaces — 533
    19.2.2 Improved measurement with aluminum surfaces — 537
    19.2.3 Precision measurement using gold surfaces — 540
    19.2.4 Dynamic measurement — 548
  19.3 Experiments with a micromechanical torsional oscillator — 549
    19.3.1 Experimental setup and measurement scheme — 550
    19.3.2 Static and dynamic measurements — 552
    19.3.3 Improved dynamic measurement — 556
    19.3.4 More precise dynamic measurement, and conclusive test for some models of the thermal Casimir force — 563
    19.3.5 Experimental test of proximity force approximation — 571
  19.4 Experiment using a configuration of two parallel plates — 573
  19.5 Related experiments — 574
    19.5.1 Thin metal layers — 574
    19.5.2 Ambient measurements — 575
    19.5.3 Measurements in liquids — 576
    19.5.4 Dynamic holography techniques — 578
  19.6 Prospects for future measurements — 578

20 Measurements of the Casimir force with semiconductors — 581
  20.1 Experiment with gold-coated sphere and silicon plate — 582

	20.1.1 Calibration of the setup	583
	20.1.2 Measurement results and experimental errors	586
	20.1.3 Comparison between experiment and theory	588
20.2	Experiment on the difference Casimir force for samples with different charge carrier densities	593
20.3	Experiment on optically modulated Casimir forces	600
	20.3.1 Experimental setup and sample preparation	601
	20.3.2 Calibration and excited-carrier lifetime measurement	603
	20.3.3 Experimental results and error analysis	606
	20.3.4 Theoretical Casimir force differences and comparison with experiment	611
	20.3.5 Tests for the effect of charge carriers in dielectrics	616
20.4	Proposed experiments with semiconductor surfaces	620
	20.4.1 The dielectric–metal transition	621
	20.4.2 Casimir force between a sphere and a patterned plate	621
	20.4.3 Pulsating Casimir force	623

21 Measurements of the Casimir force in configurations with corrugated boundaries — 625

21.1	Experiment with a sphere above a corrugated plate	625
21.2	Measurement of the lateral Casimir force	627
21.3	Calculation of the lateral Casimir force in the configuration of a sphere above a plate	632
21.4	Control of the lateral Casimir force	636
21.5	Experiment with a sphere above rectangular trenches	638

22 Measurements of the Casimir–Polder force — 643

22.1	Measurement of the thermal Casimir–Polder force	643
	22.1.1 Measurement scheme and technique	643
	22.1.2 Comparison with theory in thermal equilibrium	645
	22.1.3 Comparison with theory out of thermal equilibrium	647
22.2	Experiments on quantum reflection	649
	22.2.1 Main experimental results	649
	22.2.2 Accuracy of phenomenological potential	651

23 Applications of the Casimir force in nanotechnology — 655

23.1	Combined role of electrostatic and Casimir forces in MEMS and NEMS	655
	23.1.1 Modeling of the combined role of electrostatic and Casimir forces in MEMS and NEMS	656
	23.1.2 Experimental investigation of the stability of MEMS	659
23.2	Actuation of MEMS by the Casimir force	661
23.3	Nonlinear micromechanical Casimir oscillator	663
23.4	The Casimir–Polder interaction between atoms and carbon nanostructures	666

		23.4.1 Lifshitz-type formulas for the interaction of an atom with a multiwalled carbon nanotube	666
		23.4.2 Lifshitz-type formulas for graphene and single-walled carbon nanotubes	669
		23.4.3 Computational results for atom–nanotube interaction	674
	23.5	Prospective applications	679

24 Constraints on hypothetical interactions from the Casimir effect — 682

   24.1 Long-range forces and constraints on them from gravitational experiments — 682
      24.1.1 Light particles and extra-dimensional physics — 682
      24.1.2 Eötvos- and Cavendish-type experiments — 685
   24.2 Constraints from older measurements of the Casimir force — 687
      24.2.1 Constraints from measurements between dielectric test bodies — 688
      24.2.2 Constraints from torsion pendulum experiment — 689
      24.2.3 Constraints from ambient experiment with two crossed cylinders — 691
   24.3 Constraints from experiment with gold surfaces using an atomic force microscope — 692
   24.4 Constraints from experiment using a micromachined oscillator — 693
      24.4.1 Constraints from Casimir pressure measurement — 694
      24.4.2 Constraints from Casimir-less experiment — 696

25 Conclusions and outlook — 698

References — 703

Index — 745

1

INTRODUCTION

1.1 Zero-point oscillations and their manifestations

The Casimir effect, discovered more than 60 years ago in the seminal paper by Casimir (1948), is one of the most direct manifestations of the existence of zero-point vacuum oscillations. For a long time Casimir's paper remained relatively unknown, but starting from the 1970s it has rapidly received increasing attention and in the last few years has become highly admired.

The Casimir effect, in its simplest form, is the attraction between two electrically neutral, infinitely large, parallel conducting planes placed in a vacuum. This is an entirely quantum effect because in classical electrodynamics the force acting between two neutral planes is equal to zero. So, it is only the vacuum of the quantized electromagnetic field, i.e. the ground state of quantum electrodynamics, which causes the planes to attract each other. According to Casimir's prediction, the attractive force per unit area, i.e. the pressure between two infinitely large, neutral parallel planes made of an ideal metal at zero temperature, is given by

$$P(a) \equiv P_{\rm IM}(a) = -\frac{\pi^2}{240}\frac{\hbar c}{a^4}. \tag{1.1}$$

Here a is the separation distance between the planes, $\hbar$ is the Planck constant, and c is the velocity of light (below, the index IM is used where needed to distinguish between results obtained for ideal metals and for real materials).

Below, we shall repeatedly derive eqn (1.1) in different formalisms (this will be done for the first time in Section 2.5) and present the far-reaching generalizations of this equation for the cases of real materials at nonzero temperature and for bodies of various geometrical shapes. As an example, for two planes separated by a relatively large (on the atomic scale) distance of $a = 1\,\mu$m, the Casimir pressure (1.1) is $P \approx 1.3\,\text{mPa}$, a macroscopic value. It is remarkable that a macroscopic effect is caused by the quantum vacuum.

In fact, the roots of the Casimir effect date back to the introduction by Planck (1911) of half-quanta. According to quantum mechanics, a harmonic oscillator has discrete energy levels

$$E_n = \hbar\omega\left(n + \frac{1}{2}\right), \tag{1.2}$$

where ω is the angular frequency of the oscillator, and $n = 0, 1, 2, \ldots$ is the number of energy quanta. From eqn (1.2) it follows that the energy of the ground (vacuum) state which contains a number $n = 0$ of energy quanta is

$$E_0 = \frac{\hbar\omega}{2}, \tag{1.3}$$

i.e. it is not equal to zero. This is the energy of a zero-point oscillation with frequency ω.

The canonical quantization procedure of quantum mechanics relates the ground state energy to the arbitrariness of the operator ordering in the definition of the Hamiltonian operator $\hat{H} = \hat{H}(\hat{p}, \hat{q})$ by replacing the dynamical variables p and q in the classical Hamiltonian $H(p, q)$ with the corresponding operator quantities $\hat{p}$ and $\hat{q}$. It must be underlined that the energy E_0 of a vacuum state containing a zero number of energy quanta cannot be observed by measurements within the quantum system under consideration, i.e. in transitions between different quantum states, or, for instance, in scattering experiments. It may happen, however, that the frequency ω of the oscillator depends on some classical parameter (or parameters) external to the quantum system. It was as early as 1919 that the ground state (vacuum) energy was successfully used to explain the vapor pressures of different isotopes. In this case the mass of the isotope plays the role of the external parameter, leading to different oscillator frequencies for isotopes of different masses [a historical review was presented by Milonni (1994) and Rechenberg (1999)].

In the framework of quantum field theory any quantized field, the electromagnetic field for example, is considered as a set of oscillators of all frequencies. Then, in accordance with eqn (1.3), the energy of the ground state of a field is given by the sum of the energies of zero-point oscillations

$$E_0 = \frac{\hbar}{2} \sum_J \omega_J, \tag{1.4}$$

where the collective index J labels the quantum numbers of the field modes. For instance, for the electromagnetic field in free Minkowski space, the modes are labeled by a three-dimensional wave vector k with continuous components and a two-valued discrete index fixing the polarization state. In bounded regions of space, however, some of the wave vector components become discrete. As an example, the tangential component of the electric field vanishes on a metal surface, leading to a discrete component of the wave vector in the perpendicular direction. Note that for a quantized spinor field, the right-hand side of eqn (1.4) is negative (see Section 3.3 for more details). The sum (1.4) is clearly infinite, as always happens in quantum field theory when one tries to assign a ground state (vacuum) energy to each mode of the field. This is one of the manifestations of the problem of ultraviolet divergences.

It was Casimir who first subtracted from the infinite vacuum energy of the quantized electromagnetic field in the presence of ideal-metal planes the infinite vacuum energy of the same field in free Minkowski space. Both infinite energies were regularized, and after subtraction, the regularization was removed, leaving a finite energy per unit area,

$$E(a) \equiv E_{\rm IM}(a) = -\frac{\pi^2}{720}\frac{\hbar c}{a^3}, \tag{1.5}$$

which depends on the separation distance a between the planes. The Casimir pressure (1.1) was then obtained as

$$P(a) = -\frac{\partial E(a)}{\partial a}. \tag{1.6}$$

The removal of the infinite energy of vacuum oscillations in free Minkowski space performed by Casimir is presently the standard procedure in textbooks on quantum field theory. It is motivated by the fact that in all fields of physics, with the exception of Einstein's gravitational theory, energy is defined only up to an additive constant. Thus it is generally assumed that all physical energies should be measured starting from the top of the infinite vacuum energy in free Minkowski space. As a result, effectively the physical energy of free space is set to zero rather than being equal to infinity. Mathematically, the removal of the infinite energy of the zero-point oscillations in free space is achieved by the so-called normal ordering procedure. This operation is applied to the operators of all physical observables, defined in free Minkowski space and prewritten in a symmetrical form with respect to the creation and annihilation operators. It puts all creation operators to the left of annihilation operators as if they commute or anticommute depending on the spin of the field (Milonni 1994; Itzykson and Zuber 2005; Bogoliubov and Shirkov 1982; Weinberg 1995).

It would be incorrect, however, to neglect the infinite zero-point energy found in the presence of material boundaries, for example parallel metallic planes. In that case the frequencies of field oscillators depend on the separation distance between the planes and there is an infinite set of different vacuum states for different separations. These vacuum states change continuously with adiabatic changes in the separation distance between the planes. Thus, it is incorrect to preassign zero energy to several states between which transitions are possible. Here, in quantum field theory, the state of affairs is in perfect analogy to that discussed above in quantum mechanics. In the presence of metallic planes, there is an external parameter (the separation distance) which is similar to the mass of an isotope and leads to different frequencies of oscillators of the quantized field for different separation distances between the planes. Because of this, the finite difference between the infinite zero-point energy in the presence of metallic planes and that in free Minkowski space is an observable and gives rise to the Casimir effect.

In the remainder of this section, we briefly discuss the relation of the Casimir effect to other effects in quantum field theory connected with the existence of zero-point oscillations. It is well known that classical external fields (the Coulomb field, for instance) polarize the quantum vacuum (Itzykson and Zuber 2005). The effect of polarization of the vacuum by an external field is described by some nonzero vacuum energy depending on the field strength. The strength of the external field is the classical parameter which plays the same role as the isotope mass or the separation distance above. In fact, material boundaries can be considered as concentrated external fields. In this respect the Casimir effect, which results in a vacuum energy such as that in eqn (1.5) in quantization volumes restricted by material boundaries, is analogous to the polarization of the vacuum by an external field. We can say, then, with reasonable accuracy that material boundaries polarize the vacuum of a quantized field, and the Casimir force acting on a boundary is a result of this polarization.

Another quantum vacuum effect connected with the existence of zero-point oscillations is the creation of particles from the vacuum by an external field (Greiner *et al.* 1985; Grib *et al.* 1994). In this effect, energy is transferred from the external field to the zero-point oscillations (they are often referred to as virtual particles) transforming them into real particles. As an example, a nonstationary classical electric field can create electron–positron pairs from the vacuum. Although a material boundary is somewhat analogous to an external field, there is no particle creation from the vacuum in the case of static boundaries. However, if the boundaries are nonstationary, and the boundary conditions depend on time, there is particle creation from the vacuum in addition to the Casimir force. This effect is often called the nonstationary or dynamical Casimir effect (see Section 7.7 for further discussion).

Zero-point oscillations of the quantized electromagnetic field also contribute to many other effects of quantum electrodynamics in unbounded Minkowski space that are not only vacuum processes, but also involve real particles. Examples are spontaneous emission from atoms, the Lamb shift, and the anomalous magnetic moment of an electron (Milonni 1994). The contributions of zero-point oscillations to such processes are usually called "radiative corrections". These processes are usually considered in textbooks on quantum electrodynamics. It is significant that in free Minkowski space zero-point oscillations of quantized fields may give rise to an observable effect only if real physical particles are involved in the process. If there are material boundaries in Minkowski space, two types of effects caused by the zero-point oscillations are possible. The first type includes purely vacuum effects, such as the Casimir effect at zero temperature. In the second type, real particles are present in addition to the boundaries.

Quantum processes in the presence of boundaries are studied by quantum field theory with boundary conditions. This includes quantum field theory at nonzero temperature in the Matsubara formulation, where, in order to introduce the concept of temperature in quantum field theory, one must impose a boundary-type "identification" condition in the Euclidean time variable (see Section 5.1 for more details). In this book we shall consider quantum field theory with external conditions only in application to the Casimir effect at both zero and nonzero temperature. The Casimir effect in spaces with a non-Euclidean topology will also be discussed. Similarly to the Matsubara formulation, there are no boundaries in spaces with a non-Euclidean topology, but there are identification conditions which play the same role as the boundary conditions.

There are many other effects studied in quantum field theory with boundary conditions, where zero-point oscillations lead to important contributions. One example is provided by an atom whose spontaneous emission is changed in a cavity. Another example is the so-called apparatus correction to the anomalous magnetic moment of an electron. Here, the zero-point oscillations of the electromagnetic field (i.e. the virtual photons) that are responsible for the anomalous contribution to the magnetic moment of an electron are affected by the boundaries. In both cases a real particle is involved in the process, and the quantity to be considered is the expectation value of the energy operator in one particle state instead of the vacuum. The same holds for the cavity shift of the energy levels of a hydrogen atom. These topics, together with a number of related ones, are called "cavity quantum electrodynamics" (Dutra 2005). They are outside the scope of the present book.

1.2 Connection between van der Waals and Casimir forces

The Casimir force is closely related to the familiar phenomenon of the van der Waals force (Parsegian 2005). The van der Waals attraction acts between two nearby atoms or molecules even if neither has a permanent dipole moment (i.e., it is nonpolar). As a consequence, two neutral macrobodies separated by a short distance of a few nanometers are also attracted by the van der Waals force. Similarly to the Casimir effect, the phenomenon of the van der Waals force is entirely of quantum origin. Although atoms, molecules, and neutral macrobodies have zero net charge, they consist of moving charged particles producing a fluctuating electromagnetic field in the interatomic (or intermolecular) space, in close proximity to the surface of a macrobody (Kardar and Golestanian 1999).

The quantum theory of the van der Waals interaction was developed by London (1930). The expectation values of the operators of the dipole moment are zero for nonpolar atoms and molecules. However, the fluctuating electromagnetic field induces instantaneous dipole moments in atoms and molecules. As a result, the dispersion of the operator of the dipole moment is not equal to zero. London obtained his expression for the interatomic (or intermolecular) interaction potential in fourth-order perturbation theory for the interaction of a dipole operator with a fluctuating electric field (the interatomic potentials are discussed in Section 16.1). The result obtained is entirely quantum (because it depends on $\hbar$), but it does not contain c, i.e. it is nonrelativistic. In fact the fluctuating electromagnetic field can be considered as a model for zero-point oscillations. For closely spaced separate atoms or two atoms belonging to different macrobodies, a virtual photon emitted by one atom reaches the other atom during its lifetime. The resulting correlated oscillations of instantaneously induced dipole moments in both atoms give rise to the nonretarded (i.e. not dependent on c) van der Waals force.

Let us now consider larger separation distances between the two atoms so that a virtual photon emitted by one atom cannot reach the other during its lifetime. This case was considered for the first time by Casimir and Polder (1948), who investigated van der Waals forces in colloids. At such large distances, the usual nonretarded van der Waals force is absent. However, the correlation of the quantized electromagnetic field in the vacuum state calculated at the two spatial points where the atoms are situated is not equal to zero. Once again, this leads to correlated oscillations of the induced atomic dipole moments and results in an attractive interaction between the two atoms. This interaction, named after Casimir and Polder, is not only quantum but also relativistic. It depends on both $\hbar$ and c, and also on the atomic polarizability (see Section 16.1). Sometimes it is referred to as the retarded van der Waals interaction. The existence of interatomic (or intermolecular) retarded forces leads to similar forces acting between an atom (or molecule) and a macrobody and between two macrobodies. The role of the relativistic effects increases with separation distance and becomes dominant at separations of the order of hundred nanometers. The generic name for both the van der Waals and Casimir interactions is *dispersion forces*, because both of them are caused by dispersions of the operator of the dipole moment (Mahanty and Ninham 1976).

The work by Casimir and Polder (1948) opened the way for the development of a unified theory of van der Waals and Casimir forces between real materials. This was done by Lifshitz (1956) in the case of plane parallel dielectric plates described by a frequency-dependent dielectric permittivity. As is demonstrated below, this theory reproduces all of the results of Casimir, London, and Casimir and Polder in their respective limiting cases, and also provides smooth transitions between them.

From the above discussion, it can be seen that the Casimir force between material boundaries can be considered as simply the retarded van der Waals force. The universality of eqn (1.1), which depends only on the fundamental constants $\hbar$ and c and does not depend on charges or other interaction constants, is explained by the ideal-metal approximation used for the planes. On the surfaces of real metals, the tangential component of an electric field is not precisely equal to zero. As a result, there arise some corrections to eqn (1.1) for ideal-metal planes which depend on the electron charge and other parameters. This prompted Jaffe (2005) to argue that the Casimir effect should be tackled by the same approaches as used for the vacuum polarization contribution to the Lamb shift.

A slightly different situation arises, however, for the Casimir effect in spaces with a non-Euclidean topology, i.e. in the closed Friedmann model used in cosmology. In this case there are no material boundaries and no moving charged particles producing the fluctuating electromagnetic field. However, there are identification conditions imposed on field operators owing to the nontrivial topology of space–time, which play the same role as boundary conditions due to material boundaries. The role of a classical parameter is played by the scale factor of the metric. As a result, a universal Casimir energy density similar to (1.5) arises, which depends only on the fundamental constants and on the scale factor of the metric (see Chapter 11). Thus, this kind of Casimir effect cannot be considered as a close relative of the van der Waals forces.

1.3 The Casimir effect as a multidisciplinary subject

From the above discussion it follows that the Casimir effect is a quantum and relativistic phenomenon caused by the zero-point oscillations of quantized fields. In a general way, it can be characterized as a specific type of vacuum polarization which depends on some external classical parameters and arises owing to the presence of material boundaries or the non-Euclidean topology of a quantization volume. The electromagnetic Casimir effect in the presence of real material boundaries is a subset of the general dispersion forces and is closely connected with subtle aspects of condensed matter physics. At the same time, the Casimir effect for quantized fields of different spin in topologically nontrivial spaces goes to the heart of gravitation, cosmology, and modern unification theories beyond the Standard Model, including string theory. The quantum vacuum is the most basic type of physical reality. Thus it is not surprising that the Casimir effect is found to be important in practically all fields of modern physics. Quite recently, nanotechnological applications of the Casimir effect have also become the subject of intensive study.

Many fundamental results on the Casimir effect have been obtained using quantum field theory. The calculation of the Casimir force is a particularly complicated theoretical problem. In the simplest case of flat boundaries, the vacuum energy approaches infinity at large momentum, similarly to that in free Minkowski space. Thus, for Casimir, it was sufficient to subtract the contribution of free Minkowski space in order to obtain a finite physical result. This is, however, not the case for arbitrary domains bounded by curved surfaces (for example, for the interior of a sphere). For curved boundaries, in addition to the highest-order infinity (which is proportional to the fourth power of the cutoff momentum and is the only one present for flat boundaries), there exist other, lower-order infinities. An understanding of the general structure of these infinities for arbitrary domains with ideal boundary conditions has been obtained by using a combination of zeta function regularization and heat kernel expansion. Remarkably, for closed configurations, i.e. for the Casimir effect for one body instead of two, the Casimir force can be not only attractive but also repulsive. As was shown by Boyer (1968), the latter is true for an ideal metal spherical shell.

Investigation of the Casimir effect with quantum field theory has resulted in three main applications which will be considered in the book. It has been shown that the Casimir energy makes an important contribution to the total energy of a nucleon. In multidimensional Kaluza–Klein theories, the Casimir effect provides a mechanism for spontaneous compactification of extra spatial dimensions. Furthermore, measurements of the Casimir force in the laboratory help us to obtain constraints on the parameters of the light hypothetical particles predicted by many extensions of the Standard Model and on corrections to Newton's gravitational law predicted by extra-dimensional physics with a low-energy compactification scale.

In condensed matter physics, the Casimir effect leads to both attractive and repulsive forces in layered systems. It contributes to the interaction of a surface with the tip of an atomic force microscope and should be taken into account in the investigation of various properties of thin films, surface tension, and latent heat. The Casimir effect plays a role in both bulk and surface critical phenomena and depends on the concentration of free carriers in semiconductors.

Experimental and theoretical investigations of the Casimir force between both metal and semiconductor test bodies have helped to formulate and solve some important problems in thermodynamics and statistical physics related to the interaction of a fluctuating electromagnetic field with real materials.

In atomic physics, the Casimir effect is important for the understanding of atom–atom and atom–wall interactions. The Casimir force influences physical processes in quantum reflection and Bose–Einstein condensation. Both Casimir and van der Waals forces play a role in the absorption of atoms by various microstructures, and specifically by carbon nanotubes.

In astrophysics, gravitation, and cosmology, the Casimir effect arises in space–times with a nontrivial topology. The polarization of the vacuum due to the Casimir effect plays a role in the resolution of the problem of the cosmological constant. In some cosmological scenarios of the early Universe before the Big Bang, this polarization drives the inflation process. The theory of structure formation in the Universe employs the concept of topological defects, such as cosmic strings, which produce a Casimir-type polarization of the vacuum.

In mathematical physics, the Casimir effect has stimulated the development of powerful regularization methods based on the use of the Riemann and Epstein zeta functions and the heat kernel expansion.

In addition to fundamental physics, the Casimir effect is quickly becoming a part of nanoscience. Given the shrinking of microdevice dimensions to nanometers, the important role of the Casimir force in the performance, fabrication, and function of devices is now well recognized. Recent advances in the application of the Casimir force to nanotechnology show that it is possible to exert control over the sign of the force and its magnitude by optical modification of the charge carrier density with laser light. This opens up prospects for a new generation of nanodevices driven by the Casimir effect.

1.4 A guide to this book

The journal literature on the Casimir effect is quite extensive and contains many hundreds of papers. However, there are only a few books devoted to this subject. The first book dedicated to the Casimir effect was published in Russia by Mostepanenko and Trunov (1990). It covers all aspects of the theory, including the Casimir interaction between real bodies, before 1989 but contains only a very brief presentation of the preceding experiments. A slightly enlarged translation of this book into English was published later (Mostepanenko and Trunov 1997). The book by Milonni (1994) is partially devoted to the Casimir effect and contains a detailed investigation of Casimir's discovery in the context of quantum electrodynamics and the concept of the quantum vacuum. This book has played an important role in drawing attention to the subject. A more specialized book by Krech (1994) concentrates on the role of the Casimir effect in critical systems. A more recent book by Milton (2001) is primarily devoted to the field-theoretical aspects of the Casimir effect in ideal configurations and covers the state of knowledge in 2000.

Since the publication of the previous books, the volume of scientific information on the Casimir effect has more than doubled. In addition, new fundamental methods have been developed and some basic concepts revised. This is true for both experiment and theory. The present book sums up the state of the art in Casimir research, including fundamental field-theoretical results and their adaptation to real material bodies, measurements and their comparison with theory, and nanotechnological applications. The presentation of the three main lines (the fundamental theory, real material bodies, and experiment) is performed in three respective parts of the book, which are closely connected to each other and contain a number of cross-references. Each part is based on the previous part; nevertheless, individual parts of the book, with some obvious limitations, can be used separately by experts in the respective areas.

Part I of the book presents the physical and mathematical foundations of the Casimir effect in ideal configurations. In this part, all boundary surfaces are assumed to be perfect, and Dirichlet, Neumann, Robin, semitransparent, or identification-type boundary conditions are used. Chapter 2 presents simple models to illustrate some key points in the theory of the Casimir effect. The elementary approach to the Casimir force between two parallel ideally conducting planes is also contained here. Chapters 3 and 4 are central to the developments in Part I. In Chapter 3, field quantization in the presence of boundaries is performed, and various representations of the vacuum energy are considered. Propagators with boundary conditions are introduced. Chapter 4 contains the general theory of regularization and renormalization in the case of the Casimir effect. The regularization schemes presented here are repeatedly used in other chapters of the book. The divergent part of the vacuum energy is found using the heat kernel expansion. The finiteness of the Casimir force acting between two solid bodies is also proved here.

The foundations of the Casimir effect at nonzero temperature are considered using the Matsubara approach in Chapter 5. Here, both high- and low-temperature asymptotic expansions are discussed. In Chapter 6, several approximate methods applicable to the calculation of the Casimir energy in general geometries are discussed. In some cases, the method used does not by itself allow estimation of its accuracy (e.g. the proximity force approximation and semiclassical approaches). However, by comparison with exact results in cases where these are available, we can obtain reliable quantitative values for the accuracy and justify the use of such methods in the comparison of experiment with theory.

Chapters 7 and 8 are devoted to the Casimir effect for the configurations of two parallel planes and a rectangular box, respectively, with ideal boundary conditions. The important results for various fields, using various regularizations at zero and nonzero temperature, are discussed. The case of mixed boundary conditions is considered. Nonparallel planes (a wedge) and moving planes (the dynamical Casimir effect) are also briefly considered. Special attention is paid to the repulsive Casimir force arising in a rectangular box with particular ratios of the sides. Novel results on the Casimir piston are also discussed.

Chapter 9 presents important results on the Casimir effect for spherical and cylindrical shells with various boundary conditions. These configurations present good and informative examples for the general methods developed in Chapters 3 and 4. Chapter 9 also includes the Casimir effect for a dielectric ball. This finds applications in the bag model of quantum chromodynamics. In Chapter 10 a new, powerful description of the Casimir energy based on functional determinants is presented. This description allows one to make exact calculations of the Casimir energies and forces in general geometries. Special attention is paid to a spherical and a cylindrical shell above a plane. In both cases, exact solutions have recently been obtained. These solutions can be compared with the approximate results and thus can be used for determination of the accuracy of these results. In Chapter 11, a few examples of the Casimir effect in spaces with a non-Euclidean topology are presented. Here, the Casimir effect arises not because of the presence of material boundaries but because of identification conditions. The interactions of cosmic strings, along with applications to cosmology and the compactification of extra dimensions, are briefly discussed.

The primary purpose of Part I is to prepare the reader for the investigations of the Casimir effect between real bodies. Because of this, many results of purely mathematical character considered in the literature (such as the Casimir effect in multidimensional boxes, for automorphic fields, and for numerous topologies of space) are not covered. Here, we provide only selected references.

In Part II of the book, we concentrate on the Casimir force between real material bodies. This subject is in fact intermediate between the general theory and the experimental investigation of the Casimir effect. Experiments deal with real bodies that have a nonzero skin depth and are bounded by rough surfaces, not with perfectly shaped surfaces made of an ideal metal. Thus, to compare experiment with theory, the properties of real material boundaries must be taken into account. The theoretical methods presented in Part II were mostly developed during the last ten years in response to experimental advances. As will be clear from Part II of the book, surprisingly, the adaptation of the general theory of the Casimir effect between ideal boundaries to the experimental conditions is physically nontrivial. This presents an important theoretical challenge to some fundamental physical principles and has given rise to controversial opinions.

In Chapter 12, we present the main results of the Lifshitz theory, giving a unified description of both the van der Waals and the Casimir force between plane dielectrics. Various formulations of the well-known Lifshitz formula at zero and nonzero temperature are considered in terms of both real and imaginary Matsubara frequencies. The asymptotic expansions of the Lifshitz formula at low temperature are found, and the consistency of the Lifshitz theory with the third law of thermodynamics (which depends on the conductivity properties of the dielectric) is investigated. The results of numerical computations of the Casimir free energy per unit area and the pressure in a configuration of two dielectric semispaces are presented. The version of the Lifshitz formula for anisotropic plates is also provided. Some attention is paid to the closely related Lifshitz-type formula for radiative heat transfer through a vacuum gap. At the end of Chapter 12, we discuss the application region of the Lifshitz theory in connection with the effects of spatial dispersion.

Chapters 13 and 14 are devoted to the Casimir interaction between two parallel plates made of real metals at zero and nonzero temperature, respectively. Here, a perturbation theory in terms of the relative skin depth of the electromagnetic oscillations and the relative temperature is developed. The Kramers–Kronig relations and tabulated optical data are applied to find the dielectric permittivity of a real metal along the imaginary frequency axis. The computational results at zero temperature are shown to be in good agreement with the analytic perturbation theory. The plasma model, the Drude model, and a generalized plasma-like model that takes into account interband transitions of core electrons are considered. Special attention is paid in Chapter 14 to the problem of the zero-frequency term in the Lifshitz formula. Several approaches to the determination of this term proposed in the literature are analyzed. Computational results for the Casimir free energy, pressure, and entropy are obtained in the framework of each approach. The approach based on the use of the Drude model is shown to be in violation of the third law of thermodynamics in the case of perfect crystal lattices. It also violates the classical limit. Possible physical reasons for this are discussed. The Lifshitz formula, in application to real metals, is reformulated in terms of the Leontovich surface impedance. The generalized plasma-like model and the impedance approach are shown to be in agreement with both the third law of thermodynamics and the classical limit. The role of evanescent and traveling waves in the Casimir interaction between metals is considered. The subjects of Chapters 13 and 14 are used extensively in Part III of the book in the analysis of measurements of the Casimir force between metal test bodies.

In Chapter 15, the Casimir interaction between a metallic plate and a dielectric plate is considered. The results for the free energy, pressure, and entropy are obtained both analytically, using perturbation theory, and numerically, with the use of optical data. These results are compared with those for two dielectric plates. The consistency of the Lifshitz formula with the third law of thermodynamics is demonstrated for the interaction between a metal and a dielectric only when the static dielectric permittivity of the dielectric is finite. If it is infinite, the third law of thermodynamics is violated. The results of Chapter 15 are used in Part III of the book in the interpretation of the experiments on measuring the Casimir force between metal and semiconductor test bodies.

Chapter 16 deals with the application of the Lifshitz theory to atom–wall interactions. It starts from the derivation of the van der Waals and Casimir–Polder interatomic potentials. Then the Lifshitz formula for an atom near a cavity wall is obtained. The atom–wall interaction is investigated for the cases of metal and dielectric walls. Various approaches to the inclusion of the dc conductivity of a dielectric wall are considered in connection with the requirements of thermodynamics. The impact of magnetic properties on the atom–wall interaction is briefly presented. The atom–wall interaction out of thermal equilibrium is discussed for use in the interpretation of experiments on Bose–Einstein condensation (Chapter 22). The interaction of hydrogen atoms with a graphite wall is also calculated, keeping in mind the application to carbon nanostructures to be discussed in Chapter 23 of the book.

Chapter 17 is devoted to the calculation of the Casimir force between corrugated surfaces and in the presence of surface roughness. This subject is very important for all applications of the Casimir effect. The approximate method of pairwise summation is the simplest method, though not always an exact one, for taking account of the effect of corrugations and surface roughness described by analytic functions for real bodies of finite conductivity. This method is developed in the chapter and compared with the proximity force approximation. The corrugations and surface roughness are described by using perturbation theory in terms of relative roughness (or corrugation) amplitudes. Stochastic roughness is also considered in the same way. The consideration of corrugated surfaces leads to an important prediction about the existence of a lateral Casimir force. The experimental confirmation of this prediction is presented in Part III. The application region of the pairwise summation method for rough and corrugated surfaces is determined by the more fundamental path integral approach (which is valid for ideal metal boundaries) and by the statistical approach, taking the roughness correlation length and the nonideality of the metal into account. The role of surface roughness in the atom-wall interaction is discussed at the end of the chapter.

The most striking developments in the Casimir effect during the last ten years are new, more precise measurements of the Casimir force using modern technology. These measurements allow a quantitative comparison between experiment and theory. They have opened up promising opportunities for the use of the Casimir effect in nanotechnology and as a test for fundamental physical theories. Part III of the book covers all of these subjects.

In Chapter 18, the general requirements for Casimir force measurements are considered. Here, a brief survey of older experiments is presented and the experience from these experiments is summarized. Special attention is paid to the determination of the experimental errors, theoretical errors, and their combination. The methods for comparison of experiment and theory for measurements of the Casimir force are described.

Chapter 19 is devoted to measurements of the Casimir force between metal surfaces. The presentation starts with a discussion of an experiment using a torsion pendulum and a series of experiments with an atomic force microscope. The experimental configurations used were those of a spherical lens and a sphere above a plate, respectively. The experiments with the atomic force microscope demonstrated for the first time the corrections to the Casimir force due to the nonzero skin depth and surface roughness. The discussion of all of the experiments in this and the following chapters of Part III is combined with a careful comparison with the theory of the Casimir effect between real bodies presented in Part II of the book. Next, Chapter 19 presents experiments with a micromechanical torsional oscillator. In these experiments, which are the most precise ones to date, both the Casimir force between a plate and a sphere and the equivalent Casimir pressure between two parallel plates were measured. The comparison of the experimental results with theory leads to important conclusions concerning the validity of the various theoretical approaches to the thermal Casimir force discussed in Part II. In particular, the approach using the Drude dielectric function to determine the zero-frequency term in the Lifshitz formula is experimentally excluded at the 99.9% confidence level. Then, an experiment using a linear piezoelectric transducer is presented. This experiment exploits the original Casimir configuration of two parallel plates. The chapter ends with a discussion of several related experiments.

Chapters 20 and 21 contain presentations of experiments on measuring the Casimir force between a metal and a semiconductor and of the force in configurations with corrugated surfaces, respectively. Three experiments on the Casimir interaction between a metallized sphere and a semiconductor plate, considered in Chapter 20, allowed one to measure the change in the magnitude of the force due to a change in the semiconductor charge carrier density and to demonstrate modulation of the Casimir force with laser light. A comparison of the optical-modulation experiment with theory shows that an approach taking into account the zero-frequency conductivity of dielectric materials is experimentally excluded at a 95% confidence level. The use of corrugated surfaces allows one to study the nontrivial boundary properties of the Casimir force and to demonstrate for the first time the physical phenomenon of the lateral Casimir force, which was previously predicted theoretically (see Chapter 17).

Chapter 22 discusses measurements of the Casimir–Polder force in the experiments on Bose–Einstein condensation and quantum reflection. Special attention is paid to an experiment where the thermal Casimir–Polder force was measured for the first time (Obrecht et al. 2007). The experimental data are shown to be in disagreement with a theoretical approach taking into account the dc conductivity of dielectric materials.

In Chapter 23, the applications of the Casimir effect in nanotechnology are considered. This is a new and a rapidly developing subject, driven by the focus on miniaturization in modern technologies. When the characteristic sizes of the elements of microdevices and/or the surface separations shrink below a micrometer, the Casimir force becomes comparable to the characteristic electrostatic forces and must be taken into account in device design, operation, and fabrication. We discuss the actuation of microelectromechanical systems by the Casimir force and its influence on the oscillatory behavior of microdevices. The role of the Casimir effect in carbon nanostructures is also analyzed.

The last chapter in Part III, Chapter 24, is devoted to the constraints on non-Newtonian gravity which follow from the Casimir effect. Many extensions of the Standard Model, including supersymmetry, supergravity, and string theory, predict corrections to Newton's law of gravitation. These corrections follow from the exchange of light elementary particles between atoms of macrobodies and from extra-dimensional physics with a low-energy compactification scale. The measurements of the Casimir force and the extent of the agreement between the experimental data and theory lead to the strongest constraints on the corrections to Newtonian gravitation over a wide interaction range. In this chapter we present the constraints on the Yukawa-type corrections to Newton's law following from older measurements of the Casimir force between dielectric plates and from all modern measurements between metal plates. The so-called *Casimir-less experiment*, where the influence of the Casimir force is canceled, is also discussed.

The book ends with Chapter 25, containing our conclusions and outlook. The main conclusion is that we have already achieved very good agreement between the theory, adapted to the case of real material boundaries, which is presented in Part II of the book, and the measurements of the Casimir force. A generalization of this theory to the case of materials with spatial dispersion and a more fundamental approach to the Casimir effect at nonzero temperature are expected in the near future. Applications of the Casimir effect in both fundamental physics and nanotechnology appear very promising and may have an unexpected impact on basic scientific concepts and technological approaches.

The main notation used in this book is as follows. In relativistically covariant expressions, Greek indices $\alpha, \beta, \ldots, \mu, \nu, \ldots$ take the values $0, 1, 2, 3$, and Latin indices $i, k, l, \ldots$ take the values $1, 2, 3$. Capital Latin letters (J, J' etc.) are used as collective indices to denote a collection of quantum numbers. The scalar product of the 4-vectors a and b is written as

$$a_\mu b^\mu = a^\mu b_\mu = g_{\mu\nu} a^\mu b^\nu = g^{\mu\nu} a_\mu b_\nu = a_0 b_0 - \boldsymbol{a} \cdot \boldsymbol{b},$$

where $g_{\mu\nu}$ is the metric tensor having the signature $(+, -, -, -)$. The repetition of an index in the lower and upper positions implies a summation over this index.

In Part I of the book (with exception of Chapter 2), we use a system of units in which $\hbar = c = 1$. However, in some final expressions of major importance, the usual units are restored. In Parts II and III, which deal with real materials and experiments, the fundamental constants in all mathematical expressions are explicitly indicated. The electromagnetic equations are written in the Gaussian system of units. Some values of experimentally measured quantities and simple formulas are given in the International System (SI) of units. This is indicated in the text. When this does not create confusion, operators and c-functions are notated in a similar way. Other special notations particular to a chapter are introduced where necessary.

PART I

PHYSICAL AND MATHEMATICAL FOUNDATIONS OF THE CASIMIR EFFECT FOR IDEAL BOUNDARIES

2
SIMPLE MODELS OF THE CASIMIR EFFECT

In this chapter we discuss several basic ideas and methods related to the calculation of the Casimir energies and forces in some simple models. The simplicity of these models allows one to avoid cumbersome mathematical calculations and to demonstrate the basic problems that will be repeatedly considered in the following chapters of this book in a more sophisticated context. Such important procedures as regularization and renormalization of infinite quantities are illustrated here both physically and mathematically in a manner readily accessible to all physicists, not just to experts in quantum field theory [see also the review papers by Plunien et al. (1986) and Mostepanenko and Trunov (1988)]. The complete field quantization procedure in the presence of boundaries will be covered in Chapter 3. Despite the elementary character of the present chapter, we discuss the main physical situations where the Casimir effect arises (i.e., in regions with boundaries and in spaces with a nontrivial topology). We consider also local and global approaches to the Casimir effect and derive well-known formulas (1.1) and (1.5) for the electromagnetic Casimir pressure and energy per unit area between two parallel ideal-metal planes. A more detailed derivation and far-reaching generalizations of these formulas can be found in the following chapters of the book.

2.1 The scalar Casimir effect on an interval

We start with a scalar field $\varphi(t,x)$ which depends on the time t and one coordinate $x = x^1$, obeying the Klein–Fock–Gordon equation in two-dimensional space–time

$$\Box_2\,\varphi(t,x) + \frac{m^2 c^2}{\hbar^2}\,\varphi(t,x) = 0. \tag{2.1}$$

Here m is the mass of the field and the two-dimensional d'Alembert operator is defined by

$$\Box_2\,\varphi(t,x) = \frac{1}{c^2}\frac{\partial^2 \varphi(t,x)}{\partial t^2} - \frac{\partial^2 \varphi(t,x)}{\partial x^2}. \tag{2.2}$$

Note that this scalar field in two-dimensional space–time is dimensionless.

Let us consider the properties of the scalar field defined on an interval $0 < x < a$ with Dirichlet boundary conditions imposed at its ends,

$$\varphi(t,0) = \varphi(t,a) = 0. \tag{2.3}$$

Next we shall consider the scalar field along the entire axis $-\infty < x < \infty$. In both cases our primary goal is to find the spectrum of scalar oscillations.

For the case of the interval $[0, a]$, the scalar product of the two (in general complex) solutions of eqn (2.1), f and g, is

$$(f, g) = i \int_0^a dx \left(f^* \frac{\partial g}{\partial x_0} - \frac{\partial f^*}{\partial x_0} g \right), \tag{2.4}$$

where $x_0 = x^0 = ct$. From eqn (2.1), it follows that (f, g) does not depend on time. One may readily check that the complete orthonormal set of the positive- and negative-frequency solutions of eqn (2.1), obeying the boundary conditions (2.3) and satisfying the equalities

$$\left(\varphi_n^{(\pm)}, \varphi_{n'}^{(\pm)} \right) = \pm \delta_{nn'}, \qquad \left(\varphi_n^{(\pm)}, \varphi_{n'}^{(\mp)} \right) = 0, \tag{2.5}$$

is given by

$$\varphi_n^{(\pm)}(t, x) = \left(\frac{c}{a \omega_n} \right)^{1/2} e^{\mp i \omega_n t} \sin k_n x. \tag{2.6}$$

Here, the discrete oscillation frequencies and the wave numbers are given by

$$\omega_n = \left(\frac{m^2 c^4}{\hbar^2} + c^2 k_n^2 \right)^{1/2}, \qquad k_n = \frac{\pi n}{a}, \qquad n = 1, 2, 3, \ldots \tag{2.7}$$

and $\delta_{nn'}$ is the Kronecker delta symbol.

In Section 2.4 we shall consider the quantization procedure of the field $\varphi(t, x)$, define the vacuum state $|0\rangle$, and ensure that the quantities ω_n in eqn (2.7) are exactly the frequencies of the zero-point oscillations entering eqn (1.4) in the Introduction. Thus, the energy of the ground state (i.e. the vacuum energy) of a field $\varphi(t, x)$ on an interval is given by

$$E_0(a, m) = \frac{\hbar}{2} \sum_{n=1}^{\infty} \omega_n = \frac{\hbar}{2} \sum_{n=1}^{\infty} \left(\frac{m^2 c^4}{\hbar^2} + \frac{c^2 \pi^2 n^2}{a^2} \right)^{1/2}. \tag{2.8}$$

If the field $\varphi(t, x)$ is defined on the entire axis $-\infty < x < \infty$, the scalar product of the two solutions of eqn (2.1) takes the form

$$(f, g) = i \int_{-\infty}^{\infty} dx \left(f^* \frac{\partial g}{\partial x_0} - \frac{\partial f^*}{\partial x_0} g \right), \tag{2.9}$$

which is similar to eqn (2.4). In this case the complete orthonormal set of solutions of eqn (2.1) obeying the boundary conditions (2.3) satisfies the equalities

$$\left(\varphi_k^{(\pm)}, \varphi_{k'}^{(\pm)} \right) = \pm \delta(k - k'), \qquad \left(\varphi_k^{(\pm)}, \varphi_{k'}^{(\mp)} \right) = 0, \tag{2.10}$$

where the positive- and negative-frequency solutions are the traveling waves

$$\varphi_k^{(\pm)}(t, x) = \left(\frac{c}{4 \pi \omega_k} \right)^{1/2} e^{\mp i (\omega_k t - kx)}. \tag{2.11}$$

Here, the continuous oscillation frequencies are defined as

$$\omega_k = \left(\frac{m^2 c^4}{\hbar^2} + c^2 k^2\right)^{1/2}, \qquad (2.12)$$

with a continuous wave number $-\infty < k < \infty$, and $\delta(k - k')$ is the one-dimensional delta function.

The energy of the vacuum state of the field $\varphi(t, x_1)$ on the entire axis is given by

$$E_{0\mathrm{M}}(m) = \frac{\hbar}{2} \int_{-\infty}^{\infty} \frac{dk}{2\pi} \omega_k L = \frac{\hbar}{2\pi} \int_{0}^{\infty} dk \left(\frac{m^2 c^4}{\hbar^2} + c^2 k^2\right)^{1/2} L. \qquad (2.13)$$

In this case the sum (1.4) is interpreted as an integral with the measure $dk/(2\pi)$, and $L \to \infty$ is the length of the axis (referred to as the *normalization length*). The lower index M in the vacuum energy (2.13) labels the case of an unbounded one-dimensional space and one-dimensional time, which is a simple analogue of the free Minkowski space–time.

The expressions (2.8) and (2.13) for the vacuum energy of the scalar field in two-dimensional space–time on an interval and on the entire axis, respectively, are both infinite. They diverge at large values of n and k. Such expressions are the standard starting point in the theory of the Casimir effect. To deal with infinite quantities in a meaningful way, one must first make them finite. This is achieved by using what is referred to as a *regularization procedure*. There are many different regularization procedures that have been proposed in the literature, and some of them are discussed in this book. Here, we use the most simple one, which introduces an exponential cutoff function of the forms $\exp(-\delta c k_n)$ and $\exp(-\delta c k)$ after the summation and integration signs in eqns (2.8) and (2.13), respectively, where $\delta > 0$ is a parameter. After all of the operations with the regularized finite quantities have been performed, the regularization is removed by putting $\delta \to 0$. It is necessary to prove that the result obtained does not depend on the specific form of the cutoff function employed in the regularization procedure.

Now we apply the regularization procedure to eqn (2.8). For simplicity, we consider only the massless field with $m = 0$ and omit the argument m in the vacuum energies. From eqn (2.8), we obtain the regularized vacuum energy of a massless field on the interval $(0, a)$,

$$E_0^{(\delta)}(a) \equiv \frac{\hbar}{2} \sum_{n=1}^{\infty} \frac{c\pi n}{a} \exp\left(-\delta \frac{c\pi n}{a}\right) = \frac{\pi \hbar c}{8a} \sinh^{-2} \frac{\delta c \pi}{2a}. \qquad (2.14)$$

This quantity is finite, but it diverges when δ goes to zero. In the limit of small δ, one obtains from eqn (2.14)

$$E_0^{(\delta)}(a) = \frac{\hbar a}{2\pi c \delta^2} - \frac{\pi \hbar c}{24a} + O(\delta^2). \qquad (2.15)$$

This equation represents the vacuum energy as the sum of a singular term and a finite contribution. The latter contains a term $E(a)$ that does not depend on the regularization parameter δ.

Next we apply the regularization procedure to eqn (2.13), i.e. to the vacuum energy of a scalar field on the entire axis. Keeping $m = 0$, we obtain

$$E_{0M}^{(\delta)} \equiv \frac{\hbar c}{2\pi} \int_0^\infty k\,dk\,e^{-\delta ck} L = \frac{\hbar L}{2\pi c\delta^2}. \tag{2.16}$$

This is the regularized vacuum energy for the entire axis.

Let us now separate out the interval $(0, a)$ of the entire axis without imposing any boundary conditions at $x = 0$ and $x = a$. According to eqn (2.16), the vacuum energy for such an interval is

$$E_{0M}^{(\delta)}(a) = \frac{F_{0M}^{(\delta)}}{L} a = \frac{\hbar a}{2\pi c\delta^2}. \tag{2.17}$$

This result should be compared with eqn (2.15), obtained for the same interval *with* the boundary conditions (2.3). It is notable that eqn (2.17) coincides with the first term on the right-hand side of eqn (2.15), diverging when δ goes to zero.

Following Section 1.1, in order to obtain a finite physical result one must subtract the infinite vacuum energy (2.17) for a field on an unconstrained interval (with no boundary conditions) from the infinite vacuum energy (2.15) for an interval constrained by the boundaries. This leads to the **finite quantity**

$$E^{(\delta)}(a) \equiv E_0^{(\delta)}(a) - E_{0M}^{(\delta)}(a) = -\frac{\pi\hbar c}{24a} + O(\delta^2). \tag{2.18}$$

By removing the regularization, we obtain the Casimir energy for the scalar field on an interval,

$$E(a) = \lim_{\delta \to 0} E^{(\delta)}(a) = -\frac{\pi\hbar c}{24a}. \tag{2.19}$$

This result is analogous to eqn (1.5), obtained for the vacuum energy of the electromagnetic field between ideal-metal planes. In the next section, we shall show that eqn (2.19) does not depend on the form of cutoff function used. We shall see that eqn (2.19) retains its validity for any regularization function satisfying some general requirements. The magnitude of the Casimir energy $E(a)$ increases monotonically as the boundary points approach each other. From eqn (2.19), the Casimir force acting between the boundary points of the interval is

$$F(a) = -\frac{\partial E(a)}{\partial a} = -\frac{\pi\hbar c}{24a^2}. \tag{2.20}$$

This is similar to the electromagnetic Casimir pressure (1.1) between two parallel, ideal-metal planes.

Equation (2.18) is a typical example of what are commonly referred to as *subtraction procedures*, used in quantum field theories in order to remove infinities from divergent expectation values of physical quantities. Usually, the subtraction of any infinite quantity is interpreted as a *renormalization* of some physical constant in the bare effective action (see Chapter 4 for more details). In the simplest case considered in this section, we have subtracted a quantity equal to the vacuum energy for an unbounded axis in a given interval. Below, we shall demonstrate that the subtraction of the vacuum energy density and pressure of an unbounded Minkowski space can be formally interpreted as a renormalization of the cosmological constant.

The above results are easily generalized to other types of boundary conditions. It is of interest to consider a Dirichlet boundary condition at $x = 0$ and the Neumann boundary condition at $x = a$,

$$\varphi(t,0) = \left.\frac{\partial \varphi(t,x)}{\partial x}\right|_{x=a} = 0. \tag{2.21}$$

Such conditions are sometimes called *unusual*, *hybrid*, or *mixed*. The complete orthonormal set of solutions of eqn (2.1) has the same form as eqn (2.6), with

$$k_n = \frac{\pi}{a}\left(n + \frac{1}{2}\right), \quad n = 0, 1, 2, \ldots. \tag{2.22}$$

For a massless field, the regularized vacuum energy is given by

$$E_0^{(\delta)}(a) = \frac{\hbar}{2}\sum_{n=0}^{\infty}\frac{c\pi}{a}\left(n+\frac{1}{2}\right)e^{-\delta c\pi(2n+1)/(2a)}$$
$$= \frac{\hbar c \pi}{8a}\coth\frac{\delta c\pi}{2a}\operatorname{csch}\frac{\delta c\pi}{2a}. \tag{2.23}$$

In the limit of small δ, we obtain

$$E_0^{(\delta)}(a) = \frac{\hbar a}{2\pi c \delta^2} + \frac{\pi \hbar c}{48a} + O(\delta^2). \tag{2.24}$$

Importantly, the divergent term has the same form as in eqn (2.15) and is, thus, equal to the contribution of free space (2.17). As a result, the Casimir energy of a scalar field on an interval with the boundary conditions (2.21) is positive (Fulling *et al.* 2007a), and the respective Casimir force is repulsive:

$$E(a) = \frac{\pi \hbar c}{48a}, \quad F(a) = \frac{\pi \hbar c}{48a^2}. \tag{2.25}$$

Below, we shall discuss many situations where the Casimir force can be both attractive and repulsive.

2.2 The Abel–Plana formula and regularization

Discrete sums and integrals with respect to a continuous variable, such as those in eqns (2.8) and (2.13), respectively, are of frequent occurrence in calculations of the Casimir effect. In some cases the handling of such quantities can be simplified with the help of the Abel–Plana formula (Erdélyi et al. 1981),

$$\sum_{n=0}^{\infty} F(n) - \int_0^{\infty} F(t)\,dt = \frac{1}{2}F(0) + i\int_0^{\infty} \frac{dt}{e^{2\pi t} - 1}\left[F(it) - F(-it)\right], \quad (2.26)$$

where $F(z)$ is an analytic function in the right half-plane. This formula was first applied to the theory of the Casimir effect by Mamayev et al. (1976).

To illustrate the utility of eqn (2.26), we apply it to the massive scalar Casimir effect on an interval. We start by setting

$$F(n) = \frac{\hbar}{2}\omega_n f(\omega_n, \delta), \quad (2.27)$$

where ω_n is determined in eqn (2.7). Here, $f(\omega, \delta)$ is some cutoff function which decreases monotonically sufficiently fast with increasing ω that both the sum and the integral in eqn (2.26) converge:

$$f(\omega, \delta) \to 0 \quad \text{when} \quad \omega \to \infty \quad \text{for all} \quad \delta \neq 0. \quad (2.28)$$

This function must also satisfy the conditions

$$f(\omega, \delta) \leq 1, \quad f(\omega, 0) = 1. \quad (2.29)$$

It is evident that in the limiting case $\delta \to 0$, the integral on the right-hand side of eqn (2.26) does not depend on the specific form of $f(\omega, \delta)$. This follows from the exponentially fast convergence of this integral, which permits taking the limit $\delta \to 0$ under the integral. Thus, one can simply omit the cutoff function in all calculations, as we do below. At the same time, the independence of the results obtained of the form of the cutoff function is automatically guaranteed. This is true for all applications of the Abel–Plana formula, and is not limited to the scalar Casimir effect on an interval.

As a result, by separating the term with $n = 0$, we obtain

$$\sum_{n=0}^{\infty} F(n) = \frac{mc^2}{2} + E_0(a, m), \quad (2.30)$$

where $E_0(a, m)$ is defined in eqn (2.8). In a similar manner, taking into account the change of variable $ak = \pi t$, we find

$$\int_0^{\infty} dt\, F(t) = \frac{E_{0M}(m)a}{L} = E_{0M}(a, m), \quad (2.31)$$

where $E_{0M}(m)$ is defined in eqn (2.13). Then the Casimir energy

$$E(a, m) \equiv E_0(a, m) - E_{0M}(a, m) \quad (2.32)$$

is found from the Abel–Plana formula (2.26) with the regularization already removed:

$$E(a, m) = -\frac{mc^2}{4} + i\frac{\pi \hbar c}{2a} \int_0^\infty \frac{dt}{e^{2\pi t} - 1} \left[G_A(it) - G_A(-it)\right]. \quad (2.33)$$

Here, the function $G_A(t)$ is defined by

$$G_A(t) \equiv \left(A^2 + t^2\right)^{1/2}, \qquad A \equiv \frac{mca}{\pi \hbar}. \quad (2.34)$$

It is useful to consider the more general function $G_A^{(\alpha)}(z)$, which is defined by

$$G_A^{(\alpha)}(z) = e^{\alpha \ln(A^2 + z^2)}. \quad (2.35)$$

This has branch points $z_{1,2} = \pm iA$. By going around the branch points, one can prove the equality

$$G_A^{(\alpha)}(it) - G_A^{(\alpha)}(-it) = 2i e^{\alpha \ln(t^2 - A^2)} \sin \pi \alpha \, \theta(t - A), \quad (2.36)$$

where $\theta(x)$ is the step function. For $\alpha = 1/2$, one obtains eqn (2.34) from eqn (2.35), and

$$G_A(it) - G_A(-it) = 2i \left(t^2 - A^2\right)^{1/2} \theta(t - A) \quad (2.37)$$

from eqn (2.36).

Substituting eqn (2.37) in eqn (2.33), one arrives at

$$E(a, m) = -\frac{mc^2}{4} - \frac{\hbar c}{4\pi a} \int_{2\mu}^\infty \frac{\sqrt{y^2 - 4\mu^2}}{e^y - 1} dy, \quad (2.38)$$

where $2\pi t \equiv y$ and $\pi A = mca/\hbar \equiv \mu$ (the latter parameter has the meaning of a dimensionless mass). The first contribution on the right-hand side of eqn (2.38) is associated with the total energy of the boundary points. It does not depend on a and hence does not contribute to the Casimir force.

For $\mu = 0$ ($m = 0$), eqn (2.38) leads to

$$E(a, 0) = E(a) = -\frac{\hbar c}{4\pi a} \int_0^\infty \frac{y \, dy}{e^y - 1} = -\frac{\pi \hbar c}{24a}, \quad (2.39)$$

in agreement with eqn (2.19). In the opposite case of large masses, $\mu \gg 1$, we get

$$E(a, m) \approx -\frac{mc^2}{4} - \frac{\hbar c \sqrt{\mu}}{4\sqrt{\pi} a} e^{-2\mu}, \quad (2.40)$$

i.e. the distance-dependent term is exponentially small. The same is obtained for a configuration of two parallel planes in three-dimensional space–time for massive fields with spins 0, 1/2, and 1. This is valid, however, only for plane boundaries. If some curvature is present, either in the boundary or in space–time, the Casimir energy may depend on the mass of the field in accordance with powers of some geometrical characteristic, such as the curvature radius.

For scalar fields with mixed or antiperiodic boundary conditions (see the previous and the next section, respectively) and also for spinor fields, a modification of the Abel–Plana formula is useful for summation over half-integer numbers:

$$\sum_{n=0}^{\infty} F\left(n+\frac{1}{2}\right) - \int_0^{\infty} F(t)\,dt = -\mathrm{i}\int_0^{\infty} \frac{dt}{\mathrm{e}^{2\pi t}+1}\left[F(\mathrm{i}t) - F(-\mathrm{i}t)\right]. \qquad (2.41)$$

Further generalizations of the Abel–Plana formula are discussed by Mostepanenko and Trunov (1997) and by Saharian (2006a).

2.3 The scalar Casimir effect on a circle

As noted in Section 1.1, when the topology of the space is nontrivial (i.e., non-Euclidean), identification conditions may be imposed on fields. These are similar to boundary conditions for classical material boundaries. The simplest example is provided by the interval $0 \leq x \leq a$ whose initial and end points are identified by means of the following periodic conditions:

$$\varphi(t,0) = \varphi(t,a), \qquad \partial_x \varphi(t,x)|_{x=0} = \partial_x \varphi(t,x)|_{x=a}. \qquad (2.42)$$

The geometrical image of an interval with the identification conditions (2.42) is a circle of circumference a. Both manifolds are flat, but their topologies are different. Interval I [Fig. 2.1(a)] possesses a Euclidean topology, whereas the same interval with the conditions (2.42) possesses the topology of a circle S^1 [Fig. 2.1(b)]. In all cases the scalar field satisfies eqns (2.1) and (2.2). Here, for S^1, in contrast to eqn (2.3), new solutions are allowed, such that $\varphi \neq 0$ at the points $x = 0, a$.

The complete orthonormal set of positive- and negative-frequency solutions of eqns (2.1) and (2.2) with the identification conditions (2.42) can be written in the form

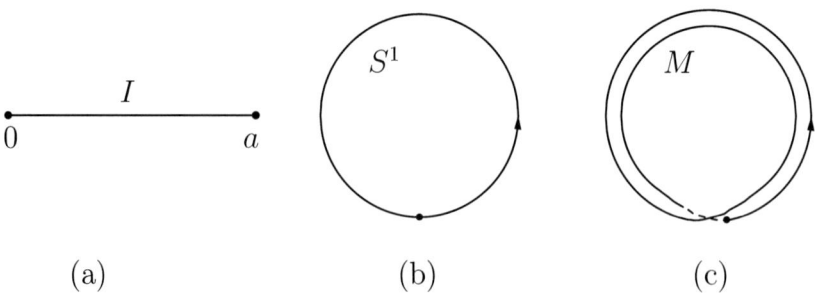

(a)   (b)   (c)

FIG. 2.1. Three one-dimensional flat manifolds, with (a) Euclidean topology, and the topology of a circle with (b) periodic and (c) antiperiodic identification conditions.

$$\varphi_n^{(\pm)}(t,x) = \left(\frac{c}{2a\omega_n}\right)^{1/2} e^{\mp i(\omega_n t - k_n x)}, \tag{2.43}$$

$$\omega_n = \left(\frac{m^2 c^4}{\hbar^2} + c^2 k_n^2\right)^{1/2}, \quad k_n = \frac{2\pi n}{a}, \quad n = 0, \pm 1, \pm 2, \ldots.$$

These solutions satisfy eqn (2.5). The only difference from eqn (2.7) is in the values of the wave numbers k_n. This leads to different oscillator frequencies than for the scalar field on the interval considered in Section 2.1.

The vacuum energy of the field $\varphi(t,x)$ on a circle S^1 is given by

$$E_0(a,m) = \frac{\hbar}{2} \sum_{n=-\infty}^{\infty} \omega_n = \hbar \sum_{n=0}^{\infty} \omega_n - \frac{\hbar}{2}\omega_0 = \hbar \sum_{n=0}^{\infty} \omega_n - \frac{mc^2}{2} \tag{2.44}$$

(here we have used the evenness of ω_n in n). In the same way, as in Section 2.2, the Casimir energy is obtained by subtracting the contribution of free Minkowski space within the length of the interval

$$E(a,m) = E_0(a,m) - E_{0M}(a,m), \tag{2.45}$$

where $E_{0M}(a,m)$ is defined in eqns (2.13) and (2.31).

Substituting eqns (2.44), (2.13), and (2.31) in eqn (2.45), we obtain

$$\begin{aligned} E(a,m) &= -\frac{mc^2}{2} + \hbar \left[\sum_{n=0}^{\infty} \omega_n - \frac{a}{2\pi}\int_0^\infty \omega_k \, dk\right] \\ &= -\frac{mc^2}{2} + \frac{2\pi \hbar c}{a}\left[\sum_{n=0}^{\infty} \sqrt{B^2 + n^2} - \int_0^\infty \sqrt{B^2 + t^2}\, dt\right], \end{aligned} \tag{2.46}$$

where $B \equiv mac/(2\pi\hbar)$ and $ak = 2\pi t$. The Casimir energy (2.46) can be calculated by using the Abel–Plana formula (2.26) and eqns (2.34) and (2.37), leading to the following result:

$$E(a,m) = -\frac{\hbar c}{\pi a}\int_\mu^\infty \frac{\sqrt{y^2 - \mu^2}}{e^y - 1}\, dy. \tag{2.47}$$

Here, $2\pi t \equiv y$ as in eqn (2.38) and the dimensionless mass is $\mu = mca/\hbar = 2\pi B$. It is notable that the term

$$\frac{1}{2}F(0) = \frac{2\pi \hbar c}{a}B = \frac{mc^2}{2} \tag{2.48}$$

on the right-hand side of the Abel–Plana formula cancels out the first term on the right-hand side of eqn (2.46). Because of this, eqn (2.47), in contrast to eqn (2.38), does not contain a contribution linear in the mass. The physical explanation for this fact is that the space with the topology of a circle does not contain boundary points and hence the vacuum energy does not contain their energy.

For a massless field, $\mu = 0$ and eqn (2.47) leads to (Mamayev and Trunov 1979a)

$$E(a,0) = E(a) = -\frac{\hbar c}{\pi a} \int_0^\infty \frac{y\,dy}{e^y - 1} = -\frac{\pi \hbar c}{6a}. \tag{2.49}$$

In the case of a large mass, $\mu \gg 1$, we obtain from eqn (2.47)

$$E(a,m) \approx -\frac{\hbar c \sqrt{\mu}}{\sqrt{2\pi a}} e^{-\mu}, \tag{2.50}$$

i.e. the Casimir energy is exponentially small.

At the end of this section, we briefly discuss what are referred to as *antiperiodic* conditions imposed on a scalar field,

$$\varphi(t, x+a) = -\varphi(t,x). \tag{2.51}$$

In the massless case, the allowed oscillator frequencies take the form

$$\omega_n = \frac{2\pi c}{a}\left(n + \frac{1}{2}\right), \quad n = 0, \pm 1, \pm 2, \ldots \tag{2.52}$$

[compare with eqn (2.22)]. The application of the Abel–Plana formula (2.41), adapted for summation over semi-integer numbers, results in the Casimir energy

$$E(a) = \frac{\pi \hbar c}{12a}. \tag{2.53}$$

We emphasize that for the antiperiodic conditions (2.51), the sign of the Casimir energy changes, similarly to the case for mixed boundary conditions. The periodic conditions (2.42) discussed at the beginning of this section can be presented as one equation,

$$\varphi(t, x+a) = \varphi(t,x), \tag{2.54}$$

similar to eqn (2.51). They were described geometrically by a circle of circumference a. If the antiperiodic conditions (2.51) are imposed on the field, one returns to the same field value

$$\varphi(t,x) = \varphi(t, x+2a) \tag{2.55}$$

only after two round trips, i.e. after traveling a distance $2a$ [see Fig. 1(c)]. Such a continuous line can be drawn on a Möbius strip. It is notable that a spinor wave function is antiperiodic and takes its initial value after two round trips, i.e. after a rotation by an angle 4π. Fields satisfying the condition (2.51) are often called *twisted*.

2.4 Local and global descriptions of the Casimir effect

In previous sections, the Casimir effect was characterized by the difference between the total vacuum energy in the presence of boundaries (or in a topologically nontrivial space) and the free, topologically trivial Minkowski space. Such an approach is called *global* because it deals with total energies. In this section we discuss another, *local*, approach to the Casimir effect, which starts from vacuum energy densities. In this case the total energy of the vacuum is obtained by the integration of the energy density over the quantization volume. The vacuum energy density is defined as the expectation value of the energy density operator of the quantized field in the vacuum state. Here, we present only the most elementary aspects of field quantization for a scalar field in two-dimensional space–time. We save the more general discussion of field quantization in the presence of boundaries for Chapter 3.

The quantization of a real scalar field on an interval $0 \leq x \leq a$ with boundary conditions (2.3) is performed through the replacement of a c-function field $\varphi(t,x)$ with the field operator

$$\varphi(t,x) = \sum_n \left[\varphi_n^{(+)}(t,x) a_n + \varphi_n^{(-)}(t,x) a_n^+ \right], \tag{2.56}$$

where the positive- and negative-frequency solutions of the field equation (2.1) are defined in eqns (2.6) and (2.7). The operators a_n and a_n^+ are the annihilation and creation operators of a scalar particle with quantum number n. They obey the standard commutation relations

$$\left[a_n, a_{n'}^+\right] = \delta_{nn'}, \qquad [a_n, a_{n'}] = \left[a_n^+, a_{n'}^+\right] = 0. \tag{2.57}$$

The vacuum state $|0\rangle$ of the field on an interval is defined by

$$a_n |0\rangle = 0. \tag{2.58}$$

The energy density operator of the scalar field in two-dimensional space–time is given by the 00-component of the energy–momentum tensor

$$T_{00}^{(0)}(t,x) = \frac{\hbar c}{2} \left\{ \frac{1}{c^2} [\partial_t \varphi(t,x)]^2 + [\partial_x \varphi(t,x)]^2 + \frac{m^2 c^2}{\hbar^2} \varphi^2(t,x) \right\}. \tag{2.59}$$

The infinite vacuum energy density of the field on an interval is given by the expectation value of the operator $T_{00}^{(0)}(t,x)$ in the vacuum state $|0\rangle$. It is calculated by the substitution of eqn (2.56) into eqn (2.59) using eqns (2.6), (2.57), and (2.58):

$$\langle 0 | T_{00}^{(0)}(t,x) | 0 \rangle = \frac{\hbar}{2a} \sum_{n=1}^{\infty} \omega_n - \frac{m^2 c^4}{2a\hbar} \sum_{n=1}^{\infty} \frac{\cos 2k_n x}{\omega_n}. \tag{2.60}$$

The total vacuum energy of the field φ on an interval is obtained by the integration of eqn (2.60):

$$E_0(a,m) = \int_0^a \langle 0|T_{00}(t,x)|0\rangle\, dx = \frac{\hbar}{2}\sum_{n=1}^{\infty}\omega_n. \qquad (2.61)$$

This is in agreement with eqn (2.8) [note that the second, oscillating, term on the right-hand side of eqn (2.60) does not contribute to the result].

We now consider the quantization of a scalar field on the entire axis $-\infty < x < \infty$. The field operator is given by

$$\varphi(t,x) = \int_{-\infty}^{\infty} dk\, \left[\varphi_k^{(+)}(t,x)a_k + \varphi_k^{(-)}(t,x)a_k^+\right], \qquad (2.62)$$

where the positive- and negative-frequency solutions are defined in eqn (2.11) and the commutation relations are as follows:

$$[a_k, a_{k'}^+] = \delta(k-k'), \qquad [a_k, a_{k'}] = [a_k^+, a_{k'}^+] = 0. \qquad (2.63)$$

The vacuum state of the scalar field on an unbounded axis is defined by the equality

$$a_k|0_M\rangle = 0. \qquad (2.64)$$

Substituting eqn (2.62) into eqn (2.59) and using eqn (2.63), we find the infinite vacuum energy density of the scalar field on the axis:

$$\langle 0_M|T_{00}^{(0)}(t,x)|0_M\rangle = \frac{\hbar}{4\pi}\int_{-\infty}^{\infty} dk\, \omega_k. \qquad (2.65)$$

Then the total vacuum energy for the whole axis is

$$E_{0M}(m) = \langle 0_M|T_{00}^{(0)}(t,x)|0_M\rangle L = \frac{\hbar}{2}\int_{-\infty}^{\infty}\frac{dk}{2\pi}\omega_k L, \qquad (2.66)$$

in agreement with eqn (2.13). We recall that L is the infinite length of the axis.

As a result, the Casimir energy density on an interval can be found using the local version of eqn (2.32),

$$\epsilon(x) = \langle 0|T_{00}^{(0)}(t,x)|0\rangle - \langle 0_M|T_{00}^{(0)}(t,x)|0_M\rangle, \qquad (2.67)$$

where the vacuum expectation values on the right-hand side of eqn (2.67) are given by eqns (2.60) and (2.65). From eqn (2.67) we obtain

$$\epsilon(x) = \frac{E(a,m)}{a} - \frac{m^2 c^4}{2a\hbar}\sum_{n=1}^{\infty}\frac{\cos 2k_n x}{\omega_n}, \qquad (2.68)$$

where $E(a,m)$ is defined in eqn (2.32) and, using eqn (2.38), we finally find

$$\epsilon(x) = -\frac{mc^2}{4a} - \frac{\hbar c}{4\pi a^2}\int_{2\mu}^{\infty}\frac{\sqrt{y^2-4\mu^2}}{e^y-1}dy - \frac{m^2 c^4}{2a\hbar}\sum_{n=1}^{\infty}\frac{\cos 2k_n x}{\omega_n}. \qquad (2.69)$$

Integration of eqn (2.69) with respect to x leads to

$$\int_0^a \epsilon(x)\,dx = E(a,m), \tag{2.70}$$

where the total Casimir energy on an interval, $E(a,m)$, is given in eqn (2.38). This is, however, true only for flat boundaries. In the case of curved boundary surfaces, there may be nonintegrable singularities in the Casimir energy density when the boundary surface is approached (see e.g. Chapter 9).

2.5 Elementary approach to the Casimir force between two parallel planes

As discussed in the Introduction, the Casimir effect is the attractive force acting between parallel ideal-metal planes which arises from vacuum oscillations of the electromagnetic field. In Chapter 7, we shall present the detailed theory of the Casimir effect between ideal-metal planes for various fields at both zero and nonzero temperature. However, it is appropriate to include an elementary derivation of eqns (1.1) and (1.5) in this chapter, which is devoted to simple models.

A configuration of two parallel planes of very large area S spaced a distance a apart is shown in Fig. 2.2. Mathematically, the area S of each plane is supposed to be infinitely large. However, the results obtained are applicable for the condition $a \ll \sqrt{S}$. From classical electrodynamics, the electric field and the magnetic induction, of both polarizations, satisfy the following boundary conditions on the surface of an ideal metal:

$$\boldsymbol{E}_\mathrm{t}(t,\boldsymbol{r})|_S = \boldsymbol{B}_\mathrm{n}(t,\boldsymbol{r})|_S = 0. \tag{2.71}$$

Here, $\boldsymbol{r}$ is the radius vector of any point, $\boldsymbol{n}$ is the unit vector normal to the surface, and the index "t" denotes the tangential component, which is parallel to

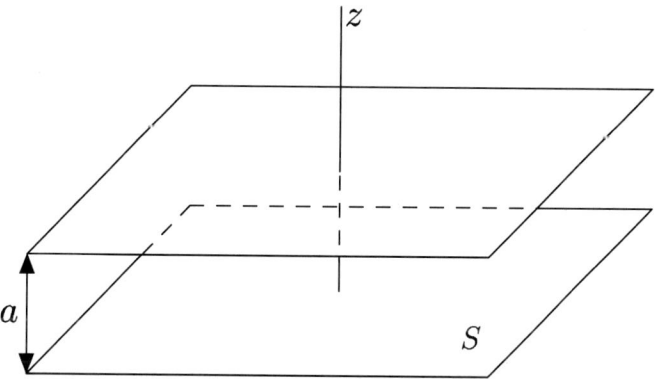

FIG. 2.2. Two parallel ideal-metal planes of area S spaced a distance a apart.

the planes. The conditions (2.71) imply that an electromagnetic field can exist only outside an ideal conductor.

The boundary conditions (2.71) can be viewed as an ideal case of the interaction of real metal surfaces with an electromagnetic field. In general, this interaction is much more complicated and is modified by the finite conductivity of a metal, i.e. by the penetration of the electromagnetic field into the metal to a characteristic length scale given by the *skin depth*. This problem becomes particularly involved at nonzero temperature (see Chapters 12 and 14).

We consider the electromagnetic field as an infinite set of harmonic oscillators with frequencies ω_J and vacuum energy (1.4), where J is the photon wave vector. In free space (i.e. without boundaries), $J = \boldsymbol{k} = (k^1, k^2, k^3)$, where all k^i are continuous. In the presence of metal planes, this is not so, however. Here and below, we shall also use the notation $x^1 \equiv x$, $x^2 \equiv y$, $x^3 \equiv z$ and $k^1 \equiv k_x$, $k^2 \equiv k_y$, $k^3 \equiv k_z$ where there is no cause for confusion with four-dimensional quantities. Let us choose Cartesian coordinates with the z-axis perpendicular to the planes. Then the components k_x, k_y remain continuous, but the component $k_z = k_{zn} = \pi n/a$, $n = 0, \pm 1, \pm 2$, becomes discrete. Note that in contrast to eqn (2.7) for a scalar field on an interval, here n can be a negative integer, which takes the two photon polarizations into account. The wave vector of the form $(k_x, k_y, 0)$ also leads to a nonzero contribution (see Section 7.2, where the complete orthonormal set of solutions of the wave equation between the two parallel planes is explicitly presented).

As a result, the vacuum energy of the electromagnetic field between the ideal-metal planes can be presented in the form

$$E_0(a) = \frac{\hbar}{2} \int_{-\infty}^{\infty} \frac{dk_x}{2\pi} \int_{-\infty}^{\infty} \frac{dk_y}{2\pi} \sum_{n=-\infty}^{\infty} \omega_{k_\perp, n} S. \qquad (2.72)$$

Here, $\boldsymbol{k}_\perp = (k_x, k_y)$ is the projection of the wave vector onto the metal planes (it is perpendicular to the z-direction, i.e. to the direction of the Casimir force), $k_\perp = |\boldsymbol{k}_\perp| = (k_x^2 + k_y^2)^{1/2}$, and the oscillator frequencies are given by

$$\omega_J = \omega_{k_\perp, n} = c\sqrt{k_\perp^2 + \left(\frac{\pi n}{a}\right)^2}. \qquad (2.73)$$

By introducing polar coordinates on the planes and noting that $\omega_{k_\perp, n}$ is an even function of n, we can rearrange eqn (2.72) to

$$E_0(a) = \frac{\hbar}{2} \int_0^\infty \frac{k_\perp dk_\perp}{2\pi} \left(2 \sum_{n=0}^{\infty} \omega_{k_\perp, n} - ck_\perp \right) S. \qquad (2.74)$$

The respective vacuum energy of the electromagnetic field in the free Minkowski space in the volume between the planes but with no boundary conditions is given by

… Elementary approach to the Casimir force between two parallel planes …

$$E_{0M}(a) = \hbar a \int_{-\infty}^{\infty} \frac{dk_x}{2\pi} \int_{-\infty}^{\infty} \frac{dk_y}{2\pi} \int_{-\infty}^{\infty} \frac{dk_z}{2\pi} \omega_k S, \qquad (2.75)$$

where

$$\omega_k = c|\mathbf{k}| = c\sqrt{k_x^2 + k_y^2 + k_z^2} \qquad (2.76)$$

and a factor of 2 has been used to take into account the two polarizations of the electromagnetic field. Equation (2.75) can be rearranged to

$$E_{0M}(a) = \frac{\hbar a}{\pi} \int_0^{\infty} \frac{k_\perp dk_\perp}{2\pi} \int_0^{\infty} dk_z\, \omega_k S. \qquad (2.77)$$

From eqns (2.74) and (2.77) the Casimir energy per unit area of the ideal-metal planes is defined as

$$E(a) \equiv \frac{E_0(a)}{S} - \frac{E_{0M}(a)}{S} \qquad (2.78)$$

$$= \hbar \int_0^{\infty} \frac{k_\perp dk_\perp}{2\pi} \left(\sum_{n=0}^{\infty} \omega_{k_\perp,n} - \frac{ck_\perp}{2} - \frac{a}{\pi}\int_0^{\infty} dk_z\, \omega_k \right).$$

Substituting eqns (2.73) and (2.76) here and introducing a new variable $t = ak_z/\pi$, we arrive at

$$E(a) = \frac{\pi \hbar c}{a} \int_0^{\infty} \frac{k_\perp dk_\perp}{2\pi} \left(\sum_{n=0}^{\infty} \sqrt{\frac{k_\perp^2 a^2}{\pi^2} + n^2} \right. \qquad (2.79)$$

$$\left. - \int_0^{\infty} dt\, \sqrt{\frac{k_\perp^2 a^2}{\pi^2} + t^2} - \frac{ak_\perp}{2\pi} \right).$$

The difference between the infinite sum and the infinite integral in eqn (2.79) is calculated using the Abel–Plana formula (2.26) with $F = G_A(t)$ defined in eqn (2.34) and $A \equiv k_\perp^2 a^2/\pi^2$. The application of this formula leads to the Casimir energy between the planes,

$$E(a) = -\frac{\pi^2 \hbar c}{a^3} \int_0^{\infty} y\, dy \int_y^{\infty} \frac{\sqrt{t^2-y^2}}{e^{2\pi t} - 1} dt, \qquad (2.80)$$

where the dimensionless variable $y = k_\perp a/\pi$ has been introduced instead of $k_\perp$. It is notable that $E(a)$ is finite. As in Section 2.2, the result (2.80) does not depend on the specific form of the regularization that might be applied to make the quantities $E_0(a)/S$ and $E_{0M}(a)/S$ in eqn (2.78) finite.

To evaluate eqn (2.80), it is sufficient to change the order of the integration:

$$E(a) = -\frac{\pi^2 \hbar c}{a^3} \int_0^{\infty} \frac{dt}{e^{2\pi t} - 1} \int_0^t y\sqrt{t^2 - y^2}\, dy$$

$$= -\frac{\pi^2 \hbar c}{3a^3} \frac{1}{(2\pi)^4} \int_0^\infty \frac{v^3\, dv}{e^v - 1}, \qquad (2.81)$$

where one more new variable, $v = 2\pi t$, has been introduced. After integration of eqn (2.81), we finally obtain (Gradshtein and Ryzhik 1994)

$$E(a) \equiv E_{\text{IM}}(a) = -\frac{\pi^2}{720} \frac{\hbar c}{a^3}, \qquad (2.82)$$

in agreement with eqn (1.5). We can use eqn (1.6) to reproduce the Casimir result (1.1) for the Casimir pressure.

Thus, in the case of the electromagnetic vacuum confined between two parallel ideal-metal planes, the final result is obtained by deleting the energy of free Minkowski space using the Abel–Plana formula. This is appropriate for the case under consideration. In more complicated configurations, however, even after the removal of the contribution from free Minkowski space, the result will in general be infinite, so that some additional renormalization has to be carried out.

3

FIELD QUANTIZATION AND VACUUM ENERGY IN THE PRESENCE OF BOUNDARIES

From the standpoint of quantum field theory, the Casimir effect is related to the vacuum polarization arising in quantization volumes restricted by boundaries or in spaces with nontrivial topology. Both boundaries and a nontrivial topology of space–time can be considered as classical external conditions, on which background the field quantization should be performed. In this chapter we briefly present some basic facts related to the quantization procedure for fields of various spin obeying boundary (or identification) conditions. We start with the classical wave equations and then consider various boundary conditions that may be imposed on their solutions. The rest of the chapter is devoted to both the canonical and the path-integral field quantization procedure in the presence of boundaries, with stress put on several different representations of the vacuum energy. Propagators with boundary conditions are also introduced. Although we touch on fields of different spin in both Minkowski and curved space–time, our presentation is primarily devoted to the case of the electromagnetic field in the presence of material boundaries, which is the main subject in the second and third parts of this book.

We recall that, starting from this chapter and throughout Part I of the book, we use units with $\hbar = c = 1$.

3.1 Field equations for fields of various spins

In the preceding chapter, we have already considered a scalar field in two-dimensional space–time. Here, all fields are defined in four-dimensional space–time, with 4-vector arguments $x \equiv x^\mu = \{x^0, x^1, x^2, x^3\} = \{x^0, \boldsymbol{r}\}$ and $x_\mu = g_{\mu\nu} x^\nu = \{x_0, x_1, x_2, x_3\} = \{x^0, -\boldsymbol{r}\}$. The Klein–Fock–Gordon equation for a free real scalar field $\varphi(x)$ is given by

$$\left(\Box + m^2\right) \varphi(x) = 0, \qquad (3.1)$$

where the four-dimensional d'Alembert operator is defined by

$$\Box = \Box_4 \equiv g^{\mu\nu} \frac{\partial^2}{\partial x^\mu \partial x^\nu} = \frac{\partial^2}{\partial x^0 \partial x^0} - \frac{\partial^2}{\partial x^1 \partial x^1} - \frac{\partial^2}{\partial x^2 \partial x^2} - \frac{\partial^2}{\partial x^3 \partial x^3} = \partial^\nu \partial_\nu \quad (3.2)$$

and m is the mass of the field. If some external source $\Upsilon(x)$ of the scalar field is present (this is not the case for scalar zero-point oscillations), eqn (3.1) is generalized to

$$\left(\Box + m^2\right) \varphi(x) = \Upsilon(x). \qquad (3.3)$$

Both eqn (3.1) and eqn (3.3) can be obtained from the Euler–Lagrange equations from the action of a scalar field

$$S[\varphi] = \int d^4x \, \mathcal{L}^{(0)}(x) = \int d^4x \left(\frac{1}{2} \partial^\nu \varphi \partial_\nu \varphi - \frac{m^2}{2} \varphi^2 + \Upsilon \varphi \right) \tag{3.4}$$

with $\Upsilon = 0$ and $\Upsilon \neq 0$, respectively [$\mathcal{L}^{(0)}(x)$ is the Lagrangian density]. Using integration by parts in eqn (3.4), one can present the action in the form

$$S[\varphi] = \int d^4x \left(-\frac{1}{2} \varphi K \varphi + \Upsilon \varphi \right), \tag{3.5}$$

where the operator

$$K \equiv K(x) = \Box + m^2 \tag{3.6}$$

is the kernel of the action.

The scalar product of the two (in general complex) solutions of eqn (3.1) is defined as

$$(\varphi_1, \varphi_2) = i \int_V dr \left(\varphi_1^* \frac{\partial \varphi_2}{\partial x_0} - \frac{\partial \varphi_1^*}{\partial x_0} \varphi_2 \right), \tag{3.7}$$

where V is the quantization volume (free infinite space or some region restricted by boundaries).

From Noether's theorem, the canonical energy–momentum tensor of the scalar field (without sources) is

$$T^{(0)}_{\mu\nu} = \partial_\mu \varphi \partial_\nu \varphi - g_{\mu\nu} L^{(0)}, \tag{3.8}$$

where $L^{(0)}$ is defined in eqn (3.4), with $\Upsilon = 0$. Sometimes what is referred to as the *metrical* energy–momentum tensor of the scalar field is also used (Chernikov and Tagirov 1968, Callan *et al.* 1970),

$$\tilde{T}^{(0)}_{\mu\nu} = T^{(0)}_{\mu\nu} - \xi \left[\partial_\mu \partial_\nu - g_{\mu\nu} \partial_\rho \partial^\rho \right] \varphi^2, \tag{3.9}$$

where $\xi = (D-2)/4(D-1)$, which differs from eqn (3.8) by a 4-divergence (D is the dimensionality of space–time).

Equation (3.1) can be generalized to the case of curved space–time in the form

$$\left(\Box + \xi R + m^2 \right) \varphi(x) = 0, \tag{3.10}$$

where $\Box = \nabla_\mu \nabla^\mu$, ∇_μ is the covariant derivative, the scalar curvature $R = R^\mu_\mu$ is the trace of the Ricci tensor, and ξ is the coupling coefficient. Equation (3.10) is conformally invariant in the limit of zero mass if ξ in the above expression is given by $\xi = 1/6$ (for $D = 4$). The case of curved space–times will be considered only in Chapter 11 and will not be discussed here.

The most important case for us is the electromagnetic field because the electromagnetic Casimir effect is experimentally observable. Here, we restrict the discussion to only the electromagnetic field in the vacuum. The case of material media is dealt with in the second part of the book. We start with the Maxwell equations for the electric field $\boldsymbol{E}(x)$ and magnetic induction $\boldsymbol{B}(x)$ (with $c = 1$):

$$\nabla \cdot \boldsymbol{E} = 4\pi\rho, \qquad \nabla \times \boldsymbol{E} + \frac{\partial \boldsymbol{B}}{\partial t} = 0, \qquad (3.11)$$

$$\nabla \times \boldsymbol{B} - \frac{\partial \boldsymbol{E}}{\partial t} = 4\pi\boldsymbol{j}, \qquad \nabla \cdot \boldsymbol{B} = 0,$$

where $\boldsymbol{j}$ and ρ are current and charge densities satisfying the local charge conservation law

$$\frac{\partial \rho}{\partial t} + \mathrm{div}\boldsymbol{j} = 0. \qquad (3.12)$$

Later, when we consider the electromagnetic zero-point energy, we shall put $\rho = 0$ and $\boldsymbol{j} = 0$.

Electrodynamics is a relativistically invariant theory. Because of this, it is often convenient to represent the physical fields $\boldsymbol{E}$ and $\boldsymbol{B}$ in terms of the 4-potential $A^\mu = (A^0, \boldsymbol{A})$ such that

$$\boldsymbol{E} = -\nabla A^0 - \frac{\partial \boldsymbol{A}}{\partial t}, \qquad \boldsymbol{B} = \nabla \times \boldsymbol{A}, \qquad (3.13)$$

and to introduce the antisymmetric field tensor

$$F^{\mu\nu} = \partial^\mu A^\nu - \partial^\nu A^\mu. \qquad (3.14)$$

The components of this tensor are the components of the electric field and magnetic induction:

$$F^{0i} = -E^i, \qquad F^{12} = -B^3, \qquad F^{13} = B^2, \qquad F^{23} = -B^1. \qquad (3.15)$$

Introducing also the 4-current $j^\mu = (\rho, \boldsymbol{j})$, one can rearrange Maxwell's equations (3.11) into the covariant form

$$\partial_\mu F^{\mu\nu} = 4\pi j^\nu, \qquad \partial_\mu \tilde{F}^{\mu\nu} = 0, \qquad (3.16)$$

where the dual tensor is

$$\tilde{F}^{\mu\nu} = \frac{1}{2} \varepsilon^{\mu\nu\beta\gamma} F_{\beta\gamma}, \qquad (3.17)$$

and $\varepsilon^{\mu\nu\beta\gamma}$ is the antisymmetric tensor equal to $+1$ or -1 depending on whether $(\mu, \nu, \beta, \gamma)$ is an even or odd transposition of the indices $(0, 1, 2, 3)$. The representation in terms of the 4-potential A^μ is especially useful for the quantization of the electromagnetic field.

Equations (3.13) and (3.14) do not define the 4-potential A^μ in a unique way, leaving the freedom for a gauge transformation

$$A^\mu(x) \to A^\mu(x) + \partial^\mu \phi(x), \qquad (3.18)$$

where $\phi(x)$ is some arbitrary smooth function. In terms of the 4-potential, the Maxwell equations (3.11) or (3.16) can be rewritten as

$$\Box A^\mu - \partial^\mu (\partial_\nu A^\nu) = 4\pi j^\mu. \tag{3.19}$$

The form of these equations is preserved under the gauge transformation (3.18). However, under some conditions that fix the gauge, a certain term in eqn (3.19) may vanish. The only relativistically invariant condition is

$$\partial_\nu A^\nu = 0, \tag{3.20}$$

which is called the *Lorentz gauge*. It is always possible to find a set of functions $\phi(x)$ in eqn (3.18) such that the condition (3.20) is satisfied. Functions belonging to this set differ from one another by a function ϕ_0 satisfying the equation $\Box \phi_0 = 0$. In the Lorentz gauge, eqn (3.19) takes the simplest form,

$$\Box A^\mu = 4\pi j^\mu. \tag{3.21}$$

For us, a particular case of these equations with the source equal to zero is most important,

$$\Box A^\mu = 0. \tag{3.22}$$

The Maxwell equations (3.19) can be obtained from the action of the electromagnetic field

$$S = \int d^4x \, \mathcal{L}^{(1)}(x) = \int d^4x \left[-\frac{1}{16\pi} F_{\mu\nu} F^{\mu\nu} - \frac{\lambda}{8\pi} (\partial_\mu A^\mu)^2 - A_\mu j^\mu \right], \tag{3.23}$$

where λ is a coefficient of what is referred to as the *gauge-fixing term*. The Euler–Lagrange equations following from eqn (3.23) are

$$\Box A^\mu - (1-\lambda) \partial^\mu (\partial_\nu A^\nu) = 4\pi j^\mu. \tag{3.24}$$

If $\lambda = 0$, eqn (3.24) coincides with eqn (3.19). The choice of $\lambda = 1$ is equivalent to the fixing of the Lorentz gauge (3.20). In this case eqn (3.24) coincides with eqn (3.21). After integration by parts, eqn (3.23) can be rearranged into the form

$$S = \int d^4x \left(\frac{1}{8\pi} A_\mu K^{\mu\nu} A_\nu - A_\mu j^\mu \right), \tag{3.25}$$

where the kernel of the action is given by the differential operator

$$K^{\mu\nu} \equiv K^{\mu\nu}(x) = g^{\mu\nu} \Box - (1-\lambda) \partial^\mu \partial^\nu. \tag{3.26}$$

The energy–momentum tensors of the electromagnetic field obtained from different forms of the action may differ by a 4-divergence. For example, by applying Noether's theorem to the action (3.23) with $\lambda = 0$, $j^\mu = 0$, we obtain

$$\tilde{T}^{(1)}_{\mu\nu} = -\frac{1}{4\pi} F_{\mu\beta} \partial_\nu A^\beta + \frac{1}{16\pi} g_{\mu\nu} F_{\beta\gamma} F^{\beta\gamma}. \tag{3.27}$$

This expression is not gauge invariant and not symmetric, i.e. $\tilde{T}^{(1)}_{\mu\nu} \neq \tilde{T}^{(1)}_{\nu\mu}$. By adding to $\tilde{T}^{(1)}_{\mu\nu}$ terms having the form of the 4-divergence $\partial_\beta (F_\mu{}^\beta A_\nu)$, it is possible to obtain a gauge-invariant, symmetric energy–momentum tensor of the electromagnetic field (Itzykson and Zuber 2005)

$$T^{(1)}_{\mu\nu} = \frac{1}{4\pi}\left(F_{\mu\beta}F^\beta{}_\nu + \frac{1}{4}g_{\mu\nu}F_{\beta\gamma}F^{\beta\gamma}\right). \tag{3.28}$$

For the 00-component of this tensor, using eqn (3.15), we obtain the familiar expression in terms of the electric field and the magnetic induction

$$T^{(1)}_{00} = \frac{\boldsymbol{E}^2 + \boldsymbol{B}^2}{8\pi}. \tag{3.29}$$

At the end of this section, we mention briefly the main facts related to a spinor field of mass m with spin one-half. This field obeys the Dirac equation

$$(i\gamma^\mu \partial_\mu - m)\psi(x) = 0, \tag{3.30}$$

where the γ^μ are 4×4 Dirac matrices satisfying the condition

$$\gamma^\mu\gamma^\nu + \gamma^\nu\gamma^\mu \equiv \{\gamma^\mu, \gamma^\nu\} = 2g^{\mu\nu} \tag{3.31}$$

and $\psi(x)$ is a 4-component bispinor. Hereafter, we shall use the representation of Dirac matrices where the matrix γ_0 is diagonal, i.e.

$$\gamma_0 = \begin{pmatrix} I & 0 \\ 0 & -I \end{pmatrix}, \quad \gamma_k = \begin{pmatrix} 0 & \sigma_k \\ -\sigma_k & 0 \end{pmatrix}, \tag{3.32}$$

and the σ_k are the Pauli matrices (I is a 2×2 unit matrix).

The action for the Dirac equation is given by

$$S[\psi] = \int d^4x\, \mathcal{L}^{(1/2)}(x), \tag{3.33}$$

where the Lagrangian density is

$$\mathcal{L}^{(1/2)} = \frac{i}{2}\left[\bar\psi\gamma^\mu\partial_\mu\psi - (\partial_\mu\bar\psi)\gamma^\mu\psi\right] - m\bar\psi\psi. \tag{3.34}$$

Here, $\bar\psi = \psi^+\gamma^0$ is the Dirac conjugate bispinor.

The scalar product of two solutions of the Dirac equation takes the form

$$(\psi_1, \psi_2) = \int_V d\boldsymbol{r}\, \psi^+\psi, \tag{3.35}$$

where V is the quantization volume (the whole space or some finite region restricted by boundaries).

The energy–momentum tensor of the spinor field is given by

$$T^{(1/2)}_{\mu\nu} = \frac{i}{2}\left(\bar\psi\gamma_\mu\partial_\nu\psi - \partial_\nu\bar\psi\gamma_\mu\psi\right). \tag{3.36}$$

It is notable that all of the energy–momentum tensors obtained above in the absence of sources satisfy the conservation law

$$\partial^\mu T^{(s)}_{\mu\nu} = 0, \tag{3.37}$$

where $s = 0, 1/2$, or 1 is the spin of the field.

3.2 Various boundaries and boundary conditions

In Part I of the book, we consider the Casimir effect in regions of space restricted by ideally smooth boundaries. Mathematically, for scalar and electromagnetic fields, the problem includes consideration of the field equation (3.1) in the scalar case and the field equation (3.11) with $\boldsymbol{j} = 0$ and $\rho = 0$, or (3.22), in the electromagnetic case, with appropriate boundary conditions. The boundaries S are considered to be stationary. This allows the separation of the time variable. For example, for the scalar field

$$\varphi_J^{(+)}(x) = \frac{1}{\sqrt{2\omega_J}} e^{-i\omega_J t} \Phi_J(\boldsymbol{r}) \tag{3.38}$$

using eqn (3.1) we get

$$-\boldsymbol{\nabla}^2 \Phi_J(\boldsymbol{r}) = \Lambda_J \Phi_J(\boldsymbol{r}), \qquad \Lambda_J \equiv \omega_J^2 - m^2. \tag{3.39}$$

Here, $\boldsymbol{\nabla}^2 = \Delta$ is the Laplace operator

$$\boldsymbol{\nabla}^2 \equiv \boldsymbol{\nabla}_{(3)}^2 = \frac{\partial^2}{\partial x_1^2} + \frac{\partial^2}{\partial x_2^2} + \frac{\partial^2}{\partial x_3^2}, \tag{3.40}$$

J is a collective index for the generalized wave vector, and $\Phi_J(\boldsymbol{r})$ satisfies some boundary conditions on the surfaces S. Equation (3.39) together with the boundary conditions imposed on the function $\Phi_J(\boldsymbol{r})$ defines a standard elliptic problem for the self-adjoint operator Δ.

The most frequently used boundary condition is the Dirichlet one,

$$\Phi_J(\boldsymbol{r})|_S = 0. \tag{3.41}$$

Physically, this means that the boundary surface is totally impermeable to the field. The Dirichlet boundary condition can be imposed on several surfaces, i.e. on two parallel planes or on the four sides of a parallelepiped. In order to solve the Dirichlet problem (3.39), (3.41), one must find an explicit expression for the eigenfrequencies ω_J.

If the normal derivative of the function $\Phi_J(\boldsymbol{r})$ on the boundary surface vanishes, i.e.

$$\left.\frac{\partial \Phi_J(\boldsymbol{r})}{\partial n}\right|_S = 0, \tag{3.42}$$

we are dealing with Neumann boundary condition. Together, eqns (3.39) and (3.42) are called the Neumann boundary problem.

A combination of the Dirichlet and Neumann boundary conditions

$$\left[u\Phi_J(\boldsymbol{r}) + \frac{\partial \Phi_J(\boldsymbol{r})}{\partial n}\right]\bigg|_S = 0, \tag{3.43}$$

where u is some parameter or a function of the radius vector, is called a Robin boundary condition.

For the electromagnetic field in the Lorentz gauge (3.20), the field equation (3.21) has the form of eqn (3.1) with $m = 0$. After separation of the time variable, i.e.

$$A^{(+)}_{J,\mu}(x) = \frac{1}{\sqrt{\omega_J}} e^{-i\omega_J t} \mathcal{A}_{J,\mu}(r), \tag{3.44}$$

we arrive at the same equation as in eqn (3.39),

$$-\nabla^2 \mathcal{A}_{J,\mu}(r) = \Lambda_J \mathcal{A}_{J,\mu}(r), \qquad \Lambda_J \equiv \omega_J^2. \tag{3.45}$$

If the boundary surfaces S are made of an ideal metal, the boundary conditions (2.71) must be satisfied at each point of S. These conditions can be equivalently rewritten as

$$n_\mu \tilde{F}^{\mu\nu}(t,r)\Big|_S = 0, \tag{3.46}$$

where $n_\mu = (0, -\boldsymbol{n})$, $\boldsymbol{n}$ is the external normal to the surface at a point $\boldsymbol{r}$, and the dual tensor $\tilde{F}^{\mu\nu}$ is defined in eqn (3.17).

For electromagnetic fields without sources, it is always possible to fix the vector potential such that

$$\mathcal{A}_J^0(r) = 0, \qquad \operatorname{div} \mathcal{A}_J(r) = 0. \tag{3.47}$$

This is usually referred to as the *Coulomb gauge*. In the Coulomb gauge, the first of the boundary conditions (2.71), $\boldsymbol{E}_t(t,\boldsymbol{r})|_S = 0$, results in

$$\mathcal{A}_{Jt}(r)|_S = 0, \tag{3.48}$$

where the index "t" marks the components of $\boldsymbol{E}$ and $\mathcal{A}_J$ tangential to the surface. To obtain eqn (3.48), we have used the first equality in eqn (3.13), and eqn (3.44).

From eqn (3.48), using the second equality in eqn (3.13), it follows that $B_n(t,\boldsymbol{r})|_S = 0$, which is the second boundary condition in eqn (2.71). Here, we have assumed that both the electric field and the magnetic induction vary sinusoidally in time as $\exp(-i\omega_J t)$, which is always true for any static configuration of boundary surfaces.

The boundary condition (3.48) is of Dirichlet type. Thus, the same elliptic boundary problem as in the case of a scalar field is relevant to the electromagnetic field. In the next chapters of Part I, the solutions of various boundary problems will be presented for a number of configurations of boundary surfaces. We shall also discuss cases where a complete solution of such problems for the electromagnetic field is not yet known.

At the end of this section, we note that for a spinor field the Dirichlet boundary condition is not meaningful, because it is in contradiction with the Dirac equation (3.30). Instead, what are referred to as *bag boundary conditions* are used, which prevent a current from flowing through the boundary (see Sections 7.5 and 9.4 for details).

In this section, we restrict ourselves to the case of ideal boundaries. In the second part of the book, nonideal boundary surfaces consisting of real dielectrics, semiconductors, and metals will be considered and more complicated boundary conditions will be discussed.

3.3 Canonical quantization and the vacuum energy as a mode expansion

The quantization procedure for a scalar field in two-dimensional space–time has already been illustrated in Section 2.4. Here, we begin with a scalar field in four-dimensional space–time satisfying the Klein–Fock–Gordon equation (3.1). For smooth static boundaries of any geometrical shape, it is always possible to introduce the positive- and negative-frequency solutions of the Klein–Fock–Gordon equation

$$\varphi_J^{(+)}(t,\mathbf{r}) = \frac{1}{\sqrt{2\omega_J}} e^{-i\omega_J t} \Phi_J(\mathbf{r}), \qquad \varphi_J^{(-)}(t,\mathbf{r}) = \left[\varphi_J^{(+)}(t,\mathbf{r})\right]^*, \qquad (3.49)$$

where $\Phi_J(\mathbf{r})$ is the solution of the elliptic boundary problem [i.e. of eqn (3.39) with one of the boundary conditions (3.41)–(3.43)]. The functions (3.49) satisfy the normalization conditions

$$\left(\varphi_J^{(\pm)}(x), \varphi_{J'}^{(\pm)}(x)\right) = \pm \delta_{JJ'}, \qquad \left(\varphi_J^{(\pm)}(x), \varphi_{J'}^{(\mp)}(x)\right) = 0, \qquad (3.50)$$

where the scalar product is defined in eqn (3.7). From eqn (3.7), we also obtain the normalization condition for the solutions $\Phi_J(\mathbf{r})$ of the boundary problem,

$$\int_V d\mathbf{r} \, \Phi_J^*(\mathbf{r}) \Phi_{J'}(\mathbf{r}) = \delta_{JJ'}. \qquad (3.51)$$

Following the procedure of canonical quantization, we present the field operator as the sum of the modes

$$\varphi(x) = \sum_J \left[\varphi_J^{(+)}(x) a_J + \varphi_J^{(-)}(x) a_J^+\right], \qquad (3.52)$$

where a_J and a_J^+ are the annihilation and creation operators of a particle with quantum numbers indicated by the collective index J. The summation over J may also mean integration if some (or all) of the quantum numbers are continuous. The annihilation and creation operators satisfy the commutation relations

$$[a_J, a_{J'}^+] = \delta_{JJ'}, \qquad [a_J, a_{J'}] = [a_J^+, a_{J'}^+] = 0. \qquad (3.53)$$

The vacuum state of the field is defined by

$$a_J |0\rangle = 0. \qquad (3.54)$$

The states with particles are obtained by applying the creation operators to the vacuum state. For example, the state with one particle is

$$|1\rangle = a_J^+|0\rangle. \tag{3.55}$$

In the case where there are no boundary conditions (i.e. when we consider the quantized scalar field in free Minkowski space), the index J coincides with the wave vector, i.e. $J \equiv \boldsymbol{k} = (k^1, k^2, k^3)$, the oscillator frequencies are given by $\omega_J = \omega_{\boldsymbol{k}} = (m^2 + \boldsymbol{k}^2)^{1/2}$, and

$$\Phi_J(\boldsymbol{r}) = \Phi_{\boldsymbol{k}}(\boldsymbol{r}) = \frac{e^{i\boldsymbol{k}\cdot\boldsymbol{r}}}{(2\pi)^{3/2}}. \tag{3.56}$$

In this case the symbol $\delta_{JJ'}$ in eqns (3.50), (3.51), and (3.53) should be understood as $\delta^3(\boldsymbol{k}-\boldsymbol{k}')$. The vacuum state of the scalar field in free Minkowski space is defined by

$$a_{\boldsymbol{k}}|0_\mathrm{M}\rangle = 0. \tag{3.57}$$

The vacuum energy density of the scalar field in the presence of boundaries is the mean value of the 00-component of the energy–momentum tensor (3.8) in the vacuum state,

$$\langle 0|T_{00}^{(0)}(x)|0\rangle = \frac{1}{2}\left\langle 0\left|\left[\sum_{\mu=0}^{3}\left(\frac{\partial\varphi}{\partial x^\mu}\right)^2 + m^2\varphi^2\right]\right|0\right\rangle. \tag{3.58}$$

Substituting eqn (3.52) in eqn (3.58) and using eqns (3.49), (3.53), and (3.54), we obtain

$$\langle 0|T_{00}^{(0)}(x)|0\rangle = \sum_J \frac{1}{4\omega_J}\left[\left(\omega_J^2 + m^2\right)\Phi_J(\boldsymbol{r})\Phi_J^*(\boldsymbol{r}) + \sum_{k=1}^{3}\frac{\partial\Phi_J(\boldsymbol{r})}{\partial x^k}\frac{\partial\Phi_J^*(\boldsymbol{r})}{\partial x^k}\right]. \tag{3.59}$$

This energy density is divergent and, in the general case, depends on the spatial point $\boldsymbol{r}$.

Now we consider the total vacuum energy of the scalar field in the quantization volume V. We assume that the functions $\Phi_J(\boldsymbol{r})$ satisfy the Dirichlet or Neumann boundary condition (3.41) or (3.42) on the boundary surface S. By integrating eqn (3.59) over V using eqns (3.39) and (3.51), the following result is obtained:

$$E_0 = \int_V d\boldsymbol{r}\, \langle 0|T_{00}^{(0)}(x)|0\rangle = \frac{1}{2}\sum_J \omega_J. \tag{3.60}$$

Note that in obtaining this result, the last term on the right-hand side of eqn (3.59) has been integrated by parts. Thus, the integration of the vacuum energy density in the presence of boundaries leads to the general result (1.4), which was discussed in Section 1.1.

For free Minkowski space without boundaries, the vacuum energy density is obtained from eqns (3.56) and (3.59):

$$\langle 0_M | T_{00}^{(0)}(x) | 0_M \rangle = \frac{1}{2} \int \frac{d\mathbf{k}}{(2\pi)^3} \omega_k. \tag{3.61}$$

The total vacuum energy in the volume V is

$$E_0 = \int_V d\mathbf{r} \, \langle 0_M | T_{00}^{(0)}(x) | 0_M \rangle = \frac{1}{2} \int \frac{d\mathbf{k}}{(2\pi)^3} \omega_k V. \tag{3.62}$$

Equations (3.61) and (3.62) are analogous to eqns (2.65) and (2.66), obtained in two-dimensional space–time. Any manipulations with them assume the use of some regularization (see Section 2.2).

We consider now the quantization of an electromagnetic field in the presence of boundaries. This is a nontrivial problem, owing to the existence of different polarization states of the photon. The point is that the physical boundary conditions (2.71) are formulated in terms of the electric field and the magnetic induction, i.e. gauge-invariant quantities. However, it is preferable to perform the quantization of the electromagnetic field in terms of the vector potential A_μ. First, we choose the Coulomb gauge, where $A_0(x) = 0$ and $\mathrm{div}\,\mathbf{A}(x) = 0$, as was done in the previous section devoted to the discussion of boundary conditions. In this gauge, eqn (3.22) takes the form

$$\frac{\partial^2 \mathbf{A}(x)}{\partial x_0^2} - \nabla^2 \mathbf{A}(x) = 0. \tag{3.63}$$

Thereafter, we separate the time variable and present the positive- and negative-frequency solutions of the wave equation (3.63) in the form

$$\mathbf{A}_J^{(+)}(x) = \frac{1}{\sqrt{\omega_J}} e^{-i\omega_J t} \mathbf{A}_J(\mathbf{r}), \qquad \mathbf{A}_J^{(-)}(x) = \left[\mathbf{A}_J^{(+)}(x) \right]^*, \tag{3.64}$$

where $\mathbf{A}_J(\mathbf{r})$ satisfies the equation

$$-\nabla^2 \mathbf{A}_J(\mathbf{r}) = \omega_J^2 \mathbf{A}_J(\mathbf{r}), \tag{3.65}$$

following from eqn (3.45) with appropriate boundary conditions. The functions $\mathbf{A}_J(\mathbf{r})$ in eqn (3.64) are orthonormal, satisfying the equation

$$\int_V d\mathbf{r} \, \mathbf{A}_J^*(\mathbf{r}) \mathbf{A}_{J'}(\mathbf{r}) = 4\pi \delta_{JJ'}. \tag{3.66}$$

Now we assume that the boundaries under consideration allow the definition of two orthonormal polarization vectors $\boldsymbol{\epsilon}_J^{(\lambda)}$:

$$\boldsymbol{\epsilon}_J^{(\lambda)} \cdot \boldsymbol{\epsilon}_J^{(\lambda')} = \delta_{\lambda\lambda'}, \qquad \lambda, \lambda' = 1, 2. \tag{3.67}$$

These are also perpendicular to the generalized wave vector defined by the quantum number J. The vector function $\boldsymbol{A}_J(\boldsymbol{r})$ in eqn (3.64) can be expanded in terms of different polarizations:

$$\boldsymbol{A}_J(\boldsymbol{r}) = \sum_{\lambda=1}^{2} \boldsymbol{A}_J^{(\lambda)}(\boldsymbol{r}) = \sum_{\lambda=1}^{2} \tilde{A}_J^{(\lambda)}(\boldsymbol{r})\boldsymbol{\epsilon}_J^{(\lambda)}, \qquad (3.68)$$

where the expansion coefficients are given by the scalar products

$$\tilde{A}_J^{(\lambda)}(\boldsymbol{r}) = \boldsymbol{A}_J(\boldsymbol{r}) \cdot \boldsymbol{\epsilon}_J^{(\lambda)}. \qquad (3.69)$$

These coefficients satisfy the normalization condition following from eqn (3.66)

$$\sum_{\lambda=1}^{2} \int_V d\boldsymbol{r}\, \tilde{A}_J^{(\lambda)*}(\boldsymbol{r})\tilde{A}_{J'}^{(\lambda)}(\boldsymbol{r}) = 4\pi\delta_{JJ'}. \qquad (3.70)$$

The two polarizations are said to be separable if eqns (3.68) and (3.69) are valid. In this case the expansion coefficients $\tilde{A}_J^{(\lambda)}(\boldsymbol{r})$ satisfy the boundary conditions.

The quantization of the electromagnetic field can be performed in the following way:

$$\boldsymbol{A}(x) = \sum_J \sum_{\lambda=1}^{2} \frac{1}{\sqrt{\omega_J}} \boldsymbol{\epsilon}_J^{(\lambda)} \left[e^{-i\omega_J t}\tilde{A}_J^{(\lambda)}(\boldsymbol{r})a_J^{(\lambda)} + e^{i\omega_J t}\tilde{A}_J^{(\lambda)*}(\boldsymbol{r})a_J^{(\lambda)+} \right], \qquad (3.71)$$

where $a_J^{(\lambda)}$ and $a_J^{(\lambda)+}$ are the annihilation and creation operators of a photon with a generalized momentum J and in the polarization state λ. They satisfy the commutation relations

$$\left[a_J^{(\lambda)}, a_{J'}^{(\lambda')+}\right] = \delta_{JJ'}\delta_{\lambda\lambda'}, \quad \left[a_J^{(\lambda)}, a_{J'}^{(\lambda')}\right] = \left[a_J^{(\lambda)+}, a_{J'}^{(\lambda')+}\right] = 0. \qquad (3.72)$$

The photon vacuum state is defined by

$$a_J^{(\lambda)}|0\rangle = 0. \qquad (3.73)$$

We note that in the simplest cases of free Minkowski space and plane boundaries, the polarization vectors perpendicular to the wave vector $\boldsymbol{k} = (k^1, k^2, k^3)$ have the form

$$\boldsymbol{\epsilon}_k^{(1)} = \frac{1}{k_\perp}\begin{pmatrix} k^2 \\ -k^1 \\ 0 \end{pmatrix}, \quad \boldsymbol{\epsilon}_k^{(2)} = \frac{1}{kk_\perp}\begin{pmatrix} k^1 k^3 \\ k^2 k^3 \\ -k_\perp^2 \end{pmatrix}, \qquad (3.74)$$

where $k = |\boldsymbol{k}|$. If there are no boundary surfaces, the coefficients $\tilde{A}_J^{(\lambda)}(\boldsymbol{r})$ in eqn (3.68) depend only on $\boldsymbol{r}$ and do not depend on the polarization state λ. In fact, in the case of curved boundary surfaces even the polarization vectors may become position-dependent and the polarizations may not be separable. Then, one is left with the general formulation (2.71) or (3.46).

Substituting eqn (3.71) into eqns (3.13) (with $A_0 = 0$) and (3.29) and using eqn (3.73), we obtain the vacuum energy density of the electromagnetic field in the presence of boundaries,

$$\langle 0|T_{00}^{(1)}(x)|0\rangle = \frac{1}{8\pi}\sum_J \sum_{\lambda=1}^{2} \frac{1}{\omega_J}\left\{\omega_J^2 \tilde{A}_J^{(\lambda)}(r)\tilde{A}_J^{(\lambda)*}(r) \right. $$
$$\left. + \left[\nabla \times \left(\tilde{A}_J^{(\lambda)}(r)\epsilon_J^{(\lambda)}\right)\right] \cdot \left[\nabla \times \left(\tilde{A}_J^{(\lambda)*}(r)\epsilon_J^{(\lambda)}\right)\right]\right\}. \quad (3.75)$$

The total energy of the electromagnetic field in the volume V is obtained from eqn (3.75), taking account of the ideal-metal boundary conditions (3.48) on the boundary surface and the Coulomb gauge (3.47). If both polarizations contribute to the result at all J, one obtains

$$E_0 = \int_V d\boldsymbol{r}\, \langle 0|T_{00}^{(1)}(x)|0\rangle = \frac{1}{2}\sum_{\lambda=1}^{2}\sum_J \omega_J = \sum_J \omega_J. \quad (3.76)$$

This is different by a factor of 2 from eqn (3.60) and reflects the existence of two polarization states of the photon. If, for some J, one of the polarizations does not contribute to the result, the respective term in eqn (3.76) is missing and there is no doubling of the scalar-field contribution for this J (see Section 7.2 for more details and an example). Similarly, in free Minkowski space we get

$$\langle 0_M|T_{00}^{(1)}(x)|0_M\rangle = \int \frac{d\boldsymbol{k}}{(2\pi)^3}k, \quad (3.77)$$

$$E_{0M} = \int_V d\boldsymbol{r}\, \langle 0_M|T_{00}^{(1)}(x)|0_M\rangle = \int \frac{d\boldsymbol{k}}{(2\pi)^3}kV,$$

where, in this case, $\omega_k = k$.

For a spinor field, the positive- and negative-frequency solutions of the Dirac equation (3.30) with appropriate boundary conditions,

$$\psi_{J,\alpha}^{(+)}(x) = e^{-i\omega_J t}\chi_{J,\alpha}^{(+)}(\boldsymbol{r}), \qquad \psi_{J,\alpha}^{(-)}(x) = e^{i\omega_J t}\chi_{J,\alpha}^{(-)}(\boldsymbol{r}), \quad (3.78)$$

are orthonormal:

$$\left(\psi_{J,\alpha}^{(\pm)}(x), \psi_{J',\alpha'}^{(\pm)}(x)\right) = \delta_{JJ'}\delta_{\alpha\alpha'}, \qquad \left(\psi_{J,\alpha}^{(\pm)}(x), \psi_{J',\alpha'}^{(\mp)}(x)\right) = 0, \quad (3.79)$$

where J is the generalized wave vector and $\alpha = 1, 2$ is the spin index. The scalar product is defined in eqn (3.35). The field operator and the Dirac conjugate operator can be presented in the form

$$\psi(x) = \sum_J \sum_{\alpha=1,2}\left[\psi_{J,\alpha}^{(+)}(x)b_{J,\alpha} + \psi_{J,\alpha}^{(-)}(x)d_{J,\alpha}^+\right], \quad (3.80)$$

$$\bar{\psi}(x) = \sum_J \sum_{\alpha=1,2} \left[\overline{\psi_{J,\alpha}^{(+)}}(x) b_{J,\alpha}^+ + \overline{\psi_{J,\alpha}^{(-)}}(x) d_{J,\alpha} \right].$$

Here, $b_{J,\alpha}$ and $b_{J,\alpha}^+$ are the annihilation and creation operators of particles, and $d_{J,\alpha}$ and $d_{J,\alpha}^+$ are the creation and annihilation operators of antiparticles. These operators satisfy the anticommutation relations

$$\left\{ b_{J,\alpha}, b_{J',\alpha'}^+ \right\} = \left\{ d_{J,\alpha}, d_{J',\alpha'}^+ \right\} = \delta_{JJ'} \delta_{\alpha\alpha'}, \tag{3.81}$$

$$\{b_{J,\alpha}, b_{J',\alpha'}\} = \left\{ b_{J,\alpha}^+, b_{J',\alpha'}^+ \right\} = \{d_{J,\alpha}, d_{J',\alpha'}\} = \left\{ d_{J,\alpha}^+, d_{J',\alpha'}^+ \right\} = 0.$$

The vacuum state of the Dirac field is defined by the operations

$$b_{J,\alpha} |0\rangle = d_{J,\alpha} |0\rangle = 0. \tag{3.82}$$

To find the vacuum energy density of the spinor field, we substitute eqn (3.80) into the 00-component of the energy–momentum tensor (3.36) and use eqns (3.81) and (3.82):

$$\langle 0 | T_{00}^{(1/2)}(x) | 0 \rangle = -\sum_{J,\alpha} \omega_J \chi_{J,\alpha}^{(-)+}(r) \chi_{J,\alpha}^{(-)}(r). \tag{3.83}$$

It is notable that the vacuum energy of the spinor field enters with a minus sign. Within the formalism of canonical quantization, this follows from the use of anticommutators in eqn (3.81) instead of the commutators for the scalar and electromagnetic fields, and it is really a consequence of the spin–statistics theorem. By integrating over the quantization volume with the use of eqn (3.79), we obtain the total vacuum energy of the spinor field,

$$E_0 = \int_V d\mathbf{r} \, \langle 0 | T_{00}^{(1/2)}(x) | 0 \rangle = -\sum_{J,\alpha} \omega_J = -2 \sum_J \omega_J. \tag{3.84}$$

The magnitude of this result differs from the similar result (3.60) for a scalar field by a factor of 4. This is due to the two types of particles described by the spinor field (particles and antiparticles) and the two possible spin states for each of them.

In free Minkowski space, the vacuum energy density and the total vacuum energy of the spinor field are given by

$$\langle 0_M | T_{00}^{(1/2)}(x) | 0_M \rangle = -2 \int \frac{d\mathbf{k}}{(2\pi)^3} \omega_k, \tag{3.85}$$

$$E_{0M} = \int_V d\mathbf{r} \, \langle 0_M | T_{00}^{(1/2)}(x) | 0_M \rangle = -2 \int \frac{d\mathbf{k}}{(2\pi)^3} \omega_k V.$$

These differ from eqns (3.61) and (3.62), obtained for a scalar field, by a factor of 4 and are opposite in sign.

3.4 Vacuum energy in terms of Green's functions

In the preceding section, we considered the easiest approach to a representation of the vacuum energy in terms of a mode expansion. In fact, there are other approaches and representations which are useful. One such case is the representation of the vacuum energy in terms of Green's functions. These are solutions of the wave equations with a delta function on the right-hand side of the equation. For a scalar field, the appropriate equation is

$$\left(\Box_x + m^2\right) G(x, x') = \delta^4(x - x'). \tag{3.86}$$

The Green's function $G(x, x')$ can be interpreted as an inhomogeneous solution of the wave equation (3.3) with a point-like source at x'. The Green's functions are not uniquely defined by the equation, since an arbitrary solution of the homogeneous equation (3.1) can be added. This arbitrariness is used to define different types of Green's functions (below, we shall use the causal one) and to satisfy the boundary conditions.

The Green's functions can be represented in terms of the solutions (3.38),

$$G(x, x') = \int_{-\infty}^{\infty} \frac{d\omega}{2\pi} \sum_J \frac{\Phi_J(\boldsymbol{r})\Phi_J^*(\boldsymbol{r}')}{-\omega^2 + \omega_J^2 - i0}\, e^{-i\omega(t-t')}. \tag{3.87}$$

In this integral, the integration over ω has poles at $\omega = \pm\omega_J$ resulting from the zeros in the denominator. The arbitrariness can be removed by defining a rule about how to go around the poles. In eqn (3.87), we have defined the causal Green's function by means of the infinitesimal addition $-i0$.

In order to establish the relation between the vacuum energy and the Green's function, we first consider the vacuum expectation value of a product of two field operators,

$$\langle 0|\varphi(x)\varphi(x')|0\rangle = \sum_J \frac{1}{2\omega_J} e^{-i\omega_J(t-t')} \Phi_J(\boldsymbol{r})\Phi_J^*(\boldsymbol{r}'), \tag{3.88}$$

which itself is not yet a Green's function. To obtain a Green's function, we need to consider the vacuum expectation value of the time-ordered product of the two field operators

$$T\varphi(x)\varphi(x') = \theta(t-t')\varphi(x)\varphi(x') + \theta(t'-t)\varphi(x')\varphi(x), \tag{3.89}$$

which results in

$$i\langle 0|T\varphi(x)\varphi(x')|0\rangle = G(x, x'). \tag{3.90}$$

The latter relation follows from carrying out the frequency integration in eqn (3.87) according to the pole-bypassing rule (for the causal Green's function, the poles are located in $\omega = \pm\sqrt{\omega_J^2 - i0}$), to obtain

$$iG(x, y) = \sum_J \frac{1}{2\omega_J}\, e^{i\omega_J|t-t'|}\, \Phi_J(\boldsymbol{r})\Phi_J^*(\boldsymbol{r}'), \tag{3.91}$$

and applying eqn (3.88) to eqn (3.89).

In order to establish the relation to the vacuum energy density as defined in eqn (3.58), we note that the energy–momentum tensor contains the product of two field operators at coincident arguments. Such expressions are, by their mathematical nature, singular and need to be regularized. This will be discussed in detail in Section 4.1; here we adopt the following procedure. We take $T^{(0)}_{\mu\nu}(x)$, eqn (3.8), with the field operators at separate points, $x \neq x'$, and consider the limit of coincidence of these arguments following the rules for the T-product of operators (3.89). In this way, we represent the vacuum expectation value of the energy–momentum tensor in the form

$$\langle 0|T^{(0)}_{00}(x)|0\rangle = -\frac{i}{2}\left(\sum_{\mu=0}^{3}\frac{\partial}{\partial x^\mu}\frac{\partial}{\partial x'^\mu} + m^2\right)G(x,x')\Big|_{x'=x}. \quad (3.92)$$

Here, we need to mention that the definition of the vacuum energy density (and of the vacuum expectation values of other observables) is not unique. In addition to the freedom to add a gradient term to $T^{(0)}_{\mu\nu}(x)$, there is another nonuniqueness following from taking the limit in eqn (3.92), which can be done in many different ways. We restrict ourselves here to arguing that eqn (3.92), after insertion of the expression (3.90) for the Green's function in terms of solutions, coincides with the definition (3.59) at least on the formal level.

The global vacuum energy (3.60) can also be expressed in terms of Green's functions by taking the spatial integral of eqn (3.92). Under the integral sign, we integrate the derivatives $\partial/\partial x'^i$ by parts. Using eqn (3.86) and making use of the fact that, owing to the time stationarity, the time dependence enters the Green's function as a difference $x_0 - x'_0$, we obtain

$$E_0 = i\int_V d\mathbf{r}\,\frac{\partial^2 G(x,x')}{\partial x_0^2}\Big|_{x'=x}. \quad (3.93)$$

A further necessary remark on the derivation of this representation is that we have performed the transformations mentioned above for separated arguments, i.e. for $x \neq x'$. In this way, the delta function on the right-hand side of eqn (3.86) does not contribute.

For an electromagnetic field, the Green's function corresponding to the vector potentials used in Section 3.3 for quantization is a tensor with components $G_{ij}(x,x')$. It is defined by the equation

$$\Box_x G_{ij}(x,x') = \delta(t-t')\Big[\delta_{ij}\,\delta^3(\mathbf{r}-\mathbf{r}')$$
$$-\partial_{x^i}\int d\mathbf{r}''\,(\nabla^2)^{-1}(\mathbf{r},\mathbf{r}'')\partial_{x'''j}\delta(\mathbf{r}''-\mathbf{r}')\Big], \quad (3.94)$$

with a unit tensor in the space of transverse functions $\partial_{x^i}A_{J,i}(x) = 0$ and the inverse of the Laplace operator

$$\nabla_r^2 (\nabla^2)^{-1}(r, r') = \delta^3(r - r') \tag{3.95}$$

on the right-hand side. The following steps are analogous to the case of a scalar field. Using the solutions of eqn (3.65), the corresponding representation is

$$G_{ij}(x, x') = \int_{-\infty}^{\infty} \frac{d\omega}{2\pi} \sum_J e^{-i\omega(t-t')} \frac{\mathcal{A}_{J,i}(r)\mathcal{A}_{J,j}^*(r')}{-\omega^2 + \omega_J^2 - i0}. \tag{3.96}$$

In the final representation of the vacuum energy,

$$E_0 = i \int_V dr \sum_{i=1}^{3} \left. \frac{\partial^2 G_{ii}(x, x')}{\partial x_0^2} \right|_{x'=x}, \tag{3.97}$$

one has to take the trace over the indices of this Green's function.

3.5 Path-integral quantization

One of the most beautiful methods in quantum field theory is path-integral quantization. The idea (Feynman 1948) is to consider the classical paths (trajectories) that a system may follow and to sum over these paths with suitable weights. In this way, quantities defined in quantum theory such as transition amplitudes can be formulated completely in terms of classical quantities. As a result, however, infinite-dimensional integrations appear, which are in general not well defined mathematically. There are nevertheless many applications where path integrals are very useful. With respect to the vacuum energy, we are usually concerned with free-field theories (the nontrivial content results from boundaries or background fields). For these, the path integral is well defined and the problem of its calculation is reduced to a Gaussian integration, which, in turn, ends up with functional determinants. The latter can be calculated if, for example, the solutions of the underlying wave equation are known. In this way, the path-integral formulation results in problems of the same difficulty as in canonical quantization. Its advantage is that in many cases it allows a much more elegant and transparent formulation. In addition, in recent years several methods have been developed within the path-integral approach which allow more direct calculations and also numerical computations.

In this section, we restrict ourselves to the case of a scalar field in order to focus on the representation of the vacuum energy, and postpone any discussion of the peculiarities related to the gauge freedom in electrodynamics. Also, we do not consider a spinor field, and restrict ourselves to the remark that the minus sign entering its vacuum energy is related to the Grassmann variables which one must use in its path-integral representation.

As the basic quantity to be represented in the path-integral approach, we take the generating functional $Z[\Upsilon]$ (we use square brackets here to denote a functional dependence) of the Green's functions for a scalar field. These can be obtained by taking functional derivatives with respect to the source $\Upsilon(x)$. For the propagator, which in this context is a two-point Green's function, we get

$$i\frac{\delta}{\delta\Upsilon(x)}\frac{\delta}{\delta\Upsilon(x')}Z[\Upsilon]|_{\Upsilon=0} = G(x,x'). \tag{3.98}$$

Here, the *functional derivative* $\delta/\delta\Upsilon(x)$ is defined as

$$\frac{\delta}{\delta\Upsilon(x)}\Upsilon(x') = \delta^4(x-x'). \tag{3.99}$$

In terms of a path integral, the generating functional is given by

$$Z[\Upsilon] = C\int D\varphi\, e^{iS[\varphi]}, \tag{3.100}$$

with the action $S[\varphi] = S[\varphi, \Upsilon]$ given by eqn (3.5), i.e. including the source $\Upsilon(x)$. In eqn (3.100), the integration goes over all fields in a suitably defined space. A general discussion of this space can be found, for example, in the book by Vasiliev (1998). Here and below, we shall be interested only in restrictions imposed on this space by boundary conditions. Concerning the constant C in front of the integral in eqn (3.100), it should be mentioned that in quantum field theory, path integrals are usually defined up to a constant (in general, infinite), which does not influence the final results. In application to the vacuum energy, this constant is, for instance, independent of external parameters such as the plate separation. We choose $C = 1$.

Another important quantity frequently used in the formulation of the vacuum energy is the *effective action*,

$$iW_{\text{eff}}[\Upsilon] = \ln Z[\Upsilon]. \tag{3.101}$$

For vanishing sources and time-independent boundaries and backgrounds, this is proportional to the total time T and the vacuum energy (Peskin and Schroeder 1995),

$$W_{\text{eff}}[0] = -TE_0 \tag{3.102}$$

such that the latter can be represented by means of

$$E_0 = \frac{i}{T}\ln Z[0] \tag{3.103}$$

in terms of a path integral. It must be mentioned that this representation does not coincide completely with the vacuum energy defined in the preceding section. However, as is discussed in Section 4.2.2, the difference between the two representations does not influence physical quantities.

From the point of view of how to perform the integration in a path integral, one has to observe that the action (3.5) is quadratic in the fields and hence the integrations are Gaussian. The finite-dimensional analogue is an integral over $\mathbb{R}^n$,

$$\int_{\mathbb{R}^n} d^n x\, e^{-x^\top \mathbf{K} x/2 + x^\top h} = (2\pi)^{n/2} (\det \mathbf{K})^{-1/2}\, e^{-h^\top \mathbf{K}^{-1} h}, \qquad (3.104)$$

where $x, h \in \mathbb{R}^n$, $\mathbf{K}$ is a real $n \times n$ matrix, and $\mathbf{K}^{-1}$ is its inverse. Written in components, the quadratic form can be represented as

$$-\frac{1}{2} x^\top \mathbf{K} x + x^\top h = -\frac{1}{2} \sum_{i,j} x_i K_{ij} x_j + \sum_i x_i h_i. \qquad (3.105)$$

The formula (3.104) assumes that all eigenvalues of $\mathbf{K}$ are positive, otherwise the integral would not converge. It is obvious that $\mathbf{K}$ can be assumed to be symmetric. By making the substitution $x = x' + \mathbf{K}^{-1} h$, i.e. by "completing the square", the linear term can be removed, and the remaining integration results in a determinant. The direct generalization of eqn (3.104) to the infinite-dimensional case can be done by the formal substitution $x_i \to \varphi(x)$ and $h_i \to \Upsilon(x)$, where the argument x takes the place of the index i. The matrix K_{ij} becomes some function of two arguments, $K(x, y)$, and the sums turn into corresponding integrations. In this way, the exponential in eqn (3.104) is transformed as follows:

$$-\frac{1}{2} \sum_{i,j} x_i K_{ij} x_j + \sum_i x_i h_i$$
$$\to -\frac{1}{2} \int d^4 x\, d^4 x'\, \varphi(x) K(x, x') \varphi(x') + \int d^4 x\, \varphi(x) \Upsilon(x). \qquad (3.106)$$

Here $K(x, x')$ is the kernel of an integral operator

$$\mathcal{K} f(x) = \int d^4 x'\, K(x, x') f(x'), \qquad (3.107)$$

where $f(x)$ is a test function. The kernel of an integral operator can be equivalently represented in the form

$$K(x, x') = \langle x' | \mathcal{K} | x \rangle. \qquad (3.108)$$

All subsequent constructions can be done with such a kernel. The differential operator $K(x)$ in eqn (3.6) is a special case of the integral operator with a local kernel $K(x, x') = \delta(x - x') K(x')$. It should be mentioned that all of these constructions can be done for very general fields and space–times. Keeping in mind the application to the Casimir effect, we restrict ourselves here to the simplest formulation.

The generalization of eqn (3.104) in the case of a scalar field leads to the generating functional

$$Z[\Upsilon] = C \left(\det \mathcal{K}\right)^{-1/2} \exp\left[\frac{i}{2}\int d^4x\, d^4x'\, \Upsilon(x) K^{-1}(x,x') \Upsilon(x')\right], \quad (3.109)$$

where $K^{-1}(x,x')$ is the inverse of $K(x,x')$, i.e.

$$\int d^4x'\, K(x,x') K^{-1}(x',x'') = \delta^4(x-x'') \quad (3.110)$$

must hold. Comparing eqn (3.109) with eqn (3.98,) we arrive at

$$K^{-1}(x,x') = G(x,x'), \quad (3.111)$$

i.e. the operator K^{-1} is the propagator of the scalar field.

From this formula, the vacuum energy is given by

$$E_0 = \frac{i}{T} \ln \left(\det \mathcal{K}\right)^{-1/2} = -\frac{i}{2T} \operatorname{Tr} \ln \mathcal{K} \quad (3.112)$$

(where we have made use of the well-known formula $\ln \det \mathcal{K} = \operatorname{Tr} \ln \mathcal{K}$). Here we have put the source $\Upsilon = 0$ and have dropped the contribution from the constant C since it gives only an irrelevant additive constant.

3.6 Propagators with boundary conditions

In the presence of boundaries, the propagator can, in general, be constructed in terms of the mode functions, for example by use of eqn (3.91). In this case it is assumed that the mode functions $\Phi_J(r)$ satisfy the corresponding boundary conditions, eqn (3.41) or eqn (3.42) for example. This representation is not always convenient. So, for a boundary of general shape, when the variables in the wave equation do not separate, explicit expressions are not available, neither for the mode functions nor for the corresponding eigenvalues. We consider here another general representation of the Green's function and of the vacuum energy which does not rely on the mode expansion and which highlights some general properties.

In order to derive this representation, we start from the expression (3.100) for the generating functional of the Green's functions $Z[\Upsilon]$ in terms of the path integral. For simplicity, we consider a scalar field $\varphi(x)$ fulfilling Dirichlet boundary conditions (3.41) on a surface S given by some functions $x_0 = u_0(\eta_0) \equiv \eta_0$, $\mathbf{r} = \mathbf{u}(\eta_1, \eta_2)$ or, in a more compact notation, $x = u(\eta)$, where $u = (u_0, \mathbf{u})$ and $\eta = (\eta_0, \eta_1, \eta_2)$.

We denote the corresponding Green's function by $^S G(x,x')$. It has to obey eqn (3.86) for $x \notin S$, i.e. outside the surface, and to obey the boundary conditions

$$^S G(x,x') = 0 \quad \text{for } x \in S \text{ or } x' \in S. \quad (3.113)$$

This Green's function can be obtained from the path-integral representation (3.100) by choosing the integration space to consist of those fields $\varphi(x)$ which fulfill the boundary conditions. By means of eqns (3.109) and (3.111), this is then reduced to the above-mentioned problem and the corresponding methods can be applied.

The idea of the new method is to start from a path integral in empty space, i.e. without boundary conditions, and to restrict the integration space by use of the corresponding functional delta functions. This method was developed for quantum electrodynamics with conductor boundary conditions by Bordag et al. (1985) and, independently, by Li and Kardar (1992) for fluctuations in a fluid. In this way, the path integral goes over fields which are free of boundary conditions. The necessary restriction of the integration space is achieved by the insertion of the functional delta function

$$\prod_{x \in S} \delta(\varphi(x)) \tag{3.114}$$

into the path integral (3.100) for the generating functional,

$$Z[\Upsilon] = \int D\varphi \prod_{x \in S} \delta(\varphi(x))\, e^{iS[\varphi]}. \tag{3.115}$$

By construction, it is clear that this is another representation of the same path integral as in the case when one integrates over a field φ fulfilling the boundary conditions.

The next steps are technical. We represent the functional delta function by a Fourier representation,

$$\prod_{x \in S} \delta(\varphi(x)) = C \int Db\, e^{i \int_S d\mu(\eta)\, b(\eta)\varphi(u(\eta))}, \tag{3.116}$$

where C is an irrelevant constant like that in eqn (3.109), $d\mu(\eta)$ is the volume element on S, and the *variable of integration*, $b(\eta)$, is an auxiliary field defined on the surface S. It is useful to rewrite the exponential in this equation in the form

$$\int_S d\mu(\eta)\, b(\eta)\varphi(u(\eta)) = \int_S d\mu(\eta) \int d^4x\, b(\eta) H(\eta, x)\varphi(x), \tag{3.117}$$

with a kernel $H(\eta, x)$, which in this case is simply $H(\eta, x) = \delta^4(x - u(\eta))$. It is clear that this construction can easily be generalized to include derivatives or to carry indices. Now we insert this representation into the path integral (3.115). The resulting integral is now over two fields, $\varphi(x)$ and $b(\eta)$. Since it is bilinear in these fields, the integration is Gaussian. It can be carried out after diagonalization of the quadratic form, which is achieved by

$$-\frac{1}{2}\int d^4x\, d^4x'\, \varphi(x) K(x, x')\varphi(x') + \int d^4x\, \varphi(x)\Upsilon(x)$$
$$+ \int_S d\eta \int d^4x\, b(\eta) H(\eta, x)\varphi(x) \tag{3.118}$$

$$= \frac{1}{2}\left[-\int d^4x\, d^4x'\, (\varphi(x) - \varphi_0(x))\, K(x,x')\, (\varphi(x') - \varphi_0(x'))\right.$$
$$\left.+ \int_S d\eta \int_S d\eta'\, b(\eta)\tilde{K}^{-1}(\eta,\eta')b(\eta') + \int d^4x \int d^4x'\, \Upsilon(x)^S G(x,x')\Upsilon(x')\right],$$

where

$$\varphi_0(x) = \int d^4x'\, K^{-1}(x,x')\left(\Upsilon(x') + \int_S d\eta\, H(\eta,x')b(\eta)\right). \tag{3.119}$$

The kernel $\tilde{K}(\eta,\eta')$ is defined by

$$\tilde{K}(\eta,\eta') = \int d^4x \int d^4x'\, H(\eta,x)G(x,x')H(\eta',x') = G(u(\eta),u(\eta')). \tag{3.120}$$

In the case of Dirichlet boundary conditions, this is just a restriction of the empty-space propagator $G(x,x')$ with both of its arguments on the surface S. The inverse, $\tilde{K}^{-1}(\eta,\eta')$, must be taken on S, i.e.

$$\int_S d\mu(\eta'')\, \tilde{K}(\eta,\eta'')\tilde{K}^{-1}(\eta'',\eta') = \delta^3(\eta - \eta'). \tag{3.121}$$

Finally, in eqn (3.118) we have introduced

$$^S G(x,x') = G(x,x') - \int d^4x'' \int d^4x''' \int_S d\mu(\eta) \int_S d\mu(\eta')$$
$$\times G(x,x'')H(\eta,x'')\tilde{K}^{-1}(\eta,\eta')H(\eta',x''')G(x''',x')$$
$$= G(x,x') - \int_S d\mu(\eta) \int_S d\mu(\eta')\, G(x,u(\eta))\tilde{K}^{-1}(\eta,\eta')G(u(\eta'),x')$$
$$\equiv G(x,x') - \overline{G}(x,x'). \tag{3.122}$$

Now the functional integration can be carried out by first shifting $\varphi(x) \to \varphi(x) + \varphi_0(x)$ and applying eqn (3.109). Subsequently, the integration over $b(\eta)$ can be carried out in the same manner. The generating functional then takes the form

$$Z[\Upsilon] = C\, (\det \mathcal{K})^{-1/2} \left(\det \tilde{\mathcal{K}}\right)^{-1/2} \exp\left[\frac{i}{2}\int d^4x\, d^4x'\, \Upsilon(x)^S G(x,x')\Upsilon(x')\right], \tag{3.123}$$

where the factor $\left(\det \tilde{\mathcal{K}}\right)^{-1/2}$ results from the integration over $b(\eta)$. In this way, we have obtained another representation for the generating functional in the presence of boundaries. The energy according to eqn (3.112) is

$$E_0 = -\frac{i}{2T}\operatorname{Tr}\ln\tilde{\mathcal{K}} = \frac{i}{T}\ln\left(\det\tilde{\mathcal{K}}\right)^{-1/2}. \tag{3.124}$$

Here, along with the constant C, we have also dropped the contribution from $\det \mathcal{K}$. This is possible since $\mathcal{K}$ results from empty space and thus does not depend on the geometry of the boundaries. So it also delivers an irrelevant constant. In fact, this constant, being the contribution from empty space, carries the divergences associated with it such that in the E_0 given by eqn (3.124), only subleading divergences remain.

From a comparison of eqns (3.122) and (3.123), it follows that $^S G(x,x')$ as defined by eqn (3.122) is just the propagator in the presence of boundaries. It has a specific representation. According to the last line of eqn (3.122), it is given by the difference between the free-space part $G(x,x')$ and a boundary-dependent part $\overline{G}(x,x')$. It can be easily checked that it fulfills eqn (3.86) and the boundary conditions. Indeed, let $\boldsymbol{r} = \boldsymbol{u}(\eta_1,\eta_2) \in S$. Then, in the second term on the right-hand side of the first line in eqn (3.122), we can apply eqn (3.121) and the resulting expression just cancels the first term. Also, it is easy to check eqn (3.86). For the first term, this is obvious. For the second term, i.e. for the boundary-dependent addition, one has to notice that, after applying the wave operator to $\overline{G}(x,x')$ within the resulting expression, a delta function $\delta^4(x-u(\eta))$ appears so that outside the surface, i.e. for $x \notin S$, this expression vanishes.

We remark that in terms of homogeneous and inhomogeneous solutions of the wave equation, the first term is just an inhomogeneous solution and the second term is a homogeneous solution which is chosen in such a way as to ensure that the boundary conditions are satisfied. A further remark concerns an alternative derivation of the representation (3.122). This can be obtained from considering the boundary conditions as constraints when one is solving the wave equation. In that case one may use Lagrange multipliers. These just correspond to the auxiliary fields $b(\eta)$ introduced in eqn (3.116). In the end, one again arrives at the representation (3.122).

A last remark concerns the structure of the propagator $^S G(x,x')$. In its derivation from the path integral, we did not specify on which side of the surface S the points x or x' are located. In fact, the representation (3.122) is valid for any choice. For instance, for a surface S dividing the space into two disconnected parts this implies that if x and x' are taken in different parts, the propagator must vanish. This property can easily be checked.

4

REGULARIZATION AND RENORMALIZATION OF THE VACUUM ENERGY

As previously discussed, the vacuum energy is a divergent quantity. In the vacuum, quantum field theory assigns half a quantum to each of the infinitely many degrees of freedom. These divergences are of an ultraviolet nature similar to that known from higher loop expansions. Their treatment, however, requires special approaches because of the presence of boundaries. Powerful methods are available for this. The most general one is the heat kernel expansion, which can be considered as the standard method and the natural language to represent these divergences. From a mathematical point of view, it is closely related to *spectral geometry* [see e.g. the book by Gilkey (1995)]. The heat kernel expansion is also related to zeta function regularization, which can be considered as the most elegant among the many different regularization schemes. In the present chapter, we use this method together with cutoff regularization to separate the divergent part of the vacuum energy.

After having regularized the vacuum energy, we consider the procedure of renormalization. We start with the case where some smooth background fields are present. To some extent this is not central to this book, but it is necessary for an understanding of the renormalization procedure. In a smooth background field, the renormalization procedure is the same as that known in quantum field theory. The divergent contributions have a structure which allows their removal by a redefinition of the parameters in the "noninteracting theory" (including the parameters of a classical background field if one is present). However, this procedure is not always possible. For the case of background fields (if these are singular or if one uses some limiting process which makes them singular), these questions are not completely settled, and we shall discuss them briefly in Section 4.3.

For the Casimir energy resulting from the boundary of a single body, geometric characteristics such as the volume, surface area, and curvature should be used for renormalization. If such characteristics are not available, the vacuum energy cannot be given a physical meaning except when the divergences are absent. The same also holds for other quantities such as the Casimir pressure and force. We shall discuss some examples below.

The most important case is the Casimir force between separate objects. Here the situation is completely different. In general, this force is always finite, as opposed to the interaction energy, which becomes finite when the contribution of the vacuum energy of free space is removed. This will be discussed in the last section of this chapter.

4.1 Regularization schemes

Regularization is a method to change an infinite quantity into a finite one. A regularization parameter is introduced such that, in the appropriate limit, the original expression is restored. Of course, this procedure is not unique, and different schemes are possible, some of which will be discussed below. Beyond this formal definition, regularizations sometimes have a direct physical meaning. For instance, since ideal conductors do not exist in nature, one has in all real applications some natural frequency, usually of the order of the plasma frequency, beyond which the reflectivity rapidly decreases. However, this decrease might not provide a regularization for some systems. The most important regularization schemes are *frequency cutoff*, *point splitting*, and *zeta function regularization*.

In a frequency cutoff regularization, one introduces some cutoff function in the mode expansion which makes the corresponding sum/integral converge. Equation (3.60) defines the nonregularized (infinite) vacuum energy E_0. In this case we introduce the regularized vacuum energy $E_0(\delta)$, and in place of eqn (3.60) we get

$$E_0(\delta) = \frac{1}{2} \sum_J \omega_J \, e^{-\delta \omega_J}. \tag{4.1}$$

The regularization is removed in the limit $\delta \to 0$, restoring eqn (3.60). Obviously, the sum in eqn (4.1) converges for any $\delta > 0$.

This regularization was used in the original work by Casimir and also in Section 2.1. A modification of this scheme would be a sharp frequency cutoff in place of the exponential in eqn (4.1). In addition, any other sufficiently fast-decreasing function of ω_J (such as a sufficiently fast-decreasing frequency-dependent permittivity) or a momentum-dependent decreasing function (a momentum cutoff) can be used.

In a point splitting regularization, one starts from the representation (3.93) of the vacuum energy in terms of a Green's function without carrying out the coincidence limit. Then

$$E_0(\epsilon) = i \int_V dr \, \frac{\partial^2}{\partial x_0^2} \, G(x, x')|_{x'=x+\epsilon}, \tag{4.2}$$

where the regularization parameter ϵ is a four-dimensional vector. To obtain a finite $E_0(\epsilon)$, it is frequently sufficient to keep only the time component nonzero [$\epsilon = (\epsilon_0, \mathbf{0})$ with $\epsilon_0 \neq 0$]. The point-splitting technique emerged from quantum field theory. It was used in operator product expansions and for quantum fields in curved backgrounds. Moretti (1999) has shown that it is equivalent to zeta function regularization.

In a zeta function regularization, one temporarily changes the power of the frequency ω_J in the mode sum (3.60), leading to

$$E_0(s) = \frac{\mu^{2s}}{2} \sum_J \omega_J^{1-2s}. \qquad (4.3)$$

This converges for $\operatorname{Re} s > (d+1)/2$, where d is the dimensionality of the space. The factor μ^{2s}, where μ has the dimension of a mass, is arbitrary. It is introduced in order to keep the dimension of E_0. It disappears on removing the regularization in the limit $s \to 0$. This regularization is called *zeta function regularization* because the vacuum energy $E_0(s)$ is given by

$$E_0(s) = \frac{\mu^{2s}}{2} \zeta_P\left(s - \frac{1}{2}\right), \qquad (4.4)$$

which is expressed in terms of the generalized zeta function

$$\zeta_P(s) = \sum_J \frac{1}{\omega_J^{2s}}. \qquad (4.5)$$

The zeta function $\zeta_P(s)$ is associated with an elliptic boundary value problem. In our case this problem is specified by eqn (3.39) with the operator $-\Delta + m^2$ along with some boundary conditions, such as eqn (3.41) or (3.42). This zeta function can be viewed as a generalization of the Riemann zeta function

$$\zeta_R(z) = \sum_{n=1}^{\infty} \frac{1}{n^z} \qquad (4.6)$$

and shares most of its beautiful properties (Elizalde et al. 1994, Elizalde 1995). For instance, $\zeta_R(z)$ is meromorphic with a single pole at $z = 1$ on the real axis and $\zeta_P(s)$ is meromorphic with a finite number of poles on the real axis. The special case of the operator $-d^2/dx^2$ on the interval $x \in [0, \pi]$ with Dirichlet boundary conditions leads to $\zeta_P(s) = \zeta_R(2s)$.

It must be mentioned that, owing to the analytic properties of $\zeta_P(s)$, the vacuum energy in this regularization is defined on the entire complex plane for the parameter s with the exception of the poles. In three-dimensional space, its sum representation (4.5) is valid for $\operatorname{Re} s > 2$ only, but still serves as a starting point for analytic continuation.

4.2 The divergent part of the vacuum energy

The regularizations (4.1), (4.2), and (4.3) were introduced to have a finite representation of the vacuum energy. Here, we consider that part of this representation which becomes singular in the limit of removing the regularization.

4.2.1 The divergent part in the cutoff regularization

We consider the vacuum energy for a scalar field in a finite volume V enclosed by a surface S where boundary conditions, such as those discussed in Section 3.2, are imposed. In this case the eigenvalues in eqn (3.39) can be labeled by an integer $n = 1, 2, \ldots$. The frequencies $\omega_J \to \omega_n = \sqrt{\Lambda_n + m^2}$ also become labeled by the same integer n. Note that in some specific cases (e.g. the rectangular boxes considered in Chapter 8) it is convenient to use several integer indices. However, the values of these indices can be renumbered and represented by one index with integer values. For $\delta \to 0$, the divergent part of the vacuum energy (4.1) results from the asymptotic behavior of the eigenvalues for $n \to \infty$,

$$\Lambda_n = C n^{2/3} \left(1 + \frac{c_1}{n^{1/3}} + \frac{c_2}{n^{2/3}} + \frac{c_3}{n} + \frac{c_4}{n^{4/3}} + \ldots \right), \quad C = \left(\frac{6\pi^2}{V}\right)^{2/3}, \quad (4.7)$$

as was first obtained by Weyl (1912). The coefficients c_i depend on the area and other geometric characteristics of S. The easiest way to calculate the asymptotic expansion of $E_0(\delta)$ in eqn (4.1) for $\delta \to 0$ is to consider its Mellin transform

$$\tilde{E}_0(s) = \int_0^\infty \frac{d\delta}{\delta} \delta^{2s} E_0(\delta) = \frac{\Gamma(2s)}{2} \sum_{n=1}^\infty \omega_n^{1-2s}, \quad \operatorname{Re} s > s_0, \quad (4.8)$$

where, in the last equality, we have integrated the individual terms of the sum over n in eqn (4.1). $\tilde{E}_0(s)$ is defined by eqn (4.8) for $\operatorname{Re} s > s_0$ (s_0 must be sufficiently large to ensure the convergence of the integral). It is a meromorphic function with poles on the real axis for $\operatorname{Re} s < s_0$. The inverse Mellin transform is

$$E_0(\delta) = \int_{-i\infty}^{i\infty} \frac{ds}{\pi i} \delta^{-2s} \tilde{E}_0(s), \quad (4.9)$$

where the integration goes parallel to the imaginary axis with $\operatorname{Re} s > s_0$. In this representation, the behavior for small δ follows from the residues of $\tilde{E}_0(s)$ at the poles situated to the left of s_0. The poles of $\tilde{E}_0(s)$ can be found by inserting the asymptotic expansion (4.7) of the eigenvalues into eqn (4.8). We include the mass term by substituting $c_2 \to \tilde{c}_2 = c_2 + m^2/C$ and obtain

$$\tilde{E}_0(s) = \frac{\Gamma(2s)}{2} C^{(1-2s)/2} \sum_{n=1}^\infty n^{(1-2s)/3} \quad (4.10)$$

$$\times \left[1 + \frac{b_1(s)}{n^{1/3}} + \frac{b_2(s)}{n^{2/3}} + \frac{b_3(s)}{n} + \frac{b_4(s)}{n^{4/3}} + \ldots \right],$$

$$b_1(s) = \frac{1-2s}{2} c_1, \quad b_2(s) = \frac{1-2s}{8} \left[4\tilde{c}_2 - (1+2s)c_1^2\right],$$

$$b_3(s) = \frac{1-2s}{48} \left[24 c_3 - 12(1+2s) c_1 \tilde{c}_2 + (3+8s+4s^2) c_1^3\right],$$

$$b_4(s) = \frac{1-2s}{384} \Big\{ 192 c_4 - (1+2s) \left[48 \tilde{c}_2^2 + 96 c_1 c_3 - 24(3+2s) c_1^2 \tilde{c}_2 \right.$$

$$\left. + (15 + 16s + 4s^2) c_1^4 \right] \Big\}.$$

Equation (4.10) can be further transformed using eqn (4.6) into

$$\tilde{E}_0(s) = \frac{\Gamma(2s)}{2} C^{(1-2s)/2} \left[\zeta_R \left(\frac{2s-1}{3} \right) + b_1(s)\zeta_R \left(\frac{2s}{3} \right) + b_2(s)\zeta_R \left(\frac{2s+1}{3} \right) \right.$$
$$\left. + b_3(s)\zeta_R \left(\frac{2s+2}{3} \right) + b_4(s)\zeta_R \left(\frac{2s+3}{3} \right) + \ldots \right]. \quad (4.11)$$

From the pole of the Riemann zeta function (4.6) at $z = 1$, it follows that $\tilde{E}_0(s)$ has simple poles at $s = 2, 3/2, 1, 1/2$ and double poles at $s = 0, -1/2, -1, \ldots$ because of the poles of the gamma function. From the residues at these poles, the divergent part of the vacuum energy is

$$E_0^{\text{div}}(\delta) = \frac{3V}{2\pi^2} \frac{1}{\delta^4} + 3b_1(1.5) \left(\frac{V}{6\pi^2} \right)^{2/3} \frac{1}{\delta^3} + \frac{3b_2(1)}{2} \left(\frac{V}{6\pi^2} \right)^{1/3} \frac{1}{\delta^2}$$
$$- \frac{3b_4(0)}{2} \left(\frac{6\pi^2}{V} \right)^{1/3} \ln \delta. \quad (4.12)$$

The highest-order divergence is $1/\delta^4$. This is proportional to the volume and corresponds to the contribution of empty space. The next-order divergence, $1/\delta^3$, is proportional to the surface area. The weakest divergence is proportional to $\ln \delta$, which comes from the first double pole.

We note that these are the contributions which must be subtracted from the vacuum energy in order to get a finite expression when the regularization is removed. We postpone discussion of the justification and interpretation of the subtraction procedure. From eqn (4.12), we see that the first five terms in the asymptotic expansion (4.7) of the eigenvalues contribute to the divergent part of the vacuum energy. Thus, for an arbitrary surface S, direct numerical approaches to the calculation of the vacuum energy as a sum over the eigenvalues have not yet been successful.

4.2.2 The divergent part in the zeta function regularization and the heat kernel expansion

The powers of the frequency in eqn (4.3) can be identically represented as an integral,

$$\omega_J^{1-2s} = \int_0^\infty \frac{dt}{t} \frac{t^{s-\frac{1}{2}}}{\Gamma(s-\frac{1}{2})} e^{-t\omega_J^2}. \quad (4.13)$$

Interchanging the order of the summation and integration, the vacuum energy in the zeta function regularization can be expressed as

$$E_0(s) = \frac{\mu^{2s}}{2} \int_0^\infty \frac{dt}{t} \frac{t^{s-\frac{1}{2}}}{\Gamma(s-\frac{1}{2})} K(t) e^{-tm^2}, \quad (4.14)$$

where

$$K(t) = \sum_J e^{-t\Lambda_J} \quad (4.15)$$

is called the *heat kernel*. This is the spatial trace over the *local heat kernel* (Seeley 1969a, 1969b),

$$K(t) = \int_V d\mathbf{r}\ K(\mathbf{r},\mathbf{r}'|t)|_{\mathbf{r}'=\mathbf{r}}. \tag{4.16}$$

The local heat kernel obeys the heat conduction equation

$$\left(\frac{\partial}{\partial t} - \nabla_{\mathbf{r}}^2\right) K(\mathbf{r},\mathbf{r}'|t) = 0 \tag{4.17}$$

with the initial condition $K(\mathbf{r},\mathbf{r}'|t=0) = \delta^3(\mathbf{r}-\mathbf{r}')$. It must also fulfill the same boundary conditions as the field $\varphi(x)$.

In general, the heat kernel is the key object in the theory of heat conduction. It is important for the Casimir effect because its behavior for small t describes the divergences in the vacuum energy. An important feature of the heat kernel is that it has an asymptotic expansion for small t,

$$K(t) = \frac{1}{(4\pi t)^{3/2}} \left(a_0 + a_{1/2}\sqrt{t} + a_1 t + a_{3/2} t^{3/2} + \ldots\right), \tag{4.18}$$

where the $a_{k/2}$ ($k = 0, 1, 2, \ldots$) are the *heat kernel coefficients*. In this expansion, the term in front of the parentheses is universal. It depends only on the dimensionality of the space [it is $(4\pi t)^{-d/2}$ in a d-dimensional space]. For an elliptic differential operator, such as the Laplace operator, the expansion is in powers of $\sqrt{t}$.

The heat kernel coefficients have a very long history. They were introduced independently several times and are known under different names such as the Minakshisundaram–Pleijel coefficients and the Seeley or Seeley–deWitt coefficients. The heat kernel coefficients have been very well investigated [recently, an excellent review was given by Vassilevich (2003)]. They are universal in the sense that they depend only on the geometric characteristics of the volume V and its enclosing surface S, such as the curvature and its derivatives, and on the type of the boundary conditions.

In the following, we consider a volume V with a background field $U(\mathbf{r})$, which can be introduced as a position-dependent mass density by the substitution $m^2 \rightarrow m^2 + U(\mathbf{r})$ in the operator (3.6). We postpone considering a curvature such as in eqn (3.10) for later. Furthermore, we assume that the volume V is bounded by a surface S. The geometric properties of the surface can be expressed in terms of its second fundamental form at a point $\mathbf{r}$,

$$L_{11}\,d\eta_1^2 + 2L_{12}\,d\eta_1\,d\eta_2 + L_{22}\,d\eta_2^2, \tag{4.19}$$

where η_1, η_2 are coordinates on a surface in three-dimensional space and

$$L_{11} = \mathbf{n}\cdot\frac{\partial^2}{\partial\eta_1^2}\mathbf{r},\quad L_{12} = L_{21} = \mathbf{n}\cdot\frac{\partial}{\partial\eta_1}\frac{\partial}{\partial\eta_2}\mathbf{r},\quad L_{22} = \mathbf{n}\cdot\frac{\partial^2}{\partial\eta_2^2}\mathbf{r}. \tag{4.20}$$

Here n is the outward-pointing normal vector to the surface at a point r (Gray 1997). The heat kernel coefficients are represented as a sum of two local integrals, one over the volume (bulk part) and the other over the surface (surface part),

$$a_{k/2} = \int_V d\boldsymbol{r}\, b_{k/2}(\boldsymbol{r}) + \int_S d\mu(\eta)\, c_{k/2}(\eta). \tag{4.21}$$

Here, we use the same parametrization for S as in Section 3.6. The surface part is absent if the volume V has no boundary (for example, an interval with periodic conditions). We must warn readers that several different notations for the heat kernel coefficients are used in the literature. Sometimes the factor $(4\pi)^{-3/2}$ is included in their definition. Sometimes the enumeration is done with integer numbers, i.e. $a_{k/2} \to a_k$. In the notation of eqns (4.18) and (4.21) and for Dirichlet boundary conditions (upper entry in the curly brackets) and Neumann boundary conditions (lower entry in the curly brackets), the first few coefficients read

$$b_0 = 1, \quad c_0 = 0, \quad b_{1/2} = 0, \quad c_{1/2} = \left\{ \begin{array}{c} -1 \\ 1 \end{array} \right\} \frac{\sqrt{\pi}}{2}, \quad b_1 = U(\boldsymbol{r}), \quad c_1 = \frac{1}{3} L_{aa},$$

$$b_{3/2} = 0, \quad c_{3/2} = \frac{\sqrt{\pi}}{192} \left(\left\{ \begin{array}{c} -1 \\ 1 \end{array} \right\} 96 U(\boldsymbol{u}(\eta)) + \left\{ \begin{array}{c} -7 \\ 13 \end{array} \right\} L_{aa}^2 + \left\{ \begin{array}{c} 10 \\ 2 \end{array} \right\} L_{ab}L_{ab} \right),$$

$$b_2 = \frac{1}{2} U^2(\boldsymbol{r}), \quad c_2 = \frac{1}{360} \left[\left\{ \begin{array}{c} -120 \\ 240 \end{array} \right\} \partial_n U(\boldsymbol{r}) + 120 U(\boldsymbol{r}) L_{aa} + 24 L_{aa;bb} \right.$$

$$\left. + \frac{1}{21} \left(\left\{ \begin{array}{c} 40 \\ 280 \end{array} \right\} L_{aa}^3 + \left\{ \begin{array}{c} -264 \\ 168 \end{array} \right\} L_{ab}L_{ab}L_{cc} + \left\{ \begin{array}{c} 320 \\ 224 \end{array} \right\} L_{ab}L_{bc}L_{ca} \right) \right]. \tag{4.22}$$

Here, there is a summation over the repeated indices $a, b = 1, 2$, i.e.

$$L_{aa} = L_{11} + L_{22}, \quad L_{ab}L_{ab} = L_{11}^2 + 2L_{12}^2 + L_{22}^2, \quad \text{etc.,} \tag{4.23}$$

and the following notation is used:

$$L_{aa;bb} = \frac{\partial^2}{\partial \eta_1^2} L_{aa} + \frac{\partial^2}{\partial \eta_2^2} L_{aa}. \tag{4.24}$$

When inserted into eqn (4.21), the coefficient b_0 leads to the volume of V, $a_0 = V$, and $c_{1/2}$ is proportional to the area S of the surface. It should be noted that the coefficients with half-integer numbers result only from the boundary.

Below, we shall consider a sphere without a background field. Here the second fundamental form is simply $L_{ab} = \delta_{ab}/R$, such that $L_{aa} = 2/R$ and the coefficients become

$$a_0 = \frac{4\pi}{3} R^3, \quad a_{1/2} = \left\{ \begin{array}{c} -1 \\ 1 \end{array} \right\} 2\pi^{3/2} R^2, \quad a_1 = \frac{8\pi}{3} R,$$

$$a_{3/2} = \left\{ \begin{array}{c} -1 \\ 7 \end{array} \right\} \frac{\pi^{3/2}}{6}, \quad a_2 = \left\{ \begin{array}{c} -1 \\ 35 \end{array} \right\} \frac{16\pi}{315 R}. \tag{4.25}$$

Thus, in terms of the heat kernel expansion, complete information about the divergences of the vacuum energy is available. This is contained in the poles of

$E_0(s)$. These poles follow from eqn (4.14) together with eqn (4.18), from the integration in the vicinity of $t = 0$. We divide the integration over t into $t \in [0, 1]$ and $t \in [1, \infty)$. The integral over the second interval gives a regular expression. The contribution to $E_0(s)$ from the interval $[0, 1]$, denoted by $\tilde{E}_0(s)$, can be calculated after a power series expansion of the exponent in eqn (4.14) has been performed. The result is

$$\tilde{E}_0(s) = \frac{\mu^{2s}}{2(4\pi)^{3/2}\Gamma\left(s - \frac{1}{2}\right)} \left[\sum_{n=0}^{\infty} \frac{\tilde{a}_n}{s - 2 + n} + \sum_{k=0}^{\infty} \frac{\tilde{a}_{(2k+1)/2}}{s - 2 + (2k+1)/2}\right], \quad (4.26)$$

where the mass has been included in the redefined coefficients

$$\tilde{a}_n = \sum_{l=0}^{n} \frac{(-1)^l}{l!} a_{n-l} m^{2l}, \quad \tilde{a}_{(2k+1)/2} = \sum_{l=0}^{k} \frac{(-1)^l}{l!} a_{(2k+1-2l)/2} m^{2l}. \quad (4.27)$$

As we are interested in the pole at $s = 0$, we separate the pole part of $\tilde{E}_0(s)$:

$$E_0^{\mathrm{P}}(s) = -\frac{\tilde{a}_2}{32\pi^2 s} = -\frac{2a_2 - 2a_1 m^2 + a_0 m^4}{64\pi^2 s}. \quad (4.28)$$

This is the part of the vacuum energy which diverges when the regularization is removed. It contains the coefficients up to and including a_2. Higher-order coefficients do not contribute.

For massive fields, the heat kernel expansion provides an expansion in inverse powers of m. This can be obtained by inserting the heat kernel expansion (4.18) into eqn (4.14) and performing the integrations in each term of the sum:

$$E_0(s) = \frac{\mu^{2s}}{16\pi^{3/2}} \sum_{k=0}^{\infty} \frac{\Gamma\left(s + \frac{k}{2} - 2\right)}{\Gamma\left(s - \frac{1}{2}\right)} m^{4-2s-k} a_{k/2}. \quad (4.29)$$

It must be stressed that this is an asymptotic expansion for $E_0(s)$.

The terms of this expansion for $0 \leq k \leq 4$ diverge when $s \to 0$. These terms contain nonnegative powers of the mass. Expanding the terms with $k \leq 4$ in eqn (4.29) in powers of s around the point $s = 0$, we arrive at

$$E_0^{\mathrm{div}}(s) = -\frac{m^4}{64\pi^2}\left(\frac{1}{s} + \ln\frac{4\mu^2}{m^2} - \frac{1}{2}\right) a_0 - \frac{m^3}{24\pi^{3/2}} a_{1/2} \quad (4.30)$$

$$+ \frac{m^2}{32\pi^2}\left(\frac{1}{s} + \ln\frac{4\mu^2}{m^2} - 1\right) a_1 + \frac{m}{16\pi^{3/2}} a_{3/2} - \frac{1}{32\pi^2}\left(\frac{1}{s} + \ln\frac{4\mu^2}{m^2} - 2\right) a_2.$$

We call this the divergent part of the vacuum energy in zeta function regularization, although it also contains some finite contributions. This definition makes sense only for a theory with a nonzero mass m or, equivalently, with a gap in the spectrum. For a massless field or a gapless spectrum, one must return to the divergent pole part, eqn (4.28), which is also meaningful for $m = 0$.

For comparison, it is instructive to express the vacuum energy using cutoff regularization, eqn (4.1), in terms of the heat kernel coefficients. As can be seen from eqns (4.8) and (4.3), the Mellin transform $\tilde{E}_0(s)$ (4.8) of the vacuum energy in cutoff regularization is related to the vacuum energy in zeta function regularization by means of

$$\tilde{E}_0(s) = \frac{\Gamma(2s)}{\mu^{2s}} E_0(s). \tag{4.31}$$

Substituting the series (4.26) into eqn (4.31) and using the inverse Mellin transform (4.9), we obtain the divergent part of the vacuum energy in cutoff regularization in terms of the heat kernel coefficients,

$$E_0^{\text{div}}(\delta) = \frac{3a_0}{2\pi^2} \frac{1}{\delta^4} + \frac{a_{1/2}}{4\pi^{3/2}} \frac{1}{\delta^3} + \frac{\tilde{a}_1}{8\pi^2} \frac{1}{\delta^2} + \frac{\tilde{a}_2}{16\pi^2} \ln \delta. \tag{4.32}$$

The coefficients $\tilde{a}_1, \tilde{a}_2$ are defined in eqn (4.27). We mention that a comparison of this formula with eqn (4.12) allows one to establish a connection between the heat kernel coefficients and the coefficients c_i in the Weyl expansion of the eigenvalues (4.7).

Now we consider the representation (3.112) of the vacuum energy which follows from the effective action. Using eqn (3.111), we represent the effective vacuum energy in the form

$$E_0 = E_{0,\text{eff}} = \frac{i}{2T} \operatorname{Tr} \ln G(x, x')|_{x=x'}. \tag{4.33}$$

Here "Tr" is understood as the sum of all of the diagonal matrix elements calculated with the functions

$$\phi_{\omega' J'}(x) = \frac{1}{\sqrt{2\pi}} e^{-i\omega' t} \Phi_{J'}(\boldsymbol{r}). \tag{4.34}$$

It is easily seen that the matrix of the Green's function (3.87) is diagonal in the basis (4.34). Because of this, we obtain

$$\ln G(x, x') = -\int_{-\infty}^{\infty} \frac{d\omega}{2\pi} \sum_J \Phi_J(\boldsymbol{r}) \Phi_J^*(\boldsymbol{r}') \ln(-\omega^2 + \omega_J^2) e^{-i\omega(t-t')}. \tag{4.35}$$

Calculating Tr of eqn (4.35) using the orthonormality of the basis functions (4.34), we get

$$\operatorname{Tr} \ln G(x, x') = -\frac{1}{2\pi} \int_{-\infty}^{\infty} d\omega' \sum_{J'} \int_{-\infty}^{\infty} dt \int d\boldsymbol{r} \int_{-\infty}^{\infty} dt' \int d\boldsymbol{r}' e^{i\omega'(t-t')}$$
$$\times \Phi_{J'}^*(\boldsymbol{r}) \int_{-\infty}^{\infty} \frac{d\omega}{2\pi} \sum_J \Phi_J(\boldsymbol{r}) \Phi_J^*(\boldsymbol{r}') \ln(-\omega^2 + \omega_J^2) e^{-i\omega(t-t')} \Phi_{J'}(\boldsymbol{r}')$$

$$= -T \int_{-\infty}^{\infty} \frac{d\omega}{2\pi} \sum_J \ln(-\omega^2 + \omega_J^2). \tag{4.36}$$

Note that the integration with respect to t in eqn (4.36) results in $2\pi\delta(\omega' - \omega)$ and the subsequent integration with respect to ω' and t' gives a total time T. Substituting eqn (4.36) into eqn (4.33) and using eqn (3.39), we arrive at

$$E_{0,\text{eff}} = -\frac{i}{2} \int_{-\infty}^{\infty} \frac{d\omega}{2\pi} \sum_J \ln(-\omega^2 + \Lambda_J + m^2). \tag{4.37}$$

Again we are faced with an infinite expression. Its zeta function regularization is

$$E_{0,\text{eff}}(s) = \frac{i}{2} \frac{\partial}{\partial s} \mu^{2s} \int_{-\infty}^{\infty} \frac{d\omega}{2\pi} \sum_J \left(-\omega^2 + \Lambda_J + m^2\right)^{-s}, \tag{4.38}$$

where μ, which has the dimension of a mass, is again an arbitrary parameter. In fact, by introducing this parameter we have added a constant in $E_{0,\text{eff}}(s)$ which arises from the differentiation of the factor μ^{2s}. However, this constant does not depend on the boundary conditions or on the background in the limit of removing the regularization $s \to 0$. The derivative of the integral in eqn (4.38) restores the logarithm and, on removal of the regularization, we return to eqn (4.38). The sum on the right-hand side is also a zeta function, but it is different from eqn (4.5). Introducing a new variable $\omega = t\sqrt{\Lambda_J + m^2}$, we obtain from eqn (4.38)

$$E_{0,\text{eff}}(s) = \frac{i}{4\pi} \frac{\partial}{\partial s} \mu^{2s} \sum_J (\Lambda_J + m^2)^{-(2s-1)/2} \int_{-\infty}^{\infty} dt \, (1 - t^2)^{-s}. \tag{4.39}$$

Using the definition of the generalized zeta function (4.5) and calculating the integral (Gradshtein and Ryzhik 1994), we get

$$E_{0,\text{eff}}(s) = -\frac{1}{4\sqrt{\pi}} \frac{\partial}{\partial s} \mu^{2s} \frac{\Gamma(s - \frac{1}{2})}{\Gamma(s)} \zeta_P\left(s - \frac{1}{2}\right). \tag{4.40}$$

Comparison with eqn (4.4) allows one to establish the relationship with the vacuum energy in zeta function regularization, eqn (4.3),

$$E_{0,\text{eff}}(s) = -\frac{1}{2\sqrt{\pi}} \frac{\partial}{\partial s} \frac{\Gamma(s - \frac{1}{2})}{\Gamma(s)} E_0(s). \tag{4.41}$$

In this representation, a remarkable property of the vacuum energy defined by eqn (3.112) follows. This energy is not singular if the zeta-function-regularized vacuum energy has at most a simple pole in $s = 0$. Indeed, representing this energy as

$$E_0(s) = E_0^{\text{P}}(s) + E_0^{\text{reg}}(s), \tag{4.42}$$

i.e. as a sum of the pole part (4.28) and a regular part for $s \to 0$, we get from eqn (4.41)

$$E_{0,\text{eff}}(0) = (\ln 2 - 1)\frac{\tilde{a}_2}{16\pi^2} + E_0^{\text{reg}}(0). \tag{4.43}$$

We note that this expression does not contain a singularity for $s \to 0$, and, as a result, $E_{0,\text{eff}}(s)$ in eqn (4.38) is finite for $s \to 0$ (provided $E_0(s)$ has only a single pole at $s = 0$). It should be mentioned that sometimes, because of this property, the zeta function regularization resulting in eqn (4.43) has been interpreted as zeta function renormalization. This interpretation, however, has limited applicability since it does not provide a unique definition of the vacuum energy. This follows from the fact that the quantity $E_0^{\text{reg}}(s)$ contains terms depending on μ [see eqn (4.30)],

$$E_0^{\text{reg}}(s) = -\frac{\tilde{a}_2}{16\pi^2}\ln\frac{\mu}{m} + \ldots \tag{4.44}$$

(unless the heat kernel coefficients in $\tilde{a}_2$ are zero). In Section 4.3, we shall treat the renormalization of $E_{0,\text{eff}}(s)$ in the same manner as in the other representations.

We conclude this section with a definition of the divergent part

$$E_{0,\text{eff}}^{\text{div}}(s) = -\frac{1}{2\sqrt{\pi}}\frac{\partial}{\partial s}\frac{\Gamma(s-\frac{1}{2})}{\Gamma(s)}E_0^{\text{div}}(s) \tag{4.45}$$

which uses eqn (4.41) and $E_0^{\text{div}}(s)$, found in eqn (4.30). Performing the differentiation in eqn (4.45) and then expanding in powers of s, we obtain

$$E_{0,\text{eff}}^{\text{div}}(s) = -\frac{\tilde{a}_2}{32\pi^2}\ln\frac{\mu^2}{m^2} \tag{4.46}$$
$$+ \frac{1}{128\pi^2}\left(-3m^4 a_0 - \frac{16\sqrt{\pi}}{3}m^3 a_{1/2} + 4m^2 a_1 + 8\sqrt{\pi} m a_{3/2}\right) + O(s).$$

Again, similarly to eqn (4.43), this expression is finite for $s \to 0$, and the notation has been chosen in uniformity with eqns (4.30) and (4.32). As expected, the first term in eqn (4.46) coincides with that in eqn (4.44).

4.3 Renormalization of the vacuum energy

After we have obtained regularized expressions for the vacuum energy, it is necessary to remove the divergences, give an interpretation of this procedure, and address the key question about its uniqueness. Nonunique features are always present owing to the choice of the regularization scheme and parameters such as μ in eqn (4.3).

The simplest case of renormalization is that of a quantum field coupled to a smooth background field. We start with this case, where one can follow the well-known procedures from quantum field theory. Next, complications to this can be added in two ways, either by adding a boundary or by making the background field singular. We conclude this section with the easiest (from the renormalization point of view) case of forces between two separate bodies, which are always finite.

4.3.1 Smooth background fields

Here we consider the vacuum energy of a quantum field of mass m with a background of a smooth classical field of mass M in unbounded space-time. Some physical examples are the quantum fields for matter and radiation in a gravitational or electrodynamic background. However, there are no smooth background fields in Casimir problems, and we discuss this case only to illustrate the renormalization procedure. Thus, it is reasonable not to deal with electromagnetic or gravitational fields but instead to choose a technically simpler example of a classical scalar field $\phi(x)$ (the background field) and a quantum field $\varphi(x)$ with the action

$$S = -\frac{1}{2}\int d^4x\,\phi(x)\left[\Box + M^2 + \lambda\phi^2(x)\right]\phi(x)$$
$$- \frac{1}{2}\int d^4x\,\varphi(x)\left[\Box + m^2 + \tilde\lambda\phi^2(x)\right]\varphi(x). \tag{4.47}$$

Here we have included a self-interaction term for the background field with some constant λ. In the action of the quantum field, the interaction term $\tilde\lambda\phi^2(x)\varphi^2(x)$ can be viewed as an additional position-dependent mass density. We assume the background field to be static, i.e. $\phi(x) \to \phi(\mathbf{r})$. Thus, when the quantum field is in the vacuum state, the system has a definite energy,

$$E = E_{\text{class}} + E_0, \tag{4.48}$$

where

$$E_{\text{class}} = \frac{1}{2}\int d\mathbf{r}\,\phi(\mathbf{r})\left[-\nabla^2 + M^2 + \lambda\phi^2(\mathbf{r})\right]\phi(\mathbf{r}) \tag{4.49}$$

is the classical part of the energy. As to the vacuum energy E_0, we take this in the form given by eqn (3.60) and use zeta function regularization (4.3),

$$E_0(s) = \frac{\mu^{2s}}{2}\sum_J \omega_J^{1-2s}. \tag{4.50}$$

The eigenvalues in eqn (4.50), after the replacement $m^2 \to m^2 + \tilde\lambda\phi^2(\mathbf{r})$, are subject to the equation

$$\left[-\nabla^2 + \tilde\lambda\phi^2(\mathbf{r})\right]\Phi_J(\mathbf{r}) = \Lambda_J\Phi_J(\mathbf{r}), \tag{4.51}$$

a generalization of eqn (3.39). The divergent part of the vacuum energy is given by eqn (4.30) with the heat kernel coefficients

$$a_1 = \tilde\lambda\int d\mathbf{r}\,\phi^2(\mathbf{r}), \qquad a_2 = \frac{\tilde\lambda^2}{2}\int d\mathbf{r}\,\phi^4(\mathbf{r}). \tag{4.52}$$

These coefficients follow from the bulk part in eqn (4.22), where we have inserted $U(\mathbf{r}) = \tilde\lambda\phi^2(\mathbf{r})$ ($a_{1/2}$ and $a_{3/2}$ are zero since we have no boundary). We drop the contribution to eqn (4.30) from $a_0 = V$, which is infinitely large in unbounded space (see the discussion at the end of this subsection). As a result,

$$E_0^{\text{div}}(s) = \frac{m^2 \tilde{\lambda}}{32\pi^2} \left(\frac{1}{s} + \ln \frac{4\mu^2}{m^2} - 1 \right) \int d\bm{r}\, \phi^2(\bm{r})$$
$$- \frac{\tilde{\lambda}^2}{64\pi^2} \left(\frac{1}{s} + \ln \frac{4\mu^2}{m^2} - 2 \right) \int d\bm{r}\, \phi^4(\bm{r}). \qquad (4.53)$$

It should be noted that the divergent part repeats the structures present in the classical energy.

Representing the complete energy (4.48) in the form

$$E = E_{\text{class}} + E_0^{\text{div}}(s) + E_0(s) - E_0^{\text{div}}(s), \qquad (4.54)$$

we can absorb $E_0^{\text{div}}(s)$ into the classical energy by introducing the renormalized parameters of the classical field

$$M_{\text{ren}}^2 = \lim_{s \to 0} \left[M^2 + \frac{m^2 \tilde{\lambda}}{32\pi^2} \left(\frac{1}{s} + \ln \frac{4\mu^2}{m^2} - 1 \right) \right],$$
$$\lambda_{\text{ren}} = \lim_{s \to 0} \left[\lambda - \frac{\tilde{\lambda}^2}{64\pi^2} \left(\frac{1}{s} + \ln \frac{4\mu^2}{m^2} - 2 \right) \right]. \qquad (4.55)$$

Thus, the renormalized classical part of the energy is given by

$$E_{\text{class}}^{\text{ren}} = \lim_{s \to 0} \left[E_{\text{class}} + E_0^{\text{div}}(s) \right]$$
$$= \frac{1}{2} \int d\bm{r}\, \phi(\bm{r}) \left[-\bm{\nabla}^2 + M_{\text{ren}}^2 + \lambda_{\text{ren}} \phi^2(\bm{r}) \right] \phi(\bm{r}). \qquad (4.56)$$

The renormalized vacuum energy is then given by

$$E_0^{\text{ren}} = \lim_{s \to 0} \left[E_0(s) - E_0^{\text{div}}(s) \right]. \qquad (4.57)$$

The same approach can be used in other regularization schemes. For instance, we can consider the cutoff regularization. When using, instead of eqn (4.50), the vacuum energy in cutoff regularization (4.1), one needs to use the divergent part given in eqn (4.32). The heat kernel coefficients are the same as before [eqn (4.52)], and the only change in the above scheme will be slightly different formulas for the renormalized mass M_{ren} and the self-interaction constant λ_{ren},

$$M_{\text{ren}}^2 = M^2 + \frac{\tilde{\lambda}}{16\pi^2} \left(\frac{2}{\delta^2} - m^2 \ln \delta \right), \qquad \lambda_{\text{ren}} = \lambda + \frac{\tilde{\lambda}^2}{32\pi^2} \ln \delta. \qquad (4.58)$$

Finally, we consider this procedure for the vacuum energy (3.112),

$$E_{0,\text{eff}} = -\frac{i}{2T} \text{Tr} \ln \left(\Box + m^2 + \tilde{\lambda} \phi^2(\bm{r}) \right), \qquad (4.59)$$

which follows from the effective action in Section 3.5 with the replacement $m^2 \to m^2 + \tilde{\lambda} \phi^2(\bm{r})$.

In the presence of the background field, the divergent part $E_{0,\text{eff}}^{\text{div}}$ can be defined in the same way as in Section 4.2.2. Inserting the heat kernel coefficients (4.52) into eqn (4.46), we obtain

$$E_{0,\text{eff}}^{\text{div}}(s) = \frac{m^2 \tilde{\lambda}}{32\pi^2}\left(\ln\frac{\mu^2}{m^2}+1\right)\int d\boldsymbol{r}\,\phi^2(\boldsymbol{r}) \qquad (4.60)$$
$$-\frac{\tilde{\lambda}^2}{64\pi^2}\ln\frac{\mu^2}{m^2}\int d\boldsymbol{r}\,\phi^4(\boldsymbol{r}) + O(s).$$

In this case the renormalized parameters of the classical field are

$$M_{\text{ren}}^2 = M^2 + \frac{m^2\tilde{\lambda}}{32\pi^2}\left(\ln\frac{\mu^2}{m^2}+1\right), \quad \lambda_{\text{ren}} = \lambda - \frac{\tilde{\lambda}^2}{64\pi^2}\ln\frac{\mu^2}{m^2}. \qquad (4.61)$$

Here, in contrast with eqns (4.55) and (4.58), both M_{ren}, λ_{ren} and M, λ are finite, but, as mentioned above, this is only a peculiarity of the representation used.

We thus obtain a finite vacuum energy (4.57) which must be added to the classical energy (4.56). As explained above, the parameters of the classical energy have been renormalized. This is, however, not a problem, since the renormalized values must be determined independently anyway (usually experimentally). This is the general scheme of renormalization known in quantum field theory.

It must be mentioned that in this model, renormalization requires the self-interaction term in the classical part, and this is in agreement with the standard counting of the superficial degrees of divergence. Furthermore, we remark that there is no renormalization of the term containing the derivatives in the classical energy. This follows from the absence of a corresponding structure in the heat kernel coefficients and implies the nonrenormalization of the classical field $\Phi(\boldsymbol{r})$.

Also, note that a renormalization scheme such as that suggested by eqn (4.55) or (4.58) is not unique. This is due to the fact that with an infinite renormalization, we can always include a finite renormalization and still remove the singularities. This is similar to a change in the definition of the divergent part of the vacuum energy. Also, the parameter μ and the choice of the regularization lead to nonuniqueness.

A discussion of this nonuniqueness involves deeply the particular model considered. For instance, within the model given by eqn (4.47), it would be natural to look for a minimum of the complete energy E in eqn (4.48). It is clear that one may perform a redistribution of the energy between the two parts of eqn (4.48). This can be viewed as an additional finite renormalization. Consequently, here, the vacuum energy does not have an independent meaning.

Another method of proceeding is to impose a normalization condition on the renormalized vacuum energy such that it becomes uniquely defined after the regularization is removed (regardless of the regularization scheme). One of these conditions is the so-called *no-tadpole* condition introduced by Graham *et* al. (2004). A second approach follows from a consideration of the mass of the quantum field together with the large-mass expansion (4.29), by demanding that

$$E_0^{\text{ren}} \to 0 \quad \text{for} \quad m \to \infty. \tag{4.62}$$

The motivation for this condition is that an infinitely heavy field should not have quantum fluctuations and hence should not produce a vacuum energy. This condition, as follows from the heat kernel expansion, is equivalent to the subtraction of all contributions involving the heat kernel coefficients a_0 through a_2 because these enter the large-mass expansion with nonnegative powers of the mass. In this manner, one can give the vacuum energy a unique meaning independent of a classical model. The definitions (4.30), (4.32), and (4.46) of the divergent part are given in such a way that the corresponding renormalized vacuum energy fulfills the normalization condition (4.62). As a consequence, the renormalized vacuum energy given by eqn (4.57) is unique, i.e. it does not depend either on the regularization chosen or on the parameter μ. This normalization condition was discussed by Bordag (2000). However, as mentioned there, this condition is meaningful for massive fields only. In the case of a massless field, this approach is not applicable, and there is as yet no known way to give a satisfactory renormalization condition independent of the classical model (unless, of course, the corresponding heat kernel coefficients are zero).

Next, we draw special attention to the divergent contribution resulting from the heat kernel coefficient a_0. On the one hand, in this simple model, we do not have a classical counterpart that has the same structure (proportional to the volume V, which is infinite here). On the other hand, this contribution does not depend on the background field. This is clearly the contribution which would be present in empty space, i.e. in the absence of the background field. Because of this, we do not relate it to the vacuum energy resulting from the background, and therefore drop it. This is the same case as when one considers only the response of the vacuum energy to a change in the background field. There is, however, one scenario where this is not possible. Namely, when we consider quantum fluctuations in a gravitational background, we cannot drop this contribution, because it is the source of the gravitational field. But in that case there exists a structure for renormalization, namely the term containing the cosmological constant, which needs to be renormalized in the same way as the gravitational constant.

4.3.2 Singular background fields and boundary conditions

The situation described in the preceding subsection changes when nonsmooth background fields are considered. From a formal point of view, one first observes that the heat kernel coefficients $a_{k/2}$ become infinite starting from some k. This is because the coefficients contain powers of the background field and its derivatives in increasing order. For instance, when the interaction potential in the model (4.47) becomes proportional to a delta function on a sphere of radius R, i.e.

$$\phi^2(\mathbf{r}) = \alpha \delta(r - R), \tag{4.63}$$

a_2 becomes singular since it contains the delta function squared, which is not a well-defined object. Mathematically, the problem is related to the noncommutativity of the two limits, one arising from the asymptotic expansion of the heat kernel for small argument t and the other from making the background singular. The physical meaning can be seen from the model (4.47) considered in the preceding subsection. If one takes the classical part in order to accommodate the renormalization, then it must contain the $\lambda\phi^4$ self-interaction term. But this term gives an infinite contribution to the classical energy in the limit (4.63). Thus one would need an infinite amount of energy in order to make the background field singular. For these reasons, the scheme of letting the background become singular does not seem very natural, as observed by Graham et al. (2002, 2003, 2004).

The situation is different when one starts from an already singular background. Here, the only case which has been investigated so far in any detail is a one-dimensional delta function potential on a spherical surface. In that case the spectral problem for the fluctuations is well defined and all the heat kernel coefficients exist. For example, with eqn (4.63), the heat kernel coefficients are (Bordag et al. 1999a)

$$a_{k/2} = 4\pi R^2 c_{k/2} \quad \text{with} \quad c_{1/2} = 0, \quad c_1 = -\alpha,$$

$$c_{3/2} = \frac{\sqrt{\pi}}{4}\alpha^2, \quad c_2 = -\frac{1}{6}\alpha^3. \tag{4.64}$$

In the above, we have written down only those coefficients which are relevant to the renormalization. It is a characteristic of this singular background that the coefficient $a_{3/2}$, with a half-integer number, appears [for a more general discussion see Bordag and Vassilevich (1999, 2004)].

Finally, we consider the vacuum energy in the presence of boundary conditions. This case is the most relevant for the Casimir effect. For simplicity, we restrict ourselves to the case of a sphere with Dirichlet or Neumann boundary conditions. The heat kernel coefficients are given by eqn (4.25). Since no background field exists, one needs to introduce other classical parameters in order to accommodate the renormalization. Blau et al. (1988) suggested the geometric structure

$$E_{\text{class}} = pV + \sigma S + h_1 R + h_2 + h_3 \frac{1}{R}, \tag{4.65}$$

where p has the meaning of a pressure and σ of a surface tension. But h_1, h_2, and h_3 do not appear to have standard meanings. Now, taking the vacuum energy in any regularization, the divergent part can be removed by a corresponding renormalization of the parameters p, σ, h_1, h_2, and h_3. This procedure is completely parallel to that in the preceding subsection, done for smooth background fields. It is also clear that it can be directly generalized to a surface of a generic shape using the heat kernel coefficients in eqn (4.22).

However, a situation may arise where there is no classical system available for the justification of the renormalization. In that case one could take the normalization condition (4.62), provided the quantum field has a mass. If the field is massless and the corresponding heat kernel coefficients do not vanish, one cannot give the vacuum energy a satisfactory interpretation. The most important examples are that of a conducting sphere of finite thickness and that of a dielectric ball. These will be discussed in Sections 9.3.3 and 9.3.4.

4.3.3 Finiteness of the Casimir force between separate bodies

We have seen in Section 4.2.2 that the divergent part of the vacuum energy follows from the heat kernel coefficients a_0 through a_2. These coefficients are represented by eqn (4.21) as integrals over local quantities: the background potential and the coefficients of the second fundamental form (4.19), including their derivatives and powers. This locality is a fundamental property of the heat kernel coefficients that holds under very general assumptions. It is believed that it is related to the locality of the ultraviolet divergences in quantum field theory.

With respect to the Casimir effect, the local nature of the coefficients determining the divergent part of the vacuum energy has a far-reaching consequence. The definition of the heat kernel coefficients presented in Section 4.2.2 is of a rather general character. It refers both to simply connected manifolds (a compact body with some finite volume restricted by a boundary surface S) and to nonsimply connected manifolds. As an example of the latter, let us consider two separate, i.e. nonintersecting bodies with volumes V_1 and V_2 and surfaces S_1 and S_2 having no common points. We also assume that there is no background field. It follows from the latter that all of the local heat kernel coefficients $b_{k/2}$ in eqn (4.22), excepting b_0, are equal to zero. Thus, none of the global heat kernel coefficients $a_{k/2}$ in eqn (4.21) with $k \geq 1$ contain a volume contribution. They are given by

$$a_{k/2} = \int_{S_1} d\mu(\eta)\, c^{(1)}_{k/2}(\eta) + \int_{S_2} d\mu(\eta)\, c^{(2)}_{k/2}(\eta). \tag{4.66}$$

Here the local coefficients $c^{(1)}_{k/2}$ and $c^{(2)}_{k/2}$ are defined by eqn (4.22) for the respective parts S_1 and S_2 of the surface S. These coefficients need not be the same. Even the boundary conditions on S_1 and S_2 may be different.

In the case of two separate interacting bodies, it is reasonable to consider the spatial region $V - V_1 - V_2$, where V is the infinite volume of the entire three-dimensional space, restricted by the boundary surface S consisting of S_1 and S_2. In doing so, we change the sign of the direction of the local normal vector $\boldsymbol{n}$ to the surface. This leads to an opposite sign for the coefficients of the second fundamental form given in eqn (4.19). For example, in the case of the Casimir interaction between two spheres with radii R_1 and R_2, $L^{(1)}_{aa} = -2/R_1$ and $L^{(2)}_{aa} = -2/R_2$. As a result, from eqns (4.21) and (4.22) we obtain

$$a_0 = V - V_1 - V_2. \tag{4.67}$$

We drop the contribution proportional to this coefficient in the divergent part of the vacuum energy (4.30). This is equivalent to subtraction of the zero-point energy arising from the free space between the bodies.

The divergent part of the vacuum energy (4.30) is then determined by the coefficients (4.66) with $1 \leq k \leq 4$. As is seen from the structure of eqn (4.66), the $a_{k/2}$ do not contain any information about the relative location of the parts S_1 and S_2 of the boundary surface S. In other words, the heat kernel coefficients do not depend on the distance between the interacting bodies under consideration. If we now insert the $a_{k/2}$ into the divergent part (4.30) or (4.32) or the pole part (4.28) of the vacuum energy in any regularization, that part is also found to be independent of the distance. The distance dependence is contained only in the finite renormalized vacuum energy E_0^{ren} defined in eqn (4.57). We emphasize that information about the distance dependence of E_0^{ren} cannot be obtained from the heat kernel expansion. It is contained in the finite part of the energy remaining after subtraction of the divergent part. As the divergent part is independent of the separation, the force between two separate bodies is always finite.

5
THE CASIMIR EFFECT AT NONZERO TEMPERATURE

So far, we have limited our discussion to the Casimir effect resulting from the energy of the vacuum state of a quantum field in the presence of boundaries. All excitations were neglected. In practice, the appropriate state of the quantum field is a state containing real particles. The typical situation is a state containing particles in thermal equilibrium. In fact, one has to consider an ensemble of states characterized by a temperature T and a probability distribution. The energy of such an ensemble in the presence of spatial boundaries is then considered as the Casimir energy at nonzero temperature.

So, let us consider a quantum system at nonzero temperature T in thermal equilibrium. It is characterized by a Gibbs distribution and a partition function

$$Z = \sum_n e^{-E_n/(k_B T)}, \tag{5.1}$$

where k_B is the Boltzmann constant. The sum is taken over all states n, and E_n is the energy of the state n. From the partition function, all thermodynamic quantities, such as the free energy,

$$\mathcal{F} = -k_B T \ln Z, \tag{5.2}$$

the pressure

$$P = -\left(\frac{d\mathcal{F}}{dV}\right)_T, \tag{5.3}$$

and the entropy

$$S = -\frac{\partial \mathcal{F}}{\partial T}, \tag{5.4}$$

can be derived.

In quantum field theory, there exist several methods to treat a system at nonzero temperature. The easiest and most frequently used method is the imaginary-time Matsubara formalism, which is applicable to a system at thermal equilibrium (Matsubara 1955). For time-dependent and nonequilibrium processes, the real time formalism may be used. But this and other related approaches have not played a significant role in the study of the Casimir effect.

5.1 The Matsubara formulation

In the Matsubara formalism, one uses a Euclidean field theory, considered as a continuation of the theory in Minkowski space–time by a rotation of the time coordinate $t \to -i\tau$. The Euclidean time τ is confined to the interval $\tau \in [0, \beta]$, where $\beta = 1/(k_B T)$ is the equivalent dimension corresponding to the inverse temperature. The fields must obey periodicity conditions on this interval; bosonic fields must be periodic, $\varphi(\tau + \beta, \boldsymbol{r}) = \varphi(\tau, \boldsymbol{r})$, and fermionic fields (which we do not consider at nonzero temperature) must be antiperiodic, $\varphi(\tau + \beta, \boldsymbol{r}) = -\varphi(\tau, \boldsymbol{r})$, in accordance with their statistics. In the limit of zero temperature one reobtains the theory on the whole time axis.

In the Matsubara formalism, the partition function Z in eqn (5.1) has the following representation in terms of a functional integral:

$$Z = C \int D\varphi\, e^{-S_E[\varphi]}, \tag{5.5}$$

where $S_E[\varphi]$ is the Euclidean action. It can be obtained from the corresponding action in Minkowski space–time (3.100) by the replacement of S with iS_E. For example, using eqn (3.5), for a scalar field with $\Upsilon = 0$, we have

$$S_E[\phi] = \frac{1}{2}\int_0^\beta d\tau \int d\boldsymbol{r}\, \varphi K_E\, \varphi, \tag{5.6}$$

where

$$K_E = \left(-\Box_E + m^2\right). \tag{5.7}$$

The *Euclidean wave operator*

$$\Box_E = \frac{\partial^2}{\partial \tau^2} + \boldsymbol{\nabla}^2 \tag{5.8}$$

is the continuation of the d'Alembertian (3.2), which in fact is the four-dimensional Laplacian. In the functional integral (5.5), the field to be integrated over must fulfill the corresponding periodicity conditions.

In general, in the Matsubara formalism, the construction of the theory, to a large extent, goes in parallel to the zero-temperature case. In this way, most of the formulas for the vacuum energy in Chapter 3 may be directly translated to the case of nonzero temperature. This is true, for instance, in the calculation of the functional integral. Since we are continuing to consider free-field theories, the functional integral is Gaussian and can be calculated directly. Using an approach similar to that in Section 3.5 and the infinite-dimensional analogue of eqn (3.104), we obtain the following from eqn (5.5) for the partition function:

$$Z = C\left(\det K_E\right)^{-1/2}, \tag{5.9}$$

where C is an irrelevant constant, which will be dropped below. Further, for the free energy we get

$$\mathcal{F} = \frac{1}{2\beta}\mathrm{Tr}\ln K_E, \tag{5.10}$$

which is analogous to eqn (3.112). The trace in this expression is taken over the space of fields to be integrated over in the functional integral (5.5).

Since we assume thermal equilibrium, it is always possible to separate the Euclidean time variable from the spatial variables. Assuming for the spatial part an eigenfunction expansion as in eqn (3.39),

$$-\nabla^2 \Phi_J(\boldsymbol{r}) = \Lambda_J \Phi_J(\boldsymbol{r}), \tag{5.11}$$

we obtain a basis in the space of fields φ,

$$\Phi_{l,J}(\tau, \boldsymbol{r}) = \frac{e^{-i\xi_l \tau}}{\sqrt{2\pi}} \Phi_J(\boldsymbol{r}), \tag{5.12}$$

which contains the *Matsubara frequencies*

$$\xi_l = 2\pi k_\mathrm{B} T l, \qquad l = 0, \pm 1, \pm 2, \ldots. \tag{5.13}$$

These functions are periodic in τ (for a fermionic field, one has to take half-integer values of l), and these are eigenfunctions of K_E,

$$K_\mathrm{E} \Phi_{l,J} = \left(\xi_l^2 + \Lambda_J + m^2\right) \Phi_{l,J}. \tag{5.14}$$

As a consequence, the trace in the free energy becomes a sum over the logarithms of the eigenvalues,

$$\mathcal{F}_0 = \frac{1}{2} k_\mathrm{B} T \sum_{l=-\infty}^{\infty} \sum_J \ln\left(\xi_l^2 + \Lambda_J + m^2\right). \tag{5.15}$$

As before for the energy, the lower index 0 stands for the nonrenormalized quantity. This formula generalizes eqn (3.112) to the case of nonzero temperature. Note that if the field is massless and all quantum numbers in the collective index J are discrete, it is assumed in both eqn (3.112) and eqn (5.15) that there are no physical states with $\Lambda_J = 0$.

In eqn (3.112), E_0 is the energy of the vacuum of the corresponding quantum field. In eqn (5.15), $\mathcal{F}_0$ is the energy (more exactly, the free energy in the thermodynamic sense) of an ensemble of states containing particles at the temperature T. In the special case of $T \to 0$, the time interval stretches over the whole axis and the sum over the Matsubara frequencies becomes an integral over the frequency ξ:

$$k_\mathrm{B} T \sum_{l=-\infty}^{\infty} f(\xi_l) \to \int_{-\infty}^{\infty} \frac{d\xi}{2\pi} f(\xi) \tag{5.16}$$

[here $f(\xi_l)$ is a function which must allow an analytic continuation from discrete values to continuous ones]. In this way, $\mathcal{F}_0$ defined in eqn (5.15) turns into the vacuum energy E_0 in eqn (3.112) [to be exact, in eqn (3.112) one should in addition pass to the Euclidean time variable].

The free energy, as given by eqn (5.15), still contains ultraviolet divergences and one has to introduce a regularization. For this, all cases considered in Chapter 4 apply. In zeta function regularization, the free energy becomes

$$\mathcal{F}_0(s) = -\frac{1}{2}\frac{\partial}{\partial s}\mu^{2s} k_\mathrm{B} T \sum_{l=-\infty}^{\infty} \sum_J (\xi_l^2 + \Lambda_J + m^2)^{-s}. \tag{5.17}$$

The regularization is removed for $s \to 0$, and μ is an arbitrary parameter with the dimension of mass. The separation of the ultraviolet divergences can be done quite easily because these are the same as at zero temperature. There are two ways to proceed. In the first method, one has to apply the Abel–Plana formula to the frequency sum in eqn (5.17). This way has the advantage that it can also be applied to the case where there is an additional dependence on l, for example through a dielectric permittivity entering in the form $\xi^2 \to \varepsilon(i\xi)\xi^2$. Another, to some extent easier, way is through the application of the Poisson summation formula (Titchmarsh 1948). According to this formula, if $c(\alpha)$ is the Fourier transform of a function $b(x)$, i.e.

$$c(\alpha) = \frac{1}{2\pi} \int_{-\infty}^{\infty} b(x) e^{-i\alpha x}\, dx, \tag{5.18}$$

then it follows that

$$\sum_{l=-\infty}^{\infty} b(l) = 2\pi \sum_{l=-\infty}^{\infty} c(2\pi l). \tag{5.19}$$

By putting

$$b(x) = e^{-zx^2}, \qquad c(\alpha) = \frac{1}{2\sqrt{\pi z}} e^{-\alpha^2/(4z)}, \tag{5.20}$$

we obtain from eqn (5.19) the following equality:

$$\sum_{l=-\infty}^{\infty} e^{-zl^2} = \sqrt{\frac{\pi}{z}} \sum_{n=-\infty}^{\infty} e^{-\pi^2 n^2/z}, \tag{5.21}$$

where $\operatorname{Re} z > 0$ is assumed. In order to use this equality, we represent eqn (5.17) as a parametric integral,

$$\mathcal{F}_0(s) = -\frac{1}{2}\frac{\partial}{\partial s}\mu^{2s} \int_0^\infty \frac{dt}{t} \frac{t^s}{\Gamma(s)} k_\mathrm{B} T \sum_{l=-\infty}^{\infty} \sum_J e^{-t(\xi_l^2+\Lambda_J+m^2)}, \tag{5.22}$$

and apply eqn (5.21) with $z = (2\pi k_\mathrm{B} T)^2 t = (2\pi/\beta)^2 t$. The result is

$$\mathcal{F}_0(s) = -\frac{1}{2}\frac{\partial}{\partial s}\mu^{2s} \sum_{n=-\infty}^{\infty} \int_0^\infty \frac{dt}{t} \frac{t^s}{\Gamma(s)\sqrt{4\pi t}} \sum_J e^{-[n^2\beta^2+4t^2(\Lambda_J+m^2)]/(4t)}. \tag{5.23}$$

The n-dependent factor in the exponential provides convergence for the t-integration at $t \to 0$ for all terms in the sum over n except for $n = 0$. The latter is just the zero-temperature contribution. This can be seen by applying eqn (5.16) to the frequency sum in eqn (5.22):

$$k_B T \sum_{l=-\infty}^{\infty} e^{-t\xi_l^2} \xrightarrow[T\to 0]{} \int_{-\infty}^{\infty} \frac{d\xi}{2\pi} e^{-t\xi^2} = \frac{1}{\sqrt{4\pi t}}. \quad (5.24)$$

As a consequence, we can split the free energy into a zero-temperature part and a temperature-dependent addition $\Delta_T \mathcal{F}_0$ as

$$\mathcal{F}_0(s) = E_{0,\text{eff}}(s) + \Delta_T \mathcal{F}_0(s). \quad (5.25)$$

Here, the vacuum energy at zero temperature is

$$E_{0,\text{eff}}(s) = -\frac{1}{2}\frac{\partial}{\partial s}\mu^{2s} \int_{-\infty}^{\infty} \frac{d\xi}{2\pi} \sum_J \left(\xi^2 + \Lambda_J + m^2\right)^{-s}. \quad (5.26)$$

This representation coincides with eqn (4.38) after the inverse Wick rotation $\xi = -i\omega$ is performed. The temperature-dependent addition (*thermal correction*) is given by

$$\Delta_T \mathcal{F}_0(s) = -\frac{\partial}{\partial s}\mu^{2s} \sum_{n=1}^{\infty} \int_0^{\infty} \frac{dt}{t}\frac{t^s}{\Gamma(s)\sqrt{4\pi t}} \sum_J e^{-\left[n^2\beta^2 + 4t^2(\Lambda_J + m^2)\right]/(4t)}. \quad (5.27)$$

Note that eqn (5.25) has a transparent physical interpretation only for the temperature-independent boundary conditions considered here (see Chapter 12 for further discussion).

The ultraviolet divergences are contained in $E_{0,\text{eff}}(s)$ and can be dealt with in the same way as described in Section 4.3. This results in the replacement of $E_{0,\text{eff}}(s)$ with E_0^{ren} in eqn (5.25). In $\Delta_T \mathcal{F}_0(s)$ in eqn (5.27), the integration over t is convergent and we can remove the regularization, i.e. we can put $s = 0$ using the equality

$$\lim_{s\to 0} \frac{\partial}{\partial s}\frac{f(s)}{\Gamma(s)} = f(0), \quad (5.28)$$

where $f(s)$ is any regular function at $s = 0$. Following this, the integration over t and the summation over n can be carried out explicitly:

$$\Delta_T \mathcal{F}_0 = k_B T \sum_J \ln\left(1 - e^{-\beta\sqrt{\Lambda_J + m^2}}\right). \quad (5.29)$$

In this formula, we see the sum of the T-dependent contributions to the free energies of the individual degrees of freedom, or modes, Λ_J of the system considered. Taking $\Lambda_J + m^2 = \omega_J^2$ into account, the total free energy of all of the oscillator modes appears to be

$$\mathcal{F}_0 = E_0^{\text{ren}} + k_B T \sum_J \ln\left(1 - e^{-\beta\omega_J}\right), \quad (5.30)$$

where the zero-temperature contribution E_0 given by eqn (3.60) has already been replaced with E_0^{ren}. For instance, if we take the volume V to be a volume of empty space, the modes are plane waves, the index J becomes the wave vector $\boldsymbol{k}$, and the sum over J turns into the corresponding momentum integration with respect to $d\boldsymbol{k}/(2\pi)^3$. As a result, from eqn (5.29) we obtain the free-energy density of black-body radiation

$$f_{\text{bb}}(T) = k_{\text{B}}T \int \frac{d\boldsymbol{k}}{(2\pi)^3} \ln\left(1 - e^{-\beta|\boldsymbol{k}|}\right) = -\frac{\pi^2 (k_{\text{B}}T)^4}{90}. \tag{5.31}$$

We note that eqn (5.31) holds for a scalar field. For an electromagnetic field, one would have to multiply it by a factor of 2. In fact, for empty space, f_{bb} defined in eqn (5.31) is the complete free energy. This is because in this case the zero-temperature part is the vacuum energy of empty space, which we have to disregard. Using the thermodynamic connection between the energy at a temperature T and the free energy

$$U(T) = -T^2 \frac{\partial}{\partial T} \frac{\mathcal{F}(T)}{T}, \tag{5.32}$$

it is evident that for $\mathcal{F}(T) = f_{\text{bb}}^{\text{em}}(T) = 2f_{\text{bb}}(T)$, the respective energy density is in agreement with Planck's black-body radiation density

$$u = \frac{\pi^2 (k_{\text{B}}T)^4}{15}. \tag{5.33}$$

If we consider the free energy in a restricted volume V, then we have to keep the zero-temperature part E_0^{ren}. For the temperature-dependent part, we have to note that we are interested in the change in energy which comes from the volume V. Therefore we need to subtract from the temperature-dependent part $\Delta_T \mathcal{F}_0$ of the free energy the corresponding amount related to empty space, i.e. the black-body radiation density f_{bb} multiplied by the volume V. As a result, we arrive at the following expression for the renormalized free energy associated with a finite volume V,

$$\tilde{\mathcal{F}}_0 = E_0^{\text{ren}} + \Delta_T \mathcal{F}_0 - V f_{\text{bb}}. \tag{5.34}$$

In the general case, however, eqn (5.34) cannot be considered as the physical Casimir free energy associated with the volume V. As we shall see in Section 5.2, the asymptotic expression for the quantity $\Delta_T \mathcal{F}_0$ at high temperatures (large separations) contains the following terms:

$$\alpha_0 \frac{(k_{\text{B}}T)^4}{(\hbar c)^3} \equiv -V\frac{\pi^2 (k_{\text{B}}T)^4}{90(\hbar c)^3}, \qquad \alpha_1 \frac{(k_{\text{B}}T)^3}{(\hbar c)^2}, \qquad \alpha_2 \frac{(k_{\text{B}}T)^2}{\hbar c}. \tag{5.35}$$

Here, we have returned to the usual units in order to underline that all of these terms are of quantum character. Note that the first of these terms is just equal to $V f_{bb}$. The coefficients α_1 and α_2 depend on geometrical parameters of the configuration (e.g. the surface area and the sum of edge lengths, see Section

8.5). They can be expressed in terms of the heat kernel coefficients $\tilde{a}_{1/2}$ and $\tilde{a}_1$ (see the next section). The presence of the terms (5.35) in the free energy would lead to forces of quantum nature which do not vanish with an increasing characteristic size of the body. At the same time, the next expansion term in the high-temperature limit of the free energy has the form of $\alpha_3 k_B T$, with a dimensionless coefficient α_3. It is of classical origin and does not contribute to the Casimir force.

In fact, the geometrical structure of the coefficients α_1 and α_2 is just the same as that in the respective infinite terms to be subtracted from the zero-temperature Casimir energy E_0 in the cutoff regularization to make it equal to a finite value E_0^{ren} (see the explicit examples in Section 8.5). Thus, all of the terms in eqn (5.35) can be absorbed by means of an additional, finite renormalization of the free energy. As a result, the physical Casimir free energy takes the form

$$\mathcal{F} = E_0^{\text{ren}} + \Delta_T \mathcal{F}, \tag{5.36}$$

where the physical thermal correction is given by (Geyer et al. 2008c)

$$\begin{aligned}\Delta_T \mathcal{F} &= \Delta_T \mathcal{F}_0 - V f_{bb} - \alpha_1 (k_B T)^3 - \alpha_2 (k_B T)^2 \\ &= \Delta_T \mathcal{F}_0 - \alpha_0 (k_B T)^4 - \alpha_1 (k_B T)^3 - \alpha_2 (k_B T)^2.\end{aligned} \tag{5.37}$$

The respective Casimir force obviously vanishes when the characteristic sizes of the volume V along all three coordinate axes go to infinity.

Below, we shall use eqns (5.36) and (5.37) to investigate the thermal Casimir effect in various configurations (see e.g. Sections 7.4.1 and 8.5 for the cases of two parallel ideal-metal planes and rectangular boxes, respectively). All other thermodynamic quantities, such as the pressure and entropy, can be derived from these formulas using eqns (5.3) and (5.4). For configurations containing translationally invariant directions, such as parallel planes or a cylinder, one must bear in mind that all quantities in eqn (5.34) must be divided by corresponding parameters such as the area of a plate so that they become the respective densities.

5.2 The Casimir effect at low and high temperature

Specific calculations of the Casimir free energy for real bodies will be considered in subsequent chapters (in Chapter 12 for dielectrics, and Chapter 14 for metals). Here we consider the low- and high-temperature asymptotic expansions of the free energy in general terms. We start with the low-temperature case. The leading contribution is, of course, the zero-temperature part, i.e. the vacuum energy E_0^{ren} in eqn (5.36). The correction $\Delta_T \mathcal{F}_0$ in eqn (5.37) is given by eqn (5.29) in terms of the eigenvalues Λ_J of the spatial part of the system under investigation. Basically, its behavior depends on the general properties of the spectrum Λ_J, particularly on whether it has a gap or is continuous. Further, it depends on the number of translationally invariant directions. The easiest example is the case where all directions are translationally invariant, i.e. empty space, which resulted in eqn (5.31). Some more complicated examples are provided by parallel planes, having two translationally invariant directions, or a cylinder, having one. So, let us assume that we have r ($r = 1, 2, \ldots$) translationally invariant directions. Then the index J becomes $J = (k_1, \ldots, k_r, n)$ and the eigenvalues Λ_J become $\Lambda_J = k_1^2 + \ldots + k_r^2 + \lambda_n$, where the λ_n are the eigenvalues in the remaining directions, which take on discrete values only. The temperature-dependent part of the nonrenormalized free energy for this configuration reads

$$\Delta_T \mathcal{F}_0 = k_\mathrm{B} T \int \frac{d^r k}{(2\pi)^r} \sum_n \ln\left(1 - e^{-\beta\sqrt{k^2 + \lambda_n + m^2}}\right). \quad (5.38)$$

We consider the asymptotic expansion of this expression for $T \to 0$, i.e. for $\beta \to \infty$, and for a massless field $m = 0$. It is clear that the dominating contribution comes from small k and the smallest eigenvalue λ_0. The next step depends on whether the smallest eigenvalue is zero or not, i.e. on whether the spectrum of λ_n has a gap. Both situations are possible. For example, for parallel planes we have $\lambda_n = (\pi n/a)^2$ ($n = 0, 1, \ldots$), with $\lambda_0 = 0$, in the electromagnetic case. For a cylinder we have $\lambda_0 \sim 1/a$, where a is the radius of the cylinder.

We first consider the gapless case, i.e. $\lambda_0 = 0$. Here, the leading contribution to the sum comes from $n = 0$, and all higher n result in exponentially suppressed contributions. After integrations, we arrive at

$$\Delta_T \mathcal{F}_0 = C_r \varrho_r (k_\mathrm{B} T)^{r+1} + \ldots, \qquad C_r = -\frac{2\Gamma(r)\zeta_\mathrm{R}(r+1)}{(4\pi)^{r/2}\Gamma(r/2)}, \quad (5.39)$$

where ϱ_r is a length (for $r = 1$), an area (for $r = 2$), or a volume (for $r = 3$). Equation (5.39) shows a power-like behavior of the thermal corrections. For $r = 3$, we get back to the black-body radiation free energy density (5.31). For $r = 1$ and $r = 2$, we get

$$C_1 = -\frac{\pi}{6}, \qquad C_2 = -\frac{\zeta_\mathrm{R}(3)}{2\pi}. \quad (5.40)$$

Then, according to eqns (5.36) and (5.37), the renormalized Casimir free energy associated with a volume V is given by

$$\mathcal{F} = E_0^{\mathrm{ren}} + C_r \varrho_r (k_\mathrm{B} T)^{r+1} - C_3 (k_\mathrm{B} T)^4 V - \alpha_1 (k_\mathrm{B} T)^3 - \alpha_2 (k_\mathrm{B} T)^2. \quad (5.41)$$

For $r = 1, 2$, this equation represents the low-temperature behavior of the Casimir free energy (for the specific example of two parallel planes, $\alpha_1 = \alpha_2 = 0$; see Section 7.4.2). For $r = 3$, $E_0^{\mathrm{ren}} = 0$, $\alpha_1 = \alpha_2 = 0$, and eqn (5.41) results in a zero Casimir free energy, as necessary in empty space.

Next we consider the case of a nonzero smallest eigenvalue $\lambda_0 \neq 0$. We expand the square root in eqn (5.38) for small k and replace $\ln(1 + x)$ with x. Then, after integration, we arrive at

$$\Delta_T \mathcal{F}_0 = -\left(\frac{\sqrt{\lambda_0}}{2\pi}\right)^{r/2} \varrho_r (k_\mathrm{B} T)^{(r+2)/2} e^{-\sqrt{\lambda_0}/(k_\mathrm{B} T)} + \ldots, \quad (5.42)$$

i.e. an exponentially suppressed contribution. Thus we arrive at the result that the low-temperature asymptotic behavior of $\Delta_T \mathcal{F}_0$ is determined by the lower part of the spectrum of the spatial operator. The power-like or exponentially suppressed behavior of the first-order correction depends on the presence or absence of a gap in the spectrum.

In contrast, the high-temperature behavior is determined by the upper part of the spectrum and can be expressed in terms of the heat kernel coefficients. In order to derive this property, we represent eqn (5.22) for the regularized free energy in the form

$$\mathcal{F}_0(s) = -\frac{1}{2}\frac{\partial}{\partial s}\mu^{2s}\int_0^\infty \frac{dt}{t}\frac{t^s}{\Gamma(s)} K_T(t)\, K(t)\, e^{-tm^2}, \qquad (5.43)$$

where

$$K_T(t) = k_B T \sum_{l=-\infty}^{\infty} e^{-t\xi_l^2} \qquad (5.44)$$

is the heat kernel associated with the temporal part of the operator (5.8) and $K(t)$ is the heat kernel of the spatial part as given by eqn (4.15). We proceed by separating the term in eqn (5.44) with $l = 0$:

$$K_T(t) = k_B T + 2k_B T \sum_{l=1}^{\infty} e^{-t\omega_l^2}. \qquad (5.45)$$

Since this term does not depend on t, its contribution to eqn (5.43) can be expressed in terms of the zeta function $\zeta_P(s)$ of the spatial part, defined in eqn (4.5). Using eqn (4.13) with $1 - 2s$ replaced by $-2s$, we obtain

$$\int_0^\infty \frac{dt}{t}\frac{t^s}{\Gamma(s)} K(t)\, e^{-tm^2} = \zeta_P(s). \qquad (5.46)$$

Noting that for elliptic problems $\zeta_P(s)$ is regular at $s = 0$, we have

$$\left.\frac{\partial}{\partial s}\mu^{2s}\zeta_P(s)\right|_{s=0} = \zeta_P'(0) + \zeta_P(0) \ln \mu^2. \qquad (5.47)$$

Substituting eqns (5.45)–(5.47) into eqn (5.43), we arrive at

$$\mathcal{F}_0(s) = -k_B T \left[\frac{\zeta_P'(0) + \zeta_P(0) \ln \mu^2}{2} + \frac{\partial}{\partial s}\mu^{2s}\sum_{l=1}^{\infty}\int_0^\infty dt \frac{t^{s-1}}{\Gamma(s)} e^{-t\xi_l^2} K(t) e^{-tm^2}\right]. \qquad (5.48)$$

The behavior of the integral with respect to t in eqn (5.48) as $T \to \infty$ is completely determined by the behavior of the heat kernel $K(t)$ at $t \to 0$. Thus, to find the behavior of $\mathcal{F}_0(s)$ at $T \to \infty$, we can use the heat kernel expansion (4.18). It should be mentioned that this statement is equivalent to the corresponding statement for the expansion in inverse powers of the large mass in eqn

(4.29). Similarly, we can hide the mass in the heat kernel coefficients by passing from $a_{n/2}$ to $\tilde{a}_{n/2}$ with the help of eqn (4.27). Thus, inserting eqn (4.18) into eqn (5.48) and carrying out the integration with respect to t and the summation over l, which results in Riemann zeta functions, we obtain

$$\mathcal{F}_0(s) = -\frac{k_B T}{2}\left[\zeta_P'(0) + \zeta_P(0) \ln \mu^2\right] \tag{5.49}$$
$$-\frac{k_B T}{(4\pi)^{3/2}} \frac{\partial}{\partial s} \mu^{2s} \sum_{n=0}^{\infty} \tilde{a}_{n/2} \frac{\Gamma(s + \frac{n}{2} - \frac{3}{2})}{\Gamma(s)} (2\pi k_B T)^{3-2s-n} \zeta_R(2s + n - 3),$$

which is an asymptotic expansion. Here, the limit $s \to 0$ can be performed. For this purpose we need to use the reflection property of the Riemann zeta function,

$$\Gamma\left(\frac{z}{2}\right)\zeta_R(z) = \pi^{(2z-1)/2}\Gamma\left(\frac{1-z}{2}\right)\zeta_R(1-z), \tag{5.50}$$

which provides the analytic continuation of this function to $\operatorname{Re} z < 0$. First we perform the differentiation in eqn (5.49) with respect to s by considering this equation as a product of two factors: $[\mu/(2\pi k_B T)]^{2s}$ and a sum from 0 to ∞ containing all other quantities. The derivative of the first factor in the limit $s \to 0$ results in $2\ln[\mu/(2\pi k_B T)]$, whereas only the two terms with $n = 3$ and 4, containing poles in the numerator, survive in the sum (one pole in the gamma function and one in the zeta function). When we differentiate the sum, there is no dependence on μ in the limit $s \to 0$ and all terms give a nonzero contribution. As a result, we arrive at the representation

$$\mathcal{F}_0 = -\frac{k_B T}{2}\left[\zeta_P'(0) + \zeta_P(0) \ln \mu^2\right] - \frac{\pi^2}{90}\tilde{a}_0(k_B T)^4 - \frac{\zeta_R(3)}{4\pi^{3/2}}\tilde{a}_{1/2}(k_B T)^3$$
$$-\frac{1}{24}\tilde{a}_1(k_B T)^2 + \frac{\tilde{a}_{3/2}}{(4\pi)^{3/2}} k_B T \ln \frac{\mu}{k_B T} - \frac{1}{16\pi^2}\left(\gamma + \ln \frac{\mu}{4\pi k_B T}\right)\tilde{a}_2$$
$$-\sum_{n=5}^{\infty} \frac{(2\pi)^{(3-2n)/2}}{2\sqrt{2}} \Gamma\left(\frac{n-3}{2}\right) \zeta_R(n-3)\tilde{a}_{n/2}(k_B T)^{4-n}. \tag{5.51}$$

Here, $\gamma = 0.577216$ is Euler's constant. It can be shown that the terms in eqn (5.51) linear in $k_B T$ do not depend on μ. To see this, we substitute eqn (4.18) into eqn (5.46) and calculate integrals with respect to t. In the limiting case $s \to 0$, with the help of eqn (4.27), this results in

$$\zeta_P(0) = \frac{1}{(4\pi)^{3/2}}(-a_{1/2}m^2 + a_{3/2}) = \frac{\tilde{a}_{3/2}}{(4\pi)^{3/2}}. \tag{5.52}$$

Using eqn (5.52), we rewrite eqn (5.51) in the form

$$\mathcal{F}_0 = -\frac{\pi^2}{90}\tilde{a}_0(k_B T)^4 - \frac{\zeta_R(3)}{4\pi^{3/2}}\tilde{a}_{1/2}(k_B T)^3 - \frac{1}{24}\tilde{a}_1(k_B T)^2$$

$$-\frac{k_{\mathrm{B}}T}{2}\left[\zeta'_{\mathrm{P}}(0) + \zeta_{\mathrm{P}}(0)\ln(k_{\mathrm{B}}T)^2\right] - \frac{1}{16\pi^2}\left(\gamma + \ln\frac{\mu}{4\pi k_{\mathrm{B}}T}\right)\tilde{a}_2$$
$$-\sum_{n=5}^{\infty}\frac{(2\pi)^{(3-2n)/2}}{2\sqrt{2}}\Gamma\left(\frac{n-3}{2}\right)\zeta_{\mathrm{R}}(n-3)\tilde{a}_{n/2}(k_{\mathrm{B}}T)^{4-n}. \quad (5.53)$$

It should be noted that after the substitution of $\zeta'_{\mathrm{P}}(0)$ and $\zeta_{\mathrm{P}}(0)$ for a specific configuration, the quantity $k_{\mathrm{B}}T$ in the logarithm is multiplied by a factor which makes the argument of the logarithm dimensionless (see the example given in Section 9.5.2). The expression (5.53) was first derived by Dowker and Kennedy (1978).

In order to get the final form of the high-temperature expansion for the free energy, one has to subtract from eqn (5.53) the divergent part of the vacuum energy $E_{0,\mathrm{eff}}^{\mathrm{div}}(0)$ given by eqn (4.46) and the three terms presented in eqn (5.35). Because $\tilde{a}_0 = a_0 = V$, the subtraction of the first term in eqn (5.35) just cancels the contribution of order T^4 in eqn (5.53). From this, one concludes that $\alpha_0 = -\pi^2 a_0/90$. The subtraction of the second term in eqn (5.35) cancels the contribution of order T^3. Keeping in mind that from eqns (4.21), (4.22), and (4.27), $\tilde{a}_{1/2} = a_{1/2} = -\sqrt{\pi}S/2$, for a scalar field with Dirichlet boundary conditions, one arrives at

$$\alpha_1 = -\frac{\zeta_{\mathrm{R}}(3)}{4\pi^{3/2}}\tilde{a}_{1/2} = \frac{\zeta_{\mathrm{R}}(3)}{8\pi}S. \quad (5.54)$$

This expression for the coefficient α_1 is obtained independently for a rectangular box in Section 8.5. Finally, the subtraction of the third term in eqn (5.35) from the right-hand side of eqn (5.53) cancels the contribution of order T^2 if we take into account the fact that $\alpha_2 = -\tilde{a}_1/24$. In Section 8.5, it is shown that for a rectangular box $\tilde{a}_1$ is proportional to the sum of the side lengths of the box. The ultraviolet renormalization concerns only the term proportional to $\tilde{a}_2$, which becomes independent of the arbitrary parameter μ. Note that the contribution linear in $k_{\mathrm{B}}T$ on the right-hand side of eqn (5.53) has the meaning of the classical limit. It will be discussed repeatedly in the following chapters.

6
APPROXIMATE AND NUMERICAL APPROACHES

The calculation of the vacuum energy for nontrivial geometries is a complicated task. Thus it took 20 years from Casimir's original work on plane parallel plates for the first calculation for a spherical shell to appear (see Chapter 9). The main technical obstacle was the separation of ultraviolet divergences, which required a detailed knowledge of the asymptotics of the Bessel functions involved. It took approximately another 20 years before this problem was solved in an efficient way in terms of the heat kernel expansion, together with zeta function regularization. Still, the progress was limited to geometries which allow a separation of variables, i.e. a reduction to a one-dimensional problem. Beyond that, the divergences remained a problem, especially for numerical approaches. To solve the problem, one could consider a mode sum representation of the vacuum energy. The calculation of the eigenvalues of a wave operator in a cavity to some given accuracy is a manageable problem. However, this does not provide a direct way to subtract the ultraviolet divergences. The general structure of the divergences is known in terms of the heat kernel coefficients together with the high-frequency expansion of the eigenvalues (see Section 4.2.2). But, after the subtraction of the first few asymptotic contributions in a numerical calculation, numerical precision is lost and the problem becomes intractable.

Because of this, before the recent exact methods appeared (see Chapter 10), several approximate methods were developed. One of them is the multiple-reflection expansion introduced by Balian and Bloch (1970). This allows an iterative calculation of the corresponding Green's function. This expansion works best for the high-frequency contribution, and it was used mainly to investigate the dependence of the divergent contributions on the geometry of the boundaries. Another set of expansions is the semiclassical expansions. These are based on the idea of the WKB approximation in quantum mechanics, or, equivalently, the eikonal approximation in optics. In application to the Casimir effect (Schaden and Spruch 1998, Jaffe and Scardicchio 2004), such methods should work best if the separation between the interacting bodies is small and the main contributions come from high frequencies. The multiple-reflection expansion and the semiclassical methods are briefly considered in Sections 6.1 and 6.2, respectively.

Since 2001, the *numerical world line approach* has been developed (Schubert 2001). Inspired by string theory, it uses the Feynman path-integral representation of the transition amplitudes. The contributions from the paths are calculated numerically and the result appears as a sum over a large number of paths (*clouds of paths*). This method was originally developed for the calculation of the one-loop effective action in background fields. Soon it was also applied to Dirichlet boundary conditions and some interesting results were obtained, which will be discussed below (Section 6.3).

Historically, the first two simple approximate methods were pairwise summation, which goes back to Lennard-Jones (1932), and the proximity force approximation (Derjaguin 1934). The method of pairwise summation, considered in Section 6.4, permits one to calculate the Casimir force between two macroscopic bodies as the sum of the forces between the microparticles belonging to them. In doing so, all effects of nonadditivity are taken into account approximately by means of a special normalization procedure. The proximity force approximation, considered in Section 6.5, is most important for the purpose of applications. According to this method, the surfaces of the interacting bodies are divided into small plane plates, to which the force per unit area known from the case of infinitely extended parallel plates is applied. Then one must sum the energies and forces from all pairs of plates. This method works well for very small separations and, so far, it has been sufficient for experimental configurations. In cases where both of the methods of pairwise summation and the proximity force approximation are applicable, they lead to coincident leading contributions to the Casimir energy and force. However, the application region of the proximity force approximation is much wider.

6.1 The multiple-reflection expansion

The multiple-reflection expansion can be interpreted as the propagation of the field under consideration from a source point r to a drain point r', which occurs freely in between the boundaries and has multiple reflections at the boundaries. These reflections are not specular, and the reflection point must be integrated over the whole surface. The multiple-reflection expansion, as initially considered (Balian and Bloch 1970), is written in terms of the Green's function

$$G_\omega(r,r') = \sum_J \frac{\Phi_J(r)\Phi_J^*(r')}{-\omega^2 + \omega_J^2 - i0} \tag{6.1}$$

of the Helmholtz equation

$$\left(-\omega^2 - \nabla^2\right) G_\omega(r,r') = \delta(r - r'). \tag{6.2}$$

The function $G_\omega(r,r')$ is related to the Green's function defined in eqn (3.87) by

$$G(x,x') = \int_{-\infty}^{\infty} \frac{d\omega}{2\pi} G_\omega(r,r')\, e^{-i\omega(t-t')}. \tag{6.3}$$

We assume a static boundary S and denote points on the boundary, $\alpha \in S$, by Greek letters, in contrast to points in the bulk, for which we keep Latin letters. In this notation, Dirichlet boundary conditions are denoted by

$$G_\omega(\alpha, r') = 0 \quad \text{for} \quad \forall\, \alpha \in S. \tag{6.4}$$

The starting point for the multiple-reflection expansion is a representation of the Green's function in terms of the potential $\nu(\boldsymbol{\beta}, \boldsymbol{r}')$ of a double layer known from electrostatics,

$$G_\omega(\boldsymbol{r}, \boldsymbol{r}') = G_\omega^{(0)}(\boldsymbol{r}, \boldsymbol{r}') + \int_S d\mu(\boldsymbol{\beta}) \frac{\partial G_\omega^{(0)}(\boldsymbol{r}, \boldsymbol{\beta})}{\partial n_\beta} \nu(\boldsymbol{\beta}, \boldsymbol{r}'), \tag{6.5}$$

where $d\mu(\boldsymbol{\beta})$ is the measure on S and $G_\omega^{(0)}(\boldsymbol{r}, \boldsymbol{r}')$ is the Green's function (6.1) with no boundary conditions, i.e. the free-space Green's function. Its well-known explicit expression is

$$G_\omega^{(0)}(\boldsymbol{r}, \boldsymbol{r}') = \frac{e^{i\omega R}}{4\pi R}, \tag{6.6}$$

where $R = |\boldsymbol{r} - \boldsymbol{r}'|$ is the distance between the points $\boldsymbol{r}$ and $\boldsymbol{r}'$. The derivative in eqn (6.5) is the normal derivative towards the interior region and must be taken before the second argument of the Green's function is put on the surface:

$$\frac{\partial G_\omega^{(0)}(\boldsymbol{r}, \boldsymbol{\beta})}{\partial n_\beta} = \boldsymbol{n}_\beta \boldsymbol{\nabla}_{\boldsymbol{r}'} \left. G_\omega^{(0)}(\boldsymbol{r}, \boldsymbol{r}') \right|_{\boldsymbol{r}'=\boldsymbol{\beta}}, \tag{6.7}$$

where $\boldsymbol{n}_\beta$ is the inward-pointing normal to the surface S at a point $\boldsymbol{\beta}$.

In eqn (6.5), the Green's function obeying the boundary conditions is represented as the sum of the free-space Green's function and the potential of a double layer $\nu(\boldsymbol{\beta}, \boldsymbol{r}')$. The latter is still unknown. It is determined by imposing the boundary conditions (6.4) on eqn (6.5). In order to do this, we remind the reader of a basic property of the potential of a double layer. Namely, the limit of putting the point $\boldsymbol{r}$ on the surface S, $\boldsymbol{r} \to \boldsymbol{\alpha}$, and the integration over $\boldsymbol{\beta}$ do not commute. Instead, the following formula holds:

$$\lim_{\boldsymbol{r} \to \boldsymbol{\alpha}} \int d\mu(\boldsymbol{\beta}) \frac{\partial G_\omega^{(0)}(\boldsymbol{r}, \boldsymbol{\beta})}{\partial n_\beta} \nu(\boldsymbol{\beta}, \boldsymbol{r}') = \int d\mu(\boldsymbol{\beta}) \frac{\partial G_\omega^{(0)}(\boldsymbol{r}, \boldsymbol{\beta})}{\partial n_\beta} \nu(\boldsymbol{\beta}, \boldsymbol{r}') + \frac{1}{2}\nu(\boldsymbol{\alpha}, \boldsymbol{r}'). \tag{6.8}$$

Here it is assumed that $\boldsymbol{r}$ approaches the point $\boldsymbol{\alpha}$ from the interior region (otherwise an additional contribution would appear with an opposite sign). The integral on the right-hand side converges.

Using eqn (6.8), we get from eqns (6.4) and (6.5) for $\boldsymbol{r} \to \boldsymbol{\alpha}$

$$0 = G_\omega^{(0)}(\boldsymbol{\alpha}, \boldsymbol{r}') + \int d\mu(\boldsymbol{\beta}) \frac{\partial G_\omega^{(0)}(\boldsymbol{r}, \boldsymbol{\beta})}{\partial n_\beta} \nu(\boldsymbol{\beta}, \boldsymbol{r}') + \frac{1}{2}\nu(\boldsymbol{\alpha}, \boldsymbol{r}'). \tag{6.9}$$

This is an integral equation for the double-layer potential

$$\nu(\boldsymbol{\alpha}, \boldsymbol{r}') = -2G_\omega^{(0)}(\boldsymbol{\alpha}, \boldsymbol{r}') - 2 \int d\mu(\boldsymbol{\beta}) \frac{\partial G_\omega^{(0)}(\boldsymbol{r}, \boldsymbol{\beta})}{\partial n_\beta} \nu(\boldsymbol{\beta}, \boldsymbol{r}'), \tag{6.10}$$

and the multiple-reflection expansion emerges from its iterative solution. The first step is to put $\nu(\boldsymbol{\beta}, \boldsymbol{r}') = \nu^{(0)}(\boldsymbol{\beta}, \boldsymbol{r}') = 0$ on the right-hand side. This gives

$$\nu^{(1)}(\boldsymbol{\alpha}, \boldsymbol{r}') = -2G_\omega^{(0)}(\boldsymbol{\alpha}, \boldsymbol{r}'), \qquad (6.11)$$

which is the first iteration for the potential. The next order is obtained by inserting eqn (6.11) into the right-hand side of eqn (6.10), and so on.

From these iterations for the potential, the respective expansion of the Green's function emerges,

$$G_\omega(\boldsymbol{r}, \boldsymbol{r}') = G_\omega^{(0)}(\boldsymbol{r}, \boldsymbol{r}') + G_\omega^{(1)}(\boldsymbol{r}, \boldsymbol{r}') + G_\omega^{(2)}(\boldsymbol{r}, \boldsymbol{r}') + \dots, \qquad (6.12)$$

which starts from the free-space Green's function (6.6). The latter appears as the contribution from zero reflections. The higher-order contributions are

$$G_\omega^{(p)}(\boldsymbol{r}, \boldsymbol{r}') = \int d\mu(\boldsymbol{\beta}) \frac{\partial G_\omega^{(0)}(\boldsymbol{r}, \boldsymbol{\beta})}{\partial n_\beta} \left[\nu^{(p)}(\boldsymbol{\beta}, \boldsymbol{r}') - \nu^{(p-1)}(\boldsymbol{\beta}, \boldsymbol{r}') \right], \quad p = 1, 2, \dots. \qquad (6.13)$$

For instance,

$$G_\omega^{(1)}(\boldsymbol{r}, \boldsymbol{r}') = -2 \int d\mu(\boldsymbol{\beta}) \frac{\partial G_\omega^{(0)}(\boldsymbol{r}, \boldsymbol{\beta})}{\partial n_\beta} G_\omega^{(0)}(\boldsymbol{\beta}, \boldsymbol{r}') \qquad (6.14)$$

is the contribution from one reflection. From the above, the physical interpretation can be seen. We have a free propagation from $\boldsymbol{r}$ to $\boldsymbol{\beta}$, some kind of operator there $(-2\partial/\partial n_\beta)$, and a free propagation from $\boldsymbol{\beta}$ to $\boldsymbol{r}'$. The point $\boldsymbol{\beta}$ must be integrated over the whole surface S. The reflection is not specular, and should be extended over the whole surface, even over shadowed regions.

This expansion was originally derived by Balian and Bloch (1970), who investigated the asymptotic distribution of the eigenvalues of the Laplace operator obeying boundary conditions. In fact, in that work, using a different notation, a systematic way was shown for how to calculate the heat kernel coefficients, and the coefficient c_1 in eqn (4.22) was derived. Several years later, this method was applied to the Casimir effect (Balian and Duplantier 1978) and the divergent surface contribution to the vacuum energy for a conducting sphere was obtained. In the following years, the multiple-reflection expansion was investigated in detail in a number of papers [for instance, by Balian and Bloch (1971) and by Hansson and Jaffe (1983a, 1983b)]. Several reformulations of this expansion are possible; for example, in the representation (6.5), a monolayer potential may be added. It is possible also to consider all kinds of boundary conditions and fields. Specifically, the multiple-reflection expansion for spinor fields was investigated by Hansson and Jaffe (1983a).

The convergence of the multiple-reflection expansion is a much-discussed question. Obviously, this method delivers at least an asymptotic expansion of the Green's functions for high frequency. This property makes it possible to obtain any heat kernel coefficient of given order from a finite number of reflections (Bordag et al. 2001b). The multiple-reflection expansion also proved useful for obtaining the heat kernel coefficients in singular background fields and in similar applications (Bordag et al. 2001b). However, the examples where it was used to obtain approximate results for the Casimir effect did not go beyond simple geometries. This was connected primarily with the increasing technical difficulties. Also, we must mention that in the application of the multiple-reflection expansion to the Casimir effect, there is no natural small expansion parameter (unless one considers small or large separations).

6.2 Semiclassical approaches

The idea of all semiclassical approaches is to consider situations where the quantum system is in some sense close to some classical ones and where classical trajectories dominate in a path-integral representation. This idea is the same as that which relates wave and ray optics. A necessary condition for the applicability of such ideas is that the phenomena under consideration are dominated by short-wavelength contributions. With respect to the Casimir effect this applies, for example, at small separations between the interacting bodies. In view of the big success that semiclassical methods have had in quantum mechanics and quantum field theory, much effort has been put into this direction, starting with the work of Schaden and Spruch (1998). Here the vacuum energy was represented as a sum over paths. Using the well-known Gutzwiller trace formula, attention was focused on the periodic paths which dominate in the semiclassical limit (Schaden 2006). More recently, a new semiclassical approach has been proposed (Jaffe and Scardicchio 2004, Scardicchio and Jaffe 2005, 2006, Schröder et al. 2005), which is inspired by the eikonal approximation of classical optics. This so-called *optical approach* includes a larger number of paths than the previous approach.

It should be mentioned that the notion of a semiclassical approximation is misleading in applications to the Casimir effect (Scardicchio and Jaffe 2005). In general, the semiclassical expansions are expansions in powers of the Planck constant $\hbar$. However, the Casimir energy and force are simply proportional to $\hbar$ [see eqns (1.1) and (1.5)]. For massless fields, no other dependence on $\hbar$ enters, since the boundaries are classical objects. Thus, there can be no expansion in powers of $\hbar$. Nevertheless, the notion of a semiclassical expansion of the Casimir energy is widely used, and it is justified insofar as one takes over the corresponding ideas from quantum mechanics.

Now we describe the basic ideas of the optical approach. The starting point is the Green's function of a Schrödinger equation related to the spatial equation (3.39) with a fictitious time variable t,

$$\left(-\mathrm{i}\frac{\partial}{\partial t} - \boldsymbol{\nabla}^2\right) G(\boldsymbol{r}, \boldsymbol{r}'; t) = \delta(\boldsymbol{r} - \boldsymbol{r}'). \tag{6.15}$$

This is connected with the Green's function $G_\omega(\boldsymbol{r}, \boldsymbol{r}')$ in eqn (6.1) by means of

$$G(\boldsymbol{r}, \boldsymbol{r}'; t) = \int_{-\infty}^{\infty} \frac{dE}{2\pi\mathrm{i}} \; G_{\sqrt{E}}(\boldsymbol{r}, \boldsymbol{r}') \, \mathrm{e}^{-\mathrm{i}Et}. \tag{6.16}$$

It should be mentioned that this Green's function is quite analogous to the local heat kernel introduced in eqn (4.17). This can be seen by inserting eqn (6.1) into eqn (6.16) and carrying out the integration over E:

$$G(\mathbf{r}, \mathbf{r}'; t) = K(\mathbf{r}, \mathbf{r}'|\mathrm{i}t)\, \theta(t). \tag{6.17}$$

Here $\theta(t)$ is the step function.

The optical approximation for this Green's function introduced by Jaffe and Scardicchio (2004) reads

$$G^{\mathrm{opt}}(\mathbf{r}, \mathbf{r}'; t) = \sum_n D_n(\mathbf{r}, \mathbf{r}'; t) \exp[\mathrm{i}S_n(\mathbf{r}, \mathbf{r}'; t)]. \tag{6.18}$$

Here the sum is over all classical paths from a point $\mathbf{r}$ to a point $\mathbf{r}'$ obeying the laws of ray optics, and n is the number of reflections that these paths undergo on the boundary. If the space between the boundaries is empty, the paths are straight lines. In eqn (6.18), the function $S_n(\mathbf{r}, \mathbf{r}'; t)$ is the action along such a path and is given by

$$S_n(\mathbf{r}, \mathbf{r}'; t) = \frac{l_n(\mathbf{r}, \mathbf{r}')}{4t}, \tag{6.19}$$

where $l_n(\mathbf{r}, \mathbf{r}')$ is the length of the path. Further, in eqn (6.18),

$$D_n(\mathbf{r}, \mathbf{r}'; t) = \det\left(\frac{\partial^2 l_n^2(\mathbf{r}, \mathbf{r}'; t)}{\partial x_i \partial x_j'}\right) \tag{6.20}$$

is the Van Vleck determinant. In this case it takes the simple form (Scardicchio and Jaffe 2005)

$$D_n(\mathbf{r}, \mathbf{r}'; t) = \frac{(-1)^n l_n(\mathbf{r}, \mathbf{r}')}{(4\pi \mathrm{i} t)^{3/2}} \sqrt{\Delta_n}, \tag{6.21}$$

where

$$\Delta_n = \frac{d\Omega_{\mathbf{r}}}{dA_{\mathbf{r}'}} \tag{6.22}$$

is the enlargement factor known from geometrical optics. Here, $dA_{\mathbf{r}'}$ is the area of the rays originating from an infinitesimal area $d\Omega_{\mathbf{r}}$. This approach can be compared with the eikonal approximation in optics or the WKB approximation in quantum mechanics. However, it goes a step beyond as it also includes the effects of the Van Vleck determinant.

In this approximation, the inverse Fourier transform of eqn (6.16),

$$G_\omega(\mathbf{r}, \mathbf{r}') = \mathrm{i} \int_0^\infty dt\, e^{\mathrm{i}\omega^2 t} G(\mathbf{r}, \mathbf{r}'; t), \tag{6.23}$$

gives

$$G_\omega^{\mathrm{opt}}(\mathbf{r}, \mathbf{r}') = \sum_n \frac{(-1)^n \sqrt{\Delta_n}}{4\pi}\, \exp[\mathrm{i}\omega l_n(\mathbf{r}, \mathbf{r}')]. \tag{6.24}$$

This function can be used in eqn (3.93) to obtain the optical approximation to the vacuum energy. Carrying out the frequency integration, one arrives at (Scardicchio and Jaffe 2005)

$$E_0^{\text{opt}} = \int d\mathbf{r} \sum_n \frac{(-1)^{n+1}\sqrt{\Delta_n}}{2\pi^2 l_n^3(\mathbf{r},\mathbf{r})}. \qquad (6.25)$$

This is the final general formula for the vacuum energy in the optical approximation with Dirichlet boundary conditions. For Neumann boundary conditions one needs to remove the sign factor $(-1)^n$.

The next step is to find all classical paths with coincident start and end points, and their enlargement factors, for some specific geometry. Then one needs to integrate over the point $\mathbf{r}$ and to sum over the number of reflections. This task involves, in general, quite complicated geometrical considerations. So here we restrict ourselves to the simple example of two parallel planes. In this case the paths are straight lines perpendicular to the planes. We take the planes perpendicular to the z-axis with a distance a between them. Paths with an even number of reflections $n = 2k$ ($k = 1, 2, \ldots$) do not depend on the position of the starting point:

$$l_{2k}(\mathbf{r},\mathbf{r}) = 2ka. \qquad (6.26)$$

Paths with an odd number $n = 2k+1$ ($k = 0, 1, \ldots$) do depend on position:

$$l_{2k+1}(\mathbf{r},\mathbf{r}) = 2ka + 2z. \qquad (6.27)$$

Accordingly, the energy becomes a sum of two contributions,

$$E^{\text{opt}} = E^{\text{opt}}_{\text{odd}} + E^{\text{opt}}_{\text{even}}. \qquad (6.28)$$

The contribution from the even number of reflections to the energy is

$$E^{\text{opt}}_{\text{even}} = -2\sum_{k=1}^{\infty} \frac{1}{2\pi^2} \int_0^a dz \frac{1}{(2ka)^4} = -\frac{\pi^2}{1440 a^3} \qquad (6.29)$$

This is finite and coincides with the known result for the scalar Casimir energy per unit area.

The contribution from the odd reflections can be written in the form

$$E^{\text{opt}}_{\text{odd}} = 2\sum_{k=0}^{\infty} \frac{1}{2\pi^2} \int_{ka}^{(k+1)a} dz \frac{1}{(2z)^4} = \frac{1}{\pi^2} \int_0^{\infty} dz \frac{1}{(2z)^4}. \qquad (6.30)$$

This contribution contains the ultraviolet divergence. This results from the zero-reflection path. If one introduces some length Λ as a regularization, all odd reflection paths add up to a distance-independent expression which does not contribute to the Casimir force and can be discarded.

Thus, the optical approach becomes exact for two parallel plates. Much work has been done to apply it to more complicated geometries aimed at checking the precision of the proximity force approximation, which is another approximate method to calculate the Casimir energy and force, discussed in Section 6.5. Specifically, the optical approach has been applied to a sphere in front of a plane (Scardicchio and Jaffe 2005). Up to four reflections were taken into account, and their contribution to the scalar Casimir energy was calculated numerically. For small separations, the result can be written in the form

$$\frac{E^{\text{opt}}}{E^{\text{PFA}}} = 1 + 0.05\frac{a}{R} + O\left(\frac{a^2}{R^2}\right), \tag{6.31}$$

where E^{PFA} is the corresponding energy in the proximity force approximation. Here a is the separation between the plane and the sphere and R is the radius of the sphere. The ratio a/R is assumed to be small. In this case the coefficient 0.05 found by optical methods, which is the first correction beyond the proximity force approximation, can be compared with the analytical result, which is 1/3 [see eqn (10.157)]. However, one can expect that if all of the reflections are summed, the optical approach will come closer to the analytical result. Similarly to the multiple-scattering expansion, the optical approach has no internal means to verify and control the accuracy of the results obtained.

The optical approach has also been applied to a number of other geometries. Specifically, it has been applied to a hyperboloid in front of a plane, where up to six reflections were included (Schröder et al. 2005), and to a tilted plane in front of another plane. Using the optical approach, local energy densities were considered and nonzero temperature was taken into consideration (Scardicchio and Jaffe 2006).

6.3 World line numerical methods

The methods considered in this section use a path-integral representation of a fictitious particle, which, in the end, is evaluated numerically. We start from eqn (3.112) representing the vacuum energy as a trace,

$$E_0 = -\frac{\text{i}}{2T}\,\text{Tr}\ln\mathcal{K}, \tag{6.32}$$

and take the operator in the form

$$\mathcal{K} = -\Box + \lambda U(\boldsymbol{r}), \tag{6.33}$$

where $U(\boldsymbol{r})$ is some background potential and λ is a coupling constant with the dimension of mass. This method was initially developed for the calculation of one-loop effective actions in background potentials (Schubert 2001) and was later applied to boundary conditions as well (Gies et al. 2003).

We restrict ourselves to an introduction to the basic ideas of the approach and present some of the results obtained. The first step is to switch to an exponential representation of the logarithm. Technically, this is the same procedure as that used in zeta function regularization. So, we write

$$E_0(s) = -\frac{1}{2T}\frac{\partial}{\partial s}\int_0^\infty \frac{dv}{v}\frac{v^s}{\Gamma(s)}\,\mathrm{Tr}\,e^{-v\mathcal{K}}, \qquad (6.34)$$

where s must be set to zero in the end. In this formula, v is an auxiliary parameter with the dimension of the second power of length, which can be interpreted as a fictitious time variable, and the trace is calculated as a space–time integral,

$$\mathrm{Tr}\,e^{-v\mathcal{K}} = \int d^4x \; <x|e^{-v\mathcal{K}}|x>. \qquad (6.35)$$

For the trace in eqn (6.34), we use the path-integral representation (Gies and Klingmüller 2006b)

$$\mathrm{Tr}\,e^{-v\mathcal{K}} = \int_{x(0)=x(v)} Dx(\tau)\,\exp\left\{-\int_0^v d\tau\left[\frac{1}{4}\dot{x}^2(\tau) + \lambda U(x(\tau))\right]\right\}. \qquad (6.36)$$

Here, the integration goes over the paths with coincident start and end points, and the Wick rotation has already been done. The dot denotes the derivative with respect to the fictitious variable v. Substitution of eqn (6.36) into eqn (6.34) leads to

$$E_0(s) = -\frac{1}{2T}\frac{\partial}{\partial s}\int_0^\infty \frac{dv}{v}\frac{v^s}{\Gamma(s)}\int_{x(0)=x(v)} Dx(\tau)\,\exp\left\{-\int_0^v d\tau\left[\frac{\dot{x}^2(\tau)}{4} + \lambda U(x(\tau))\right]\right\}. \qquad (6.37)$$

But this is still not the final expression. For a time-independent background potential, one has to separate the total time T, which drops out. Also, it is necessary to separate the divergences. Afterwards, the derivative can be calculated using eqn (5.28). Thus, one arrives at a representation where the path integral can be evaluated numerically. To do so it is necessary to generate a large number of random loops, called a *cloud of loops*, and to evaluate the integral on these loops.

Computations of Casimir energies by means of world line numerics were initiated by Gies *et al.* (2003), who developed the necessary algorithm to create the clouds of loops. Boundary conditions were implemented by selecting from the cloud those loops which fulfill the boundary conditions. In this procedure, it is quite easy to implement Dirichlet boundary conditions. However, implementing Neumann or other boundary conditions involving a derivative is technically more complicated. This circumstance has prevented a more general use of the method beyond Dirichlet boundary conditions.

Later, the numerical precision was increased considerably, and the distance dependence of the Casimir force acting between a sphere or a cylinder and a plane were calculated for a scalar field obeying Dirichlet boundary conditions (Gies and Klingmüller 2006b). The known limiting cases for large and small separations were reproduced. Particularly, at short separations, the first analytic corrections beyond the proximity force approximation found for a cylinder in front of a plane (Bordag 2006a) and for a sphere in front of a plane (Bordag and Nikolaev 2008) were reproduced (see Sections 10.3.3 and 10.4.3 for details of these corrections).

Valuable information was also obtained for a semi-infinite plane ($x \geq 0$) parallel to an infinite plane (a) or another semi-infinite plane (b) (one stacked exactly above the other) (Gies and Klingmüller 2006c). In both cases, for dimensional reasons, the force can be written in the form

$$F = -\gamma_0 \frac{S}{a^4} - \gamma_{a,b} \frac{L}{a^3}. \tag{6.38}$$

Here, the first term is the contribution from the force density for two infinitely extended plates, i.e. the contribution without any edge effect. It is proportional to the area S of the upper plate, and

$$\gamma_0 = \frac{\pi^2}{480} = 2.056 \times 10^{-2} \tag{6.39}$$

is the value known for the Casimir effect for a scalar field between such plates (see Section 7.1.1). For dimensional reasons, the edge contribution is proportional to the length L of the edge, and it is assumed that the separation a of the planes is the smallest parameter, i.e. $a \ll L$, and that $L \ll \sqrt{S}$. Under these assumptions, the numerical results for the coefficients in eqn (6.38) are

$$\gamma_a = 5.23(2) \times 10^{-3}, \qquad \gamma_b = 2.30(1) \times 10^{-3}. \tag{6.40}$$

Here the digits in brackets indicate the numerical error. These results allow one, at least for a scalar field, to estimate the contribution of the edge effects.

6.4 Pairwise summation

In this section, we consider a simple approximate method which allows calculation of the Casimir force between two bodies as a sum of the forces acting between their constituents (atoms or molecules). Although the Casimir force is not an additive quantity, the effects of nonadditivity can be partially taken into account with the help of a special normalization procedure which relates the case under consideration to a similar configuration where both the additive and the exact results are available. The additive method has been widely used in the theory of disperion forces, following Lennard-Jones (1932). This is a simple calculation for many configurations of experimental interest. Under certain conditions, the results obtained turn out to be very accurate, although the method of pairwise summation (PWS) does not contain any internal means for the determination of their accuracy.

To illustrate the method, we start with a configuration of two thick plates (semispaces) at a sufficiently large separation a, described by constant dielectric permittivities $\varepsilon_0^{(1)}$ and $\varepsilon_0^{(2)}$. This also includes the case of ideal metals in the limit where $\varepsilon_0^{(1,2)} \to \infty$. Let the boundary plane of the lower semispace be $z = 0$ and let that of the upper semispace be $z = a$. We assume that two atoms (one in the lower semispace at a point r_1 and the other in the upper semispace at a point r_2) are characterized by an interaction energy

$$E^{AA}(r) = -\frac{B}{r^7}, \qquad (6.41)$$

where B is some constant and $r = |r_2 - r_1|$. After integration of eqn (6.41) over the lower semispace, we find the additive interaction energy of an atom at a point r_2 with the lower semispace,

$$E_A^{\text{add}}(z_2) = -2\pi N_1 B \int_0^\infty \rho \, d\rho \int_{-\infty}^0 \frac{dz_1}{\left[(z_2 - z_1)^2 + \rho^2\right]^{7/2}} = -\frac{\pi N_1 B}{10 z_2^4}. \qquad (6.42)$$

Here N_1 is the number of atoms per unit volume in the lower semispace. Integrating eqn (6.42) over the volume of the upper semispace, we find the additive Casimir energy of the two plates (semispaces),

$$E_{\text{pp}}^{\text{add}}(a) = -\frac{\pi N_1 N_2 BS}{10} \int_a^\infty \frac{dz_2}{z_2^4} = -\frac{\pi N_1 N_2 BS}{30 a^3}, \qquad (6.43)$$

where N_2 is the density of atoms in the upper semispace and S is the infinite area of the boundary surface. Note that the same dependence on separation as in eqn (6.43) is obtained from the Lifshitz formula for two dilute semispaces sufficiently far apart from each other (see Section 16.1). There, the constant B is related to the static atomic polarizabilities. This justifies our choice of the atom–atom interaction energy in the form (6.41).

As mentioned above, results such as eqn (6.43) do not take the effects of nonadditivity into account. The role of these effects in a configuration of two semispaces can be characterized by the *normalization constant*

$$K_E = \frac{E_{\text{pp}}^{\text{add}}(a)}{E(a)S} = \frac{24 B N_1 N_2}{\pi \Psi(\varepsilon_0^{(1)}, \varepsilon_0^{(2)})}. \qquad (6.44)$$

Here, the exact expression for the Casimir energy per unit area of the semispaces at sufficiently large separations, $E(a)$, is obtained from the Lifshitz formula [see eqn (16.8)]. The latter takes the nonadditivity effects into account. Generally, it is true that $K_E > 1$. For ideal metals $\varepsilon_0^{(1)}, \varepsilon_0^{(2)} \to \infty$, $\Psi(\varepsilon_0^{(1)}, \varepsilon_0^{(2)}) \to 1$, and $E(a) = E_{\text{IM}}(a)$ as defined in eqn (1.5).

We now deal with two arbitrarily shaped bodies V_1 and V_2. In this case the additive interaction energy takes the form

$$E^{\text{add}}(a) = -B N_1 N_2 \int_{V_1} dr_1 \int_{V_2} dr_2 \, |r_2 - r_1|^{-7}. \qquad (6.45)$$

By assuming that for two arbitrary bodies the effects of nonadditivity play approximately the same role as for two thick parallel plates, one can define the *normalized* interaction energy as (Mostepanenko and Sokolov 1988)

$$E^{\text{tot}}(a) = \frac{E^{\text{add}}(a)}{K_E} = -\frac{\pi}{24}\Psi(\varepsilon_0^{(1)},\varepsilon_0^{(2)})\int_{V_1}d\boldsymbol{r}_1\int_{V_2}d\boldsymbol{r}_2\,|\boldsymbol{r}_2-\boldsymbol{r}_1|^{-7}. \quad (6.46)$$

For ideal-metal plates, the function $\Psi(\varepsilon_0^{(1)},\varepsilon_0^{(2)})$ is replaced by unity. In this case eqn (6.46) represents the Casimir energy of two ideal-metal bodies in the framework of the PWS method.

As a first application of the PWS method, we consider the configuration of a sphere of radius R above a plate. This configuration is often used in the measurement of the Casimir force (see Chapters 19–21). Let both bodies be made of an ideal metal. The application of eqn (6.45) leads to

$$E^{\text{add}}_{\text{sp}}(a) = -BN_1N_2\int_a^{a+2R}dz_2\,\pi\rho_2^2(z_2)\int_{-\infty}^0 dz_1\int_0^\infty\frac{2\pi\rho_1\,d\rho_1}{[\rho_1^2+(z_2-z_1)^2]^{7/2}}, \quad (6.47)$$

where

$$\rho_2^2(z_2) = R^2 - (R+a-z_2)^2. \quad (6.48)$$

All integrations in eqn (6.47) can be performed explicitly, with the result

$$E^{\text{add}}_{\text{sp}}(a) = -\pi^2 BN_1N_2\frac{R^3(R+2a)}{30a^2(R+a)^3} = -\frac{\pi^2}{30}BN_1N_2\frac{R}{a^2}\left[1+O\left(\frac{a}{R}\right)\right]. \quad (6.49)$$

The respective additive Casimir force is given by

$$F^{\text{add}}_{\text{sp}}(a) = -\frac{\pi^2}{15}BN_1N_2\frac{R}{a^3}\left[1+O\left(\frac{a}{R}\right)\right]. \quad (6.50)$$

By dividing the additive energy (6.49) and force (6.50) into the normalization constant (6.44) with $\Psi(\varepsilon_0^{(1)},\varepsilon_0^{(2)})=1$, one obtains the following approximate expressions for the Casimir energy and force in the sphere–plate configuration, as given by the PWS method:

$$E^{\text{sp}}_{\text{IM}}(a) = -\frac{\pi^3 R}{720a^2}, \qquad F^{\text{sp}}_{\text{IM}}(a) = -\frac{\pi^3 R}{360a^3}. \quad (6.51)$$

The same results can be obtained directly from eqn (6.46).

Note that in eqns (6.49) and (6.50) and in the final equation (6.51), we do not include corrections of order a/R and of higher orders in this parameter. The point is that the PWS method uses a normalization to the case of two parallel plates. So, the results obtained are meaningful only if the geometrical region that gives the major contribution to the force corresponds closely to a configuration of two parallel plates, as happens for very small a/R. In fact, corrections to eqn (6.51) in powers of a/R obtained from eqn (6.46) are physically meaningless. At the same time, for three-dimensional bodies V_1 and V_2 at arbitrary separation, the additive result (6.45) correctly reproduces the dependence of the interaction energy on the separation distance (Barash 1988).

Another configuration of experimental interest is a cylinder of radius R above a plane plate (a semispace). In the ideal-metal case, the electromagnetic Casimir force between a cylindrical shell and a plane was recently calculated analytically starting from the first principles of quantum field theory (see Section 10.3). However, the approximate methods preserve their usefulness because they are applicable to nonideal materials also (see the next section).

Let the cylinder be parallel to the (x, y) coordinate plane at the shortest distance a. The Casimir energy is then given by eqn (6.46) with $\Psi(\varepsilon_0^{(1)}, \varepsilon_0^{(2)}) = 1$,

$$E_{\text{IM}}^{\text{cp}}(a) = -\frac{\pi}{12} \int_a^{a+2R} dz_2\, \rho_2(z_2) \int_{-\infty}^0 dz_1 \int_0^\infty \frac{2\pi \rho_1\, d\rho_1}{[\rho_1^2 + (z_2 - z_1)^2]^{7/2}}, \qquad (6.52)$$

where $\rho_2(z_2)$ is defined in eqn (6.48). Calculating all integrals in eqn (6.52) explicitly, one obtains

$$E_{\text{IM}}^{\text{cp}}(a) = -\frac{\pi^3 R^2}{240 a^{5/2}} \frac{a+R}{(a+2R)^{5/2}}. \qquad (6.53)$$

As in the configuration of a sphere above a plane, only the leading contribution in the small parameter a/R is physically meaningful. Thus, eqn (6.53) leads to the following approximate expressions for the Casimir energy and force in the ideal-metal cylinder–plane configuration:

$$E_{\text{IM}}^{\text{cp}}(a) = -\frac{\pi^3}{960 a^2} \sqrt{\frac{R}{2a}}, \qquad F_{\text{IM}}^{\text{cp}}(a) = -\frac{\pi^3}{384 a^3} \sqrt{\frac{R}{2a}}. \qquad (6.54)$$

In Section 10.3, these results are reproduced using exact methods, and the first corrections to them of the order of a/R are also presented.

Finally, we briefly review the application of the PWS method to the case of a spherical lens of thickness H and radius R at a shortest separation a above a finite disk of radius L and thickness D (Bezerra et al. 1997). Both the lens and the plate are supposed to be made of dielectric materials with permittivities $\varepsilon_0^{(1)}$ and $\varepsilon_0^{(2)}$, respectively. The results obtained are useful for the estimation of corrections to the Casimir force due to the finiteness of the plate (see Section 19.2.3). It is assumed that the conditions $a \ll H, D, R$ and $D, H \ll R, L$ are satisfied, which are usually fulfilled in experimental configurations.

As shown by Bezerra et al. (1997), the application of eqn (6.46) to the configuration of a spherical lens above a finite plate results in

$$E_{\text{fin}}(a) = -\frac{\pi^3}{120} \Psi(\varepsilon_0^{(1)}, \varepsilon_0^{(2)}) \int_Q^1 t\, dt \int_{R-H}^{Rt} du \int_{R+a}^{R+D+a} dv \frac{(v+u)^2}{(v-u)^5}, \qquad (6.55)$$

where $Q \equiv \max(R/\sqrt{R^2 + L^2}, (R-H)/R)$. Thus, the magnitude of Q depends on the relative sizes of the lens and the plate. If $L \leq \sqrt{2RH}$, then

$Q = R/\sqrt{R^2 + L^2}$. However, if $L > \sqrt{2RH}$, then $Q = (R - H)/R$ and the value of the Casimir force, owing to the rapid decrease of the potential $1/r^7$ with separation, does not depend on any further increase in L.

To calculate the Casimir force, we remove one integration in eqn (6.55) and arrive at

$$F_{\text{fin}}(a) = -\frac{\pi^3}{120}\Psi(\varepsilon_0^{(1)}, \varepsilon_0^{(2)}) \int_Q^1 t\, dt \int_{R-H}^{Rt} du \left[\frac{(R+a+u)^2}{(R+a-u)^5} - \frac{(R+a+D+u)^2}{(R+a+D-u)^5}\right]. \tag{6.56}$$

In the limiting case of an infinite plate, $L \to \infty$, in the lowest order of the small parameter a/R, eqn (6.56) results in

$$F(a) = -\Psi(\varepsilon_0^{(1)}, \varepsilon_0^{(2)})\frac{\pi^3 R}{360 a^3}. \tag{6.57}$$

This equation is a generalization of the second equality in eqn (6.51), which applies to the case of a dielectric sphere above a dielectric plate. In the limit $\varepsilon_0^{(1)}, \varepsilon_0^{(2)} \to \infty$, eqn (6.57) coincides with the respective equation in eqn (6.51).

Now we perform the remaining integrations in eqn (6.56) explicitly. The main contribution to the result, which depends on the size of the plate, appears in the third order in the small parameter a/R and has the form

$$F_{\text{fin}}(a) \approx \left[1 + \frac{a^3}{R^3}\frac{1}{(1-Q)^3}\right] F(a), \tag{6.58}$$

where $F(a)$ is defined in eqn (6.57). For typical configurations used in an experiment, the correction to the value of unity in this equation is very small (see Section 19.2.3).

6.5 The proximity force approximation

Another approximate method for the calculation of the Casimir force between bodies of arbitrary shape is the proximity force approximation (PFA). This powerful method, which allows generalization to the case of bodies made of real materials and to forces of different physical nature, was suggested by Derjaguin (1934). It was applied to the interpretation of measurements of the Casimir force in the sphere–plate configuration by Derjaguin et al. (1956) and reconsidered for application to various forces by Blocki et al. (1977).

We start from the most general formulation of the prescription given by the PFA for the calculation of the interaction energy between two arbitrarily shaped bodies V_1 and V_2. Let the top surface of the lower body and the bottom surface of the upper body be described by the equations $z_1 = z_1(x, y)$ and $z_2 = z_2(x, y) > z_1(x, y)$ in an appropriate coordinate system. The separation distance between these surfaces along the z-axis is given by

$$z(x, y) = z_2(x, y) - z_1(x, y). \tag{6.59}$$

This is the variable width of the gap between the two interacting bodies. We use a to denote the smallest value of $z(x, y)$. Now we consider an arbitrary point

(x, y) and replace the elements of the curved surfaces dS_1 and dS_2 around the points $z_1(x,y)$ and $z_2(x,y)$, respectively, with parallel surface elements $dx\, dy$. In so doing, we simultaneously replace the unknown interaction energy between the elements of the curved surfaces dS_1 and dS_2 with a known energy $E(z(x,y))$ per unit area of plane surface elements spaced at a separation $z(x,y)$, defined in eqn (6.59). Then the interaction energy between the bodies V_1 and V_2 is obtained by the integration of interaction energies between the plane surface elements. The integration region σ is the part of the (x,y) plane where both surfaces are defined:

$$E_{\text{PFA}}(a) = \iint_\sigma dx\, dy\, E(z(x,y)). \tag{6.60}$$

If $E(z(x,y))$ represents the Casimir energy per unit area of ideal-metal planes, the E_{PFA} obtained is also related to ideal-metal surfaces $z_1(x,y)$ and $z_2(x,y)$. If, however, $E(z(x,y))$ takes into account real material properties of the interacting bodies (see Chapters 12–15), so does E_{PFA}. Note that in this case E may depend on the properties of a real body at a point (x,y): $E = E(x, y, z(x,y))$.

Equation (6.60) is a universal one. It can be applied to arbitrary bodies, leading to results of varying precision when compared with the respective exact analytical results where those are available. This precision is usually determined by the separation distance a in comparison with the geometrical parameters of the bodies. From eqn (6.60), the force acting between the interacting bodies is given by

$$F_{\text{PFA}}(a) = -\frac{\partial E_{\text{PFA}}(a)}{\partial a}. \tag{6.61}$$

Equations (6.60) and (6.61) can be further simplified in the case where the surfaces of the interacting bodies are described by continuous functions having continuous derivatives up to an arbitrary order. If one also assumes that there is a single point $x = y = 0$ where the width of the gap $z(x,y)$ reaches a minimum, one can use a Taylor expansion in the form

$$z(x,y) = a + \frac{z''_{xx}(0,0)}{2}x^2 + \frac{z''_{yy}(0,0)}{2}y^2 + \ldots = a + \frac{x^2}{2R_x} + \frac{y^2}{2R_y} + \ldots. \tag{6.62}$$

Here, the directions x, y are chosen along the principal axes of the quadratic form of the function $z(x,y)$. Because of this, there is no cross term in xy in eqn (6.62). R_x and R_y are the principal radii of curvature at the point $(0,0)$.

Substituting eqn (6.62) into eqn (6.60) and introducing new variables $\xi = x/\sqrt{2R_x}$, $\eta = y/\sqrt{2R_y}$, one obtains

$$E_{\text{PFA}}(a) = 2\sqrt{R_x R_y} \iint_\sigma d\xi\, d\eta\, E(a + \xi^2 + \eta^2). \tag{6.63}$$

In terms of the polar coordinates φ and $\zeta = \sqrt{\xi^2 + \eta^2}$, this can be rearranged as

$$E_{\text{PFA}}(a) = 2\sqrt{R_x R_y} \int_0^\infty 2\pi\zeta\, d\zeta\, E(a+\zeta^2). \tag{6.64}$$

Finally, taking into account that $a + \zeta^2 = z$, we arrive at (Blocki et al. 1977)

$$E_{\text{PFA}}(a) = 2\pi \bar{R}\, \mathcal{P}(a), \quad \mathcal{P}(a) \equiv \int_a^\infty E(z)\, dz, \quad \bar{R} = \sqrt{R_x R_y}. \tag{6.65}$$

Using eqn (6.61), one obtains an expression for the force in the framework of the PFA,

$$F_{\text{PFA}}(a) = 2\pi \bar{R}\, E(a), \tag{6.66}$$

where $E(a)$ is the energy per unit area in the configuration of two parallel plates. Note that the geometrical mean of the two principal radii, $\bar{R}$, is connected with the Gaussian curvature at the origin $K(0,0)$ by $\bar{R} = 1/\sqrt{K(0,0)}$.

We emphasize that eqns (6.65) and (6.66) do not have as wide an application range as eqn (6.60). In fact, they are applicable only to compact gaps that have a single point of minimum width and are characterized by a finite mean curvature. Two typical examples are a sphere above a plane, and two spheres of radii R_1 and R_2. However, this formulation of the PFA in the form of eqns (6.65) and (6.66) is not applicable to cases where at least one radius of curvature is infinitely large (for instance, for a cylinder above a plane) or when a gap cannot be characterized by a single mean curvature radius. The latter happens, for instance, when the boundary surfaces are described by periodic functions. Equations (6.65) and (6.66) are also not applicable when the characteristic size of the upper body in the z-direction is smaller than or of the order of a. In fact, for the applicability of eqn (6.65) it is required that this size goes beyond the range of z where the interaction energy $E(z)$ drops to zero, so that the integral becomes independent of its upper limit. In all cases when this condition is not satisfied, the formulation of the PFA given by eqn (6.60) works well.

The first example of the application of the PFA is an ideal-metal sphere of radius R above an ideal-metal plane [see Fig. 6.1(a)]. The gap is restricted by the plane $z_1(x,y) = 0$ and by the lower hemisphere

$$z_2(x,y) = a + R - \sqrt{R^2 - x^2 - y^2}. \tag{6.67}$$

Replacing the surface elements of the sphere dS by the plane plate $dx\,dy$ parallel to the plane $z_1 = 0$ and introducing the polar coordinates $x = \rho\cos\varphi, y = \rho\sin\varphi$, we rearrange eqn (6.60) into the form

$$E_{\text{PFA}} \equiv E_{\text{IM}}^{\text{sp}}(a) = -\int_0^R 2\pi\rho\, d\rho\, \frac{\pi^2}{720 z_2^3(\rho)}, \tag{6.68}$$

where $z_2(\rho)$ is obtained from eqn (6.67), and $E(z_2)$ for ideal-metal plates is taken from eqn (1.5). The integration in eqn (6.68) results in

$$E_{\text{IM}}^{\text{sp}}(a) = -\frac{\pi^3}{720}\frac{R^2}{a^2(R+a)} = -\frac{\pi^3 R}{720 a^2}\left[1 + O\left(\frac{a}{R}\right)\right]. \tag{6.69}$$

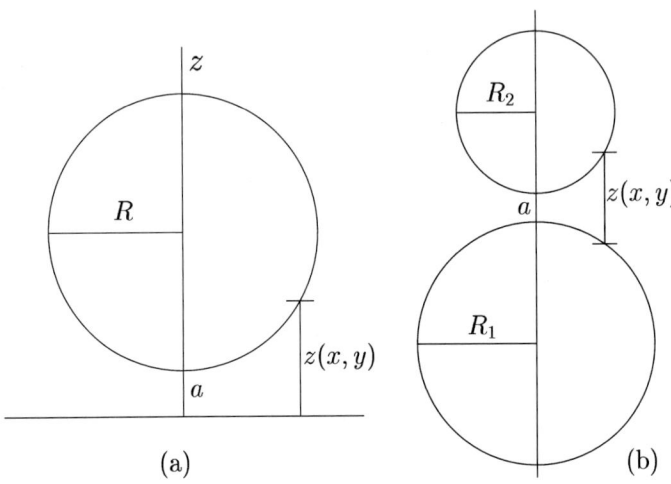

FIG. 6.1. Configurations (a) of a sphere above a plane and (b) two spheres of different radii.

Here, only the leading order is meaningful, and the correction of the order of a/R should be omitted. The negative differentiation of this expression with respect to the separation returns us to eqn (6.51) for the Casimir force between an ideal-metal sphere and an ideal-metal plane obtained using the PWS method. The same result also follows from eqn (6.66), which is readily applicable to the sphere–plate configuration. In this case one obtains $R_x = R_y = R$ from eqn (6.67), so that

$$F_{\text{PFA}} \equiv F_{\text{IM}}^{sp}(a) = 2\pi R E_{\text{IM}}(a) = -\frac{\pi^3 R}{360 a^3}. \tag{6.70}$$

Equation (6.70) allows a far-reaching generalization to the case of a sphere and a plate made of real materials and kept at nonzero temperature. In this case $E_{\text{IM}}(a)$ must be replaced with the free energy per unit area, $\mathcal{F}(a,T)$, for a configuration of two plane parallel plates made of a real dielectric or real metal (see Chapters 12–15). Then, according to eqn (6.70), the Casimir force between a large sphere and a plate is given by

$$F^{\text{sp}}(a,T) = 2\pi R \mathcal{F}(a,T). \tag{6.71}$$

This equation is widely used in Part III of the book for comparison of experiment and theory in relation to measurements of the Casimir force. Qualitatively, the error introduced by the use of the approximate eqn (6.71) is of the order of the terms neglected above, i.e. of the order of a/R. More exact, quantitative conclusions concerning the accuracy of the PFA can be found in Sections 10.3.3 and

10.4.3, where some exact analytical results valid beyond the PFA are presented. Experimentally, however, a/R is usually of order 10^{-3}, so that theoretical results based on the use of eqn (6.71) are highly reliable.

The configuration of two ideal-metal spheres of radii R_1 and $R_2 \leq R_1$ can also be considered using both eqn (6.60) and eqn (6.66). Let the sphere centers lie on the z-axis [see Fig. 6.1(b)]. The width of the gap between the lower and the upper sphere is described by the function

$$z(x,y) = z(\rho) = a + R_1 - \sqrt{R_1^2 - \rho^2} + R_2 - \sqrt{R_2^2 - \rho^2}, \quad (6.72)$$

where $\rho^2 = x^2 + y^2$. Using eqn (1.5) for the energy per unit area of ideal-metal planes, we obtain from eqn (6.60)

$$E_{\text{IM}}^{\text{ss}}(a) = -\frac{\pi^3}{360} \int_0^{R_2} \frac{\rho \, d\rho}{z^3(\rho)}. \quad (6.73)$$

Calculating this integral under the condition $a/R_2 \ll 1$, we arrive at

$$E_{\text{PFA}}(a) \equiv E_{\text{IM}}^{\text{ss}}(a) = -\frac{\pi^3}{720 a^2} \frac{R_1 R_2}{R_1 + R_2} \left[1 + O\left(\frac{a}{R_2}\right) \right]. \quad (6.74)$$

Neglecting the terms of order of a/R_2 and using eqn (6.61), we obtain the Casimir force acting between two ideal-metal spheres,

$$F_{\text{PFA}}(a) \equiv F_{\text{IM}}^{\text{ss}}(a) = -\frac{\pi^3}{360 a^3} \frac{R_1 R_2}{R_1 + R_2}. \quad (6.75)$$

The same result can be obtained from eqn (6.66). For this purpose, using eqn (6.72), we find

$$R_x = \frac{1}{z''_{xx}(0,0)} = \frac{R_1 R_2}{R_1 + R_2}, \quad R_y = \frac{1}{z''_{yy}(0,0)} = \frac{R_1 R_2}{R_1 + R_2}. \quad (6.76)$$

Then eqn (6.66) leads to

$$F_{\text{IM}}^{\text{ss}}(a) = 2\pi \frac{R_1 R_2}{R_1 + R_2} E_{\text{IM}}(a) = -\frac{\pi^3}{360 a^3} \frac{R_1 R_2}{R_1 + R_2}. \quad (6.77)$$

We now come to the case of an ideal-metal cylinder of radius R above an ideal metal plane. In this case one of the principal radii becomes infinite. Because of this, we use the formulation of the PFA in the form of eqn (6.60). The minimal separation distance between the plane and cylindrical surface is given by a. The gap between the two bodies is restricted by the surfaces $z_1(x,y) = 0$ and

$$z_2(y) = R + a - \sqrt{R^2 - y^2}. \quad (6.78)$$

In this case the cylinder axis coincides with the x-axis. In accordance with eqn (6.60), the Casimir energy per unit length of the cylinder is given by

$$E_{\text{PFA}}(a) \equiv E_{\text{IM}}^{\text{cp}}(a) = -2\int_0^R dy \frac{\pi^2}{720 z_2^3(y)}. \tag{6.79}$$

Calculating this integral under the condition $a/R \ll 1$, we obtain

$$E_{\text{IM}}^{\text{cp}}(a) = -\frac{\pi^3}{960 a^2}\sqrt{\frac{R}{2a}}\left[1+O\left(\frac{a}{R}\right)\right], \tag{6.80}$$

in accordance with the first equality in eqn (6.54), derived using the PWS method. As usual, only the leading contribution is meaningful and the terms of the order of a/R should be omitted. Then the Casimir force per unit length is given by the second equality in eqn (6.54). By analogy with eqns (6.70) and (6.77), the Casimir force per unit length between a cylinder and a plane can be represented using the Casimir energy per unit area of two parallel planes,

$$F_{\text{IM}}^{\text{cp}}(a) = \frac{15\pi}{8}\sqrt{\frac{R}{2a}}E_{\text{IM}}(a) = -\frac{\pi^3}{384 a^3}\sqrt{\frac{R}{2a}}. \tag{6.81}$$

In this form, the result obtained can be generalized to a cylinder and a plate made of any real materials.

One more example, where the formulation (6.66) of the PFA does not work, is the case of a paraboloid of sufficiently small height H, above a plane. For this configuration, $z_1(x, y) = 0$ and the surface of the paraboloid can be represented in the form

$$z_2 = a + \frac{H}{L^2}(x^2 + y^2) = a + \frac{H}{L^2}\rho^2, \tag{6.82}$$

where L is the radius of the top cross section. Substituting this in eqn (6.60), one obtains

$$E_{\text{PFA}}(a) \equiv E_{\text{IM}}^{\text{par}}(a) = -\int_0^L 2\pi\rho\, d\rho \frac{\pi^2}{720 z_2^3(\rho)}$$
$$= -\frac{\pi^3}{720}\frac{L^2}{H}\int_0^H \frac{dt}{(a+t)^3} = -\frac{\pi^3}{1440}\frac{L^2}{H}\left[\frac{1}{a^2} - \frac{1}{(a+H)^2}\right]. \tag{6.83}$$

From eqn (6.61), the respective Casimir force is given by

$$F_{\text{PFA}}(a) \equiv F_{\text{IM}}^{\text{par}}(a) = -\frac{\pi^3}{720}\frac{L^2}{H}\left[\frac{1}{a^3} - \frac{1}{(a+H)^3}\right]. \tag{6.84}$$

This result is applicable when H is of the order of a. For paraboloids of large height $H \gg a$, the formulation of the PFA (6.66) is also applicable. This leads to a Casimir force equal to the first term on the right-hand side of eqn (6.84).

7

THE CASIMIR EFFECT FOR TWO IDEAL-METAL PLANES

In the present chapter, we consider the simple, yet most important configuration of two parallel ideal-metal planes. The original Casimir effect is based on this configuration (Casimir 1948). First we present the theory of the scalar and electromagnetic Casimir effects between parallel planes. In comparison with Chapter 2, some basic facts are added concerning the relation between local and global approaches and the polarizations of the electromagnetic field. The radiative corrections to the Casimir force are considered.

The configuration of two parallel ideal-metal planes is the first configuration where we investigate the Casimir effect at nonzero temperature. Here, we present general analytical formulas for the Casimir free energy, entropy, and pressure and consider the limits of low and high temperature. The agreement with thermodynamics of the results obtained is analyzed. This is the starting point for the thermal Casimir force between real materials, considered in Part II of the book. The spinor Casimir effect between planes and the Casimir effect for a wedge are also discussed.

At the end of the chapter, we briefly consider the dynamic Casimir effect connected with uniformly moving or oscillating planes.

7.1 The scalar Casimir effect for parallel planes

Here, we consider the Casimir vacuum energy of a scalar field in a configuration of two parallel planes in three dimensions (see Fig. 2.2) with Dirichlet or mixed boundary conditions.

7.1.1 Dirichlet boundary conditions

We start from local quantities and pay special attention to the regions of space external to the plane–plane configuration. Let the two planes be at $x^3 \equiv z = 0$ and $z = a$.

In the region between the planes, the complete orthonormal set of solutions to eqn (3.1) satisfying the Dirichlet boundary conditions is given by eqn (3.49), where

$$\Phi_J(\boldsymbol{r}) = \Phi_{k_\perp, n}(\boldsymbol{r}) = \frac{1}{\pi\sqrt{2a}}\, e^{i(k_x x + k_y y)} \sin k_{zn} z, \qquad (7.1)$$

$$\omega_{k_\perp, n}^2 = m^2 + k_\perp^2 + k_{zn}^2, \qquad k_{zn} = \frac{\pi n}{a}.$$

Substituting these solutions into eqn (3.59), we obtain the vacuum energy density

$$\langle 0|T_{00}^{(0)}(z)|0\rangle = \frac{1}{2a}\int_0^\infty \frac{k_\perp \, dk_\perp}{2\pi} \sum_{n=1}^\infty \left[\omega_{k_\perp,n} - \frac{m^2 + k_\perp^2}{\omega_{k_\perp,n}} \cos(2k_{zn}z)\right], \quad (7.2)$$

where the last term is oscillating and does not contribute to the total energy. In free Minkowski space, the complete orthonormal set of solutions is given by eqn (3.56), leading to

$$\langle 0_M|T_{00}^{(0)}|0_M\rangle = \frac{1}{2}\int \frac{d\mathbf{k}}{(2\pi)^3}\omega_k, \quad (7.3)$$

where $d\mathbf{k} = dk_x\, dk_y\, dk_z$, $\omega_k = (m^2 + \mathbf{k}^2)^{1/2}$, and the vacuum states $|0\rangle$ and $|0_M\rangle$ have been defined in eqns (3.54) and (3.57), respectively.

The Casimir energy density

$$\epsilon(z) = \langle 0|T_{00}^{(0)}(z)|0\rangle - \langle 0_M|T_{00}^{(0)}|0_M\rangle \quad (7.4)$$

is found from eqns (7.2) and (7.3) using the Abel–Plana formula (2.26) in the same manner as in Section 2.5, with the result

$$\epsilon(z) = -\frac{1}{2a}\int_0^\infty \frac{k_\perp\, dk_\perp}{2\pi}\left[\frac{\sqrt{m^2+k_\perp^2}}{2} + \frac{2\pi}{a}\int_A^\infty \frac{\sqrt{t^2-A^2}}{e^{2\pi t}-1}dt \quad (7.5)\right.$$
$$\left. + (m^2+k_\perp^2)\sum_{n=1}^\infty \frac{\cos 2k_{zn}z}{\omega_{k_\perp,n}}\right],$$

where

$$A \equiv \frac{a\sqrt{m^2+k_\perp^2}}{\pi}, \qquad t \equiv \frac{ak_z}{\pi}. \quad (7.6)$$

Note that the first term on the right-hand side of eqn (7.5) is connected with the energy of the boundary planes (see below), whereas the third term is oscillating and, as explained previously, does not contribute to the Casimir energy. As a result, the Casimir energy per unit area of the planes is given by

$$E(a) = \int_0^a dz\, \epsilon(z) = -\frac{1}{2}\int_0^\infty \frac{k_\perp dk_\perp}{2\pi}\left(\frac{\sqrt{m^2+k_\perp^2}}{2} + \frac{2\pi}{a}\int_A^\infty \frac{\sqrt{t^2-A^2}}{e^{2\pi t}-1}dt\right). \quad (7.7)$$

For a massless field ($m = 0$), we have $A = ak_\perp/\pi$ and the integrals can be calculated simply, in the same way as at the end of Section 2.5. The result is

$$E(a) = -\frac{\pi^2}{1440a^3} - \frac{1}{8\pi}\int_0^\infty k_\perp^2\, dk_\perp. \quad (7.8)$$

The first term on the right-hand side of this equation is just one-half of eqn (2.82), obtained for the electromagnetic Casimir effect. From eqn (7.8), the Casimir

pressure between two parallel planes due to scalar field oscillations takes the form
$$P(a) = -\frac{\pi^2}{480a^4}. \tag{7.9}$$

The second, infinite, term in eqn (7.8) does not depend on the separation distance and does not contribute to the force. Below, we discuss its physical meaning.

We start by noting that so far, only the region of space $0 < z < a$ between the parallel planes has been considered. Now let us find the vacuum energy of a semispace $z < 0$ with a Dirichlet boundary condition on the plane $z = 0$. In this case the complete orthonormal set of solutions of eqn (3.1) can be presented in the form

$$\varphi_k^{(\pm)}(t, \boldsymbol{r}) = \frac{1}{\pi\sqrt{2\pi\omega_k}} e^{\mp i(\omega_k t - k_x x - k_y y)} \sin k_z z, \tag{7.10}$$

where $0 \leq k_z < \infty$ and $\omega_k^2 = m^2 + |\boldsymbol{k}|^2$. These solutions are normalized in terms of the scalar product (3.7), where the integration is over the semispace $z \leq 0$. The set of solutions of eqn (3.1) in free Minkowski space is given in eqn (3.56). It can be presented in a form similar to eqn (7.10),

$$\varphi_{kj}^{(\pm)}(t, \boldsymbol{r}) = \frac{1}{2\pi\sqrt{2\pi\omega_k}} e^{\mp i(\omega_k t - k_x x - k_y y)} \psi_{k_z j}(z), \tag{7.11}$$

where

$$\psi_{k_z 1}(z) = \cos k_z z, \qquad \psi_{k_z 2}(z) = \sin k_z z, \tag{7.12}$$

and $0 \leq k_z < \infty$. These functions are normalized for the entire volume of free Minkowski space.

Calculating the Casimir energy density in the same way as for the region in between the planes, we arrive at the result

$$\begin{aligned}\epsilon(z) &= -\frac{1}{2\pi} \int_0^\infty \frac{k_\perp dk_\perp}{2\pi} \int_0^\infty dk_z \frac{m^2 + k_\perp^2}{\omega_k} \cos 2k_z z \\ &= -\frac{1}{2\pi} \int_0^\infty \frac{k_\perp dk_\perp}{2\pi} (m^2 + k_\perp^2) K_0\left(2\sqrt{m^2 + k_\perp^2} |z|\right), \end{aligned} \tag{7.13}$$

where the $K_\nu(z)$ are the Bessel functions of imaginary argument.

The total Casimir energy in the region $z \leq 0$ per unit area of the boundary plane $z = 0$ is given by

$$E = \int_{-\infty}^0 \epsilon(z)\, dz = -\frac{1}{8} \int_0^\infty \frac{k_\perp dk_\perp}{2\pi} \sqrt{m^2 + k_\perp^2}. \tag{7.14}$$

Note that the first term on the right-hand side of eqn (7.7) is just twice the expression (7.14). This is because eqn (7.14) represents the Casimir energy arising from one side of a plane, whereas the respective term in eqn (7.7) originates from the external sides of two different planes.

To conclude, when the local approach to the Casimir effect is used for plane boundary surfaces, the energy densities obtained may contain some constant and position-dependent terms. This is true for both interior and exterior regions of the configuration under consideration. The energy densities obtained are not defined uniquely and can be changed by adding a 4-divergence to the energy–momentum tensor (Itzykson and Zuber 2005). This does not influence the total Casimir energy obtained by the integration of the energy density over the volume of the system. After integration over the volume, there also remain terms (generally infinite) that do not depend on the separation between the interacting surfaces. In particular, this is the case for the exterior regions outside the boundary planes. Such constant terms can be interpreted as the proper energies of these planes, and they do not influence measurable quantities such as the Casimir force or pressure that are defined as derivatives of the energy with respect to the separation. Because of this, in the subsequent text we concentrate only on those contributions to the Casimir energy that are dependent on the separation between the interacting surfaces. The problem of uniqueness in the definition of the vacuum energy density will be discussed further in Chapter 11, devoted to spaces with non-Euclidean topologies.

7.1.2 Mixed boundary conditions

Now we consider a scalar field for a configuration of two parallel planes with a Dirichlet boundary condition on one plane, $z = 0$, and a Neumann boundary condition on the other plane, $z = a$:

$$\varphi(t, x, y, 0) = \left.\frac{\partial \varphi(t, x, y, z)}{\partial z}\right|_{z=a} = 0. \tag{7.15}$$

The complete orthonormal set of solutions of eqn (3.1) satisfying these conditions is given by eqns (3.49) and (7.1), where $k_{zn} = \pi n/a$ is replaced with

$$k_{zn} = \frac{\pi}{a}\left(n + \frac{1}{2}\right), \quad n = 0, 1, 2, \ldots. \tag{7.16}$$

The vacuum energy of the scalar field (for simplicity, we consider the massless case) is expressed as

$$E_0(a) = \frac{1}{2}\int_0^\infty \frac{k_\perp dk_\perp}{2\pi} \sum_{n=0}^\infty \sqrt{k_\perp^2 + \frac{\pi^2}{a^2}\left(n + \frac{1}{2}\right)^2}\, S, \tag{7.17}$$

where S is the area of the planes.

Now we subtract from eqn (7.17) the vacuum energy of free Minkowski space for the volume between the planes. This is done using a version of the Abel–Plana formula (2.41) adapted for summation over half-integers. The resulting finite Casimir energy per unit area is

$$E(a) = \frac{\pi}{a}\int_0^\infty \frac{k_\perp dk_\perp}{2\pi}\int_{k_\perp a/\pi}^\infty \frac{\sqrt{t^2 - (k_\perp^2 a^2/\pi^2)}}{e^{2\pi t} + 1}dt. \quad (7.18)$$

Substituting the new variable $y = k_\perp a/\pi$ and changing the order of the integrations, we obtain

$$E(a) = \frac{\pi^2}{2a^3}\int_0^\infty \frac{dt}{e^{2\pi t} + 1}\int_0^t \sqrt{t^2 - y^2}\, y\, dy = \frac{7}{8}\frac{\pi^2}{1440 a^3}. \quad (7.19)$$

As is seen from eqn (7.19), the Casimir energy for the case of mixed boundary conditions is different only by a factor of $-7/8$ from the first term in eqn (7.8), related to Dirichlet boundary conditions on both planes. It is also notable that for mixed boundary conditions there is no separation-independent contribution to the Casimir energy, such as found in eqn (7.8), which describes the energy of the boundary planes. This is because, for mixed boundary conditions, the summation starts from zero instead of unity.

From eqn (7.19), the repulsive Casimir pressure between the planes is equal to

$$P(a) = -\frac{\partial E(a)}{\partial a} = \frac{7}{8}\frac{\pi^2}{480 a^4}, \quad (7.20)$$

i.e. a factor $-7/8$ different from the Casimir pressure in eqn (7.9). This is another example where the subtraction of the infinite vacuum energy of free space leaves us with a positive Casimir energy. The scalar Casimir effect with Robin boundary conditions on two parallel planes was considered by Romeo and Saharian (2002).

7.2 The electromagnetic Casimir effect between parallel planes

In this section, we present a more detailed picture of the electromagnetic Casimir effect between two ideal-metal parallel planes and between an ideal-metal plane and an infinitely permeable plane. This includes the complete orthonormal set of solutions and a justification of eqn (3.76) for the summation of modes. We also illustrate the method of zeta function regularization. Our presentation is based on the canonical quantization of the electromagnetic field in Section 3.3.

7.2.1 Ideal-metal planes

In the case of parallel planes at $z = 0, a$, the set of solutions of the Dirichlet boundary problem introduced in eqns (3.64) and (3.65) takes the form

$$\mathcal{A}_{k_\perp,n}(\mathbf{r}) = \begin{pmatrix} b_x \cos k_x x \sin k_y y \sin k_{zn} z \\ b_y \sin k_x x \cos k_y y \sin k_{zn} z \\ b_z \sin k_x x \sin k_y y \cos k_{zn} z \end{pmatrix}. \quad (7.21)$$

Here, $k_{zn} = \pi n/a$, $n = 0, 1, 2, \ldots$, and thus the first and second components of the vector potential vanish on the planes. This is equivalent to the boundary condition (2.71) written in terms of the electric field and magnetic induction, as discussed in Section 3.2. It is easily seen that the normalization condition (3.66), with

$$\delta_{JJ'} = \delta(k_x - k'_x)\delta(k_y - k'_y)\delta_{nn'}, \quad \omega^2_{k_\perp,n} = k^2_\perp + k^2_{zn}, \qquad (7.22)$$

$$\int_V d\mathbf{r} \equiv \int_{-\infty}^{\infty} dx \int_{-\infty}^{\infty} dy \int_0^a dz,$$

and the gauge-fixing condition $\mathrm{div}\mathbf{A}_{k_\perp,n} = 0$, leads to the following values of the coefficients b_x and b_z:

$$b_x = b_y = \frac{4\sqrt{2\pi}}{\sqrt{a}} \frac{k_{zn}}{\sqrt{(k_x + k_y)^2 + 2k^2_{zn}}},$$

$$b_z = -\frac{4\sqrt{2\pi}}{\sqrt{a}} \frac{k_x + k_y}{\sqrt{(k_x + k_y)^2 + 2k^2_{zn}}}. \qquad (7.23)$$

Now we introduce the polarizations of the electromagnetic field between the planes, which will be repeatedly used throughout the book. The plane formed by the wave vector $\mathbf{k} = (k_x, k_y, k_{zn})$ and the normal to the boundary plane $\mathbf{n}$ is called the plane of incidence. The two polarization vectors introduced in eqn (3.74) can be rewritten in the form

$$\boldsymbol{\epsilon}_{\mathbf{k}}^{(1)} = \frac{1}{k_\perp}\begin{pmatrix} k_y \\ -k_x \\ 0 \end{pmatrix}, \quad \boldsymbol{\epsilon}_{\mathbf{k}}^{(2)} = \frac{1}{kk_\perp}\begin{pmatrix} k_x k_{zn} \\ k_y k_{zn} \\ -k^2_\perp \end{pmatrix}. \qquad (7.24)$$

The vector $\boldsymbol{\epsilon}_{\mathbf{k}}^{(1)}$ is perpendicular to $\mathbf{k}$ and to the plane of incidence, whereas $\boldsymbol{\epsilon}_{\mathbf{k}}^{(2)}$ is perpendicular to $\mathbf{k}$ but parallel to the plane of incidence. The electromagnetic wave with $\mathbf{E}$ parallel to $\boldsymbol{\epsilon}_{\mathbf{k}}^{(1)}$ is called the transverse electric (TE) mode. For the TE mode, the magnetic induction $\mathbf{B}$ is in the plane of incidence. The electromagnetic wave with $\mathbf{E}$ parallel to $\boldsymbol{\epsilon}_{\mathbf{k}}^{(2)}$ is called the transverse magnetic (TM) mode. For the TM mode, $\mathbf{B}$ is perpendicular to the plane of incidence.

From eqns (7.21) and (7.24), we can find the coefficients (3.69) of the expansion of the vector potential (7.21) in the polarization vectors,

$$\tilde{A}^{(1)}_{k_\perp,n}(\mathbf{r}) = \frac{b_x}{k_\perp}\left(\frac{\partial}{\partial x} - \frac{\partial}{\partial y}\right) \cos k_x x \cos k_y y \sin k_{zn} z, \qquad (7.25)$$

$$\tilde{A}^{(2)}_{k_\perp,n}(\mathbf{r}) = -\frac{1}{kk_\perp}\left(b_x \frac{\partial^2}{\partial x \partial z} + b_x \frac{\partial^2}{\partial y \partial z} + b_z k^2_\perp\right) \sin k_x x \sin k_y y \cos k_{zn} z.$$

We emphasize that $\tilde{A}^{(1)}_{k_\perp,0}(\mathbf{r}) = 0$. Thus, for all $n \geq 1$ there are two different polarizations of the electromagnetic field confined between the parallel planes, but at $n = 0$ only one polarization survives.

Now we can use the general equations (3.75) and (3.76) to find the total Casimir energy of the electromagnetic vacuum:

$$E_0(a) = \left(\frac{1}{2} \int_0^\infty \frac{k_\perp dk_\perp}{2\pi} \omega_{k_\perp,0} + \int_0^\infty \frac{k_\perp dk_\perp}{2\pi} \sum_{n=1}^\infty \omega_{k_\perp,n} \right) S. \qquad (7.26)$$

Here we have accounted for the fact that there is only one polarization state at $n = 0$. Equation (7.26), derived using the complete orthonormal set of solutions of the wave equation (3.63), coincides with eqn (2.72), formally obtained by the summation of the oscillator frequencies over both negative and positive n. In Section 2.5, a finite Casimir energy per unit area of the planes was obtained from eqn (7.26) by subtracting the energy of free Minkowski space and using the Abel–Plana formula. It was shown that the result obtained does not depend on the form of the cutoff function used to regularize eqn (7.26) and the respective expression in free space to be subtracted from eqn (7.26).

Here, we demonstrate the application of zeta function regularization, discussed in Section 4.1, to the case of two parallel planes. To begin, we disregard the first term on the right-hand side of eqn (7.26) because it does not depend on the separation distance (see the previous section). Next, we introduce the regularization parameter s and rewrite eqn (7.26) in the regularized form

$$E_0^{(s)}(a) = \sum_{n=1}^\infty \int_0^\infty \frac{k_\perp dk_\perp}{2\pi} \left(k_\perp^2 + \frac{\pi^2 n^2}{a^2} \right)^{(1-2s)/2} S. \qquad (7.27)$$

Note that we have dropped the factor μ^{2s} in eqn (4.3). By making the change of variable $k_\perp = \pi n y/a$, we obtain

$$E_0^{(s)}(a) = \frac{1}{2\pi} \left(\frac{\pi}{a} \right)^{3-2s} \sum_{n=1}^\infty \frac{1}{n^{2s-3}} \int_0^\infty y\, dy \, (y^2 + 1)^{(1-2s)/2} S. \qquad (7.28)$$

The sum in eqn (7.28) reduces to the Riemann zeta function (4.6) with $z = 2s - 3$. As was explained in Chapter 4, this function is defined by eqn (4.6) for $\operatorname{Re} z > 1$, i.e. for $\operatorname{Re} s > 2$. We, however, need the value of $\zeta_R(z)$ at $z = -3$ in the limit of removing the regularization, $s \to 0$. If we use the definition of $\zeta_R(z)$ according to eqn (4.6), the value of $\zeta_R(-3)$ evidently diverges. To obtain from $E_0^{(s)}(a)$ the physical Casimir energy per unit area $E(a)$, the method of zeta function regularization suggests the use of the analytic continuation of the Riemann zeta function. As was discussed in Section 4.1, there exists a meromorphic function with a simple pole at $z = 1$ which can be obtained by analytic continuation of the right-hand side of eqn (4.6) to the entire complex plane. Such an analytic continuation is unique and well defined, for instance, at the point $z = -3$. Needless to say, the values of this analytic continuation for $\operatorname{Re} z < 1$ are not represented by the right-hand side of eqn (4.6). For $\operatorname{Re} z < 0$, these values can be obtained from the reflection relation (5.50).

Equation (5.50) results in $\zeta_R(-3) = 1/120$. The integral in eqn (7.28) also can be calculated at $\operatorname{Re} s > 3/2$:

$$\int_0^\infty y\, dy\, (y^2+1)^{(1-2s)/2} = -\frac{1}{3-2s}. \tag{7.29}$$

Substituting the regularized values of both the sum and the integral in eqn (7.28) and replacing $E_0^{(s)}(a)/S$ with $E(a)$ in the limiting case $s \to 0$, we obtain

$$E(a) \equiv E_{\mathrm{IM}}(a) = -\frac{\pi^2}{720 a^3}, \tag{7.30}$$

i.e. just the Casimir energy per unit area (2.82). From this, the electromagnetic Casimir pressure is equal to that given by eqn (1.1).

It should be noted that with the use of analytic continuation of the zeta function at $z = -3$ (i.e. in the limit $s \to 0$), not only did the vacuum energy remain finite, but this procedure also made this energy equal to the physical value (7.30) obtained in Section 2.5 after the subtraction of the vacuum energy of free Minkowski space. Thus, the application of this method to two parallel planes is sometimes referred to as renormalization by zeta function regularization. For more complicated configurations, the final result in the limit of removing the regularization will in general be infinite and some additional renormalization might be needed.

As can be seen from the above, the zeta function regularization method for the configuration of parallel planes is not as physically transparent as regularization using a cutoff function. The latter makes the vacuum energy in the presence of boundary planes and the respective energy in the free Minkowski space individually finite. In that case, the Abel–Plana formula allows one to find a finite difference between the two quantities when the regularization is removed. However, as was shown in Chapter 4, zeta function regularization has some mathematical advantages and helps one to find the general structure of the ultraviolet divergences.

7.2.2 An ideal-metal plane and an infinitely permeable plane

The ideal-metal planes considered in Sections 2.5 and 7.2.1 are idealized thin plates made of a material with an infinitely large magnitude of the dielectric permittivity (more realistic models of metals will be considered in Part II). It was H. B. G. Casimir himself who raised a question, in a letter to T. H. Boyer, regarding magnetic boundary conditions in his effect. Stimulated by this letter, Boyer (1974) solved the problem of the Casimir interaction between an ideal-metal plane and an infinitely permeable plane characterized by an infinitely large magnetic permeability. On the infinitely permeable plane, the tangential component of the magnetic induction vanishes:

$$\boldsymbol{B}_{\mathrm{t}}(t, \boldsymbol{r})|_S = 0. \tag{7.31}$$

We assume that the plane $z = 0$ is made of an ideal metal and the plane $z = a$ is infinitely permeable.

It is easily seen that the complete orthonormal set of solutions (7.21) satisfying Dirichlet boundary conditions on the plane at $z = 0$,

$$\mathcal{A}_{x;k_\perp,n}|_{z=0} = \mathcal{A}_{y;k_\perp,n}|_{z=0} = 0, \tag{7.32}$$

also satisfies eqn (7.31) on the plane at $z = a$ if we replace $k_{zn} = \pi n/a$ with the k_{zn} defined in eqn (7.16). This follows from the second equality in eqn (3.13) and from the fact that the vector potential (7.22) with the above replacement of k_{zn} satisfies the Neumann boundary conditions

$$\left.\frac{\partial \mathcal{A}_{x;k_\perp,n}}{\partial z}\right|_{z=a} = \left.\frac{\partial \mathcal{A}_{x;k_\perp,n}}{\partial z}\right|_{z=a} = 0. \tag{7.33}$$

With the replacement of k_{zn}, we get

$$\mathcal{A}_{z;k_\perp,n}|_{z=a} = 0, \tag{7.34}$$

and eqn (7.33) becomes equivalent to eqn (7.31). Thus, the configuration of an ideal-metal plane and an infinitely permeable plane becomes equivalent to the simpler Casimir problems for a scalar field with mixed boundary conditions considered in Sections 2.1 and 7.1.2.

In analogy to eqn (7.27), the regularized Casimir energy takes the form

$$E_0^{(s)}(a) = \sum_{n=0}^{\infty} \int_0^\infty \frac{k_\perp \, dk_\perp}{2\pi} \left[k_\perp^2 + \frac{\pi^2}{a^2}\left(n + \frac{1}{2}\right)^2 \right]^{(1-2s)/2} S. \tag{7.35}$$

By making the change of variable $k_\perp = \pi y(n+1/2)/a$, this can be rearranged as

$$E_0^{(s)}(a) = \frac{1}{2\pi}\left(\frac{\pi}{a}\right)^{3-2s} \sum_{n=0}^\infty \left(\frac{2}{2n+1}\right)^{2s-3} \int_0^\infty y \, dy \, (y^2+1)^{(1-2s)/2} S. \tag{7.36}$$

The sum can be expressed as $\zeta(2s-3, 1/2)$ in terms of the Hurwitz zeta function

$$\zeta(z,q) \equiv \sum_{n=0}^\infty \frac{1}{(n+q)^z}, \tag{7.37}$$

which is well defined for $\mathrm{Re}\, z > 1$ and $q \neq 0, -1, -2, \ldots$. We, however, need $\zeta(-3, 1/2)$ in the limit when the regularization is removed. This value can be obtained from the following analytic continuation (Gradshteyn and Ryzhik 1994):

$$\zeta(z,q) = \frac{2\Gamma(1-z)}{(2\pi)^{1-z}}\left[\sin\frac{\pi z}{2}\sum_{n=1}^\infty \frac{\cos(2\pi qn)}{n^{1-z}} + \cos\frac{\pi z}{2}\sum_{n=1}^\infty \frac{\sin(2\pi qn)}{n^{1-z}}\right]. \tag{7.38}$$

This continuation applies to the region $\mathrm{Re}\, z < 0$ and $0 < q \leq 1$.

In our case, eqn (7.38) leads to

$$\zeta\left(-3, \frac{1}{2}\right) = -\frac{7}{8}\frac{1}{200}. \tag{7.39}$$

Using eqns (7.29) and (7.39) with the regularization removed, we obtain from eqn (7.36) the Casimir energy per unit area (Boyer 1974)

$$E(a) = \frac{7}{8}\frac{\pi^2}{720a^3}. \tag{7.40}$$

This is the positive Casimir energy, leading to the repulsive Casimir pressure

$$P(a) = \frac{7}{8}\frac{\pi^2}{240a^4}, \tag{7.41}$$

which is equal to a factor of $-7/8$ times the classical Casimir result for ideal-metal planes. Note that the change from attraction to repulsion occurs because of the mixed boundary conditions. If two infinitely permeable planes were considered, we would return to the results (7.30) and (1.1) for ideal-metal planes.

There are various generalizations of eqn (7.41) to the case of nonzero temperature (Santos et al. 1999). In Part II of the book, we shall discuss the influence of the magnetic properties of real materials on the Casimir force.

7.3 The radiative corrections to the Casimir force

The interaction of the electromagnetic field with the electron–positron field gives additional contributions to the Casimir effect. These are the radiative corrections which occur for all quantum electrodymanic processes. The vacuum energy, as we have considered it so far, can be represented in terms of Feynman graphs in the lowest order with respect to this interaction. This is referred to as a *one-loop* contribution. The radiative corrections are two- and higher-loop contributions. Their relative magnitude depends strongly on the configuration considered. First we consider the physically relevant configuration of conductor boundary conditions and a massive spinor field. For two parallel planes at a separation a, the vacuum energy, including the first radiative correction, can be written in the form

$$E(a) = -\frac{\pi^2}{720a^3}\left[1 - \alpha\, G_{\mathrm{p}}\left(\frac{a}{\lambda_{\mathrm{C}}}\right) + \ldots\right], \tag{7.42}$$

where the Compton wavelength of the electron λ_{C} and the fine structure constant α (in the Gaussian system of units) are defined as

$$\lambda_{\mathrm{C}} = \frac{\hbar}{m_e c} \approx 3.86 \times 10^{-13}\,\mathrm{m}, \qquad \alpha = \frac{e^2}{\hbar c} \approx \frac{1}{137}, \tag{7.43}$$

and $G_{\mathrm{p}}(a/\lambda_{\mathrm{C}})$ is a dimensionless function. The representation (7.42) follows simply from dimensional considerations.

The smallness of the radiative corrections is determined by two factors. The first is the smallness of the coupling constant α. A similar expression in, for example, quantum chromodynamics would have a bigger coupling constant in place of α. The second factor is of geometrical nature, and its smallness follows from the magnitude of the function $G_{\rm p}(a/\lambda_{\rm C})$. This function was derived by Bordag et al. (1985). In the region of interest, the plate separation a is much larger than $\lambda_{\rm C}$ in eqn (7.43). For instance, if the separation is about a micrometer, we have $\lambda_{\rm C}/a \sim 10^{-7}$. Thus, we need to know the behavior of the function $G_{\rm p}$ for large arguments,

$$G_{\rm p}\left(\frac{a}{\lambda_{\rm C}}\right) = \frac{9}{128\pi} \frac{\lambda_{\rm C}}{a} + O\left(\frac{\lambda_{\rm C}^2}{a^2}\right). \tag{7.44}$$

Thus, in quantum electrodynamics, for a separation of $1\,\mu$m, the radiative correction is suppressed by approximately ten orders of magnitude. It is too small to be measurable.

Although this chapter is devoted to the configuration of two ideal-metal planes, we briefly mention here the radiative corrections to the Casimir energy arising in the configurations of an ideal-metal spherical shell and a rectangular box. The dominant, one-loop, contributions to the Casimir energy in these configurations are considered in Chapters 9 and 8, respectively. Thus, for a conducting spherical shell, a similar calculation (Bordag and Lindig 1998) results in

$$E(R) = \frac{0.0461766}{R}\left[1 - \alpha\, G_{\rm s}\left(\frac{R}{\lambda_{\rm C}}\right) + \ldots\right], \tag{7.45}$$

where R is the radius of the sphere. The leading-order contribution to the Casimir energy of a spherical shell is considered in Section 9.3.3. The function $G_{\rm s}(R/\lambda_{\rm C})$ was calculated for a large argument and found to have an expansion similar to eqn (7.44),

$$G_{\rm s}\left(\frac{R}{\lambda_{\rm C}}\right) = \left(-0.001306\,\ln\frac{\lambda_{\rm C}}{R} + 0.01117\right)\frac{\lambda_{\rm C}}{R} + O\left(\frac{\lambda_{\rm C}^2}{R^2}\right). \tag{7.46}$$

Here, a logarithmic contribution has appeared, which is a result of the curvature. The smallness of this radiative correction is similar to that for parallel planes.

Thus, at present, the interest in radiative corrections to the Casimir effect is only theoretical. However, there are a number of interesting problems in this area of research. First, there is the question of the ultraviolet divergences. From the calculations by Bordag et al. (1985) and Bordag and Lindig (1998), it can be seen that the renormalization of the additional loop can be done by charge renormalization in the same way as in quantum electrodynamics without boundary conditions. None of the remaining divergences affect the Casimir pressure. However, a general investigation of the divergences resulting from higher loops in the presence of boundaries is still missing.

In the remaining part of this section, we discuss some extensions of the above results and some questions of more theoretical interest. We start with the function $G_\mathrm{p}(a/\lambda_\mathrm{C})$. It was derived in the form of a double integral, and its asymptotic expansion for large argument was found (Bordag et al. 1985). It was also calculated numerically for all arguments (Milton 2001). It is a smooth, monotonic function and its value at zero argument is $G_\mathrm{p}(0) = 4.016$. This is the radiative correction in the case of a massless spinor field. A similar result can be expected for the spherical case. So one can conclude that for massless spinor fields, there is no geometrical suppression of the radiative correction. The radiative corrections to the Casimir effect between parallel planes at nonzero temperature were investigated by Robaschik et al. (1987).

For further discussion of the above results, it is necessary to consider some details of their derivation. In parallel, we shall discuss the corresponding formulas for a scalar field. The starting point is the functional-integral representation of the vacuum energy, eqn (3.103). It should be mentioned that the derivation of this representation in Section 3.5 is also valid for interacting fields. The generating functional $Z[\Upsilon]$ is given by eqn (3.100) and the action now consists of three parts,

$$S_\mathrm{QED} = S_\mathrm{em} + S_\mathrm{spinor} + S_\mathrm{int}, \tag{7.47}$$

where the action of the electromagnetic field S_em is given by eqn (3.23) and the action of the spinor field S_spinor is given by eqns (3.33) and (3.34). The interaction term,

$$S_\mathrm{int} = -e \int d^4x\, \bar\psi \gamma^\mu \psi A_\mu, \tag{7.48}$$

is the usual one following from the covariant derivative

$$D_\mu = \partial_\mu + \mathrm{i}e A_\mu. \tag{7.49}$$

In case of a scalar field, we would take

$$S_\mathrm{scalar} = S[\varphi] + \frac{\lambda}{4!} \int d^4x\, \varphi^4(x), \tag{7.50}$$

where $S[\varphi]$ is given by eqn (3.5) and λ is the coupling constant.

In both cases, the general Feynman rules for calculating higher-loop corrections retain their validity in the presence of boundary conditions which enter through the propagators. According to these rules, the effective action $W_\mathrm{eff}[0]$ in eqn (3.101) for a vanishing source is the sum of all connected vacuum graphs, i.e. all connected graphs with no external legs. Then, from eqn (3.101), the effective action for a vanishing source can be expressed through

$$\ln Z_\mathrm{scalar}[0] = \ln Z^{(0)}_\mathrm{scalar}[0] + \frac{1}{8}\,\infty + \ldots \tag{7.51}$$

for the scalar field and

$$\ln Z_{\text{QED}}[0] = \ln Z_{\text{QED}}^{(0)}[0] + \begin{array}{c}\includegraphics{}\end{array} + \ldots \qquad (7.52)$$

for quantum electrodynamics. Here, the wavy line represents the photon propagator and the solid lines represent the spinor propagator. In eqns (7.51) and (7.52), the superscript (0) denotes the contribution following from the noninteracting parts of the actions and gives rise to the vacuum energies that we have considered so far. For example, for parallel plates we get the electromagnetic Casimir energy (1.5) or (2.82), and for the scalar field we get one half of that. These are the one-loop contributions. The radiative corrections are given by the graphs in eqns (7.51) and (7.52) and have two loops. Using the corresponding Feynman rules, and inserting the two-loop contribution for the scalar field into the vacuum energy (3.103), one obtains

$$E_{0,\text{scalar}}^{(\text{2-loop})} = \frac{i\lambda}{2T} \int d^4x \left[{}^S G(x,x) \right]^2, \qquad (7.53)$$

where ${}^S G(x, x')$ is the scalar propagator obeying the corresponding boundary conditions. This might be given, for example, by eqn (3.122). It should be mentioned that any representation of the propagator can be inserted into eqn (7.53), for example a representation obtained using mirror images.

In quantum electrodynamics, the one-loop part consists of two contributions, one from the electromagnetic field and one from the spinor field. The former one, as in the scalar case, just gives the Casimir effect that we have considered so far: that is, eqn (1.5) or (2.82) for parallel planes with conducting boundary conditions. For the spinor contribution, the situation is different. Here, the only local boundary condition is the bag boundary condition [see eqn (7.102)]. However, in the presence of the interaction (7.48), this condition is not compatible with the conductor boundary conditions (3.46). The only compatible way is to keep the spinor field without boundary conditions. Thus, the spinor field must be considered for the whole space. As a consequence, the electromagnetic field must also be considered for the whole space, i.e. on both sides of the boundary. In this configuration, the radiative correction to the energy (3.103) following from the graph in eqn (7.52) is

$$E_{0,\text{QED}}^{(\text{2-loop})} = \frac{i}{2T} \int d^4x \int d^4y \, {}^S G_{\nu\mu}(x,y) \, \Pi^{\mu\nu}(y-x). \qquad (7.54)$$

Here, ${}^S G_{\mu\nu}(x, y)$ is the photon propagator obeying conductor boundary conditions, which can be constructed as a direct generalization from eqn (3.122), and

$$\Pi^{\mu\nu}(x-y) = -ie^2 \, \text{Tr} \, \gamma^\mu S(x-y) \gamma^\nu S(y-x) \qquad (7.55)$$

is the quantum electrodynamical polarization tensor for the spinor propagator $S(x - y)$ in free space. The physical picture for this setup is that we have an infinitely thin conducting surface and a spinor field which freely penetrates this surface. One could think of a metallic mesh whose spacing is bigger than the Compton wavelength of the electron. Exactly this configuration was used to derive eqs (7.42) and (7.45). The consistency of this configuration was reconfirmed later (Bordag and Scharnhorst 1998) by considering semitransparent mirrors, which reproduce the result (7.44) in the limit of becoming ideal conductors.

Another configuration where the boundary conditions are compatible with respect to the interaction is obtained by taking bag boundary conditions for both the electromagnetic and the spinor field. These boundary conditions can be used in both quantum electrodynamics and quantum chromodynamics. However, in this scheme the radiative corrections to the Casimir effect have not been considered so far.

The radiative corrections for a scalar field with a self-interaction given by eqn (7.50) have been calculated more than once. The first calculation was done by Ford (1979), who, however, did not obtain a finite result for parallel planes. The finite result, obtained by Kay (1979) and Toms (1980b), is

$$E(a) = -\frac{\pi^2}{1440a^3}\left[1 + \frac{5}{64}\lambda + O\left(\lambda^2\right)\right]. \tag{7.56}$$

The number 5/64 is the scalar analogue of $G_p(0)$ in eqn (7.42). This result has been later reconfirmed several times, and various boundary conditions and massive fields have been considered (Barone et al. 2004).

For the electrodynamics of photons interacting with a massless scalar field in a cube of size a, the first radiative correction has been calculated (Peterson et al. 1982), and results in

$$E(a) = \frac{0.09166}{a}\left[1 + 8.07\,\alpha + O\left(\alpha^2\right)\right], \tag{7.57}$$

where the leading contribution is the energy (8.64) for the electromagnetic field in a cube, and α is the fine structure constant.

For parallel planes with periodic boundary conditions, the Casimir energy is the same as the free energy in a finite-temperature theory with no boundary conditions. This was first discussed by Toms (1980a). There and in a large number of other papers [e.g. Langfeld et al. (1995)], mass generation due to the vacuum energy was investigated. This includes spontaneous symmetry breaking and phase transitions which occur when the radiative corrections given by the graph in eqn (7.51) are taken into account. It should be mentioned that these studies, strictly speaking, go beyond the Casimir effect as far as it is considered in this book.

We conclude this section by mentioning some controversies concerning the leading order of the radiative corrections. Using an effective-action approach, Kong and Ravndal (1997) and Scharnhorst (1998) found that the leading term in the radiative correction is of the second order in the fine structure constant and that, in addition, it is suppressed by geometrical smallness. However, later it was recognized (Bordag and Scharnhorst 1998, Ravndal and Thomassen 2001) that the first-order term is indeed present and does dominate. This was reconfirmed recently by Aghababaie and Burgess (2004), who identified the correct procedure to be used in the effective-action framework.

7.4 Two parallel planes at nonzero temperature

As explained in Chapter 5, temperature is an important parameter in the Casimir effect. Here, we consider the thermal Casimir force in the configuration of two parallel ideal-metal planes at $z = 0, a$ (Fierz 1960, Mehra 1967, Brown and Maclay 1969). These planes are supposed to be in thermal equilibrium with the environment at a temperature T. We start with the general case and then consider the limiting cases of low and high temperature.

7.4.1 General case

In Section 7.2, eqn (7.22), we have already determined the eigenfrequencies $\omega_{k_\perp, n}$ of the electromagnetic oscillations between ideal-metal planes. At nonzero temperature, instead of the energy (7.26), one must consider the free energy. For the electromagnetic case in the configuration of parallel planes, eqns (5.29) and (7.26) result in

$$\mathcal{F}_0(a, T) = \int_0^\infty \frac{k_\perp dk_\perp}{2\pi} \left\{ \frac{1}{2}\omega_{k_\perp, 0} + k_B T \ln\left(1 - e^{-\omega_{k_\perp,0}/k_B T}\right) \right.$$
$$\left. + 2\sum_{n=1}^\infty \left[\frac{1}{2}\omega_{k_\perp, n} + k_B T \ln\left(1 - e^{-\omega_{k_\perp, n}/k_B T}\right)\right]\right\} S, \quad (7.58)$$

where S is the (infinite) area of the planes. Equation (7.58) reflects the existence of two polarization states of the photon for all $n \geq 1$, and of only one polarization state at $n = 0$. This equation can be identically rewritten as

$$\mathcal{F}_0(a, T) = k_B T \int_0^\infty \frac{k_\perp dk_\perp}{2\pi} \left[\ln\left(2\sinh\frac{\omega_{k_\perp, 0}}{2k_B T}\right) + 2\sum_{n=1}^\infty \ln\left(2\sinh\frac{\omega_{k_\perp, n}}{2k_B T}\right)\right] S$$
$$\equiv \mathcal{F}_0^{(n=0)} + \mathcal{F}_0^{(n\geq 1)}(a). \quad (7.59)$$

In the limit $T \to 0$, the value of $\mathcal{F}_0$ leads to the zero-point energy defined in eqn (7.26). We notice that $\mathcal{F}_0^{(n=0)}$ does not depend on the separation, and concentrate our attention on rearrangements of $\mathcal{F}_0^{(n\geq 1)}$.

First we use the argument principle (Ahlfors 1979)

$$\sum_n g(a_n) - \sum_m g(b_m) = \frac{1}{2\pi i} \oint_{C_1} g(z) \, d\ln \Delta(z), \quad (7.60)$$

where $g(z)$ is some analytic function and $\Delta(z)$ is a meromorphic function inside a closed contour C_1, with a_n and b_m being the zeros and poles of $\Delta(z)$ located inside C_1 [$\Delta(z)$ is assumed to be analytic on C_1]. We put

$$g(\omega) = \ln\left(2\sinh\frac{\omega}{2k_B T}\right), \quad (7.61)$$

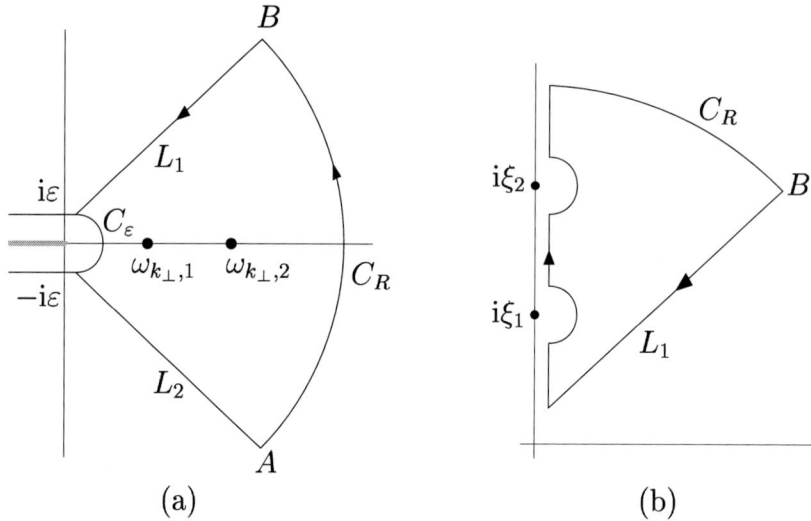

FIG. 7.1. Integration paths (a) C_1 and (b) C_2 in the plane of the complex frequency. The Matsubara frequencies are ξ_l and the photon eigenfrequencies are $\omega_{k_\perp,n}$ (Geyer et al. 2003).

$$\Delta(\omega) = e^{-aq(\omega)} \sin a\sqrt{\omega^2 - k_\perp^2}, \quad q^2(\omega) = k_\perp^2 - \omega^2$$

and choose the contour C_1 in the plane of the complex frequency ω to go around the poles in the counterclockwise direction, as shown in Fig. 7.1(a). Here, the two arcs have an infinitely small radius ε and an infinitely large radius R, and the two straight lines L_1, L_2 are inclined at angles of ± 45 degrees to the real axis. It is clear that $\Delta(\omega_{k_\perp,n}) = (-1)^n \sin \pi n = 0$ and that $\Delta(\omega)$ has no poles (a function with these properties is called a *mode-generating function*; see Chapter 9). Using eqns (7.59)–(7.61), we arrive at

$$\mathcal{F}_0^{(n\geq 1)}(a,T) = k_\mathrm{B}T \int_0^\infty \frac{k_\perp dk_\perp}{\pi} \frac{S}{2\pi i} \oint_{C_1} \ln\left(2\sinh\frac{\omega}{2k_\mathrm{B}T}\right) d\ln\Delta(\omega). \quad (7.62)$$

Note that the function $g(\omega)$ in eqn (7.61) has branch points at imaginary frequencies $\omega_l = i\xi_l$, where the $\xi_l = 2\pi k_\mathrm{B}Tl$ with $l = 0, \pm 1, \pm 2, \ldots$ are the Matsubara frequencies (see Section 5.1). The contour C_1 in Fig. 7.1(a) is chosen so as to avoid all of the branch points and to enclose all oscillation frequencies $\omega_{k_\perp,n}$ with $n \geq 1$. To calculate the countor integral in eqn (7.62), we represent it as a sum of four integrals along the contours C_R, L_1, C_ε, and L_2. It is easy to check that the integral along C_R vanishes when $R \to \infty$. Then we integrate the remaining three integrals by parts. It is seen that all of the terms, other than the integrals obtained after integration by parts, cancel each other or are equal to zero [such as at the points A and B in Fig. 7.1(a)]. The integral along L_1 can be calculated using the Cauchy theorem applied to the closed contour C_2 [see Fig. 7.2(b)], inside which the function under consideration is analytic:

$$-\int_{L_1} \coth \frac{\omega}{2k_B T} \ln \Delta(\omega)\, d\omega = \int_{i\varepsilon}^{i\infty} \coth \frac{\omega}{2k_B T} \ln \Delta(\omega)\, d\omega. \qquad (7.63)$$

The path $(i\varepsilon, i\infty)$ contains semicircles of radius ε about the poles $i\xi_l$ of the function $\coth(\omega/2k_B T)$. The integral along the line L_2 is calculated in a similar way (details are presented by Geyer et al. 2003). As a result of the integration by parts, we obtain only poles instead of branch points and represent $\mathcal{F}_0^{(n \geq 1)}(a, T)$ in the form

$$\mathcal{F}_0^{(n \geq 1)}(a, T) = \frac{1}{2} \int_0^\infty \frac{k_\perp dk_\perp}{\pi} \frac{S}{2\pi i} \int_{-i\infty}^{i\infty} \coth \frac{\omega}{2k_B T} \ln \Delta(\omega)\, d\omega. \qquad (7.64)$$

The integration of this equation, involving poles at the points $i\xi_l$, leads to

$$\mathcal{F}_0^{(n \geq 1)}(a, T) = -\frac{iS}{4\pi} \int_0^\infty \frac{k_\perp dk_\perp}{\pi} \int_{-\infty}^\infty \cot \frac{\xi}{2k_B T} \ln \Delta(i\xi)\, d\xi$$

$$+ \frac{S}{4} \int_0^\infty \frac{k_\perp dk_\perp}{\pi} \sum_{l=-\infty}^\infty \text{res}\left[\coth \frac{\omega}{2k_B T} \ln \Delta(\omega); i\xi_l\right], \qquad (7.65)$$

where

$$\Delta(i\xi) = e^{-a\sqrt{k_\perp^2 + \xi^2}} \sinh a\sqrt{k_\perp^2 + \xi^2}. \qquad (7.66)$$

Noting that $\Delta(i\xi)$ is an even function of ξ, we conclude that the apparently pure imaginary integral on the right-hand side of eqn (7.65) vanishes. After the calculation of the residues, and using the evenness of the function $\Delta(i\xi_l)$, we arrive at the result

$$\mathcal{F}_0^{(n \geq 1)}(a, T) = \frac{k_B T S}{\pi} \int_0^\infty k_\perp dk_\perp \sum_{l=0}^\infty{}' \ln \Delta(i\xi_l), \qquad (7.67)$$

where the prime on the summation sign means that the term for $l = 0$ has to be multiplied by $1/2$.

The free energy (7.59), (7.67) is infinite. The finite Casimir energy per unit area of the planes is obtained by subtraction from $\mathcal{F}_0(a, T)$ of the free energy for infinitely separated planes,

$$\mathcal{F}(a, T) = \frac{\mathcal{F}_0(a, T)}{S} - \lim_{a \to \infty} \frac{\mathcal{F}_0(a, T)}{S}. \qquad (7.68)$$

This is equivalent to the replacement of $\Delta(i\xi_l)$ in eqn (7.67),

$$\Delta(i\xi_l) = \frac{1}{2}\left(1 - e^{-2a\sqrt{k_\perp^2 + \xi_l^2}}\right) \qquad (7.69)$$

[see eqn (7.66)], with Δ/Δ_∞, where $\Delta_\infty = 1/2$. The result is

$$\mathcal{F}(a,T) = \frac{k_B T}{\pi} \sum_{l=0}^{\infty}{}' \int_0^\infty k_\perp dk_\perp \ln\left(1 - e^{-2a\sqrt{k_\perp^2 + \xi_l^2}}\right). \tag{7.70}$$

From eqn (7.70), the thermal Casimir pressure between the parallel planes takes the form

$$P(a,T) = -\frac{\partial \mathcal{F}(a,T)}{\partial a} = -\frac{2k_B T}{\pi} \sum_{l=0}^{\infty}{}' \int_0^\infty k_\perp dk_\perp \frac{\sqrt{k_\perp^2 + \xi_l^2}}{e^{2a\sqrt{k_\perp^2 + \xi_l^2}} - 1}. \tag{7.71}$$

In Part II of the book, the Casimir entropy, defined as

$$S(a,T) = -\frac{\partial \mathcal{F}(a,T)}{\partial T}, \tag{7.72}$$

plays an important role in thermodynamic tests of the various approaches to the calculation of the Casimir force between real materials. From eqns (7.70) and (7.72), one obtains

$$S(a,T) = -\frac{1}{T}\mathcal{F}(a,T) + \frac{k_B}{\pi}\sum_{l=1}^{\infty} \xi_l^2 \ln\left(1 - e^{-2a\xi_l}\right). \tag{7.73}$$

To calculate the Casimir free energy, pressure, and entropy for two ideal-metal planes, it is convenient to introduce a dimensionless variable and a parameter

$$y = 2a\sqrt{k_\perp^2 + \xi_l^2}, \qquad \tau = 2\pi\frac{T}{T_{\text{eff}}} = 4\pi a k_B T, \tag{7.74}$$

respectively, where $k_B T_{\text{eff}} = 1/(2a)$. Expressed in terms of these, the Casimir free energy per unit area and the pressure are

$$\mathcal{F}(a,T) = \frac{\tau}{16\pi^2 a^3} \sum_{l=0}^{\infty}{}' \int_{\tau l}^\infty y\, dy \ln(1 - e^{-y}), \tag{7.75}$$

$$P(a,T) = \frac{\tau}{16\pi^2 a^4} \sum_{l=0}^{\infty}{}' \int_{\tau l}^\infty dy \frac{y^2}{e^y - 1}.$$

It is very convenient to rewrite eqn (7.75) with the help of the Poisson summation formula (5.18), (5.19). This formula was first used in the theory of the Casimir effect by Mehra (1967), Brown and Maclay (1969), and Schwinger et al. (1978). In the case of the free energy, we substitute

$$b(l) = \frac{\tau}{32\pi^2 a^3} \int_{\tau|l|}^\infty y\, dy \ln\left(1 - e^{-y}\right) \tag{7.76}$$

and obtain

$$c(\alpha) = \frac{1}{\pi} \int_0^\infty b(x) \cos \alpha x \, dx \tag{7.77}$$

because $b(x)$ is an even function of x. Then, according to eqn (5.19),

$$\mathcal{F}(a,T) = \sum_{l=-\infty}^{\infty} b(l) = 4\pi \sum_{l=0}^{\infty}{}' c(2\pi l), \tag{7.78}$$

where

$$c(2\pi l) = \frac{\tau}{32\pi^3 a^3} \int_0^\infty dx \cos 2\pi l x \int_{\tau x}^{\infty} y \, dy \ln\left(1 - e^{-y}\right) \tag{7.79}$$

$$= \frac{1}{32\pi^3 a^3} \int_0^\infty dv \cos ltv \int_v^\infty y \, dy \ln\left(1 - e^{-y}\right).$$

Here, a new variable $v = \tau x$ has been introduced and $t \equiv T_{\text{eff}}/T$.

Substituting eqn (7.79) in eqn (7.78) and changing the order of the integration, we get

$$\mathcal{F}(a,T) = \frac{1}{8\pi^2 a^3} \sum_{l=0}^{\infty}{}' \int_0^\infty y \, dy \ln\left(1 - e^{-y}\right) \int_0^y dv \cos ltv. \tag{7.80}$$

Calculating the integrals (Gradshteyn and Ryzhik 1994) and separating the $l = 0$ term, we finally obtain

$$\mathcal{F}(a,T) = -\frac{\pi^2}{720 a^3} \left\{ 1 + \frac{45}{\pi^3} \sum_{l=1}^{\infty} \left[\frac{\coth(\pi l t)}{t^3 l^3} + \frac{\pi}{t^2 l^2 \sinh^2(\pi t l)} \right] - \frac{1}{t^4} \right\}. \tag{7.81}$$

As is seen from eqn (7.81), the contribution with $l = 0$ is just the Casimir energy per unit area at $T = 0$ obtained in eqn (7.30), whereas the other terms correspond to the thermal correction.

In a similar way, by applying the Poisson formula to the second equality in eqn (7.75), we obtain an expression for the thermal Casimir pressure,

$$P(a,T) = -\frac{\pi^2}{240 a^4} \left\{ 1 + \frac{30}{\pi^4} \sum_{l=1}^{\infty} \left[\frac{1}{t^4 l^4} - \frac{\pi^3}{tl} \frac{\cosh(\pi l t)}{\sinh^3(\pi t l)} \right] \right\}. \tag{7.82}$$

As with the free energy, the first term on the right-hand side of eqn (7.82) is the Casimir pressure at $T = 0$ (1.1), and the other terms represent the thermal correction.

In perfect analogy to the Casimir entropy (7.72), (7.73), the entropy per unit area is given by (Mitter and Robaschik 2000)

$$S(a,T) = \frac{3k_B}{8\pi a^2} \left\{ \sum_{l=1}^{\infty} \left[\frac{\coth(\pi l t)}{t^2 l^3} + \frac{\pi}{t l^2 \sinh^2(\pi t l)} + \frac{2\pi^2 \cosh(\pi l t)}{3 l \sinh^3(\pi t l)} \right] - \frac{4\pi^3}{135 t^3} \right\}. \tag{7.83}$$

Note that eqn (7.81) can be obtained directly from the general equations (5.36) and (5.37), which represent the renormalized free energy associated with a volume V. In fact, for two parallel planes, $\alpha_1 = \alpha_2 = 0$ and eqns (5.36), (5.37), and (5.29) represent the electromagnetic Casimir free energy per unit area in the form

$$\mathcal{F}(a,T) = -\frac{\pi^2}{720a^3} + \frac{k_B T}{\pi} \int_0^\infty k_\perp dk_\perp \sum_{n=0}^\infty{}' \ln\left(1 - e^{-\omega_{k_\perp,n}/k_B T}\right) + \frac{\pi^2 (k_B T)^4 a}{45}. \quad (7.84)$$

By introducing the new variable

$$z = \frac{1}{k_B T}\omega_{k_\perp,n} = \frac{1}{k_B T}\sqrt{k_\perp^2 + \left(\frac{\pi n}{a}\right)^2} \quad (7.85)$$

and changing the order of summation and integration, we can rearrange eqn (7.84) as

$$\mathcal{F}(a,T) = -\frac{\pi^2}{720a^3} + \frac{(k_B T)^3}{\pi} \sum_{n=0}^\infty{}' \int_{2\pi nt}^\infty z\, dz \ln\left(1 - e^{-z}\right) + \frac{\pi^2}{720a^3 t^4}. \quad (7.86)$$

The integral entering eqn (7.86) can be evaluated using the series expansion

$$\int_{2\pi nt}^\infty z\, dz \ln\left(1 - e^{-z}\right) = -\sum_{l=1}^\infty \frac{1}{l} \int_{2\pi nt}^\infty z\, dz\, e^{-lz}$$

$$= -\frac{1}{8\pi a^3} \sum_{l=1}^\infty \frac{1}{(lt)^3}(1 + 2\pi nlt) e^{-2\pi nlt}. \quad (7.87)$$

Substituting this into eqn (7.86) and performing the summation in n, we again obtain eqn (7.81).

The limiting cases for low and high temperature of the above expressions for the free energy, pressure, and entropy are considered below.

7.4.2 The limit of low temperature

Here, we consider the asymptotic behavior of the Casimir free energy, pressure and entropy at low temperatures, i.e. under the condition $T \ll T_{\text{eff}}$. This is equivalent to $t \equiv T_{\text{eff}}/T \gg 1$. Terms in powers of the small parameter $1/t$ in eqn (7.81) result only from those containing $\coth(\pi tl)$ and $-1/t^4$. They are given by

$$\mathcal{F}(a,T) = -\frac{\pi^2}{720a^3}\left[1 + \frac{45\zeta_R(3)}{\pi^3}\left(\frac{T}{T_{\text{eff}}}\right)^3 - \left(\frac{T}{T_{\text{eff}}}\right)^4\right], \quad (7.88)$$

where $\zeta_R(3) \approx 1.202$. Equation (7.88) coincides with the low-temperature behavior of the Casimir free energy obtained in eqn (5.41) for configurations with

$r = 2$ translationally invariant directions. All corrections to eqn (7.88) are exponentially small. The leading exponentially small correction to be added to the terms in the square brackets in eqn (7.88) is

$$\frac{180}{\pi^2}\left(\frac{T}{T_{\text{eff}}}\right)^2 e^{-2\pi T_{\text{eff}}/T}. \tag{7.89}$$

For example, at a separation of $a = 1\,\mu\text{m}$, the effective temperature is $T_{\text{eff}} \approx 1145\,\text{K}$ and the asymptotic expression (7.88) is clearly applicable at room temperature.

In a similar way, the low-temperature behavior of the Casimir pressure (7.82) is given by

$$P(a, T) = -\frac{\pi^2}{240 a^4}\left[1 + \frac{1}{3}\left(\frac{T}{T_{\text{eff}}}\right)^4\right]. \tag{7.90}$$

It is notable that if one obtains eqn (7.90) from eqn (7.88) using the first equality in eqn (7.71), the second term on the right-hand side of eqn (7.88) does not contribute to the result, because it does not depend on a. The leading exponentially small correction to be added to the terms in the square brackets in eqn (7.90) has the form

$$-\frac{120}{\pi}\frac{T}{T_{\text{eff}}} e^{-2\pi T_{\text{eff}}/T}. \tag{7.91}$$

Now we deal with the low-temperature behavior of the Casimir entropy per unit area (7.83). Terms in powers of the small parameter $1/t$ arise from the first and the fourth term in the curly brackets in eqn (7.83):

$$S(a, T) = \frac{3\zeta_{\text{R}}(3) k_{\text{B}}}{8\pi a^2}\left(\frac{T}{T_{\text{eff}}}\right)^2 \left[1 - \frac{4\pi^3}{135\zeta_{\text{R}}(3)}\frac{T}{T_{\text{eff}}}\right]. \tag{7.92}$$

The leading exponentially small correction inside the square brackets in eqn (7.92) is

$$\frac{8\pi^2}{3\zeta_{\text{R}}(3)}\left(\frac{T_{\text{eff}}}{T}\right)^2 e^{-2\pi T_{\text{eff}}/T}. \tag{7.93}$$

Equation (7.92) can also be obtained from eqn (7.88) using eqn (7.72).

The properties of the Casimir entropy are important as a test of the consistency of the Matsubara quantum field theory with thermodynamics. As can be observed from eqn (7.83), $S(a, T) > 0$ at any a and any T. From eqn (7.92), we also get

$$S(a, T) \to 0 \quad \text{when} \quad T \to 0, \tag{7.94}$$

i.e. the third law of thermodynamics (the Nernst heat theorem) is satisfied. Thus, the Matsubara formulation is in agreement with the thermodynamic test when applied to ideal-metal planes. This test will be an important guide when we deal with plates made of real materials in Part II of the book.

We conclude this section with a remark about units. If the fundamental constants $\hbar$ and c were restored in eqns (7.81)–(7.88) and (7.90) above, the right-hand side of each equation would be multiplied by $\hbar c$. The Casimir free energy, pressure, and entropy also depend on $\hbar c$ through the definition of the effective temperature: $k_B T_{\rm eff} = \hbar c/(2a)$. In the low-temperature limit considered in this section, all of the above quantities are of quantum and relativistic character because they depend on both $\hbar$ and c.

7.4.3 The limit of high temperature

Now we consider the Casimir free energy, pressure, and entropy in the configuration of two parallel ideal-metal planes under the opposite condition $T \gg T_{\rm eff}$, i.e. $t \ll 1$. It is easier to obtain the respective asymptotic expressions from eqn (7.75) rather than from eqns (7.81)–(7.83).

We consider the contribution to the first equation in eqn (7.75) with $l = 0$, i.e. the zero-Matsubara-frequency term:

$$\mathcal{F}(a,T) = \frac{\tau}{32\pi^2 a^3} \int_0^\infty y\,dy\,\ln\left(1 - e^{-y}\right) = -\frac{k_B T}{8\pi a^2}\zeta_R(3). \tag{7.95}$$

It is easy to see that in the limit of high temperature, the neglected contributions of all terms with $l \geq 1$ are exponentially small:

$$\frac{\tau}{16\pi^2 a^3}\int_{\tau l}^\infty y\,dy\,\ln\left(1 - e^{-y}\right) = -\frac{\tau}{16\pi^2 a^3}\sum_{n=1}^\infty \frac{1}{n}\int_{\tau l}^\infty y e^{-ny}\,dy$$

$$= -\frac{\tau}{16\pi^2 a^3}\sum_{n=1}^\infty \frac{1}{n^3}(1 + \tau ln)e^{-\tau ln}. \tag{7.96}$$

Thus, the leading correction to eqn (7.95) is equal to

$$-\frac{k_B T}{2a^2}\frac{T}{T_{\rm eff}}e^{-2\pi T/T_{\rm eff}}. \tag{7.97}$$

In a similar way, from the second equation in eqn (7.75), one obtains the following for the Casimir pressure at $T \gg T_{\rm eff}$:

$$P(a,T) = -\frac{k_B T}{4\pi a^3}\zeta_R(3). \tag{7.98}$$

The leading correction to eqn (7.98) is given by

$$-\frac{\pi k_B T}{a^3}\left(\frac{T}{T_{\rm eff}}\right)^2 e^{-2\pi T/T_{\rm eff}}. \tag{7.99}$$

For the Casimir entropy per unit area at high temperature, one obtains

$$S(a,T) = \frac{k_B}{8\pi a^2}\zeta_R(3), \tag{7.100}$$

where the leading exponentially small correction is

$$-\frac{k_B \pi}{a^2}\left(\frac{T}{T_{\text{eff}}}\right)^2 e^{-2\pi T/T_{\text{eff}}}. \tag{7.101}$$

In fact, at room temperature the asymptotic expressions (7.95), (7.98), and (7.100) work well at separations larger than 6 μm.

We note that eqns (7.95), (7.98), and (7.100) have the same form in units where $\hbar = c = 1$ and where $\hbar$ and c are indicated explicitly, because the results obtained at high temperature do not depend on $\hbar$ and c. This is the so-called *classical limit* (Feinberg *et al.* 2001, Scardicchio and Jaffe 2006). As is known from quantum statistical physics, in this limit the Bose–Einstein and Fermi–Dirac quantum distribution functions reduce to the Maxwell–Boltzmann distribution. In Part II of this book, agreement with the classical limit is also considered as a test of consistency for any theory of the thermal Casimir force between real material bodies.

It is also worth noting that the asymptotic behaviors of the thermal Casimir pressure between ideal-metal planes at low and high temperature are connected by means of inversion symmetry (Brown and Maclay 1969).

7.5 The spinor Casimir effect between parallel planes

A spinor field of mass m is described by the Dirac equation (3.30) and the energy–momentum tensor (3.36). As was noted at the end of Section 3.2, a Dirichlet boundary condition cannot be imposed on a bispinor ψ, because this would be in contradiction with the Dirac equation. Therefore we use the *bag boundary condition* (Johnson 1975)

$$\left(i\boldsymbol{\gamma}\cdot\boldsymbol{n}+1\right)\psi(x)\big|_S = 0, \tag{7.102}$$

where $\boldsymbol{n}$ is the unit vector normal to the surface. It can be easily observed that if eqn (7.102) is satisfied, the current of Dirac particles flowing through the surface in the direction $\boldsymbol{n}$ is equal to zero. To make sure that this is really the case, we multiply eqn (7.102) by the Dirac conjugate bispinor $\bar{\psi}$ from the left and obtain

$$\boldsymbol{j}(x)\cdot\boldsymbol{n}\big|_S = i\bar{\psi}(x)\psi(x)\big|_S, \tag{7.103}$$

where $\boldsymbol{j} = \bar{\psi}\boldsymbol{\gamma}\psi$ is the current of Dirac particles. By considering the Hermitian conjugate of eqn (7.102), we obtain

$$\left[-i\psi^+(x)\boldsymbol{\gamma}^+\cdot\boldsymbol{n}+\psi^+(x)\right]\big|_S = 0. \tag{7.104}$$

Multiplication of eqn (7.104) from the right by the Dirac matrix γ^0 and use of the anticommutation relations (3.31) leads to

$$\left[-i\bar{\psi}(x)\boldsymbol{\gamma}\cdot\boldsymbol{n}+\bar{\psi}(x)\right]\big|_S = 0. \tag{7.105}$$

Finally, multiplication of the above equation by $\psi(x)$ from the right results in

$$\boldsymbol{j}(x)\cdot\boldsymbol{n}|_S = -i\bar{\psi}(x)\psi(x)|_S. \quad (7.106)$$

From a comparison of eqns (7.103) and (7.106), we conclude that

$$\boldsymbol{j}(x)\cdot\boldsymbol{n}|_S = i\bar{\psi}(x)\psi(x)|_S = 0. \quad (7.107)$$

Now we consider two parallel planes $z = 0, a$, with the boundary condition (7.102) on both of them. The normal vector is given by $\boldsymbol{n} = (0,0,1)$ and $\boldsymbol{n} = (0,0,-1)$ on the planes $z = 0$ and $z = a$, respectively. Solutions of the Dirac equation (3.30) can be obtained in the usual manner:

$$\psi(x) = e^{-i\omega t}\begin{pmatrix}\varphi(\boldsymbol{r})\\\chi(\boldsymbol{r})\end{pmatrix}, \quad \chi(\boldsymbol{r}) = -\frac{i\boldsymbol{\sigma}\cdot\boldsymbol{\nabla}\varphi(\boldsymbol{r})}{m+\omega}, \quad (7.108)$$

where φ and χ are two-component spinors and we use the standard representation of the Dirac matrices

$$\gamma^i = \begin{pmatrix}0 & \sigma^i\\-\sigma^i & 0\end{pmatrix}, \quad \gamma^0 = \begin{pmatrix}I & 0\\0 & -I\end{pmatrix} \quad (7.109)$$

[I is the 2×2 unit matrix and $\boldsymbol{\sigma} = (\sigma_x,\sigma_y,\sigma_z)$ is the Pauli matrices].

In the configuration under consideration, the upper spinor takes the form

$$\varphi(\boldsymbol{r}) = e^{i(k_x x + k_y y)}\left(u e^{ik_z z} + v e^{-ik_z z}\right), \quad (7.110)$$

where u and v are constant spinors. Substituting eqns (7.108) and (7.110) into the Dirac equation (3.30), we obtain

$$\nabla^2\varphi + (\omega^2 - m^2)\varphi = 0, \quad \omega^2 = m^2 + k_\perp^2 + k_z^2. \quad (7.111)$$

Equation (7.108) can be also used to eliminate the lower spinor χ from the boundary condition (7.102):

$$[(m+\omega)\boldsymbol{\sigma}\cdot\boldsymbol{n}\varphi + \boldsymbol{\sigma}\cdot\boldsymbol{\nabla}\varphi]|_S = 0. \quad (7.112)$$

The substitution of eqn (7.110) into eqn (7.112) at $z = 0$ and $z = a$ leads to two equations for the spinors u and v, which are compatible only when the eigenvalues $k_z = k_{zn}$ satisfy the following equality (Mamayev and Trunov 1980, Mostepanenko and Trunov 1997):

$$f(k_{zn}a) = ma\sin k_{zn}a + k_{zn}a\cos k_{zn}a = 0. \quad (7.113)$$

Note that eqn (7.113) is obtained with the condition $k_z \neq 0$. In fact, $k_z = 0$ does not satisfy the boundary condition (7.112). The respective eigenfrequencies ω_n are found from eqn (7.111) by replacing k_z with k_{zn}.

The vacuum energy of the spinor field, defined as

$$E_0(a) = \int_{-\infty}^{\infty} dx \int_{-\infty}^{\infty} dy \int_0^a dz \, \langle 0|T_{00}^{(1/2)}(z)|0\rangle, \qquad (7.114)$$

where $T_{\mu\nu}^{(1/2)}$ is presented in eqn (3.36), takes the form

$$E_0(a) = -2 \int_0^{\infty} \frac{k_\perp dk_\perp}{2\pi} \sum_{n=1}^{\infty} \omega_n S. \qquad (7.115)$$

It is easy to explicitly find the finite Casimir energy of a spinor field between the planes in the massless case $m = 0$. In this case eqn (7.113) can be solved with the result

$$k_{zn} = \frac{\pi}{a}\left(n + \frac{1}{2}\right), \qquad n = 0, 1, 2, \ldots. \qquad (7.116)$$

Subtracting from eqn (7.115) the vacuum energy of the spinor field in the volume between the planes in free Minkowski space

$$E_{0M}(a) = -\frac{2a}{\pi} \int_0^{\infty} \frac{k_\perp dk_\perp}{2\pi} \int_0^{\infty} dk_z \, \omega_k S \qquad (7.117)$$

[compare this with eqn (2.75) for the electromagnetic field], and dividing by the area of the planes, we arrive at the Casimir energy per unit area,

$$E(a) = -\frac{1}{a} \int_0^{\infty} k_\perp dk_\perp \left[\sum_{n=0}^{\infty} \sqrt{A^2 + \left(n + \frac{1}{2}\right)^2} - \int_0^{\infty} dt \sqrt{A^2 + t^2} \right]. \qquad (7.118)$$

Here, $A \equiv ak_\perp/\pi$ and $t \equiv ak_z/\pi$. The application of the Abel–Plana formula (2.41) adapted for summation over half-integers leads to

$$E(a) = -\frac{7\pi^2}{2880a^3} \qquad (7.119)$$

(see the end of Section 2.5 for calculation details, which are the same as for the electromagnetic case). The respective Casimir pressure is given by

$$P(a) = -\frac{7\pi^2}{960a^4}. \qquad (7.120)$$

To change these equations to the usual units, $\hbar c$ must be added to the numerators of eqns (7.119) and (7.120).

For a massive spinor field, one must perform a summation over the roots of eqn (7.113). This can be done using the argument principle (7.60). As a result, for $ma \ll 1$, small corrections to eqn (7.119) are obtained. For $ma \gg 1$, $E(a)$ is exponentially small in the parameter ma (Mostepanenko and Trunov 1997).

7.6 The Casimir effect for a wedge

In this section we calculate the Casimir energy density for an ideal-metal wedge, i.e. for two planes that are inclined at a given angle φ_0. Let the wedge axis coincide with the z-axis and let the polar coordinates in the coordinate plane (x, y) be (ρ, φ). For a massless scalar field, this problem was solved by Dowker and Kennedy (1978) using zeta function regularization. The electromagnetic case was considered by Deutsch and Candelas (1979) using Green's functions and point-splitting regularization techniques. Later these results were rederived by means of Schwinger's source theory (Brevik and Lygren 1996) and in terms of a local zeta function (Nesterenko et al. 2002). Here we consider only an realistic case of an electromagnetic field, bearing in mind the application of the results obtained in Part II of the book to the estimation of the effect of nonparallelity of the plates.

After separation of the time variable in accordance with eqn (3.64), the set of solutions of the boundary problem (3.48), (3.65) for the two independent polarizations of the electromagnetic field can be presented in the form

$$\boldsymbol{A}_J^{(1)}(\boldsymbol{r}) = -\beta_J e^{ik_z z} \begin{pmatrix} (\alpha n/\rho) J_{\alpha n}(k_\rho \rho) \sin \alpha n \varphi \\ k_\rho J'_{\alpha n}(k_\rho \rho) \cos \alpha n \varphi \\ 0 \end{pmatrix},$$

$$\boldsymbol{A}_J^{(2)}(\boldsymbol{r}) = \beta_J \frac{e^{ik_z z}}{\omega_J} \begin{pmatrix} -k_z k_\rho J'_{\alpha n}(k_\rho \rho) \sin \alpha n \varphi \\ -k_z (\alpha n/\rho) J_{\alpha n}(k_\rho \rho) \cos \alpha n \varphi \\ i k_\rho^2 J_{\alpha n}(k_\rho \rho) \sin \alpha n \varphi \end{pmatrix}.$$
(7.121)

We note that the Coulomb gauge (3.47) has been used in the above. Here, $J_\nu(z)$ is a Bessel function and $J'_\nu(z)$ is the derivative of it with respect to its argument z. The collective quantum number is $J = (k_\rho, n, k_z)$, where $0 \leq k_\rho < \infty$, $-\infty < k_z < \infty$, $n = 0, 1, 2, \ldots$, and the oscillator frequency is equal to $\omega_J^2 = k_z^2 + k_\rho^2$. The parameter α is defined by $\alpha \equiv \pi/\varphi_0$, and the normalization factor is given by

$$\beta_J = \frac{\alpha}{\pi^{3/2} k_\rho} \left(1 - \frac{1}{2} \delta_{n0}\right).$$
(7.122)

Note that $\boldsymbol{A}_J^{(1)}(\boldsymbol{r})$ is the transverse electric mode and $\boldsymbol{A}_J^{(2)}(\boldsymbol{r})$ is the transverse magnetic mode introduced in Section 7.2. The tangential components of both modes vanish on the wedge faces, at $\varphi = 0$, φ_0, as is required by the boundary conditions [allowance must be made for the fact that the tangential components of the vector potentials (7.121) are defined by their ρ and z components, i.e. by the first and the third component instead of the first and the second component, as was the case in the configuration of two parallel planes perpendicular to the z-axis]. A set of vector potentials similar to eqn (7.121) was used by Bezerra de Mello et al. (2007) in an investigation of the electromagnetic Casimir effect inside an ideal-metal cylindrical shell in a cosmic-string space–time.

Representing the field operator in the form

$$\boldsymbol{A}(x) = \sum_J \sum_{\lambda=1}^{2} \frac{1}{\sqrt{\omega_J}} \left[e^{-i\omega_J t} \boldsymbol{\mathcal{A}}_J^{(\lambda)}(\boldsymbol{r}) a_J^{(\lambda)} + e^{i\omega_J t} \boldsymbol{\mathcal{A}}_J^{(\lambda)*}(\boldsymbol{r}) a_J^{(\lambda)+} \right], \quad (7.123)$$

substituting this in eqn (3.29), and using eqn (3.13) with $A_0 = 0$, eqn (3.72), and eqn (3.73), we obtain the following expression for the electromagnetic vacuum energy density inside a wedge:

$$\langle 0|T_{00}^{(1)}(\rho)|0\rangle = \frac{\alpha}{4\pi^2} \sum_{n=0}^{\infty}{}' \int_{-\infty}^{\infty} dk_z \int_0^{\infty} dk_\rho \frac{k_\rho^3}{\sqrt{k_\rho^2 + k_z^2}} \quad (7.124)$$

$$\times \left\{ \left(1 + 2\frac{k_z^2}{k_\rho^2}\right) \left[J_{\alpha n}'^2(k_\rho\rho) + \frac{\alpha^2 n^2}{k_\rho^2 \rho^2} J_{\alpha n}^2(k_\rho\rho) \right] + J_{\alpha n}^2(k_\rho\rho) \right\}.$$

The respective expression in cylindrical coordinates in free Minkowski space is

$$\langle 0_M|T_{00}^{(1)}|0_M\rangle = \frac{1}{4\pi^2} \int_{-\infty}^{\infty} dk_z \int_0^{\infty} k_\rho \, dk_\rho \, \omega_J \quad (7.125)$$

and does not depend on position.

For the case of integer α, eqn (7.124) can be simplified using the following summation formulas (Prudnikov et al. 1986, Bezerra de Mello et al. 2007):

$$\sum_{n=0}^{\infty}{}' J_{\alpha n}^2(k_\rho\rho) = \frac{1}{2\alpha} \sum_{l=0}^{\alpha-1} J_0\left(2k_\rho\rho \sin\frac{\pi l}{\alpha}\right), \quad (7.126)$$

$$\sum_{n=0}^{\infty}{}' \left[J_{\alpha n}'^2(k_\rho\rho) + \frac{\alpha^2 n^2}{k_\rho^2 \rho^2} J_{\alpha n}^2(k_\rho\rho) \right] = \frac{1}{2\alpha} \sum_{l=0}^{\alpha-1} \cos\frac{2\pi l}{\alpha} J_0\left(2k_\rho\rho \sin\frac{\pi l}{\alpha}\right).$$

Substituting eqn (7.126) in eqn (7.124), we obtain

$$\langle 0|T_{00}^{(1)}(\rho)|0\rangle = \frac{1}{8\pi^2} \sum_{l=0}^{\alpha-1} \int_{-\infty}^{\infty} dk_z \int_0^{\infty} dk_\rho \frac{k_\rho}{\sqrt{k_\rho^2 + k_z^2}} \quad (7.127)$$

$$\times \left[(k_\rho^2 + 2k_z^2) \cos\frac{2\pi l}{\alpha} + k_\rho^2 \right] J_0\left(2k_\rho\rho \sin\frac{\pi l}{\alpha}\right).$$

The term with $l = 0$ in this equation is equal to the vacuum energy density in free Minkowski space (7.125). Thus it is canceled in the Casimir energy density

$$\varepsilon(\rho) = \langle 0|T_{00}^{(1)}(\rho)|0\rangle - \langle 0_M|T_{00}^{(1)}|0_M\rangle. \quad (7.128)$$

Bearing in mind that both of the quantities (7.125) and (7.127) are divergent, we introduce a cutoff function $\exp(-\delta\omega)$ in the integrand and change to polar coordinates (ω, θ) on the plane (k_ρ, k_z). As a result, eqn (7.128) takes the form

$$\varepsilon(\rho) = \frac{1}{4\pi^2} \lim_{\delta\to 0} \sum_{l=1}^{\alpha-1} \int_0^{\pi/2} d\theta \cos\theta \left[(1 + \sin^2\theta) \cos\frac{2\pi l}{\alpha} + \cos^2\theta \right]$$

$$\times \int_0^\infty d\omega\, \omega^3 e^{-\delta\omega} J_0\left(2\omega\rho\cos\theta \sin\frac{\pi l}{\alpha}\right). \tag{7.129}$$

Now we introduce a new variable $v = \sin\theta$ and use the substitution $y_l = 1/\sin(\pi l/\alpha)$. Then eqn (7.129) can be rearranged to

$$\varepsilon(\rho) = \frac{1}{2\pi^2} \lim_{\delta\to 0} \sum_{l=1}^{\alpha-1} \frac{1}{y_l^2} \int_0^1 dv\,(y_l^2 - 1 - v^2) \int_0^\infty d\omega\, \omega^3 e^{-\delta\omega} J_0\left(\omega\frac{2\rho\sqrt{1-v^2}}{y_l}\right). \tag{7.130}$$

Using the formula (Prudnikov et al. 1986)

$$\int_0^\infty d\omega\, \omega^3 e^{-\delta\omega} J_0\left(\omega\frac{2\rho\sqrt{1-v^2}}{y_l}\right) = -\frac{\partial^3}{\partial\delta^3}\frac{y_l}{\sqrt{\delta^2 y_l^2 + 4\rho^2(1-v^2)}}, \tag{7.131}$$

we rewrite eqn (7.130) in the form

$$\varepsilon(\rho) = -\frac{1}{4\pi^2\rho} \lim_{\delta\to 0} \sum_{l=1}^{\alpha-1} \frac{1}{y_l}\frac{\partial^3}{\partial\delta^3} \int_0^1 dv\,\frac{y_l(y_l^2 - 1 - v^2)}{\sqrt{\delta^2 y_l^2 + 4\rho^2(1-v^2)}}. \tag{7.132}$$

By performing the integration, differentiation, and limiting transition in eqn (7.132), we obtain

$$\varepsilon(\rho) = -\frac{1}{16\pi^2\rho^4} \sum_{l=1}^{\alpha-1} y_l^4, \tag{7.133}$$

and after the summation we finally find

$$\varepsilon(\rho) = -\frac{1}{720\pi^2\rho^4}(\alpha^2 - 1)(\alpha^2 + 11). \tag{7.134}$$

This result has been obtained for an integer α (i.e. for $\varphi_0 = \pi/2, \pi/3, \ldots$). However, the formula (7.134) obtained can be analytically continued to any non-integer value of α.

It is interesting to note that for $\alpha = 1$ ($\varphi_0 = \pi$) we have $\varepsilon(\rho) = 0$, as would be expected for the Casimir energy density of a massless field near a single plane. If we perform the limiting transition $\varphi_0 \to 0$, $\rho \to \infty$ under the condition $\varphi_0\rho \equiv a = \text{const}$, the wedge is transformed into two parallel planes with a separation distance a between them. In agreement with this, it can be seen in eqn (7.134) that $a\varepsilon$ goes to $-\pi^2/(720a^3)$, i.e. the Casimir energy per unit area of the planes (7.30). Thus, the results obtained for both configurations are consistent.

In a similar way, it is not difficult to obtain the Casimir energy density of a massless scalar field described by the metrical energy–momentum tensor (3.9) (Deutsch and Candelas 1979),

$$\varepsilon(\rho) = -\frac{1}{1440\pi^2\rho^4}(\alpha^4 - 1). \tag{7.135}$$

7.7 The dynamical Casimir effect

There are various dynamical Casimir effects which arise from the movement of planes. For a uniformly moving plane, the Casimir force acquires a velocity-dependent correction. For an accelerated plane, the Casimir force is accompanied by the creation of particles from the vacuum.

7.7.1 Uniformly moving plane

Here, we briefly discuss the simplest case where one of the ideal-metal planes, at $z = 0$, is stationary and the other one, at $z = a(t) = a_0 + vt$, moves with a constant velocity v in the positive direction of the z-axis. For a massless scalar field, this situation was considered by Bordag et al. (1984) using the Green's function method. The vacuum energy density was obtained from eqn (3.92), in each of the three domains [below the plane $z = 0$, in between the planes, and above the plane $z = a(t)$]. After subtraction of the contribution from free Minkowski space, the Casimir pressure between the two planes was obtained under the condition that $v \ll 1$ ($v \ll c$ in the usual units):

$$P[a(t)] = -\frac{\pi^2}{480 a^4(t)} \left[1 + \frac{8}{3} v^2 + O(v^4) \right]. \tag{7.136}$$

It is seen that the first contribution on the right-hand side of eqn (7.136) agrees with eqn (7.9) if we replace a with $a(t)$. The second contribution is a nontrivial correction due to the movement of the upper plane.

Similar results were obtained using the same method for the electromagnetic Casimir effect (Bordag et al. 1986). For $v \ll 1$, we get

$$P[a(t)] = -\frac{\pi^2}{240 a^4(t)} \left[1 - \left(\frac{10}{\pi^2} - \frac{2}{3} \right) v^2 + O(v^4) \right]. \tag{7.137}$$

It is notable that for $v \ll 1$ the velocity-dependent correction to the Casimir pressure has opposite signs in the scalar and electromagnetic cases. Formally, one may consider also the case of large velocities, $1 - v \ll 1$. In this case even the leading term cannot be obtained from eqn (1.1) by the substitution $a \to a(t)$. The result is

$$P[a(t)] = -\frac{3}{8\pi^2 a^4(t)} \left\{ 1 + \frac{(1-v^2)^2}{16} + O\left[(1-v^2)^4\right] \right\}. \tag{7.138}$$

Here we shall leave our discussion of the dynamic Casimir effect due to a uniformly moving plane, as it has been discussed previously in the literature (Mostepanenko and Trunov 1997, Bordag et al. 2001a).

7.7.2 Particle creation from an accelerated plane

Another, more interesting, modification of the dynamical Casimir effect is the creation of particles from the vacuum by accelerated boundaries (Moore 1970,

Fulling and Davies 1976). The creation of particles from the vacuum by nonstationary external fields is a well-explored area (Greiner et al. 1985, Grib et al. 1994). Bearing in mind that a material boundary can be considered as a kind of concentrated external field, it is not surprising that moving boundaries act in the same way as a nonstationary external field. We shall outline the main ideas of particle creation by moving boundaries with the example of a massless scalar field on an interval $[0, a(t)]$, where $a(t) = a_0 \equiv a(0)$ for $t \leq 0$ and $a(t)$ is some function of t for $t > 0$.

The boundary conditions (2.3) now read

$$\varphi(t, 0) = \varphi(t, a(t)) = 0. \tag{7.139}$$

We assume that $|\varphi'(t)| < 1$. This allows one to consider the boundary point (a mirror) as some material body. The original papers (Moore 1970, Fulling and Davies 1976) reduced the problem of nonstationary boundary conditions to a static one by means of a conformal transformation. This, however, is possible only in two-dimensional space–time. Here, we follow another approach (Razavy and Terning 1985, Law 1995) applicable both in two and in four dimensions.

At $t < 0$, the complete orthonormal set of solutions of eqn (2.1) with $m = 0$ and boundary conditions (7.139) is given by eqn (2.6), where we replace a with a_0. The complete orthonormal set of solutions of the same boundary problem at $t \geq 0$, $\chi_n^{(\pm)}(t, x)$, should satisfy the initial conditions

$$\chi_n^{(\pm)}(0, x) = \varphi_n^{(\pm)}(0, x) = \frac{1}{\sqrt{\pi n}} \sin \frac{\pi n x}{a_0}, \tag{7.140}$$

$$\left. \frac{\partial \chi_n^{(\pm)}(t, x)}{\partial t} \right|_{t=0} = \left. \frac{\partial \varphi_n^{(\pm)}(t, x)}{\partial t} \right|_{t=0} = \mp i \frac{\pi n}{a_0} \varphi_n^{(\pm)}(0, x).$$

The functions $\chi_n^{(+)}(t, x)$, as yet unknown, can be found in the form of a series

$$\chi_n^{(+)}(t, x) = \frac{1}{\sqrt{\pi n}} \sum_k Q_{nk}(t) \sqrt{\frac{a_0}{a(t)}} \sin \frac{\pi k x}{a(t)}, \tag{7.141}$$

where the initial conditions for Q_{nk} are given by

$$Q_{nk}(0) = \delta_{nk}, \quad Q'_{nk}(0) = -i \frac{\pi n}{a_0} \delta_{nk}. \tag{7.142}$$

Note that $\chi_n^{(-)}(t, x) = \chi_n^{(+)*}(t, x)$. It is obvious that both of the boundary conditions in eqn (7.139) are identically satisfied in eqn (7.141). Now we substitute eqn (7.141) into the field equation (2.1) with $m = 0$ and, after some rearrangement, arrive at an infinite coupled system of differential equations with respect to the functions $Q_{nk}(t)$ (Law 1995),

$$Q''_{nk}(t) + \frac{\pi^2 k^2}{a^2(t)} Q_{nk}(t) \tag{7.143}$$

$$= \sum_j \left[2\nu(t) h_{kj} Q'_{nj}(t) + \nu'(t) h_{kj} Q_{nj}(t) + \nu^2(t) \sum_l h_{jk} h_{jl} Q_{nl}(t) \right].$$

Here, the following notation has been introduced:

$$\nu(t) = \frac{a'(t)}{a(t)}, \qquad h_{kj} = -h_{jk} = (-1)^{k-j} \frac{2kj}{j^2 - k^2}, \quad j \neq k. \tag{7.144}$$

Let the boundary point $a(t)$ return to its initial position a_0 after some time T and remain at rest. For $t > T$, we have $\nu(t) = 0$ and the right-hand side of eqn (7.143) vanishes so that the two linearly independent solutions become the same as at $t < 0$:

$$Q_{nk}^{(1)}(t) = e^{-i\pi kt/a_0}, \qquad Q_{nk}^{(2)}(t) = e^{i\pi kt/a_0}. \tag{7.145}$$

Thus, at $t > T$ the solution of eqn (7.143) with the initial conditions (7.142) can be represented in the form

$$Q_{nk}(t) = \alpha_{nk} e^{-i\pi kt/a_0} + \beta_{nk} e^{i\pi kt/a_0}, \tag{7.146}$$

where α_{nk} and β_{nk} are the Bogoliubov coefficients.

This is a familiar situation in the S-matrix theory of particle creation from the vacuum by a nonstationary external field. The operator of a scalar field called the *in* field (i.e. the field defined for $t < 0$ when the boundary point a_0 is at rest) is given by eqn (2.56), where the functions $\varphi_n^{(\pm)}(t,x)$ are defined in eqn (2.6) with $\omega_n = k_n = \pi n/a_0$. The annihilation and creation operators a_n, a_n^+ are called the operators of the *in* particles. The *in* vacuum state is defined by

$$a_n |0_{\text{in}}\rangle = 0. \tag{7.147}$$

The field operator at any moment $t \geq 0$, expressed in terms of the creation and annihilation operators a_n^+, a_n, is given by

$$\varphi(t,x) = \sum_n \left[\chi_n^{(+)}(t,x) a_n + \chi_n^{(-)}(t,x) a_n^+ \right]. \tag{7.148}$$

However, at $t > T$ it is possible to reexpand this operator in terms of the solutions $\varphi_k^{(\pm)} = \exp(\mp i\pi kt/a_0) \sin(\pi kx/a_0)/\sqrt{\pi k}$. For this purpose we substitute eqns (7.141) and (7.146) into eqn (7.148), with the result

$$\varphi(t,x) = \sum_k \frac{1}{\sqrt{\pi k}} \sin \frac{\pi k x}{a_0} \left(e^{-i\pi kt/a_0} b_k + e^{i\pi kt/a_0} b_k^+ \right), \tag{7.149}$$

where the annihilation and creation operators are given by the equality

$$b_k = \sum_n \sqrt{\frac{k}{n}} \left(\alpha_{nk} a_n + \beta_{nk}^* a_n^+ \right) \tag{7.150}$$

and the Hermitian conjugate of this equality. The operators b_k and b_k^+ are called the operators of the *out* particles.

Equation (7.150), connecting the creation and annihilation operators of the *in* and *out* particles, is called the *Bogoliubov transformation*. The coefficients of the Bogoliubov transformation satisfy the equality

$$\sum_k \frac{k}{n} \left(|\alpha_{nk}|^2 - |\beta_{nk}|^2 \right) = 1, \qquad (7.151)$$

which is a consequence of the unitarity condition (Grib et al. 1994, Birrell and Davies 1982). The vacuum state at $t > T$ (the *out* vacuum) is defined by

$$b_k |0_{\text{out}}\rangle = 0. \qquad (7.152)$$

The number of particles with the quantum number k created from the vacuum state $|0_{\text{in}}\rangle$ during a time T is given by the following matrix element calculated using eqn (7.150):

$$n_k = \langle 0_{\text{in}} | b_k^+ b_k | 0_{\text{in}} \rangle = k \sum_{n=1}^{\infty} \frac{1}{n} |\beta_{nk}|^2. \qquad (7.153)$$

The total number of particles created in all modes during the time T is

$$N = \sum_{k=1}^{\infty} n_k = \sum_{k=1}^{\infty} k \sum_{n=1}^{\infty} \frac{1}{n} |\beta_{nk}|^2. \qquad (7.154)$$

An approximate solution of eqn (7.143) can be found in the case where the boundary point oscillates harmonically with a small amplitude under the condition of parametric resonance,

$$a(t) = a_0 \left[1 + \varepsilon \sin(2\omega_1 t) \right], \qquad (7.155)$$

where $\omega_1 = \pi/a_0$, $\varepsilon \ll 1$. Using the theory of parametrically excited systems (Bogoliubov and Mitropolsky 1985), the coefficients α_{nk} and β_{nk} in eqn (7.146) can be treated as slowly varying functions of time. Substitution of eqn (7.146) into eqn (7.143), after averaging over fast oscillations with frequencies $\omega_k = k\omega_1$, where $k = 2, 3, \ldots$, leads in the first order in powers of ε to a simplified system of equations (Dodonov and Klimov 1996):

$$\frac{d\alpha_{n1}}{d\tau} = -\beta_{n1} + 3\alpha_{n3}, \qquad \frac{d\alpha_{nk}}{d\tau} = (k+2)\alpha_{n,k+2} - (k-2)\alpha_{n,k-2},$$

$$\qquad (7.156)$$

$$\frac{d\beta_{n1}}{d\tau} = -\alpha_{n1} + 3\beta_{n3}, \qquad \frac{d\beta_{nk}}{d\tau} = (k+2)\beta_{n,k+2} - (k-2)\beta_{n,k-2}.$$

Here, the term *slow time*, represented by $\tau = \varepsilon\omega_1 t/2$, is introduced. Under the conditions $\tau \ll 1$ and $\tau \gg 1$, approximate solutions of eqn (7.156) can be found, resulting in (Dodonov and Klimov 1996)

$$n_1(t) \approx \frac{1}{4}(\varepsilon\omega_1 t)^2, \qquad n_1(t) \approx \frac{4\varepsilon\omega_1 t}{\pi^2} + \frac{2}{\pi^2}\ln 4 - \frac{1}{2}, \qquad (7.157)$$

respectively. The total number of created particles with all quantum numbers is $N \approx n_1(t)$ if $\tau \ll 1$, and $N \sim \tau^2 \gg n_1(t)$ if $\tau \gg 1$.

The energy of the first mode is $\omega_1 n_1(t)$. The total energy in all modes is found from eqn (7.153):

$$E(t) = \sum_k \omega_1 k n_k(t) = \omega_1 \sum_{k=1}^{\infty} \sum_{n=1}^{\infty} \frac{k^2}{n} |\beta_{nk}|^2. \tag{7.158}$$

The calculation of this double sum does not require explicit expressions for the coefficients β_{nk}, and leads to (Dodonov and Klimov 1996)

$$E(t) = \frac{\omega_1}{4} \sinh^2(2\tau). \tag{7.159}$$

Thus, the total energy of the created particles increases faster than their number, indicating a rapid pumping of energy into the high-frequency modes at the expense of the low-frequency ones.

The above results can be generalized to the case of an electromagnetic field in a three-dimensional oscillating cavity. In this case both the total number and the total energy of the created photons grow exponentially with time. Note that the periodically oscillating boundary (7.155) is mathematically equivalent to an external electric field periodic in time. The number of bosonic particles created by such a field from the vacuum depends exponentially on time if the condition of parametric resonance is satisfied (Narozhnyi and Nikishov 1973, Mostepanenko and Frolov 1974, Mostepanenko 2003). This concept is promising for the creation of photons from the vacuum by use of the dynamical Casimir effect.

There is extensive literature on various aspects of the creation of photons due to the dynamical Casimir effect using various theoretical methods. For example, a Hamiltonian approach to the description of photon creation was suggested by Haro and Elizalde (2006). Cavities with the insertion of dispersive mirrors or a slab with a time-dependent dielectric permittivity were considered (Schaller et al. 2002, Uhlmann et al. 2004). Multiple-scale analysis was applied to the calculation of the flux of created particles (Crocce et al. 2002). Photon creation in a harmonically oscillating one-dimensional cavity with mixed boundary conditions (see Sections 7.1.2 and 7.2.2) has been analyzed (Alves et al. 2006). A collection of papers on the subject has been compiled and edited by Barton et al. (2005).

The experimental observation of the dynamical Casimir effect is a complicated problem because the internal mechanical properties of the oscillating wall do not permit oscillations in the GHz region (Dodonov and Dodonov 2006). Because of this, instead of a real moving metallic surface, it has been proposed to use an effective electron–hole plasma mirror created on the surface of a semiconductor by illuminating it with laser pulses of appropriate frequency (Yablonovitch 1989, Braggio et al. 2004). In this case, the conducting layer is created periodically on the surface, simulating mechanical oscillations. The present status of a proposed experiment which aims at measuring the dynamical Casimir effect by using the effective motion of a wall of a superconducting microwave resonant cavity has been reported by Agnesi et al. (2008).

8
THE CASIMIR EFFECT IN RECTANGULAR BOXES

As was mentioned in Chapters 1 and 2, the Casimir energy and force may change sign depending on the geometry of the configuration and the type of boundary conditions. A dramatic example of this situation, which has given rise to many discussions in the literature for several decades, is the case of a rectangular box with sides a, b, and c. Lukosz (1971) noticed that the electromagnetic Casimir energy inside an ideal-metal box may change sign depending on side lengths a, b, and c. A detailed investigation of the Casimir energy for fields of different spins, where it may again be either positive or negative, inside a rectangular box as a function of the box dimensions was performed by Mamayev and Trunov (1979a, 1979b). In particular, analytical results for two- and three-dimensional boxes were obtained by repeated application of the Abel–Plana formula (2.26). Ambjørn and Wolfram (1983) used the Epstein zeta function to calculate the Casimir energy for a scalar and an electromagnetic field in hypercuboidal regions in n-dimensional space–time. The problem of isolation of the divergent terms in the vacuum energy and their interpretation received the most attention. In recent years, this problem has been reformulated in terms of a **rectangular box divided into two sections by an ideal-metal movable partition (piston)** (Cavalcanti 2004, Hertzberg *et al.* 2005). It was shown that the Casimir force acting on a piston with Dirichlet boundary conditions attracts it to the nearest wall. Based on this, some doubts about the results previously obtained demonstrating Casimir repulsion in cubes have been raised. Below, we present both the old classical results on the Casimir effect in ideal-metal rectangular boxes and the recent results related to boxes with a piston. We demonstrate that the two sets of results are in mutual agreement, and the attraction (or repulsion for a piston with Neumann boundary conditions) of a piston to one of the box faces does not negate the Casimir repulsion for boxes with some appropriate ratio between a, b, and c.

8.1 The scalar Casimir effect in a rectangle

In this section, we consider the simplest geometry where the problems connected with the change of the sign of the Casimir energy and of the force arise. This is the case of a massless scalar field in a rectangle $a \times b$. For this configuration, we provide two different regularizations (one using the Abel–Plana formula and the other using the Epstein zeta function) which lead to coincident results for the Casimir energy. Then we discuss the Casimir force acting on a partition (piston) which divides the rectangle into two rectangles.

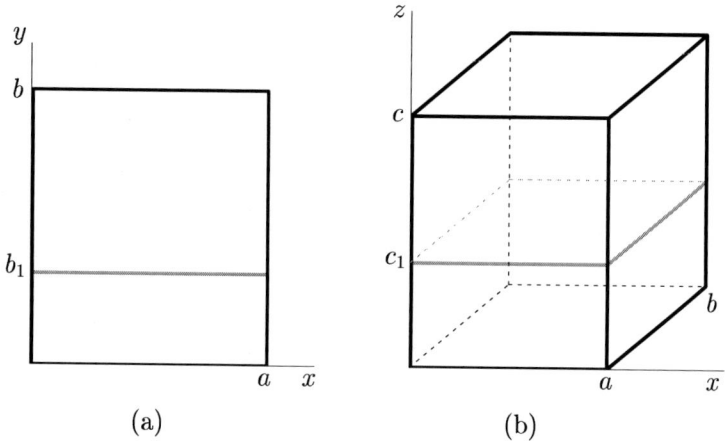

FIG. 8.1. Configurations of (a) a rectangle and (b) a rectangular box, with pistons at $y = b_1$ and $z = c_1$, respectively.

8.1.1 Regularization using the Abel–Plana formula

Let the massless scalar field $\varphi(t, x, y)$ be defined in a rectangle $0 \leq x \leq a$, $0 \leq y \leq b$ with Dirichlet boundary conditions on all sides, i.e. $x = 0, a$ and $y = 0, b$ [see Fig. 8.1(a)]. The complete orthonormal set of solutions of the equation

$$\Box_3 \varphi(t, x, y) = 0 \tag{8.1}$$

with these boundary conditions is given by

$$\varphi_{nl}^{(\pm)}(t, x, y) = \sqrt{\frac{2}{ab\omega_{nl}}} e^{\mp i\omega_{nl} t} \sin k_n x \sin k_l y, \tag{8.2}$$

where

$$k_n = \frac{\pi n}{a}, \quad k_l = \frac{\pi l}{b}, \quad \omega_{nl}^2 = k_n^2 + k_l^2, \quad n, l = 1, 2, 3, \ldots. \tag{8.3}$$

Note that n and l cannot be equal to zero because in that case the solution (8.2) vanishes.

Repeating the same calculations as in Section 3.3, but in three-dimensional space–time, we obtain the total vacuum energy of the scalar field inside a rectangle,

$$E_0(a, b) = \frac{\pi}{2} \sum_{n,l=1}^{\infty} \sqrt{\left(\frac{n}{a}\right)^2 + \left(\frac{l}{b}\right)^2}. \tag{8.4}$$

In order to perform the summation, we apply the Abel–Plana formula (2.26) twice. As was explained in Section 2.2, in so doing some cutoff function is introduced which makes all of the results finite. However, it is not necessary to write it out explicitly, because the result obtained after removing the regularization does not depend on its specific form. First, we apply the Abel–Plana formula to perform the summation over l, with the result

$$E_0(a,b) = \frac{\pi}{2} \sum_{n=1}^{\infty} \left[-\frac{n}{2a} + \int_0^{\infty} dt \sqrt{\left(\frac{n}{a}\right)^2 + \left(\frac{t}{b}\right)^2} \right.$$
$$\left. -2 \int_{bn/a}^{\infty} \sqrt{\left(\frac{t}{b}\right)^2 - \left(\frac{n}{a}\right)^2} \frac{dt}{e^{2\pi t}-1} \right]. \qquad (8.5)$$

Note that in obtaining eqn (8.5), eqn (2.37) has been used.

Next, we apply the Abel–Plana formula (2.26) to perform the first two summations over n on the right-hand side of eqn (8.5):

$$-\frac{1}{2a} \sum_{n=1}^{\infty} n = -\frac{1}{2a} \int_0^{\infty} t\, dt + \frac{1}{24a},$$

$$\sum_{n=1}^{\infty} \int_0^{\infty} dt \sqrt{\left(\frac{n}{a}\right)^2 + \left(\frac{t}{b}\right)^2} = -\frac{1}{2b} \int_0^{\infty} t\, dt \qquad (8.6)$$

$$+ \int_0^{\infty} dt \int_0^{\infty} dv \sqrt{\left(\frac{v}{a}\right)^2 + \left(\frac{t}{b}\right)^2} - \frac{b}{8\pi^2 a^2} \zeta_{\mathrm{R}}(3).$$

Now we substitute eqn (8.6) into eqn (8.5). We also introduce a new variable $u = at/(nb)$ in the last integral on the right-hand side of eqn (8.5). The result is

$$E_0 = \frac{\pi}{2} \left[-\frac{1}{2}\left(\frac{1}{a} + \frac{1}{b}\right) \int_0^{\infty} t\, dt + \int_0^{\infty} dt \int_0^{\infty} dv \sqrt{\left(\frac{v}{a}\right)^2 + \left(\frac{t}{b}\right)^2} \right.$$
$$\left. + \frac{1}{24a} - \frac{b}{8\pi^2 a^2} \zeta_{\mathrm{R}}(3) + \frac{2}{a} G\left(\frac{b}{a}\right) \right], \qquad (8.7)$$

where

$$G(z) = -z \int_1^{\infty} du \sqrt{u^2 - 1} \sum_{n=1}^{\infty} \frac{n^2}{e^{2\pi n u z} - 1}. \qquad (8.8)$$

Representing $G(z)$ in the form

$$G(z) = -z \sum_{n=1}^{\infty} n^2 \sum_{l=1}^{\infty} \int_1^{\infty} du \sqrt{u^2 - 1}\, e^{-2\pi n l u z}, \qquad (8.9)$$

and using the representation for the Bessel functions of imaginary argument (Gradshteyn and Ryzhik 1994)

$$K_\nu(z) = \frac{(z/2)^\nu \Gamma(1/2)}{\Gamma\left(\nu + \frac{1}{2}\right)} \int_1^{\infty} e^{-zt} (t^2 - 1)^{(2\nu-1)/2}\, dt, \qquad (8.10)$$

we can rearrange eqn (8.8) as

$$G(z) = -\frac{1}{2\pi}\sum_{n=1}^{\infty}\sum_{l=1}^{\infty}\frac{n}{l}K_1(2\pi nlz). \qquad (8.11)$$

Equation (8.7) contains two infinite integrals on the right-hand side. The second one is proportional to the vacuum energy of the free unbounded two-dimensional space contained within a rectangle of area ab. As was discussed in Section 1.1, the physical energies are counted from the top of the vacuum energy in free space, and thus this integral should be omitted. The first integral on the right-hand side of eqn (8.7) is proportional to the perimeter of the rectangle $2(a+b)$, which plays the role of the boundary surface in the two-dimensional case. Omission of this integral is equivalent to a renormalization of the geometrical object inherent in the configuration under consideration, as we discussed in Chapter 4. We shall return to the physical meaning of this omission below when we discuss the Casimir force acting on a piston.

As a result, the renormalized Casimir energy of a rectangle is

$$E(a,b) = \frac{\pi}{48a} - \frac{b}{16\pi a^2}\zeta_R(3) + \frac{\pi}{a}G\left(\frac{b}{a}\right). \qquad (8.12)$$

This is, in fact, symmetric with respect to the interchange of a and b. This symmetry is, however, implicit. The order of performing the summations chosen above is advantageous when $b \geq a$. In this case, as follows from eqn (8.11), $G(b/a)$ is of order $\exp(-2\pi b/a)$, i.e. is exponentially small. For example, even for $a = b$, the contribution from $G(b/a)$ to the Casimir energy is only about 1%. Numerical computations using the full equation (8.12) show that the Casimir energy E is positive if

$$0.36537 \leq \frac{b}{a} \leq 2.73686 \qquad (8.13)$$

and negative if $b > 2.73686$ or $b < 0.36537$. The Casimir forces acting on opposite sides of the rectangle

$$F_a(a,b) = -\frac{\partial E(a,b)}{\partial a} = \frac{\pi}{48a^2} - \frac{b}{8\pi a^3}\zeta_R(3) + \frac{\pi}{a^2}G\left(\frac{b}{a}\right) + \frac{\pi b}{a^3}G'\left(\frac{b}{a}\right),$$

$$F_b(a,b) = -\frac{\partial E(a,b)}{\partial b} = \frac{1}{16\pi a^2}\zeta_R(3) - \frac{\pi}{a^2}G'\left(\frac{b}{a}\right) \qquad (8.14)$$

are repulsive and attractive, correspondingly, in the same intervals (a prime is used here to denote a derivative with respect to the argument).

8.1.2 Regularization using the Epstein zeta function

The Epstein zeta function and its analytic continuation are very convenient tools for the investigation of the analytic properties of multiple summations. The Epstein zeta function can be defined as (Erdélyi et al. 1981)

$$Z_p(a_1, a_2, \ldots, a_p; s) = \sum_{n_1,\ldots,n_p=-\infty}^{\infty} \left[(n_1 a_1)^2 + \ldots + (n_p a_p)^2 \right]^{-s/2}$$
$$\times (1 - \delta_{n_1 0} \cdots \delta_{n_p 0}). \tag{8.15}$$

The inclusion of the negative product of δ-symbols in eqn (8.15) is equivalent to the condition that the term with all $n_i = 0$ is omitted.

The series in eqn (8.15) is convergent when $\operatorname{Re} s > p$. It can be analytically continued over the entire complex plane except for a pole at $s = p$, using the reflection formula

$$a_1 \cdots a_p \Gamma\left(\frac{s}{2}\right) \pi^{-s/2} Z_p(a_1, \ldots, a_p; s) \tag{8.16}$$
$$= \Gamma\left(\frac{p-s}{2}\right) \pi^{(s-p)/2} Z_p\left(\frac{1}{a_1}, \ldots, \frac{1}{a_p}; p-s\right).$$

We regularize the expression for the total vacuum energy of the scalar field in the rectangle (8.4) by introducing a regularization parameter s:

$$E_0^{(s)}(a,b) = \frac{\pi}{2} \sum_{n,l=1}^{\infty} \left[\left(\frac{n}{a}\right)^2 + \left(\frac{l}{b}\right)^2 \right]^{(1-s)/2}. \tag{8.17}$$

Equation (8.17) can be expressed in terms of the Epstein and Riemann zeta functions in the following way:

$$E_0^{(s)}(a,b) = \frac{\pi}{8} \Bigg\{ \sum_{n,l=-\infty}^{\infty} \left[\left(\frac{n}{a}\right)^2 + \left(\frac{l}{b}\right)^2 \right]^{-(s-1)/2} (1 - \delta_{n0}\delta_{l0})$$
$$- \frac{2}{a} \sum_{n=1}^{\infty} n^{-(s-1)} - \frac{2}{b} \sum_{l=1}^{\infty} l^{-(s-1)} \Bigg\}$$
$$= \frac{\pi}{8} \left[Z_2\left(\frac{1}{a}, \frac{1}{b}; s-1\right) - 2\left(\frac{1}{a} + \frac{1}{b}\right) \zeta_R(s-1) \right]. \tag{8.18}$$

In the limiting case $s \to 0$ (i.e. when the regularization is removed), the quantity (8.18) is divergent. By using the reflection relations (5.50) for the Riemann zeta function and (8.16) for the Epstein zeta function, we obtain the analytic continuation of $E_0^{(s)}$,

$$E_0^{(s)}(a,b) = -\frac{ab}{32\pi} Z_2(a,b; 3-s) + \frac{1}{8\pi}\left(\frac{1}{a} + \frac{1}{b}\right) \zeta_R(2-s). \tag{8.19}$$

After removing the regularization ($s \to 0$), we finally obtain the finite Casimir energy for a rectangle,

$$E(a,b) = -\frac{ab}{32\pi} Z_2(a,b;3) + \frac{\pi}{48}\left(\frac{1}{a} + \frac{1}{b}\right). \tag{8.20}$$

Let us now compare eqn (8.20) with the Casimir energy for the rectangle (8.12) derived using repeated application of the Abel–Plana formula. For this purpose, it is convenient to express the Epstein zeta function Z_2 in terms of the auxiliary function

$$S(\eta, \kappa; q) \equiv \pi^{-q/2} \Gamma\left(\frac{q}{2}\right) \sum_{k=-\infty}^{\infty} \left[\left(\frac{\eta}{\pi}\right)^2 + \left(\frac{k}{\kappa}\right)^2\right]^{-q/2}, \quad (8.21)$$

where $\eta \neq 0$ and the series is convergent for $\operatorname{Re} q > 1$. The result is

$$Z_2(a, b; 3) = \frac{2\pi^{3/2}}{\Gamma(3/2)} \sum_{l=1}^{\infty} S\left(\pi b l, \frac{1}{a}; 3\right) + \frac{2\zeta_R(3)}{a^3}. \quad (8.22)$$

There is an integral representation for the function S (Ambjørn and Wolfram 1983),

$$S(\eta, \kappa; q) = \kappa \left\{ \Gamma\left(\frac{q-1}{2}\right) \left(\frac{\eta}{\sqrt{\pi}}\right)^{1-q} \right. \quad (8.23)$$

$$\left. + \int_0^\infty dx\, x^{-(q+1)/2} e^{-\eta^2/(\pi x)} \left[\vartheta(0; \kappa^2 x) - 1\right] \right\},$$

where

$$\vartheta(z; x) = \sum_{n=-\infty}^{\infty} e^{-\pi n^2 x} e^{2\pi n z} \quad (8.24)$$

is the Jacobi theta function.

Substituting eqn (8.24) into eqn (8.23) and performing the integration with respect to x, we arrive at

$$S(\eta, \kappa; q) = \kappa \left(\frac{\eta}{\sqrt{\pi}}\right)^{1-q} \left[\Gamma\left(\frac{q-1}{2}\right) + 4 \sum_{n=1}^{\infty} (\eta \kappa n)^{(q-1)/2} K_{(q-1)/2}(2\eta \kappa n)\right]. \quad (8.25)$$

As a result, eqn (8.22) takes the form

$$Z_2(a, b; 3) = \frac{2\pi^2}{3ab^2} + \frac{16\pi}{a^2 b} \sum_{n,l=1}^{\infty} \frac{n}{l} K_1\left(2\pi n l \frac{b}{a}\right) + \frac{2\zeta_R(3)}{a^3}$$

$$= \frac{2\pi^2}{3ab^2} + \frac{2\zeta_R(3)}{a^3} - \frac{32\pi^2}{a^2 b} G\left(\frac{b}{a}\right). \quad (8.26)$$

The substitution of eqn (8.26) into eqn (8.20) brings us back to eqn (8.12), with $G(z)$ defined in eqn (8.11). Thus, the methods based on the Abel–Plana formula and the Epstein zeta function are in agreement. They lead to the same finite result for the Casimir energy of a scalar field in a rectangle.

8.1.3 A Casimir piston in a rectangle

Let us now consider a partition (a piston) $y = b_1$ with a Dirichlet boundary condition on it which divides the rectangle into two rectangles [see Fig. 8.1(a)]. This piston may take any position $0 < b_1 < b$. Because of this, it is often called a *movable* piston. We shall calculate the Casimir force acting on the piston, including the infinite integrals in eqn (8.7), making sure that it is finite (Cavalcanti 2004). For this purpose, we explicitly introduce the cutoff function

$$f(\omega_{nl}\delta) = \exp\left[-\delta\sqrt{\left(\frac{n}{a}\right)^2 + \left(\frac{l}{b}\right)^2}\right] \tag{8.27}$$

when applying the Abel–Plana formula, and rewrite eqn (8.7) before removing the regularization in the form

$$E_0^{(\delta)}(a,b) = I_0(a,b) + I_1(a,b) + E(a,b), \tag{8.28}$$

where $E(a,b)$ is the finite Casimir energy defined in eqn (8.12) and the regularized infinite integrals are

$$I_0(a,b) = \frac{2\pi^2}{\delta^3}ab, \qquad I_1(a,b) = -\frac{\pi}{4\delta^2}(a+b). \tag{8.29}$$

Equations (8.28) and (8.29) have been derived in the absence of the piston. The total regularized energy of the two boxes $a \times b_1$ and $a \times (b - b_1)$ in the presence of a piston is given by

$$E_0^{(\delta)}(a, b_1) + E_0^{(\delta)}(a, b - b_1) = -\frac{\pi}{4\delta^2}(2a + b) + \frac{2\pi^2}{\delta^3}ab + \frac{\pi}{24a} - \frac{b}{16\pi a^2}\zeta_R(3)$$
$$+ \frac{\pi}{a}\left[G\left(\frac{b_1}{a}\right) + G\left(\frac{b-b_1}{a}\right)\right]. \tag{8.30}$$

As can be seen from eqn (8.30), the divergent terms, and the finite terms outside the square brackets containing the function G, do not depend on b_1. Thus, the force acting on the piston,

$$F(a,b,b_1) = -\frac{\partial}{\partial b_1}\left[E_0^{(\delta)}(a,b_1) + E_0^{(\delta)}(a,b-b_1)\right]$$
$$= -\frac{\pi}{a^2}\left[G'\left(\frac{b_1}{a}\right) - G'\left(\frac{b-b_1}{a}\right)\right], \tag{8.31}$$

is uniquely defined regardless of the regularization and renormalization procedures used.

It can be shown that $F(a,b,b_1)$ is always negative. This means that the piston is attracted to the nearer side of the rectangle. One can also say that the piston is repelled from the more remote side. If the forces acting on the piston from the top and bottom sides are equal in magnitude (this is the case, for instance, for $b = 2a$, $b_1 = a$), the resulting force is equal to zero, i.e. we have a state of equilibrium. This equilibrium is, however, unstable, and its violation in any direction would lead to movement of the piston to the nearer side of the rectangle.

In particular, for $b \gg a$ and $b_1 \sim a$, eqn (8.31) results in

$$F(a,b,b_1) = \frac{\pi}{a^2} \sum_{n,l=1}^{\infty} K_1' \left(2\pi n l \frac{b_1}{a} \right) \approx -\frac{\pi}{2(a^3 b_1)^{1/2}} \exp\left(-2\pi \frac{b_1}{a}\right). \quad (8.32)$$

In this case the piston is attracted to the bottom side a even if, for instance, $b_1 = a$, i.e. the rectangle $a \times b_1$ satisfies the inequality (8.13), under which the Casimir energy should be positive and the force should be repulsive. This, however, does not mean that the regularization-independent results obtained for the piston raise questions about eqns (8.12) and (8.13), obtained after omitting the two infinite integrals I_1 and I_2 from eqn (8.29). The reason is that eqn (8.12) is relevant to the case of an empty space outside the rectangle $a \times b$. In this case the vacuum energy outside the rectangle does not depend on a and b, and there is no force acting on the rectangle from the outside, whereas eqn (8.32) is relevant to the case where there is a long additional rectangle $a \times (b - b_1)$ above the rectangle $a \times b_1$, with Dirichlet boundary conditions on its sides. The vacuum energy inside the rectangle $a \times (b - b_1)$ depends on b_1 and gives rise to an additional force which would be absent for an isolated rectangle $a \times b_1$. This changes the physical situation and leads to a total Casimir force attracting the piston to the nearer side of the rectangle.

One can conclude that although the regularization-independent results may be considered as somewhat more transparent than the regularization-dependent ones, the above consideration of a piston inside a rectangle neither adds to nor diminishes the reliability of the classical result (8.12) for the Casimir energy inside an empty rectangle in free space.

8.2 The scalar Casimir effect in a three-dimensional box

In this section we consider the problem of the attractive and repulsive Casimir forces for a massless scalar field in the more realistic geometry of a three-dimensional box with sides a, b, and c arranged along the x, y, and z axes, respectively [see Fig. 8.1(b)]. It will be shown that in spite of additional technical difficulties, the qualitative results obtained for a rectangle are preserved for a three-dimensional box [see the review by Actor (1995); a scalar field confined by *soft* boundaries represented as a harmonic-oscillator potential was considered by Actor and Bender (1995)].

The complete orthonormal set of solutions of eqn (3.1) with $m = 0$ and with Dirichlet boundary conditions on all six faces of the box takes the form

$$\varphi_{nlp}^{(\pm)}(t, \mathbf{r}) = \sqrt{\frac{4}{abc\omega_{nlp}}} e^{\mp i\omega_{nlp} t} \sin k_n x \sin k_l y \sin k_p z, \quad (8.33)$$

where
$$k_p = \frac{\pi p}{c}, \quad \omega_{nlp}^2 = k_n^2 + k_l^2 + k_p^2, \quad p = 1, 2, 3, \ldots, \tag{8.34}$$

and the other notation has been introduced in eqn (8.3).

The total vacuum energy of the scalar field inside the box is given by

$$E_0(a,b,c) = \frac{\pi}{2} \sum_{n,l,p=1}^{\infty} \sqrt{\left(\frac{n}{a}\right)^2 + \left(\frac{l}{b}\right)^2 + \left(\frac{p}{c}\right)^2}. \tag{8.35}$$

All desired results can be obtained by repeated application of the Abel–Plana formula (three times in this case) or by using the Epstein zeta function regularization technique. In this case we begin with the Epstein zeta function approach.

First, we identically rearrange the regularized eqn (8.35) for the vacuum energy in order to separate the contributions of the various Epstein and Riemann zeta functions in the sums from unity to infinity:

$$E_0^{(s)}(a,b,c) = \frac{\pi}{8} \Bigg\{ \frac{1}{2} \sum_{n,l,p=-\infty}^{\infty} (1 - \delta_{n0}\delta_{l0}\delta_{p0}) \left[\left(\frac{n}{a}\right)^2 + \left(\frac{l}{b}\right)^2 + \left(\frac{p}{c}\right)^2\right]^{-(s-1)/2}$$

$$- \frac{1}{2} \sum_{l,p=-\infty}^{\infty} (1 - \delta_{l0}\delta_{p0}) \left[\left(\frac{l}{b}\right)^2 + \left(\frac{p}{c}\right)^2\right]^{-(s-1)/2}$$

$$- \frac{1}{2} \sum_{n,p=-\infty}^{\infty} (1 - \delta_{n0}\delta_{p0}) \left[\left(\frac{n}{a}\right)^2 + \left(\frac{p}{c}\right)^2\right]^{-(s-1)/2}$$

$$- \frac{1}{2} \sum_{n,l=-\infty}^{\infty} (1 - \delta_{n0}\delta_{l0}) \left[\left(\frac{n}{a}\right)^2 + \left(\frac{l}{b}\right)^2\right]^{-(s-1)/2}$$

$$+ \sum_{n=1}^{\infty} \left(\frac{n}{a}\right)^{1-s} + \sum_{l=1}^{\infty} \left(\frac{l}{b}\right)^{1-s} + \sum_{p=1}^{\infty} \left(\frac{p}{c}\right)^{1-s} \Bigg\}. \tag{8.36}$$

Equation (8.35) is obtained from eqn (8.36) when $s \to 0$. Using the definitions of the Epstein and Riemann zeta functions in eqns (8.15) and (4.6), respectively, we represent eqn (8.36) in the form

$$E_0^{(s)}(a,b,c) = \frac{\pi}{16} \Bigg[Z_3\left(\frac{1}{a},\frac{1}{b},\frac{1}{c}; s-1\right) - Z_2\left(\frac{1}{b},\frac{1}{c}; s-1\right) \tag{8.37}$$
$$- Z_2\left(\frac{1}{a},\frac{1}{c}; s-1\right) - Z_2\left(\frac{1}{a},\frac{1}{b}; s-1\right) + 2\left(\frac{1}{a}+\frac{1}{b}+\frac{1}{c}\right)\zeta_R(s-1) \Bigg].$$

All terms on the right-hand side of eqn (8.37) are evidently divergent in the limiting case $s \to 0$ when the regularization is removed. To perform renormalization by means of the zeta function, we use the reflection relations (8.16) (for the

Epstein zeta function) and (5.50) (for the Riemann zeta function). As a result, the finite Casimir energy of the massless scalar field inside the box is

$$E(a,b,c) = -\frac{abc}{32\pi^2}Z_3(a,b,c;4) + \frac{bc}{64\pi}Z_2(b,c;3) \qquad (8.38)$$
$$+ \frac{ac}{64\pi}Z_2(a,c;3) + \frac{ab}{64\pi}Z_2(a,b;3) - \frac{\pi}{96}\left(\frac{1}{a} + \frac{1}{b} + \frac{1}{c}\right).$$

Equation (8.26) provides an explicit expression for $Z_2(a,b;3)$ and, with appropriate permutations of the arguments, for $Z_2(b,c;3)$ and $Z_2(a,c;3)$. Now we shall obtain a similar expression for $Z_3(a,b,c;4)$.

For this purpose, we employ once more the function $S(\eta,\kappa;q)$ defined in eqn (8.21). From the definitions of the functions $S(\eta,\kappa;q)$ and $Z_3(a,b,c;4)$, it follows that

$$Z_3(a,b,c;4) = \pi^2 \sum_{l,p=-\infty}^{\infty} (1 - \delta_{l0}\delta_{p0})S(\eta_{lp},\kappa;4) + \frac{\zeta_R(4)}{a^4}, \qquad (8.39)$$

where

$$\kappa \equiv \frac{1}{a}, \quad \eta_{lp} \equiv \pi\sqrt{(bl)^2 + (cp)^2}. \qquad (8.40)$$

Now we substitute the expression (8.25) for S in eqn (8.39) and arrive at

$$Z_3(a,b,c;4) = \frac{\pi^4}{45a^4} + \frac{\sqrt{\pi}}{a}\sum_{l,p=-\infty}^{\infty}(1-\delta_{l0}\delta_{p0})\left[(bl)^2+(cp)^2\right]^{-3/2} \qquad (8.41)$$

$$\times \left\{\frac{\sqrt{\pi}}{2} + 4\sum_{j=1}^{\infty}\frac{K_{3/2}\left[2\pi j\sqrt{(bl)^2+(cp)^2}/a\right]}{\left[\pi j\sqrt{(bl)^2+(cp)^2}/a\right]^{-3/2}}\right\}.$$

Equation (8.41) can be identically rewritten as

$$Z_3(a,b,c;4) = \frac{\pi^4}{45a^4} + \frac{\pi}{2a}Z_2(b,c;3) + \frac{32\pi^2}{a^2bc}R\left(\frac{b}{a},\frac{c}{a}\right), \qquad (8.42)$$

where the following notation has been introduced:

$$R(z_1,z_2) = \frac{z_1 z_2}{8}\sum_{l,p=-\infty}^{\infty}(1-\delta_{l0}\delta_{p0}) \qquad (8.43)$$

$$\times \sum_{j=1}^{\infty}\left(\frac{j}{\sqrt{l^2z_1^2+p^2z_2^2}}\right)^{3/2} K_{3/2}\left(2\pi j\sqrt{l^2z_1^2+p^2z_2^2}\right).$$

Substituting eqn (8.42) in eqn (8.38), we notice that the contributions of the form $Z_2(b,c;3)$ cancel each other. Using eqn (8.26), we represent the final result in the form (Edery 2007)

$$E(a,b,c) = -\frac{\pi^2 bc}{1440a^3} + \frac{\zeta_R(3)(b+c)}{32\pi a^2} - \frac{\pi}{96a} \qquad (8.44)$$

$$-\frac{\pi}{2a}\left[G\left(\frac{b}{a}\right)+G\left(\frac{c}{a}\right)\right]-\frac{1}{a}R\left(\frac{b}{a},\frac{c}{a}\right),$$

where G is defined in eqn (8.11). After division of this equation by bc, we obtain in the limit $b,\,c\to\infty$ the scalar Casimir energy per unit area [see the first term in eqn (7.8)] for a configuration of two plane parallel plates.

This equation contains both positive and negative contributions. However, it is easily seen that for any relationship between a, b, and c, $E(a,b,c)<0$ holds. Thus, for a massless scalar field inside an ideal-metal box, all opposite faces attract each other independently of the magnitudes of a, b, and c. It is simple to check analytically that this is really so under the conditions $b/a,\,c/a\geq 1$. In this case the energy is approximately given by the first three terms on the right-hand side of eqn (8.44), the other contributions containing the Bessel functions of imaginary argument being exponentially small. Note that eqn (8.44) is in fact symmetric relative to cyclic permutations of the arguments a, b, and c. This symmetry is, however, implicit. In particular, if the above conditions $b/a,\,c/a\geq 1$ are violated, it is worthwhile to use another order of arguments in eqn (8.44). The absence of a Casimir repulsion for a scalar field in a three-dimensional box is qualitatively different from the case of a rectangle, where opposite sides repel each other if the condition (8.13) is fulfilled.

The result (8.44) for the Casimir energy of a scalar field in a box can be obtained by a three-fold application of the Abel–Plana formula, in analogy to Section 8.1, where it was applied twice. The advantage of the Abel–Plana formula in comparison with the Epstein zeta function is that the former permits explicit separation of divergent terms that are hidden in the analytic continuation of the latter. To apply the Abel–Plana formula, we introduce the cutoff function

$$f(\omega_{nlp}\delta)=e^{-\delta\sqrt{(n/a)^2+(l/b)^2+(p/c)^2}}. \tag{8.45}$$

As a result, we obtain the regularized Casimir energy in the form

$$E_0^{(\delta)}(a,b,c)=E(a,b,c)+I_0(a,b,c)+I_1(a,b,c)+I_2(a,b,c), \tag{8.46}$$

where the finite Casimir energy $E(a,b,c)$ is defined in eqn (8.44) and the regularized integrals are given by

$$I_0(a,b,c)=\frac{\pi}{2}\int_0^\infty dt\int_0^\infty dv\int_0^\infty du\sqrt{\left(\frac{v}{a}\right)^2+\left(\frac{t}{b}\right)^2+\left(\frac{u}{c}\right)^2}$$
$$\times e^{-\delta\sqrt{(v/a)^2+(t/b)^2+(u/c)^2}},$$

$$I_1(a,b,c)=-\frac{\pi}{4}\int_0^\infty dt\int_0^\infty dv\left[\sqrt{\left(\frac{v}{a}\right)^2+\left(\frac{t}{b}\right)^2}e^{-\delta\sqrt{(v/a)^2+(t/b)^2}}\right. \tag{8.47}$$
$$\left.+\sqrt{\left(\frac{v}{a}\right)^2+\left(\frac{t}{c}\right)^2}e^{-\delta\sqrt{(v/a)^2+(t/c)^2}}+\sqrt{\left(\frac{v}{b}\right)^2+\left(\frac{t}{c}\right)^2}e^{-\delta\sqrt{(v/b)^2+(t/c)^2}}\right],$$

$$I_2(a,b,c) = \frac{\pi}{8}\int_0^\infty t\,dt\left[\frac{1}{a}e^{-\delta t/a} + \frac{1}{b}e^{-\delta t/b} + \frac{1}{c}e^{-\delta t/c}\right].$$

The analytic continuations of the Epstein and Riemann zeta functions used above are effectively equivalent to the omission of the integrals (8.47).

A physical interpretation of this omission can be obtained from an explicit integration in eqn (8.47), which leads to

$$I_0(a,b,c) = \frac{12\pi^2 abc}{\delta^4}, \qquad I_1(a,b,c) = -\frac{\pi^2(ab+ac+bc)}{\delta^3},$$
$$I_2(a,b,c) = \frac{\pi(a+b+c)}{8\delta^2}. \tag{8.48}$$

Note that I_0 is proportional to the box volume, I_1 is proportional to the total surface area of the box, and I_2 is proportional to the sum of the sides. Thus the omission of all of these integrals, which is done implicitly in the zeta function regularization and explicitly in the regularization using the Abel–Plana formula, can be interpreted as deletion of the vacuum energy of free space in the volume of the box and renormalization of the respective geometrical objects.

Now we consider a Casimir piston for a massless scalar field in a three-dimensional box. Let the piston be at a point $z = c_1$, as shown in Fig. 8.1(b). The total regularized energy of the two three-dimensional boxes $a \times b \times c_1$ and $a \times b \times (c-c_1)$ in the presence of the piston is given by

$$\begin{aligned}E_0^{(\delta)}(a,b,c_1) + E_0^{(\delta)}(a,b,c-c_1) &= \frac{\pi(2a+2b+c)}{8\delta^2} - \frac{\pi^2(2ab+ac+bc)}{\delta^3} \\ &+ \frac{12\pi^2 abc}{\delta^4} - \frac{\pi^2 bc}{1440a^3} + \frac{\zeta_R(3)(2b+c)}{32\pi a^2} - \frac{\pi}{48a} \\ &- \frac{\pi}{2a}\left[2G\!\left(\frac{b}{a}\right) + G\!\left(\frac{c_1}{a}\right) + G\!\left(\frac{c-c_1}{a}\right) + \frac{2}{\pi}R\!\left(\frac{b}{a},\frac{c_1}{a}\right) + \frac{2}{\pi}R\!\left(\frac{b}{a},\frac{c-c_1}{a}\right)\right].\end{aligned} \tag{8.49}$$

As can be seen from eqn (8.49), none of the singular terms depend on the piston position c_1. The same is true for all finite terms outside the square brackets. Thus the Casimir force acting on the piston does not depend on whether the divergent integrals are omitted. It is given by

$$\begin{aligned}F(a,b,c,c_1) &= -\frac{\partial}{\partial c_1}\left[E_0^{(\delta)}(a,b,c_1) + E_0^{(\delta)}(a,b,c-c_1)\right] \\ &= \frac{\pi}{2a^2}\left[G'\!\left(\frac{c_1}{a}\right) - G'\!\left(\frac{c-c_1}{a}\right) + \frac{2}{\pi}R'\!\left(\frac{b}{a},\frac{c_1}{a}\right) - \frac{2}{\pi}R'\!\left(\frac{b}{a},\frac{c-c_1}{a}\right)\right],\end{aligned} \tag{8.50}$$

where the prime on R denotes the derivative with respect to the second argument.

As in the case of the rectangle, eqn (8.50) determines the Casimir attraction of a piston to the nearest face of the box. As an example, we consider a box with

$c_1 \ll a = b$ and $c \to \infty$. In this case eqn (8.50) leads to the following attractive force (Hertzberg et al. 2005):

$$F(a, a, \infty, c_1) = -\frac{\pi^2 a^2}{480 c_1^4} + \frac{\zeta_R(3) a}{8\pi c_1^3} - \frac{\pi}{96 c_1^2} + \frac{A}{a^2}. \tag{8.51}$$

Here

$$A = \frac{2\pi \zeta_R(3) - Z_2(1, 1; 4)}{32\pi^2} \approx 0.00483. \tag{8.52}$$

Another example is that of a cube $a = b = c_1$, preserving the condition $c \to \infty$ (Edery 2007). In this case the attractive force on the piston originating entirely from this cube is $F_1 = -0.005244/a^2$. However, the force $F_2 = 0.004832/a^2$ acting on the piston from the box $a \times b \times (c - c_1)$ repels it from the cube bottom at $z = 0$ (Edery 2007). The total force acting on the piston, $F = F_1 + F_2 = -0.000412/a^2$, attracts it to the cube bottom at $z = 0$ and simultaneously repels it from the top of the box $a \times b \times (c - c_1)$ at $z = c$. This example demonstrates that the force depends on whether there is empty space outside the box $a \times b \times (c - c_1)$. As was shown above, the Casimir force for a scalar field inside any isolated box (including a box of dimensions $a \times b \times (c - c_1)$ with $c \to \infty$) is attractive. However, the presence of an adjacent cube $a \times b \times c_1$ changes the attraction to a repulsion. The Dirichlet piston problem for a massless scalar field and an arbitrary rectangular box with dimensions $a \times b \times c$ was solved by Edery (2007) and by Hertzberg et al. (2007).

8.3 The electromagnetic Casimir effect in a three-dimensional box

The most realistic configuration related to the Casimir effect in a rectangular box is the case of an electromagnetic vacuum confined in a three-dimensional box with arbitrary sides $a \times b \times c$. As a first approximation, it is possible to consider walls made of an ideal metal and impose on them the boundary conditions (2.71).

The complete set of solutions of eqn (3.65) in the Coulomb gauge with the boundary condition (3.48) on the walls is given by

$$\mathcal{A}_{nlp}(\mathbf{r}) = \frac{4\sqrt{2\pi}}{\sqrt{abc}} \begin{pmatrix} b_x \cos k_{xn} x \sin k_{yl} y \sin k_{zp} z \\ b_y \sin k_{xn} x \cos k_{yl} y \sin k_{zp} z \\ b_z \sin k_{xn} x \sin k_{yl} y \cos k_{zp} z \end{pmatrix}. \tag{8.53}$$

Here, $k_{xn} = \pi n/a$, $k_{yl} = \pi l/b$, and $k_{zp} = \pi p/c$, where $n, l, p = 0, 1, 2, \ldots$ but only one of the three indices may have a value zero (otherwise the vector potential would be equal to zero). As is seen from eqn (8.53), on each wall of the box the respective pair of components of the vector potential forming a tangential vector vanishes. The normalization condition (3.66) must be considered with

$$\delta_{JJ'} = \delta_{nn'} \delta_{ll'} \delta_{pp'}, \quad \omega_J^2 = \omega_{nlp}^2 = k_{xn}^2 + k_{yl}^2 + k_{zp}^2,$$

$$\int_V d\boldsymbol{r} \equiv \int_0^a dx \int_0^b dy \int_0^c dz. \tag{8.54}$$

Together with the gauge-fixing condition $\text{div}\boldsymbol{A}_{nlp} = 0$, this leads to

$$b_x = b_y = \frac{k_{zp}}{\sqrt{(k_{xn} + k_{yl})^2 + 2k_{zp}^2}},$$

$$b_z = -\frac{k_{xn} + k_{yl}}{\sqrt{(k_{xn} + k_{yl})^2 + 2k_{zp}^2}}. \tag{8.55}$$

The total regularized energy of the electromagnetic vacuum inside a box can be written using eqn (3.76) in the form

$$E_0^{(s)}(a,b,c) = \frac{1}{2}\left(2\sum_{n,l,p=1}^{\infty} \omega_{nlp}^{1-s} + \sum_{l,p=1}^{\infty} \omega_{0lp}^{1-s} + \sum_{n,p=1}^{\infty} \omega_{n0p}^{1-s} + \sum_{n,l=1}^{\infty} \omega_{nl0}^{1-s}\right). \tag{8.56}$$

In order to express this quantity in terms of the zeta function, we do a transformation similar to that in the previous section and obtain

$$E_0^{(s)}(a,b,c) = \frac{\pi}{8}\left[Z_3\left(\frac{1}{a}, \frac{1}{b}, \frac{1}{c}; s-1\right) - 2\left(\frac{1}{a} + \frac{1}{b} + \frac{1}{c}\right)\zeta_R(s-1)\right]. \tag{8.57}$$

If we compare this with eqn (8.37), the electromagnetic case appears somewhat simpler than the scalar one.

To perform the renormalization by means of the zeta function, we use the reflection relations (8.16) and (5.50) and arrive at

$$E(a,b,c) = -\frac{abc}{16\pi^2}Z_3(a,b,c;4) + \frac{\pi}{48}\left(\frac{1}{a} + \frac{1}{b} + \frac{1}{c}\right). \tag{8.58}$$

The Epstein zeta function $Z_3(a,b,c;4)$ contained in this equation has already been calculated in eqns (8.42) and (8.26). Substituting these equations in eqn (8.58), we obtain the electromagnetic Casimir energy inside a three-dimensional box,

$$E(a,b,c) = -\frac{\pi^2 bc}{720a^3} - \frac{\zeta_R(3)c}{16\pi b^2} + \frac{\pi}{48}\left(\frac{1}{a} + \frac{1}{b}\right) + \frac{\pi}{b}C\left(\frac{c}{b}\right) - \frac{2}{a}R\left(\frac{b}{a}, \frac{c}{a}\right). \tag{8.59}$$

After dividing both sides by bc, we obtain in the limit $b, c \to \infty$ the electromagnetic Casimir energy per unit area of two plane parallel plates obtained in eqns (2.82) and (7.30).

The same expression has been obtained using the Abel–Plana formula (Mamayev and Trunov 1979a, 1979b). All of the calculations are similar to those presented in Section 8.1.1. As was mentioned above, the advantage of the Abel–Plana formula is that it allows clear identification of the divergent contributions that are effectively omitted in the analytic continuation of the zeta functions. Introducing the cutoff function (8.45) and applying the Abel–Plana formula three times, we arrive at the following regularized vacuum energy of the electromagnetic field inside a box:

$$E_0^{(\delta)}(a,b,c) = E(a,b,c) + \tilde{I}_0(a,b,c) + \tilde{I}_2(a,b,c). \qquad (8.60)$$

Here, the finite electromagnetic Casimir energy $E(a,b,c)$ has been presented in eqn (8.59) and the regularized divergent integrals are

$$\tilde{I}_0(a,b,c) = 2I_0(a,b,c), \qquad \tilde{I}_2(a,b,c) = -2I_2(a,b,c), \qquad (8.61)$$

where $I_1(a,b,c)$ and $I_3(a,b,c)$ are defined in eqn (8.47). From eqn (8.48), we obtain

$$\tilde{I}_0(a,b,c) = \frac{24\pi^2 abc}{\delta^4}, \qquad \tilde{I}_2(a,b,c) = -\frac{\pi(a+b+c)}{4\delta^2}. \qquad (8.62)$$

Thus, in the calculation of the electromagnetic Casimir energy inside a box, we have discarded the vacuum energy of free space inside the volume of the box and renormalized the constant proportional to the total size of the sides. We emphasize that eqn (8.60) does not contain a divergent integral proportional to the total surface area of the box [notated as $I_2(a,b,c)$ in eqn (8.48)], which is present for a scalar field. For the electromagnetic field, the divergent contributions to the vacuum energy proportional to the areas of the various faces cancel each other.

The electromagnetic Casimir energy inside a box (8.59) can be both negative and positive depending on the relationship between the sides a, b, and c. The respective Casimir forces between the opposite faces of a box

$$F_x = -\frac{\partial E(a,b,c)}{\partial a}, \qquad F_y = -\frac{\partial E(a,b,c)}{\partial b}, \qquad F_z = -\frac{\partial E(a,b,c)}{\partial c} \qquad (8.63)$$

can be both attractive and repulsive. Thus, for a cube with $a = b = c$, computations using eqn (8.59) result in

$$E(a) = \frac{0.09166}{a} > 0. \qquad (8.64)$$

Because of this, the opposite faces of an ideal-metal cube would attract each other. For a box with $a = b$ but arbitrary c, the Casimir energy $E(a,a,c)$ is positive under the condition

$$0.2942 < \frac{c}{a} < 3.429 \qquad (8.65)$$

and negative if this condition is not satisfied. Maclay (2000) performed numerical computations of Casimir energies and forces for boxes of various dimensions and found the regions in (a,b,c) space where the energy is negative or positive.

The electromagnetic Casimir effect in three-dimensional boxes was considered by Hacyan et al. (1993) using the Hertz potentials.

Next, we consider the electromagnetic Casimir force acting on a piston at $z = c_1$, shown in Fig. 8.1(b). This was first considered by Hertzberg et al. (2005). The regularized electromagnetic Casimir energy of the boxes $a \times b \times c_1$ and $a \times b \times (c - c_1)$ is given by

$$E_0^{(\delta)}(a,b,c_1) + E_0^{(\delta)}(a,b,c-c_1) = -\frac{\pi(2a+2b+c)}{4\delta^2} + \frac{24\pi^2 abc}{\delta^4}$$
$$- \frac{\pi^2 bc}{720a^3} - \frac{\zeta_R(3)c}{16\pi b^2} + \frac{\pi}{24}\left(\frac{1}{a}+\frac{1}{b}\right) + \frac{\pi}{b}\left[G\left(\frac{c_1}{b}\right) + G\left(\frac{c-c_1}{b}\right)\right]$$
$$- \frac{2}{a}\left[R\left(\frac{b}{a},\frac{c_1}{a}\right) + R\left(\frac{b}{a},\frac{c-c_1}{a}\right)\right]. \tag{8.66}$$

Just as for a scalar field, the divergent terms do not influence the force acting on the piston, and the force is equal to

$$F(a,b,c,c_1) = -\frac{\pi}{b^2}\left[G'\left(\frac{c_1}{b}\right) - G'\left(\frac{c-c_1}{b}\right)\right]$$
$$+ \frac{2}{a^2}\left[R'\left(\frac{b}{a},\frac{c_1}{a}\right) - R'\left(\frac{b}{a},\frac{c-c_1}{a}\right)\right]. \tag{8.67}$$

Here, the prime stands for differentiation with respect to the argument (the second argument in the case of R). For a box with $c_1 \ll a = b$ and $c \to \infty$, the asymptotic behavior of the force on the piston is given by (Hertzberg et al. 2005)

$$F(a,a,\infty,c_1) = -\frac{\pi^2 a^2}{240 c_1^4} + \frac{\pi}{48 c_1^2} - \frac{Z_2(1,1;4)}{16\pi^2 a^2} < 0, \tag{8.68}$$

where $Z_2(1,1;4) \approx 6.027$. Thus, this force is attractive. Note that the last term on the right-hand side of eqn (8.68) originates from the rectangular box $a \times b \times (c - c_1)$. This term does not depend on c_1, and demonstrates the influence on the piston of the box $a \times b \times (c - c_1)$ adjacent to the box $a \times b \times c_1$.

In the electromagnetic case, further generalizations of configurations with pistons have been considered in the literature. Thus, Barton (2006) found, using other boundary conditions (weakly reflecting dielectrics) that the force on a circular piston inside a cylinder can change sign as the distance from the top of the cylinder increases. In the case of two ideal-metal pistons inside an ideal-metal cylinder of arbitrary cross section, the force acting on each of the pistons was shown to be attractive (Marachevsky 2007).

We conclude from the above that the examples of a piston introduced into a rectangle or a three-dimensional box do not add direct information about the problem of the reality of the Casimir repulsion in rectangular boxes. So far, all results on this subject have been obtained using a model of ideal-metal walls. However, a conclusive resolution of the problem requires consideration of real material walls of finite conductivity.

8.4 Rectangular boxes with different boundary conditions

In the previous sections, we have considered Dirichlet boundary conditions imposed on scalar and electromagnetic fields at the internal edges and faces of two- and three-dimensional boxes. Similar results can be obtained for different types of boundary conditions. As a simple case, let us begin with a massless scalar field in a rectangle whose opposite sides are identified with each other (we have the topology of a 2-torus, which can be symbolically written as $S^1 \times S^1$). In this case $k_n = 2\pi n/a$, $k_l = 2\pi l/a$ [see eqn (2.43)], and the vacuum energy is given by

$$E_0(a,b) = \pi \sum_{n,l=-\infty}^{\infty} \left[\left(\frac{n}{a}\right)^2 + \left(\frac{l}{b}\right)^2\right]^{1/2}. \tag{8.69}$$

Bearing in mind that the term with $n = l = 0$ does not contribute in eqn (8.69), the regularized vacuum energy can be presented in the form

$$E_0^{(s)}(a,b) = \pi \sum_{n,l=-\infty}^{\infty} (1 - \delta_{n0}\delta_{l0}) \left[\left(\frac{n}{a}\right)^2 + \left(\frac{l}{b}\right)^2\right]^{-(s-1)/2}$$

$$= \pi Z_2\left(\frac{1}{a}, \frac{1}{b}; s-1\right). \tag{8.70}$$

This should be compared with the more complicated result (8.18) valid for Dirichlet boundary conditions.

Using the reflection formula (8.16) and removing the regularization, we arrive at the finite result

$$E(a,b) = -\frac{ab}{4\pi} Z_2(a,b;3). \tag{8.71}$$

An explicit expression for $Z_2(a,b;3)$ has already been obtained in eqn (8.26).

A massless scalar field in a three-dimensional box with identified opposite faces (which has the topology $S^1 \times S^1 \times S^1$ of a 3-torus) has a vacuum energy

$$E_0(a,b,c) = \pi \sum_{n,l,p=-\infty}^{\infty} \left[\left(\frac{n}{a}\right)^2 + \left(\frac{l}{b}\right)^2 + \left(\frac{p}{c}\right)^2\right]^{1/2}. \tag{8.72}$$

The regularized value of this energy is given by

$$E_0^{(s)}(a,b,c) = \pi \sum_{n,l,p=-\infty}^{\infty} (1 - \delta_{n0}\delta_{l0}\delta_{p0}) \left[\left(\frac{n}{a}\right)^2 + \left(\frac{l}{b}\right)^2 + \left(\frac{p}{c}\right)^2\right]^{-(s-1)/2}$$

$$= \pi Z_3\left(\frac{1}{a}, \frac{1}{b}, \frac{1}{c}; s-1\right). \tag{8.73}$$

This expression is much simpler than the corresponding eqn (8.37) for Dirichlet boundary conditions. The finite physical result is obtained using the reflection formula (8.16) and removing the regularization:

$$E(a,b,c) = -\frac{abc}{2\pi^2} Z_3(a,b,c;4). \tag{8.74}$$

An explicit expression for $Z_3(a,b,c;4)$ is contained in eqns (8.42) and (8.26).

It is of some interest to consider a hybrid situation, for example an identification condition on one pair of sides of a rectangle, and a Dirichlet boundary condition on the other (we have the topology of $S^1 \times I$, where I is a Euclidean interval). The vacuum energy of a massless scalar field takes the form

$$E_0(a,b) = \pi \sum_{n=-\infty}^{\infty} \sum_{l=1}^{\infty} \left[\left(\frac{n}{a}\right)^2 + \left(\frac{l}{2b}\right)^2\right]^{1/2}$$

$$= \frac{\pi}{2} \sum_{n=-\infty}^{\infty} \sum_{l=-\infty}^{\infty} (1-\delta_{l0}) \left[\left(\frac{n}{a}\right)^2 + \left(\frac{l}{2b}\right)^2\right]^{1/2}. \quad (8.75)$$

This can be identically rearranged to

$$E_0(a,b) = \frac{\pi}{2} \sum_{n,l=-\infty}^{\infty} (1-\delta_{n0}\delta_{l0}) \left[\left(\frac{n}{a}\right)^2 + \left(\frac{l}{2b}\right)^2\right]^{1/2} - \frac{\pi}{a} \sum_{n=1}^{\infty} n. \quad (8.76)$$

Then the regularized vacuum energy in the hybrid configuration is

$$E_0^{(s)}(a,b) = \frac{\pi}{2}\left[Z_2\left(\frac{1}{a},\frac{1}{2b};s-1\right) - \frac{2}{a}\zeta_R(s-1)\right], \quad (8.77)$$

which should be compared with eqn (8.18). After application of the reflection relations (5.50) and (8.16), the following finite result is obtained:

$$E(a,b) = -\frac{ab}{4\pi} Z_2(a,2b;3) + \frac{\pi}{12a}, \quad (8.78)$$

where $Z_2(a,2b;3)$ can be expressed using eqn (8.26).

The calculational technique used above can be applied in other cases; for example, scalar and electromagnetic fields in configurations $S^1 \times I \times I$, $S^1 \times S^1 \times I$, etc. can be considered. Mamayev and Trunov (1979a, 1979b) obtained the Casimir energy in these configurations using the Abel–Plana formula. Corresponding results for multidimensional rectangular cavities, including the case of Neumann boundary conditions, were obtained by Edery (2006, 2007). Multidimensional rectangular cavities were also considered by Caruso et al. (1991, 1999) and Li et al. (1997).

We now pursue our discussion with an interesting example of the role of mixed boundary conditions in rectangular cavities. We demonstrate that if a Neumann boundary condition is imposed on the piston, and Dirichlet boundary conditions are imposed on the other sides, the piston is *repelled* from the nearest side (Zhai and Li 2007). Let us begin with the simplest case of a rectangle $a \times b$ containing no piston, with Dirichlet boundary conditions on the sides of length b. For the sides with length a, we impose a Dirichlet boundary condition at $0 \leq x \leq a$, $y = 0$ and a Neumann boundary condition at $0 \leq x \leq a$, $y = b$.

Using the oscillation frequencies obtained when we studied mixed boundary conditions in Sections 2.1, 7.1.2, and 7.2.2, the nonregularized Casimir energy of a massless scalar field in such a rectangle is given by

$$E_0(a,b) = \frac{\pi}{2} \sum_{n=1}^{\infty} \sum_{l=0}^{\infty} \sqrt{\left(\frac{n}{a}\right)^2 + \left(l+\frac{1}{2}\right)^2 \frac{1}{b^2}}$$

$$= \frac{\pi}{2} \sum_{n,l=1}^{\infty} \left[\sqrt{\left(\frac{n}{a}\right)^2 + \left(\frac{l}{2b}\right)^2} - \sqrt{\left(\frac{n}{a}\right)^2 + \left(\frac{l}{b}\right)^2}\right]. \quad (8.79)$$

The regularized energy can be rearranged in the following way:

$$E_0^{(s)}(a,b) = \frac{\pi}{2} \sum_{n,l=1}^{\infty} \left\{\left[\left(\frac{n}{a}\right)^2 + \left(\frac{l}{2b}\right)^2\right]^{-(s-1)/2} - \left[\left(\frac{n}{a}\right)^2 + \left(\frac{l}{b}\right)^2\right]^{-(s-1)/2}\right\}$$

$$= \frac{\pi}{8} \sum_{n,l=-\infty}^{\infty} (1-\delta_{n0}\delta_{l0}) \left\{\left[\left(\frac{n}{a}\right)^2 + \left(\frac{l}{2b}\right)^2\right]^{-(s-1)/2} - \left[\left(\frac{n}{a}\right)^2 + \left(\frac{l}{b}\right)^2\right]^{-(s-1)/2}\right\}$$

$$- \frac{\pi}{4b} \sum_{l=0}^{\infty} \left(l+\frac{1}{2}\right)^{-(s-1)}$$

$$= \frac{\pi}{8}\left[Z_2\left(\frac{1}{a},\frac{1}{2b};s-1\right) - Z_2\left(\frac{1}{a},\frac{1}{b};s-1\right)\right] - \frac{\pi}{4b}\zeta\left(s-1,\frac{1}{2}\right), \quad (8.80)$$

where the Hurwitz zeta function $\zeta(z,q)$ is defined in eqn (7.37).

Using the reflection relations (7.38) and (8.16), we obtain, after removing the regularization, the following finite Casimir energy for a scalar field in a rectangle with one pair of mixed boundary conditions and one pair of Dirichlet conditions (Zhai and Li 2007):

$$E(a,b) = -\frac{ab}{32\pi}\left[2Z_2(a,2b;3) - Z_2(a,b;3)\right] - \frac{\pi}{96b}. \quad (8.81)$$

Using eqn (8.26) for $Z_2(a,b;3)$, we can represent eqn (8.81) in the final form

$$E(a,b) = -\frac{\zeta_R(3)b}{16\pi a^2} + \frac{\pi}{a}\left[G\left(\frac{2b}{a}\right) - G\left(\frac{b}{a}\right)\right], \quad (8.82)$$

where the function $G(z)$ is defined in eqn (8.11).

The same result can be obtained by repeated application of the Abel–Plana formulas (2.26) and (2.31). In so doing, the cutoff function

$$f(\omega_{nl}\delta) = e^{-\delta\sqrt{(n/a)^2+(2l+1)^2/(4b^2)}} \quad (8.83)$$

is used. This method not only allows one to find $E(a,b)$ but also specifies explicit expressions for the divergent integrals in the regularized vacuum energy

$$E_0^{(\delta)}(a,b) = I_0(b) + I_1(a,b) + E(a,b), \tag{8.84}$$

where

$$I_2(a,b) = \frac{2\pi^2}{\delta^3}ab, \qquad I_1(b) = -\frac{\pi}{4\delta^2}b. \tag{8.85}$$

The first integral in eqn (8.85) equals that in eqn (8.29). Its omission is equivalent to the subtraction of the energy of free space corresponding to the area of the rectangle. The second integral depends only on b. Thus, the use of mixed boundary conditions on the sides of size a removes the infinity proportional to a from the vacuum energy.

We now introduce a piston into the rectangle at $y = b_1 < b$ [see Fig. 8.1(a)]. Let the boundary condition on the piston be of Neumann type and let that on the four other sides be of Dirichlet type. Using eqns (8.82), (8.84), and (8.85), we find that the divergent terms in the energy for the two boxes $a \times b_1$ and $a \times (b - b_1)$ do not depend on the piston's position b_1:

$$E_0^{(\delta)}(a,b_1) + E_0^{(\delta)}(a,b-b_1) = -\frac{\pi}{2\delta^2}b + \frac{2\pi^2}{\delta^3}ab - \frac{\zeta_R(3)b}{16\pi a^2} \tag{8.86}$$
$$+ \frac{\pi}{a}\left[G\left(\frac{2b_1}{a}\right) - G\left(\frac{b_1}{a}\right) + G\left(\frac{2b-2b_1}{a}\right) - G\left(\frac{b-b_1}{a}\right)\right].$$

Thus, the force acting on the piston is given by

$$F(a,b,b_1) = \frac{\pi}{a^2}\left[-2G'\left(\frac{2b_1}{a}\right) + G'\left(\frac{b_1}{a}\right) + 2G'\left(\frac{2b-2b_1}{a}\right) - G'\left(\frac{b-b_1}{a}\right)\right]. \tag{8.87}$$

In the limiting case $b \to \infty$, eqn (8.87) reduces to

$$F(a,\infty,b_1) = \frac{\pi}{a^2}\left[-2G'\left(\frac{2b_1}{a}\right) + G'\left(\frac{b_1}{a}\right)\right]. \tag{8.88}$$

Numerical calculations show that this force is positive for all values of a and b_1. Thus, a piston with a Neumann boundary condition is repelled from the nearer side of the rectangle. The same results are obtained for a massless scalar field inside a three-dimensional box $a \times a \times c$ with Dirichlet boundary conditions on all sides, divided into two sections by an infinitely permeable piston at $z = c_1$ (Zhai and Li 2007). The force acting on the piston is again repulsive. For example, in the limiting case $c \to \infty$ and $c_1 = a$ (the configuration of a cube), the force F on the piston is equal to $0.000412/a^2$.

The above examples demonstrate that rectangular configurations with pistons leave room for the existence of repulsive Casimir forces. The final resolution of this problem has to include real material properties of the boundary surfaces.

8.5 Rectangular boxes at nonzero temperature

An independent problem is the thermal Casimir effect in a rectangular box. The first calculations on this subject (Ambjørn and Wolfram 1983) resulted in a divergent free energy after removing the regularization. More recent results appear to be either infinite (Santos and Tort 2000) or ambiguous (Jáuregui et al. 2006). Lim and Teo (2007) reconsidered the derivation of the Casimir free energy for massless scalar and electromagnetic fields using zeta function regularization. However, the renormalization was not complete. This led to irregular conclusions about the behavior of the free energy as a function of temperature. Here we consider the thermal Casimir effect in a rectangular box starting from the general equations (5.36) and (5.37) for the renormalized free energy associated with a finite volume V. These equations lead to physically meaningful results for the Casimir free energy and force for rectangular boxes with arbitrary sides $a \times b \times c$.

8.5.1 The scalar Casimir effect

We start with the case of a massless scalar field with Dirichlet boundary conditions. Equations (5.36), (5.37), and (5.29) can be written in the form

$$\mathcal{F}(a,b,c,T) = E(a,b,c) + k_\mathrm{B} T \sum_{n,l,p=1}^{\infty} \ln\left(1 - e^{-\beta \omega_{nlp}}\right) - \alpha_0 (k_\mathrm{B} T)^4$$
$$- \alpha_1 (k_\mathrm{B} T)^3 - \alpha_2 (k_\mathrm{B} T)^2, \tag{8.89}$$

where $\beta = 1/(k_\mathrm{B} T)$, $\alpha_0 = -\pi^2 abc/90$, ω_{nlp} is defined in eqn (8.34), and the finite Casimir energy at zero temperature is given by eqn (8.44).

Equation (8.89) represents the finite renormalized value of the Casimir free energy for a scalar field with Dirichlet boundary conditions in a box with sides a, b, and c, valid at any temperature. However, the values of α_1 and α_2 remain unknown. To determine them, one must find an asymptotic expression for the nonrenormalized thermal correction $\Delta_T \mathcal{F}_0$ at high temperature (or large separation). To do this, we rearrange $\Delta_T \mathcal{F}_0$ to the form

$$\Delta_T \mathcal{F}_0(a,b,c,T) = k_\mathrm{B} T X(\beta_a, \beta_b, \beta_c), \tag{8.90}$$

$$X(\beta_a, \beta_b, \beta_c) \equiv \sum_{n,l,p=1}^{\infty} \ln\left(1 - e^{-\sqrt{\beta_a^2 n^2 + \beta_b^2 l^2 + \beta_c^2 p^2}}\right),$$

where

$$\beta_a = \frac{\pi \beta}{a}, \quad \beta_b = \frac{\pi \beta}{b}, \quad \beta_c = \frac{\pi \beta}{c}. \tag{8.91}$$

Note that the quantity X does not depend on a, b, c, and T separately, but only through the products aT, bT, and cT. Below, we find an asymptotic expression for X under the conditions β_a, β_b, $\beta_c \ll 1$. This can be done by repeated application of the Abel–Plana formula (2.26).

First we put

$$F(n) = \sum_{l,p=1}^{\infty} \ln\left(1 - e^{-\sqrt{\beta_a^2 n^2 + \beta_b^2 l^2 + \beta_c^2 p^2}}\right). \tag{8.92}$$

Then the application of eqn (2.26) leads to

$$X(\beta_a, \beta_b, \beta_c) = -\frac{1}{2} \sum_{l,p=1}^{\infty} \ln\left(1 - e^{-\sqrt{\beta_b^2 l^2 + \beta_c^2 p^2}}\right) \quad (8.93)$$

$$+ \int_0^\infty dt \sum_{l,p=1}^{\infty} \ln\left(1 - e^{-\sqrt{\beta_a^2 t^2 + \beta_b^2 l^2 + \beta_c^2 p^2}}\right) + O(\ln\beta_a, \ln\beta_b, \ln\beta_c).$$

Here, we take into account the fact that the last term on the right-hand side of eqn (2.26) with F defined in eqn (8.92) is of order $\ln\beta_a$, $\ln\beta_b$, and $\ln\beta_c$.

Applying the Abel–Plana formula to each of the sums in eqn (8.93), we get

$$X(\beta_a, \beta_b, \beta_c) = \frac{1}{4} \sum_{p=1}^{\infty} \ln\left(1 - e^{-\beta_c p}\right) \quad (8.94)$$

$$- \frac{1}{2}\left(\frac{1}{\beta_a} + \frac{1}{\beta_b}\right) \int_0^\infty dy \sum_{p=1}^\infty \ln\left(1 - e^{-\sqrt{y^2 + \beta_c^2 p^2}}\right)$$

$$+ \frac{1}{\beta_a \beta_b} \int_0^\infty dy \int_0^\infty dv \sum_{p=1}^\infty \ln\left(1 - e^{-\sqrt{y^2 + v^2 + \beta_c^2 p^2}}\right) + O(\ln\beta_a, \ln\beta_b, \ln\beta_c).$$

Bearing in mind that

$$\sum_{p=1}^\infty \ln\left(1 - e^{-\beta_c p}\right) = -\frac{\pi^2}{6\beta_c} + O(\ln\beta_c) \quad (8.95)$$

and applying the Abel–Plana formula to the remaining two sums, the following result is obtained:

$$X(\beta_a, \beta_b, \beta_c) = -\frac{\pi^2}{24\beta_c} + \frac{1}{4}\left(\frac{1}{\beta_a} + \frac{1}{\beta_b}\right) \int_0^\infty dy \ln\left(1 - e^{-y}\right)$$

$$- \frac{1}{2}\left(\frac{1}{\beta_a \beta_c} + \frac{1}{\beta_b \beta_c} + \frac{1}{\beta_a \beta_b}\right) \int_0^\infty dy \int_0^\infty dv \ln\left(1 - e^{-\sqrt{y^2 + v^2}}\right)$$

$$+ \frac{1}{\beta_a \beta_b \beta_c} \int_0^\infty dy \int_0^\infty dv \int_0^\infty dw \ln\left(1 - e^{-\sqrt{y^2 + v^2 + w^2}}\right)$$

$$+ O(\ln\beta_a, \ln\beta_b, \ln\beta_c). \quad (8.96)$$

Calculating all integrals and using eqn (8.90) and the notation in eqn (8.91), we arrive at the asymptotic expression for the nonrenormalized thermal correction at high temperature (or large separation):

$$\Delta_T \mathcal{F}_0(a,b,c,T) = -\frac{\pi}{24}(k_B T)^2 (a+b+c) + \frac{\zeta_R(3)}{4\pi}(k_B T)^3 (ac+bc+ab)$$

$$- \frac{\pi^2}{90}(k_B T)^4 abc + O(k_B T \ln\beta_a, k_B T \ln\beta_b, k_B T \ln\beta_c). \quad (8.97)$$

Thus, it has been demonstrated that at high temperature (or large separation) the nonrenormalized thermal correction really contains terms of the form (5.35).

From a comparison of eqn (8.97) with eqn (5.35), it follows (Geyer et al. 2008c) that

$$\alpha_1 = \frac{\zeta_R(3)}{4\pi}(ac+bc+ab), \quad \alpha_2 = -\frac{\pi}{24}(a+b+c), \qquad (8.98)$$

i.e. α_1 is proportional to the total surface area of the box, and α_2 is proportional to the sum of the sides. These geometrical objects (together with the box volume) are usually renormalized when $E_{0,\text{eff}}$ is replaced with E_0^{ren} (see Sections 8.1.1, 8.2, and 8.3). Thus the subtraction of the last three terms on the right-hand side of eqn (8.89) can be interpreted as an additional finite renormalization, giving a physically meaningful temperature-dependent contribution to the Casimir energy.

As mentioned in Section 5.1, α_0, α_1, and α_2 can be expressed in terms of heat kernel coeffcients. Thus, keeping in mind that the heat kernel coefficient $a_{1/2} = -\sqrt{\pi}S/2$ [this follows from eqn (4.25) and has also been shown in the review by Vassilevich (2003)], where S is the surface area of box, and comparing eqn (8.97) with the general asymptotic expression for the free energy (5.53), one arrives at

$$\alpha_1 = -\frac{\zeta_R(3)}{4\pi^{3/2}}a_{1/2} = \frac{\zeta_R(3)}{8\pi}S \qquad (8.99)$$

in agreement with eqn (8.98). A similar comparison of the general expression (5.53) with eqn (8.97) leads to $\alpha_2 = -a_1/24$. The heat kernel coefficient a_1 can be calculated from the known expression for the heat kernel coefficient for an angle θ (Nesterenko et al. 2003),

$$c_1(\theta) = \frac{\pi^2 - \theta^2}{6\theta}. \qquad (8.100)$$

In the case of a rectangle, $\theta = \pi/2$ and $c_1 = \pi/4$. In the three-dimensional case, this must be multiplied by the lengths of all sides, leading to

$$a_1 = 4c_1(a+b+c) = \pi(a+b+c). \qquad (8.101)$$

From this, the expression for α_2 in eqn (8.98) is again obtained.

The Casimir force acting between opposite faces of the box is obtained from (8.89) and (8.98):

$$F_x(a,b,c,T) = -\frac{\partial \mathcal{F}(a,b,c,T)}{\partial a} = F_x(a,b,c) + \frac{\pi^2}{a^3}\sum_{n,l,p=1}^{\infty}\frac{n^2}{\omega_{nlp}(e^{\beta\omega_{nlp}}-1)}$$

$$-\frac{\pi^2(k_BT)^4}{90}bc + \frac{\zeta_R(3)}{4\pi}(k_BT)^3(b+c) - \frac{\pi}{24}(k_BT)^2. \qquad (8.102)$$

It is well known that the scalar Casimir energy $E(a,b,c)$ is negative and the respective force $F_x(a,b,c) = -\partial E(a,b,c)/\partial a$ is attractive for any ratio of a, b, and c (see Section 8.2). Because of this, we restrict ourselves to the consideration of a cube $a = b = c$.

In the limit of low temperature, $T \ll T_{\text{eff}}$, the leading terms in eqn (8.89) are

$$\mathcal{F}(a,T) = E(a) - k_B T e^{-\pi\sqrt{3}/(ak_B T)} \tag{8.103}$$
$$+ \frac{\pi^2 (k_B T)^4}{90} a^3 - \frac{3\zeta_R(3)(k_B T)^3}{4\pi} a^2 + \frac{\pi(k_B T)^2}{24} a,$$

where the scalar Casimir energy for a cube can be calculated numerically using eqns (8.44), (8.11), and (8.43), with the result

$$E(a) = -\frac{0.0102}{a}. \tag{8.104}$$

At an arbitrary temperature, it is convenient to represent both the free energy and the force in terms of the dimensionless variable $t = T_{\text{eff}}/T$:

$$\mathcal{F}(a,T) = E_0^{\text{ren}}(a) + \frac{1}{2at} \sum_{n,l,p=1}^{\infty} \ln\left(1 - e^{-2\pi t \sqrt{n^2+l^2+p^2}}\right)$$
$$+ \frac{\pi^2}{1440 a t^4} - \frac{3\zeta_R(3)}{32\pi} \frac{1}{at^3} + \frac{\pi}{32} \frac{1}{at^2}, \tag{8.105}$$

$$F_x(a,T) = F_x(a) + \frac{\pi}{a^2} \sum_{n,l,p=1}^{\infty} \frac{n^2}{\sqrt{n^2+l^2+p^2}} \frac{1}{e^{2\pi t \sqrt{n^2+l^2+p^2}} - 1}$$
$$- \frac{\pi^2}{1440 a^2 t^4} + \frac{\zeta_R(3)}{16\pi} \frac{1}{a^2 t^3} - \frac{\pi}{96} \frac{1}{a^2 t^2},$$

where the force at zero temperature, in agreement with eqn (8.104), is given by

$$F_x(a) = -\frac{0.0102}{3a^2}. \tag{8.106}$$

In Fig. 8.2(a), the scalar Casimir free energy in a cube, obtained from (8.105), is plotted as a function of the side length a at $T = 300$ K (solid line). As can be

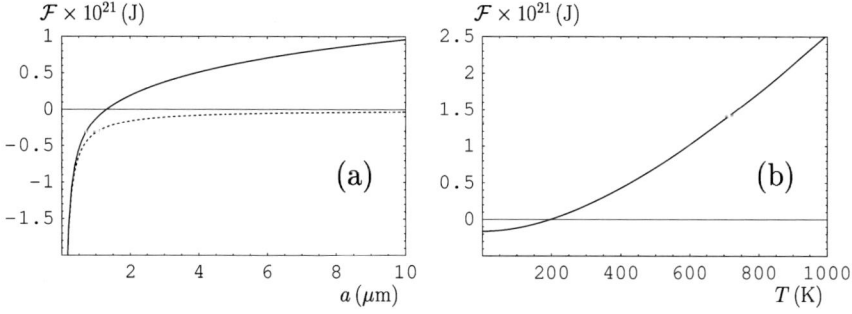

FIG. 8.2. The scalar Casimir free energy for a cube as a function of (a) side length a at $T = 300$ K (solid line; the dashed line shows the energy at $T = 0$) and (b) temperature at $a = 2\,\mu\text{m}$ (Geyer et al. 2008c).

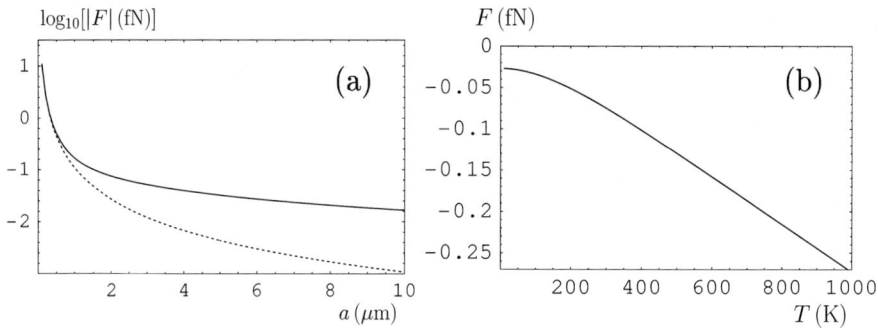

FIG. 8.3. The scalar Casimir force between opposite faces of a cube as a function of (a) side length a at $T = 300\,\text{K}$ (solid line; the dashed line shows the force at $T = 0$) and (b) temperature at $a = 2\,\mu\text{m}$ (Geyer et al. 2008c).

seen from this figure, the free energy increases monotonically with increasing a. At large separations (not shown in the figure), it approaches to a constant. The Casimir energy $E = E(a)$ at zero temperature is shown in the same figure by the dashed line. In Fig. 8.2(b), the scalar Casimir free energy is plotted as a function of temperature for a cube with $a = 2\,\mu\text{m}$. It is seen that at large temperatures $\mathcal{F}$ is proportional to the temperature in accordance with the classical limit.

The magnitude of the scalar Casimir force obtained from eqn (8.105) (on a logarithmic scale) as a function of a at $T = 300\,\text{K}$ is shown in Fig. 8.3(a) by the solid line. The force is attractive for cubes of any size and its magnitude goes to zero with increasing a. In the same figure, the dashed line shows the magnitude of the Casimir force (on a logarithmic scale) acting on opposite faces of the cube at $T = 0$. Figure 8.3(b) shows the Casimir force as a function of temperature for a cube with $a = 2\,\mu\text{m}$. It can be seen that for both negative and positive values of the free energy, the Casimir force is attractive.

8.5.2 The electromagnetic Casimir effect

Now we consider the electromagnetic thermal Casimir effect in a rectangular box with sides of lengths a, b, and c. For an electromagnetic field, the renormalized free energy (5.36), (5.37), and (5.29) is specified as

$$\mathcal{F}(a,b,c,T) = E(a,b,c) + k_\text{B}T\left[\sum_{l,p=1}^{\infty} \ln\!\left(1 - e^{-\beta\omega_{lp}}\right)\right. \tag{8.107}$$

$$\left. + \sum_{n,l=1}^{\infty} \ln\!\left(1 - e^{-\beta\omega_{nl}}\right) + \sum_{n,p=1}^{\infty} \ln\!\left(1 - e^{-\beta\omega_{np}}\right) + 2\sum_{n,l,p=1}^{\infty} \ln\!\left(1 - e^{-\beta\omega_{nlp}}\right)\right]$$

$$- \tilde{\alpha}_0(k_\text{B}T)^4 - \tilde{\alpha}_1(k_\text{B}T)^3 - \tilde{\alpha}_2(k_\text{B}T)^2.$$

Here, $\tilde{\alpha}_0 = -\pi^2 abc/45$, ω_{nlp} is defined in eqn (8.34), $\omega_{nl} = \omega_{nl0}$, and $\tilde{\alpha}_1$, $\tilde{\alpha}_2$ have to be determined. The electromagnetic Casimir energy at $T = 0$, $E(a,b,c)$, is given by eqn (8.59).

In order to find the coefficients $\tilde{\alpha}_1$ and $\tilde{\alpha}_2$, one must determine the asymptotic behavior of $\Delta_T \mathcal{F}_0$ in the limit of high temperature (or large separation). In the electromagnetic case, the nonrenormalized thermal correction can be identically rearranged to the form

$$\Delta_T \mathcal{F}_0(a,b,c,T) = k_B T\, Y(\beta_a, \beta_b, \beta_c), \qquad (8.108)$$

$$Y(\beta_a,\beta_b,\beta_c) = 2X(\beta_a,\beta_b,\beta_c) + \sum_{l,p=1}^{\infty} \ln\left(1 - e^{-\sqrt{\beta_b^2 l^2 + \beta_c^2 p^2}}\right)$$

$$+ \sum_{n,l=1}^{\infty} \ln\left(1 - e^{-\sqrt{\beta_a^2 n^2 + \beta_b^2 l^2}}\right) + \sum_{n,p=1}^{\infty} \ln\left(1 - e^{-\sqrt{\beta_a^2 n^2 + \beta_c^2 p^2}}\right),$$

where the asymptotic behavior of $X(\beta_a, \beta_b, \beta_c)$ at small β_a, β_b, and β_c has been determined above. Taking eqns (8.90) and (8.97) into account, this is given by

$$X(\beta_a,\beta_b,\beta_c) = -\frac{\pi^2}{24}\left(\frac{1}{\beta_a} + \frac{1}{\beta_b} + \frac{1}{\beta_c}\right) + \frac{\pi \zeta_R(3)}{4}\left(\frac{1}{\beta_a \beta_b} + \frac{1}{\beta_a \beta_c} + \frac{1}{\beta_b \beta_c}\right)$$

$$- \frac{\pi^5}{90} \frac{1}{\beta_a \beta_b \beta_c} + O(\ln \beta_a, \ln \beta_b, \ln \beta_c). \qquad (8.109)$$

The asymptotic behavior of $Y(\beta_a, \beta_b, \beta_c)$ at small β_a, β_b, β_c is obtained, in perfect analogy with the case of a scalar field, by repeated application of the Abel–Plana formula (2.26) to the remaining three summations in eqn (8.108). The result obtained, taking account of eqn (8.109), is

$$Y(\beta_a,\beta_b,\beta_c) = \frac{\pi^2}{12}\left(\frac{1}{\beta_a} + \frac{1}{\beta_b} + \frac{1}{\beta_c}\right) - \frac{\pi^5}{45}\frac{1}{\beta_a \beta_b \beta_c}$$
$$+ O(\ln \beta_a, \ln \beta_b, \ln \beta_c). \qquad (8.110)$$

Substituting this into eqn (8.108) and using the notation in eqn (8.91), one obtains an asymptotic expression for the thermal correction at high temperature (or large separation)

$$\Delta_T \mathcal{F}_0(a,b,c,T) = \frac{\pi}{12}(k_B T)^2 (a+b+c) - \frac{\pi^2}{45}(k_B T)^4 abc$$
$$+ O(k_B T \ln \beta_a, k_B T \ln \beta_b, k_B T \ln \beta_c). \qquad (8.111)$$

It is notable that eqn (8.111) does not contain a contribution proportional to the surface area of the box [in contrast to eqn (8.97) for the scalar field]. In the electromagnetic case, such a contribution is also absent in the divergent Casimir energy of the box E_0 at zero temperature (see Section 8.3). Thus, from eqns (8.111) and (5.35), one arrives at (Geyer et al. 2008c)

$$\tilde{\alpha}_1 = 0, \quad \tilde{\alpha}_2 = \frac{\pi}{12}(a+b+c). \qquad (8.112)$$

Hence, similarly to the scalar case, $\tilde{\alpha}_0$, $\tilde{\alpha}_1$, and $\tilde{\alpha}_2$ have the same geometrical structure as the infinite expressions in eqn (8.62).

The Casimir force acting between opposite faces of a box is obtained as the negative derivative of eqn (8.107) with respect to a:

$$F_x(a,b,c,T) = F_x(a,b,c) + \frac{\pi^2}{a^3}\left[\sum_{n,l=1}^{\infty} \frac{n^2}{\omega_{nl}(e^{\beta\omega_{nl}} - 1)} + \sum_{n,p=1}^{\infty} \frac{n^2}{\omega_{np}(e^{\beta\omega_{np}} - 1)}\right.$$

$$\left. + 2\sum_{n,l,p=1}^{\infty} \frac{n^2}{\omega_{nlp}(e^{\beta\omega_{nlp}} - 1)}\right] - \frac{\pi^2(k_BT)^4}{45}bc + \frac{\pi}{12}(k_BT)^2. \tag{8.113}$$

It is known that the electromagnetic Casimir energy inside a box at $T = 0$, $E(a,b,c)$, can be both positive and negative and that the Casimir force, $F_x(a,b,c) = -\partial E(a,b,c)/\partial a$, can be both attractive and repulsive depending on the ratio of the sides a, b, and c (see Section 8.3). Here we consider in more detail the thermal electromagnetic Casimir effect for a cube $a = b = c$, where the electromagnetic Casimir energy at zero temperature (8.64) is positive and the force is repulsive.

For a cube, the electromagnetic Casimir free energy (8.107) and force (8.113) are given by

$$\mathcal{F}(a,T) = E(a) + \frac{3}{2at}\sum_{n,l=1}^{\infty} \ln\left(1 - e^{-2\pi t\sqrt{n^2+l^2}}\right)$$

$$+ \frac{1}{at}\sum_{n,l,p=1}^{\infty} \ln\left(1 - e^{-2\pi t\sqrt{n^2+l^2+p^2}}\right) + \frac{\pi^2}{720at^4} - \frac{\pi}{16at^2},$$

$$F_x(a,T) = F_x(a) + \frac{2\pi}{a^2}\left[\sum_{n,l=1}^{\infty} \frac{n^2}{\sqrt{n^2+l^2}} \frac{1}{e^{2\pi t\sqrt{n^2+l^2}} - 1}\right. \tag{8.114}$$

$$\left. + \sum_{n,l,p=1}^{\infty} \frac{n^2}{\sqrt{n^2+l^2+p^2}} \frac{1}{e^{2\pi t\sqrt{n^2+l^2+p^2}} - 1}\right] - \frac{\pi^2}{720a^2t^4} + \frac{\pi}{48a^2t^2},$$

where the force at $T = 0$ is

$$F_x(a) = \frac{0.09166}{3a^2}. \tag{8.115}$$

In Fig. 8.4(a), we plot the electromagnetic Casimir free energy in a cube as a function of a at $T = 300\,\text{K}$ (solid line). The Casimir energy at $T = 0$ is shown in the same figure by the dashed line. As can be seen in this figure, the electromagnetic Casimir free energy decreases with increasing separation. Similarly to the scalar case, at large separations $\mathcal{F}$ approaches a constant. In Fig. 8.4(b), the electromagnetic Casimir free energy is shown as a function of temperature for a cube with $a = 2\,\mu\text{m}$. The free energy decreases with increasing T. At high temperature, $\mathcal{F}$ approaches the classical limit. The respective thermal

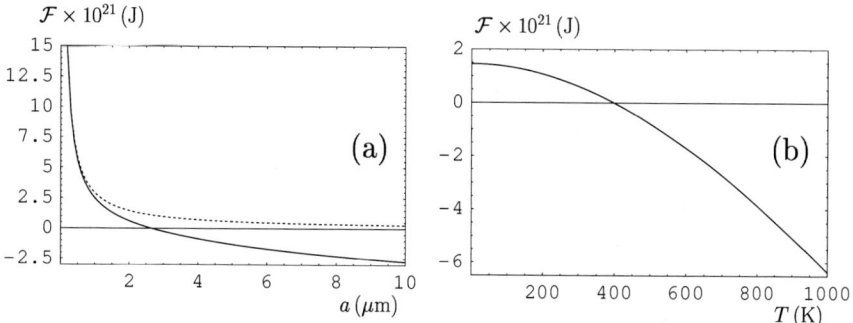

FIG. 8.4. The electromagnetic Casimir free energy for a cube as a function of (a) side length a at $T = 300\,\text{K}$ (solid line; the dashed line shows the energy at $T = 0$) and (b) temperature at $a = 2\,\mu\text{m}$ (Geyer et al. 2008c).

electromagnetic Casimir force at $T = 300\,\text{K}$, as a function of a, is shown in Fig. 8.5(a) by the solid line. It is positive (i.e. repulsive) for cubes of any size. Thus, thermal effects for cubes in the electromagnetic case increase the strength of the Casimir repulsion. The dashed line in Fig. 8.5(a) shows the electromagnetic Casimir force at $T = 0$ as a function of a. This force is given by eqn (8.115), i.e. it is always repulsive. Figure 8.5(b) shows the electromagnetic Casimir force in a cube of side $a = 2\,\mu\text{m}$ as a function of temperature. It is seen that the force increases with increasing temperature.

Note that the results presented here differ from those found by Lim and Teo (2007), where the terms of order $(k_B T)^4$ and of lower orders in the Casimir free energy were obtained in the high-temperature regime. Also, the Casimir free energy obtained by Lim and Teo (2007) for both the scalar and the electromagnetic field is always a decreasing function of temperature, opposite to the result in Fig. 8.2(b). This is due to the fact that Lim and Teo (2007) did not perform subtraction of the contributions from black-body radiation and of the

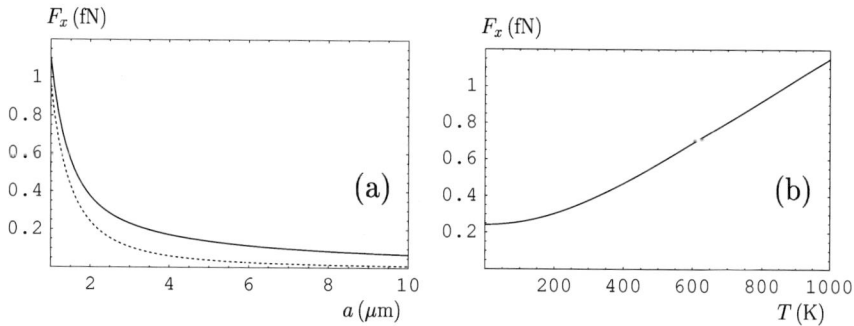

FIG. 8.5. The electromagnetic Casimir force between the opposite faces of cube as a function of (a) side length a at $T = 300\,\text{K}$ (solid line; the dashed line shows the force at $T = 0$) and (b) temperature at $a = 2\,\mu\text{m}$ (Geyer et al. 2008c).

terms proportional to the surface area of the box and the sum of its sides.

The above equations (8.89), (8.102), (8.107), and (8.113) can be used to compute the scalar and electromagnetic free energy and force for boxes with arbitrary sides a, b, and c. Specifically, it follows that the temperature-dependent contribution to the electromagnetic Casimir force [which is obtained as $-\partial \Delta_T \mathcal{F}/\partial a$ from the physical thermal correction defined in eqn (5.37)] can be both positive and negative depending on the sides a, b, and c. On the one hand, as shown above (see Fig. 8.5), for a cube $a \times a \times a$ the temperature-dependent contribution to the Casimir force is positive, and computations show that this is preserved for any box $a \times b \times b$ with $a > b$. On the other hand, for a box with $b = c = 10\,\mu$m and $a_1 = 2.942\,\mu$m or $a_2 = 34.29\,\mu$m, the Casimir energy at $T = 0$ is equal to zero (see Section 8.3). Computations using eqn (8.113) show that for a box $a_1 \times b \times b$ the temperature-dependent contribution to the force is negative, whereas for a box $a_2 \times b \times b$ it is positive.

The thermal corrections to the Casimir energy and force acting on a piston were investigated by Hertzberg et al. (2007) for a scalar field with Dirichlet or Neumann boundary conditions using the definition (5.30). The electromagnetic Casimir free energy and force acting on a piston were found in the case of ideal-metal rectangular boxes and a cavities with a general cross section (Hertzberg et al. 2007). In the limit of low temperature, the thermal correction to the Casimir force on a piston was shown to be exponentially small. In the case of an intermediate temperature $a \ll \beta \ll b, c$, Hertzberg et al. (2007) obtained terms of order $(k_\mathrm{B}T)^4$ and of order $(k_\mathrm{B}T)^2$ in the electromagnetic Casimir free energy. In the scalar Casimir free energy, a term of order $(k_\mathrm{B}T)^3$ was also obtained. This results in a contribution to the force which does not depend on the position of the piston. The same results for the thermal correction to the Casimir force acting on a piston are obtained if the free energy is defined in accordance with eqn (5.37). This is because the contribution of black-body radiation to the energy of the entire box in Fig. 8.1 is equal to

$$-abc_1 f_{bb} - ab(c - c_1) f_{bb} = -abc\, f_{bb}, \qquad (8.116)$$

i.e. it does not depend on the position of the piston. This is also true for the terms of order $(k_\mathrm{B}T)^3$ and $(k_\mathrm{B}T)^2$, which are proportional to the surface area of each section of the box and to the sum of its sides.

Notwithstanding the fact that the two expressions for the Casimir free energy (5.30) and (5.37) lead to a common Casimir force acting on the piston, the latter should be considered as preferable on physical grounds. The reason is that from a thermodynamic point of view (Kubo 1968), any equilibrium system can be characterized by both the free energy (5.2) and the respective pressure (5.3). From this point of view, it would not be consistent to allow consideration of the force acting on the piston but to exclude from consideration those forces which act on the faces of the box where this piston serves as a partition. Thus, it is important to use the definition of the Casimir free energy in a box (5.37) which leads to physically meaningful forces acting not only on the piston, but on all box faces as well. The point of view (Hertzberg *et al.* 2005, 2007) that the definition of a force acting on a box face requires elastic deformations of a single body that is treated as perfect is not accurate. To define a force and a pressure in a static configuration, one need not involve elastic deformations. This is simply done using the principle of virtual work, and virtual displacements through the action of real forces (Charlton 1973).

The above results were obtained for rectangular boxes with Dirichlet boundary conditions (in the scalar case) and for ideal-metal boxes (in the electromagnetic case). In the same way as for zero temperature, the consideration of the thermal Casimir effect in a rectangular box has to incorporate real material properties of the boundary surfaces. At present, this problem remains unsolved.

9
SINGLE SPHERICAL AND CYLINDRICAL BOUNDARIES

In this chapter we consider the Casimir effect for simple, single bodies having a spherical or cylindrical shape. As we have seen in Section 4.3.3, the Casimir force and the corresponding interaction energy for separate bodies are always finite. Thus renormalization is not an issue. However, for single bodies, such as the rectangular boxes considered in the previous chapter, this is not the case, because the divergent contributions to the vacuum energy depend on the lengths of the sides. For a sphere, the divergent contributions usually depend on its radius and cannot be neglected. Therefore renormalization schemes, as discussed in Section 4.3.2, must be employed. This requires some classical system for the interpretation of the renormalization. While such schemes are known to work well in quantum field theory, for example in the case of quantum corrections to the mass of solitons, for the Casimir effect there is presently a lack of physical examples.

It must be mentioned that the interest in the Casimir effect for single bodies such as a sphere is enormous. This is due to a number of reasons. Historically, the first and most intriguing follows from Casimir's model for the electron (Casimir 1953). He assumed the charge of the electron to be distributed over a conductive spherical shell. The resulting electrostatic self-repulsion would be balanced by an attractive Casimir force (in analogy with the case of parallel metal planes). If this model has been proved to be correct, it would have revolutionized physics by allowing calculation of the fine structure constant. However, the question of what balances the electrostatic self-repulsion of the electron and other charged particles has still not been answered and is probably poorly formulated. As Casimir himself said, "It would have been embarrassing if you really got 137 because such a model can never be very close to reality..." (Casimir 1999). But Boyer (1968) showed that the Casimir effect for a conducting sphere is repulsive. Recalculations using different methods (Davies 1972, Balian and Duplantier 1977, 1978, DeRaad and Milton 1981) confirmed this result. The interest in this configuration has continued over the decades. While the initial calculation was very laborious, the use of advanced computational tools has simplified such calculations considerably. The basic methods used, are the multiple-reflection expansion (Balian and Duplantier 1977), the Green's function method (DeRaad and Milton 1981), and the zeta function method. As we shall see in Section 9.3.3, there is a complete understanding of the cancellation of the divergences.

The second reason for the interest in the Casimir effect for a sphere is the phenomenological bag model of hadrons (Chodos *et al.* 1974), which is intended to describe their properties until a complete solution of the confinement problem is available. Here, boundary conditions are imposed on the quark and gluon fields in order to prevent the escape of color from some region in space (the bag, in the simplest case, is a sphere). For gluon fields, because of duality, this is equivalent to the Casimir energy of a conducting sphere. For quark fields, a special type of boundary condition appears, which will be considered in Section 9.4. However, within this model, the renormalization issue is not really settled and the model remains physically unsatisfactory. An attempt to collect different phenomenological approaches together within the bag model was recently undertaken by Milton (2001). Here we restrict ourselves to a representation of the Casimir effect for quark fields in a bag and focus on the calculation procedure.

Thirdly, the Casimir effect has received much attention as a possible explanation for sonoluminescence. The latter is the production of short flashes of light from air bubbles in a liquid driven by an ultrasonic wave. This phenomenon has been experimentally investigated [see the review by Barber *et al.* (1997)] but a complete understanding of the underlying physics has still not been reached. Schwinger (1992, 1993) attempted to relate the energy release in the light flashes to differences in the vacuum energy of a collapsing dielectric ball. Eberlein (1996) adopted the dynamical Casimir effect as an explanation. Work in this direction was continued by Liberati *et al.* (2000), but there is as yet no final answer. At this point it should be mentioned that a spherical geometry is also being used in connection with attempts to consider the vacuum energy in black holes.

A fourth reason for the interest in spherical geometries is driven by the repulsive character of the force acting on a conducting spherical shell (the same holds for some rectangular cavities; see Chapter 8). It would be exciting to find a repulsive Casimir force between real bodies in contrast to the attractive van der Waals forces between atoms and molecules. If found, this would provide a means to reduce stiction in nanomechanical systems.

In this chapter, we focus on the technical methods necessary to calculate the Casimir effect in spherical and cylindrical geometries and on the analysis of the ultraviolet divergences. In general, most of the technical matter is well known from classical electrodynamics, and the mathematics involved does not go much beyond Bessel functions. Nevertheless, we feel it worthwhile to collect the relevant formulas together in one place and to present them from a common point of view. Our representation is completely in terms of global quantities. For a local treatment, especially for a representation in terms of Green's dyadics, we refer the reader to the book by Milton (2001), where the relevant literature is collected together. The physical content of the Casimir effect for a sphere depends much on the ultraviolet divergences. We discuss them in detail using the approach of heat kernel coefficients. Adopting the philosophy of renormalization used in Section 4.3, we discuss for which system (the interior or exterior of a sphere or the total space) a meaningful definition of the vacuum energy is possible. We shall come to the conclusion that this is possible only for a few selected cases.

9.1 Separation of variables and mode summation

Problems with boundary or matching conditions on a sphere or on a cylinder allow a separation of variables in the corresponding wave equation (3.1), (3.22), or (3.30). This reduces the problem to one-dimensional radial equations. However, because of the necessary summation over the orbital momenta, the details of the renormalization are quite different from the one-dimensional case. As a consequence, these problems require their own technical tools. The most important ones are the Green's function method, especially that based on dyadic Green's functions (Milton 2001), the multiple-reflection method (Balian and Duplantier 1977, 1978), and the mode summation method. Thus different regularization schemes have been employed, which are sometimes referred to as separate methods. In fact, all these methods are equivalent, at least for the calculation of global quantities such as the vacuum energy. For the calculation of local quantities such as the energy density or the charge density, the Green's function method is preferable. However, since we are interested only in global quantities, in order to avoid unnecessary technical details, we use only the mode summation method below.

9.1.1 *Spherical symmetry*

We start from the representations of the vacuum energy in eqn (3.60) for a scalar field, eqn (3.76) for an electromagnetic field, and eqn (3.84) for a spinor field. The corresponding eigenvalue problems are eqns (3.39) and (3.45). For the scalar field, using separation of variables in spherical coordinates (r, θ, φ), the solution to the Laplace equation is

$$\Phi_J(\boldsymbol{r}) = f_{l,n}(r) Y_{lM}(\theta, \varphi). \tag{9.1}$$

The index $J = (n, l, M)$ is now composed of the orbital momentum $l = 0, 1, \ldots$, its azimuthal component M with $|M| \leq l$, and the radial index $n = 1, 2, \ldots$.

For the electromagnetic field, instead of eqn (3.68), the separation of variables and the separation of polarizations are achieved by use of a representation of the vector potential in the form

$$\boldsymbol{\mathcal{A}}_J(\boldsymbol{r}) = \boldsymbol{E}^{\text{TE}} f_{l,n}^{\text{TE}}(r) Y_{lM}(\theta, \varphi) + \boldsymbol{E}^{\text{TM}} f_{l,n}^{\text{TM}}(r) Y_{lM}(\theta, \varphi), \tag{9.2}$$

with the known (nonnormalized) operator polarization vectors

$$\boldsymbol{E}^{\text{TE}} = \boldsymbol{L}, \qquad \boldsymbol{E}^{\text{TM}} = \boldsymbol{\nabla} \times \boldsymbol{L}. \tag{9.3}$$

Here, $\boldsymbol{L}$ is the operator of the orbital angular momentum, having the property

$$\boldsymbol{L}^2 Y_{lM}(\theta, \varphi) = l(l+1) Y_{lM}(\theta, \varphi) \tag{9.4}$$

The spherical harmonics $Y_{lM}(\theta, \varphi)$ are defined as

$$Y_{lM}(\theta, \varphi) = \frac{1}{\sqrt{2\pi}} \sqrt{\frac{2l+1}{2}} \sqrt{\frac{(l-M)!}{(l+M)!}} \, e^{iM\varphi} P_l^M(\cos\theta), \tag{9.5}$$

and the $P_l^M(z)$ are the associated Legendre functions. The polarization vectors (9.3) are orthogonal to each other, i.e. $\boldsymbol{E}^{\text{TE}} \cdot \boldsymbol{E}^{\text{TM}} = 0$, and satisfy the equalities

$$\boldsymbol{\nabla} \cdot \boldsymbol{E}^{\text{TE}} = \boldsymbol{\nabla} \cdot \boldsymbol{E}^{\text{TM}} = 0, \quad \boldsymbol{\nabla} \times \boldsymbol{E}^{\text{TE}} = \boldsymbol{E}^{\text{TM}}, \quad \boldsymbol{\nabla} \times \boldsymbol{E}^{\text{TM}} = -\boldsymbol{\nabla}^2 \boldsymbol{E}^{\text{TE}}. \tag{9.6}$$

These polarization vectors are conveniently chosen in spherical geometry instead of the $\boldsymbol{\epsilon}_j^{(\lambda)}$ in eqn (3.74), which were used in the plane parallel geometry. It must be mentioned that in the decomposition (9.2) the angular momentum $l = 0$, i.e. the s-wave, is absent. This is because the orbital-momentum operator, when acting on a function which does not depend on the angles such as $Y_{00}(\theta, \varphi)$, leads to zero. Therefore, in contrast to the scalar case, the orbital momenta have values $l = 1, 2, \ldots$. Note that in accordance with the definition (9.3), $\boldsymbol{E}^{\text{TE}}$ is dimensionless and the dimension of $\boldsymbol{E}^{\text{TM}}$ is 1/cm.

Using eqn (3.13), and the time dependence $\sim \exp(-i\omega t)$ in eqn (3.44), the electric field and magnetic induction are

$$\boldsymbol{E}_J(\boldsymbol{r}) = i\omega \left[\boldsymbol{E}^{\text{TE}} f_{l,n}^{\text{TE}}(r) + \boldsymbol{E}^{\text{TM}} f_{l,n}^{\text{TM}}(r) \right] Y_{lM}(\theta, \varphi),$$
$$\boldsymbol{B}_J(\boldsymbol{r}) = \left[-\boldsymbol{\nabla}^2 \boldsymbol{E}^{\text{TE}} f_{l,n}^{\text{TM}}(r) + \boldsymbol{E}^{\text{TM}} f_{l,n}^{\text{TE}}(r) \right] Y_{lM}(\theta, \varphi). \tag{9.7}$$

The notation "TE", denoting a transverse electric field, follows from this representation since the vector $\boldsymbol{E}^{\text{TE}}$ is perpendicular to the wavefront of a spherical wave, i.e. $\boldsymbol{n} \cdot \boldsymbol{E}^{\text{TE}} = 0$, where $\boldsymbol{n}$ is the unit vector in the radial direction. The case of a transverse magnetic field, denoted by "TM", is similar owing to duality. This can be seen in eqn (9.7) where, except for some factors, the two lines differ only by an interchange of the radial functions $f_{l,n}^{\text{TE}}(r)$ and $f_{l,n}^{\text{TM}}(r)$. The radial function $f_{l,n}^{\text{TE}}(r)$ has the dimension of 1/cm, whereas $f_{l,n}^{\text{TM}}(r)$ is dimensionless. Thus, the physical fields (9.7) have the correct dimension of 1/cm². Similar formulas can be written down for the spinor case. Since these are technically more involved, we postpone their discussion to Section 9.4.

In all cases, the wave equations (3.39) and (3.45) translate into the radial equation

$$\left[-\frac{\partial^2}{\partial r^2} - \frac{2}{r} \frac{\partial}{\partial r} + \frac{l(l+1)}{r^2} \right] f_{l,n}(r) = \Lambda_{l,n} f_{l,n}(r) \tag{9.8}$$

for each of the functions $f_{l,n}(r)$ in eqn (9.1) or $f_{l,n}^{\text{TE}}(r)$ or $f_{l,n}^{\text{TM}}(r)$ in eqn (9.2), as the polarization vectors (9.3) commute with the Laplace operator. In eqn (9.8), we have used the Laplace operator in spherical coordinates,

$$\boldsymbol{\nabla}^2 = \frac{\partial^2}{\partial r^2} + \frac{2}{r} \frac{\partial}{\partial r} - \frac{\boldsymbol{L}^2}{r^2}, \tag{9.9}$$

and the eigenvalues of the angular-momentum operator from eqn (9.4). The solutions of eqn (9.8) are the spherical Bessel functions.

Further progress depends on the boundary conditions. We assume them to be compatible with the separation of variables such that they translate into conditions on the radial functions. We assume further that these conditions, together with eqn (9.8), constitute a well-posed eigenvalue problem. This may be on an interval $r \in [0, R]$ (the interior of a sphere) or on an infinite interval $r \in [R, \infty)$ (the exterior region). In the following, we consider these two cases separately.

9.1.2 Mode summation for the interior problem

For the interior of the sphere, we have to consider eqn (9.8) on the interval $r \in [0, R]$. For an arbitrary k, the solutions are spherical Bessel functions $j_l(z)$ obeying

$$\left[-\frac{\partial^2}{\partial r^2} - \frac{2}{r}\frac{\partial}{\partial r} + \frac{l(l+1)}{r^2} \right] j_l(kr) = k^2 \, j_l(kr), \tag{9.10}$$

which, by means of

$$j_l(z) = \sqrt{\frac{\pi}{2z}} J_{l+\frac{1}{2}}(z), \tag{9.11}$$

are related to the ordinary Bessel functions. The functions $j_l(z)$ are chosen because they are regular at the origin, i.e.

$$j_l(z) \underset{r \to 0}{\sim} \sqrt{\frac{\pi}{2}} \left(\frac{z}{2}\right)^l \frac{1}{\Gamma\left(l + \frac{3}{2}\right)}, \tag{9.12}$$

ensuring the renormalizability of the wave function.

The discrete eigenvalues of the radial problem appear as solutions of the corresponding boundary condition. For a scalar field with Dirichlet boundary conditions, from eqns (3.41) and (9.1), the radial functions $f_{n,l}(r)$ must vanish at $r = R$. This is achieved for discrete radial momenta $k_{n,l}$ obtained from

$$j_l(kR) = 0, \qquad k = \frac{j_{\nu,n}}{R}, \qquad n = 1, 2, \dots, \tag{9.13}$$

where the $j_{\nu,n}$ are the zeros of the Bessel functions, i.e.

$$J_\nu(j_{\nu,n}) = 0, \qquad \nu \equiv l + \frac{1}{2}. \tag{9.14}$$

In this way, the solutions of eqn (9.8) are given by

$$f_{l,n}(r) = c \, j_l\left(\frac{j_{\nu,n}}{R} r\right), \tag{9.15}$$

where c is a constant, and the eigenvalues are

$$\Lambda_{l,n} = \left(\frac{j_{\nu,n}}{R}\right)^2. \tag{9.16}$$

At this point we introduce a *mode-generating function* $\Delta_l(k)$, which is defined as a function whose zeros are just the eigenvalues $\Lambda_{n,l}$ (9.16). For Dirichlet boundary conditions, we define

$$\Delta_l^{D,i}(k) = (kR)^{-(2l+1)/2} J_{l+\frac{1}{2}}(kR). \tag{9.17}$$

The superscript "D" indicates Dirichlet boundary conditions and the "i" refers to the interior. The definition of this function is not unique; for instance, it can always be multiplied by an arbitrary constant. We have introduced the factor $(kR)^{-\nu}$ in order to give the function positive real values along the imaginary frequency axis, preserving the finiteness at $k = 0$. Other examples of mode-generating functions will be given in subsequent sections.

Now we insert the eigenvalues $\Lambda_{l,n}$, which are given by eqn (9.16) or can be obtained from a mode-generating function $\Delta_l(k)$, into the regularized expression for the vacuum energy. In zeta function regularization (4.3), we then have

$$E_0(s) = \frac{\mu^{2s}}{2} \sum_{n=1}^{\infty} \sum_{l=0}^{\infty} (2l+1) \left(\Lambda_{l,n} + m^2\right)^{(1-2s)/2}. \tag{9.18}$$

The summation in eqn (4.3) was over a generic index J, which in this case has the form $J = (n, l, M)$. The summation over M can be carried out owing to the azimuthal symmetry of the mode-generating function, and this summation results in the factor $(2l + 1)$.

Equation (9.18) was derived for a massive scalar field. It is clear that such a representation holds for any problem with spherical symmetry in an interior region. In principle, using an asymptotic expansion such as eqn (4.7) for the eigenvalues, one can use such a representation to construct the analytic continuation to $s = 0$. This method is attractive because it is in terms of eigenvalues, i.e. in terms of the physical spectrum. However, it is technically quite complicated and is not used in practice. An easier way to proceed is to transform the sum over n into an integral over the radial momentum using the argument principle (7.60):

$$E_0(s) = \frac{\mu^{2s}}{2} \sum_{l=0}^{\infty} (2l+1) \int_\gamma \frac{dk}{2\pi i} \left(k^2 + m^2\right)^{(1-2s)/2} \frac{\partial}{\partial k} \ln \Delta_l(k). \tag{9.19}$$

Here, a closed contour γ contains all of the zeros of the mode-generating function $k_{l,n} = \sqrt{\Lambda_{l,n}}$. It must be mentioned that the same contour can be chosen for all l without any loss in convergence of the sum over l. The next step is to deform γ towards the imaginary axis as was done in Section 7.4.1. Thus we use the fact that the mode-generating function does not have a pole at $k = 0$, which is ensured by the behavior of the Bessel functions for small argument in eqn (9.17). We assume also that there are no poles in the upper complex half-plane, which holds at least for all mode-generating functions following from the boundary conditions. Finally, we assume $\Delta_l(-k) = \Delta_l(k)$ to hold. With this, we turn the upper half of the contour γ towards the positive imaginary axis and the lower half towards the negative imaginary axis, i.e. make substitutions $k \to \pm ik$

with $k \geq 0$. Then we introduce the function $G_m^{(\alpha)}$, as defined in eqn (2.35), with $\alpha = (1/2) - s$. As a result, eqn (9.19) takes the form

$$E_0(s) = -\frac{\mu^{2s}}{2} \sum_{l=0}^{\infty} (2l+1) \int_m^{\infty} \frac{dk}{2\pi i} \left[G_m^{(\alpha)}(ik) - G_m^{(\alpha)}(-ik) \right] \frac{\partial}{\partial k} \ln \Delta_l(ik). \quad (9.20)$$

Using the equality (2.36), we arrive at

$$E_0(s) = -\mu^{2s} \frac{\cos(\pi s)}{2\pi} \sum_{l=0}^{\infty} (2l+1) \int_m^{\infty} dk \, (k^2 - m^2)^{(1-2s)/2} \frac{\partial}{\partial k} \ln \Delta_l(ik). \quad (9.21)$$

It should be mentioned that this expression is real since $\Delta_l(ik)$ is real. For example, for Dirichlet boundary conditions, from eqn (9.17),

$$\Delta_l^{D,i}(ik) = (kR)^{-(2l+1)/2} I_{l+\frac{1}{2}}(kR) \quad (9.22)$$

follows, where $I_\nu(z)$ is a modified Bessel function related to the usual Bessel functions by $J_\nu(iz) = i^\nu I_\nu(z)$.

The representation (9.21) for the regularized vacuum energy in an interior region is the most convenient form for further calculations. This is because an explicit knowledge of the eigenvalues is not necessary, and the use of the integral representation instead of the sum over eigenvalues is convenient for analytic continuation.

9.1.3 Mode summation for the exterior problem

For the exterior of the sphere, the mode summation method is slightly more complicated since one has, at least partly, a continuous spectrum. This is because the space is unbounded. At the same time, there is a contribution proportional to the volume of the space which is infinite, and which must be eliminated because of the definition of the vacuum energy as the change with respect to empty space. In general, there are several ways to handle this problem. One is to use the local energy density and to integrate it over the space, dropping the unwanted volume contribution. Another way, which we follow here, is to enclose the whole system inside a large sphere and to take the limit where its radius R_c tends to infinity. Of course, the two approaches are equivalent, but in the latter it is not necessary to introduce local quantities.

In a large sphere of radius R_c, the spectrum is discrete and we start from eqn (9.19) with an appropriate mode-generating function $\Delta_l^{R_c}(k)$. We have to impose boundary conditions on that sphere. In the end, the vacuum energy will not depend on the type of these boundary conditions, and we take Dirichlet ones for simplicity. In this way, we return to an interior problem with an additional smaller sphere inside.

The next task is to remove the large sphere, i.e. to consider the limit $R_c \to \infty$. To do so, we consider the general problem of scattering on a hard sphere given by the boundary conditions at $r = R$. The corresponding radial equation for this problem reads

$$\left[-\frac{\partial^2}{\partial r^2} - \frac{2}{r}\frac{\partial}{\partial r} + \frac{l(l+1)}{r^2} + V(r)\right]\phi_{l,k}(r) = k^2\phi_{l,k}(r), \qquad (9.23)$$

with $k \in [0, \infty)$. We take $\phi_{l,k}(r)$ to be the regular scattering solution, i.e. the one which becomes the free solution for a vanishing scattering center ($R \to 0$),

$$\phi_{l,k}(r) \underset{k \to 0}{\sim} j_{l+\frac{1}{2}}(kr). \qquad (9.24)$$

Details of these definitions can be found, for example, in the books by Newton (1966) and Taylor (1972).

For $r \to \infty$, the scattering solution becomes a superposition of incoming and outgoing spherical waves,

$$\phi_{l,k}(r) \underset{r \to \infty}{\sim} \frac{1}{2}\left[f_l(k)h_l^{(2)}(kr) + f_l^*(k)h_l^{(1)}(kr)\right], \qquad (9.25)$$

where

$$h_l^{(1,2)}(z) = \sqrt{\frac{\pi}{2z}}\, H_{l+\frac{1}{2}}^{(1,2)}(z) \qquad (9.26)$$

are the spherical Hankel functions of the first and second kind, which also satisfy eqn (9.10). The coefficients in this superposition are the Jost functions $f_l(k)$ and their complex conjugates. These are uniquely determined by the equation, the boundary conditions, and the potential $V(r)$. In general, the Jost functions are commonly used in potential scattering (for details, see the above-mentioned books). Here we need only their relation to the scattering phase shifts $\delta_l(k)$,

$$\frac{f_l(k)}{f_l^*(k)} = e^{-2i\delta_l(k)}, \qquad (9.27)$$

and some basic properties. The Jost functions are meromorphic functions of the variable k in the upper half of the complex plane. Their continuation to the real axis is a continuous function and the only poles in the upper half-plane are on the imaginary axis (the corresponding solutions describe bound states). In addition, we should mention their reflection property $f_l(-k) = f_l^*(k)$ for real k.

Now we apply the boundary condition on the large sphere to the scattering solution and define the mode-generating function

$$\Delta_l^{R_c}(k) = \phi_{l,k}(R_c). \qquad (9.28)$$

We insert this function into the representation (9.19) of the vacuum energy and, in order to perform the limit $R_c \to \infty$, we divide the integration contour into two parts γ_1 and γ_2, one above the real axis and the other below. For large R_c, we use the asymptotic expression (9.25). Resulting from the asymptotic behavior of the Hankel functions,

$$H^{(1,2)}_{l+\frac{1}{2}}(kR_c) \sim \exp(\pm ikR_c), \qquad (9.29)$$

the primary contribution on γ_1 is

$$\ln \Delta_l^{R_c}(k) = \ln f_l(k) + \ldots, \qquad (9.30)$$

where we have neglected contributions which vanish when $R_c \to \infty$ or which do not depend on R. On the other part, γ_2, we have

$$\ln \Delta_l^{R_c}(k) = \ln f_l^*(k) + \ldots. \qquad (9.31)$$

Now we return the integration path to the real axis. This is possible because the poles have been removed together with the dropped contributions. From eqn (9.19), with the use of eqn (9.27), we get

$$E_0(s) = \frac{\mu^{2s}}{2} \sum_{l=0}^{\infty} (2l+1) \int_0^{\infty} \frac{dk}{\pi} \left(k^2 + m^2\right)^{(1-2s)/2} \frac{\partial}{\partial k} \delta_l(k). \qquad (9.32)$$

From eqn (9.32), we obtain a representation of the vacuum energy for the exterior problem. It is parallel to the representation (9.21) for the interior problem. It has the advantage of being in terms of the physical spectrum and the scattering phase shifts. Its disadvantage is the oscillating behavior of the phase shifts and their complicated asymptotic behavior, which makes both the numerical evaluation and the analytic continuation in s difficult. For these reasons, it is preferable to rotate the integration to the imaginary axis. For that, with eqn (9.27), we return to the Jost functions. In the contribution containing $f_l(k)$, we rotate the γ_1 part of the contour by means of $k \to ik$ and, in the other contribution, we use $f_l^*(k) = f_l(-k)$ and rotate γ_2 in the opposite direction. In this way, eqn (9.32) leads to

$$E_0(s) = -\frac{\mu^{2s}}{2} \sum_{l=0}^{\infty} (2l+1) \int_0^{\infty} \frac{dk}{2\pi i} \left[G_m^{(\alpha)}(ik) - G_m^{(\alpha)}(-ik)\right] \frac{\partial}{\partial k} \ln f_l(ik), \qquad (9.33)$$

where $G_m^{(\alpha)}$ is defined in eqn (2.35) and $\alpha = (1/2)-s$. Then, by using the equality (2.36), we obtain

$$E_0(s) = -\mu^{2s} \frac{\cos(\pi s)}{2\pi} \sum_{l=0}^{\infty} (2l+1) \int_m^{\infty} dk \left(k^2 - m^2\right)^{(1-2s)/2} \frac{\partial}{\partial k} \ln f_l(ik). \qquad (9.34)$$

This is one more representation of the vacuum energy for the exterior region. Formally, it is identical to eqn (9.21) for the interior region. This allows one to unify the notation by defining a mode-generating function $\Delta_l(k)$ for the exterior

region. From eqns (9.34) and (9.21), it follows that we can use the Jost function for this purpose:
$$\Delta_l(k) = f_l(k). \tag{9.35}$$
We remark that this can be multiplied by an arbitrary constant. Now eqn (9.21) is a representation of the regularized vacuum energy valid in both regions, inside and outside the sphere.

9.1.4 Cylindrical symmetry

In this subsection, we collect together the formulas for a cylinder with a circular section. We follow the same method as in the case of a sphere, and elaborate on the differences. We separate the variables in the wave equation in cylindrical coordinates (ρ, φ, z), and represent the solutions for a scalar field in the form
$$\Phi_J(\boldsymbol{r}) = f_{l,n}(\rho)\, e^{il\varphi}\, e^{ik_z z}. \tag{9.36}$$
Now the index $J = (n, l, k_z)$ is composed of the radial index n, the angular momentum $l = 0, \pm 1, \pm 2, \ldots$, and the momentum k_z parallel to the axis of the cylinder. We have translational invariance in the z-direction for both the interior and the exterior problems.

The electromagnetic potential can be represented by the decomposition
$$\boldsymbol{\mathcal{A}}_J(\boldsymbol{r}) = \left[\boldsymbol{E}^{\mathrm{TE}} f^{\mathrm{TE}}_{l,n}(\rho) + \boldsymbol{E}^{\mathrm{TM}} f^{\mathrm{TM}}_{l,n}(\rho)\right] e^{il\varphi}\, e^{ik_z z} \tag{9.37}$$
with the operator polarization vectors
$$\boldsymbol{E}^{\mathrm{TE}} = (\boldsymbol{e}_z \times \boldsymbol{\nabla})\,, \qquad \boldsymbol{E}^{\mathrm{TM}} = \boldsymbol{\nabla} \times (\boldsymbol{e}_z \times \boldsymbol{\nabla})\,. \tag{9.38}$$
Here,
$$\boldsymbol{e}_z = \begin{pmatrix} 0 \\ 0 \\ 1 \end{pmatrix} \tag{9.39}$$
is the unit vector pointing along the axis of the cylinder. These polarization vectors are orthogonal to each other, i.e. $\boldsymbol{E}^{\mathrm{TE}} \cdot \boldsymbol{E}^{\mathrm{TM}} = 0$, and satisfy the equations $\boldsymbol{\nabla} \cdot \boldsymbol{E}^{\mathrm{TE}} = \boldsymbol{\nabla} \cdot \boldsymbol{E}^{\mathrm{TM}} = 0$ and the other properties in eqn (9.6). According to eqn (9.38), the operator polarization vectors $\boldsymbol{E}^{\mathrm{TE}}$ and $\boldsymbol{E}^{\mathrm{TM}}$ have the dimensions 1/cm and 1/cm^2, respectively.

Using eqns (3.13) and (3.47), and the time dependence $\sim \exp(-i\omega t)$ in eqn (3.44), the decomposition of the field strengths is
$$\boldsymbol{E}_J(\boldsymbol{r}) = i\omega \left[\boldsymbol{E}^{\mathrm{TE}} f^{\mathrm{TE}}_{l,n}(\rho) + \boldsymbol{E}^{\mathrm{TM}} f^{\mathrm{TM}}_{l,n}(\rho)\right] e^{il\varphi}\, e^{ik_z z},$$
$$\boldsymbol{B}_J(\boldsymbol{r}) = \left[-\boldsymbol{\nabla}^2 \boldsymbol{E}^{\mathrm{TE}} f^{\mathrm{TM}}_{l,n}(\rho) + \boldsymbol{E}^{\mathrm{TM}} f^{\mathrm{TE}}_{l,n}(\rho)\right] e^{il\varphi}\, e^{ik_z z}. \tag{9.40}$$
Because of
$$\boldsymbol{E}^{\mathrm{TE}} = -\boldsymbol{e}_\rho \frac{1}{\rho}\frac{\partial}{\partial \varphi} + \boldsymbol{e}_\varphi \frac{\partial}{\partial \rho}, \tag{9.41}$$
where $\boldsymbol{e}_\rho$ and $\boldsymbol{e}_\varphi$ are the unit vectors in the radial and azimuthal directions, the amplitude $f^{\mathrm{TE}}_{l,n}(\rho)$ describes an electric field orthogonal to the axis of the cylinder.

By duality, the amplitude $f_{l,n}^{TM}(\rho)$ describes a magnetic induction orthogonal to the axis of the cylinder. The function $f_{l,n}^{TE}(\rho)$ is dimensionless and the function $f_{l,n}^{TM}(\rho)$ has the dimension of length. As in the spherical case, the polarization vectors commute with the Laplacian, and eqns (3.39) and (3.45) result in the radial equation

$$\left(-\frac{\partial^2}{\partial \rho^2} - \frac{1}{\rho}\frac{\partial}{\partial \rho} + \frac{l^2}{\rho^2} + k_z^2\right) f_{l,n,k_z}(\rho) = \Lambda_{l,n}\, f_{l,n,k_z}(\rho). \tag{9.42}$$

The dependence on the momentum k_z can be separated, and by defining $\Lambda_{l,n} = \tilde{\Lambda}_{l,n} + k_z^2$ we get the pure radial equation

$$\left(-\frac{\partial^2}{\partial \rho^2} - \frac{1}{\rho}\frac{\partial}{\partial \rho} + \frac{l^2}{\rho^2}\right) f_{l,n}(\rho) = \tilde{\Lambda}_{l,n}\, f_{l,n}(\rho). \tag{9.43}$$

The vacuum energy takes the form

$$E_0(s) = \frac{\mu^{2s}}{2} \int_{-\infty}^{\infty} \frac{dk_z}{2\pi} \sum_{n=1}^{\infty} \sum_{l=-\infty}^{\infty} \left(\tilde{\Lambda}_{l,n} + k_z^2 + m^2\right)^{(1-2s)/2}, \tag{9.44}$$

which is the general representation for the cylindrical geometry in terms of the discrete eigenvalues $\tilde{\Lambda}_{l,n}$ in the plane perpendicular to the axis of the cylinder.

We proceed in parallel to the spherical case and define mode-generating functions. We consider first the interior region, $\rho \in [0, R]$, and take as an example Dirichlet boundary conditions. The solutions of eqn (9.43) are Bessel functions obeying

$$\left(-\frac{\partial^2}{\partial \rho^2} - \frac{1}{\rho}\frac{\partial}{\partial \rho} + \frac{l^2}{\rho^2}\right) J_l(k\rho) = k^2\, J_l(k\rho), \tag{9.45}$$

where we have chosen functions which are regular at $\rho = 0$. The eigenvalues $\tilde{\Lambda}_{l,n}$ follow as solutions of the boundary conditions:

$$J_l(kR) = 0, \quad k^2 = \tilde{\Lambda}_{l,n} = \frac{j_{l,n}}{R}, \quad n = 1, 2, \ldots. \tag{9.46}$$

The solutions of eqn (9.43) obeying the boundary conditions are

$$f_{l,n}(\rho) = c\, J_l\left(\frac{j_{l,n}}{R}\rho\right), \tag{9.47}$$

where c is some constant. Finally we define the mode-generating function,

$$\Delta_l^{D,i}(k) = (kR)^{-l} J_l(kR), \tag{9.48}$$

again multiplied by a factor which ensures real values along the imaginary frequency axis. We see that in contrast to the spherical case, described by eqn (9.17), the orbital momentum is an integer.

Separation of variables and mode summation

The procedure used to transform the summation over the radial index n into an integral is identical to that in the spherical case, and from eqn (9.44) we arrive at the representation

$$E_0(s) = -\mu^{2s}\frac{\cos(\pi s)}{2\pi}\sum_{l=-\infty}^{\infty}\int_{-\infty}^{\infty}\frac{dk_z}{2\pi}$$
$$\times \int_{\sqrt{k_z^2+m^2}}^{\infty} dk \,(k^2 - k_z^2 - m^2)^{(1-2s)/2}\frac{\partial}{\partial k}\ln\Delta_l(ik). \quad (9.49)$$

Sometimes a mode-generating function does not depend on the momentum k_z. In that case the integration over k_z can be carried out and the representation of the vacuum energy is simplified. For this purpose, we change the order of integrations with respect to k_z and k and introduce a new variable $t = k_z/\sqrt{k^2 - m^2}$. The result is

$$E_0(s) = \mu^{2s}\frac{1}{4\sqrt{\pi}\Gamma\left(s-\frac{1}{2}\right)\Gamma(2-s)}\sum_{l=-\infty}^{\infty}\int_m^{\infty} dk \,(k^2 - m^2)^{1-s}\frac{\partial}{\partial k}\ln\Delta_l(ik). \quad (9.50)$$

A similar formula holds for the exterior region also. However, we first have to define the mode-generating function. For this purpose, we consider the scattering problem in the plane perpendicular to the cylinder,

$$\left(-\frac{\partial^2}{\partial\rho^2} - \frac{1}{\rho}\frac{\partial}{\partial\rho} + \frac{l^2}{\rho^2}\right)\phi_{l,k}(\rho) = k^2\,\phi_{l,k}(\rho). \quad (9.51)$$

The regular scattering solution $\phi_{l,k}(\rho)$ is the one which for $R \to 0$ turns into the free solution. In this case it is the Bessel function $J_l(k\rho)$. For large ρ, it becomes a superposition of incoming and outgoing cylindrical waves,

$$\phi_{l,k}(\rho) \underset{\rho\to\infty}{\sim} \frac{1}{2}\left[f_l(k)\,H_l^{(2)}(k\rho) + f_l^*(k)\,H_l^{(1)}(k\rho)\right], \quad (9.52)$$

and defines the corresponding Jost functions $f_l(k)$. These have the same general properties as in the spherical case. Now we use a large cylinder with Dirichlet boundary conditions at $\rho = R_c$ and define the eigenvalues inside this cylinder by

$$\phi_{l,k}(R_c) = 0, \qquad k^2 = \tilde{\Lambda}_{l,n}. \quad (9.53)$$

The corresponding mode-generating function is given by the formula (9.28) from the preceding subsection with the function $\phi_{l,k}(r)$ defined here in eqn (9.51). Next, we follow the same steps as in the spherical case. We perform the limit $R_c \to \infty$ by separately considering the two parts of the contour γ in a representation parallel to eqn (9.19) and, using eqn (9.29), we arrive at formulas identical to eqns (9.49) and (9.50) with a mode-generating function now defined by the Jost function of the cylindrical scattering problem (9.52),

$$\Delta_l(ik) = f_l(ik). \tag{9.54}$$

Thus we obtain the basic formulas for the calculation of the vacuum energy in the cylindrical geometry.

It should be mentioned that on the level of these formulas, the difference with respect to the spherical geometry is just in the indices of the Bessel functions involved and in the k_z integration. In general, the cylindrical problem is less symmetric than the spherical one. For the scalar problem, in the case of a mode-generating function which is independent of k_z, this difference in symmetry does not show up in the formulas. The same holds for the electromagnetic field with conductor boundary conditions. However, for a dielectric cylinder, the mode-generating function does depend on k_z and, even more importantly, the polarizations are not separable (see Section 9.6.2).

9.2 The scalar Casimir effect for a spherical shell

Studying the Casimir effect for a scalar field on a spherical shell is a useful exercise as the mathematical methods developed are important for both electromagnetic and spinor fields. In this section we consider a scalar field obeying eqn (3.39) and certain boundary conditions. First we consider a massive field and, in the last subsection, the massless case. We collect the relevant formulas together, especially the heat kernel coefficients. Using the representation of the vacuum energy derived in Section 9.1, we construct the analytic continuation of the vacuum energy in zeta function regularization. Then we discuss the relevant models with respect to their renormalization and, finally, represent the known results.

9.2.1 Boundary conditions and mode-generating functions

As discussed in Section 3.2, there are independent Dirichlet [eqn (3.41)] and Neuman [eqn (3.42)] boundary conditions for the scalar field $\Phi_J(\boldsymbol{r})$ which, with eqn (9.1), imply for the radial function $f_{l,n}(r)$ the conditions

$$f_{l,n}(r)|_{r=R} = 0 \quad \text{or} \quad \left. \frac{\partial}{\partial r} f_{l,n}(r) \right|_{r=R} = 0. \tag{9.55}$$

The Robin boundary condition (3.43) implies

$$\left[r \frac{\partial}{\partial r} f_{l,n}(r) + u f_{l,n}(r) \right]_{r=R} = 0, \tag{9.56}$$

where u is a parameter or a function of the radius r. For $u = 0$, eqn (9.56) is the Neumann boundary condition. For $u = 1$, it is the boundary condition which appears in the next section for the TM mode of the electromagnetic field.

In the remaining part of this section, we consider only the Dirichlet and Neumann conditions. It should be mentioned that the separation of variables holds independently for any set of the indices.

In the following, we consider three cases for each set of boundary conditions. These are labeled as follows:

- (i), the interior region of the sphere, $r \in [0, R]$;
- (e), the exterior region of the sphere, $r \in [R, \infty)$;
- (b), both regions together, $r \in [0, \infty)$.

Since the boundaries determined by the conditions (9.55) and (9.56) are impenetrable, i.e. the fields in both regions are completely independent of each other, the corresponding spectral problems are independent. The third case is mathematically the sum of the first two. As a result, the vacuum energy or, more exactly, the regularized vacuum energy of the case (b), is the sum of the energies of the cases (i) and (e). We also note that the heat kernel coefficients for (b) are the sum of the coefficients for (i) and (e). However, regarding the physical interpretation, and especially the renormalization, the third case may behave quite differently and therefore should be considered independently.

As seen in Section 9.1, the solutions of the wave equation are spherical Bessel functions. In the interior region, we take those which are regular at the origin,

$$f_{l,n}(r) = j_l(kr), \tag{9.57}$$

with k following from the boundary conditions. In the Dirichlet case, eqn (9.13) is satisfied and the mode-generating function can be defined by

$$\Delta_l^{D,i}(k) = (kR)^{-\nu} J_\nu(kR), \qquad \nu \equiv l + \frac{1}{2}. \tag{9.58}$$

This is the same as that already discussed in Section 9.1.1. We shall use the notation ν below.

For Neumann boundary conditions, we have to demand

$$[j_l(kR)]' = 0, \tag{9.59}$$

where, here and below, the prime denotes the following differentiation with respect to the argument kr:

$$\left. \frac{dF(kr)}{d(kr)} \right|_{r=R} \equiv [F(kR)]'. \tag{9.60}$$

For this condition, the mode-generating function can be taken in the form

$$\Delta_l^{N,i}(k) = (kR)^{1-l} \left[\frac{1}{\sqrt{kR}} J_\nu(kR) \right]'. \tag{9.61}$$

Again, the factor in front is chosen appropriately using eqns (9.11) and (9.12).

Now we consider the exterior region. Here we have to use the regular scattering solution defined in Section 9.1, which for $r \geq R$ is

$$\phi_{l,k}(r) = \frac{1}{2}\left[f_l(k)h_l^{(2)}(kr) + f_l^*(k)h_l^{(1)}(kr)\right], \tag{9.62}$$

i.e. it coincides with the asymptotic expression (9.25). The Jost functions must be chosen such that eqn (9.62) satisfies the boundary conditions. In the Dirichlet case this is achieved by

$$f_l^D(k) = i\frac{2\sqrt{\pi}}{\Gamma(\nu)}\left(\frac{kR}{2}\right)^{l+1} h_\nu^{(1)}(kR). \tag{9.63}$$

The factors in front are chosen in order to satisfy eqn (9.24), where we have used eqn (9.26) and

$$H_\nu^{(1,2)}(z) \underset{z\to 0}{\approx} \mp i\frac{\Gamma(\nu)}{\pi}\left(\frac{z}{2}\right)^{-\nu}. \tag{9.64}$$

Now, from eqn (9.35), we can define the mode-generating function for the Dirichlet boundary conditions in the exterior region with

$$\Delta_l^{D,e}(k) = (kR)^\nu H_\nu^{(1)}(kR), \tag{9.65}$$

where we have used eqn (9.26) and have omitted some constants.

For the Neumann boundary conditions (9.55), we proceed in the same way. The expression (9.62) satisfies the second equality in eqn (9.55) if the Jost function is chosen to be

$$f_l^N(k) = -i\frac{4\sqrt{\pi}}{(l+1)\Gamma(\nu)}\left(\frac{kR}{2}\right)^{l+2}[h_l^{(1)}(kR)]'. \tag{9.66}$$

From this, we define the mode-generating function by

$$\Delta_l^{N,e}(k) = -(kR)^{l+2}\left[\frac{1}{\sqrt{kR}}H_\nu^{(1)}(kR)\right]', \tag{9.67}$$

which ensures the necessary properties along the imaginary frequency axis.

9.2.2 Analytic continuation for regularized vacuum energy and divergent contributions

In the preceding subsection we derived the mode-generating functions, eqns (9.58), (9.61), (9.65), and (9.67), which are used in the representation (9.21) of the vacuum energy in the zeta function regularization. In this subsection we perform the analytic continuation in the regularization parameter s. In order to simplify the notation, we represent the vacuum energy (9.21) in the form

$$E_0(s) = -\frac{(\mu R)^{2s}}{R}\frac{\cos(\pi s)}{\pi}\sum_{l=0}^\infty \nu \int_{mR}^\infty dk\, [k^2 - (mR)^2]^{(1-2s)/2}\frac{\partial}{\partial k}\ln \tilde{\Delta}_l(ik), \tag{9.68}$$

where we have substituted $k \to k/R$ and introduced the notation $\tilde{\Delta}_l(ik) \equiv \Delta_l(ik/R)$. Note that the mode-generating functions derived in the preceding subsection depend only on the argument kR. Thus, the generating functions $\tilde{\Delta}_l(ik)$ depend only on k, and they are used in this form in eqn (9.68). The explicit form for the functions $\tilde{\Delta}_l(ik)$ follows from the relations for the Bessel functions of imaginary argument

$$J_\nu(iz) = i^\nu I_\nu(z), \qquad H_\nu^{(1)}(iz) = i^{-\nu} \frac{2}{i\pi} K_\nu(z). \qquad (9.69)$$

The results are displayed in Table 9.1.

The need to perform the analytic continuation follows from the fact that for a decreasing regularization parameter s, the sum and the integral in eqn (9.68) become divergent. Since the logarithm of the mode-generating function is a complicated object, a direct analytic continuation is impossible. A way out is to use the uniform asymptotic expansion of that logarithm for large ν and k with fixed $z \equiv k/\nu$. This expansion can be constructed using the uniform asymptotic expansion of the modified Bessel functions in the following general form:

$$\ln \tilde{\Delta}_l(ik) \simeq \nu D_{-1}(z) + D_0(z) + \frac{1}{\nu} D_1(z) + \ldots \qquad (9.70)$$

[the functions $D_i(z)$ are discussed below]. Now we define the function

$$h_{as}(\nu, z) = \sum_{i=-1}^{3} \frac{D_i(z)}{\nu^i} \qquad (9.71)$$

from the first five terms of the expansion (9.70). Then, in eqn (9.68), we first subtract and then add $h_{as}(\nu, z)$ from the $\tilde{\Delta}_l(ik)$, such that we obtain two parts of the regularized vacuum energy:

$$E_0(s) = E_0^{\text{fin}} + E_0^{\text{as}}(s). \qquad (9.72)$$

Here, the *finite part* is

$$E_0^{\text{fin}} = -\frac{1}{\pi R} \sum_{l=0}^{\infty} \nu \int_{mR}^{\infty} dk \sqrt{k^2 - (mR)^2} \frac{\partial}{\partial k} \left[\ln \tilde{\Delta}_l(ik) - h_{as}(\nu, z) \right], \qquad (9.73)$$

where we have put $s = 0$. The *asymptotic part* is given by

TABLE 9.1. The mode-generating functions $\tilde{\Delta}_l(ik)$ for a scalar field.

Region	Dirichlet boundary condition	Neumann boundary condition
Interior	$\tilde{\Delta}_l^{D,i}(ik) = k^{-\nu} I_\nu(k)$	$\tilde{\Delta}_l^{N,i}(ik) = k^{1-l} \left[I_\nu(k)/\sqrt{k} \right]'$
Exterior	$\tilde{\Delta}_l^{D,e}(ik) = k^\nu K_\nu(k)$	$\tilde{\Delta}_l^{N,e}(ik) = -k^{2+l} \left[K_\nu(k)/\sqrt{k} \right]'$

$$E_0^{\text{as}}(s) = -\frac{(\mu R)^{2s}}{R} \frac{\cos(\pi s)}{\pi} \sum_{l=0}^{\infty} \nu \int_{mR}^{\infty} dk \left[k^2 - (mR)^2 \right]^{(1-2s)/2} \frac{\partial}{\partial k} h_{\text{as}}(\nu, z). \tag{9.74}$$

In fact, the finiteness of E^{fin} imposes some conditions on $h_{\text{as}}(\nu, z)$. Note that it is not defined uniquely. For example, it is possible to include more than five terms in the expansion (9.71). This would be unnecessary but might be useful for numerical evaluations. Also, it is possible to obtain the expansion in terms of the orbital momentum l instead of ν. However, it can be shown that this leads to unnecessarily complicated intermediate steps. In general, the finite part can be calculated only numerically. This is, however, an easy task. We finish the discussion of the finite part with the remark that for the numerical evaluation, it is useful to integrate by parts and to use the representation

$$E_0^{\text{fin}} = \frac{1}{\pi R} \sum_{l=0}^{\infty} \nu \int_{mR}^{\infty} \frac{dk\, k}{\sqrt{k^2 - (mR)^2}} \left[\ln \tilde{\Delta}_l(ik) - h_{\text{as}}(\nu, z) \right]. \tag{9.75}$$

Surface contributions do not appear provided that, in $h_{\text{as}}(\nu, k)$, all constant contributions, shown below in Table 9.2, are preserved.

In order to treat the asymptotic part $E_0^{\text{as}}(s)$ of the vacuum energy, we need the uniform asymptotic expansions of the mode-generating functions. These follow from the corresponding expansions of the modified Bessel functions (Abramowitz and Stegun 1972),

$$\left.\begin{array}{c} I_\nu(\nu z) \\ K_\nu(\nu z) \end{array}\right\} \simeq \frac{\pi^{\mp 1/2}}{\sqrt{2\nu}} \frac{1}{(1+z^2)^{1/4}} e^{\pm \nu \eta(z)} \left[1 + \sum_{k=1}^{\infty} (\pm 1)^k \frac{u_k(t)}{\nu^k} \right], \tag{9.76}$$

$$\left.\begin{array}{c} I'_\nu(\nu z) \\ K'_\nu(\nu z) \end{array}\right\} \simeq \pm \frac{\pi^{\mp 1/2}}{\sqrt{2\nu}} \frac{(1+z^2)^{1/4}}{z} e^{\pm \nu \eta(z)} \left[1 + \sum_{k=1}^{\infty} (\pm 1)^k \frac{v_k(t)}{\nu^k} \right]$$

with the following notation:

$$\eta(z) = \sqrt{1+z^2} + \ln \frac{z}{1+\sqrt{1+z^2}}, \qquad t = \frac{1}{\sqrt{1+z^2}}. \tag{9.77}$$

The Debye polynomials $u_k(t)$ and $v_k(t)$ are given by the recurrence relations

$$u_0(t) = 1, \qquad v_0(t) = 1, \tag{9.78}$$

$$u_{k+1}(t) = \frac{1}{2} t^2 (1-t^2) u'_k(t) + \frac{1}{8} \int_0^t d\tau\, (1 - 5\tau^2) u_k(\tau), \qquad k = 0, 1, \ldots,$$

$$v_k(t) = u_k(t) - t(1-t^2) \left[\frac{1}{2} u_{k-1}(t) + t u'_{k-1}(t) \right], \qquad k = 1, 2, \ldots.$$

Inserting eqn (9.76) into the $\tilde{\Delta}_l(ik)$ given in Table 9.1 and expanding the logarithm, we find the functions $D_i(z)$. For $i = -1$, the function $D_{-1}(z)$ is common to both the Dirichlet and the Neumann boundary conditions. It is given by

$$D_{-1}(z) = \pm [\eta(z) - \ln z] \tag{9.79}$$

for the interior region (+) and the exterior region (−). The functions $D_0(z)$ are shown in Table 9.2. Starting from $i = 1$, the $D_i(z)$ are polynomials in t and can be represented in the form

$$D_i(z) = \sum_{a=i}^{3i} x_{i,a} t^a, \quad i = 1, 2, 3. \tag{9.80}$$

The coefficients $x_{i,a}$ are real numbers shown in Tables 9.3 and 9.4. The symmetry between the interior and exterior cases results in the signs shown. A symmetry between the Dirichlet and Neumann boundary conditions is observed only in $D_{-1}(z)$ and $D_0(z)$.

For future discussion, it is useful to define individual contributions to the asymptotic part of the vacuum energy by

$$E_0^{\mathrm{as}}(s) = \sum_{i=-1}^{3} A_i(s), \tag{9.81}$$

where

$$A_i(s) = -\frac{(\mu R)^{2s}}{R} \frac{\cos(\pi s)}{\pi} \sum_{l=0}^{\infty} \nu \int_{mR}^{\infty} dk \left[k^2 - (mR)^2\right]^{(1-2s)/2} \frac{\partial}{\partial k} \frac{D_i(z)}{\nu^i}. \tag{9.82}$$

It should be mentioned that the analytic continuation requires careful treatment in the complex plane. Details are presented by Bordag et al. (1997). First we consider the Dirichlet boundary condition in the interior region. The results for $i = -1, 0$ are

$$A_{-1}(s) = -\frac{(\mu R)^{2s}}{R} \frac{1}{4\sqrt{\pi} \Gamma\left(s - \frac{1}{2}\right)} \tag{9.83}$$
$$\times \sum_{j=0}^{\infty} \frac{(-i)^j}{j!} \frac{2\Gamma(s+j-1)}{1-2s-2j} (mR)^{2j} \zeta\left(2s + 2j - 3, \frac{1}{2}\right),$$

$$A_0(s) = -\frac{(\mu R)^{2s}}{R} \frac{1}{4\Gamma\left(s - \frac{1}{2}\right)} \tag{9.84}$$

TABLE 9.2. The functions $D_0(z)$ entering the asymptotic expansion $h_{\mathrm{as}}(\nu, z)$ in eqn (9.71).

Region	Dirichlet boundary condition	Neumann boundary condition
Interior	$\frac{1}{2}\ln t - \frac{1}{2}\ln(2\pi\nu) - \nu \ln \nu$	$-\frac{1}{2}\ln t + \frac{1}{2}\ln \frac{\pi}{2\nu} - \nu \ln \nu$
Exterior	$\frac{1}{2}\ln t + \frac{1}{2}\ln \frac{\pi}{2\nu} + \nu \ln \nu$	$-\frac{1}{2}\ln t + \frac{1}{2}\ln \frac{\pi\nu}{2} + \nu \ln \nu$

TABLE 9.3. The coefficients $x_{i,a}$ for Dirichlet boundary conditions. The upper sign is for the interior region and the lower sign is for the exterior region.

					a				
i	1	2	3	4	5	6	7	8	9
1	$\pm\frac{1}{8}$	0	$\mp\frac{5}{24}$	0	0	0	0	0	0
2	0	$\frac{1}{16}$	0	$-\frac{3}{8}$	0	$\frac{5}{16}$	0	0	0
3	0	0	$\pm\frac{25}{384}$	0	$\mp\frac{531}{640}$	0	$\pm\frac{221}{128}$	0	$\mp\frac{1105}{1152}$

TABLE 9.4. The coefficients $x_{i,a}$ for Neumann boundary conditions. The upper sign is for the interior region and the lower sign is for the exterior region.

					a				
i	1	2	3	4	5	6	7	8	9
1	$\mp\frac{7}{8}$	0	$\pm\frac{7}{24}$	0	0	0	0	0	0
2	0	$-\frac{9}{16}$	0	$\frac{7}{8}$	0	$-\frac{7}{16}$	0	0	0
3	0	0	$\mp\frac{199}{384}$	0	$\pm\frac{1349}{640}$	0	$\mp\frac{371}{128}$	0	$\pm\frac{1463}{1152}$

$$\times \sum_{j=0}^{\infty} \frac{(-1)^j}{j!} \Gamma\left(s+j-\frac{1}{2}\right) (mR)^{2j} \zeta\left(2s+2j-2, \frac{1}{2}\right).$$

For $i = 1, 2, 3$, the contributions $A_i(s)$ are

$$A_i(s) = -\frac{(\mu R)^{2s}}{R} \sum_{a=i}^{3i} \frac{x_{i,a}}{\Gamma\left(\frac{a}{2}\right)\Gamma\left(s-\frac{1}{2}\right)} \qquad (9.85)$$
$$\times \sum_{j=0}^{\infty} \frac{(-1)^j}{j!} \Gamma\left(s+j+\frac{a-1}{2}\right) (mR)^{2j} \zeta\left(2s+2j+i-2, \frac{1}{2}\right),$$

where $\zeta(z, q)$ is the Hurwitz zeta function defined in eqn (7.37). It results from a summation over the orbital momenta.

With these formulas, the functions $A_i(s)$ can be represented as a power series in mR, which can be shown to converge for $mR < 1$. There is also an alternative representation which is valid for all values of mR (Bordag et al. 1997). From the representations (9.83)–(9.85), the analytic continuation of E_0^{as}, eqn (9.81), to $s = 0$ can be found easily. We have poles in $E_0^{\text{as}}(s)$ originating from the poles of the gamma functions and the Hurwitz zeta functions, and the corresponding finite parts. The finite parts result in rather cumbersome formulas, so here we restrict ourselves to the pole part,

$$E_0^{\text{as}}(s) = \frac{1}{s\pi R}\left[\frac{1}{630} + \frac{1}{12}(mR)^2 - \frac{1}{48}(mR)^4\right] + O(s^0). \tag{9.86}$$

We note that these formulas have been derived for Dirichlet boundary conditions in the interior region. Using eqn (9.79) and Tables 9.2–9.4, similar formulas can be derived for the other cases also.

Since the finite part of the vacuum energy E_0^{fin}, eqn (9.75), by construction does not have a pole contribution, eqn (9.86) is the complete pole contribution of the vacuum energy in zeta function regularization. In fact, eqn (9.86) coincides with the pole part in $E_0^{\text{div}}(s)$, eqn (4.30), with the corresponding heat kernel coefficients inserted from eqn (4.25). It is also possible to calculate the heat kernel coefficients starting from $E_0^{\text{as}}(s)$ in eqn (9.74), as is done below in eqn (9.131). For instance, the coefficients for the outside region follow from those for the inside region by multiplication by $(-1)^{k+1}$, where $k \neq 0$.

9.2.3 The renormalized vacuum energy for a massive scalar field

In this subsection, we consider the vacuum energy for a massive scalar field with boundary conditions on a sphere. We shall concentrate on the mass dependence, and study some general properties and various contributions to the vacuum energy. This gives a deeper understanding of the structure of the vacuum energy and puts the case of massless fields into a broader context.

We start by dividing the renormalized vacuum energy (4.57) into two parts,

$$E_0^{\text{ren}} = E_0^{\text{fin}} + E_0^{\text{an}}, \tag{9.87}$$

where E_0^{fin} is the finite part, given in eqn (9.73), and the *analytical part* is defined by

$$E_0^{\text{an}} = \lim_{s\to 0}\left[E_0^{\text{as}}(s) - E_0^{\text{div}}(s)\right]. \tag{9.88}$$

The asymptotic part is contained in eqn (9.86) and the divergent part in eqn (4.30), with the corresponding heat kernel coefficients inserted. The limit $s\to 0$ in eqn (9.88) is finite since the ultraviolet divergences are contained in $E_0^{\text{as}}(s)$ and they are subtracted out by means of $E_0^{\text{div}}(s)$.

With eqn (9.87), we have a representation of the renormalized vacuum energy which is suitable for a discussion of its general properties and for numerical evaluation. First we consider some general properties related to the dependence on the mass m. These properties can be formulated in terms of the heat kernel coefficients. For large m, we use the property of the heat kernel formalism that it provides an asymptotic expansion in the inverse powers of the mass. This can be seen directly from eqn (4.29). With our renormalization prescription (4.57), we have subtracted all contributions containing nonnegative powers of the mass such that E_0^{ren} vanishes for $m\to\infty$. In fact, this is the *large-mass normalization condition* discussed in Section 4.3.1. Therefore, the leading order for $m\to\infty$ in E_0^{ren} comes from the first nonvanishing heat kernel coefficient $a_{k/2}$ with $k\geq 5$,

$$E_0^{\text{ren}} \underset{m\to\infty}{\sim} -\frac{\Gamma(\frac{k}{2}-2)}{32\pi^2}\frac{a_{k/2}}{m^{2k-4}}. \tag{9.89}$$

For a sphere, the first nonvanishing coefficient is $a_{5/2}$. It was first calculated by Kennedy (1978) [see also Bordag et al. (1996a)]:

$$a_{5/2} = \frac{\pi^{3/2}}{120R^2} \quad \text{and} \quad a_{5/2} = \frac{47\pi^{3/2}}{60R^2} \tag{9.90}$$

for Dirichlet and Neumann boundary conditions, respectively. It is the same for both interior and exterior regions. As a consequence, in all of the cases considered in Section 9.2.1 this coefficient is nonzero and the vacuum energy decreases according to

$$E_0^{\text{ren}} \underset{m \to \infty}{\sim} -\frac{a_{5/2}}{32\pi^{3/2}m}. \tag{9.91}$$

Inserting eqn (9.90) here, we obtain the behavior of E_0^{ren} for $R \to \infty$:

$$E_0^{\text{ren}} \sim -\frac{1}{3840mR^2} \quad \text{and} \quad E_0^{\text{ren}} \sim -\frac{47}{1920mR^2} \tag{9.92}$$

for Dirichlet and Neumann boundary conditions, respectively.

The behavior for the opposite case, i.e. for $m \to 0$, is determined by $E_0^{\text{div}}(s)$ which is subtracted in the process of renormalization. This is because the regularized energy $E_0(s)$ has a finite limit at $m \to 0$ and this limit can be interchanged with the limit at $s \to 0$. In $E_0^{\text{div}}(s)$, given in eqn (4.30), at $m \to 0$ the logarithmic contribution dominates such that the renormalized vacuum energy becomes

$$E_0^{\text{ren}} \underset{m \to 0}{\sim} -\frac{\ln(mR)}{16\pi^2} a_2. \tag{9.93}$$

Therefore, this limit is determined by the heat kernel coefficient a_2. It must be mentioned that this is a result of our normalization prescription (4.62). However, if we were to change this prescription by excluding the logarithmic contribution from $E_0^{\text{div}}(s)$, which would be equivalent to a finite renormalization, the renormalized vacuum energy would diverge logarithmically at $m \to \infty$. In this way, the renormalized vacuum energy, as a function of m, can never be finite in both limits, i.e. for both large and small masses. Here, it is assumed that the heat kernel coefficient a_2 is nonzero, otherwise the logarithmic contribution is absent and E_0^{ren} has a finite limit for $m \to 0$. An example of this is the case (b), where the interior and exterior regions are considered together.

In eqn (9.93), the radius R appears in the argument of the logarithm. It enters automatically for dimensional reasons, since a_2 does not depend on m and the argument of the logarithm must be dimensionless. Therefore this formula gives the dependence of the renormalized vacuum energy on the radius for $R \to 0$.

Now, having studied the general properties of the renormalized vacuum energy, we consider the two parts in eqn (9.87) separately. The finite part is given by eqn (9.73) or (9.75), where the latter is more convenient for numerical evaluation. Since the mode-generating functions and the asymptotic part h_{as} are given explicitly or in terms of Bessel functions, the numerical evaluation might appear

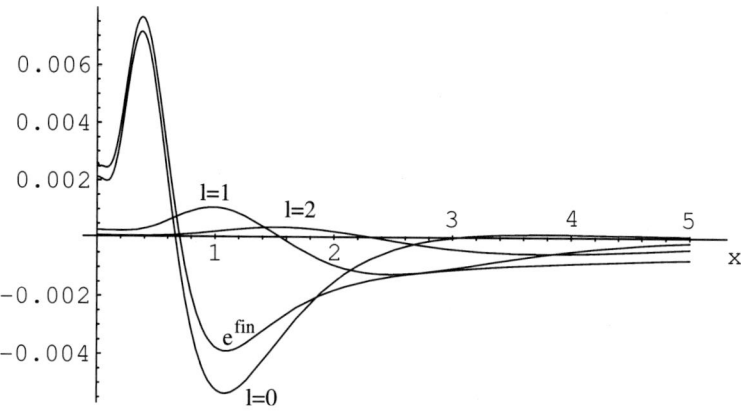

FIG. 9.1. The function $e^{\mathrm{fin}}(x)$ and the contributions to it from the first three orbital momenta entering eqn (9.94).

simple. However, in taking the difference between the logarithm of the mode-generating function and its asymptotic expansion, a small value results from the difference of two large terms in some regions of parameter space, leading to a loss of numerical precision. Also, the convergence of the orbital-momentum sum goes as $\sum \nu^{-2}$, so a precise numerical evaluation is not trivial.

We take as an example the Dirichlet boundary condition in the interior region and represent eqn (9.75) in the form

$$E_0^{\mathrm{fin}} = \frac{e^{\mathrm{fin}}(mR)}{R} \equiv \frac{1}{R} \sum_{l=0}^{\infty} e^{\mathrm{f}}_{l+\frac{1}{2}}(mR) \qquad (9.94)$$

in terms of a dimensionless function $e^{\mathrm{fin}}(mR)$ and its contributions $e^{\mathrm{fin}}_\nu(mR)$ from individual orbital momenta. These functions have a finite value for $mR = 0$ and decrease for $mR \to \infty$. Examples are shown in Fig. 9.1 as functions of the argument $x = mR$. It is seen that the dominating contribution comes from $l = 0$, i.e. from the s-wave, in accordance with expectations.

Finally we consider the analytical part E_0^{an}, given in eqn (9.88). An explicit expression can be obtained from eqn (9.81) using eqns (9.83) and (9.85) and subtracting the divergent part of the vacuum energy $F_0^{\mathrm{div}}(s)$. As already mentioned and as can be seen explicitly from eqn (9.74), the pole contributions cancel and the limit $s \to 0$ in eqn (9.88) is finite. From the above, we obtain E_0^{an} as a convergent series, which is an alternative representation (Bordag et al. 1997). Since the corresponding formulas are too cumbersome to be displayed here, we restrict ourselves to a graphical representation for Dirichlet boundary conditions in the interior region. For this purpose, in addition to eqn (9.94), we introduce also the dimensionless functions $e^{\mathrm{ren}}(mR)$ and $e^{\mathrm{an}}(mR)$:

$$E_0^{\mathrm{ren}} = \frac{e^{\mathrm{ren}}(mR)}{R}, \qquad E_0^{\mathrm{an}} = \frac{e^{\mathrm{an}}(mR)}{R}. \qquad (9.95)$$

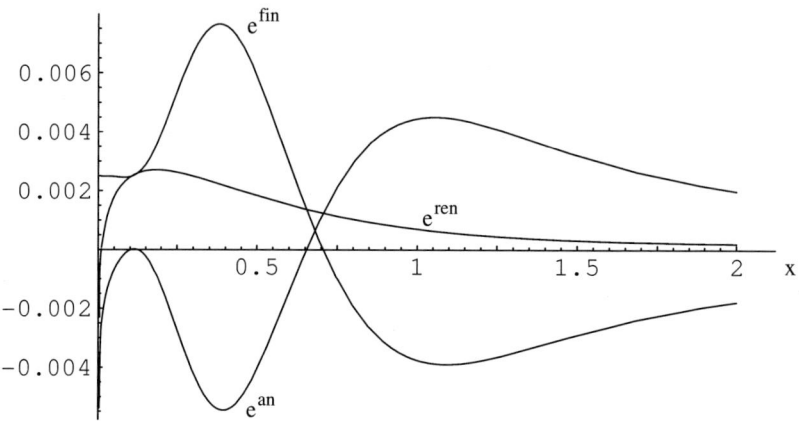

FIG. 9.2. The function $e^{\text{ren}}(x)$ and the contributions to it $e^{\text{fin}}(x)$ and $e^{\text{an}}(x)$.

These functions are plotted versus $x = mR$ in Fig. 9.2. For $x \to \infty$, all of them decrease no more slowly than what is given by eqn (9.92). For $x \to 0$, the functions $e^{\text{an}}(mR)$ and $e^{\text{ren}}(mR)$ diverge according to eqn (9.93), whereas $e^{\text{fin}}(mR)$ has a finite limit. It should be mentioned that this behavior depends on the region considered (interior, exterior, or both). The logarithmic behavior is present for the interior and the exterior cases considered separately, as the coefficient a_2 is nonzero. In contrast, if one considers both regions together, i.e. the case (b) defined in Section 9.2.1, the situation is different. Since the coefficients a_2 in the interior and exterior regions differ by a sign only, the logarithmic contributions cancel in the sum of the energies, and the vacuum energy has a finite limit at $m \to 0$.

From Fig. 9.2, the relative contributions of the finite and analytic parts can be observed. For small x, the latter dominates. For large x, both are of the same order and partly compensate each other. In the intermediate region, they are mostly of the same order.

In this subsection we considered, as an example, a Dirichlet boundary condition in the interior region. Using the methods and formulas discussed here, other cases can also be considered (Bordag et al. 1997, Bordag et al. 2001a, Kirsten 2000). The e^{ren} obtained are smooth functions similar to the e^{ren} in Fig. 9.2, interpolating between the asymptotic behaviors for small and large mR.

9.2.4 The vacuum energy for a massless scalar field

In contrast to the massive scalar field considered in the preceding subsection, here we lack any a priori normalization condition. Therefore, as explained in Section 4.3, we choose a classical model whose parameters will accommodate a renormalization. Before considering such a model, we remark that the boundary conditions (9.55) do not introduce any dimensional parameter. Hence the vacuum energy in zeta function regularization (4.3) or (9.68) can be represented as

$$E_0(s) = \frac{(\mu R)^{2s}}{R} e_0(s), \tag{9.96}$$

where $e_0(s)$ is a dimensionless function depending only on s. On the other hand, from the heat kernel expansion and from eqn (4.28) for $m=0$, the pole contribution is

$$e_0(s) = -\frac{a_2 R}{32\pi^2}\frac{1}{s} + \tilde{e}_0 + O(s), \tag{9.97}$$

where we have introduced the notation $\tilde{e}_0$ for that part of $e_0(s)$ which is independent of s. We note that for dimensional reasons $a_2 R$ is a number [see eqn (4.25)]. Combining (9.96) and (9.97), and using $(\mu R)^{2s} \approx 1 + 2s\ln(\mu R)$, we get

$$E_0(s) = -\frac{a_2}{32\pi^2}\left[\frac{1}{s} + 2\ln(\mu R)\right] + \frac{\tilde{e}_0}{R} + O(s). \tag{9.98}$$

Now we consider the classical model (4.65) together with the vacuum energy (9.98). For the needs of renormalization, it is sufficient to keep only the last term in eqn (4.65). Then, for the total energy, we get

$$E^{\text{tot}} = \frac{h_3}{R} - \frac{a_2}{32\pi^2}\left[\frac{1}{s} + 2\ln(\mu R)\right] + \frac{\tilde{e}_0}{R} + O(s). \tag{9.99}$$

Obviously, we can accommodate the ultraviolet divergence by a redefinition of the parameter h_3

$$h_3 \to h_3^{\text{ren}} = h_3 - \frac{a_2 R}{32\pi^2 s}. \tag{9.100}$$

Here, we are left with the remaining, finite renormalization of h_3^{ren}. From this it follows that the contribution $\tilde{e}_0/R$, which one could naively consider as the vacuum energy left after removing the ultraviolet divergence, is indistinguishable from the classical contribution h_3^{ren}/R. In this way, the only meaningful part of the vacuum energy is that containing the logarithm of R in eqn (9.99). It should be mentioned that this is a rather trivial contribution, since it depends only on the heat kernel coefficient. We conclude this discussion with the remark that any further attempt to give the energy (9.99) a physical meaning requires additional information from the classical model considered, i.e. from outside the pure quantum part.

The above considerations have assumed a nonvanishing heat kernel coefficient $a_2 \neq 0$. This holds for all cases, i.e. for Dirichlet and Neumann boundary conditions in both the interior and the exterior region. In the case of the sum of the two regions, $a_2 = 0$. In Section 9.4, we shall have another example of a vanishing a_2. In such cases, the vacuum energy in zeta function regularization is finite. More exactly, it has a finite continuation to $s = 0$ and no renormalization is required. Also, the arbitrariness arising from the parameter μ drops out. A similar situation holds in the cutoff regularization (4.32). In the presence of boundary conditions, for symmetry reasons, it also follows from $a_2 = 0$ that

$a_1 = 0$. Then one is left with the divergent contribution from $a_{1/2}$, which is proportional to the surface area of the sphere. In this case one could consider the vacuum energy together with the classical model and remove the divergence by a redefinition of the surface tension σ in eqn (4.65).

Thus, there is an important difference between the cases defined in Section 9.2.1. For (i) or (e), because $a_2 \neq 0$, only the logarithmic contribution from the vacuum energy can have a physical meaning. As regards the case (b), where $a_2 = 0$, the vacuum energy

$$E_0^{\text{ren}} = \frac{\tilde{e}_0}{R} \tag{9.101}$$

can be considered as having a physical meaning, similar to the Casimir interaction energy for two parallel ideally conducting planes. This meaning, however, should be understood only in the context of the total energy (9.99). Note that the scalar field considered here does not have a direct physical application. It is merely the simplest model for study. However, the conclusions connected with the renormalization might hold for other fields also.

The calculation of the vacuum energy (9.72) in the massless case proceeds along the same lines as in the massive case discussed in the preceding subsection. The corresponding results can be obtained by putting $m = 0$ in formulas such as eqns (9.75) and (9.83)–(9.85). For $m = 0$, these formulas simplify in such a way that it is instructive to show them in some detail. The starting point is the separation (9.72) of the vacuum energy into finite and asymptotic parts. Whereas the calculation of the finite part does not change much, the asymptotic part simplifies. We define, as before, the asymptotic part $E_0^{\text{as}}(s)$ by eqn (9.81). The functions $A_i(s)$ in eqn (9.82) now read

$$A_i(s) = -\frac{(\mu R)^{2s}}{R} \frac{\cos(\pi s)}{\pi} \sum_{l=0}^{\infty} \nu^{2-2s} \int_0^{\infty} dz\, z^{1-2s} \frac{\partial}{\partial z} \frac{D_i(z)}{\nu^i}, \tag{9.102}$$

where we have substituted the variable of integration using $k = \nu z$. We see that the dependence on the orbital momentum factorizes so that the sum over l can be carried out directly using eqn (7.37). The remaining integrations are simple. Using

$$\int_0^{\infty} dz\, z^{1-2s} \frac{\partial}{\partial z} [\eta(z) - \ln z] = -\frac{\Gamma\left(\frac{1}{2} - s\right) \Gamma(s-1)}{4\sqrt{\pi}},$$

$$\int_0^{\infty} dz\, z^{1-2s} \frac{\partial}{\partial z} \ln t = \frac{\pi}{2\cos(\pi s)}, \tag{9.103}$$

$$\int_0^{\infty} dz\, z^{1-2s} \frac{\partial}{\partial z} t^a = -\frac{\Gamma\left(\frac{3}{2} - s\right) \Gamma\left(s + \frac{a-1}{2}\right)}{\Gamma\left(\frac{a}{2}\right)},$$

we get the functions

$$A_{-1}(s) = \frac{(\mu R)^{2s}}{R} \frac{1}{4\sqrt{\pi}} \frac{\Gamma(s-1)}{\Gamma\left(s + \frac{1}{2}\right)} \zeta\left(2s - 3, \frac{1}{2}\right),$$

$$A_0(s) = -\frac{(\mu R)^{2s}}{4R}\zeta\left(2s-2,\frac{1}{2}\right), \tag{9.104}$$

$$A_i(s) = -\frac{(\mu R)^{2s}}{R}\sum_{a=i}^{3i}\frac{x_{i,a}}{\Gamma\left(\frac{a}{2}\right)}\frac{\Gamma\left(s+\frac{a-1}{2}\right)}{\Gamma\left(s-\frac{1}{2}\right)}\zeta\left(2s-2+i,\frac{1}{2}\right), \quad i=1,2,3.$$

As discussed above, these formulas also follow from eqns (9.83)–(9.85) for $m = 0$.

From this representation, using eqn (9.81) and Table 9.3 or 9.4, we immediately get the asymptotic part of the vacuum energy. For example, for Dirichlet boundary conditions in the interior region, this part is

$$E_0^{\text{as,D}}(s) = \frac{1}{630\pi R}\left[\frac{1}{s} + 2\ln(\mu R) + \frac{1385}{384} - \frac{229\gamma}{64} - \frac{561}{64}\ln 2\right.$$
$$\left. - \frac{315}{8}\zeta_R'(-1) + \frac{2205}{8}\zeta_R'(-3)\right] + O(s), \tag{9.105}$$

where γ is Euler's constant. The corresponding formula for Neumann boundary conditions is

$$E_0^{\text{as,N}}(s) = -\frac{35}{630\pi R}\left[\frac{1}{s} + 2\ln(\mu R) - \frac{1411}{384} + \frac{451\gamma}{320} + \frac{267}{64}\ln 2\right.$$
$$\left. - \frac{63}{8}\zeta_R'(-1) - \frac{63}{8}\zeta_R'(-3)\right] + O(s). \tag{9.106}$$

For the exterior region, we get the same expressions with an opposite sign. This is due to the symmetry between the asymptotic expansions of the Bessel functions involved and the equality $\zeta(0, 1/2) = \zeta(-2, 1/2) = 0$. As a consequence, in the case (b) the asymptotic part of the vacuum energy is zero.

Therefore, the remaining task is to calculate E_0^{fin}, which is most conveniently done using eqn (9.75) in the form

$$E_0^{\text{fin}} = \frac{1}{\pi R}\sum_{l=0}^{\infty}\nu^2\int_0^{\infty}dz\left[\ln\tilde{\Delta}_l(i\nu z) - h_{\text{as}}(\nu, z)\right]. \tag{9.107}$$

This can only be performed numerically. Note that one might attempt to calculate E_0^{fin} analytically from eqn (9.73) with $m = 0$ and eqns (9.70) and (9.71). This leads to a series of the type

$$E_0^{\text{fin}} \simeq -\frac{1}{\pi R}\sum_{l=0}^{\infty}\nu^2\int_0^{\infty}dz\, z\frac{\partial}{\partial z}\sum_{i=3}^{\infty}\frac{D_i(z)}{\nu^i}, \tag{9.108}$$

which does not converge and, thus, cannot be used for the calculation of E_0^{fin}.

TABLE 9.5. Zeta functions and the Casimir energy of a massless scalar field for a spherical shell with Dirichlet boundary conditions.

d	ζ_P interior	ζ_P exterior	$RE_0(s)$
2	$+0.0098540$	-0.0084955	$+0.0006792$
	$-0.0039062/s$	$-0.0039062/s$	$-0.0039062/s$
3	$+0.0088920$	-0.0032585	$+0.0028168$
	$+0.0010105/s$	$-0.0010105/s$	
4	-0.0017939	$+0.0004544$	-0.0006698
	$+0.0002670/s$	$+0.0002670/s$	$+0.0002670/s$
5	-0.0009450	$+0.0003739$	-0.0002856
	$-0.0001343/s$	$+0.0001343/s$	
6	$+0.0002699$	-0.0000611	$+0.0001044$
	$-0.0000335/s$	$-0.0000335/s$	$-0.0000335/s$
7	$+0.0001371$	-0.0000555	$+0.0000408$
	$+0.0000214/s$	$0.0000214/s$	
8	-0.0000457	$+0.0000101$	-0.0000178
	$+5.228\times 10^{-6}/s$	$+5.228\times 10^{-6}/s$	$+5.228\times 10^{-6}/s$
9	-0.0000230	$+0.0000094$	-0.0000068
	$-3.769\times 10^{-6}/s$	$+3.769\times 10^{-6}/s$	

Calculations of E_0^{fin} and E_0^{as} have been done repeatedly. The best compilation has been given by Kirsten (2000). The results are represented in terms of the generalized zeta function, which is related to the vacuum energy by means of

$$\frac{1}{R}\zeta_P\left(s-\frac{1}{2}\right) = 2E_0(s). \tag{9.109}$$

Computations were performed in d-dimensional space. This is instructive since the vacuum energy, especially its divergences, is different in odd and even dimensions. Such calculations were pioneered by Bender and Milton (1994). The results for Dirichlet boundary conditions are presented in Table 9.5 (Kirsten 2000). The regular contribution to the zeta function ζ_P and its pole part at $s = 0$ are presented for the interior region (first column) and the exterior region (second column). The third column contains the vacuum energy $RE_0(s)$ for both regions taken together. In odd dimensions, the latter is a finite quantity. Similar results for Neumann boundary conditions are presented Table 9.6 (Kirsten 2000).

For the case of $d = 3$ spatial dimensions, the vacuum energies of the scalar field in the interior and exterior regions taken together are

$$E^D(R) \equiv E_0^{D,\text{ren}} = \frac{0.0028168}{R}, \qquad E^N(R) \equiv E_0^{N,\text{ren}} = -\frac{0.2238216}{R}, \tag{9.110}$$

TABLE 9.6. Zeta functions and the Casimir energy of a massless scalar field for a spherical shell with Neumann boundary conditions.

d	ζ_P interior	ζ_P exterior	$RE_0(s)$
2	-0.3446767	-0.0215672	-0.1831220
	$-0.0195312/s$	$-0.0195312/s$	$-0.0195312/s$
3	-0.4597174	$+0.0120743$	-0.2238216
	$-0.0353678/s$	$+0.0353678/s$	
4	-0.5153790	-0.00603940	-0.2607092
	$-0.0447159/s$	$-0.0447159/s$	$-0.0447159/s$
5	-0.5552071	$+0.0030479$	-0.2760796
	$-0.0489213/s$	$+0.0489213/s$	
6	-0.5949395	-0.0128321	-0.3038858
	$-0.0513727/s$	$-0.0513727/s$	$-0.0513727/s$

and their sum is negative.

9.3 The electromagnetic Casimir effect for a spherical shell and for a dielectric ball

In this section, we study the electromagnetic Casimir energy for a perfectly conducting spherical shell and for a dielectric ball. These two cases have attracted great attention since the publication of the paper by Boyer (1968). We start with a derivation of the boundary and matching conditions, using a complete separation of variables and polarizations, and then derive the mode-generating functions. An important point is a discussion of the ultraviolet divergences. Finally, we display numerical and analytical results for the Casimir energy.

9.3.1 Boundary conditions and separation of polarizations

The boundary conditions for an electromagnetic field on a conductor were considered in Section 3.2. For a sphere of radius R, they are given by

$$\boldsymbol{E}_t|_{r=R} = 0, \qquad \boldsymbol{B}_n|_{r=R} = 0. \qquad (9.111)$$

For a dielectric ball, we have to consider the more general problem where one medium, inside the sphere, has a permittivity ε_1 and a permeability μ_1, and the other medium, outside, has a permittivity ε_2 and a permeability μ_2. We need to consider the permeability of the material in order to discuss the special case where the speeds of light, $c_i = 1/\sqrt{\varepsilon_i \mu_i}$, are equal inside and outside the ball. Nevertheless, we continue to refer to the ball as a dielectric ball.

From classical electrodynamics, the continuity boundary conditions across the surface of the dielectric ball at $r = R$ require that the quantities

$$\varepsilon \boldsymbol{E}_n, \quad \boldsymbol{E}_t, \quad \boldsymbol{B}_n, \quad \frac{1}{\mu}\boldsymbol{B}_t \qquad (9.112)$$

are continuous. Sometimes these are referred to as *matching conditions*. In the case of a conducting spherical shell, the eigenvalue equation (3.45) remains unchanged. However, for the dielectric ball, because of the different speeds of light inside and outside, the eigenvalues in eqn (3.45) take the form

$$\Lambda_J = -\varepsilon_i \mu_i \omega_J^2. \tag{9.113}$$

For the boundary and matching conditions in eqns (9.111) and (9.112), the polarizations can be separated using the polarization vectors (9.3) and (9.6). Then we need to apply the conditions (9.111) or (9.112) to the fields (9.7). The normal component can be obtained by multiplying by the normal vector $\boldsymbol{n}$:

$$E_\text{n} = \boldsymbol{n} \cdot \boldsymbol{E}, \qquad B_\text{n} = \boldsymbol{n} \cdot \boldsymbol{B}. \tag{9.114}$$

For the two tangential components, it is convenient to take the projections

$$\begin{aligned} E_\text{t}^{(1)} &= \boldsymbol{L} \cdot \boldsymbol{E}, & E_\text{t}^{(2)} &= (\boldsymbol{L} \times \boldsymbol{n}) \cdot \boldsymbol{E}, \\ B_\text{t}^{(1)} &= \boldsymbol{L} \cdot \boldsymbol{B}, & B_\text{t}^{(2)} &= (\boldsymbol{L} \times \boldsymbol{n}) \cdot \boldsymbol{B}. \end{aligned} \tag{9.115}$$

In order to insert eqn (9.7) into eqns (9.114) and (9.115), we need the following properties of the operator polarization vectors (9.3):

$$\boldsymbol{L} \cdot \boldsymbol{E}^\text{TE} = \boldsymbol{L}^2, \qquad \boldsymbol{L} \cdot \boldsymbol{E}^\text{TM} = 0, \qquad \boldsymbol{n} \cdot \boldsymbol{E}^\text{TE} = 0, \qquad \boldsymbol{n} \cdot \boldsymbol{E}^\text{TM} = \frac{\text{i}}{r} \boldsymbol{L}^2,$$

$$(\boldsymbol{L} \times \boldsymbol{n}) \cdot \boldsymbol{E}^\text{TE} = 0, \qquad (\boldsymbol{L} \times \boldsymbol{n}) \cdot \boldsymbol{E}^\text{TM} = -\frac{1}{r} \frac{\partial}{\partial r} \boldsymbol{L}^2 r. \tag{9.116}$$

These can be easily verified using identities such as

$$\boldsymbol{n} \cdot \boldsymbol{L} = 0, \quad \boldsymbol{n} \cdot (\boldsymbol{\nabla} \times \boldsymbol{L}) = \frac{\text{i}}{r} \boldsymbol{L}^2, \quad (\boldsymbol{L} \times \boldsymbol{n}) \cdot (\boldsymbol{\nabla} \times \boldsymbol{L}) = -\frac{1}{r} \frac{\partial}{\partial r} \boldsymbol{L}^2 r. \tag{9.117}$$

With these relations, the projections of the field strengths are

$$\boldsymbol{E}_\text{t}^{(1)} = \text{i}\omega \boldsymbol{L}^2 f_l^\text{TE}(r) Y_{lM}(\theta,\varphi), \quad \boldsymbol{B}_\text{t}^{(1)} = -\boldsymbol{\nabla}^2 \boldsymbol{L}^2 f_l^\text{TM}(r) Y_{lM}(\theta,\varphi), \tag{9.118}$$

$$\boldsymbol{E}_\text{t}^{(2)} = -\frac{\text{i}\omega}{r} \boldsymbol{L}^2 \frac{\partial}{\partial r} r f_l^\text{TM}(r) Y_{lM}(\theta,\varphi), \quad \boldsymbol{B}_\text{t}^{(2)} = -\frac{\text{i}}{r} \boldsymbol{L}^2 \frac{\partial}{\partial r} r f_l^\text{TE}(r) Y_{lM}(\theta,\varphi),$$

$$E_\text{n} = \frac{\text{i}}{r} \boldsymbol{L}^2 f_l^\text{TE}(r) Y_{lM}(\theta,\varphi), \quad B_\text{n} = \frac{\text{i}}{r} \boldsymbol{L}^2 f_l^\text{TM}(r) Y_{lM}(\theta,\varphi).$$

Then the conductor boundary conditions (9.111) imply

$$\left. f_l^\text{TE}(r) \right|_{r=R} = 0, \qquad \left. \left[r f_l^\text{TM}(r) \right]' \right|_{r=R} = 0, \tag{9.119}$$

where the prime denotes differentiation with respect to r. For the TE mode, this is a Dirichlet boundary condition. For the TM mode, this is a Robin condition (9.56) with $u = 1$. For the dielectric ball, eqn (9.112) results in the following terms being continuous across $r = R$:

$$f_l^\text{TE}(r), \quad \frac{1}{\mu} \left[r f_l^\text{TE}(r) \right]', \quad \varepsilon f_l^\text{TM}(r), \quad \left[r f_l^\text{TM}(r) \right]'. \tag{9.120}$$

In the above, simpler matching conditions follow if we set $\mu = 1$.

9.3.2 The mode-generating functions

In the electromagnetic case, the mode-generating functions follow from the boundary conditions similarly to the scalar case. For the TE mode, this is obvious, since the boundary condition is of Dirichlet type. For the TM mode, there is an additional radius in the condition. All discussions related to the interior and exterior regions are the same as in Section 9.2.1 and we can write down the mode-generating functions immediately. For the TE mode, we have in the interior $\Delta_l^{\text{TE,i}}(k) = (kR)^{-\nu} J_\nu(kR)$, which is the same as eqn (9.58). In the exterior, $\Delta_l^{\text{TE,e}}(k) = (kR)^{\nu} H_\nu^{(1)}(kR)$, which is the same as eqn (9.65). Correspondingly, for the TM mode we have

$$\Delta_l^{\text{TM,i}}(k) = (kR)^{-l} \left[\sqrt{kR} J_\nu(kR)\right]', \quad \Delta_l^{\text{TM,e}}(k) = -(kR)^{l+1} \left[\sqrt{kR} H_\nu^{(1)}(kR)\right]' \tag{9.121}$$

instead of eqns (9.61) and (9.67). We have modified the factors in front of the Bessel functions to ensure all of the required properties at $k = 0$ and along the imaginary frequency axis. The next step is to rotate the momentum k to the imaginary axis. For this it is common to use the modified Riccati–Bessel functions, which, by means of

$$s_l(z) = \sqrt{\frac{\pi z}{2}} I_\nu(z), \quad e_l(z) = \sqrt{\frac{2z}{\pi}} K_\nu(z), \tag{9.122}$$

are related to the modified Bessel functions. The mode-generating functions for the electromagnetic field are shown in terms of these functions in Table 9.7. Here we have substituted $k \to k/R$, used the notation $\tilde{\Delta}_l(ik) \equiv \Delta_l(ik/R)$, and omitted unimportant numerical factors.

For the dielectric ball with the matching conditions (9.120), we have to consider the problem on the whole axis $r \in [0, \infty)$. It can be treated similarly to the case of the exterior region in Section 9.1.3. We have to write down the regular scattering solutions to eqn (9.23) satisfying the matching conditions. These solutions can be written in the form

$$\phi_{l,k}^{(T)}(r) = j_l(qr)\theta(R-r) + \frac{1}{2}\left[f_l^{(T)}(k) h_l^{(2)}(kr) + f_l^{(T)*}(k)\, h_l^{(1)}(kr)\right]\theta(r-R), \tag{9.123}$$

where the index "(T)" stands for either TE or TM. The radial momenta are defined according to eqn (9.113):

TABLE 9.7. The mode-generating functions for the electromagnetic field with conductor boundary conditions.

Region	TE mode	TM mode
Interior	$\tilde{\Delta}_l^{\text{TE,i}}(ik) = k^{-l-1} s_l(k)$	$\tilde{\Delta}_l^{\text{TM,i}}(ik) = k^{-l} s_l'(k)$
Exterior	$\tilde{\Delta}_l^{\text{TE,e}}(ik) = k^l e_l(k)$	$\tilde{\Delta}_l^{\text{TM,e}}(ik) = -k^{l+1} e_l'(k)$

$$q = \sqrt{\varepsilon_1\mu_1}\,\omega, \quad k = \sqrt{\varepsilon_2\mu_2}\,\omega. \tag{9.124}$$

This is a free solution in the interior and it coincides with the asymptotic form (9.25) in the exterior. The Jost functions can be found by inserting eqn (9.123) into the matching conditions (9.120). The resulting equations can be solved. Then, using the Wronskian

$$h_l^{(1)}(z)\,[h_l^{(2)}(z)]' - [h_l^{(1)}(z)]'\,h_l^{(2)}(z) = -2\frac{\mathrm{i}}{z^2}, \tag{9.125}$$

the Jost functions can be obtained in the form

$$f_l^{\mathrm{TE}}(k) = -\mathrm{i}kR\sqrt{\frac{\varepsilon_2}{\mu_2}}\left\{\mu_1 j_l(qR)\left[kR\,h_l^{(1)}(kR)\right]' - \mu_2\left[qR\,j_l(qR)\right]'h_l^{(1)}(kR)\right\}, \tag{9.126}$$

$$f_l^{\mathrm{TM}}(k) = -\mathrm{i}kR\sqrt{\frac{\mu_2}{\varepsilon_2}}\left\{\varepsilon_1 j_l(qR)\left[kR\,h_l^{(1)}(kR)\right]' - \varepsilon_2\left[qR\,j_l(qR)\right]'h_l^{(1)}(kR)\right\}.$$

Note that this spectral problem has a purely continuous spectrum, i.e. there are no bound states and no surface plasmons. Accordingly, the Jost functions (9.126) have no zeros on the positive imaginary k-axis. By rotating the argument to the imaginary axis, we arrive at the final expressions which define the mode-generating functions for the dielectric ball,

$$\tilde{\Delta}^{\mathrm{TE}}(\mathrm{i}k) = \sqrt{\varepsilon_1\mu_2}\,s_l'(q)e_l(k) - \sqrt{\varepsilon_2\mu_1}\,s_l(q)e_l'(k),$$
$$\tilde{\Delta}^{\mathrm{TM}}(\mathrm{i}k) = \sqrt{\mu_1\varepsilon_2}\,s_l'(q)e_l(k) - \sqrt{\mu_2\varepsilon_1}\,s_l(q)e_l'(k). \tag{9.127}$$

Here, we have used the same change of variable $k \to k/R$ as before.

9.3.3 The electromagnetic Casimir effect for a conducting spherical shell

The mode-generating functions obtained in the preceding subsection allow one to write the regularized vacuum energy (9.68) in the form

$$E_0(s) = -\frac{(\mu R)^{2s}}{R}\frac{\cos(\pi s)}{\pi}\sum_{l=1}^{\infty}\nu\int_0^{\infty}dk\,k^{1-2s}\frac{\partial}{\partial k}\ln\tilde{\Delta}_l(\mathrm{i}k). \tag{9.128}$$

Here, we taken into account the fact that the electromagnetic field is massless and that the values of the orbital momentum start from $l = 1$. For $\tilde{\Delta}_l(\mathrm{i}k)$, we have to insert only one generating function or a product of the mode-generating functions listed in Table 9.7, depending on the case considered. As before for the scalar field, we define the asymptotic part of the vacuum energy by eqn (9.74). For this purpose, we expand the mode-generating function in a series (9.70) and determine $h_{\mathrm{as}}(\nu, z)$ according to eqn (9.71). The asymptotic part $E_0^{\mathrm{as}}(s)$ is represented by eqn (9.81), where, in the massless case, the functions $A_i(s)$ are

$$A_i(s) = -\frac{(\mu R)^{2s}}{R}\frac{\cos(\pi s)}{\pi}\sum_{l=1}^{\infty}\nu^{2-2s}\int_0^{\infty}dz\,z^{1-2s}\frac{\partial}{\partial z}\frac{D_i(z)}{\nu^i}. \tag{9.129}$$

The functions $D_i(z)$ can be obtained in the same way as in the scalar case. For $i = -1$, they coincide with those given in eqn (9.79). For $i = 0$, the functions $D_0(z)$ are shown in Table 9.8. The functions $D_i(z)$ with $i = 1, 2, 3$ are given by eqn (9.80) with the coefficients $x_{i,a}$ presented in Table 9.9. Using the integrals in eqns (9.103), we arrive at

$$A_{-1}(s) = \frac{(\mu R)^{2s}}{R} \frac{\Gamma(s-1)}{\Gamma(s+\frac{1}{2})} \zeta\left(2s - 3, \frac{3}{2}\right),$$

$$A_0(s) = -\frac{(\mu R)^{2s}}{4R} \zeta\left(2s - 2, \frac{3}{2}\right), \quad (9.130)$$

$$A_i(s) = -\frac{(\mu R)^{2s}}{R} \sum_{a=i}^{3i} \frac{x_{i,a}}{\Gamma\left(\frac{a}{2}\right)} \frac{\Gamma\left(s + \frac{a-1}{2}\right)}{\Gamma\left(s - \frac{1}{2}\right)} \zeta\left(2s - 2 + i, \frac{3}{2}\right), \quad i = 1, 2, 3$$

instead of eqn (9.104). In comparison with eqns (9.83)–(9.85), here the coefficients $x_{i,a}$ must be taken from Table 9.9 instead of Table 9.3, and the parameters of the Hurwitz zeta functions are different. Now, we can substitute eqn (9.130) into eqn (9.81) and get the asymptotic part of the vacuum energy $E_0^{\text{as}}(s)$. In order to discuss the ultraviolet divergences, we calculate the first five heat kernel coefficients. These can be obtained from the vacuum energy in zeta function regularization (4.26) with $m = 0$ by taking the corresponding residues,

$$a_{k/2} = 16\pi^{3/2} \text{res}\left[\Gamma\left(s - \frac{1}{2}\right) E_0(s); 2 - \frac{k}{2}\right]. \quad (9.131)$$

Note that E_0^{fin} does not contribute to the poles in eqn (9.131). Thus, according to eqn (9.72), we can use $E_0^{\text{as}}(s)$ for the calculation of $a_{k/2}$. The results are

$$a_0 = \frac{4\pi}{3}R^3, \quad a_{1/2} = \begin{Bmatrix} -1 \\ 1 \end{Bmatrix} 2\pi^{3/2}R^2, \quad a_1 = -\begin{Bmatrix} 2 \\ 14 \end{Bmatrix}\frac{2\pi}{3}R,$$

$$a_{3/2} = \begin{Bmatrix} 23 \\ 7 \end{Bmatrix}\frac{\pi^{3/2}}{6}, \quad a_2 = -\begin{Bmatrix} 1 \\ 7 \end{Bmatrix}\frac{16\pi}{315R}. \quad (9.132)$$

The upper entries in the curly brackets are for the TE mode and the lower are for the TM mode. These heat kernel coefficients are different from those for a scalar field given in eqn (4.25). They were calculated for a general surface by Bernasconi et al. (2003).

TABLE 9.8. The functions $D_0(z)$ entering the asymptotic expansion (9.71) in the electromagnetic case.

Region	Dirichlet boundary condition	Neumann boundary condition
Interior	$\frac{1}{2}\ln t - \left(\nu + \frac{1}{2}\right)\ln\nu - \ln 2$	$-\frac{1}{2}\ln t + \left(-\nu + \frac{1}{2}\right)\ln\nu + \ln 2$
Exterior	$\frac{1}{2}\ln t + \left(\nu - \frac{1}{2}\right)\ln\nu$	$-\frac{1}{2}\ln t + \left(\nu + \frac{1}{2}\right)\ln\nu$

TABLE 9.9. The coefficients $x_{i,a}$ for the TM mode. The upper signs are for the interior region, and the lower signs are for the exterior region.

					a				
i	1	2	3	4	5	6	7	8	9
1	$\mp\frac{1}{8}$	0	$\pm\frac{7}{24}$	0	0	0	0	0	0
2	0	$-\frac{1}{16}$	0	$\frac{3}{8}$	0	$-\frac{7}{16}$	0	0	0
3	0	0	$\mp\frac{23}{384}$	0	$\pm\frac{549}{640}$	0	$\mp\frac{259}{128}$	0	$\pm\frac{1463}{1152}$

The coefficients $a_{k/2}$ in eqn (9.132) are for the interior region. Those for the exterior region follow, as in the scalar case, by multiplication by $(-1)^{k+1}$ ($k \neq 0$), except for $a_{3/2}$. The latter, in the exterior region, is given by $a_{3/2} = -\pi^{3/2}/6$ for the TE mode and $a_{3/2} = -17\pi^{3/2}/6$ for the TM mode. Note that the coefficients $a_{1/2}$ for the TE and TM modes differ only in sign, as a consequence of conformal symmetry. Therefore they cancel when the two modes are added together.

With these coefficients, we are in a position to discuss the ultraviolet divergences of the Casimir energy for an ideally conducting spherical shell. Since $a_2 \neq 0$, the vacuum energies for the interior and exterior regions taken separately are divergent and cannot be uniquely defined. This holds for each mode and also for their sum. However, if one considers the interior and exterior regions together, there is a cancellation in the coefficients a_2 and a_1 such that their combined contribution vanishes. Similarly to the case of a massless scalar field, here also the vacuum energy can be uniquely defined, and in zeta function regularization one obtains a finite result. In other regularization schemes there is, in general, a surface divergence which must be removed by some renormalization. In addition, in the electromagnetic case, if the TE and TM modes are taken together, this divergence disappears because of the above-mentioned symmetry in the coefficients $a_{1/2}$. The only remaining potentially divergent contribution comes from the coefficients $a_{3/2}$. In zeta function regularization (4.3) and in cutoff regularization (4.1), this contribution does not show up. Hence, in these regularization schemes the Casimir energy for a conducting sphere is finite when the interior and exterior and both polarizations are taken together. In other regularization schemes, a divergence arising from $a_{3/2}$ may show up. However, since $a_{3/2}$, for dimensional reasons, does not depend on the radius, this divergence is also independent of the radius (provided that the regularization does not introduce any artificial dependence on the radius). Therefore it may be dropped, as it does not contribute to the Casimir force acting on the spherical shell. It must be remarked that, historically, just these fortunate compensations enabled Boyer (1968) to obtain his well-known result for a conducting sphere.

At this point it may be useful to make a comment on the physical interpretation. We have seen that the Casimir effect for a spherical shell can give a finite result if the interior and exterior regions are taken together. However, it must be underlined that this implies an infinitely thin conducting shell. Any attempt to consider a shell with a finite thickness will necessarily destroy the symmetry. Specifically, the coefficient a_2 becomes proportional to the difference between the two radii (Bordag et al. 1999a). Thus, it becomes impossible to define the Casimir energy in a meaningful way.

We now continue discussing the case of an infinitely thin, ideally conducting spherical shell. Substituting $A_i(s)$ from eqn (9.130) into eqn (9.81), we obtain the asymptotic part of the vacuum energy in the form

$$E_0^{as,TE,i}(s) = \frac{1}{630\pi R}\left[\frac{1}{s} + 2\ln(\mu R) - \frac{561\ln 2}{64} + \frac{5355\pi}{128} - \frac{229\gamma}{64} + \frac{52013}{384}\right.$$
$$\left. - \frac{315}{8}\zeta_R'(-1) + \frac{2205}{8}\zeta_R'(-3)\right] + O(s),$$

$$E_0^{as,TM,i}(s) = \frac{1}{630\pi R}\left[\frac{1}{s} + 2\ln(\mu R) + \frac{1743\ln 2}{64} - \frac{3465\pi}{128} + \frac{539\gamma}{64} - \frac{18067}{384}\right.$$
$$\left. - \frac{315}{8}\zeta_R'(-1) + \frac{2205}{8}\zeta_R'(-3)\right] + O(s). \quad (9.133)$$

These expressions are for the interior region. If we add the corresponding expressions for the exterior region to eqn (9.133), the result is

$$E_0^{as,TE,b}(s) = \frac{17}{128R} + O(s), \quad E_0^{as,TM,b}(s) = -\frac{11}{128R} + O(s). \quad (9.134)$$

In the derivation of these formulas, a number of cancellations have occurred, which are the same as in the scalar case [however, here we have nonzero contributions from $\zeta(2, 3/2) = -1/4$ and $\zeta(0, 3/2) = -1$]. Thus, in the case (b), i.e. when the interior and exterior regions are taken together, the asymptotic part of the vacuum energy is finite and nonzero. It is also nonzero if the TE and TM contributions are added:

$$E_0^{as,TE,b}(s) + E_0^{as,TM,b}(s) = \frac{3}{64R}. \quad (9.135)$$

It is equal to the analytic part defined in eqn (9.88), since the divergent part is zero. The finite part can be calculated in the same way as in Section 9.2.4. These calculations have been carried out many times and reported in the literature for different spatial dimensions [the results compiled in the book by Kirsten (2000) are presented in Table 9.10]. Specifically, in three dimensions, the electromagnetic Casimir energy for an ideally conducting spherical shell,

$$E(R) \equiv E_0^{ren} = \frac{0.0461766}{R}, \quad (9.136)$$

is repulsive.

TABLE 9.10. Zeta functions and electromagnetic Casimir energy for a perfectly conducting spherical shell.

d	ζ_P interior	ζ_P exterior	$RE_0(s)$
2	−0.3446767	−0.0215672	−0.1831220
	− 0.0195312/s	− 0.0195312/s	− 0.0195312/s
3	+0.1678471	−0.0754938	+0.0461766
	+ 0.0080841/s	− 0.0080841/s	
4	+0.5008593	−0.1942082	+0.1533255
	+ 0.0231719/s	− 0.0564056/s	− 0.0332337/s
5	+1.0463255	−0.2981425	+0.3740915
	+ 0.1838665/s	− 0.1838665/s	

9.3.4 The Casimir effect for a dielectric ball

As mentioned in Section 9.3.1, we consider a dielectric ball where the inside medium has a permittivity ε_1 and a permeability μ_1 and the outside medium has a permittivity ε_2 and a permeability μ_2. Such a configuration may describe scenarios such as a dielectric ball made of an insulator, a bubble in a liquid, or a hadronic bag. The electromagnetic field couples through the macroscopic Maxwell equations and its vacuum energy depends on the radius of the ball and on the parameters ε_i and μ_i ($i = 1, 2$). Since such a dielectric can be considered as a reasonable idealization of a real physical body, one could expect that the vacuum energy of the electromagnetic field (or at least the corresponding pressure) would be finite. However, as discussed below, this is not the case, as the ultraviolet divergences cannot be satisfactorily removed. In fact, this is still an unresolved puzzle which continues to stimulate interest in this configuration.

The dielectric ball was originally considered in connection with the Casimir model of an electron (Milton 1980, Brevik and Kolbenstvedt 1982). Soon it was realized that this configuration has unremovable ultraviolet divergences which do not allow one to obtain a finite result like that for an ideal-metal spherical shell. An exception is the case for equal speeds of light $c_i = 1/\sqrt{\varepsilon_i \mu_i}$ (in relation to the value in vacuum) in the inside and outside regions, as found by Brevik and Kolbenstvedt (1982). These authors discussed this case in connection with the hadronic bag model.

A decade later, this configuration became popular owing to attempts to explain sonoluminescence as the release of vacuum energy from the collapse of air bubbles in water (Schwinger 1992, 1993). Although this phenomenon can probably be explained by different physical processes (Liberati *et al.* 2000), the vacuum energy of a dielectric ball was intensively investigated. In terms of heat kernel coefficients, it was shown (Bordag *et al.* 1999a) that beyond the dilute approximation, ultraviolet divergences are present independent of the regularization scheme used. This prevents any conclusive result. At the same time, in the dilute approximation, after removal of the residual divergences, there are regularization schemes which lead to a unique result. Here, the *dilute approximation* is understood as the contribution to the vacuum energy up to the second order of some small parameter characterizing the diluteness of the ball material. Definite results for $E(R)$ are obtained in the following two cases. In the first case, it is assumed that $|c_1 - c_2| \ll c_1, c_2$ and, thus, $|c_1 - c_2|$ plays the role of a small parameter. In this case it is usually also assumed that $\mu_1 = \mu_2 = 1$ but $\varepsilon_1 \neq \varepsilon_2$ ($\varepsilon_{1,2} \approx 1$). In the second case, it is assumed that $c_1 = c_2$ but there is a nonzero parameter

$$\xi = \frac{\varepsilon_1 - \varepsilon_2}{\varepsilon_1 + \varepsilon_2} = \frac{\mu_2 - \mu_1}{\mu_1 + \mu_2}. \tag{9.137}$$

In this case, the results are obtained for small ξ and for $\xi = 1$. Both cases have been solved by several authors, who obtained agreement using different approaches.

We start with the consideration of the heat kernel coefficients of the electromagnetic field in the presence of a dielectric ball. These can be obtained using eqn (9.131) from the vacuum energy in zeta function regularization, where the asymptotic part given by eqns (9.81) and (9.82) must be inserted. The functions $D_i(z)$ follow from the uniform asymptotic expansion of the mode-generating functions (9.127). Changing the factor in front for convenience, we denote them by

$$\tilde{\Delta}_l^{(\rho)}(ik) = \alpha^\rho s_l'(q) e_l(k) - s_l(q) e_l'(k), \quad \alpha = \sqrt{\frac{\varepsilon_1 \mu_2}{\varepsilon_2 \mu_1}}, \tag{9.138}$$

with $\rho = \pm 1$ for the TE and TM modes, respectively. With this notation, we calculate the heat kernel coefficients (9.131) using the formulas (9.128) and (9.129) and the mode-generating functions (9.138). The only difference from the preceding sections is that the functions $D_i(z)$ are now more complicated. With eqns (9.70) and (9.76), we get

$$D_{-1}^{(\rho)}(z) = \eta\left(\frac{z}{c_1}\right) - \eta\left(\frac{z}{c_2}\right), \quad D_0^{(\rho)}(z) = \ln\left(\frac{\alpha^\rho c_1 t_2 c_2 t_1}{2\sqrt{c_1 c_2 t_1 t_2}}\right), \tag{9.139}$$

where $\eta(z)$ is defined in eqn (9.77) and

$$t_{1,2} = \frac{1}{\sqrt{1 + c_{1,2} z^2}}. \tag{9.140}$$

The expressions for $D_i^{(\rho)}$ ($i = 1, 2, 3$) have been presented by Bordag et al. (1999a). As before, the sum over the orbital momenta can be carried out, leading to the Hurwitz zeta function. For $i = -1$, using the first line in eqn (9.103), we obtain

$$A_{-1}(s) = \frac{(\mu R)^{2s}}{R} \frac{\Gamma\left(s - \frac{1}{2}\right)}{4\sqrt{\pi}\Gamma(s+1)} \zeta\left(2s - 3, \frac{3}{2}\right) \left(c_1^{1-2s} - c_2^{1-2s}\right). \tag{9.141}$$

For $i = 0, 1, 2, 3$, the integration over z cannot be carried out and we are left with

$$A_i(s) = -\frac{(\mu R)^{2s}}{R}\frac{\cos(\pi s)}{\pi}\zeta\left(2s+i-2,\frac{3}{2}\right)\int_0^\infty dz\, z^{1-2s}\frac{\partial}{\partial z}D_i^{(\rho)}(z). \qquad (9.142)$$

Let us consider, as an example, the calculation of the heat kernel coefficient a_2. For this purpose, according to eqn (9.131), we need the residue of $E_0^{\text{as}}(s)$ at $s = 0$. For $i = -1$, the pole comes from the gamma function in eqn (9.141) and we get

$$\text{res}\,[A_{-1}(s); 0] = \frac{127}{1920\pi R}(c_1 - c_2). \qquad (9.143)$$

For $i = 0$, there is no pole contribution, since the z-integration is convergent at $s = 0$. The same holds for $i = 2$. For $i = 1$, the z-integration is divergent in the upper bound because $D_1^{(\rho)}(z) = (c_1 - c_2)/(8z) + O(z^{-3})$. This results in a pole contribution with

$$\text{res}\,[A_1(s); 0] = -\frac{11}{192\pi R}(c_1 - c_2). \qquad (9.144)$$

The last contribution is for $i = 3$. Here the pole comes from the Hurwitz zeta function and the residue is

$$\text{res}\,[A_3(s); 0] = \frac{1}{\pi R}\int_0^\infty dz\, D_3^{(\rho)}(z), \qquad (9.145)$$

where an integration by parts has been done. This remaining integration cannot be performed analytically. However, eqn (9.145) is a smooth function of the parameter α and of the speed of light c_i. This function can be expanded in powers of $(c_1 - c_2)$. When this is done, the zeroth and all even orders are absent. To get the heat kernel coefficient a_2, we take the sum of eqns (9.143), (9.144), and (9.145) [multiplied by $16\pi^{3/2}\Gamma(-1/2) = -32\pi^2$ according to eqn (9.131)] for both values $\rho = \pm 1$. It so happens that in this sum the first-order contributions cancel whereas the third order and higher are present:

$$a_2 = -\frac{2656\pi}{5005 R}\frac{(c_1-c_2)^3}{c_2^2} + O\left[(c_1-c_2)^4\right]. \qquad (9.146)$$

The other coefficients have been calculated using the same methods (Bordag et al. 1999a):

$$a_0 = \frac{8\pi}{3}R^3\left(c_1^{-3} - c_2^{-3}\right), \qquad a_{1/2} = -2\pi^{3/2}R^2\frac{(c_1^2-c_2^2)^2}{c_1^2 c_2^2(c_1^2+c_2^2)^2},$$

$$a_1 = -\frac{22\pi}{3}R\left(c_1^{-1} - c_2^{-1}\right) + 8\pi R\int_0^\infty \frac{dz}{z}\frac{\partial}{\partial z}\left[D_1^{(1)}(z) + D_1^{(-1)}(z)\right],$$

$$a_{3/2} = \pi^{3/2}\frac{(c_1^2 - c_2^2)^2}{(c_1^2+c_2^2)^2}. \qquad (9.147)$$

Equations (9.146) and (9.147) determine the ultraviolet divergences for the dielectric ball for any regularization. For instance, in zeta function regularization, where we have a divergent contribution from a_2 only, it follows that in the dilute approximation the vacuum energy $E_0(s)$ has a finite and unique continuation to $s = 0$. This is because eqn (9.146) starts from the third-order term. However, the vacuum energy is not finite and cannot be uniquely defined beyond the dilute approximation. In the cutoff regularization, from eqn (9.147), inserted into eqn (4.32), divergences follow which are proportional to $1/\delta^4$ and $1/\delta^3$. They are linear and quadratic in $c_1 - c_2$. However, by removing them in some way, one can be sure that because of eqn (9.146), the result for the energy in the second order will always be the same.

Thus, the vacuum energy for the dielectric ball is uniquely defined in the dilute approximation but not beyond it. The latter case is still an unresolved problem. As discussed, the case of a zero-point electromagnetic field interacting with a material body is a physical one and should result in finite measurable quantities, such as the surface stress. It has been argued that the permittivity cannot be considered as a constant but should be a function of the frequency which approaches unity at high frequencies. Because of this, the divergences which result from the high frequencies might disappear. To investigate this possibility, a frequency-dependent dielectric permittivity inside the ball

$$\varepsilon_1(\omega) = 1 - \frac{w_1}{\omega^2} + \frac{w_2}{\omega^4} - \ldots \tag{9.148}$$

and $\varepsilon_2 = 1$ outside were considered (Bordag and Kirsten 2002). As a result, a nonvanishing contribution to the heat kernel coefficient a_2 was found,

$$a_2 = \frac{4\pi}{3} w_1^2 R^3 + \frac{16\pi}{3} w_2 R^3. \tag{9.149}$$

Thus, the divergence is weakened by dispersion, as expected, but not sufficiently to resolve the problem. The reason is that the permittivity does not decrease faster than $\sim \omega^{-2}$ (see Section 12.1). Another attempt to resolve the problem considers a smooth function $\varepsilon(r)$ instead of a function with a jump at $r = R$. As shown by Bordag et al. (1998a), however, for a smooth function the heat kernel coefficient a_2 is nonzero even in the dilute approximation. Thus, this approach does not work either.

In the remainder of this subsection, we consider the vacuum energy in the dilute approximation, for which reliable results can be obtained. As explained, depending on the regularization scheme, either the vacuum energy is finite or the divergent contributions can be discarded, leaving a unique result. We start with the first case, putting $\mu_1 = \mu_2 = 1$, and keep contributions up to the second order in $c_1 - c_2$. It is assumed that both of the dielectric permittivities $\varepsilon_{1,2}$ are approximately unity. In this case, the vacuum energy is

$$E(R) = E_0^{\text{ren}} = \frac{23}{384\pi R}(c_1 - c_2)^2 + O\left[(c_1 - c_2)^3\right]. \tag{9.150}$$

This has been found independently using several methods.

Barton (1999) treated the dielectric ball as a small perturbation with respect to empty space and arrived at eqn (9.150) in second-order perturbation theory. A completely different approach is based on the integration of the Casimir-Polder forces acting between the molecules of the ball in a vacuum (Milton and Ng 1998). The starting point is the physical picture of a ball composed of a finite number of polarizable molecules. In Section 16.1, it will be shown that any two molecules at points r_1 and r_2 in a dilute medium interact through the Casimir-Polder potential

$$V(r_1, r_2) = -\frac{23}{(4\pi)^3 N^2} \frac{(\varepsilon_1 - 1)^2}{|r_1 - r_2|^7}, \tag{9.151}$$

where N is the density of molecules. Integrating with respect to both coordinates in eqn (9.151) over the volume of the ball, taking into account the density of molecules (see Section 6.4) and dividing by 2 to avoid double counting, the interaction energy becomes

$$E = -\frac{23}{8\pi} \frac{(\varepsilon_1 - 1)^2}{(4\pi)^2} \int_V dr_1 \int_V dr_2 \frac{1}{|r_1 - r_2|^\gamma}. \tag{9.152}$$

Here we have changed the power in the denominator to γ in order to make the integrations convergent. This is a regularization and, to make the integrals convergent, one needs to make $\gamma < 3$. At the end, one has to perform a continuation to $\gamma = 7$. The integrations can be carried out explicitly (Milton and Ng 1998), and one arrives at

$$E = -\frac{23}{2^{2+\gamma}\pi} (\varepsilon_1 - 1)^2 \frac{\Gamma\left(2 - \frac{\gamma}{2}\right)}{\Gamma\left(4 - \frac{\gamma}{2}\right)(3 - \gamma)} \frac{1}{R^{\gamma-6}}. \tag{9.153}$$

This expression has a unique analytic continuation to $\gamma = 7$, where it takes a finite value coinciding with eqn (9.150), where c_2 is set equal to unity. The regularization used in eqn (9.152) is not the only possible one. Another choice could be to use point splitting, i.e. the replacement $r_1 - r_2 \to r_1 - r_2 + \delta$ in the denominator, with $\delta \to 0$ at the end. In that case one finds divergent contributions proportional to the inverse powers of δ. Disregarding them, one arrives at the same result, i.e. eqn (9.150) with $c_2 = 1$.

It is interesting that the energy (9.150) is positive although it is represented in eqn (9.152) as an integral over pairwise energies which are negative. The change in sign comes from the analytic continuation in eqn (9.153) and demonstrates that a finite, physically meaningful quantity, such as the vacuum energy for a dilute dielectric ball, may have a sign contrary to expectation.

There is also a way to calculate the same vacuum energy from the mode-generating functions that we have obtained so far. One has to insert eqn (9.138) into eqn (9.128) and take the expansion in $2\delta_c \equiv c_2 - c_1$ up to the second order. In this case it is convenient to start from eqn (9.127) and to write $\sqrt{\varepsilon_{1,2}} = 1 \pm \delta_c$. Then the product of the mode-generating functions can be written in the form

$$\tilde{\Delta}_l^{\text{ball}}(ik) \equiv \tilde{\Delta}_l^{\text{TE}}(ik) \tilde{\Delta}_l^{\text{TM}}(ik) \tag{9.154}$$

$$= [s'_l(q)e_l(k) - s_l(q)e'_l(k)]^2 - \delta_c^2 [s'_l(q)e_l(k) + s_l(q)e'_l(k)]^2.$$

The Riccati–Bessel functions in eqn (9.154) can be expanded in powers of a small parameter δ_c. Then, using the above procedure, the vacuum energy E_0 in eqn (9.128) can be calculated (Lambiase et al. 2001), leading to the result (9.150). This result has also been confirmed numerically with high precision (Brevik et al. 1999).

Finally, we consider the second case discussed above, i.e. a dielectric–diamagnetic ball where the speeds of light inside and outside are equal. In this case the arguments of the Riccati–Bessel functions in the mode-generating functions are equal, and from eqn (9.127) we get (Brevik et al. 1999)

$$\tilde{\Delta}^{(c_1=c_2)}_l(ik) \equiv \tilde{\Delta}^{\mathrm{TE}}_l(ik)\tilde{\Delta}^{\mathrm{TM}}_l(ik) = \sqrt{\varepsilon_1\mu_1\varepsilon_2\mu_2}\left\{[s_l(k)e'_l(k)]^2 + [s'_l(k)e_l(k)]^2\right\}$$
$$-(\varepsilon_1\mu_2 + \varepsilon_2\mu_1)s_l(k)s'_l(k)e_l(k)e'_l(k) = \frac{\mu_1}{4\varepsilon_1}(\varepsilon_1 + \varepsilon_2)^2\left(1 - \xi^2\sigma_l^2\right). \quad (9.155)$$

Here, a simplification has been done using $\varepsilon_1\mu_1 = \varepsilon_2\mu_2$, the definition (9.137) of the parameter ξ, and the identity

$$\sigma_l^2 = 4s_l(k)s'_l(k)e_l(k)e'_l(k) + 1, \quad \text{where} \quad \sigma_l = [s_l(k)e_l(k)]'. \quad (9.156)$$

Dropping a constant, the mode-generating function can be rewritten as

$$\tilde{\Delta}^{(c_1=c_2)}_l(ik) = 1 - \xi^2\sigma_l^2. \quad (9.157)$$

Note that for $\xi = 1$ this is just the mode-generating function of a conducting sphere when both polarizations and the interior and exterior regions are taken together, i.e. it is the product of the functions in Table 9.7. In this way, for the dielectric–diamagnetic ball with equal speeds of light, the vacuum energy for $\xi = 1$ is known. This has also been calculated for small ξ in the perturbation order to ξ^2. In that case the vacuum energy is

$$E_0(s) = \xi^2 \frac{(\mu R)^{2s}}{R} \frac{\cos(\pi s)}{\pi} \sum_{l=1}^{\infty} \nu \int_0^{\infty} dk\, k^{1-2s} \frac{\partial}{\partial k}\sigma_l^2. \quad (9.158)$$

As explained above, this expression has a finite analytic continuation to $s = 0$. This continuation can be constructed in the usual way by subtracting and adding back the corresponding part of the asymptotic expansion for the logarithm of the mode-generating function, as we did for the scalar field in Section 9.2.2.

Sometimes the mode-generating function (9.157) is presented in the form of an asymptotic series, and the higher orders of this series have been discussed (Brevik et al. 1998). The resulting series representation for the vacuum energy allows the limit $s \to 0$. However, this series diverges and cannot be used to get reliable approximations for the vacuum energy. It should be mentioned that the same also holds for scalar, electromagnetic, and other fields and for all boundary conditions.

There is a much more elegant way to calculate the vacuum energy (9.158) explicitly (Klich 2000). For this purpose, the summation over the orbital momenta in eqn (9.158) is performed using the addition theorem for the Bessel functions (Abramowitz and Stegun 1972)

$$\sum_{l=0}^{\infty}(2l+1)s_l(k)e_l(q)P_l(v) = \frac{kq}{Q}e^{-Q}, \tag{9.159}$$

where $Q = \sqrt{k^2 + q^2 - 2kqv}$, $|v| \leq 1$. Putting $q = k$ and taking the derivative with respect to k, we arrive at

$$\sum_{l=0}^{\infty}(2l+1)\sigma_l P_l(v) = \frac{1 - k\tilde{Q}}{\tilde{Q}}e^{-k\tilde{Q}}, \tag{9.160}$$

where $\tilde{Q} = \sqrt{2 - 2v}$. Now we square this equation and integrate over v using the orthogonality relation of the Legendre polynomials,

$$\int_{-1}^{1} dv\, P_l(v)P_{l'}(v) = \frac{2\delta_{ll'}}{2l+1}. \tag{9.161}$$

Again taking the derivative with respect to k, we obtain the relation

$$\sum_{l=0}^{\infty}(2l+1)\left(\sigma_l^2\right)' = \int_{-1}^{1} dv \frac{(1-k\tilde{Q})(-2+k\tilde{Q})}{\tilde{Q}} e^{-2k\tilde{Q}} = \frac{1}{2k}\left[(1+2k)^2 e^{-4k} - 1\right]. \tag{9.162}$$

It remains to subtract the contribution from $l = 0$ on the right-hand side of this equation. Using $s_0(k) = \sinh(k)$ and $e_0(k) = \exp(-k)$, we arrive at

$$\sum_{l=1}^{\infty}(2l+1)\left(\sigma_l^2\right)' = \frac{1}{2k}\left[(1-2k)^2 e^{-4k} - 1\right]. \tag{9.163}$$

Since the sum here converges, this formula can be substituted into eqn (9.158) and we arrive at

$$E_0(s) = \frac{\xi^2}{4}\frac{(\mu R)^{2s}}{R}\frac{\cos(\pi s)}{\pi}\int_0^{\infty} dk\, k^{-2s}\left[(1-2k)^2 e^{-4k} - 1\right]. \tag{9.164}$$

The integral converges for $1/2 < s < 3/2$. When we integrate by parts, it becomes

$$E_0(s) = \frac{\xi^2}{4}\frac{(\mu R)^{2s}}{R}\frac{\cos(\pi s)}{\pi}\frac{1}{2s-1}\int_0^{\infty} dk\, k^{1-2s}\frac{\partial}{\partial k}\left[(1-2k)^2 e^{-4k} - 1\right]. \tag{9.165}$$

In this representation, we can put $s = 0$ in the integral. Now the integration can be carried out and we arrive at

$$E_0^{\text{ren}} = \frac{5}{32\pi}\frac{\xi^2}{R} + O\left(\xi^4\right). \tag{9.166}$$

To conclude, the only results known so far for the Casimir effect for a dielectric ball are eqn (9.166), which is for $c_1 = c_2$ and small ξ; the limiting case of it for

$\xi = 1$, which is for a conducting sphere; and eqn (9.150), for the dilute case. It is also known that beyond the dilute approximation there are unremovable (at least so far) ultraviolet divergences. This is a rather unsatisfactory situation, and further work is necessary.

9.4 The spinor Casimir effect for a sphere

In this section we consider the Casimir effect for a spinor field obeying bag boundary conditions on a sphere. The physical motivation is its relevance to the bag model in QCD which describes hadrons. Although this model is phenomenologically quite successful, all attempts to include the vacuum energy of gluons and quarks have remained unsatisfactory owing to ultraviolet divergences. One might expect that a renormalization scheme, as outlined in Section 4.3.2, should be applicable. However, the ultraviolet divergences calculated in terms of the heat kernel coefficients have not yet been incorporated into the bag model in a satisfactory way (Milton 2001). Note that this problem is more likely to be connected with the formulation of the bag model than with the calculation of the vacuum energy. As we shall see below, the divergences of the spinor vacuum energy can be isolated in terms of the heat kernel coefficients similarly to other fields. The finite parts of the vacuum energy can be calculated numerically, whereas the interpretation of these results within the bag model remains an open question.

Here we focus on the spinor field. As to the gluon field, the results for the vacuum energy are the same as for conductor boundary conditions if one ignores the higher-loop corrections. In this case the corresponding results from Section 9.3.3 apply. We allow the spinor field to be massive, in which case additional ultraviolet divergences are present. As is seen below, the methods previously applied in this chapter for a scalar field work in the spinor case also.

We start from the Dirac equation (3.30) with the bag boundary conditions (7.102). The solutions of the Dirac equation in spherical coordinates have the form

$$\psi_{j,M,\sigma,k}(\boldsymbol{r}) = \frac{1}{\sqrt{r}} \begin{pmatrix} \sqrt{k_0+m}\, J_{(2j+1-\sigma)/2}(kr)\, i\Omega^l_{j,M}(\theta,\varphi) \\ -\sigma\sqrt{k_0-m}\, J_{(2j+1+\sigma)/2}(kr)(\boldsymbol{n}\cdot\boldsymbol{\sigma})\Omega^l_{j,M}(\theta,\varphi) \end{pmatrix}, \quad (9.167)$$

where $\boldsymbol{n} = \boldsymbol{r}/r$, and $\boldsymbol{\sigma}$ is the Pauli matrices. The Pauli spinors

$$\Omega^l_{j,M}(\theta,\varphi) = \begin{pmatrix} \sqrt{\frac{2l+2\sigma M+1}{2(2l+1)}}\, Y_{l,M-\frac{1}{2}}(\theta,\varphi) \\ \sigma\sqrt{\frac{2l-2\sigma M+1}{2(2l+1)}}\, Y_{l,M+\frac{1}{2}}(\theta,\varphi) \end{pmatrix} \quad (9.168)$$

are the eigenstates of the total momentum $\boldsymbol{J} = \boldsymbol{L} + \frac{1}{2}\boldsymbol{\sigma}$. The eigenvalues of the operators $\boldsymbol{J}^2$, $\boldsymbol{L}^2$, $\boldsymbol{J}_z$, and $I + \boldsymbol{\sigma}\cdot\boldsymbol{L}$ are $j(j+1)$, $l(l+1)$, M, and $\sigma(2j+1)/2$, respectively. These eigenvalues are related by $j = (2l+\sigma)/2$ with spin projections $\sigma = \pm 1$. Each solution (9.167) describes two states with $k_0 = \pm\sqrt{m^2+k^2}$, corresponding to a particle and an antiparticle. The radial functions in eqn (9.167) are the Bessel functions. The solution is regular at $r = 0$, as is required for the interior region of a sphere. The corresponding solutions for the exterior region, i.e. the scattering solutions, can be obtained by replacement of the Bessel functions with Hankel functions.

The matrix on the left-hand side of the bag boundary condition (7.102) can be written in the form

$$i\boldsymbol{\gamma}\cdot\boldsymbol{n} + 1 = \begin{pmatrix} I & i(\boldsymbol{\sigma}\cdot\boldsymbol{n}) \\ -i(\boldsymbol{\sigma}\cdot\boldsymbol{n}) & I \end{pmatrix}. \tag{9.169}$$

When applied to eqn (9.167), this leads to the condition

$$\left[\sqrt{k_0 + m}\, J_{(2j+1-\sigma)/2}(kr) - \sigma\sqrt{k_0 - m}\, J_{(2j+1+\sigma)/2}(kr)\right]\Big|_{r=R} = 0, \tag{9.170}$$

where we have used $(\boldsymbol{\sigma}\cdot\boldsymbol{n})(\boldsymbol{\sigma}\cdot\boldsymbol{n}) = 1$. In fact, two conditions are obtained, one from the upper two components of the bispinor and another one from the lower two. They are identical up to a constant factor. The solutions of eqn (9.170) with respect to k are the discrete eigenvalues. Therefore we can take the left-hand side of it as the mode-generating function. However, the formulas are simplified if one takes the product of the mode-generating functions for particles and antiparticles, i.e. for the two signs of k_0. One has to be careful with the signs. From this the mode-generating function, up to a constant factor, is

$$\Delta_j(k) = \sigma\left[J^2_{(2j+1+\sigma)/2}(kR) - J^2_{(2j+1-\sigma)/2}(kR)\right]$$
$$- \frac{2m}{k} J_{(2j+1+\sigma)/2}(kR) J_{(2j+1-\sigma)/2}(kR). \tag{9.171}$$

The function (9.171) takes the same value for $\sigma = 1$ and $\sigma = -1$. Because of this, one can take either value of σ. Finally, we turn to the imaginary radial momentum $k \to ik/R$ and, again dropping the constant factor, the mode-generating function for the bag boundary condition becomes

$$\tilde{\Delta}_j(ik) = I^2_{j+1}(k) + I^2_j(k) + \frac{2mR}{k} I_{j+1}(k) I_j(k). \tag{9.172}$$

The regularized vacuum energy can be derived in the same way as for the scalar field [see eqn (9.68)], and it becomes

$$E_0(s) = 4\frac{(\mu R)^{2s}}{R}\frac{\cos(\pi s)}{\pi}\sum_{n=1}^{\infty} n\int_{mR}^{\infty} dk\,\left[k^2 - (mR)^2\right]^{(1-2s)/2}\frac{\partial}{\partial k}\ln\tilde{\Delta}_j(ik). \tag{9.173}$$

Here, $j = (2n - 1)/2$ and we have taken into account the fact that the vacuum energy of a spinor field enters with the opposite sign and that we have an additional factor of 4 from the two kinds of particles and two spin projections (see Section 3.3). There is also another difference with respect to the scalar field;

namely, in the spinor case, the mode-generating function (9.172) depends on the mass m.

Next we consider the asymptotic expansion of the mode-generating function. For convenience, we rewrite it in the form

$$\tilde{\Delta}_j(ik) = I_j^2(k)\left(1 + \frac{j^2}{k^2} - \frac{2mRj}{k^2}\right) + [I_j'(k)]^2 + 2\frac{mR-j}{k}I_j(k)I_j'(k) \quad (9.174)$$

using the recurrence relation

$$I_{j+1}(k) = I_j'(k) - \frac{j}{k}I_j(k). \quad (9.175)$$

From eqn (9.76), we represent the logarithm of the mode-generating function (9.174) in the form (9.70) with $z \equiv k/j$. Then, similarly to eqn (9.71), we define

$$h_{as}(j,z) = \sum_{i=-1}^{3} \frac{D_i(z)}{j^i}. \quad (9.176)$$

Here, the functions $D_i(z)$ are given by

$$D_{-1}(z) = 2[\eta(z) - \ln z], \quad D_0(z) = \ln\frac{2}{1+\sqrt{1+z^2}} - \ln(2\pi j),$$

$$D_i(z) = \sum_{a=i}^{3i} x_{i,a} t^a, \quad i = 1, 2, 3. \quad (9.177)$$

The coefficients $x_{i,a}$ are shown in Table 9.11.

The corresponding formulas for the exterior region are obtained by multiplication of the functions $D_i(z)$ by $(-1)^{2i+1}$. In addition, one must reverse the sign of the mass, i.e. $m \to -m$, because of the different sign in the recursion formula

$$K_{j+1}(k) = -K_j'(k) + \frac{j}{k}K_j(k). \quad (9.178)$$

Now we insert $h_{as}(j,z)$ (9.176) into the vacuum energy (9.173) and follow the same calculation procedure that was used in Section 9.2.2 in the case of the massive scalar field. The result is (Elizalde et al. 1998)

TABLE 9.11. The coefficients $x_{i,a}$ for a spinor field, using the notation $x = mR$.

i	1	2	3	4	5	6	7	8	9
1	$\frac{4x-1}{4}$	0	$\frac{1}{12}$	0	0	0	0	0	0
2	0	$-\frac{x^2}{12}$	$\frac{1}{8} - \frac{x}{2}$	$\frac{1}{8} - \frac{x}{2}$	$-\frac{1}{8}$	$-\frac{1}{8}$	0	0	0
3	0	0	$\frac{5}{192} - \frac{x}{8} + \frac{x^3}{3}$	$\frac{1}{8} - \frac{x}{2} + \frac{x^2}{2}$	$\frac{9}{320}$	$-\frac{x}{4} + \frac{x^2}{2}$	$\frac{x-1}{2}$	$\frac{56x-23}{64}$	$\frac{3}{8}$ $\frac{179}{576}$

$$A_{-1}(z) = \frac{(\mu R)^{2s}}{R^2\sqrt{\pi}\,\Gamma\left(s-\frac{1}{2}\right)} \sum_{k=0}^{\infty} \frac{(-1)^k}{k!} (mR)^{2k} \frac{\Gamma(k+s-1)}{2k+2s-1}$$

$$\times \left[2\zeta\left(2k+2s-3,\frac{1}{2}\right) + \zeta\left(2k+2s-2,\frac{1}{2}\right)\right],$$

$$A_0(z) = -\frac{(\mu R)^{2s}}{R^2\sqrt{\pi}\,\Gamma\left(s-\frac{1}{2}\right)} \sum_{k=0}^{\infty} \frac{(-1)^k}{k!} (mR)^{2k} \frac{\Gamma(k+s)}{2k+2s-1} \quad (9.179)$$

$$\times \left[2\zeta\left(2k+2s-2,\frac{1}{2}\right) + \zeta\left(2k+2s-1,\frac{1}{2}\right)\right],$$

$$A_i(z) = -\frac{(\mu R)^{2s}}{R^2\sqrt{\pi}\,\Gamma\left(s-\frac{1}{2}\right)} \sum_{k=0}^{\infty} \frac{(-1)^k}{k!} (mR)^{2k} \sum_{a=i}^{3i} \frac{\Gamma\left(k+s+\frac{a-1}{2}\right)}{\Gamma(a/2)}$$

$$\times \left[2\zeta\left(2k+2s+i-2,\frac{1}{2}\right) + \zeta\left(2k+2s+i-1,\frac{1}{2}\right)\right]$$

for $i = 1, 2, 3$. This is a power series representation which converges for $mR < 1$. A representation which is valid for $mR \geq 1$ was also given by Elizalde et al. (1998).

From eqn (9.179), with eqn (9.131), it is possible to find the heat kernel coefficients. These are given by

$$a_0 = -\frac{16\pi R^3}{3}, \quad a_{1/2} = 0, \quad a_1 = \frac{16\pi R}{3}\left[1 + 3mR + (mR)^2\right],$$

$$a_{3/2} = 8\pi^{3/2}\left[mR + (mR)^2\right], \quad (9.180)$$

$$a_2 = \frac{16}{63\pi R}\left[1 + \frac{21}{5}mR - 21(mR)^2 + 21(mR)^3 + \frac{21}{2}(mR)^4\right].$$

The magnitude of the coefficient a_0 represents the volume of the sphere multiplied by a factor of 4, accounting for particles, antiparticles, and the two spin projections. The negative sign is due to the Fermi statistics. There is no simple explanation for why the next coefficient, $a_{1/2}$, vanishes. The higher coefficients depend on the mass of the field. This is the mass dependence which appears through the mode-generating function (9.172). Thus, all parameters in the renormalization procedure, as discussed in Section 4.3.2, except for σ, are needed. This is the case for a massless spinor field also. Note that if one takes the interior and exterior regions of the sphere together, the situation for the massive spinor field is worse in comparison with the scalar and electromagnetic cases, where the heat kernel coefficients a_1 and a_2 from both sides of the surface compensated each other. This happens only partly in the spinor case. Owing to the mass dependence of the mode-generating functions, there are remaining contributions. In this case the heat kernel coefficients are

$$a_0 = 0, \quad a_{1/2} = 0, \quad a_1 = -32\pi mR^2,$$

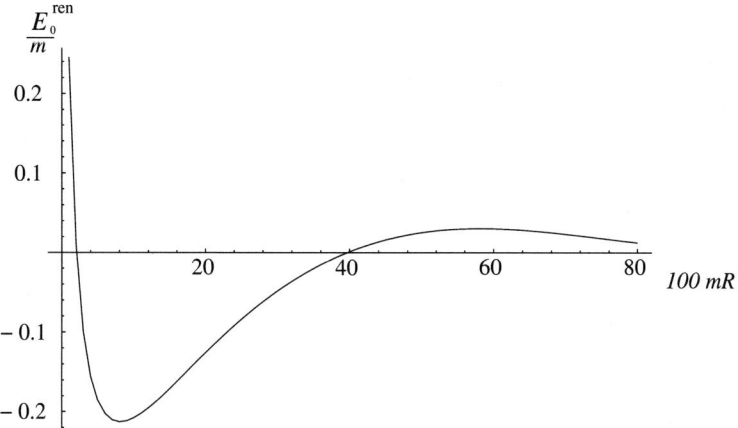

FIG. 9.3. The vacuum energy of a spinor field of mass m in the interior of a bag as a function of mR (Elizalde et al. 2008).

$$a_{3/2} = 16\pi^{3/2}(mR)^2, \quad a_2 = \frac{32\pi m}{15}\left[1 + 5(mR)^2\right]. \quad (9.181)$$

As a result, there are additional divergences which are proportional to powers of the mass.

As already mentioned, it is not clear if the formulation of the bag model can accommodate the renormalization of these divergences. If one ignores this problem, it is possible to calculate the finite part of the vacuum energy (Elizalde et al. 1998). The finite part was defined using the large-mass normalization condition (4.62). However, its application in this case is more complicated than for a massive scalar field, where the large-mass behavior was given by the heat kernel expansion. For the spinor field, an additional mass dependence is contained in the mode-generating function. As a result, it is not possible to determine the large-mass behavior analytically. Elizalde et al. (1998) implemented the renormalization condition (4.62) after a numerical analysis of the large-mass behavior. Within this approach, the renormalized vacuum energy in the interior region of the bag was calculated numerically (see Fig. 9.3). Note that this is a demonstration of the ability to calculate the finite part of the vacuum energy rather than a physical result related to the bag model, because the renormalization issues are not settled.

For a massless spinor field in the interior and exterior regions taken together, the heat kernel coefficient a_2 is zero. In this case a unique finite result can be calculated. The corresponding energy is (Milton 1983, Elizalde et al. 1998)

$$E(R) \equiv E_0^{\text{ren}} = \frac{0.0204}{R}. \quad (9.182)$$

9.5 Spherical shell at nonzero temperature

The calculation of the free energy and other thermodynamic quantities for a spherical shell follows the general lines of Chapter 5. The limiting cases of high and low temperature can be obtained analytically. In this section, we focus on these cases.

9.5.1 Low-temperature expansion

The low-temperature expansion of the Casimir free energy was discussed in Section 5.2. As with all configurations with a finite volume, a gap in the spectrum is present for a sphere. In this case the physical thermal correction for the interior of the sphere is defined by eqns (5.37) and (5.29) with $m = 0$:

$$\Delta_T \mathcal{F}_i = k_B T \sum_{n=1}^{\infty} \sum_l (2l+1) \ln\left(1 - e^{-\beta \Lambda_{l,n}}\right) \qquad (9.183)$$
$$- \alpha_{0,i}(k_B T)^4 - \alpha_{1,i}(k_B T)^3 - \alpha_{2,i}(k_B T)^2.$$

Here, the $\Lambda_{l,n}$ are the eigenvalues of the Laplace operator [for instance eqn (9.16) for Dirichlet boundary conditions]. For $T \to 0$ ($\beta \to \infty$), the first term on the right-hand side of eqn (9.183) decreases exponentially as $\exp(-\beta/R)$ according to the scale set by the radius R of the sphere. A similar suppression takes place for a massive field.

Now we consider a massless field in the exterior region of the sphere. Here, we have a gapless continuous spectrum. In this case the asymptotic expansion for low temperature can be derived from eqn (9.183) using the same methods as in Section 9.1.3. We place the whole system into a large sphere with radius R_c and mode-generating function $\phi_{l,k}(R_c)$. Then the radial sum in the temperature-dependent part of the free energy (9.183) can be written as an integral,

$$\Delta_T \mathcal{F}_e = k_B T \sum_l (2l+1) \int_\gamma \frac{dk}{2\pi i} \ln\left(1 - e^{-\beta k}\right) \frac{\partial}{\partial k} \ln \phi_{l,k}(R_c) \qquad (9.184)$$
$$- \alpha_{0,e}(k_B T)^4 - \alpha_{1,e}(k_B T)^3 - \alpha_{2,e}(k_B T)^2.$$

Here, the limit $R_c \to \infty$ can be performed by the same sequence of steps as that which resulted in eqn (9.32). In so doing, the contribution depending only on R_c was dropped. Now we explicitly write out this contribution, which originates from the black-body radiation inside a sphere with radius R_c. The result is

$$\Delta_T \mathcal{F}_e = k_B T \sum_l (2l+1) \int_0^\infty \frac{dk}{\pi} \ln\left(1 - e^{-\beta k}\right) \frac{\partial}{\partial k} \delta_l(k) \qquad (9.185)$$
$$- \frac{4\pi}{3} R_c^3 \frac{\pi^2}{90}(k_B T)^4 - \alpha_{0,e}(k_B T)^4 - \alpha_{1,e}(k_B T)^3 - \alpha_{2,e}(k_B T)^2,$$

where the $\delta_l(k)$ are the scattering phase shifts defined in eqn (9.27).

The expansion of the first term on the right-hand side of eqn (9.185) at low temperature can be obtained by expanding the phase shifts for small argument into a Taylor series. Taking into account also the fact that $\alpha_{0,e} = -V_e \pi^2/90$, where V_e is the difference between the volumes of spheres with radii R_c and R, one arrives at

$$\Delta_T \mathcal{F}_e = -\frac{1}{\pi} \sum_{j=1}^{\infty} \zeta_R(j+1)(k_B T)^{j+1} \sum_l (2l+1) \delta_l^{(j)}(0) \quad (9.186)$$
$$+ \alpha_{0,i}(k_B T)^4 - \alpha_{1,e}(k_B T)^3 - \alpha_{2,e}(k_B T)^2.$$

The Riemann zeta functions appear from the integration over k.

As examples, we consider a massless scalar field and an electromagnetic field with boundary conditions on a spherical shell of radius R. According to Sections 9.2.4 and 9.3.3, a finite vacuum energy at zero temperature, E_0^{ren}, can be obtained by considering the interior and exterior regions together. The physical thermal correction can also be obtained as a sum of eqns (9.183) and (9.186). In the calculation of this sum, one must take into account the proportionalities $\alpha_1 \sim a_{1/2}$ and $\alpha_2 \sim a_1$ (see the end of Section 5.2) and the symmetry properties of the heat kernel coefficients $a_{1/2,i} = a_{1/2,e}$, $a_{1,i} = -a_{1,e}$ (see the end of Section 9.2.2). As a result,

$$\Delta_T \mathcal{F} = -\frac{1}{\pi} \sum_{j=1}^{\infty} \zeta_R(j+1)(k_B T)^{j+1} \sum_l (2l+1) \delta_l^{(j)}(0) - 2\alpha_{1,i}(k_B T)^3. \quad (9.187)$$

To calculate the derivatives in eqn (9.187), it is convenient to express the phase shifts in terms of the mode-generating functions using eqns (9.27) and (9.35). The expressions for the mode-generating functions can be taken from eqns (9.65) and (9.67) for the scalar field and from Section 9.3.2 for the electromagnetic field. The lowest-order nonzero contributions for the scalar field are

$$\delta_0'(0) = -R, \qquad \delta_1'''(0) = -2R^3 \quad (9.188)$$

for the Dirichlet boundary condition and

$$\delta_0'''(0) = -2R^3, \qquad \delta_1'''(0) = R^3 \quad (9.189)$$

for the Neumann boundary condition. For the electromagnetic field, where the orbital momenta start from $l = 1$, we have

$$\delta_1'''(0) = -2R^3, \qquad \delta_1'''(0) = 4R^3 \quad (9.190)$$

for the TE and TM modes, respectively.

From these expressions, using eqn (9.187), the low-temperature behavior of the thermal correction follows. For the scalar field, using also eqn (5.54) and the expressions for the heat kernel coefficient $a_{1/2}$ in eqn (4.25), we obtain

$$\Delta_T \mathcal{F}^D = \frac{\pi^3 R}{6}(k_B T)^2 - \zeta_R(3) R^2 (k_B T)^3,$$

$$\Delta_T \mathcal{F}^N = \zeta_R(3) R^2 (k_B T)^3 - \frac{\pi^3 R^3}{90} (k_B T)^4. \tag{9.191}$$

For the electromagnetic field, we get

$$\Delta_T \mathcal{F}^{TE} = -\zeta_R(3) R^2 (k_B T)^3 + \frac{\pi^3 R^3}{15} (k_B T)^4,$$

$$\Delta_T \mathcal{F}^{TM} = \zeta_R(3) R^2 (k_B T)^3 - \frac{2\pi^3 R^3}{15} (k_B T)^4. \tag{9.192}$$

Note that above expressions hold for $k_B T R \ll 1$ such that the limit $R \to \infty$ cannot be performed here.

The total thermal correction for the electromagnetic field can be obtained as a sum of the two contributions in eqn (9.192):

$$\Delta_T \mathcal{F}(R, T) = -\frac{\pi^3 R^3}{15} (k_B T)^4. \tag{9.193}$$

Combining eqn (9.193) with the energy at zero temperature (9.136) we get (Balian and Duplantier 1977, 1978)

$$\mathcal{F}(R, T) = \frac{0.0461766}{R} - \frac{\pi^3 R^3}{15} (k_B T)^4. \tag{9.194}$$

The corresponding radial force acting on the spherical shell is repulsive:

$$F(R, T) = \frac{0.0461766}{R^2} + \frac{\pi^3 R^2}{5} (k_B T)^4. \tag{9.195}$$

Thus, for a spherical shell, the thermal correction is of the same sign as the electromagnetic Casimir force at zero temperature, i.e. it leads to additional repulsion.

9.5.2 High-temperature expansion

For the high-temperature expansion of the free energy, we use eqn (5.53) with $m = 0$. Equation (5.53) contains the derivative of the zeta function in addition to the heat kernel coefficients. Using the definition (4.5), this derivative can be expressed through the determinant of the operator P, which in our case is the negative Laplace operator:

$$\zeta_P'(0) = -\ln \det P. \tag{9.196}$$

The determinant of the operator P entering eqn (9.196) is of wider interest and can be found in the literature for various configurations. For a scalar field in the interior region of a spherical shell with a Dirichlet boundary condition, one obtains (Bordag et al. 1996a)

$$\zeta_P'(0) = -\frac{3}{32} - \frac{\ln 2}{12} - \frac{\ln R}{24} + \frac{3\zeta_R(3)}{16\pi^2} + \frac{1}{2} \zeta_R'(-1). \tag{9.197}$$

Note that $\det P$ in eqn (9.196) and R in eqn (9.197) are understood mathematically as dimensionless quantities. After the substitution of these equations and the heat kernel coefficients in eqn (5.53), the correct dimensions are restored. For the exterior region of a sphere with Dirichlet boundary conditions, we get

$$\zeta'_{\rm P}(0) = -\frac{3}{32} + \frac{\ln 2}{6} - \frac{\ln R}{24} - \frac{3\zeta_{\rm R}(3)}{16\pi^2} + \frac{1}{2}\zeta'_{\rm R}(-1). \qquad (9.198)$$

For an electromagnetic field with the interior and exterior regions taken together, the following result is obtained (Bordag et al. 2002):

$$\zeta'_{\rm P}(0) = -\frac{1}{8} + \frac{13\ln 2}{6} + \frac{\ln R}{2} + 6\zeta'_{\rm R}(-1). \qquad (9.199)$$

According to the discussion at the end of Section 5.2, the contributions proportional to $(k_{\rm B}T)^4$, $(k_{\rm B}T)^3$, and $(k_{\rm B}T)^2$ should be subtracted [see also eqns (9.183) and (9.186)]. Using eqn (9.197) for a scalar field with a Dirichlet boundary condition in the interior region, we obtain

$$\mathcal{F}_{\rm i}^{\rm D}(R,T) = \left[\frac{3}{64} + \frac{\ln 2}{24} - \frac{3\zeta_{\rm R}(3)}{32\pi^2} - \frac{\zeta'_{\rm R}(-1)}{4} + \frac{\ln(Rk_{\rm B}T)}{48}\right] k_{\rm B}T + O(T^0). \qquad (9.200)$$

Now we consider the interior and exterior regions together. In this case the high-temperature expansion of the free energy of the scalar field is given by

$$\mathcal{F}^{\rm D}(R,T) = \left[\frac{3}{32} - \frac{\ln 2}{24} - \frac{\zeta'(-1)}{2} + \frac{\ln(Rk_{\rm B}T)}{24}\right] k_{\rm B}T + O(T^{-1}). \qquad (9.201)$$

Note that the term $O(T^0)$ in eqn (9.200) contains the vacuum energy at $T = 0$ and contributions proportional to $a_{n/2}$ with $n \geq 4$. As to eqn (9.201), here the term $O(T^0)$ contains only the vacuum energy at $T = 0$ and contributions proportional to $a_{n/2}$ with odd n.

A similar formula holds for an electromagnetic field. Taking the TE and TM modes together, the high-temperature expansion of the free energy in the entire space is

$$\mathcal{F}(R,T) = \left[\frac{1}{16} - \frac{13\ln 2}{12} - 3\zeta'(-1) - \frac{\ln(Rk_{\rm B}T)}{4}\right] k_{\rm B}T + O(T^0). \qquad (9.202)$$

We conclude with the remark that the dependence on Planck's constant can be restored by the substitution $T \to T/(\hbar c)$ in the logarithms in eqns (9.200)–(9.202). The terms of $\mathcal{F}$ which are linear in $k_{\rm B}T$ do not depend on $\hbar$. This is the classical limit (see Section 7.4.3), or the *entropic contribution*. It does not enter the internal energy (5.32).

9.6 The Casimir effect for a cylinder

Other than for a sphere, there are only a few cases where a separation of variables is possible and explicit calculations of the Casimir energy can be done. Among them, the case of a cylinder is the only one without corners or edges. For this reason, it has attracted attention as the next object of study. However, because of the lower symmetry it is technically more involved. The first results for a cylinder were published only at the beginning of the 1980's (DeRaad and Milton 1981). For a conducting cylindrical shell, the electromagnetic Casimir force is attractive. This is somewhat analogous to the case of a sufficiently long rectangular box (see Section 8.3). This result was confirmed by Gosdzinsky and Romeo (1998), who also considered a scalar field. A dielectric cylinder was discussed by Brevik and Nyland (1994). Divergences similar to those for a dielectric sphere were noted. Later it was found that the heat kernel coefficient a_2 for a dielectric cylinder behaves in the same way as for a dielectric sphere, i.e. it vanishes in the dilute approximation and is nonzero beyond it (Bordag and Pirozhenko 2001). As a consequence, a meaningful result can be obtained in the dilute approximation only. The same two descriptions of the dilute approximation as in Section 9.4.3 can be considered. In the cases of both different and equal speeds of light, the Casimir energy calculated up to the second order of the respective small parameter was found to be equal to zero (Milton et al. 1999). These results were analytically reconfirmed by Klich and Romeo (2000) for equal speeds of light and by Barton (2001) for the dilute approximation using a perturbative expansion in $(c_2 - c_1)$. Further reconfirmations have been obtained by several authors (Romeo and Milton 2005, 2006, Cavero-Pelaez and Milton 2005, 2006).

In the first subsection we derive the basic formulas in the case of a cylindrical geometry for a conducting shell and a dielectric cylinder. For the latter, the polarizations do not separate. As a result, the formulas are more involved and a generalization of the mode summation method is required. We sketch the derivation of the heat kernel coefficients, highlighting the differences from the spherical case. The differences arise from the need to consider zero orbital momentum separately. Then we collect together the expressions necessary for the numerical evaluation of the vacuum energy and present the known results. In the second subsection, the case of a dielectric cylinder is considered. Again, we present some basic formulas and sketch the calculation of the heat kernel coefficients.

9.6.1 Conducting cylindrical shell

In the cylindrical case, the vacuum energy is given by eqn (9.50) with $m = 0$. We use the mode decomposition (9.40) of the field strengths in a cylindrical geometry and apply the general boundary condition (3.46) in the form (9.111), where r is replaced with ρ. In this case the azimuthal and z-components play the role of the tangential components, whereas the normal component is the radial one. Thus eqns (9.40) and (9.41) lead to the conditions

$$\left.\frac{\partial}{\partial \rho} f_{l,n}^{\text{TE}}(\rho)\right|_{\rho=R} = 0, \qquad \left.f_{l,n}^{\text{TM}}(\rho)\right|_{\rho=R} = 0. \qquad (9.203)$$

In the cylindrical configuration all solutions to the wave equation are Bessel and Hankel functions, following the particular case (9.48). We can write down all of

TABLE 9.12. The mode-generating functions $\tilde{\Delta}_l(ik)$ for the two polarizations of an electromagnetic field for a conducting cylindrical shell.

Region	TE mode	TM mode
Interior	$\tilde{\Delta}_l^{\text{TE},i}(ik) = k^{1-l} I_l'(k)$	$\tilde{\Delta}_l^{\text{TM},i}(ik) = k^{-l} I_l(k)$
Exterior	$\tilde{\Delta}_l^{\text{TE},e}(ik) = -k^{1+l} K_l'(k)$	$\tilde{\Delta}_l^{\text{TM},e}(ik) = k^l K_l(k)$

the mode-generating functions immediately. These mode-generating functions, after rotation of the momentum to the imaginary axis and with the usual substitution $k \to k/R$, are shown in Table 9.12. Here again, multiples have been chosen to satisfy all of the requirements imposed on the mode-generating functions. To make progress, one hopes that the corresponding steps for the spherical case can be followed directly. In fact, this is possible to a large extent. Especially, the discussion of the heat kernel coefficients and of the divergences is similar. For the cases of the interior and exterior regions taken separately, the heat kernel coefficient a_2 is nonzero. Consequently, as the field is massless, we do not have a natural renormalization condition and the vacuum energy cannot be defined uniquely. For the case where both regions are taken together, the sum of the coefficients a_2 is zero (the contributions from inside and outside the shell cancel each other) and the vacuum energy can be uniquely defined. Since the coefficient a_1 is zero for the same reasons, a divergent contribution may arise only from the coefficients $a_{1/2}$ and $a_{3/2}$. Further, just as in the spherical case, the contributions to the coefficient $a_{1/2}$ resulting from the two polarizations cancel. The remaining coefficient, $a_{3/2}$, for dimensional reasons, delivers a divergent contribution which does not depend on the radius of the cylinder and may be omitted as not physical. Because of these close similarities with the spherical case, we restrict ourselves to the case where the interior and exterior regions are taken together. Here, the vacuum energy for the two polarizations is uniquely defined and takes finite values in zeta function regularization. There is, however, a difference with respect to the spherical case because one needs to consider the orbital momentum $l = 0$ separately (see below).

In order to simplify the notation, we define the mode-generating functions for the interior and exterior regions taken together,

$$\tilde{\Delta}_l^{\text{TE}} = -k^2 I_l'(k) K_l'(k), \qquad \tilde{\Delta}_l^{\text{TM}} = I_l(k) K_l(k), \qquad (9.204)$$

Further, we divide the vacuum energy into two contributions,

$$E_0(s) = E_{l=0}(s) + E_{l \neq 0}(s). \qquad (9.205)$$

Here, $E_{l=0}(s)$ is the term in the sum (9.50) with $l = 0$ and $E_{l \neq 0}(s)$ is the remaining part. In the latter contribution, we act exactly as in the spherical case. First we define the asymptotic part $h_{\text{as}}(l, z)$ from the first several terms of the uniform asymptotic expansion of the mode-generating functions for large k and l, while $z = k/l$ is fixed. We arrive at the general formula

$$h_{\mathrm{as}}(l, z) = \sum_{i=-1}^{3} \frac{D_i(z)}{l^i}, \qquad (9.206)$$

which is similar to eqn (9.71). Using eqn (9.76), we find the following for the nonzero coefficients $D_i(z)$ in the case of the interior and exterior regions taken together:

$$D_0^{\mathrm{TE}} = -\ln t, \qquad D_0^{\mathrm{TM}} = \ln t, \qquad D_2^{\mathrm{TE,TM}} = \sum_{j=1}^{3} x_{2j}^{\mathrm{TE,TM}} t^{2j}, \qquad (9.207)$$

$$x_2^{\mathrm{TE}} = -\frac{3}{8}, \quad x_4^{\mathrm{TE}} = \frac{5}{4}, \quad x_6^{\mathrm{TE}} = -\frac{7}{8}, \quad x_2^{\mathrm{TM}} = \frac{1}{8}, \quad x_4^{\mathrm{TM}} = -\frac{3}{4}, \quad x_6^{\mathrm{TM}} = \frac{5}{8},$$

where t is defined in eqn (9.77).

To continue, we represent $E_{l\neq 0}(s)$ as a sum of finite and asymptotic parts similarly to eqn (9.72). The asymptotic part is represented as in eqn (9.81), where the functions $A_i(s)$ are obtained from eqn (9.50):

$$A_i(s) = \frac{(\mu R)^{2s}}{R^2} \frac{1}{2\sqrt{\pi}\Gamma(s-\frac{1}{2})\Gamma(2-s)} \sum_{l=1}^{\infty} l^{2-2s-i} \int_0^{\infty} dz\, z^{2-2s} \frac{\partial}{\partial z} D_i(z). \qquad (9.208)$$

The sum over l results in a Riemann zeta function. Integrating over z with the use of eqn (9.103), we get

$$A_0 = \mp \frac{(\mu R)^{2s}}{R^2} \frac{\sqrt{\pi}}{4\Gamma(s-\frac{1}{2})\Gamma(2-s)\sin(\pi s)} \zeta_R(2s-2),$$

$$A_2 = -\frac{(\mu R)^{2s}}{R^2} \frac{1}{2\sqrt{\pi}\Gamma(s-\frac{1}{2})} \sum_{j=1}^{3} x_{2j} \frac{\Gamma(s+j-1)}{\Gamma(j)} \zeta_R(2s), \qquad (9.209)$$

where the signs in A_0 denote the TE and TM modes, respectively, and the x_{2j} are defined in eqn (9.207). We note that the quantity $E_{l\neq 0}^{\mathrm{as}}(s) = A_0 + A_2$ has a pole at $s=0$ arising from the term with $j=1$ in A_2:

$$E_{l\neq 0}^{\mathrm{as,TE}}(s) = \frac{3}{64\pi R^2 s} + O(1), \qquad E_{l\neq 0}^{\mathrm{as,TM}}(s) = -\frac{1}{64\pi R^2 s} + O(1). \qquad (9.210)$$

Now we consider the contribution to $E_0(s)$ from $l=0$ which will compensate this pole part. This contribution consists only of an integration and no summation:

$$E_{l=0}(s) = \frac{(\mu R)^{2s}}{R^2} \frac{1}{4\sqrt{\pi}\Gamma(s-\frac{1}{2})\Gamma(2-s)} \int_0^{\infty} dk\, k^{2-2s} \frac{\partial}{\partial k} \ln \tilde{\Delta}_0(ik). \qquad (9.211)$$

Here, in order to get the asymptotic expansion of the mode-generating function, we need the asymptotic expansions of the modified Bessel functions for large argument (Abramowitz and Stegun 1972),

$$\left.\begin{array}{c} I_l(k) \\ K_l(k) \end{array}\right\} \simeq \frac{\pi^{\mp 1/2}}{\sqrt{2k}} e^{\pm k} \left[1 + \sum_{i=1}^{\infty} (\pm 1)^i \frac{\tilde{u}_i}{k^i} \right], \quad (9.212)$$

$$\left.\begin{array}{c} I'_l(k) \\ K'_l(k) \end{array}\right\} \simeq \pm \frac{\pi^{\mp 1/2}}{\sqrt{2k}} e^{\pm k} \left[1 + \sum_{i=1}^{\infty} (\pm 1)^i \frac{\tilde{v}_i}{k^i} \right].$$

The quantities $\tilde{u}_i$ and $\tilde{v}_i$ are numbers which can be obtained as a limiting case $\tilde{u}_i = \lim_{t \to 0} t^{-i} u_i(t)$ from the Debye polynomials (9.78) (and similarly for $\tilde{v}_i$). Inserting these formulas into the mode-generating functions (9.204), we get

$$\ln \tilde{\Delta}_0^{\rm TE} = \ln k - \ln 2 - \frac{3}{8k^2} + \ldots,$$

$$\ln \tilde{\Delta}_0^{\rm TM} = -\ln k - \ln 2 + \frac{1}{8k^2} + \ldots. \quad (9.213)$$

The higher expansion orders are not needed for the determination of the divergences which appear in $E_{l=0}(s)$ in the limit $s \to 0$. However, eqn (9.213) cannot be inserted directly into eqn (9.211) because the integral would diverge at $k = 0$. For this reason we define the asymptotic functions $\tilde{h}_{\rm as}(k)$ as

$$\tilde{h}_{\rm as}^{\rm TE}(k) = \ln \sqrt{1+k^2} - \frac{7}{8(1+k^2)}, \quad \tilde{h}_{\rm as}^{\rm TM}(k) = -\ln \sqrt{1+k^2} + \frac{5}{8(1+k^2)}. \quad (9.214)$$

These functions differ from eqn (9.213) in higher orders in k [not shown in eqn (9.213)] and by a constant, which does not contribute to eqn (9.211) because of the derivative. Inserting eqn (9.214) into eqn (9.211), we get the respective contributions to $E_{l \neq 0}^{\rm as}(s)$,

$$E_{l=0}^{\rm as, TE}(s) = \frac{(\mu R)^{2s}}{R^2} \frac{(-3 + 7s)\Gamma(s-1)}{32\sqrt{\pi}\Gamma\left(s - \frac{1}{2}\right)},$$

$$E_{l=0}^{\rm as, TM}(s) = \frac{(\mu R)^{2s}}{R^2} \frac{(1 - 5s)\Gamma(s-1)}{32\sqrt{\pi}\Gamma\left(s - \frac{1}{2}\right)}. \quad (9.215)$$

It can be checked that the pole part in eqn (9.215) cancels the poles in eqn (9.210). It can be shown that the same holds for the pole at $s = 1$, i.e. for the heat kernel coefficient a_1. Now the nonzero heat kernel coefficients can be easily calculated. Using eqn (9.131), we get

$$a_{1/2}^{\rm TE} = 2\pi^{5/2} R, \qquad a_{3/2}^{\rm TE} = \frac{5}{16} \pi^{5/2} \frac{1}{R}, \quad (9.216)$$

$$a_{1/2}^{\rm TM} = -2\pi^{5/2} R, \qquad a_{3/2}^{\rm TM} = -\frac{1}{16} \pi^{5/2} \frac{1}{R}.$$

A compensation between the TE and TM modes in $a_{1/2}$ can be observed.

Finally, in order to calculate the finite part of the vacuum energy, we define

$$E_0^{\text{fin}} = E_{l=0}^{\text{fin}} + E_{l\neq 0}^{\text{fin}}. \tag{9.217}$$

The expression for $E_{l=0}^{\text{fin}}$ (and a similar one for $E_{l\neq 0}^{\text{fin}}$) is obtained from eqn (9.211) using integration by parts:

$$E_{l=0}^{\text{fin,TE}} = \frac{1}{4\pi R^2} \int_0^\infty dk\, k\, \left[\ln \tilde{\Delta}_0^{\text{TE}}(ik) - \tilde{h}_{\text{as}}^{\text{TE}}(k)\right],$$

$$E_{l\neq 0}^{\text{fin,TE}} = \frac{1}{2\pi R^2} \sum_{l=1}^\infty \int_0^\infty dk\, k\, \left[\ln \tilde{\Delta}_l^{\text{TE}}(ik) - h_{\text{as}}^{\text{TE}}(l,k)\right]. \tag{9.218}$$

Similar equations can be written for the TM mode. Further, we need to add to eqn (9.218) the regular part of E_0^{as}, which in this case coincides with E_0^{an} defined in eqn (9.88). This is the sum of the two contributions. From eqn (9.209), we get the contribution to the regular part with $l \neq 0$:

$$E_{l\neq 0}^{\text{an,TE}} = \frac{1}{\pi R^2} \left[-\frac{25}{128} + \frac{3}{32}\ln(4\pi) - \frac{1}{16\pi^2}\zeta_R(3)\right],$$

$$E_{l\neq 0}^{\text{an,TM}} = \frac{1}{\pi R^2} \left[\frac{11}{128} - \frac{1}{32}\ln(4\pi) + \frac{1}{16\pi^2}\zeta_R(3)\right]. \tag{9.219}$$

The contribution with $l = 0$ can be obtained from eqn (9.215):

$$E_{l=0}^{\text{an,TE}} = \frac{5 - \ln 8}{32\pi R^2}, \qquad E_{l=0}^{\text{an,TM}} = \frac{-3 + \ln 2}{32\pi R^2}. \tag{9.220}$$

The complete vacuum energy for the TE mode is given by

$$E_0^{\text{ren,TE}} = E_{l=0}^{\text{fin,TE}} + E_{l=0}^{\text{an,TE}} + E_{l\neq 0}^{\text{fin,TE}} + E_{l\neq 0}^{\text{an,TE}}, \tag{9.221}$$

and similarly for the TM mode.

A numerical evaluation of $E_0^{\text{ren, TE}}$ and $E_0^{\text{ren, TM}}$ was done by Gosdzinsky and Romeo (1998). The results for the renormalized vacuum energy are

$$E_0^{\text{ren,TE}} = -\frac{0.002256}{R^2}, \qquad E_0^{\text{ren,TM}} = \frac{0.000098}{R^2}. \tag{9.222}$$

The sum of these results gives the vacuum energy for the electromagnetic field in the case of a conducting cylindrical shell,

$$E(R) \equiv E_0^{\text{ren}} = -\frac{0.002158}{R^2}, \tag{9.223}$$

in accordance with the original result of DeRaad and Milton (1981). It is seen that the TE mode dominates in the energy and that the force is attractive.

9.6.2 Dielectric cylinder

For the case of a dielectric cylinder, the polarizations of an electromagnetic field do not separate, and we need to generalize the derivation of the mode sum representation of the vacuum energy. We start from the expansions (9.40) of the field strengths and define the cylindrical components E_a and B_a, where $a = (\rho, \varphi, z)$, by

$$\boldsymbol{E} = \sum_a \boldsymbol{e}_a E_a \equiv \sum_a \boldsymbol{e}_a \tilde{E}_a \, e^{il\varphi} e^{ik_z z}, \qquad \boldsymbol{B} = \sum_a \boldsymbol{e}_a B_a \equiv \sum_a \boldsymbol{e}_a \tilde{B}_a \, e^{il\varphi} e^{ik_z z}. \tag{9.224}$$

Using eqn (9.41) and the corresponding formula for $\boldsymbol{E}^{\text{TM}}$,

$$\boldsymbol{E}^{\text{TM}} = -\boldsymbol{e}_\rho \frac{\partial}{\partial \rho} \frac{\partial}{\partial z} - \boldsymbol{e}_\varphi \frac{1}{\rho} \frac{\partial}{\partial \varphi} \frac{\partial}{\partial z} + \boldsymbol{e}_z \left(\frac{\partial^2}{\partial \rho^2} + \frac{1}{\rho} \frac{\partial}{\partial \rho} + \frac{1}{\rho^2} \frac{\partial^2}{\partial \varphi^2} \right), \tag{9.225}$$

we get

$$\tilde{E}_\rho = \frac{\omega l}{\rho} f_{l,n}^{\text{TE}}(\rho) + \omega k_z \left[f_{l,n}^{\text{TM}}(\rho) \right]', \quad \tilde{E}_\varphi = i\omega \left[f_{l,n}^{\text{TE}}(\rho) \right]' + i \frac{\omega l k_z}{\rho} f_{l,n}^{\text{TM}}(\rho),$$

$$\tilde{E}_z = -i\omega(\varepsilon_i \mu_i \omega^2 - k_z^2) f_{l,n}^{\text{TM}}(\rho), \quad \tilde{B}_\rho = -ik_z \left[f_{l,n}^{\text{TE}}(\rho) \right]' - i\varepsilon_i \mu_i \omega^2 \frac{\omega l}{\rho} f_{l,n}^{\text{TM}}(\rho),$$

$$\tilde{B}_\varphi = \frac{k_z l}{\rho} f_{l,n}^{\text{TE}}(\rho) + \varepsilon_i \mu_i \omega^2 \left[f_{l,n}^{\text{TM}}(\rho) \right]', \quad \tilde{B}_z = -(\varepsilon_i \mu_i \omega^2 - k_z^2) f_{l,n}^{\text{TE}}(\rho). \tag{9.226}$$

Here we have also used eqn (9.113), taking into account the fact that there are two different dielectric media inside and outside the cylinder.

In terms of these cylindrical components, the matching conditions at $\rho = R$ require that

$$\varepsilon E_\rho, \quad E_\varphi, \quad E_z, \quad B_\rho, \quad \frac{1}{\mu} B_\varphi, \quad \frac{1}{\mu} B_z \tag{9.227}$$

should be continuous. These conditions define a scattering problem on the entire axis $\rho \in [0, \infty)$, and for this case the scattering solutions are

$$\phi_{l,k}^{\text{TE}}(\rho) = \alpha J_l(q\rho) \theta(R - \rho)$$
$$+ \frac{1}{2} \left[f_l^{\text{TE}}(k, k_z) H_l^{(2)}(k\rho) + f_l^{\text{TE}*}(k, k_z) H_l^{(1)}(k\rho) \right] \theta(\rho - R),$$
$$\phi_{l,k}^{\text{TM}}(\rho) = \beta J_l(q\rho) \theta(R - \rho) \tag{9.228}$$
$$+ \frac{1}{2} \left[f_l^{\text{TM}}(k, k_z) H_l^{(2)}(k\rho) + f_l^{\text{TM}*}(k, k_z) H_l^{(1)}(k\rho) \right] \theta(\rho - R).$$

Here the notation $q = \sqrt{\varepsilon_1 \mu_1 \omega^2 - k_z^2}$ and $k = \sqrt{\varepsilon_2 \mu_2 \omega^2 - k_z^2}$ has been used and we have inserted two arbitrary constants α and β, where α is dimensionless and β has the dimension of length. This is necessary because the two solutions are not independent. The Jost functions $f_l^{\text{TE}}(k, k_z)$ and $f_l^{\text{TM}}(k, k_z)$ must be determined from the matching conditions (9.227). In fact, only four of the six conditions

(9.227) are independent, and we impose continuity boundary conditions on the following four combinations:

$$E_\varphi + \frac{k_z l}{\rho k^2} E_z, \quad E_z, \quad B_\varphi + \frac{k_z l}{\rho k^2} B_z, \quad B_z. \tag{9.229}$$

The resulting equations are

$$\text{Re}\left[f_l^{\text{TE}}(k,k_z) H_l^{(2)\prime}(kR)\right] = \frac{q}{k} J_l'(qR)\alpha + \frac{k_z l}{R} \frac{k^2-q^2}{k^3} J_l(qR)\beta,$$

$$\text{Re}\left[f_l^{\text{TM}}(k,k_z) H_l^{(2)}(kR)\right] = \frac{q^2}{k^2} J_l(qR)\beta, \tag{9.230}$$

$$\text{Re}\left[f_l^{\text{TM}}(k,k_z) H_l^{(2)\prime}(kR)\right] = \frac{k_z l}{R} \frac{k^2-q^2}{\varepsilon_2 \mu_1 \omega^2 k^3} J_l(qR)\alpha + \frac{\varepsilon_1 q}{\varepsilon_2 k} J_l'(qR)\beta,$$

$$\text{Re}\left[f_l^{\text{TE}}(k,k_z) H_l^{(2)}(kR)\right] = \frac{\mu_2 q^2}{\mu_1 k^2} J_l(qR)\alpha.$$

The important difference with respect to the spherical case is that these equations cannot be separated into TE and TM parts. This can be observed on using the Jost functions following from eqn (9.230),

$$f_l^{\text{TE}}(k,k_z) = \frac{i\pi kR}{2}\left(w_{11}\alpha + w_{12}\beta\right), \quad f_l^{\text{TM}}(k,k_z) = \frac{i\pi kR}{2}\left(w_{21}\alpha + w_{22}\beta\right), \tag{9.231}$$

where

$$w_{11} = \frac{q}{k} J_l'(qR) H_l^{(1)}(kR) - \frac{\mu_2 q^2}{\mu_1 k^2} J_l(qR) H_l^{(1)\prime}(kR),$$

$$w_{12} = \frac{k_z l}{R} \frac{k^2-q^2}{k^3} J_l(qR) H_l^{(1)}(kR), \tag{9.232}$$

$$w_{21} = \frac{k_z l}{R} \frac{k^2-q^2}{\varepsilon_2 \mu_1 \omega^2 k^3} J_l(qR) H_l^{(1)}(kR),$$

$$w_{22} = \frac{\varepsilon_1 q}{\varepsilon_2 k} J_l'(qR) H_l^{(1)}(kR) - \frac{q^2}{k^2} J_l(qR) H_l^{(1)\prime}(kR).$$

With these Jost functions, the scattering solutions (9.228) are determined as functions of the parameters α and β, which describe, up to a common factor, the relative weights of the two polarizations. The polarizations do not decouple, because a solution with a single polarization on one side of the boundary (say with $\beta = 0$, i.e. a pure TE wave inside the cylinder) has both polarizations outside.

However, this mixing does not create a problem in the calculation of the vacuum energy. As before, we place the dielectric cylinder inside a larger concentric cylinder with radius $R_c > R$. On that cylinder, we impose conductor boundary conditions. In this way, we get a discrete spectrum which includes both polarizations. We apply the boundary conditions to the field strengths in the same way as in the preceding subsection. So we can use eqn (9.203), where now we have to insert the solutions (9.228) at $\rho = R_c > R$. The resulting equations are

$$f_l^{\text{TE}}(k,k_z)H_l^{(2)'}(kR_c) + f_l^{\text{TE}*}(k,k_z)H_l^{(1)'}(kR_c) = 0,$$
$$f_l^{\text{TM}}(k,k_z)H_l^{(2)}(kR_c) + f_l^{\text{TM}*}(k,k_z)H_l^{(1)}(kR_c) = 0. \quad (9.233)$$

Then we substitute eqn (9.231), for the Jost functions, into eqn (9.233). The result is

$$\text{Re}\left[w_{11}H_l^{(2)'}(kR_c)\right]\alpha + \text{Re}\left[w_{12}H_l^{(2)'}(kR_c)\right]\beta = 0, \quad (9.234)$$
$$\text{Re}\left[w_{21}H_l^{(2)}(kR_c)\right]\alpha + \text{Re}\left[w_{22}H_l^{(2)}(kR_c)\right]\beta = 0.$$

This is a homogeneous system of equations for the coefficients α and β. It has nontrivial solutions only if its determinant, which we denote by $\Delta_l(k,k_z,R,R_c)$, is equal to zero. The solutions of the equation $\Delta_l(k,k_z,R,R_c) = 0$ with respect to k give discrete frequencies inside the large conducting cylinder. So we can take $\Delta_l(k,k_z,R,R_c)$ as a mode-generating function, as defined in Section 9.1.2. The next step is to remove the large cylinder, and we can do this in the same way as in Section 9.1.3. We represent the sum over the discrete eigenvalues as an integral as in eqn (9.19) and divide the integration contour into two parts, one above the real axis and the other below. On the upper part, for $R_c \to \infty$, the contributions which contain a product of two Hankel functions of the second kind dominate [see eqn (9.29)], and we keep them. On the lower part, we keep the contributions which contain a product of two Hankel functions of the first kind. All other parts of the mode-generating function deliver vanishing contributions and can be omitted. We denote the part of the mode-generating function which we keep on the upper part by $\Delta_l(k,k_z)$. From eqn (9.234), this is equal to

$$\Delta_l(k,k_z) = w_{11}w_{22} - w_{12}w_{21}. \quad (9.235)$$

On the lower part of the contour, we take the complex conjugate. The remaining steps, for instance the deformation of the integration contour towards the imaginary axis, proceed in the same way as in Section 9.1.4 and we end with eqn (9.49) for $m = 0$, with eqn (9.235) inserted. Note that in this case the integration with respect to k_z cannot be done, since $\Delta_l(k,k_z)$ depends on k_z.

Finally, after a rotation $k \to ik$, the mode-generating function can be expressed in terms of modified Bessel functions. Now we change the variables by the substitutions $k \to k/R$ and $k_z \to k_z/R$. Omitting the constant factor, we rewrite the mode-generating function of the dimensionless variables in the form

$$\tilde{\Delta}_l(ik,k_z) = D_l^{\text{TE}}(q,k)D_l^{\text{TM}}(q,k) + \left[\frac{c_1^2 - c_2^2}{c_1^2 c_2^2}\frac{\omega k_z l}{qk}I_l(q)K_l(k)\right]^2, \quad (9.236)$$

where $c_i = 1/\sqrt{\varepsilon_i\mu_i}$ and

$$D_l^{\text{TE}}(q,k) = \mu_1 k I_l'(q)K_l(k) - \mu_2 q I_l(q)K_l'(k),$$

$$D_l^{\text{TM}}(q,k) = \varepsilon_1 k I_l'(q) K_l(k) - \varepsilon_2 q I_l(q) K_l'(k). \tag{9.237}$$

The second term in eqn (9.236) results from the coupling between the two polarizations. It is absent, for example, for the s-wave, i.e. for $l = 0$, and also for equal speeds of light inside and outside. In the absence of polarization coupling, $D_l^{\text{TE}}(q,k)$ and $D_l^{\text{TM}}(q,k)$ are the mode-generating functions for the corresponding polarizations.

Now we are in a position to obtain the mode sum representation of the Casimir energy for a dielectric cylinder. For this purpose we make the same change of variables ($k \to k/R$, $k_z \to k_z/R$) in eqn (9.49) with $m = 0$, and substitute the mode-generating function from eqn (9.236). The result is

$$E_0(s) = -\frac{(\mu R)^{2s}}{R^2} \frac{\cos(\pi s)}{2\pi} \sum_{l=-\infty}^{\infty} \int_{-\infty}^{\infty} \frac{dk_z}{2\pi} \int_{k_z}^{\infty} dk \, (k^2 - k_z^2)^{(1-2s)/2} \frac{\partial}{\partial k} \ln \tilde{\Delta}_l(ik, k_z). \tag{9.238}$$

This equation can be used for the calculation of the heat kernel coefficients. The calculation procedure follows the same lines as for the dielectric ball. However, it is more involved technically (Bordag and Pirozhenko 2001). The reasons are that it is necessary to investigate the asymptotics of double integrals (since the integration with respect to k_z cannot be carried out explicitly) and the contribution with $l = 0$ requires a separate treatment. The calculations are rather cumbersome, but straightforward. The starting point is eqn (9.238). The heat kernel coefficients can be calculated using eqn (9.131) and, as before, we can replace the mode-generating function with its asymptotic expansion. We have to consider large l, k, and k_z with fixed $z = k/l$ and $\eta = k_z/(zl)$. Using eqn (9.76), the expansion of the mode-generating function (9.236) with $l \neq 0$ takes the form

$$\ln \tilde{\Delta}_l(ik, k_z) \simeq \sum_{i=-1}^{\infty} \frac{D_i(z,\eta)}{l^i}. \tag{9.239}$$

For $l = 0$, we consider large k and k_z with $\tilde{\eta} = k_z/k$ fixed, and the asymptotic expansion is

$$\ln \tilde{\Delta}_0(ik, k_z) \simeq \sum_{i=-1}^{\infty} \frac{\tilde{D}_i(\tilde{\eta})}{k^i}, \tag{9.240}$$

where we have used eqn (9.212). Explicit formulas for the functions $D_i(z,\eta)$ and $\tilde{D}_i(\tilde{\eta})$ are given by Bordag and Pirozhenko (2001). For the calculation of the asymptotic part, it is sufficient to take $i \leq 3$ in eqn (9.239) and $i \leq 2$ in eqn (9.240). Then the asymptotic part of the vacuum energy is given as a sum of the corresponding contributions

$$E_0^{\text{as}}(s) = \sum_{i=-1}^{3} A_i(s) + \tilde{A}(s). \tag{9.241}$$

The functions $A_i(s)$ are found with the help of eqns (9.238) and (9.239). In doing so, we change the order of integrations with respect to k and k_z and use the relation between variables $k_z = \eta z l$. The summation over all $l \neq 0$ results in the Riemann zeta function:

$$A_i(s) = -\frac{(\mu R)^{2s}}{R^2}\frac{\cos(\pi s)}{\pi^2}\zeta_R(2s+i-2) \qquad (9.242)$$

$$\times \int_0^\infty dk\, k^{2-2s}\int_0^1 d\eta\,(1-\eta^2)^{(1-2s)/2}\left(\frac{\partial}{\partial k} - \frac{\eta}{k}\frac{\partial}{\partial \eta}\right)D_i(z,\eta).$$

The function $\tilde{A}(s)$ is found similarly by using eqn (9.240) instead of eqn (9.239), and the relation between the variables $k_z = \tilde{\eta}k$:

$$\tilde{A}(s) = -\frac{(\mu R)^{2s}}{R^2}\frac{\cos(\pi s)}{2\pi^2} \qquad (9.243)$$

$$\times \int_0^\infty dk\, k^{2-2s}\int_0^1 d\tilde{\eta}\,(1-\tilde{\eta}^2)^{(1-2s)/2}\left(\frac{\partial}{\partial k} - \frac{\tilde{\eta}}{k}\frac{\partial}{\partial \tilde{\eta}}\right)\left[\frac{\tilde{D}_1(\tilde{\eta})}{k} + \frac{\tilde{D}_2(\tilde{\eta})}{k^2}\right].$$

The calculation of the heat kernel coefficients is a repeat of that for the diclectric ball. Again, for a_2 there is a compensation, this time between $A_2(s)$ and $\tilde{A}(s)$. The total contribution to a_2 comes from the pole of the zeta function in $A_3(s)$:

$$a_2 = \frac{2c_2}{R^2}\int_0^\infty dk\, k^{2-2s}\int_0^1 d\eta\sqrt{1-\eta^2}\left(\frac{\partial}{\partial k} - \frac{\eta}{k}\frac{\partial}{\partial \eta}\right)D_3(z,\eta). \qquad (9.244)$$

It is not possible to calculate this expression explicitly, but it has been shown that the expansion of a_2 in powers of the difference of speeds of light starts from the third order:

$$a_2 = O\left[(c_1-c_2)^3\right]. \qquad (9.245)$$

The values of a_2 in eqn (9.244) have been calculated numerically and plotted as a function of c_1/c_2 (Bordag and Pirozhenko 2001). Also, the lower coefficients have been calculated. For example,

$$a_0 = -2\pi R^2\left(c_1^{-3} - c_2^{-3}\right), \quad a_{1/2} = -2\pi^{3/2}R\frac{(c_1^2-c_2^2)}{c_1^4 c_2^4(\mu_1+\mu_2)(\varepsilon_1+\varepsilon_2)}, \qquad (9.246)$$

which are similar to eqn (9.147).

As mentioned at the beginning of this subsection, in the dilute approximation the vacuum energy is defined uniquely. This is now confirmed by eqn (9.245), which states that the coefficient a_2 is equal to zero when defined in the perturbation order of $(c_1-c_2)^2$. The respective vacuum energy has been calculated using various methods. As an example, we mention here the pairwise summation of the Casimir-Polder potentials (9.151) in the dilute approximation. In place of eqn (9.152), we now have the formula

$$E(\gamma) = -\frac{23}{8\pi}\frac{(\varepsilon_1-1)^2}{(4\pi)^2} \qquad (9.247)$$

$$\times \int d^3r_1 \, d^3r_2 \left[(z_1 - z_2)^2 + \rho_1^2 + \rho_2^2 - 2\rho_1\rho_2 \cos(\varphi_1 - \varphi_2)\right]^{-\gamma/2}.$$

Here, γ is a regularization parameter and we have to set $\gamma \to 7$ at the end. The energy (9.247) is equal to (Milton et al. 1999)

$$E(\gamma) = -\frac{23}{2^{3+\gamma}\pi}(\varepsilon_1 - 1)^2 \frac{\Gamma\left(\frac{\gamma-1}{2}\right)\Gamma\left(2 - \frac{\gamma}{2}\right)}{(3-\gamma)\Gamma\left(\frac{\gamma}{2}\right)\Gamma\left(\frac{7-\gamma}{2}\right)} \frac{L}{R^{\gamma-5}}, \tag{9.248}$$

where L is the infinite length of the cylinder. For $\gamma = 7$, one obtains $E(7) = 0$. This result was confirmed by Barton (2001) using the perturbative method and by Cavero-Pelaez and Milton (2005) by first taking the orbital-momentum sum. It should be mentioned that the latter calculation, though it follows the same lines as that for the ball starting from eqn (9.159) (see Section 9.3.4), is considerably more involved, and it cannot be represented here.

Finite results have been obtained also for the case of equal speeds of light inside and outside the cylinder. In this case the mode-generating function (9.236) can be written in a simpler form,

$$\Delta_l(ik, k_z) = D_l^{\text{TE}}(k) D_l^{\text{TM}}(k) = \frac{(\varepsilon_1 + \varepsilon_2)(\mu_1 + \mu_2)}{4}\left[1 - \xi^2 k^2 \sigma_l^2(k)\right]. \tag{9.249}$$

Here, we have used the notation $\sigma_l(k) = [I_l(k) K_l(k)]'$ and the identity

$$\sigma_l^2(k) = \frac{1}{k^2} + 4I_l(k)I_l'(k)K_l(k)K_l'(k). \tag{9.250}$$

For $\xi = 1$, this mode-generating function coincides, up to a factor, with the product of the functions (9.204) for a conducting cylindrical shell. We can denote the corresponding vacuum energy by $E(\xi^2)$. As has been found numerically (Milton et al. 1999) and analytically (Klich and Romeo 2000), this quantity is equal to zero for small ξ to the order ξ^2. Higher orders in the expansion for small ξ have also been calculated (Nesterenko and Pirozhenko 1999, Klich and Romeo 2000). Numerical results for the energy, calculated as a function of ξ^2 for $0 \le \xi \le 1$, interpolate smoothly between $E(0) = 0$ and the ideal-conductor value $E(1) = E_0^{\text{ren}}$ in eqn (9.223) (Nesterenko and Pirozhenko 1999). Thus, the vacuum energy for a dilute cylinder has been calculated in those cases where a unique result can be obtained.

10

THE CASIMIR FORCE BETWEEN OBJECTS OF ARBITRARY SHAPE

In Chapter 3, the vacuum energy of quantum fields with boundary conditions on bodies of arbitrary shape was expressed in general terms. The force acting between two bodies can be obtained as a derivative of the energy with respect to the distance between them. As argued in Section 4.3.3, this force is always finite, in agreement with what is expected on physical grounds, since force is a directly measurable quantity. However, calculation of the Casimir force using the general expressions of Chapter 3 is plagued by ultraviolet divergences at intermediate steps. None of the expressions given there represents the force in terms of convergent sums and integrals. All of them contain regularizations which cannot be removed in general expressions. As a consequence, even a direct numerical approach is nearly impossible. For instance, when working with mode expansion, one has to calculate the eigenvalues for a general geometry with a very high precision, since it is necessary to subtract several terms of the asymptotic expansion of the eigenvalues. To date, there has been no successful attempt in the literature to numerically calculate the Casimir force for a complicated geometry in this manner.

A new approach has been found only recently. As shown by Bulgac *et al.* (2006) and Emig *et al.* (2006), it is possible to rewrite a representation of the vacuum energy in terms of a functional determinant such as eqn (3.124) which does not contain any ultraviolet divergences and which is finite at all intermediate steps. The key idea is to subtract the vacuum energy of each of the interacting bodies separately. These are the only contributions which contain divergences. The remaining expression can be rewritten in a compact form in terms of convergent sums and integrals. The first applications of this method were to the force between two parallel cylinders or between two spheres, which are considered below. Here, reliable numerical results were obtained together with asymptotic expansions to high orders for large separation. Also, the technically more involved asymptotic expansion for small separation was obtained. To leading order, this reproduces the proximity force approximation (PFA) and, from the next-to-leading order, the first corrections beyond the PFA were calculated (Bordag 2006a).

An independent and different derivation of the above-mentioned representation was performed in a remarkable paper by Kenneth and Klich (2006) using the block structure of the functional determinants for potentials with disjoint support. Using general assumptions, it was shown that the Casimir force between two bodies which are mirror images of each other is always attractive. As we shall see, this is quite illuminating for the understanding of the representation of the vacuum energy in terms of functional determinants. We note also that a microscopic derivation of the Lifshitz formula in terms of functional determinants was performed by Renne (1971).

It should be mentioned that the technical tools necessary for the new approach were developed earlier, in applications to corrugated planes (Li and Kardar 1992, Büscher and Emig 2004, 2005). The calculations for corrugated surfaces are more complicated than those for a sphere or a cylinder in front of a plane. The reason is that inversion of matrices for a single plane with corrugations is not possible in an explicit way. In this chapter, we provide only a brief summary of the results obtained for corrugated surfaces.

We need to add a remark about the notation. In this chapter we use a variety of different operators, infinite-dimensional matrices, and expansions. The most appropriate mathematical language for this is in terms of operators and mappings. This is, however, quite abstract and would hide the structure and the physical meaning of the mathematical expressions. Therefore we use more specific notations which, we hope, provide more physical intuition.

10.1 Various approaches to the calculation of the Casimir energy

The various techniques used for obtaining a finite representation of the interaction energy in the form of simple formulas are quite different from each other. We divide them into two groups. One group applies to boundary conditions, i.e. to the case where the interaction of the quantum field with the bodies is limited to their surface. The other is relevant to background fields. Here the interaction takes place in the whole volume of the bodies; this group can be applied, for example, to the case of the Casimir force acting between dielectrics.

The general starting point is the representation in eqns (3.111) and (3.112) for the vacuum energy in terms of the trace of the logarithm of the Green's function or, equivalently, in terms of a functional determinant. This is the complete vacuum energy, in the sense that it still includes the contribution from empty space which must be removed at a later step. Here, the Green's function includes the complete interaction with the background. We elaborate on this representation in Section 10.1.2 and derive the *T-matrix approach*. Although the latter approach is also applicable to the case of boundaries, for these we start in the next section from eqn (3.124), where the contribution of empty space has already been removed, and which is formulated in terms of the boundary surface. It should be mentioned that an alternative derivation of this representation has recently been given in terms of source theory (Emig *et al.* 2007, 2008).

10.1.1 *Functional-determinant representation for the case of boundary conditions on separate bodies*

In this section, we consider the vacuum energy of a scalar field obeying Dirichlet or Neumann boundary conditions and that of an electromagnetic field obeying ideal-metal boundary conditions on the surfaces of two separate bodies. For these, we introduce the following notation. Let S be the boundary surface given by

$$\boldsymbol{r} = \boldsymbol{u}(\boldsymbol{\eta}) = \boldsymbol{u}(\eta_1, \eta_2), \tag{10.1}$$

where $\boldsymbol{r}$ is a radius vector in the initial space. The surface is assumed to be static, such that it can be defined by functions $\boldsymbol{u}(\boldsymbol{\eta})$ with parameters $\boldsymbol{\eta} = (\eta_1, \eta_2)$, which can be viewed as coordinates on S. Further, we assume the surface to consist of two nonintersecting parts, S_A and S_B, with

$$S = S_A \cup S_B, \qquad S_A \cap S_B = 0. \tag{10.2}$$

Accordingly, we have to define two parametrizations, $\boldsymbol{u}_A(\boldsymbol{\eta}_A)$ and $\boldsymbol{u}_B(\boldsymbol{\eta}_B)$. Some specific examples are considered below.

For the vacuum energy, we start from the representation (3.124) in terms of the trace of the logarithm of the operator $\tilde{\mathcal{K}}$. Since we consider only static boundary surfaces here it is convenient to first perform the Wick rotation $x_0 \to ix_4$ and to calculate the trace over the time variable,

$$\operatorname{Tr} \ln \tilde{\mathcal{K}} = i \int_{-\infty}^{\infty} dx_4 \, \langle x_4 | \operatorname{Tr} \ln \tilde{\mathcal{K}} | x_4 \rangle, \tag{10.3}$$

where "Tr" on the right-hand side of eqn (10.3) and below in this section is calculated over only the spatial variables. Taking into account the translational invariance with respect to the time variable, we represent the expression on the right-hand side of eqn (10.3) in the form

$$\begin{aligned}
i \int_{-\infty}^{\infty} dx_4 \, \langle x_4 | \operatorname{Tr} \ln \tilde{\mathcal{K}} | x_4 \rangle &= i \int_{-\infty}^{\infty} dx_4 \int_{-\infty}^{\infty} \frac{d\xi}{2\pi} \operatorname{Tr} \ln \tilde{\mathcal{K}}_\xi \\
&= 2iT \int_{-\infty}^{\infty} \frac{d\xi}{2\pi} \operatorname{Tr} \ln \tilde{\mathcal{K}}_\xi,
\end{aligned} \tag{10.4}$$

where $\operatorname{Tr} \ln \tilde{\mathcal{K}}_\xi$ is the Fourier image of $\operatorname{Tr} \ln \tilde{\mathcal{K}}$.

Substituting eqns (10.3) and (10.4) into eqn (3.124), we arrive at the expression for the vacuum energy,

$$E_0 = \frac{1}{2\pi} \int_0^{\infty} d\xi \, \operatorname{Tr} \ln \tilde{\mathcal{K}}_\xi. \tag{10.5}$$

Considering that the translational invariance with respect to the time variable is equivalent to the diagonality of the operators $\tilde{\mathcal{K}}$ and $\ln \tilde{\mathcal{K}}$ in the basis $|x_4\rangle = \exp(i\xi x_4)/\sqrt{2\pi}$, the integral kernel of the operator $\tilde{\mathcal{K}}_\xi$ is given by [see eqn (3.120)]

$$\tilde{K}_\xi(\boldsymbol{\eta}, \boldsymbol{\eta}') = \int d\boldsymbol{r} \int d\boldsymbol{r}' \, H(\boldsymbol{\eta}, \boldsymbol{r}) \, G_\xi(\boldsymbol{r}, \boldsymbol{r}') H(\boldsymbol{\eta}', \boldsymbol{r}'). \tag{10.6}$$

For the scalar field, the Green's function $G_\xi(\boldsymbol{r}, \boldsymbol{r}')$ is

$$G_\xi(\boldsymbol{r},\boldsymbol{r}') = \int \frac{d\boldsymbol{k}}{(2\pi)^3} \frac{e^{i\boldsymbol{k}\cdot(\boldsymbol{r}-\boldsymbol{r}')}}{\xi^2 + k^2}. \tag{10.7}$$

This is the Fourier image with respect to the time variable of the Green's function (3.87) for a scalar field in free space. For an electromagnetic field (see Section 3.2), we take the Coulomb gauge, and the relevant field is the vector potential $\boldsymbol{A}(x)$. For this field, $\nabla \cdot \boldsymbol{A} = 0$, and the corresponding Green's function is

$$G_\xi(\boldsymbol{r},\boldsymbol{r}')_{ij} = \int \frac{d\boldsymbol{k}}{(2\pi)^3} \left(\delta_{ij} - \frac{k_i k_j}{k^2}\right) \frac{e^{i\boldsymbol{k}\cdot(\boldsymbol{r}-\boldsymbol{r}')}}{\xi^2 + k^2}. \tag{10.8}$$

In Section 3.6, we introduced the projector $H(\eta, x)$ for a scalar field with Dirichlet boundary conditions. With the notation used in this section, this is

$$H(\boldsymbol{\eta},\boldsymbol{r}) = \delta^3(\boldsymbol{r} - \boldsymbol{u}(\boldsymbol{\eta})). \tag{10.9}$$

By integration over $\boldsymbol{r}$, it puts the argument of the Green's function onto the surface, as in eqn. (3.120). This projector can be generalized in an obvious way to the case of Neumann boundary conditions:

$$H(\boldsymbol{\eta},\boldsymbol{r}) = \delta^3(\boldsymbol{r} - \boldsymbol{u}(\boldsymbol{\eta}))(\boldsymbol{n} \cdot \nabla_{\boldsymbol{r}}). \tag{10.10}$$

Another generalization is that for the case of an electromagnetic field. Here the boundary conditions are for a generic boundary surface, given by eqn (3.46). In fact, these are two independent conditions, and it is always possible to define two projection vectors $H_i^{(s)}$ with $s = 1, 2$ such that these boundary conditions are equivalent to

$$H_i^{(s)}(\boldsymbol{\eta},\boldsymbol{r}) = H_i^{(s)} \delta^3(\boldsymbol{r} - \boldsymbol{u}(\boldsymbol{\eta})). \tag{10.11}$$

In the following, we do not use the most general form of $H_i^{(s)}$. We restrict ourselves to special cases. The first is that of a plane surface perpendicular to the z-axis, for which an obvious choice is

$$\boldsymbol{H}^{(1)} = \boldsymbol{e}_z \times \nabla, \qquad \boldsymbol{H}^{(2)} = \boldsymbol{e}_z \times (\boldsymbol{e}_z \times \nabla). \tag{10.12}$$

The second example is that of a spherical surface. Here we define

$$\boldsymbol{H}^{(1)} = \boldsymbol{L}, \qquad \boldsymbol{H}^{(2)} = \boldsymbol{L} \times \boldsymbol{n}, \tag{10.13}$$

where $\boldsymbol{L}$ is the orbital-momentum operator and $\boldsymbol{n}$ is the normal vector (see Section 9.1.1). Using eqn (9.6), the fulfillment of the boundary conditions (9.111) can be checked immediately. It must be mentioned that the kernel (10.6) now carries additional indices indicating the two polarizations and, in place of eqn (10.6), we have to write

$$\tilde{K}_\xi^{s,t}(\boldsymbol{\eta},\boldsymbol{\eta}') = \int d\boldsymbol{r} \int d\boldsymbol{r}' \, H_i^{(s)}(\boldsymbol{\eta},\boldsymbol{r}) \, [G_\xi(\boldsymbol{r},\boldsymbol{r}')]_{ij} \, H_j^{(t)+}(\boldsymbol{\eta}',\boldsymbol{r}'). \tag{10.14}$$

Here, the second factor is taken in Hermitian conjugation. For this, we note that $(\boldsymbol{L} \times \boldsymbol{n})^+ = -\boldsymbol{n} \times \boldsymbol{L}$. When inserting this kernel into the representation (10.5) for the vacuum energy, one needs to take the trace over the indices s and t also.

After having defined all necessary objects on a generic surface, we consider the division (10.2) of the surface into two parts and represent the kernel of $\tilde{\mathcal{K}}$ as a block matrix with respect to the two surfaces,

$$\tilde{K}_\xi^{(b)} = \begin{pmatrix} \tilde{K}_{\xi,AA}(\boldsymbol{\eta}_A, \boldsymbol{\eta}'_A) \,, & \tilde{K}_{\xi,AB}(\boldsymbol{\eta}_A, \boldsymbol{\eta}'_B) \\ \tilde{K}_{\xi,BA}(\boldsymbol{\eta}_B, \boldsymbol{\eta}'_A) \,, & \tilde{K}_{\xi,BB}(\boldsymbol{\eta}_B, \boldsymbol{\eta}'_B) \end{pmatrix}. \tag{10.15}$$

It must be mentioned that the kernels $\tilde{K}_{\xi,AA}(\boldsymbol{\eta}_A, \boldsymbol{\eta}'_A)$ and $\tilde{K}_{\xi,BB}(\boldsymbol{\eta}_B, \boldsymbol{\eta}'_B)$, if viewed as matrices, are square matrices whereas, in general, the off-diagonal kernels are rectangular. Also, the parameters $\boldsymbol{\eta}_A$ and $\boldsymbol{\eta}_B$ may have completely different structures, since the two parts of the surface S may be different.

The multiplication of two matrices such as those in eqn (10.15) must involve a summation and integration over the corresponding variables. This becomes important if, for example, we consider the inverse of one of the kernels on the diagonal. The square matrices can be inverted using

$$\int d\mu(\boldsymbol{\eta}_{A''}) \, \tilde{K}_{\xi,AA}(\boldsymbol{\eta}_A, \boldsymbol{\eta}''_A) \, \tilde{K}_{\xi,AA}^{-1}(\boldsymbol{\eta}''_A, \boldsymbol{\eta}'_A) = \delta^2(\boldsymbol{\eta}_A - \boldsymbol{\eta}'_A). \tag{10.16}$$

The measure is given by the induced metric,

$$d\mu(\boldsymbol{\eta}_A) = \sqrt{g_A} \, d\boldsymbol{\eta}_A, \tag{10.17}$$

where

$$g_A = \det\left(\frac{\partial \mathbf{r}}{\partial \eta_i} \cdot \frac{\partial \mathbf{r}}{\partial \eta_j}\right). \tag{10.18}$$

Similar equations hold with $A \to B$.

With eqns (10.5) and (10.15), we have a representation for the complete vacuum energy related to the two surfaces S_A and S_B. It also contains the vacuum self-energies of the individual surfaces, which are infinite. The most important step is the separation of these energies into a distance-independent part carrying all the divergences. This can be done by rewriting eqn (10.15) as

$$\tilde{K}_\xi^{(b)} = \begin{pmatrix} \tilde{K}_{\xi,AA} & 0 \\ 0 & 1 \end{pmatrix} \begin{pmatrix} 1 & 0 \\ 0 & \tilde{K}_{\xi,BB} \end{pmatrix} \begin{pmatrix} 1 & \tilde{K}_{\xi,AA}^{-1}\tilde{K}_{\xi,AB} \\ \tilde{K}_{\xi,BB}^{-1}\tilde{K}_{\xi,BA} & 1 \end{pmatrix}. \tag{10.19}$$

Here, in order to simplify the notation, we have dropped the arguments. This is a product of three matrices. The first two are diagonal and depend only on the corresponding individual surfaces, whereas the third matrix includes the nondiagonal contributions. The matrix element "1" in all of the matrices must be understood as the unit matrix in the corresponding subspace, and the inverse matrices are the same as those defined in eqn (10.16). The next step is to note that in eqn (10.5), $\operatorname{Tr} \ln \mathcal{K} = \ln \det \mathcal{K}$. Thus, from eqn (10.19), we get three additive contributions to the vacuum energy. The first two of these contributions can be dropped, since they do not depend on the distance between the surfaces.

In so doing, we have replaced the operator $\mathcal{K}$ in eqn (10.5) with its integral kernel. This is possible if one includes all integrations with respect to the kernel arguments in the definition of the trace. The last step is to rewrite the resulting expression using the following simple relation which holds for block matrices:

$$\det \begin{pmatrix} 1 & B \\ C & 1 \end{pmatrix} = \det(1 - BC) = \det(1 - CB). \tag{10.20}$$

We then arrive at the following representation for the contribution to the vacuum energy depending on the distance between the two surfaces,

$$E = \frac{1}{2\pi} \int_0^\infty d\xi \, \mathrm{Tr} \ln \left(1 - \tilde{K}_{\xi,AA}^{-1} \tilde{K}_{\xi,AB} \tilde{K}_{\xi,BB}^{-1} \tilde{K}_{\xi,BA}\right). \tag{10.21}$$

Here, the arguments have been dropped and we emphasize once again that the integration over them is included in the definition of the "Tr".

It must be mentioned that the contributions to the energy from the first two factors on the right-hand side of eqn (10.19) are just what one would get if each of the surfaces S_A and S_B were present alone. Hence, these are the energies associated with a single surface and therefore they do not depend on the distance between the two surfaces. Keeping in mind the general statement in Section 4.3.3 that the force between two bodies must be finite, it is the subtraction of these energies which removes the infinities. To indicate that the expression for the vacuum energy (10.21) is finite, we have dropped the lower index 0.

Equation (10.21) is a representation of the vacuum interaction energy between two surfaces S_A and S_B which does not contain ultraviolet divergences. In fact, all integrations and summations in this expression are convergent. Below, we shall use it in specific calculations.

10.1.2 T-matrix approach for potentials with disjoint support

In the preceding subsection, we considered the vacuum energy in the presence of boundaries. There the role of the interaction was played by some boundary conditions. In this subsection, we consider a different scenario, where the quantum field interacts with a background potential $V(\mathbf{r})$. For this, we have in mind primarily the interaction of an electromagnetic field with a dielectric body. In contrast to the case of ideal-metal boundary conditions, here the quantum field penetrates into the body and the interaction takes place throughout the whole volume, not just at the surface.

For the vacuum energy, we use the representation (3.112). As in Section 10.1.1, we assume a static background, and, similarly to eqn (10.5), arrive at

$$E_0 = -\frac{1}{2\pi} \int_0^\infty d\xi \, \mathrm{Tr} \ln \mathcal{G}_\xi^{(V)}. \tag{10.22}$$

Here, $\mathcal{G}_\xi^{(V)}$ is the operator whose integral kernel is the Green's function of the field in the background potential $V(r)$, obeying the equation

$$[\xi^2 - \nabla^2 + V(r)] G_\xi^{(V)}(r, r') = \delta^3(r - r'). \tag{10.23}$$

Note that in eqn (10.5), the operator $\tilde{\mathcal{K}}_\xi$ is used, whereas eqn (10.22) is expressed in terms of the operator $\mathcal{G}_\xi^{(V)}$. Keeping in mind eqn (3.111), this resulted in the negative sign on the right-hand side of eqn (10.22). A specific example of the potential would be provided by a region D given by a characteristic function $\chi(r)$ defined as $\chi(r) = 1$ for $r \in D$ and $\chi(r) = 0$ for $r \notin D$, filled with a dielectric of permittivity ε, so that

$$V(r) = (\varepsilon - 1)\chi(r)\xi^2. \tag{10.24}$$

However, the following derivations are valid for any background potential $V(r)$. Note that the frequency dependence of this potential does not pose any difficulty, since we consider only a static region D.

We continue by introducing the T-matrix operator. For this, it is necessary to perform a number of transformations, which are done at the operator level. For all quantities involved, we introduce corresponding operators and denote them by calligraphic letters. Thus, we rewrite eqn (10.23) in the form

$$[\xi^2 - \nabla^2 + \mathcal{V}] \mathcal{G}_\xi^{(V)} = \mathbf{1}, \tag{10.25}$$

where $\mathbf{1}$ is the unit operator and $\mathcal{V}$ is the operator for multiplication by the function $V(r)$. The operator $\mathcal{G}_\xi^{(0)}$ for the Green's function without a potential is defined by

$$(\xi^2 - \nabla^2) \mathcal{G}_\xi^{(0)} = \mathbf{1}. \tag{10.26}$$

The integral kernel of $\mathcal{G}_\xi^{(0)}$ is given by eqn (10.7). Using eqn (10.26), we rewrite eqn (10.25) in the form of an integral equation for the Green's function,

$$\mathcal{G}_\xi^{(V)} = \mathcal{G}_\xi^{(0)} - \mathcal{G}_\xi^{(0)} \mathcal{V} \mathcal{G}_\xi^{(V)}. \tag{10.27}$$

In scattering theory, this is known as the Lippmann–Schwinger equation. Another expression for the same quantity is

$$\mathcal{G}_\xi^{(V)} = \left(1 + \mathcal{G}_\xi^{(0)} \mathcal{V}\right)^{-1} \mathcal{G}_\xi^{(0)}, \tag{10.28}$$

which is a formal solution of eqn (10.25). This also gives a perturbative solution if expanded in powers of the potential. Equation (10.27) can be rewritten with the help of eqn (10.28) in the form

$$\mathcal{G}_\xi^{(V)} = \mathcal{G}_\xi^{(0)} - \mathcal{G}_\xi^{(0)} \mathcal{T} \mathcal{G}_\xi^{(0)}, \tag{10.29}$$

where

$$\mathcal{T} = \mathcal{V} \left(1 + \mathcal{G}_\xi^{(0)} \mathcal{V}\right)^{-1} \tag{10.30}$$

is called the *T-matrix*. This is widely used in the theory of light scattering, where it is the basic object for expressing the properties of scatterers [see e.g. the book by Bohren and Huffmann (1998)]. Using

$$\mathcal{S} = 1 - \mathcal{G}_\xi^{(0)} \mathcal{T}, \tag{10.31}$$

the T-matrix can be related to the scattering matrix (S-matrix). Using eqn (10.28), the T-matrix (10.30) can also be related to the Green's function:

$$\mathcal{T} = \mathcal{V} - \mathcal{V}\mathcal{G}_\xi^{(V)}\mathcal{V}. \tag{10.32}$$

Below, we shall also use the relation

$$\mathcal{V}\mathcal{G}_\xi^{(V)} = \mathcal{T}\mathcal{G}_\xi^{(0)}, \tag{10.33}$$

which follows from eqn (10.28) by multiplication by $\mathcal{V}$ using eqn (10.30).

In this section, we are interested in the vacuum interaction of two disjoint bodies. Therefore we consider a potential consisting of the sum

$$V(\mathbf{r}) = V_A(\mathbf{r}) + V_B(\mathbf{r}). \tag{10.34}$$

We define the corresponding operators $\mathcal{G}_\xi^{V_A}$ and $\mathcal{G}_\xi^{V_B}$ using eqn (10.25) for the individual potentials $V_A(\mathbf{r})$ and $V_B(\mathbf{r})$, respectively. In a similar way, we define the individual T-matrices $\mathcal{T}^A$ and $\mathcal{T}^B$. Now we separate the contributions from the individual potentials. For this purpose, we rewrite the following expression entering eqn (10.28):

$$1 + \mathcal{G}_\xi^{(0)}(\mathcal{V}_A + \mathcal{V}_B) = \left(1 + \mathcal{G}_\xi^{(0)}\mathcal{V}_A\right)\left(1 + \mathcal{G}_\xi^{(0)}\mathcal{V}_B\right) - \mathcal{G}_\xi^{(0)}\mathcal{V}_A\mathcal{G}_\xi^{(0)}\mathcal{V}_B$$
$$= \left(1 + \mathcal{G}^{(0)}\mathcal{V}_A\right)(1 - \mathcal{M}_\xi)\left(1 + \mathcal{G}^{(0)}\mathcal{V}_B\right). \tag{10.35}$$

Here we have introduced

$$\mathcal{M}_\xi = \left(1 + \mathcal{G}_\xi^{(0)}\mathcal{V}_A\right)^{-1}\mathcal{G}_\xi^{(0)}\mathcal{V}_A\mathcal{G}_\xi^{(0)}\mathcal{V}_B\left(1 + \mathcal{G}_\xi^{(0)}\mathcal{V}_B\right)^{-1}, \tag{10.36}$$

which can be expressed, using eqn (10.30) applied to the potentials V_A and V_B, in terms of the T-matrices:

$$\mathcal{M}_\xi = \mathcal{G}_\xi^{(0)}\mathcal{T}^A\mathcal{G}_\xi^{(0)}\mathcal{T}^B. \tag{10.37}$$

To obtain this equation, we have also made use of

$$(1 + \mathcal{G}_\xi^{(0)}\mathcal{V})^{-1}\mathcal{G}_\xi^{(0)} = \mathcal{G}_\xi^{(0)}(1 + \mathcal{V}\mathcal{G}_\xi^{(0)})^{-1},$$
$$\mathcal{V}(1 + \mathcal{G}_\xi^{(0)}\mathcal{V})^{-1} = (1 + \mathcal{V}\mathcal{G}_\xi^{(0)})^{-1}\mathcal{V}. \tag{10.38}$$

Inserting eqn (10.35) into eqn (10.28) and substituting the result into $\mathrm{Tr} \ln \mathcal{G}_\xi^V$ in eqn (10.22), we arrive at

$$\mathrm{Tr} \ln \mathcal{G}_\xi^{(V_A + V_B)} = -\mathrm{Tr} \ln \mathcal{G}_\xi^{(0)} + \mathrm{Tr} \ln \mathcal{G}_\xi^{(V_A)} + \mathrm{Tr} \ln \mathcal{G}_\xi^{(V_B)} - \mathrm{Tr} \ln(1 - \mathcal{M}_\xi). \tag{10.39}$$

Note that eqn (10.39) holds for any background potential. On a formal level, it can also be justified by an expansion in powers of the potentials. In this sense, eqn (10.39) can be viewed as a resummation of the perturbative expansion. With respect to the vacuum energy, the first three contributions on the right-hand side of eqn (10.39) do not depend on the distance between the two bodies described by the potentials $V_A(r)$ and $V_B(r)$. This is because the first term does not depend on the potentials, and the second and the third terms depend individually on only one of them. Therefore the vacuum interaction between the two bodies is contained solely in the last term. Dropping the first three terms in eqn (10.39), we arrive at a finite expression for the part of the vacuum energy (10.22) depending on the separation distance:

$$E = \frac{1}{2\pi} \int_0^\infty d\xi \, \text{Tr} \ln (1 - \mathcal{M}_\xi) = \frac{1}{2\pi} \int_0^\infty d\xi \, \text{Tr} \ln \left(1 - \mathcal{G}_\xi^{(0)} T^A \mathcal{G}_\xi^{(0)} T^B \right). \quad (10.40)$$

This equation is sometimes referred to as the *TGTG representation* or the *T-matrix representation*. Note that it is possible to perform cyclic permutations of the operators within a trace.

Now we take the final step in the derivation using the fact that the two potentials have a disjoint support. We denote by

$$A = \text{supp} \, V_A(r), \qquad B = \text{supp} \, V_B(r) \qquad (10.41)$$

the regions in space where the potentials take nonvanishing values, and we assume that these regions do not overlap, i.e. $A \cap B = 0$. We are going to use this property in the representation (10.37) of the operator $\mathcal{M}_\xi$. For this, we note that the integral kernel $T(r, r')$ of a T-matrix obtained using eqn (10.32) [or, equivalently, eqn (10.30)] is nonvanishing only if both arguments are in one support region. As a consequence, the kernel of the operator $\mathcal{M}_\xi$ in eqn (10.40), with all arguments and integrations shown explicitly, is given by

$$M_\xi(r, r') = \int_A dr'' \int_B d\tilde{r} \int_B d\tilde{r}' \, T^{V_A}(r, r'')$$
$$\times G_\xi^{(0)}(r'', \tilde{r}) T^{V_B}(\tilde{r}, \tilde{r}') G_\xi^{(0)}(\tilde{r}', r'), \qquad (10.42)$$

where $T(r, r')$ is the integral kernel of the operator T. Here we have marked the arguments from region B with a tilde; those from region A are without a tilde. We note that after insertion of $\mathcal{M}_\xi$ into the expansion of the logarithm, the arguments r and r' will take values from region A only. This structure makes it meaningful to change the notation and to write T_{AA} for the T-matrix whose integral kernel is $T^{V_A}(r, r'')$, with both arguments belonging to region A, and to write T_{BB} for the kernel $T^{V_B}(\tilde{r}, \tilde{r}')$. In a similar way, we introduce $\mathcal{G}_{\xi,AB}^{(0)}$ and $\mathcal{G}_{\xi,BA}^{(0)}$. These are the operators for the free-space Green's functions with arguments in the corresponding regions. With this notation, again dropping the arguments for brevity, and performing cyclic permutations of the operators, we represent the vacuum energy (10.40) in the form (Kenneth and Klich 2006, 2008)

$$E = \frac{1}{2\pi} \int_0^\infty d\xi \, \text{Tr} \ln \left(1 - T_{AA} \mathcal{G}_{\xi,AB}^{(0)} T_{BB} \mathcal{G}_{\xi,BA}^{(0)} \right). \qquad (10.43)$$

This is the final form for the vacuum interaction energy for two background potentials with nonoverlapping supports. It can be used, for example, with the potential (10.24) for two dielectric bodies. The expression (10.43) does not contain ultraviolet divergences, which go away after the first three terms in eqn (10.39) are dropped in the case of nonoverlapping support regions. The latter is necessary, since any overlap of the two background potentials may result in a nonvanishing contribution to some heat kernel coefficients, which are connected with corresponding divergences (these coefficients, however, would not depend on the relative location of the two potentials).

The representation (10.43) of the vacuum energy for background potentials is complementary to the representation (10.21) for boundary conditions. The structures of the two are similar. The main difference is in the integration regions: the surfaces of the interacting bodies in eqn (10.21) and the volumes bounded by these surfaces in eqn (10.43). The operator $\tilde{\mathcal{K}}_{AB}$ in eqn (10.21) is in fact the same free-space Green's function as $\mathcal{G}^{(0)}_{\xi,AB}$ in eqn (10.43), and the inverse kernels $\tilde{\mathcal{K}}^{-1}_{AA}$, $\tilde{\mathcal{K}}^{-1}_{BB}$ are the T-matrices for scattering on an impermeable surface. In this sense, the representation (10.21) is the limiting case of eqn (10.43).

It should be mentioned that other equivalent representations have been independently derived, using the block structure of the functional determinant that appears for nonintersecting support regions (Kenneth and Klich 2006) and using the multi-scattering approach (Bulgac et al. 2006).

10.2 Casimir attraction between two bodies

The sign of the Casimir force continues to be a topic of great interest. Historically, it first appeared in connection with Boyer's result for a conducting sphere (see Section 9.3.3). At present, the interest in the sign of the Casimir force is generated by the search for nonadhesive configurations in nanomechanical applications. The following general rules for this sign can be formulated. For dielectric bodies, in the dilute approximation, one may start with the pairwise summation of attractive forces. This results in an attractive force between two distinct bodies, since the corresponding integration is convergent. The sign may change if the integration is divergent. For example, this happens for a single dielectric ball in the dilute approximation (see Section 9.3.4). To what extent these statements are valid beyond the dilute approximation is, in general, not known. A sign change of the Casimir force can also occur owing to the boundary conditions. The simplest cases are the repulsion between two planes, when one of them has Dirichlet and the other has Neumann boundary conditions (see Section 7.2.2), and the repulsion between dielectric layers sandwiching a third body such that the permittivities are $\varepsilon_1 > \varepsilon_2 > \varepsilon_3$ (see Section 19.5.3). Except for above situations, there is currently only one general statement about the sign of the Casimir force [which has been formulated as a theorem by Kenneth and Klich (2006)]: *For two nonmagnetic bodies which are mirror images of each other, the electromagnetic and scalar Casimir forces are attractive.* An immediate consequence of this statement is that the force between a single body and a plane is attractive provided that the same boundary conditions are imposed on both of them (see below).

The proof of the above statement gives a deeper insight into the structure of general representations of the vacuum energy as in eqns (10.21) and (10.43). We start with the definition of an operator J which performs a reflection in the symmetry plane $z = 0$:

$$J\mathbf{r} \equiv J \begin{pmatrix} x \\ y \\ z \end{pmatrix} = \begin{pmatrix} x \\ y \\ -z \end{pmatrix}. \tag{10.44}$$

Let the two regions A and B be connected by a mirror symmetry relative to the $z = 0$ symmetry plane:

$$JA = B. \tag{10.45}$$

We assume the region A to be entirely in the half-space $z < 0$. The mapping (10.44) leads to an unitary operator $\mathcal{J}$ that transforms the wave functions and Green's functions by means of

$$\mathcal{J}\psi(\mathbf{r}) = \psi(J\mathbf{r}). \tag{10.46}$$

For a vector field, we also have to reflect its third component:

$$\mathcal{J} \begin{pmatrix} A_x \\ A_y \\ A_z \end{pmatrix} = \begin{pmatrix} A_x \\ A_y \\ -A_z \end{pmatrix}. \tag{10.47}$$

Now we consider the representation (10.43) of the vacuum energy for two interacting bodies given by background potentials in the regions A and B, which are related by eqn (10.45). The two potentials introduced in eqn (10.34) are assumed to be related accordingly:

$$V_B(\mathbf{r}) = V_A(J\mathbf{r}). \tag{10.48}$$

Under these assumptions, the operators of the free-space Green's functions and T-matrices are connected by

$$\mathcal{G}^{(0)}_{\xi,AB} = \mathcal{J}^+ \mathcal{G}^{(0)}_{\xi,BA} \mathcal{J}^+, \qquad T_{AA} = \mathcal{J}^+ T_{BB} \mathcal{J}, \tag{10.49}$$

respectively. Using cyclic permutations within the trace, this allows one to represent the vacuum energy (10.43) in the form

$$E = \frac{1}{2\pi} \int_0^\infty d\xi \, \mathrm{Tr} \ln \left(1 - \mathcal{Y}_\xi^2\right), \tag{10.50}$$

where

$$\mathcal{Y}_\xi = \sqrt{T_{BB}} \mathcal{G}^{(0)}_{\xi,BA} \mathcal{J}^+ \sqrt{T_{BB}}. \tag{10.51}$$

The proof continues by showing that, in the operator sense, the following inequalities hold:

$$\mathcal{G}^{(0)}_{\xi,BA}\mathcal{J}^+ > 0, \qquad \frac{\partial}{\partial a}\mathcal{G}^{(0)}_{\xi,BA}\mathcal{J}^+ < 0. \tag{10.52}$$

Here a is the minimal distance between points in A and their mirror images. To prove eqn (10.52), we assume that $\psi(\mathbf{r})$ is a function with support in B and consider

$$I(a) = \int_B d\mathbf{r} \int_B d\mathbf{r}' \; \psi^*(\mathbf{r})\mathcal{G}^{(0)}_{\xi,BA}\mathcal{J}^+\psi(\mathbf{r}'). \tag{10.53}$$

Using the representation (10.7) for the Green's function, we arrive at

$$I(a) = \int_B d\mathbf{r} \int_B d\mathbf{r}' \; \psi^*(\mathbf{r}) \int \frac{d\mathbf{k}_\perp}{(2\pi)^2} \frac{1}{2\gamma} e^{i\mathbf{k}_\perp \cdot (\mathbf{r}_\perp - \mathbf{r}'_\perp) - \gamma(z+z')}\psi(\mathbf{r}'), \tag{10.54}$$

where the subscript $\perp$ denotes two-dimensional vectors perpendicular to the z axis (i.e. parallel to the symmetry plane). In eqn (10.54), we have carried out the integration over k_3 making use of the fact that z and z' take only positive values, i.e. $z > 0$ and $z' > 0$, and have defined $\gamma = \sqrt{\xi^2 + k_\perp^2}$. This expression can be rewritten in the form

$$I(a) = \int \frac{d\mathbf{k}_\perp}{(2\pi)^2} \frac{1}{2\gamma} \left| \int_B d\mathbf{r} \; \psi^*(\mathbf{r}) e^{i\mathbf{k}_\perp \cdot \mathbf{r}_\perp - \gamma z} \right|^2, \tag{10.55}$$

demonstrating the first statement in eqn (10.52).

To prove the second statement, we replace the integration variable $\mathbf{r} = (\mathbf{r}_\perp, z)$ with $\mathbf{r}_1 = (\mathbf{r}_\perp, z_1)$, where $z_1 = z - a/2$. Then eqn (10.55) takes the form

$$I(a) = \int \frac{d\mathbf{k}_\perp}{(2\pi)^2} \frac{1}{2\gamma} e^{-\gamma a} \left| \int_B d\mathbf{r}_1 \; \psi^*(\mathbf{r}_1) e^{i\mathbf{k}_\perp \cdot \mathbf{r}_\perp - \gamma z_1} \right|^2. \tag{10.56}$$

The derivative of eqn (10.56) with respect to a is evidently negative. Now we note that $\mathcal{Y}_\xi > 0$ follows from eqns (10.52) and (10.51), since the operator for the T-matrix is Hermitian. Therefore one can apply a Feynman–Hellman argument and conclude that the eigenvalues of the operator $\mathcal{Y}_\xi$, λ_n, satisfy the inequality $0 \leq \lambda_n < 1$ [the right-hand inequality follows from the unitarity of the scattering matrix (10.31)]. As a consequence, the logarithm $\ln(1-\lambda_n^2)$ is a monotonically decreasing function of a. Finally, the absolute convergence of the sums and integrations involved in the trace in eqn (10.50) transfers these properties to the vacuum energy which, thus, is shown to be a monotonic function of a. Obviously, it takes negative values and provides an attractive force. Note that from the above formulas it is also clear that the logarithm and, thus, the energy vanish for $a \to \infty$. Hence, after dropping the distance-independent terms, as was done in the preceding sections, only a correctly normalized interaction energy emerges.

The above derivation was done for a scalar field. However, essentially the same arguments allow one to extend these results to an electromagnetic field (Kenneth and Klich 2006). It should also be noted that although we have used the T-matrix approach here, the same statement can be proved using the representation (10.21) for the vacuum energy in the presence of boundary conditions. Here one needs to assume that the surfaces S_A and S_B are related by mirror symmetry and that the same boundary conditions are imposed on both. A generalization of the above results by Kenneth and Klich (2006) was given by Bachas (2007).

The derivation presented in this section leads to more insights. Making use of the cyclic property of the trace, we can change the representation (10.50) of the vacuum energy to

$$E = \frac{1}{2\pi} \int_0^\infty d\xi \, \text{Tr} \ln\left(1 - \mathcal{N}_\xi^2\right), \tag{10.57}$$

where

$$\mathcal{N}_\xi = \mathcal{T}_{BB} \mathcal{G}_{\xi,BA}^{(0)} \mathcal{J}^+. \tag{10.58}$$

This representation allows us to split the vacuum energy into two terms:

$$E = E^{(+)} + E^{(-)}, \qquad E^{(\pm)} = \frac{1}{2\pi} \int_0^\infty d\xi \, \text{Tr} \ln\left(1 \pm \mathcal{N}_\xi\right). \tag{10.59}$$

Here, $E^{(+)}$ and $E^{(-)}$ are the vacuum energies for when the symmetry plane has Neumann and Dirichlet boundary conditions, respectively, for the body B. This follows from general symmetry considerations, since the corresponding wave functions, owing to the mirror symmetry, separate into symmetric and antisymmetric ones. Thus, the corresponding boundary condition on the plane is automatically fulfilled. The same can also be shown by applying the formalism developed in the two preceding sections to the body B and a plane.

By means of eqn (10.59), the vacuum energy for two mirror-symmetric bodies becomes the sum of two vacuum energies, each between one of these bodies and the symmetry plane. For one of them, we have to use Dirichlet boundary conditions and, for the other, Neumann boundary conditions on the plane. It is not difficult to show that the same holds for an electromagnetic field. Therefore the body may be a dielectric or an ideal metal. The two boundary conditions on the plane are in one case ideal-metal boundary conditions and in the other case, for duality reasons, bag boundary conditions (see Section 7.5).

10.3 Application to cylindrical geometry

Here we consider some specific applications of the general results obtained in the preceding sections, starting with the simplest cases (other than the configuration of two parallel planes). These are the cases of parallel cylinders and a cylinder parallel to a plane. We restrict ourselves to a scalar field obeying Dirichlet or Neumann boundary conditions on the surfaces. Since these are waveguide geometries, the results cover the electromagnetic case also. We start with a specification of the general formulas for the given geometry. Then we consider the case of large separations. The technically more involved case of short separations is also considered. From this we obtain the PFA and calculate the first correction to it.

10.3.1 *Two parallel cylinders and a cylinder parallel to a plane*

We start with a scalar field obeying Dirichlet boundary conditions on two cylinders with radii R_A and R_B (see Fig. 10.1). The cylinder axes are parallel to the z-axis, $x = 0$ is the symmetry plane of the cylinder axes, and $L = 2b$ is the distance between the centers of the two cylinders. We employ the representation (10.21) for the vacuum energy and make use of the translational invariance along the z-axis. In fact, we have to consider the energy per unit length of the cylinders. For this purpose, we explicitly perform the integration with respect to z contained inside the "Tr" in eqn (10.21). Taking into account eqns (10.6) and (10.7) and introducing the polar coordinates $\gamma = \sqrt{\xi^2 + k_z^2}$ and $-\pi/2 \leq \varphi_\gamma \leq \pi/2$ on the right half-plane (ξ, k_z), we obtain an expression for the Casimir energy per unit length of the cylinders,

$$E = \frac{1}{4\pi} \int_0^\infty d\gamma\, \gamma\, \mathrm{Tr} \ln\left(1 - \tilde{K}_{\gamma,AA}^{-1} \tilde{K}_{\gamma,AB} \tilde{K}_{\gamma,BB}^{-1} \tilde{K}_{\gamma,BA}\right). \tag{10.60}$$

Here, the trace does not contain an integration with respect to z. The kernels $\tilde{K}_\gamma$ are represented similarly to eqn (10.14), where the Green's function $\mathcal{G}_\xi(\boldsymbol{r}, \boldsymbol{r}')$ must be replaced with

$$G_\gamma(\boldsymbol{r}_\perp - \boldsymbol{r}'_\perp) = \int \frac{d\boldsymbol{k}_\perp}{(2\pi)^2} \frac{e^{i\boldsymbol{k}_\perp \cdot (\boldsymbol{r}_\perp - \boldsymbol{r}'_\perp)}}{\gamma^2 + k_\perp^2}, \tag{10.61}$$

where the vectors $\boldsymbol{r}_\perp = (x, y)$ and $\boldsymbol{r}'_\perp = (x', y')$ describe points belonging to the cylindrical shells A and B, respectively. It is convenient to describe the surfaces of the cylinders in polar coordinates centered on the cylinder axes. Then the cylindrical shell A is described by a vector $\boldsymbol{\rho} = (R_A \cos\varphi, R_A \sin\varphi)$ defined in a coordinate system centered at the point $(-b, 0, 0)$. In a similar way, the surface of the cylinder B is described by a vector $\boldsymbol{\rho}' = (R_B \cos\varphi', R_B \sin\varphi')$ defined relative to the origin $(b, 0, 0)$. According to eqns (10.6) and (10.9), one can write

$$\tilde{K}_{\gamma,AB} = G_\gamma(\boldsymbol{\rho} - \boldsymbol{L} - \boldsymbol{\rho}')|_{\rho=R_A,\, \rho'=R_B}, \tag{10.62}$$

where $\boldsymbol{L}$ is the vector connecting the points $(-b, 0, 0)$ and $(b, 0, 0)$. For the coordinate η on the surfaces we can take an angle, i.e. $\eta \to \varphi$, and the corresponding integration is over the interval $[0, 2\pi]$. The measure (10.17) amounts to a constant on which the vacuum energy does not depend. In this way, all quantities entering eqn (10.60) are defined. In order to start the calculations, it is useful to write the trace in a convenient basis. We take

$$|l\rangle = \frac{e^{il\varphi}}{\sqrt{2\pi}}, \tag{10.63}$$

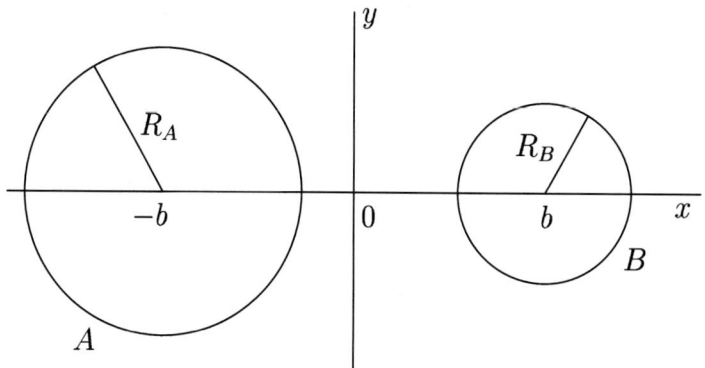

FIG. 10.1. A configuration of two cylinders or of one cylinder parallel to a plane.

where l is an integer. Then all quantities $\tilde{K}_\gamma$ become matrices with matrix elements enumerated by the indices l and l',

$$\left(\tilde{K}_{\gamma,AB}\right)_{ll'} = \langle l|\tilde{K}_{\gamma,AB}|l'\rangle. \tag{10.64}$$

It is convenient to represent the propagator (10.61) in the basis

$$u_l(k_\perp;\boldsymbol{\rho}) = J_l(k_\perp\rho)\frac{e^{il\varphi}}{\sqrt{2\pi}}, \tag{10.65}$$

where the $J_l(z)$ are Bessel functions. For this purpose, we express the difference $\boldsymbol{r}_\perp - \boldsymbol{r}'_\perp$ and $\boldsymbol{k}_\perp = (k_\perp\cos\varphi_k, k_\perp\sin\varphi_k)$ in polar coordinates with the origin at the point $(-b,0,0)$:

$$\boldsymbol{r}_\perp - \boldsymbol{r}'_\perp = \boldsymbol{\rho} - \boldsymbol{\rho}'_1 = \boldsymbol{\rho} - (\boldsymbol{L} + \boldsymbol{\rho}'). \tag{10.66}$$

Using the expansion of two-dimensional plane waves into cylindrical waves

$$e^{i\boldsymbol{k}_\perp\cdot\boldsymbol{\rho}} = \sum_{l=-\infty}^{\infty} i^l J_l(k_\perp\rho)e^{il(\varphi-\varphi_k)}, \tag{10.67}$$

we arrive at the representation

$$G_\gamma(\boldsymbol{\rho}-\boldsymbol{\rho}'_1) = \int_0^\infty \frac{dk_\perp\, k_\perp}{\gamma^2+k_\perp^2} \sum_{l=-\infty}^{\infty} u_l(k_\perp;\boldsymbol{\rho})u_l^*(k_\perp;\boldsymbol{\rho}'_1). \tag{10.68}$$

Instead of the functions $u_l(k_\perp;\boldsymbol{\rho}'_1)$, it is convenient to use basis functions in polar coordinates with the origin at the point $(b,0,0)$. To do so, we apply the *translation formula*

$$u_l(k_\perp; \boldsymbol{L}+\boldsymbol{\rho}') = \sum_{l'=-\infty}^{\infty} A_{ll'}(k_\perp; \boldsymbol{L}) u_{l'}(k_\perp; \boldsymbol{\rho}'), \qquad (10.69)$$

where

$$A_{ll'}(k_\perp; \boldsymbol{L}) = J_{l-l'}(k_\perp L) e^{i(l-l')\varphi_L} \qquad (10.70)$$

(in our case we have to put $\varphi_L = 0$). Inserting eqn (10.68), with account of eqns (10.69) and (10.70), into eqn (10.62), we can calculate the matrix element (10.64):

$$\left(\tilde{K}_{\gamma,AB}\right)_{ll'} = \int_0^\infty \frac{dk_\perp\, k_\perp}{\gamma^2 + k_\perp^2} J_l(k_\perp R_A) J_{l'}(k_\perp R_B) J_{l-l'}(k_\perp L). \qquad (10.71)$$

For kernels with both arguments on the same surface, as a special case of the above expression, we find

$$\left(\tilde{K}_{\gamma,AA}\right)_{ll'} = \delta_{ll'} \int_0^\infty \frac{dk_\perp\, k_\perp}{\gamma^2 + k_\perp^2} J_l(k_\perp R_A) J_l(k_\perp R_A), \qquad (10.72)$$

and similarly for $A \to B$. This formula can be made more explicit by carrying out the integration with respect to $k_\perp$ (Gradshteyn and Ryzhik 1994):

$$\left(\tilde{K}_{\gamma,AA}\right)_{ll'} = \delta_{ll'}\, I_l(\gamma R_A) K_l(\gamma R_A). \qquad (10.73)$$

This is a well-known representation of the propagator in terms of the modified Bessel functions. From eqn (10.73), it follows that

$$\left(\tilde{K}_{\gamma,AA}^{-1}\right)_{ll'} = \delta_{ll'} \frac{1}{I_l(\gamma R_A) K_l(\gamma R_A)}. \qquad (10.74)$$

In a similar way, we can perform the integration with respect to $k_\perp$ in eqn (10.71). For this purpose $J_{l-l'}(k_\perp L)$ must be represented as a sum of two Hankel functions:

$$J_{l-l'}(k_\perp L) = \frac{1}{2}\left[H^{(1)}_{l-l'}(k_\perp L) + H^{(2)}_{l-l'}(k_\perp L)\right] \qquad (10.75)$$
$$= \frac{1}{2}\left[H^{(1)}_{l-l'}(k_\perp L) - (-1)^{l-l'} H^{(1)}_{l-l'}(-k_\perp L)\right].$$

Substituting eqn (10.75) into eqn (10.71), we can represent the matrix element $(\tilde{K}_{\gamma,AB})_{ll'}$ as a sum of two integrals. A change of the variable $k_\perp \to -k_\perp$ in the second of these integrals leads to

$$(\tilde{K}_{\gamma,AB})_{ll'} = \frac{1}{2} \int_0^\infty \frac{k_\perp\, dk_\perp}{k_\perp^2 + \gamma^2} J_l(k_\perp R_A) J_{l'}(k_\perp R_B) H^{(1)}_{l-l'}(k_\perp L) \qquad (10.76)$$

$$+ \frac{1}{2}(-1)^{l-l'} \int_{-\infty}^{0} \frac{k_\perp dk_\perp}{k_\perp^2 + \gamma^2} J_l(-k_\perp R_A) J_{l'}(-k_\perp R_B) H_{l-l'}^{(1)}(k_\perp L).$$

Note that both integrals on the right-hand side are convergent. Using the properties of Bessel functions, we get

$$(\tilde{K}_{\gamma,AB})_{ll'} = \frac{1}{2} \int_{-\infty}^{\infty} \frac{k_\perp dk_\perp}{k_\perp^2 + \gamma^2} J_l(k_\perp R_A) J_{l'}(k_\perp R_B) H_{l-l'}^{(1)}(k_\perp L). \quad (10.77)$$

Now we integrate in the upper half-plane of the complex variable $k_\perp$, over the real axis and an infinitely remote semicircle bypassed in the counterclockwise direction. Calculating the residue at the simple pole $k_\perp = i\gamma$, we obtain

$$\left(\tilde{K}_{\gamma,AB}\right)_{ll'} = (-1)^{l'} I_l(\gamma R_A) K_{l-l'}(\gamma L) I_{l'}(\gamma R_B). \quad (10.78)$$

Using these formulas, the vacuum energy (10.60) for a scalar field with Dirichlet boundary conditions on two parallel cylinders can be represented in the form

$$E = \frac{1}{4\pi} \int_0^\infty d\gamma\, \gamma\, \text{Tr} \ln(1 - \mathcal{M}_\gamma). \quad (10.79)$$

The matrix elements of $\mathcal{M}$ are

$$\mathcal{M}_{\gamma,ll'} = \frac{I_{l'}(\gamma R_A)}{K_l(\gamma R_A)} \sum_{l''} K_{l-l''}(\gamma L) I_{l''}(\gamma R_B) \frac{1}{K_{l''}(\gamma R_B)} K_{l''-l'}(\gamma L). \quad (10.80)$$

Here, we take into account the fact that for the quantity $\tilde{K}_{\gamma,BA}$, the matrix elements (10.64) are calculated with the basis functions (10.65) defined in polar coordinates centered at the point $(b,0,0)$. In this case, in the transition matrix (10.70), we have $\varphi_L = \pi$. As a result, the matrix element $(\tilde{K}_{\gamma,BA})_{ll'}$ is defined by eqn (10.78), where $(-1)^{l'}$ is replaced with $(-1)^l$. Note that in eqn (10.80) and below, the summations with respect to indices l are performed from $-\infty$ to ∞.

Now we consider two cylinders of equal radii $R = R_A = R_B$, and take into account the mirror symmetry given by $\mathcal{J}\varphi = \pi - \varphi$. The symmetry operator $\mathcal{J}$ in the basis (10.63) is

$$\mathcal{J}_{ll'} = (-1)^l \delta_{l,-l'}. \quad (10.81)$$

Using this operator, we define, in analogy with eqn (10.58), the operator $\mathcal{N}$ with the following matrix elements:

$$\mathcal{N}_{\gamma,ll'}^D = \sum_{l'',l'''} \left(\tilde{K}_{\gamma,AA}^{-1}\right)_{l,l''} \left(\tilde{K}_{\gamma,AB}\right)_{l'',l'''} \mathcal{J}_{l''',l'}$$

$$= (-1)^{l'} \left(\tilde{K}_{\gamma,AA}^{-1}\right)_{l,l} \left(\tilde{K}_{\gamma,AB}\right)_{l,-l'}. \quad (10.82)$$

This expression can be simplified using eqns (10.74) and (10.78), with the result

$$N^D_{\gamma,ll'} = \frac{1}{K_l(\gamma R)} K_{l+l'}(\gamma L) I_{l'}(\gamma R). \tag{10.83}$$

Then the final formula for the vacuum energy per unit length of two parallel cylinders of equal radii, for a scalar field with Dirichlet boundary conditions, is

$$E = \frac{1}{4\pi} \int_0^\infty d\gamma\, \gamma\, \mathrm{Tr} \ln\left(1 - \mathcal{N}_\gamma^2\right). \tag{10.84}$$

This energy can also be split into a sum of two terms as in eqn (10.59), where now

$$E^{(\pm)} = \frac{1}{4\pi} \int_0^\infty d\gamma\, \gamma\, \mathrm{Tr} \ln\left(1 \pm \mathcal{N}_\gamma\right) \tag{10.85}$$

is the vacuum energy for a cylinder with Dirichlet boundary conditions and a plane with Neumann (upper sign) or Dirichlet (lower sign) boundary conditions. Note that the distance from the plane to the center of the cylinder is $L/2$.

Now we discuss the changes arising for two cylinders with Neumann boundary conditions. In this case the projector is defined by eqn (10.10). The normal derivative is now the derivative with respect to the variable ρ. Then, instead of eqn (10.62), in accordance with eqn (10.14), we must take

$$\tilde{K}_{\gamma,AB} = \frac{\partial}{\partial \rho} \frac{\partial}{\partial \rho'} G_\gamma(\boldsymbol{\rho} - \boldsymbol{L} - \boldsymbol{\rho}')\bigg|_{\rho=R_A,\,\rho'=R_B}. \tag{10.86}$$

We have to add these derivatives to all of the quantities $\tilde{K}_\gamma$. Therefore for two cylinders of equal radii with Neumann boundary conditions, the vacuum energies per unit length are given by eqns (10.84) and (10.85) with

$$N^N_{\gamma,ll'} = \frac{1}{K'_l(\gamma R)} K_{l+l'}(\gamma L) I'_{l'}(\gamma R). \tag{10.87}$$

Here, the prime denotes differentiation with respect to the argument of the Bessel functions.

This approach was first developed by Emig et al. (2006). It is interesting to remark that the above formulas show a very transparent origin of the different signs of the Casimir energy in its dependence on the boundary conditions. For example, in the case of a cylinder parallel to a plane, for Dirichlet conditions on the cylinder and Dirichlet or Neumann conditions on the plane, $E^{(+)} > 0$ and $E^{(-)} < 0$, corresponding to repulsion and attraction, respectively. However, the magnitude of the repulsive contribution is smaller than the magnitude of the attractive one. For Neumann conditions on the cylinder, we have to use eqn (10.87) and take into account the fact that the derivative of the modified Bessel function $K_l(z)$ is negative. So we get the same picture, i.e. for similar boundary conditions on the cylinder and the plane we have attraction, and for different boundary conditions we have repulsion, which is, however, weaker than the attraction.

10.3.2 Cylinder parallel to a plane at large separation

We consider Dirichlet boundary conditions on both the cylinder and the plane and, expanding the logarithm, rewrite the vacuum energy (10.85) in the form

$$E^D \equiv E^{(-)} = -\frac{1}{4\pi} \int_0^\infty d\gamma\, \gamma \sum_{s=0}^\infty \frac{1}{s+1} \sum_{l_1, l_2, \ldots, l_{s+1}} N_{\gamma, l_1 l_2} N_{\gamma, l_2 l_3} \cdots N_{\gamma, l_{s+1} l_1}. \tag{10.88}$$

The matrix elements $N_{\gamma, ll'}$ are given by eqn (10.83). At large separations, the inequality $R \ll L$ is true and one can use the asymptotic expansion for small arguments for the Bessel functions containing R. The leading contribution to the Casimir energy per unit length (10.88) at large separations is given by the terms with $l_i = 0$. Taking into account the fact that

$$K_0(z) = -\ln\frac{z}{2} - C + O(z^2), \qquad I_0(z) = 1 + O(z^2), \tag{10.89}$$

where C is Euler's constant, we can rewrite the matrix element (10.83) in the form

$$N_{\gamma, ll'}^D = \frac{\delta_{l0}\delta_{l'0} K_0(\gamma L)}{-\ln(\gamma R/2) - C} + O\left[(\gamma R)^2\right]. \tag{10.90}$$

Now we substitute eqn (10.90) into eqn (10.88) and introduce the new variable $y = \gamma L$:

$$E^D = -\frac{1}{4\pi L^2} \int_0^\infty dy\, y \sum_{s=0}^\infty \frac{1}{s+1} \frac{K_0^{s+1}(y)}{[-\ln(R/L)]^{s+1}} \left[1 + \frac{\ln(y/2) + C}{\ln(R/L)}\right]^{-s-1}$$

$$+ O\left(\frac{R^2}{L^2}\right). \tag{10.91}$$

Expanding this in powers of the small parameter $1/\ln(R/L)$ and taking into account the fact that the main term is obtained with $s = 0$, we arrive at (Emig et al. 2006)

$$E^D = \frac{1}{4\pi L^2} \frac{1}{\ln(R/L)} + O\left(\ln^{-2}\frac{R}{L}\right). \tag{10.92}$$

The respective leading contribution to the Casimir force at large separations between the cylinder and the plate is

$$F^D = -\frac{\partial E^D}{\partial a} = -2\frac{\partial E^D}{\partial L} = \frac{1}{\pi L^3} \frac{1}{\ln(R/L)}. \tag{10.93}$$

The origin of this logarithmic behavior is connected with the properties of the two-dimensional Green's function. Thus, the vacuum energy for a circle in front of a line, i.e. the same problem without the third spatial dimension, shows this logarithmic behavior too (Bordag 2006a).

The corresponding energy per unit length for Neumann boundary conditions on the plane and on the cylinder follows from eqn (10.87). In this case the leading contribution to the Casimir energy per unit length (10.88) is given by terms with $l = 0$ and $l = \pm 1$. Using the integration variable y, we have

$$E^{\mathrm{N}} = -\frac{1}{8\pi} \frac{R^2}{L^4} \int_0^\infty y^3 dy \, [K_0(y) + 2K_2(y)] + O\left(\frac{R^4}{L^4} \ln \frac{R}{L}\right). \tag{10.94}$$

Here, the first term under the integral results from $l = 0$ and the second one from $l = \pm 1$. Performing the integration in eqn (10.94), we obtain (Emig et al. 2006)

$$E^{\mathrm{N}} = -\frac{5R^2}{2\pi L^4} + O\left(\frac{R^4}{L^4} \ln \frac{R}{L}\right). \tag{10.95}$$

The respective main contribution to the Casimir force per unit length is given by

$$F^{\mathrm{N}} = -\frac{20R^2}{\pi} \frac{1}{L^5}. \tag{10.96}$$

The energy and force per unit length obtained decrease faster than those for Dirichlet boundary conditions. Therefore the electromagnetic case, which is the sum of the cases with Dirichlet and Neumann conditions, is given by eqns (10.92) and (10.93).

10.3.3 The limit of short separations, and corrections beyond the proximity force approximation

For short separations between the cylinder and the plane, we introduce a small parameter describing this limit,

$$\varepsilon = \frac{a}{R} \ll 1, \qquad a = \frac{L}{2} - R. \tag{10.97}$$

We start from eqn (10.88) and introduce a new integration variable $\tilde{\gamma} = \gamma R$ and new summation indices $l_1 = l$, $l_{i+1} = l + \tilde{l}_i$ ($i \geq 1$). The result is

$$E^{\mathrm{D}} = -\frac{1}{4\pi R^2} \sum_{s=0}^\infty \frac{1}{s+1} \int_0^\infty d\tilde{\gamma}\, \tilde{\gamma} \sum_l \sum_{\tilde{l}_1} \cdots \sum_{\tilde{l}_s} Q_{\tilde{\gamma}; l, \tilde{l}_1, \ldots, \tilde{l}_s}, \tag{10.98}$$

where

$$Q_{\tilde{\gamma}; l, \tilde{l}_1, \ldots, \tilde{l}_s} = N_{\tilde{\gamma}, l, l+\tilde{l}_1} N_{\tilde{\gamma}, l+\tilde{l}_1, l+\tilde{l}_2} \cdots N_{\tilde{\gamma}, l+\tilde{l}_s, l} \tag{10.99}$$

and

$$N^{\mathrm{D}}_{\tilde{\gamma}, ll'} = \frac{1}{K_l(\tilde{\gamma})} K_{l+l'}\left(2\tilde{\gamma}(1+\varepsilon)\right) I_{l'}(\tilde{\gamma}). \tag{10.100}$$

In eqn (10.98), the summations are done over all integers.

In this representation, it is not possible to perform the limit $\varepsilon \to 0$ directly. The reason is that with decreasing separation, larger and larger momenta $\tilde{\gamma}$ and orbital momenta become important. For example, the convergence of the integration in $\tilde{\gamma}$ may be lost, as can be seen from eqn (10.100).

Since, in the limit $\varepsilon \to 0$, large values of $\tilde{\gamma}$, l, and $\tilde{l}_i$ give the main contribution to the energy, we replace the Bessel functions with their uniform asymptotic expansions and approximate the sums by integrals. Further, we use the symmetry under $l \to -l$ and represent the Casimir energy (10.98) as

$$E^D = -\frac{1}{2\pi R^2} \sum_{s=0}^{\infty} \frac{1}{s+1} \quad (10.101)$$

$$\times \int_0^\infty d\tilde{\gamma}\,\tilde{\gamma} \int_0^\infty dl \int_{-\infty}^\infty d\tilde{l}_1 \int_{-\infty}^\infty d\tilde{l}_2 \ldots \int_{-\infty}^\infty d\tilde{l}_s \, Q_{\tilde{\gamma};l,\tilde{l}_1,\ldots,\tilde{l}_s}.$$

Here, Q is still given by eqn (10.99); however, the Bessel functions in eqn (10.100) are replaced with their uniform asymptotic expansions (9.76). First we consider the exponential factors in $N_{ll'}$. These combine into

$$N_{\tilde{\gamma},ll'} \sim e^{-\tilde{\eta}}, \quad (10.102)$$

where

$$\tilde{\eta} = (l+l')\,\eta\!\left(\frac{2\tilde{\gamma}(1+\varepsilon)}{l+l'}\right) - l\,\eta\!\left(\frac{\tilde{\gamma}}{l}\right) - l'\,\eta\!\left(\frac{\tilde{\gamma}}{l'}\right) \quad (10.103)$$

and $\eta(z)$ is defined in eqn (9.77). From this result it can be observed that the dominating contribution to eqn (10.101) comes from $\tilde{\gamma} \sim l \sim 1/\varepsilon$, $\tilde{l}_i \sim 1/\sqrt{\varepsilon}$. Hence we substitute

$$\tilde{\gamma} = \frac{t\sqrt{1-\tau^2}}{\varepsilon}, \quad l = \frac{t\tau}{\varepsilon}, \quad \tilde{l}_i = n_i\sqrt{\frac{4t}{\varepsilon}}, \quad (10.104)$$

where the factor $\sqrt{4t}$ has been introduced for the sake of convenience. With these substitutions, the energy becomes

$$E^D = -\frac{1}{2\pi R^2 \varepsilon^3} \sum_{s=0}^{\infty} \frac{1}{s+1} \quad (10.105)$$

$$\times \int_0^\infty dt\, t^2 \int_0^1 d\tau \int_{-\infty}^\infty dn_1 \int_{-\infty}^\infty dn_2 \ldots \int_{-\infty}^\infty dn_s \left(\frac{4t}{\varepsilon}\right)^{s/2} Q(t,\tau,n_1,\ldots,n_s).$$

Next we consider eqn (10.100) with the asymptotic expansions of the Bessel functions and the substitutions (10.104) inserted. After changing the notation to

$$N_{\tilde{\gamma},ll'} \to N^{\mathrm{as}}_{nn'}, \quad (10.106)$$

we get

$$N^{\mathrm{as}}_{nn'} = \sqrt{\frac{\varepsilon}{4\pi t}}\, e^{-\tilde{\eta}^{\mathrm{as}}}\left(1 + a^{(1/2)}_{nn'}(t,\tau)\sqrt{\varepsilon} + a^{(1)}_{nn'}(t,\tau)\varepsilon + \ldots\right), \quad (10.107)$$

where

$$\tilde{\eta}^{\mathrm{as}} = 2t + (n-n')^2. \quad (10.108)$$

Details, including the explicit form of the functions $a^{(1/2)}_{nn'}$ and $a^{(1)}_{nn'}$, are presented by Bordag (2006a).

Now we insert eqn (10.107) into the energy (10.105):

$$E^D = -\frac{1}{2\pi a^2}\sqrt{\frac{R}{a}}\sum_{s=0}^{\infty}\frac{1}{s+1}\int_0^{\infty}\frac{dt}{t}\frac{t^{5/2}e^{-2(s+1)t}}{\sqrt{4\pi}} \quad (10.109)$$

$$\times \int_0^1 d\tau \int_{-\infty}^{\infty}\frac{dn_1}{\sqrt{\pi}}\cdots\int_{-\infty}^{\infty}\frac{dn_s}{\sqrt{\pi}}Q^{as}(t,\tau,n_1,\ldots,n_s).$$

Here the asymptotic expansion of the function Q^{as} is

$$Q^{as}(t,\tau,n_1,\ldots,n_s) = e^{-\eta_1}\left[1 + \sqrt{\varepsilon}\sum_{i=0}^{s}a^{(1/2)}_{n_i,n_{i+1}}(t,\tau)\right. \quad (10.110)$$

$$\left. + \varepsilon\left(\sum_{0\leq i<j\leq s}a^{(1/2)}_{n_i,n_{i+1}}(t,\tau)a^{(1/2)}_{n_j,n_{j+1}}(t,\tau) + \sum_{i=0}^{s}a^{(1)}_{n_i,n_{i+1}}(t,\tau)\right) + \ldots\right],$$

where

$$\eta_1 = \sum_{i=0}^{s}(n_i - n_{i+1})^2. \quad (10.111)$$

We note that in Q^{as}, a reexpansion has been performed. Now, with eqn (10.109), we have an expression which delivers an expansion for small ε and has a remarkably simple structure. First of all, the integrations over the n_i are Gaussian and can be carried out in a closed form. For example, in the leading order, we note that

$$\int_{-\infty}^{\infty}\frac{dn_1}{\sqrt{\pi}}\cdots\int_{-\infty}^{\infty}\frac{dn_s}{\sqrt{\pi}}e^{-\eta_1} = \frac{1}{\sqrt{s+1}}. \quad (10.112)$$

For symmetry reasons, the contributions proportional to $\sqrt{\varepsilon}$ in eqn (10.110) drop out and the contributions of order ε become the leading-order corrections. Then, the integrations over t and τ can also be carried out, with the result (Bordag 2006a)

$$E^D = -\frac{\pi^3}{1920a^2}\sqrt{\frac{R}{2a}}\left[1 + \frac{7}{36}\frac{a}{R} + O\left(\frac{a^2}{R^2}\right)\right], \quad (10.113)$$

where we have substituted ε using eqn (10.97). In this formula, the leading term coincides with the PFA (see Section 6.5), giving it an independent confirmation. The next-to-leading order is the first correction beyond the PFA. It should be mentioned that eqn (10.113) was the first analytical result for a correction beyond the PFA. Shortly after this result was obtained, it was confirmed by the numerical world line methods considered in Section 6.3 (Gies and Klingmüller 2006a, 2006b).

The calculation for Neumann boundary conditions is completely parallel to the above Dirichlet case, and the result is (Bordag 2006a)

$$E^N = -\frac{\pi^3}{1920a^2}\sqrt{\frac{R}{2a}}\left[1 + \left(\frac{7}{36} - \frac{40}{3\pi^2}\right)\frac{a}{R} + O\left(\frac{a^2}{R^2}\right)\right]. \quad (10.114)$$

Finally, because in the geometry considered above the polarizations of the electromagnetic field separate, the sum of eqns (10.113) and (10.114) is the vacuum energy for the electromagnetic field (Bordag 2006a):

$$E^{\text{em}}(a) = -\frac{\pi^3}{960a^2}\sqrt{\frac{R}{2a}}\left[1 + \left(\frac{7}{36} - \frac{20}{3\pi^2}\right)\frac{a}{R} + O\left(\frac{a^2}{R^2}\right)\right]. \quad (10.115)$$

Note that the numerical coefficient in the correction of order a/R to the PFA is equal to -0.48103. From eqn (10.115), the electromagnetic Casimir force between an ideal-metal cylinder and an ideal-metal plane in close proximity per unit length of the cylinder is given by (Mostepanenko 2008)

$$F^{\text{em}}(a) = -\frac{\pi^3}{384a^3}\sqrt{\frac{R}{2a}}\left[1 - \left(\frac{4}{\pi^2} - \frac{7}{60}\right)\frac{a}{R} + O\left(\frac{a^2}{R^2}\right)\right]. \quad (10.116)$$

Here, the coefficient to the correction of order a/R to the PFA is equal to -0.288618, demonstrating that for the force, the PFA is even more precise than for the energy.

There are a few other exact results for configurations involving cylinders. Thus, the electromagnetic Casimir energy per unit length for a configuration of two concentric cylinders with radii $R_A < R_B$ was obtained by Mazzitelli et al. (2003). Here, in contrast to the configuration of a cylinder above a plate, the variables do separate. Specifically, for a small gap $a = R_B - R_A$ between the cylinders, where $a/R_A \ll 1$, both the first- and the second-order corrections to the result given by the PFA for this configuration were obtained (Lombardo et al. 2008):

$$E^{\text{em}}(a) = -\frac{\pi^3 R_A}{360a^3}\left[1 + \frac{1}{2}\frac{a}{R_A} - \left(\frac{2}{\pi^2} + \frac{1}{10}\right)\frac{a^2}{R_A^2} + O\left(\frac{a^3}{R_A^3}\right)\right]. \quad (10.117)$$

An exact result for the Casimir energy per unit length for two eccentric cylinders was obtained by Dalvit et al. (2006) and Mazzitelli et al. (2006). Exact results for the Casimir energy for various configurations involving cylinders and plates were found by Rahi et al. (2008).

10.4 Applications to spherical geometry

From the point of view of experimental applications, the configuration of a (large) sphere in front of a plane is the most important one. In this section, we apply the method of Section 10.1.1 to a scalar field obeying boundary conditions on two spheres and on a sphere in front of a plane. This case is largely parallel to the cylindrical case considered in the preceding section. However, it is technically somewhat more complicated, since we have to deal with one more orbital momentum, and with $3j$-symbols, which enter the mathematical expressions for the Casimir energy. Although this should not create any special problem, the calculation turns out to be not so trivial. As before, we start from the case of two spheres of different radii and then consider a sphere in front of a plane using the symmetry arguments of Section 10.2. As an application, we consider the Casimir force between a sphere and a plane at large and small separations, where we present the analytical correction beyond the PFA.

10.4.1 General formulas for two spheres and for a sphere in front of a plane

We start from a scalar field obeying Dirichlet boundary conditions on two spheres of radii R_A and R_B separated by a distance L. Figure 10.1, which was previously used as a representation of two cylinders or of a cylinder near a plane, will now be used as a schematic of two spheres or of a sphere in front of a plane. We parametrize each surface by the spherical coordinates (r, θ, φ) and (r', θ', φ'), with $r = R_A$ and $r' = R_B$, respectively (the angles θ and θ' are measured from the y-axis, which goes through the centers of both spheres; see Fig. 10.1). We take the vacuum energy as given by eqn (10.21), and we have to specify the kernels $\tilde{K}_{AB}$. For the parameters η on the surfaces, we take the angular variables (θ, φ) such that, similarly to (10.62),

$$\tilde{K}_{\xi,AB}(\boldsymbol{\eta}, \boldsymbol{\eta}') = G_\xi(\boldsymbol{r} - \boldsymbol{L} - \boldsymbol{r}')|_{r=R_A, r'=R_B}, \tag{10.118}$$

where $\boldsymbol{L}$ is the vector pointing from the center of sphere A to the center of B. Considering one sphere, we get

$$\tilde{K}_{\xi,AA}(\boldsymbol{\eta}, \boldsymbol{\eta}') = G_\xi(\boldsymbol{r} - \boldsymbol{r}')|_{r=R_A, r'=R_A}, \tag{10.119}$$

and similarly with $A \to B$. The integrations and the trace in (10.21) go over the solid angle Ω in the corresponding variables. The measure is $d\mu(\boldsymbol{\eta}) = d\Omega \equiv d\varphi\, d\theta \sin\theta$, where we have dropped the radius since it gives only a constant factor which will cancel in eqn (10.21).

With these formulas, we have defined all quantities entering the energy (10.21). The trace must be calculated in an appropriate basis. Owing to the geometry considered, this basis is given by the spherical harmonics $|l, M\rangle = Y_{lM}(\theta, \varphi)$ defined in eqn (9.5). Thus, the kernels are

$$\begin{aligned}\left(\tilde{K}_{\xi,AB}\right)_{lM,l'M} &= \langle l, M \,|\, \tilde{K}_{\xi,AB} \,|\, l', M\rangle \\ &= \int_0^{2\pi} d\varphi \int_0^\pi d\theta \sin\theta \int_0^{2\pi} d\varphi' \int_0^\pi d\theta' \sin\theta' \\ &\quad \times Y_{lM}^*(\theta, \varphi)\, G_\xi(\boldsymbol{r} - \boldsymbol{L} - \boldsymbol{r}')|_{r=R_A, r'=R_B}\, Y_{l'M}(\theta', \varphi').\end{aligned} \tag{10.120}$$

Note that in the coordinates chosen, the configuration of two spheres has an azimuthal symmetry. For this reason, all matrix elements are diagonal in the magnetic quantum number M. The trace in eqn (10.21) includes a sum over M, and in the remaining trace, the orbital momenta take values $l \geq |M|$. In addition, the matrices in eqn (10.21) depend only on $|M|$.

Applications to spherical geometry

For the calculation of the matrix elements in eqn (10.120), it is useful to introduce the basis

$$u_{lM}(k;\mathbf{r}) = j_l(kr)Y_{lM}(\theta,\varphi), \tag{10.121}$$

where the $j_l(z)$ are the spherical Bessel functions (9.11). Using the expansion of plane waves into spherical waves

$$e^{i\mathbf{k}\cdot\mathbf{r}} = 4\pi \sum_{l=0}^{\infty} \sum_{M=-l}^{l} i^l j_l(kr) Y_{lM}^*(\theta_k,\varphi_k) Y_{lM}(\theta,\varphi), \tag{10.122}$$

where $k = |\mathbf{k}|$, we can represent the Green's function (10.7) in the form

$$G_\xi(\mathbf{r}-\mathbf{r}') = \frac{2}{\pi} \int_0^\infty \frac{dk\, k^2}{\xi^2 + k^2} \sum_{l,M} u_{lM}(k;\mathbf{r}) u_{lM}^*(k;\mathbf{r}'). \tag{10.123}$$

This formula can be used to calculate the matrix elements of the kernel (10.119), where both arguments are defined in the same coordinate system. In order to use different coordinate systems, in the case of the kernel (10.118), we must apply the translation formula, which, for the basis functions (10.121), is given by

$$u_{lM}(k;\mathbf{r}'+\mathbf{L}) = \sum_{l',M'} A_{lM,l'M'}(k;\mathbf{L})\, u_{l'M'}(k;\mathbf{r}'). \tag{10.124}$$

The translation coefficients $A_{lM,l'M'}$ can be obtained by the repeated application of eqn (10.122):

$$A_{lM,l'M'}(k;\mathbf{L}) = i^{l'-l}(-1)^{M'} \int d\Omega_k\, Y_{lM}(\theta_k,\varphi_k) Y_{l',-M'}(\theta_k,\varphi_k) e^{i\mathbf{k}\cdot\mathbf{L}}$$

$$= 4\pi i^{l'-l}(-1)^{M'} \sum_{l'',M''} i^{l''} j_{l''}(kL) Y_{l''M''}(\theta_L,\varphi_L) \tag{10.125}$$

$$\times (-1)^{M''} \int d\Omega_k\, Y_{lM}(\theta_k,\varphi_k) Y_{l',-M'}(\theta_k,\varphi_k) Y_{l'',-M''}(\theta_k,\varphi_k).$$

Here we have used $Y_{lM}^*(\theta,\varphi) = (-1)^M Y_{l,-M}(\theta,\varphi)$. Now we take into consideration the fact that the vector $\mathbf{L}$ is directed along the axis connecting the centers of our spheres. As a consequence, $\theta_L \equiv 0$ and

$$Y_{l''M''}(\theta_L,\varphi_L) = \delta_{0M''} \sqrt{\frac{2l''+1}{4\pi}}. \tag{10.126}$$

Using the integral over three spherical harmonics

$$\int d\Omega\, Y_{lM}(\theta,\varphi) Y_{l'M'}(\theta,\varphi) Y_{l''M''}(\theta,\varphi)$$

$$= \sqrt{\frac{(2l+1)(2l'+1)(2l''+1)}{4\pi}} \begin{pmatrix} l & l' & l'' \\ M & M' & M'' \end{pmatrix} \begin{pmatrix} l & l' & l'' \\ 0 & 0 & 0 \end{pmatrix}, \quad (10.127)$$

and the standard properties of 3j-symbols (Varshalovich et al. 1988), the translation coefficients (10.125) can be represented in the form

$$A_{lM,l'M'}(k;\boldsymbol{L}) = \delta_{MM'} i^{l'-l}(-1)^M \sum_{l''=|l-l'|}^{l+l'} i^{l''} j_{l''}(kL) H_{ll'M}^{l''}, \quad (10.128)$$

where

$$H_{ll'M}^{l''} = \sqrt{(2l+1)(2l'+1)}(2l''+1) \begin{pmatrix} l & l' & l'' \\ 0 & 0 & 0 \end{pmatrix} \begin{pmatrix} l & l' & l'' \\ M & -M & 0 \end{pmatrix}. \quad (10.129)$$

Note that the first 3j-symbol on the right-hand side of eqn (10.129) is not equal to zero only when $l+l'+l''$ is even. There is also an alternative derivation for the translation coefficients, which may be useful for computations (Wittmann 1988).

Using eqns (10.123), (10.124), and (10.128), we get the following for the matrix element (10.120):

$$\left(\tilde{K}_{\xi,AB}\right)_{lM,l'M} = \frac{2}{\pi} \int_0^\infty \frac{dk\, k^2}{\xi^2 + k^2} j_l(kR_A) j_{l'}(kR_B) A_{lM,l'M}^*(k;\boldsymbol{L}) \quad (10.130)$$

$$= \frac{2}{\pi} \int_0^\infty \frac{dk\, k^2}{\xi^2 + k^2} j_l(kR_A) j_{l'}(kR_B) i^{l-l'}(-1)^M \sum_{l''} i^{-l''} j_{l''}(kL) H_{ll'M}^{l''}.$$

Here and below, the summation over l'' is performed in the same limits as in eqn (10.128).

For the matrix element of the kernel (10.119), we get the following expression from eqn (10.123):

$$\left(\tilde{K}_{\xi,AA}\right)_{lM,l'M} = \delta_{ll'} \frac{2}{\pi} \int_0^\infty \frac{dk\, k^2}{\xi^2 + k^2} j_l(kR_A) j_{l'}(kR_A), \quad (10.131)$$

which is diagonal in the indices l and l'. Substituting the definition (9.11) of the spherical Bessel functions here and integrating similarly to eqn (10.72), we arrive at

$$\left(\tilde{K}_{\xi,AA}\right)_{lM,l'M} = \delta_{ll'} \frac{1}{R_A} I_\nu(\xi R_A) K_\nu(\xi R_A), \quad (10.132)$$

where $\nu \equiv l + 1/2$. In fact, eqn (10.132) is a well-known representation of the free-space propagator in spherical coordinates. A similar formula results from $A \to B$. For the inverse operator, we get

$$\left(\tilde{K}_{\xi,AA}^{-1}\right)_{lM,l'M} = \delta_{ll'} \frac{R_A}{I_\nu(\xi R_A) K_\nu(\xi R_A)}. \quad (10.133)$$

Applications to spherical geometry

In the matrix element (10.130), the k-integration can also be carried out. For this purpose, we substitute eqn (9.11) into eqn (10.130) and obtain

$$\left(\tilde{K}_{\xi,AB}\right)_{lM,l'M} = \sqrt{\frac{\pi}{2R_AR_BL}} i^{l-l'}(-1)^M \sum_{l''} i^{-l''} H_{ll'M}^{l''}$$
$$\times \int_0^\infty \frac{dk\sqrt{k}}{\xi^2+k^2} J_\nu(kR_A)J_{\nu'}(kR_B)J_{\nu''}(kL). \tag{10.134}$$

The integral on the right-hand side of eqn (10.134) is calculated in perfect analogy to the integral (10.71), leading to

$$\left(\tilde{K}_{\xi,AB}\right)_{lM,l'M} = (-1)^{l+M}\sqrt{\frac{\pi}{2\xi R_AR_BL}} I_\nu(\xi R_A)I_{\nu'}(\xi R_B)$$
$$\times \sum_{l''} (-1)^{l''} K_{\nu''}(\xi L) H_{ll'M}^{l''}. \tag{10.135}$$

The vacuum energy for a scalar field obeying Dirichlet boundary conditions on two spheres can now be represented in the form

$$E = \frac{1}{2\pi}\int_0^\infty d\xi\, \text{Tr} \ln(1-\mathcal{M}_\xi). \tag{10.136}$$

Here the matrix $\mathcal{M}_\xi$ is a product of two matrices,

$$\mathcal{M}_{\xi,lM,l'M'} = \delta_{MM'} \sum_{\tilde l} (P_{\xi,AB})_{lM,\tilde lM} (P_{\xi,BA})_{\tilde lM,l'M}, \tag{10.137}$$

where

$$(P_{\xi,AB})_{lM,\tilde lM} \equiv \left(\tilde K_{\xi,AA}^{-1}\right)_{lM,lM} \left(\tilde K_{\xi,AB}\right)_{lM,\tilde lM} \tag{10.138}$$
$$= (-1)^{l+M}\sqrt{\frac{\pi R_A}{2\xi L R_B}} \frac{I_{\tilde\nu}(\xi R_B)}{K_\nu(\xi R_A)} \sum_{l''} (-1)^{l''} K_{\nu''}(\xi L) H_{l\tilde lM}^{l''}.$$

The matrix elements of $P_{\xi,BA}$ can be obtained from this equation by interchanging the radii R_A and R_B and taking into account the fact that in eqn (10.126), $\theta_L = \pi$. In eqn (10.137), we have explicitly indicated that the matrix is diagonal in the magnetic quantum numbers. The trace in eqn (10.136) is over all of the orbital quantum numbers.

The vacuum energy (10.136) is for the case of two different spheres. Now we consider two similar spheres with $R = R_A = R_B$. They are symmetric under the mirror symmetry $\theta \to \pi - \theta$, $\varphi \to \varphi$. The corresponding operator $\mathcal{J}$ in the (ll') representation is given by

$$\mathcal{J}_{lM,l'M} = (-1)^{l+M}\delta_{ll'}. \tag{10.139}$$

Following eqn (10.58), this allows us to define

$$N_{\xi,lM,l'M}^D = \sum_{\tilde l} \left(\tilde K_{\xi,AA}^{-1}\right)_{lM,lM} \left(\tilde K_{\xi,AB}\right)_{lM,\tilde lM'} \mathcal{J}_{\tilde lM,l'M} \tag{10.140}$$

$$= \sqrt{\frac{\pi}{2\xi L}} \frac{I_{\nu'}(\xi R)}{K_\nu(\xi R)} \sum_{l''} K_{\nu''}(\xi L) H_{ll'M}^{l''}.$$

Here, we have taken into account the fact that $H_{ll'M}^{l''}$ in eqn (10.129) is not equal to zero only for even $l+l'+l''$. With the superscript D, we have indicated that Dirichlet boundary conditions are imposed on the sphere. With this definition, following eqn (10.57), the matrix elements of the operator $\mathcal{M}$ become a product of two matrices:

$$M_{\xi,lM,l'M'} = \delta_{MM'} \sum_{\tilde{l}} N_{\xi,lM,\tilde{l}M} N_{\xi,\tilde{l}M,l'M}. \tag{10.141}$$

As a consequence, the vacuum energy of the two spheres becomes a sum of the vacuum energies:

$$E^{(\pm)} = \frac{1}{2\pi} \int_0^\infty d\xi \, \mathrm{Tr} \ln\left(1 \pm \mathcal{N}_\xi\right). \tag{10.142}$$

Here, if the matrix element (10.140) is used, $E^{(+)}$ is the vacuum energy for a sphere with Dirichlet boundary conditions in front of a plane with Neumann boundary conditions. In this case, $E^{(-)}$ corresponds to Dirichlet boundary conditions both on the sphere and on the plane. We note that the distance between the center of the sphere and the plane is $L/2$.

Here we add some necessary remarks for generalizing these formulas to the case of Neumann boundary conditions on the sphere. We have to go back to eqn (10.10) to define the projection of the free-space Green's function. The normal derivative here is the derivative with respect to the radial variable, which must be taken into account in eqns (10.118) and (10.119). After these changes, all other steps can be done in the same way as for Dirichlet boundary conditions, and we can write down the matrix elements of the operator $\mathcal{N}_\xi$ to be inserted into eqn (10.142) as

$$N_{\xi,lM,l'M}^{\mathrm{N}} = \sqrt{\frac{\pi}{2\xi L}} \frac{(I_{\nu'}(z)/\sqrt{z})'\big|_{z=\xi R}}{(K_\nu(z)/\sqrt{z})'\big|_{z=\xi R}} \sum_{l''} K_{\nu''}(\xi L) H_{ll'M}^{l''}. \tag{10.143}$$

Here, the derivatives act on the arguments of the Bessel functions, and we have taken into account the relation between the spherical Bessel functions and the ordinary bessel functions. Finally, we mention that all remarks concerning the sign of the energy made at the end of Section 10.3.1 apply to this case also.

10.4.2 A sphere and a plane at large separation

We consider a sphere in front of a plane with Dirichlet boundary conditions. Following eqn (10.142), the vacuum energy is given by

$$E^{\mathrm{D}} \equiv E^{(-)} = -\frac{1}{\pi} \int_0^\infty d\xi \sum_{s=0}^\infty \frac{1}{s+1} {\sum_{M=0}^\infty}' \sum_{l_1=M}^\infty \cdots \sum_{l_{s+1}=M}^\infty N_{\xi,l_1 M,l_2 M} \cdots N_{\xi,l_{s+1} M,l_1 M}. \tag{10.144}$$

Here, the prime denotes an extra factor of $1/2$ in the sum over M for $M = 0$. The energy $E^N \equiv E^{(+)}$ for Neumann boundary conditions on the plane can be obtained by replacement of $1/(s+1)$ with $(-1)^{s+1}/(s+1)$. The boundary conditions on the sphere are determined by the matrix elements in eqn (10.140) or eqn (10.143), which we insert into eqn (10.144).

First we consider large separations, or, equivalently, a small sphere, i.e. $R \ll L/2 = b$ (see Fig. 10.1). In this case the Bessel functions depending on R can be expanded directly under the integration sign and in the summations in eqn (10.144). Similarly to Section 10.3.2, the leading contribution comes from $l = l' = 0$ and $s = 0$ (this is the first order of the expansion of the logarithm). Higher orders can also be generated. For Dirichlet boundary conditions on both the sphere and the plane, we get

$$E^{DD} = -\frac{R}{8\pi b^2}\left[1 + \frac{5}{8}\frac{R}{b} + \frac{421}{144}\frac{R^2}{b^2} + O\left(\frac{R^3}{b^3}\right)\right]. \tag{10.145}$$

In the same way for Neumann boundary conditions on the plane and Dirichlet boundary conditions on the sphere, we get

$$E^{ND} = \frac{R}{8\pi b^2}\left[1 + \frac{3}{8}\frac{R}{b} + \frac{403}{144}\frac{R^2}{b^2} + O\left(\frac{R^3}{b^3}\right)\right]. \tag{10.146}$$

A slightly different picture appears for Neumann boundary conditions on the sphere. Because, for $l = 0$, the expansion of the Bessel function $K_{1/2}(z) \sim \sqrt{z}$ in eqn (10.143) starts from the square root of the radius, the leading order of the expansion inside the derivative in eqn (10.143) is a constant and drops out. As a consequence, for Neumann conditions on the plane, we find

$$E^{NN} = -\frac{17R^3}{96\pi b^4}\left[1 + \frac{379}{170}\frac{R^2}{b^2} + \frac{801}{4352}\frac{R^3}{b^3} + O\left(\frac{R^4}{b^4}\right)\right] \tag{10.147}$$

and, for Dirichlet boundary conditions on the plane, the result is

$$E^{DN} = \frac{17R^3}{96\pi b^4}\left[1 + \frac{379}{170}\frac{R^2}{b^2} + \frac{479}{4352}\frac{R^3}{b^3} + O\left(\frac{R^4}{b^4}\right)\right]. \tag{10.148}$$

The results in eqns (10.145)–(10.148) were obtained by Emig (2008). The energy for two spheres with Dirichlet boundary conditions on both is $E^{DD} + E^{ND}$, and for Neumann boundary conditions on both it is $E^{DN} + E^{NN}$. It can be seen that for two spheres with Dirichlet boundary conditions, the Casimir energy decreases faster than for a sphere with Dirichlet boundary conditions in front of a plane. A similar statement is valid for two spheres with Neumann boundary conditions.

10.4.3 *Corrections beyond the proximity force approximation at small separations*

The limit of small separations $a \ll R$ between a sphere and a plane is the most interesting case from an experimental point of view (see e.g. Section 18.1.2). The small parameter ε in this problem is defined in eqn (10.97). The spherical case can be considered in a similar way to the cylindrical one (see Section 10.3.3), with the exception that an asymptotic expansion of the $3j$-symbols is needed.

We start from the representation (10.144) for the vacuum energy. We introduce a new integration variable $\tilde{\xi} = \xi R$ and new summation indices according to $l_1 = l$, $l_{i+1} = l + \tilde{l}_i$ ($i \geq 1$). Then the vacuum energy is given by

$$E^D = -\frac{1}{\pi R} \int_0^\infty d\tilde{\xi} \sum_{s=0}^\infty \frac{1}{s+1} \sum_{M=0}^\infty {\sum_{l=M}^\infty}' \sum_{\tilde{l}_1=-l}^\infty \cdots \sum_{\tilde{l}_s=-l}^\infty Z_{\tilde{\xi};M,l,\tilde{l}_1,\ldots,\tilde{l}_s}, \qquad (10.149)$$

where

$$Z_{\tilde{\xi};M,l,\tilde{l}_1,\ldots,\tilde{l}_s} = N_{\tilde{\xi},lM,l+\tilde{l}_1,M} N_{\tilde{\xi},l+\tilde{l}_1,M,l+\tilde{l}_2,M} \cdots N_{\tilde{\xi},l+\tilde{l}_s,M,lM}. \qquad (10.150)$$

For Dirichlet boundary conditions on the sphere, the matrix elements are given by

$$N^D_{\tilde{\xi},lM,l'M} = \frac{\sqrt{\pi}}{2\sqrt{\tilde{\xi}(1+\varepsilon)}} \frac{I_{\nu'}(\tilde{\xi})}{K_\nu(\tilde{\xi})} \sum_{l''} K_{\nu''}\left(2\tilde{\xi}(1+\varepsilon)\right) H^{l''}_{ll'M}. \qquad (10.151)$$

For Neumann boundary conditions, the matrix elements are

$$N^N_{\tilde{\xi},lM,l'M} = \frac{\sqrt{\pi}}{2\sqrt{\tilde{\xi}(1+\varepsilon)}} \frac{\left(I_{\nu'}(z)/\sqrt{z}\right)'\big|_{z=\tilde{\xi}}}{\left(K_\nu(z)/\sqrt{z}\right)'\big|_{z=\tilde{\xi}}} \sum_{l''} K_{\nu''}\left(2\tilde{\xi}(1+\varepsilon)\right) H^{l''}_{ll'M}. \qquad (10.152)$$

Now, similarly to Section 10.3.3, we take into account the fact that at small separations the dominating contributions come from the large frequencies $\tilde{\xi}$ and the large summation indices involved in eqn (10.149). As we are interested in the asymptotic expansion for small ε, we replace all sums with integrals. The error introduced is assumed to be exponentially small. After that, we make the following substitutions in eqn (10.149),

$$\tilde{\xi} = \frac{t}{\varepsilon}\sqrt{1-\tau^2}, \quad l = \frac{t\tau}{\varepsilon}, \quad M = \sqrt{\frac{t}{\varepsilon}}\tau\mu, \quad \tilde{l}_i = n_i\sqrt{\frac{4t}{\varepsilon}}, \qquad (10.153)$$

as the main contributions come from $\tilde{\xi} \sim 1/\varepsilon$, $l \sim 1/\varepsilon$, $\tilde{l}_i \sim 1/\sqrt{\varepsilon}$ ($i = 1, \ldots, s$), and $M \sim 1/\sqrt{\varepsilon}$. The variable τ is the cosine of the angle in the (ξ, l) plane. After this, we expand the matrix elements for $\varepsilon \to 0$ and calculate the remaining integrals.

With the new variables, the expression for the energy reads

$$E^D = -\frac{1}{4\pi R\varepsilon^2} \sum_{s=0}^\infty \frac{1}{s+1} \int_0^\infty dt\, t \int_0^1 \frac{d\tau\,\tau}{\sqrt{1-\tau^2}} \int_{-\infty}^\infty \frac{d\mu}{\sqrt{\pi}} \qquad (10.154)$$

$$\times \int_{-\infty}^{\infty} \frac{dn_1}{\sqrt{\pi}} \cdots \int_{-\infty}^{\infty} \frac{dn_s}{\sqrt{\pi}} \left(\frac{4\pi t}{\varepsilon}\right)^{(s+1)/2} Z^{as}(t,\tau,\mu,n_1,\ldots,n_s),$$

where Z^{as} is given by eqn (10.150), under the condition that in eqns (10.151) and (10.152) the Bessel functions and the $3j$-symbols are replaced with their asymptotic expansions. The key difference in comparison with the cylindrical case is the asymptotic expansion of the $3j$-symbols for large l and M. It is different from the known semiclassical expansions and can be obtained using the saddle point method in an integral representation. The result is (Bordag and Nikolaev 2008)

$$N_{\xi,l+l_i,M,l+l_{i+1},M}^{D,as} = \sqrt{\frac{\varepsilon}{4\pi t}} e^{-\tilde{\eta}^{as}} e^{-\mu^2} \tag{10.155}$$
$$\times \left[1 + a_{n_i,n_{i+1}}^{(1/2)}(t,\tau,\mu)\sqrt{\varepsilon} + a_{n_i,n_{i+1}}^{(1)}(t,\tau,\mu)\varepsilon + \ldots\right],$$

where $\tilde{\eta}^{as}$ is defined in eqn (10.108). The functions $a_{nn'}^{(1/2)}$ and $a_{nn'}^{(1)}$ are polynomials in n and n'.

Substituting eqn (10.155) into eqn (10.150), we obtain Z^{as} in the form

$$Z^{as}(t,\tau,\mu,n_1,\ldots,n_s) = e^{-\eta_1}\left[1 + \sqrt{\varepsilon}\sum_{i=0}^{s} a_{n_i,n_{i+1}}^{(1/2)}(t,\tau,\mu) \right. \tag{10.156}$$
$$\left. +\varepsilon\left(\sum_{0\leq i<j\leq s} a_{n_i,n_{i+1}}^{(1/2)}(t,\tau,\mu)a_{n_j,n_{j+1}}^{(1/2)(t,\tau,\mu)} + \sum_{i=0}^{s} a_{n_i,n_{i+1}}^{(1)}(t,\tau,\mu)\right) + \ldots\right],$$

where η_1 is defined in eqn (10.111).

With this expression, the integrations over the n_i in eqn (10.154) are Gaussian and can be carried out [see e.g. eqn (10.112)]. The remaining integrations can also be carried out. Finally, we arrive at the result (Bordag and Nikolaev 2008)

$$E^{D}(a) = -\frac{\pi^3}{1440}\frac{R}{a^2}\left[1 + \frac{1}{3}\frac{a}{R} + O\left(\frac{a^2}{R^2}\right)\right] \tag{10.157}$$

for the Casimir energy with Dirichlet boundary conditions on both the sphere and the plane. In eqn (10.157), the leading order is the known result from the PFA. The contribution of order $\sqrt{\varepsilon}$ has dropped out in the integrations over n_i, because of symmetry. Thus the first correction beyond the PFA is of order ε.

Note that the coefficient $1/3$ in eqn (10.157) has been confirmed numerically by world line methods (Gies and Klingmüller 2006b) and by extrapolation from direct numerical calculation at medium separations (Emig 2008).

The corresponding calculations for Neumann boundary conditions are analogous. One has to take into account the derivatives in eqn (10.152). The result is (Bordag and Nikolaev 2008, 2009)

$$E^{\mathrm{N}}(a) = -\frac{\pi^3}{1440}\frac{R}{a^2}\left[1 + \left(\frac{1}{3} - \frac{10}{\pi^2}\right)\frac{a}{R} + O\left(\frac{a^2}{R^2}\right)\right], \quad (10.158)$$

where the structure is very similar to that in the cylindrical case [eqn (10.115)].

10.5 Corrugated planes

An experimentally important configuration is that of two planes covered by corrugations. These corrugations may be of various shapes and sizes, ranging from corrugations of small amplitude (compared with the separation between the planes) to ones of large amplitude. The structure of the corrugations can also vary, from stochastic to periodic, with various periods. From a theoretical point of view, the general formulas of Section 10.1 apply to all cases and provide a representation of the Casimir energy which is free of divergences and is accessible to direct numerical computation. However, in practice this has turned out to be a complicated task and, at present, corrugations of various shapes remain an active area of research.

For many applications it is realistic to consider the corrugations as small perturbations of a flat surface. In this section, we briefly present the theoretical results for corrugated planes in the lowest perturbation order. For corrugations of arbitrary shape, only a scalar field with Dirichlet or Neumann boundary conditions can be considered. This is because, in general, the polarizations of an electromagnetic field do not separate. We consider two parallel planes $z_n^{(0)} = \pm a/2$ ($n = 1, 2$) perpendicular to the z-axis, covered with small corrugations. The functions describing the corrugations, $h_n(\boldsymbol{r}_\perp)$, where $\boldsymbol{r}_\perp$ is a vector in the plane perpendicular to the z-axis, satisfy the conditions

$$\int d\boldsymbol{r}_\perp\, h_n(\boldsymbol{r}_\perp) = 0, \qquad \max|h_n(\boldsymbol{r}_\perp)| \ll a. \quad (10.159)$$

The surfaces of the corrugated planes are described by the equations

$$z_n(\boldsymbol{r}_\perp) = z_n^{(0)} + h_n(\boldsymbol{r}_\perp). \quad (10.160)$$

If we restrict ourselves to second-order perturbation theory in the corrugation functions, the vacuum energy per unit area takes the following form:

$$E(a) = E^{(0)}(a) + \frac{1}{2}\sum_{n,n'}\int d\boldsymbol{r}_\perp \int d\boldsymbol{r}'_\perp\, h_n(\boldsymbol{r}_\perp) h_{n'}(\boldsymbol{r}'_\perp) R_{nn'}(\boldsymbol{r}_\perp - \boldsymbol{r}'_\perp). \quad (10.161)$$

Here, $E^{(0)}(a)$ is the Casimir energy per unit area between flat planes with the corresponding boundary conditions, and the kernel $R_{nn'}$ does not depend on the corrugations. It is determined by the type of boundary conditions and is symmetric in n and n' (Li and Kardar 1992).

In the case of periodic uniaxial corrugations, the polarizations of an electromagnetic field do separate. For corrugations of sinusoidal form on only one plane, one has the functions

$$h_1(\mathbf{r}_\perp) = A \sin \frac{2\pi x}{\Lambda}, \qquad h_2(\mathbf{r}_\perp) = 0. \tag{10.162}$$

Emig et al. (2003) obtained an explicit expression for the electromagnetic Casimir energy per unit area (10.161) in the ideal-metal case:

$$E^{\text{em}}(a) = -\frac{\pi^2}{720 a^3} - \frac{A^2}{a^5} \left[G_{\text{TM}}\left(\frac{a}{\Lambda}\right) + G_{\text{TE}}\left(\frac{a}{\Lambda}\right) \right]. \tag{10.163}$$

The explicit form of the functions $G_{\text{TM}}(\xi)$ and $G_{\text{TE}}(\xi)$ is

$$G_{\text{TM}}(\xi) = \frac{\pi^3 \xi}{480} - \frac{\pi^2 \xi^4}{30} \ln(1-u) + \frac{\pi}{1920 \xi} \text{Li}_2(1-u) + \frac{\pi \xi^3}{24} \text{Li}_2(u)$$
$$+ \frac{\xi^2}{24} \text{Li}_3(u) + \frac{\xi}{32\pi} \text{Li}_4(u) + \frac{1}{64 \pi^2} \text{Li}_5(u) + \frac{1}{256 \pi^3 \xi} \left[\text{Li}_6(u) - \frac{\pi^6}{945} \right],$$

$$G_{\text{TE}}(\xi) = \frac{\pi^3 \xi}{1440} - \frac{\pi^2 \xi^4}{30} \ln(1-u) + \frac{\pi}{1920 \xi} \text{Li}_2(1-u) - \frac{\pi \xi}{48}(1+2\xi^2) \text{Li}_2(u)$$
$$+ \left(\frac{\xi^2}{48} - \frac{1}{64} \right) \text{Li}_3(u) + \frac{5\xi}{64\pi} \text{Li}_4(u) + \frac{7}{128 \pi^2} \text{Li}_5(u)$$
$$+ \frac{1}{256 \pi^3 \xi} \left[\frac{7}{2} \text{Li}_6(u) - \pi^2 \text{Li}_4(u) + \frac{\pi^6}{135} \right]. \tag{10.164}$$

Here, $\text{Li}_n(u)$ is the polylogarithm function and $u \equiv \exp(-4\pi\xi)$.

The behavior of these functions for small ξ, i.e. for $a \ll \Lambda$, is

$$G_{\text{TM}}(\xi) = \frac{\pi^2}{480} + O(\xi^2), \qquad G_{\text{TE}}(\xi) = \frac{\pi^2}{480} + O(\xi^2). \tag{10.165}$$

For large ξ, i.e. for $a \gg \Lambda$, one obtains

$$G_{\text{TM}}(\xi) = \frac{\pi^3 \xi}{480} \left(1 + \frac{5}{126 \xi^2} \right) + O\left(e^{-4\pi\xi} \right),$$
$$G_{\text{TE}}(\xi) = \frac{\pi^3 \xi}{1440} \left(1 + \frac{1}{6 \xi^2} \right) + O\left(e^{-4\pi\xi} \right). \tag{10.166}$$

As can be seen from these formulas, for corrugations with a large wavelength Λ, i.e. for $\xi \to 0$, the energy is

$$E^{\text{em}}(a) = -\frac{\pi^2}{720 a^3} \left(1 + 3 \frac{A^2}{a^2} \right), \tag{10.167}$$

where both polarizations give the same contribution. For short-wavelength corrugations, i.e. for $\xi \to \infty$, there are only two contributions from each mode all the way up to exponentially small terms, which together give the expression

$$E^{\text{em}}(a) = -\frac{\pi^2}{720a^3}\left(1 + 2\pi\frac{A^2}{a\Lambda}\right). \tag{10.168}$$

Now we consider the case of uniaxial sinusoidal corrugations of equal amplitudes and periods on both planes,

$$h_1(\mathbf{r}_\perp) = A\sin\frac{2\pi x}{\Lambda}, \qquad h_2(\mathbf{r}_\perp) = A\sin\frac{2\pi(x+x_0)}{\Lambda}, \tag{10.169}$$

where x_0 is the relative phase shift between the corrugations. In this case, instead of eqn (10.163), the Casimir energy per unit area is given by (Emig et al. 2003)

$$E^{\text{em}}(a) = -\frac{\pi^2}{720a^3} - \frac{2A^2}{a^5}\left[G_{\text{TM}}\left(\frac{a}{\Lambda}\right) + G_{\text{TE}}\left(\frac{a}{\Lambda}\right)\right]$$
$$+ \frac{A^2}{a^5}\cos\frac{2\pi x_0}{\Lambda}\left[J_{\text{TM}}\left(\frac{a}{\Lambda}\right) + J_{\text{TE}}\left(\frac{a}{\Lambda}\right)\right]. \tag{10.170}$$

The explicit expressions for the functions $J_{\text{TM}}(\xi)$ and $J_{\text{TE}}(\xi)$ are

$$J_{\text{TM}}(\xi) = \frac{\pi^2}{120}(16\xi^4 - 1)\operatorname{arctanh}(\sqrt{u}) + \sqrt{u}\left[\frac{\pi}{12}\left(\xi^3 - \frac{1}{80\xi}\right)\Phi\left(u, 2, \frac{1}{2}\right)\right.$$
$$+ \frac{\xi^2}{12}\Phi\left(u, 3, \frac{1}{2}\right) + \frac{\xi}{16\pi}\Phi\left(u, 4, \frac{1}{2}\right) + \frac{1}{32\pi^2}\Phi\left(u, 5, \frac{1}{2}\right)$$
$$+ \left.\frac{1}{128\pi^3\xi}\Phi\left(u, 6, \frac{1}{2}\right)\right],$$

$$J_{\text{TE}}(\xi) = \frac{\pi^2}{120}(16\xi^4 - 1)\operatorname{arctanh}(\sqrt{u}) + \sqrt{u}\left[-\frac{\pi}{12}\left(\xi^3 + \frac{\xi}{2} + \frac{1}{80\xi}\right)\Phi\left(u, 2, \frac{1}{2}\right)\right.$$
$$+ \frac{1}{24}\left(\xi^2 - \frac{3}{4}\right)\Phi\left(u, 3, \frac{1}{2}\right) + \frac{5}{32\pi}\left(\xi - \frac{1}{20\xi}\right)\Phi\left(u, 4, \frac{1}{2}\right)$$
$$+ \left.\frac{7}{64\pi^2}\Phi\left(u, 5, \frac{1}{2}\right) + \frac{7}{256\pi^3\xi}\Phi\left(u, 6, \frac{1}{2}\right)\right]. \tag{10.171}$$

Here, the Lerch transcendental function is defined as

$$\Phi(z, s, p) = \sum_{n=0}^{\infty}\frac{z^n}{(n+p)^s}. \tag{10.172}$$

The behavior of the functions (10.171) for small argument ξ, i.e. for long-wavelength corrugations with $\Lambda \gg a$, is

$$J_{\text{TM}}(\xi) + J_{\text{TE}}(\xi) = \frac{\pi^2}{120} + O(\xi^2). \tag{10.173}$$

For large ξ, i.e. in the limit of short wavelengths of the corrugations, $\Lambda \ll a$, the contribution of the functions (10.171) in the energy (10.170) is exponentially small:

$$J_{\text{TM}}(\xi) + J_{\text{TE}}(\xi) = \frac{4\pi^2}{15} e^{-2\pi\xi} \left[\xi^4 + O(\xi^2) \right]. \tag{10.174}$$

Equations (10.163) and (10.170) are used in Section 17.5.3 for the calculation of the normal and lateral Casimir force and for estimation of the accuracy of various approximate methods.

Note that in the case of ideal-metal plates with rectangular periodic corrugations, exact expressions for the Casimir energy have been obtained (Emig 2003, Büscher and Emig, 2004, 2005) which do not use perturbative expansions in powers of the corrugation amplitudes. We discuss some of the results obtained in Section 21.5 in connection with an experiment on the measurement of the Casimir force between a sphere and a plate covered with rectangular trenches.

11
SPACES WITH NON-EUCLIDEAN TOPOLOGY

In Section 2.3, we considered the simplest example of a case where the Casimir effect arises owing to identification conditions, signifying the topology of a circle S^1 in this case. This is in fact the single topologically nontrivial one-dimensional spatial manifold in two-dimensional space–time. Here, we briefly review some more complicated spaces with nontrivial topology (both flat and curved) with respect to the Casimir effect. As an important application of the numerous results obtained in this field, we consider in more detail the vacuum energy–momentum tensor due to the Casimir effect in the closed Friedmann model. A related subject is the role of the Casimir effect in multidimensional Kaluza–Klein theories, where it provides one of the mechanisms for compactification of extra spatial dimensions. This is also reflected in the present chapter. We conclude with a brief discussion of the Casimir effect for topological defects, such as cosmic strings and domain walls. Some grand unification theories predict the formation of such defects in the early Universe. Although cosmic strings are not considered anymore as a primary source of primordial density perturbations, there are astrophysical effects where they play a large role.

11.1 Topologically nontrivial flat spaces

Here, we consider examples of the Casimir effect of topological origin in three- and four-dimensional space–times.

11.1.1 Three-dimensional space–time

Following S^1 in complexity are the topologically nontrivial flat spaces in two-dimensional space (three-dimensional space–time). In fact, we have already considered one such example in Section 8.4. This is the case of a massless scalar field in a rectangle with identified opposite sides (with the topology of a 2-torus $S^1 \times S^1$). In fact, a 2-torus can be described as a Euclidean plane R^2 with identified points having coordinates (x, y) and $(x + na, y + lb)$, where n and l take any integer values from $-\infty$ to $+\infty$. The manifold obtained is compact and a scalar field defined on it satisfies periodic identification conditions in both coordinates:

$$\varphi(t, 0, y) = \varphi(t, a, y), \qquad \varphi(t, x, 0) = \varphi(t, x, b),$$

(11.1)

$$\left.\frac{\partial \varphi(t, x, y)}{\partial x}\right|_{x=0} = \left.\frac{\partial \varphi(t, x, y)}{\partial x}\right|_{x=a}, \qquad \left.\frac{\partial \varphi(t, x, y)}{\partial y}\right|_{y=0} = \left.\frac{\partial \varphi(t, x, y)}{\partial y}\right|_{y=b}.$$

An explicit expression for the vacuum energy density in this case is obtained from eqns (8.71) and (8.26),

$$\varepsilon(a,b) = \frac{E(a,b)}{ab} = -\frac{\pi}{6ab^2} - \frac{\zeta_R(3)}{2\pi a^3} + \frac{8\pi}{a^2 b} G\left(\frac{b}{a}\right), \qquad (11.2)$$

where the function $G(z)$ is defined in eqn (8.11).

Another topologically nontrivial two-dimensional flat space has the topology of a cylinder $S^1 \times R^1$; i.e. points with the coordinates $(x+na,y)$, where $n = 0, \pm 1, \pm 2, \ldots$, are identified. A scalar field $\varphi(t,x,y)$ defined on such a noncompact manifold satisfies the boundary conditions

$$\varphi(t,0,y) = \varphi(t,a,y), \qquad \left.\frac{\partial \varphi(t,x,y)}{\partial x}\right|_{x=0} = \left.\frac{\partial \varphi(t,x,y)}{\partial x}\right|_{x=a}. \qquad (11.3)$$

The Casimir energy density for a massless scalar field defined on a plane with the topology of a cylinder can be obtained from eqn (11.2) in the limiting case $b \to \infty$:

$$\varepsilon(a) = \lim_{b \to \infty} \varepsilon(a,b) = -\frac{\zeta_R(3)}{2\pi a^3}. \qquad (11.4)$$

Note that the flat manifolds with the topologies of a 2-torus and a cylinder are homogeneous. As a result, the energy densities $\varepsilon(a,b)$ and $\varepsilon(a)$ defined in eqns (11.2) and (11.4) do not depend on the coordinates.

It has been known that there are only four complete two-dimensional flat spaces with topologies different from the topology of the Euclidean plane R^2 (Efimov 1980). Two of them, $S^1 \times S^1$ and $S^1 \times R^1$, have already been considered. The third is a plane with the topology of the Klein surface. This manifold is inhomogeneous. It is obtained by the identification of all points with the coordinates $[x+na, (-1)^n y + lb]$, where $n, l = 0, \pm 1, \pm 2, \ldots$. This results in the following periodic identification conditions imposed on the scalar field:

$$\varphi(t,x,0) = \varphi(t,x,b), \qquad \varphi(t,0,y) = \varphi(t,a,b-y),$$

$$\left.\frac{\partial \varphi(t,x,y)}{\partial y}\right|_{y=0} = \left.\frac{\partial \varphi(t,x,y)}{\partial y}\right|_{y=b}, \qquad \left.\frac{\partial \varphi(t,x,y)}{\partial x}\right|_{x=0} = \left.\frac{\partial \varphi(t,x,b-y)}{\partial x}\right|_{x=a}. \qquad (11.5)$$

Here, finite expressions for the vacuum energy density can be found not by using the canonical energy–momentum tensor of the scalar field (3.8) but by using the metrical energy–momentum tensor (3.9) with $\xi = 1/8$ and $D = 3$. The final result is obtained by the application of the Abel–Plana formulas (2.26) and (2.41) (Mamayev and Mostepanenko 1985):

$$\tilde{\varepsilon}(a,b;y) = -\frac{\pi}{12ab^2} - \frac{\pi}{8a^2 b} - \frac{\zeta_R(3)}{16\pi a^3} + \frac{2\pi}{a^2 b} G\left(\frac{b}{2a}\right) + \frac{4\pi}{b^3} H(a,b;y), \qquad (11.6)$$

where

$$H(a,b;y) = -\sum_{k=1}^{\infty} k^2 \cos\frac{4\pi k y}{b} \int_1^\infty \frac{ds}{\sinh(2\pi k s a/b)} \frac{2s^2-1}{\sqrt{s^2-1}}. \qquad (11.7)$$

As can be seen from eqn (11.7),

$$\int_0^b H(a,b;y)\,dy = 0. \tag{11.8}$$

The last complete two-dimensional manifold with a non-Euclidean topology is the Möbius strip of infinite width. It is obtained from a plane by the identification of all points with coordinates $[x + na, (-1)^n y]$. A scalar field defined on this noncompact manifold must satisfy the identification conditions

$$\varphi(t,0,y) = \varphi(t,a,-y), \qquad \left.\frac{\partial \varphi(t,x,y)}{\partial x}\right|_{x=0} = \left.\frac{\partial \varphi(t,x,-y)}{\partial x}\right|_{x=a}. \tag{11.9}$$

Using the metrical energy–momentum tensor (3.9) with $\xi = 1/8$, the following Casimir energy density is obtained:

$$\tilde{\varepsilon}(a;y) = -\frac{\zeta_R(3)}{16\pi a^3} - \frac{1}{16\pi^2 a^3}\int_0^\infty dq\, q^2 \cos\left(\frac{qy}{a}\right)\int_1^\infty \frac{ds}{\sinh(qs/2)}\frac{2s^2-1}{\sqrt{s^2-1}}. \tag{11.10}$$

It is easily seen that eqn (11.10) is the limiting case of eqns (11.6) and (11.7), found for the Klein surface, when $b \to \infty$.

The above four configurations with different topologies are complete flat manifolds. As an example of an incomplete flat manifold, we consider a Möbius strip of finite width b. This is a one-sided nonorientable surface and can be obtained by giving a paper strip a half-twist, and then gluing the ends of the strip together to form a single strip. For this topology the identification conditions (11.9), imposed on a scalar field, must be supplemented with

$$\varphi(t,x,0) = \varphi(t,x,b) = 0. \tag{11.11}$$

Using the same technique based on the Abel–Plana formulas, and omitting the exponentially small terms for $b \geq a$ and the oscillating terms, we obtain (Mamayev and Trunov 1979c)

$$\tilde{\varepsilon}(a,b) = -\frac{\zeta_R(3)}{16\pi a^3} + \frac{\pi}{24a^2 b} - \frac{\pi}{24ab^2}. \tag{11.12}$$

11.1.2 Four-dimensional space–time

Now we turn to the Casimir energy density in three-dimensional flat spaces with non-Euclidean topologies (i.e. in four-dimensional space–time). There are many complete flat manifolds in this case, and we shall restrict ourselves to only one example, briefly mentioned in Section 8.4. This is the massless scalar field in a three-dimensional box with identified opposite faces (with the topology of a 3-torus $S^1 \times S^1 \times S^1$). From eqns (8.74), (8.42), and (8.26), we obtain the following for the Casimir energy density (Mamayev and Trunov 1979a):

$$\varepsilon(a,b,c) = -\frac{\pi^2}{90a^4} - \frac{\pi^2}{6abc^2} - \frac{\zeta_R(3)}{2\pi ab^3} + \frac{8\pi}{ab^2 c}G\left(\frac{c}{b}\right) - \frac{16}{a^2 bc}R\left(\frac{b}{a},\frac{c}{a}\right), \tag{11.13}$$

where $G(z)$ and $R(z_1, z_2)$ are defined in eqns (8.11) and (8.43), respectively.

In the limiting case $c \to \infty$, the 3-torus $S^1 \times S^1 \times S^1$ transforms into $S^1 \times S^1 \times R^1$ and the energy density takes the form

$$\varepsilon(a,b,\infty) = -\frac{\pi^2}{90a^4} - \frac{\zeta_R(3)}{2\pi ab^3}. \tag{11.14}$$

From this, for $a = b$, one obtains (Dowker and Critchley 1976, Goncharov 1982)

$$\varepsilon(a,a,\infty) = -\frac{0.301}{a^4}. \tag{11.15}$$

If, in addition to $c \to \infty$, the limiting case $b \to \infty$ is considered, our manifold transforms into $S^1 \times R^2$ and, from eqn (11.14), the corresponding Casimir energy density is

$$\varepsilon(a,\infty,\infty) = -\frac{\pi^2}{90a^4}. \tag{11.16}$$

Flat spaces with non-Euclidean topology give rise to Casimir effects for fields of various spin. As an example, we consider a massless spinor field in a space with a topology of a 3-torus $S^1 \times S^1 \times S^1$. For $a \leq b \leq c$, one obtains a positive energy density (Mamayev and Trunov 1980)

$$\varepsilon(a,b,c) \approx \frac{2\pi^2}{45a^4} + \frac{2\pi}{3abc^2} + \frac{2\zeta_R(3)}{\pi ab^3}, \tag{11.17}$$

where the exponentially small terms have been omitted.

Topologically nontrivial spaces with a nonzero vacuum energy density are of interest both for cosmology and for multidimensional physics with spontaneously compactified extra dimensions. In both cases, considered below, not only the non-Euclidean topology but also the curvature of the space plays an important role. For topologically nontrivial flat spaces, many results for the Casimir energy density have been obtained for both scalar and spinor twisted fields (DeWitt *et al.* 1979; Isham 1978a, 1978b). Twisted fields have already been mentioned in Section 2.3. For a space with the topology $S^1 \times R^2$, Casimir energy densities for twisted spinor fields were found by Ford (1980). The intermediate case between the usual type of fields and that of twisted fields is the case of *automorphic* fields. The Casimir energy densities for such fields in topologically nontrivial flat spaces were studied by Banach and Dowker (1979a, 1979b). All these questions are, however, outside the scope of this book; they have been considered in more detail by Mostepanenko and Trunov (1997).

11.2 Topologically nontrivial curved spaces

Let us consider a curved Riemann space–time of arbitrary dimensionality, keeping in mind the applications to cosmology and extra-dimensional physics. In general, the metric takes the form

$$ds^2 = g_{\mu\nu}(x)\, dx^\mu\, dx^\nu, \tag{11.18}$$

where $g_{\mu\nu}$ is the metric tensor. A generalization of the scalar field equation to curved space–time was presented in eqn (3.10), where all of the necessary notation was also introduced. The respective energy–momentum tensor (below we consider cases where only the diagonal components are nonzero) is obtained by the variation of the action with respect to $g^{\mu\nu}$ (Grib et al. 1994, Birrell and Davies 1982):

$$T^{(0)}_{\mu\mu} = (1 - 2\xi)\partial_\mu\varphi\partial_\mu\varphi + \left(2\xi - \frac{1}{2}\right)g_{\mu\mu}\partial_\nu\varphi\partial^\nu\varphi \qquad (11.19)$$
$$- \xi(\varphi\nabla_\mu\nabla_\mu\varphi + \nabla_\mu\nabla_\mu\varphi \cdot \varphi) + \left[\left(\frac{1}{2} - 2\xi\right)m^2 g_{\mu\mu} - \xi G_{\mu\mu} - 2\xi^2 R g_{\mu\mu}\right]\varphi^2.$$

Here

$$G_{\mu\nu} = R_{\mu\nu} - \frac{1}{2}R g_{\mu\nu} \qquad (11.20)$$

is the Einstein tensor, and $R_{\mu\nu}$ is the Ricci tensor. The generalization of the Dirac equation to the case of curved space–time requires the introduction of local orthonormal tetrad vectors. This is considered in the books mentioned above.

11.2.1 Three-dimensional space–time

An example of a manifold with a non-Euclidean topology and curvature is the surface of a 2-sphere S^2 with a constant radius a_0. The metric of the corresponding space–time is given by

$$ds^2 = dt^2 - a_0^2(d\theta^2 + \sin^2\theta\, d\varphi^2). \qquad (11.21)$$

This is the three-dimensional analogue of Einstein space–time. Equation (3.10) with $\xi = 1/8$ and $R = 2/a_0^2$ takes the form

$$\frac{\partial^2 \varphi(x)}{\partial t^2} - \frac{1}{a_0^2}\nabla^2_{(2)}\varphi(x) + \left(m^2 + \frac{1}{4a_0^2}\right)\varphi(x) = 0. \qquad (11.22)$$

The complete orthonormal set of positive- and negative-frequency solutions of eqn (11.22) can be expressed in terms of the spherical harmonics:

$$\varphi^{(+)}_{lM}(t,\theta,\varphi) = \frac{1}{a_0\sqrt{2\omega_l}}e^{-i\omega_l t}Y_{lM}(\theta,\varphi), \qquad (11.23)$$
$$\varphi^{(-)}_{lM} = \varphi^{(+)*}_{lM}, \quad l = 0, 1, 2, \ldots, \quad M = 0, \pm 1, \ldots \pm l,$$
$$\omega_l^2 = m^2 + \frac{1}{a_0^2}\left(l + \frac{1}{2}\right)^2.$$

The spherical harmonics $Y_{lM}(\theta,\varphi)$ are defined in eqn (9.5). Bearing in mind the applications to gravitational theory, we consider below all of the components of the vacuum energy–momentum tensor, and not just the energy density.

Substituting the field operator (3.52) expressed in terms of the set of solutions (11.23) into the energy–momentum tensor (11.19) and calculating the mean values in the vacuum state, we arrive at

$$\langle 0|T_0{}^0|0\rangle = \frac{1}{4\pi a_0^2}\sum_{l=0}^{\infty}\left(l + \frac{1}{2}\right)\omega_l, \qquad (11.24)$$

$$-\langle 0|T_i{}^k|0\rangle = \frac{1}{8\pi a_0^4} \sum_{l=0}^{\infty} \left(l+\frac{1}{2}\right)^3 \frac{1}{\omega_l} \delta_i{}^k.$$

To obtain these expressions, summation theorems for the spherical functions have been used, which follow from the equality (Varshalovich et al. 1988)

$$\sum_{M=-l}^{l} Y_{lM}(\theta_1,\varphi_1) Y_{lM}^*(\theta_2,\varphi_2) = \frac{2l+1}{4\pi} P_l(\cos\omega), \qquad (11.25)$$

$$\cos\omega = \cos\theta_1 \cos\theta_2 + \sin\theta_1 \sin\theta_2 \cos(\varphi_1-\varphi_2),$$

where the $P_l(z)$ are the Legendre polynomials.

The expressions (11.24) are, as usual, divergent. In the case of flat space–time, we made them finite by subtracting the contributions of free Minkowski space with Euclidean topology. Here we proceed in the same spirit and subtract from the expressions (11.24) the corresponding expressions in the Minkowski space–time tangential at any fixed point to the Riemann space–time under consideration. This can be most easily done by introducing a cutoff function to make all quantities finite, and then using the Abel–Plana formula (2.41) adapted for summation over half-integers. The result is independent of the form of the cutoff function, as explained in Section 2.2. It is given by

$$\varepsilon = \frac{m^3}{2\pi} \int_0^1 \frac{d\xi\, \xi\sqrt{1-\xi^2}}{\exp(2\pi m a_0 \xi)+1}, \qquad (11.26)$$

$$P = -\frac{m^3}{4\pi} \int_0^1 \frac{d\xi\, \xi^3}{\sqrt{1-\xi^2}\left[\exp(2\pi m a_0 \xi)+1\right]}.$$

Recall that the quantity $-T_i{}^i$ (with no summation over the three-space index i) has the physical meaning of a pressure. The Casimir energy density in eqn (11.26) is positive and nonzero only for a massive field. Together, the energy density and pressure satisfy the thermodynamic relation

$$P = -\frac{\partial E}{\partial S}, \qquad (11.27)$$

where $E = \varepsilon S$ and $S = 4\pi a_0^2$ is the area of the sphere.

Under the condition $a_0 \gg m^{-1}$, eqn (11.26) leads to

$$\varepsilon \approx \frac{m}{96\pi a_0^2}\left[1-\frac{7}{40(ma_0)^2}\right], \quad P \approx -\frac{7}{3840\pi m a_0^4}. \qquad (11.28)$$

Under the opposite condition $a_0 \ll m^{-1}$, one obtains

$$\varepsilon \approx -P \approx \frac{m^3}{12\pi}. \qquad (11.29)$$

11.2.2 Four-dimensional space–time

The most interesting example of a topologically nontrivial three-dimensional space with nonzero curvature is a three-dimensional sphere S^3 with a constant radius a_0. In cosmology, the related static space–time $R^1 \times S^3$ is known as the Einstein model. The metric of this space–time is

$$ds^2 = dt^2 - a_0^2 [dr^2 + \sin^2 r \, (d\theta^2 + \sin^2\theta \, d\varphi^2)] \qquad (11.30)$$
$$= a_0^2 [d\eta^2 - dr^2 - \sin^2 r \, (d\theta^2 + \sin^2\theta \, d\varphi^2)],$$

where r, θ, φ are dimensionless coordinates on a three-space of constant curvature $+1$ and $\eta = t/a_0$ is what is usually referred to as the *conformal time*. The field equation (3.10), with $\xi = 1/6$ and $R = 6/a_0^2$, is transformed to the form

$$\frac{\partial^2 \varphi(x)}{\partial t^2} - \frac{1}{a_0^2} \nabla^2_{(3)} \varphi(x) + \left(m^2 + \frac{1}{a_0^2}\right)\varphi(x) = 0. \qquad (11.31)$$

The complete orthonormal set of solutions of this equation can be represented as

$$\varphi^{(+)}_{klM}(t, r, \theta, \varphi) = \frac{1}{\sqrt{2\omega_l a_0^3}} e^{-i\omega_l t} \Phi_{klM}(r, \theta, \varphi), \qquad (11.32)$$

$$\varphi^{(-)}_{klM} = \varphi^{(+)\,*}_{klM}, \quad k = 1, 2, 3, \ldots, \quad l = 0, 1, 2, \ldots, k-1,$$

$$M = 0, \pm 1, \ldots \pm l, \qquad \omega_k^2 = m^2 + \frac{k^2}{a_0^2}.$$

The four-dimensional spherical functions are given by

$$\Phi_{klM}(r, \theta, \varphi) = \frac{1}{\sqrt{\sin r}} \sqrt{\frac{k(k+l)!}{(k-l+1)!}} \, P^{-l-1/2}_{k-1/2}(\cos r) Y_{lM}(\theta, \varphi), \qquad (11.33)$$

where the $P^\nu_\mu(z)$ are the associated Legendre functions defined on the cut.

The vacuum energy–momentum tensor of a scalar field is calculated in a manner analogous to eqn (11.24), using summation formulas for the four-dimensional spherical functions following from the equality (Grib *et al.* 1994)

$$\sum_{l=0}^{k-1} \sum_{M=-l}^{l} \Phi_{klM}(r_1, \theta_1, \varphi_1) \Phi^*_{klM}(r_2, \theta_2, \varphi_2) = \frac{k}{\pi^2} \frac{\sin k\rho}{\sin \rho}, \qquad (11.34)$$

$$\cos\rho = \cos r_1 \cos r_2 + \sin r_1 \sin r_2 [\cos\theta_1 \cos\theta_2 + \sin\theta_1 \sin\theta_2 \cos(\varphi_1 - \varphi_2)],$$

where ρ is the geodesic distance between the points $(r_1, \theta_1, \varphi_1)$ and $(r_2, \theta_2, \varphi_2)$. The result is given by

$$\langle 0|T_0{}^0|0\rangle = \frac{1}{4\pi^2 a_0^3} \sum_{k=1}^{\infty} k^2 \omega_k, \qquad (11.35)$$

$$-\langle 0|T_i{}^k|0\rangle = \frac{1}{12\pi^2 a_0^5}\sum_{k=1}^{\infty}\frac{k^4}{\omega_k}\delta_i{}^k.$$

Subtracting from eqn (11.35) the respective contributions in the tangential Minkowski space–time and using the Abel–Plana formula (2.26), we obtain the finite Casimir energy density and the pressure of a scalar field in the Einstein model:

$$\varepsilon = \frac{1}{2\pi^2 a_0^4}\int_{ma_0}^{\infty}\frac{\xi^2\,d\xi}{\exp(2\pi\xi)-1}\sqrt{\xi^2-m^2 a_0^2}, \qquad (11.36)$$

$$P = \frac{1}{6\pi^2 a_0^4}\int_{ma_0}^{\infty}\frac{\xi^4\,d\xi}{[\exp(2\pi\xi)-1]\sqrt{\xi^2-m^2 a_0^2}}.$$

The thermodynamic relation between the Casimir energy and pressure now takes the form

$$P = -\frac{\partial E}{\partial V}, \qquad (11.37)$$

where $E=\varepsilon V$, and $V=2\pi^2 a_0^3$ is the volume of the spherical space S^3. Here, $\varepsilon>0$ as in the case of the two-dimensional sphere S^2.

In the massless case eqn (11.36) leads to (Ford 1975, 1976)

$$\varepsilon = \frac{1}{2\pi^2 a_0^4}\int_0^{\infty}\frac{\xi^3\,d\xi}{\exp(2\pi\xi)-1} = \frac{1}{480\pi^2 a_0^4}, \qquad (11.38)$$

$$P = \frac{\varepsilon}{3} = \frac{1}{1440\pi^2 a_0^4}.$$

Here, the spectral density of the vacuum energy density has the Planck form, with the temperature defined from the equality $k_B T = 1/(2\pi a_0)$.

For a massive field with $ma_0 \gg 1$, we obtain the following from eqn (11.36) (Mamayev et al. 1976):

$$\varepsilon \approx \frac{(ma_0)^{5/2}}{8\pi^3 a_0^4}e^{-2\pi ma_0}, \qquad P \approx \frac{(ma_0)^{7/2}}{12\pi^2 a_0^4}e^{-2\pi ma_0}. \qquad (11.39)$$

It is notable that the Casimir energy density and pressure of a massive field on S^3 are exponentially small. This is not typical for spaces with nonzero curvature [compare with eqn (11.28) for S^2].

Similar results can be obtained for the Casimir effect for spinor and electromagnetic fields on S^3. For example, for a massive spinor field, the Casimir energy density and pressure are given by (Grib et al. 1980, 1994)

$$\varepsilon = \frac{1}{2\pi^2 a_0^4}\int_{ma_0}^{\infty}\frac{(4\xi^2+1)\,d\xi}{\exp(2\pi\xi)+1}\sqrt{\xi^2-m^2 a_0^2}, \qquad (11.40)$$

$$P = \frac{1}{6\pi^2 a_0^4} \int_{ma_0}^{\infty} \frac{\xi^2(4\xi^2+1)\,d\xi}{[\exp(2\pi\xi)+1]\sqrt{\xi^2-m^2a_0^2}}.$$

In the massless case eqn (11.40) leads to

$$\varepsilon = \frac{1}{2\pi^2 a_0^4} \int_0^{\infty} \frac{\xi(4\xi^2+1)}{\exp(2\pi\xi)+1}\,d\xi = \frac{17}{960\pi^2 a_0^4}, \tag{11.41}$$

$$P = \frac{\varepsilon}{3} = \frac{17}{2880\pi^2 a_0^4}.$$

Finally, for the electromagnetic Casimir effect on S^3,

$$\varepsilon = 3P = \frac{11}{240\pi^2 a_0^4}. \tag{11.42}$$

It is significant that for all of the examples considered in Sections 11.1 and 11.2, the Casimir energy densities and pressures do not depend on the energies attributed to material walls. Here, the Casimir effect arises owing to the nontrivial topology, and not to material boundaries, which are not present. This allows one to uniquely define not only the separation-dependent global energies but also the energy densities.

11.3 Nontrivial topologies in cosmology

The most important physical system where spaces with nontrivial topology play a role is our Universe. The cosmological models that describe the structure and evolution of the Universe are the solutions of Einstein's equations

$$G_{\mu\nu} = -8\pi G T_{\mu\nu}, \tag{11.43}$$

where Einstein's tensor $G_{\mu\nu}$ is defined in eqn (11.20), G is the gravitational constant, and $T_{\mu\nu}$ is the energy–momentum tensor of matter. For example, in the case of a spatially homogeneous and isotropic closed space–time, the metric is given by eqn (11.30) with the replacement of $a_0 \to a(t)$, where $a(t)$ is the *scale factor*. This gives rise to the closed Friedmann model, which expands from a singular state to some maximum radius and then contracts to a singular state when the energy–momentum tensor of classical matter (dust and radiation) is substituted on the right-hand side of eqn (11.43). Like the Einstein model considered in the previous section, the closed Friedmann model has the topology $R^1 \times S^3$, but it is not static. From this it follows that eqns (11.36)–(11.42) for the Casimir energy density and pressure of various fields in the Einstein model preserve their validity in the closed Friedmann model provided that a_0 is replaced by $a(t)$.

In a nonstatic homogeneous isotropic space, however, the vacuum energy-momentum tensor contains contributions not only from the terms of Casimir origin discussed above, but also terms describing vacuum polarization and production of particles from the vacuum that occur for nonstationary external fields. Both of these latter effects are determined by the nonstationarity of the metric and depend on the derivatives of the scale factor with respect to time. They are not of topological origin and are also present in open and quasi-Euclidean models, which retain the same topology as flat Minkowski space–time with no identification conditions. In the closed model, the scalar curvature is given by

$$R = \frac{6}{a^3}(\ddot{a} + a), \tag{11.44}$$

where the dot stands for the derivative with respect to the conformal time η, defined by $d\eta = dt/a(\eta)$ [compare with eqn (11.30)]. The scalar-field equation (3.10) with $\xi = 1/6$ takes the form

$$\frac{\partial^2 \varphi}{\partial \eta^2} + 2\frac{\dot{a}}{a}\frac{\partial \varphi}{\partial \eta} - \nabla^2_{(3)}\varphi + \left(m^2 a^2 + \frac{\ddot{a}}{a} + 1\right)\varphi = 0, \tag{11.45}$$

which should be compared with (11.31). It can be transformed to an oscillator equation with a time-dependent frequency [equation (11.31) is an oscillator equation with a constant frequency],

$$\frac{d^2 g(\eta)}{d\eta^2} + [k^2 + m^2 a^2(\eta)]g(\eta) = 0, \tag{11.46}$$

by the substitution

$$\varphi(x) = \frac{1}{a(\eta)} g_k(\eta) \Phi_{klM}(r, \theta, \varphi) \tag{11.47}$$

and using the eigenvalue equation

$$-\nabla^2_{(3)} \Phi_{klM}(r, \theta, \varphi) = (k^2 - 1)\Phi_{klM}(r, \theta, \varphi). \tag{11.48}$$

Regarding the effect of particle production from the vacuum, only massive particles are created in homogeneous, isotropic cosmological models. This is because these models are conformally flat, and the field equations are conformally invariant in the limit of zero mass [with the exception of the scalar field equation (3.10) with an arbitrary value of the coupling coefficient ξ]. From eqn (11.46), it is seen that for $m = 0$ the oscillator frequency does not depend on time and, thus, particle production does not occur. Details of particle production from the vacuum in a gravitational field can be found in the books by Birrell and Davies (1982) and by Grib et al. (1994).

The contribution to the energy–momentum tensor of a scalar field due to the vacuum polarization of non-Casimir origin is given by (Mamayev 1980, Grib et al. 1994)

$$\varepsilon^{\text{pol}} = \frac{1}{960\pi^2 a^4}(2\ddot{c}c - \dot{c}^2 - 2c^4), \tag{11.49}$$

$$P^{\text{pol}} = \frac{1}{2880\pi^2 a^4}(-2\dddot{c} + 2\ddot{c}c - \dot{c}^2 + 8\dot{c}c - 2c^4),$$

where $c \equiv \dot{a}/a$ [note that the above references deal with a complex scalar field and, thus, the energy–momentum tensor there is twice that in eqn (11.49)]. If the scale factor of the metric $a(\eta)$ is determined by classical radiation, the Casimir energy density (11.38) [with the replacement of a_0 by $a(\eta)$] is much smaller than the polarization energy (11.49) due to the nonstationarity of the metric. Thus, at the Compton time $t = m^{-1}$ the Casimir energy density in the closed Friedmann model is smaller than the polarization energy density by a factor of 10^{-74}. By the time the classical background matter becomes dust-like ($t \sim 10^{14}$ s), the Casimir energy density is as small as 10^{-6} of the polarization energy density (11.49) originating from the nonstationarity of space–time. Note also that sometimes, in the literature, any vacuum polarization energy in a nonstationary gravitational background is called a gravitational Casimir effect by analogy with the dynamical Casimir effect (see Section 7.7). Such terminology is, however, not justified. The reason is that neither the vacuum polarization of a massless field (11.49) (which is often called the *conformal anomaly*) nor the creation of massive particles from the vacuum is related to the periodicity (identification) conditions giving rise to the Casimir effect. In particular, both the conformal anomaly and the effect of particle creation from the vacuum occur also in the quasi-Euclidean and open Friedmann models, where no identification conditions are imposed on the quantum field and the Casimir effect does not arise.

In eqn (11.43), classical matter is the source of the gravitational field. In the Planck epoch, when classical matter was absent, some kind of semiclassical Einstein equations can be considered,

$$G_{\mu\nu} = -8\pi G \langle 0|T_{\mu\nu}|0\rangle^{\text{ren}}, \qquad (11.50)$$

where the renormalized vacuum energy–momentum tensor of the quantized fields plays the role of the source. The equations (11.50) are called *self-consistent*. For a massless scalar field in a closed homogeneous, isotropic model, the quantity $\langle 0|T_{\mu\nu}|0\rangle^{\text{ren}}$ is determined by the sum of the Casimir contribution (11.38) [with a_0 replaced by $a(\eta)$] and the polarization contribution (11.49). The 00-component of eqn (11.50) reduces to the following (the spatial components are connected with this component by the conservativity condition):

$$3(c^2 + 1) = \frac{G}{120\pi a^2}(2\ddot{c}c - \dot{c}^2 - 2c^4 + 2). \qquad (11.51)$$

Here, the last contribution on the right-hand side is due to the Casimir effect. The following nonsingular solution of eqn (11.51) was obtained by Mamayev and Mostepanenko (1980) and by Starobinsky (1980):

$$a = \sqrt{\frac{G}{360\pi}}\frac{1}{\cos\eta} = \sqrt{\frac{G}{360\pi}}\cosh\left(t\sqrt{\frac{360\pi}{G}}\right). \qquad (11.52)$$

Note that in the absence of classical background matter, the vacuum polarization energies of Casimir and non-Casimir origin are of the same order of magnitude. The solution (11.52) describes a scale factor exponentially increasing with time. Similar solutions were obtained also for spinor and electromagnetic fields. Later, such scale factors were used in the theory of inflation driven by the classical inflaton field (Guth 1981, Linde 1990).

As is seen in eqn (11.50), the vacuum energy–momentum tensor of a quantized field serves as a source of a gravitational field. This makes it unique and particularly important. The advantage of spaces with non-Euclidean topology, as compared with regions restricted by material boundaries, is that in the former case the identification conditions should be understood in a literal sense. By contrast, for material boundaries the boundary conditions are usually idealizations. This creates difficulties in interpretation, as discussed in Chapter 4. In any case, the finite, renormalized Casimir energy density gravitates in the same way as all other types of energy, i.e. as is required by the equivalence principle (Bimonte et al. 2007, Fulling et al. 2007b).

From a theoretical standpoint, on cosmological scales, space–time may possess different possible topologies [see the review by Lachièze-Rey and Luminet (1995)]. It is common knowledge that the metric tensor does not fix the topological structure. For example, one may consider a quasi-Euclidean space–time with a metric

$$ds^2 = dt^2 - a^2(t)(dx^2 + dy^2 + dz^2), \tag{11.53}$$

where the coordinates x, y, z are supposed to be dimensionless and $a(t)$ is the scale factor measured in meters. For a quasi-Euclidean metric, the curvature of three-space is equal to zero. If the three-space has a Euclidean topology, the Casimir energy density also is equal to zero. However, one may introduce in a three-space the topology of a 3-torus, i.e. identify all points with coordinates $(x+lL, y+nL, z+pL)$, where $l, n, p = 0, \pm 1, \pm 2, \ldots$ and L is the identification scale. In this case, the Casimir energy density is obtained from eqn (11.13) by the replacement of a, b, and c with $a(t)L$:

$$\varepsilon(a) = -\frac{A}{a^4(t)L^4}, \qquad A = 0.8375. \tag{11.54}$$

In recent formulations, the self-consistent Einstein equations usually include the cosmological term:

$$G_{\mu\nu} + \Lambda g_{\mu\nu} = -8\pi G \langle 0|T_{\mu\nu}|0\rangle^{\text{ren}}. \tag{11.55}$$

It is of interest that the 00-component of these equations with the source (11.54),

$$-\frac{3}{a^3}\left(\frac{da}{dt}\right)^2 + \Lambda = \frac{8\pi G A}{L^4 a^4}, \tag{11.56}$$

has a nonsingular solution (Zel'dovich and Starobinsky 1984)

$$a(t) = \frac{1}{L}\left(\frac{8\pi G A}{\Lambda}\right)^{1/4}\left[\cosh\left(2\sqrt{\frac{\Lambda}{3}}t\right)\right]^{1/2}. \tag{11.57}$$

Thus, for spatially flat, homogeneous isotropic models of the Universe with the topology $S^1 \times S^1 \times S^1$, the Casimir energy density (11.54) can drive the inflation process [recall that the asymptotic behavior of the scale factor (11.57) at large t is the same as in (11.52)]. Note that, in the literature, the cosmological term $\Lambda g_{\mu\nu}$ in eqn (11.55) is identified on occasion with some energy–momentum tensor of Casimir origin. This is, however, not supported by the above results, because the Casimir energy–momentum tensor, as a rule, does not have a vacuum-type but a radiation-type equation of state.

Phenomenologically, the nontrivial topology of space–time may lead to some observable effects if the identification scale is smaller than the horizon. This was first discussed by Ellis (1971) and by Sokolov and Shvartsman (1974). Nontrivial topologies result in multiple images of some cosmic sources [several specific examples were analyzed by Fagundes (1989), Müller et al. (2002), and Mota et al. (2005)]. Numerous calculations of the Casimir energy density in spherical and cylindrical models of the Universe with various dimensionalities can be found in the book by Elizalde et al. (1994) and in the review by Bytsenko et al. (1996).

11.4 Compactification of extra dimensions

The Casimir effect in spaces with a non-Euclidean topology provides one of the mechanisms for the compactification of extra spatial dimensions in multidimensional Kaluza–Klein theories. These theories are widely used in modern schemes for the unification of all fundamental interactions, including the gravitational interaction. The main idea of the Kaluza–Klein approach is that the true dimensionality of space–time is $D = 4 + N$, where the additional N spatial dimensions are compactified and form a compact space. Originally this idea was used in the 1920s for the unification of gravitation and electromagnetism. For this purpose a five-dimensional space–time with a topology $M^4 \times S^1$ was employed, where M^4 is a flat four-dimensional Minkowski space–time. The geometrical size of S^1 was supposed to be of the order of the Planck length $l_{\text{Pl}} = \sqrt{G} \sim 10^{-33}$ cm (Wesson 2006).

According to present concepts, a promising unified theory of fundamental interactions is superstring theory (Polchinski 1998). String theory usually deals with a ten-dimensional space–time of the form $M^4 \times K^6$, where K^6 is a six-dimensional compact space. It was common to assume that the extra dimensions were compactified on the Planck scale, as in the old Kaluza–Klein theory. At the present time, however, theoretical schemes with low-energy compactification scale of the order of 1 TeV have become popular (Antoniadis et al. 1998, Arkani-Hamed et al. 1999). In these schemes the compactification length can be as large as 1 nm or even 1 μm. The important problem of string theories is to find the compactification mechanism responsible for the stability of a compact manifold K^6. It is remarkable that one such mechanism is provided by the Casimir effect.

To illustrate how the Casimir effect compactifies extra dimensions, we consider the multidimensional Einstein equations of the form

$$G_{AB} + \Lambda_D g_{AB} = -8\pi G_D \langle 0|T_{AB}|0\rangle^{\text{ren}}. \tag{11.58}$$

Here, G_{AB} and T_{AB} are the Einstein tensor and the energy–momentum tensor, respectively, in $D = (4 + N)$-dimensional space–time. The indices A and B take values $0, 1, \ldots, D - 1$, and Λ_D and G_D are the cosmological and gravitational constants in D dimensions. The space–time is assumed to be $M^4 \times K^N$, where K^N is an N-dimensional compact manifold (the internal space). The vacuum energy–momentum tensor on the right-hand side of eqn (11.58) is of Casimir origin. The question arises whether or not the equations (11.58) have a solution describing an internal compact space K^N with reasonable physical properties.

This question is the start of a large body of research. However, a satisfactory solution has not been found yet. Because of this, we shall only list briefly some of the results obtained. Birmingham et al. (1988) computed Casimir energies for scalar and spinor fields in even-dimensional Kaluza–Klein spaces of the form $M^4 \times S^{N_1} \times S^{N_2} \times \cdots$. In the massless case, they found a stable solution of eqn (11.58) in a space–time of the form $M^4 \times S^2 \times S^2$ (i.e. for a four-dimensional compact internal space). The Casimir energy densities arising in a multidimensional space–time $M^4 \times T^N$, where T^N is an N-dimensional torus, were discussed by Buchbinder and Odintsov (1989). This includes the case $N = 6$, which is of interest in superstring theory. A space–time of the form $M^4 \times B$, where B is the Klein bottle, was considered by Blau et al. (1984). Solutions of eqn (11.58) for a space–time with a spatial section $S^3 \times S^3$ were obtained by Dowker (1984) at any nonzero temperature. It was found that the Casimir energy–momentum tensor of a massless spinor field determines a definite value for the radius of the sphere. The literature on the subject is quite extensive, and many more complicated Kaluza–Klein geometries have been studied (Elizalde et al. 1994).

As an example, we consider Casimir energy densities and the corresponding solutions of eqn (11.58) in the space–time $M^4 \times S^N$ (Chodos and Myers 1984, Candelas and Weinberg 1984). To preserve the Poincaré invariance in Minkowski space–time M^4, we look for solutions with block diagonal metric and Ricci tensors

$$g_{AB} = \begin{pmatrix} g_{\mu\nu} & 0 \\ 0 & g_{ab}(\boldsymbol{u}) \end{pmatrix}, \qquad R_{AB} = \begin{pmatrix} 0 & 0 \\ 0 & R_{ab}(\boldsymbol{u}) \end{pmatrix}. \qquad (11.59)$$

Here, $g_{\mu\nu}$, with $\mu, \nu = 0, 1, 2, 3$, is the usual diagonal metric tensor in Minkowski space–time M^4, and $g_{ab}(\boldsymbol{u})$ is the metric tensor on a manifold S^N with coordinates $\boldsymbol{u}$ ($a, b = 4, 5, \ldots, D-1$). The Ricci tensor R_{ab} on a sphere and the scalar curvature R_D coincide with those calculated using the metric tensor $g_{ab}(\boldsymbol{u})$. The Casimir energy momentum tensor also has the block structure

$$\langle 0|T_{\mu\nu}|0\rangle^{\text{ren}} = T_1 g_{\mu\nu}, \qquad \langle 0|T_{ab}(\boldsymbol{u})|0\rangle^{\text{ren}} = T_2 g_{ab}(\boldsymbol{u}), \qquad (11.60)$$

where T_1 and T_2 do not depend on $\boldsymbol{u}$, owing to the homogeneity of space.

Candelas and Weinberg (1984) expressed T_1 and T_2 for massless scalar and spinor fields in terms of the effective potential $V(a) = C_N/a^4$:

$$T_1 \Omega_N = -V(a), \qquad T_2 \Omega_N = -\frac{2a^2}{N} \frac{dV(a)}{d(a^2)}. \qquad (11.61)$$

Here, a is the sphere radius, and

$$\Omega_N = \int d^N u \sqrt{|\det g_{ab}|} \qquad (11.62)$$

is the sphere volume. As a result, $T_{1,2} \sim a^{-D}$ as expected from dimensional considerations in D-dimensional space–time. Explicit values of C_N were calculated for odd N (i.e. for odd-dimensional internal spaces) using dimensional regularization and the generalized zeta function. For a massless scalar field with conformal coupling and for a massless spinor field, $C_N > 0$ for $N = 3$ and 7 but $C_N < 0$ for $N = 5$ and 9. It was shown also that in these cases $|C_N| \ll 1$. For example, $C_3 = 0.7146 \times 10^{-5}$ for the conformally coupled scalar field and $C_3 = 19.45 \times 10^{-5}$ for the spinor field. For scalar fields with minimal coupling [$\xi = 0$ in eqn (3.10)], $C_N > 0$ for any odd N. In this case, for example, $C_3 = 7.569 \times 10^{-5}$.

Substitution of eqns (11.59)–(11.62) in eqn (11.58) leads to the following consistency conditions on the sphere radius and on the cosmological constant:

$$a^2 = \frac{8\pi C_N (N+4) G}{N(N-1)}, \qquad \Lambda_D = -\frac{N^2(N-1)^2(N+2)}{16\pi C_N (N+4)^2 G}, \qquad (11.63)$$

where G is the usual gravitational constant connected with the multidimensional one by the equality $G = G_D/\Omega_N$. As is seen from eqn (11.63), the Casimir energy density of a massless conformally coupled scalar field or a massless spinor field defines a consistent sphere radius for $N = 3$ and 7, where C_N is positive. The Casimir energy density of a minimally coupled scalar field defines a consistent sphere radius for any odd N. In all cases the cosmological constant remains negative.

The semiclassical Einstein equations (11.58) are written in one-loop approximations, i.e. under the assumption that the gravitational field remains classical. Thus, the result obtained for a compactification radius

$$a = \sqrt{\frac{8\pi C_N (N+4)}{N(N-1)}} \, l_{\text{Pl}} \qquad (11.64)$$

is meaningful only if $C_N \geq 1$ and, consequently, $a > l_{\text{Pl}}$. This, however, demands a large number of fields. Depending on the field type, this number may be of order 10^2 or 10^3. A large number of required fields is a characteristic feature of spontaneous-compactification schemes using the Casimir effect.

11.5 Topological defects

A vacuum polarization of the Casimir type arises in the space–time of the *topological defects* which may be formed in phase transitions in the very early Universe (Vilenkin and Shellard 1994). One of the most popular topological defects is the cosmic string, which can be formed owing to the spontaneous breaking of an

axial symmetry. The space–time of an infinitely thin cosmic string is described by the metric
$$ds^2 = dt^2 - d\rho^2 - \rho^2 d\varphi^2 - dz^2, \tag{11.65}$$
where the spatial points (ρ, φ, z) and $(\rho, \varphi + \beta, z)$ are to be identified (Dowker 1987). This leads to a periodicity of quantized fields in the variable φ with a period β and results in a nonzero vacuum energy density. The metric (11.65) describes a locally flat space–time for all $\rho \neq 0$ that has a cone-like singularity at $\rho = 0$. Frolov and Serebriany (1987) determined the general structure of the renormalized vacuum energy–momentum tensor for conformally invariant massless fields in the space–time (11.65) as

$$\langle 0|T_\mu{}^\nu|0\rangle^{\mathrm{ren}} = -\frac{1}{1440\pi^2 \rho^4} f(\delta) \operatorname{diag}(1, 1, -3, 1), \tag{11.66}$$

where $f(\delta)$ depends on the spin of the field and $\delta = 2\pi/\beta$. From the theory of cosmological phase transitions, $\delta^{-1} = 1 - 4\mu^*$, where $\mu^* = G\mu \sim 10^{-6}$ and $\mu \sim 10^{22}$ g/cm is the mass per unit length of the string.

For a conformally invariant scalar field [eqn (3.10) with $m = 0$ and $\xi = 1/6$], we get $f(\delta) = \delta^4 - 1$ and eqn (11.66) leads to the result obtained earlier by Helliwell and Konkowski (1986),

$$\langle 0|T_\mu{}^\nu|0\rangle^{\mathrm{ren}} = -\frac{\delta^4 - 1}{1440\pi^2 \rho^4} \operatorname{diag}(1, 1, -3, 1). \tag{11.67}$$

The Casimir energy density following from this equation,

$$\varepsilon(\rho) = \langle 0|T_0{}^0|0\rangle^{\mathrm{ren}} = -\frac{\delta^4 - 1}{1440\pi^2 \rho^4}, \tag{11.68}$$

coincides with the Casimir energy density of a conformally invariant scalar field inside a wedge (7.135) if one replaces α with δ. For an electromagnetic field, $f(\delta) = 2(\delta^2 - 1)(\delta^2 + 11)$ and the Casimir energy density, as given by eqn (11.66) with $\mu = \nu = 0$, coincides with the electromagnetic Casimir energy density inside a wedge (7.134) if the same replacement is made. The case of a massive scalar field was considered by Linet (1987).

As can be seen in eqns (11.66)–(11.68), the local vacuum energy density has a singularity proportional to ρ^{-4}, where ρ is the distance from the string. Therefore the global vacuum energy of the cosmic string diverges. One may avoid this problem by considering a string of finite thickness. Some partial results in this direction were obtained by Khusnutdinov and Bordag (1999). The vacuum energy of quantized fields at nonzero temperature in the space–time of a cosmic string has been investigated by, for example, Linet (1992) and Frolov et al. (1995).

Casimir-type effects can also be expected in the interaction between two parallel cosmic strings. Here, the attractive force arises because of vacuum fluctuations analogously to the force acting between two metal planes. The metric of two parallel cosmic strings can be represented in the form (Letelier 1987)

$$ds^2 = dt^2 - e^{-2\Lambda(\boldsymbol{r}_\perp)}(dx^2 + dy^2) - dz^2, \tag{11.69}$$

where

$$\Lambda(\boldsymbol{r}_\perp) = 4 \sum_{k=1}^{2} \mu_k^* \ln \frac{|\boldsymbol{r}_\perp - \boldsymbol{a}_k)|}{\rho_0}. \tag{11.70}$$

Here, $\boldsymbol{r}_\perp = (x,y)$ is a vector in the (x,y) plane, which is perpendicular to the strings; the $\boldsymbol{a}_k$ are the positions of the strings in that plane, and ρ_0 is the unit of length.

The calculation of the Casimir energy and force in the metric (11.69) is complicated because the variables do not separate. However, perturbative calculation in powers of the small parameters μ_k^* is possible. The result can be represented in the form

$$E(a) = -\sigma \frac{\mu_1^* \mu_2^*}{a^2}, \qquad F(a) = -2\sigma \frac{\mu_1^* \mu_2^*}{a^3}, \tag{11.71}$$

where a is the distance between the strings and σ is a number. The values of σ were calculated by Bordag (1990) and by Gal'tsov et al. (1995) for a massless scalar field, by Aliev et al. (1997) for a massless spinor field, and by Aliev (1997) for an electromagnetic field. The same Casimir-type attraction occurs for magnetic strings (Bordag 1991) and for cosmic strings carrying magnetic flux. Heat kernel coefficients in space–times with cone-like singularities have been calculated by Fursaev (1994), Cognola et al. (1994), and Bordag et al. (1996b). Similar methods have been applied also to other topological defects, for instance, global monopoles (Bezerra de Mello et al. 1999), which are formed when spherical symmetry is broken at a phase transition, and domain walls, which are formed owing to the violation of some discrete symmetry.

Very recently, what are referred to as *braneworld* cosmological scenarios have become more and more popular. These scenarios are connected with superstring theory. They assume that the four-dimensional Universe that we are living in is positioned on a brane embedded in a bulk multidimensional space. It is also assumed that the known elementary particles and their interactions described by the Standard Model are confined to the brane. The quantum fields that can propagate in the bulk result in a Casimir-type contribution to the vacuum energy and in forces acting on branes. These forces must be taken into account in investigations of world brane dynamics (Fabinger and Hořava 2000, Nojiri et al. 2000). Many different models have been suggested to describe braneworlds, and here we shall not discuss the extensive literature on the subject, which would demand a separate book. It is worth noting only that most frequently used brane models were suggested by Randall and Sundrum (1999a). The surface and bulk Casimir energy densities and the interaction forces in models of this type were investigated by Elizalde (2006) and Saharian (2004, 2006b, 2006c).

PART II

THE CASIMIR FORCE BETWEEN REAL BODIES

12

THE LIFSHITZ THEORY OF THE VAN DER WAALS AND CASIMIR FORCES BETWEEN PLANE DIELECTRICS

Starting with this chapter, we deal with the Casimir effect for real bodies made of various materials rather than idealized boundaries. This means that the electrical, optical, and mechanical properties of real materials will be properly accounted for to provide a consistent theory of the Casimir force. The great diversity of the available materials and their properties makes this problem rather challenging. However, some general theoretical results of broad interest can be obtained [see the review by Klimchitskaya et al. (2009a)]. These results will be used for comparing experiment and theory in Part III of the book.

The basic theory providing a unified description of both the van der Waals interaction and the Casimir interaction between planar dielectrics was given by Lifshitz (1956). This theory is the subject of the present chapter. Some results of general character presented here will be also used in later chapters devoted to metals. In condensed matter physics, all materials can be separated into dielectrics or metals based on their conductivity properties at zero temperature (Mott 1990). Dielectrics have zero conductivity at $T = 0$. There are different kinds of dielectrics in nature. They are classified by the behavior of the electron density of states N as a function of the electron energy E_e at $T = 0$. If $N(E_F) = 0$, where E_F is the Fermi energy, and the bandgap between the filled valence band and empty conduction band is wide in comparison with the thermal energy, then the material is called an *insulator*. When $N(E_F) = 0$ but the bandgap is not as wide, a dielectric is called an *intrinsic semiconductor*. There are also dielectrics where $N(E_F) \neq 0$ but the conductivity at $T = 0$ is equal to zero owing to electron correlations (Mott 1990). These are called *Mott–Hubbard dielectrics*. All doped semiconductors with a doping concentration n below the critical value ($n < n_{cr}$) are also dielectrics. For them $N(E_F) \neq 0$, but the conductivity at $T = 0$ is equal to zero because the charge carriers are localized in the vicinity of the impurity centers (Shklovskii and Efros 1984). At $T \neq 0$ all dielectrics possess some nonzero conductivity, which may be extremely small (as for insulators) or relatively large (as for doped semiconductors with $n < n_{cr}$). Metals have a nonzero conductivity at zero temperature. Semiconductors with a doping concentration above critical behave like metals.

In this chapter, we present the Lifshitz theory as applied to dielectric plates at both zero and nonzero temperature, leaving the case of metals for Chapters 13 and 14. Various planar configurations are considered. The consistency of the theory with the requirements of thermodynamics, and its application region are discussed. Note that in Parts II and III of the book, we explicitly write out the fundamental constants $\hbar$ and c in all formulas.

12.1 The Lifshitz formula for two semispaces at zero temperature

Here, we consider the simplest configuration of two identical parallel dielectric semispaces, separated by a vacuum gap of width a, at zero temperature. Let the boundary planes (x, y) of the semispaces be described by the equations $z = \pm a/2$. Considering only nonmagnetic materials in the absence of charge and current densities, the Maxwell equations take the form

$$\nabla \cdot \boldsymbol{D}(t, \boldsymbol{r}) = 0, \qquad \nabla \times \boldsymbol{E}(t, \boldsymbol{r}) + \frac{1}{c} \frac{\partial \boldsymbol{B}(t, \boldsymbol{r})}{\partial t} = 0,$$

$$\nabla \times \boldsymbol{B}(t, \boldsymbol{r}) - \frac{1}{c} \frac{\partial \boldsymbol{D}(t, \boldsymbol{r})}{\partial t} = 0, \qquad \nabla \cdot \boldsymbol{B}(t, \boldsymbol{r}) = 0, \qquad (12.1)$$

where $\boldsymbol{D}$ is the electric displacement, and $\boldsymbol{E}$ and $\boldsymbol{B}$ are the electric field and magnetic induction, respectively [compare with eqn (3.11)]. The oscillation frequencies of the electromagnetic field between and inside the dielectric semispaces must be found from the solution of eqn (12.1), supplemented by the standard continuity boundary conditions of classical electrodynamics imposed at the planes $z = \pm a/2$ (Jackson 1999, Landau et al. 1984),

$$E_{1t}(t, \boldsymbol{r}) = E_{2t}(t, \boldsymbol{r}), \qquad D_{1n}(t, \boldsymbol{r}) = D_{2n}(t, \boldsymbol{r}),$$
$$B_{1n}(t, \boldsymbol{r}) = B_{2n}(t, \boldsymbol{r}), \qquad B_{1t}(t, \boldsymbol{r}) = B_{2t}(t, \boldsymbol{r}). \qquad (12.2)$$

Here, $\boldsymbol{n}$ is the normal to the boundary plane directed inside the dielectric, and the subscripts "n" and "t" refer to the normal and tangential components, respectively (the subscript 1 refers to the vacuum and the subscript 2 to the semispace).

The Lifshitz theory ignores spatial dispersion. In this case the material equation connecting $\boldsymbol{D}$ and $\boldsymbol{E}$ takes the form

$$\boldsymbol{D}(t, \boldsymbol{r}) = \int_{-\infty}^{t} \varepsilon(t - t', \boldsymbol{r}) \boldsymbol{E}(t', \boldsymbol{r}) \, dt', \qquad (12.3)$$

where the kernel of the integral operator is $\varepsilon(\tau, \boldsymbol{r})$ (here we consider an isotropic dielectric with time-independent properties; the generalization to the anisotropic case is straightforward). According to eqn (12.3), the electric displacement $\boldsymbol{D}(t, \boldsymbol{r})$ at position $\boldsymbol{r}$ and time t is determined by the electric field $\boldsymbol{E}$ at the same point $\boldsymbol{r}$ but at different times $t' \leq t$. This indicates the role of temporal dispersion. Representing fields as Fourier transforms,

$$\boldsymbol{E}(t, \boldsymbol{r}) = \int_{-\infty}^{\infty} \boldsymbol{E}(\omega, \boldsymbol{r}) e^{-i\omega t} \, d\omega, \qquad \boldsymbol{D}(t, \boldsymbol{r}) = \int_{-\infty}^{\infty} \boldsymbol{D}(\omega, \boldsymbol{r}) e^{-i\omega t} \, d\omega, \qquad (12.4)$$

and substituting this in eqn (12.3), we arrive at

$$\boldsymbol{D}(\omega, \boldsymbol{r}) = \varepsilon(\omega, \boldsymbol{r})\boldsymbol{E}(\omega, \boldsymbol{r}). \tag{12.5}$$

Here, the frequency-dependent dielectric permittivity,

$$\varepsilon(\omega, \boldsymbol{r}) = \int_0^\infty \varepsilon(\tau, \boldsymbol{r}) e^{i\omega\tau} \, d\tau, \tag{12.6}$$

a concept central to the Lifshitz theory, is introduced. For the case of two similar homogeneous dielectric plates separated by a vacuum gap of width a, we get

$$\varepsilon(\omega, \boldsymbol{r}) = \begin{cases} \varepsilon(\omega), & |z| \geq a/2, \\ 1, & |z| < a/2. \end{cases} \tag{12.7}$$

The standard analytical representation of the frequency-dependent dielectric permittivity of a dielectric material is given by (Parsegian 2005)

$$\varepsilon(\omega) = 1 + \sum_{j=1}^{K} \frac{g_j}{\omega_j^2 - \omega^2 - i\gamma_j\omega}, \tag{12.8}$$

where the $\omega_j \neq 0$ are the oscillator frequencies, the g_j are the oscillator strengths, the γ_j are the damping parameters, and K is the number of oscillators. Usually, $\gamma_j \ll \omega_j$. From eqn (12.8), the dielectric permittivity of a dielectric at zero frequency takes the form

$$\varepsilon_0 \equiv \varepsilon(0) = 1 + \sum_{j=1}^{K} \frac{g_j}{\omega_j^2} < \infty. \tag{12.9}$$

For polar dielectrics consisting of molecules having an intrinsic dipole moment (see Section 12.7), an additional term must be added on the right-hand side of eqn (12.8).

There are many different derivations of the well-known Lifshitz formula for the Casimir force per unit area of the semispaces present in the literature. The original derivation by Lifshitz (1956) was based on the assumption that dielectric materials in a state of thermal equilibrium are characterized by randomly fluctuating sources of long-wavelength electromagnetic fields. The force acting between the semispaces was found as the zz-component of the Maxwell energy–momentum tensor statistically averaged using the fluctuation–dissipation theorem. Later, the correlation functions of the fluctuating electromagnetic field were expressed in terms of the Green's functions of the Maxwell equations (Dzyaloshinskii et al. 1961, Lifshitz and Pitaevskii 1980). Another approach to the derivation of the Lifshitz formula is based on scattering theory (Bordag et al. 2001a, Genet et al. 2003b). In this approach, an electromagnetic wave propagating in the dielectric material is scattered in the empty gap between the two semispaces and the transmitted and reflected waves are considered. The final results for the energy and force per unit area of the plates are expressed in terms of the reflection coefficients.

12.1.1 Representation in terms of imaginary frequencies

Here, we present another derivation of the Lifshitz formula, starting directly from the zero-point energy of the electromagnetic field in the presence of dielectric semispaces. The main ideas of such a derivation were formulated by van Kampen et al. (1968) and later elaborated by Ninham et al. (1970), Schram (1973), Langbein (1973), Zhou and Spruch (1995), and Klimchitskaya et al. (2000). Let us consider a monochromatic electromagnetic field

$$\boldsymbol{E}(t,\boldsymbol{r}) = \boldsymbol{E}(\boldsymbol{r})\,\mathrm{e}^{-\mathrm{i}\omega t}, \qquad \boldsymbol{B}(t,\boldsymbol{r}) = \boldsymbol{B}(\boldsymbol{r})\,\mathrm{e}^{-\mathrm{i}\omega t} \tag{12.10}$$

[it is assumed that the physical fields are equal to the real parts of eqn (12.10)]. Substituting eqn (12.10) into the Maxwell equations (12.1) and using eqns (12.5) and (12.7), we obtain equations inside the dielectric semispaces in the form

$$\boldsymbol{\nabla}\cdot\boldsymbol{E}(\boldsymbol{r}) = 0, \qquad \boldsymbol{\nabla}\times\boldsymbol{E}(\boldsymbol{r}) - \mathrm{i}\frac{\omega}{c}\boldsymbol{B}(\boldsymbol{r}) = 0,$$

$$\boldsymbol{\nabla}\times\boldsymbol{B}(\boldsymbol{r}) + \mathrm{i}\varepsilon(\omega)\frac{\omega}{c}\boldsymbol{E}(\boldsymbol{r}) = 0, \qquad \boldsymbol{\nabla}\cdot\boldsymbol{B}(\boldsymbol{r}) = 0, \tag{12.11}$$

where $\boldsymbol{D}(\boldsymbol{r}) = \varepsilon(\omega)\boldsymbol{E}(\boldsymbol{r})$ has been used. In the gap between the semispaces, $\varepsilon(\omega)$ in eqn (12.11) must be replaced with unity. From eqn (12.11), second-order equations for the fields follow:

$$\nabla^2\boldsymbol{E}(\boldsymbol{r}) + \varepsilon(\omega)\frac{\omega^2}{c^2}\boldsymbol{E}(\boldsymbol{r}) = 0, \qquad \nabla^2\boldsymbol{B}(\boldsymbol{r}) + \varepsilon(\omega)\frac{\omega^2}{c^2}\boldsymbol{B}(\boldsymbol{r}) = 0. \tag{12.12}$$

The complete orthonormal set of solutions of eqn (12.12) can be found to be

$$\boldsymbol{E}_J(\boldsymbol{r}) = \boldsymbol{e}_p(z,\boldsymbol{k}_\perp)\exp(\mathrm{i}\boldsymbol{k}_\perp\cdot\boldsymbol{r}_\perp), \qquad \boldsymbol{B}_J(\boldsymbol{r}) = \boldsymbol{g}_p(z,\boldsymbol{k}_\perp)\exp(\mathrm{i}\boldsymbol{k}_\perp\cdot\boldsymbol{r}_\perp). \tag{12.13}$$

Here, $\boldsymbol{r} = (x,y,z) = (\boldsymbol{r}_\perp, z)$, $\boldsymbol{k}_\perp = (k_x, k_y)$, and the collective index $J = \{p, \boldsymbol{k}_\perp, \omega\}$, where $p = \mathrm{TM, TE}$ labels the two independent polarizations of the electromagnetic field (transverse magnetic and transverse electric) defined in Section 7.2.1.

Substituting eqn (12.13) into eqn (12.12), we obtain the oscillator equations for the functions $\boldsymbol{e}_p$ and $\boldsymbol{g}_p$,

$$\boldsymbol{e}_p''(z,\boldsymbol{k}_\perp) - k^2\boldsymbol{e}_p(z,\boldsymbol{k}_\perp) = 0, \qquad \boldsymbol{g}_p''(z,\boldsymbol{k}_\perp) - k^2\boldsymbol{g}_p(z,\boldsymbol{k}_\perp) = 0, \tag{12.14}$$

where

$$k^2 = k^2(\omega, \boldsymbol{k}_\perp) = k_\perp^2 - \varepsilon(\omega)\frac{\omega^2}{c^2} \tag{12.15}$$

and a prime denotes a derivative with respect to z. In the vacuum gap between the semispaces, k^2 must be replaced with

$$q^2 = q^2(\omega, \boldsymbol{k}_\perp) = k_\perp^2 - \frac{\omega^2}{c^2}. \tag{12.16}$$

In a similar way, the first and fourth Maxwell equations in eqn (12.11) lead to equations for the projections of the vectors $\boldsymbol{e}_p$ and $\boldsymbol{g}_p$ on the x, y, and z axes:

$$e_{p,z}'(z,\boldsymbol{k}_\perp) + \mathrm{i}k_x e_{p,x}(z,\boldsymbol{k}_\perp) + \mathrm{i}k_y e_{p,y}(z,\boldsymbol{k}_\perp) = 0,$$

$$g'_{p,z}(z,k_\perp) + ik_x g_{p,x}(z,k_\perp) + ik_y g_{p,y}(z,k_\perp) = 0. \tag{12.17}$$

Now let us satisfy the continuity boundary conditions (12.2). We start by considering the electric field and electric displacement. The first condition in eqn (12.2), for the electric field, is satisfied if $e_{p,x}$ and $e_{p,y}$ are continuous when crossing the boundary planes $z = \pm a/2$. From the first equation in eqn (12.17), this is equivalent to the continuity of $e'_{p,z}$. The second equation in eqn (12.2), for the electric displacement, is satisfied if $\varepsilon e_{p,z}$ is continuous when crossing the boundary planes. Note that $e_{p,z}$ is not equal to zero for $p =$ TM only, so that we now consider the transverse magnetic mode. By using the exponentially decreasing solutions of the first equation in eqn (12.14) inside the semispaces, we can represent $e_{\mathrm{TM},z}$ in the form

$$e_{\mathrm{TM},z}(z,k_\perp) = \begin{cases} A\exp(kz), & z < -a/2, \\ B\exp(qz) + C\exp(-qz), & |z| < a/2, \\ D\exp(-kz), & z > a/2. \end{cases} \tag{12.18}$$

Then the continuity of $e'_{\mathrm{TM},z}$ and $\varepsilon(\mathbf{r},\omega)e_{\mathrm{TM},z}$ at $z = \pm a/2$ implies the following system of equations:

$$\begin{aligned} Ak\exp(-ka/2) &= Bq\exp(-qa/2) - Cq\exp(qa/2), \\ -Dk\exp(-ka/2) &= Bq\exp(qa/2) - Cq\exp(-qa/2), \\ A\varepsilon\exp(-ka/2) &= B\exp(-qa/2) + C\exp(qa/2), \\ D\varepsilon\exp(-ka/2) &= B\exp(qa/2) + C\exp(-qa/2). \end{aligned} \tag{12.19}$$

This is a linear homogeneous system of algebraic equations with unknown quantities A, B, C, and D. It has a nontrivial solution under the condition that the determinant of its known coefficients is equal to zero,

$$\Delta^{\mathrm{TM}}(\omega,k_\perp) \equiv e^{-ka}\left[(\varepsilon q + k)^2 e^{qa} - (\varepsilon q - k)^2 e^{-qa}\right] = 0. \tag{12.20}$$

In a similar way, the third and fourth continuity conditions in eqn (12.2), for the magnetic induction, are satisfied if all of the components $g_{p,x}$, $g_{p,y}$, and $g_{p,z}$ are continuous when crossing the boundary planes $z = \pm a/2$. Owing to the second equation in (12.17), the continuity of $g_{p,x}$ and $g_{p,y}$ is equivalent to the continuity of $g'_{p,z}$. The component $g_{p,z}$ is not equal to zero for the transverse electric mode $p =$ TE (see Section 7.2.1). Representing $g_{\mathrm{TE},z}(z,k_\perp)$ in the form of eqn (12.18) with some coefficients E, F, G, H, we obtain from the third and fourth conditions in eqn (12.2) the same system of equations as in eqn (12.19), where ε is replaced with unity. The condition that this system has a nontrivial solution is

$$\Delta^{\mathrm{TE}}(\omega,k_\perp) \equiv e^{-ka}\left[(q+k)^2 e^{qa} - (q-k)^2 e^{-qa}\right] = 0. \tag{12.21}$$

The solutions of the transcendental equations (12.20) and (12.21) are the photon eigenfrequencies $\omega_{k_\perp,n}^{\mathrm{TM}}$ and $\omega_{k_\perp,n}^{\mathrm{TE}}$ for the configuration of two dielectric semispaces separated by a vacuum gap. Thus Δ^{TM} and Δ^{TE} are the mode-generating functions, as discussed in Section 9.3.2. Note that in reality these eigenfrequencies are complex [because the dielectric permittivity (12.8) is complex], with positive imaginary parts which are much smaller than the real parts. But we now continue the derivation of the Lifshitz formula by neglecting the imaginary parts of the photon eigenfrequencies. The justification for this approach was provided by Barash and Ginzburg (1975). It is discussed below.

In terms of the as yet unknown photon eigenfrequencies, the vacuum energy of the electromagnetic field in thermal equilibrium with the dielectric plates at zero temperature is given by

$$E_0(a) = \frac{\hbar}{4\pi} \int_0^\infty k_\perp dk_\perp \sum_n (\omega_{k_\perp,n}^{TM} + \omega_{k_\perp,n}^{TE}) S. \qquad (12.22)$$

The sums over the solutions of eqns (12.20) and (12.21) can be calculated by using the argument principle (7.60). We consider a closed contour consisting of a semicircle C_+ of infinite radius in the right half of the complex plane ω with its center at the origin, and the imaginary axis $[i\infty, -i\infty]$ bypassed in a counterclockwise manner. Given that the functions Δ^{TM} and Δ^{TE} do not have poles inside this contour, we obtain

$$\sum_n \omega_{k_\perp,n}^{TM,TE} = \frac{1}{2\pi i} \left[\int_{i\infty}^{-i\infty} \omega \, d\ln \Delta^{TM,TE}(\omega, k_\perp) + \int_{C_+} \omega \, d\ln \Delta^{TM,TE}(\omega, k_\perp) \right]. \qquad (12.23)$$

The second integral on the right-hand side of eqn (12.23) is calculated simply, under the natural assumption

$$\lim_{\omega \to \infty} \varepsilon(\omega) = 1, \qquad \lim_{\omega \to \infty} \frac{d\varepsilon(\omega)}{d\omega} = 0 \qquad (12.24)$$

satisfied by the dielectric permittivity (12.8). The result is infinite and does not depend on a:

$$\int_{C_+} \omega \, d\ln \Delta^{TM,TE}(\omega, k_\perp) = 4 \int_{C_+} d\omega. \qquad (12.25)$$

Introducing a new variable $\xi = -i\omega$, we rearrange eqn (12.23) to the form

$$\sum_n \omega_{k_\perp,n}^{TM,TE} = \frac{1}{2\pi} \int_\infty^{-\infty} \xi \, d\ln \Delta^{TM,TE}(i\xi, k_\perp) + \frac{2}{\pi} \int_{C_+} d\xi, \qquad (12.26)$$

where the last term on the right-hand side does not depend on a.

The energy given by eqns (12.22) and (12.26) is infinite. The finite Casimir energy per unit area of the boundary planes $z = \pm a/2$ is obtained by the subtraction from $E_0(a)$ of the energy for infinitely separated semispaces:

$$E(a) = \frac{E_0(a)}{S} - \lim_{a \to \infty} \frac{E_0(a)}{S}. \qquad (12.27)$$

This is equivalent to the dropping of the integral along C_+ and the replacement of $\Delta^{\mathrm{TM,TE}}(i\xi, k_\perp)$ with $\Delta^{\mathrm{TM,TE}}/\Delta_\infty^{\mathrm{TM,TE}}$, where

$$\Delta_\infty^{\mathrm{TM}}(i\xi, k_\perp) = e^{(q-k)a}(\varepsilon q + k)^2, \qquad \Delta_\infty^{\mathrm{TE}}(i\xi, k_\perp) = e^{(q-k)a}(q + k)^2,$$

$$q^2 = q^2(i\xi, k_\perp) = k_\perp^2 + \frac{\xi^2}{c^2}, \qquad k^2 = k^2(i\xi, k_\perp) = k_\perp^2 + \varepsilon(i\xi)\frac{\xi^2}{c^2}. \quad (12.28)$$

The result is

$$E(a) = \frac{\hbar}{8\pi^2} \int_0^\infty k_\perp dk_\perp \int_\infty^{-\infty} \xi\, d\left[\ln\frac{\Delta^{\mathrm{TM}}(i\xi, k_\perp)}{\Delta_\infty^{\mathrm{TM}}(i\xi, k_\perp)} + \ln\frac{\Delta^{\mathrm{TE}}(i\xi, k_\perp)}{\Delta_\infty^{\mathrm{TE}}(i\xi, k_\perp)}\right]. \quad (12.29)$$

Integrating eqn (12.29) by parts and using the explicit expressions (12.20), (12.21), and (12.28), we arrive at the Lifshitz formula at zero temperature,

$$E(a) = \frac{\hbar}{4\pi^2} \int_0^\infty k_\perp dk_\perp \int_0^\infty d\xi \left\{\ln\left[1 - r_{\mathrm{TM}}^2(i\xi, k_\perp)e^{-2qa}\right] \right. \quad (12.30)$$
$$\left. + \ln\left[1 - r_{\mathrm{TE}}^2(i\xi, k_\perp)e^{-2qa}\right]\right\}.$$

In the above, the following notation has been introduced:

$$r_{\mathrm{TM}}(i\xi, k_\perp) = \frac{\varepsilon(i\xi)q(i\xi, k_\perp) - k(i\xi, k_\perp)}{\varepsilon(i\xi)q(i\xi, k_\perp) + k(i\xi, k_\perp)}, \quad r_{\mathrm{TE}}(i\xi, k_\perp) = \frac{q(i\xi, k_\perp) - k(i\xi, k_\perp)}{q(i\xi, k_\perp) + k(i\xi, k_\perp)}, \quad (12.31)$$

where q and k are defined in eqn (12.28).

The quantities r_{TM} and r_{TE} are the familiar Fresnel reflection coefficients for the transverse magnetic and electric waves (Landau *et al.* 1984),

$$r_{\mathrm{TM}}(\theta_0) = \frac{\varepsilon \cos\theta_0 - \sqrt{\varepsilon - \sin^2\theta_0}}{\varepsilon \cos\theta_0 + \sqrt{\varepsilon - \sin^2\theta_0}}, \quad r_{\mathrm{TE}}(\theta_0) = \frac{\cos\theta_0 - \sqrt{\varepsilon - \sin^2\theta_0}}{\cos\theta_0 + \sqrt{\varepsilon - \sin^2\theta_0}}, \quad (12.32)$$

where θ_0 is the angle of incidence from the vacuum gap onto the semispace. Considering that for real electromagnetic waves in vacuum $\sin\theta_0 = k_\perp c/\omega$, we see that $r_{\mathrm{TM}}(\theta_0)$ and $r_{\mathrm{TE}}(\theta_0)$ coincide with $r_{\mathrm{TM}}(\omega, k_\perp)$ and $r_{\mathrm{TE}}(\omega, k_\perp)$, respectively. However, an important property of the coefficients (12.31) is that they are calculated along the imaginary frequency axis and depend on the two *independent* variables ξ and $k_\perp$. There are no constraints such as $k_\perp c \leq \xi$ in the two integrations with respect to $k_\perp$ and ξ in the Lifshitz formula (12.30). Using eqn (1.6), the Lifshitz formula for the Casimir pressure is obtained:

$$P(a) = -\frac{\hbar}{2\pi^2} \int_0^\infty k_\perp dk_\perp \int_0^\infty d\xi\, q \left\{\left[r_{\mathrm{TM}}^{-2}(i\xi, k_\perp)e^{2qa} - 1\right]^{-1} \right. \quad (12.33)$$
$$\left. + \left[r_{\mathrm{TE}}^{-2}(i\xi, k_\perp)e^{2qa} - 1\right]^{-1}\right\}.$$

In the above derivation, the small imaginary parts of the photon eigenfrequencies were neglected. If the complex nature of the photon eigenfrequencies is taken into account, the vacuum energy of the electromagnetic field is not given by eqn (12.22), which is already clear from the complexity of the right-hand side of this equation. In fact, a dielectric medium with a dielectric permittivity (12.8) is represented as a set of oscillators

$$\frac{d^2x(t)}{dt^2} + \gamma_j \frac{dx(t)}{dt} + \omega_j^2 x(t) = f(t), \qquad (12.34)$$

where $f(t) = f_\omega \exp(-i\omega t)$ is a harmonically varying external force modeling the influence of electromagnetic fluctuations. If the solutions of eqns (12.20) and (12.21), $\omega_{k_\perp,n}^{TM}$ and $\omega_{k_\perp,n}^{TE}$, are complex, the respective eigenfunctions are not orthogonal. They can, however, be expanded in terms of the orthogonal eigenfunctions of some auxiliary electrodynamic system that has real eigenfrequencies. This leads to the result that although the representation of the energy as a sum (12.22) of complex eigenfrequencies is invalid, the representation (12.30) remains correct for complex eigenfrequencies. The qualitative reason for the validity of this statement is that the energy (12.30) depends only on the dielectric permittivity along the imaginary frequency axis, which is always real. A rigorous proof of the validity of eqns (12.30) and (12.33) in the case of complex photon eigenfrequencies, and an exact formulation of the auxiliary electrodynamic problem can be found in the review by Barash and Ginzburg (1975) and in the book by Milonni (1994). Note that in the case of complex photon eigenfrequencies, the correct expression for the Casimir energy must be written as an integral with respect to ξ with limits from 0 to ∞ [see eqn (12.30)]. Although an integration from $-\infty$ to ∞ [as in eqn (12.29)] can be equivalently used for real eigenfrequencies, it is not permitted for complex eigenfrequencies because the functions $\Delta^{TM,TE}$ and the reflection coefficients are not, in this case, even functions of ξ (Barash and Ginzburg 1984).

12.1.2 Representation in terms of real frequencies

Equations (12.30) and (12.33) represent the Casimir energy and pressure as functions of the separation in terms of integrals over imaginary frequencies. Equivalently, these equations can be expressed in terms of real frequencies. To do so, we notice that for any function $f(\omega)$ that is analytic in the first quadrant of the complex ω plane,

$$\int_{C_q} d\omega \, f(\omega) = 0, \qquad (12.35)$$

where the contour C_q consists of the positive half $(0, \infty)$ of the real axis, a 90° arc of infinitely large radius, and the half $(i\infty, 0)$ of the imaginary axis. By assuming that $f(\omega)$ vanishes on the arc of infinitely large radius and that $f(i\xi)$ is real, from eqn (12.35) we get

$$\mathrm{Im} \int_0^\infty d\omega \, f(\omega) = \mathrm{Im}\left[i \int_0^\infty d\xi \, f(i\xi)\right] = \int_0^\infty d\xi \, f(i\xi). \qquad (12.36)$$

Now we apply eqn (12.36) to eqn (12.30). This is possible because, in the renormalized energy per unit area and the pressure, the respective integrals along the arcs of infinitely large radius vanish. Then eqn (12.30) takes the form

$$E(a) = \frac{\hbar}{4\pi^2} \int_0^\infty k_\perp dk_\perp \int_0^\infty d\omega \,\text{Im}\, \{\ln[1 - r_{\text{TM}}^2(\omega, k_\perp)e^{-2qa}] \quad (12.37)$$
$$+ \ln[1 - r_{\text{TE}}^2(\omega, k_\perp)e^{-2qa}]\},$$

where $q = q(\omega, k_\perp)$ has been defined in eqn (12.16), and r_{TM} and r_{TE} are obtained from eqn (12.31) by the replacement of $i\xi$ with ω. Note that in contrast to eqn (12.30), where $q \equiv q(i\xi, k_\perp)$ is real and positive, in eqn (12.37) $q \equiv q(\omega, k_\perp)$ takes both real and pure imaginary values depending on whether $k_\perp \geq \omega/c$ or $k_\perp < \omega/c$. As a result, eqn (12.37) contains integrals of rapidly oscillating functions. This makes the representation (12.30) more convenient for numerical computations.

In a similar way, eqn (12.33) for the Casimir pressure can be rearranged to a form containing an integration along the real frequency axis,

$$P(a) = -\frac{\hbar}{2\pi^2} \int_0^\infty k_\perp dk_\perp \int_0^\infty d\omega \,\text{Im}\, \{q[r_{\text{TM}}^{-2}(\omega, k_\perp)e^{2qa} - 1]^{-1} \quad (12.38)$$
$$+ q[r_{\text{TE}}^{-2}(\omega, k_\perp)e^{2qa} - 1]^{-1}\}.$$

We note that pure imaginary values of $q(\omega, k_\perp)$ in eqns (12.37) and (12.38) (which are taken at $\omega > ck_\perp$) correspond to the contributions of *propagating* waves, while real values of $q(\omega, k_\perp)$ (taken at $\omega \leq ck_\perp$) describe what are referred to as *evanescent* waves.

12.1.3 The limiting cases of small and large separations

Computational results for the Casimir energy and pressure obtained using eqns (12.30) and (12.33) for some typical dielectrics are presented in Section 12.6.2. Here, we consider the characteristic behavior of the pressure $P(a)$ at small and large separations in comparison with the characteristic absorption wavelength of the dielectric material λ_0. At $a \ll \lambda_0$, the most important region, $qa \sim 1$, in the integral with respect to $k_\perp$ in eqn (12.33) corresponds to a rather large q. This is the region of high frequencies ξ, where $q \approx k$ and, thus, $r_{\text{TE}} \ll r_{\text{TM}}$. By introducing a new variable $y = 2aq$ instead of $k_\perp$, and changing the order of integration with respect to ξ and to $k_\perp$, we get

$$P(a) = -\frac{\hbar}{16\pi^2 a^3} \int_0^\infty d\xi \int_{\xi/\omega_c}^\infty y^2 \, dy \, [r_{\text{TM}}^{-2}(i\xi, y)e^y - 1]^{-1}, \quad (12.39)$$

where $\omega_c = c/(2a)$ is the *characteristic frequency*. Now, in the same approximation, it is possible to replace the lower integration limit ξ/ω_c with zero and cancel the q and k in the numerator and denominator in the definition (12.31) of the reflection coefficient $r_{\rm TM}$. As a result (Lifshitz and Pitaevskii 1980)

$$P(a) = -\frac{H}{6\pi a^3}, \quad H = \frac{3\hbar}{8\pi}\int_0^\infty d\xi \int_0^\infty y^2\, dy \left\{\left[\frac{\varepsilon(i\xi)+1}{\varepsilon(i\xi)-1}\right]^2 e^y - 1\right\}^{-1}, \tag{12.40}$$

where H is called the *Hamaker constant*. This is the nonretarded van der Waals pressure in the configuration of two semispaces. It is of nonrelativistic character and thus does not depend on the velocity of light c.

Now we consider the opposite limit of large separation, satisfying the condition $a \gg \lambda_0$. At the same time, it is assumed that the separation is small enough to neglect thermal effects (see Section 12.3). In this case, low frequencies provide the dominant contribution to the pressure. As a result, the dielectric permittivity along the imaginary frequency axis can be replaced with its static value ε_0 defined in eqn (12.9). Introducing dimensionless variables

$$y = 2aq, \quad \zeta = \frac{\xi}{\omega_c}, \tag{12.41}$$

and using eqn (12.31), we represent the pressure at large separations in the form

$$P(a) = -\frac{\pi^2}{240}\frac{\hbar c}{a^4}\Psi(\varepsilon_0). \tag{12.42}$$

Here, the function $\Psi(\varepsilon_0)$ is defined as

$$\Psi(\varepsilon_0) = \frac{15}{2\pi^4}\int_0^\infty d\zeta \int_\zeta^\infty y^2\, dy \left\{\left[\left(\frac{\varepsilon_0 y + \sqrt{y^2 + \zeta^2(\varepsilon_0-1)}}{\varepsilon_0 y - \sqrt{y^2 + \zeta^2(\varepsilon_0-1)}}\right)^2 e^y - 1\right]^{-1} \right.$$
$$\left. + \left[\left(\frac{y + \sqrt{y^2 + \zeta^2(\varepsilon_0-1)}}{y - \sqrt{y^2 + \zeta^2(\varepsilon_0-1)}}\right)^2 e^y - 1\right]^{-1}\right\}. \tag{12.43}$$

Equation (12.42) includes retardation effects, an essential factor at large separations. The relativistic character of eqn (12.42) is reflected in the explicit dependence on the velocity of light c. In the limiting case $\varepsilon_0 \to \infty$, the function $\Psi(\varepsilon_0) \to 1$ and one obtains the case of ideal-metal plates.

12.2 The Lifshitz formula for stratified and magnetic media

The formalism presented in the previous section can be easily generalized to the case of two dissimilar semispaces, to stratified media consisting of several plane parallel layers, to two plates of finite thickness, etc. We start with the case of two semispaces made of dissimilar dielectric materials with dielectric permittivities

$\varepsilon^{(1)}(\omega)$ and $\varepsilon^{(2)}(\omega)$. A straightforward repetition of the derivation in Section 12.1.1 results in the following expression for the Casimir energy:

$$E(a) = \frac{\hbar}{4\pi^2} \int_0^\infty k_\perp dk_\perp \int_0^\infty d\xi \left\{ \ln\left[1 - r_{\text{TM}}^{(1)}(i\xi, k_\perp) r_{\text{TM}}^{(2)}(i\xi, k_\perp) e^{-2qa}\right] \right.$$
$$\left. + \ln\left[1 - r_{\text{TE}}^{(1)}(i\xi, k_\perp) r_{\text{TE}}^{(2)}(i\xi, k_\perp) e^{-2qa}\right] \right\}, \qquad (12.44)$$

where the reflection coefficients are given by

$$r_{\text{TM}}^{(n)}(i\xi, k_\perp) = \frac{\varepsilon^{(n)}(i\xi) q(i\xi, k_\perp) - k^{(n)}(i\xi, k_\perp)}{\varepsilon^{(n)}(i\xi) q(i\xi, k_\perp) + k^{(n)}(i\xi, k_\perp)}, \qquad (12.45)$$

$$r_{\text{TE}}^{(n)}(i\xi, k_\perp) = \frac{q(i\xi, k_\perp) - k^{(n)}(i\xi, k_\perp)}{q(i\xi, k_\perp) + k^{(n)}(i\xi, k_\perp)}, \quad k^{(n)}(i\xi, k_\perp) = \left[k_\perp^2 + \varepsilon^{(n)}(i\xi) \frac{\xi^2}{c^2}\right]^{1/2}.$$

In a similar manner, the generalization of the Casimir pressure to the case of two dissimilar semispaces reads

$$P(a) = -\frac{\hbar}{2\pi^2} \int_0^\infty k_\perp dk_\perp \int_0^\infty d\xi \, q \left\{ \left[\frac{e^{2qa}}{r_{\text{TM}}^{(1)}(i\xi, k_\perp) r_{\text{TM}}^{(2)}(i\xi, k_\perp)} - 1\right]^{-1} \right.$$
$$\left. + \left[\frac{e^{2qa}}{r_{\text{TE}}^{(1)}(i\xi, k_\perp) r_{\text{TE}}^{(2)}(i\xi, k_\perp)} - 1\right]^{-1} \right\}. \qquad (12.46)$$

Next we consider a stratified medium consisting of three layers of finite thickness a and $d_{\pm 1}$ with dielectric permittivities $\varepsilon^{(n)}(\omega)$ ($n = 0, \pm 1$) enclosed between two semispaces with dielectric permittivities $\varepsilon^{(\pm 2)}(\omega)$ (see Fig. 12.1). For

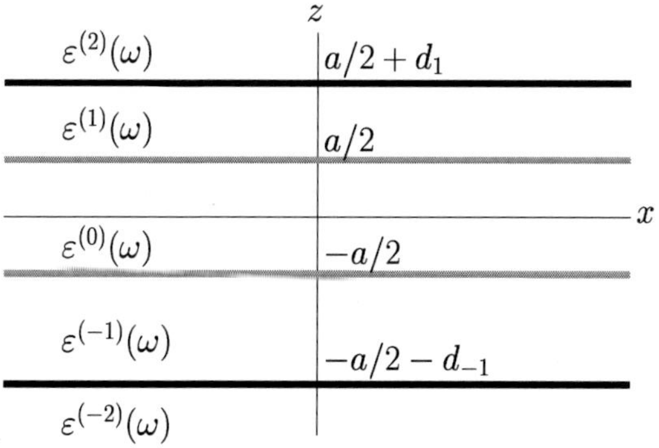

FIG. 12.1. Stratified medium consisting of three layers of finite thickness d_{-1}, a, and d_1 with dielectric permittivities $\varepsilon^{(-1)}(\omega)$, $\varepsilon^{(0)}(\omega)$, and $\varepsilon^{(1)}(\omega)$, enclosed between two semispaces $z \leq -a/2 - d_{-1}$ and $z \geq a/2 + d_1$ with dielectric permittivities $\varepsilon^{(-2)}(\omega)$ and $\varepsilon^{(-2)}(\omega)$, respectively.

this purpose, the electrodynamic boundary conditions (12.2) are imposed on the four boundary planes $z = \pm a/2$, $z = a/2 + d_1$, and $z = -a/2 - d_{-1}$ for the solutions of the Maxwell equations (12.11) written for each of the five strata. This leads to two homogeneous algebraic systems, each containing eight linear equations. By setting the determinants of these systems equal to zero, we arrive at two transcendental equations for the determination of the photon eigenfrequencies in the stratified medium. Then, using eqn (12.22) and applying the argument principle under the renormalization condition that the physical energy should vanish when $a \to \infty$, we arrive at the Casimir energy per unit area [see Bordag et al. (2001a) for details]:

$$E(a) = \frac{\hbar}{4\pi^2} \int_0^\infty k_\perp dk_\perp \int_0^\infty d\xi \left\{ \ln\left[1 - R_{\mathrm{TM}}^{(+)}(i\xi, k_\perp) R_{\mathrm{TM}}^{(-)}(i\xi, k_\perp) e^{-2k^{(0)}a}\right] \right. $$
$$\left. + \ln\left[1 - R_{\mathrm{TE}}^{(+)}(i\xi, k_\perp) R_{\mathrm{TE}}^{(-)}(i\xi, k_\perp) e^{-2k^{(0)}a}\right] \right\}. \tag{12.47}$$

Here,

$$k^{(n)} \equiv k^{(n)}(i\xi, k_\perp) = \left[k_\perp^2 + \varepsilon^{(n)}(i\xi)\frac{\xi^2}{c^2}\right]^{1/2}, \quad n = 0, \pm 1, \pm 2, \tag{12.48}$$

and the reflection coefficients in the region $|z| < a/2$ for electromagnetic waves incident on the planes $z = \pm a/2$ are given by

$$R_{\mathrm{TM}}^{(\pm)}(i\xi, k_\perp) = \frac{r_{\mathrm{TM}}^{(0,\pm1)} + r_{\mathrm{TM}}^{(\pm1,\pm2)} e^{-2k^{(\pm1)}d_{\pm1}}}{1 + r_{\mathrm{TM}}^{(0,\pm1)} r_{\mathrm{TM}}^{(\pm1,\pm2)} e^{-2k^{(\pm1)}d_{\pm1}}},$$

$$R_{\mathrm{TE}}^{(\pm)}(i\xi, k_\perp) = \frac{r_{\mathrm{TE}}^{(0,\pm1)} + r_{\mathrm{TE}}^{(\pm1,\pm2)} e^{-2k^{(\pm1)}d_{\pm1}}}{1 + r_{\mathrm{TE}}^{(0,\pm1)} r_{\mathrm{TE}}^{(\pm1,\pm2)} e^{-2k^{(\pm1)}d_{\pm1}}}. \tag{12.49}$$

In these formulas, all upper signs or, alternatively, all lower signs must be chosen. The reflection coefficients $r_{\mathrm{TM,TE}}^{(n,n')}$ on the various boundary planes are defined as follows:

$$r_{\mathrm{TM}}^{(n,n')} \equiv r_{\mathrm{TM}}^{(n,n')}(i\xi, k_\perp) = \frac{\varepsilon^{(n')}(i\xi)k^{(n)} - \varepsilon^{(n)}(i\xi)k^{(n')}}{\varepsilon^{(n')}(i\xi)k^{(n)} + \varepsilon^{(n)}(i\xi)k^{(n')}},$$

$$r_{\mathrm{TE}}^{(n,n')} \equiv r_{\mathrm{TE}}^{(n,n')}(i\xi, k_\perp) = \frac{k^{(n)} - k^{(n')}}{k^{(n)} + k^{(n')}}. \tag{12.50}$$

Note that the coefficients given by eqn (12.50) coincide with eqn (12.31), obtained for the case of two semispaces, if $n = 0$, $n' = 1$, and $\varepsilon^{(0)}(i\xi) = 1$, i.e. there is a vacuum in the region $|z| < a/2$.

For the Casimir pressure in the region $|z| < a/2$, the following expression is obtained from eqn (12.47):

$$P(a) = -\frac{\hbar}{2\pi^2} \int_0^\infty k_\perp dk_\perp \int_0^\infty d\xi \, k^{(0)} \left\{ \left[\frac{e^{2k^{(0)}a}}{R_{\mathrm{TM}}^{(+)}(i\xi, k_\perp) R_{\mathrm{TM}}^{(-)}(i\xi, k_\perp)} - 1\right]^{-1} \right.$$

$$+ \left[\frac{e^{2k^{(0)}a}}{R_{\text{TE}}^{(+)}(i\xi, k_\perp) R_{\text{TE}}^{(-)}(i\xi, k_\perp)} - 1 \right]^{-1} \right\}. \tag{12.51}$$

Equations (12.47) and (12.51) can be easily generalized to an arbitrary number of plane parallel layers [see e.g. Tomaš (2002) and Raabe et al. (2003)].

Using the reflection coefficients (12.49) for the planes $z = \pm a/2$, it is possible to consider various configurations which are of interest for the experimental applications in Part III of the book. In particular, one can investigate the effect of thin layers deposited on thick plates. It is also simple to obtain an expression for the Casimir energy and pressure in a configuration of two plates of finite thickness d with a dielectric permittivity $\varepsilon(\omega)$. For this purpose we put in the above equations $d_1 = d_{-1} = d$, $\varepsilon^{(0)}(i\xi) = \varepsilon^{(\pm 2)}(i\xi) = 1$, and $\varepsilon^{(\pm 1)}(i\xi) = \varepsilon(i\xi)$. Then the Casimir energy per unit area and the Casimir pressure are given by eqns (12.30) and (12.33), where the reflection coefficients must be replaced with

$$r_{\text{TM}}(i\xi, k_\perp) = \frac{\varepsilon^2(i\xi)q^2 - k^2}{\varepsilon^2(i\xi)q^2 + k^2 + 2qk\varepsilon(i\xi)\coth(kd)},$$

$$r_{\text{TE}}(i\xi, k_\perp) = \frac{q^2 - k^2}{q^2 + k^2 + 2qk\coth(kd)}. \tag{12.52}$$

The quantities q and k are defined in eqn (12.28). Another configuration of interest in applications is a system consisting of two thick plates (semispaces) with a layer of thickness a sandwiched between them. In this case the Casimir energy per unit area and the Casimir pressure are given by eqns (12.47) and (12.51). The reflection coefficients are obtained from eqn (12.49) in the limit $d_1, d_{-1} \to \infty$. The result is

$$R_{\text{TM}}^{(\pm)}(i\xi, k_\perp) = r_{\text{TM}}^{(0,\pm 1)}(i\xi, k_\perp), \qquad R_{\text{TE}}^{(\pm)}(i\xi, k_\perp) = r_{\text{TE}}^{(0,\pm 1)}(i\xi, k_\perp), \tag{12.53}$$

where $r_{\text{TM}}^{(0,\pm 1)}$ and $r_{\text{TE}}^{(0,\pm 1)}$ are defined in eqn (12.50).

In the above, we have considered nonmagnetic dielectrics. The results obtained, however, can easily be generalized to the case of magnetodielectrics with a frequency-dependent magnetic permeability $\mu(\omega)$. We consider a configuration of two magnetodielectric semispaces at a separation a. In this case the third Maxwell equation in eqn (12.11) inside the dielectrics must be replaced with

$$\nabla \times \boldsymbol{H}(\boldsymbol{r}) + i\varepsilon(\omega)\frac{\omega}{c}\boldsymbol{E}(\boldsymbol{r}) = 0, \tag{12.54}$$

where $\boldsymbol{H}(\boldsymbol{r}) = \boldsymbol{B}(\boldsymbol{r})/\mu(\omega)$ is the magnetic field. The first three continuity boundary conditions imposed at the boundary planes $z = \pm a/2$ in eqn (12.2) remain the same, whilst the fourth is replaced with

$$\boldsymbol{H}_{1t}(t, \boldsymbol{r}) = \boldsymbol{H}_{2t}(t, \boldsymbol{r}). \tag{12.55}$$

By repeating all calculations performed in Section 12.1.1, we once more arrive at the Lifshitz formulas for the Casimir energy per unit area (12.30) and the pressure (12.33). However, the reflection coefficients for the transverse magnetic and transverse electric fields must be replaced with

$$r_{\text{TM},\mu}(i\xi, k_\perp) = \frac{\varepsilon(i\xi)q(i\xi, k_\perp) - k_\mu(i\xi, k_\perp)}{\varepsilon(i\xi)q(i\xi, k_\perp) + k_\mu(i\xi, k_\perp)},$$

$$r_{\text{TE},\mu}(i\xi, k_\perp) = \frac{\mu(i\xi)q(i\xi, k_\perp) - k_\mu(i\xi, k_\perp)}{\mu(i\xi)q(i\xi, k_\perp) + k_\mu(i\xi, k_\perp)},$$

(12.56)

where

$$k_\mu^2(i\xi, k_\perp) = k_\perp^2 + \varepsilon(i\xi)\mu(i\xi)\frac{\xi^2}{c^2}. \quad (12.57)$$

For nonmagnetic dielectrics, $\mu(\omega) = 1$ and the reflection coefficients (12.56) coincide with eqn (12.31). The Lifshitz formulas can be generalized to an arbitrary number of plane parallel layers of magnetodielectrics (Buhmann et al. 2005, Tomaš 2005).

Kenneth et al. (2002) have shown that, for the case of materials with frequency-independent ε and μ, there exists a range of values for ε and μ for which the Casimir energy (free energy) is positive and, thus, the Casimir force between the two magnetodielectric semispaces is repulsive. Later, it was shown, however, that in the range of frequencies which contributes most to the Casimir force, the magnetic permeability μ of real materials is nearly equal to unity, and its magnitude is always far away from the values needed for achieving Casimir repulsion (Iannuzzi and Capasso 2003).

12.3 Two semispaces at nonzero temperature

Now we return to the simplest configuration of two semispaces at a separation a described by a frequency-dependent dielectric permittivity. In contrast to Section 12.1, where temperature was equal to zero, we now assume that the semispaces are at a temperature T in thermal equilibrium with the environment. In this section we consider two different representations for the Casimir free energy and pressure in the Lifshitz theory.

12.3.1 *Representation in terms of Matsubara frequencies*

The material of the semispaces is described by the dielectric permittivity (12.8), which may depend slightly on T through the temperature dependence of the oscillator parameters. However, for most dielectrics whose permittivity can be represented in the form of eqn (12.8), the dependence of the oscillator parameters on the temperature is negligible and the photon eigenfrequencies for both the TM and the TE polarizations satisfy eqns (12.20) and (12.21) derived in Section 12.1.1. By considering the free energy of each oscillator mode instead of the energy (as was done in Section 7.4.1 for two ideal-metal planes), we get the following expression for the free energy, analogous to eqn (7.59):

$$\mathcal{F}_0(a,T) = k_B T \int_0^\infty \frac{k_\perp dk_\perp}{2\pi} \sum_n \left[\ln\left(2\sinh\frac{\hbar\omega_{k_\perp,n}^{TM}}{2k_B T}\right) + \ln\left(2\sinh\frac{\hbar\omega_{k_\perp,n}^{TE}}{2k_B T}\right) \right] S. \tag{12.58}$$

As in Section 12.1.1, we first consider the case of real eigenfrequencies $\omega_{k_\perp,n}^{TM,TE}$.

The summation over the roots of eqns (12.20) and (12.21) can be performed by using the argument principle (7.60). This results in

$$\mathcal{F}_0(a,T) = k_B T \int_0^\infty \frac{k_\perp dk_\perp}{2\pi} \frac{1}{2\pi i} \oint_{C_1} \ln\left(2\sinh\frac{\hbar\omega}{2k_B T}\right)$$
$$\times d\left[\ln\Delta^{TM}(\omega,k_\perp) + \ln\Delta^{TE}(\omega,k_\perp)\right] S, \tag{12.59}$$

where the contour C_1 is shown in Fig. 7.1(a) (note that the functions Δ^{TM} and Δ^{TE} have no poles). The integral along the contour C_1 is calculated similarly to what was done in Section 7.4.1. The contour C_1 in Fig. 7.1(a) is chosen so as to avoid all of the branch points of the logarithm at the imaginary frequencies $\omega_l = i\xi_l$, where

$$\xi_l = 2\pi \frac{k_B T}{\hbar} l \tag{12.60}$$

are the Matsubara frequencies, written here with all fundamental constants in contrast to Chapter 5 ($l = 0, \pm 1, \pm 2, \ldots$). The integral along C_1 can be presented as a sum of four integrals along the contours L_1, C_ϵ, L_2, and the arc C_R [see Fig. 7.1(a)]. The integrals along L_1, C_ϵ, and L_2 can be integrated by parts. The terms besides the integrals cancel each other at the points $\pm i\epsilon$ and contribute to a separation-independent infinite constant at points A and B [compare with eqn (12.26)]. The integral along L_1 is calculated by the application of the Cauchy theorem to the closed contour C_2 in Fig. 7.1(b):

$$-\int_{L_1} \coth\frac{\hbar\omega}{2k_B T} \ln\Delta^{TM,TE}(\omega,k_\perp)\, d\omega = \int_{i\epsilon}^{i\infty} \coth\frac{\hbar\omega}{2k_B T} \ln\Delta^{TM,TE}(\omega,k_\perp)\, d\omega, \tag{12.61}$$

where the infinite separation-independent contribution from the arc C_R has been omitted. Note that the integration path $(i\epsilon, i\infty)$ contains semicircles of infinitely small radius ϵ about the singular points $i\xi_l$ (poles) of the function $\coth(\hbar\omega/2k_B T)$. The analogous formula for the integral along L_2 is

$$-\int_{L_2} \coth\frac{\hbar\omega}{2k_B T} \ln\Delta^{TM,TE}(\omega,k_\perp)\, d\omega = \int_{-i\infty}^{-i\epsilon} \coth\frac{\hbar\omega}{2k_B T} \ln\Delta^{TM,TE}(\omega,k_\perp)\, d\omega. \tag{12.62}$$

As a result, the Casimir free energy without the contribution from the complete semicircle C_R is given by

$$\mathcal{F}_0(a,T) = \frac{\hbar S}{8\pi^2 i} \int_0^\infty k_\perp dk_\perp \int_{-i\infty}^{i\infty} \coth\frac{\hbar\omega}{2k_B T} \left[\ln\Delta^{TM}(\omega,k_\perp) + \ln\Delta^{TE}(\omega,k_\perp)\right] d\omega, \tag{12.63}$$

where the integration with respect to ω runs along the segments of the imaginary axis between the Matsubara frequencies ξ_l and around infinitely small semicircles with centers at ξ_l [see Fig. 7.1(b)].

The expression (12.63) is still infinite. To remove the divergence, we use the renormalization condition that the free energy should be equal to zero in the case of infinitely separated interacting bodies. Thus, the finite Casimir free energy per unit area is defined by eqn (7.68). This definition is equivalent to the replacement of $\Delta^{\mathrm{TM,TE}}(\omega, k_\perp)$ with the ratio $\Delta^{\mathrm{TM,TE}}(\omega, k_\perp)/\Delta_\infty^{\mathrm{TM,TE}}(\omega, k_\perp)$, where $\Delta_\infty^{\mathrm{TM,TE}}(\omega, k_\perp)$ is defined in eqn (12.28), with $i\xi = \omega$. The result is

$$\mathcal{F}(a,T) = \frac{\hbar}{8\pi^2 i} \int_0^\infty k_\perp dk_\perp \int_{-i\infty}^{i\infty} d\omega \coth \frac{\hbar\omega}{2k_B T} \left\{ \ln\left[1 - r_{\mathrm{TM}}^2(\omega, k_\perp) e^{-2aq}\right] \right.$$
$$\left. + \ln\left[1 - r_{\mathrm{TE}}^2(\omega, k_\perp) e^{-2aq}\right] \right\}, \tag{12.64}$$

where the reflection coefficients are defined in eqn (12.31) with $i\xi = \omega$ and q is defined in (12.16).

The integration in eqn (12.64), involving poles at the Matsubara frequencies $i\xi_l$, leads to

$$\mathcal{F}(a,T) = \frac{\hbar}{8\pi^2 i} \int_0^\infty k_\perp dk_\perp \int_{-\infty}^{\infty} d\xi \cot \frac{\hbar\xi}{2k_B T} \left\{ \ln\left[1 - r_{\mathrm{TM}}^2(i\xi, k_\perp) e^{-2aq}\right] \right.$$
$$\left. + \ln\left[1 - r_{\mathrm{TE}}^2(i\xi, k_\perp) e^{-2aq}\right] \right\}$$
$$+ \frac{\hbar}{8\pi} \int_0^\infty k_\perp dk_\perp \sum_{l=-\infty}^{\infty} \operatorname{res} \left\{ \coth \frac{\hbar\omega}{2k_B T} \left[\ln\left(1 - r_{\mathrm{TM}}^2(\omega, k_\perp) e^{-2aq}\right)\right.\right.$$
$$\left.\left. + \ln\left(1 - r_{\mathrm{TE}}^2(\omega, k_\perp) e^{-2aq}\right)\right] ; i\xi_l \right\}. \tag{12.65}$$

In the case of real photon eigenfrequencies under consideration here, r_{TM} and r_{TE} are even functions of both ω and ξ. As a consequence, the seemingly pure imaginary integral on the right-hand side of eqn (12.65) vanishes. Using the evenness of the functions r_{TM} and r_{TE} with respect to ξ and calculating the residues, we obtain

$$\mathcal{F}(a,T) = \frac{k_B T}{2\pi} \sideset{}{'}\sum_{l=0}^{\infty} \int_0^\infty k_\perp dk_\perp \left\{ \ln\left[1 - r_{\mathrm{TM}}^2(i\xi_l, k_\perp) e^{-2aq_l}\right] \right.$$
$$\left. + \ln\left[1 - r_{\mathrm{TE}}^2(i\xi_l, k_\perp) e^{-2aq_l}\right] \right\}. \tag{12.66}$$

The explicit expressions for the reflection coefficients computed at the Matsubara frequencies (12.60) are

$$r_{\mathrm{TM}}(i\xi_l, k_\perp) = \frac{\varepsilon_l q_l - k_l}{\varepsilon_l q_l + k_l}, \qquad r_{\mathrm{TE}}(i\xi_l, k_\perp) = \frac{q_l - k_l}{q_l + k_l}, \tag{12.67}$$

where

$$q_l^2 = k_\perp^2 + \frac{\xi_l^2}{c^2}, \qquad k_l^2 = k_\perp^2 + \varepsilon_l \frac{\xi_l^2}{c^2}, \qquad \varepsilon_l = \varepsilon(i\xi_l). \tag{12.68}$$

It should be noted that eqn (12.66) for the Casimir free energy can be obtained from eqn (12.30) for the energy at $T = 0$ (and vice versa) by the formal substitution

$$\frac{\hbar}{2\pi} \int_0^\infty d\xi \longleftrightarrow k_B T \sum_{l=0}^\infty {}' \tag{12.69}$$

and the replacement of the continuous frequencies ξ with the discrete Matsubara frequencies ξ_l. With the same replacement, the Casimir pressure at zero temperature (12.33) can be transformed into the Casimir pressure at nonzero temperature following from eqn (12.66),

$$P(a,T) = -\frac{k_B T}{\pi} \sum_{l=0}^\infty {}' \int_0^\infty q_l k_\perp dk_\perp \left\{ \left[r_{\text{TM}}^{-2}(i\xi_l, k_\perp) e^{2aq_l} - 1 \right]^{-1} \right.$$
$$\left. + \left[r_{\text{TE}}^{-2}(i\xi_l, k_\perp) e^{2aq_l} - 1 \right]^{-1} \right\}. \tag{12.70}$$

We note, however, that the substitution (12.69) is somewhat formal because, for real materials, the dielectric properties may be different at $T = 0$ and $T \neq 0$.

In the same way as in Section 12.1.1, the Lifshitz formulas (12.66) and (12.70) derived in the case of real photon eigenfrequencies can be generalized to the case of complex eigenfrequencies with small imaginary parts. Similarly to the case of zero temperature, for complex eigenfrequencies the free energy is not given by eqn (12.58), but the auxiliary electrodynamic problem results in eqn (12.66), which becomes applicable for both real and complex dielectric permittivities of the form (12.8). For this to be applicable in thermal equilibrium, net heat losses should be absent on average (Barash and Ginzburg 1975, Ginzburg 1989). In other words, the allowable dissipative processes in dielectrics lead to a nonzero imaginary part for the photon eigenfrequencies, but each act of absorption must be followed by a corresponding emission. This condition is explained in more detail in Section 14.3.4.

The above derivation can be easily generalized to the case of arbitrary stratified media considered in Section 12.2. In particular, the Casimir free energy per unit area and the Casimir pressure for two dissimilar semispaces are given by eqns (12.44) and (12.46), where the replacement (12.69) is made:

$$\mathcal{F}(a,T) = \frac{k_B T}{2\pi} \sum_{l=0}^\infty {}' \int_0^\infty k_\perp dk_\perp \left\{ \ln\left[1 - r_{\text{TM}}^{(1)}(i\xi_l, k_\perp) r_{\text{TM}}^{(2)}(i\xi_l, k_\perp) e^{-2aq_l}\right] \right.$$
$$\left. + \ln\left[1 - r_{\text{TE}}^{(1)}(i\xi_l, k_\perp) r_{\text{TE}}^{(2)}(i\xi_l, k_\perp) e^{-2aq_l}\right] \right\}, \tag{12.71}$$

$$P(a,T) = -\frac{k_B T}{\pi} \sum_{l=0}^\infty {}' \int_0^\infty q_l k_\perp dk_\perp \left\{ \left[\frac{e^{2aq_l}}{r_{\text{TM}}^{(1)}(i\xi_l, k_\perp) r_{\text{TM}}^{(2)}(i\xi_l, k_\perp)} - 1 \right]^{-1} \right.$$

$$+ \left[\frac{e^{2aq_l}}{r_{\text{TE}}^{(1)}(i\xi_l, k_\perp) r_{\text{TE}}^{(2)}(i\xi_l, k_\perp)} - 1 \right]^{-1} \Bigg\}.$$

The Casimir free energy and pressure at nonzero temperature for the three-layer system enclosed between two semispaces considered in Section 12.2 are obtained from eqns (12.47) and (12.51) with the same replacement. The use in eqns (12.66) and (12.70) of the reflection coefficients (12.52), where $\varepsilon(i\xi)$, q, and k are replaced with ε_l, q_l, and k_l, allows one to calculate the Casimir free energy per unit area and the pressure at nonzero temperature in the configuration of two dielectric plates of finite thickness. Computational results obtained using eqns (12.66) and (12.70) for some specific materials are presented in Section 12.6.2. Note also that by using the proximity force approximation (Section 6.5), one can convert computational results for the free energy (12.66) into a force acting between a plate and a sphere of radius $R \gg a$. For this purpose, the values of $\mathcal{F}(a, T)$ must be multiplied by $2\pi R$. This will be used in Part III of the book for the comparison between experiment and theory.

12.3.2 Representation in terms of real frequencies

Here, we present another form of the Lifshitz formulas for the Casimir free energy per unit area and the Casimir pressure at nonzero temperature, using integration over real frequencies rather than a summation over the Matsubara frequencies. To do so we start from eqn (12.64), where the integration path of the integral with respect to ω consists of segments of the imaginary frequency axis between the imaginary Matsubara frequencies $i\xi_l$, and semicircles centered on $i\xi_l$. Using the evenness of the functions r_{TM}, r_{TE}, and q with respect to ω, we can rearrange eqn (12.64) as

$$\mathcal{F}(a, T) = -\frac{\hbar i}{4\pi^2} \int_0^\infty k_\perp dk_\perp \int_0^{i\infty} d\omega \coth \frac{\hbar \omega}{2 k_B T} \left\{ \ln \left[1 - r_{\text{TM}}^2(\omega, k_\perp) e^{-2aq} \right] \right.$$
$$\left. + \ln \left[1 - r_{\text{TE}}^2(\omega, k_\perp) e^{-2aq} \right] \right\}. \quad (12.72)$$

Similarly to Section 12.1.2, we now consider a function $-if(\omega)$ such that its integral along the imaginary frequency axis is real. The function $-if(\omega)$ is supposed to be analytic inside a contour C_q which consists of the positive half $(0, \infty)$ of the real axis, a 90° arc of infinitely large radius, and the half $(i\infty, 0)$ of the imaginary axis, with infinitely small semicircles centered on the imaginary Matsubara frequencies. For the renormalized quantities, the infinitely remote arc does not contribute, and thus we assume the same property for the function $-if(\omega)$. Then the application of the Cauchy theorem results in

$$\text{Im} \int_0^\infty f(\omega) \, d\omega = \text{Re} \left[-i \int_0^{i\infty} f(\omega) \, d\omega \right] = -i \int_0^{i\infty} f(\omega) \, d\omega. \quad (12.73)$$

The application of the equality (12.73) to eqn (12.72) leads to

$$\mathcal{F}(a,T) = \frac{\hbar}{4\pi^2}\int_0^\infty k_\perp dk_\perp \int_0^\infty d\omega \coth\frac{\hbar\omega}{2k_B T}\mathrm{Im}\left\{\ln\left[1-r_{\mathrm{TM}}^2(\omega,k_\perp)e^{-2aq}\right]\right.$$
$$\left.+\ln\left[1-r_{\mathrm{TE}}^2(\omega,k_\perp)e^{-2aq}\right]\right\}. \quad (12.74)$$

This is the Lifshitz formula for the Casimir free energy along the real frequency axis. In a similar way, the Casimir pressure (12.70) can be represented in the form

$$P(a,T) = -\frac{\hbar}{2\pi^2}\int_0^\infty k_\perp dk_\perp \int_0^\infty d\omega \coth\frac{\hbar\omega}{2k_B T}\mathrm{Im}\left\{q\left[r_{\mathrm{TM}}^{-2}(\omega,k_\perp)e^{2aq}-1\right]^{-1}\right.$$
$$\left.+q\left[r_{\mathrm{TE}}^{-2}(\omega,k_\perp)e^{2aq}-1\right]^{-1}\right\}. \quad (12.75)$$

In the limit $T \to 0$, eqns (12.74) and (12.75) coincide with the respective equations (12.37) and (12.38) obtained for the case of zero temperature. The contributions of propagating waves to eqns (12.74) and (12.75) are obtained for $\omega > ck_\perp$ (pure imaginary q), and the contributions of evanescent waves for $\omega \leq ck_\perp$ (real q).

12.4 Correlation of energy and free energy

At nonzero temperature T, the physical quantities describing the Casimir effect are determined by both the zero-point oscillations of the electromagnetic field and thermal photons. For real physical bodies, however, there may be a nontrivial interplay between the influences of the virtual and thermal photons. Here, we discuss this problem in a general form applicable not only to dielectrics but also to metals. We define the important concept of the thermal correction in relation to real materials.

The Casimir free energy per unit area (12.66) and the Casimir pressure (12.70) are conventionally represented in the form of the two contributions

$$\mathcal{F}(a,T) = \mathcal{E}(a,T) + \Delta\mathcal{F}(a,T),$$
$$P(a,T) = \mathcal{P}(a,T) + \Delta P(a,T), \quad (12.76)$$

using the Abel–Plana formula (2.26) (Bordag et al. 2000a). The first terms on the right-hand sides of eqn (12.76), $\mathcal{E}(a,T)$ and $\mathcal{P}(a,T)$, have the same form as the right-hand sides of eqns (12.30) and (12.33) for the Casimir energy and pressure at $T=0$, $E(a)$ and $P(a)$, respectively. The only difference is that in $E(a)$ and $P(a)$, the dielectric permittivity $\varepsilon(\omega)$ at $T=0$ is used in the reflection coefficients, whereas in $\mathcal{E}(a,T)$ and $\mathcal{P}(a,T)$ the dielectric permittivity $\varepsilon = \varepsilon(\omega,T)$ is substituted. The second terms on the right-hand sides of eqn (12.76), with the Abel–Plana formula, lead to

$$\Delta\mathcal{F}(a,T) = \frac{ik_B T}{2\pi}\int_0^\infty dt\,\frac{F(i\xi_1 t) - F(-i\xi_1 t)}{e^{2\pi t}-1}, \quad (12.77)$$

$$\Delta P(a,T) = -\frac{\mathrm{i}k_B T}{\pi}\int_0^\infty dt\,\frac{\Phi(\mathrm{i}\xi_1 t)-\Phi(-\mathrm{i}\xi_1 t)}{e^{2\pi t}-1},$$

where the functions $F(x)$ and $\Phi(x)$ are defined as follows:

$$F(x)=\int_0^\infty k_\perp dk_\perp \left\{\ln\left[1-r_{\mathrm{TM}}^2(\mathrm{i}x,k_\perp)e^{-2aq}\right]+\ln\left[1-r_{\mathrm{TE}}^2(\mathrm{i}x,k_\perp)e^{-2aq}\right]\right\},$$

$$\Phi(x)=\int_0^\infty k_\perp dk_\perp\, q\left\{\left[r_{\mathrm{TM}}^{-2}(\mathrm{i}x,k_\perp)e^{2aq}-1\right]^{-1}+\left[r_{\mathrm{TE}}^{-2}(\mathrm{i}x,k_\perp)e^{2aq}-1\right]^{-1}\right\},$$

$$q\equiv q(\mathrm{i}x,k_\perp)=\sqrt{k_\perp^2+\frac{x^2}{c^2}}. \qquad (12.78)$$

These functions are also calculated using the dielectric permittivity $\varepsilon(\omega,T)$.

Sometimes, in the literature, the argument T in the quantities $\mathcal{E}(a,T)$ and $\mathcal{P}(a,T)$ is omitted and they are identified with the Casimir energy per unit area, $E(a)$, and the Casimir pressure, $P(a)$, respectively, at zero temperature, and the quantities $\Delta\mathcal{F}(a,T)$ and $\Delta P(a,T)$ are called the *thermal corrections*. This is, however, not always justified. Physically, a thermal correction is the difference between the value of some quantity at a temperature T and at $T=0$. Because of this, the thermal corrections to the zero-temperature Casimir energy per unit area, $E(a)$, and to the Casimir pressure, $P(a)$, should be rigorously defined as

$$\Delta_T \mathcal{F}(a,T) = \mathcal{F}(a,T) - E(a), \qquad \Delta_T P(a,T) = P(a,T) - P(a). \qquad (12.79)$$

Importantly, for some materials

$$\Delta_T \mathcal{F}(a,T) \neq \Delta \mathcal{F}(a,T), \qquad \Delta_T P(a,T) \neq \Delta P(a,T), \qquad (12.80)$$

where $\Delta\mathcal{F}(a,T)$ and $\Delta P(a,T)$ are defined in eqns (12.76) and (12.77). This is the case for all materials whose permittivity depends on the temperature, i.e. $\varepsilon=\varepsilon(\omega,T)$. As a result, the quantities $\mathcal{E}(a,T)$ and $\mathcal{P}(a,T)$ in eqn (12.76) do not have a clearly defined physical meaning at any nonzero temperature. At zero temperature, the equalities $\mathcal{E}(a,0)=E(a)$ and $\mathcal{F}(a,0)=E(a)$ are valid. However, at $T>0$ the quantity $\mathcal{E}(a,T)$ is not equal to the Casimir energy at nonzero temperature as defined in thermodynamics,

$$E(a,T) = -T^2\frac{\partial}{\partial T}\frac{\mathcal{F}(a,T)}{T} \qquad (12.81)$$

[compare with eqn (5.32) for the energy from black-body radiation]. By using the Lifshitz formula (12.66), the Casimir energy at nonzero temperature can be expressed as

$$E(a,T) = \frac{k_B T^2}{2\pi}\int_0^\infty k_\perp dk_\perp \sum_{l=0}^\infty{}' \left(\frac{2\pi k_B l}{\hbar}\frac{\partial}{\partial \xi_l}+\frac{\partial\varepsilon}{\partial T}\frac{\partial}{\partial\varepsilon}\right) \qquad (12.82)$$
$$\times\left\{\ln\left[1-r_{\mathrm{TM}}^2(\mathrm{i}\xi_l,k_\perp)e^{-2aq_l}\right]+\ln\left[1-r_{\mathrm{TE}}^2(\mathrm{i}\xi_l,k_\perp)e^{-2aq_l}\right]\right\}.$$

If ε is temperature-independent, i.e. $\partial\varepsilon/\partial T=0$, the differentiation with respect to ε does not contribute to $E(a,T)$. In this case only, $\mathcal{E}$ does not depend on T and is equal to the Casimir energy per unit area at $T = 0$: $\mathcal{E} = \mathcal{E}(a) = E(a)$. For a temperature-independent ε, $\mathcal{P} = \mathcal{P}(a) = P(a)$ is also valid. For the case when $\partial \varepsilon / \partial T = 0$, eqn (12.82) allows an obvious interpretation as the combined energy due to the zero-point and thermal photons. To confirm that this is the case, we consider the additive sum of the photon eigenfrequencies between the semispaces and the Planck photons (Barash 1988),

$$E_0(a,T) = \hbar \int_0^\infty \frac{k_\perp dk_\perp}{2\pi} \sum_n \left[\omega^{TM}_{k_\perp,n} \left(\frac{1}{2} + \frac{1}{e^{\hbar \omega^{TM}_{k_\perp,n}/(k_B T)} - 1} \right) \right.$$
$$\left. + \omega^{TE}_{k_\perp,n} \left(\frac{1}{2} + \frac{1}{e^{\hbar \omega^{TE}_{k_\perp,n}/(k_B T)} - 1} \right) \right] S. \qquad (12.83)$$

This can be rearranged to give

$$E_0(a,T) = \frac{\hbar}{2} \int_0^\infty \frac{k_\perp dk_\perp}{2\pi} \sum_n \left(\omega^{TM}_{k_\perp,n} \coth \frac{\hbar \omega^{TM}_{k_\perp,n}}{2k_B T} + \omega^{TE}_{k_\perp,n} \coth \frac{\hbar \omega^{TE}_{k_\perp,n}}{2k_B T} \right). \qquad (12.84)$$

The sum in eqn (12.84) can be calculated using the argument principle (7.60). The integration path in the complex plane of ω consists of a semicircle of infinite radius in the right half-plane and the segments of the imaginary frequency axis between the imaginary Matsubara frequencies, with infinitely small semicircles centered at $i\xi_l$. By calculating the residuals of the function $\omega \coth[\hbar\omega/(2k_B T)]$ at the points $\omega_l = i\xi_l$ and subtracting the energy of infinitely separated semispaces, we arrive at a finite Casimir energy per unit area at nonzero temperature, $E(a,T)$, as given in eqn (12.82) with $\partial \varepsilon / \partial T = 0$ [see Bezerra et al. (2002b) for details]. This quantity depends on T through the Matsubara frequencies only.

Thus, in the case of a temperature-independent dielectric permittivity, the effects due to zero-point and thermal photons can be considered as additive. This is, however, not so for real materials with dielectric properties which depend on the temperature. For such materials, a complicated interplay between the effects of zero-point and thermal photons can be expected. Nevertheless, a representation of the Casimir free energy per unit area and the Casimir pressure in the form of eqn (12.76) may be formally useful, although the quantities $\Delta \mathcal{F}$ and $\Delta \mathcal{P}$ have no physical meaning with regard to the thermal corrections in this case.

12.5 Asymptotic properties of the Lifshitz formula at low and high temperature

It is useful to investigate the behavior of the Casimir free energy per unit area and the Casimir pressure for dielectrics, as given by the Lifshitz formulas (12.66) and (12.70), at both low and high temperature. This allows one to find the Casimir entropy and to check the consistency of the Lifshitz theory with thermodynamics, for various models of the dielectric response.

12.5.1 Finite static dielectric permittivity

As explained in Section 7.4, the regions of low and high temperature are defined with respect to the *effective temperature*, which, in the usual units, takes the form

$$k_B T_{\text{eff}} = \frac{\hbar c}{2a}. \tag{12.85}$$

From eqn (12.85), it becomes clear that the regions of low and high temperature ($T \ll T_{\text{eff}}$ and $T \gg T_{\text{eff}}$, respectively) can be understood as the regions of small and large separations between the plates. This makes the asymptotic properties of the Lifshitz formula experimentally important. The theoretical meaningfulness of the asymptotic properties is connected with requirements on the behavior of the Casimir free energy at low temperature imposed by the Nernst heat theorem, and with the validity of the classical limit at high temperature. As we shall see below, thermodynamic constraints serve as a useful guide for clarifying the question of what properties of real materials should be included in the model of the dielectric response in the Lifshitz theory.

We start with the Casimir free energy per unit area (12.66) in a configuration of two dielectric semispaces. The dielectric permittivity along the imaginary frequency axis is obtained from eqn (12.8),

$$\varepsilon_l = \varepsilon(i\xi_l) = 1 + \sum_{j=1}^{K} \frac{g_j}{\omega_j^2 + \xi_l^2 + \gamma_j \xi_l}. \tag{12.86}$$

Note that in the case of dielectrics, both of the reflection coefficients $r_{\text{TM}}(i\xi, k_\perp)$ and $r_{\text{TE}}(i\xi, k_\perp)$ in eqn (12.31) are discontinuous at the point $(0,0)$ when considered as functions of the two variables ξ and $k_\perp$. This is because their limiting values at the point $(0,0)$ depend on the path chosen to approach $(0,0)$ in the $(\xi, k_\perp)$ plane. Thus, if one approaches the point $(0,0)$ along a straight line $k_\perp = 0$ (the case of normal incidence) the limiting values of the reflection coefficients (12.31),

$$r_{\text{TM}}(0,0) = \frac{\sqrt{\varepsilon_0} - 1}{\sqrt{\varepsilon_0} + 1}, \qquad r_{\text{TE}}(0,0) = -\frac{\sqrt{\varepsilon_0} - 1}{\sqrt{\varepsilon_0} + 1}, \tag{12.87}$$

are equal to the reflection coefficients of physical electromagnetic waves on a dielectric surface (Landau et al. 1984). If, however, one approaches the point $(0,0)$ along a line $\xi = 0$, the limiting values of the reflection coefficients are

$$r_{\text{TM}}(0,0) = \frac{\varepsilon_0 - 1}{\varepsilon_0 + 1}, \qquad r_{\text{TE}}(0,0) = 0. \tag{12.88}$$

The latter result follows from eqn (12.31) and describes the zero-frequency contribution in the Lifshitz formula.

To find the low-temperature behavior of the free energy, we exploit the representation (12.76). Fortunately, the oscillator parameters are almost independent of the temperature for most dielectrics, in which case $\mathcal{E}(a,T) \approx E(a)$ and $\mathcal{P}(a,T) \approx P(a)$. In terms of the dimensionless variables

$$y = 2q_l a, \qquad \zeta_l = \frac{\xi_l}{\omega_c} = \tau l, \qquad \tau = 4\pi \frac{k_B a T}{\hbar c} = 2\pi \frac{T}{T_{\text{eff}}} \tag{12.89}$$

[recall that $\omega_c = c/(2a)$ was introduced in Section 12.1.3], the quantities $E(a)$ and $\Delta\mathcal{F}(a,T)$ defined in eqns (12.30) and (12.77) take the form

$$E(a) = \frac{\hbar c}{32\pi^2 a^3} \int_0^\infty d\zeta \int_\zeta^\infty dy\, f(\zeta, y),$$

$$f(\zeta, y) = y \left\{ \ln\left[1 - r_{\text{TM}}^2(i\zeta, y) e^{-y}\right] + \ln\left[1 - r_{\text{TE}}^2(i\zeta, y) e^{-y}\right] \right\},$$

$$\Delta\mathcal{F}(a,T) = \frac{i\hbar c\tau}{32\pi^2 a^3} \int_0^\infty dt\, \frac{F(it\tau) - F(-it\tau)}{e^{2\pi t} - 1},$$

$$F(x) \equiv \int_x^\infty dy\, f(x, y). \tag{12.90}$$

The reflection coefficients (12.31), expressed in terms of the same variables, are

$$r_{\text{TM}}(i\zeta, y) = \frac{\varepsilon y - \sqrt{y^2 + \zeta^2(\varepsilon - 1)}}{\varepsilon y + \sqrt{y^2 + \zeta^2(\varepsilon - 1)}}, \tag{12.91}$$

$$r_{\text{TE}}(i\zeta, y) = \frac{y - \sqrt{y^2 + \zeta^2(\varepsilon - 1)}}{y + \sqrt{y^2 + \zeta^2(\varepsilon - 1)}}, \qquad \varepsilon \equiv \varepsilon(i\omega_c \zeta).$$

In this representation, they depend on the separation through the characteristic frequency ω_c. The dielectric permittivity (12.86), as a function of the continuous dimensionless variables, is given by

$$\varepsilon(i\omega_c \zeta) = 1 + \sum_{j=1}^K \frac{\tilde{g}_j}{1 + \alpha_j \zeta^2 + \beta_j \zeta}, \tag{12.92}$$

where

$$\tilde{g}_j = \frac{g_j}{\omega_j^2}, \qquad \alpha_j = \alpha_j(a) = \frac{\omega_c^2}{\omega_j^2}, \qquad \beta_j = \beta_j(a) = \frac{\omega_c \gamma_j}{\omega_j^2}. \tag{12.93}$$

Now we substitute $\varepsilon(i\omega_c \zeta)$ from eqn (12.92) into the reflection coefficients (12.91). We are seeking the asymptotic behavior of $\Delta\mathcal{F}(a,T)$ at low temperature or, equivalently, at $\tau \ll 1$. To find this, we expand the function $f(x,y)$ defined in eqn (12.90) in powers of $x = t\tau$. The subsequent integration of this expansion with respect to y from x to infinity results in

$$F(\mathrm{i}x) - F(-\mathrm{i}x) = \frac{8\mathrm{i}bx}{\varepsilon_0^2 - 1}\mathrm{Li}_2(r_0^2) + \frac{\mathrm{i}\pi}{2}r_0^2(\varepsilon_0 + 1)x^2 - 240\mathrm{i}C_4 x^3 + O(x^4). \quad (12.94)$$

Here, $\mathrm{Li}_n(z)$ is the polylogarithm function, and the following notation has been used:

$$b = b(a) = \sum_{j=1}^{K} \tilde{g}_j \beta_j(a), \qquad r_0 = \frac{\varepsilon_0 - 1}{\varepsilon_0 + 1}. \quad (12.95)$$

We note that $\varepsilon_0 = \varepsilon(0)$ and that the dependence of b on a comes from the definition of the characteristic frequency ω_c. Substituting eqn (12.94) into eqn (12.90) and integrating with respect to t from zero to infinity, we obtain (Klimchitskaya and Geyer 2008)

$$\mathcal{F}(a,T) = E(a) - \frac{\hbar c}{32\pi^2 a^3}\left[\frac{b(a)\mathrm{Li}_2(r_0^2)}{3(\varepsilon_0^2 - 1)}\tau^2 + \frac{\zeta_R(3)r_0^2(\varepsilon_0 + 1)}{8\pi^2}\tau^3 - C_4\tau^4 + O(\tau^5)\right]. \quad (12.96)$$

Note that the expression (12.96) was obtained by Geyer et al. (2005b) and Klimchitskaya et al. (2006c) for a simplified model of a dielectric with $\gamma_j = 0$ and, consequently, $b(a) = 0$. In that model the perturbative expansion starts from the term of order τ^3, which does not contribute to the Casimir pressure. The coefficient of τ^3 in the simplified model is the same as for the complete dielectric permittivity (12.8). In the simplified model, the coefficient C_4 is given by

$$C_4 = \frac{1}{720}(\sqrt{\varepsilon_0} - 1)(\varepsilon_0^2 + \varepsilon_0\sqrt{\varepsilon_0} - 2). \quad (12.97)$$

Now we find the first terms in the asymptotic expansion at $\tau \ll 1$ for the Casimir pressure. In terms of dimensionless variables, the quantity $\Delta P(a,T)$ defined in eqns (12.77) and (12.78) takes the form

$$\Delta P(a,T) = -\frac{\mathrm{i}\hbar c\tau}{32\pi^2 a^4}\int_0^\infty dt\,\frac{\Phi(\mathrm{i}t\tau) - \Phi(-\mathrm{i}t\tau)}{e^{2\pi t} - 1}, \quad (12.98)$$

and the function $\Phi(x) \equiv \Phi_{\mathrm{TM}}(x) + \Phi_{\mathrm{TE}}(x)$ is given by

$$\Phi_\lambda(x) = \int_x^\infty \frac{y^2\,dy\,r_\lambda^2(\mathrm{i}x,y)}{e^y - r_\lambda^2(\mathrm{i}x,y)}, \quad (12.99)$$

where $\lambda = \mathrm{TM}$ or TE.

First, let us determine the leading term of the expansion of $\Phi_{\mathrm{TE}}(x)$ in powers of x. By introducing a new variable $v = y/x$, we arrive at

$$\Phi_{\mathrm{TE}}(x) = x^3\int_1^\infty \frac{v^2\,dv\,r_{\mathrm{TE}}^2(\mathrm{i}x,v)}{e^{vx} - r_{\mathrm{TE}}^2(\mathrm{i}x,v)}. \quad (12.100)$$

Note that the reflection coefficient $r_{TE}(ix, v)$ depends on x only through the frequency dependence of ε according to eqn (12.92). Expanding in powers of x in eqn (12.100), we obtain

$$\Phi_{TE}(ix) - \Phi_{TE}(-ix) = -i\left[1 - \frac{\sqrt{\varepsilon_0}}{2}(3 - \varepsilon_0)\right]\frac{x^3}{2} + O(x^5). \quad (12.101)$$

The expansion of $\Phi_{TM}(x)$ from eqn (12.99) in powers of x is somewhat more cumbersome. Following a procedure presented by Geyer et al. (2005b), we arrive at

$$\Phi_{TM}(ix) - \Phi_{TM}(-ix) = -i\frac{32b(a)x}{\varepsilon_0^2 - 1}\text{Li}_2(r_0^2) - i\left\{1 + \frac{\sqrt{\varepsilon_0}}{2}(2\varepsilon_0^2 - 3\varepsilon_0 - 1)\right.$$
$$\left. + \frac{3b(a)}{2(\varepsilon_0^2 - 1)}\left[\frac{(\varepsilon_0 - 1)^2}{\varepsilon_0^2}(1 + 5\varepsilon_0 + 2\varepsilon_0^2) + \frac{32b(a)}{\varepsilon_0 - 1}\frac{\ln(1 - r_0^2)}{r_0}\right]\right\}\frac{x^3}{3}$$
$$+ O(x^5). \quad (12.102)$$

Adding eqns (12.101) and (12.102), substituting the result obtained into eqn (12.98), and integrating with respect to t, we find the asymptotic expression for the Casimir pressure at low temperature as

$$P(a,T) = P(a) - \frac{\hbar c}{32\pi^2 a^4}\left\{\frac{2b(a)\text{Li}_2(r_0^2)}{3(\varepsilon_0^2 - 1)}\tau^2 \right. \quad (12.103)$$
$$+ \frac{\tau^4}{240}\left[\frac{1}{3}(\sqrt{\varepsilon_0} - 1)(\varepsilon_0^2 + \varepsilon_0\sqrt{\varepsilon_0} - 2) + \frac{b(a)r_0}{2\varepsilon_0^2}(1 + 5\varepsilon_0 + 2\varepsilon_0^2)\right.$$
$$\left.\left. + \frac{16b^2(a)}{(\varepsilon_0 - 1)^3}\ln(1 - r_0^2)\right] + O(\tau^5)\right\}.$$

Note that the term of order τ^2 can be also obtained by differentiation of the term of order τ^2 in the free energy (12.96), taking into account the fact that $b = b(a)$ (recall that $b(a) \sim 1/a$ and $\tau \sim a$).

Now we are in a position to calculate the Casimir entropy and to perform a thermodynamic test of the Lifshitz theory. Using the definition of the entropy (5.4), we get

$$S(a,T) = \frac{k_B\tau}{8\pi a^2}\left[\frac{2b(a)\text{Li}_2(r_0^2)}{3(\varepsilon_0^2 - 1)} + \frac{3\zeta_R(3)r_0^2(\varepsilon_0 + 1)}{8\pi^2}\tau - 4C_4\tau^2 + O(\tau^3)\right]. \quad (12.104)$$

As is seen from eqn (12.104), the Casimir entropy goes to zero when T goes to zero (note that $\tau \sim T$), i.e. the Nernst heat theorem is satisfied when the dielectric permittivity is given by eqn (12.8) with a finite value of ε_0 defined in eqn (12.9). In this limiting case, the entropy depends on $\hbar$ through the definition of τ in eqn (12.89).

The results obtained for the Casimir free energy, pressure, and entropy can be generalized to the case of two dissimilar semispaces described by dielectric permittivities $\varepsilon^{(1)}(\omega)$ and $\varepsilon^{(2)}(\omega)$, each having the form of eqn (12.8) but with different values of the oscillator parameters. The respective static permittivities are $\varepsilon_0^{(1)} = \varepsilon^{(1)}(0)$ and $\varepsilon_0^{(2)} = \varepsilon^{(2)}(0)$. Restricting ourselves to the lower perturbation orders, the generalized expressions are the following:

$$\Delta \mathcal{F}(a,T) = -\frac{\hbar c}{32\pi^2 a^3} \left\{ \frac{\mathrm{Li}_2(r_0^{(1)} r_0^{(2)})}{6} \left[\frac{b^{(1)}(a)}{(\varepsilon_0^{(1)})^2 - 1} + \frac{b^{(2)}(a)}{(\varepsilon_0^{(2)})^2 - 1} \right] \tau^2 \right.$$
$$\left. + \frac{\zeta_R(3) r_0^{(1)} r_0^{(2)}}{8\pi^2} \frac{\varepsilon_0^{(1)} + \varepsilon_0^{(2)} + 2\varepsilon_0^{(1)} \varepsilon_0^{(2)}}{\varepsilon_0^{(1)} + \varepsilon_0^{(2)}} \tau^3 \right\}, \quad (12.105)$$

$$\Delta P(a,T) = -\frac{\hbar c}{32\pi^2 a^4} \frac{\mathrm{Li}_2(r_0^{(1)} r_0^{(2)})}{3} \left[\frac{b^{(1)}(a)}{(\varepsilon_0^{(1)})^2 - 1} + \frac{b^{(2)}(a)}{(\varepsilon_0^{(2)})^2 - 1} \right] \tau^2,$$

$$S(a,T) = \frac{k_B \tau}{8\pi a^2} \left\{ \frac{\mathrm{Li}_2(r_0^{(1)} r_0^{(2)})}{3} \left[\frac{b^{(1)}(a)}{(\varepsilon_0^{(1)})^2 - 1} + \frac{b^{(2)}(a)}{(\varepsilon_0^{(2)})^2 - 1} \right] \right.$$
$$\left. + \frac{3\zeta_R(3) r_0^{(1)} r_0^{(2)}}{8\pi^2} \frac{\varepsilon_0^{(1)} + \varepsilon_0^{(2)} + 2\varepsilon_0^{(1)} \varepsilon_0^{(2)}}{\varepsilon_0^{(1)} + \varepsilon_0^{(2)}} \tau \right\}.$$

Here,

$$r_0^{(n)} = \frac{\varepsilon_0^{(n)} - 1}{\varepsilon_0^{(n)} + 1}, \qquad b^{(n)}(a) = \sum_{j=1}^{K} \tilde{g}_j^{(n)} \beta_j^{(n)}(a). \quad (12.106)$$

It is interesting to note that if $\varepsilon_0^{(n)}$ decreases to unity (i.e. when rarefied materials are considered), some of the denominators in the above formulas go to zero. This does not mean, however, that the respective physical quantities, such as the free energy, pressure, or entropy, are larger for more rarefied bodies, because the respective magnitudes of the oscillator parameters decrease more rapidly.

To end this subsection, we consider the case of high temperature ($\tau \gg 1$) or, equivalently, of a large separation distance between the plates. In this case the approximation of a static dielectric permittivity is applicable, and only the zero-frequency term of the Lifshitz formula determines the total result (similarly to Section 7.4.3, where all terms with Matsubara frequencies ξ_l, $l \geq 1$, are exponentially small). At zero frequency, we obtain the following from eqn (12.91) for the reflection coefficients:

$$r_{\mathrm{TM}}^{(n)}(0,y) = r_0^{(n)}, \qquad r_{\mathrm{TE}}^{(n)}(0,y) = 0. \quad (12.107)$$

The Lifshitz formula (12.71), written in terms of the dimensionless variables (12.89), for a configuration of two dissimilar semispaces takes the form

$$\mathcal{F}(a,T) = \frac{k_B T}{8\pi a^2} \sum_{l=0}^{\infty}{}' \int_{\zeta_l}^{\infty} y\, dy \left\{ \ln\left[1 - r_{\mathrm{TM}}^{(1)}(i\zeta_l, y) r_{\mathrm{TM}}^{(2)}(i\zeta_l, y) e^{-y}\right] \right.$$

$$+ \ln\left[1 - r_{\text{TE}}^{(1)}(i\zeta_l, y) r_{\text{TE}}^{(2)}(i\zeta_l, y) e^{-y}\right]\right\}. \tag{12.108}$$

Neglecting all terms with $l \geq 1$, we obtain

$$\mathcal{F}(a, T) = \frac{k_B T}{16\pi a^2} \int_0^\infty y \, dy \ln\left[1 - r_0^{(1)} r_0^{(2)} e^{-y}\right], \tag{12.109}$$

where $r_0^{(n)}$ is defined in eqn (12.106). The integration with respect to y in this equation results in

$$\mathcal{F}(a, T) = -\frac{k_B T}{16\pi a^2} \text{Li}_3[r_0^{(1)} r_0^{(2)}]. \tag{12.110}$$

In a similar manner, at $\tau \gg 1$ the Casimir pressure is given by

$$P(a, T) = -\frac{k_B T}{8\pi a^3} \text{Li}_3[r_0^{(1)} r_0^{(2)}]. \tag{12.111}$$

The respective Casimir entropy,

$$S(a, T) = \frac{k_B}{16\pi a^2} \text{Li}_3[r_0^{(1)} r_0^{(2)}], \tag{12.112}$$

is positive and independent of the temperature, and does not depend on $\hbar$, i.e. the classical limit is achieved (Feinberg et al. 2001).

12.5.2 Static conductivity of the dielectric material and the third law of thermodynamics

As remarked in the introduction to this chapter, at nonzero temperature all dielectrics possess some nonzero conductivity. The value of this conductivity may be different for different types of dielectrics, but in all cases it is several orders of magnitude smaller than the conductivity of metals. In the calculations in the preceding section we have ignored the conductivity properties of dielectric materials at $T \neq 0$ by assuming that the dielectric permittivity is independent of temperature and has a finite value at zero frequency. One might expect that the inclusion of a negligible or relatively small conductivity for a dielectric at nonzero temperature in the model of the dielectric response would not lead to theoretical results significantly different from those obtained neglecting this conductivity. What is more, one would expect that computations using the Lifshitz theory with the conductivity included at $T \neq 0$ would be more exact. However, as we demonstrate below, the results of such computations are simply invalid, as they are in contradiction with thermodynamics.

Now we repeat the calculation of the Casimir free energy and entropy for two dissimilar dielectrics at low temperature, taking into account the conductivity that arises for $T > 0$ at zero frequency. To do so, we must replace the dielectric permittivity $\varepsilon(\omega)$ defined in eqn (12.8) with (Palik 1985)

$$\tilde{\varepsilon}^{(n)}(\omega, T) = \varepsilon^{(n)}(\omega) + i \frac{4\pi \sigma_0^{(n)}(T)}{\omega}, \quad n = 1, 2, \tag{12.113}$$

where $\sigma_0^{(n)}(T)$ is the static conductivity. There is a universal behavior for the conductivity of dielectric materials as a function of temperature. For pure insulators and intrinsic semiconductors, it is given by

$$\sigma_0^{(n)}(T) \sim \exp\left[-\frac{\Delta^{(n)}}{2k_\mathrm{B}T}\right], \qquad (12.114)$$

where Δ is the bandgap. By convention, the material is called an *insulator* if $\Delta \geq 2\text{--}3\,\mathrm{eV}$ and an *intrinsic semiconductor* if $\Delta < 2\text{--}3\,\mathrm{eV}$ (Mott 1990). For Mott–Hubbard dielectrics and doped semiconductors with a doping concentration below the critical value (at sufficiently low T), a similar dependence of the conductivity on the temperature occurs:

$$\sigma_0^{(n)}(T) \sim \exp\left[-\frac{C^{(n)}}{2k_\mathrm{B}T}\right], \qquad (12.115)$$

where $C^{(n)}$ is a parameter (Mott 1990, Shklovskii and Efros 1984).

The dielectric permittivity (12.113) at the imaginary Matsubara frequencies takes the form

$$\tilde{\varepsilon}_l^{(n)} \equiv \tilde{\varepsilon}^{(n)}(i\xi_l, T) = \varepsilon^{(n)}(i\xi_l) + \frac{\beta^{(n)}(T)}{l}, \qquad (12.116)$$

where $\varepsilon_l^{(n)} = \varepsilon^{(n)}(i\xi_l)$ is defined in eqn (12.86) and $\beta^{(n)}(T) = 2\hbar\sigma_0^{(n)}(T)/(k_\mathrm{B}T)$. We note that the parameter β is very small. For example, for SiO_2 at $T = 300\,\mathrm{K}$, $\beta \sim 10^{-12}$ (Shackelford and Alexander 2001). Substituting the dielectric permittivity (12.116) into the reflection coefficients (12.91), we obtain

$$\tilde{r}_{\mathrm{TM}}^{(n)}(0, y) = 1, \qquad \tilde{r}_{\mathrm{TE}}^{(n)}(0, y) = 0. \qquad (12.117)$$

Comparing this with eqn (12.107), we see that the magnitude of $\tilde{r}_{\mathrm{TM}}$ at zero frequency is different when the conductivity is included, regardless of whether it is high or low [for the nonrelativistic van der Waals force, this was noticed by Davies and Ninham (1972)]. In fact, the inclusion of the static conductivity leads to a discontinuity in the transverse-magnetic reflection coefficient at zero frequency. Bearing in mind that the conductivity is nonzero only when $T \neq 0$, we conclude that $\tilde{r}_{\mathrm{TM}}$, as a function of ξ and T, is discontinuous at the point $\xi = 0$, $T = 0$.

The free energy $\tilde{\mathcal{F}}(a, T)$ calculated with the dielectric permittivities $\tilde{\varepsilon}_l^{(n)}$ is expressed by eqn (12.108), where the reflection coefficients denoted with a tilde are used. We separate the term with $l = 0$ from the terms with $l \geq 1$ and subtract and add the term

$$\frac{k_\mathrm{B}T}{16\pi a^2}\int_0^\infty y\,dy\,\ln\left[1 - r_0^{(1)}r_0^{(2)}e^{-y}\right]. \qquad (12.118)$$

As a result, the free energy takes the form

$$\tilde{\mathcal{F}}(a, T) = \frac{k_\mathrm{B}T}{16\pi a^2}\int_0^\infty y\,dy\left\{\ln(1 - e^{-y}) - \ln\left[1 - r_0^{(1)}r_0^{(2)}e^{-y}\right]\right\}$$

$$+\frac{k_{\rm B}T}{16\pi a^2}\int_0^\infty y\,dy\,\ln\left[1-r_0^{(1)}r_0^{(2)}{\rm e}^{-y}\right] \qquad (12.119)$$

$$+\frac{k_{\rm B}T}{16\pi a^2}\sum_{l=1}^\infty\int_{\zeta_l}^\infty y\,dy\,\left\{\ln\left[1-\tilde{r}_{\rm TM}^{(1)}({\rm i}\zeta_l,y)\tilde{r}_{\rm TM}^{(2)}({\rm i}\zeta_l,y){\rm e}^{-y}\right]\right.$$
$$\left.+\ln\left[1-\tilde{r}_{\rm TE}^{(1)}({\rm i}\zeta_l,y)\tilde{r}_{\rm TE}^{(2)}({\rm i}\zeta_l,y){\rm e}^{-y}\right]\right\}.$$

Now we expand the last, third term on the right-hand side of eqn (12.119), containing a summation over l, in powers of the small parameters $\beta^{(n)}/l$. Combining the zero-order contribution in this expansion with the second term on the right-hand side of eqn (12.119), we obtain the free energy $\mathcal{F}(a,T)$ calculated with the dielectric permittivities $\varepsilon_l^{(n)}$, which neglects the conductivity (see Section 12.5.1). Calculating the first integral on the right-hand side of eqn (12.119), we arrive at

$$\tilde{\mathcal{F}}(a,T)=\mathcal{F}(a,T)-\frac{k_{\rm B}T}{16\pi a^2}\left[\zeta_R(3)-{\rm Li}_3\left(r_0^{(1)}r_0^{(2)}\right)\right]+R(a,T). \qquad (12.120)$$

Here, $R(a,T)=O(\beta^{(n)})$ stands for the first and higher-order contributions in the expansion of the third term on the right-hand side of eqn (12.119) in powers of $\beta^{(n)}/l$. It can be proven that both $R(a,T)$ and $\partial R(a,T)/\partial T$ go to zero exponentially fast when T goes to zero [see Geyer et al. (2005b) for details].

Equation (12.120) leads to an important conclusion about the thermodynamic inconsistency of the Lifshitz theory for dielectrics when one includes the dc conductivity in the model of the dielectric response. Substituting eqn (12.120) into eqn (5.4), we obtain the entropy per unit area for the two semispaces described by the dielectric permittivities (12.116),

$$\tilde{S}(a,T)=S(a,T)+\frac{k_{\rm B}}{16\pi a^2}\left[\zeta_R(3)-{\rm Li}_3\left(r_0^{(1)}r_0^{(2)}\right)\right]-\frac{\partial R(a,T)}{\partial T}, \qquad (12.121)$$

where $S(a,T)$, given by eqn (12.105), is the entropy for plates with dielectric permittivities $\varepsilon^{(n)}$.

In the limit of $T\to 0$, we obtain from eqn (12.121)

$$\tilde{S}(a,0)=\frac{k_{\rm B}}{16\pi a^2}\left[\zeta_R(3)-{\rm Li}_3\left(r_0^{(1)}r_0^{(2)}\right)\right]>0. \qquad (12.122)$$

The right-hand side of this equation depends on the parameters of the system under consideration (the separation distance a) and implies a violation of the third law of thermodynamics (the Nernst heat theorem). This conclusion is general for any material that has zero conductivity at zero temperature. Such materials include not only simple insulators, but also intrinsic semiconductors, Mott–Hubbard dielectrics, solids with ionic conductivity, and doped semiconductors with doping concentrations below the critical value. The formal reason for the violation of Nernst's theorem, as was noticed by Intravaia and Henkel (2008), is connected with the discontinuity of the transverse-magnetic reflection coefficient $\tilde{r}_{\rm TM}$ indicated above, as a function of ξ and T at the point $\xi = 0$, $T = 0$. Physically, the disagreement with the Nernst theorem is explained by the drift current which arises in dielectric semispaces described by the dielectric permittivity (12.113) and leads to Joule heating. As explained in detail in Section 14.3.4, this violates thermal equilibrium, which is the basic application condition for the Lifshitz theory.

Thus, in the framework of the Lifshitz theory, one should neglect the dc conductivity of dielectrics at nonzero temperature. Violation of this rule leads to theoretical results significantly different from those obtained with the neglect of conductivity at nonzero temperature. In Section 12.6.3, we compute the Casimir free energy and pressure between specific dielectric materials and show that the inclusion of the dc conductivity would lead to an enormously large, nonphysical thermal correction. Recent experiments by Chen *et al.* (2007a, 2007b) demonstrated that theoretical results obtained by neglecting the dc conductivity of doped semiconductors with a concentration of charge carriers below critical were consistent with measurement data, whereas results obtained by taking the dc conductivity into account were excluded by the data. A detailed analysis of these experiments and a comparison with theory are contained in Section 20.3.

12.6 Computational results for typical dielectrics

In this section, we present computational results for the Casimir free energy per unit area and the Casimir pressure for a configuration of two semispaces at different separations and temperatures, with different models of the dielectric response.

12.6.1 *Dielectric permittivity along the imaginary frequency axis*

To perform computations using the Lifshitz formulas (12.66) and (12.70) with the reflection coefficients (12.67), one needs to know the dielectric permittivity $\varepsilon(i\xi_l)$ at the imaginary Matsubara frequencies. Computations at zero temperature using eqns (12.30) and (12.33) require a knowledge of $\varepsilon(i\xi)$ at all imaginary frequencies from zero to infinity.

The calculation of the dielectric permittivity can be performed using tabulated optical data for the complex index of refraction $n(\omega) = n_1(\omega) + i n_2(\omega)$. The most complete set of such data, for various materials over wide frequency regions, has been provided by Palik (1985). Using the data for $n_1(\omega)$ and $n_2(\omega)$, one can obtain both the real and the imaginary part of the dielectric permittivity:

$$\operatorname{Re} \varepsilon(\omega) = n_1^2(\omega) - n_2^2(\omega), \qquad \operatorname{Im} \varepsilon(\omega) = 2 n_1(\omega) n_2(\omega). \tag{12.123}$$

The permittivity of a dielectric material is characterized by a finite static value of ε_0, defined in eqn (12.9). For dielectric permittivities $\varepsilon(\omega)$ which are regular at $\omega = 0$, the standard Kramers–Kronig relations apply (Landau *et al.* 1984):

$$\mathrm{Re}\,\varepsilon(\omega) = 1 + \frac{1}{\pi}\mathrm{P}\int_{-\infty}^{\infty}\frac{\mathrm{Im}\,\varepsilon(\xi)}{\xi-\omega}\,d\xi, \quad \mathrm{Im}\,\varepsilon(\omega) = -\frac{1}{\pi}\mathrm{P}\int_{-\infty}^{\infty}\frac{\mathrm{Re}\,\varepsilon(\xi)}{\xi-\omega}\,d\xi, \quad (12.124)$$

where the integrals are understood as principal values. From these equations, one can obtain the Kramers–Kronig relation for the dielectric permittivity along the imaginary frequency axis,

$$\varepsilon(i\xi) = 1 + \frac{2}{\pi}\int_0^\infty \frac{\omega\,\mathrm{Im}\,\varepsilon(\omega)}{\omega^2+\xi^2}\,d\omega \qquad (12.125)$$

[here we have used the fact that $\mathrm{Im}\,\varepsilon(-\omega) = -\mathrm{Im}\,\varepsilon(\omega)$].

There are two procedures, which lead to practically the same result, for determining $\varepsilon(i\xi)$ using optical data for the complex index of refraction. One procedure is to numerically integrate the data for $\mathrm{Im}\,\varepsilon(\omega)$, as given by eqn (12.123), in accordance with eqn (12.125). Another procedure is to make a fit between the optical data for $\mathrm{Im}\,\varepsilon(\omega)$ and the oscillator analytical expression following from eqn (12.8),

$$\mathrm{Im}\,\varepsilon(\omega) = \sum_{j=1}^{K}\frac{g_j\gamma_j\omega}{(\omega_j^2-\omega^2)^2+\gamma_j^2\omega^2}. \qquad (12.126)$$

In this fit, the number of oscillators K and the oscillator parameters g_j, ω_j, and γ_j are determined. Then the dielectric permittivity along the imaginary frequency axis is given by the analytic expression (12.86). The use of the imaginary part of ε is preferred over the real part for the oscillator fit because, in the region of absorption bands, $n_1 \sim n_2$ and $\mathrm{Re}\,\varepsilon \ll \mathrm{Im}\,\varepsilon$.

Let us illustrate both procedures for two typical dielectrics. First we consider silicon (Si), which is the main dielectric material (an intrinsic semiconductor in the above classification) used in the semiconductor industry. The tabulated optical data for the complex refractive index of Si extend from 0.00496 eV to 2000 eV (Palik 1985). Let us use the first procedure to obtain $\varepsilon(i\xi)$ for Si. The tabulated optical data have been measured across such a wide frequency region that there is no need to use any extrapolation of the data to smaller frequencies when using eqn (12.125) in order to find the dielectric permittivity at all contributing imaginary Matsubara frequencies. Computational results for Si are presented in Fig. 12.2(a), where $\varepsilon(i\xi)$ is plotted as a function of $\log_{10}\xi$ (ξ is measured in rad/s; 1 eV = 1.51927×10^{15} rad/s). The first Matsubara frequency at $T = 300\,\mathrm{K}$, $\xi_1 = 2.47 \times 10^{14}$ rad/s, is indicated by the dashed line. As is seen in the figure, $\varepsilon(i\xi)$ is practically constant below approximately 10^{15} rad/s and decreases rapidly at higher frequencies. The static dielectric permittivity of Si is equal to $\varepsilon_0 = \varepsilon(0) = 11.66$. Silicon is a material which is characterized by an electronic polarization only. Thus, there is only one step in Fig. 12.2(a) in the behavior of ε along the imaginary frequency axis.

Another dielectric material, with a different behavior of the dielectric permittivity as a function of $i\xi$, is vitreous SiO_2. We shall obtain $\varepsilon(i\xi)$ for SiO_2 using the second procedure. For many materials, including SiO_2, the dielectric

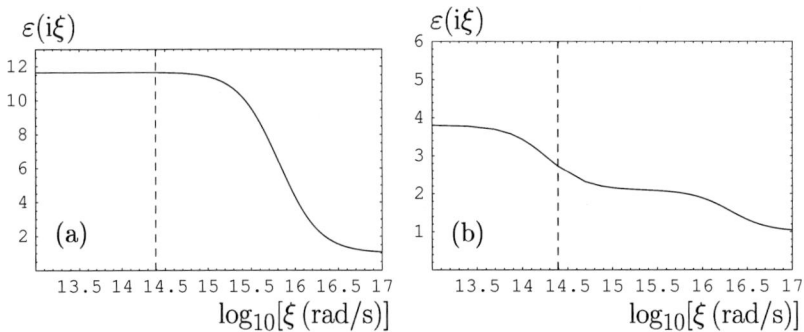

FIG. 12.2. The dielectric permittivities along the imaginary frequency axis as functions of the logarithm of the frequency for (a) Si and (b) SiO$_2$. The dashed lines indicate the position of the first Matsubara frequency at $T = 300$ K.

permittivity along the imaginary frequency axis can be approximated to a high accuracy by the Ninham–Parsegian representation (Mahanty and Ninham 1976)

$$\varepsilon(i\xi) = 1 + \frac{C_{\rm UV}\,\omega_{\rm UV}^2}{\xi^2 + \omega_{\rm UV}^2} + \frac{C_{\rm IR}\,\omega_{\rm IR}^2}{\xi^2 + \omega_{\rm IR}^2}, \qquad (12.127)$$

which is obtained from eqn (12.92) with $K = 2$, $\beta_j = 0$, $g_1 = C_{\rm UV}\omega_{\rm UV}^2$, $g_2 = C_{\rm IR}\omega_{\rm IR}^2$, $\omega_1 = \omega_{\rm UV}$, and $\omega_2 = \omega_{\rm IR}$. The Ninham–Parsegian approximation represents the effects of electronic polarization by one effective oscillator with a frequency in the ultraviolet spectrum. For dielectrics with ionic polarization, one more oscillator term is present, with a frequency in the infrared spectrum. For SiO$_2$, the values of the parameters $C_{\rm UV} = 1.098$, $C_{\rm IR} = 1.703$, $\omega_{\rm UV} = 2.033 \times 10^{16}$ rad/s, and $\omega_{\rm IR} = 1.88 \times 10^{14}$ rad/s have been determined from a fit to optical data (Hough and White 1980). The tabulated data for the complex refractive index of SiO$_2$ extend from 0.0025 eV to 2000 eV. The dependence of $\varepsilon(i\xi)$ on $\log_{10}\xi$ for SiO$_2$, as given by eqn (12.127), is shown in Fig. 12.2(b). A characteristic feature of SiO$_2$ is that the dependence of $\varepsilon(i\xi)$ on ξ in Fig. 12.2(b) contains two steps. One of them (to the right of the first Matsubara frequency, indicated by the dashed vertical line) is due to electronic polarization, and the second is due to ionic polarization. The static dielectric permittivity of SiO$_2$ is equal to $\varepsilon_0 = \varepsilon(0) = 3.81$. The two dielectric permittivities in Fig. 12.2 can be used to compute the Casimir free energy per unit area and the Casimir pressure for a configuration of two dielectric semispaces.

12.6.2 Free energy and pressure as functions of separation and temperature

We start with the computation of the Casimir free energy as a function of separation distance in a configuration of two semispaces (thick plates) made of Si and SiO$_2$. For this purpose, the dielectric permittivities computed at the imaginary Matsubara frequencies (see Fig. 12.2) have been substituted into the Lifshitz formula (12.66). The computational results for the logarithm of the magnitude of the free energy at $T = 300$ K are shown in Fig. 12.3(a). As is seen in Fig. 12.3(a),

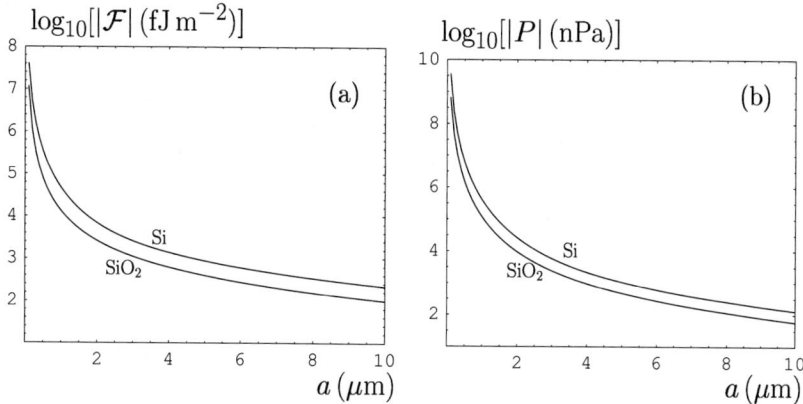

FIG. 12.3. Logarithm of the magnitude of (a) the Casimir free energy per unit area and (b) the Casimir pressure for a configuration of two semispaces made of Si and SiO$_2$ at $T = 300\,\text{K}$, as a function of the separation distance.

the magnitude of the free energy for Si is larger than for SiO$_2$ at all separations. At large separations, the magnitude of the free energy per unit area is very small (for example, at $a = 10\,\mu\text{m}$, it is only $100\,\text{fJ}\,\text{m}^{-2}$ for SiO$_2$; recall that $1\,\text{fJ} = 10^{-15}\,\text{J}$). However, at separations below $1\,\mu\text{m}$, the magnitude of the free energy per unit area can be millions of times larger.

In Table 12.1, the magnitude of the free energy per unit area for Si (column 3) and for SiO$_2$ (column 5) computed at $T = 300\,\text{K}$ is presented for separation

TABLE 12.1. Magnitude of the Casimir free energy per unit area $(\text{nJ}\,\text{m}^{-2})$ at two different temperatures, $T = 0$ and $T = 300\,\text{K}$, as a function of separation.

a	Si		Vitreous SiO$_2$	
(μm)	$T = 0$	$T = 300\,\text{K}$	$T = 0$	$T = 300\,\text{K}$
0.1	120.39	120.43	22.954	23.020
0.2	15.930	15.958	3.3234	3.3473
0.3	4.7850	4.8070	1.0884	1.1032
0.4	2.0292	2.0471	0.49558	0.50662
0.5	1.0416	1.0566	0.26964	0.27858
0.6	0.6036	0.6165	0.16400	0.17157
0.7	0.3804	0.3916	0.10766	0.11425
0.8	0.2550	0.2649	0.74717	0.80556
0.9	0.1792	0.1880	0.05409	0.05934
1.0	0.1306	0.1386	0.04048	0.04525

distances from 100 nm to $1\,\mu$m. For the purpose of comparison, the magnitude of the Casimir energy per unit area has been computed by using the Lifshitz formula (12.30) derived for $T=0$. The results for Si are presented in column 2 and for SiO_2 in column 4. As can be seen from a comparison of columns 3 and 2, and also columns 5 and 4, the relative thermal correction

$$\delta_T \mathcal{F}(a,T) = \frac{\mathcal{F}(a,T) - E(a)}{E(a)} \qquad (12.128)$$

increases with an increase in separation distance. For example, at $a=100$ nm we have $\delta_T \mathcal{F}(a, 300\,\text{K}) = 0.033\%$ for Si and 0.29% for SiO_2. However, at $a=1\,\mu$m, the relative thermal correction $\delta_T \mathcal{F}(a, 300\,\text{K})$ is 6.12% for Si and 11.8% for SiO_2. This means that in the Casimir regime, the thermal correction is not significant at the shortest separations but it must be taken into account at separations larger than several hundred nanometers. We note that the absolute thermal correction $\Delta_T \mathcal{F}(a,T)$ has the same sign (negative) as the Casimir free energy, and increases its magnitude.

In Fig. 12.3(b), the computational results for the logarithm of the magnitude of the Casimir pressure at $T=300$ K are presented. They were computed using the Lifshitz formula (12.70). The magnitude of the pressure for Si is larger than for SiO_2. At separations below $1\,\mu$m, the magnitude of the Casimir pressure reaches measurable levels. In Table 12.2, the magnitude of the pressure for Si (column 3) and for SiO_2 (column 5) computed at $T=300$ K is presented for various separations below $1\,\mu$m. In columns 2 (for Si) and 4 (for SiO_2), the

TABLE 12.2. Magnitude of the Casimir pressure (mPa) at two different temperatures, $T=0$ and $T=300$ K, as a function of separation.

a (μm)	Si		Vitreous SiO_2	
	$T=0$	$T=300$ K	$T=0$	$T=300$ K
0.1	3457.95	3458.07	643.896	644.985
0.2	235.24	235.32	45.966	46.119
0.3	47.478	47.523	9.9445	9.9978
0.4	15.149	15.183	3.3818	3.4088
0.5	6.2306	6.2553	1.4703	1.4869
0.6	3.0116	3.0304	0.74571	0.75713
0.7	1.6279	1.6425	0.42032	0.42878
0.8	0.9551	0.9667	0.25583	0.26240
0.9	0.5966	0.6060	0.16507	0.17037
1.0	0.3916	0.3993	0.11151	0.11588

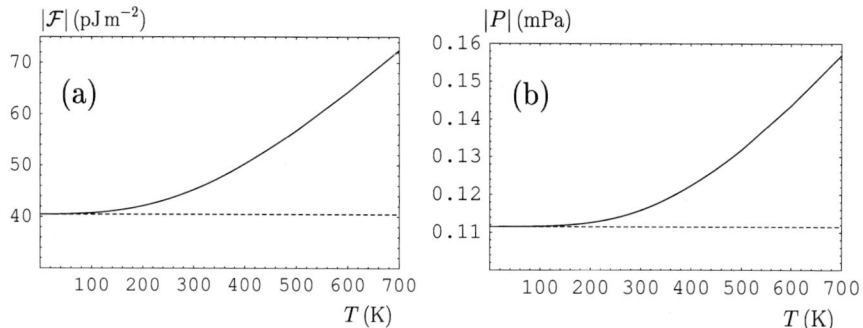

FIG. 12.4. Magnitude of (a) the Casimir free energy per unit area and (b) the Casimir pressure for a configuration of two semispaces made of SiO_2 at a separation of $a = 1\,\mu m$, as a function of temperature. The dashed line indicates the magnitude of (a) the Casimir energy and (b) the Casimir pressure at zero temperature.

magnitude of the Casimir pressure at $T = 0$ computed using eqn (12.33) is given. As can be seen from a comparison of columns 3 and 2 for Si and 5 and 4 for SiO_2, the relative thermal correction to the pressure, defined similarly to eqn (12.128), plays an important role at $a = 1\,\mu m$. Thus, for Si we have $\delta_T P(a, 300\,\text{K}) = 1.96\%$ and, for SiO_2, $\delta_T P(a, 300\,\text{K}) = 3.92\%$. However, these are of smaller magnitude than the relative thermal correction for the free energy. Keeping in mind that in the proximity force approximation (see Section 6.5), the Casimir force for the experimental configuration of a sphere above a plate is proportional to the Casimir free energy in the configuration of two plates, the large magnitude of the thermal correction to the free energy is significant (see Part III for discussion on this subject).

Next we turn our attention to the dependence of the Casimir free energy per unit area and the Casimir pressure on the temperature when the separation is fixed. The computations were performed using the Lifshitz formulas (12.66) for the free energy and (12.70) for the pressure in a configuration of two SiO_2 semispaces at a separation $a = 1\,\mu m$. The computational results for the magnitude of the Casimir free energy as a function of temperature are presented in Fig. 12.4(a). For the purpose of comparison, the Casimir energy computed at $T = 0$ using eqn (12.30) is shown in the same figure by a dashed line. It is seen that at high temperatures the thermal correction reaches tens of percent. However, the role of the thermal correction is noticeable even at $T = 100\,\text{K}$ and thus cannot be neglected at $T = 300\,\text{K}$, where it contributes more than 10% (as was discussed above).

In Fig. 12.4(b), the computational results for the magnitude of the Casimir pressure between the two SiO_2 semispaces at $a = 1\,\mu m$ apart are presented as a function of temperature. The Casimir pressure at zero temperature between the same semispaces was computed using the Lifshitz formula (12.33). This is shown by the dashed line. Similarly to the Casimir free energy, the thermal correction becomes noticeable at $T = 100\,\text{K}$, reaches several percent at $T = 300\,\text{K}$, and increases rapidly with an increase in temperature.

Additional results related to SiO_2 and Si can be found in Chapter 16 in connection with the atom–wall interaction and in Chapter 20 in connection with experimental investigation of the Casimir force.

12.6.3 The inclusion of dc conductivity

As was shown in Section 12.5.2, the inclusion of the dc conductivity always present in dielectric materials at nonzero temperature leads to a situation out of thermal equilibrium, which is the basic application condition for the Lifshitz theory, and results in a contradiction with thermodynamics. Thus, such an inclusion is theoretically not acceptable. Because of this, the dc conductivity was ignored in the computations in Section 12.6.2.

Are there significant differences between the results obtained using the Lifshitz formula with the dielectric permittivity $\varepsilon(i\xi)$, ignoring the conductivity at nonzero temperature, and with the permittivity $\tilde{\varepsilon}(i\xi)$, taking this conductivity into account? To answer this question, here we describe computations of the relative thermal correction $\delta_T \mathcal{F}$ to the Casimir energy per unit area [see eqn (12.128)] and the corresponding correction $\delta_T P$ to the Casimir pressure using the Lifshitz formulas (12.66), (12.70), (12.30), and (12.33). Computations for Si ignoring the conductivity were performed with the dielectric permittivity $\varepsilon(i\xi)$, as given in Fig. 12.2(a). The results for $\delta_T \mathcal{F}$, as a function of separation, are presented in Fig. 12.5(a) by the solid line. Corresponding computations for Si were performed with the dielectric permittivity

$$\tilde{\varepsilon}(i\xi) = \varepsilon(i\xi) + \frac{4\pi\sigma_0(T)}{\xi} \qquad (12.129)$$

[see eqn (12.113)], where, at $T = 300\,\text{K}$, a typical value of the conductivity of Si, $\sigma_0 = 1.4 \times 10^7\,\text{s}^{-1}$, was used. Importantly, the calculation results including the dc conductivity of Si do not depend on the value of σ_0 over a wide range $0 < \sigma_0 < 10^{13}\,\text{s}^{-1}$. This is because the difference between the calculational results using $\tilde{\varepsilon}(i\xi)$ and $\varepsilon(i\xi)$ is determined mostly by contributions to the zero-frequency term in the Lifshitz formula. As with the contributions from all Matsubara frequencies ξ_l with $l \geq 1$, these contributions are approximately the same in both cases if $\sigma_0 < 10^{13}\,\text{s}^{-1}$. The calculational results for $\delta_T \mathcal{F}$ for Si with the conductivity included are presented in Fig. 12.5(a) by the dashed line.

As is seen in Fig. 12.5(a), the inclusion of the dc conductivity of Si leads to an enormous increase in the thermal correction at all separations. Thus, when the conductivity is ignored (solid line), the thermal correction is equal to 1.44%, 6.13%, 14.1%, and 25.7% at separations $a = 0.5\,\mu\text{m}$, $1\,\mu\text{m}$, $1.5\,\mu\text{m}$, and $2\,\mu\text{m}$, respectively. When the conductivity is included (dashed line), the thermal correction is equal to 39.5%, 81.9%, 127.7%, and 177.2% at the same separations.

Similar results are obtained for the thermal correction to the Casimir pressure between Si semispaces. Calculational results obtained using the Lifshitz formulas

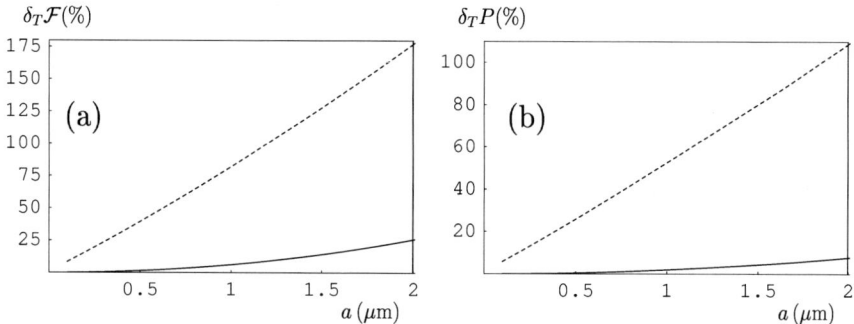

FIG. 12.5. Relative thermal correction (a) to the Casimir energy per unit area and (b) to the Casimir pressure as a function of separation. Computations were performed for Si at $T = 300\,\text{K}$ by using a dielectric permittivity with a finite static value (solid lines) and taking into account the conductivity at nonzero temperature (dashed lines).

with the dielectric permittivities $\varepsilon(i\xi_l)$ and $\tilde{\varepsilon}(i\xi_l)$ are shown in Fig. 12.5(b) by the solid and dashed lines, respectively. It can be clearly seen that the inclusion of the dc conductivity leads to a large increase in the thermal correction. Thus, at separations of $1\,\mu\text{m}$ and $2\,\mu\text{m}$, the thermal correction computed using $\varepsilon(i\xi_l)$ is equal to 1.96% and 7.7%, whereas the thermal correction computed using $\tilde{\varepsilon}(i\xi_l)$ with the conductivity included is equal to 52.5% and 108.7%, respectively.

An even greater increase in the thermal correction due to the inclusion of conductivity in the model of the dielectric response takes place for SiO_2. For SiO_2, computations were performed with the help of the same Lifshitz formulas, using the dielectric permittivity $\varepsilon(i\xi)$ shown in Fig. 12.2(b), with the neglect of the dc conductivity, and using the permittivity $\tilde{\varepsilon}(i\xi)$ given in eqn (12.129), which includes the dc conductivity. The conductivity of SiO_2 at $T = 300\,\text{K}$ is equal to $\sigma_0 = 29.7\,\text{s}^{-1}$ (Shackelford and Alexander 2001). The calculational results for $\delta_T\mathcal{F}$ and $\delta_T P$ are shown in Fig. 12.6(a,b), respectively. In both part (a) and part (b), the solid line indicates results calculated with the dc conductivity neglected and the dashed line indicates results with the dc conductivity included. As is seen in Fig. 12.6(a), the thermal correction to the Casimir energy increases from 11.8% and 41.5% to 256% and 464% at separations of $a = 1\,\mu\text{m}$ and $2\,\mu\text{m}$, respectively, when the conductivity of SiO_2 is included in the model of the dielectric response. In Fig. 12.6(b), the calculational results demonstrate a dramatic increase in the thermal correction to the Casimir pressure. Thus, at separations of $1\,\mu\text{m}$ and $2\,\mu\text{m}$, the thermal corrections are only 3.9% and 15.4%, respectively, when the conductivity of SiO_2 is not included. If the conductivity of SiO_2 at $T = 300\,\text{K}$ is taken into account, $\delta_T P$ is equal to 182% and 314% at the same separation distances.

It is instructive also to compare the behavior of the Casimir free energy and the Casimir pressure at large separations (high temperatures) for the cases where

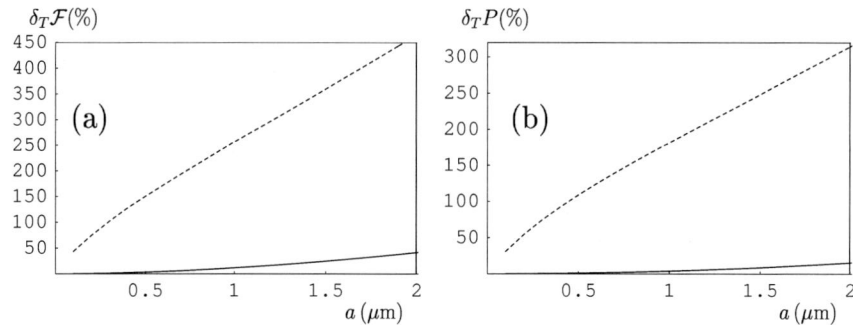

FIG. 12.6. Relative thermal correction (a) to the Casimir energy per unit area and (b) to the Casimir pressure as a function of separation. Computations were performed for SiO_2 at $T = 300\,\text{K}$ by using a dielectric permittivity with a finite value at zero frequency (solid lines) and taking into account the dc conductivity at nonzero temperature (dashed lines).

the dc conductivity of the dielectric materials is neglected or included. In the absence of dc conductivity, the analytical asymptotic behavior of the free energy and pressure for a configuration of two dissimilar dielectric semispaces is given in eqns (12.110) and (12.111), respectively. If the dc conductivity is included in the model of the dielectric response, the reflection coefficients at zero frequency are given by eqn (12.117) and the asymptotic behaviors of the free energy and pressure are given by the zero-frequency terms of the Lifshitz formulas (12.66) and (12.70), respectively:

$$\tilde{\mathcal{F}}(a,T) = -\frac{k_B T}{16\pi a^2}\zeta_R(3), \qquad \tilde{P}(a,T) = -\frac{k_B T}{8\pi a^3}\zeta_R(3). \qquad (12.130)$$

For two Si semispaces, $r_0^{(1)} = r_0^{(2)} = r_0^{Si} = 0.842$. This leads to

$$\frac{\tilde{\mathcal{F}}(a,T)}{\mathcal{F}(a,T)} = \frac{\tilde{P}(a,T)}{P(a,T)} = \frac{\zeta_R(3)}{\text{Li}[(r_0^{Si})^2]} = 1.52. \qquad (12.131)$$

For two SiO_2 semispaces, $r_0^{(1)} = r_0^{(2)} = r_0^{SiO_2} = 0.584$ and the corresponding ratios are equal to 3.36. Thus, the inclusion of the static conductivity leads to a large increase in the magnitudes of both the Casimir free energy per unit area and the Casimir pressure in the asymptotic limit of large separation or high temperature.

Such large thermal corrections and magnitudes of the free energy and pressure, as calculated in this section with inclusion of the static conductivities of dielectrics, are inconsistent with the basic principles of thermodynamics. As mentioned in Section 12.5.2 and considered in detail in Section 20.3, these corrections have already been excluded experimentally.

12.7 Problems with polar dielectrics

In the above, we have considered dielectric permittivities in the form of eqn (12.8). This form is commonly used for the description of nonpolar dielectrics (see Section 12.6.1). Equation (12.8) includes the electronic polarization which is inherent in all dielectrics. The respective oscillator frequencies ω_j belong to the ultraviolet spectrum. Some dielectrics, however, have an ionic component (typical examples are SiO_2 and Al_2O_3). These dielectrics possess *ionic polarization*. Their dielectric permittivity can also be represented in the form of eqn (12.8) but with oscillator frequencies in the infrared spectrum [see e.g. the Ninham–Parsegian representation (12.127)]. In both cases the molecules do not possess intrinsic dipole moments, but only induced dipole moments due to the influence of the fluctuating electromagnetic field. Another type of dielectric is the *polar dielectrics*, whose molecules possess intrinsic dipole moments which are oriented in an external electromagnetic field. In general, the dielectric permittivity of a dielectric with all three types of polarization can be approximately represented, on the imaginary frequency axis, in the form (Parsegian 2005)

$$\varepsilon(i\xi) = 1 + \frac{C_{UV}\,\omega_{UV}^2}{\xi^2 + \omega_{UV}^2} + \frac{C_{IR}\,\omega_{IR}^2}{\xi^2 + \omega_{IR}^2} + \frac{d}{1 + \xi\tau_d}. \qquad (12.132)$$

Here, we have included for simplicity only one oscillator term describing the electronic polarization and one oscillator term describing the ionic polarization (the Ninham–Parsegian model). These terms can be obtained from eqn (12.86) with $\gamma_j = 0$. The last term on the right-hand side of eqn (12.132), with temperature-dependent parameters d and τ_d, is the *Debye term*, which describes the dipole orientation polarization. The typical values of $1/\tau$ are in the microwave region of the spectrum.

Let us consider mica as an example of a polar dielectric which possesses all three types of polarization. The dielectric permittivity of mica along the imaginary frequency axis is plotted in Fig. 12.7(a). It corresponds to the following values of the parameters in eqn (12.132): $\omega_{UV} = 10.33$ eV, $C_{UV} = 1.48$, $\omega_{IR} = 3.95 \times 10^{-2}$ eV, $C_{UV} = 2.0$, and, at room temperature, $\tau_d = 5 \times 10^{-8}$ s and $d = 0.4$ (Parsegian 2005). As is seen in Fig. 12.7(a), there are three horizontal steps in the functional dependence of $\varepsilon(i\xi)$ on $\log_{10}\xi$ due to the three types of polarization. The step due to the electronic polarization is in the frequency region around 10^{15} rad/s. If it were extrapolated to zero frequency, this step would lead to $\varepsilon_0^e = 2.45$. The step due to both the electronic and the ionic polarization is in a region of order 10^{11}–10^{12} rad/s. Extrapolation of this step to zero frequency leads to $\varepsilon_0^{ei} = 4.45$. Finally, there is a third step at frequencies below 10^8 rad/s due to the electronic, ionic, and orientation polarizations together. As a result, the static permittivity of mica due to all three types of polarization, ε_0^p, is equal to 4.85.

At separations below 1 µm the Casimir energy at zero temperature, $E(a)$, is mostly determined by the electronic polarization. It is instructive to compare the

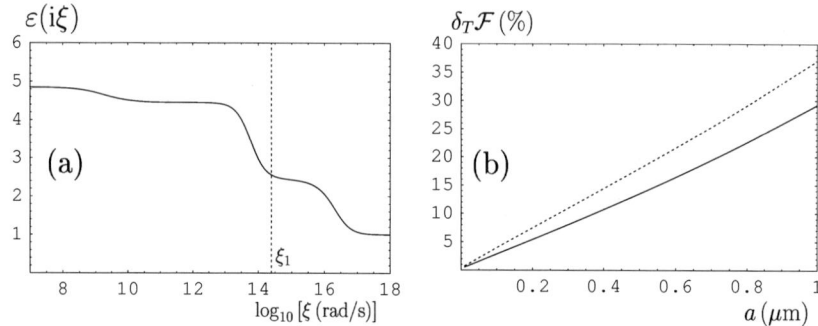

FIG. 12.7. (a) Dielectric permittivity of mica along the imaginary frequency axis. The dashed line indicates the first Matsubara frequency at $T = 300\,\text{K}$ (Klimchitskaya and Geyer 2008). (b) The relative thermal correction to the Casimir energy of mica plates as a function of separation at $T = 300\,\text{K}$. The solid line was computed with inclusion of the electronic and ionic polarizations, whereas the dashed line includes the orientation polarization as well.

values of the thermal correction $\delta_T \mathcal{F}(a, T)$ defined in eqn (12.128) calculated including the various types of polarization. Thus, for Si, which possesses electronic polarization only, we obtain from Table 12.1 $\delta_T \mathcal{F}(a, T) = 1.44\%$ at $T = 300\,\text{K}$ and $a = 500\,\text{nm}$. If we disregard both the ionic and the orientation polarization of mica and take into account only its electronic polarization, the relative thermal correction is $\delta_T \mathcal{F}(a, T) = 1.25\%$ at the same T and a. Thus, the roles of the electronic polarization for Si and mica are in fairly good agreement. However, in mica, the ionic and orientation polarizations are also present. In Fig. 12.7(b), we plot the relative thermal correction $\delta_T \mathcal{F}(a, T)$ for mica versus the separation at $T = 300\,\text{K}$. The solid line was computed taking the electronic and ionic polarizations into account, i.e. using a dielectric permittivity with $\varepsilon_0^{\text{ei}} = 4.45$. The dashed line was computed by using the complete dielectric permittivity (12.132), i.e. including the orientation polarization also. As is seen in Fig. 12.7(b) (solid line), at $a = 500\,\text{nm}$ the relative thermal correction reaches 13.5% (compare with the 1.25% found above when only the electronic polarization was included). Thus, inclusion of the ionic polarization leads to a marked increase in the relative thermal correction.

In Fig. 12.7(b), it is also seen that the role of the orientation polarization increases with an increase in separation distance. Thus, at $a = 100\,\text{nm}$ the inclusion of the orientation polarization leads to a 1% increase in the relative thermal correction, but at $a = 1\,\mu\text{m}$ it leads to an 8% increase. We should emphasize that the Debye term in the dielectric permittivity (12.132) leads to problems in the Lifshitz theory. This term influences only the zero-frequency contribution to the Casimir free energy (12.66). As a result, the thermal correction (12.79) with the inclusion of the orientation polarization is given by (Klimchitskaya and Geyer

2008)

$$\Delta_T \mathcal{F}^{(\mathrm{p})}(a,T) = \Delta_T \mathcal{F}(a,T) - \frac{k_\mathrm{B} T}{16\pi a^2}\left[\mathrm{Li}_3(r_{0,\mathrm{p}}^2) - \mathrm{Li}_3(r_{0,\mathrm{ei}}^2)\right]. \qquad (12.133)$$

Here, $\Delta_T \mathcal{F}(a,T)$ is the thermal correction due to the electronic and ionic polarizations only, and

$$r_{0,\mathrm{p}} = \frac{\varepsilon_0^\mathrm{p} - 1}{\varepsilon_0^\mathrm{p} + 1}, \qquad r_{0,\mathrm{ei}} = \frac{\varepsilon_0^\mathrm{ei} - 1}{\varepsilon_0^\mathrm{ei} + 1}. \qquad (12.134)$$

Note that $r_{0,\mathrm{p}}$ depends on the temperature through the parameter d in eqn (12.132). At temperatures of about 300 K, eqn (12.133) contains a contribution approximately linear in the temperature. The first term on the right-hand side of eqn (12.133), $\Delta_T \mathcal{F}(a,T)$, has a standard form, considered in Section 12.5.1. Hough and White (1980) questioned whether or not the Debye term should be included in the model of dielectric response used in the Lifshitz theory. According to them, the inclusion of the orientation degrees of freedom that come into play at very low frequencies much below the first Matsubara frequency is not justified. This problem calls for further investigation.

12.8 The Lifshitz formula for anisotropic plates

In the above, only isotropic plates were considered. In this case the dielectric permittivity of a plate material can be described by one function $\varepsilon(\omega)$. The results obtained, however, can be generalized simply to anisotropic plates described by a diagonal tensor with the components $\varepsilon_{xx}(\omega)$, $\varepsilon_{yy}(\omega)$, and $\varepsilon_{zz}(\omega)$. A derivation of the Lifshitz formula following the lines of Section 12.1.1 (at zero temperature) or 12.3.1 (at nonzero temperature) returns us to eqn (12.30) or (12.66), respectively. However, the expressions for the reflection coefficients must be modified to take into account the character of the crystal anisotropy and the orientation of the optical axis.

12.8.1 Uniaxial crystals

Here, we consider two semispaces or two plates of finite thickness d made of a uniaxial crystal (graphite, for example) which is characterized by two dissimilar dielectric permittivities $\varepsilon_x(\omega) = \varepsilon_y(\omega)$ and $\varepsilon_z(\omega)$. Bearing in mind the applications to graphite plates and carbon nanostructures in Part III of the book, we assume that the boundary planes of the semispaces are in the plane (x,y) and the crystal optical axis z is perpendicular to it. For the two semispaces, the Casimir free energy per unit area is given by eqn (12.66) and the Casimir pressure by eqn (12.70), with the following generalized reflection coefficients (Greenaway et al. 1969):

$$r_{\mathrm{TM}}^{(\mathrm{u})}(i\xi_l, k_\perp) = \frac{\sqrt{\varepsilon_{xl}\varepsilon_{zl}}q_l - k_{zl}}{\sqrt{\varepsilon_{xl}\varepsilon_{zl}}q_l + k_{zl}}, \qquad r_{\mathrm{TE}}^{(\mathrm{u})}(i\xi_l, k_\perp) = \frac{q_l - k_{xl}}{q_l + k_{xl}}. \qquad (12.135)$$

Here, the following notation has been introduced:

$$k_{xl}^2 = k_\perp^2 + \varepsilon_{xl}\frac{\xi_l^2}{c^2}, \quad k_{zl}^2 = k_\perp^2 + \varepsilon_{zl}\frac{\xi_l^2}{c^2}, \quad \varepsilon_{xl} = \varepsilon_x(i\xi_l), \quad \varepsilon_{zl} = \varepsilon_z(i\xi_l). \quad (12.136)$$

For an isotropic crystal, $\varepsilon_x = \varepsilon_y = \varepsilon_z$ and eqn (12.135) coincides with eqn (12.67).

For two plates of finite thickness d made of a uniaxial crystal, the Casimir free energy per unit area and the Casimir pressure are once more given by eqns (12.66) and (12.70), with the following generalized reflection coefficients:

$$r_{\text{TM}}^{(u)}(i\xi_l, k_\perp) = \frac{\varepsilon_{xl}\varepsilon_{zl}q_l^2 - k_{zl}^2}{\varepsilon_{xl}\varepsilon_{zl}q_l^2 + k_{zl}^2 + 2\sqrt{\varepsilon_{xl}\varepsilon_{zl}}q_l k_{zl} \coth(k_{zl}d)},$$

$$r_{\text{TE}}^{(u)}(i\xi_l, k_\perp) = \frac{q_l^2 - k_{xl}^2}{q_l^2 + k_{xl}^2 + 2q_l k_{xl} \coth(k_{xl}d)}. \quad (12.137)$$

In the limiting case of thick plates ($d \to \infty$), eqn (12.137) coincides with eqn (12.135). For an isotropic material, eqn (12.137) coincides with the previously derived eqn (12.52).

12.8.2 Casimir torque

We now consider a plate made of a uniaxial crystal described by the dielectric permittivities $\varepsilon_x(\omega)$ and $\varepsilon_y(\omega) = \varepsilon_z(\omega)$. Thus, the optical axis of one of the plates is aligned with the x-axis. Let the optical axis of the second plate [which is also in the (x,y) plane] be rotated by an angle φ with respect to the x-axis. As was shown by Parsegian and Weiss (1972) in the nonretarded case and by Barash (1973) with retardation effects taken into account, a torque arises between the two plates. This torque is due to the modification of the electromagnetic zero-point energy in the presence of the plates. It leads to a rotation of the plates until their optical axes are aligned. The free energy $\mathcal{F} = \mathcal{F}(a, \varphi, T)$ can be expressed by the Lifshitz formula, but the formulas for the reflection coefficients are rather cumbersome [they have been presented by Munday et al. (2005)]. The torque is given by

$$M(a, \varphi, T) = -\frac{\partial \mathcal{F}(a, \varphi, T)}{\partial \varphi} S. \quad (12.138)$$

In the nonretarded limit ($a \ll \lambda_0$, see Section 12.1.3) and for small anisotropies $|\varepsilon_x - \varepsilon_y|/\varepsilon_y \ll 1$, one obtains (Munday et al. 2005)

$$M(a, \varphi, T) = -\frac{\bar{\omega} S}{64\pi^2 a^2} \sin(2\varphi), \quad (12.139)$$

where $\bar{\omega}$ is the effective frequency, calculated using the dielectric permittivities of both plates along the imaginary frequency axis. In the relativistic limit ($a \gg \lambda_0$), $M \sim -S\sin(2\varphi)/a^3$ (Mostepanenko and Trunov 1997).

A Casimir torque also arises in a configuration consisting of asymmetric bodies. In the same way as for the case of similar bodies with anisotropic properties, the torque tends to change the mutual orientation of the bodies. Typical examples are those of two finite plates with uniaxial sinusoidal corrugations and of a corrugated sphere above a corrugated plate. These configurations will be considered in more detail in Chapter 21 in connection with the measurements of the lateral Casimir force by Chen *et al.* (2002a, 2002b).

12.9 Lifshitz-type formula for radiative heat transfer

There are a number of physical phenomena apart from the Casimir effect that are caused by zero-point and thermal fluctuations of the electromagnetic field. One example is the radiative heat transfer between two plane, parallel plates (semispaces) at different temperatures $T_1 > T_2$, separated by an empty gap of width a. This phenomenon was studied by Rytov (1959), and later on reconsidered by Polder and van Hove (1971), by Loomis and Maris (1994), and by Volokitin and Persson (2001, 2004) [see also the review by Volokitin and Persson (2007)]. The approach followed by Polder and van Hove is closely related to the Lifshitz theory of the van der Waals and Casimir interaction between macroscopic bodies. In this approach the heat transfer is regarded as occurring via fluctuating electromagnetic fields radiated by the two bodies, whose sources are the random thermal electric currents that are present inside the plates. The statistical properties of the random currents are determined by using the fluctuation–dissipation theorem. The same assumptions about the properties of dielectric materials as in the Lifshitz theory are used to describe the radiative heat transfer. Specifically, it is supposed that the materials can be described by a frequency-dependent dielectric permittivity $\varepsilon(\omega)$. Keeping in mind that in this case the plates are kept at different temperatures, it is assumed that each of the plates is in local thermal equilibrium.

This phenomenon is characterized by the power G per unit area of the heat transfer from plate one to plate two. An expression for G was found by Loomis and Maris (1994) using the average value of the Poynting vector in the gap between the two plates. In terms of the reflection coefficients (12.45) along the real axis (i.e. iξ must be replaced with ω), the power of the heat transfer is given by (Volokitin and Persson, 2001, 2007, Bezerra *et al.* 2007)

$$G(a, T_1, T_2) = \frac{\hbar}{4\pi^2} \int_0^\infty \omega\, d\omega \left[\frac{1}{\exp(\hbar\omega/k_B T_1) - 1} - \frac{1}{\exp(\hbar\omega/k_B T_2) - 1} \right]$$

$$\times \left\{ \int_0^{\omega/c} k_\perp dk_\perp \left[\frac{\left[1 - |r^{(1)}_{\rm TM}(\omega, k_\perp)|^2\right]\left[1 - |r^{(2)}_{\rm TM}(\omega, k_\perp)|^2\right]}{|1 - r^{(1)}_{\rm TM}(\omega, k_\perp) r^{(2)}_{\rm TM}(\omega, k_\perp) e^{-2aq}|^2} \right. \right.$$

$$\left. + \frac{\left[1 - |r^{(1)}_{\rm TE}(\omega, k_\perp)|^2\right]\left[1 - |r^{(2)}_{\rm TE}(\omega, k_\perp)|^2\right]}{|1 - r^{(1)}_{\rm TE}(\omega, k_\perp) r^{(2)}_{\rm TE}(\omega, k_\perp) e^{-2aq}|^2} \right]$$

$$+ 4 \int_{\omega/c}^\infty k_\perp dk_\perp e^{-2qa} \left[\frac{\operatorname{Im} r^{(1)}_{\rm TM}(\omega, k_\perp) \operatorname{Im} r^{(2)}_{\rm TM}(\omega, k_\perp)}{|1 - r^{(1)}_{\rm TM}(\omega, k_\perp) r^{(2)}_{\rm TM}(\omega, k_\perp) e^{-2aq}|^2} \right.$$

$$\left. \left. + \frac{\operatorname{Im} r^{(1)}_{\rm TE}(\omega, k_\perp) \operatorname{Im} r^{(2)}_{\rm TE}(\omega, k_\perp)}{|1 - r^{(1)}_{\rm TE}(\omega, k_\perp) r^{(2)}_{\rm TE}(\omega, k_\perp) e^{-2aq}|^2} \right] \right\}.$$

(12.140)

Equation (12.140) is similar to the Lifshitz formula (12.74) expressed in terms of reflection coefficients along the real frequency axis. It is presented as a sum of contributions of propagating waves (the integration with limits from 0 to ω/c) and evanescent waves (the integration with limits from ω/c to infinity).

Note that for $r_{\text{TM}}^{(2)} = r_{\text{TE}}^{(2)} = 0$, the contribution of the evanescent waves vanishes. For $r_{\text{TM}}^{(2)} = r_{\text{TE}}^{(2)} = 0$ and $T_2 = 0$ (i.e. in the absence of the second semispace), the contribution of the propagating waves reduces to the well-known Kirchhoff formula for the flux of radiation from a single surface with reflection coefficients $r_{\text{TM}} = r_{\text{TM}}^{(1)}$ and $r_{\text{TE}} = r_{\text{TE}}^{(1)}$ at a temperature $T = T_1$:

$$G(T) = \frac{\hbar}{4\pi^2 c^2} \int_0^\infty d\omega \frac{\omega^3}{\exp(\hbar\omega/k_{\text{B}}T) - 1} \qquad (12.141)$$

$$\times \int_0^1 p\, dp \left[2 - |r_{\text{TM}}(\omega, k_\perp)|^2 - |r_{\text{TE}}(\omega, k_\perp)|^2 \right],$$

where $p = k_\perp c/\omega$. In Section 14.5, we present the results of calculations for the power of radiative heat transfer between metallic surfaces using eqn (12.140).

12.10 Application region of the Lifshitz formula

The application region of the Lifshitz formula is a challenging question. On the one hand, it was derived in Sections 12.1.1.and 12.3.1 for dielectric media described by real dielectric permittivities depending only on the frequency; under some conditions, including the condition of thermal equilibrium, that derivation was generalized to the case of complex $\varepsilon(\omega)$. The validity of this generalization was independently confirmed in the original derivation of the Lifshitz formula using the fluctuation–dissipation theorem and the scattering approach (see the references in Section 12.1.1). We emphasize that the formulas obtained are valid only in a state of thermal equilibrium. On the other hand, as was demonstrated in Section 12.5.2, the application of the Lifshitz formula to complex permittivities that include the dc conductivity of dielectric materials at nonzero temperature leads to a violation of the third law of thermodynamics and to contradictions with experimental data (see Section 20.3). Such anomalies occur because in the presence of a drift current, the condition of thermal equilibrium is violated. As a result, the fluctuation–dissipation theorem becomes inapplicable. It was concluded that for all materials that have zero conductivity at zero temperature, the conductivity arising at nonzero temperature must be neglected in the framework of the Lifshitz theory. A related rule for metals is formulated in Section 14.6.3.

An important assumption used in all derivations of the Lifshitz formula is that the dielectric permittivity depends only on the frequency, i.e. only the temporal dispersion is taken into account. However, the final representation of the Lifshitz formula in terms of the reflection coefficients for various systems [see, e.g. eqns (12.30), (12.33), (12.66), and (12.70)] tempts one to apply these formulas to a wider class of media possessing not only temporal but also spatial dispersion. When this is done, the dielectric permittivity $\varepsilon(\omega)$ is replaced with $\varepsilon(\omega, \boldsymbol{k})$ and the modified reflection coefficients for reflection of electromagnetic waves from a medium with spatial dispersion are substituted into the usual Lifshitz formulas. Such an approach was used long ago (Heinrichs 1973, Kleinman and Landman 1974). The results obtained were criticized by Barash and Ginzburg (1975) as not reliable. In the last few years, however, some others have applied the Lifshitz formula to media with spatial dispersion [see e.g. Sernelius (2005), Svetovoy and Esquivel (2005), Esquivel and Svetovoy (2004), Esquivel et al. (2003), and Contreras-Reyes and Mochán (2005)]. Below, we demonstrate why the case of media with spatial dispersion is outside the application region of the standard Lifshitz theory (Klimchitskaya and Mostepanenko 2007).

As was shown in Section 12.1.1, the starting point for the derivation of the photon eigenfrequencies and reflection coefficients (12.31) is the set of continuity boundary conditions (12.2). These conditions form the basis for all of the derivations of the Fresnel reflection coefficients available in the literature. However, if the material in the semispaces possesses spatial dispersion, these boundary conditions do not apply. In electrodynamics with spatial dispersion, the physical fields $\boldsymbol{E}$ and $\boldsymbol{B}$ at the interface are usually finite, though the electric displacement $\boldsymbol{D}$ can tend to infinity (Agranovich and Ginzburg 1984). Taking this fact into account and integrating the Maxwell equations (12.1) over the thickness of the boundary layer [see e.g. Stratton (1941)], one reproduces the first and third boundary conditions in eqn (12.2), but arrives at modified second and fourth conditions at $z = \pm a/2$ (Agranovich and Ginzburg 1984, Ginzburg 1985):

$$E_{1t}(t,\boldsymbol{r}) = E_{2t}(t,\boldsymbol{r}), \qquad D_{2n}(t,\boldsymbol{r}) - D_{1n}(t,\boldsymbol{r}) = 4\pi\sigma(t,\boldsymbol{r}), \qquad (12.142)$$

$$B_{1n}(t,\boldsymbol{r}) = B_{2n}(t,\boldsymbol{r}), \qquad [\boldsymbol{n} \times (\boldsymbol{B}_2(t,\boldsymbol{r}) - \boldsymbol{B}_1(t,\boldsymbol{r}))] = \frac{4\pi}{c}\boldsymbol{j}(t,\boldsymbol{r}).$$

Here, the induced surface charge and current densities are given by

$$\sigma(t,\boldsymbol{r}) = \frac{1}{4\pi}\int_1^2 \boldsymbol{\nabla}\cdot[\boldsymbol{n}\times[\boldsymbol{D}(t,\boldsymbol{r})\times\boldsymbol{n}]]dl, \quad \boldsymbol{j}(t,\boldsymbol{r}) = \frac{1}{4\pi}\int_1^2 \frac{\partial \boldsymbol{D}(t,\boldsymbol{r})}{\partial t}dl$$
(12.143)

[all of the notation used here was described immediately after eqn (12.2)]. If one uses eqn (12.3), which is valid in the absence of spatial dispersion, eqn (12.143) leads to $\sigma = 0$ and $\boldsymbol{j} = 0$, and eqn (12.142) coincides with the standard continuity boundary conditions (12.2). If, however, spatial dispersion is present, it is necessary to use the boundary conditions (12.142).

Now we turn to the possibility of using the permittivity $\varepsilon(\omega,\boldsymbol{k})$ and related reflection coefficients in the Lifshitz theory. If the material of the semispaces possesses both temporal and spatial dispersion, eqn (12.3) should be generalized to

$$D_k(t,\boldsymbol{r}) = \int_{-\infty}^t dt' \int d\boldsymbol{r}\, \hat{\varepsilon}_{kl}(t-t',\boldsymbol{r},\boldsymbol{r}')E_l(t',\boldsymbol{r}') \qquad (12.144)$$

(this equation also includes anisotropic materials). If the medium is uniform in space (i.e. all points are equivalent), the kernel $\hat{\varepsilon}$ does not depend on $\boldsymbol{r}$ and $\boldsymbol{r}'$

separately, but only on the difference $\boldsymbol{R} \equiv \boldsymbol{r} - \boldsymbol{r}'$. In this case, by performing the Fourier transformation

$$\boldsymbol{E}(t, \boldsymbol{r}) = \int_{-\infty}^{\infty} d\omega \int d\boldsymbol{k}\, \boldsymbol{E}(\omega, \boldsymbol{k}) e^{i(\boldsymbol{k} \cdot \boldsymbol{r} - \omega t)} \qquad (12.145)$$

(and similarly for $\boldsymbol{D}$) and substituting the result into eqn (12.144), one can introduce dielectric permittivities

$$\varepsilon_{ml}(\omega, \boldsymbol{k}) = \int_0^{\infty} d\tau \int d\boldsymbol{R}\, \hat{\varepsilon}_{ml}(\tau, \boldsymbol{R}) e^{-i(\boldsymbol{k} \cdot \boldsymbol{R} - \omega \tau)}. \qquad (12.146)$$

These permittivities depend on both the wave vector and the frequency and bring eqn (12.144) into a form analogous to eqn (12.5),

$$D_m(\omega, \boldsymbol{k}) = \varepsilon_{ml}(\omega, \boldsymbol{k}) E_l(\omega, \boldsymbol{k}). \qquad (12.147)$$

However, for a Casimir configuration of two semispaces with a separation a, the system is not uniform, owing to the presence of a gap. Here, it is incorrect to assume that the kernel $\hat{\varepsilon}$ depends only on $\boldsymbol{R}$ and τ, and hence it is not possible to introduce $\varepsilon_{ml}(\omega, \boldsymbol{k})$. Specifically, owing to the translational invariance in the plane of the plates, one can introduce $\varepsilon_x(\omega, \boldsymbol{k}_\perp) = \varepsilon_y(\omega, \boldsymbol{k}_\perp)$, but one cannot introduce $\varepsilon_z(\omega, \boldsymbol{k})$, because the translational invariance along the z-axis is violated. Thus, the Lifshitz formulas cannot be used with the generalized reflection coefficients (12.135) derived for uniaxial crystals. In the presence of boundaries, the kernel $\hat{\varepsilon}$ for systems with spatial dispersion depends not only on $\boldsymbol{R}$ and τ but also on the distance from the boundary (Agranovich and Ginzburg 1984). An approximate phenomenological approach to dealing with this case, applicable to some physical problems other than the Casimir effect, has been described by Agranovich and Ginzburg (1984). The main features of this approach are the following. For electromagnetic waves with a wavelength λ, the kernel $\hat{\varepsilon}(\tau, \boldsymbol{r}, \boldsymbol{r}')$ is significantly large only in the vicinity of the point $\boldsymbol{r}$, in a region with a characteristic dimension $l \ll \lambda$ (in fact, for nonmetallic condensed media, l is of the order of the lattice constant). One can then assume that $\hat{\varepsilon}$ is a function of $\boldsymbol{R} = \boldsymbol{r} - \boldsymbol{r}'$, except for a layer of thickness l adjacent to the boundary surface. If one is not interested in this subsurface layer, the quantity $\varepsilon_{ml}(\omega, \boldsymbol{k})$ can be used to describe the remainder of the medium.

Note that the approximate phenomenological approach outlined above, is widely used in the theory of the anomalous skin effect for the investigation of bulk physical phenomena involving electromagnetic fields (Kliewer and Fuchs 1968). To take the boundary into account, some fictitious infinite system is introduced and the electromagnetic fields in this system are discontinuous on the boundary surface. This discontinuity should not be confused with the discontinuity in the physical fields of the real system in the presence of spatial dispersion given by eqns (12.142) and (12.143). There is, however, another approach to the anomalous skin effect in polycrystals (Kaganova and Kaganov 2001), based on the use of the local Leontovich impedance, taking into account the shape of the Fermi surface (the concept of the Leontovich impedance is discussed in connection with the Lifshitz formula in Sections 13.4 and 14.4). The frequency- and wave-vector-dependent dielectric permittivity in the presence of boundaries has also been successfully applied in some other problems, for example the study of electromagnetic interactions of molecules with a surface (Ford and Weber 1984). However, from a fundamental theoretical point of view, for a bounded medium, it is incorrect to use the same kernel in eqn (12.144), depending on $\boldsymbol{R} = \boldsymbol{r} - \boldsymbol{r}'$, as for an unbounded medium. According to Foley and Devaney (1975), this would lead to nonconservation of the number of particles and, consequently, to a violation of the law of conservation of energy (Barash and Ginzburg 1975). Nevertheless, this approach has recently been used by Svetovoy (2008). Additionally, specular reflection of charge carriers from the boundary planes was assumed. It has been proven, however, that for spatially dispersive materials the scattering of carriers is neither specular nor diffuse (Foley and Devaney 1975).

Thus, it is unlikely that a phenomenological approach using a frequency- and wave-vector-dependent dielectric permittivity $\varepsilon(\omega, \boldsymbol{k})$ for the spatially nonuniform configuration of two parallel plates separated by a gap could be applicable in combination with the usual Lifshitz formula to calculating the Casimir force for cases with spatial dispersion. The Casimir force is very sensitive to the behavior of the dielectric permittivity in the layer of thickness l adjacent to the boundary surface, i.e. where the approximate description by means of $\varepsilon(\omega, \boldsymbol{k})$ is not applicable. A consistent theory of the Casimir force taking spatial dispersion into account should start from the boundary conditions (12.142) and (12.143) and the general connection (12.144) between the electric field and electric displacement. Some steps towards a theory of the van der Waals and Casimir force including spatial dispersion were taken by Barash and Ginzburg (1975), where a general expression for the free energy was obtained in terms of the thermal Green's function of the electromagnetic field and the polarization operator. The generalization of the Lifshitz formula in terms of the scattering matrices (10.43) opens up opportunities to include the effect of spatial dispersion. To do so, one must find the corresponding T-matrices of the operators $\mathcal{T}_{AA}$ and $\mathcal{T}_{BB}$. In principle, this could be done by solving Maxwell's equations for a given Casimir configuration with nonlocally responding materials. As regards the standard Lifshitz formulas considered in this book, they are applicable only in the absence of spatial dispersion.

13

THE CASIMIR INTERACTION BETWEEN REAL-METAL PLATES AT ZERO TEMPERATURE

In contrast to dielectrics, metals are materials that have a nonzero conductivity at zero temperature. This conductivity is determined by the presence of free electrons. For metals, the valence band is the conduction band as well (i.e. this band is half-filled). The ideal-metal boundary conditions considered in the first part of the book do not allow any penetration of the electromagnetic field inside the metal. In reality, however, there is some nonzero penetration depth (the skin depth) of an electromagnetic wave into the metal interior which depends on the quality of the metal and on the wave frequency. Thus, there should be corrections to the original Casimir expressions (1.1) and (1.5) for the pressure and energy per unit area due to the nonzero skin depth.

The effect of the skin depth on the Casimir force can be investigated in the framework of the Lifshitz theory. To do this, it is necessary to model the properties of the metal by a dielectric function which depends only on the frequency. Traditionally, the effect of the skin depth has been studied at zero temperature using the free-electron plasma model. A first-order correction to the Casimir result (1.1) in the relative penetration depth of electromagnetic waves into the metal was obtained in this way by Dzyaloshinskii *et al.* (1961), with an error in the numerical coefficient which was corrected by Hargreaves (1965).

In this chapter, we consider both analytical calculations of the Casimir energies and forces between real metal plates and numerical computations using tabulated optical data for the complex index of refraction of metals. Comparison between the results of the analytical and numerical computations permits one to infer the main properties of metals affecting the Casimir force and how they enter into the Lifshitz theory. We also introduce the concept of the Leontovich surface impedance and related boundary condition and indicate the application region of the impedance approach. This chapter should be considered as a preparation for Chapter 14, where the complicated problem of the thermal Casimir force between real metal plates is considered.

13.1 Perturbation theory in the relative skin depth, and the plasma model

We consider two parallel metal semispaces separated by a gap of width a at zero temperature. It is reasonable to consider separation distances from about $0.1\,\mu$m to $1\,\mu$m. The contributing frequencies in this region are those of visible light and infrared optics. In addition, thermal corrections cannot exceed a few percent. This justifies our assumption that the temperature is equal to zero.

At the relatively high frequencies of infrared optics, a good approximation for the dielectric properties of metals is provided by the free-electron plasma model

$$\varepsilon_{\mathrm{p}}(\omega) = 1 - \frac{\omega_{\mathrm{p}}^2}{\omega^2}, \tag{13.1}$$

where the plasma frequency is given by

$$\omega_{\mathrm{p}}^2 = \frac{4\pi n_e e^2}{m^*}. \tag{13.2}$$

Here, n_e is the density of electrons, m^* is their effective mass, and e is the electron charge. The problem can easily be framed in terms of the relative skin depth based on the plasma wavelength,

$$\frac{\delta_0}{a} = \frac{\lambda_{\mathrm{p}}}{2\pi a} = \frac{c}{\omega_{\mathrm{p}} a} \ll 1, \tag{13.3}$$

where δ_0 is the skin depth and λ_{p} is the plasma wavelength.

In terms of the dimensionless variables (12.41), the Casimir energy (12.30) takes the form

$$E(a) = \frac{\hbar c}{32\pi^2 a^3} \int_0^\infty y\, dy \int_0^y d\zeta \left\{ \ln\left[1 - r_{\mathrm{TM}}^2(i\zeta, y) e^{-y}\right] \right.$$
$$\left. + \ln\left[1 - r_{\mathrm{TE}}^2(i\zeta, y) e^{-y}\right] \right\}, \tag{13.4}$$

equivalent to that in the first two lines of eqn (12.90). The reflection coefficients are expressed in terms of dimensionless variables in eqn (12.91). The dielectric permittivity (13.1) along the imaginary frequency axis reads

$$\varepsilon_{\mathrm{p}}(i\xi) = \varepsilon_{\mathrm{p}}(i\omega_c \zeta) = 1 + \frac{\omega_{\mathrm{p}}^2}{\omega_c^2 \zeta^2} = 1 + \frac{4a^2}{\delta_0^2 \zeta^2}, \tag{13.5}$$

where $\omega_c = c/(2a)$ is the characteristic frequency.

Now we introduce the quantity

$$A(y) = \int_0^y d\zeta \left\{ \ln\left[1 - r_{\mathrm{TM}}^2(i\zeta, y) e^{-y}\right] + \ln\left[1 - r_{\mathrm{TE}}^2(i\zeta, y) e^{-y}\right] \right\} \tag{13.6}$$

and expand it in powers of the small parameter δ_0/a up to the fourth order,

$$A(y) = 2y\ln(1 - e^{-y}) + \frac{8}{3}\frac{\delta_0}{a}\frac{y^2}{e^y - 1} - \frac{12}{5}\left(\frac{\delta_0}{a}\right)^2 \frac{y^3 e^y}{(e^y - 1)^2} \tag{13.7}$$
$$+ \frac{8}{105}\left(\frac{\delta_0}{a}\right)^3 \frac{y^4(-1 + 22e^y + 19e^{2y})}{(e^y - 1)^3}$$

$$-\frac{1}{135}\left(\frac{\delta_0}{a}\right)^4 \frac{y^5 e^y (179 + 1042 e^y + 179 e^{2y})}{(e^y - 1)^4}.$$

Substituting eqn (13.7) into eqn (13.4) and integrating with respect to y, we finally find

$$E_\mathrm{p}(a) = -\frac{\pi^2 \hbar c}{720 a^3} \left[1 - 4\frac{\delta_0}{a} + \frac{72}{5}\left(\frac{\delta_0}{a}\right)^2 \right. \tag{13.8}$$

$$\left. -\frac{320}{7}\left(1 - \frac{\pi^2}{210}\right)\left(\frac{\delta_0}{a}\right)^3 + \frac{400}{3}\left(1 - \frac{163\pi^2}{7350}\right)\left(\frac{\delta_0}{a}\right)^4 \right],$$

where the subscript "p" indicates that the Casimir energy is obtained using the plasma model. The zeroth-order term in eqn (13.8) coincides with eqn (2.82), obtained for ideal-metal planes, as expected. Note that multiplication of eqn (13.8) by $2\pi R$ leads to the Casimir force acting between a plate and a large sphere of radius R (see Section 6.5).

In a similar way, the Casimir pressure (12.33) is given in terms of the dimensionless variables (12.41) by

$$P(a) = -\frac{\hbar c}{32\pi^2 a^4} \int_0^\infty y^2 \, dy \, B(y), \tag{13.9}$$

$$B(y) = \int_0^y d\zeta \left\{ [r_\mathrm{TM}^{-2}(i\zeta, y) e^y - 1]^{-1} + [r_\mathrm{TE}^{-2}(i\zeta, y) e^y - 1]^{-1} \right\}.$$

Expanding the integrand in $B(y)$ up to the fourth power in δ_0/a and integrating with respect to ζ, we obtain

$$B(y) = \frac{2y}{e^y - 1} - \frac{8}{3}\frac{\delta_0}{a}\frac{y^2 e^y}{(e^y - 1)^2} + \frac{12}{5}\left(\frac{\delta_0}{a}\right)^2 \frac{y^3 e^y (1 + e^y)}{(e^y - 1)^3}$$

$$- \frac{8}{105}\left(\frac{\delta_0}{a}\right)^3 \frac{y^4 e^y (19 + 82 e^y + 19 e^{2y})}{(e^y - 1)^4} \tag{13.10}$$

$$+ \frac{1}{135}\left(\frac{\delta_0}{a}\right)^4 \frac{y^5 e^y (179 + 2621 e^y + 2621 e^{2y} + 179 e^{3y})}{(e^y - 1)^5}.$$

After substitution of this expansion in eqn (13.9) and integration with respect to y, the result is

$$P_\mathrm{p}(a) = -\frac{\pi^2 \hbar c}{240 a^4} \left[1 - \frac{16}{3}\frac{\delta_0}{a} + 24\left(\frac{\delta_0}{a}\right)^2 \right. \tag{13.11}$$

$$\left. -\frac{640}{7}\left(1 - \frac{\pi^2}{210}\right)\left(\frac{\delta_0}{a}\right)^3 + \frac{2800}{9}\left(1 - \frac{163\pi^2}{7350}\right)\left(\frac{\delta_0}{a}\right)^4 \right].$$

The coefficient of δ_0/a was calculated by Hargreaves (1965) [see also Schwinger et al. (1978) and Milonni (1994)]. The coefficient of $(\delta_0/a)^2$ was found by Mostepanenko and Trunov (1985). The coefficients of the third and fourth powers of δ_0/a

were calculated by Bezerra et al. (2000a). Bezerra et al. (2001) have also found the coefficients of the fifth and sixth powers of δ_0/a.

Equations (13.8) and (13.11) were generalized by Geyer et al. (2002) to the case of two semispaces made of dissimilar metals with plasma frequencies $\omega_p^{(1)}$ and $\omega_p^{(2)}$. In this case there are two small parameters δ_1/a and δ_2/a, defined as in eqn (13.3) via the respective plasma wavelengths and plasma frequencies. The resulting Casimir energy per unit area and Casimir pressure are given by (Geyer et al. 2002)

$$E_p(a) = -\frac{\pi^2 \hbar c}{720 a^3} \left\{ 1 - 4\frac{\delta}{a} + \frac{72}{5}\left(\frac{\delta}{a}\right)^2 - \frac{320}{7}\left(\frac{\delta}{a}\right)^3 \left[1 - \frac{2\pi^2}{105}(1-3\kappa)\right] \right.$$
$$\left. + \frac{400}{3}\left(\frac{\delta}{a}\right)^4 \left[1 - \frac{326\pi^2}{3675}(1-3\kappa)\right] \right\}, \qquad (13.12)$$

$$P_p(a) = -\frac{\pi^2 \hbar c}{240 a^4} \left\{ 1 - \frac{16}{3}\frac{\delta}{a} + 24\left(\frac{\delta}{a}\right)^2 - \frac{640}{7}\left(\frac{\delta}{a}\right)^3 \left[1 - \frac{2\pi^2}{105}(1-3\kappa)\right] \right.$$
$$\left. + \frac{2800}{9}\left(\frac{\delta}{a}\right)^4 \left[1 - \frac{326\pi^2}{3675}(1-3\kappa)\right] \right\}.$$

Here, the effective quantities δ and κ are defined as

$$\delta = \frac{\delta_1 + \delta_2}{2}, \qquad \kappa = \frac{\delta_1 \delta_2}{(\delta_1 + \delta_2)^2}. \qquad (13.13)$$

In the case of similar semispaces, $\delta_1 = \delta_2 = \delta_0$, $\delta = \delta_0$, $\kappa = 1/4$, and eqn (13.12) coincides with eqns (13.8) and (13.11).

In what follows, we compare computational results obtained using eqns (13.8) and (13.11) with those obtained by other methods used in the literature at zero temperature.

13.2 Drude model and the Lifshitz formula at zero temperature

Here, we introduce the dielectric permittivity in the Drude model, which has been the subject of many discussions in connection with the thermal Casimir force. We derive the Drude dielectric permittivity starting from the Maxwell equations and present some computational results for the Casimir energy per unit area and the Casimir pressure obtained using the Drude and plasma models.

13.2.1 The Drude dielectric permittivity

As was discussed in Section 12.6.1, one method to describe the behavior of the dielectric permittivity along the imaginary frequency axis is the numerical integration of the optical data for $\text{Im}\,\varepsilon(\omega)$ using the Kramers–Kronig relations. This approach will be used for metals in next section. Here, we note only that for metals, the optical data are usually not known in the required frequency region. Thus, for gold (the most important metal in the experimental investigations of the Casimir force; see Chapters 19 and 20), data are available only at $\omega \geq 0.125\,\text{eV}$. Because of this, it is necessary to extrapolate the available data for $\text{Im}\,\varepsilon(\omega)$ to lower frequencies using some reasonable procedure. It was suggested (Lamoreaux 1999, Lambrecht and Reynaud 2000a, Boström and Sernelius 2000a) that this extrapolation should be performed using the imaginary part of the Drude dielectric permittivity (or dielectric function)

$$\varepsilon_{\text{D}}(\omega) = 1 - \frac{\omega_{\text{p}}^2}{\omega(\omega + i\gamma)}, \tag{13.14}$$

where γ is the relaxation parameter. In contrast to the plasma model, the Drude model takes into account the relaxation processes of conduction electrons.

The Drude dielectric function (or the Drude model) plays an outstanding role in the theory of the Casimir effect between real metals. Because of this, it is necessary to dwell on this subject in more detail and discuss the origin and application region of eqn (13.14). Let us derive the Drude dielectric permittivity (13.14) starting from the Maxwell equations (12.1) for an unbounded nonmagnetic medium without spatial dispersion. As mentioned at the beginning of this chapter, the most important property of metals is their nonzero conductivity at zero temperature. For metals, the electric displacement is determined by

$$\frac{\partial \boldsymbol{D}(t,\boldsymbol{r})}{\partial t} = \frac{\partial \boldsymbol{E}(t,\boldsymbol{r})}{\partial t} + \frac{4\pi}{c}\boldsymbol{j}(t,\boldsymbol{r}), \tag{13.15}$$

where $\boldsymbol{j}(t,\boldsymbol{r}) = \sigma_0 \boldsymbol{E}(t,\boldsymbol{r})$ is the electric current density induced in the metal, and σ_0 is the conductivity at zero frequency. Physically, the condition that the metal is unbounded means that it should be much larger than the size of the wavefront of the electromagnetic waves under consideration. Representing the solutions of the Maxwell equations (12.1) and (13.15) in the form of monochromatic waves (12.10), we obtain

$$\nabla^2 \boldsymbol{E}(\boldsymbol{r}) + k^2 \boldsymbol{E}(\boldsymbol{r}) = 0, \qquad \nabla^2 \boldsymbol{B}(\boldsymbol{r}) + k^2 \boldsymbol{B}(\boldsymbol{r}) = 0, \tag{13.16}$$

where

$$k^2 = \frac{\omega^2}{c^2} + i\frac{4\pi\sigma_0\omega}{c^2} \equiv \varepsilon_{\text{n}}(\omega)\frac{\omega^2}{c^2}. \tag{13.17}$$

Here, the dielectric permittivity for the *normal skin effect* has been introduced,

$$\varepsilon_{\text{n}}(\omega) = 1 + i\frac{4\pi\sigma_0}{\omega}. \tag{13.18}$$

This dielectric permittivity is applicable at sufficiently low frequencies (the region of the normal skin effect), where the relation $\boldsymbol{j} = \sigma_0 \boldsymbol{E}$ is valid. These frequencies are lower than the frequencies of infrared optics. In fact, a permittivity of the form (13.18) has already been used in Chapter 12 for the description of the conductivity of dielectrics at nonzero temperature [see eqn (12.129)].

For the frequency range of interest, one could hope to interpolate between eqn (13.1), applicable at the relatively high frequencies of infrared optics up to ω_p and eqn (13.18), applicable at the relatively low frequencies of the normal skin effect. This interpolation is provided by the replacement

$$\sigma_0 \to \sigma(\omega) = \frac{\sigma_0[1+\mathrm{i}(\omega/\gamma)]}{1+(\omega^2/\gamma^2)}. \tag{13.19}$$

Substituting eqn (13.19) into eqn (13.18) and taking into account the relationship between the static conductivity and the relaxation parameter (Ashcroft and Mermin 1976)

$$\sigma_0 = \frac{\omega_p^2}{4\pi\gamma}, \tag{13.20}$$

we recover the dielectric permittivity of the Drude model (13.14). Thus, the Drude model (13.14) is approximately applicable in a wide frequency region from quasistatic frequencies (where relaxation processes are important) to infrared optics (where relaxation does not play any role). At $\omega \ll \gamma$, one can neglect ω in comparison with $\mathrm{i}\gamma$ in eqn (13.14), and $\varepsilon_D(\omega) \approx \varepsilon_n(\omega)$, whereas at $\omega \gg \gamma$ one can neglect $\mathrm{i}\gamma$ in comparison with ω, and $\varepsilon_D(\omega) \approx \varepsilon_p(\omega)$ (here we disregard the region of frequencies of the anomalous skin effect, which will be considered in Section 13.4 in connection with the Leontovich surface impedance).

The Drude dielectric function has a simple pole at zero frequency. Because of this, the second of the Kramers–Kronig relations (12.124), derived for functions that are regular at $\omega = 0$, does not apply in this case. It is replaced by (Landau et al. 1984)

$$\operatorname{Im}\varepsilon(\omega) = -\frac{1}{\pi}\mathrm{P}\int_{-\infty}^{\infty}\frac{\operatorname{Re}\varepsilon(\xi)}{\xi-\omega}d\xi + \frac{4\pi\sigma_0}{\omega}. \tag{13.21}$$

This, however, does not affect the Kramers–Kronig relation expressing $\varepsilon(\mathrm{i}\xi)$ via $\operatorname{Im}\varepsilon(\omega)$, so that eqn (12.125) remains valid.

It should be stressed that the Drude dielectric permittivity (13.14) is in fact a function of temperature, i.e. $\varepsilon_D \equiv \varepsilon_D(\omega,T)$, through the temperature dependence of the relaxation parameter $\gamma = \gamma(T)$. In Section 14.3.2, this point will be considered in detail, including the functional form of this dependence. However, for the purpose of extrapolation of the tabulated optical data to lower frequencies, the value of the relaxation parameter at room temperature, $T = 300\,\mathrm{K}$, is commonly used (Lamoreaux 1999, Lambrecht and Reynaud 2000a, Boström and Sernelius 2000a). The Casimir energy per unit area and the Casimir pressure at zero temperature calculated in this way have a somewhat dubious physical meaning. On the one hand, the Lifshitz formulas used [eqns (13.4) and (13.9)] describe the Casimir effect at zero temperature. On the other hand, the Drude dielectric function substituted into these formulas is related to the temperature $T = 300\,\mathrm{K}$.

13.2.2 Computations using the plasma and Drude models

Now we consider some computational results obtained using the dielectric permittivities of the plasma and Drude models at zero temperature (with the caveat that the relaxation parameter of the Drude model has been determined at $T = 300$ K). We have used the following values of the Drude parameters for gold: $\omega_\text{p} = 9.0$ eV, and $\gamma = 0.035$ eV (Palik 1985, Lambrecht and Reynaud 2000a). The choice of these values is explained in the next section. We start with the plasma model. Bearing in mind that the Casimir energy per unit area and pressure vary rapidly with a change in separation distance, we present the computational results in the form of correction factors

$$\eta_E(a) = \frac{E_\text{p}(a)}{E(a)}, \qquad \eta_P(a) = \frac{P_\text{p}(a)}{P(a)}. \tag{13.22}$$

Here, $E_\text{p}(a)$ and $P_\text{p}(a)$ are computed by the substitution of the dielectric permittivity of the plasma model (13.1) into the Lifshitz formulas (13.4) and (13.9), respectively. The quantities $E(a)$ and $P(a)$ are the Casimir energy per unit area and pressure for ideal-metal planes defined in eqns (1.5) and (1.1), respectively.

The computational results for the correction factor η_E are presented in Fig. 13.1(a) by the solid line labeled 1 and those for the correction factor η_P by the solid line labeled 2, within the range of separations from 100 nm to 1 μm. The dashed lines 1 and 2 in Fig. 13.1(a) show the same correction factors where $E_\text{p}(a)$ and $P_\text{p}(a)$ are calculated using the perturbative expansions (13.8) and (13.11), respectively. As can be seen in Fig. 13.1(a), the perturbation expansions reproduce the results obtained using the Lifshitz formulas very accurately over a wide range of separations. Relatively large deviations between the dashed and solid lines occur only at separations below the plasma wavelength (for gold, $\lambda_\text{p} = 2\pi c/\omega_\text{p} \approx 137$ nm). Thus, at $a = 130$ nm the relative deviation between the correction factor $\eta_E^{(\text{per})}$ computed using the perturbative formulas and $\eta_E^{(L)}$ computed using the Lifshitz formula, $[\eta_E^{(\text{per})} - \eta_E^{(L)}]/\eta_E^{(L)}$, reaches 3.4%. The relative deviation between the computational results for η_E decreases rapidly with increasing separation. For example, at separations of 150 nm, 200 nm, and 1 μm, this deviation is equal to 1.6%, 0.36%, and 0.01%, respectively. Similarly, at $a = 130$ nm the relative deviation of the correction factors for the Casimir pressure, $[\eta_P^{(\text{per})} - \eta_P^{(L)}]/\eta_P^{(L)}$, is equal to 10.4% (solid and dashed lines labeled 2). At separations of 150 nm, 200 nm, and 1 μm, this deviation is equal to 4.85%, 1.1%, and 0.06%, respectively.

Now we consider the computational results obtained from the substitution of the Drude model (13.14) into the Lifshitz formulas (13.4) for the Casimir energy per unit area and (13.9) for the Casimir pressure. They are presented in the form of correction factors η_E and η_P in Fig. 13.1(b) by the dashed line 1 (for the energy per unit area) and the dashed line 2 (for the pressure) as functions of separation distance. For comparison purposes, the same correction factors as in Fig. 13.1(a), computed using the plasma model, are shown by the solid lines. In Fig. 13.1(b),

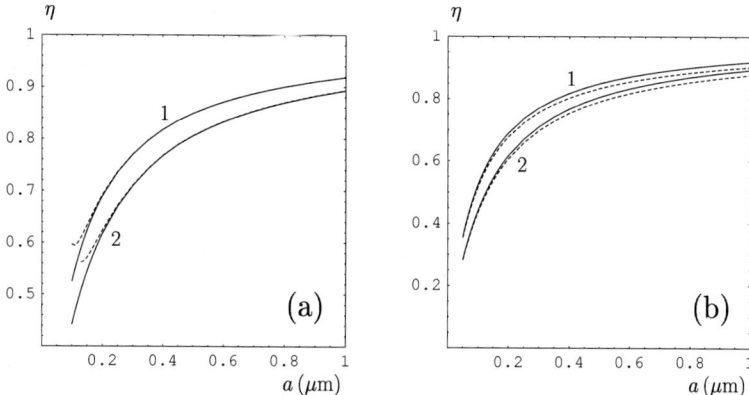

FIG. 13.1. (a) Correction factors to the Casimir energy per unit area, η_E (solid line 1), and to the Casimir pressure, η_P (solid line 2), computed using the Lifshitz formula and the plasma model as functions of separation. The dashed lines 1 and 2 show the correction factors η_E and η_P calculated using the perturbative expansions in the relative skin depth. (b) Correction factors to the Casimir energy per unit area (dashed line 1) and to the Casimir pressure (dashed line 2), computed using the Lifshitz formula and the Drude model. The solid lines 1 and 2 show the correction factors η_E and η_P calculated using the plasma model.

a wider range of separations than in Fig. 13.1(a) has been chosen, i.e. from 50 nm to 1 μm. As is seen in Fig. 13.1(b), at the shortest separations, from 50 nm to 100 nm, the computational results obtained using the Drude and plasma models for both the Casimir energy per unit area and the Casimir pressure are very close. With increasing separation distance, there are slightly increasing deviations between the results obtained with the use of the Drude and plasma models. Thus, for the Casimir energy per unit area the deviation between the correction factors $\eta_E^{(L)}$ computed using the plasma and Drude models normalized to the plasma value is equal to only 1.1% and 1.5% at separations $a = 50$ nm and 100 nm, respectively. At separations $a = 200$ nm and 1 μm, this deviation increases to 1.75% and 1.8%, respectively. For the Casimir pressure, the respective quantity is equal to 0.88% and 1.3% at separations $a = 50$ nm and 100 nm, respectively. At $a = 200$ nm and 1 μm, it reaches 1.7% and 1.8%, respectively.

Thus, at separations below 1 μm, where one might expect that thermal effects are not important, the Drude and plasma models lead to results differing by less than 2%, and this difference decreases with a decrease in separation distance.

13.3 Computations using tabulated optical data

Both the plasma and the Drude model disregard important physical processes determined by interband transitions of core electrons. As was mentioned in Section 13.2.1, the dielectric permittivity along the imaginary frequency axis $\varepsilon(i\xi)$

can be obtained using the Kramers–Kronig relation (12.125) through the numerical integration of tabulated optical data for $\operatorname{Im}\varepsilon(\omega)$. As an example, we shall illustrate this procedure using gold semispaces described by the tabulated optical data collected by Palik (1985).

At high frequencies the available data extend up to $\omega = 10\,000$ eV. This is quite sufficient to compute highly accurate values of $\varepsilon(i\xi)$ at all frequencies contributing to the Casimir energy per unit area (13.4) and to the Casimir pressure (13.9). Note that the characteristic frequency at a typical separation $a = 0.5\,\mu$m is equal to $\omega_c = c/(2a) = 3 \times 10^{14}$ rad/s $= 0.197$ eV. Numerical analysis of the Lifshitz formulas demonstrates that highly accurate results (up to five significant figures) can be obtained if the upper integration limit with respect to ξ is chosen equal to a value from $10\omega_c$ to $15\omega_c$ instead of infinity. A further increase of the upper integration limit does not noticeably influence the results obtained.

The situation at low frequencies is more complicated. Here, optical data are available only for $\omega \geq 0.125$ eV, i.e. starting from a frequency of the same order as ω_c. This does not permit one to obtain $\varepsilon(i\xi)$ from the Kramers–Kronig relation using the tabulated data alone. At the same time, numerical analysis of the Lifshitz formulas shows that to obtain highly accurate results, one must take the lower integration limit with respect to ξ as low as $10^{-3}\omega_c$ instead of zero (a further decrease of the lower integration limit has only a negligible influence on the results obtained). Thus, values of $\varepsilon(i\xi)$ at frequencies $\xi \geq 10^{-3}\omega_c$ must be available in order to obtain highly accurate values of the Casimir energy per unit area and the Casimir pressure. For this reason, it is necessary to extrapolate the tabulated optical data for gold to frequencies $\omega < 0.125$ eV using some reasonable analytical procedure. Lamoreaux (1999) suggested extrapolation of the data to lower frequencies using $1/\omega$, the expected behavior for a metal. This is the same behavior as that given by the Drude model (13.14). Extrapolation of the optical data for $\operatorname{Im}\varepsilon(\omega)$ to low frequencies by use of the imaginary part of the Drude dielectric permittivity,

$$\operatorname{Im}\varepsilon_\mathrm{D}(\omega) = \frac{\omega_\mathrm{p}^2 \gamma}{\omega(\omega^2 + \gamma^2)}, \qquad (13.23)$$

was done in papers by Lambrecht and Reynaud (2000a,b), Boström and Sernelius (2000a), and Klimchitskaya et al. (2000). The results of these four papers are in mutual agreement. It was also shown that the computations by Lamoreaux (1999) contained errors in the interpolation and extrapolation procedures. On the basis of the optical data collected by Palik (1985), Lambrecht and Reynaud (2000a) arrived at the following values of the Drude parameters at room temperature:

$$\omega_\mathrm{p} = 9.0\,\mathrm{eV}, \qquad \gamma = 0.035\,\mathrm{eV}. \qquad (13.24)$$

Different sets of the Drude parameters and their impact on the calculated values of $\varepsilon(i\xi)$ and the Casimir pressure have been discussed by Pirozhenko et al. (2006). We shall return to this point in Section 18.3 in connection with the comparison between experiment and theory. It should be emphasized that only the room

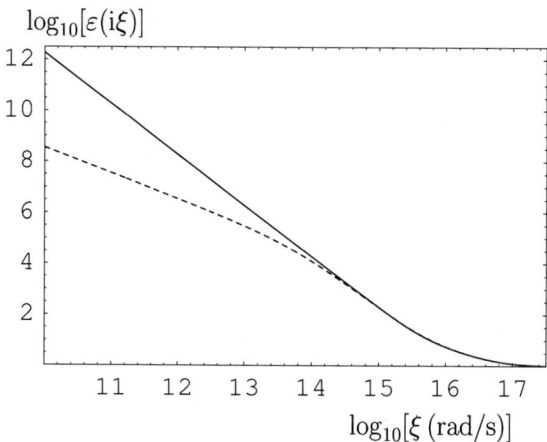

FIG. 13.2. The dashed line shows the dielectric permittivity of gold along the imaginary frequency axis computed using the Kramers–Kronig relations and tabulated optical data extrapolated to low frequencies by use of the Drude dielectric function, as discussed in the text. The solid line shows the same permittivity computed using the generalized plasma model (see Section 13.5).

temperature (300 K) value of the relaxation parameter and the optical data were used.

In Fig. 13.2 (dashed line), we present computational results for $\varepsilon(i\xi)$, as a function of ξ, over a wide frequency region from 10^{10} rad/s to 3.16×10^{17} rad/s. These results were obtained from the Kramers–Kronig relation (12.125) using the tabulated optical data for $\mathrm{Im}\,\varepsilon(\omega)$ (Palik 1985), extrapolated to low frequencies by use of eqn (13.23) with the Drude parameters (13.24). The $\varepsilon(i\xi)$ obtained can be used for computations of the Casimir effect at separations $a > 5$ nm, where the characteristic frequencies are less than 3×10^{16} rad/s. The dielectric function $\varepsilon(i\xi)$ was substituted into the Lifshitz formulas at zero temperature (13.4) and (13.9) for the Casimir energy per unit area and the pressure, written in terms of dimensionless variables, which are more convenient for numerical integration. The results were computed over a range of separations from $a = 50$ nm to $a = 1\,\mu$m (note that computations at larger separations using the Lifshitz formula at $T = 0$ would be meaningless because of the role played by thermal effects).

In Fig. 13.3, the correction factor to the Casimir energy per unit area, η_E, calculated using the tabulated optical data extrapolated by use of the Drude model to low frequencies, is shown by the solid line labeled 1. The dashed line labeled 1 shows the same correction factor calculated using the Drude model (13.14) over the entire frequency range from zero to infinity. As is seen in Fig. 13.3, at large separations the deviations between the solid and dashed lines labeled 1 are minimal. Thus, at $a = 1\,\mu$m the deviation between the correction factors η_E computed using the optical data and the Drude model, normalized to

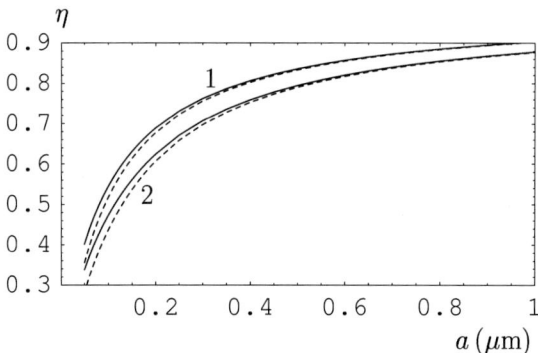

FIG. 13.3. Correction factors to the Casimir energy per unit area, η_E (solid line 1), and to the Casimir pressure, η_P (solid line 2), as functions of separation computed using the Lifshitz formula and the tabulated optical data extrapolated to low frequencies by use of the Drude model. The dashed lines 1 and 2 show the correction factors η_E and η_P computed using just the Drude model.

η_E computed using the optical data, is equal to only 0.08%. At lower separations of $a = 200$ nm, 100 nm, and 50 nm, the relative deviation between the correction factors computed in the two different ways is equal to 1.7%, 5%, and 12%, respectively. This deviation is connected with the fact that the Drude model does not take into account the interband transitions of core electrons occurring at high frequencies that contribute more and more to the Casimir energy as the separation distance decreases.

Similar calculations can be performed for the Casimir pressure. The correction factor η_P calculated using the tabulated optical data extrapolated by use of the Drude model to low frequencies is shown by the solid line labeled 2 in Fig. 13.3. The dashed line labeled 2 indicates the factor η_P computed using the Drude model (13.14). At the largest separation $a = 1\,\mu\text{m}$, the relative deviation between the factors calculated by the two different methods is equal to 0.14%. However, this deviation increases rapidly with decreasing the separation distance. As an illustration, at separations of $a = 200$ nm, 100 nm, and 50 nm this deviation is equal to 2.8%, 7.7%, and 16%, respectively. What this means is that the interband transitions of core electrons neglected by the Drude model but taken into account in the optical data are an even more important factor in the Casimir pressure than in the Casimir energy. At separations below 50 nm, the deviation between the computational results obtained using the optical data and the simple Drude model increases further. In fact, the Lifshitz formula at zero temperature using the tabulated optical data extrapolated by use of the Drude model provides an adequate approach to the calculation of van der Waals forces in the range of separations from a few nanometers to several tens of nanometers. Its applicability at larger separations will be discussed in Chapter 14.

13.4 Surface impedance approach

The reflection properties of electromagnetic waves on metal surfaces can be described in terms of the Leontovich surface impedance (Landau et al. 1984), as an alternative concept to the frequency-dependent permittivity. In this section we present the analytic expressions for the Leontovich impedance in different frequency regions and discuss the impedance representations of the reflection coefficients.

13.4.1 The concept of the Leontovich impedance

The possibility of introducing the Leontovich impedance Z is based on the fact that the angle of transmission for an electromagnetic wave inside a metal near its surface is approximately zero, i.e. the plane wave propagates perpendicular to the surface. As a consequence, the fields $\boldsymbol{E}_t$ and $\boldsymbol{B}_t$ are related by

$$\boldsymbol{E}_t = Z[\boldsymbol{H}_t \times \boldsymbol{n}], \qquad (13.25)$$

where $\boldsymbol{n}$ is a unit vector normal to the surface directed inside the metal. According to Leontovich, eqn (13.25) can be used as a boundary condition imposed on a metal surface for the determination of the field outside the metal. Equation (13.25) allows one to determine the electromagnetic field outside the metal, without considering the propagation of the electromagnetic field in the interior of the metal. It is close in spirit to the original Casimir approach for ideal metals and to a special reformulation of the standard continuity boundary conditions (12.2), which also does not require consideration of field propagation inside a metal (Emig and Büscher 2004). For an ideal metal, $Z = 0$. The boundary condition (13.25) is valid when $|Z| \ll 1$. For good conductors, this inequality is satisfied over a wide frequency region.

The asymptotic form of the Leontovich impedance in various frequency regions can be obtained from the solution of the kinetic equations (Lifshitz and Pitaevskii 1981). We consider a conducting semispace in an external electromagnetic field of frequency ω. At low frequencies (the region of the normal skin effect), the following conditions hold:

$$l \ll \delta_n(\omega) = \frac{c}{\sqrt{2\pi\sigma_0\omega}}, \qquad l \ll \frac{v_F}{\omega}, \qquad (13.26)$$

where l is the mean free path of the conduction electrons, $\delta_n(\omega)$ is the penetration depth of the field inside the metal, σ_0 is the static conductivity, and v_F is the Fermi velocity. In the frequency region determined by the inequalities (13.26), the external field leads to the initiation of a real current of the conduction electrons. The normal skin effect is characterized by a volume relaxation described by the relaxation parameter $\gamma = \gamma(T)$ (see Section 13.2.1). As a result, the mean free path of the conduction electrons is also temperature-dependent, $l = l(T) = v_F/\gamma(T)$, and increases with a decrease in temperature (this is discussed in more detail in Section 14.4). In the domain of the normal skin effect, the surface impedance takes the form (Landau et al. 1984)

$$Z = Z_\mathrm{n}(\omega) = (1-\mathrm{i})\sqrt{\frac{\omega}{8\pi\sigma_0}}. \tag{13.27}$$

Note that Re Z describes the relaxation processes and the inequality Re $Z > 0$ is always satisfied. As discussed in Section 13.2.1, in the region of the normal skin effect the connection between the electric field and electric current has the form

$$\boldsymbol{j}(t,\boldsymbol{r}) = \sigma_0 \boldsymbol{E}(t,\boldsymbol{r}), \tag{13.28}$$

i.e. it is local.

At higher frequencies or larger l, eqn (13.28) is violated and the connection between the electric field and the current becomes nonlocal. This is what is referred to as the *anomalous skin effect*. This occurs when the inequalities

$$\delta_\mathrm{a}(\omega) \ll l, \qquad \delta_\mathrm{a}(\omega) \ll \frac{v_\mathrm{F}}{\omega}, \tag{13.29}$$

are obeyed, where for an isotropic metal with a spherical Fermi surface, the skin depth is given by (Kaganova and Kaganov 2001)

$$\delta_\mathrm{a}(\omega) = \left(\frac{4\pi c^2 \hbar^3}{\omega e^2 S_\mathrm{F}}\right)^{1/3}, \tag{13.30}$$

where S_F is the total area of the Fermi surface. Owing to the spatial nonlocality, the concept of a dielectric permittivity depending only on the frequency is incorrect for the description of a metal in the frequency region of the anomalous skin effect. The Leontovich impedance in the frequency region (13.29) is given by (Kaganova and Kaganov 2001)

$$Z_\mathrm{a}(\omega) = \frac{2(1-\mathrm{i}\sqrt{3})}{3\sqrt{3}} \frac{\omega \delta_\mathrm{a}(\omega)}{c}, \tag{13.31}$$

where $\delta_\mathrm{a}(\omega)$ is defined in (13.30). Note that for some metals, especially for alloys, instead of eqn (13.29), the inequalities $v_\mathrm{F}/\omega \ll l \ll \delta$ hold, specifying the *relaxation domain* (Wooten 1972, Casimir and Ubbink 1967); here, spatial relaxation is also important.

On further increase of the frequency, the inequalities

$$\frac{v_\mathrm{F}}{\omega} \ll \delta_0 \ll l \tag{13.32}$$

hold, where the penetration depth δ_0 is defined in eqn (13.3). These inequalities determine the frequency region of *infrared optics* (note that the condition $\hbar\omega \ll \varepsilon_\mathrm{F} = \hbar\omega_\mathrm{p}$ is also obeyed here). In this domain, volume relaxation does not play any role and spatial dispersion is absent. The respective impedance function is pure imaginary (Lifshitz and Pitaevskii 1981):

$$Z_i(\omega) = -i\frac{\omega}{\sqrt{\omega_p^2 - \omega^2}}. \qquad (13.33)$$

In the frequency regions of the normal skin effect and of infrared optics, where spatial locality is preserved, we get

$$Z(\omega) = \frac{1}{\sqrt{\varepsilon(\omega)}}. \qquad (13.34)$$

However, in the frequency region of the anomalous skin effect, spatial locality is violated and the metal cannot be characterized by a dielectric permittivity depending only on the frequency. In this frequency region, eqn (13.34) is meaningless but, nevertheless, the impedance boundary condition (13.25) is appropriate.

13.4.2 The Lifshitz formula with the Leontovich impedance

The derivation of the photon eigenfrequencies between two metallic semispaces $z \geq a/2$ and $z \leq -a/2$ is performed in the same way as in Section 12.1.1. Now the electromagnetic field $\boldsymbol{E}(t, \boldsymbol{r})$, $\boldsymbol{B}(t, \boldsymbol{r})$ is considered only in the free space between the semispaces. The electric field and magnetic induction are represented in the form of eqns (12.10) and (12.13). This leads to eqn (12.14), following from the Maxwell equations, where the k^2 defined in eqn (12.15) must be replaced with q^2 defined in eqn (12.16). The equations (12.17) remain unchanged. Now we need to satisfy the impedance boundary conditions (13.25) imposed at $z = \pm a/2$. Expressing the magnetic induction using the Maxwell equations and taking the direction of the normal vector into account [$\boldsymbol{n} = (0, 0, 1)$ at the plane $z = a/2$ and $\boldsymbol{n} = (0, 0, -1)$ at $z = -a/2$], we find

$$e_{p,x}\left(\pm\frac{a}{2}, k_\perp\right) = \pm\frac{iZc}{\omega}\left[ik_x e_{p,z}\left(\pm\frac{a}{2}, k_\perp\right) - e'_{p,x}\left(\pm\frac{a}{2}, k_\perp\right)\right],$$

$$e_{p,y}\left(\pm\frac{a}{2}, k_\perp\right) = \pm\frac{iZc}{\omega}\left[ik_y e_{p,z}\left(\pm\frac{a}{2}, k_\perp\right) - e'_{p,y}\left(\pm\frac{a}{2}, k_\perp\right)\right]. \qquad (13.35)$$

Similar equations also hold for $\boldsymbol{g}_p$. The equations (12.14) (where k^2 is replaced with q^2) and (12.17) together with eqn (13.35) lead to the following equations for the determination of the photon eigenfrequencies (Geyer et al. 2003):

$$\Delta^{\mathrm{TM}}(\omega, k_\perp) = e^{-aq}(1 - \eta^2)\left(\sinh aq - \frac{2i\eta}{1 - \eta^2}\cosh aq\right) = 0,$$

$$\Delta^{\mathrm{TE}}(\omega, k_\perp) = e^{-aq}(1 - \kappa^2)\left(\sinh aq - \frac{2i\kappa}{1 - \kappa^2}\cosh aq\right) = 0, \qquad (13.36)$$

where

$$\eta = \eta(\omega) = \frac{Z\omega}{cq}, \qquad \kappa = \kappa(\omega) = \frac{Zcq}{\omega}. \tag{13.37}$$

The calculation of the Casimir energy $E_0(a)$ is performed analogously to that in Section 12.1.1 using the argument principle. One must take into account that in this case, if the impedance functions satisfy the conditions

$$\lim_{\omega\to\infty} Z(\omega) = \text{const}, \qquad \lim_{\omega\to\infty} \frac{dZ(\omega)}{d\omega} = 0, \tag{13.38}$$

the integral along the semicircle of infinitely large radius C_+ vanishes. As in Section 12.1.1, the finite Casimir energy per unit area $E(a)$ is obtained by the subtraction from $E_0(a)$ of the energy for infinitely separated plates [see eqn (12.27)]. Thus, $E(a)$ is represented by eqn (12.29) with

$$\Delta_\infty^{\text{TM}}(\mathrm{i}\xi, k_\perp) = \frac{1}{2}\left[1 + \frac{Z(\mathrm{i}\xi)\xi}{cq(\mathrm{i}\xi, k_\perp)}\right]^2, \quad \Delta_\infty^{\text{TE}}(\mathrm{i}\xi, k_\perp) = \frac{1}{2}\left[1 + \frac{Z(\mathrm{i}\xi)cq(\mathrm{i}\xi, k_\perp)}{\xi}\right]^2, \tag{13.39}$$

where q is defined in eqn (12.28). Finally, the Casimir energy per unit area is given by the Lifshitz formula (12.30), where new expressions for the reflection coefficients are obtained from eqns (13.36) and (13.39) as follows:

$$r_{\text{TM}}(\mathrm{i}\xi, k_\perp) = \frac{cq(\mathrm{i}\xi, k_\perp) - Z(\mathrm{i}\xi)\xi}{cq(\mathrm{i}\xi, k_\perp) + Z(\mathrm{i}\xi)\xi}, \quad r_{\text{TE}}(\mathrm{i}\xi, k_\perp) = \frac{cq(\mathrm{i}\xi, k_\perp)Z(\mathrm{i}\xi) - \xi}{cq(\mathrm{i}\xi, k_\perp)Z(\mathrm{i}\xi) + \xi}. \tag{13.40}$$

The Casimir pressure expressed in terms of the Leontovich impedance is given by eqn (12.33) with the reflection coefficients (13.40). A Lifshitz formula with the impedance reflection coefficients (13.40) was obtained by Kats (1977).

The Casimir energy per unit area and the Casimir pressure, as given by the Lifshitz formulas with the impedance reflection coefficients (13.40), were computed by Bezerra et al. (2002c). The results obtained were compared with those obtained from the Lifshitz formulas with the reflection coefficients (12.31) expressed in terms of the dielectric permittivity. In the range of separations from 100 nm to 1 μm, the dielectric permittivity (13.1) and the Leontovich impedance (13.33) for infrared optics are applicable. It was shown that the difference between the Casimir energies computed using ε_{p} and Z_{i} normalized to the energy computed using ε_{p} was less than 0.3% in the range of separations from 100 nm to 400 nm. In the range from 400 nm to 1 μm, the computational results obtained from ε_{p} and Z_{i} were practically coincident. Similar computational results were obtained for the Casimir pressure. The largest relative deviation, of 0.5%, between the Casimir pressures computed using ε_{p} and Z_{i} was reached at $a = 100$ nm. With increasing separation, this deviation quickly goes to zero. The application regions of the impedances for the anomalous skin effect (13.31) and for the normal skin effect (13.27) are at separations $a > 1$ μm, where thermal effects are important. That is why these impedances are considered in Chapter 14, devoted to the Casimir effect between real metals at nonzero temperature.

Sometimes, in the literature, the so-called *exact surface impedances* are considered, along with the related boundary conditions (Esquivel et al. 2003, Brevik et al. 2005)

$$\boldsymbol{E}_\mathrm{t} = Z_\mathrm{TE}(\omega,k_\perp)[\boldsymbol{H}_\mathrm{t} \times \boldsymbol{n}], \qquad Z_\mathrm{TM}(\omega,k_\perp)\boldsymbol{H}_\mathrm{t} = \boldsymbol{n} \times \boldsymbol{E}_\mathrm{t}. \qquad (13.41)$$

If the connection between the electric field and electric current is local, i.e. is given by eqn (13.28), the Maxwell equations lead to

$$Z_\mathrm{TM}(\omega,k_\perp) = \frac{1}{\sqrt{\varepsilon(\omega)}}\left[1 - \frac{c^2 k_\perp^2}{\varepsilon(\omega)\omega^2}\right]^{1/2},$$

$$Z_\mathrm{TE}(\omega,k_\perp) = \frac{1}{\sqrt{\varepsilon(\omega)}}\left[1 - \frac{c^2 k_\perp^2}{\varepsilon(\omega)\omega^2}\right]^{-1/2}. \qquad (13.42)$$

In terms of these impedances, the reflection coefficients take the form

$$r_\mathrm{TM}(\mathrm{i}\xi,k_\perp) = \frac{cq - Z_\mathrm{TM}\xi}{cq + Z_\mathrm{TM}\xi}, \qquad r_\mathrm{TE}(\mathrm{i}\xi,k_\perp) = \frac{cqZ_\mathrm{TE} - \xi}{cqZ_\mathrm{TE} + \xi}. \qquad (13.43)$$

The impedances (13.42) coincide with the Leontovich impedance (13.34) under the condition $c^2 k_\perp^2/(|\varepsilon|\omega^2) \ll 1$. However, the Leontovich impedance and the Leontovich boundary condition (13.25) should not be considered as only approximations to the impedances (13.42) and boundary conditions (13.41). A point to be noted is that the impedances (13.42) are obtained from the Maxwell equations by using the local equation (13.28), whereas the Leontovich impedance is applicable when the connection between the electric field and the current becomes nonlocal. Thus, the physical meaning of the Leontovich impedance and the Leontovich boundary condition (13.25) is different from that of the impedances (13.42) and the "exact" impedance boundary conditions (13.41). In fact, the substitution of the impedances (13.42) in eqn (13.43) simply returns us back to the reflection coefficients (12.31) expressed in terms of the frequency-dependent dielectric permittivity. In this sense, the use of the "exact" impedances is equivalent to the use of the dielectric permittivity, while the use of the Leontovich surface impedance is not. In Chapter 14, we shall see that this leads to important consequences for the theoretical description of the thermal Casimir force.

13.5 The generalized plasma-like dielectric permittivity

In the above, we have considered several models for the dielectric response of a metal used in the literature to calculate the Casimir force at zero temperature in the framework of the Lifshitz theory. These are the plasma model (13.1), the Drude model (13.14), the dielectric function obtained using tabulated optical data extrapolated to low frequencies by use of the Drude model, and the Leontovich surface impedance. Although each model leads to different results for the Casimir energy per unit area and for the Casimir pressure, the approach based on the use of real optical data may seem preferable because it takes more complete account of all of the available information about the dielectric properties of the metal. As was shown in Section 12.5.2, however, some real properties of dielectrics and semiconductors (i.e. the conductivity at nonzero temperature) must not be included in the model of the dielectric response if we wish to preserve the agreement between the Lifshitz theory and thermodynamics. As we shall see in Chapter 14, the relaxation properties of conduction electrons play the same crucial role in the theory of the thermal Casimir force, leading to its inconsistency with thermodynamics. Because of this, here we consider a model of a metal in terms of a generalized plasma-like permittivity taking into account all of its properties with the exception of the relaxation properties of the conduction electrons. These properties are primarily due to the electron–phonon interaction and also due to the scattering of electrons on other electrons, on impurities, and on grain boundaries. In reality, the relaxation properties are almost absent at $T = 0$, but they enter into computations performed using the tabulated optical data and the parameters of the Drude model determined at $T = 300\,\text{K}$ (this inconsistency in the formalism used was discussed at the end of Section 13.2.1). Here, we consider the formulation of the generalized plasma-like model, derive the generalized Kramers–Kronig relations, and present computational results.

13.5.1 Generalized plasma-like permittivity and optical data

A generalized plasma-like dielectric permittivity was considered, for instance, by Jackson (1999) for the description of a metal at frequencies much larger than the Drude relaxation parameter. In the theory of the Casimir effect, it was first used by Klimchitskaya *et al.* (2007a). We represent it in a form similar to the permittivity of a dielectric material (12.8),

$$\varepsilon_{\text{gp}}(\omega) = 1 - \frac{\omega_{\text{p}}^2}{\omega^2} + \sum_{j=1}^{K} \frac{g_j}{\omega_j^2 - \omega^2 - i\gamma_j\omega}. \tag{13.44}$$

We recall that the $\omega_j \neq 0$ are the resonant frequencies of the oscillators describing the core electrons, the γ_j are the relaxation frequencies, and the g_j are the oscillator strengths. The term $-\omega_{\text{p}}^2/\omega^2$ takes the conduction electrons into account with their relaxation properties being ignored.

For any metal, the parameters of the oscillators ω_j, g_j, and γ_j and the number of oscillators K must be determined from a fit to the available optical data. As was explained in Sections 12.6.1 and 13.3, it is convenient to perform a fit to the imaginary part of the dielectric permittivity (13.44) given by eqn (12.126). Before making a fit, we must subtract from the optical data the contribution from the relaxation properties of the conduction electrons. This can be approximately done by subtraction from $\text{Im}\,\varepsilon(\omega)$ in eqn (12.123), as determined from the complete optical data, of the imaginary part of the Drude function (13.23). The resulting expression for $\text{Im}\,\varepsilon^{\text{s}}(\omega)$ to be used in the fit is

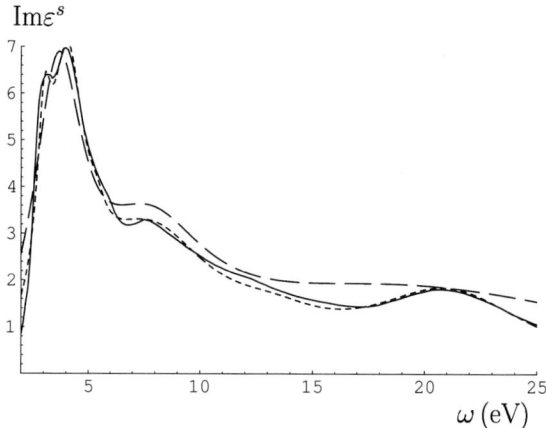

FIG. 13.4. Tabulated optical data for the imaginary part of the dielectric permittivity for gold with the contribution from the relaxation of conduction electrons subtracted, shown by the solid line, as a function of frequency (Decca et al. 2007b). The oscillator fits are shown by the long-dashed line (three oscillators) and the short-dashed line (six oscillators).

$$\mathrm{Im}\,\varepsilon^{\mathrm{s}}(\omega) = 2n_1(\omega)n_2(\omega) - \frac{\omega_{\mathrm{p}}^2 \gamma}{\omega(\omega^2 + \gamma^2)}, \qquad (13.45)$$

where n_1 and n_2 are the real and imaginary parts of the complex index of refraction.

As an example, we have performed a six-oscillator fit of eqn (12.126) to eqn (13.45) for gold by using the tabulated optical data from the handbook by Palik (1985), and $\omega_{\mathrm{p}} = 9.0\,\mathrm{eV}$ and $\gamma = 0.035\,\mathrm{eV}$ (Lambrecht and Reynaud 2000a). In Fig. 13.4, we plot the quantity $\mathrm{Im}\,\varepsilon^{\mathrm{s}}(\omega)$, as defined in eqn (13.45), within the frequency region from 2.0 eV to 25 eV (solid line). For $\omega < 2\,\mathrm{eV}$, the imaginary part of the dielectric permittivity (13.45) vanishes rapidly because we have subtracted the contribution from the relaxation properties of the conduction electrons. For $\omega > 2.5\,\mathrm{eV}$, there is only a negligible contribution from the conduction electrons, so that $\mathrm{Im}\,\varepsilon(\omega) \approx \mathrm{Im}\,\varepsilon^{\mathrm{s}}(\omega)$. The upper limit of the frequency region under consideration (25 eV) is sufficiently high to consider separation distances above $a = 100\,\mathrm{nm}$. The resulting set of oscillator parameters ω_j, γ_j, and g_j is presented in Table 13.1. Also, in Fig. 13.4, the imaginary part (12.126) of the permittivity (13.44), computed with the oscillator parameters in Table 13.1, is shown by the short-dashed line. In the same figure, the three-oscillator fit by Parsegian and Weiss (1981) is shown by the long-dashed line. As is seen in Fig. 13.4, the short-dashed line based on the six-oscillator fit reproduces the actual data better than does the long-dashed line using the three-oscillator fit.

Using the oscillator parameters in Table 13.1, the dielectric permittivity along the imaginary frequency axis can be found from eqn (13.44):

TABLE 13.1. The oscillator parameters for Au in eqn (13.44) found from the six-oscillator fit to the tabulated optical data (Palik 1985).

j	ω_j (eV)	γ_j (eV)	g_j (eV^2)
1	3.05	0.75	7.091
2	4.15	1.85	41.46
3	5.4	1.0	2.700
4	8.5	7.0	154.7
5	13.5	6.0	44.55
6	21.5	9.0	309.6

$$\varepsilon_{\text{gp}}(i\xi) = 1 + \frac{\omega_p^2}{\xi^2} + \sum_{j=1}^{6} \frac{g_j}{\omega_j^2 + \xi^2 + \gamma_j \xi}. \tag{13.46}$$

This correctly reproduces $\varepsilon_{\text{gp}}(i\xi)$ for the generalized plasma-like permittivity, as a function of imaginary frequency, up to $\xi \approx 15\,\text{eV}$. This corresponds to approximately $15\omega_c$, where the characteristic frequency ω_c is computed for a separation $a = 100\,\text{nm}$. At shorter separations, a more precise determination of $\varepsilon_{\text{gp}}(i\xi)$ at higher ξ is needed. This can be done by numerical integration of the optical data with the contribution of the relaxation of conduction electrons subtracted, as given in eqn (13.45), by using the Kramers–Kronig relations. However, the standard Kramers–Kronig relations (12.124), (12.125), and (13.21) are not applicable because they do not take into account the presence of the second-order pole at $\omega = 0$ in eqn (13.44).

13.5.2 Generalized Kramers–Kronig relations for the plasma and plasma-like permittivities

The derivation of the Kramers–Kronig relations for the dielectric permittivity (13.44) is direct, but one must take care of the fact that there is a pole of second order at $\omega = 0$. We consider the contour C_1 presented in Fig. 13.5(a). According to the Cauchy theorem,

$$\int_{C_1} \frac{\varepsilon(\omega) - 1}{\omega - \omega_0} d\omega = 0, \tag{13.47}$$

where $\varepsilon(\omega)$ is equal to either $\varepsilon_p(\omega)$ or $\varepsilon_{\text{gp}}(\omega)$, because inside C_1 the function under the integral is analytic. At infinity, $\epsilon(\omega) \to 1$ and therefore the function under the integral tends to zero more rapidly than $1/\omega$. Thus, the integral along the semicircle of infinite radius is equal to zero. We pass around the points 0 and ω_0 along the semicircles C_ρ and C_δ with radii ρ and δ, respectively. It can be easily seen that

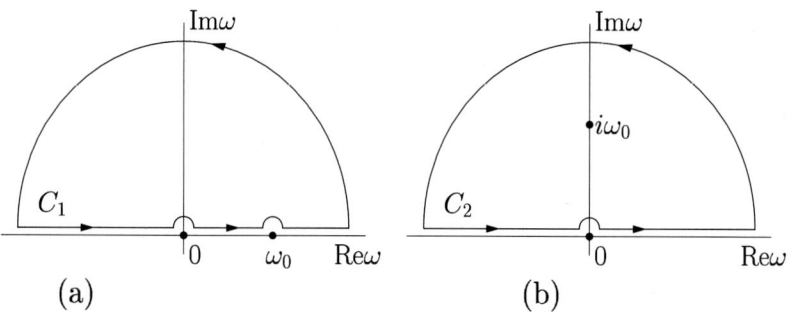

FIG. 13.5. The integration contours in (a) eqn (13.47) and (b) eqn (13.58), consisting of the real frequency axis, semicircles C_ρ and C_δ around the points 0 and ω_0 of infinitely small radii ρ and δ, respectively, and semicircles of infinitely large radii centered at $\omega = 0$ (Klimchitskaya et al. 2007a).

$$\int_{C_\delta} \frac{\varepsilon(\omega) - 1}{\omega - \omega_0} d\omega = -\pi i \operatorname{Res} \left. \frac{\varepsilon(\omega) - 1}{\omega - \omega_0} \right|_{\omega=\omega_0} = -\pi i [\varepsilon(\omega_0) - 1]. \quad (13.48)$$

The similar integral around the point $\omega = 0$ is more involved. Using eqn (13.44), we represent it as the sum of the integral

$$\int_{C_\rho} \frac{A(\omega)}{\omega - \omega_0} d\omega = -\frac{2A(0)}{\omega_0} \rho, \quad (13.49)$$

where

$$A(\omega) \equiv \sum_{j=1}^{K} \frac{g_j}{\omega_j^2 - \omega^2 - i\gamma_j \omega}, \quad (13.50)$$

and

$$-\omega_p^2 \int_{C_\rho} \frac{d\omega}{\omega^2(\omega - \omega_0)} \equiv \frac{\omega_p^2}{\omega_0^2} \int_{C_\rho} \left[\frac{\omega_0}{\omega^2} - \frac{1}{\omega - \omega_0} + \frac{1}{\omega} \right] d\omega. \quad (13.51)$$

It is clear that eqn (13.49) vanishes when $\rho \to 0$. The second and third integrals on the right-hand side of eqn (13.51) are calculated as follows:

$$\int_{C_\rho} \frac{d\omega}{\omega - \omega_0} = -\frac{2}{\omega_0} \rho, \qquad \int_{C_\rho} \frac{d\omega}{\omega} = -\pi i. \quad (13.52)$$

The first integral can be represented as the principal value of an integral of another function along the real frequency axis:

$$\int_{C_\rho} \frac{d\omega}{\omega^2} = \frac{2}{\rho} = \omega_0 P \int_{-\infty}^{\infty} \frac{d\omega}{\omega^2(\omega - \omega_0)}. \quad (13.53)$$

This integral is divergent. To be exact, in eqn (13.47) it must be considered together with the integral

$$\text{P} \int_{-\infty}^{\infty} \frac{\varepsilon(\omega) - 1}{\omega - \omega_0} d\omega. \tag{13.54}$$

By doing accordingly and substituting eqns (13.48), (13.49), and (13.51)–(13.53) into eqn (13.47), we arrive at

$$-\frac{i\pi\omega_p^2}{\omega_0^2} - i\pi[\varepsilon(\omega_0) - 1] + \text{P} \int_{-\infty}^{\infty} \frac{d\omega}{\omega - \omega_0} \left[\varepsilon(\omega) - 1 + \frac{\omega_p^2}{\omega^2}\right] = 0. \tag{13.55}$$

The last integral on the left-hand side of eqn (13.55) is already finite, as expected.

Now we replace the integration variable ω with ξ, replace ω_0 with ω, and represent the function $\varepsilon(\omega)$ as $\varepsilon(\omega) = \operatorname{Re}\varepsilon(\omega) + i\operatorname{Im}\varepsilon(\omega)$. Taking into account that

$$\text{P} \int_{-\infty}^{\infty} \frac{d\omega}{\omega - \omega_0} = 0 \tag{13.56}$$

and separating the real and imaginary parts in eqn (13.55), we obtain the generalized Kramers–Kronig relations (Klimchitskaya et al. 2007a)

$$\operatorname{Re}\varepsilon(\omega) = 1 + \frac{1}{\pi}\text{P} \int_{-\infty}^{\infty} \frac{\operatorname{Im}\varepsilon(\xi)}{\xi - \omega} d\xi - \frac{\omega_p^2}{\omega^2}, \tag{13.57}$$

$$\operatorname{Im}\varepsilon(\omega) = -\frac{1}{\pi}\text{P} \int_{-\infty}^{\infty} \frac{d\xi}{\xi - \omega} \left[\operatorname{Re}\varepsilon(\xi) + \frac{\omega_p^2}{\xi^2}\right]$$

[in contrast to the standard relations (12.124) for an $\varepsilon(\omega)$ which is regular at $\omega = 0$, or to (13.21) for an $\varepsilon(\omega)$ with a simple pole at $\omega = 0$].

The dielectric permittivity along the imaginary frequency axis can be determined through the use of the integral

$$\int_{C_2} \frac{\omega[\varepsilon(\omega) - 1]}{\omega^2 + \omega_0^2} d\omega = \pi i[\varepsilon(i\omega_0) - 1] \tag{13.58}$$

along the contour C_2 in Fig. 13.5(b). By integrating along C_2, we get

$$i\pi \frac{\omega_p^2}{\omega_0^2} + \text{P} \int_{-\infty}^{\infty} \frac{\omega[\varepsilon(\omega) - 1]}{\omega^2 + \omega_0^2} d\omega = i\pi[\varepsilon(i\omega_0) - 1]. \tag{13.59}$$

Now we replace ω_0 with ξ, separate the real and imaginary parts of $\varepsilon(\omega)$ under the integral, and make use of the identities

$$\text{P} \int_{-\infty}^{\infty} \frac{\xi \, d\xi}{\xi^2 + \omega^2} = 0, \qquad \text{P} \int_{-\infty}^{\infty} \frac{\xi \operatorname{Re}\varepsilon(\xi) \, d\xi}{\xi^2 + \omega^2} = 0. \tag{13.60}$$

Thus, the result to be compared with eqn (12.125) is (Klimchitskaya et al. 2007a)

$$\varepsilon(i\xi) = 1 + \frac{2}{\pi} \int_0^{\infty} \frac{\omega \operatorname{Im}\varepsilon(\omega)}{\omega^2 + \xi^2} d\omega + \frac{\omega_p^2}{\xi^2}. \tag{13.61}$$

For the usual plasma model with $\varepsilon = \varepsilon_p$ in eqn (13.1), $\operatorname{Im}\varepsilon(\omega) = 0$ and the generalized Kramers–Kronig relations (13.57) and (13.61) are satisfied identically. Direct substitution of eqn (13.44) into eqns (13.57) and (13.61) shows that the generalized plasma-like dielectric permittivity ε_{gp} exactly satisfies the generalized Kramers–Kronig relations as well (Klimchitskaya et al. 2007a).

Now we are in a position to calculate the generalized plasma-like dielectric permittivity along the imaginary frequency axis directly from the tabulated optical data without using the analytic representation (13.44). For this purpose we substitute the imaginary part of the dielectric permittivity (13.45), as given by the optical data after subtraction of the contribution of the relaxation of conduction electrons, into the right-hand side of the Kramers–Kronig relation (13.61). The $\varepsilon_{\text{gp}}(i\xi)$ obtained, as a function of the frequency, is plotted by the solid line in Fig. 13.2 in the frequency interval from 10^{10} rad/s to 3.16×10^{17} rad/s. This interval can be easily widened using the available data at higher frequencies. In the frequency region $\xi < 2.3 \times 10^{16}$ rad/s, the same values of $\varepsilon_{\text{gp}}(i\xi)$ are obtained analytically from eqn (13.46) with the oscillator parameters in Table 13.1. At higher frequencies, there are minor deviations between the results of the numerical computations of $\varepsilon_{\text{gp}}(i\xi)$, as given by the solid line in Fig. 13.2, and the analytical eqn (13.46). To obtain a more exact analytic representation for $\varepsilon_{\text{gp}}(i\xi)$ over a wider frequency region, one must perform an oscillator fit with a larger number of oscillators valid at higher frequencies (i.e. at frequencies above $25\,\text{eV} = 3.8 \times 10^{16}$ rad/s).

Note that at frequencies larger than 10^{15} rad/s, the generalized plasma-like dielectric permittivity along the imaginary frequency axis is equal to that obtained using the optical data extrapolated to low frequencies by means of the Drude model. From this it follows that at short separation distances, the calculation results for the Casimir effect at zero temperature using the two permittivities should be almost equal.

13.5.3 Computations using the generalized plasma-like model

We have substituted the dielectric permittivity of the generalized plasma-like model along the imaginary frequency axis, $\varepsilon_{\text{gp}}(i\xi)$, as shown by the solid line in Fig. 13.2, into the Lifshitz formula (13.4) for the Casimir energy per unit area. The results obtained, $E_{\text{gp}}(a)$, were normalized by the Casimir energy between ideal-metal planes $E(a)$ defined in eqn (1.5). The resulting correction factor $\eta_E(a) = E_{\text{gp}}(a)/E(a)$, as a function of separation in the region from $a = 50$ nm to $a = 1\,\mu$m, is shown by the solid line in Fig. 13.6(a). In the same figure, the dashed line shows the correction factor $\eta_E(a)$ computed by using the tabulated optical data extrapolated to low frequencies by means of the Drude model. As expected, the largest deviation between the two factors is at the largest separation of $a = 1\,\mu$m. Here, the difference between $\eta_E(a)$ computed using the optical data extrapolated by use of the Drude model and $\eta_E(a)$ computed with the generalized plasma model, normalized by $\eta_E(a)$ found using the optical data and the Drude model, reaches -1.7%. At smaller separations, the relative difference between the correction factors computed by the two different methods decreases rapidly. Thus, at separations $a = 200$ nm, 100 nm, and 50 nm this relative difference is equal to -1.2%, -0.7%, and 0.02%, respectively. This is explained by the

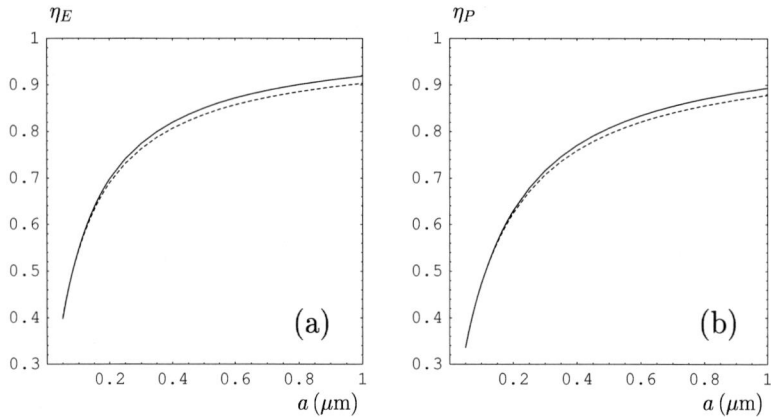

FIG. 13.6. Correction factors as functions of separation (a) to the Casimir energy per unit area and (b) to the Casimir pressure, calculated using the dielectric permittivity of the generalized plasma-like model (solid lines) and using tabulated optical data extrapolated to low frequencies by use of the Drude model (dashed lines).

decreasing role of low frequencies, where the dielectric permittivities in the two approaches differ, in the calculation results at smaller separations.

The Casimir pressure for the generalized plasma-like model, $P_{\text{gp}}(a)$, was computed by substitution of the dielectric permittivity shown by the solid line in Fig. 13.2 into the Lifshitz formula (13.9). The resulting correction factor $\eta_P = P_{\text{gp}}(a)/P(a)$, as a function of separation distance, is shown in Fig. 13.6(b) [$P(a)$ is defined in eqn (1.1)]. In the same figure, the correction factor η_P computed using the tabulated optical data extrapolated by use of the Drude model to low frequencies is shown by the dashed line. The largest relative deviation between the two values of η_P computed by the two different methods indicated above is equal to -1.7% at a separation $a = 1\,\mu\text{m}$. At separations $a = 200\,\text{nm}$, $100\,\text{nm}$, and $50\,\text{nm}$ the relative deviation between the two calculated values of η_P is equal to -0.86%, -0.18%, and 0.44%, respectively. Similarly to the case of the Casimir energy, the effect of the differences between the dielectric permittivities represented by the solid and dashed lines in Fig. 13.2 on the correction factor η_P decreases at small separations.

At the end of this section, we note that the application of the generalized plasma model at zero temperature is free of any of the ambiguity discussed in Section 13.2.1 with respect to the Drude model. This is because the generalized plasma-like dielectric permittivity ignores the temperature-dependent relaxation processes of conduction electrons and takes into account the relaxation due to core electrons alone, which depends negligibly on the temperature.

14

THE CASIMIR INTERACTION BETWEEN REAL METALS AT NONZERO TEMPERATURE

As discussed in Chapter 12, the theoretical description of the thermal Casimir force between dielectrics at nonzero temperature runs into puzzling difficulties. It appears that the conductivity of a dielectric material resulting from nonzero temperature should not be included in the model of the dielectric response if one wishes to comply with the applicability conditions of the Lifshitz theory and preserve its agreement with thermodynamics.

Conductivity is an inherent property of metals at any temperature. At nonzero temperature, conductivity is inevitably linked with the dissipation of energy. Because of this, one may expect that for real metals the difficulties experienced with the Lifshitz theory discussed in the case of dielectrics may deepen. This is indeed so, and historically such problems with the Lifshitz theory were first discovered not for dielectrics but for metals. In particular, it was shown that the substitution of the Drude dielectric function into the Lifshitz formula at nonzero temperature leads to a violation of Nernst's heat theorem in the case of metals with perfect crystal lattices (Bezerra *et al.* 2002a, 2002b, 2004). This unexpected result gave rise to several different opinions in the literature, and a consensus has not yet been reached. Because of this, we present below several different points of view and several different approaches to the resolution of the problem contained in the literature.

We show that the root of the problem is in the discontinuity of the zero-frequency term of the Lifshitz formula for the TE mode, as a function of frequency and temperature, which arises when the Drude model is used to characterize the low-frequency dielectric permittivity of metals. The thermal Casimir free energy and pressure are computed using both the plasma and the Drude model, and using tabulated optical data for the complex index of refraction extrapolated by use of the Drude model to low frequencies. The thermodynamic consistency of the plasma and Drude models is analyzed. It is shown that the plasma model combined with the Lifshitz formula is in agreement with thermodynamics, while the Drude model is not if the metal crystal lattice is perfect. For lattices with impurities the Drude model satisfies the Nernst theorem, although the Casimir entropy is negative over a wide temperature region. We present physical arguments for why the Drude model is outside the application region of the Lifshitz formula. The approximate approach based on the Leontovich surface impedance is considered and its consistency with thermodynamics is investigated. The role of evanescent and traveling waves in the Casimir effect between metals is also

14.1 The problem associated with the zero-frequency term in the Lifshitz formula

As was discussed in Section 12.5.2, for dielectrics, the problem concerning the physical role of the dc conductivity is connected with the discontinuity of the zero-frequency transverse magnetic coefficient r_{TM}, as a function of frequency and temperature. With the dc conductivity disregarded, this coefficient is given by eqn (12.107), but with the dc conductivity included at $T \neq 0$, it is equal to unity in accordance with eqn (12.117). A similar situation arises for metals, but the roles of the reflection coefficients for the two different polarizations of the electromagnetic field are reversed.

We first recall that eqns (12.66)–(12.68), representing the Casimir free energy per unit area for dielectric semispaces, are also applicable to metals with an appropriately chosen dielectric permittivity. It is useful to begin with the case of ideal metals and compare eqn (7.70), obtained in this case, with the Lifshitz formula (12.66). From this comparison, we arrive at the conclusion that the Casimir interaction between ideal metals is described by the Lifshitz formula with reflection coefficients equal to plus or minus unity at any frequency:

$$r_{\text{TM}}(i\xi_l, k_\perp) = 1, \qquad r_{\text{TE}}(i\xi_l, k_\perp) = -1. \tag{14.1}$$

This conclusion is in agreement with eqn (12.67), expressing the reflection coefficients under the assumption that for an ideal metal ε_l is infinitely large at any l. There is, however, a problem concerning the value of $r_{\text{TE}}(i\xi_l, k_\perp)$ at $l = 0$. It contains a term $\varepsilon_l \xi_l^2 / c^2$ through the definition of k_l in eqn (12.68) which makes the value of k_0 indefinite when $\varepsilon_0 = \infty$ and $\xi_0 = 0$. To avoid this difficulty and achieve agreement between the definition of the reflection coefficients and the Lifshitz formula on the one hand, and the independently obtained Casimir free energy between ideal metals on the other hand, Schwinger et al. (1978) proposed a special prescription (in the literature, this is often referred to as *Schwinger's prescription*). According to this prescription, the limit $\varepsilon \to \infty$ should be taken first, which provides agreement with both of the equalities in eqn (14.1) at all $\xi > 0$, and the limit $\xi \to 0$, to obtain the zero-frequency term of the Lifshitz formula (12.66), should be taken afterwards.

The Schwinger prescription works well when we are dealing with ideal metals. For real metals, however, the dielectric permittivity is finite at any $\xi > 0$. The functional form of its approach to infinity when ξ goes to zero is fixed. This converts the limiting value of

$$\frac{1}{c^2} \lim_{\xi \to 0} \varepsilon(i\xi) \xi^2 \tag{14.2}$$

into a subject for calculation and does not leave room for any special prescriptions.

From this point of view, let us consider the reflection coefficients at $\xi = 0$ and the zero-frequency term of the Lifshitz formula for the various models of real metals used in Chapter 13. Substituting the dielectric permittivity of the plasma model (13.1) or the generalized plasma-like model (13.46) into the reflection coefficients (12.66), we arrive at

$$r_{\text{TM}}(0, k_\perp) = 1, \qquad r_{\text{TE}}(0, k_\perp) = \frac{k_\perp c - \sqrt{k_\perp^2 c^2 + \omega_{\text{p}}^2}}{k_\perp c + \sqrt{k_\perp^2 c^2 + \omega_{\text{p}}^2}}. \qquad (14.3)$$

The case of an ideal metal results from ε_{p} in the limit $\omega_{\text{p}} \to \infty$. In this limiting case, eqn (14.3) coincides with eqn (14.1). Thus, descriptions of real metals using the plasma model or the generalized plasma-like model in the Lifshitz theory are in agreement with the independently investigated case of ideal metals.

This conclusion is supported by a consideration of large separation distances (or, alternatively, high temperatures), where, as discussed in Section 12.5.1, the zero-frequency term of the Lifshitz formula determines the total result. Substituting eqn (14.3) into the zero-frequency term of eqns (12.66) and (12.70), we obtain asymptotic expressions for the Casimir free energy per unit area and the Casimir pressure, as given by the plasma model,

$$\mathcal{F}_{\text{p}}(a, T) = -\frac{k_B T \zeta_R(3)}{16\pi a^2} + \frac{k_B T}{16\pi a^2} \int_0^\infty y\, dy \ln\left[1 - \frac{\omega_c^4}{\omega_{\text{p}}^4}\left(y - \sqrt{y^2 + \frac{\omega_{\text{p}}^2}{\omega_c^2}}\right)^4 e^{-y}\right],$$

$$(14.4)$$

$$P_{\text{p}}(a, T) = -\frac{k_B T \zeta_R(3)}{8\pi a^3} - \frac{k_B T}{16\pi a^3} \int_0^\infty y^2\, dy \left[\frac{\omega_c^4}{\omega_{\text{p}}^4}\left(y + \sqrt{y^2 + \frac{\omega_{\text{p}}^2}{\omega_c^2}}\right)^4 e^y - 1\right]^{-1},$$

where $\omega_c = c/(2a)$. In the limiting case $\omega_{\text{p}} \to \infty$, eqn (14.4) leads to

$$\mathcal{F}_{\text{p}}(a, T) = -\frac{k_B T \zeta_R(3)}{16\pi a^2} + \frac{k_B T}{16\pi a^2} \int_0^\infty y\, dy \ln\left(1 - e^{-y}\right) = -\frac{k_B T \zeta_R(3)}{8\pi a^2},$$

$$(14.5)$$

$$P_{\text{p}}(a, T) = -\frac{k_B T \zeta_R(3)}{8\pi a^3} - \frac{k_B T}{16\pi a^3} \int_0^\infty \frac{y^2\, dy}{e^y - 1} = -\frac{k_B T \zeta_R(3)}{4\pi a^3},$$

i.e. it returns us to eqns (7.95) and (7.98) obtained for ideal metals. Thus, real metals described by the plasma model or the generalized plasma-like model satisfy the *classical limit* discussed in Section 7.4.3 (Feinberg et al. 2001, Scardicchio and Jaffe 2006).

A different situation arises when real metals are characterized by the dielectric permittivity of the Drude model (13.14) or any other dielectric permittivity inversely proportional to the first power of the frequency [see e.g. the dielectric permittivity for the normal skin effect in eqn (13.18)]. This case is very important, because the same behavior of the permittivity follows from the Maxwell equations in the region of quasistatic frequencies (see Section 13.2.1). Substituting the dielectric permittivity of the Drude model (13.14) into the reflection coefficients (12.67), we obtain

$$r_{\mathrm{TM}}(0, k_\perp) = 1, \qquad r_{\mathrm{TE}}(0, k_\perp) = 0. \tag{14.6}$$

The difference between the second equations in eqns (14.3) and (14.6) is caused by the fact that for the plasma and plasma-like models the limit in eqn (14.2) is equal to ω_p^2/c^2, but for the Drude model this limit is equal to zero.

The second equality in eqn (14.6) is in contrast to the second equality in eqn (14.1) given for the case of ideal metals. This shows that the description of metals using the Drude model is in direct disagreement with the idealized case of ideal metals. This can be clearly seen from the asymptotic limit of large separations (high temperatures), where the zero-frequency term of the Lifshitz formula alone determines the total result. Substituting eqn (14.6) into the terms with $l = 0$ in eqns (12.66) and (12.70), we obtain asymptotic expressions for the Casimir free energy and pressure, as given by the Drude model,

$$\mathcal{F}_{\mathrm{D}}(a, T) = -\frac{k_{\mathrm{B}} T}{16 \pi a^2} \zeta_{\mathrm{R}}(3), \qquad P_{\mathrm{D}}(a, T) = -\frac{k_{\mathrm{B}} T}{8 \pi a^3} \zeta_{\mathrm{R}}(3). \tag{14.7}$$

These expressions are just one-half of the respective expressions (7.95) and (7.98) obtained in the case of ideal metals. In fact, the results (14.7) are counterintuitive because at large separation distances the corrections due to the finite conductivity of a metal do not play any role, and real metals should behave like ideal ones. Moreover, eqn (14.7) does not make any distinction between real metals with different relaxation parameters γ. The relaxation parameter may be arbitrarily small but, surprisingly, the asymptotic values for the Casimir free energy and pressure (14.7) are equal to one-half of those obtained for ideal metals. Thus, the theoretical description using the Drude model violates the classical limit given by eqns (7.95) and (7.98). The missing contribution from the transverse electric mode at zero frequency in the Lifshitz formula combined with the Drude model was first mentioned by Barash (1988). Recently, attention to this problem has been rekindled in relation to the work by Boström and Sernelius (2000b), who predicted a large thermal correction to the Casimir force at short separations using the Drude model (see Section 14.3.1 below). At large separations, eqn (14.7) was also reobtained using a microscopic statistical approach to the Casimir force (Buenzli and Martin 2008). In this approach, plates are considered as a set of nonrelativistic point particles of several types confined by walls.

The peculiarities in the behavior of the reflection coefficient $r_{\mathrm{TE}}(i\xi, k_\perp)$ in the case of the Drude dielectric permittivity are connected with the fact that this coefficient is discontinuous at the point $\xi = 0$, $k_\perp = 0$ as a function of the two variables ξ and $k_\perp$. To be a continuous function of $(\xi, k_\perp)$ at the point $(0,0)$, $r_{\rm TE}(i\xi, k_\perp)$ must have the same limiting value when the point $(\xi, k_\perp)$ approaches the point $(0,0)$ along any path. This is, however, not the case. Substituting the Drude dielectric function (13.14) into eqn (12.31), one obtains

$$r_{\rm TE}(i\xi, k_\perp) = \frac{(c^2 k_\perp^2 + \xi^2)^{1/2} - \{c^2 k_\perp^2 + \xi^2 + \omega_p^2[\xi/(\xi+\gamma)]\}^{1/2}}{(c^2 k_\perp^2 + \xi^2)^{1/2} + \{c^2 k_\perp^2 + \xi^2 + \omega_p^2[\xi/(\xi+\gamma)]\}^{1/2}}. \tag{14.8}$$

If the point $(\xi, k_\perp)$ approaches the point $(0,0)$ along the path $k_\perp = 0$, we obtain $r_{\rm TE} \to -1$ from eqn (14.8). The same holds if $(\xi, k_\perp)$ approaches $(0,0)$ along any path $k_\perp = \alpha\xi$ with $\alpha \neq 0$. In contrast, if the point $(\xi, k_\perp)$ approaches the point $(0,0)$ along the path $\xi = 0$, as happens in the zero-frequency term of the Lifshitz formula, we get $r_{\rm TE} \to 0$. Note that for real electromagnetic waves satisfying the mass-shell equation $c^2 k^2 = \omega^2$ incident on real metals, the reflection coefficients always satisfy eqn (14.1) in the limit of zero frequency and $k_\perp$ (the Hagen–Rubens law). The same holds for the reflection coefficients (14.3) determined by the plasma model. Actually, from eqn (14.3), one obtains $r_{\rm TM} \to 1$ and $r_{\rm TE} \to -1$ when $k_\perp \to 0$, as it should be. This is connected with the fact that both of the reflection coefficients $r_{\rm TM}(i\xi, k_\perp)$ and $r_{\rm TE}(i\xi, k_\perp)$ determined by the permittivity of the plasma model are continuous functions of the two variables ξ and $k_\perp$ at the point $(0,0)$. Below, we consider the physical consequences which follow when different dielectric permittivities are used to compute the thermal Casimir force. The zero-frequency term of the Lifshitz formula calculated using the Leontovich impedance is considered separately in Section 14.4.

14.2 Perturbation theory for metals described by the plasma model

Here, we consider in more detail the *plasma model approach* to the thermal Casimir force, based on the use of the Lifshitz formulas at nonzero temperature (12.66) and (12.70) combined with the permittivity of the plasma model (13.1). This approach was suggested independently by Genet *et al.* (2000) and by Bordag *et al.* (2000a). In the paper by Genet *et al.* (2000), numerical computations were performed and analytic espressions were obtained up to first order in the relative skin depth of electromagnetic oscillations in the metal defined in eqn (13.3). In the paper by Bordag *et al.* (2000a), corrections to the Casimir force due to both nonzero temperature and the skin depth were treated perturbatively to several perturbation orders. The analytic results were found to be in good agreement with the results of the numerical computations. We present numerical computations using the plasma model in Section 14.3.1, where they are compared with corresponding computations using the Drude model. In this section, we restrict ourselves to the use of perturbation theory and demonstrate the agreement of the results obtained with the Nernst heat theorem.

14.2.1 Casimir free energy per unit area and Casimir pressure

We start with the Lifshitz formula for the Casimir free energy per unit area (12.108) expressed in terms of the dimensionless variables (12.89). For simplicity, we consider the case of similar semispaces, so that $r_{\text{TM}}^{(1)} = r_{\text{TM}}^{(2)}$ and $r_{\text{TE}}^{(1)} = r_{\text{TE}}^{(2)}$. Applying the Poisson summation formulas (5.18) and (5.19), we obtain from eqn (12.108)

$$\mathcal{F}(a,T) = \frac{\hbar c}{16\pi^2 a^3} \sum_{l=0}^{\infty}{}' \int_0^\infty y\,dy \int_0^y d\zeta \cos(lt\zeta) \qquad (14.9)$$
$$\times \left\{ \ln\left[1 - r_{\text{TM}}^2(i\zeta, y)e^{-y}\right] + \ln\left[1 - r_{\text{TE}}^2(i\zeta, y)e^{-y}\right] \right\}.$$

In accordance with eqn (12.90), the term of the sum with $l = 0$ is the Casimir energy per unit area at zero temperature. The terms with $l \geq 1$ represent the temperature corrections to it, $\Delta_T \mathcal{F}(a,T)$ (recall that the permittivity of the plasma model is independent of the temperature). In accordance with eqn (14.9), the temperature correction can be represented as a sum

$$\Delta_T \mathcal{F}(a,T) = \Delta_T^{\text{TM}} \mathcal{F}(a,T) + \Delta_T^{\text{TE}} \mathcal{F}(a,T), \qquad (14.10)$$

where

$$\Delta_T^{\text{TM}} \mathcal{F}(a,T) = \frac{\hbar c}{16\pi^2 a^3} \sum_{l=1}^{\infty} \int_0^\infty y\,dy \int_0^y d\zeta \cos(lt\zeta) \ln\left[1 - r_{\text{TM}}^2(i\zeta, y)e^{-y}\right], \qquad (14.11)$$

and similarly for $\Delta_T^{\text{TE}} \mathcal{F}(a,T)$ with the index TM replaced with TE. The reflection coefficients are defined in eqn (12.91).

We are now going to account fully for the thermal effects. The effect of the skin depth will be treated perturbatively up to the second power of the small parameter δ_0/a defined in eqn (13.3). We substitute the dielectric permittivity of the plasma model (13.5) into eqn (14.11) (and into a similar equation for the TE mode) and use the following expansions:

$$\ln\left[1 - r_{\text{TM}}^2(i\zeta, y)e^{-y}\right] = \ln(1 - e^{-y}) + 2\frac{\delta_0}{a}\frac{\zeta^2}{y(e^y - 1)} - 2\left(\frac{\delta_0}{a}\right)^2 \frac{\zeta^4 e^y}{y^2(e^y - 1)^2}, \qquad (14.12)$$

$$\ln\left[1 - r_{\text{TE}}^2(i\zeta, y)e^{-y}\right] = \ln(1 - e^{-y}) + 2\frac{\delta_0}{a}\frac{y}{e^y - 1} - 2\left(\frac{\delta_0}{a}\right)^2 \frac{y^2 e^y}{(e^y - 1)^2}.$$

The integrations of eqn (14.12) with respect to ζ in eqn (14.11) are trivial. All further integrations with respect to y as required in eqn (14.11) can be performed with the help of the table by Gradshteyn and Ryzhik (1994). Putting together the contributions from both modes, the total thermal correction for two semispaces made of a metal described by the plasma model takes the form [see Geyer et al. (2001) for details]

$$\Delta_T \mathcal{F}_p(a,T) = -\frac{\hbar c}{8\pi^2 a^3} \sum_{l=1}^{\infty} \left\{ \frac{\pi}{2(lt)^3} \coth(\pi lt) - \frac{1}{(lt)^4} + \frac{\pi^2}{2(lt)^2} \frac{1}{\sinh^2(\pi lt)} \right.$$
$$+ \frac{\delta_0}{a} \left[\frac{\pi}{(lt)^3} \coth(\pi lt) - \frac{4}{(lt)^4} + \frac{\pi^2}{(lt)^2} \frac{1}{\sinh^2(\pi lt)} + \frac{2\pi^3}{lt} \frac{\coth(\pi lt)}{\sinh^2(\pi lt)} \right] \quad (14.13)$$
$$- \left(\frac{\delta_0}{a}\right)^2 \left[\frac{\pi}{(lt)^5} + \frac{2\pi^4}{\sinh^2(\pi lt)} \left(1 - 3\coth^2(\pi lt) + \frac{\coth(\pi lt)}{\pi lt} - \frac{1}{(\pi lt)^2} \right) \right.$$
$$\left. + \frac{12}{(lt)^4} \ln(1 - e^{-2\pi lt}) - \frac{12\pi}{(lt)^3 (e^{2\pi lt} - 1)} - \frac{6}{\pi (lt)^5} \text{Li}_2(e^{-2\pi lt}) \right] \Bigg\}.$$

Here, $t = T_{\text{eff}}/T$ and the effective temperature T_{eff} is defined in eqn (12.85). The zeroth-order terms in δ_0/a on the right-hand side of eqn (14.13) coincide with the thermal correction for an ideal metal in eqn (7.81) if one takes into account that $\zeta_R(4) = \pi^4/90$.

Now we consider the limiting cases of eqn (14.13) at low and high temperature (or small and large separation, respectively). At low temperature, $t \gg 1$, and, keeping only the largest of the exponentially small contributions, we obtain

$$\Delta_T \mathcal{F}_p(a,T) = -\frac{\hbar c}{8\pi a^3} \left\{ \frac{\zeta_R(3)}{2t^3} - \frac{\pi^3}{90 t^4} + \frac{2\pi}{t^2} e^{-2\pi t} \right. \quad (14.14)$$
$$\left. + \frac{\delta_0}{a} \left[\frac{\zeta_R(3)}{t^3} - \frac{2\pi^3}{45 t^4} + \frac{8\pi^2}{t} e^{-2\pi t} \right] - \left(\frac{\delta_0}{a}\right)^2 \left[\frac{\zeta_R(5)}{t^5} - 16\pi^3 e^{-2\pi t} \right] \right\}.$$

The power-type corrections of zeroth order in δ_0/a in eqn (14.14) coincide with the thermal correction for an ideal metal at low temperature in eqn (7.88), and the exponentially small term in it coincides with the respective result (7.89) obtained for an ideal metal. We remark that the second-order term in δ_0/a in eqn (14.14) contains term in powers of t, not just the exponentially small term. It is, however, of order t^{-5} in agreement with the result of Bordag et al. (2000a), who proved the absence of thermal corrections with powers lower than $1/t^5$ in the terms of order $(\delta_0/a)^k$ with $k = 2, 3, 4, 5$, and 6.

Now we consider the limiting case of high temperature (or large separation). It is easier to extract it directly from eqns (14.10)–(14.12) than from eqn (14.13). To do this, we must perform the integration with respect to ζ in the same manner as above and then change the order of summation and integration with respect to y. Owing to the small value of the parameter t, all summations can be performed by the use of the formula (Gradshteyn and Ryzhik 1994)

$$\sum_{l=1}^{\infty} \frac{\sin(lty)}{l} = \frac{\pi - ty}{2}, \quad (14.15)$$

which is valid under the condition $0 < ty < 2\pi$. For the integration with respect to y, all functions under the integrals decrease with y as $\exp(-y)$, so the infinity in the upper integration limit can be replaced with the required accuracy by $\tilde{y} = 2\pi/t - \alpha$, where $\alpha > 0$. As a result, the high-temperature limit of the thermal correction to the Casimir energy per unit area with neglect of exponentially small corrections is given by

$$\Delta_T \mathcal{F}_p(a,T) = -\frac{\hbar c}{8\pi a^3} \left\{ \frac{\zeta_R(3)}{2t} - \frac{\pi^3}{90} \right. \tag{14.16}$$

$$\left. + \frac{\delta_0}{a} \left[-\frac{\zeta_R(3)}{t} + \frac{2\pi^3}{45} \right] + \left(\frac{\delta_0}{a}\right)^2 \left[\frac{3\zeta_R(3)}{t} - \frac{4\pi^3}{25} \right] \right\}.$$

The terms of zeroth order in δ_0/a on the right-hand side of eqn (14.16) coincide with the high-temperature limit of the thermal correction for an ideal metal, $\mathcal{F}(a,T) - E(a)$, where $\mathcal{F}(a,T)$ and $E(a)$ are defined in eqns (7.95) and (1.5), respectively.

Similar analytic expressions can be obtained for the Casimir pressure calculated using the Lifshitz formula (12.70) and the plasma dielectric function (13.5). In terms of the dimensionless variables (12.89), the Casimir pressure takes the form

$$P(a,T) = -\frac{\hbar c}{16\pi^2 a^4} \sum_{l=0}^{\infty}{}' \int_{\zeta_l}^{\infty} y^2 \, dy \left\{ [r_{\text{TM}}^{-2}(i\zeta_l, y)e^y - 1]^{-1} + [r_{\text{TE}}^{-2}(i\zeta, y)e^y - 1]^{-1} \right\}, \tag{14.17}$$

where the reflection coefficients are given in eqn (12.91). Applying the Poisson summation formula (5.18), (5.19), we rearrange eqn (14.17) to the form

$$P(a,T) = -\frac{\hbar c}{16\pi^2 a^4} \sum_{l=0}^{\infty}{}' \int_0^{\infty} y^2 \, dy \int_0^y d\zeta \cos(lt\zeta) \tag{14.18}$$

$$\times \left\{ [r_{\text{TM}}^{-2}(i\zeta,y)e^y - 1]^{-1} + [r_{\text{TE}}^{-2}(i\zeta,y)e^y - 1]^{-1} \right\}.$$

In the same way as in the case of the Casimir free energy (14.9), the term with $l = 0$ is the Casimir pressure at zero temperature, $P(a)$, whereas the terms with $l \geq 1$ represent the thermal correction

$$\Delta_T P(a,T) = -\frac{\hbar c}{16\pi^2 a^4} \sum_{l=1}^{\infty} \int_0^{\infty} y^2 \, dy \int_0^y d\zeta \cos(lt\zeta) \tag{14.19}$$

$$\times \left\{ [r_{\text{TM}}^{-2}(i\zeta,y)e^y - 1]^{-1} + [r_{\text{TE}}^{-2}(i\zeta,y)e^y - 1]^{-1} \right\}.$$

To obtain an analytic expression for the thermal correction to the Casimir pressure, we expand the terms in eqn (14.19) containing the reflection coefficients up to the second power of the small parameter δ_0/a:

$$[r_{\text{TM}}^{-2}(i\zeta,y)e^y - 1]^{-1} = \frac{1}{e^y - 1} - 2\frac{\delta_0}{a}\frac{\zeta^2 e^y}{y(e^y - 1)^2} + 2\left(\frac{\delta_0}{a}\right)^2 \frac{\zeta^4 e^y(e^y + 1)}{y^2(e^y - 1)^3},$$

$$\left[r_{\text{TE}}^{-2}(i\zeta, y)e^y - 1\right]^{-1} = \frac{1}{e^y - 1} - 2\frac{\delta_0}{a}\frac{ye^y}{(e^y - 1)^2} + 2\left(\frac{\delta_0}{a}\right)^2 \frac{y^2 e^y (e^y + 1)}{(e^y - 1)^3}. \quad (14.20)$$

Substituting eqn (14.20) into eqn (14.19) and calculating all integrals with respect to ζ and y, we obtain the thermal correction to the Casimir pressure,

$$\Delta_T P_{\text{p}}(a, T) = -\frac{\hbar c}{8\pi^2 a^4} \sum_{l=1}^{\infty} \left\{ \frac{1}{(lt)^4} - \frac{\pi^3}{lt}\frac{\cosh(\pi lt)}{\sinh^3(\pi lt)} \right. \quad (14.21)$$

$$+ \frac{\delta_0}{a}\frac{\pi^3}{lt \sinh 2(\pi lt)}\left[\frac{1}{(\pi lt)^2}\sinh(\pi lt)\cosh(\pi lt)\right.$$

$$\left. + 4\coth(\pi lt) + 2\pi lt - 6\pi lt \coth^2(\pi lt) + \frac{1}{\pi lt}\right]$$

$$+ 3\left(\frac{\delta_0}{a}\right)^2 \frac{\pi^3}{lt \sinh^2(\pi lt)}\left[-4\pi lt + 5(\pi lt)^2 \coth(\pi lt)\right.$$

$$\left. \left. + 12\pi lt \coth^2(\pi lt) - 8(\pi lt)^2 \coth^3(\pi lt) - 4\coth(\pi lt)\right]\right\}.$$

Note that the first two terms on the right-hand side of this equation, which are of zeroth order in δ_0/a, coincide with the thermal correction to the Casimir pressure for ideal-metal planes in eqn (7.82).

In the case of low temperature (or small separation), where $t \gg 1$, we can replace the hyperbolic functions with their asymptotic expressions. Keeping the exponentially small contributions and performing summations, we obtain from eqn (14.21)

$$\Delta_T P_{\text{p}}(a, T) = -\frac{\hbar c}{8\pi a^4}\left\{ \frac{\pi^3}{90 t^4} - \frac{4\pi^2}{t}e^{-2\pi t} \right. \quad (14.22)$$

$$\left. + \frac{\delta_0}{a}\left[\frac{\zeta_R(3)}{t^3} - 16\pi^3 e^{-2\pi t}\right] - 36\pi^4 t \left(\frac{\delta_0}{a}\right)^2 e^{-2\pi t}\right\}.$$

The leading term of zeroth order in δ_0/a in this equation coincides with the thermal correction for the case of an ideal metal in the asymptotic expression (7.90). The exponentially small correction to it coincides with that obtained for an ideal metal in eqn (7.91).

The case of high temperature (or large separation) can be considered in the same way as for the Casimir energy. We start directly from eqns (14.21) and (14.20) and use eqn (14.15). The result is

$$\Delta_T P_{\text{p}}(a, T) = -\frac{\hbar c}{8\pi a^4}\left\{ \frac{\zeta_R(3)}{t} - \frac{\pi^3}{30} \right. \quad (14.23)$$

$$+ \frac{\delta_0}{a}\left[-\frac{3\zeta_R(3)}{t} + \frac{8\pi^3}{45}\right] + \left(\frac{\delta_0}{a}\right)^2\left[\frac{12\zeta_R(3)}{t} - \frac{4\pi^3}{5}\right]\right\}.$$

Here, the first two terms of zeroth order in δ_0/a coincide with $P(a,T) - P(a)$, where $P(a,T)$ and $P(a)$ are defined in eqns (7.98) and (1.1), respectively, i.e. they coincide with the thermal correction to the Casimir pressure in the case of an ideal metal.

Equations (14.13) and (14.21) take account exactly of the thermal effects but are restricted to a second-order expansion with respect to the parameter δ_0/a. Note that at small separation (or low temperature) the perturbation orders $\sim (\delta_0/a)^k$ with $k = 2, 3, 4, 5, 6$ contain a contribution from thermal corrections only to order t^{-5} and higher (Bordag et al. 2000a). Because of this, eqns (14.14) and (14.22) obtained at $T \ll T_{\text{eff}}$ can be made more exact by adding higher-order corrections in the relative skin depth as discussed in Section 13.1.

14.2.2 Agreement with the Nernst heat theorem

In Sections 12.5.1 and 12.5.2, the third law of thermodynamics (the Nernst heat theorem) was used as a test of consistency for the model of the dielectric response. In this chapter we shall repeatedly apply Nernst's theorem to verify if the models under consideration are thermodynamically consistent. The asymptotic expression (14.14) for the Casimir free energy per unit area leads to the following expression for the Casimir entropy in the plasma model (Bezerra et al. 2002b, 2004)

$$S_p(a,T) = -\frac{\partial \mathcal{F}_p(a,T)}{\partial T} = -\frac{\partial \Delta_T \mathcal{F}_p(a,T)}{\partial T} \tag{14.24}$$

$$= \frac{k_B \tau^2}{16\pi^3 a^2}\left\{\frac{3}{2}\zeta_R(3) - \frac{\pi^2}{45}\tau + \frac{\delta_0}{a}\left[3\zeta_R(3) - \frac{4\pi^2}{45}\tau\right] - \left(\frac{\delta_0}{a}\right)^2 \frac{5\zeta_R(5)}{4\pi^2}\tau^2\right\}.$$

Here, we have omitted terms that vanish as $\exp(-4\pi^2/\tau)$ when τ vanishes. For convenience when we compare this result with the results obtained in previous chapters, we have returned to the variable $\tau = 2\pi/t = 2\pi T/T_{\text{eff}}$ defined in eqn (12.89), which is linear in the temperature. The zeroth-order contribution in the parameter δ_0/a on the right-hand side of eqn (14.24) coincides with the Casimir entropy of an ideal metal found in eqn (7.92). The terms of order δ_0/a and $(\delta_0/a)^2$ are the corrections to this result due to the skin depth. Equation (14.24) should be compared also with the Casimir entropy between dielectric semispaces given in eqn (12.104). It can be seen that the main contribution in eqn (12.104) is of order τ, whereas for metals described by the plasma model it is of order τ^2.

As can be seen in eqn (14.24), the Casimir entropy goes to zero when the temperature vanishes. This means that the Nernst heat theorem is satisfied. Thus, the Lifshitz theory in combination with the plasma model is thermodynamically consistent. The Casimir entropy, as given by the Lifshitz formula and the plasma model, is positive at all temperatures. To illustrate this, in Fig. 14.1(a) we present

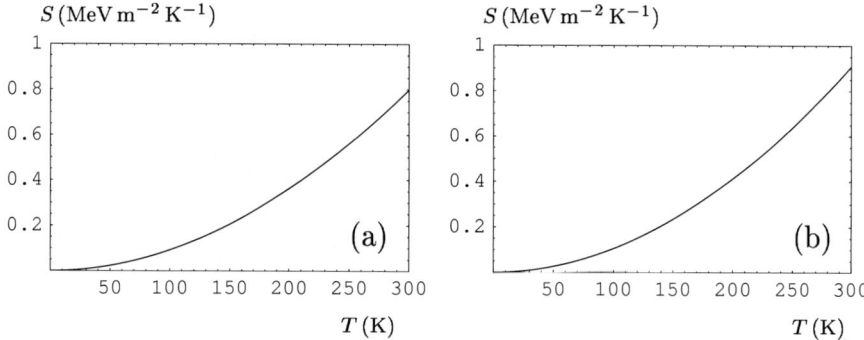

FIG. 14.1. The Casimir entropy per unit area as a function of temperature computed for (a) two ideal-metal planes and (b) two metallic semispaces described by the plasma model. The separation distance is $a = 300$ nm (Mostepanenko and Geyer 2008).

computational results for the Casimir entropy, as a function of temperature, in a configuration of two ideal-metal planes at a separation $a = 300$ nm. In Fig. 14.1(b), the Casimir entropy computed by using the Lifshitz formula for metallic plates described by the plasma model with $\omega_p = 9.0$ eV and spaced at the same separation is presented. As can be seen from a comparison of the two figures, qualitatively the behavior of the Casimir entropy in the two cases is very similar. The asymptotic result (14.24) is in good agreement with the results of numerical computations. Of particular interest is the fact that the analytic expressions obtained for the Casimir free energy, pressure, and entropy are applicable at arbitrarily low temperatures and provide a rigorous proof of the consistency of the Casimir entropy with the third law of thermodynamics, whereas numerical computations are valid at some small but not arbitrarily low temperature.

14.3 Metals described by the Drude model

Systematic computations of the thermal Casimir force using the Drude model approach were started by Boström and Sernelius (2000b). These authors found that the use of the Drude model results in an anomalously large thermal correction, in qualitative disagreement with the result for ideal-metal planes and also for real metal plates described by the plasma model. The predicted thermal correction becomes appreciable even at relatively small separations below 1 μm. In addition, the thermal correction is opposite in sign to the main contribution to the Casimir force within a wide range of separations, i.e. it decreases the magnitude of the total force. This effect pertains equally to the Drude model and to the dielectric permittivity obtained from optical data extrapolated by use of the Drude model to low frequencies (the latter procedure was discussed in Section 13.3). It is due to the fact that the large magnitude of the correction is determined solely by the zero-frequency term of the Lifshitz formula. According to eqn (14.6), the reflection coefficient for the transverse electric mode at zero frequency vanishes if the Drude model is used, and this leads to the large thermal correction. After presenting computational results, we shall show that the Drude model approach violates the Nernst heat theorem for metals with perfect crystal lattices. Then we consider the role of impurities and discuss why the Drude model is not applicable in the Lifshitz theory. We arrive at the conclusion that the Drude current violates thermal equilibrium, which is the basic applicability condition of the Lifshitz theory. At the end of the section, various attempts to modify the reflection coefficients are considered.

14.3.1 Prediction of large thermal corrections below 1 μm

To illustrate this situation, we first consider the results of numerical computations obtained by using the Drude model alone and then present some results obtained using tabulated optical data extrapolated by use of the Drude model. The most informative characteristic is the relative thermal correction to the Casimir energy per unit area defined in eqn (12.128). For comparison purposes, we have computed this correction for both the Drude and the plasma model, given in eqns (13.14) and (13.1), using the Lifshitz formulas (12.108) [with $r^{(1)} = r^{(2)}$] and (13.4). Such a comparison between the two models was performed by Klimchitskaya and Mostepanenko (2001). The same values of the parameters for gold, $\omega_p = 9.0\,\text{eV}$ and $\gamma = 0.035\,\text{eV}$, as in Section 13.3, were used in the computations. In Fig. 14.2(a), the computational results for the relative thermal correction at $T = 300\,\text{K}$, as a function of separation, are shown by the solid line (the Drude model) and by the dashed line (the plasma model) within the range of separations from 100 nm to 10 μm. In the figure, a dramatic difference between the two lines is immediately apparent. The dashed line obtained using the plasma model is in qualitative agreement with the case of an ideal metal. It demonstrates only a small thermal correction at short separations. At separations of 100 nm, 500 nm, and 1 μm the relative thermal correction, as given by the plasma

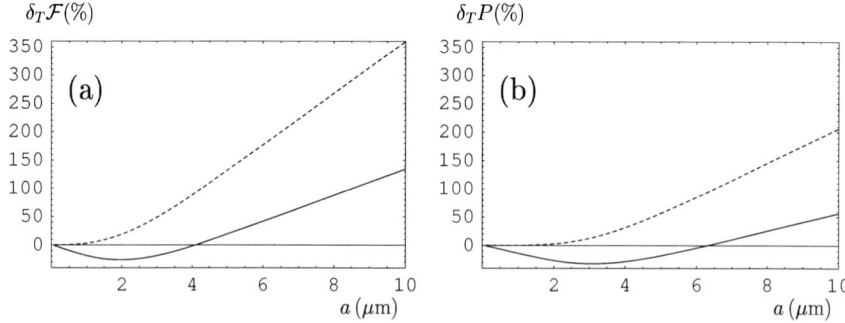

FIG. 14.2. The relative thermal correction to (a) the Casimir energy per unit area and (b) the Casimir pressure as a function of separation for two metallic semispaces at $T = 300\,\text{K}$. The solid lines indicate results computed using the Drude model. The dashed lines show results computed using the plasma model (see text for details).

model, is equal to 0.18%, 0.67%, and 3.22%, respectively. However, the solid line, obtained using the Drude model, is qualitatively different from the case of an ideal metal. Within the range of separations from 100 nm to 4.1 μm, the thermal correction shown by the solid line is negative. This means that thermal effects decrease the magnitude of the Casimir energy. At a separation $a \approx 4.1\,\mu$m, the solid line demonstrates zero thermal effect. This is counterintuitive because, at a separation of 4 μm, the effective temperature $T_{\text{eff}} = 286$ K is only a little lower than the temperature for which the computation was performed ($T = 300$ K). Thus, there should be a considerable thermal effect. The most striking feature of the solid line is the prediction of a large thermal correction at separations below 1 μm. Thus, at separations of 300 nm, 500 nm, 700 nm, and 1 μm the Drude model predicts thermal corrections equal to –5.8%, –10%, –13.9%, and –18.9%, respectively, of the Casimir energy at zero temperature.

A similar comparison between the Drude and plasma models was performed by calculating the Casimir pressure (Klimchitskaya and Mostepanenko 2001). The pressure was computed at $T = 300$ K using eqn (12.70) and at zero temperature using eqn (12.46). The relative thermal correction was defined as in eqn (12.128):

$$\delta_T P_{\text{D(p)}}(a, T) = \frac{P_{\text{D(p)}}(a, T) - P_{\text{D(p)}}(a)}{P_{\text{D(p)}}(a, T)}. \tag{14.25}$$

In Fig. 14.2(b), the computational results obtained by using the Drude model are shown by the solid line, and those obtained by using the plasma model by the dashed line. The dashed line demonstrates a positive thermal correction, in qualitative agreement with the case of an ideal metal. At separations below one micrometer, the thermal effect predicted by the dashed line is relatively small. Thus, at $a = 100$ nm, 500 nm, and 1 μm the relative thermal correction to the Casimir pressure is equal to 0.0016%, 0.058%, and 0.29%, respectively. A different behavior of the thermal correction to the Casimir pressure is predicted by the solid line in Fig. 14.2(b), computed using the Drude model. This thermal correction is negative over a wide range of separations from $a = 100$ nm to $a = 6.3\,\mu$m. At 6.3 μm the thermal correction becomes equal to zero, and it changes its sign to positive at larger separations. The magnitude of the thermal correction at separations below 1 μm is quite large. At separations of 300 nm, 500 nm, 700 nm, and 1 μm it takes values of –3.5%, –6.4%, –9.4%, and –13.8%, respectively.

The large magnitudes of the thermal correction predicted by the Drude model at short separation distances make it possible to test this experimentally. Related precision measurements of the Casimir pressure were performed in several experiments by Decca et al. (2003a, 2003b, 2004, 2005b, 2007a, 2007b). The large thermal corrections predicted by the Drude model were excluded at the 99.9% confidence level. These experiments are discussed in detail in Section 19.3. Note that according to Yampol'skii et al. (2008), a decreasing magnitude of the Casimir pressure occurs only for a thin film described by the Drude model near a semispace, whereas for bulk samples the magnitude of the Casimir pressure increases with temperature (or separation). Geyer et al. (2008e) demonstrated that this statement was incorrect. What actually happens is that for two semispaces described by the Drude model, the decrease of the magnitude of the Casimir pressure is much larger than for a film near a semispace.

Now we consider some computational results obtained using the tabulated optical data for gold extrapolated to low frequencies by using the Drude model. The extrapolation procedure was considered in Section 13.3. It results in the dielectric permittivity $\varepsilon(i\xi)$ shown by the dashed line in Fig. 13.2. The results of a calculation using this permittivity practically coincide with results obtained using the permittivity of the Drude model $\varepsilon_\mathrm{D}(i\xi)$ at separations a larger than 300 or 400 nm. Relatively large deviations (25% and more) are obtained at separations below 50 nm. In Fig. 14.3(a), we plot the correction factor to the Casimir energy, $\eta_E = \mathcal{F}_\mathrm{D}(a,T)/E(a)$, as a function of the separation (solid line), where $\mathcal{F}_\mathrm{D}(a,T)$ was computed at $T = 300\,\mathrm{K}$ with the tabulated optical data extrapolated by use of the Drude model, and $E(a)$ is defined in eqn (1.5). In the same figure, we reproduce the correction factor to the Casimir energy also computed using the optical data extrapolated by use of the Drude model, but at zero temperature (see solid line 1 in Fig. 13.3). Here, the same calculation procedure as described in Section 12.6 was used. The separation range from 50 nm to 1 μm has been chosen to illustrate the anomalously large thermal correction predicted by the Drude model approach at short separations.

From a comparison of the solid and dashed lines in Fig. 14.3(a), the relative thermal correction to the Casimir energy, $\delta_T \mathcal{F}_\mathrm{D}(a,T)$, is equal to –0.54%, –1.5%, and –3.6% at separations of 50 nm, 100 nm, and 200 nm, respectively. At separations larger than 300 or 400 nm, the magnitudes of the thermal correction coincide with those listed above, obtained from computations using the simple Drude model. It is important to note that even at short separations the Casimir energy at zero temperature, shown by the dashed line, deviates significantly from the Casimir free energy, shown by the solid line. This demonstrates that the Casimir energy at zero temperature computed using the tabulated data extrapolated by use of the Drude model is in fact physically meaningless even at relatively short separations in the region from 50 nm to 1 μm.

The dielectric permittivity obtained from the optical data and extrapolated to low frequencies by means of the Drude model can be used also to compute the Casimir pressure. In Fig. 14.3(b), we plot the correction factor to the Casimir pressure, $\eta_P = P_\mathrm{D}(a,T)/P(a)$, as a function of the separation (solid line), where $P_\mathrm{D}(a,T)$ was computed at $T = 300\,\mathrm{K}$ and $P(a)$ is defined in eqn (1.5). The dashed line reproduces the correction factor to the Casimir pressure computed at zero temperature using the tabulated optical data extrapolated by use of the Drude model. The relative thermal correction, $\delta_T P_\mathrm{D}(a,T)$, is equal to –0.2%, –0.71%, and −2.0% at separations of 50 nm, 100 nm, and 200 nm, respectively. Similarly to the case considered above, the Casimir pressure at nonzero temperature [solid line in Fig. 14.3(b)] deviates significantly from the dashed line. Thus,

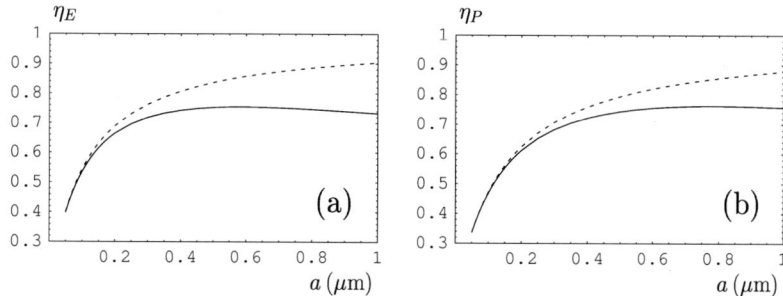

FIG. 14.3. The correction factor to (a) the Casimir energy and (b) the Casimir pressure as a function of separation for two semispaces at $T = 300\,\text{K}$. The solid lines indicate results computed using the optical data extrapolated by use of the Drude model. The dashed lines show results computed in the same way but at $T = 0$ (see text for details).

the Casimir pressure at $T = 0$, computed using the optical data extrapolated to low frequencies by means of the Drude model, has no definite physical meaning in the separation region $a \geq 50\,\text{nm}$.

The Drude model approach to the thermal Casimir force, as suggested by Boström and Sernelius (2000b), was developed further and used in computations by Brevik et al. (2002), Høye et al. (2003), Milton (2004), and Brevik et al. (2005). The last paper led to extended discussions [see the Comment by Bezerra et al. (2006) and the Reply by Høye et al. (2006)]. The important theoretical question about this approach is whether it is consistent with thermodynamics. This problem is analyzed below in detail.

14.3.2 Violation of Nernst's theorem for Drude metals with perfect crystal lattices

Bezerra et al. (2002a, 2002b) first demonstrated that the Drude model combined with the Lifshitz formula results in a nonzero (negative) Casimir entropy at zero temperature which is a function of the separation distance between the plates and, thus, violates the third law of thermodynamics (the Nernst heat theorem). Later, Bezerra et al. (2004) presented a rigorous proof of this statement for metals with perfect crystal lattices having no impurities. Keeping in mind the fundamental importance of the agreement between the Lifshitz theory and thermodynamics, demonstrated already in the case of dielectrics in Chapter 12, we now consider this subject for metals.

We start from the Lifshitz formula for the Casimir free energy per unit area, eqn (12.108), expressed in terms of the dimensionless variables (12.89), in the case of two similar plates, so that $r_{\text{TM}}^{(1)} = r_{\text{TM}}^{(2)}$ and $r_{\text{TE}}^{(1)} = r_{\text{TE}}^{(2)}$. The reflection coefficients are given in eqn (12.91). We consider metals described by the dielectric permittivity of the Drude model (13.14). We replace ω in eqn (13.14) with the imaginary Matsubara frequencies and rearrange this equation in terms of the dimensionless variables:

$$\varepsilon_D(i\omega_c \zeta_l) = 1 + \frac{\tilde{\omega}_p^2}{\zeta_l[\zeta_l + \tilde{\gamma}(T)]}. \tag{14.26}$$

Here, $\tilde{\omega}_p = \omega_p/\omega_c$ and $\tilde{\gamma}(T) = \gamma(T)/\omega_c$. We have also indicated explicitly that the relaxation parameter of the Drude model depends on the temperature. This dependence is very important in the proof given below. We recall that in the absence of relaxation, $\tilde{\gamma}(T) = 0$ and ε_D coincides with the dielectric permittivity of the plasma model (13.1), (13.5). In this section, the reflection coefficients obtained by the substitution of eqn (14.26) into eqn (12.91) are denoted by $r_{TM}^{(D)}(i\zeta_l, y)$ and $r_{TE}^{(D)}(i\zeta_l, y)$. The reflection coefficients obtained by the substitution of the plasma model (13.5) into (12.91) are notated as $r_{TM}^{(p)}(i\zeta_l, y)$ and $r_{TE}^{(p)}(i\zeta_l, y)$. The superscripts "D" and "p" will be used for the respective Casimir energies. In terms of dimensionless variables, we obtain from eqn (12.91) that $r_{TE}^{(D)}(0, y) = 0$, whereas

$$r_{TE}^{(p)}(0, y) = r_{TE}^{(p)}(i\zeta_l, y) = \frac{y - \sqrt{\tilde{\omega}_p^2 + y^2}}{y + \sqrt{\tilde{\omega}_p^2 + y^2}} \neq 0. \tag{14.27}$$

What this means is that there is no smooth transition from $\mathcal{F}_D$ to $\mathcal{F}_p$ when $\gamma(T) \to 0$. This nonanalyticity originates from the vanishing zero-frequency contribution of the transverse electric mode to the free energy $\mathcal{F}_D(a, T)$. When we consider the limiting case $T \to 0$ for a perfect crystal lattice [i.e. with $\gamma(T) \to 0$; see below], the reflection coefficient $r_{TE}^{(D)}(i\zeta, y)$ with $\zeta \neq 0$ goes to the frequency-independent value $r_{TE}^{(p)}$ given in eqn (14.27). If the limiting case $\zeta \to 0$ is then considered, we arrive at the same value (14.27), but not at the zero value, $r_{TE}^{(D)}(0, y) = 0$. Thus, the reflection coefficient $r_{TE}^{(D)}(i\zeta, y)$ is discontinuous as a function of ζ and T at the point $\zeta = 0$, $T = 0$.

For the calculation of the Drude free energy $\mathcal{F}_D(a, T)$, it is useful to add to and subtract from it the free energy $\mathcal{F}_p(a, T)$ calculated using the plasma model. We also write out separately the zero-frequency terms of the Drude free energy and of the subtracted plasma free energy, taking into account that

$$r_{TE}^{(D)}(0, y) = 0, \qquad r_{TM}^{(p)}(0, y) = r_{TM}^{(D)}(0, y) = 1. \tag{14.28}$$

The representation of the Drude free energy then takes the form

$$\mathcal{F}_D(a, T) = \mathcal{F}_p(a, T) - \frac{k_B T}{16\pi a^2} \int_0^\infty y\, dy \ln\left[1 - r_{TE}^{(p)\,2}(0, y)e^{-y}\right] \tag{14.29}$$
$$+ \frac{k_B T}{8\pi a^2} \sum_{l=1}^\infty \int_{\zeta_l}^\infty y\, dy \left\{ \ln\left[1 - r_{TM}^{(D)\,2}(i\zeta_l, y)e^{-y}\right] - \ln\left[1 - r_{TM}^{(p)\,2}(i\zeta_l, y)e^{-y}\right] \right\}$$

$$+ \ln\left[1 - r_{\text{TE}}^{(\text{D})\,2}(\mathrm{i}\zeta_l, y)\mathrm{e}^{-y}\right] - \ln\left[1 - r_{\text{TE}}^{(\text{p})\,2}(\mathrm{i}\zeta_l, y)\mathrm{e}^{-y}\right]\right\}.$$

An important feature of this representation is that the zero-frequency contributions are contained only in the first two terms on the right-hand side of eqn (14.29). All other terms contain integration from some nonzero lower integration limit to infinity because the summation starts from a nonzero Matsubara frequency ζ_1.

Let us find the asymptotic behavior of the Drude free energy (14.29) in the limiting case $T \to 0$. This problem is solved differently for perfect crystal lattices and for crystal lattices with impurities. In the present subsection, we consider perfect crystal lattices, postponing the discussion of impurities to the next subsection. Note that for perfect crystal lattices $\tilde{\gamma}(T)$ is the smallest of the three dimensionless parameters, $\tilde{\omega}_\text{p}$, ζ_l with $l \geq 1$, and $\tilde{\gamma}(T)$ contained in eqn (14.26), at any T. This is clear from the following.

At $T = 300\,\text{K}$, for good metals, $\gamma \sim 10^{13}$–$10^{14}\,\text{rad/s}$ (in the case of gold, for example, $\gamma = 0.035\,\text{eV} = 5.32 \times 10^{13}\,\text{rad/s}$), whereas $\xi_1 = 2\pi k_\text{B}T/\hbar = 2.46 \times 10^{14}\,\text{rad/s}$. Keeping in mind that $\xi_l = l\xi_1$, we obtain $\gamma < \xi_l$. When T decreases from room temperature to approximately $T_\text{D}/4$, where T_D is the Debye temperature, $\gamma(T) \sim T$, i.e. it decreases so that it shows the same behavior as ξ_l, preserving the inequality $\gamma < \xi_l$. Note that for gold, $T_\text{D} = 165\,\text{K}$ (Kittel 1996). At temperatures below $T_\text{D}/4$ the relaxation parameter decreases even more rapidly than ξ_l with decreasing T. At first it decreases as $\sim T^5$ according to the Bloch–Grüneisen law, taking electron–phonon collisions into account (Ashcroft and Mermin 1976). As a result, at $T = 30\,\text{K}$ $\gamma(T)/\xi_1(T) = 4.9 \times 10^{-2}$, and at $T = 10\,\text{K}$ $\gamma(T)/\xi_1(T) = 1.8 \times 10^{-3}$. At liquid helium temperatures the relaxation parameter decreases as $\sim T^2$ owing to electron–electron scattering (Kittel 1996). For perfect crystal lattices the rule $\gamma(T) \sim T^2$ is followed down to zero temperature. Because of this, at very low temperatures the condition $\gamma(T) \ll \xi_1(T)$ is satisfied.

The largest parameter of the above three is ω_p (recall that for gold we use $\omega_\text{p} = 9.0\,\text{eV} = 1.37 \times 10^{16}\,\text{rad/s}$). For example, for gold at $T = 300\,\text{K}$, $70\,\text{K}$, and $10\,\text{K}$ we have $\gamma(T)/\omega_\text{p} = \tilde{\gamma}(T)/\tilde{\omega}_\text{p} = 3.88 \times 10^{-3}$, 6.71×10^{-4}, and 1.06×10^{-6}, respectively.

Now we return to eqn (14.29) and notice that for all $l \geq 1$, the reflection coefficients (12.91) have continuous derivatives with respect to the ratio $\gamma(T)/\xi_l(T)$ at the point $\gamma(T)/\xi_l(T) = 0$. Under the condition $\gamma(T) \ll \xi_l(T)$ proven above, which is satisfied at all sufficiently low temperatures, the parameter $\gamma(T)/\xi_l(T) = \tilde{\gamma}(T)/\zeta_l(T)$ is small. So we can expand the reflection coefficients in a Taylor series around the point $\tilde{\gamma}(T)/\zeta_l(T) = 0$ keeping only the first-order terms

$$r_{\text{TM}}^{(\text{D})\,2}(\mathrm{i}\zeta_l, y) = r_{\text{TM}}^{(\text{p})\,2}(\mathrm{i}\zeta_l, y) - \frac{\tilde{\gamma}(T)}{\zeta_l(T)} R_{\text{TM}}(\mathrm{i}\zeta_l, y),$$

$$r_{\text{TE}}^{(D)\,2}(i\zeta_l, y) = r_{\text{TE}}^{(p)\,2}(i\zeta_l, y) - \frac{\tilde{\gamma}(T)}{\zeta_l(T)} R_{\text{TE}}(i\zeta_l, y), \qquad (14.30)$$

where

$$R_{\text{TM}}(i\zeta_l, y) = \frac{2\zeta_l^2 \alpha y \left[1 + \alpha^2(2y^2 - \zeta_l^2)\right] r_{\text{TM}}^{(p)}(i\zeta_l, y)}{\sqrt{1 + \alpha^2 y^2}\,[y + \alpha \zeta_l^2(\alpha y + \sqrt{1 + \alpha^2 y^2})]^2},$$

$$R_{\text{TE}}(i\zeta_l, y) = -\frac{2\alpha\, y\, r_{\text{TE}}^{(p)}(i\zeta_l, y)}{\sqrt{1 + \alpha^2 y^2}\,(\alpha y + \sqrt{1 + \alpha^2 y^2})^2}, \qquad (14.31)$$

and $\alpha \equiv 1/\tilde{\omega}_p = \delta_0/(2a) \ll 1$ (note that we keep the argument T when ζ_l participates in the expansion parameter but omit it otherwise).

Similar expansions for the logarithms which appear in eqn (14.29) are

$$\ln\left[1 - r_{\text{TM}}^{(D)\,2}(i\zeta_l, y) e^{-y}\right] = \ln\left[1 - r_{\text{TM}}^{(p)\,2}(i\zeta_l, y) e^{-y}\right]$$
$$- \frac{\tilde{\gamma}(T)}{\zeta_l(T)} \frac{R_{\text{TM}}(i\zeta_l, y)\, e^{-y}}{1 - r_{\text{TM}}^{(p)\,2}(i\zeta_l, y) e^{-y}}, \qquad (14.32)$$

$$\ln\left[1 - r_{\text{TE}}^{(D)\,2}(i\zeta_l, y) e^{-y}\right] = \ln\left[1 - r_{\text{TE}}^{(p)\,2}(i\zeta_l, y) e^{-y}\right]$$
$$- \frac{\tilde{\gamma}(T)}{\zeta_l(T)} \frac{R_{\text{TE}}(i\zeta_l, y)\, e^{-y}}{1 - r_{\text{TE}}^{(p)\,2}(i\zeta_l, y) e^{-y}}.$$

Substituting eqn (14.32) into eqn (14.29), we obtain

$$\mathcal{F}_{\text{D}}(a, T) = \mathcal{F}_{\text{p}}(a, T) - \frac{k_B T}{16 \pi a^2} \int_0^\infty y\, dy \ln\left[1 - r_{\text{TE}}^{(p)\,2}(0, y) e^{-y}\right] + \mathcal{F}_\gamma(a, T). \qquad (14.33)$$

Here, $\mathcal{F}_\gamma(a, T)$ is the contribution to the Drude free energy which depends on the relaxation parameter γ. It is given by

$$\mathcal{F}_\gamma(a, T) = \frac{\tilde{\gamma}(T)}{\tilde{\omega}_p} \frac{k_B T}{8\pi a^2} \sum_{l=1}^\infty \frac{\tilde{\omega}_p}{\zeta_l(T)} \int_0^\infty y\, dy \qquad (14.34)$$

$$\times \left[\frac{R_{\text{TM}}(i\zeta_l, y)\, e^{-y}}{1 - r_{\text{TM}}^{(p)\,2}(i\zeta_l, y) e^{-y}} + \frac{R_{\text{TE}}(i\zeta_l, y)\, e^{-y}}{1 - r_{\text{TE}}^{(p)\,2}(i\zeta_l, y) e^{-y}} \right].$$

Note that in this expression the small parameter $\tilde{\gamma}(T)/\tilde{\omega}_p = \gamma(T)/\omega_p$ is factored out, as it does not depend on the summation index.

The asymptotic expression for the plasma free energy $\mathcal{F}_{\text{p}}(a, T)$ at low temperature is given by

$$\mathcal{F}_{\text{p}}(a, T) = E_{\text{p}}(a) + \Delta_T \mathcal{F}_{\text{p}}(a, T), \qquad (14.35)$$

where $E_{\text{p}}(a)$ is the Casimir energy per unit area at $T = 0$, and the thermal correction is obtained from eqn (14.14). The respective Casimir entropy $S_{\text{p}}(a, T)$ is presented in eqn (14.24). It goes to zero as $\sim T^2$ when T vanishes.

The second term on the right-hand side of eqn (14.33) is linear in the temperature. The coefficient of T can be calculated perturbatively. For this purpose we use eqn (14.27) and expand the logarithm under the integral in powers of the relative skin depth δ_0/a. Then the integral with respect to y in eqn (14.33) is evaluated explicitly, resulting in

$$-\frac{k_\mathrm{B}T}{16\pi a^2}\int_0^\infty y\,dy\,\ln\!\left[1-r_\mathrm{TE}^{(\mathrm{p})\,2}(0,y)e^{-y}\right] = \frac{k_\mathrm{B}T\zeta_\mathrm{R}(3)}{16\pi a^2}\left\{1-4\frac{\delta_0}{a}+12\left(\frac{\delta_0}{a}\right)^2\right.$$

$$\left.-32\left(\frac{\delta_0}{a}\right)^3\!\left[1-\frac{\zeta_\mathrm{R}(5)}{16\zeta_\mathrm{R}(3)}\right]+80\left(\frac{\delta_0}{a}\right)^4\!\left[1-\frac{\zeta_\mathrm{R}(5)}{4\zeta_\mathrm{R}(3)}\right]\right\}. \qquad (14.36)$$

Now we consider the low-temperature behavior of the last term on the right-hand side of eqn (14.33), $\mathcal{F}_\gamma(a,T)$, which depends on the relaxation parameter. As pointed out above, in the case of a perfect crystal lattice $\gamma(T)\to 0$ when $T\to 0$ no slower than $\sim T^2$. We shall show that the $\mathcal{F}_\gamma(a,T)$ expressed by eqn (14.34) also goes to zero when T vanishes as $\sim T^2\ln T$. For this purpose we expand R_TM and R_TE in eqn (14.31) up to the first order in the small parameter α (recall that $\mathcal{F}_\gamma$ is already proportional to the smallest parameter in our problem, γ/ω_p):

$$R_\mathrm{TM}(i\zeta_l,y)=\frac{2\zeta_l^2\alpha}{y},\qquad R_\mathrm{TE}(i\zeta_l,y)=2y\alpha. \qquad (14.37)$$

Substituting eqn (14.36) into eqn (14.33), we obtain

$$\mathcal{F}_\gamma(a,T)=\frac{\gamma(T)}{\omega_\mathrm{p}}\frac{k_\mathrm{B}T}{4\pi a^2}\sum_{l=1}^\infty\left(\zeta_l\int_{\zeta_l}^\infty\frac{dy}{e^y-1}+\frac{1}{\zeta_l}\int_{\zeta_l}^\infty\frac{y^2\,dy}{e^y-1}\right). \qquad (14.38)$$

In the limiting case $T\to 0$, both sums in eqn (14.38) can be evaluated easily. The asymptotic form of the first sum is given by

$$\sum_{l=1}^\infty\zeta_l\int_{\zeta_l}^\infty\frac{dy}{e^y-1}=\frac{2\pi T}{T_\mathrm{eff}}\sum_{k=1}^\infty\frac{1}{k}\left[\frac{1}{e^{k\tau}-1}+\frac{1}{(e^{k\tau}-1)^2}\right]=\frac{1}{\tau}\zeta_\mathrm{R}(3)+\zeta_\mathrm{R}(2). \qquad (14.39)$$

Here, we have used the notation $\tau=2\pi T/T_\mathrm{eff}$ and kept only two terms in the expansion of $\exp(k\tau)$ in the denominators. For the second sum in eqn (14.38), we obtain

$$\sum_{l=1}^\infty\frac{1}{\zeta_l}\int_{\zeta_l}^\infty\frac{y^2\,dy}{e^y-1}=\frac{1}{\tau}\sum_{k=1}^\infty\left[-\frac{2}{k^3}\ln(1-e^{-k\tau})+\frac{2\tau}{e^{k\tau}-1}+\frac{\tau^2}{k}\frac{e^{k\tau}}{(e^{k\tau}-1)^2}\right]$$

$$=-\frac{2\zeta_\mathrm{R}(3)}{\tau}\ln\tau+\frac{3\zeta_\mathrm{R}(3)}{\tau}+2\zeta_\mathrm{R}(2). \qquad (14.40)$$

Substituting eqns (14.39) and (14.40) into eqn (14.38), we arrive at

$$\mathcal{F}_\gamma(a,T) = \frac{\gamma(T)}{\omega_p} \frac{k_B T_{\text{eff}} \zeta_R(3)}{4\pi^2 a^2} \left[-\ln \tau + 2 + \frac{3\zeta_R(2)}{2\zeta_R(3)} \tau \right]. \tag{14.41}$$

As can be seen in eqn (14.41), the leading term in $\mathcal{F}_\gamma(a,T)$, as a function of temperature, behaves as $-\gamma(T)\ln(T/T_{\text{eff}})$ and goes to zero when $T \to 0$ because for a perfect crystal lattice $\gamma(T) \sim T^2$ at liquid helium and lower temperatures.

Now we are in a position to find the low-temperature behavior of the Casimir entropy, as defined by the Drude model. It is obtained from eqn (14.33):

$$S_D(a,T) = -\frac{\partial \mathcal{F}_D(a,T)}{\partial T} \tag{14.42}$$

$$= S_p(a,T) + \frac{k_B}{16\pi a^2} \int_0^\infty y\, dy \ln\left[1 - r_{\text{TE}}^{(p)\,2}(0,y) e^{-y}\right] - \frac{\partial \mathcal{F}_\gamma(a,T)}{\partial T}.$$

Here, $S_p(a,T)$ is the low-temperature asymptotic expression (14.24) for the Casimir entropy given by the plasma model. It goes to zero when T vanishes. The second contribution on the right-hand side of eqn (14.42) is independent of the temperature. The asymptotic behavior of the last contribution in the limit of low temperature is obtained from eqn (14.41) taking into account that $\gamma(T) = \gamma_0 T^2$:

$$-\frac{\partial \mathcal{F}_\gamma(a,T)}{\partial T} = -\frac{k_B \zeta_R(3)}{2\pi a^2} \frac{\gamma(T)}{\omega_p} \frac{1}{\tau} \left[-2\ln \tau + 3 + \frac{9\zeta_R(2)}{2\zeta_R(3)} \tau \right]. \tag{14.43}$$

At low T, this contribution goes to zero as $T \ln(T/T_{\text{eff}})$. Taking the limit $T \to 0$ on both sides of eqn (14.42), we arrive at

$$S_D(a,0) = \frac{k_B}{16\pi a^2} \int_0^\infty y\, dy \ln\left[1 - r_{\text{TE}}^{(p)\,2}(0,y) e^{-y}\right]$$

$$= -\frac{k_B \zeta_R(3)}{16\pi a^2}\left[1 - 4\frac{\delta_0}{a} + 12\left(\frac{\delta_0}{a}\right)^2 - \cdots\right] < 0, \tag{14.44}$$

where the higher-order contributions are presented in eqn (14.36).

Thus, at zero temperature the Casimir entropy calculated using the Drude model for metals with perfect crystal lattices takes a nonzero negative value. This value, in violation of the Nernst heat theorem, depends on the parameters of the system, such as the separation distance between the plates and the plasma frequency. The asymptotic expression (14.44) is in good agreement with the results of numerical computations using the Lifshitz formula. In Fig. 14.4(a), we present computational results for the Casimir entropy in a configuration of two plates made of a Drude metal with a perfect crystal lattice at a separation $a = 1\,\mu$m (the same Drude parameters as above, i.e. $\omega_p = 9.0$ eV and $\gamma(T = 300\,\text{K}) = 0.035$ eV, have been used). As is seen in the figure, the Casimir entropy is negative over a wide temperature region and nonzero at $T = 0$. Keeping in mind that the Drude parameter γ describes the relaxation properties of conduction electrons, the demonstrated contradiction with thermodynamics questions the possibility

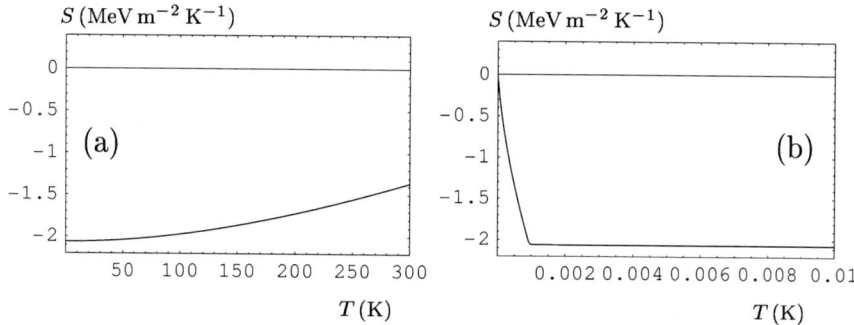

Fig. 14.4. The Casimir entropy per unit area as a function of temperature in a configuration of two metallic semispaces at a separation $a = 1\,\mu\text{m}$ described by the Drude model for (a) a metal with a perfect crystal lattice (b) a metal with a typical concentration of impurities (Mostepanenko and Geyer 2008). See text for further discussion.

to describe these properties within the framework of the Lifshitz theory by means of the Drude model. As was mentioned in Section 14.3.1, the large thermal corrections to the Casimir pressure predicted by the Drude model at separations below $1\,\mu\text{m}$ are excluded experimentally (see Section 19.3 for details). The violation of the Nernst theorem under the condition that one of the reflection coefficients is discontinuous as a function of the frequency and temperature was demonstrated on general grounds by Intravaia and Henkel (2008). Their proof is applicable to both the case of dielectrics, as discussed in Section 12.5.2, and the case of metals.

Note that the use of the Drude model in the Lifshitz formula is usually justified by the fact that it provides a smooth transition between the regions of the normal skin effect and infrared optics. This ignores the region of the anomalous skin effect, where the Lifshitz formula in terms of $\varepsilon(\omega)$ is not applicable (see Section 13.4.1). With a decrease in T, the application region of the normal skin effect becomes narrower and the application region of the anomalous skin effect widens (see Section 14.4.2). However, at any $T > 0$, there exists a frequency range of small frequencies where the normal skin effect is applicable. Bearing in mind that the violation of the Nernst heat theorem demonstrated above is completely determined by the zero-frequency term of the Lifshitz formula, it is worthwhile to use the Drude model at low T when considering the thermodynamic consistency of the Lifshitz theory (Klimchitskaya et al. 2009a). This invalidates the critical remarks (Pitaevskii 2008b) concerning the violation of the Nernst theorem in the Drude model approach proved by Bezerra et al. (2004).

14.3.3 The role of impurities

As noted in the previous subsection, for perfect crystal lattices the relaxation parameter $\gamma(T)$ is less than the other two parameters, ω_p and ξ_l, used in the theoretical description of the thermal Casimir force by means of the Drude model.

It was shown that at arbitrarily low temperatures the following inequality is satisfied:

$$\gamma(T) \ll \xi_l(T), \qquad l = 1, 2, 3, \ldots \tag{14.45}$$

This was used to demonstrate the violation of the Nernst theorem.

Boström and Sernelius (2004) noticed, however, that the Drude model can be in agreement with the Nernst theorem if the crystal lattice of the metal contains some fraction of impurities. This conclusion was supported by numerical computations of the Casimir entropy as a function of temperature, showing a negative entropy which undergoes an abrupt increase to zero at very low temperatures. The validity of the Nernst theorem for metals with impurities is not in contradiction with the above results on the violation of this theorem for a perfect crystal lattice. The reason is that for metals with impurities, the relaxation parameter reaches some minimal (residual) value $\gamma_{\text{res}} > 0$ when the temperature decreases. With a further decrease of temperature, the value γ_{res} remains unchanged and the inequality (14.45) used in Section 14.3.2 is not valid anymore. Thus, the conclusions based on eqn (14.45) are not applicable to metals with impurities.

The magnitude of the residual relaxation can be determined from the resistivity ratio of a sample. This is defined as the ratio of the resistivity of the sample at room temperature to its residual resistivity. For relatively pure samples containing only a small fraction of impurities, the resistivity ratio may be as high as 10^6 (Kittel 1996). For typical samples, the resistivity ratio is of order 10^3. Thus, for gold, with $\gamma(T = 300\,\text{K}) = 5.32 \times 10^{13}$ rad/s, the residual relaxation parameter is equal to $\gamma_{\text{res}} = 5.32 \times 10^7$ rad/s for pure samples and $\gamma_{\text{res}} = 5.32 \times 10^{10}$ rad/s for typical samples. For relatively pure samples, the inequality (14.45) is valid at $T \gg \hbar \gamma_{\text{res}}/(2\pi k_{\text{B}}) = 6.5 \times 10^{-5}$ K and not applicable at lower temperatures.

Høye et al. (2007) considered gold samples with impurities and obtained an analytic expression for the thermal correction to the Casimir energy at very low temperatures. They based their derivation on the inequality [opposite to eqn (14.45)]

$$\xi_l(T) \ll \gamma(T). \tag{14.46}$$

The resulting asymptotic expression for the thermal correction to the Casimir energy per unit area was found in the form

$$\Delta_T \mathcal{F}_{\text{D}}(a, T) = C_1 T^2 (1 - C_2 T^{1/2} + \cdots) \tag{14.47}$$

and approximated as

$$\Delta_T \mathcal{F}_{\text{D}}(a, T) = \frac{C_1 T^2}{1 + C_2 T^{1/2}}. \tag{14.48}$$

The respective Casimir entropy

$$S_{\text{D}}(a, T) = -\frac{\partial \Delta_T \mathcal{F}_{\text{D}}(a, T)}{\partial T} = -C_1 T \frac{4 + 3 C_2 T^{1/2}}{2(1 + C_2 T^{1/2})^2} \tag{14.49}$$

goes to zero when T vanishes.

Høye et al. (2007) assumed $\gamma = \gamma(T = 300\,\text{K})$ in eqn (14.46) and determined the coefficients C_1 and C_2 under this assumption. Klimchitskaya and Mostepanenko (2008a) noted that for samples with impurities, one should instead use $\gamma = \gamma_\text{res}$. As a result, for typical gold samples the following coefficients were obtained:

$$C_1 = 5.81 \times 10^{-10}\,\text{J}\,\text{m}^{-2}\,\text{K}^{-2}, \qquad C_2 = 95.75\,\text{K}^{-1/2} \qquad (14.50)$$

[to be compared with $C_1 = 5.81 \times 10^{-13}\,\text{J}\,\text{m}^{-2}\,\text{K}^{-2}$ and $C_2 = 3.03\,\text{K}^{-1/2}$ as found by Høye et al. (2007)].

For illustration, in Fig. 14.4(b) we present the Casimir entropy, as a function of temperature, computed using eqns (14.49) and (14.50) for gold plates at a separation $a = 1\,\mu\text{m}$ with a typical concentration of impurities. The asymptotic results at low temperatures in the application region of eqn (14.49) were connected smoothly to the results of numerical computations at higher temperatures. As can be seen from a comparison of Figs. 14.4(a) and (b), at a temperature above approximately $10^{-3}\,\text{K}$ the Casimir entropies for the perfect crystal lattice and for the lattice with impurities coincide. However, at $T < 10^{-3}\,\text{K}$ the Casimir entropy of the lattice with impurities undergoes an abrupt increase to zero. Based on the zero Casimir entropy at zero temperature for lattices with impurities, Høye et al. (2007) concluded that the Nernst heat theorem is not violated when a realistic Drude dispersion model is used.

It must be emphasized, however, that the vanishing of the Casimir entropy at zero temperature for lattices with impurities does not solve the problem of the inconsistency of the Drude model when combined with the Lifshitz formula with basic thermodynamic principles. The reason is that for metals with perfect crystal lattices the Drude model, as was shown above, violates the Nernst theorem. This fact alone makes the Drude approach to the thermal Casimir force unacceptable, as being in violation of quantum statistical physics. As discussed by Mostepanenko et al. (2006b), a perfect crystal lattice is a truly equilibrium system with a nondegenerate dynamical state of lowest energy. The entropy at zero temperature is proportional to the logarithm of the number of states with the lowest energy, i.e. to the logarithm of unity. Thus, in accordance with quantum statistical physics, in the case of a perfect crystal lattice the Casimir entropy must vanish. Any theoretical approach that violates this demand is thermodynamically inconsistent. It is pertinent to note also that a perfect crystal lattice is one of the basic models used in condensed matter physics. It provides a correct description of quantum states in a lattice and of many physical processes, including electron–phonon and electron–electron interactions. The model of a perfect crystal lattice describes relaxation processes leading to the dissipation of energy. Because of this, although a perfect crystal lattice is an idealization (as is any physical model), it is much more realistic than, for instance, the model of an ideal metal. If, for some reason, the model of a perfect crystal lattice were recognized to be in contradiction with thermodynamics, this would lead to the eventual abandonment of most of condensed matter physics. That is the reason why the attempt to attain agreement between the Drude model approach to the thermal Casimir force and thermodynamics at the cost of introducing impurities cannot be considered as satisfactory.

In this connection, Høye et al. (2008) have argued that the Drude model combined with the Lifshitz formula is thermodynamically consistent in the case of a perfect crystal lattice also. To justify this statement, they introduced a definition of a perfect crystal lattice as a limiting case of a lattice with impurities when $\gamma_{\text{res}} \to 0$. However, the common definition of a perfect crystal lattice in condensed matter physics (Kittel 1996) defines it as a lattice with a perfect crystal structure without impurities, i.e. with $\gamma_{\text{res}} = 0$ from the outset. This common definition allows one to develop a quantum theory of electron–phonon interactions which is based on the symmetry properties of the lattice. Specifically, the relaxation parameter γ turns out to be temperature-dependent and vanishes with vanishing temperature. The definition of a perfect crystal lattice introduced by Høye et al. (2008) violates lattice symmetry properties. For example, it would not be possible to reproduce the standard results of the theory of elementary excitations on the basis of this definition. Thus, for metals with perfect crystal lattices as commonly understood, the Lifshitz theory combined with the Drude model violates the Nernst heat theorem and is thermodynamically inconsistent.

14.3.4 Why the Drude model is not applicable in the Lifshitz theory

As shown in Section 13.2, the dielectric permittivities for the normal skin effect (13.18) and the Drude model (13.14) are the consequences of Maxwell's equations in the respective frequency regions. Because of this, it may be considered as somewhat surprising that the substitution of these permittivities into the Lifshitz formula leads to unexpected predictions (such as an anomalously large thermal correction at separations below $1\,\mu$m) and even to the violation of the basic principles of thermodynamics. Below, we elucidate the physical reasons why the Drude model is not applicable in the Lifshitz theory of thermal Casimir forces.

We begin with a consideration of real metal plates in an external (not fluctuating) field of electromagnetic plane waves of any frequency. In Section 13.2, the dielectric permittivities $\varepsilon_n(\omega)$ and $\varepsilon_D(\omega)$ were derived from the Maxwell equations for an infinite medium (a semispace or a plate of infinitely large area) with no external sources, zero induced charge density, and a nonzero induced current $\boldsymbol{j} = \sigma_0 \boldsymbol{E}$. In such a medium, there are no walls (at least in the lateral directions) limiting the flow of charges. Physically, the condition that a plate is infinite means that its planar dimensions are much larger than the size of a wavefront. In the framework of the Drude model (13.14), the total current is given by

$$\boldsymbol{j}_{\text{tot}}(t,\boldsymbol{r}) = \text{Re}\left[-\frac{i\omega}{4\pi}\varepsilon_D(\omega)\boldsymbol{E}(\boldsymbol{r})e^{-i\omega t}\right] \qquad (14.51)$$

$$= \frac{\omega}{4\pi}\left(1 - \frac{\omega_p^2}{\omega^2 + \gamma^2}\right)\text{Im}\left[\boldsymbol{E}(\boldsymbol{r})e^{-i\omega t}\right] + \frac{\sigma_0\gamma^2}{\omega^2 + \gamma^2}\text{Re}\left[\boldsymbol{E}(\boldsymbol{r})e^{-i\omega t}\right].$$

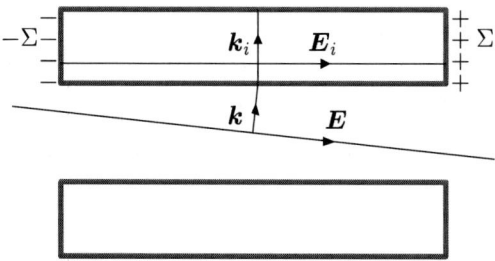

FIG. 14.5. A configuration of two metallic plates of finite size. An incident electromagnetic wave in a vacuum gap is characterized by an electric field E and a wave vector k (E_i and k_i in the interior of a plate). A short-lived current inside the metal leads to the accumulation of surface charges with densities $\pm\Sigma$ (Mostepanenko and Geyer 2008).

The first term on the right-hand side of this equation has the meaning of the displacement current, whereas the second term is proportional to the physical electric field and describes a real current of conduction electrons. Under the condition $\gamma \ll \omega < \omega_p$, i.e. in the region of infrared optics, the first term dominates. This is the displacement current of the plasma model. In accordance with eqn (13.19), it corresponds to a pure imaginary conductivity $\sigma(\omega) = i\sigma_0\gamma/\omega$. Under the opposite condition $\omega \ll \gamma$, i.e. in the region of the normal skin effect, the second term on the right-hand side of eqn (14.51) dominates; this term describes the real current of the conduction electrons.

For real metal semispaces (for plates of finite thickness and area), the applicability conditions of the Drude model are violated. The size of the wavefront of a plane wave is much larger than any conceivable metal plate (see Fig. 14.5). For plane waves of very low frequency, the electric field E_i inside the plate is almost constant and is parallel to the boundary surface (see Fig. 14.5, where k and k_i are the wave vectors outside and inside the plate, respectively). As described in textbooks (Landau et al. 1984, Jackson 1999), a constant electric field E_i in a metal creates a very short-lived current of conduction electrons leading to the formation of almost constant charge densities $\pm\Sigma$ on opposite sides of the plate (see Fig. 14.5). As a result, both the electric field and the current inside the plate are screened out. The field outside the plate becomes equal to the superposition of the incident field E and the field E_Σ produced by the charge densities $+\Sigma$ (Jackson 1999). This was considered for two wires by Bimonte (2007) in connection with Johnson noise. We emphasize that the time interval during which charges on the sides of finite plates are accumulated and the total electric field inside a metal becomes zero is extremely short. As an example, for gold with $4\pi\sigma_0 \approx 3.5 \times 10^{18}\,\text{s}^{-1}$, the electric field inside the metal vanishes after the very short time of 10^{-18} s (Geyer et al. 2007).

If the area of the plates is infinitely large, there is a drift current of conduction electrons in the longitudinal direction under the influence of the electric field $\boldsymbol{E}_i$ given by the second term on the right-hand side of eqn (14.51). The interaction of the conduction electrons with the elementary excitations of the crystal lattice (phonons) leads to Joule heating of the metal (ohmic losses). Thus, if one wishes to substitute the Drude dielectric function into the Lifshitz formula, it is necessary to introduce an interaction between the plates and a heat reservoir. This interaction is needed to preserve the constant temperature of the plates. It describes a unidirectional flux of heat from the plates to the heat reservoir (Bryksin and Petrov 2008). An interaction of this kind is an irreversible process violating time reversal symmetry (it is well known that any drift current is t-asymmetric). However, irreversible processes, in principle, cannot be considered in the state of thermal equilibrium assumed in the Lifshitz theory. According to the definition of thermal equilibrium, not only is the temperature constant, but all irreversible processes connected with the dissipation of energy to heat have been terminated (Kondepugi and Prigogine 1998; Rumer and Ryvkin 1980). In thermal equilibrium, only a zero-on-average bilateral fluctuating exchange of heat between a system and a heat reservoir is admissible. Thus, the Drude dielectric function cannot be substituted into the Lifshitz formula, because the drift current violates thermal equilibrium, which is the basic applicability condition of the Lifshitz theory. As noted by Geyer *et al.* (2003), the fluctuating electromagnetic field in thermal equilibrium with a metal plate cannot lead to the initiation of a drift current and the heating of the metal. This is strictly prohibited by thermodynamics.

In the above, we have discussed plane waves of very low frequency and the limiting case when the frequency vanishes. This is relevant only to the zero-frequency contribution to the Lifshitz formula. For electromagnetic oscillations at all other Matsubara frequencies $\xi_l = l\xi_1$ with $l \geq 1$ (recall that at room temperature $\xi_1 = 2.47 \times 10^{14}$ rad/s, i.e. a rather high frequency), the drift current of conduction electrons [given by the second term on the right-hand side of eqn (14.51)] is small in comparison with the displacement current. As explained above, the latter corresponds to a pure imaginary conductivity of the plasma (or generalized plasma-like) model. The displacement current does not lead to Joule heating of the metal and is consistent with the state of thermal equilibrium. That is the reason why the conduction electrons in metals can be incorporated into the Lifshitz theory by means of the generalized plasma-like dielectric permittivity (Mostepanenko and Geyer 2008). It is worth noting that the substitution of the Drude model into the Lifshitz formula does not create problems in the case of zero temperature considered in Chapter 13. This is connected with the fact that the frequencies around zero do not provide any contribution to the integral with respect to frequency in the Lifshitz formulas (12.30) and (12.33).

14.3.5 *Attempts at modifying the reflection coefficients*

Several attempts have been undertaken in the literature to modify the reflection coefficients in the Lifshitz formula in order to make it consistent with the Drude model. One attempt (Svetovoy and Lokhanin 2001) suggested the use of the value -1 for the transverse electric reflection coefficient at zero frequency for real metals. When this was done, both of the reflection coefficients at all nonzero Matsubara frequencies were assumed to be the same as in the framework of the Drude model approach. This is a phenomenological prescription. According to this suggestion, at zero frequency real metals have the same reflection coefficients (14.1) as given for ideal metals. Bezerra et al. (2004) checked whether the Lifshitz theory combined with the above suggestion was thermodynamically consistent. It was shown that the respective value of the Casimir entropy at zero temperature,

$$\tilde{S}_\mathrm{D}(a,0) = \frac{k_\mathrm{B}\zeta_\mathrm{R}(3)}{4\pi a^2} \frac{\delta_0}{a}\left(1 - 3\frac{\delta_0}{a}\right) > 0, \qquad (14.52)$$

is positive, i.e. the Nernst heat theorem is violated [here, $\delta_0 = \lambda_\mathrm{p}/(2\pi)$, λ_p is the plasma wavelength, and $a > \lambda_\mathrm{p}$ is assumed]. This should be compared with the negative value of the Casimir entropy at zero temperature (14.44) within the standard Drude model approach. In Section 19.3.3, we show that the theoretical Casimir pressures computed using the phenomenological prescription of Svetovoy and Lokhanin (2001) are excluded by the experimental data at a 95% confidence level.

Another attempt to modify the reflection coefficients in the Lifshitz formula takes into account screening effects and diffusion currents of free charge carriers (Dalvit and Lamoreaux 2008). The modified coefficients for the transverse magnetic and transverse electric modes at any frequency were obtained through the use of the Boltzmann transport equation, which takes into account not only the standard drift current $\boldsymbol{j}$, but also the diffusion current $eD\,\boldsymbol{\nabla} n$, where D is the diffusion constant and $\boldsymbol{\nabla} n$ is the gradient of the charge carrier density. In this approach the transverse magnetic coefficient along the imaginary frequency axis takes the form

$$r_\mathrm{TM}^\mathrm{mod}(\mathrm{i}\xi, k_\perp) = \frac{\tilde{\varepsilon}(\mathrm{i}\xi)q - k - k_\perp^2 \eta^{-1}(\mathrm{i}\xi)\varepsilon^{-1}(\mathrm{i}\xi)\left[\tilde{\varepsilon}(\mathrm{i}\xi) - \varepsilon(\mathrm{i}\xi)\right]}{\tilde{\varepsilon}(\mathrm{i}\xi)q + k + k_\perp^2 \eta^{-1}(\mathrm{i}\xi)\varepsilon^{-1}(\mathrm{i}\xi)\left[\tilde{\varepsilon}(\mathrm{i}\xi) - \varepsilon(\mathrm{i}\xi)\right]}. \qquad (14.53)$$

Here, q and k are defined in eqn (12.28), with the dielectric permittivity

$$\tilde{\varepsilon}(\mathrm{i}\xi) = \varepsilon(\mathrm{i}\xi) + \frac{\omega_\mathrm{p}^2}{\xi(\xi + \gamma)}, \qquad (14.54)$$

where $\varepsilon(\mathrm{i}\xi)$ is presented in eqn (12.86) and describes core electrons. By analogy with eqn (13.46), $\tilde{\varepsilon}(\mathrm{i}\xi)$ can be called the *generalized Drude-like permittivity*. The parameter $\eta(\mathrm{i}\xi)$ is defined as

$$\eta(\mathrm{i}\xi) = \left[k_\perp^2 + \kappa^2 \frac{\varepsilon_0}{\varepsilon(\mathrm{i}\xi)} \frac{\tilde{\varepsilon}(\mathrm{i}\xi)}{\tilde{\varepsilon}(\mathrm{i}\xi) - \varepsilon(\mathrm{i}\xi)}\right]^{1/2}, \qquad (14.55)$$

where $1/\kappa$ is the *screening length*. The transverse electric reflection coefficient $r_\mathrm{TE}^\mathrm{mod}(\mathrm{i}\xi, k_\perp)$ is given by the standard expression (12.67), where the dielectric permittivity (14.54) is substituted.

Dalvit and Lamoreaux (2008) restricted the application of their approach to intrinsic semiconductors only. They used a specific Debye–Hückel expression for the screening length,

$$\frac{1}{\kappa} = \frac{1}{\kappa_{\rm DH}} = R_{\rm DH} = \sqrt{\frac{\varepsilon_0 k_{\rm B} T}{4\pi e^2 n}}. \tag{14.56}$$

This expression is applicable to particles obeying Maxwell–Boltzmann statistics. It is obtained from the general representation for the screening length (Chazalviel 1999)

$$\frac{1}{\kappa} = R = \sqrt{\frac{\varepsilon_0 D}{4\pi \sigma_0(T)}}. \tag{14.57}$$

For this purpose, one must use the following expression for the dc conductivity (Ashcroft and Mermin 1976),

$$\sigma_0(T) = \mu |e| n, \tag{14.58}$$

and Einstein's relation for the case of Maxwell–Boltzmann statistics (Chazalviel 1999)

$$\frac{D}{\mu} = \frac{k_{\rm B} T}{|e|}. \tag{14.59}$$

Here μ is the mobility of the charge carriers.

However, there is nothing in the theory that restricts the application region of the reflection coefficients $r_{\rm TM,TE}^{\rm mod}$ with the Debye–Hückel screening length (14.56) to only intrinsic semiconductors. These coefficients are applicable to all materials where the density of charge carriers is not too large, so that they are described by Maxwell–Boltzmann statistics. This means that in the framework of the proposed approach, it is natural to apply eqns (14.53)–(14.55) to doped semiconductors with a dopant concentration below the critical value, to solids with ionic conductivity, to Mott-Habbard dielectrics, etc. (Decca et al. 2008).

Here, we consider the application of this approach to metallic plates. Metals and semiconductors of the metallic type are characterized by a rather high concentration of charge carriers, which obey the quantum Fermi–Dirac statistics. The general transport equation, however, is equally applicable to classical and quantum systems. The only difference that one must take into account is the type of statistics. Substituting the Einstein relation appropriate to the case of Fermi–Dirac statistics (Ashcroft and Mermin 1976, Chazalviel 1999),

$$\frac{D}{\mu} = \frac{2 E_{\rm F}}{3|e|}, \tag{14.60}$$

where $E_{\rm F} = \hbar \omega_{\rm p}$ is the Fermi energy, into eqn (14.57), one arrives at the following expression for the Thomas–Fermi screening length (Chazalviel 1999):

$$\frac{1}{\kappa} = \frac{1}{\kappa_{\rm TF}} = R_{\rm TF} = \sqrt{\frac{\varepsilon_0 E_{\rm F}}{6\pi e^2 n}}. \tag{14.61}$$

With this definition of the parameter κ, it is natural to apply eqns (14.53)–(14.55) to metals.

The behavior of the Casimir free energy and entropy at low temperature with the use of the modified reflection coefficients $r^{\text{mod}}_{\text{TM,TE}}(i\xi, k_\perp)$ was determined by Mostepanenko (2009) for metals and by Klimchitskaya (2009) for dielectrics [see also Mostepanenko et al. (2009)]. For all metals, the screening length (14.61) is very small. As a result, at any reasonable separation distance between the plates, the dimensionless parameter $2a\kappa_{\text{TF}}$ is very large. Using a perturbative expansion in the small parameter $(2a\kappa_{\text{TF}})^{-1}$, one can conclude (Mostepanenko 2009) that the Casimir entropy per unit area of the plates at zero temperature, $S^{\text{mod}}(a,T)$, in the case of a perfect crystal lattice is given by the same equation (14.44) as in the standard Drude model approach with the unmodified reflection coefficients (12.67). Thus, in the case of metals, the Casimir entropy at zero temperature is negative and the proposed modification of the reflection coefficients is inconsistent with thermodynamics. In Section 19.3.4, it will be shown that when applied to metals, the approach taking screening effects into account is excluded by the experimental data at a 99.9% confidence level.

For dielectric materials, the asymptotic behavior of the Casimir entropy at low temperature can be found using a perturbative expansion in powers of the small parameter $4\pi\sigma_0(T)/\xi_1$ (see Section 12.5.2). As a result (Klimchitskaya 2009), for intrinsic semiconductors and insulators [for these materials, $n(T)$ decays exponentially to zero with vanishing temperature] $\tilde{S}(a,T)$ goes to zero when $T \to 0$, i.e. the approach under consideration is in agreement with the Nernst heat theorem. However, there is a wide class of dielectric materials (such as doped semiconductors with a dopant concentration below the critical value and solids with ionic conductivity) for which n does not go to zero when T goes to zero. Although $\sigma_0(T)$ in eqn (14.58) goes to zero exponentially fast for all dielectrics when T goes to zero, for most of them this happens owing to a vanishing mobility μ. For all such materials, in accordance with eqn (14.56), $\kappa_{\text{DH}} \to \infty$ when $T \to 0$. As a result (Klimchitskaya 2009),

$$S^{\text{mod}}(a,0) = \frac{k_B}{16\pi a^2}\left[\zeta_R(3) - \text{Li}_3(r_0^2)\right] > 0, \quad (14.62)$$

where r_0 is defined in eqn (12.95). Thus, the approach taking screening effects into account is in contradiction with thermodynamics.

We emphasize that the existence of dielectric materials for which n does not go to zero but μ does go to zero when T vanishes demonstrates that the reflection coefficient (14.53) at $\xi = 0$ is ambiguous. In reality, for such materials $r^{\text{mod}}_{\text{TM}}(0, k_\perp) \to 1$ when T and μ simultaneously vanish. This is because $\kappa_{\text{DH}} \to \infty$ when $T \to 0$ (i.e. the screening length goes to zero), in disagreement with the physical intuition that there should be no screening at zero mobility. This ambiguity is connected with the break in continuity of the reflection coefficient $r^{\text{mod}}_{\text{TM}}(i\xi, k_\perp)$ at the point $\xi = 0$, $T = 0$. If one takes the limit $T \to 0$ first, keeping $\xi = \text{const} \neq 0$, the standard Fresnel reflection coefficient r_{TM} from eqn (12.67) with no screening is reproduced. This property is preserved in the subsequent limiting transition $\xi \to 0$. Once again, the violation of the Nernst heat theorem is caused just by the break in the continuity of the reflection coefficients at the origin of the (ξ, T) plane (Intravaia and Henkel 2008). In Section 20.3.5, it will be shown that the theoretical predictions using the modified reflection coefficients for dielectrics are excluded experimentally at a 70% confidence level.

In Sections 12.5.2 and 14.3.4, the contradictions with thermodynamics and with experimental data of the standard Lifshitz theory when the dc conductivity and relaxation of conduction electrons are included were explained by the violation of its applicability condition, i.e. of thermal equilibrium, in the presence of a drift current. In a similar way, the approach taking screening effects into account applies the Lifshitz theory to physical phenomena involving both drift and diffusion currents. The latter current is caused by a nonequilibrium distribution of charge carriers in an external field, i.e. by a physical situation out of thermal equilibrium. The reason is that the diffusion current is determined by a nonzero gradient of charge carrier density, whereas for homogeneous systems in thermal equilibrium the charge carrier density must be homogeneous. We emphasize that the Boltzmann transport equation used to derive the reflection coefficients $r_{\text{TM,TE}}^{\text{mod}}$ describes only nonequilibrium processes. It is not symmetric under time reversal. As a result, the Boltzmann equation describes processes which are related only to an increase of entropy (Rumer and Ryvkin 1980). Thus, screening effects and diffusion currents cannot be considered in the framework of the Lifshitz theory, as they violate its main applicability condition of thermal equilibrium (Decca et al. 2008). In fact, the fluctuating electromagnetic field can create neither drift nor diffusion currents.

14.4 Leontovich impedance approach at nonzero temperature

The description of metals at zero temperature by means of the Leontovich impedance was considered in Section 13.4. This can be easily generalized to the case of nonzero temperature (Bezerra et al. 2002c, Geyer et al. 2003). The derivation of the Lifshitz formula starts from the expression (12.58) for the free energy of all oscillator modes between two metallic semispaces. However, here, the photon eigenfrequencies do not satisfy eqns (12.20) and (12.21) found from the continuity boundary conditions (12.2). Instead, they satisfy eqn (13.36), following from the impedance boundary conditions (13.25). The additional steps in the derivation repeat those in Section 12.3.1, with the simplification that the integrals along the arcs of infinitely large radius vanish under the condition (13.38). As a result, the Casimir free energy per unit area is expressed by eqn (12.66), with the impedance reflection coefficients (13.40) calculated at imaginary Matsubara frequencies $i\xi_l$. The respective Casimir pressure is given by eqn (12.70) with the same reflection coefficients. Equivalent formulations in terms of real frequencies are presented in eqns (12.74) and (12.75), where the impedance reflection coefficients defined on the real frequency axis must be used. All generalizations of the Lifshitz formula considered in Section 12.2 refer equally to the case where the reflection coefficients are expressed not in terms of the dielectric permittivity but in terms of the Leontovich surface impedance.

In Section 13.4.2, the computational results obtained using the Lifshitz formula at zero temperature with the impedance reflection coefficients were compared with those obtained using the dielectric permittivity. It was shown that in the frequency region of infrared optics, the two approaches lead to practically coincident results. Below, we consider the thermal Lifshitz formula (12.66) with the impedance reflection coefficients (13.40) in various frequency regions. It is shown that in application to the thermal Casimir force, the impedance approach cannot be considered as only an approximate reformulation of the standard Lifshitz theory, but leads to important results that cannot be obtained using other approaches.

14.4.1 Impedance in the frequency region of the normal skin effect

As considered in Section 13.4.1, the frequency region of the normal skin effect is determined by eqn (13.26). The impedance function for this frequency region is given by eqn (13.27). According to Section 13.4.1, the normal skin effect is characterized by a volume relaxation described by a temperature-dependent relaxation parameter $\gamma = \gamma(T)$. This parameter decreases with decreasing temperature (see Section 14.3.2). As a result, the mean free path of the conduction electrons $l = v_F/\gamma(T)$ increases with decreasing temperature and the application conditions (13.26) of the normal skin effect are violated. Thus, the frequency region of the normal skin effect reduces to zero when the temperature vanishes and becomes wider when the temperature becomes higher. Typically, at room temperature, the frequency region of the normal skin effect extends from zero frequency to frequencies of order 10^{12} rad/s. The characteristic frequency of the Casimir effect $\omega_c = c/(2a)$ belongs to this region at separations larger than 100 μm. This is the asymptotic limit of large separations (high temperatures), where the zero-frequency term of the Lifshitz formula alone determines the total result.

To find the Casimir free energy in the region of the normal skin effect, we substitute the impedance function (13.27) into the reflection coefficients (13.40) and consider the limit of zero frequency. The result is

$$r_{\text{TM}}(0, k_\perp) = 1, \qquad r_{\text{TE}}(0, k_\perp) = 1. \tag{14.63}$$

Substituting this into eqn (12.66) and restricting ourselves to the contribution of the zero-frequency term, we arrive at the Casimir free energy in the region of the normal skin effect,

$$\mathcal{F}_n(a, T) = -\frac{k_B T}{8\pi a^2} \zeta_R(3). \tag{14.64}$$

This coincides with eqn (7.95), obtained for ideal-metal planes in the case of high temperatures (or large separations). Equation (14.64) is in agreement with the classical limit.

The results (14.63) and (14.64) are different from those obtained using the dielectric permittivity for the normal skin effect (13.18) and the standard reflection coefficients (12.67). For the dielectric permittivity for the normal skin effect, one obtains eqn (14.6) instead of eqn (14.63) and only one-half of the free energy in eqn (14.64). Thus, the impedance approach in the region of the normal skin effect leads to the same result as was obtained for ideal metals. The standard approach using the dielectric permittivity for the normal skin effect leads to the same result as for the Drude model. The latter is in contradiction with the classical limit (see Section 14.1).

Torgerson and Lamoreaux (2004) and Bimonte (2006a) used the impedance for the normal skin effect to compute the thermal correction to the Casimir pressure at small separations. These computations, which are in disagreement with the experimental data, are considered in Section 14.5 in connection with the role played by evanescent waves.

14.4.2 Impedance in the region of the anomalous skin effect

At higher frequencies, satisfying eqn (13.29), the impedance for the anomalous skin effect is applicable. It is given by eqn (13.31). At room temperature for gold, for instance, the frequency region of the anomalous skin effect is rather narrow. It includes frequencies of order from 10^{12} rad/s to 10^{13} rad/s. The characteristic frequency of the Casimir effect ω_c belongs to this frequency region at separations of the order of tens of micrometers. With a decrease in temperature, however, the lower application limit of the anomalous skin effect decreases. Substituting the impedance (13.31) into the reflection coefficients (13.40), one gets the same result (14.63) as for the normal skin effect. It is different only by a phase factor from the case of an ideal metal (14.1).

Svetovoy and Lokhanin (2003) verified that the Casimir free energy calculated using the impedance for the anomalous skin effect leads to zero entropy at zero temperature in accordance with the Nernst theorem. At very low temperatures satisfying the condition

$$k_B T \ll \left(\frac{\omega_c}{\omega_p}\right)^2 \frac{\hbar \omega_p v_F}{2\pi c}, \tag{14.65}$$

the asymptotic behavior of the thermal correction to the Casimir energy per unit area, $\Delta_T \mathcal{F} \sim \tau^{4/3} \ln \tau$, was obtained [the parameter τ is linear in the temperature; this parameter was defined in eqn (12.89)]. The respective Casimir entropy vanishes as $-\tau^{1/3} \ln \tau$ when the temperature goes to zero.

As explained in Section 13.4.1, in the frequency region of the anomalous skin effect the concept of a dielectric permittivity which depends only on the frequency is not applicable because of the effects of spatial nonlocality. In this frequency region, the impedance approach is the only consistent approach for the theoretical description of the Casimir force between metallic bodies. Computations of the Casimir energy and free energy for gold plates using the impedance for the anomalous skin effect were performed by Geyer et al. (2003). At separations of about 5 µm and larger (to be exact, the impedance for the anomalous skin effect is applicable at $a \gg 2.4\,\mu\mathrm{m}$), the Casimir energy obtained practically coincides with that for the ideal-metal case. The thermal correction to the Casimir energy calculated using the impedance for the anomalous skin effect in its application region is approximately equal to that computed using the impedance for infrared optics (see below). Note, however, that if the impedance for the anomalous skin effect is applied at shorter separations outside its application region, the resulting magnitude of the thermal correction is significantly overestimated. As an example, at $a = 0.15\,\mu\mathrm{m}$ the values of the relative thermal correction $\delta_T \mathcal{F}$ computed using the impedances for the anomalous skin effect and for infrared optics at $T = 300\,\mathrm{K}$ are 1.55% and 0.0182%, respectively. At $T = 70\,\mathrm{K}$, the respective values are 0.485% and 2.76×10^{-4}%. This means that outside its application region, use of the impedance for the anomalous skin effect overestimates the magnitude of the thermal correction by up to a factor of 85 at $T = 300\,\mathrm{K}$ and up to a factor of 1757 at $T = 70\,\mathrm{K}$. Such anomalously large thermal corrections at separations from 100 to 500 nm at $T \leq 70\,\mathrm{K}$ were obtained by Svetovoy and Lokhanin (2003) by using only the impedance for the anomalous skin effect outside its application region.

14.4.3 Impedance in the region of infrared optics

The frequency region of infrared optics is determined by eqn (13.32). It extends between frequencies of order 10^{14} rad/s to frequencies of order $\omega_\mathrm{p}/10$. The application region of infrared optics does not depend on the temperature. The analytic form of the Leontovich impedance in this region is given by eqn (13.33). The impedance for infrared optics leads to approximately the same results for the thermal corrections to the Casimir energy and pressure as does the dielectric permittivity of the plasma model (see the computational results in Section 14.3.1). At large separation distances (formally, in the application region of the anomalous skin effect and then the normal skin effect), the predictions from all three impedances practically coincide with that for an ideal metal.

An important question is the agreement between the Casimir entropy computed using the impedance for infrared optics, and the Nernst heat theorem. We consider this point under the assumption that eqn (13.33) can be extrapolated from the frequency region of infrared optics to all lower frequencies. As is discussed in the next section, this is justified if the characteristic frequency ω_c belongs to the region of infrared optics. We start from the Lifshitz formula (12.108) for the Casimir free energy in terms of the dimensionless variables (12.89) and represent it as the sum of the Casimir energy, $E(a)$, and the thermal correction to it, $\Delta_T \mathcal{F}(a, T)$, defined in eqn (12.90). The impedance reflection coefficients are, in terms of dimensionless variables,

$$r_\mathrm{TM}(i\zeta_l, y) = \frac{y - Z(i\zeta_l\omega_\mathrm{c})\zeta_l}{y + Z(i\zeta_l\omega_\mathrm{c})\zeta_l}, \quad r_\mathrm{TE}(i\zeta_l, y) = \frac{yZ(i\zeta_l\omega_\mathrm{c}) - \zeta_l}{yZ(i\zeta_l\omega_\mathrm{c}) + \zeta_l}. \quad (14.66)$$

Using the same variables, the impedance for infrared optics (13.33) takes the form

$$Z_{\mathrm{i}}(\mathrm{i}\zeta_l\omega_{\mathrm{c}}) = \frac{\alpha\zeta_l}{\sqrt{1+\alpha^2\zeta_l^2}}, \qquad (14.67)$$

where α, defined after eqn (14.31), is much less than unity throughout the entire application region of the impedance approach. At zero frequency, from eqns (14.66) and (14.67), we obtain

$$r_{\mathrm{TM}}(0,y) = 1, \qquad r_{\mathrm{TE}}(0,y) = -\frac{1-\alpha y}{1+\alpha y}. \qquad (14.68)$$

These reflection coefficients at zero Matsubara frequency were used when the impedance approach was compared with experiment (Decca et al. 2005b). In this case the values of the impedance function at all nonzero Matsubara frequencies, $Z_{\mathrm{i}}(\mathrm{i}\xi_l)$, were calculated by use of eqn (13.34) with $\omega = \mathrm{i}\xi_l$, where $\varepsilon(\mathrm{i}\xi_l)$ was obtained using optical data, as described in Section 13.3.

Now, we expand the function $f(x,y)$ defined in eqn (12.90) in powers of the small parameter α using eqns (14.66) and (14.67):

$$f(x,y) = 2y\ln(1-\mathrm{e}^{-y}) + 4\alpha\frac{x^2+y^2}{\mathrm{e}^y-1} - 8\alpha^2\frac{(x^4+y^4)\mathrm{e}^y}{y(\mathrm{e}^y-1)^2} + O(\alpha^3). \qquad (14.69)$$

Substituting eqn (14.69) into the definition of $F(x)$ in eqn (12.90), we obtain

$$F(x) = I_0(x) + \alpha I_1(x) + \alpha^2 I_2(x) + O(\alpha^3), \qquad (14.70)$$

where

$$I_0(x) = 2\int_x^\infty y\,dy\,\ln(1-\mathrm{e}^{-y}), \qquad I_1(x) = 4\int_x^\infty dy\,\frac{x^2+y^2}{\mathrm{e}^y-1},$$

$$I_2(x) = -8\int_x^\infty dy\,\frac{(x^4+y^4)\mathrm{e}^y}{y(\mathrm{e}^y-1)^2}. \qquad (14.71)$$

Integrating with respect to y and expanding the results in powers of x, we arrive at [details are presented by Bezerra et al. (2007)]

$$F(\mathrm{i}t\tau) - F(-\mathrm{i}t\tau) = \pi\mathrm{i}(t\tau)^2 - \frac{2}{3}\mathrm{i}(t\tau)^3 + 4\mathrm{i}\alpha\left[\pi(t\tau)^2 - \frac{4}{3}(t\tau)^3\right] + O[(t\tau)^4]. \qquad (14.72)$$

Here, we limit ourselves to terms of order τ^3, where powers of order α^2 do not contribute to the result (note that $\tau = 2\pi T/T_{\mathrm{eff}}$). Substituting this into eqn (12.90), we obtain the thermal correction to the Casimir energy and, from eqn (12.79), the Casimir free energy

$$\mathcal{F}_{\mathrm{i}}(a,T) = E_{\mathrm{i}}(a) - \frac{\hbar c}{8\pi a^3}\left\{\frac{\zeta_R(3)}{2t^3} - \frac{\pi^3}{90t^4} + \frac{\delta_0}{a}\left[\frac{\zeta_R(3)}{t^3} - \frac{2\pi^3}{45t^4}\right]\right\}, \qquad (14.73)$$

where we have introduced the notation $t = T_{\mathrm{eff}}/T$ for convenience in comparing this result with eqn (14.14). It is evident that eqn (14.73) coincides with the respective perturbation orders in eqn (14.14) and, thus, the impedance for infrared optics leads to the same low-temperature behavior of the Casimir free energy as does the dielectric permittivity of the plasma model. From eqn (14.73), the Casimir entropy at low temperature is given by

$$S_\text{i}(a,T) = \frac{k_\text{B}\tau^2}{16\pi^3 a^2}\left\{\frac{3}{2}\zeta_\text{R}(3) - \frac{\pi^2}{45}\tau + \frac{\delta_0}{a}\left[3\zeta_\text{R}(3) - \frac{4\pi^2}{45}\tau\right]\right\}. \tag{14.74}$$

This coincides with the respective perturbation orders of eqn (14.24) obtained using the dielectric permittivity of the plasma model. The Casimir entropy (14.74) goes to zero when T vanishes, in agreement with the Nernst theorem.

14.4.4 Impedance using the Drude model

In the previous subsection, we extrapolated the impedance function for infrared optics to all lower frequencies, including zero frequency. However, in the Lifshitz formula the zero Matsubara frequency does not belong to the region of infrared optics. In the literature, an approach was suggested in which different impedance functions should be substituted into the Lifshitz formula within different frequency regions in accordance with their applicability conditions [see e.g. the comment by Svetovoy (2004)]. As was shown by Geyer et al. (2004), the use of two impedance functions (that for infrared optics around the characteristic frequency and that for the anomalous skin effect at the first few Matsubara frequencies) results in a violation of the Nernst theorem.

Sometimes, in the literature, the impedance of the Drude model, Z_D, is used (Bimonte 2006b). It is defined by eqn (13.34) with the dielectric permittivity (13.14). At small frequencies satisfying the condition $\omega \ll \gamma$, the impedance Z_D coincides with the impedance for the normal skin effect Z_n. In the region of infrared optics, where $\omega \gg \gamma$, the impedance of the Drude model coincides with Z_i. Therefore Z_D provides a smooth analytic interpolation between the impedances for the normal skin effect and infrared optics.

It can be seen from direct computations that the impedance Z_D results in an anomalously large thermal correction to the Casimir pressure at separations below $1\,\mu$m. To demonstrate this, we substituted the impedance Z_D into the Lifshitz formula (14.17) with the impedance reflection coefficients (14.66) and calculated the thermal correction to the Casimir pressure. At a separation distance $a = 200$ nm at $T = 300$ K, the result for gold is $|\Delta_T P| = 14.7$ mPa. The ratio of this value to the magnitude of the thermal correction for ideal-metal planes computed under the same conditions is 7200. At a separation $a = 1\,\mu$m, the magnitude of the thermal correction to the Casimir pressure calculated using the impedance of the Drude model is larger by a factor of 20 than for ideal-metal planes. Note that the thermal correction obtained using the impedance Z_D increases the magnitude of the Casimir pressure at all separation distances. Recall also that if the dielectric permittivity of the Drude model is used in the computations, the thermal correction decreases the magnitude of the pressure over a wide separation region (see Section 14.3.1).

Bezerra et al. (2007) have demonstrated that substitution of the impedance of the Drude model into the Lifshitz formula for the free energy results in the violation of Nernst's theorem for metals with perfect crystal lattices. To do so, Z_D was presented in terms of the dimensionless variables:

$$Z_D(i\zeta_l \omega_c) = \frac{\alpha \zeta_l}{\sqrt{\alpha^2 \zeta_l^2 + [1/(1+x_l)]}}, \qquad (14.75)$$

where $x_l = \gamma(T)/\xi_l$ and α has been defined after eqn (14.67). It was then expanded in powers of the small parameter x_l:

$$Z_D(i\zeta_l \omega_c) = \frac{\alpha \zeta_l}{\sqrt{1+\alpha^2 \zeta_l^2}} + \frac{(\alpha\zeta_l)^{-2} x_l}{2[1+(\alpha\zeta_l)^{-2}]^{3/2}} + O(x_l^2)$$
$$= Z_i(i\zeta_l \omega_c) + \tilde{x}_l + O(x_l^2). \qquad (14.76)$$

[The impedance for infrared optics Z_i is expressed in this form in eqn (14.67).] The quantity $\tilde{x}_l$ can be identically represented as

$$\tilde{x}_l = \frac{\gamma(T)}{2\omega_p}(1+\alpha^2 \zeta_l^2)^{-3/2} \ll 1. \qquad (14.77)$$

The proof of the violation of Nernst's theorem follows along the same lines as in Section 14.3.2. We consider the Casimir free energy and separate the zero-frequency term:

$$\mathcal{F}_D(a,T) = \mathcal{F}_D^{(l=0)}(a,T) + \mathcal{F}_D^{(l\geq 1)}(a,T). \qquad (14.78)$$

The reflection coefficients obtained with the impedance Z_D satisfy eqn (14.63). This results in

$$\mathcal{F}_D^{(l=0)}(a,T) = \frac{k_B T}{8\pi a^2}\int_0^\infty y\, dy \ln(1-e^{-y}) = -\frac{k_B T}{8\pi a^2}\zeta_R(3). \qquad (14.79)$$

Now we expand the term $\mathcal{F}_D^{(l\geq 1)}(a,T)$ in powers of the small parameters $\tilde{x}_l$ and Z_i:

$$\mathcal{F}_D^{(l\geq 1)}(a,T) = \mathcal{F}_i^{(l\geq 1)}(a,T) + \frac{k_B T \gamma(T)}{2\pi a^2 \omega_p}\frac{\ln \tau}{\tau}\zeta_R(3)$$
$$+ O[Z_i\tilde{x}_l, \tilde{x}_l^2] + \frac{k_B T \gamma(T)}{2\pi a^2 \omega_p}O(\tau^{-1}), \qquad (14.80)$$

where $\mathcal{F}_i^{(l\geq 1)}$ is the contribution of all Matsubara frequencies with $l \geq 1$ to the Casimir free energy computed using the impedance for infrared optics. Details of the perturbative expansion leading to eqn (14.80) were presented by Bezerra et al. (2007). Taking into account that $\tau \sim T$ and at low temperature $\gamma(T) \sim T^2$, $Z_i \sim T$, and $\tilde{x}_l \sim T^2$, we conclude that not only do the last two terms on the right-hand side of eqn (14.80) vanish when the temperature goes to zero, but their derivatives with respect to temperature vanish also. Because of this, we omit these terms below.

Now we add the zero-frequency contribution $\mathcal{F}_D^{(l=0)}$, defined in eqn (14.79), on the left- and right-hand sides of eqn (14.80). In addition, on the right-hand side of eqn (14.80), we add and subtract the zero-frequency contribution $\mathcal{F}_i^{(l=0)}$ obtained using the impedance for infrared optics. Taking eqn (14.68) into account, this contribution is given by

$$\mathcal{F}_i^{(l=0)}(a,T) = -\frac{k_B T}{16\pi a^2}\zeta_R(3) + \frac{k_B T}{16\pi a^2}\int_0^\infty y\,dy\, \ln\left[1 - \left(\frac{1-\alpha y}{1+\alpha y}\right)^2 e^{-y}\right]. \quad (14.81)$$

After using the identical transformations indicated above, eqn (14.80) takes the form

$$\mathcal{F}_D(a,T) = \mathcal{F}_i(a,T) - \frac{k_B T}{16\pi a^2}\left\{\zeta_R(3) + \int_0^\infty y\,dy\,\ln\left[1 - \left(\frac{1-\alpha y}{1+\alpha y}\right)^2 e^{-y}\right]\right\}$$
$$+ \frac{k_B T \gamma(T)}{2\pi a^2 \omega_p}\frac{\ln\tau}{\tau}\zeta_R(3). \quad (14.82)$$

Here, the asymptotic behavior of the free energy obtained using the impedance for infrared optics is given in eqn (14.73). Calculating the negative derivative of both sides of eqn (14.82) with respect to temperature and putting $T = 0$, we obtain the Casimir entropy

$$S_D(a,0) = \frac{k_B}{16\pi a^2}\left\{\zeta_R(3) + \int_0^\infty y\,dy\,\ln\left[1 - \left(\frac{1-\alpha y}{1+\alpha y}\right)^2 e^{-y}\right]\right\} > 0. \quad (14.83)$$

Thus, the use of the impedance of the Drude model results in a violation of the Nernst theorem. The positive value of the entropy at $T = 0$ can be represented perturbatively by expanding it in powers of the small parameter $\alpha = \delta_0/(2a)$:

$$S_D(a,0) = \frac{k_B \zeta_R(3)}{4\pi a^2}\frac{\delta_0}{a}\left\{1 - 3\frac{\delta_0}{a} + O\left[\left(\frac{\delta_0}{a}\right)^2\right]\right\} > 0. \quad (14.84)$$

We conclude that the combination of the Lifshitz formula with the Leontovich impedance for the Drude model is thermodynamically inconsistent. The measurement data presented in Section 19.3 demonstrate that the large thermal corrections to the Casimir force predicted from the use of this impedance at short separation distances are experimentally excluded (Bezerra et al. 2007).

14.5 The role of evanescent and propagating waves

As was noted in Section 12.3.2, the representation of the Lifshitz formula in terms of real frequencies enables one to represent the Casimir free energy and pressure as a sum of contributions from propagating and evanescent waves. The propagating waves satisfy the condition $ck_\perp < \omega$, which is valid for real photons. They propagate both in the vacuum gap and inside the semispaces. The respective values of q, as defined in eqn (12.16), are pure imaginary. Evanescent waves satisfy the condition $\omega \leq ck_\perp$. They may propagate only along the boundary planes $z = \pm a/2$. The electromagnetic field of an evanescent wave decreases exponentially with distance from the interface between the vacuum gap and the semispace. For the evanescent waves, the quantity q is real.

Using these definitions, one can represent the Casimir free energy per unit area (12.74) in the form

$$\mathcal{F}(a,T) = \mathcal{F}_{\text{PW}}(a,T) + \mathcal{F}_{\text{EW}}(a,T), \tag{14.85}$$

where the contributions of propagating and evanescent waves are given by

$$\mathcal{F}_{\text{PW}}(a,T) = \frac{\hbar}{4\pi^2} \int_0^\infty k_\perp dk_\perp \int_{ck_\perp}^\infty d\omega \coth\frac{\hbar\omega}{2k_B T} \text{Im}\left\{\ln\left[1 - r_{\text{TM}}^2(\omega,k_\perp)e^{-2aq}\right]\right.$$
$$\left. + \ln\left[1 - r_{\text{TE}}^2(\omega,k_\perp)e^{-2aq}\right]\right\},$$

$$\mathcal{F}_{\text{EW}}(a,T) = \frac{\hbar}{4\pi^2} \int_0^\infty k_\perp dk_\perp \int_0^{ck_\perp} d\omega \coth\frac{\hbar\omega}{2k_B T} \text{Im}\left\{\ln\left[1 - r_{\text{TM}}^2(\omega,k_\perp)e^{-2aq}\right]\right.$$
$$\left. + \ln\left[1 - r_{\text{TE}}^2(\omega,k_\perp)e^{-2aq}\right]\right\}. \tag{14.86}$$

In a similar way, the Casimir pressure (12.75) can also be expressed as a sum of contributions from propagating and evanescent waves,

$$P(a,T) = P_{\text{PW}}(a,T) + P_{\text{EW}}(a,T). \tag{14.87}$$

By using the identity

$$\coth\frac{x}{2} = 1 + \frac{2}{\exp(x) - 1}, \tag{14.88}$$

we can present both the Casimir free energy and the Casimir pressure, as given by eqns (12.74) and (12.75), in the form of eqn (12.76), where $\mathcal{E}(a,T)$ and $\mathcal{P}(a,T)$ are contained in eqns (12.37) and (12.38), respectively, and $\Delta\mathcal{F}(a,T)$ and $\Delta P(a,T)$ are given by

$$\Delta\mathcal{F}(a,T) = \frac{\hbar}{2\pi^2}\int_0^\infty k_\perp dk_\perp \int_0^\infty \frac{d\omega}{\exp\left(\frac{\hbar\omega}{k_B T}\right) - 1} \text{Im}\left\{\ln\left[1 - r_{\text{TM}}^2(\omega,k_\perp)e^{-2aq}\right]\right.$$
$$\left. + \ln\left[1 - r_{\text{TE}}^2(\omega,k_\perp)e^{-2aq}\right]\right\},$$

$$\Delta P(a,T) = -\frac{\hbar}{\pi^2}\int_0^\infty k_\perp dk_\perp \int_0^\infty \frac{d\omega}{\exp\left(\frac{\hbar\omega}{k_B T}\right) - 1} \text{Im}\left\{q\left[r_{\text{TM}}^{-2}(\omega,k_\perp)e^{2aq} - 1\right]^{-1}\right.$$
$$\left. + q\left[r_{\text{TE}}^{-2}(\omega,k_\perp)e^{2aq} - 1\right]^{-1}\right\}. \tag{14.89}$$

The quantities $\mathcal{E}(a,T)$ and $\mathcal{P}(a,T)$ are of the same functional form as the Casimir energy and pressure, $E(a)$ and $P(a)$, at zero temperature [see eqns (12.37) and

(12.38)]. They are, however, calculated with $\varepsilon = \varepsilon(\omega, T)$ (see the discussion in Section 12.4). Equation (14.89) provides one more equivalent representation for the quantities $\Delta\mathcal{F}$ and $\Delta\mathcal{P}$ [the first representation is contained in eqns (12.77) and (12.78)]. If the dielectric permittivity (or the impedance) is independent of the temperature, these quantities have the physical meaning of thermal corrections to the Casimir energy and pressure, as discussed in Section 12.4.

All of the quantities $\mathcal{E}(a,T)$, $\mathcal{P}(a,T)$, $\Delta\mathcal{F}(a,T)$, and $\Delta\mathcal{P}(a,T)$ can be represented as sums of contributions due to propagating and evanescent waves, as was done for the complete free energy and pressure in eqns (14.85) and (14.86). In the case of the quantities $\mathcal{E}(a,T)$ and $\mathcal{P}(a,T)$, such a representation is difficult to use in numerical computations because the contributions $\mathcal{E}_{\text{PW}}$ and $\mathcal{E}_{\text{EW}}$ (and similarly for $\mathcal{P}_{\text{PW}}$ and $\mathcal{P}_{\text{EW}}$) contain integrals of rapidly oscillating functions. Representations of the form

$$\Delta\mathcal{F}(a,T) = \Delta\mathcal{F}_{\text{PW}}(a,T) + \Delta\mathcal{F}_{\text{EW}}(a,T), \qquad (14.90)$$
$$\Delta\mathcal{P}(a,T) = \Delta\mathcal{P}_{\text{PW}}(a,T) + \Delta\mathcal{P}_{\text{EW}}(a,T)$$

are much more useful. Using the definition of propagating and evanescent waves, from eqn (14.89) one obtains

$$\Delta\mathcal{F}_{\text{PW}}(a,T) = \frac{\hbar}{2\pi^2}\int_0^\infty k_\perp dk_\perp \int_{ck_\perp}^\infty \frac{d\omega}{\exp\left(\frac{\hbar\omega}{k_B T}\right) - 1} \text{Im}\left\{\ln\left[1 - r_{\text{TM}}^2(\omega, k_\perp)e^{-2aq}\right]\right.$$
$$\left. + \ln\left[1 - r_{\text{TE}}^2(\omega, k_\perp)e^{-2aq}\right]\right\},$$

$$\Delta\mathcal{F}_{\text{EW}}(a,T) = \frac{\hbar}{2\pi^2}\int_0^\infty k_\perp dk_\perp \int_0^{ck_\perp} \frac{d\omega}{\exp\left(\frac{\hbar\omega}{k_B T}\right) - 1} \text{Im}\left\{\ln\left[1 - r_{\text{TM}}^2(\omega, k_\perp)e^{-2aq}\right]\right.$$
$$\left. + \ln\left[1 - r_{\text{TE}}^2(\omega, k_\perp)e^{-2aq}\right]\right\} \qquad (14.91)$$

(and similarly for the pressure). These expressions are convenient for numerical computations because the exponents in the denominators of the integrands ensure quick convergence. Note that both $\Delta\mathcal{F}_{\text{PW}}$ and $\Delta\mathcal{F}_{\text{EW}}$ contain contributions from transverse electric and transverse magnetic waves, as do $\Delta\mathcal{P}_{\text{PW}}$ and $\Delta\mathcal{P}_{\text{EW}}$.

Torgerson and Lamoreaux (2004) calculated the contribution from transverse electric waves to the quantity $\Delta\mathcal{P}(a,T)$ using the real-frequency axis formalism and the impedance function for the normal skin effect $Z_n(\omega)$ defined in eqn (13.27). At a separation $a = 1\,\mu\text{m}$, a large thermal effect was found for the Casimir pressure between two gold plates (a factor of 30 larger than in the case of an ideal metal). This is several times smaller (and has the opposite sign) than was predicted by Boström and Sernelius (2000b) at this separation using the dielectric permittivity of the Drude model, but is still a very large effect. According to Torgerson and Lamoreaux (2004), the large thermal effect found by them is explained by the increased role of the transverse electric evanescent waves contributing to $\Delta\mathcal{P}(a,T)$ at low frequencies. Qualitatively the same result, with a corrected magnitude of the thermal effect, was found by Bimonte (2006a).

Here, we present results of numerical computations of both the transverse electric and the transverse magnetic contributions to $\Delta P(a,T)$, as functions of separation, at $T = 300\,\text{K}$ for gold plates described by the impedance Z_n. We also study the role of the propagating and evanescent waves in the significant increase in the value of $\Delta P(a,T)$ when compared with the case of an ideal metal. The static conductivity $\sigma_0 = 2.8 \times 10^{17}\,\text{s}^{-1}$ was used in these computations, as obtained from eqn (13.20) with $\omega_\text{p} = 9.0\,\text{eV}$ and $\gamma = 0.035\,\text{eV}$. The computational results are presented in Table 14.1, where column 1 contains the values of the separation distance between the plates, and column 2 the values of ΔP. These values were computed using the imaginary-axis formulas (12.33), (12.70), and (12.76) and also using the real-axis formula (14.89) with coincident results. Note that although the magnitude of ΔP increases with decreasing separation, the relative thermal correction becomes smaller at shorter separations. The same is the case for metals described by the dielectric permittivity of the plasma model (see Section 14.2). In column 3 of Table 14.1, the ratio of ΔP computed using the impedance Z_n to ΔP^IM computed for an ideal metal is presented. Columns 4 and 5 contain the relative contributions to ΔP from transverse electric evanescent waves and propagating waves, respectively. In columns 6 and 7, the relative contributions to ΔP from transverse magnetic evanescent waves and propagating waves, respectively, are presented. All computations were performed at separations $a \geq 200\,\text{nm}$ in order to remain in the application region of the impedance approach. For example, at $a = 200\,\text{nm}$ the characteristic frequency $\omega_\text{c} = c/(2a) = 7.5 \times 10^{14}\,\text{rad/s}$ and $Z_\text{n}(\omega_\text{c}) = 1.4 \times 10^{-2} \ll 1$. As can be seen from columns 2 and 3 in Table 14.1, the magnitudes of ΔP computed using the impedance for the normal skin effect Z_n are rather large. At $a = 1\,\mu\text{m}$, ΔP is larger by a factor of 16 than ΔP^IM for ideal-metal planes. This ratio quickly increases with decreasing separation. At $a = 200\,\text{nm}$, it is as large as 5900. By summation of the values presented in columns 4 and 5, in one case and 6 and 7 in the other, we find that the dominant contributions to ΔP are given

TABLE 14.1. The quantity ΔP for two Au plates at $T = 300\,\text{K}$ (column 2) and various contributions to it as a function of separation, computed using the impedance for the normal skin effect Z_n. See text for further discussion.

$a\,(\mu\text{m})$	$\Delta P\,(\text{mPa})$	$\dfrac{\Delta P}{\Delta P^\text{IM}}$	$\dfrac{\Delta P_{\text{TE,EW}}}{\Delta P}$	$\dfrac{\Delta P_{\text{TE,PW}}}{\Delta P}$	$\dfrac{\Delta P_{\text{TM,EW}}}{\Delta P}$	$\dfrac{\Delta P_{\text{TM,PW}}}{\Delta P}$
0.2	−12.1	5.9×10^3	0.998	-9×10^{-5}	2×10^{-3}	-2×10^{-4}
0.25	−5.4	2.6×10^3	0.997	-7×10^{-5}	3.6×10^{-3}	-2×10^{-4}
0.3	−2.8	1.4×10^3	0.995	-3×10^{-6}	5×10^{-3}	-3×10^{-4}
0.35	−1.6	7.8×10^2	0.994	1.5×10^{-4}	6×10^{-3}	-3×10^{-4}
0.4	−0.96	4.7×10^2	0.992	5×10^{-4}	8×10^{-3}	-2×10^{-4}
1	−0.032	16	0.91	0.03	0.03	0.02

by the transverse electric mode, whereas the contributions from the transverse magnetic mode are negligibly small. The largest value obtained for the latter at $a = 1\,\mu$m is 0.05. Similarly, comparing columns 4 and 5 in one case and 6 and 7 in the other, we conclude that in both cases the dominant contribution is given by the evanescent waves. If we consider only the contribution from the transverse electric mode discussed by Torgerson and Lamoreaux (2004), we obtain $\Delta P_{\text{TE}}/\Delta P_{\text{TE}}^{\text{IM}} = 29.6$ and 1.2×10^4 at separations $a = 1\,\mu$m and $0.2\,\mu$m, respectively.

The large thermal effect discussed above is caused by the contribution of evanescent waves at relatively low frequencies in the GHz range. The characteristic frequencies in the separation region considered (from 200 nm to $1\,\mu$m) belong to the region of infrared optics. Thus, if the impedance Z_n were used to calculate the complete Casimir free energy $\mathcal{F}(a,T)$ and pressure $P(a,T)$, the results would be in error. This is because at the frequencies of infrared optics, the impedance for the normal skin effect Z_n leads to a much smaller skin depth than the correct value, $\lambda_p/(2\pi)$, given by the impedance for infrared optics Z_i. In other words, the impedance Z_n when applied in the frequency region of infrared optics significantly underestimates the effect of the skin depth. In this situation, it is beyond reason to believe that the use of the impedance Z_n may lead to a proper theoretical description of the thermal correction. It is notable also that the quantity $\Delta P(a,T)$ computed by Torgerson and Lamoreaux (2004) is not equal to the thermal correction to the Casimir pressure $\Delta_T P(a,T)$, because the conductivity σ_0 and, consequently, the impedance function Z_n are functions of temperature (see Section 12.4 for a discussion of this subject). The experimental results presented in Section 19.3 exclude the theoretical approach using the impedance for the normal skin effect to describe the thermal Casimir force at separations of $1\,\mu$m and lower (Bezerra et al. 2007).

The role of propagating and evanescent waves has also been investigated for plates described by a frequency-dependent dielectric permittivity. Henkel et al. (2004) have shown that at short separation distances the evanescent waves provide a dominant contribution to the Casimir force between plates described by the plasma model and also plates made of weakly absorbing materials. A detailed investigation of the propagating and evanescent waves at any separation between the plates in the framework of the plasma model was performed by Intravaia and Lambrecht (2005) and Intravaia et al. (2007). Specifically, it was shown that while the propagating waves always contribute to the attractive force, evanescent waves may lead to a repulsive contribution as well. Of course, the resulting Casimir force between the plates always remains attractive, as expected.

Evanescent waves play an important role in radiative heat transfer through a vacuum gap, which was considered in Section 12.10. Equation (12.140) represents the power of heat transfer, $G(a, T_1, T_2)$, as a sum of contributions from propagating and evanescent waves. The total power G and the contributions to it from the two types of waves, G_{PW} and G_{EW}, were computed as a function of separation for two gold plates at temperatures $T_1 = 320\,\text{K}$ and $T_2 = 300\,\text{K}$

using the dielectric permittivity of the Drude model (13.14), the impedance for the normal skin effect (13.27), and the impedance of the Drude model (13.34), (13.14) (Bezerra et al. 2007). As an example, at $a = 0.3\,\mu\mathrm{m}$ the power of heat transfer calculated using the dielectric permittivity of the Drude model is equal to $G_\mathrm{D} = 89\,\mathrm{W\,m^{-2}}$. If, at the same separation, the impedances for the normal skin effect and that of the Drude model are used, the results are $185\,\mathrm{W\,m^{-2}}$ and $107\,\mathrm{W\,m^{-2}}$, respectively. The contribution of the transverse electric evanescent waves to these results is dominant. Thus, when ε_D is used in the computations, they contribute 94% of the total power. For the impedances Z_n and Z_D, the contribution of transverse electric evanescent waves is equal to 93% and 95%, respectively. These contributions decrease with increasing separation distance. For example, at $a = 0.8\,\mu\mathrm{m}$ the transverse electric waves contribute 66% of the power of heat transfer when ε_D is used in the computations and 55% and 69% when Z_n and Z_D, respectively, are used.

As can be seen from the above, the use of different models leads to different predictions for the power of radiative heat transfer through a vacuum gap. At present there is insufficient experimental information which could constrain the theoretical choice.

14.6 Metals described by the generalized plasma-like model

In the foregoing, several different approaches to the thermal Casimir force were considered using different models for the dielectric permittivity and surface impedance. Many observations were made indicating that the Lifshitz theory runs into problems when it includes the drift current of the conduction electrons. When the relaxation of the conduction electrons is taken into account, one arrives at contradictions between the Lifshitz theory and thermodynamics. We have discussed an attempt to avoid the violation of the Nernst theorem by the inclusion of impurities. This, however, cannot be considered as satisfactory because, in accordance with quantum statistical physics, the Casimir entropy must vanish at zero temperature in the case of a perfect crystal lattice. In addition, the large thermal corrections predicted when one takes the relaxation of the conduction electrons into account are excluded experimentally. Another attempt exploits the concept of the Leontovich impedance and impedance boundary conditions. This allows one to avoid consideration of the volume relaxation and provides a practical resolution of the problem, at least on a phenomenological level, because it leads to results in qualitative agreement with the case of an ideal metal and is consistent with experiment. The impedance approach, however, is not applicable at separations below the plasma wavelength and, thus, cannot be used for comparison with measurements of the Casimir force between metals at short separations.

As discussed in Section 14.3.4, a drift current violates the applicability condition of the Lifshitz theory, i.e. thermal equilibrium. Thus, in the Lifshitz theory, the relaxation properties of conduction electrons connected with the drift current should be disregarded. This is parallel to our conclusion stated in Section

12.5.2 that, in the framework of the Lifshitz theory, one should disregard the conductivity of dielectrics at nonzero temperature. Below, we show that the generalized plasma-like model, considered at zero temperature in Section 13.5, can be used to describe conduction electrons in metals in the framework of the Lifshitz theory at any temperature, in a way consistent with the basic principles of thermodynamics.

14.6.1 Computational results

As discussed in Chapters 13 and 14, the usual plasma model (13.1) disregards all relaxation processes. Because of this, at short separation distances below the plasma wavelength, predictions obtained from the Lifshitz formula using the plasma model deviate from those obtained using tabulated optical data (see Sections 13.2.2 and 13.3). Thus, the usual plasma model cannot be used to calculate the van der Waals force. The generalized plasma-like model (13.44), (13.46) eliminates this defect. It takes into account the relaxation of core electrons due to interband transitions, but disregards the relaxation of conduction electrons, just as suggested in the introduction to this section. Thanks to this property, the generalized plasma-like dielectric permittivity provides a consistent description of the thermal Casimir force between real metals in the Lifshitz theory. We begin with computational results obtained using this permittivity and consider general theoretical properties of this approach in the following subsections.

We have performed computations of the Casimir free energy per unit area and the Casimir pressure between gold semispaces at $T = 300\,\text{K}$ using the Lifshitz formulas (12.108) [with $r^{(1)} = r^{(2)}$] and (14.17) expressed in terms of dimensionless variables. Keeping in mind applications at both small and large separations, we used the dielectric permittivity of the generalized plasma model along the imaginary frequency axis, shown as a solid line in Fig. 13.2. This was obtained by using the dispersion relation (13.61) and the tabulated optical data with the contribution from the relaxation of conduction electrons subtracted, as discussed in Section 13.5.2. In Fig. 14.6(a), the solid line presents the correction factor to the Casimir energy, $\eta_E = \mathcal{F}_{\text{gp}}(a,T)/E(a)$, as a function of the separation at $T = 300\,\text{K}$, where $\mathcal{F}_{\text{gp}}(a,T)$ is computed using the generalized plasma-like permittivity, and the energy for an ideal metal $E(a)$ is defined in eqn (1.5). For comparison purposes, in the same figure the dashed line shows the correction factor when the dielectric permittivity is obtained from the optical data extrapolated to zero frequency by means of the Drude model (see Section 13.3). As is seen in Fig. 14.6(a), the solid and dashed lines demonstrate different behaviors; the dashed line is a nonmonotonic function of the separation.

It is instructive to compare Fig. 14.6(a) with Fig. 13.6(a), which presents the same correction factor at zero temperature. The solid line in Fig. 14.6(a) at separations below $1\,\mu\text{m}$ is very close to the solid line in Fig. 13.6(a), indicating that the thermal corrections predicted by the generalized plasma-like model are small at short separations. In contrast, the dashed line in Fig. 14.6(a) deviates significantly from the dashed line in Fig. 13.6(a) even at $a = 200\,\text{nm}$, demon-

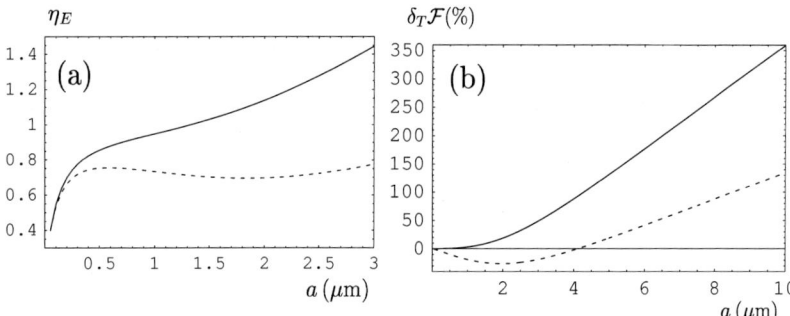

FIG. 14.6. (a) The correction factor and (b) the relative thermal correction to the Casimir free energy as a function of separation for two gold semispaces at $T = 300\,\text{K}$. The solid lines were computed by using the generalized plasma-like dielectric permittivity. The dashed lines represent results obtained using the optical data extrapolated to low frequencies by use of the Drude model.

strating a large thermal effect, as predicted when the relaxation of conduction electrons is taken into account.

In Fig. 14.6(b), we present the relative thermal correction to the Casimir energy computed using the generalized plasma model,

$$\delta_T \mathcal{F}_{\text{gp}}(a, T) = \frac{\Delta_T \mathcal{F}_{\text{gp}}(a, T)}{E_{\text{gp}}(a)}, \tag{14.92}$$

as a function of separation (solid line). In the same figure, the dashed line shows the relative thermal correction $\delta_T \mathcal{F}_{\text{D}}(a, T)$ computed using the optical data extrapolated by means of the Drude model [it practically coincides with the relative thermal correction computed using the Drude model alone, as shown by the solid line in Fig. 14.2(a)]. As is seen in Fig. 14.6(b), the thermal correction computed using the generalized plasma model is positive and increases with increasing separation. It almost coincides with the relative thermal correction computed using the usual plasma model [the dashed line in Fig. 14.2(a)].

Similar results are obtained for the Casimir pressure. In Fig. 14.7(a), the correction factor $\eta_P = P_{\text{gp}}(a, T)/P(a)$ calculated using the generalized plasma-like model as a function of separation is shown by the solid line [the Casimir pressure for ideal-metal planes, $P(a)$, is defined in eqn (1.1)]. In the same figure, the correction factor to the Casimir pressure computed using the optical data extrapolated by means of the Drude model is shown by the dashed line. The latter is a nonmonotonic function of separation. A comparison with the respective correction factors plotted at zero temperature in Fig. 13.6(b) shows that at separations below $1\,\mu\text{m}$, the thermal effects at $T = 300\,\text{K}$ only slightly influence the correction factor calculated using the generalized plasma-like model. At the same time, the dashed line in Fig. 14.7(a) deviates significantly from the dashed line

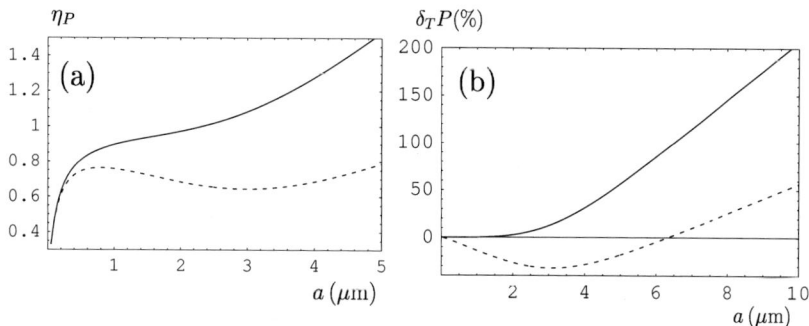

FIG. 14.7. (a) The correction factor and (b) the relative thermal correction to the Casimir pressure as a function of separation for two gold semispaces at $T = 300$ K. The solid lines were computed by using the generalized plasma-like dielectric permittivity. The dashed lines represent results obtained using the optical data extrapolated to low frequencies by use of the Drude model.

in Fig. 13.6(b) at separations above 200 nm. This demonstrates an anomalously large thermal effect, as predicted by the Drude model.

The relative thermal correction to the Casimir pressure,

$$\delta_T P_{\mathrm{gp}}(a, T) = \frac{\Delta_T P_{\mathrm{gp}}(a, T)}{P_{\mathrm{gp}}(a)}, \qquad (14.93)$$

computed using the generalized plasma-like model is shown by the solid line in Fig. 14.7(b). It almost coincides with the respective result for the usual plasma model, shown by the dashed line in Fig. 14.2(b). The Drude model predicts a nonmonotonic relative thermal correction, as shown in Fig. 14.7(b) by the dashed line.

Some analytic asymptotic results for the Casimir free energy in the framework of the generalized plasma-like model are considered below.

14.6.2 Perturbation theory for the generalized plasma model

To obtain the behavior of the free energy at low and high temperature (small and large separations), we use the analytic expression for the plasma-like dielectric permittivity given in eqns (13.44) and (13.46). Keeping in mind applications to real experimental situations, we consider the Casimir free energy in a configuration of two similar plates of finite thickness d. It is convenient to use the dimensionless variables (12.89) and start from the Casimir free energy in the form of eqn (12.108) with $r^{(1)} = r^{(2)}$. For plates of finite thickness, the reflection coefficients are contained in eqn (12.52). In terms of dimensionless variables, they are given by

$$r_{\mathrm{TM}}(i\zeta_l, y) = \frac{(\varepsilon_l^2 - 1)(y^2 - \zeta_l^2)}{(\varepsilon_l + 1)y^2 + (\varepsilon_l - 1)\zeta_l^2 + 2\varepsilon_l y h_l(y) \coth\left[h_l(y)\, d/(2a)\right]},$$

$$r_{\text{TE}}(i\zeta_l, y) = \frac{(\varepsilon_l - 1)\zeta_l^2}{2y^2 + (\varepsilon_l - 1)\zeta_l^2 + 2yh_l(y)\coth[h_l(y)\,d/(2a)]}, \quad (14.94)$$

where

$$h_l(y) = \left[y^2 + (\varepsilon_l - 1)\zeta_l^2\right]^{1/2}. \quad (14.95)$$

The generalized plasma-like dielectric permittivity along the imaginary frequency axis [eqn (13.46) with an arbitrary number of oscillators K] can also be presented in terms of the dimensionless variables:

$$\varepsilon_l = \varepsilon_{\text{gp}}(i\omega_c\zeta_l) = 1 + \frac{\tilde{\omega}_p^2}{\zeta_l^2} + A_l = 1 + \frac{1}{\alpha^2 \zeta_l^2} + A_l. \quad (14.96)$$

Here the following notation has been used:

$$A_l = A(\zeta_l) = \sum_{j=1}^{K} \frac{\tilde{g}_j}{1 + \alpha_j \zeta_l^2 + \beta_j \zeta_l}, \quad (14.97)$$

where α_j, β_j, and $\tilde{g}_j$ are defined in eqn (12.93) and $\tilde{\omega}_p = \omega_p/\omega_c = 1/\alpha$.

Now we represent the free energy in the form of eqn (12.76), where $\mathcal{E}(a) = E(a)$ and $\Delta\mathcal{F}(a,T) = \Delta_T\mathcal{F}(a,T)$ are given in eqn (12.90), and determine the asymptotic behavior of the thermal correction $\Delta_T\mathcal{F}(a,T)$ at low temperature (Geyer et al. 2007).

First, we expand the reflection coefficients (14.94), with ζ_l replaced by ζ, in powers of the small parameter α, preserving all powers, including the fourth order. Thus we consider separation distances such that $\alpha = \lambda_p/(4\pi a) \ll 1$, i.e. $a > \lambda_p$ (see Section 14.4.4). In fact, it is more convenient to expand the logarithmic functions contained in $f(\zeta, y)$, as defined in eqn (12.90). The results are as follows:

$$y \ln\left[1 - r_{\text{TM}}^2(i\zeta, y)e^{-y}\right] = y \ln(1 - e^{-y}) + \alpha \frac{4\zeta^2}{e^y - 1} - \alpha^2 \frac{8e^y \zeta^4}{y(e^y - 1)^2}$$
$$+ \alpha^3 \frac{2\zeta^2 \left\{2\zeta^4 (3e^y + 1)^2 + 3(e^y - 1)^2 y^2 \left[y^2 - 2\zeta^2 - \zeta^2 A(\zeta)\right]\right\}}{3y^2 (e^y - 1)^3}$$
$$- \alpha^4 \frac{8e^y \zeta^4 \left\{2\zeta^4 (e^y + 1)^2 + (e^y - 1)^2 y^2 \left[y^2 - 2\zeta^2 - \zeta^2 A(\zeta)\right]\right\}}{y^3 (e^y - 1)^4},$$
$$(14.98)$$

$$y \ln\left[1 - r_{\text{TE}}^2(i\zeta, y)e^{-y}\right] = y \ln(1 - e^{-y}) + \alpha \frac{4y^2}{e^y - 1} - \alpha^2 \frac{8y^3 e^y}{(e^y - 1)^2}$$
$$+ \alpha^3 \frac{2y^2 \left[-3(e^y - 1)^2 \zeta^2 A(\zeta) + y^2 (15e^{2y} + 18e^y - 1)\right]}{3(e^y - 1)^3}$$

$$-\alpha^4 \frac{8y^3 e^y \left[-(e^y-1)^2 \zeta^2 A(\zeta) + y^2 \left(e^{2y} + 6e^y + 1\right)\right]}{(e^y-1)^4}.$$

It is significant that these expansions do not depend on d (the thickness of the plates), which is contained in eqn (14.94). This is because the factor in the denominator of eqn (14.94),

$$\coth\left[\frac{d}{2a} h_l(y)\right] = \coth\left(\frac{d}{2a}\sqrt{y^2 + \frac{1}{\alpha^2} + A_l \zeta_l^2}\right) \qquad (14.99)$$

$$= \frac{1 + \exp\left[-d\sqrt{1 + \alpha^2 y^2 + \alpha^2 A_l \zeta_l^2}/(a\alpha)\right]}{1 - \exp\left[-d\sqrt{1 + \alpha^2 y^2 + \alpha^2 A_l \zeta_l^2}/(a\alpha)\right]},$$

behaves asymptotically as $1 + 2\exp[-d/(a\alpha)]$ when α goes to zero. Thus, this factor can only contribute exponentially small terms in the expansion (14.98), providing the plate thickness d is much larger than the skin depth [recall that $2a\alpha = \lambda_p/(2\pi)$]. Under this condition, the perturbative expansions (14.98) are the same for two semispaces and for two plates of finite thickness. We note also that the terms in eqn (14.98) of order α^0, α, and α^2 do not contain contributions from the core electrons. The latter are contained only in the terms of order α^3 and α^4.

Below, we consider the limit of low temperature, i.e. of the small parameter τ defined in eqn (12.89). The contribution from the terms of order α^0, α, and α^2 in eqn (14.98) to the thermal correction is the same as for the usual plasma model. It is denoted by $\Delta_T \mathcal{F}_p(a,T)$ and is given by eqn (14.14), where $t = T_{\text{eff}}/T$ and the exponentially small terms are omitted. Now we deal with the terms of order α^3 and α^4 in eqn (14.98), which contain the contributions from the core electrons. The respective functions in eqn (12.90) can be denoted by $F^{(3)}(x)$ and $F^{(4)}(x)$. They are given by

$$F^{(3)}(x) = -2\alpha^3 \left\{ [A(x) - 1] x^2 \int_x^\infty \frac{y^2 \, dy}{e^y - 1} - \frac{1}{3} \int_x^\infty \frac{y^4 \left(15 e^{2y} + 18 e^y - 1\right)}{(e^y - 1)^3} dy \right.$$

$$\left. + [A(x) + 2] x^4 \int_x^\infty \frac{dy}{e^y - 1} - \frac{2}{3} x^6 \int_x^\infty \frac{(3e^y + 1)^2}{y^3 (e^y - 1)^3} dy \right\}, \qquad (14.100)$$

$$F^{(4)}(x) = 8\alpha^4 \left\{ A(x) x^2 \int_x^\infty \frac{y^3 e^y \, dy}{(e^y - 1)^2} \right.$$

$$- \int_x^\infty \frac{y^5 \left(e^{2y} + 6 e^y + 1\right) e^y \, dy}{(e^y - 1)^4} + [A(x) + 2] x^6 \int_x^\infty \frac{e^y \, dy}{y (e^y - 1)^2}$$

$$\left. - x^4 \int_x^\infty \frac{y e^y \, dy}{y^3 (e^y - 1)^2} - 2 x^8 \int_x^\infty \frac{e^y (e^y + 1)^2}{y^3 (e^y - 1)^4} dy \right\}.$$

Calculating all of the integrals in eqn (14.100) as asymptotic expansions at small x [details are presented by Geyer et al. (2007)], we arrive at

$$F^{(3)}(i\tau t) - F^{(3)}(-i\tau t) = -2i\alpha^3 \left[2\tau^3 t^3 \zeta_R(3) \sum_{j=1}^{K} \tilde{g}_j \beta_j + \pi \tau^4 t^4 \left(\sum_{j=1}^{K} \tilde{g}_j + 2 \right) \right],$$

$$F^{(4)}(i\tau t) - F^{(4)}(-i\tau t) = 8i\alpha^4 \left[8\tau^3 t^3 \zeta_R(3) \sum_{j=1}^{K} \tilde{g}_j \beta_j + \pi \tau^4 t^4 \right]. \quad (14.101)$$

The terms omitted in eqn (14.101) are of order τ^5.

We denote the contribution from the terms of order α^3 and α^4 in the thermal correction by $\Delta \mathcal{F}_g(a,T)$. This contains both the higher-order terms of the usual plasma model that were not taken into account in eqn (14.14) and the terms specific to the generalized plasma-like model. Substituting eqn (14.101) into eqn (14.98), we obtain

$$\Delta \mathcal{F}_g(a,T) = -\frac{\hbar c}{32 \pi^2 a^3} \left\{ -\alpha^3 \left[\frac{\zeta_R(3)}{60} \sum_{j=1}^{K} \tilde{g}_j \beta_j \tau^4 + \frac{3 \zeta_R(5)}{2\pi^4} \left(\sum_{j=1}^{K} \tilde{g}_j + 2 \right) \tau^5 \right] \right.$$

$$\left. + \alpha^4 \left[\frac{4 \zeta_R(3)}{15} \sum_{j=1}^{K} \tilde{g}_j \beta_j \tau^4 + \frac{6 \zeta_R(5)}{\pi^4} \tau^5 \right] \right\}. \quad (14.102)$$

The total Casimir free energy computed using the generalized plasma-like permittivity can now be found as

$$\mathcal{F}_{gp}(a,T) = E_{gp}(a) + \Delta_T \mathcal{F}_p(a,T) + \Delta \mathcal{F}_g(a,T), \quad (14.103)$$

where $E_{gp}(a)$ is the Casimir energy at zero temperature (see Section 13.5.3). Using eqns (14.14) (with exponentially small contributions omitted) and (14.102), we get the result

$$\mathcal{F}_{gp}(a,T) = E_{gp}(a) - \frac{\hbar c \zeta_R(3)}{16 \pi a^3} \frac{1}{t^3} \left\{ 1 + 2 \frac{\delta_0}{a} \right.$$

$$- \frac{\pi^3}{45 \zeta_R(3)} \frac{1}{t} \left[1 + 4 \frac{\delta_0}{a} + \frac{3 \zeta_R(3)}{4} \left(\frac{\delta_0}{a} \right)^3 \sum_{j=1}^{K} \tilde{g}_j \beta_j - 6 \zeta_R(3) \left(\frac{\delta_0}{a} \right)^4 \sum_{j=1}^{K} \tilde{g}_j \beta_j \right]$$

$$\left. - \frac{2 \zeta_R(5)}{\zeta_R(3)} \frac{1}{t^2} \left(\frac{\delta_0}{a} \right)^2 \left[1 + \frac{3}{2} \frac{\delta_0}{a} \left(\sum_{j=1}^{K} \tilde{g}_j + 2 \right) - 3 \left(\frac{\delta_0}{a} \right)^2 \right] \right\}. \quad (14.104)$$

Here we can see that the free energy calculated using the generalized plasma-like permittivity contains a correction of order $1/t^4 = (T/T_{\text{eff}})^4$ not only in the terms of order $(\delta_0/a)^0$ and δ_0/a (as in the usual plasma model) but also in the third- and fourth-order expansion terms in δ_0/a. In the usual plasma model, the terms of order $(\delta_0/a)^3$ and $(\delta_0/a)^4$ contain thermal corrections only to the order of $(T/T_{\rm eff})^5$ and higher. To estimate the relative role of the additional terms arising from the use of the generalized plasma-like permittivity, we can employ the following values of the parameters for gold from Table 13.1:

$$\sum_{j=1}^{6}\tilde{g}_j = 6.3175, \qquad \sum_{j=1}^{6}\tilde{g}_j\beta_j = \begin{cases} 0.272, & a = 200\,{\rm nm}, \\ 0.109, & a = 500\,{\rm nm}. \end{cases} \quad (14.105)$$

Equation (14.104) is applicable at low temperatures $T \ll T_{\rm eff}$. It is easy to obtain the high-temperature asymptotic behavior of the Casimir free energy as determined by the generalized plasma-like dielectric permittivity. In this case the zero-frequency term of the Lifshitz formula alone determines the total result. However, the substitution of the generalized plasma-like permittivity (13.46) into the reflection coefficients (12.67) leads at zero frequency to the same reflection coefficients (14.3) as for the usual plasma model. Thus, in the high-temperature limit, the Casimir free energy and pressure found by using the generalized plasma-like model are given by eqn (14.4), which was found for the usual plasma model. As noted in Section 14.1, the asymptotic expressions obtained are in agreement with the case of an ideal metal and with the classical limit.

14.6.3 Agreement with the Nernst heat theorem

The asymptotic expression for the Casimir free energy in the limit of low temperatures in eqn (14.104) allows one to find the temperature behavior of the Casimir entropy and to test the consistency of the generalized plasma-like model with the third law of thermodynamics. For this purpose we calculate the Casimir entropy from eqns (5.4) and (14.104), with the result

$$S_{\rm gp}(a,T) = \frac{3\zeta_{\rm R}(3)k_{\rm B}}{32\pi^3 a^2}\tau^2 \left\{ 1 + 2\frac{\delta_0}{a} \right.$$

$$- \frac{2\pi^2}{135\zeta_{\rm R}(3)}\tau\left[1 + 4\frac{\delta_0}{a} + \frac{3\zeta_{\rm R}(3)}{4}\left(\frac{\delta_0}{a}\right)^3 \sum_{j=1}^{K}\tilde{g}_j\beta_j - 6\zeta_{\rm R}(3)\left(\frac{\delta_0}{a}\right)^4 \sum_{j=1}^{K}\tilde{g}_j\beta_j\right]$$

$$\left. - \frac{5\zeta_{\rm R}(5)}{6\zeta_{\rm R}(3)\pi^2}\tau^2\left(\frac{\delta_0}{a}\right)^2\left[1 + \frac{3}{2}\frac{\delta_0}{a}\left(\sum_{j=1}^{K}\tilde{g}_j + 2\right) - 3\left(\frac{\delta_0}{a}\right)^2\right]\right\}. \quad (14.106)$$

As can be seen from eqn (14.106), $S_{\rm gp}(a,T)$ remains positive and goes to zero when $T \to 0$ (recall that $\tau = 2\pi/t = 2\pi T/T_{\rm eff}$). This means that the Lifshitz theory combined with the generalized plasma-like dielectric permittivity is consistent with thermodynamics.

As can be seen from the foregoing, the generalized plasma-like model leads to reasonable physical results at both short and long separation distances and is consistent with thermodynamics. In Chapter 19, it will be shown also that this model is consistent with all available experimental data in the case of metal surfaces. As argued in Sections 14.3.4 and 14.3.5, the Drude model leads to a nonzero drift current of conduction electrons, which results in the violation of thermal equilibrium. This physical situation is outside the application region of the Lifshitz formula. The generalized plasma-like permittivity, which disregards the relaxation of conduction electrons, leads only to a displacement current. Thus, there is both theoretical and experimental evidence showing that in the application of the Lifshitz theory, the conduction electrons should be described in the framework of the generalized plasma-like model, so that their relaxation is disregarded. In Section 19.3, several experimental results are presented which provide convincing confirmation of this rule.

15

THE CASIMIR INTERACTION BETWEEN A METAL AND A DIELECTRIC

This chapter is devoted to the Casimir interaction between two parallel plates, one metallic and the other dielectric. This hybrid configuration of metal and dielectric plates was first considered by Geyer *et al.* (2005a), and attracted instant attention owing to its unique properties. As discussed in Chapter 12, for dielectrics the transverse electric reflection coefficient vanishes at zero frequency. Thus, owing to the mathematical structure of the Lifshitz formula, it is not important which of the proposed models of the real metal is used when we calculate the Casimir force between dielectric and metal plates. Experimentally, this opens up considerable opportunities for the investigation of the role of the conductivity properties in the Casimir effect by keeping the metal plate fixed but considering dielectrics (semiconductors) with different conductivity properties, varying from insulating to metallic. Such experiments have already been successfully performed and have yielded important new insights (see Chapter 20).

Below, we demonstrate that if the static permittivity of a dielectric plate is finite, the Lifshitz theory is thermodynamically consistent and the Nernst heat theorem is satisfied for the Casimir entropy. The remarkable feature of the dielectric–metal configuration is that the Casimir entropy takes negative values within some temperature interval while going to zero when the temperature vanishes. This is akin to the Casimir–Polder interaction of an atom with a plate (see Chapter 16). In contrast, if the dc conductivity of the dielectric material at nonzero temperature is included in the model of the dielectric response, the Nernst theorem is violated for the Casimir entropy in the configuration of a metal and a dielectric plate. This provides additional confirmation of the rules formulated in Chapters 12 and 14 on how to apply the Lifshitz theory to real materials in a thermodynamically and experimentally consistent way. Approximate analytical formulas for the Casimir energy density and pressure at zero temperature in the configuration of one metal and one dielectric plate are also presented.

15.1 An ideal-metal plate and a plate with constant permittivity

Below, we consider the model where one plate is made of an ideal metal and the other plate is a dielectric with a frequency-independent $\varepsilon = \varepsilon_0$. It appears that this simplified model correctly reflects some important features inherent in real materials. The Casimir free energy in a configuration of two thick dissimilar plates (semispaces) is given by eqn (12.108), expressed in terms of the dimensionless variables (12.89). Let the metal plate be labeled (1) and the dielectric plate

(2). The free energy can be represented exactly by eqn (12.76), where, for a model with temperature-independent parameters, $\mathcal{E}(a,T) = E(a)$ (the Casimir energy at zero temperature) and $\Delta\mathcal{F}(a,T) = \Delta_T \mathcal{F}(a,T)$ (the thermal correction to the Casimir energy). Equation (14.1) is valid for the reflection coefficients of the ideal metal, labeled (1). Then, by using the notation $r^{(2)}_{\text{TM,TE}}(i\zeta,y) \equiv r_{\text{TM,TE}}(i\zeta,y)$, we can represent the Casimir energy at zero temperature and the thermal correction to it by eqn (12.90), where $r^2_{\text{TM,TE}}$ is replaced with $r^{(1)}_{\text{TM,TE}} r^{(2)}_{\text{TM,TE}}$ and

$$f(\zeta,y) = y\left\{\ln\left[1 - r_{\text{TM}}(i\zeta,y)e^{-y}\right] + \ln\left[1 + r_{\text{TE}}(i\zeta,y)e^{-y}\right]\right\}. \tag{15.1}$$

Here, the reflection coefficients for the dielectric plate are given by eqn (12.91). The values of these coefficients at $\zeta = 0$ are contained in eqn (12.107) with the index (n) omitted, and r_0 is defined in eqn (12.95).

15.1.1 The asymptotic behavior at low and high temperature

Now we proceed with the derivation of the asymptotic behavior of the Casimir free energy at low temperature (Geyer et al. 2006). This can be done under the condition $\tau \equiv 2\pi T/T_{\text{eff}} \ll 1$ [see eqn (12.85) for the definition of T_{eff}]. The expansion of $f(x,y)$ defined in eqn (15.1) takes the form

$$f(x,y) = y\ln\left(1 - r_0 e^{-y}\right) - x^2\left(\frac{\varepsilon_0 - 1}{4y}e^{-y} - \frac{\varepsilon_0}{\varepsilon_0 + 1}\sum_{n=1}^{\infty} r_0^n \frac{e^{-ny}}{y}\right) + O(x^3). \tag{15.2}$$

To find $F(x)$ in eqn (12.90), we integrate the right-hand side of eqn (15.2) with respect to y. Note that the first term on the right-hand side of eqn (15.2) does not contribute to the first expansion orders in $F(ix) - F(-ix)$, which are in fact the quantity of interest similarly to eqn (12.90). This is because in the expression

$$\int_x^\infty y\, dy \ln(1 - r_0 e^{-y}) = \int_0^\infty v\, dv \ln(1 - r_0 e^{-v}) + O(x^2), \tag{15.3}$$

where a new variable $v = y - x$ has been introduced, the contribution from the first-order term in x vanishes. Thus, this term can contribute to $F(ix) - F(-ix)$ starting only from the third order in x. Integrating the second term on the right-hand side of eqn (15.2) with the use of the formula

$$\int_x^\infty dy \frac{e^{-ny}}{y} = -\text{Ei}(-nx), \tag{15.4}$$

where $\text{Ei}(z)$ is the exponential integral function, we finally obtain

$$F(ix) - F(-ix) = i\pi \frac{(\varepsilon_0 - 1)^2}{4(\varepsilon_0 + 1)} x^2 - 240 i K_4 x^3 + O(x^4). \tag{15.5}$$

Here, K_4 is some as yet unknown coefficient. It cannot be determined at this stage, because all terms in the expansion of $f(x,y)$ in powers of x contribute to it.

Substituting eqn (15.5) into eqn (12.90) for $\Delta \mathcal{F}$ and using eqn (12.76), we find

$$\mathcal{F}(a,T) = E(a) - \frac{\hbar c}{32\pi^2 a^3} \left[\frac{\zeta_R(3)}{16\pi^2} \frac{(\varepsilon_0 - 1)^2}{\varepsilon_0 + 1} \tau^3 - K_4 \tau^4 + O(\tau^5) \right]. \quad (15.6)$$

Then the Casimir pressure is obtained:

$$P(a,T) = -\frac{\partial \mathcal{F}(a,T)}{\partial a} = P(a) - \frac{\hbar c}{32\pi^2 a^4} \left[K_4 \tau^4 + O(\tau^5) \right]. \quad (15.7)$$

In order to determine the coefficient K_4, we now consider the Casimir pressure in the form of eqn (12.76). Here, $\mathcal{P}(a,T) \equiv P(a)$ is the Casimir pressure at zero temperature. For the case of a frequency-independent $\varepsilon = \varepsilon_0$ under consideration here, the function $f(\zeta, y)$ in eqn (15.1) does not depend on the separation, and $P(a)$ is obtained from eqn (12.90) by differentiation of the factor in front of the integral:

$$P(a) = \frac{3\hbar c}{32\pi^2 a^4} \int_0^\infty d\zeta \int_\zeta^\infty f(\zeta, y) \, dy. \quad (15.8)$$

The thermal correction $\Delta P(a,T) = \Delta_T P(a,T)$ is given by eqn (12.98), where, for our configuration, we obtain from eqn (12.99)

$$\Phi_{\mathrm{TM}}(x) = \int_x^\infty \frac{y^2 \, dy \, r_{\mathrm{TM}}(ix, y)}{e^y - r_{\mathrm{TM}}(ix, y)}, \quad \Phi_{\mathrm{TE}}(x) = -\int_x^\infty \frac{y^2 \, dy \, r_{\mathrm{TE}}(ix, y)}{e^y + r_{\mathrm{TE}}(ix, y)}. \quad (15.9)$$

To find the expansion of $\Phi(ix) - \Phi(-ix)$ in powers of x, we deal first with $\Phi_{\mathrm{TE}}(x)$. We add and subtract the asymptotic behavior of the integrand function at small x,

$$-\frac{y^2 \, r_{\mathrm{TE}}(ix,y)}{e^y + r_{\mathrm{TE}}(ix,y)} = \frac{1}{4}(\varepsilon_0 - 1)x^2 e^{-y} + O(x^3), \quad (15.10)$$

inside the integral in eqn (15.9). Then we introduce a new variable $v = y/x$ and expand the integrand in powers of x, with the result

$$\Phi_{\mathrm{TE}}(x) = \frac{1}{4}(\varepsilon_0 - 1)x^2 e^{-x} \quad (15.11)$$

$$+ x^3 \int_1^\infty dv \left[v^2 \sum_{n=1}^\infty (-1)^n r_{\mathrm{TE}}^n(iv) e^{-nvx} - \frac{1}{4}(\varepsilon_0 - 1)e^{-vx} \right]$$

$$= \frac{1}{4}(\varepsilon_0 - 1)x^2 (1 - x) - x^3 \int_1^\infty dv \left[\frac{v^2 r_{\mathrm{TE}}(iv)}{1 + r_{\mathrm{TE}}(iv)} + \frac{\varepsilon_0 - 1}{4} \right] + O(x^4).$$

By integrating on the right-hand side of eqn (15.11), we arrive at

$$\Phi_{\mathrm{TE}}(x) = \frac{\varepsilon_0 - 1}{4} x^2 - \frac{1}{6}(\varepsilon_0 \sqrt{\varepsilon_0} - 1)x^3 + O(x^4). \quad (15.12)$$

To deal with $\Phi_{\mathrm{TM}}(x)$, we add and subtract under the integral in eqn (15.9) the first two expansion terms of the integrand function in powers of x:

$$\Phi_{\mathrm{TM}}(x) = \int_x^\infty y^2 \, dy \left[\frac{r_0}{e^y - r_0} - \frac{\varepsilon_0 r_0 e^{-y} x^2}{y^2 (\varepsilon_0 + 1)(1 - r_0 e^{-y})^2} \right] \quad (15.13)$$

$$+ \int_x^\infty y^2\, dy \left[\frac{r_{\text{TM}}(ix,y)}{e^y - r_{\text{TM}}(ix,y)} - \frac{r_0}{e^y - r_0} + \frac{\varepsilon_0 r_0 e^{-y} x^2}{y^2(\varepsilon_0 + 1)(1 - r_0 e^{-y})^2} \right].$$

Calculating the asymptotic expansions of both integrals in powers of x, we find

$$\Phi_{\text{TM}}(x) = 2\text{Li}_3(r_0) - \frac{\varepsilon_0(\varepsilon_0 - 1)}{2(\varepsilon_0 + 1)} x^2 \tag{15.14}$$

$$- \frac{1}{6} \left[\varepsilon_0 - 1 + \varepsilon_0(\varepsilon_0 \sqrt{\varepsilon_0} - 1) - 3\varepsilon_0(\varepsilon_0 - 1)\sqrt{\varepsilon_0} \right] x^3 + O(x^4),$$

where $\text{Li}_n(z)$ is the polylogarithm function. After summing eqns (15.12) and (15.14), the following result is obtained:

$$\Phi(ix) - \Phi(-ix) = -\frac{2i}{3} \left(1 - 2\varepsilon_0\sqrt{\varepsilon_0} + \varepsilon_0^2\sqrt{\varepsilon_0} \right) x^3 + O(x^4). \tag{15.15}$$

Substituting this in eqn (12.98) and using eqn (12.76), we get

$$P(a,T) = P(a) - \frac{\hbar c}{32\pi^2 a^4} \left[\frac{1}{360} \left(1 - 2\varepsilon_0\sqrt{\varepsilon_0} + \varepsilon_0^2\sqrt{\varepsilon_0} \right) \tau^4 + O(\tau^5) \right]. \tag{15.16}$$

On comparing this equation with eqn (15.7), we get

$$K_4 = \frac{1}{360} \left(1 - 2\varepsilon_0\sqrt{\varepsilon_0} + \varepsilon_0^2\sqrt{\varepsilon_0} \right). \tag{15.17}$$

Thus, the explicit asymptotic expression (15.6) for the free energy has also been completely determined.

Using the definition (5.4), we obtain from eqn (15.6) the asymptotic behavior of the Casimir entropy at low temperature (Geyer et al. 2005a, 2006),

$$S(a,T) = \frac{3k_B \zeta_R(3)(\varepsilon_0 - 1)^2}{128\pi^3 a^2(\varepsilon_0 + 1)} \tau^2 \tag{15.18}$$

$$\times \left[1 - \frac{8\pi^2(\varepsilon_0 + 1)\left(1 - 2\varepsilon_0\sqrt{\varepsilon_0} + \varepsilon_0^2\sqrt{\varepsilon_0}\right)}{135\zeta_R(3)(\varepsilon_0 - 1)^2} \tau + O(\tau^2) \right].$$

As can be seen in eqn (15.18), the entropy of the Casimir interaction between metal and dielectric plates vanishes with vanishing temperature, as is required by the Nernst heat theorem. At low temperatures, the entropy remains positive and goes to zero with vanishing temperature. At the same time, at larger temperatures entropy is nonmonotonic and may take negative values (Geyer et al. 2005a). This will be illustrated with numerical computations in Section 15.3. Note that the perturbative expansions (15.6), (15.16), and (15.18) do not allow one to consider the limiting case $\varepsilon_0 \to \infty$. The mathematical reason is that in a power expansion of a function depending on ε_0 as a parameter, the limiting transitions $\varepsilon_0 \to \infty$ and $\tau \to 0$ are not interchangeable.

Now we consider the opposite limiting case $\tau \gg 1$, i.e. the limit of high temperature (large separation). Here, the main contribution to the free energy (12.108) is given by the term with $l = 0$, whereas all terms with $l \geq 1$ are exponentially small. As a result, the free energy is given by

$$\mathcal{F}(a,T) = \frac{\hbar c \tau}{64\pi^2 a^3} \int_0^\infty y\, dy \ln(1 - r_0 e^{-y}), \qquad (15.19)$$

where, owing to eqn (12.107), only the transverse magnetic mode contributes. By integrating eqn (15.19) and using the definition of τ, we obtain

$$\mathcal{F}(a,T) = -\frac{k_B T}{16\pi a^2} \text{Li}_3(r_0). \qquad (15.20)$$

For the Casimir pressure and entropy at $\tau \gg 1$, we get

$$P(a,T) = -\frac{k_B T}{8\pi a^3} \text{Li}_3(r_0), \qquad S(a,T) = \frac{k_B}{16\pi a^2} \text{Li}_3(r_0). \qquad (15.21)$$

15.1.2 The Casimir energy and pressure at zero temperature

Now we consider the Casimir energy per unit area, $E(a)$, and the pressure, $P(a)$, in the configuration of an ideal-metal plate and a plate with a constant permittivity ε_0. By using eqns (12.90) and (15.8) with the function $f(\zeta, y)$ defined in eqn (15.1), we obtain

$$E(a) = -\frac{\pi^2}{720} \frac{\hbar c}{a^3} \psi_{\text{DM}}(\varepsilon_0), \qquad P(a) = -\frac{\pi^2}{240} \frac{\hbar c}{a^4} \psi_{\text{DM}}(\varepsilon_0), \qquad (15.22)$$

where the function $\psi_{\text{DM}}(\varepsilon_0)$ is defined as

$$\psi_{\text{DM}}(\varepsilon_0) = -\frac{45}{2\pi^4} \int_0^\infty d\zeta \int_\zeta^\infty f(\zeta, y)\, dy. \qquad (15.23)$$

In fact, the function ψ_{DM} in eqns (15.22) and (15.23) has the physical meaning of a correction factor to the famous Casimir results (1.1) and (1.5), obtained for two ideal-metal plates. It is equal to the function φ_{DM} multiplied by r_0, as introduced in the textbook by Lifshitz and Pitaevskii (1980).

The function $\psi_{\text{DM}}(\varepsilon_0)$ in eqn (15.23) can be put into a simpler analytical form. For this purpose, we represent the logarithms in eqn (15.1) as a series and change the order of the integrations:

$$\psi_{\text{DM}}(\varepsilon_0) = \frac{45}{2\pi^4} \sum_{n=1}^\infty \frac{1}{n} \int_0^\infty y\, dy\, e^{-ny} \int_0^y d\zeta\, [r_{\text{TM}}^n(i\zeta, y) + (-1)^n r_{\text{TE}}^n(i\zeta, y)]. \qquad (15.24)$$

Introducing a new variable $w = \zeta/y$, we rearrange eqn (15.24) to the form

$$\psi_{\text{DM}}(\varepsilon_0) = \frac{45}{2\pi^4} \sum_{n=1}^\infty \frac{1}{n} \int_0^\infty y^2\, dy\, e^{-ny} \int_0^1 dw\, [r_{\text{TM}}^n(iw) + (-1)^n r_{\text{TE}}^n(iw)], \qquad (15.25)$$

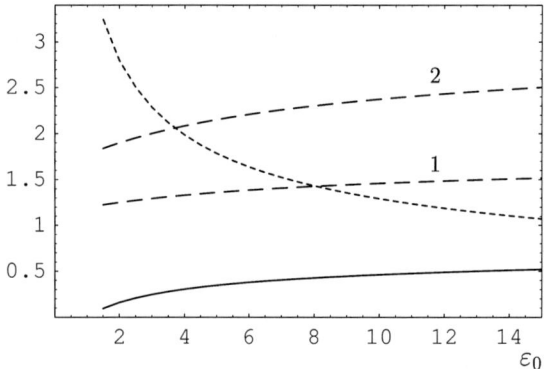

FIG. 15.1. The factor $\psi_{\rm DM}(\varepsilon_0)$ in the Casimir energy per unit area (solid line) as a function of the static dielectric permittivity. The long-dashed lines 1 and 2 and the short-dashed line show the coefficients $C_1(\varepsilon_0)$, $C_2(\varepsilon_0)$, and $B(\varepsilon_0)$, respectively, in eqn (15.46) for the Casimir energy per unit area between plates made of a real metal and a dielectric (Geyer et al. 2008a).

where

$$r_{\rm TM}(iw) = \frac{\varepsilon_0 - \sqrt{1+(\varepsilon_0-1)w^2}}{\varepsilon_0 + \sqrt{1+(\varepsilon_0-1)w^2}}, \qquad r_{\rm TE}(iw) = \frac{1-\sqrt{1+(\varepsilon_0-1)w^2}}{1+\sqrt{1+(\varepsilon_0-1)w^2}}. \tag{15.26}$$

Calculating the integral with respect to y and performing the summation with respect to n in eqn (15.25), we arrive at (Geyer et al. 2008a)

$$\psi_{\rm DM}(\varepsilon_0) = \frac{45}{2\pi^4}\int_0^1 dw\,\{{\rm Li}_4\,[r_{\rm TM}(iw)] + {\rm Li}_4\,[-r_{\rm TE}(iw)]\}. \tag{15.27}$$

In Fig. 15.1, the function $\psi_{\rm DM}(\varepsilon_0)$ given in eqn (15.27) is plotted versus ε_0 as a solid line (when $\varepsilon_0 \to 1$, it goes to zero, and when $\varepsilon_0 \to \infty$, it goes to unity, reproducing the limit of an ideal metal).

It is notable that the model used in this section (one ideal-metal plate and a second plate of constant permittivity) correctly reproduces the zero-temperature Casimir energy and pressure only in the retarded regime (i.e. at sufficiently large separations). Regarding the thermal corrections in eqns (15.6) and (15.16), the expressions obtained are also valid at small separations (the nonretarded regime) as long as the parameter τ is sufficiently small at sufficiently low temperatures.

15.2 Metal and dielectric plates with permittivities depending on frequency

In this section, analytical expressions for the Casimir energy, free energy, and pressure are presented for a configuration of two plates, one made of a metal and the other of a dielectric, taking into account the dependence of the permittivities of both the metal and the dielectric on the frequency (Geyer et al. 2008a).

The metal plate is described by the dielectric permittivity of the plasma model $\varepsilon^{(1)}(\omega) = \varepsilon_p(\omega)$ (13.1). As discussed in Chapters 13 and 14, this description is good for separations between the plates greater than the plasma wavelength.

For the dielectric plate, we use the representation of the dielectric permittivity along the imaginary frequency axis

$$\varepsilon^{(2)}(i\xi_l) = \varepsilon(i\xi_l) = 1 + \sum_j \frac{C_j}{1 + (\xi_l^2/\omega_j^2)}, \quad \sum_j C_j = \varepsilon_0 - 1, \qquad (15.28)$$

where $\varepsilon_0 = \varepsilon(0) < \infty$ [to be compared with eqn (12.132) with $d = 0$ and eqn (12.86) with $\gamma_j = 0$]. The dielectric material is assumed to be nonpolar. Equation (15.28) leads to the same values of the reflection coefficients at zero frequency, given in eqn (12.107), as were obtained from the simplified model with a constant dielectric permittivity $\varepsilon(\omega) = \varepsilon_0$ in the previous section. The representation (15.28) gives rather precise description of the dielectric permittivity for many dielectrics. It has been successfully used for a comparison of experimental data with theory (Bergström 1997). We first consider asymptotic expressions for the Casimir free energy, pressure, and entropy in the low-temperature limit and then approximate formulas for the Casimir energy and pressure at zero temperature.

15.2.1 The low- and high-temperature limits

The Casimir free energy is represented by eqn (12.108). Owing to the equality $r_{\mathrm{TE}}^{(2)}(0, y) = 0$, valid for the dielectric plate, the transverse electric mode does not contribute to the free energy (12.108), regardless of the value of the reflection coefficient $r_{\mathrm{TE}}^{(1)}(0, y)$ of the metal. Because of this, in the configuration of a metal and a dielectric, the various approaches to the definition of $r_{\mathrm{TE}}^{(1)}(0, y)$ discussed in Chapter 14 actually lead to the same result. We represent the free energy in the form of eqn (12.76), with $\mathcal{E}(a, T) = E_\omega(a)$ and the thermal correction $\Delta \mathcal{F}_\omega(a, T)$ defined in eqn (12.90). Here and below, the index ω means that plate materials with frequency-dependent permittivities are considered. In the case of different plates, the quantities $r_{\mathrm{TM,TE}}^2$ in the definition of the function $f_\omega(\zeta, y)$ in eqn (12.90) must be replaced with $r_{\mathrm{TM,TE}}^{(1)} r_{\mathrm{TM,TE}}^{(2)} = r_{\mathrm{TM,TE}}^{(\mathrm{p})} r_{\mathrm{TM,TE}}^{(\omega)}$, where the index "(p)" indicates that the metal is described by the plasma model. Now the function $f(x, y)$ can be represented in the form

$$f_\omega(x, y) = f_{\mathrm{TM}}^{(\omega)}(x, y) + f_{\mathrm{TE}}^{(\omega)}(x, y), \qquad (15.29)$$

$$f_{\mathrm{TM,TE}}^{(\omega)}(x, y) = y \ln\left[1 - r_{\mathrm{TM,TE}}^{(\mathrm{p})}(ix, y) r_{\mathrm{TM,TE}}^{(\omega)}(ix, y) e^{-y}\right].$$

To obtain the analytical expressions of interest, we develop a perturbation theory in the two small parameters τ (in the limit of low temperature) and $\alpha = \delta_0/(2a)$, where δ_0 is the skin depth defined in Section 13.1. The condition $\alpha \ll 1$ is satisfied at separations above the plasma wavelength. For the sake of simplicity, we shall consider dielectrics which can be described by eqn (15.28) with only one oscillator, i.e. with $j = 1$. High-resistivity Si is a typical example of such a material. The function $F(x)$ in eqn (12.90) can be conveniently represented in the form

$$F^{(\omega)}(x) = F_{\text{TM}}^{(\omega)}(x) + F_{\text{TE}}^{(\omega)}(x), \qquad F_{\text{TM,TE}}^{(\omega)}(x) = \int_x^\infty dy\, f_{\text{TM,TE}}^{(\omega)}(x,y). \qquad (15.30)$$

As the first step, we perform an expansion in powers of the small parameter α. This results in

$$F_{\text{TM}}^{(\omega)}(x) = \int_x^\infty y\, dy \ln\left[1 - r_{\text{TM}}^{(\omega)}(ix,y)e^{-y}\right] + 2x^2\alpha \int_x^\infty dy \frac{r_{\text{TM}}^{(\omega)}(ix,y)}{e^y - r_{\text{TM}}^{(\omega)}(ix,y)}$$

$$- 2x^4\alpha^2 \int_x^\infty dy \frac{e^y r_{\text{TM}}^{(\omega)}(ix,y)}{y[e^y - r_{\text{TM}}^{(\omega)}(ix,y)]^2} + O(\alpha^3),$$

$$F_{\text{TE}}^{(\omega)}(x) = \int_x^\infty y\, dy \ln\left[1 + r_{\text{TE}}^{(\omega)}(ix,y)e^{-y}\right] - 2\alpha \int_x^\infty y^2 dy \frac{r_{\text{TE}}^{(\omega)}(ix,y)}{e^y + r_{\text{TE}}^{(\omega)}(ix,y)}$$

$$+ 2\alpha^2 \int_x^\infty y^3 dy \frac{e^y r_{\text{TE}}^{(\omega)}(ix,y)}{[e^y + r_{\text{TE}}^{(\omega)}(ix,y)]^2} + O(\alpha^3). \qquad (15.31)$$

The reflection coefficients $r_{\text{TM,TE}}^{(\omega)}$ of the dielectric in eqn (15.31) are obtained after substitution of eqn (15.28), with $j = 1$, into eqn (12.91):

$$r_{\text{TM}}^{(\omega)}(ix,y) = \frac{[1 + (\varepsilon_0 - 1)/(1 + \chi^2 x^2)]y - \sqrt{y^2 + [(\varepsilon_0 - 1)/(1 + \chi^2 x^2)]x^2}}{[1 + (\varepsilon_0 - 1)/(1 + \chi^2 x^2)]y + \sqrt{y^2 + [(\varepsilon_0 - 1)/(1 + \chi^2 x^2)]x^2}},$$

$$r_{\text{TE}}^{(\omega)}(ix,y) = \frac{y - \sqrt{y^2 + [(\varepsilon_0 - 1)/(1 + \chi^2 x^2)]x^2}}{y + \sqrt{y^2 + [(\varepsilon_0 - 1)/(1 + \chi^2 x^2)]x^2}}, \qquad (15.32)$$

where $\chi \equiv \omega_c/\omega_1$ and $\omega_c = c/(2a)$ is the characteristic frequency.

Let us consider in sequence the contributions to $F^{(\omega)}(x)$ in eqn (15.30) from the terms of order α^0, α, and α^2 in eqn (15.31). Regarding the terms of order α^0 [the first terms on the right-hand side of the equalities in eqn (15.31)], an expansion in powers of small x (small τ) performed using eqn (15.32) leads to

$$F_{\alpha^0}^{(\omega)}(x) = F(x) - \chi^2 x^4 \left\{ \frac{3\varepsilon_0^2 + 2\varepsilon_0 - 1}{(\varepsilon_0 + 1)^2} \sum_{n=1}^\infty nr_0^n \operatorname{Ei}(-nx) \right.$$

$$\left. - r_0^2 \sum_{n=1}^\infty nr_0^n \operatorname{Ei}[-(n+1)x] + \frac{\varepsilon_0 - 1}{4} \operatorname{Ei}(-x) \right\} + O(x^5). \qquad (15.33)$$

Here, $F(x)$ has already been calculated in Section 15.1.1 and results in eqn (15.5). The additional contributions on the right-hand side of eqn (15.33) lead to a term of order τ^4 in $F_{\alpha^0}^{(\omega)}(i\tau t) - F_{\alpha^0}^{(\omega)}(-i\tau t)$ and of order τ^5 in the free energy. Thus, they can be omitted (as in Section 15.1.1, we keep only terms of order τ^3 and τ^4).

To find the contribution to $F^{(\omega)}(x)$ of order α [which is labeled $F_\alpha^{(\omega)}(x)$], we expand the following quantities under the integrals in eqn (15.31) in powers of x:

$$\frac{x^2 r_{\text{TM}}^{(\omega)}(ix,y)}{e^y - r_{\text{TM}}^{(\omega)}(ix,y)} = x^2 \frac{r_0}{e^y - r_0} - x^4 \frac{e^y r_0 \varepsilon_0}{y^2(\varepsilon_0+1)(e^y - r_0)^2} \quad (15.34)$$

$$-2\chi^2 x^4 \frac{e^y r_0}{(\varepsilon_0+1)(e^y - r_0)^2} + O(x^5),$$

$$-\frac{y^2 r_{\text{TE}}^{(\omega)}(ix,y)}{e^y + r_{\text{TE}}^{(\omega)}(ix,y)} = x^2 \frac{(\varepsilon_0-1)e^{-y}}{4} - x^4 \frac{(\varepsilon_0-1)^2 e^{-2y}(2e^y-1)}{16 y^2}$$

$$-\chi^2 x^4 \frac{(\varepsilon_0-1)e^{-y}}{4} + O(x^5).$$

By integrating the third terms on the right-hand sides of eqn (15.34) with respect to y from x to infinity, we find that their contributions to $F_\alpha^{(\omega)}(i\tau t) - F_\alpha^{(\omega)}(-i\tau t)$ are only of order τ^5 and, thus, their contributions to the free energy are of order τ^6. Thus, they can be omitted. The integration of the first two terms on the right-hand sides of eqn (15.34) leads to

$$F_{\text{TM},\alpha}^{(\omega)}(i\tau t) - F_{\text{TM},\alpha}^{(\omega)}(-i\tau t) = i\alpha\tau^3 t^3(\varepsilon_0-1)(\varepsilon_0+2), \quad (15.35)$$

$$F_{\text{TE},\alpha}^{(\omega)}(i\tau t) - F_{\text{TE},\alpha}^{(\omega)}(-i\tau t) = \frac{i}{4}\alpha\tau^3 t^3(\varepsilon_0-1)(\varepsilon_0+3).$$

From eqn (15.35), we get

$$F_\alpha^{(\omega)}(i\tau t) - F_\alpha^{(\omega)}(-i\tau t) = \frac{i}{4}\alpha\tau^3 t^3(\varepsilon_0-1)(5\varepsilon_0+11). \quad (15.36)$$

Considering the terms of order α^2 in eqn (15.31), their lowest-order contributions to $F_{\text{TM},\alpha^2}^{(\omega)}(i\tau t) - F_{\text{TM},\alpha^2}^{(\omega)}(-i\tau t)$ and $F_{\text{TE},\alpha^2}^{(\omega)}(i\tau t) - F_{\text{TE},\alpha^2}^{(\omega)}(-i\tau t)$ are of order τ^4 and τ^5, respectively. This leads to respective contributions of order τ^5 and τ^6 to the free energy, which we omit in our analysis.

Using eqn (12.90), the respective correction to the Casimir free energy takes the form

$$\Delta\mathcal{F}_\alpha^{(\omega)}(a,T) = -\frac{\hbar c}{30720\pi^2 a^3}\alpha\tau^4(\varepsilon_0-1)(5\varepsilon_0+11). \quad (15.37)$$

Remarkably, $\alpha\tau^4 \sim a^3$ and the correction (15.37) does not depend on the separation. Thus, there are no corrections to the Casimir pressure of order $\alpha\tau^q$ with $q \leq 4$ due to the nonzero skin depth of a real-metal plate. Note that in a configuration of two ideal-metal plates, the leading thermal correction to the Casimir pressure at low temperature is of order τ^4 [see eqn (7.82)]. If the nonideality of the metal is taken into account, a correction of order τ^3 arises, as shown in eqn (14.22). From this it follows that the thermal correction in a metal–dielectric configuration is less sensitive to the effect of nonzero skin depth than in a configuration of two metals.

Combining the contributions from the zero- and first-order terms in α in eqns (15.6) and (15.37), the free energy at low temperature for the configuration of a real metal and a real dielectric is (Geyer et al. 2008a)

$$\mathcal{F}^{(\omega)}(a,T) = E^{(\omega)}(a) - \frac{\hbar c}{32\pi^2 a^3}\left[\frac{\zeta_R(3)}{16\pi^2}\frac{(\varepsilon_0-1)^2}{\varepsilon_0+1}\tau^3 - K_4\tau^4\right.$$
$$\left. + \frac{1}{960}(\varepsilon_0-1)(5\varepsilon_0+11)\alpha\tau^4 + O(\tau^5)\right], \qquad (15.38)$$

where K_4 is defined in eqn (15.17). It should be noted that if the representation (15.28) for the dielectric permittivity is used, the low-temperature behavior of the free energy is not influenced by the absorption bands of the dielectric material and is determined only by the static dielectric permittivity. If, instead of eqn (15.28), the representation (12.86) is used with nonzero relaxation parameters γ_j, the perturbative expansion of $\mathcal{F}^{(\omega)}(a,T)$ starts from a term of order τ^2 depending on the absorption bands [in comparison with eqn (12.96) for the case of two dielectric plates]. For the Casimir pressure $P^{(\omega)}(a,T)$ between plates made of a real metal and a dielectric, eqn (15.16) is preserved, with the replacement of $P(a)$ by $P^{(\omega)}(a)$ given in the next subsection.

From eqns (5.4) and (15.38), we obtain the asymptotic behavior of the Casimir entropy at small τ for a metal plate and a dielectric plate made of real materials,

$$S^{(\omega)}(a,T) = \frac{3k_B \zeta_R(3)(\varepsilon_0-1)^2}{128\pi^3 a^2 (\varepsilon_0+1)}\tau^2 \left\{1 - \frac{\pi^2(\varepsilon_0+1)}{45\zeta_R(3)(\varepsilon_0-1)}\tau\right.$$
$$\left. \times \left[\frac{8(1-2\varepsilon_0\sqrt{\varepsilon_0}+\varepsilon_0^2\sqrt{\varepsilon_0})}{3(\varepsilon_0-1)} - (5\varepsilon_0+11)\alpha\right] + O(\tau^2)\right\}. \qquad (15.39)$$

As can be seen in eqn (15.39), $S^{(\omega)}(a,T)$ goes to zero when the temperature vanishes, as is required by the Nernst heat theorem. This conclusion remains valid even if the model of the dielectric includes nonzero relaxation parameters γ_j. In the latter case, however, the Casimir entropy vanishes as the first power of the temperature, as in eqn (12.104).

We complete this subsection with a consideration of the high-temperature limit. Here, the zero-frequency term (15.19) of the Lifshitz formula determines the total result. As mentioned above, in the configuration of a metal plate and a dielectric plate only the transverse magnetic mode (for which the reflection coefficient of the metal is equal to unity) contributes to the zero-frequency term. Thus, unlike the case of two plates made of real metals, there are no corrections due to nonzero skin depth at large separations (high temperatures) for the case of one metal and one dielectric plate. As a result, for metal and dielectric plates made of real materials, eqns (15.20) and (15.21), obtained for an ideal metal and a dielectric plate with a constant permittivity, remain valid.

15.2.2 Analytical results at zero temperature

Now we derive an analytical representation for the Casimir energy $E^{(\omega)}(a)$ in the configuration of a metal and a dielectric plate described by the frequency-dependent dielectric permittivities (13.1) and (15.28). Expanding the right-hand side of the first equality in eqn (12.90) in powers of α, where the function f is replaced with $f^{(\omega)}$ defined in eqn (15.29), we obtain

$$E^{(\omega)}(a) = \frac{\hbar c}{32\pi^2 a^3} \left\{ \int_0^\infty d\zeta \int_\zeta^\infty y\, dy \left[\ln\left(1 - r_{\rm TM}^{(\omega)}(i\zeta, y) e^{-y}\right) \right. \right.$$

$$\left. + \ln\left(1 + r_{\rm TE}^{(\omega)}(i\zeta, y) e^{-y}\right) \right] \quad (15.40)$$

$$+ 2\alpha \int_0^\infty d\zeta \left[\zeta^2 \int_\zeta^\infty dy \frac{r_{\rm TM}^{(\omega)}(i\zeta, y)}{e^y - r_{\rm TM}^{(\omega)}(i\zeta, y)} - \int_\zeta^\infty y^2\, dy \frac{r_{\rm TE}^{(\omega)}(i\zeta, y)}{e^y + r_{\rm TE}^{(\omega)}(i\zeta, y)} \right]$$

$$- 2\alpha^2 \int_0^\infty d\zeta \left[\zeta^4 \int_\zeta^\infty dy \frac{e^y r_{\rm TM}^{(\omega)}(i\zeta, y)}{y\left(e^y - r_{\rm TM}^{(\omega)}(i\zeta, y)\right)^2} - \int_\zeta^\infty y^3\, dy \frac{e^y r_{\rm TE}^{(\omega)}(i\zeta, y)}{\left(e^y + r_{\rm TE}^{(\omega)}(i\zeta, y)\right)^2} \right] \right\}.$$

Here, the reflection coefficients for a dielectric with a frequency-dependent dielectric permittivity are defined in eqn (15.32). For many dielectrics, in the representation (15.28) with one oscillator, the characteristic frequency at typical separations is much less than the absorption frequency, leading to $\chi = \omega_c/\omega_1 \ll 1$. In fact, the small parameter χ is of the order of another small parameter, α. The expansion of eqn (15.32) in powers of χ takes the form

$$r_{\rm TM}^{(\omega)}(ix, y) = r_{\rm TM}(ix, y) \quad (15.41)$$

$$-\chi^2 \frac{(\varepsilon_0 - 1)x^2 y\left[2y^2 + (\varepsilon_0 - 2)x^2\right]}{\sqrt{(\varepsilon_0 - 1)x^2 + y^2}\,(\varepsilon_0 y + \sqrt{(\varepsilon_0 - 1)x^2 + y^2})^2} + O(\chi^4),$$

$$r_{\rm TE}^{(\omega)}(ix, y) = r_{\rm TE}(ix, y)$$

$$+\chi^2 \frac{(\varepsilon_0 - 1)x^2 y}{\sqrt{(\varepsilon_0 - 1)x^2 + y^2}\,(y + \sqrt{(\varepsilon_0 - 1)x^2 + y^2})^2} + O(\chi^4),$$

where $r_{\rm TM,TE}(ix, y)$ are the reflection coefficients for a dielectric with a frequency-independent dielectric permittivity ε_0. Our goal is to obtain an expression for $E^{(\omega)}(a)$ up to the second power in the small parameters α and χ.

To attain this goal, we note that eqn (15.41) contains terms of zero order and second order in χ. Thus, both of these terms should be substituted in the expression for the zero-order term in α in eqn (15.40). Considering the terms of order α and α^2 in eqn (15.40), we should restrict ourselves to only the zero-order term in χ, i.e. replace $r^{(\omega)}_{\text{TM,TE}}(i x, y)$ with $r_{\text{TM,TE}}(i x, y)$. The calculation method for all of the coefficients accompanying α, α^2, and χ^2 is the same as was used in Section 15.1.2 for obtaining an analytical expression for the function $\psi_{\text{DM}}(\varepsilon_0)$. It consists of expansion of the integrands in a power series, changing the order of the integrals, and introducing a new variable $w = \zeta/y$. For the order α^0 in eqn (15.40), we obtain the contribution already calculated in eqns (15.22) and (15.27) and the contribution of order χ^2. The latter takes the form

$$E^{(\omega)}_{\chi^2}(a) = \frac{\hbar c \chi^2 (\varepsilon_0 - 1)}{32\pi^2 a^3} \sum_{n=1}^{\infty} \int_0^{\infty} dy\, y^4 e^{-ny} \int_0^1 dw\, \frac{w^2}{\sqrt{(\varepsilon_0 - 1)w^2 + 1}} \quad (15.42)$$

$$\times \left\{ \frac{[2 + (\varepsilon_0 - 2)w^2]\, r_{\text{TM}}^{n-1}(iw)}{\left[\varepsilon_0 + \sqrt{(\varepsilon_0 - 1)w^2 + 1}\right]^2} + \frac{(-1)^{n-1} r_{\text{TE}}^{n-1}(iw)}{\left[1 + \sqrt{(\varepsilon_0 - 1)w^2 + 1}\right]^2} \right\},$$

where the reflection coefficients $r_{\text{TM,TE}}(iw)$ are given in eqn (15.26). After integration with respect to y and summation, we get

$$E^{(\omega)}_{\chi^2}(a) = \frac{3\hbar c \chi^2}{4\pi^2 a^3} \int_0^1 dw\, \frac{w^2}{\sqrt{(\varepsilon_0 - 1)w^2 + 1}} \quad (15.43)$$

$$\times \left\{ \frac{2 + (\varepsilon_0 - 2)w^2}{\varepsilon_0 + 1 - w^2} \text{Li}_5[r_{\text{TM}}(iw)] + \text{Li}_5[-r_{\text{TE}}(iw)] \right\},$$

Following the same procedure, for the terms of order α in eqn (15.40) we obtain

$$E^{(\omega)}_{\alpha}(a) = \frac{3\hbar c \alpha}{8\pi^2 a^3} \int_0^1 dw\, \left\{ w^2 \text{Li}_4[r_{\text{TM}}(iw)] + \text{Li}_4[-r_{\text{TE}}(iw)] \right\}. \quad (15.44)$$

Similarly, for the terms of order α^2 in eqn (15.40), we obtain

$$E^{(\omega)}_{\alpha^2}(a) = -\frac{3\hbar c \alpha^2}{2\pi^2 a^3} \int_0^1 dw\, \left\{ w^4 \text{Li}_4[r_{\text{TM}}(iw)] + \text{Li}_4[-r_{\text{TE}}(iw)] \right\}. \quad (15.45)$$

By combining eqns (15.22) and (15.43)–(15.45), we arrive at the Casimir energy for a configuration of a metal plate and a dielectric plate made of real materials,

$$E^{(\omega)}(a) = -\frac{\pi^2 \hbar c}{720 a^3} \psi_{\text{DM}}(\varepsilon_0) \left[1 - C_1(\varepsilon_0)\frac{\delta_0}{a} + C_2(\varepsilon_0)\frac{\delta_0^2}{a^2} - B(\varepsilon_0)\frac{\omega_c^2}{\omega_1^2} \right], \quad (15.46)$$

where the positive coefficients C_1, C_2, and B are defined as

$$C_1(\varepsilon_0)\frac{\delta_0}{a} \equiv -\frac{E^{(\omega)}_{\alpha}(a)}{E(a)}, \quad C_2(\varepsilon_0)\frac{\delta_0^2}{a^2} \equiv \frac{E^{(\omega)}_{\alpha^2}(a)}{E(a)}, \quad B(\varepsilon_0)\frac{\omega_c^2}{\omega_1^2} \equiv -\frac{E^{(\omega)}_{\chi^2}(a)}{E(a)}, \quad (15.47)$$

and $E(a)$ is given in eqn (15.22).

In Fig. 15.1, the above coefficients are plotted as functions of ε_0 by the long-dashed lines 1 and 2 (C_1 and C_2, respectively) and the short-dashed line (B). From eqn (15.46), it is easy to obtain the respective analytical expression for the Casimir pressure,

$$P^{(\omega)}(a) = -\frac{\pi^2 \hbar c}{240 a^4} \psi_{\mathrm{DM}}(\varepsilon_0) \left[1 - \frac{4}{3} C_1(\varepsilon_0) \frac{\delta_0}{a} + \frac{5}{3} C_2(\varepsilon_0) \frac{\delta_0^2}{a^2} - \frac{5}{3} B(\varepsilon_0) \frac{\omega_c^2}{\omega_1^2} \right]. \quad (15.48)$$

Equations (15.46) and (15.48) allow one to find the Casimir energy per unit area and the Casimir pressure between a metal and a dielectric plate with rather high precision (see the next section).

15.3 Computational results

As mentioned in Section 15.1.1, in a configuration with one plate made of a metal and another made of a dielectric, the thermodynamic quantities may be nonmonotonic functions of the temperature (and separation). To demonstrate that this is really so, we consider the case where one plate is made of gold and the other plate is made of a real dielectric (Si or α-Al$_2$O$_3$). Both of these dielectrics possess relatively large values of the static dielectric permittivity, but they exhibit different behaviors of $\varepsilon(i\xi)$ around the characteristic frequency defined at a separation of about $1\,\mu$m. The dielectric permittivity of Si along the imaginary frequency axis is shown in Fig. 12.2(a). There is only one step in this figure. The dielectric permittivity of α-Al$_2$O$_3$ along the imaginary frequency axis can be given in the Ninham–Parsegian representation (12.127), where $\omega_{\mathrm{IR}} = 1 \times 10^{14}$ rad/s and $\omega_{\mathrm{UV}} = 2 \times 10^{16}$ rad/s are the characteristic absorption frequencies, and $C_{\mathrm{IR}} = 7.03$ and $C_{\mathrm{UV}} = 2.072$ are the corresponding absorption strengths (Bergström 1997). This material has not only an electronic but also an ionic polarization, and the corresponding dependence of $\varepsilon(i\xi)$ on ξ contains two steps. The dielectric permittivity of gold as a function of ξ is shown in Fig. 13.2 (for a metal–dielectric configuration, the computational results obtained by using the dashed and solid lines in Fig. 13.2 coincide at $T = 300$ K).

The computations were performed by using the Lifshitz formula for the free energy (12.108). In Fig. 15.2(a), the computational results for the relative thermal correction (12.128) to the Casimir energy between Au and Si plates at $T = 300$ K are shown as a function of the separation between the plates (solid line). In the same figure, the dashed line shows the relative thermal correction for the case of an ideal metal and a dielectric with a constant dielectric permittivity $\varepsilon_0 = 11.66$ (as for Si at $\xi = 0$). As can be seen in Fig. 15.2(a) (solid line), there is a wide range of separations $0.2\,\mu$m $\leq a \leq 1.3\,\mu$m where the relative thermal correction to the Casimir energy is negative (in terms of the dimensionless variable τ, this holds for $0.33 \leq \tau \leq 2.14$). The minimum value of the thermal correction $\delta_T \mathcal{F}^{(\omega)} = -0.006$ is achieved at $a = 0.95\,\mu$m ($\tau = 1.56$). This means that the Casimir entropy in the case under consideration is negative within the range of separations $0.2\,\mu$m $\leq a \leq 0.95\,\mu$m (or, in terms of τ, for $0.33 \leq \tau \leq 1.56$).

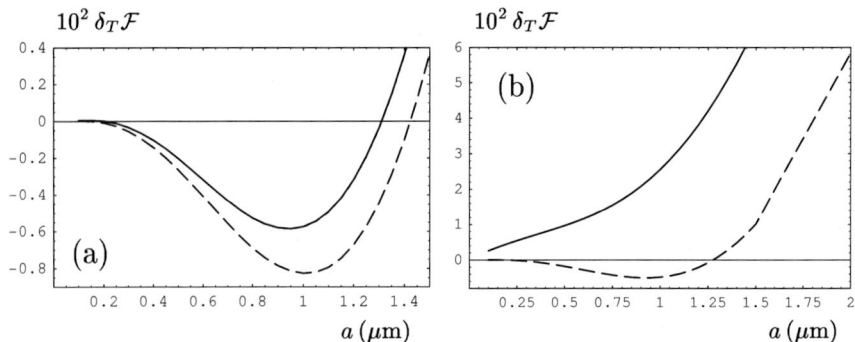

FIG. 15.2. Relative thermal correction to the Casimir energy at $T = 300\,\text{K}$ as a function of separation for two plates, one made of Au and the other of (a) Si and (b) α-Al_2O_3, shown by the solid lines. The corresponding quantity is shown by the dashed lines for an ideal metal and a dielectric with a frequency-independent dielectric permittivity (a) $\varepsilon_0 = 11.66$ and (b) $\varepsilon_0 = 10.1$ (Geyer et al. 2005a).

A comparison with the dashed line shows that for Si the simple model used in Section 15.1 leads to the same qualitative results, with only minor differences in the minimum value of $\delta_T \mathcal{F}$ and the widths of the intervals where the thermal correction and the Casimir entropy are negative (Geyer et al. 2005a).

We now consider the Casimir interaction of a gold plate with a plate made of α-Al_2O_3. As discussed above, the behavior of the dielectric permittivity of α-Al_2O_3 along the imaginary frequency axis is different from that of Si. The computational results for the relative thermal correction to the Casimir energy, as a function of separation at $T = 300\,\text{K}$, are shown in Fig. 15.2(b) by the solid line. The dashed line in Fig. 15.2(b) was computed for an ideal-metal plate and a dielectric plate with a frequency-independent dielectric permittivity $\varepsilon_0 = 10.1$. As is seen in Fig. 15.2(b), in this case the solid line represents a monotonically increasing positive function of the separation distance. The corresponding Casimir entropy is also nonnegative within the range of separations reflected in the figure. The application of the simplified model of Section 15.1 to α-Al_2O_3 leads to qualitatively different results, shown by the dashed line in Fig. 15.2(b). This line demonstrates a negative thermal correction within the range of separations from $0.25\,\mu\text{m}$ to $1.27\,\mu\text{m}$ and a negative Casimir entropy within the range from $0.25\,\mu\text{m}$ to $0.9\,\mu\text{m}$. Thus, the inclusion of realistic optical data is especially important for dielectrics with ionic polarization.

Now we compare the results of the analytical and numerical computations of the Casimir energy per unit area and the Casimir pressure for a configuration of two plates, one made of Au and the other of Si. The analytical results, $E_a^{(\omega)}(a)$ and $P_a^{(\omega)}(a)$, were calculated by using eqns (15.46), (15.48) and the plasma frequency of gold $\omega_p = 9.0\,\text{eV}$, and setting the characteristic absorption frequency

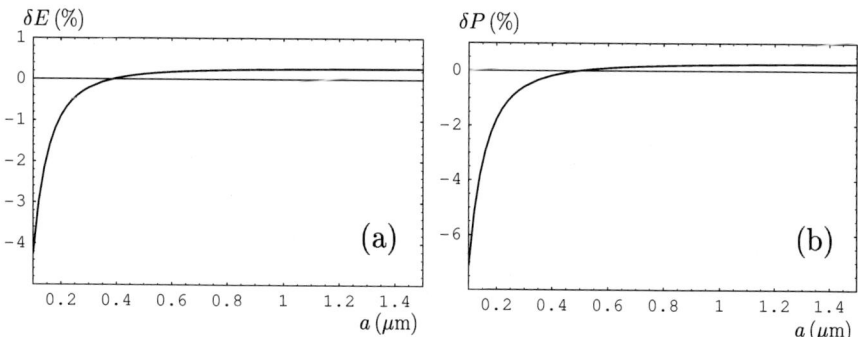

FIG. 15.3. Relative difference between analytical and numerical results for (a) the Casimir energy per unit area and (b) the Casimir pressure, at zero temperature, versus separation (Geyer et al. 2008a).

of Si equal to $\omega_1 = 4.2\,\text{eV}$ (Palik 1985). The numerical results, $E_n^{(\omega)}(a)$ and $P_n^{(\omega)}(a)$, were computed using the Lifshitz formulas (12.44) and (12.46) with the dielectric permittivity of Si shown in Fig. 12.2(a) and the dielectric permittivity of gold shown in Fig. 13.2 (at separations of about $1\,\mu\text{m}$ the contributing frequencies belong to the region of infrared optics, so that the solid and dashed lines lead to almost coincident results). In Fig. 15.3, we plot in percent the relative difference between the analytical and numerical computations versus separation for (a) the Casimir energy, $\delta E = (E_a^{(\omega)} - E_n^{(\omega)})/E_n^{(\omega)}$, and (b) the Casimir pressure, $\delta P = (P_a^{(\omega)} - P_n^{(\omega)})/P_n^{(\omega)}$. As can be seen in Fig. 15.3, the largest deviations between the analytical and numerical results (-4.3% and -7.1% for the energy and pressure, respectively) occur at the shortest separation of 100 nm. This is because the plasma model works well only at separations larger than the plasma wavelength. At shorter separations, in order to obtain precise results, one should use not the analytical representation of ε but the tabulated optical data. At separations larger than 200 and 300 nm, $|\delta E|$ is less than 0.9% and 0.25%, respectively. Regarding $|\delta P|$, this quantity is less than 0.9% and 0.25% at separations larger than 250 and 370 nm, respectively. The analytical formulas obtained for the Casimir energy per unit area and the Casimir pressure at zero temperature, between metal and dielectric plates, provide rather precise results over a wide separation range with a precision of a fraction of a percent. Thus, in some cases it is unnecessary to perform the much more cumbersome numerical computations using the Lifshitz formula together with optical data for the complex index of refraction.

15.4 Conductivity of a dielectric plate and the Nernst heat theorem

In the earlier sections of this chapter, it was assumed that at zero frequency the permittivity of the dielectric plate had some finite value $\varepsilon_0 = \varepsilon(0)$. As was discussed in Section 12.5, however, at nonzero temperature all dielectrics possess some small but physically real dc conductivity $\sigma_0 = \sigma_0(T)$, which decreases exponentially with vanishing temperature. In order to include this conductivity, the dielectric permittivity of the dielectric plate $\varepsilon(\omega)$ used above must be replaced with that in eqn (12.113) [with the upper index (n) omitted]. As a result, the permittivity of the dielectric plate along the imaginary frequency axis is given by eqn (12.116). In Section 12.5.2, it was shown that for a configuration of two dielectric plates, the inclusion of the dc conductivity in the model of the dielectric response results in a violation of the Nernst heat theorem for the Casimir entropy.

A relevant question is the possible role of the dc conductivity of the dielectric in the Casimir interaction between a metal and a dielectric. As was demonstrated by Geyer et al. (2006, 2008a), the inclusion of the conductivity of the dielectric plate leads to an inconsistency of the Lifshitz theory with thermodynamics regardless of what approach is used for the description of the metal. To show this, we substitute the dielectric permittivity (12.116) into eqn (12.108) for the Casimir free energy instead of the dielectric permittivity $\varepsilon(i\xi_l)$ from eqn (15.28) used in the previous sections of this chapter. For the metal, the dielectric permittivity of the plasma model (13.1) is used. Then we get the Casimir free energy $\tilde{\mathcal{F}}(a,T)$, which takes into account the dc conductivity of the dielectric plate. It is convenient to separate the zero-frequency term of $\tilde{\mathcal{F}}(a,T)$ and substract and add the zero-frequency term of the free energy $\mathcal{F}^{(\omega)}(a,T)$ calculated by using the dielectric permittivity $\varepsilon(i\xi_l)$:

$$\tilde{\mathcal{F}}(a,T) = \frac{k_B T}{16\pi a^2} \int_0^\infty y\, dy\, [\ln(1-e^y) - \ln(1-r_0 e^y)] + \frac{k_B T}{16\pi a^2} \int_0^\infty y\, dy \ln(1-r_0 e^y)$$
$$+ \frac{k_B T}{8\pi a^2} \sum_{l=1}^\infty \int_{\zeta_l}^\infty y\, dy \left\{ \ln\left[1 - r_{\mathrm{TM}}^{(\mathrm{p})}(i\zeta_l, y)\tilde{r}_{\mathrm{TM}}(i\zeta_l, y)e^{-y}\right] \right. \quad (15.49)$$
$$\left. + \ln\left[1 - r_{\mathrm{TE}}^{(\mathrm{p})}(i\zeta_l, y)\tilde{r}_{\mathrm{TE}}(i\zeta_l, y)e^{-y}\right] \right\}.$$

Here, the reflection coefficients $\tilde{r}_{\mathrm{TM,TE}}$ are found from eqn (12.91), where the dielectric permittivity $\varepsilon(i\xi_l)$ from eqn (15.28) is replaced with $\tilde{\varepsilon}(i\xi_l)$ from eqn (12.116).

To find the behavior of $\tilde{\mathcal{F}}(a,T)$ at low temperatures, we expand the last integral on the right-hand side of eqn (15.49) in powers of the small parameter $\beta(T)/l$. The zero-order contribution in this expansion, together with the second integral on the right-hand side of eqn (15.49), is equal to the Casimir free energy $\mathcal{F}^{(\omega)}(a,T)$ calculated by using the dielectric permittivity $\varepsilon(i\xi_l)$. Calculating explicitly the first integral on the right-hand side of eqn (15.49), we rearrange this equation to the form

$$\tilde{\mathcal{F}}(a,T) = \mathcal{F}^{(\omega)}(a,T) - \frac{k_B T}{16\pi a^2}[\zeta_R(3) - \mathrm{Li}_3(r_0)] + Q(a,T), \quad (15.50)$$

where $Q(a,T)$ contains all of the powers in the expansion of the last integral on the right-hand side of eqn (15.49) in the small parameter $\beta(T)/l$ equal to or higher than the first power. The explicit expression for the main term in $Q(a,T)$, linear in $\beta(T)/l$, is

$$Q_1(a,T) = \frac{k_B T}{8\pi a^2} \sum_{l=1}^{\infty} \frac{\beta(T)}{l} \int_{\zeta_l}^{\infty} \frac{dy\, y^2 e^{-y}}{\sqrt{y^2 + \zeta_l^2(\varepsilon_l - 1)}} \qquad (15.51)$$

$$\times \left\{ \frac{(2-\varepsilon_l)\zeta_l^2 - 2y^2}{\left[\sqrt{y^2 + \zeta_l^2(\varepsilon_l-1)} + \varepsilon_l y\right]^2} \frac{r_{\rm TM}^{(p)}(i\zeta_l, y)}{1 - r_{\rm TM}^{(p)}(i\zeta_l, y) r_{\rm TM}^{(\omega)}(i\zeta_l, y) e^{-y}} \right.$$

$$\left. + \frac{\zeta_l^2}{\left[\sqrt{y^2 + \zeta_l^2(\varepsilon_l-1)} + y\right]^2} \frac{r_{\rm TE}^{(p)}(i\zeta_l, y)}{1 - r_{\rm TE}^{(p)}(i\zeta_l, y) r_{\rm TE}^{(\omega)}(i\zeta_l, y) e^{-y}} \right\},$$

where $\varepsilon_l = \varepsilon(i\xi_l)$ is defined in eqn (15.28).

To determine the asymptotic behavior of eqn (15.51) in the limiting case where $\tau \to 0$, we expand the integrated function in powers of τ (recall that $\zeta_l = \tau l$) and consider the main contribution in this expansion at $\tau = 0$,

$$Q_1(a,T) = -\frac{k_B T r_0}{4\pi a^2(\varepsilon_l^2 - 1)} \sum_{l=1}^{\infty} \frac{\beta(T)}{l} \int_{\zeta_l}^{\infty} \frac{dy\, y e^{-y}}{1 - r_0 e^{-y}} \qquad (15.52)$$

$$= -\frac{k_B T \beta(T)}{4\pi a^2(\varepsilon_l^2 - 1)} \sum_{n=1}^{\infty} \frac{r_0^n}{n^2} \left[\sum_{l=1}^{\infty} \frac{e^{-n\tau l}}{l} + n\tau \sum_{l=1}^{\infty} e^{-n\tau l} \right].$$

Performing the summation over l, we get

$$Q_1(a,T) = -\frac{k_B T \beta(T)}{4\pi a^2(\varepsilon_l^2 - 1)} \sum_{n=1}^{\infty} \frac{r_0^n}{n^2} \left[-\ln(1 - e^{-n\tau}) + \frac{n\tau}{e^{n\tau} - 1} \right]. \qquad (15.53)$$

The right-hand side of eqn (15.53) can be rearranged with the help of the equality

$$-\ln(1 - e^{-n\tau}) + \frac{n\tau}{e^{n\tau} - 1} = -\ln\tau + 1 - \ln n + O(\tau^2). \qquad (15.54)$$

As a result, we obtain

$$Q_1(a,T) = -\frac{k_B {\rm Li}_2(r_0)}{2\pi a^2(\varepsilon_l^2 - 1)} T\beta(T) \ln\tau + T\beta(T) O(\tau^0). \qquad (15.55)$$

Taking into account the fact that $\beta(T) \sim (1/T)\exp(-{\rm const}/T)$ (see Section 12.5.2), we arrive at

$$Q_1(a,T) \sim e^{-{\rm const}/T} \ln(T/T_{\rm eff}). \qquad (15.56)$$

From eqn (15.56), it follows that both $Q_1(a,T)$ and its derivative with respect to T go to zero when T vanishes. The terms of higher order in the small parameters

$\beta(T)/l$ and τ, omitted in the above analysis, go to zero even faster than Q_1. Thus, the quantity $Q(a,T)$ in eqn (15.50) and its derivative with respect to T have limits equal to zero when the temperature goes to zero.

Now we are in a position to find the asymptotic behavior of the entropy for the metal–dielectric configuration with the inclusion of the dc conductivity of the dielectric plate. Using eqn (5.4), we obtain from eqn (15.50)

$$\tilde{S}(a,T) = S^{(\omega)}(a,T) + \frac{k_B}{16\pi a^2}[\zeta_R(3) - \text{Li}_3(r_0)] - \frac{\partial Q(a,T)}{\partial T}, \qquad (15.57)$$

where $S^{(\omega)}(a,T)$ is defined in eqn (15.39). In the limit $T \to 0$, eqn (15.57) results in

$$\tilde{S}(a,0) = \frac{k_B}{16\pi a^2}[\zeta_R(3) - \text{Li}_3(r_0)] > 0, \qquad (15.58)$$

i.e. for a metal–dielectric configuration with the inclusion of the dc conductivity of the dielectric plate, the Nernst heat theorem is violated. This result was obtained by Geyer et al. (2006) for dielectric and ideal-metal plates and by Geyer et al. (2008a) for plates made of a dielectric and a real metal. The analogous result for two dielectric plates was discussed in Section 12.5.2.

Thus, the metal–dielectric configuration provides one more confirmation of the conclusion reached in Section 12.5.2 that the Lifshitz theory is inconsistent with thermodynamics when the dc conductivity of a dielectric is included in the model of dielectric response. This happens because of the violation of thermal equilibrium by the drift current. In Chapter 20, measurements of the Casimir force between a metal sphere and a semiconductor plate are considered. For a semiconductor plate with a dopant concentration below the critical value (i.e. in the dielectric phase), the theory including the dc conductivity is shown to be inconsistent with the measurement data. The data are shown to be in agreement with theory when the dc conductivity of the dielectric plate is neglected. This confirms the rule, formulated in Section 12.5.2, that the dc conductivity of dielectric materials should be neglected. Further discussion of this subject is contained in the next chapter.

16
THE LIFSHITZ THEORY OF ATOM–WALL INTERACTIONS

The Lifshitz theory of the van der Waals and Casimir interactions between two semispaces separated by a gap can be applied to obtain the interaction potential between two atoms or molecules at a distance r or between an atom (or molecule) and a semispace. For this purpose, the substance of both semispaces (or of the one semispace) is assumed to be rarefied. As a result, the interaction energy (free energy) of two semispaces can be obtained as an additive sum of interatomic potentials between pairs of constituent atoms or, alternatively, as a sum of atom–wall potentials between the separate atoms of one semispace and another semispace. This permits one to reproduce the well-known results for atom–atom interaction potentials at small separations (London 1930) and at large separations (Casimir and Polder 1948) usually obtained by the application of quantum mechanical or quantum electrodynamic perturbation theory to the dipole–dipole interaction.

The Lifshitz-type formula for the atom–wall interaction obtained when one of the two walls is treated as a rarefied medium contains both the nonrelativistic and the relativistic limit and describes the smooth transition between them. Below, we apply this formula for various wall materials and various atoms. It is shown that the Casimir interaction between an atom and a metal wall does not depend on the model of the metal used. The respective Casimir entropy satisfies the Nernst heat theorem. In the case of an atom interacting with a dielectric wall, the Lifshitz theory is thermodynamically consistent if the dc conductivity of the dielectric material is neglected. If the dc conductivity of the dielectric wall is included in the model of the dielectric response, the Nernst heat theorem is violated. In the present chapter, we consider an attempt to remedy this situation by using a generalization of the Lifshitz theory taking spatial dispersion into account. We discuss the possible impact of the magnetic properties of an atom and a wall material on the atom–wall interaction. We also discuss the atom–wall interaction in the nonequilibrium case, when the temperatures of the wall and of the environment are different, and consider the atom–wall interaction when the wall material is anisotropic. The results presented in this chapter will be used in Part III of the book in connection with the role of the Casimir–Polder force in experiments on Bose–Einstein condensation and quantum reflection (Chapter 22), and in applications in nanotechnology (Chapter 23).

16.1 The van der Waals and Casimir–Polder interatomic potentials

We start with the case of two rarefied semispaces at a short separation a, much less than the characteristic absorption wavelength, at zero temperature. Let the dielectric permittivity of one semispace be $\varepsilon^{(1)}(\omega)$ and that of the other be $\varepsilon^{(2)}(\omega)$. For the case $\varepsilon^{(1)}(\omega) = \varepsilon^{(2)}(\omega)$, the van der Waals pressure is given in eqn (12.40). The respective van der Waals energy per unit area is obtained by the integration of eqn (12.40) with respect to a [and generalizing it to $\varepsilon^{(1)}(\omega) \neq \varepsilon^{(2)}(\omega)$]:

$$E(a) = -\frac{H}{12\pi a^2}, \qquad (16.1)$$

$$H = \frac{3\hbar}{8\pi} \int_0^\infty d\xi \int_0^\infty y^2\, dy \left\{ \frac{[\varepsilon^{(1)}(i\xi) + 1] [\varepsilon^{(2)}(i\xi) + 1]}{[\varepsilon^{(1)}(i\xi) - 1] [\varepsilon^{(2)}(i\xi) - 1]} e^y - 1 \right\}^{-1}.$$

For rarefied semispaces, $\varepsilon^{(n)}(i\xi) - 1 \ll 1$. Expanding eqn (16.1) in these two small parameters and integrating with respect to y, we obtain

$$E(a) = -\frac{\hbar}{64\pi^2 a^2} \int_0^\infty d\xi \left[\varepsilon^{(1)}(i\xi) - 1\right] \left[\varepsilon^{(2)}(i\xi) - 1\right]. \qquad (16.2)$$

Then we expand the dielectric permittivity in powers of the numbers of atoms $N^{(n)}$ per unit volume, preserving only the first-order contributions (Lifshitz 1956):

$$\varepsilon^{(n)}(i\xi) = 1 + 4\pi \alpha^{(n)}(i\xi) N_n + O(N_n^2), \qquad (16.3)$$

where the $\alpha^{(n)}(\omega)$ are the dynamic polarizabilities of the atoms (or molecules). Substituting eqn (16.3) into eqn (16.2), we obtain

$$E(a) = -\frac{\hbar N_1 N_2}{4a^2} \int_0^\infty d\xi\, \alpha^{(1)}(i\xi) \alpha^{(2)}(i\xi). \qquad (16.4)$$

Exactly this dependence of the energy per unit area between the two semispaces is obtained by a pairwise summation of the interatomic potentials $E^{AA}(r) = -A/r^6$ over the volumes of the two semispaces. To see this, we choose a coordinate plane (x, y) coinciding with the boundary surface of the lower semispace. Then the additive interaction energy of an atom in the upper semispace with the lower semispace is given by

$$E_{nr}^{A,add}(z_2) = -2\pi A N_1 \int_0^\infty \rho\, d\rho \int_{-\infty}^0 \frac{dz_1}{\left[(z_2 - z_1)^2 + \rho^2\right]^3} = -\frac{\pi N_1 A}{6 z_2^3}. \qquad (16.5)$$

Integrating this result over the volume of the upper semispace, we find the additive interaction energy between the two plates (semispaces) due to the interatomic potential $E^{AA}(r)$:

$$E_{pp,nr}^{add}(a) = -\frac{\pi N_1 N_2 A S}{6} \int_a^\infty \frac{dz_2}{z_2^3} = -\frac{\pi N_1 N_2 A S}{12 a^2}, \qquad (16.6)$$

where S is the infinite surface area of the boundary plane. Comparing the right-hand side of eqn (16.4) with $E_{pp,nr}^{add}(a)/S$ obtained from eqn (16.6), we arrive at the following explicit form of the interatomic potential:

$$E^{AA}(r) = -\frac{A}{r^6}, \qquad A = \frac{3\hbar}{\pi}\int_0^\infty d\xi\, \alpha^{(1)}(i\xi)\alpha^{(2)}(i\xi). \tag{16.7}$$

Equation (16.7) is in agreement with the standard quantum mechanical result obtained by London (1930), which does not take the relativistic retardation into account. To underline this, the quantities (16.5) and (16.6) have been marked "nr" (nonrelativistic).

Alternatively, one can consider the case of a large separation between rarefied semispaces. In this case the Casimir pressure is given by eqn (12.42). The respective Casimir energy per unit area is obtained by the negation and integration of eqn (12.42), taking into account the fact that now $\varepsilon^{(1)}(\omega) \neq \varepsilon^{(2)}(\omega)$:

$$E(a) = -\frac{\pi^2 \hbar c}{720 a^3}\Psi(\varepsilon_0^{(1)}, \varepsilon_0^{(2)}), \quad \Psi(\varepsilon_0^{(1)}, \varepsilon_0^{(2)}) = \frac{15}{2\pi^4}\int_0^\infty d\zeta \int_\zeta^\infty y^2\, dy \tag{16.8}$$

$$\times \left\{ \left[\frac{\varepsilon_0^{(1)} y + \sqrt{y^2 + \zeta^2(\varepsilon_0^{(1)} - 1)}}{\varepsilon_0^{(1)} y - \sqrt{y^2 + \zeta^2(\varepsilon_0^{(1)} - 1)}} \frac{\varepsilon_0^{(2)} y + \sqrt{y^2 + \zeta^2(\varepsilon_0^{(2)} - 1)}}{\varepsilon_0^{(2)} y - \sqrt{y^2 + \zeta^2(\varepsilon_0^{(2)} - 1)}} e^y - 1 \right]^{-1} \right.$$

$$\left. + \left[\frac{y + \sqrt{y^2 + \zeta^2(\varepsilon_0^{(1)} - 1)}}{y - \sqrt{y^2 + \zeta^2(\varepsilon_0^{(1)} - 1)}} \frac{y + \sqrt{y^2 + \zeta^2(\varepsilon_0^{(2)} - 1)}}{y - \sqrt{y^2 + \zeta^2(\varepsilon_0^{(2)} - 1)}} e^y - 1 \right]^{-1} \right\},$$

where $\varepsilon_0^{(n)} \equiv \varepsilon^{(n)}(0)$ are the static dielectric permittivities of the plate materials.

Expanding the integrated function in powers of the small parameters $\varepsilon_0^{(n)} - 1$ and using eqn (16.3), we obtain

$$\Psi(\varepsilon_0^{(1)}, \varepsilon_0^{(2)}) = \frac{15}{2\pi^2}\alpha^{(1)}(0)\alpha^{(2)}(0) N_1 N_2 \int_0^\infty d\zeta \int_\zeta^\infty y^2\, dy \left[\left(2 - \frac{\zeta^2}{y^2}\right)^2 + \frac{\zeta^4}{y^4} \right]. \tag{16.9}$$

The integration in eqn (16.9) results in

$$\Psi(\varepsilon_0^{(1)}, \varepsilon_0^{(2)}) = \frac{138}{\pi^2}\alpha^{(1)}(0)\alpha^{(2)}(0) N_1 N_2. \tag{16.10}$$

Thus, using eqn (16.8), the Casimir energy per unit area for a configuration of two dilute semispaces at a large separation is given by

$$E(a) = -\frac{23\hbar c}{120 a^3}\alpha^{(1)}(0)\alpha^{(2)}(0) N_1 N_2. \tag{16.11}$$

The same dependence on separation distance was obtained in eqn (6.43) by the pairwise summation of interatomic potentials $E^{AA}(r)$ in eqn (6.41) acting between the atoms belonging to the two semispaces. We equate the right-hand sides of eqn (6.43) divided by S and eqn (16.11) and obtain

$$E^{AA}(r) = -\frac{B}{r^7}, \qquad B = \frac{23\hbar c}{4\pi}\alpha^{(1)}(0)\alpha^{(2)}(0). \tag{16.12}$$

This is the retarded interaction potential between two sufficiently remote atoms derived by Casimir and Polder (1948) using perturbation theory in the dipole–dipole interaction.

16.2 The Lifshitz formula for an atom above a plate

The Lifshitz formula can be used to describe the van der Waals and Casimir interactions between an isolated atom or a molecule and a semispace, a material plate, or a layered structure. This is achieved by considering two dissimilar semispaces and assuming that the substance of one semispace is rarefied (Lifshitz and Pitaevskii 1980).

The free energy for a configuration of two dissimilar semispaces is given by eqn (12.71). In order to derive the free energy for an atom near the dielectric semispace labeled (1), we consider the rarefied dielectric semispace labeled (2). Then we expand the dielectric permittivity $\varepsilon^{(2)}(i\xi_l)$ in powers of the number of atoms $N_2 \equiv N$ per unit volume in accordance with eqn (16.3). Substituting $\varepsilon^{(2)}(i\xi_l)$ from eqn (16.3) with $\alpha^{(2)}(i\xi) \equiv \alpha(i\xi)$ into eqns (12.67) and (12.68), written for the medium labeled (2), and expanding the quantities obtained to the first power in N, we find

$$r^{(2)}_{\text{TM}}(i\xi_l, k_\perp) = \pi\alpha(i\xi_l)N\left(2 - \frac{\xi_l^2}{q_l^2 c^2}\right) + O(N^2),$$

$$r^{(2)}_{\text{TE}}(i\xi_l, k_\perp) = -\pi\alpha(i\xi_l)\frac{N\xi_l^2}{q_l^2 c^2} + O(N^2). \tag{16.13}$$

Using eqn (16.13), the free energy (12.71) takes the form

$$\mathcal{F}(a,T) = -\frac{k_B T N}{2}\sum_{l=0}^{\infty}{}' \alpha(i\xi_l)\int_0^\infty k_\perp dk_\perp \tag{16.14}$$

$$\times \left[\left(2 - \frac{\xi_l^2}{q_l^2 c^2}\right) r^{(1)}_{\text{TM}}(i\xi_l, k_\perp) - \frac{\xi_l^2}{q_l^2 c^2} r^{(1)}_{\text{TE}}(i\xi_l, k_\perp)\right] e^{-2aq_l} + O(N^2).$$

Using the additivity of the first-order term in the expansion of the free energy in powers of N, we can also write

$$\mathcal{F}(a,T) = N\int_a^\infty \mathcal{F}^A(z,T)\,dz + O(N^2), \tag{16.15}$$

where $\mathcal{F}^A(z,T)$ is the free energy of one atom spaced a distance z from the dielectric wall labeled (1).

Setting the right-hand sides of eqns (16.14) and (16.15) equal to each other and calculating the derivative with respect to a in the limit $N \to 0$, we arrive at

$$\mathcal{F}^A(a,T) = -k_B T\sum_{l=0}^{\infty}{}' \alpha(i\xi_l)\int_0^\infty k_\perp dk_\perp\, q_l e^{-2aq_l} \tag{16.16}$$

$$\times \left\{ 2 r_{\rm TM}^{(1)}(i\xi_l, k_\perp) - \frac{\xi_l^2}{q_l^2 c^2} \left[r_{\rm TM}^{(1)}(i\xi_l, k_\perp) + r_{\rm TE}^{(1)}(i\xi_l, k_\perp) \right] \right\}.$$

From here, we obtain a Lifshitz-type formula for the Casimir–Polder force acting on an atom [or a molecule if $\alpha(i\xi_l)$ is the molecular dynamic polarizability] near a dielectric wall:

$$F^{\rm A}(a,T) = -\frac{\partial \mathcal{F}^{\rm A}(a,T)}{\partial a} = -2k_{\rm B} T \sum_{l=0}^{\infty}{}' \alpha(i\xi_l) \int_0^\infty k_\perp dk_\perp q_l^2 e^{-2aq_l} \quad (16.17)$$

$$\times \left\{ 2 r_{\rm TM}^{(1)}(i\xi_l, k_\perp) - \frac{\xi_l^2}{q_l^2 c^2} \left[r_{\rm TM}^{(1)}(i\xi_l, k_\perp) + r_{\rm TE}^{(1)}(i\xi_l, k_\perp) \right] \right\}.$$

The interesting characteristic feature of eqns (16.16) and (16.17) describing the atom–wall interaction is that the transverse electric reflection coefficient at zero frequency, $r_{\rm TE}^{(1)}(0, k_\perp)$, does not contribute to the result, as it is multiplied by the factor $\xi_0^2 = 0$. Because of this, in the case of a metal wall the values obtained for the free energy and force do not depend on the model of the metal used. However, in the case of a dielectric wall, the results obtained depend on the transverse magnetic reflection coefficient at zero frequency, $r_{\rm TM}^{(1)}(0, k_\perp)$. We shall see later that this can be used as a test for the model of the dielectric permittivity of the wall material (Sections 16.4.1 and 16.4.3).

Below, we perform computations of the van der Waals and Casimir–Polder interactions between different atoms and semispaces made of different materials. For this purpose it is useful to express eqns (16.16) and (16.17) in terms of the dimensionless variables introduced in eqn (12.89). Thus, the Casimir–Polder free energy for an atom near a semispace is given by

$$\mathcal{F}^{\rm A}(a,T) = -\frac{k_{\rm B} T}{8a^3} \sum_{l=0}^{\infty}{}' \alpha(i\zeta_l \omega_c) \int_{\zeta_l}^\infty dy \, e^{-y} \quad (16.18)$$

$$\times \left\{ 2y^2 r_{\rm TM}^{(1)}(i\zeta_l, y) - \zeta_l^2 \left[r_{\rm TM}^{(1)}(i\zeta_l, y) + r_{\rm TE}^{(1)}(i\zeta_l, y) \right] \right\}.$$

The reflection coefficients are expressed in terms of these variables in eqn (12.91). The respective expression for the Casimir–Polder force acting on an atom or a molecule is

$$F^{\rm A}(a,T) = -\frac{k_{\rm B} T}{8a^4} \sum_{l=0}^{\infty}{}' \alpha(i\zeta_l \omega_c) \int_{\zeta_l}^\infty dy \, y e^{-y} \quad (16.19)$$

$$\times \left\{ 2y^2 r_{\rm TM}^{(1)}(i\zeta_l, y) - \zeta_l^2 \left[r_{\rm TM}^{(1)}(i\zeta_l, y) + r_{\rm TE}^{(1)}(i\zeta_l, y) \right] \right\}.$$

It is also simple to write Lifshitz-type formulas for the Casimir–Polder interaction of an atom (or molecule) with a multilayered structure or a plate of finite thickness. For this purpose, the reflection coefficients in eqns (16.16) and (16.17) must be replaced with those given in eqn (12.49) or (12.52). The generalization to the case of an atom interacting with a semispace or a plate of finite thickness made of a uniaxial crystal can be done by replacement of the reflection coefficients with those defined in eqn (12.135) or (12.137). Here, we consider atoms in the ground state and assume that thermal radiation is not strong enough to excite electrons to higher states. The case of excited atoms was discussed by Buhmann and Welsch (2007). The possible impact of virtual-photon absorption on the Casimir–Polder force was considered by Buhmann and Scheel (2008).

Similarly to Section 12.4, the Casimir–Polder free energy, as given in eqn (16.18), can be represented in the form

$$\mathcal{F}^A(a,T) = \mathcal{E}^A(a,T) + \Delta\mathcal{F}^A(a,T). \tag{16.20}$$

Here, the quantities $\mathcal{E}^A(a,T)$ and $\Delta\mathcal{F}^A(a,T)$ are obtained from eqn (16.18) by the application of the Abel-Plana formula (2.26):

$$\mathcal{E}^A(a,T) = \frac{\hbar c}{32\pi a^4} \int_0^\infty d\zeta \int_\zeta^\infty dy\, h(\zeta, y), \tag{16.21}$$

$$h(\zeta, y) = -\alpha(i\omega_c \zeta) e^{-y} \left\{ 2y^2 r_{\text{TM}}^{(1)}(i\zeta, y) - \zeta^2 \left[r_{\text{TM}}^{(1)}(i\zeta, y) + r_{\text{TE}}^{(1)}(i\zeta, y) \right] \right\},$$

$$\Delta\mathcal{F}^A(a,T) = \frac{i\hbar c\tau}{32\pi a^4} \int_0^\infty dt\, \frac{H(it\tau) - H(-it\tau)}{e^{2\pi t} - 1}, \qquad H(x) \equiv \int_x^\infty dy\, h(x, y).$$

As noted in Section 12.4, for a temperature-independent permittivity we get $\mathcal{E}^A(a,T) = E^A(a)$, where $E^A(a)$ is the Casimir–Polder energy at zero temperature, and $\Delta\mathcal{F}^A(a,T) = \Delta_T \mathcal{F}^A(a,T)$ is the thermal correction to it.

Equations (16.16) and (16.17) present the Casimir–Polder energy and force at any separation between an atom and a wall. In the nonrelativistic limit (short separations compared with the characteristic absorption wavelength of the wall material), the summation in these equations can be replaced by an integration over continuous frequencies according to eqn (12.69). Then the Casimir–Polder (or van der Waals) atom–wall interaction energy is given by

$$E^A(a) = -\frac{\hbar}{2\pi} \int_0^\infty d\xi\, \alpha(i\xi) \int_0^\infty k_\perp dk_\perp \sqrt{k_\perp^2 + \frac{\xi^2}{c^2}}\, e^{-2a\sqrt{c^2 k_\perp^2 + \xi^2}/c} \tag{16.22}$$

$$\times \left\{ 2r_{\text{TM}}^{(1)}(i\xi, k_\perp) - \frac{\xi^2}{c^2 k_\perp^2 + \xi^2} \left[r_{\text{TM}}^{(1)}(i\xi, k_\perp) + r_{\text{TE}}^{(1)}(i\xi, k_\perp) \right] \right\}$$

$$\approx -\frac{\hbar}{\pi} \int_0^\infty d\xi\, \alpha(i\xi) \int_0^\infty k_\perp^2 dk_\perp e^{-2ak_\perp} r_{\text{TM}}^{(1)}(i\xi, k_\perp).$$

In the nonrelativistic limit, we obtain the following from eqn (12.67):

$$r_{\text{TM}}^{(1)}(i\xi, k_\perp) \approx \frac{\varepsilon^{(1)}(i\xi) - 1}{\varepsilon^{(1)}(i\xi) + 1}. \tag{16.23}$$

Substituting this into eqn (16.22), we arrive at the van der Waals energy for the atom–wall interaction,

$$E^A(a) = -\frac{C_3}{a^3}, \quad C_3 = \frac{\hbar}{4\pi}\int_0^\infty d\xi\, \alpha(i\xi)\frac{\varepsilon^{(1)}(i\xi) - 1}{\varepsilon^{(1)}(i\xi) + 1}. \tag{16.24}$$

Note that this expression can be obtained directly from eqn (16.1), which represents the interaction energy between two parallel semispaces in the nonrelativistic approximation. We assume that one of the semispaces is dilute, i.e. $\varepsilon^{(2)}(\omega) - 1 \ll 1$, use eqn (16.3) with $n = 2$, and repeat the calculations done at the beginning of this section. The relativistic limit of the atom–wall interaction will be considered separately for metals and dielectrics.

16.3 Interaction of atoms with a metal wall

As mentioned above, the Lifshitz-type formulas describing the atom–wall interaction are not sensitive to the value of the TE reflection coefficient of the metal at zero frequency. This makes the results obtained to a large extent independent of the model used to describe the metal wall. At the same time, the configuration of an atom near a metallic wall suggests some interesting and unexpected features connected with the behavior of the Casimir–Polder entropy at low temperature. This can be observed even in the simplest case of an atom interacting with an ideal-metal plane.

16.3.1 Atom near an ideal-metal plane

The Casimir–Polder free energy of the atom–wall interaction is given by eqn (16.16). On substitution of the reflection coefficients (14.1) for an ideal-metal plane, we obtain

$$\mathcal{F}^A(a,T) = -2k_B T \sum_{l=0}^\infty {}' \alpha(i\xi_l) \int_0^\infty k_\perp dk_\perp q_l e^{-2aq_l}. \tag{16.25}$$

In the region of small and moderate separation distances between the atom and the metal plane, where thermal effects can be neglected, the free energy is approximately equal to the energy. The latter can be obtained from eqn (16.25) by the substitution of eqn (12.69):

$$E^A(a) = -\frac{\hbar}{\pi}\int_0^\infty d\xi\, \alpha(i\xi) \int_0^\infty k_\perp dk_\perp q e^{-2aq}. \tag{16.26}$$

Introducing the dimensionless variables (12.41) and integrating with respect to y, we can rearrange this equation to

$$E^A(a) = -\frac{\hbar c}{16\pi a^4}\int_0^\infty d\zeta\, \alpha(i\omega_c\zeta)(\zeta^2 + 2\zeta + 2)e^{-\zeta}. \tag{16.27}$$

If we consider moderate separations from about $1\,\mu$m to $3\,\mu$m only, the approximation of a static atomic polarizability, $\alpha(i\omega_c\zeta) \approx \alpha(0)$, works well. In this case eqn (16.27) leads to

$$E^A(a) \equiv E^A_{CP}(a) = -\frac{C_4}{a^4} = -\frac{3\hbar c}{8\pi a^4}\alpha(0). \tag{16.28}$$

This result was first obtained by Casimir and Polder (1948).

At shorter separations, we can use the single-oscillator model for the dynamic atomic polarizability,

$$\alpha(i\xi) = \frac{\alpha(0)}{1+(\xi^2/\omega_0^2)} = \frac{\alpha(0)}{1+(\omega_c^2/\omega_0^2)\zeta^2} \equiv \alpha(i\omega_c\zeta) \tag{16.29}$$

(we shall compare this model below with accurate results for the polarizability of some specific atoms). Here, ω_0 is the characteristic absorption frequency for the atom under consideration. We perform the calculation under the assumption $\beta_A \equiv \omega_c/\omega_0 = \lambda_0/(4\pi a) \gg 1$, where $\lambda_0 = 2\pi c/\omega_0$ is the characteristic absorption wavelength. This assumption is satisfied at atom–plane separations $a \ll \lambda_0$. Once more, we introduce the new variables (12.41) into eqn (16.26), but change the order of integrations with respect to ζ and y rather than first integrate with respect to y as was done previously. The result is

$$E^A(a) = -\frac{\hbar c}{16\pi a^4}\int_0^\infty y^2\, dy\, e^{-y}\int_0^y d\zeta\, \alpha(i\omega_c\zeta). \tag{16.30}$$

Substituting eqn (16.29) eqn into (16.30) and integrating with respect to ζ, we get

$$E^A(a) = -\frac{\hbar c}{16\pi a^4}\alpha(0)\frac{1}{\beta_A}\int_0^\infty y^2\, dy\, e^{-y}\arctan(\beta_A y). \tag{16.31}$$

Taking into account the fact that $\beta_A \gg 1$, it is possible to replace $\arctan(\beta_A y)$ with $\pi/2$ without loss of accuracy. This leads to

$$E^A(a) = -\frac{\hbar c}{4\lambda_0 a^3}\alpha(0). \tag{16.32}$$

Note that although eqn (16.32) demonstrates the same distance dependence (the inverse third power of separation) as does the nonrelativistic limit in eqn (16.24), it is quite different in nature. Particularly, eqn (16.24) does not contain the velocity of light, as appropriate for the nonrelativistic limit, whereas eqn (16.32) does. In fact, the nonrelativistic limit cannot be achieved for an ideal-metal wall. The dependence of the Casimir–Polder energy at short separation distances in eqn (16.32) on the velocity of light should be considered as a sign that at these separations the approximation of an ideal metal is not applicable.

Now we consider any separation distance larger than $1\,\mu$m, including values larger than $3\,\mu$m. Here, we can use the static atomic polarizability but must take thermal effects into account. Introducing the dimensionless variables (12.41), we rearrange the free energy (16.25) to the form

$$\mathcal{F}^A(a,T) = -\frac{k_B T}{4a^3}\alpha(0)\sum_{l=0}^{\infty}{}'\int_{\zeta_l}^\infty y^2\, dy\, e^{-y}. \tag{16.33}$$

After performing the integration and summation, we obtain

$$\mathcal{F}^A(a,T) = E^A_{CP}(a)\,\eta(a,T), \tag{16.34}$$

where the Casimir–Polder energy is given in eqn (16.28) and the correction factor is

$$\eta(a,T) = \frac{\tau}{6}\left[1 + \frac{2}{e^\tau - 1} + \frac{2\tau e^\tau}{(e^\tau - 1)^2} + \frac{\tau^2 e^\tau(e^\tau + 1)}{(e^\tau - 1)^3}\right]. \tag{16.35}$$

Note that the parameter τ, defined in eqn (12.89), is linear in the separation and the temperature. The asymptotic behavior of the correction factor (16.35) at low temperature is given by

$$\eta(a,T) = 1 - \frac{\tau^4}{2160} + \frac{\tau^6}{15120} - \frac{\tau^8}{241920} + O(\tau^{10}). \tag{16.36}$$

The Casimir–Polder force between an atom and an ideal-metal plane can also be presented in a form similar to eqn (16.34). For this purpose, we can use eqn (16.19) with the reflection coefficients (14.1) or calculate the negative derivative with respect to a of both sides of eqn (16.34). The result is

$$F^A(a,T) = F^A_{CP}(a)\kappa(a,T), \qquad F^A_{CP}(a) = -\frac{3\hbar c}{2\pi a^5}\alpha(0). \tag{16.37}$$

The correction factor to the Casimir–Polder force $F^A_{CP}(a)$ can be represented as

$$\kappa(a,T) = \frac{3}{4}\eta(a,T) + \frac{\tau^4 e^\tau(e^{2\tau} + 4e^\tau + 1)}{24(e^\tau - 1)^4}. \tag{16.38}$$

The asymptotic behavior of this correction factor at low temperature is given by

$$\kappa(a,T) = 1 - \frac{\tau^6}{30240} - \frac{\tau^8}{241920} + O(\tau^{10}). \tag{16.39}$$

It is interesting that representations like eqns (16.34) and (16.36) for the free energy and eqns (16.37) and (16.39) for the force of the atom–wall interaction can be obtained directly from the results of Chapter 15, which deals with the Casimir interaction between metallic and dielectric plates. As an example, we start with eqn (15.16) for the Casimir pressure in the configuration of an ideal-metal plane and a dielectric plate with a static permittivity ε_0. We assume that the material of the dielectric plate is dilute, so that from eqn (16.15) we obtain

$$P(a,T) = N\mathcal{F}^A(a,T) + O(N^2) \quad \text{and} \quad P(a) = NE^A_{CP}(a) + O(N^2) \tag{16.40}$$

at nonzero and zero temperature, respectively. Then eqns (15.16) and (16.3) result in

$$N\mathcal{F}^A(a,T) + O(N^2) = NE^A_{CP}(a) \tag{16.41}$$

$$-\frac{\hbar c\tau^4}{11520\pi^2 a^4}\left\{1 - 2\left[1 + 4\pi\alpha(0)N\right]^{3/2} + \left[1 + 4\pi\alpha(0)N\right]^{5/2} + O(\tau)\right\}.$$

Expanding in powers of N and taking the limit $N \to 0$, we arrive at

$$\mathcal{F}^A(a, T) = E_{\text{CP}}^A(a)\left[1 - \frac{\tau^4}{2160} + O(\tau^5)\right], \tag{16.42}$$

i.e. the same leading term at small τ as in eqn (16.36) is obtained.

Now we are in a position to find the entropy of the Casimir–Polder interaction. Calculating the negative derivative of eqn (16.34) with respect to temperature, we derive the expression

$$S(a, T) = \frac{3k_B}{2a^3}\alpha(0)\sigma(a, T), \tag{16.43}$$

$$\sigma(a, T) = \frac{1}{\tau}\eta(a, T) - \frac{\tau^3 e^\tau (e^{2\tau} + 4e^\tau + 1)}{6(e^\tau - 1)^4}.$$

It can be easily seen that the asymptotic expansion of the entropy factor σ at low temperature is given by

$$\sigma(a, T) = -\frac{\tau^3}{540} + \frac{\tau^5}{2520} + O(\tau^7). \tag{16.44}$$

Thus, the Casimir–Polder entropy goes to zero when temperature vanishes, in accordance with the Nernst heat theorem. Note, however, that at low temperatures (small τ) the entropy (16.43) takes negative values. In Fig. 16.1, we plot the entropy factor σ obtained from eqn (16.43) for the configuration of an atom near an ideal-metal plane as a function of τ. As is seen in Fig. 16.1, the Casimir–Polder entropy is negative for $0 < \tau < 3$ and positive for larger τ. This is in accordance with the corresponding results presented in Chapter 15 (Sections 15.1.1 and 15.3) for a configuration of metal and dielectric plates. Keeping in mind

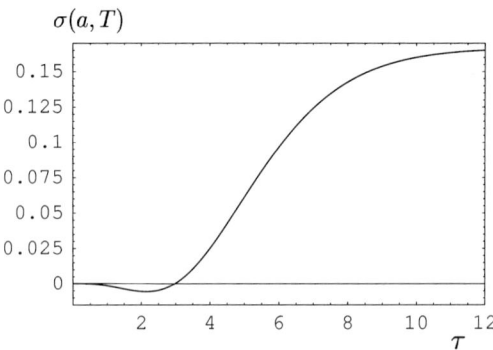

FIG. 16.1. The entropy factor σ obtained from eqn (16.43) for an atom near an ideal metal wall as a function of τ (Bezerra *et al.* 2008).

that the Lifshitz formula for an atom near a metal plate was obtained from the formula describing the case of two parallel plates, one of which is metallic and the other is a dilute dielectric, the similarity obtained in the behavior of the entropy appears quite natural.

In the high-temperature limit $T \gg T_{\text{eff}}$, only the zero-frequency term in eqn (16.33) determines the total result, whereas all terms with $l \geq 1$ are exponentially small. In this case eqn (16.33) leads to

$$\mathcal{F}^{\text{A}}(a, T) = -\frac{k_{\text{B}} T}{4a^3} \alpha(0). \qquad (16.45)$$

This is the classical limit of the Casimir–Polder free energy because the right-hand side of eqn (16.45) does not depend on $\hbar$. The respective expressions for the Casimir–Polder entropy and force are given by

$$S^{\text{A}}(a, T) = -\frac{k_{\text{B}}}{4a^3} \alpha(0), \qquad F^{\text{A}}(a, T) = -\frac{3 k_{\text{B}} T}{4a^4} \alpha(0). \qquad (16.46)$$

16.3.2 A real-metal plate and an atom

Here, we consider a metal plate made of Au, described by the plasma model (13.1) with a plasma frequency $\omega_{\text{p}} = 9.0\,\text{eV}$. This allows rather precise results at separation distances larger than the plasma wavelength $\lambda_{\text{p}} = 137\,\text{nm}$. At these separations, the dynamic polarizability of an atom can be represented using the single-oscillator model (16.29). For example, for the metastable helium atom He*, we have $\alpha(0) = 315.63$ a.u. [one atomic unit (a.u.) of polarizability is equal to $1.482 \times 10^{-31}\,\text{m}^3$] and $\omega_0 = 1.18\,\text{eV} = 1.794 \times 10^{15}\,\text{rad/s}$ (Brühl et al. 2002). Equation (16.29), with the above value of ω_0, is appropriate in the frequency region contributing to the Casimir–Polder interaction. This is demonstrated below by a comparison of computational results obtained using the single-oscillator model with results obtained using a highly accurate atomic dynamic polarizability. In this subsection, we present computations of the correction factors to the Casimir–Polder free energy and force at separations larger than λ_{p}. We also discuss the influence of various factors, such as the nonzero skin depth of the metal, the dynamic polarizability of the atom, and nonzero temperature, on the results obtained.

In Fig. 16.2(a), we present the correction factor $\eta(a, T)$ to the Casimir–Polder energy $E_{\text{CP}}^{\text{A}}(a)$ defined in eqn (16.34) as a function of the separation at $T = 300\,\text{K}$ (Babb et al. 2004). Computations of the Casimir–Polder free energy were performed using the Lifshitz formula (16.18) at separations $a \geq 150\,\text{nm}$. Line 1 was computed with the plasma dielectric function (13.1) and the dynamic atomic polarizability (16.29). This line takes into account all of the major factors which have an impact on the free energy. Other results for $\eta(a, T)$ are also plotted, omitting one or more of the above factors, in Fig. 16.2(a) for comparison. Line 2 was computed using the Lifshitz formula (16.18) with the reflection coefficients (14.1), i.e. for an ideal-metal wall, and the dynamic atomic polarizability (16.29). Line 3 was computed with the Lifshitz formula combined with the plasma model,

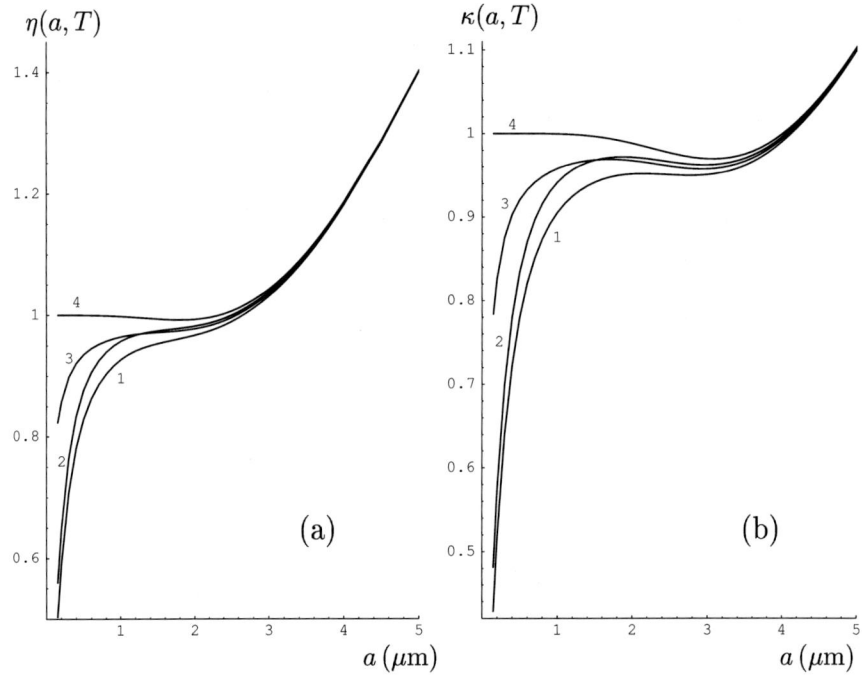

FIG. 16.2. Correction factors to the Casimir–Polder energy (a) and force (b) for an He* atom and an Au wall computed at $T = 300\,\mathrm{K}$ including the effect of the nonzero skin depth of the metal and the dynamic polarizability of the atom (line 1), with inclusion of only the dynamic polarizability (line 2), with inclusion of only the nonzero skin depth (line 3), and for an ideal metal and an atom described by a static polarizability (line 4), versus separation (Babb et al. 2004).

but with a static atomic polarizability. Finally, line 4 was computed with the Lifshitz formula for an ideal-metal wall and an atom with a static polarizability. Thus, all of the lines 1–4 take into account the effect of nonzero temperature. They can be compared with the flat line $\eta(a,T) = 1$ [not shown in Fig. 16.2(a)] representing the case of an atom described by a static polarizability near an ideal-metal wall at $T = 0$.

As can be seen from a comparison of lines 3 and 4 on the one hand and lines 1 and 2 on the other hand, at short separations the effect of the nonzero skin depth of the metal wall is much greater for an atom described by a static polarizability than for an atom described by a dynamic polarizability. In particular, for a real atom characterized by a dynamic polarizability, the corrections due to the nonzero skin depth are much less than for two metal plates. As shown in Section 13.2.2, for two parallel plates, the use of the plasma model instead of the tabulated optical data at the separations considered leads to an error of less than 2%. For the atom–wall interaction, however, the use of the plasma model leads to less than 1% error in the values of the Casimir–Polder free energy and force compared with the use of $\varepsilon(i\xi)$ obtained from the complete tabulated optical data. One can also conclude that at shorter separations, taking proper account of the dynamic atomic polarizability is more important than taking account of the nonzero skin depth. This becomes clear from a comparison of lines 2 and 3 with line 4. At intermediate separation distances from 1 to $3\,\mu\text{m}$, the dynamic atomic polarizability and the nonzero skin depth of the metal play qualitatively equal roles. With increasing a, the role of the dynamic polarizability becomes negligible, and the free energy is determined by only $\alpha(0)$. The high-temperature asymptotic expression (16.45) becomes applicable at $a > 6\,\mu\text{m}$.

Overall, Fig. 16.2(a) leads to the conclusion that at the shortest separation considered here, the total correction to the Casimir–Polder free energy due to the various factors can be as large as 50%. This should be taken into account in any comparison of measurement data with theory. At intermediate separations from 1 to $3\,\mu\text{m}$, the corrections may vary from 5% to 7%, which should be taken into account in precise experiments.

In Fig. 16.2(b), similar computational results for the correction factor $\kappa(a,T)$ to the Casimir–Polder force are presented as a function of the separation for a He* atom near an Au wall (Babb et al. 2004). Computations were performed using the Lifshitz formula (16.19). Lines 1–4 in Fig. 16.2(b) are numbered similar to those in Fig. 16.2(a). Line 1 takes into account all corrections to the Casimir–Polder force, line 2 was computed for an ideal-metal wall but for an atom with a dynamic polarizability, line 3 was computed for a real-metal wall but for an atom described by a static polarizability, and line 4 takes into account only the thermal effects. The lines in Fig. 16.2(b) demonstrate the same characteristic features as were discussed above with respect to the free energy. In particular, at short separations the effect of the nonzero skin depth is suppressed if the dynamic polarizability of the atom is taken into account. The dynamic polarizability appears to be more important at short separations than does the effect of the nonzero skin depth. At intermediate separations, the two effects lead to approximately equal contributions. The high-temperature asymptotic expression (16.46) becomes applicable at $a > 8\,\mu\text{m}$.

As can be seen in Fig. 16.2(b), in the case of the force, the correction factors play a stronger role than in the case of the free energy. For example, at the shortest separation considered in Fig. 16.2(b), the overall correction to the force reaches 57%. At intermediate separations from 1 to $3\,\mu\text{m}$, the correction varies between 5% to 9%.

Now we discuss the validity of the single-oscillator model (16.29) used here at separations $a \geq 150\,\text{nm}$. For this purpose, we have repeated some computations using highly accurate values of the nonrelativistic dynamic atomic polarizability of the He* atom (Yan and Babb 1998) with a relative error of about 10^{-6}. In Section 16.3.4, we show some results obtained using highly accurate polarizabilities of the He* and Na atoms to compute the Casimir–Polder interaction at short separations larger than a few nanometers, where the single-oscillator model leads to large errors. An accurate representation of the polarizabilities along the imaginary frequency axis can be presented in the form

$$\alpha(i\xi_l) = \sum_{j=1}^{K} \frac{g_j}{\omega_{0j}^2 + \xi_l^2}, \qquad (16.47)$$

where K is the number of oscillators and the g_j are the oscillator strengths. The static atomic polarizability is expressed as

$$\alpha(0) = \sum_{j=1}^{K} \frac{g_j}{\omega_{0j}^2}. \qquad (16.48)$$

In Fig. 16.3, the accurate dynamic polarizabilities, normalized to their static values, are shown as functions of the frequency for He* (line 1) and for Na (line 2).

The correction factor to the Casimir free energy of an He* atom interacting with an Au wall was computed using the Lifshitz formula (16.18), the plasma model (13.1), and the highly accurate polarizability shown in Fig. 16.3, at the shortest separation considered in this section, $a = 150$ nm, where the largest difference between the results obtained by use of the two models of the dynamic polarizability is expected. Here, the characteristic frequency w_c is equal to 10^{15} rad/s $\approx \xi_4$ at $T = 300$ K. The numerical data related to Fig. 16.3 show that at $\xi \leq \xi_{10}$, the difference in the relative polarizability given by the two models is less than 1%. This difference reaches 28% at $\xi = \xi_{40} = 10\,w_c$ (the higher Matsubara frequencies do not contribute to the Casimir–Polder interaction). The respective variation in the correction factor $\eta(a,T)$ at $a = 150$ nm is negligibly small. Thus, using the accurate and single-oscillator polarizabilities, we obtain $\eta(a,T) = 0.5039$ and 0.5032, respectively, i.e. only a 0.14% difference. For the correction factor to the force, the respective results are $\kappa(a,T) = 0.4298$

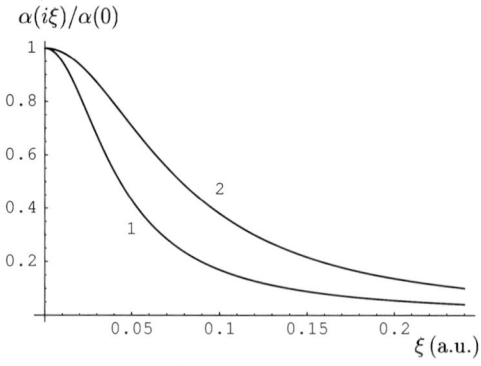

FIG. 16.3. Highly accurate normalized atomic dynamic polarizabilities for He* (line 1) and for Na (line 2) versus frequency expressed in atomic units (Babb et al. 2004). One a.u. of frequency is equal to 27.21 eV.

and 0.4284, i.e. only a 0.32% difference. With increasing separation, these differences decrease rapidly. Because of this, the single-oscillator model can be used reliably to calculate the Casimir–Polder free energy and force at separations of $a \geq 150\,\text{nm}$ with an accuracy better than 1%.

Within the range of separation distances from 1 to $3\,\mu\text{m}$, where the thermal correction is small, the role of the corrections to the Casimir–Polder energy due to the nonzero skin depth and the dynamic polarizability can be illustrated analytically. For this purpose, we can start from the plasma and single-oscillator models and use the perturbative expansion in the relative skin depth δ_0/a (as in Sections 13.1 and 14.6.2) and in the oscillator parameter β_A. Expanding the function $h(\zeta, y)$ in eqn (16.21) up to the second power in both parameters, we obtain

$$h(\zeta, y) = -\alpha(0)e^{-y}\left[2y^2 - 2\beta_A^2\zeta^2 y^2 + \left(\frac{\zeta^4}{y} - 3\zeta^2 y\right)\frac{\delta_0}{a}\right.$$
$$\left. + \frac{1}{2}\left(2\zeta^4 - \frac{\zeta^6}{y^2} + \zeta^2 y^2\right)\left(\frac{\delta_0}{a}\right)^2\right]. \tag{16.49}$$

Now we substitute eqn (16.49) into the first equality of eqn (16.21), change the order of integrations, and calculate the integrals with respect to ζ and y. The result is

$$E^A(a) = E^A_{\text{CP}}(a)\left[1 - \frac{20}{3}\beta_A^2 - \frac{8}{5}\frac{\delta_0}{a} + \frac{62}{21}\left(\frac{\delta_0}{a}\right)^2\right], \tag{16.50}$$

where $E^A_{\text{CP}}(a)$ is defined in eqn (16.28). Substituting the parameters given above for the Au wall and He* atom, we find that at $a = 1\,\mu\text{m}$ the correction to unity due to the nonzero skin depth is equal to -0.034, whereas the correction due to the dynamic polarizability is equal to -0.046. At $a = 2\,\mu\text{m}$, these corrections are -0.018 and -0.012, respectively, and they decrease further with increasing separation. From eqn (16.50), it can be seen that at a separation distance of about $1\,\mu\text{m}$, the corrections to the Casimir energy due to the nonzero skin depth and the dynamic polarizability of the atom play a qualitatively equal role, as was discussed on the basis of the numerical computations.

The respective expression for the Casimir–Polder force is obtained from the negative derivative of both sides of eqn (16.50) with respect to a:

$$F^A(a) = F^A_{\text{CP}}(a)\left[1 - 10\beta_A^2 - 2\frac{\delta_0}{a} + \frac{31}{7}\left(\frac{\delta_0}{a}\right)^2\right], \tag{16.51}$$

where $F^A_{\text{CP}}(a)$ is defined in eqn (16.37).

16.3.3 Asymptotic behavior at low temperature

We now turn to an examination of the low-temperature behavior of the Casimir–Polder free energy, entropy, and force for an atom interacting with a metallic wall made of a real metal. This allows one to solve complicated problems about the consistency of the Lifshitz theory, as adapted for atom–wall interactions, with thermodynamics. The asymptotic expressions obtained in this subsection can also serve as a test for some generalizations of the Lifshitz theory. As above, we describe a real metal by means of the plasma model, and the atom by means of a single-oscillator expression for the dynamic polarizability. Thus, separation distances larger than 150 nm are applicable.

We start once again from the function $h(x,y)$ defined in eqn (16.21) but expand it to the second power of only one parameter, δ_0/a:

$$h(x,y) = -\frac{\alpha(0)}{1+\beta_A^2 x^2} e^{-y} \left[2y^2 + \left(\frac{x^4}{y} - 3x^2 y\right) \frac{\delta_0}{a} \right.$$

$$\left. + \frac{1}{2}\left(2x^4 - \frac{x^6}{y^2} + x^2 y^2\right)\left(\frac{\delta_0}{a}\right)^2 \right] \equiv h_0(x,y) + h_1(x,y) + h_2(x,y), \quad (16.52)$$

where the $h_k(x,y)$, with $k=0,1,2$, are the contributions to $h(x,y)$ of order $(\delta_0/a)^k$. The function $H(x)$ defined in eqn (16.21) is given by

$$H(x) = H_1(x) + H_2(x) + H_3(x), \qquad H_k(x) = \int_x^\infty dy\, h_k(x,y). \quad (16.53)$$

To calculate the thermal correction to the Casimir–Polder energy defined in eqn (16.21), one needs to find the difference $H(i t\tau) - H(-i t\tau)$. This is most easily done for every $H_k(x)$ separately. Thus, for $k=0$,

$$H_0(x) = -\frac{2\alpha(0)}{1+\beta_A^2 x^2} \int_x^\infty dy\, e^{-y} y^2 = -\frac{2\alpha(0)}{1+\beta_A^2 x^2} e^{-x}(2 + 2x + x^2). \quad (16.54)$$

Expanding this in powers of x, we obtain

$$H_0(i\tau) - H_0(-i\tau) = -4i\alpha(0)\tau^3 t^3 \left[\frac{1}{3} - \left(\frac{1}{10} - \frac{\beta_A^2}{3}\right)\tau^2 t^2 \right.$$

$$\left. + \left(\frac{1}{168} - \frac{\beta_A^2}{10} + \frac{\beta_A^4}{3}\right)\tau^4 t^4 \right]. \quad (16.55)$$

Substituting eqn (16.55) into the third equality of eqn (16.21) and integrating with respect to t, we find the respective contribution to the thermal correction,

$$\Delta_T \mathcal{F}_0^A(a,T) = \frac{\hbar c \alpha(0)}{128\pi a^4} \tau^4 \left[\frac{1}{45} - \frac{\tau^2}{315}\left(1 - \frac{10}{3}\beta_A^2\right) \right.$$

$$\left. + \frac{\tau^4}{5040}\left(1 - \frac{84}{5}\beta_A^2 + 56\beta_A^4\right) \right]. \quad (16.56)$$

For $\beta_A = 0$, this is just the thermal correction for an atom described by a static polarizability near an ideal-metal plane, calculated using a different method in Section 16.3.1 and contained in eqns (16.34) and (16.36).

In a similar way, for $k = 1$ we have

$$H_1(x) = -\frac{\alpha(0)}{1+\beta_A^2 x^2}\frac{\delta_0}{a}\int_x^\infty dy\, e^{-y}\left(\frac{x^4}{y} - 3x^2 y\right) \quad (16.57)$$

$$= \frac{\alpha(0)}{1+\beta_A^2 x^2}\frac{\delta_0}{a}\left[3x^2 e^{-x}(1+x) - x^4\Gamma(0,x)\right],$$

where $\Gamma(z, x)$ is the incomplete gamma function. Expanding this in powers of x, we obtain

$$H_1(i t\tau) - H_1(-it\tau) = i\alpha(0)\tau^4 t^4 \frac{\delta_0}{a}\left(\pi + \pi\beta_A^2 \tau^2 t^2 - \frac{4}{45}\tau^3 t^3\right). \quad (16.58)$$

The contribution from this to the thermal correction is given by

$$\Delta_T \mathcal{F}_1^A(a, T) = -\frac{\hbar c\alpha(0)}{128\pi a^4}\tau^5 \frac{\delta_0}{a}\left[\frac{3\zeta_R(5)}{\pi^4} + \beta_A^2\tau^2\frac{45\zeta_R(7)}{2\pi^6} - \frac{\tau^3}{1350}\right]. \quad (16.59)$$

For $k = 2$ we arrive at

$$H_2(x) = -\frac{\alpha(0)}{2(1+\beta_A^2 x^2)}\left(\frac{\delta_0}{a}\right)^2\int_x^\infty dy\, e^{-y}\left(2x^4 - \frac{x^6}{y^2} + x^2 y^2\right) \quad (16.60)$$

$$= -\frac{\alpha(0)}{2(1+\beta_A^2 x^2)}\left(\frac{\delta_0}{a}\right)^2\left[2x^4 e^{-x} - x^6\Gamma(-1,x) + x^2 e^{-x}(2+2x+x^2)\right].$$

After an expansion in powers of x, the following equality is obtained:

$$H_2(it\tau) - H_2(-it\tau) = \frac{i\alpha(0)}{2}\tau^5 t^5\left(\frac{\delta_0}{a}\right)^2\left[\frac{20}{3} - \pi\tau t + \frac{2}{15}(1+50\beta_A^2)\tau^2 t^2\right]. \quad (16.61)$$

This results in the respective contribution to the thermal correction,

$$\Delta_T \mathcal{F}_2^A(a, T) = -\frac{\hbar c\alpha(0)}{128\pi a^4}\tau^6\left(\frac{\delta_0}{a}\right)^2\left[\frac{5}{189} - \frac{45\zeta_R(7)}{4\pi^6}\tau + \frac{1}{1800}(1+50\beta_A^2)\tau^2\right]. \quad (16.62)$$

Taking eqns (16.56), (16.59), and (16.62) together, we find the low-temperature asymptotic behavior of the Casimir–Polder free energy,

$$\Delta_T \mathcal{F}^A(a, T) = \frac{\hbar c\alpha(0)}{128\pi a^4}\tau^4\left\{\frac{1}{45} - \frac{\tau^2}{315}\left(1 - \frac{10}{3}\beta_A^2\right)\right. \quad (16.63)$$

$$+ \frac{\tau^4}{5040}\left(1 - \frac{84}{5}\beta_A^2 + 56\beta_A^4\right) - \tau\frac{\delta_0}{a}\left[\frac{3\zeta_R(5)}{\pi^4} + \beta_A^2\tau^2\frac{45\zeta_R(7)}{2\pi^6} - \frac{\tau^3}{1350}\right]$$

$$\left. - \tau^2\left(\frac{\delta_0}{a}\right)^2\left[\frac{5}{189} - \frac{45\zeta_R(7)}{4\pi^6}\tau + \frac{1}{1800}(1+50\beta_A^2)\tau^2\right]\right\}.$$

This expression includes the effect of both the nonzero skin depth of the metal plate and the dynamic polarizability of the atom. Several terms on the right-hand side of eqn (16.63) do not contribute to the Casimir–Polder force, because the quantities τ/a and $\tau\beta_A$ do not depend on the separation distance a. Calculating the negative derivative of eqn (16.63) with respect to a, we obtain the thermal correction to the Casimir–Polder force at zero temperature (16.51),

$$\Delta_T F^A(a,T) = \frac{\hbar c \alpha(0)}{128\pi a^5} \tau^6 \left[\frac{2}{315} - \frac{\tau^2}{30}\left(\frac{1}{42} - \frac{1}{5}\beta_A^2\right) \right.$$
$$\left. - 3\tau^2 \frac{\delta_0}{a} - \tau\left(\frac{\delta_0}{a}\right)^2 \left(\frac{45\zeta_R(7)}{4\pi^6} - 2\tau\right) \right]. \quad (16.64)$$

At $\delta_0 = \beta_A = 0$, this expression coincides with the thermal correction contained in eqns (16.37) and (16.39) derived for an ideal-metal wall interacting with an atom characterized by a static polarizability.

Equation (16.63) allows the calculation of the Casimir–Polder entropy at low temperature. Calculating the negative derivative with respect to the temperature of both sides of eqn (16.63), we arrive at

$$S^A(a,T) = -\frac{k_B \alpha(0)}{32 a^3} \tau^3 \left[\frac{4}{45} - \frac{2\tau^2}{105}\left(1 - \frac{10}{3}\beta_A^2\right) \right.$$
$$\left. - \tau\frac{\delta_0}{a}\frac{15\zeta_R(5)}{\pi^4} - \frac{10}{63}\tau^2\left(\frac{\delta_0}{a}\right)^2 \right]. \quad (16.65)$$

It can be seen that this entropy goes to zero when the temperature vanishes, i.e. the Nernst heat theorem is satisfied. Although we used the plasma model in the derivation of eqn (16.65), this conclusion is valid for any other approach to the description of real metals, including the Drude model approach discussed in Section 14.3. The point to note is that the TE reflection coefficient at zero frequency does not contribute to the Casimir–Polder atom–wall interaction. Regarding the contributions of all other Matsubara frequencies and the TM reflection coefficient at $\xi = 0$, several different theoretical approaches to the description of a real metal in the framework of the Lifshitz theory lead to practically coincident results (see Chapter 14). Thus the standard Lifshitz theory of atom–wall interactions in the case of a metal wall is thermodynamically consistent. At the same time, as can be seen from eqn (16.65), $S^A(a,T)$ at low temperature is negative. This property of the atom–wall configuration, discussed above for the case of an ideal-metal wall (Section 16.3.1), is also preserved for real-metal walls. Note that the asymptotic expressions (16.45) and (16.46) obtained for an atom near an ideal-metal wall at high temperature are valid for real-metal walls as well. A nonzero skin depth does not play any role at high temperatures (or large separations).

In contrast to the case of a metal wall, the interaction of an atom with a dielectric wall runs into problems when the dc conductivity of the wall material is included in the model of the dielectric response (see Section 16.4.3 below). This led to an attempt to modify the TM reflection coefficient at zero frequency by including the microscopic density of charge carriers as a characteristic of the wall material in addition to the macroscopic dielectric permittivity (Pitaevskii 2008a). Whenever feasible, this approach would recover the cases of a metallic and a dielectric wall for a large and a small density, respectively, of charge carriers. In Section 16.4.3, we consider this problem both for a dielectric wall and in the limiting case of a metallic wall. It will be shown that for metals and for many types of dielectrics, the proposed modification results in a violation of Nernst's theorem.

16.3.4 The case of short separations

In this subsection, we shall present an accurate description of the atom–wall interaction, including the dynamic polarizability of the atom and the real-metal properties of the wall, at short separation distances from 3 nm to 150 nm. Separations of a few nanometers belong to the region of van der Waals forces. Larger separations belong to the transition region between van der Waals and Casimir forces. At short separations below 150 nm, the use of the plasma model leads to large errors in the atom–wall interaction. Because of this, here we use the dielectric permittivity of an Au wall along the imaginary frequency axis presented in Fig. 13.2. This permittivity was obtained from the tabulated optical data for the complex index of refraction (see Section 13.3). As was justified in Section 14.6, the solid line in Fig. 13.2, obtained using the generalized plasma model, is theoretically preferable to the dashed line, obtained from extrapolation of the optical data by means of the Drude model. However, in the separation region below 150 nm, the two approaches to the definition of $\varepsilon(i\xi)$ lead to almost coincident computational results for the atom–wall interaction.

Computations of the Casimir–Polder free energy were performed using the Lifshitz formula (16.18). For convenience in comparison with the nonrelativistic van der Waals atom–wall energy (16.24), we represent the free energy (16.18) in the identical form

$$\mathcal{F}^A(a,T) = -\frac{C_3(a,T)}{a^3}, \tag{16.66}$$

where

$$C_3(a,T) = -\frac{k_B T}{8} \sum_{l=0}^{\infty}{}' \alpha(i\zeta_l \omega_c) \int_{\zeta_l}^{\infty} dy\, e^{-y} \tag{16.67}$$

$$\times \left\{ 2y^2 r_{\text{TM}}^{(1)}(i\zeta_l, y) - \zeta_l^2 \left[r_{\text{TM}}^{(1)}(i\zeta_l, y) + r_{\text{TE}}^{(1)}(i\zeta_l, y) \right] \right\}.$$

Here, the dependence of C_3 on a and T signifies the deviation from the standard van der Waals result (16.24). As two examples, we consider atoms of metastable He* and Na, for which highly accurate values of the dynamic polarizability are available (Yan and Babb 1998, Kharchenko et al. 1997). These values are reproduced graphically in Fig. 16.3. We shall compare the results obtained for the coefficient C_3 with results computed using the single-oscillator model (16.29). The parameters of this model for He* were presented in Section 16.3.2. For Na, the parameters are $\omega_0 = 2.14\,\text{eV}$ and $\alpha(0) = 162.68$ a.u. (Derevianko et al. 1999).

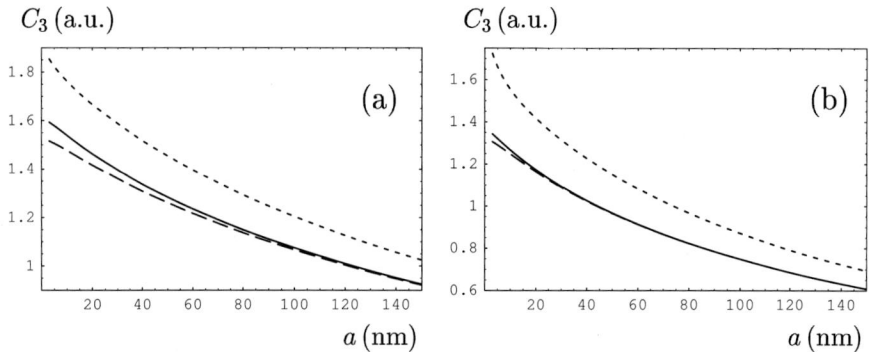

FIG. 16.4. Van der Waals coefficient C_3 versus separation for (a) metastable He* and (b) Na atoms near an Au wall, calculated by use of the complete optical data for Au and accurate atomic dynamic polarizabilities (solid lines) or by use of the single-oscillator model (long-dashed lines). The short-dashed lines were computed for an ideal metal by use of the accurate dynamic polarizabilities of the atoms (Caride et al. 2005).

The single-oscillator model was used for computations of the Casimir–Polder interaction between a hydrogen atom and a silver wall (Boström and Sernelius 2000c).

The computational results for the van der Waals coefficient C_3 in the case of an Au wall at $T = 300\,\text{K}$ are represented in Fig. 16.4 by the solid lines for (a) metastable He* and (b) Na (Caride et al. 2005). Note that one atomic unit (a.u.) for C_3 is equal to $4.032 \times 10^{-3}\,\text{eV}\,\text{nm}^3$. In the same figure, the long-dashed lines show the results obtained with a single-oscillator model for the atomic dynamic polarizability. The short-dashed lines illustrate the dependence of C_3 on a in the case of a wall made of an ideal metal but with the accurate dynamic polarizability of the atom. Note that at the separations considered, the computational results depend only slightly on the temperature.

As can be seen from a comparison of the solid and long-dashed lines with the short-dashed lines, the use of realistic properties of the wall metal is important for all separations under consideration. At the shortest separation of $a = 3\,\text{nm}$, the result for an ideal-metal wall differs from the accurate result given by the solid line by about 16% for He* and by 28% for Na. These strong deviations decrease only slightly with increasing separation. From Fig. 16.4, it can be also seen that the use of the accurate data for the polarizability (in comparison with the single-oscillator model) is of most importance at short separations below a few tens of nanometers. Thus, at $a = 3\,\text{nm}$ the relative error in C_3 given by the single-oscillator model is 4.4% for He* and 2.2% for Na. At $a = 15\,\text{nm}$, the single-oscillator model becomes more precise. For He*, it leads to an error of only 3.3%, and for Na an error of 1.6%. The values of the van der Waals coefficient C_3 for He* and Na were tabulated by Caride et al. (2005). Note that at $a = 150\,\text{nm}$, the above calculation results obtained by the use of the tabulated optical data for Au and those in Section 16.3.2 obtained using the simple plasma model differ by less than 1%. Thus, at $a = 150$ nm the value of C_3 computed using the optical data for Au and the accurate dynamic polarizability of He* is equal to 0.925 a.u. If the metal is described by the plasma model and the accurate atomic polarizability is used, the respective result is 0.918 a.u.

As shown in Section 16.3.2, the effect of the atomic dynamic polarizability (used in the single-oscillator model) strongly influences the Casimir–Polder free energy in comparison with the original result (16.28) obtained in the static approximation. We emphasize that in the case of the van der Waals atom–wall interaction, the influence of the dynamic effects is even greater than for the Casimir–Polder force. Thus, if we restrict ourselves to only the static polarizability of the He* atom, the values of C_3 are found to be 11.6 and 1.64 times greater than those given by the solid line in Fig. 16.4(a) at separations of $a = 3$ nm and $a = 150$ nm, respectively.

16.4 Interaction of atoms with a dielectric wall

The case of a dielectric wall provides a test to investigate the impact of the dc conductivity of the wall material on the Casimir–Polder interaction. The point to note is that the inclusion of the dc conductivity in the model of the dielectric response changes the TM reflection coefficient at zero frequency. In contrast to the TE reflection coefficient at zero frequency (which does not enter the zero-frequency contribution to the Lifshitz formula describing the atom–wall interaction), this leads to a change in the Casimir–Polder free energy and force. This change is most pronounced at large separation distances of a few micrometers and it is observable in experiments on Bose–Einstein condensation and quantum reflection. Thus the interaction of various atoms with dielectric walls is of special interest. As we shall see below, the properties of this interaction can be used as a validity test of the Lifshitz theory and its generalizations.

16.4.1 Asymptotic properties at low and high temperature for a finite static permittivity of the wall material

The dielectric permittivity of the wall material is described here by eqn (15.28). As mentioned in Section 15.2, this equation gives a very accurate description of the dielectric properties along the imaginary frequency axis. All parameters in eqn (15.28) are assumed to be temperature-independent in accordance with the discussion in Section 12.5.1. The dynamic polarizability of an atom is described by the single-oscillator model (16.29). As shown in Section 16.3.4, this model provides very accurate results for atom–wall interactions at separations larger than a few tens of nanometers. Under these conditions, we shall find the asymptotic behavior of the Casimir–Polder free energy and entropy at low temperature and investigate the consistency of the Lifshitz theory with thermodynamics.

We start from the representation of the Casimir–Polder free energy in eqn (16.20), with $\mathcal{E}^A(a,T) = E^A(a)$ and $\Delta \mathcal{F}^A(a,T) = \Delta_T \mathcal{F}^A(a,T)$ defined in eqn

(16.21). In this section we consider the case $\varepsilon_0 = \varepsilon(0) < \infty$, i.e. the dc conductivity of the dielectric material is disregarded. To find the primary contribution to the low-temperature asymptotic behavior of the Casimir–Polder free energy, it is sufficient to restrict ourselves to a frequency-independent permittivity $\varepsilon_l = \varepsilon_0$. This was discussed in Section 12.5.1 for the case of two dielectric plates and in Section 15.2.1 for the case of a metal and a dielectric plate. Now we use eqn (16.29) for the dynamic polarizability, and the expressions for the reflection coefficients

$$r_{\rm TM}(i\zeta_l, y) = \frac{\varepsilon_0 y - \sqrt{y^2 + \zeta_l^2(\varepsilon_0 - 1)}}{\varepsilon_0 y + \sqrt{y^2 + \zeta_l^2(\varepsilon_0 - 1)}}, \quad r_{\rm TE}(i\zeta_l, y) = \frac{y - \sqrt{y^2 + \zeta_l^2(\varepsilon_0 - 1)}}{y + \sqrt{y^2 + \zeta_l^2(\varepsilon_0 - 1)}} \tag{16.68}$$

following from eqn (12.91) in order to expand the function $H(x)$ from eqn (16.21) in powers of x. Preserving only the terms up to the third power in x, we obtain

$$H(x) = -\alpha(0)\left[4r_0 + r_0\left(4\beta_{\rm A}^2 - 2\frac{\varepsilon_0}{\varepsilon_0 + 1} - 1\right)x^2 + C_{\rm D}(\varepsilon_0)x^3\right]. \tag{16.69}$$

Here, r_0 is defined in eqn (12.95) and $\beta_{\rm A}$ is defined after eqn (16.29). The coefficient $C_{\rm D}(\varepsilon_0)$ is given by

$$C_{\rm D}(\varepsilon_0) = r_0 \frac{7\varepsilon_0 + 1}{3(\varepsilon_0 + 1)} + \frac{(\sqrt{\varepsilon_0} - 1)\left[(3\varepsilon_0^2 + 1)(2\sqrt{\varepsilon_0} + 1) + 3\varepsilon_0(\sqrt{\varepsilon_0} - 1)\right]}{3(\sqrt{\varepsilon_0} + 1)(\varepsilon_0 + 1)^2}$$
$$+ \frac{2\varepsilon_0^2}{(\varepsilon_0 + 1)^{5/2}}\left({\rm Artanh}\sqrt{\frac{\varepsilon_0}{\varepsilon_0 + 1}} - {\rm Arcoth}\sqrt{\varepsilon_0 + 1}\right). \tag{16.70}$$

In the limiting case $\varepsilon_0 \to 1$, we have $C_{\rm D}(\varepsilon_0) \to 0$ so that $H(x) \to 0$, as expected. Typical values of this coefficient are $C_{\rm D}(\varepsilon_0) = 0.585$ and 7.60 for $\varepsilon_0 = 1.5$ and 16, respectively. For the commonly used dielectrics SiO$_2$, with $\varepsilon_0 = 3.81$, and Si, with $\varepsilon_0 = 11.66$, we get $C_{\rm D}(\varepsilon_0) = 2.70$ and 6.33, respectively, from eqn (16.70).

As a result, from eqn (16.70) we obtain

$$H(i\tau t) - H(-i\tau t) = 2i\tau^3 t^3 \alpha(0) C_{\rm D}(\varepsilon_0). \tag{16.71}$$

Substituting this into eqn (16.21) for the thermal correction, we arrive at the Casimir–Polder free energy

$$\mathcal{F}^{\rm A}(a, T) = E^{\rm A}(a) - \frac{\hbar c \pi^3}{240 a^4}\alpha(0) C_{\rm D}(\varepsilon_0)\left(\frac{T}{T_{\rm eff}}\right)^4. \tag{16.72}$$

The respective low-temperature behavior of the Casimir–Polder entropy is given by (Klimchitskaya et al. 2008a)

$$S^{\rm A}(a, T) = \frac{\pi^3 k_{\rm B}}{30 a^3}\alpha(0) C_{\rm D}(\varepsilon_0)\left(\frac{T}{T_{\rm eff}}\right)^3. \tag{16.73}$$

As can be seen from eqn (16.73), the entropy goes to zero when the temperature vanishes, in accordance with the Nernst heat theorem. Thus, the Lifshitz theory of the atom–plate interaction is thermodynamically consistent if the dc conductivity of the dielectric plate is neglected. This conclusion is still true when the frequency dependence of the dielectric permittivity is taken into account, as long as $\varepsilon(0) < \infty$. Note that for an atom near a dielectric wall, the Casimir–Polder entropy is positive. This is in contrast to the case of an atom near a metal wall considered in Sections 16.3.1 and 16.3.3.

The limiting case of high temperature, $T \gg T_{\text{eff}}$, can be considered in the same way as in Section 16.3.1. In this case only the zero-frequency term gives the dominant contribution to the Lifshitz formulas (16.18) and (16.19), the other terms being exponentially small. This leads to the following asymptotic expressions for the Casimir–Polder free energy and force:

$$\mathcal{F}^A(a,T) = -\frac{k_B T}{4a^3}\alpha(0)r_0, \qquad F^A(a,T) = -\frac{3k_B T}{4a^4}\alpha(0)r_0. \tag{16.74}$$

From eqn (16.74), the high-temperature (large-separation) asymptotic expression for the Casimir–Polder entropy is given by

$$S^A(a,T) = \frac{k_B}{4a^3}\alpha(0)r_0. \tag{16.75}$$

In the limiting case $\varepsilon_0 \to \infty$, we have $r_0 \to 1$, and eqns (16.74) and (16.75) coincide with the respective expressions in eqns (16.45) and (16.46) valid for an atom near an ideal (or real) metal wall.

16.4.2 Computations of the free energy

Here, we present some typical computational results characterizing the interaction of atoms with dielectric walls at different separation distances. The free energy of the Casimir–Polder interaction was computed using the Lifshitz formula (16.18) with a frequency-dependent dielectric permittivity $\varepsilon(\omega)$. For comparison purposes, dielectric walls described by a static dielectric permittivity will be also considered. The dynamic polarizability of an atom was represented by the single-oscillator model. At the shortest separations, accurate data for the atomic dynamic polarizability were also used.

We start with the example of an Si wall interacting with an He* or Na atom at separations from 3 nm to 150 nm. The dielectric permittivity of Si along the imaginary frequency axis, obtained from the tabulated optical data for the complex index of refraction, was considered in Section 12.6.1. Computational results were presented in Fig. 12.2(a) for $T = 300$ K. The single-oscillator and accurate dynamic polarizabilities for He* and Na were considered in Sections 16.3.2 and 16.3.4. The Casimir–Polder free energy (16.18) was represented by eqn (16.66) in terms of the separation-dependent van der Waals coefficient $C_3(a,T)$. The computational results are shown in Fig. 16.5(a) for metastable helium He* and in Fig. 16.5(b) for Na. The solid lines were obtained by using the accurate atomic dynamic polarizabilities and the long-dashed lines by using the single-oscillator model (16.29). The short-dashed lines were computed with the accurate

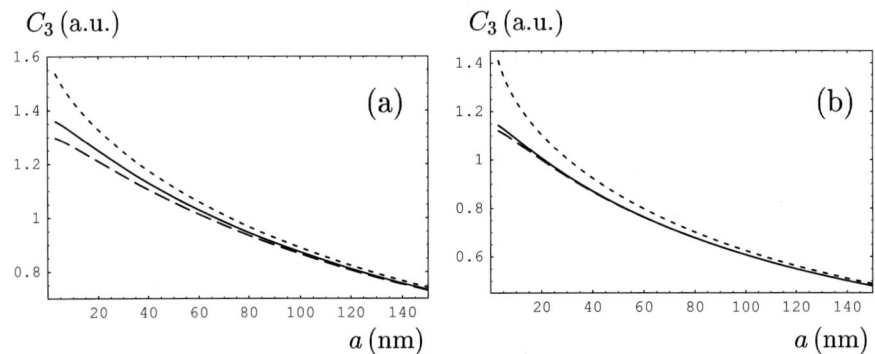

FIG. 16.5. Van der Waals coefficient C_3 versus separation for (a) metastable He* and (b) Na atoms near an Si wall calculated by use of the complete optical data for Si and accurate atomic dynamic polarizabilities (solid lines) or by use of the single-oscillator model (long-dashed lines). The short-dashed lines were computed for a semiconductor described at all frequencies by the static dielectric permittivity of Si and the accurate dynamic polarizabilities of the atoms (Caride et al. 2005).

dynamic polarizabilities but with the assumption that the dielectric permittivity of the wall material does not depend on the frequency and is equal to its static value. At the shortest separation $a = 3$ nm, the error in C_3 due to the use of the static dielectric permittivity is equal to approximately 13% for He* and 24% for Na. As is seen in Fig. 16.5, $C_3(a,T)$ is a decreasing function of distance. Even at the shortest separations considered, of about 3 nm, there is no range of separations where C_3 is constant. Thus, for Si the pure van der Waals regime (16.24) is not realized.

Figure 16.5 and the corresponding numerical data permit us to follow the influence of the characteristics of the atom and of the Si on the Casimir–Polder interaction. Thus, comparing the solid and long-dashed lines in Fig. 16.5(a), one can conclude that for He* the use of the single-oscillator model leads to a 4.4% error at $a = 3$ nm and 3.1% error at $a = 15$ nm. From Fig. 16.5(b), one can conclude that for Na, these errors are 1.8% and 1%, respectively. With an increase of the separation distance to 150 nm, the errors given by the single-oscillator model decrease to 0.4% for He* and practically to zero for Na. This confirms that at larger separations, the single-oscillator model is quite sufficient for high-precision calculations of the Casimir–Polder interaction (Caride et al. 2005, Mostepanenko et al. 2006a).

Now we consider a much wider range of separations, where relativistic effects are more pronounced and thermal effects come into play. At larger separations it is more convenient to represent the Casimir–Polder free energy not in the form of eqn (16.66) but as

$$\mathcal{F}^A(a,T) = -\frac{C_4(a,T)}{a^4}, \tag{16.76}$$

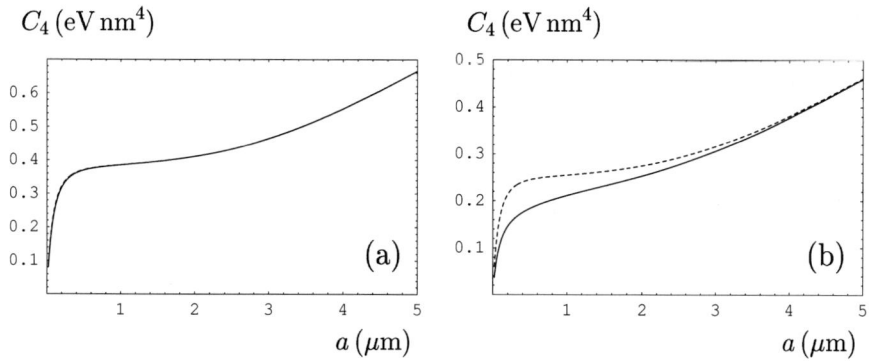

FIG. 16.6. Casimir–Polder coefficient C_4 versus separation for an Na atom interacting with (a) Si and (b) SiO$_2$ walls. The solid and dashed lines were computed by using frequency-dependent permittivities and static dielectric permittivities, respectively, for the wall material.

i.e. separating the inverse fourth power of a as in eqn (16.28), obtained at relatively large separations. We have computed the function $C_4(a,T)$ at $T = 300$ K for the Na atom and two wall materials (Si and SiO$_2$), demonstrating the different roles of the frequency dependence of the dielectric permittivity. Computations were performed in the separation region from 20 nm to 5 μm using the Lifshitz formula (16.28). The dielectric permittivity of SiO$_2$ along the imaginary frequency axis is presented in Fig. 12.2(b). As opposed to the case of Si, it has two steps owing to the two types of polarization (see Section 12.6.1). The dynamic polarizability of the Na atom was described in the single-oscillator approximation. As can be seen in Fig. 16.5(b), at separations $a > 20$ nm, the single-oscillator model for Na leads to the same results as does the accurate dynamic polarizability.

In Fig. 16.6, the coefficient C_4 is plotted as a function of separation for an Na atom interacting with (a) Si and (b) SiO$_2$ walls. The solid lines were computed using frequency-dependent dielectric permittivities. The dashed lines were computed using static dielectric permittivities for Si and SiO$_2$ ($\varepsilon_0 = 11.66$ and 3.81, respectively) at all frequencies. As can be seen in Fig. 16.6(a), for Si the solid and dashed lines practically coincide. The largest deviation between them is equal to 9%, which happens at the shortest separation $a = 20$ nm. At $a = 100$ nm, this deviation falls to 2.8%, and it is less than 1% at $a > 200$ nm. Thus, the frequency dependence of the dielectric permittivity of Si plays a minor role in the atom–wall interaction. For SiO$_2$, the dashed line deviates markedly from the solid line. The largest deviation, of 38%, is achieved at $a = 20$ nm. It decreases slowly to 32% at 200 nm, 21% at 700 nm, and 10% at 1.6 μm. Because of this, for SiO$_2$ the dependence of the dielectric permittivity on frequency must be taken into account in precise computations of the atom–wall interaction. From Fig. 16.6(a), it can be seen that within the range of separations from about 0.5 μm to 2 μm the function C_4 can be approximated by a constant. This means that in this range, the Casimir–Polder free energy falls as $1/a^4$, i.e. following the same law as eqn (16.28) for an ideal metal. At larger separations, thermal effects contribute increasingly to the Casimir–Polder free energy.

16.4.3 Various approaches to including the dc conductivity, and the Nernst theorem

As shown in Sections 16.3.3 and 16.4.1, the Lifshitz theory of atom–wall interactions is thermodynamically consistent in the cases of a metal wall and a dielectric wall when the dc conductivity is neglected. Now we reconsider the problem of the consistency of the Lifshitz theory for the interaction of an atom with a dielectric wall when the dc conductivity is included. Similarly to the cases of two dielectric plates (Section 12.5.2) and dielectric and metal plates (Section 15.4), the inclusion of the dc conductivity in the model of the dielectric response results in a violation of the Nernst heat theorem. We also discuss the consistency with thermodynamics of a generalization of the Lifshitz theory of atom–wall interactions (Pitaevskii 2008a) taking into account the effect of Debye screening. It is shown that this generalization also violates Nernst's theorem for a wide class of dielectric materials.

The permittivity of a dielectric plate with the dc conductivity included is given by eqn (12.113), where the upper index (n) is omitted. As shown in Section 12.5.2, the inclusion of the dc conductivity leads to only negligible additions to all terms of the Lifshitz formula with $l \geq 1$. These additions decay exponentially to zero with vanishing temperature. However, the term with $l = 0$ is modified because, according to eqn (16.68), the inclusion of the dc conductivity leads to the replacement of the reflection coefficient $r_{\rm TM}(0, y) = r_0$ with $\tilde{r}_{\rm TM}(0, y) = 1$. As a result, with the dc conductivity included, the free energy of the atom–wall interaction at low temperature takes the form

$$\tilde{\mathcal{F}}^{\rm A}(a, T) = \mathcal{F}^{\rm A}(a, T) - \frac{k_{\rm B} T}{4a^3}(1 - r_0)\alpha(0) + N(a, T). \qquad (16.77)$$

Here, $\mathcal{F}^{\rm A}(a, T)$ is given by eqn (16.72), and $N(a, T)$ goes exponentially fast to zero together with its derivative when the temperature goes to zero. From eqn (16.77), one can immediately arrive at a violation of the Nernst heat theorem (Klimchitskaya et al. 2008a) by taking the derivative with respect to T:

$$\tilde{S}^{\rm A}(a, T) = \frac{k_{\rm B}\alpha(0)}{4a^3}(1 - r_0) > 0. \qquad (16.78)$$

This is analogous to the corresponding results obtained for the configurations of two dielectric plates and of a metal and a dielectric plate (see Sections 12.5.2 and 15.4, respectively). It provides one more confirmation of the fact that the Lifshitz theory with inclusion of the dc conductivity of the dielectric material is thermodynamically inconsistent. Thus, in the corresponding applications of the Lifshitz theory, the dc conductivity must be disregarded. In the case of the atom–plate interaction, this conclusion was confirmed experimentally in a measurement of the Casimir–Polder interaction between ^{87}Rb atoms and an SiO$_2$ wall (Obrecht et al. 2007). It was shown that with the dc conductivity neglected, the Lifshitz theory was in excellent agreement with the data (Obrecht et al. 2007). By contrast, with the dc conductivity of SiO$_2$ included, the predictions of the Lifshitz theory were found to be inconsistent with the measurement data (Klimchitskaya and Mostepanenko 2008b). A discussion of this experiment and a comparison of the data with theory is contained in Section 22.1.

An interesting attempt to generalize the Lifshitz theory and to solve the problem of the zero-frequency contribution to eqn (16.18) describing the atom–wall interaction was undertaken by Pitaevskii (2008a). According to the proposed generalization, the electric field in the conductor can be screened owing to the presence of free charge carriers with a density n. As a result, the potential around a point charge e takes the Yukawa-type form $e\exp(-\kappa r)/r$. If the charge carriers can be described by classical Maxwell–Boltzmann statistics, the Debye–Hückel approximation is valid, where $\kappa = \kappa_{\rm DH}$ is defined by eqn (14.56). The effect of screening, which is taken into account only for the static field, leads to the replacement of the transverse magnetic reflection coefficient at zero frequency, $r_{\rm TM}(0, y) = r_0$, as given by eqn (16.68), with

$$r_{\rm TM}^{\rm mod}(0, y) = \frac{\varepsilon_0 \sqrt{4a^2\kappa^2 + y^2} - y}{\varepsilon_0 \sqrt{4a^2\kappa^2 + y^2} + y}. \tag{16.79}$$

At the same time, all of the coefficients $r_{\rm TM,TE}(i\zeta_l, y)$ with $l \geq 1$ remain unchanged (Pitaevskii 2008a). The reflection coefficient (16.79) is a particular case of the reflection coefficient (14.53) defined at any frequency in the approach of Dalvit and Lamoreaux (2008). To see this, we can put $\xi = 0$ and $k_\perp = y/(2a)$ in eqn (14.53). When the total density of charge carriers n is zero, eqn (16.79) leads to the same result as eqn (16.68). For $n \to \infty$, at fixed $T \neq 0$, $r_{\rm TM}^{\rm mod}(0, y) = \tilde{r}_{\rm TM}(0, y) = 1$, as in the case of the standard Lifshitz theory when the dc conductivity is included in the model of the dielectric response.

The proposed generalization (Pitaevskii 2008a) includes the effect of the conductivity properties of the material in the zero-frequency term of the Lifshitz formula by introducing the microscopic quantity n rather than by adding a contribution from conduction electrons to the frequency-dependent dielectric permittivity as in eqn (12.113). However, this procedure can be formulated identically as the inclusion of spatial dispersion in the standard Lifshitz theory. For this purpose, the plate material is characterized by two dissimilar dielectric permittivities $\varepsilon_x = \varepsilon_y$ and ε_z, as was done in Section 12.8.1 for uniaxial crystals. If the two dielectric permittivities ε_x and ε_z depend only on the frequency, the reflection coefficients are given by eqn (12.135) with the notation in eqn (12.136). In this case the standard Lifshitz formulas for the free energy and force are valid. To obtain the suggested generalization (Pitaevskii 2008a), one puts

$$\varepsilon_{x0} = \varepsilon_x(0) = \varepsilon_0, \qquad \varepsilon_{z0} = \varepsilon_z(0) = \varepsilon_0\left(1 + \frac{\kappa^2}{k_\perp^2}\right). \tag{16.80}$$

We emphasize that ε_{z0} in eqn (16.80) depends on the wave vector in the plane of the plate, i.e. it is incorrect to substitute it into the reflection coefficient $r_{\rm TM}^{(u)}(i\xi_l, k_\perp)$ in eqn (12.135) and into the Lifshitz formula (see the discussion in Section 12.10). If, however, one disregards this warning and substitutes eqn (16.80) in eqn (12.135) at zero frequency, the following result is obtained:

$$r_{\rm TM}^{(u)}(0, k_\perp) = \frac{\varepsilon_0 \sqrt{k_\perp^2 + \kappa^2} - k_\perp}{\varepsilon_0 \sqrt{k_\perp^2 + \kappa^2} + k_\perp}. \tag{16.81}$$

Taking into account the relation (12.89) between dimensional and dimensionless variables $y = 2ak_\perp$, one arrives at

$$r_{\rm TM}^{(u)}(0, k_\perp) \equiv r_{\rm TM}^{\rm mod}(0, y), \tag{16.82}$$

i.e. the reflection coefficients (16.79) and (16.81) coincide precisely. This shows that the generalization of the Lifshitz theory under consideration (Pitaevskii 2008a) is in fact the standard Lifshitz theory, with the plate material described by two dissimilar dielectric permittivities where one of them is allowed to depend on the wave vector (see Section 12.10).

Now we check if the proposed generalization is thermodynamically consistent. The calculation of the Casimir–Polder free energy at low temperature with the modified TM reflection coefficient at zero frequency (16.79) results in

$$\mathcal{F}^{\rm A,mod}(a, T) = \mathcal{F}^{\rm A}(a, T) - \frac{k_B T \alpha(0)}{8a^3} \int_0^\infty r_{\rm TM}^{\rm mod}(0, y) e^{-y} y^2 \, dy + \frac{k_B T}{4a^3} \alpha(0) r_0, \tag{16.83}$$

where the exponentially small terms have been omitted and $\mathcal{F}^{\rm A}(a, T)$ is defined in eqn (16.72). The respective Casimir–Polder entropy is given by

$$S^{\rm A,mod}(a, T) = S^{\rm A}(a, T) + \frac{k_B \alpha(0)}{4a^3} \left[\frac{1}{2} \int_0^\infty r_{\rm TM}^{\rm mod}(0, y) e^{-y} y^2 \, dy - r_0 \right]$$
$$+ \frac{k_B T}{8a^3} \alpha(0) \int_0^\infty \frac{\partial r_{\rm TM}^{\rm mod}(0, y)}{\partial T} e^{-y} y^2 \, dy, \tag{16.84}$$

where $S^{\rm A}(a, T)$ has been defined in eqn (16.73). It is easily seen that the last term on the right-hand side of eqn (16.84) goes to zero when the temperature vanishes, regardless of the temperature dependence of n. The second term is more involved. We discuss it first for various kinds of dielectric materials.

If $n(T)$ decays exponentially to zero with vanishing temperature (as is true for pure insulators and intrinsic semiconductors), then so does $\kappa(T)$. As a result, $r_{\rm TM}^{\rm mod}(0, y) \to r_0$ and the Casimir–Polder entropy $S^{\rm A,mod}(a, 0)$ is equal to zero, in accordance with the Nernst theorem. However, if n does not go to zero when T goes to zero (this is true, for instance, for dielectric materials such as semiconductors doped below the critical doping concentration, semimetals with strong electronic correlation, and solids with ionic conductivity), $\kappa \to \infty$ with vanishing temperature and $r_{\rm TM}^{\rm mod}(0, y) \to 1$ when $T \to 0$. In this case we obtain from eqn (16.84) (Klimchitskaya et al. 2008a)

$$S^{\rm A,mod}(a,0) = \tilde{S}^{\rm A}(a,0) = \frac{k_{\rm B}\alpha(0)}{4a^3}(1 - r_0) > 0, \qquad (16.85)$$

i.e. the proposed generalization violates the Nernst heat theorem. Precisely the same result for the entropy of the atom–wall interaction appears (Klimchitskaya et al. 2009b) in the approach by Dalvit and Lamoreaux (2008), where the TM reflection coefficient is modified at all Matsubara frequencies (see Section 14.3.5). As can be seen from the comparison of eqns (16.78) and (16.85), both theoretical approaches with the modified reflection coefficients violate the Nernst heat theorem in the same way as does the standard Lifshitz theory with the inclusion of the dc conductivity. In fact, the conductivity is given by $\sigma_0(T) = n|e|\mu$, where μ is the mobility of the charge carriers (Ashcroft and Mermin 1976). Although $\sigma_0(T)$ goes to zero exponentially fast for all dielectrics when T goes to zero, for most of them this happens owing to a vanishing μ. For instance, the conductivity of SiO_2 is ionic in nature and is determined by the concentration of impurities (alkali ions), which are always present as trace constituents. According to the above result, for this material the generalization of the Lifshitz theory under consideration violates the Nernst theorem (note that here the entropy of the fluctuating field is nonzero at $T = 0$ and depends on the separation, whereas the entropy of the plates is separation-independent). This is in agreement with the fact that the extension of the Lifshitz theory with the inclusion of spatial dispersion is controversial (see Section 12.10). Note that Pitaevskii (2008b) ignored the contradiction of his theory with the Nernst theorem for semiconductors doped below the critical value and dielectric-type semimetals.

According to Svetovoy (2008), the nonlocal generalization of the Lifshitz theory satisfies the Nernst theorem; this was stated specifically for ionic conductors possessing an activation-type conductivity. To prove this, the thermal dependence of the mobility, $\sim \exp(-C/k_{\rm B}T)$, where C is the activation energy, was arbitrarily separated and attributed to the "effective density of charges, which are able to move". This transfer of the temperature dependence from μ to n is incorrect (Klimchitskaya et al. 2008c) because the commonly used density of charge carriers n that produce the effect of screening in ionic conductors and their mobility are independently measured quantities (Schütt and Gerdes 1992). In Section 20.3.5, it will be shown that the proposed generalization is also inconsistent with experimental data on the measurement of the Casimir force between an Au sphere and an Si plate illuminated with laser pulses.

The reason for the contradictions between the generalization of the Lifshitz theory by Pitaevskii (2008a) and thermodynamics is the same as in the approach by Dalvit and Lamoreaux (2008). Physically, this generalization includes the effect of screening, i.e. nonzero gradients of n. This situation is out of thermal equilibrium, which is the basic applicability condition of the Lifshitz theory (Geyer et al. 2008d). As was recognized by Pitaevskii (2008a), "It is not clear if the fields with the very low frequencies... are in the thermodynamic equilibrium with the bodies. The problem is worth experimental investigation."

Now we consider the thermodynamic consistency of a theoretical approach which uses electrostatic screening for the description of atom–wall interactions in the case of metal walls. In Section 16.3.3, it was shown that the standard Lifshitz theory of the interaction of an atom with a metal wall is thermodynamically consistent. To check the thermodynamic consistency of an approach with a modified TM reflection coefficient given by (16.79) or (16.81), we separate the zero-frequency contribution in eqn (16.16) and represent the modified free energy in the form

$$\mathcal{F}^{A,\mathrm{mod}}(a,T) = \mathcal{F}^{A}(a,T) - k_B T \alpha(0) \int_0^\infty k_\perp^2 \, dk_\perp e^{-2ak_\perp} \left[r_{\mathrm{TM}}^{(u)}(0, k_\perp) - 1 \right], \tag{16.86}$$

where $\mathcal{F}^A(a,T)$ is the free energy of the atom–wall interaction in the standard Lifshitz theory with $r_{\mathrm{TM}}(0,k_\perp) = 1$ for a metal wall [in contrast to eqn (16.83), dimensional variables are used in eqn (16.86)]. The asymptotic behavior of $\mathcal{F}^A(a,T)$ at low T is given in eqn (16.64).

To find the asymptotic behavior of the second term on the right-hand side of eqn (16.86) at low temperature, one must take into account the fact that in this case, the electrons obey quantum statistics. As a result, instead of the Debye–Hückel κ defined in eqn (14.56), applicable when the electrons in a metal obey Maxwell–Boltzmann statistics, one must use the expression (14.61) for κ_{TF} derived in the Thomas–Fermi approximation. Thus, using eqn (16.81), we obtain the result that the asymptotic behavior of the modified Casimir–Polder free energy at low temperature is given by

$$\mathcal{F}^{A,\mathrm{mod}}(a,T) = \mathcal{F}^{A}(a,T) + 2k_B T \alpha(0) \int_0^\infty dk_\perp \frac{k_\perp^3 e^{-2ak_\perp}}{\varepsilon_0 \sqrt{k_\perp^2 + \kappa_{\mathrm{TF}}^2} + k_\perp}, \tag{16.87}$$

where ε_0 is the contribution of core electrons to the dielectric permittivity of the metal.

Bearing in mind that the integral on the right-hand side of eqn (16.87) is temperature-independent, we obtain the asymptotic behavior of the modified Casimir–Polder entropy at low temperature,

$$S^{A,\mathrm{mod}}(a,T) = S^{A}(a,T) - 2k_B \alpha(0) \int_0^\infty dk_\perp \frac{k_\perp^3 e^{-2ak_\perp}}{\varepsilon_0 \sqrt{k_\perp^2 + \kappa_{\mathrm{TF}}^2} + k_\perp}, \tag{16.88}$$

where $S^A(a,T)$ is presented in eqn (16.65). In the limiting case $T \to 0$, we obtain

$$S^{A,\mathrm{mod}}(a,0) = -2k_B \alpha(0) \int_0^\infty dk_\perp \frac{k_\perp^3 e^{-2ak_\perp}}{\varepsilon_0 \sqrt{k_\perp^2 + \kappa_{\mathrm{TF}}^2} + k_\perp} < 0, \tag{16.89}$$

i.e. the generalization of the Lifshitz theory of the atom–wall interaction by the inclusion of spatial dispersion due to electrostatic screening, as suggested by

Pitaevskii (2008a), violates the Nernst heat theorem. Note that the approach by Dalvit and Lamoreaux (2008) satisfies the Nernst heat theorem in the case of an atom interacting with a metal wall (Klimchitskaya et al. 2009b).

16.5 The impact of magnetic properties on atom–wall interaction

Previous sections were devoted to a pure electrically polarizable atom near a metallic or dielectric wall described by the frequency-dependent dielectric permittivity ε and a magnetic permeability μ equal to unity. However, the role of the magnetic properties of an atom and the wall material has also been discussed in the literature. Magnetic properties have received much attention due to an expectation of a repulsive atom–wall interaction based on the work of Boyer (1974) that an ideal-metal plane repels an infinitely permeable plane (see section 7.2.2). Keeping in mind that both the atoms and the walls used in cavity quantum electrodynamics may possess magnetic proiperties, the impact of these properties on the atom–wall interaction deserves consideration.

Safari et al. (2008) developed the theory of atom–wall interaction for the case of both a polarizable and a (para)magnetizable atom near a magnetodielectric macrobody. This theory was applied to the case of an atom near a semispace (thick magnetodielectric wall described by the frequency-dependent ε and μ). It was shown that the resulting potential of the atom–wall interaction is very similar to the well-known potential of a polarizable atom interacting with a dielectric wall. It is pertinent to note that Safari et al. (2008) deal with paramagnetic atoms which are magnetizable but have no intrinsic magnetic moment. This is usually referred to as *Van Vleck paramagnetism* (Van Vleck 1932). It is caused by the deformation of the electronic structure of the atom by an external field which creates the induced magnetic moment. Usually such deformation leads to a diamagnetic effect. However, in some specific cases paramagnetism results. Thus, Van Vleck paramagnetism is of induced origin and the respective magnetic susceptibility is temperature-independent.

Here, we present the Lifshitz-type formulas describing the impact of magnetic properties on atom–wall interaction for both magnetizable atoms and for atoms possessing an intrinsic (permanent) magnetic moment. Such atoms (for instance H or Rb) participate in different physical processes involving atom–surface interaction (see e.g. Sections 22.1 and 23.4). The case of atoms possessing a permanent magnetic moment is interesting in two aspects. First, the magnitude of a permanent magnetic moment is much larger than that of an induced one. Second, the resulting magnetic susceptibility is temperature-dependent.

We start from the Lifshitz formula for the free energy per unit area in the configuration of two parallel dissimilar magnetodielectric semispaces separated by a distance a, in thermal equilibrium at temperature T [see the first equality in eqn (12.71)]. The reflection coefficients $r_{\rm TM,TE}^{(n)}$ for each one of the two magnetodielectric semispaces ($n = 1, 2$) are given by eqn (12.56), where ε and μ are replaced with $\varepsilon^{(1)}$ and $\mu^{(1)}$ or $\varepsilon^{(2)}$ and $\mu^{(2)}$.

In order to obtain the Lifshitz-type formula for the free energy of a magnetic atom near a magnetodielectric semispace we use the same method as was used in Section 16.2 for the case of an electrically polarizable atom near a dielectric semispace. For this purpose we leave the first semispace ($n = 1$) unchanged, but we replace the second semispace by a rarefied magnetodielectric medium. Upon expanding the dielectric permittivity and the magnetic permeability of the latter medium in powers of the number N_2 of atoms per unit volume, and keeping only contributions of order N_2, we obtain eqn (16.3) with $n = 2$ and

$$\mu^{(2)}(i\xi) = 1 + 4\pi\beta^{(2)}(i\xi)N_2 + O\left(N_2^2\right). \tag{16.90}$$

Here, $\beta^{(2)}(i\xi)$ is the dynamic magnetic susceptibility for an atom of the rarefied material (the semispace with $n = 2$). Similar to Section 16.2 we use the notation $N_2 \equiv N$ and $\beta^{(2)}(i\xi) \equiv \beta(i\xi)$. It should be remembered that the quantity $\alpha(i\xi)$ is usually temperature-independent whereas $\beta(i\xi)$, for paramagnetic materials displaying orientation polarization, is proportional to the inverse of the temperature.

By repeating the derivation of Section 16.2, we arrive at the Lifshitz-type formula for the Casimir-Polder free energy for a magnetic atom near a magnetodielectric semispace (Bimonte et al. 2009)

$$\mathcal{F}^A(a,T) = -k_B T \sum_{l=0}^{\infty}{}' \int_0^\infty k_\perp dk_\perp q_l e^{-2aq_l} \tag{16.91}$$

$$\times \left\{ 2\left[\alpha(i\xi_l)r^{(1)}_{\rm TM}(i\xi_l, k_\perp) + \beta(i\xi_l)r^{(1)}_{\rm TE}(i\xi_l, k_\perp)\right] \right.$$

$$\left. -\frac{\xi_l^2}{q_l^2 c^2}[\alpha(i\xi_l) + \beta(i\xi_l)]\left[r^{(1)}_{\rm TM}(i\xi_l, k_\perp) + r^{(1)}_{\rm TE}(i\xi_l, k_\perp)\right] \right\}.$$

At zero temperature a similar formula for the energy of a magnetizable atom was obtained by Safari et al. (2008) using the Green's function method. For a nonmagnetic atom, $\beta(i\xi_l) = 0$, near a dielectric wall, $\mu(i\xi_l) = 1$, eqn (16.91) coincides with eqn (16.16).

Starting from eqn (16.91) it is straightforward to derive the expression for the force acting on a magnetic atom placed near a magnetodielectric wall

$$F^A(a,T) = -\frac{\partial \mathcal{F}^A(a,T)}{\partial a} = -2k_B T \sum_{l=0}^{\infty}{}' \int_0^\infty k_\perp dk_\perp q_l^2 e^{-2aq_l} \tag{16.92}$$

$$\times \left\{ 2\left[\alpha(i\xi_l)r^{(1)}_{\rm TM}(i\xi_l, k_\perp) + \beta(i\xi_l)r^{(1)}_{\rm TE}(i\xi_l, k_\perp)\right] \right.$$

$$\left. -\frac{\xi_l^2}{q_l^2 c^2}[\alpha(i\xi_l) + \beta(i\xi_l)]\left[r^{(1)}_{\rm TM}(i\xi_l, k_\perp) + r^{(1)}_{\rm TE}(i\xi_l, k_\perp)\right] \right\}.$$

Computations show (Bimonte et al. 2009) that the largest influence of the magnetic properties on the atom–wall interaction holds for paramagnetic atoms possessing a permanent magnetic moment and for a wall made of ferromagnetic dielectrics. For such atoms the dynamic magnetic susceptibility along the imaginary frequency axis is given by (Morrish 1965, Vonsovskii 1974)

$$\beta(i\xi_l) = \frac{g^2\mu_B^2 J(J+1)}{3k_B T} \frac{1}{1+\tau\xi_l}, \quad (16.93)$$

where g is the Lande factor, $\mu_B = e\hbar/(2m_e c)$ is the Bohr magneton, m_e is the electron mass, J is the total momentum and τ is the relaxation time. As an example, ground state atoms of H and ^{87}Rb have approximately equal magnetic moments (Valberg and Ramsey 1971). For both atoms $g = 1$ and $J = 1/2$ [the magnetic moment of H atoms was determined by Phipps and Taylor (1927); relativistic and radiative corrections are discussed by Faustov (1970)]. For various atoms at $T = 300$ K, τ varies in the range from 10^{-10} to 10^{-4} s and it increases when the temperature decreases.

It can be seen (Bimonte et al. 2009) that for Rb atoms the influence of magnetic properties on the Casimir–Polder force is negligibly small, as compared to H atoms. Bimonte et al. (2009) have performed numerical computations of the Casimir–Polder force acting between H atoms with frequency-dependent electric polarizability and magnetic susceptibility and walls made of either an ideal metal, Au, Fe or a ferromagnetic dielectric. In the first three cases inclusion of the atomic magnetic moment was shown to lead to a decrease of the force magnitude, while in the fourth case to an increase of it. Although the impact of the permanent magnetic moment of an atom on the atom–wall interaction was found to be always equal to only a fraction of a percent, it is nevertheless larger than the effect of the induced (para)magnetic moment (Safari et al. 2008). The smallness of the corrections to the atom–wall interaction from the magnetic properties allows one to disregard them in related experiments and theoretical approaches.

16.6 Atom–wall interactions in the nonequilibrium case

In previous sections it was supposed that an atom interacts with a plate which is in thermal equilibrium with an environment. If the substance of the plate is at a temperature T_P but the environment (remote bodies) is at temperature T_E, the system is out of thermal equilibrium. This case, which is beyond the scope of standard Casimir problems, was investigated by Antezza et al. (2005). It opens up interesting opportunities for the measurement of thermal Casimir forces in experiments on Bose–Einstein condensation and quantum reflection, considered in Chapter 22. Because of this, here we provide a brief summary of the results obtained, which is needed for a comparison of experiment with theory.

It is assumed that the plate is locally in thermal equilibrium and the atom is in its ground state. Thus, the thermal radiation between the plate and the environment is not of sufficient strength to excite atomic electrons to higher states. In this case the Casimir–Polder force consists of three contributions. One of them, $F^A(a, T_E)$, is given by the standard expression (equilibrium) for the Lifshitz formula (16.17) at the environmental temperature. The two additional contributions have a common form specific to the nonequilibrium situation. The first of the two, $F_\text{n}^\text{A}(a, T_\text{P})$, is taken at the plate temperature T_P and the second, $-F_\text{n}^\text{A}(a, T_\text{E})$, is taken at the environmental temperature T_E. The total force acting between the atom and the plate is given by (Antezza *et al.* 2005)

$$F^\text{A}(a, T_\text{P}, T_\text{E}) = F^\text{A}(a, T_\text{E}) + F_\text{n}^\text{A}(a, T_\text{P}) - F_\text{n}^\text{A}(a, T_\text{E}). \tag{16.94}$$

When the temperatures of the plate and environment are equal, the last two contributions on the right-hand side of eqn (16.94) cancel each other and we return to the Lifshitz expression (16.17). The explicit form of the nonequilibrium contribution is given by an integral along the real frequency axis (Antezza *et al.* 2005):

$$F_\text{n}^\text{A}(a, T) = -K \int_0^\infty d\omega \int_0^\infty dx\, f(\omega, x) \exp\left(-\frac{2\omega x a}{c}\right), \tag{16.95}$$

$$f(\omega, x) = \frac{\omega^4 x^2}{\exp\left(\frac{\hbar\omega}{k_\text{B} T}\right) - 1} \left[|p(\omega, x)| + \operatorname{Re}\varepsilon(\omega) - 1 - x^2\right]^{1/2}$$

$$\times \left[\frac{1}{|\sqrt{p(\omega, x)} + ix|^2} + \frac{(2x^2 + 1)(x^2 + 1 + |p(\omega, x)|)}{|\sqrt{p(\omega, x)} + i\varepsilon(\omega)x|^2}\right],$$

$$K \equiv \frac{2\sqrt{2}\hbar\alpha(0)}{\pi c^4}, \qquad p(\omega, x) \equiv \varepsilon(\omega) - 1 - x^2.$$

Note that the nonequilibrium contribution is derived in the approximation of a static atomic polarizability. Because of this, the equilibrium contribution $F^\text{A}(a, T_\text{E})$ must be calculated in the same approximation (as shown in Section 16.3.2, at separations larger than 2 μm the correction to the Casimir–Polder energy due to the dynamic atomic polarizability is less than 1.2%). Equation (16.94) in the absence of environmental radiation ($T_\text{E} = 0$) was obtained by Henkel *et al.* (2002). As was shown by Antezza *et al.* (2005), at separations of a few micrometers the theoretical effects of conditions out of thermal equilibrium influence the force significantly and provide promising opportunities for the measurement of atom–wall interactions (see Chapter 22). Thus, by increasing T_P while keeping the environment at room temperature, one can increase the magnitude of the attractive force. If $T_\text{P} < T_\text{E}$, the force changes sign from attraction to repulsion with increasing separation distance.

In the limiting case of large separations, the main contribution to the integral with respect to t in eqn (16.95) is given by $t \ll 1$. Expanding $f(\omega, t)$ in powers of t and integrating, one obtains (Antezza *et al.* 2005, Antezza 2006)

$$F_\text{n}^\text{A}(a, T) = -\frac{Kc^3}{4a^3} \int_0^\infty \frac{\omega\, d\omega}{\exp\left(\frac{\hbar\omega}{k_\text{B} T}\right) - 1} g(\omega), \tag{16.96}$$

$$g(\omega) = \left[|\varepsilon(\omega) - 1| + \operatorname{Re}\varepsilon(\omega) - 1\right]^{1/2} \frac{2 + |\varepsilon(\omega) - 1|}{\sqrt{2}\,|\varepsilon(\omega) - 1|}.$$

For temperatures such that $\omega_T = k_B T/\hbar$ is much less than the characteristic absorption frequency of the plate material, one may replace $\varepsilon(\omega)$ in eqn (16.96) with the static permittivity ε_0. Then, when the integration with respect to ω is performed and the definition of K in eqn (16.95) is used, the nonequilibrium contribution (16.96) reduces to (Antezza et al. 2005, Antezza 2006)

$$F_n^A(a,T) = -\frac{\pi \alpha(0)(k_B T)^2}{6c\hbar a^3} \frac{\varepsilon_0 + 1}{\sqrt{\varepsilon_0 - 1}}. \qquad (16.97)$$

This expression is obtained under the assumption that the dielectric material is described by a finite permittivity at zero frequency, i.e. the dc conductivity is ignored. We emphasize that the substitution of $\varepsilon(\omega) = \varepsilon_0 + 4\pi i \sigma_0/\omega$ in eqn (16.96) leads to practically the same values of $F_n^A(a,T)$ as are given by eqn (16.97), if the condition $\sigma_0 \ll \omega_T$ is satisfied. The latter condition is easily fulfilled for typical dielectric materials. Thus, as opposed to the equilibrium contribution to the Casimir–Polder force $F^A(a,T)$, given by the standard Lifshitz formula, the nonequilibrium contribution $F_n^A(a,T)$ is not sensitive to the inclusion of a small dc conductivity in the model of the dielectric response.

For metals, $\sigma_0 > \omega_T$. Because of this, at all frequencies contributing to the integral (16.96), the real part of $\varepsilon(\omega)$ can be neglected in comparison with the imaginary part. As a result, $g(\omega) \approx \sqrt{2\pi\sigma_0/\omega}$ and the integration in eqn (16.96) leads to (Antezza et al. 2005)

$$F_n^A(a,T) = -\frac{\alpha(0)\,\zeta_R(3/2)\,\sqrt{\sigma_0}\,(k_B T)^{3/2}}{c\sqrt{2\hbar}\,a^3}. \qquad (16.98)$$

Thus, at large separations the nonequilibrium contribution to the atom–wall force possesses very different temperature dependences in the cases of dielectric and metal wall materials.

16.7 Anisotropic materials: interaction of hydrogen atoms with graphite

In this section, we consider the interaction of atoms with a wall made of an anisotropic material. This is interesting as an application of the Lifshitz formula for anisotropic plates discussed in Section 12.8.1 and in connection with the carbon nanostructures to be discussed in Section 23.4. Bearing these applications in mind, here we choose a hydrogen atom or molecule to be the microparticle interacting with a graphite wall. We pay special attention to the behavior of the dielectric permittivity of graphite along the imaginary frequency axis. The van der Waals coefficient is calculated as a function of the separation for the interaction of hydrogen atoms and molecules with a graphite semispace and with a plate of finite thickness.

16.7.1 Dielectric permittivity of graphite along the imaginary frequency axis

Graphite is an example of the uniaxial crystals discussed in Section 12.8.1. Thus, it can be characterized by two dissimilar dielectric permittivities $\varepsilon_x(\omega) = \varepsilon_y(\omega)$

and $\varepsilon_z(\omega)$. Below, we consider a microparticle located near a graphite semispace or a plate of finite thickness d restricted by a boundary $z = \text{const}$. The optical axis z is perpendicular to the boundary plane of the semispace. This configuration is useful for applications of the theory to nanostructures such as carbon nanotubes. The Casimir–Polder free energy for the interaction of an atom with a graphite semispace or a wall of finite thickness is given by eqns (16.16) and (16.18) with the reflection coefficients (12.135) (for a semispace) or (12.137) (for a plate of thickness d). To calculate the free energy and the force of the atom–graphite wall interaction, one needs to know the dielectric permittivities $\varepsilon_x(i\xi_l)$ and $\varepsilon_z(i\xi_l)$ at the imaginary Matsubara frequencies.

These quantities can be computed with the help of the Kramers-Kronig relation (12.125), applied separately to ε_x and ε_z. The imaginary parts of the respective dielectric permittivities along the real frequency axis are expressed through the real and imaginary parts of the complex refractive index of graphite, $n_{x,z}(\omega)$, for ordinary and extraordinary rays in the same way as in Section 12.8.1.

The handbook by Palik (1991) contains measurement data for both the real and the imaginary parts of $n_{x,z}(\omega)$, obtained by various authors in the frequency region from $\Omega_1 = 0.02\,\text{eV}$ to $\Omega_2 = 40\,\text{eV}$. The use of these data to calculate $\varepsilon_{x,z}(i\xi)$ with eqn (12.125) is, however, plagued by two problems. First, the interval $[\Omega_1, \Omega_2]$ is too narrow to calculate $\varepsilon_{x,z}(i\xi)$ at all Matsubara frequencies contributing to the Casimir–Polder force (by comparison, for Au the complex refractive index has been measured up to $10000\,\text{eV}$; see Section 13.3). Second, although for n_x the data from the various authors are in agreement, in the case of n_z there are contradictory data in the literature at $\omega \leq 15.5\,\text{eV}$.

The first problem can be solved by the use of extrapolation. At high frequencies $\omega \geq \Omega_2$, the imaginary parts of the dielectric permittivities of graphite can be presented analytically in the form (Palik 1991)

$$\text{Im}\,\varepsilon_{x,z}^{(\text{h})}(\omega) = \frac{A_{x,z}}{\omega^3}. \tag{16.99}$$

Here, the values of the constants $A_x = 9.60 \times 10^3\,\text{eV}^3$ and $A_z = 3.49 \times 10^4\,\text{eV}^3$ have been determined by the condition of a smooth overlap with the values of $\text{Im}\,\varepsilon_{x,z}$ given by the tabulated data at $\omega = \Omega_2$.

At low frequencies $\omega \leq \Omega_1$, one may extrapolate $\text{Im}\,\varepsilon_x$ with the help of the Drude model,

$$\text{Im}\,\varepsilon_x^{(\text{l})}(\omega) = \frac{\omega_p^2 \gamma}{\omega(\omega^2 + \gamma^2)}, \tag{16.100}$$

where the plasma frequency $\omega_p = 1.226\,\text{eV}$ and the relaxation parameter $\gamma = 0.04\,\text{eV}$ have again been determined from the condition of a smooth overlap with the tabulated data at $\omega = \Omega_1$.

The extrapolation of the tabulated data for $\text{Im}\,\varepsilon_z$ to the region of low frequencies is connected with the second problem discussed above, i.e. the contradictory measurement results from different authors. Thus, the measurement data for

$n_z(\omega)$ of Klucker et al. (1974) differ considerably from the data of Venghaus (1975) in the frequency region $\omega \leq 15.5\,\text{eV}$. According to both of these papers, the imaginary part of $\varepsilon_z(\omega)$ can be extrapolated to low frequencies $\omega \leq \Omega_1$ by taking it to be a constant:

$$\operatorname{Im}\varepsilon_z^{(l)}(\omega) \equiv \varepsilon''_{z0} = \text{const.} \tag{16.101}$$

The values of this constant, however, were found to be different: $\varepsilon''_{z0} = 3$ (Venghaus 1975) and $\varepsilon''_{z0} = 0$ (Klucker et al. 1974).

As a result, the calculation of the dielectric permittivities of graphite along the imaginary frequency axis using eqn (12.125) was performed as follows:

$$\varepsilon_{x,z}(i\xi) = 1 + \frac{2}{\pi}\int_0^{\Omega_1} d\omega \frac{\omega\,\operatorname{Im}\varepsilon_{x,z}^{(l)}}{\omega^2+\xi^2} + \frac{2}{\pi}\int_{\Omega_1}^{\Omega_2} d\omega \frac{\omega\,\operatorname{Im}\varepsilon_{x,z}^{(t)}}{\omega^2+\xi^2} + \frac{2}{\pi}\int_{\Omega_2}^\infty d\omega \frac{\omega\,\operatorname{Im}\varepsilon_{x,z}^{(h)}}{\omega^2+\xi^2}, \tag{16.102}$$

where $\operatorname{Im}\varepsilon_{x,z}^{(t)}$ were found from the tabulated optical data and $\operatorname{Im}\varepsilon_{x,z}^{(h,l)}$ were given by eqns (16.99)–(16.101). Substituting eqns (16.99)–(16.101) in eqn (16.102), one finds (Blagov et al. 2005)

$$\varepsilon_x(i\xi) = 1 + \frac{2\omega_p^2}{\pi\xi(\xi^2-\gamma^2)}\left(\xi\arctan\frac{\Omega_1}{\gamma} - \gamma\arctan\frac{\Omega_1}{\xi}\right) \tag{16.103}$$

$$+ \frac{2}{\pi}\int_{\Omega_1}^{\Omega_2} d\omega \frac{\omega\,\operatorname{Im}\varepsilon_x^{(t)}}{\omega^2+\xi^2} + \frac{2A_x}{\pi\xi^2}\left(\frac{1}{\Omega_2} - \frac{1}{\xi}\arctan\frac{\xi}{\Omega_2}\right),$$

$$\varepsilon_z(i\xi) = 1 + \frac{\varepsilon''_{z0}}{\pi}\ln\frac{\xi+\Omega_1}{\xi} + \frac{2}{\pi}\int_{\Omega_1}^{\Omega_2} d\omega \frac{\omega\,\operatorname{Im}\varepsilon_z^{(t)}}{\omega^2+\xi^2} + \frac{2A_z}{\pi\xi^2}\left(\frac{1}{\Omega_2} - \frac{1}{\xi}\arctan\frac{\xi}{\Omega_2}\right).$$

The calculation results for $\varepsilon_x(i\xi)$ and $\varepsilon_z(i\xi)$ obtained from eqn (16.103) by the use of the tabulated optical data (Palik 1991, Klucker et al. 1974, Venghaus 1975) are shown in Figs. 16.7(a) and 16.7(b), respectively, in the frequency range from $\xi_1 = 2.47\times 10^{14}\,\text{rad/s}$ to ξ_{2000} at $T = 300\,\text{K}$ (Blagov et al. 2005). These results allow a precise calculation of the Casimir–Polder interaction in the range of separations $a \geq 3\,\text{nm}$ (as usual, with increasing separation, the number of Matsubara frequencies giving a nonnegligible contribution to the result decreases). At zero Matsubara frequency $\xi_0 = 0$, eqn (16.103) leads to $r_{\text{TM}}^{(u)}(0,k_\perp) = 1$ for both of the reflection coefficients defined in eqns (12.135) and (12.137), which follows from the fact that $\varepsilon_x(i\xi) \to \infty$ when $\xi \to 0$. As always in this chapter devoted to the atom–wall configuration, the other reflection coefficient $r_{\text{TE}}^{(u)}(0,k_\perp)$ does not contribute to the result, owing to the multiplier ξ_0^2 on the right-hand side of eqns (16.16)–(16.19).

The dependence of $\varepsilon_x(i\xi)$ on ξ in Fig. 16.7(a) is typical of good conductors (compare with Fig. 13.2 for Au). In Fig. 16.7(b), the solid line was obtained with $\varepsilon''_{z0} = 3$, as found by Venghaus (1975) and reviewed by Palik (1991). The dashed line in Fig. 16.7(b) was obtained with $\varepsilon''_{z0} = 0$ (Klucker et al. 1974, Palik

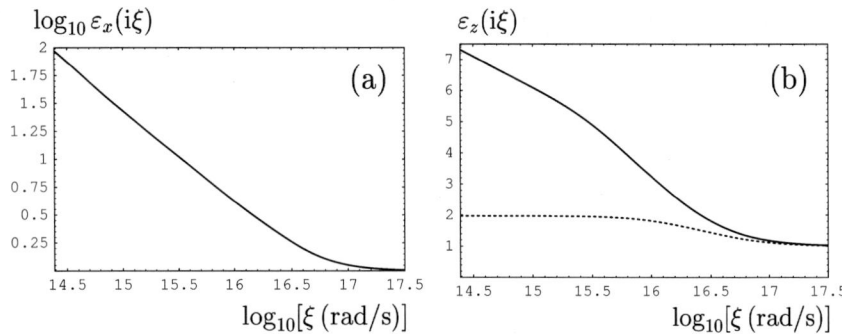

FIG. 16.7. Dielectric permittivities of graphite along the imaginary frequency axis (a) for directions parallel to the hexagonal layers and (b) for the direction perpendicular to the layers. The solid and dashed lines in part (b) were obtained by using different sets of optical data (Blagov et al. 2005). See text for discussion.

1991). It is seen that the dashed line differs markedly from the solid line in the frequency region $\xi < 10^{17}$ rad/s. The respective differences in the free energy are discussed in the next section. There are physical arguments, however, to prefer the solid line in Fig. 16.7(b) as giving the correct behavior of ε_z along the imaginary frequency axis. The point to note is that the difference between the two lines is due to the absence of absorption bands near the frequencies of 5 eV and 11 eV in the data of Klucker et al. (1974) related to ε_z [note that in the data for ε_x, there are absorption bands in both sets of data (Klucker et al. 1974, Venghaus 1975)]. This raises doubts about the former measurement data related to ε_z because, from the theory of the band structure of graphite (Johnson and Dresselhaus 1973), it follows that corresponding absorption bands must be present simultaneously in sets of data for ε_x and ε_z.

16.7.2 Computational results for plates of different thickness

To perform computations, one needs sufficiently precise information on the behavior of the dynamic polarizability of hydrogen atoms and molecules along the imaginary frequency axis. A highly accurate expression for the atomic dynamic polarizability of hydrogen is given by eqn (16.47) with $K = 10$. The respective parameters of the oscillators were found by Johnson et al. (1967). They are listed in Table 16.1 (note that one a.u. of energy $= 4.3597 \times 10^{-18}$ J $= 27.11$ eV). It is worthwhile to mention that before substitution of eqn (16.47) into the Lifshitz formula, the atomic polarizability must be expressed in cubic meters.

In addition to the accurate expression (16.47), the atomic dynamic polarizability of a hydrogen atom can be expressed in terms of the simpler single-oscillator model (16.29) with $\omega_0 \equiv \omega_{0A} = 11.65$ eV and $\alpha(0) \equiv \alpha_A(0) = 4.50$ a.u. (Rauber et al. 1982). Below, it is demonstrated that the substitution of the expressions (16.29) and (16.47) into the Lifshitz formula leads to equal results

TABLE 16.1. The values of the strengths and eigenenergies of the oscillators for a hydrogen atom in the framework of the ten-oscillator model.

j	g_j (a.u.)	ω_{0j} (a.u.)
1	0.41619993	0.37500006
2	0.08803654	0.44533064
3	0.08993244	0.48877611
4	0.10723836	0.56134416
5	0.10489786	0.68364018
6	0.08700329	0.89169023
7	0.06013601	1.2698693
8	0.03259492	2.0478339
9	0.01199044	4.0423429
10	0.00197021	12.194172

within the limits of the required accuracy. This allows one to use the simpler eqn (16.29) in the computations.

For a hydrogen molecule, the single-oscillator model for the dynamic polarizability is even more precise than for the atom. For this reason, it is acceptable to represent the dynamic polarizability of a hydrogen molecule by eqn (16.29) with parameters $\omega_0 \equiv \omega_{0M} = 14.09$ eV and $\alpha(0) \equiv \alpha_M(0) = 5.439$ a.u. (Rauber et al. 1982).

Computations of the Casimir–Polder free energy (16.16) were performed for a hydrogen atom or molecule at a separation a from a hexagonal-plane boundary of a graphite semispace or a plate of thickness d (Blagov et al. 2005, Klimchitskaya et al. 2006a). Note that the separation distance between two hexagonal layers in graphite is approximately 0.336 nm. Because of this, all computations were performed for $a \geq 3$ nm, where one can neglect the atomic structure of graphite and describe it in terms of the dielectric permittivities $\varepsilon_x(\omega)$ and $\varepsilon_z(\omega)$. The free energy was represented in the form of eqn (16.66) with a separation- and temperature-dependent van der Waals coefficient $C_3(a, T)$. The expression for $C_3(a, T)$ is contained in eqn (16.18), written in terms of the dimensionless variables (12.89). In these variables, the reflection coefficients for a graphite semispace (12.135) are rearranged as follows:

$$r_{\text{TM}}^{(u)}(i\zeta_l, y) = \frac{\sqrt{\varepsilon_{xl}\varepsilon_{zl}}y - f_z(y,\zeta_l)}{\sqrt{\varepsilon_{xl}\varepsilon_{zl}}y + f_z(y,\zeta_l)}, \qquad r_{\text{TE}}^{(u)}(i\zeta_l, y) = \frac{y - f_x(y,\zeta_l)}{y + f_x(y,\zeta_l)}, \qquad (16.104)$$

where

$$f_z^2(y,\zeta_l) = y^2 + \zeta_l^2(\varepsilon_{zl} - 1), \qquad f_x^2(y,\zeta_l) = y^2 + \zeta_l^2(\varepsilon_{xl} - 1). \qquad (16.105)$$

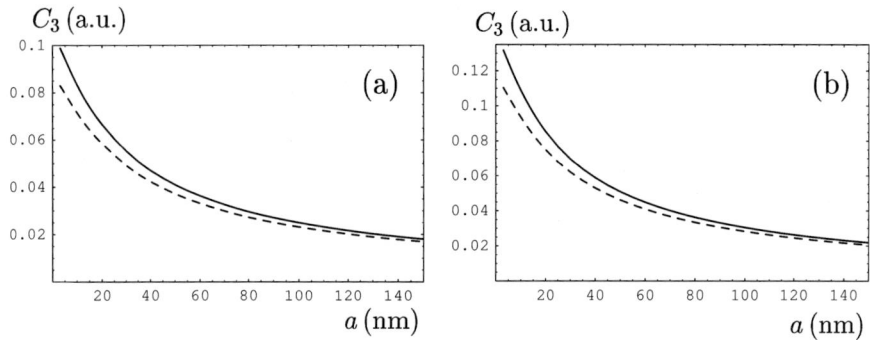

FIG. 16.8. Van der Waals coefficient C_3 versus separation for (a) a hydrogen atom and (b) a molecule near a graphite semispace (Blagov et al. 2005). The solid and dashed lines were obtained by using different sets of optical data [the solid and dashed lines, respectively, in Fig. 16.7(b)].

Analogously, the reflection coefficients (12.137) for a graphite plate of thickness d take the form

$$r_{\rm TM}^{(u)}(i\zeta_l, y) = \frac{\varepsilon_{xl}\varepsilon_{zl}y^2 - f_z^2(y,\zeta_l)}{\varepsilon_{xl}\varepsilon_{zl}y^2 + f_z^2(y,\zeta_l) + 2\sqrt{\varepsilon_{xl}\varepsilon_{zl}}yf_z(y,\zeta_l)\coth[f_z(y,\zeta_l)d/(2a)]},$$

$$r_{\rm TE}^{(u)}(i\zeta_l, y) = \frac{y^2 - f_x^2(y,\zeta_l)}{y^2 + f_x^2(y,\zeta_l) + 2yf_x(y,\zeta_l)\coth[f_x(y,\zeta_l)d/(2a)]}. \quad (16.106)$$

We then substituted into the equation for $C_3(a,T)$ the reflection coefficients for a semispace (16.104), the accurate atomic dynamic polarizability (16.47) with the parameters in Table 16.1, and the data in Fig. 16.7(a) for ε_x and Fig. 16.7(b) (solid line) for ε_z. The computational results for C_3 as a function of separation are shown in Fig. 16.8(a) by the solid line. For comparison, the dashed line in Fig. 16.8(a) shows the computational results obtained with the use of the alternative data for ε_z [the dashed line in Fig. 16.7(b)]. As can be seen in Fig. 16.8(a), at the shortest separation $a = 3$ nm the use of the alternative data for ε_z leads to a 15% error in the value of C_3, which decreases with an increase in separation (Blagov et al. 2005).

The computation of C_3 was repeated using the same procedure as described above but using the single-oscillator model (16.29) for the atomic dynamic polarizability instead of the accurate expression (16.47). The computational results were found to be practically coincident with those in Fig. 16.8(a) (the maximum deviation within the range of separations from 3 to 150 nm was less than 0.2%). This shows that for hydrogen atoms, the single-oscillator model is more exact than for He* or Na (see Section 16.3.4). Thus, the single-oscillator approximation is quite sufficient for an investigation of the Casimir–Polder interaction of hydrogen atoms (and, consequently, molecules) with a graphite surface.

Similarly to the above, the van der Waals coefficient $C_3(a,T)$ for the interaction of an H molecule with a graphite semispace can be calculated (Blagov

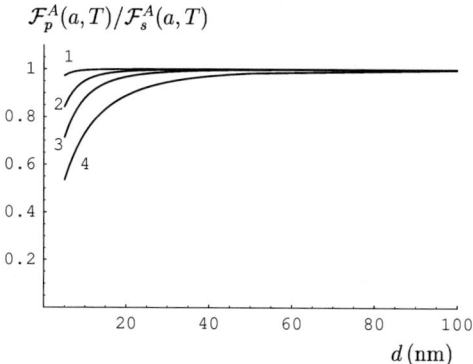

FIG. 16.9. Ratio of the free energy of the Casimir–Polder atom–plate interaction to that for the atom–semispace interaction as a function of plate thickness for a hydrogen atom located at different separations from a graphite surface (Blagov et al. 2005). Lines 1, 2, 3, and 4 correspond to separations $a = 3, 10, 20$, and $50\,\mathrm{nm}$, respectively.

et al. 2005). The only difference is the use of the dynamic polarizability of a molecule instead of that for an atom. The computational results are presented in Fig. 16.8(b) by the solid line (the dashed line was calculated with the alternative data for the dielectric permittivity ε_z discussed in Section 16.7.1). A comparison of Figs. 16.7(a) and 16.7(b) leads to the conclusion that the magnitudes of the van der Waals coefficient for the H_2 molecule are larger than for the atom.

Now we consider a hydrogen atom at a separation a from a graphite plate of thickness d. It is interesting to investigate the dependence of the Casimir–Polder free energy as a function of d and compare it with the case of an atom at the same separation from a semispace. Computations of the free energy were performed by use of eqn (16.18) with the reflection coefficients (16.104) (for a semispace) and (16.106) (for a plate of thickness d). The values of the dielectric permittivities along the imaginary frequency axis were taken from Fig. 16.7 (solid lines), and the single-oscillator model for the atomic dynamic polarizability was used. In Fig. 16.9, the ratio of the free energy $\mathcal{F}_p^A(a,T)$ for an atom near a plate of thickness d to $\mathcal{F}_s^A(a,T)$ for an atom near a semispace is plotted as a function of d for different atom–semispace (or atom–plate) separations (lines 1, 2, 3, and 4 are for separations $a = 3, 10, 20$, and $50\,\mathrm{nm}$, respectively). As is seen in Fig. 16.9, at a separation $a = 3\,\mathrm{nm}$ the finite thickness of the plate has a noticeable effect on the free energy (more than 1% change) only for thicknesses $d < 8\,\mathrm{nm}$. At separations $a = 10, 20$, and $50\,\mathrm{nm}$, the finite thickness of the plate leads to a smaller magnitude of the van der Waals free energy, as compared with a semispace, of more than 1% if the thickness of the plate is less than 19, 32, and $61\,\mathrm{nm}$, respectively. For example, if the separation between the atom and the plate is $a = 3\,\mathrm{nm}$, then plates of thickness larger than $8\,\mathrm{nm}$ can be considered as semispaces with good accuracy.

17

THE CASIMIR FORCE BETWEEN ROUGH SURFACES AND CORRUGATED SURFACES

In the preceding chapters of Part II of this book, devoted to the Casimir force between real bodies, we have discussed many effects due to the presence of charge carriers at both zero and nonzero temperature. In doing so, we supposed the surfaces of the bodies under consideration (parallel plates) to be perfectly smooth. It was also mentioned that many of the results obtained can be approximately extended to the case of a material ball of large radius in close proximity to a flat plate. The surface of the ball was assumed to be perfectly spherical.

The surfaces of real bodies, however, are not characterized by a perfect geometrical shape. Even if special efforts are made to avoid large-scale deviations from a planar or spherical shape, any real surface is invariably covered with disorder called *surface roughness*. This disorder may be on a relatively large or small scale (in relation to the separation distance between the two bodies), and may be formed by native atoms or molecules or may consist of defects and clusters of them or foreign inclusions. In some cases the roughness profile can be described mathematically by a regular function, but in other cases the roughness can be considered as stochastic. In the Lifshitz theory, dispersion forces are determined by the reflection of electromagnetic waves from a surface. Thus, scattering processes on rough surfaces become important for investigations of corrections to the Casimir force due to surface roughness. In the case of surfaces with stochastic roughness, the scattering of both classical waves and quantum particles was considered by Brown *et al.* (1985), Tutov *et al.* (1999), and Leskova *et al.* (2005) and in many other papers. It was concluded that surface roughness might have a significant effect on the van der Waals force at short separations by increasing its magnitude (Maradudin and Mazur 1980, Mazur and Maradudin 1981, Rabinovich and Churaev 1989).

The calculation of roughness corrections to the Casimir force between real bodies is an extremely difficult problem. As shown in Chapters 6 and 10, even for perfectly shaped bodies different from plane parallel plates, exact calculations have been performed only recently and only for a few simple cases. Because of this, various approximate methods have been developed. In many cases it is possible to consider the roughness as *small* in the sense that the characteristic roughness amplitude is much less than the separation distance between the two bodies. Then perturbative methods can be applied. An important characteristic of roughness is the correlation length, which is the average distance between adjacent peaks and valleys. The ratio between the roughness period (for periodic roughness) or the correlation length and the separation distance shows when a simple phenomenological calculation of the roughness correction to the Casimir force is possible and when a more fundamental theory is necessary. In the case of stochastic roughness, the root-mean-square deviation of the surface from flatness (i.e. the *dispersion*) plays the role of the amplitude. It is important to remember that the role of roughness should be considered not in isolation, but in combination with the previously investigated corrections to the Casimir force between real bodies due to the dielectric properties, nonzero skin depth, and nonzero temperature.

In this chapter we consider all of the above-mentioned problems, emphasizing those approaches that can be used for comparison between theory and the measurement data presented in Part III of the book. Thus, the method of pairwise summation (PWS) allows one to calculate roughness corrections for large-scale roughness of both the regular and the stochastic type (van Bree et al. 1974). The method of geometrical averaging is based on the proximity force approximation (PFA). This was discussed in Section 6.5. Here, it is considered in connection with the limits of its validity. The relationship between these two phenomenological methods is discussed with the help of the example of two nonparallel plates which, in itself, is of interest for the understanding of experiments. The more fundamental approaches suggested in the literature which are needed for the description of short-scale roughness are also considered. At the end of the chapter, we present various approaches to the theoretical description of both the normal and the lateral Casimir force between sinusoidally corrugated surfaces. This configuration has already been successfully used in the first experiment on the measurement of the lateral Casimir force (Chen *et al.* 2002a, 2002b). It presents interesting opportunities to study the application regions of various methods and demonstrates the nontrivial geometry dependence of the Casimir force. The influence of sinusoidal corrugations on atom-wall interactions is also considered.

17.1 Method of pairwise summation for real bodies with rough surfaces

Here, we apply the approximate phenomenological method of PWS formulated in Section 6.4 to describe the roughness corrections to the Casimir force between real material bodies. After the formulation of the method, a perturbation theory in the relative roughness amplitude is developed for the configurations of two parallel plates and a sphere above a plate. The results obtained are applied to the case of large-scale roughness in accordance with the validity region of the PWS method. A description of the role of stochastic roughness in the framework of the PWS method is also provided.

17.1.1 Formulation of the method

The PWS method was formulated in Section 6.4 for perfectly shaped bodies made of ideal materials (dielectrics with constant permittivities $\varepsilon_0^{(1)}$ and $\varepsilon_0^{(2)}$

and the limiting case when $\varepsilon_0^{(i)} \to \infty$) It suggests that the additive interaction energy of two bodies with volumes V_1 and V_2 can be obtained as an integral of the interatomic potentials of the retarded Casimir-Polder interaction (16.12) over the volumes of both bodies. This energy is expressed by eqn (6.45). After division by the normalization factor (6.44) which takes approximate account of the nonadditivity, the resulting interaction energy takes the form (6.46). In this chapter we consider bodies with roughness, and both the total interaction energy and the energy per unit area of the plates are denoted by an index "R":

$$E_R^{\text{tot}}(a) = -\frac{\pi\hbar c}{24}\Psi(\varepsilon_0^{(1)},\varepsilon_0^{(2)})\int_{V_1} d\mathbf{r}_1 \int_{V_2} d\mathbf{r}_2\, |\mathbf{r}_1 - \mathbf{r}_2|^{-7}, \qquad (17.1)$$

where the function $\Psi(\varepsilon_0^{(1)},\varepsilon_0^{(2)})$ is defined in eqn (16.8). Recall that the PWS method is formulated in such a way that in the particular case of two perfectly shaped parallel semispaces, it leads to the exact result. Thus, for two perfectly shaped semispaces,

$$\int_{V_1} d\mathbf{r}_1 \int_{V_2} d\mathbf{r}_2\, |\mathbf{r}_1 - \mathbf{r}_2|^{-7} = \frac{\pi}{30 a^3} S, \qquad (17.2)$$

where S is the infinite area of the boundary plane of a semispace. Substituting this into eqn (17.1), we return to the same energy per unit area $E(a) = E_R^{\text{tot}}(a)/S$ as in eqn (16.8).

Now we consider two parallel plates of thickness D with sides of length $2L$, and with surface roughness described by regular functions. Let us first assume that the plate materials are described by constant dielectric permittivities $\varepsilon_0^{(1)}$ and $\varepsilon_0^{(2)}$. This assumption means that the separation distance between the plates is sufficiently large. The choice of the coordinate system (see Fig. 17.1) permits the description of the rough surfaces of the first and second plates by the equations

$$z_1^{(s)} = A_1 f_1(x,y), \qquad z_2^{(s)} = a + A_2 f_2(x,y), \qquad (17.3)$$

where a is the mean distance between the plates. The amplitudes A_1 and A_2 are defined in such a way that $\max |f_i(x,y)| = 1$. The origin of the z-axis is chosen such that

$$\langle z_1^{(s)}\rangle \equiv A_1\langle f_1(x,y)\rangle = \frac{A_1}{(2L)^2}\int_{-L}^{L} dx \int_{-L}^{L} dy\, f(x,y) = 0,$$

$$\langle z_2^{(s)}\rangle \equiv a + A_2\langle f_2(x,y)\rangle = a. \qquad (17.4)$$

Corrections to the Casimir force between the plates due to the surface roughness described by eqn (17.4) can be calculated perturbatively under the assumptions $A_i \ll a$, $a \ll D$, and $a \ll L$. In addition, in all of the experimental situations considered in Part III of the book, $a/D, a/L \ll A_i/a$. Because of this, the perturbation theory used in the PWS approach for the description of the surface

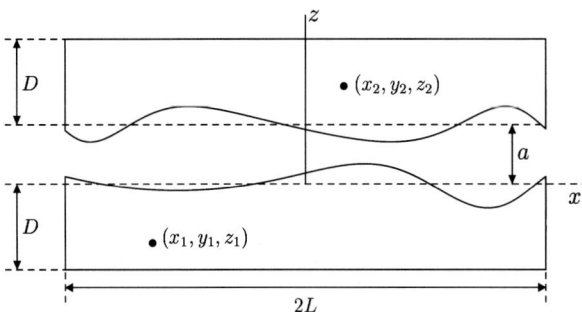

FIG. 17.1. Two parallel plates with rough surfaces.

roughness is based on a perturbative expansion in powers of A_i/a but is restricted to the zeroth order in a/D and a/L.

The above method uses pairwise summation of the retarded Casimir–Polder potentials (16.12). Because of this, the results obtained are approximately applicable only at sufficiently large separations. Alternatively, one could use the same additive method but with the nonretarded interatomic potential (16.7). In this case the additive interaction energy between the two bodies V_1 and V_2 is given by

$$E_{R,nr}^{add}(a) = -AN_1N_2 \int_{V_1} d\mathbf{r}_1 \int_{V_2} d\mathbf{r}_2 \, |\mathbf{r}_1 - \mathbf{r}_2|^{-6}, \qquad (17.5)$$

where N_1 and N_2 are the numbers of atoms per unit volume of the first and the second body. The normalization constant, $K_{E,nr}$, can be obtained in the same way as in Section 6.4, devoted to the relativistic case. By applying eqn (17.5) to two perfectly shaped plates (or semispaces) at short separations a, one obtains the additive interaction energy $E_{pp,nr}^{add}(a)$ defined in eqn (16.6). At the same time, the exact van der Waals interaction energy per unit area, $E(a)$, in this case is given by eqn (16.1). By dividing eqn (16.6) by eqn (16.1), the normalization factor is found to be

$$K_{E,nr} = \frac{E_{pp,nr}^{add}(a)}{E(a)S} = \frac{\pi^2 A N_1 N_2}{H}, \qquad (17.6)$$

where H is the Hamaker constant, defined in eqn (16.1). As a result, a normalized interaction energy between the bodies V_1 and V_2 taking partial account of nonadditivity is obtained from eqn (17.5):

$$E_{R,nr}^{tot}(a) \equiv \frac{E_{R,nr}^{add}(a)}{K_{E,nr}} = -\frac{H}{\pi^2} \int_{V_1} d\mathbf{r}_1 \int_{V_2} d\mathbf{r}_2 \, |\mathbf{r}_1 - \mathbf{r}_2|^{-6}. \qquad (17.7)$$

For two perfectly shaped semispaces, this equation leads to the exact result (16.1). It can be applied to plates with rough surfaces described by eqn (17.3) by performing a perturbative expansion in the small parameters A_i/a.

It should be kept in mind, however, that the application region of the van der Waals interaction potential is rather narrow. In fact, eqn (17.5) is valid only at separations r below a few nanometers. At larger separations, relativistic effects come into play. This is the beginning of a wide transition region between the nonrelativistic and pure relativistic regimes. At the end of this region (typically at separations of a few hundred nanometers), the relativistic Casimir–Polder potential (16.12) becomes fully applicable. If one takes into account the fact that it is hard to decrease roughness amplitudes much below 1 nm, it becomes clear that the application region of the perturbation theory based on eqn (17.7) is very narrow.

From the above, it may appear that the PWS method developed below for the treatment of surface roughness is applicable only in the pure relativistic regime, where the role of roughness is in fact very small. For the transition region of separations from a few nanometers to a few hundred nanometers, where the roughness corrections are really important from the experimental point of view, this method seems to provide no more than an order-of-magnitude estimate of the role of roughness. However, the application region of the PWS method based on the Casimir-Polder potential (16.12) and the perturbative expansion obtained below can be widened to shorter separations. For this purpose, one can use a separation-dependent normalization factor, obtained as the ratio of the additive interaction energy between two perfectly shaped semispaces (16.1) and the exact Casimir energy per unit area (13.4) in this configuration calculated using frequency-dependent permittivities. The latter energy can be represented in the form

$$E(a) = -\frac{\pi^2 \hbar c}{720 a^3} \eta_E(a), \tag{17.8}$$

where, for two dissimilar plates, $r^2_{\text{TM,TE}}$ in eqn (13.4) is replaced with a product $r^{(1)}_{\text{TM,TE}} r^{(2)}_{\text{TM,TE}}$. As a result, the correction factor $\eta_E(a)$ is given by

$$\eta_E(a) = -\frac{45}{2\pi^4} \int_0^\infty y\, dy \int_0^y d\zeta \left\{ \ln\left[1 - r^{(1)}_{\text{TM}}(i\zeta, y) r^{(2)}_{\text{TM}}(i\zeta, y) e^{-y}\right] \right.$$
$$\left. + \ln\left[1 - r^{(1)}_{\text{TE}}(i\zeta, y) r^{(2)}_{\text{TE}}(i\zeta, y) e^{-y}\right] \right\} > 0. \tag{17.9}$$

The reflection coefficients entering eqn (17.9) can be found from eqn (12.91), with $\varepsilon = \varepsilon^{(1)}$ for the first plate (or semispace) and $\varepsilon = \varepsilon^{(2)}$ for the second plate (or semispace). Note that at sufficiently large a, when one can neglect the dependence of the dielectric permittivity on frequency, $\eta_E(a)$ becomes separation-independent and coincides with the function $\Psi(\varepsilon_0^{(1)}, \varepsilon_0^{(2)})$ defined in eqn (16.8). Thus, the separation-dependent normalization factor is equal to

$$\tilde{K}_E(a) = \frac{E_{\text{pp}}^{\text{add}}(a)}{E(a)S} = \frac{24 B N_1 N_2}{\pi \hbar c \eta_E(a)}. \tag{17.10}$$

After division of the additive energy between two rough bodies $E_R^{\text{add}}(a)$, obtained from eqn (6.45), by this factor, the refined version of the PWS method results in

$$E_{\rm R}^{\rm tot}(a) = \frac{E_{\rm R}^{\rm add}(a)}{\tilde{K}_E(a)} = -\frac{\pi \hbar c}{24}\eta_E(a) \int_{V_1} d\boldsymbol{r}_1 \int_{V_2} d\boldsymbol{r}_2 \, |\boldsymbol{r}_1 - \boldsymbol{r}_2|^{-7}. \qquad (17.11)$$

As shown below, eqn (17.11) provides a basis for the *multiplicative* approach to taking account of the corrections due to surface roughness and frequency-dependent dielectric properties. The latter is important for metals, where the corrections due to the nonzero skin depth depend on the separation (see Section 13.1). Because of this, we shall often discuss the multiplicative approach below in relation to metals. In this case the multiplier $\eta_E(a)$ incorporates corrections due to the nonzero skin depth and the integral over V_1 and V_2 due to the surface roughness. In Section 17.2, the multiplicative approach is compared with another method for taking account of roughness and the skin depth based on geometrical averaging. Using eqn (17.2), it is evident that for two perfectly shaped semispaces, eqn (17.11) leads to the exact result (13.4).

The PWS method in the version given by eqn (17.11) has some ambiguity when applied to the Casimir pressure. On the one hand, the Casimir pressure between two rough bodies can be obtained from eqn (17.11) as

$$\tilde{P}_{\rm R}(a) = \frac{\pi \hbar c}{24S}\frac{\partial}{\partial a}\left[\eta_E(a) \int_{V_1} d\boldsymbol{r}_1 \int_{V_2} d\boldsymbol{r}_2 \, |\boldsymbol{r}_1 - \boldsymbol{r}_2|^{-7}\right], \qquad (17.12)$$

where S is the surface area and both the function η_E and the integrals depend on a. Using eqn (17.2), it can be seen that for perfectly shaped semispaces, eqn (17.12) results in exactly the same pressure as eqn (17.8). On the other hand, it is natural to define a nonnormalized additive pressure by the equation below, which follows from eqn (6.45):

$$P_{\rm R}^{\rm add}(a) = \frac{BN_1 N_2}{S}\frac{\partial}{\partial a} \int_{V_1} d\boldsymbol{r}_1 \int_{V_2} d\boldsymbol{r}_2 \, |\boldsymbol{r}_1 - \boldsymbol{r}_2|^{-7}. \qquad (17.13)$$

The additive pressure between two perfectly shaped semispaces is obtained from eqn (6.43):

$$P^{\rm add}(a) = -\frac{BN_1 N_2 \pi}{10 a^4}. \qquad (17.14)$$

The corresponding exact Casimir pressure can be represented with the help of eqn (13.9) as

$$P(a) = -\frac{\pi^2 \hbar c}{240 a^4}\eta_P(a), \qquad (17.15)$$

$$\eta_P(a) = \frac{15}{2\pi^4}\int_0^\infty y^2\, dy \int_0^y d\zeta \left\{\left[r_{\rm TM}^{(1)}{}^{-1}(i\zeta, y) r_{\rm TM}^{(2)}{}^{-1}(i\zeta, y)e^y - 1\right]^{-1} \right.$$
$$\left. + \left[r_{\rm TE}^{(1)}{}^{-1}(i\zeta, y) r_{\rm TE}^{(2)}{}^{-1}(i\zeta, y)e^y - 1\right]^{-1}\right\}.$$

Following the same lines as above for the energy, we can define a separation-dependent normalization factor for the pressure according to

$$\tilde{K}_P(a) = \frac{P^{\text{add}}(a)}{P(a)} = \frac{24 B N_1 N_2}{\pi \hbar c \eta_P(a)}. \qquad (17.16)$$

Then the normalized Casimir pressure between two rough bodies is given by

$$P_R(a) = \frac{P_R^{\text{add}}(a)}{\tilde{K}_P(a)} = \frac{\pi \hbar c}{24 S} \eta_P(a) \frac{\partial}{\partial a} \int_{V_1} d\mathbf{r}_1 \int_{V_2} d\mathbf{r}_2 \, |\mathbf{r}_1 - \mathbf{r}_2|^{-7}. \qquad (17.17)$$

This equation also coincides exactly with the Lifshitz result (17.15) in the particular case of two perfectly shaped semispaces. Equation (17.17) is analogous to eqn (17.11). It allows the roughness and skin depth corrections to the Casimir pressure to be taken into account multiplicatively.

It can be easily seen that at sufficiently large separation distances, where one can neglect the frequency dependence of the dielectric permittivity, eqns (17.12) and (17.17) lead to a common result. In this case $\eta_E(a) = \Psi(\varepsilon_0^{(1)}, \varepsilon_0^{(2)})$ and $\eta_P(a) = \eta_P$ (i.e. the two quantities do not depend on the separation), and $\eta_P = \Psi(\varepsilon_0^{(1)}, \varepsilon_0^{(2)})$ also. However, at relatively short separations, eqns (17.12) and (17.17) lead to slightly different results. The difference between the two results can be estimated analytically for metals described by the plasma model using the perturbative description of surface roughness (see Section 17.1.3).

All of the above approaches using the PWS method were developed for zero temperature. This is justified because the surface roughness is the most important factor at relatively short separation distances, where the temperature corrections are very small and can be disregarded.

17.1.2 Perturbation theory in the roughness amplitudes for two parallel plates

Now we perform the perturbative expansion of the integral (17.2) containing the Casimir–Polder interatomic potential over the volumes of the rough plates shown in Fig. 17.1 (Bordag et al. 1994, 1995a). The same expansion can be used in both eqn (17.1) and eqn (17.11), which present the two versions of the PWS method, and also in eqns (17.17) and (17.12) for the Casimir pressure. We start with the nonnormalized energy of an atom at a point (x_2, y_2, z_2) belonging to the upper plate. This energy is obtained by the additive summation of the Casimir–Polder potentials (16.12) over the volume of the lower plate

$$E_A^{\text{add}}(x_2, y_2, z_2) = -B N_1 \int_{-L}^{L} dx_1 \int_{-L}^{L} dy_1 \int_{-D}^{A_1 f_1(x_1, y_1)} dz_1 \qquad (17.18)$$
$$\times \left[(x_2 - x_1)^2 + (y_2 - y_1)^2 + (z_2 - z_1)^2 \right]^{-7/2}.$$

We expand this expression in powers of the parameter A_1/z_2, which is small owing to the inequalities $A_1 \ll a \le z_2$. In doing these calculations, we neglect the corrections of order z_2/L and z_2/D, as explained in Section 17.1.1. The result of the expansion up to the fourth order can be presented in the form

$$E_A^{\text{add}}(x_2, y_2, z_2) = -B N_1 \left\{ \frac{\pi}{10 z_2^4} + \int_{-L}^{L} dx_1 \int_{-L}^{L} dy_1 \left[\frac{z_2 f_1(x_1, y_1)}{X^{7/2}} \frac{A_1}{z_2} \right. \right.$$

$$+ \frac{7z_2^3 f_1^2(x_1,y_1)}{2X^{9/2}} \left(\frac{A_1}{z_2}\right)^2 + \frac{7z_2^3}{6X^{9/2}}\left(\frac{9z_2^2}{X}-1\right) f_1^3(x_1,y_1)\left(\frac{A_1}{z_2}\right)^3$$

$$+ \frac{21z_2^5}{8X^{11/2}}\left(\frac{11z_2^2}{X}-3\right) f_1^4(x_1,y_1)\left(\frac{A_1}{z_2}\right)^4 \Bigg]\Bigg\}, \quad (17.19)$$

where $X = (x_1 - x_2)^2 + (y_1 - y_2)^2 + z_2^2$. Note that we have performed the limit $L \to \infty$ in the first term on the right-hand side of eqn (17.19), which describes the contribution from a perfectly flat plate surface with no roughness [a similar perturbative expansion can be performed starting from the van der Waals potential (16.7) (Bordag et al. 1995b)].

The interaction energy between the rough plates can be obtained by the integration of eqn (17.19) over the volume V_2 of the upper plate, including the shape of its boundary surface:

$$E_R^{\text{add}}(a) = N_2 \int_{-L}^{L} dx_2 \int_{-L}^{L} dy_2 \int_{a+A_2 f_2(x_2,y_2)}^{a+D} dz_2 \, E_A(x_2, y_2, z_2). \quad (17.20)$$

The nonnormalized Casimir pressure between the plates is given by

$$P_R^{\text{add}}(a) = -\frac{1}{(2L)^2} \frac{\partial E_R^{\text{add}}(a)}{\partial a}. \quad (17.21)$$

Substituting eqn (17.20) into eqn (17.21), we obtain

$$P_R^{\text{add}}(a) = \frac{N_2}{(2L)^2} \int_{-L}^{L} dx_2 \int_{-L}^{L} dy_2 \, E_A[x_2, y_2, a + A_2 f_2(x_2, y_2)]. \quad (17.22)$$

Here we have omitted the contribution of the upper integration limit in eqn (17.20) using the condition $a \ll D$.

Using eqn (17.19), the quantity E_A in eqn (17.22) can be presented as an expansion up to the fourth order in the small parameter A_2/a. After the substitution of this expansion into eqn (17.22), the Casimir pressure takes the form

$$P_R^{\text{add}}(a) = P^{\text{add}}(a) \sum_{k=0}^{4} \sum_{l=0}^{4-k} c_{kl} \left(\frac{A_1}{a}\right)^k \left(\frac{A_2}{a}\right)^l, \quad (17.23)$$

where $P^{\text{add}}(a)$ is the pressure (17.14) between two semispaces with flat surfaces obtained by the additive summation of the interatomic potentials. In the zeroth-order expansion (i.e. for $k = l = 0$), $P_R^{\text{add}}(a) = P_{\text{pp}}^{\text{add}}(a)$. It follows from this that $c_{00} = 1$. From our choice (17.4), we also obtain $c_{01} = c_{10} = 0$. The coefficients whose first index is zero are the simplest ones:

$$c_{02} = 10\langle f_2^2\rangle, \quad c_{03} = -20\langle f_2^3\rangle, \quad c_{04} = 35\langle f_2^4\rangle. \quad (17.24)$$

The remaining coefficients in eqn (17.23) are

$$c_{20} = \frac{35}{\pi} a^7 \langle\langle f_1^2 Y^{-9}\rangle\rangle, \quad c_{30} = \frac{35}{\pi} a^7 \langle\langle f_1^3 \varphi_1(Y)\rangle\rangle, \quad (17.25)$$

$$c_{40} = \frac{105}{4\pi}a^9\langle\langle f_1^4 \varphi_2(Y)\rangle\rangle, \quad c_{11} = -\frac{70}{\pi}a^7\langle\langle f_1 f_2 Y^{-9}\rangle\rangle,$$

$$c_{12} = \frac{35}{\pi}a^7\langle\langle f_1 f_2^2 \varphi_1(Y)\rangle\rangle, \quad c_{21} = -\frac{35}{\pi}a^7\langle\langle f_1^2 f_2 \varphi_1(Y)\rangle\rangle,$$

$$c_{13} = -\frac{105}{\pi}a^9\langle\langle f_1 f_2^3 \varphi_2(Y)\rangle\rangle, \quad c_{31} = -\frac{105}{\pi}a^9\langle\langle f_1^3 f_2 \varphi_2(Y)\rangle\rangle,$$

$$c_{22} = \frac{315}{\pi}a^9\langle\langle f_1^2 f_2^2 \varphi_2(Y)\rangle\rangle.$$

Here, the following notation has been used:

$$Y = \left[(x_1 - x_2)^2 + (y_1 - y_2)^2 + a^2\right]^{1/2}, \tag{17.26}$$

$$\varphi_1(Y) = 9a^2 Y^{-11} - Y^{-9}, \quad \varphi_2(Y) = 11a^2 Y^{-13} - 3Y^{-11},$$

and the averaging procedure for a function of four variables

$$\langle\langle \Phi(x_1, y_1; x_2, y_2)\rangle\rangle = \frac{1}{(2L)^2} \int_{-L}^{L} dx_2 \int_{-L}^{L} dy_2 \int_{-L}^{L} dx_1 \int_{-L}^{L} dy_1 \Phi(x_1, y_1; x_2, y_2). \tag{17.27}$$

For brevity, we have used $f_1 \equiv f_1(x_1, y_1)$, $f_2 \equiv f_2(x_2, y_2)$.

The calculation of the expansion coefficients (17.25) is rather cumbersome. The simplest ones are c_{l0}, which depend only on one roughness function. Integrating over x_2 and y_2 according to eqn (17.27) with $L = \infty$, we obtain the results which are valid in the zeroth order of the small parameter a/L,

$$c_{20} = 10\langle f_1^2\rangle, \quad c_{30} = -20\langle f_1^3\rangle, \quad c_{40} = 35\langle f_1^4\rangle. \tag{17.28}$$

These results are in agreement with eqn (17.24) and can also be obtained from symmetry considerations. For the calculation of the mixed coefficients depending on both roughness functions f_1 and f_2, these functions and their powers, f_1^i and f_2^k, should be considered as periodic functions with a period $2L$ in both coordinates. The Fourier coefficients can be denoted as $[g_{00}^{(i)}, g_{l,mn}^{(i)}]$ and $[h_{00}^{(k)}, h_{l,mn}^{(k)}]$ ($l = 1, 2, 3, 4$; $m, n = 0, 1, 2, \ldots$) for the functions f_1^i and f_2^k, respectively. Then the mixed expansion coefficients c_{ik} take the following form [details were presented by Bordag et al. (1995a), although several misprints in the original publication have been corrected here]:

$$c_{11} = -\frac{4}{3}\sqrt{\frac{2}{\pi}} \sum_{m,n=0}^{\infty} G_{mn}^{(1,1)} z_{nm}^{7/2} K_{7/2}(z_{mn}), \tag{17.29}$$

$$c_{12} = \frac{2}{3}\sqrt{\frac{2}{\pi}} \sum_{m,n=0}^{\infty} G_{mn}^{(1,2)} z_{nm}^{7/2} \left[z_{mn} K_{9/2}(z_{mn}) - K_{7/2}(z_{mn})\right],$$

$$c_{13} = -\frac{1}{6}\sqrt{\frac{2}{\pi}} \sum_{m,n=0}^{\infty} G_{mn}^{(1,3)} z_{nm}^{9/2} \left[z_{mn} K_{11/2}(z_{mn}) - K_{9/2}(z_{mn})\right],$$

$$c_{22} = 210 g_{00}^{(2)} h_{00}^{(2)} + \frac{1}{4}\sqrt{\frac{2}{\pi}} \sum_{m,n=0}^{\infty} G_{mn}^{(2,2)} z_{nm}^{9/2} \left[z_{mn} K_{11/2}(z_{mn}) - K_{9/2}(z_{mn})\right],$$

where

$$G_{mn}^{(i,k)} = \frac{1}{4}(1 + \delta_{m0} + \delta_{n0}) \sum_{l=1}^{4} g_{l.mn}^{(i)} h_{l,mn}^{(k)}, \quad z_{mn} = \frac{\pi a}{L}\sqrt{m^2 + n^2}, \qquad (17.30)$$

and $K_\nu(z)$ are the Bessel functions of imaginary argument. Note that c_{21} is obtained from c_{12} in eqn (17.29) by the replacement of $G_{mn}^{(1,2)}$ with $-G_{mn}^{(2,1)}$. The coefficient c_{31} is obtained from c_{13} by the replacement of $G_{mn}^{(1,3)}$ with $G_{mn}^{(3,1)}$.

As mentioned above, the approximate approach based on the PWS is applicable to the case of large-scale roughness with a characteristic uniaxial length scale $\Lambda \gg a$. Under the stronger condition $\Lambda \gg 2\pi a$, the expressions (17.29) for the expansion coefficients can be simplified considerably. For this purpose, we first perform an identical transformation of eqn (17.29) using the explicit expressions for the Bessel functions of imaginary argument and obtain (Bordag et al. 1995a)

$$c_{11} = -20 \sum_{m,n=0}^{\infty} G_{mn}^{(1,1)} e^{-z_{nm}} \left(1 + z_{mn} + \frac{2}{5}z_{mn}^2 + \frac{1}{15}z_{mn}^3\right),$$

$$c_{12} = 60 \sum_{m,n=0}^{\infty} G_{mn}^{(1,2)} e^{-z_{nm}} \left(1 + z_{mn} + \frac{13}{30}z_{mn}^2 + \frac{1}{10}z_{mn}^3 + \frac{1}{90}z_{mn}^4\right),$$

$$c_{13} = -140 \sum_{m,n=0}^{\infty} G_{mn}^{(1,3)} e^{-z_{nm}} \left[1 + z_{mn}\Pi(z_{mn})\right], \qquad (17.31)$$

$$c_{22} = 210 \left\{ g_{00}^{(2)} h_{00}^{(2)} + \sum_{m,n=0}^{\infty} G_{mn}^{(2,2)} e^{-z_{nm}} \left[1 + z_{mn}\Pi(z_{mn})\right]\right\},$$

where

$$\Pi(z) = 1 + \frac{25}{56}z + \frac{19}{168}z^2 + \frac{1}{168}z^3 + \frac{1}{840}z^4. \qquad (17.32)$$

For roughness characterized by a lateral scale Λ (such as the period or correlation length), the main contribution to c_{kl} is given by the Fourier coefficients of the roughness functions $f_{1,2}$ and of their powers with $m, n \sim 2L/\Lambda$. For the harmonics with much larger values of m and n, the respective quantities $G_{mn}^{(i,k)}$ decrease rapidly. From the definition (17.30), we obtain the result that $z_{mn} \sim 2\pi a/\Lambda$. Under the condition $2\pi a/\Lambda \ll 1$, we can put $z_{mn} = 0$ in eqn (17.31) without loss of accuracy, and arrive at

$$c_{11} = -20 \sum_{m,n=0}^{\infty} G_{mn}^{(1,1)}, \quad c_{12} = 60 \sum_{m,n=0}^{\infty} G_{mn}^{(1,2)}, \qquad (17.33)$$

$$c_{13} = -140 \sum_{m,n=0}^{\infty} G_{mn}^{(1,3)}, \quad c_{22} = 210 \left[g_{00}^{(2)} h_{00}^{(2)} + \sum_{m,n=0}^{\infty} G_{mn}^{(2,2)} \right].$$

Taking into account eqn (17.30) for $G_{mn}^{(i,k)}$ and elementary properties of Fourier expansions, we can rewrite eqn (17.33) in the form

$$c_{11} = -20\langle f_1 f_2 \rangle, \quad c_{12} = 60\langle f_1 f_2^2 \rangle, \quad c_{13} = -140\langle f_1 f_2^3 \rangle, \qquad (17.34)$$
$$c_{22} = 210\langle f_1^2 f_2^2 \rangle, \quad c_{21} = -60\langle f_1^2 f_2 \rangle, \quad c_{31} = -140\langle f_1^3 f_2 \rangle.$$

Here and below, $f_1 = f_1(x,y)$ and $f_2 = f_2(x,y)$, i.e. both functions are already expressed in terms of the same coordinates.

Substituting eqns (17.24), (17.28), and (17.34) into eqn (17.23), we obtain the perturbative expansion for the Casimir pressure

$$P_R^{\text{add}}(a) = P^{\text{add}}(a)\kappa_P(a), \qquad (17.35)$$

where

$$\kappa_P(a) = 1 + 10 \left[\langle f_1^2 \rangle \frac{A_1^2}{a^2} - 2\langle f_1 f_2 \rangle \frac{A_1 A_2}{a^2} + \langle f_2^2 \rangle \frac{A_2^2}{a^2} \right] \qquad (17.36)$$

$$+ 20 \left[\langle f_1^3 \rangle \frac{A_1^3}{a^3} - 3\langle f_1^2 f_2 \rangle \frac{A_1^2 A_2}{a^3} + 3\langle f_1 f_2^2 \rangle \frac{A_1 A_2^2}{a^3} - \langle f_2^3 \rangle \frac{A_2^3}{a^3} \right]$$

$$+ 35 \left[\langle f_1^4 \rangle \frac{A_1^4}{a^4} - 4\langle f_1^3 f_2 \rangle \frac{A_1^3 A_2}{a^4} + 6\langle f_1^2 f_2^2 \rangle \frac{A_1^2 A_2^2}{a^4} - 4\langle f_1 f_2^3 \rangle \frac{A_1 A_2^3}{a^4} + \langle f_2^4 \rangle \frac{A_2^4}{a^4} \right].$$

We emphasize that eqns (17.35) and (17.36) are obtained under the condition $2\pi a/\Lambda \ll 1$. However, the PWS method is applicable in a wider range of separations. Thus, if the condition $2\pi a/\Lambda \ll 1$ is violated, one should use not eqn (17.35) but eqn (17.23) with the coefficients (17.31) (see Section 17.2.2). The mixed terms in eqn (17.36) which contain both f_1 and f_2 are evidence of interference-like behavior. For example, for $f_2 = \mp f_1$, eqns (17.35) and (17.36) result in

$$P_R^{\text{add}} = P^{\text{add}} \left[1 + 10\langle f_1^2 \rangle \frac{(A_1 \pm A_2)^2}{a^2} + 20\langle f_1^3 \rangle \frac{(A_1 \pm A_2)^3}{a^3} + 35\langle f_1^4 \rangle \frac{(A_1 \pm A_2)^4}{a^4} \right]. \qquad (17.37)$$

From eqns (17.35) and (17.36), we easily obtain the respective Casimir energy for two rough plates of area $S = (2L)^2$,

$$E_R^{\text{add}}(a) = -(2L)^2 \int_{-\infty}^{a} P_R^{\text{add}}(z)\,dz = E_{\text{pp}}^{\text{add}}(a)\kappa_E(a), \qquad (17.38)$$

where

$$\kappa_E(a) = 1 + 6 \left[\langle f_1^2 \rangle \frac{A_1^2}{a^2} - 2\langle f_1 f_2 \rangle \frac{A_1 A_2}{a^2} + \langle f_2^2 \rangle \frac{A_2^2}{a^2} \right]$$

$$+10\left[\langle f_1^3\rangle\frac{A_1^3}{a^3}-3\langle f_1^2 f_2\rangle\frac{A_1^2 A_2}{a^3}+3\langle f_1 f_2^2\rangle\frac{A_1 A_2^2}{a^3}-\langle f_2^3\rangle\frac{A_2^3}{a^3}\right] \quad (17.39)$$

$$+15\left[\langle f_1^4\rangle\frac{A_1^4}{a^4}-4\langle f_1^3 f_2\rangle\frac{A_1^3 A_2}{a^4}+6\langle f_1^2 f_2^2\rangle\frac{A_1^2 A_2^2}{a^4}-4\langle f_1 f_2^3\rangle\frac{A_1 A_2^3}{a^4}+\langle f_2^4\rangle\frac{A_2^4}{a^4}\right],$$

and where $E_{\text{pp}}^{\text{add}}(a)$ is the Casimir energy (6.43) obtained by additive summation for the configuration of two perfectly shaped semispaces.

Both the Casimir pressure (17.35) and the energy (17.38) can be normalized, as discussed in Section 17.1.1, in order to take approximate account of the effects of nonadditivity. Thus, for the Casimir energy per unit area $E_R(a) = E_R^{\text{tot}}(a)/S$, we divide both sides of eqn (17.38) by the normalization factor (17.10) and obtain

$$E_R(a) = \frac{E_R^{\text{add}}(a)}{S\tilde{K}_E(a)} = -\frac{\pi^2\hbar c}{720 a^3}\eta_E(a)\kappa_E(a), \quad (17.40)$$

where $\eta_E(a)$ and $\kappa_E(a)$ are defined in eqns (17.9) and (17.39), respectively. In this equation, the factor $\eta_E(a)$ takes into account the corrections to the ideal-metal Casimir energy due to the nonzero skin depth. The factor $\kappa_E(a)$ takes into account the corrections due to surface roughness.

To normalize the Casimir pressure, we divide both sides of eqn (17.35) by the factor (17.16). The result is

$$P_R(a) = \frac{P_R^{\text{add}}(a)}{\tilde{K}_P(a)} = -\frac{\pi^2\hbar c}{240 a^4}\eta_P(a)\kappa_P(a), \quad (17.41)$$

where $\eta_P(a)$ is defined in eqn (17.15) and $\kappa_P(a)$ in eqn (17.36). Similarly to the energy, eqn (17.41) represents the pressure as an ideal-metal result with two correction factors, one due to the nonzero skin depth and the other due to surface roughness.

An alternative procedure for obtaining the Casimir pressure using PWS is given by eqn (17.12). Calculating the negative derivative of both sides of eqn (17.40) with respect to a, we arrive at

$$\tilde{P}_R(a) = \frac{\pi^2\hbar c}{720}\frac{\partial}{\partial a}\left[\frac{1}{a^3}\eta_E(a)\kappa_E(a)\right]. \quad (17.42)$$

As can be seen in eqn (17.42), the pressure $\tilde{P}_R(a)$ is not represented as a product of two correction factors. However, numerically, $\tilde{P}_R(a) \approx P_R(a)$ for a wide range of parameters (see the next subsection), and for practical purposes it is more convenient to use $P_R(a)$ as defined in eqn (17.41).

17.1.3 Applications to large-scale roughness

The perturbative equations (17.40) and (17.41) obtained in the previous subsection can be applied to various kinds of large-scale roughness. The most typical

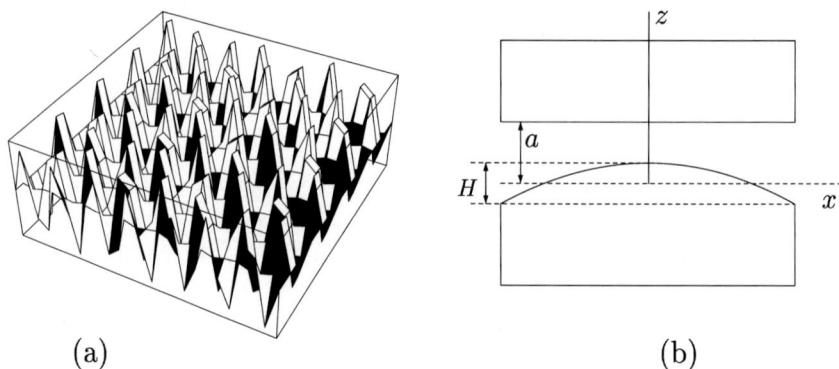

FIG. 17.2. (a) An example of surface roughness periodic in the x and y directions. (b) Side view of two finite-size cylinders where the top surface of the lower cylinder, of height H, is of paraboloidal shape.

example is roughness with a sinusoidal profile under the condition that the separation distance and the roughness period satisfy the condition $2\pi a \ll \Lambda$. For sinusoidal roughness, we also assume that the roughness period is much less than the plate size $2L$. In this case one can replace the integration from $-L$ to L over the plate in all quantities of the form $\langle f_1^k f_2^l \rangle$ with an integration over the period of the function. The case of nonperiodic functions $f_{1,2}$ will also be considered, and various definitions of the Casimir pressure in eqns (17.41) and (17.42) will be compared.

Let the surfaces of the plates be covered by roughness with a sinusoidal profile, such that functions $f_{1,2}$ in eqn (17.3) are given by (Klimchitskaya and Shabaeva 1996)

$$f_1(x,y) = \sin\alpha x \sin\beta y, \quad f_2(x,y) = \sin(\alpha x + \phi_1)\sin(\beta y + \phi_2). \quad (17.43)$$

These functions describe roughness with two different periods $\Lambda_x = 2\pi/\alpha$ and $\Lambda_y = 2\pi/\beta$ in the x and y directions, respectively. A typical example is shown in Fig. 17.2(a). The Casimir pressure can be calculated using eqn (17.41), applicable under the conditions $2\pi a \ll \Lambda_x, \Lambda_y$. Using the functions (17.43) and performing the integration from 0 to Λ_x with respect to x and from 0 to Λ_y with respect to y, we obtain the following average values:

$$\langle f_1^2 \rangle = \langle f_2^2 \rangle = \frac{1}{4}, \quad \langle f_1 f_2 \rangle = \frac{1}{4}\cos\phi_1 \cos\phi_2, \quad \langle f_i^3 \rangle = \langle f_i f_j^2 \rangle = 0,$$
$$\langle f_1^4 \rangle = \langle f_2^4 \rangle = \frac{9}{64}, \quad \langle f_1^2 f_2^2 \rangle = \frac{1}{64}(2+\cos\phi_1)(2+\cos\phi_2),$$
$$\langle f_1 f_2^3 \rangle = \langle f_1^3 f_2 \rangle = \frac{3}{64}\cos\phi_1 \cos\phi_2. \quad (17.44)$$

Now we substitute eqn (17.44) into eqns (17.35) and (17.36). The result is

$$P_R(a) = -\frac{\pi^2 \hbar c}{240 a^4} \eta_P(a) \left\{ 1 + \frac{5}{2} \left(\frac{A_1^2 + A_2^2}{a^2} - 2\cos\phi_1 \cos\phi_2 \frac{A_1 A_2}{a^2} \right) \right.$$
$$+ \frac{315}{64} \left[\frac{A_1^4 + A_2^4}{a^4} - \frac{4}{3} \cos\phi_1 \cos\phi_2 \frac{A_1^3 A_2 + A_1 A_2^3}{a^4} \right.$$
$$\left. \left. + \frac{2}{3}(2 + \cos\phi_1)(2 + \cos\phi_2) \frac{A_1^2 A_2^2}{a^4} \right] \right\}. \tag{17.45}$$

For equal roughness amplitudes $A_1 = A_2 = A$, eqn (17.45) simplifies to

$$P_R(a) = -\frac{\pi^2 \hbar c}{240 a^4} \eta_P(a) \left[1 + 5(1 - \cos\phi_1 \cos\phi_2) \frac{A^2}{a^2} \right. \tag{17.46}$$
$$\left. + \frac{105}{32}(7 - 3\cos\phi_1 \cos\phi_2 + 2\cos\phi_1 + 2\cos\phi_2) \frac{A^4}{a^4} \right].$$

All of the terms in eqns (17.45) and (17.46) containing $\cos\phi_i$ are of interference-like character. If $\cos\phi_1 = 0$ or, alternatively, $\cos\phi_2 = 0$, the interference is absent in the second perturbation order. Note that eqns (17.45) and (17.46) were obtained for roughness functions whose periods along each of the coordinates x and y on both plates are equal, i.e. $\Lambda_{x,1} = \Lambda_{x,2}$ and $\Lambda_{y,1} = \Lambda_{y,2}$. If these periods were different along at least one coordinate axis, i.e. $\Lambda_{x,1} \neq \Lambda_{x,2}$, all interference terms in eqns (17.45) and (17.46) would disappear. The presence of interference terms when the roughness functions have equal periods along the respective coordinate axes leads to the interesting phenomenon of the lateral Casimir force (Golestanian and Kardar 1997, 1998). To define the lateral force in a configuration of two plates with the roughness profiles (17.43), we first obtain the Casimir energy per unit area using eqns (17.40), (17.39), and (17.44):

$$E_R(a) = -\frac{\pi^2 \hbar c}{720 a^3} \eta_E(a) \left\{ 1 + \frac{3}{2} \left(\frac{A_1^2 + A_2^2}{a^2} - 2\cos\phi_1 \cos\phi_2 \frac{A_1 A_2}{a^2} \right) \right.$$
$$+ \frac{135}{64} \left[\frac{A_1^4 + A_2^4}{a^4} - \frac{4}{3} \cos\phi_1 \cos\phi_2 \frac{A_1^3 A_2 + A_1 A_2^3}{a^4} \right.$$
$$\left. \left. + \frac{2}{3}(2 + \cos\phi_1)(2 + \cos\phi_2) \frac{A_1^2 A_2^2}{a^4} \right] \right\}. \tag{17.47}$$

Let the phase shift ϕ_2 be fixed and let the Casimir energy $E_R(a)$ be considered as a function of $\phi_1 \equiv 2\pi x_0/\Lambda_x$. Then, from eqn (17.47), we derive the following expression for the lateral Casimir force:

$$F_R^{\text{lat}}(a) = -\frac{\partial E_R(a) S}{\partial x_0} = \frac{\pi^3 \hbar c S}{120 a^3 \Lambda_x} \eta_E(a) \sin\phi_1 \frac{A_1 A_2}{a^2} \tag{17.48}$$
$$\times \left\{ \cos\phi_2 + \frac{15}{32} \left[2\cos\phi_2 \frac{A_1^2 + A_2^2}{a^2} - (2 + \cos\phi_2) \frac{A_1 A_2}{a^2} \right] \right\}.$$

The lateral force is discussed in more detail in Section 17.5, devoted to the case of plates with uniaxial sinusoidal corrugations.

Now we consider the case of deviations of the plate surfaces from flatness with the same scale as the plate size. As a first example, let both plates be cylinders of some finite height, with the top surface of the lower cylinder being of paraboloidal shape [see Fig. 17.2(b)]. It is assumed that the height of the paraboloid is small, i.e. $H \ll a$. In this case the deviation amplitudes are $A_1 = H/2$ and $A_2 = 0$, and the functions describing the deviation from flatness in the cylindrical coordinates take the form

$$f_1(\rho, \varphi) = 1 - \frac{2\rho^2}{L^2}, \qquad f_2(\rho, \varphi) = 0. \tag{17.49}$$

Note that in cylindrical coordinates, the averaging defined in eqn (17.4) is understood as

$$\langle f(\rho, \varphi) \rangle = \frac{1}{\pi L^2} \int_0^{2\pi} d\varphi \int_0^L \rho \, d\rho \, f(\rho, \varphi). \tag{17.50}$$

Using the functions (17.49), the following average values can easily be calculated:

$$\langle f_1^2 \rangle = \frac{1}{3}, \qquad \langle f_1^3 \rangle = 0, \qquad \langle f_1^4 \rangle = \frac{1}{5}, \tag{17.51}$$

and all of the average values containing the function f_2 are equal to zero. Substituting eqn (17.51) into eqn (17.41), we obtain the Casimir pressure in the configuration of Fig. 17.2(b) (Bordag et al. 1995a),

$$P_R(a) = -\frac{\pi^2 \hbar c}{240 a^4} \eta_P(a) \left[1 + \frac{5}{6} \left(\frac{H}{a} \right)^2 + \frac{7}{16} \left(\frac{H}{a} \right)^4 \right]. \tag{17.52}$$

This result is in agreement with the corresponding result in Chapter 6 obtained by additive summation for a paraboloid of any height H. To see this, we can consider the limiting case of eqn (6.84) under the condition $H \ll a$ and take into account the fact that the separation distance in eqn (6.84) is defined as $a - H/2$, where a is defined as in eqn (17.52). The same Casimir pressure (17.52) is obtained if the convex paraboloid shown in Fig. 17.2(b) is replaced with a concave one described by the equation

$$f_1(\rho, \varphi) = -1 + \frac{2\rho^2}{L^2}. \tag{17.53}$$

If we consider the configurations with two convex or two concave paraboloidal surfaces for both plates with equal amplitudes $A_1 = A_2 = H/2$, the nonzero average values of the functions $f_{1,2}$ and their products and powers are

$$\langle f_1^2 \rangle = \langle f_2^2 \rangle = -\langle f_1 f_2 \rangle = \frac{1}{3}, \tag{17.54}$$

$$\langle f_1^4 \rangle = \langle f_2^4 \rangle = \langle f_1^2 f_2^2 \rangle = -\langle f_1^3 f_2 \rangle = -\langle f_1 f_2^3 \rangle = \frac{1}{5}.$$

Substituting these expressions into eqns (17.41) and (17.36), we obtain

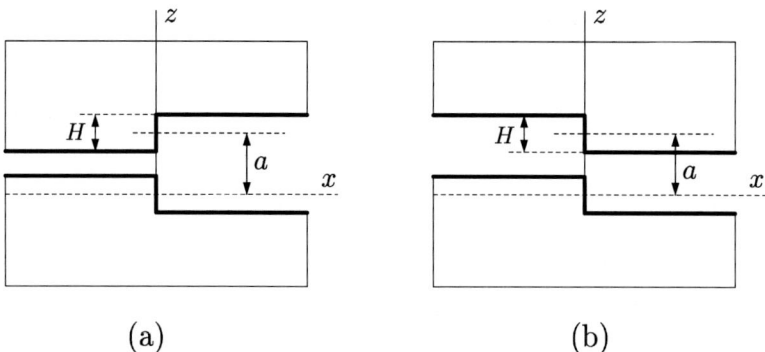

FIG. 17.3. Two parallel plates with a step of height H on their boundary surfaces. (a) The separations between the two halves of the plates are different. (b) The separations between the two halves of the plates are the same.

$$P_R(a) = -\frac{\pi^2 \hbar c}{240 a^4} \eta_P(a) \left[1 + \frac{10}{3} \left(\frac{H}{a}\right)^2 + 7 \left(\frac{H}{a}\right)^4 \right]. \qquad (17.55)$$

Note that in contrast to eqn (17.52), this Casimir pressure contains contributions from interference-like mixed terms.

Another example of a deviation from flatness is shown in Fig. 17.3(a), where the surfaces of both plates contain a step of height $H \ll a$ and width L. Here, the amplitudes of the functions describing the two surfaces are equal, i.e. $A_1 = A_2 = H/2$, and the functions themselves are given by

$$f_1(x,y) = \begin{cases} 1, & -L \le x < 0, \\ -1, & 0 \le x \le L, \end{cases} \quad f_2(x,y) = -f_1(x,y). \qquad (17.56)$$

Using eqn (17.37) with the first condition, we arrive at

$$P_R(a) = -\frac{\pi^2 \hbar c}{240 a^4} \eta_P(a) \left[1 + 10 \left(\frac{H}{a}\right)^2 + 35 \left(\frac{H}{a}\right)^4 \right]. \qquad (17.57)$$

For small H/a, this result coincides with half the sum of the exact Casimir pressures between two pairs of plates at separation distances $a_{1,2} = a \pm H$. For the similar configuration of two plates with steps on their surfaces shown in Fig. 17.3(b), we have $f_1 = f_2$. Using eqn (17.37) with the second condition, all corrections due to the presence of the steps vanish. As a result, the Casimir pressure coincides with that for two parallel plates with flat surfaces at a separation a.

In the above calculations, we have used the multiplicative expression (17.41) for the Casimir pressure $P_R(a)$ obtained within the PWS method. It is interesting to compare the results obtained with those found from the alternative expression for the Casimir pressure (17.42). To compare the two sets of results analytically, we shall consider metallic plates and describe the metal by means of the plasma model (13.1). In this case the functions $\eta_E(a)$ and $\eta_P(a)$ in eqns (17.40)–(17.42) can be represented in the following form:

$$\eta_E(a) = \sum_{i=0}^{4} \frac{d_i}{a^i}, \qquad \eta_P(a) = \sum_{i=0}^{4} \frac{3+i}{3}\frac{d_i}{a^i}, \qquad (17.58)$$

where the coefficients d_i are defined in eqn (13.8); for example, $d_0 = 1$, $d_1 = -4\delta_0$, $d_2 = 72\delta_0^2/5$, etc. where δ_0 is the skin depth. In a similar way, the functions $\kappa_E(a)$ and $\kappa_P(a)$ entering eqns (17.40)–(17.42) can be represented as

$$\kappa_E(a) = \sum_{k=0}^{4} \frac{h_k}{a^k}, \qquad \kappa_P(a) = \sum_{k=0}^{4} \frac{3+k}{3}\frac{h_k}{a^k}, \qquad (17.59)$$

where the coefficients h_k are defined in eqn (17.39).

Now we are in a position to estimate the difference between the Casimir pressures (17.41) and (17.42). For this purpose we substitute eqns (17.58) and (17.59) into eqns (17.41) and (17.42), and arrive at

$$\tilde{P}_R(a) - P_R(a) = \frac{\pi^2 \hbar c}{2160 a^4} \sum_{i=1}^{4} \sum_{k=1}^{4-i} ik \frac{d_i h_k}{a^{i+k}}. \qquad (17.60)$$

Taking into account the fact that, in accordance with eqn (17.39), $h_1 = 0$, we obtain

$$\tilde{P}_R(a) - P_R(a) = \frac{\pi^2 \hbar c}{2160 a^4} \left(2\frac{d_1 h_2}{a^3} + 3\frac{d_1 h_3}{a^4} + 4\frac{d_2 h_2}{a^4} \right), \qquad (17.61)$$

i.e. the difference under consideration is a quantity of the third order in the small parameters (of the first order in the relative skin depth and of the second order in the relative roughness amplitude). We note also that $h_2, d_2 > 0$, whereas $d_1 < 0$. This leads to a partial compensation of the terms contributing to eqn (17.61). As an example, we can estimate the relative contribution from the difference of pressures (17.61) to the Casimir pressure. For this purpose, we consider the case of sinusoidal roughness on gold plates described by eqn (17.43) with equal roughness amplitudes $A_1 = A_2 = 20\,\text{nm}$ and $\phi_1 = \phi_2 = \pi/2$. The Casimir pressure calculated using the multiplicative equation (17.41) is presented in eqn (17.46), where we put $\cos\phi_1 = \cos\phi_2 = 0$. At a separation distance $a = 200\,\text{nm}$, the relative difference of the pressures $[\tilde{P}_R(a) - P_R(a)]/|P_R(a)|$ is only 0.01%. At the same time, the relative contribution from the term of order $(A/a)^4$ to the Casimir pressure at $a = 200\,\text{nm}$ is about 22%. From this we can conclude that it may be meaningful to calculate the Casimir pressure taking account of the roughness correction up to the fourth order in A/a in the framework of the multiplicative approach. Further discussion of the various approaches to surface roughness is contained in Sections 17.2–17.4. In Part III of the book, these approaches are used for comparison between experiment and theory.

17.1.4 Perturbation theory for a sphere above a plate

The configuration of a sphere (or a spherical lens) of large radius of curvature R, in comparison with the separation distance to a plate, is very important for experiments on measuring the Casimir force. Because of this, we consider here the perturbative description of the roughness corrections to the Casimir force in this configuration. In fact, only a relatively thin spherical section of height h and diameter $2r$ contributes to the Casimir force. The parameters of the plate (thickness D and length $2L$) are the same as those shown in Fig. 17.1. Note that $h \sim D$ so that $a/h \ll A_i/a$.

The roughness of the plate is described by the function $z_1^{(s)}$ given in eqn (17.3). The rough surface of the spherical lens can be described in polar coordinates ρ and φ:

$$z_2^{(s)} = a + R - \sqrt{R^2 - \rho^2} + A_2 f_2(\rho, \varphi). \tag{17.62}$$

The perturbative calculation of the Casimir force acting between the rough surfaces of the sphere and plate is performed in the same way as for two plates in Section 17.1.2. The unnormalized energy of one atom of the sphere located at a point (ρ, φ, z_2) is given by

$$E_A^{\text{add}}(\rho, \varphi, z_2) = -BN_1 \int_{-L}^{L} dx_1 \int_{-L}^{L} dy_1 \int_{-D}^{z_1^{(s)}(x_1, y_1)} dz_1 \tag{17.63}$$

$$\times \left[(x_1 - \rho \cos \varphi)^2 + (y_1 - \rho \sin \varphi)^2 + (z_1 - z_2)^2\right]^{-7/2}.$$

This function can be expanded in powers of A_1/z_2 according to eqn (17.19), where (x_2, y_2, z_2) is now replaced with (ρ, φ, z_2) and

$$X = (x_1 - \rho \cos \varphi)^2 + (y_1 - \rho \sin \varphi)^2 + z_2^2. \tag{17.64}$$

The additive interaction energy between the sphere and the plate takes the form

$$E_R^{\text{add}}(a) = N_2 \int_0^{2\pi} d\varphi \int_0^r \rho\, d\rho \int_{z_2^{(s)}(\rho,\varphi)}^{h+a} dz_2\, E_A^{\text{add}}(\rho, \varphi, z_2). \tag{17.65}$$

Calculating the negative derivative of both sides of this equation, we obtain the unnormalized Casimir force acting between the sphere and the plate,

$$F_R^{\text{add}}(a) = -N_2 \int_0^{2\pi} d\varphi \int_0^r \rho\, d\rho\, E_A^{\text{add}}[\rho, \varphi, z_2^{(s)}(\rho, \varphi)], \tag{17.66}$$

where the function $z_2^{(s)}$ is defined in eqn (17.62).

Below, we restrict ourselves to second-order perturbation theory in each roughness amplitude. Then we keep only the first three terms on the right-hand side of eqn (17.19). Expanding the $E_A^{\text{add}}(\rho, \varphi, z_2)$ obtained up to the second power in the small parameter A_2/a, we get

$$F_R^{\text{add}}(a) = F_{\text{sp}}^{\text{add}}(a) \sum_{k=0}^{2} \sum_{l=0}^{2-k} C_{kl} \left(\frac{A_1}{a}\right)^k \left(\frac{A_2}{a}\right)^l. \quad (17.67)$$

Here, $F_{\text{sp}}^{\text{add}}(a)$ is the Casimir force between a perfectly shaped sphere and a plate (or semispace) calculated by integration of the interatomic Casimir–Polder potential (16.12). It is given in eqn (6.50), obtained under the condition $a \ll R$. From the definition of the functions describing the roughness, we have $C_{00} = 1$. The other coefficients in eqn (17.67) can be found in a manner analogous to those for rough plates. The results are (Klimchitskaya and Pavlov 1996)

$$C_{01} = -\frac{6}{\pi Ra} \int_0^{2\pi} d\varphi \int_0^\infty \rho \, d\rho \, f_2(\rho, \varphi) \left(1 + \frac{\rho^2}{2aR}\right)^{-5}, \quad (17.68)$$

$$C_{02} = \frac{15}{\pi Ra} \int_0^{2\pi} d\varphi \int_0^\infty \rho \, d\rho \, f_2^2(\rho, \varphi) \left(1 + \frac{\rho^2}{2aR}\right)^{-6},$$

$$C_{10} = \frac{15a^4}{\pi^2 R} \langle\langle f_1 Y^{-7} \rangle\rangle, \qquad C_{20} = \frac{105a^5}{4\pi^2 R^2} \langle\langle (2aR + \rho^2) f_1^2 Y^{-9} \rangle\rangle,$$

$$C_{20} = -\frac{105a^5}{4\pi^2 R^2} \langle\langle (2aR + \rho^2) f_1 f_2 Y^{-9} \rangle\rangle.$$

In these expressions, we have used the notation

$$Y = \left[(x_1 - \rho \cos\varphi)^2 + (y_1 - \rho \sin\varphi)^2 + \left(a + \frac{\rho^2}{2R}\right)^2\right]^{1/2} \quad (17.69)$$

and the following averaging procedure for functions depending on four variables:

$$\langle\langle \Phi(\rho, \varphi; x_1, y_1) \rangle\rangle = \int_0^{2\pi} d\varphi \int_0^\infty \rho \, d\rho \int_{-\infty}^\infty dx_1 \int_{-\infty}^\infty dy_1 \, \Phi(\rho, \varphi; x_1, y_1). \quad (17.70)$$

For the sake of brevity, it has also been assumed that $f_1 = f_1(x_1, y_1)$, $f_2 = f_2(\rho, \varphi)$.

The calculation of the expansion coefficients (17.68) can be performed by considering the function $f_1 = f_1(x_1, y_1)$ as a periodic one with a period $2L$ in both variables. The function $f_2(\rho, \varphi)$ is considered as a periodic function of ρ with a period r. By applying the properties of the respective Fourier expansions to large-scale roughness on the plate and the sphere with periods much larger than the separation distance but much smaller than $\sqrt{aR}$, we can express C_{kl} in terms of the average quantities $\langle f_1^k f_2^l \rangle$, where $f_1 = f_1(\rho\cos\varphi, \rho\sin\varphi)$ and

$f_2 = f_2(\rho, \varphi)$ (Klimchitskaya and Pavlov 1996). Then the Casimir force can be represented in the form (up to fourth-order perturbation theory)

$$F_R^{\text{add}}(a) = F_{\text{sp}}^{\text{add}}(a)\kappa_E(a), \tag{17.71}$$

where the factor $\kappa_E(a)$ coincides with that defined in eqn (17.39). Thus, direct application of perturbation theory to calculating the force in the configuration of a sphere above a plate, under certain conditions, leads to the same correction factor as for the energy in the configuration of two parallel plates.

Equation (17.71) can be normalized in order to take approximate account of the effects of nonadditivity, similarly to what was done in Section 17.1.2. For the configuration of a perfectly shaped sphere and a semispace, a very accurate expression for the force can be obtained using the PFA and eqn (17.8),

$$F(a) = 2\pi R E(a) = -\frac{\pi^3 \hbar c R}{360 a^3} \eta_E(a). \tag{17.72}$$

Note that here the PFA has been used to establish the connection between the case of a perfectly smooth large sphere near a plate and the case of two plates but not for the description of surface roughness. The normalization factor for the force acting between a sphere and a plate can then be defined from eqns (6.50) and (17.72) as

$$\tilde{K}_F(a) = \frac{F_{\text{sp}}^{\text{add}}(a)}{F(a)} = \frac{24 B N_1 N_2}{\pi \hbar c \eta_E(a)} = \tilde{K}_E(a). \tag{17.73}$$

It can be seen that this factor coincides with the factor (17.10) for the energy in the configuration of two parallel plates. By normalizing the force (17.71) with the help of the factor (17.73), we obtain

$$F_R(a) = \frac{F_R^{\text{add}}(a)}{\tilde{K}_F(a)} = -\frac{\pi^3 \hbar c R}{360 a^3} \eta_E(a) \kappa_E(a). \tag{17.74}$$

In similarity to eqn (17.41), the correction factor $\kappa_E(a)$ takes into account corrections due to surface roughness and the correction factor $\eta_E(a)$ due to the nonzero skin depth. In Sections 17.2, 17.3, and 17.5 the PWS method will be applied in several configurations and compared with the PFA and exact results where they are available.

Although eqn (17.74) was obtained for large-scale roughness, the respective periods were assumed to be much less than $\sqrt{aR}$. Now we consider deviations from a planar and a spherical shape described by the functions

$$f_1(x,y) = \cos\left(\frac{2\pi x}{\Lambda_x} + \phi_1\right), \quad f_2(\rho, \varphi) = \cos\left(\frac{2\pi \rho}{\Lambda_\rho} + \phi_2\right), \tag{17.75}$$

i.e. uniaxial corrugations on the plate and concentric corrugations on the sphere, respectively. Here, it is assumed that the periods are of the order of the size of the plate and the spherical lens: $\Lambda_x \sim L$, $\Lambda_\rho \sim r$. In this case eqn (17.74) with $\kappa_E(a)$ defined in eqn (17.39) is not applicable. Instead, direct calculation using eqn (17.68) results in (Klimchitskaya and Pavlov 1996)

$$F_R(a) = -\frac{\pi^3 \hbar c R}{360 a^3} \eta_E(a) \left[1 + 3\cos\phi_1 \frac{A_1}{a} - 3\cos\phi_2 \frac{A_2}{a} \right.$$
$$\left. + 3\left(\frac{A_1}{a}\right)^2 + 3\left(\frac{A_2}{a}\right)^2 - 12\cos\phi_1 \frac{A_1 A_2}{a^2} \right]. \quad (17.76)$$

The important difference between this expression and the cases considered above is the presence of first-order terms in the relative distortion amplitudes. The leading terms of the first order, which can arise when large-scale deviations from perfect geometry are present, depend on the phase shifts and change their magnitude and sign depending on the position of the sphere.

17.1.5 *Stochastic roughness with large correlation length*

In the preceding subsections, the roughness of the plate surfaces was described by regular functions f_1 and f_2. However, the roughness of real material surfaces is often extremely irregular and can be better modeled by stochastic functions. We consider once more the configuration shown in Fig. 17.1, where the surface roughness is now considered as stochastic. The surfaces of the plates are described by

$$z_1^{(s)} = \delta_1 f_1(x, y), \qquad z_2^{(s)} = a + \delta_2 f_2(x, y), \quad (17.77)$$

where the two stochastic functions $\{\delta_i f_i(x, y)\}$, $i = 1, 2$, have variances δ_i and zero mean values $\langle \delta_i f_i(x, y) \rangle_i = 0$. The symbol $\langle \ \rangle_i$ denotes an average over the ensemble of all particular realizations $\delta_i f_i(x, y)$ of the corresponding stochastic function. In the framework of PWS, the energy of an atom belonging to the upper plate can be represented by eqn (17.18), where the amplitude A_1 in the upper integration limit is replaced with the variance δ_1. The perturbative expansion of this energy in eqn (17.19) also preserves its validity with this replacement.

In our further discussion, we assume a normal distribution at each point of the surface and that the correlation lengths of the stochastic functions under consideration are sufficiently large in comparison with the separation distance between the plates. In this situation, any correlation of the roughness at two different points of the surface may occur for sufficiently remote points only, and the following mean values are obtained:

$$\langle f_i \rangle_i = \langle f_i^3 \rangle_i = 0, \quad \langle f_i^2 \rangle_i = 1, \quad \langle f_i^4 \rangle_i = 3. \quad (17.78)$$

By averaging eqn (17.19) over the realizations $\delta_1 f_1(x, y)$ of the stochastic function $\{\delta_1 f_1(x, y)\}$ with the use of eqn (17.78), we arrive at

$$\langle E_A^{\text{add}}(x_2, y_2, z_2) \rangle_1 = -BN_1 \left\{ \frac{\pi}{10 z_2^4} \right. \quad (17.79)$$

$$+ \int_{-L}^{L} dx_1 \int_{-L}^{L} dy_1 \left[\frac{7z_2^3}{2X^{9/2}} \frac{\delta_1^2}{z_2^2} + \frac{63 z_2^5}{8 X^{11/2}} \left(\frac{11 z_2^2}{X} - 3 \right) \frac{\delta_1^4}{z_2^4} \right] \bigg\}.$$

If the deviations of the plate surfaces from planes are described by stationary stochastic functions with $\delta_i = $ const, eqn (17.79) in the limiting case $L \to \infty$ leads to

$$\langle E_A^{\text{add}}(x_2, y_2, z_2) \rangle_1 = -B N_1 \frac{\pi}{10 z_2^4} \left(1 + 10 \frac{\delta_1^2}{z_2^2} + 105 \frac{\delta_1^4}{z_2^4} \right). \tag{17.80}$$

In order to obtain the Casimir energy in the configuration of two rough plates, we integrate eqn (17.80) over the volume V_2 of the upper plate, taking into account the fact that its boundary surface is described by the stochastic function $z_2^{(s)}$ defined in eqn (17.77). Then we calculate the mean value over all realizations of the stochastic function $\{\delta_2 f_2(x, y)\}$ and obtain

$$\langle \langle E_A^{\text{add}}(x_2, y_2, z_2) \rangle_1 \rangle_2 = N_2 \left\langle \int_{-L}^{L} dx_2 \int_{-L}^{L} dy_2 \int_{a + \delta_2 f_2(x_2, y_2)}^{a+D} dz_2 \, \langle E_A^{\text{add}}(x_2, y_2, z_2) \rangle_1 \right\rangle_2. \tag{17.81}$$

Calculating the negative derivative with respect to a on both sides of eqn (17.81) and dividing by $(2L)^2$ in accordance with eqn (17.21), we obtain the Casimir pressure

$$P_R^{\text{add}}(a) = \frac{N_2}{(2L)^2} \int_{-L}^{L} dx_2 \int_{-L}^{L} dy_2 \, \langle \langle E_A^{\text{add}}[x_2, y_2, a + \delta_2 f_2(x_2, y_2)] \rangle_1 \rangle_2. \tag{17.82}$$

Now we expand the quantity $\langle E_A^{\text{add}} \rangle_1$ in powers of the small parameter δ_2/a using eqn (17.80). Calculating the mean value over the realizations $\delta_2 f_2(x_2, y_2)$ of the stochastic function $\{\delta_2 f_2(x_2, y_2)\}$ with the help of eqn (17.78) and dividing the result obtained by the normalization factor (17.16), we find the Casimir pressure with partial account for the effects of nonadditivity (Bordag et al. 1995c),

$$P_R(a) = -\frac{\pi^2 \hbar c}{240 a^4} \eta_P(a) \left[1 + 10 \frac{\delta_1^2 + \delta_2^2}{a^2} + 105 \frac{(\delta_1^2 + \delta_2^2)^2}{a^4} \right]. \tag{17.83}$$

It can be seen that the correction factor due to stochastic roughness obtained here depends on $\delta_1^2 + \delta_2^2$ and does not depend on the correlation lengths of the stochastic functions. This is because we have considered stochastic roughness with a sufficiently large correlation length. The case where this condition is not satisfied is considered in Section 17.4. The above result, up to the second-order term, was obtained by van Bree et al. (1974). For a typical value of $\delta_{1,2}/a = 0.1$, the correction due to stochastic roughness given by eqn (17.83) is equal to 24% of the pressure between plates with plane surfaces, of which 4% results from the fourth-order contribution.

The same calculation procedure as presented above can be applied to the configuration of a sphere (or a spherical lens) of large radius above a plate with stochastic roughness. For stationary stochastic functions, the Casimir force acting between a sphere and a plate is given by

$$F_R(a) = -\frac{\pi^3 \hbar c}{360 a^3} \eta_E(a) \left[1 + 6\frac{\delta_1^2 + \delta_2^2}{a^2} + 45\frac{(\delta_1^2 + \delta_2^2)^2}{a^4} \right]. \quad (17.84)$$

The case of surface roughness described by nonstationary stochastic functions is more complicated. Some results for the Casimir pressure in this case have been obtained under the condition that the mean values are constant but the variances depend on coordinates, i.e. $\delta_i = \delta_i(x, y)$ (Bordag et al. 1995c, Klimchitskaya and Pavlov 1996).

17.2 The proximity force approximation for real rough bodies

The PFA introduced in Section 6.5 can be used for the case of two perfectly shaped macroscopic bodies, such as a sphere and a plate or a cylinder and a plate, but it can also be used for two bodies with surface roughness. In this case, the roughness profiles on two opposite plates are represented by a set of pairs of parallel plates at corresponding separations, and the sum of the results obtained for each pair is considered. This approximate approach to including surface roughness has some advantages in comparison with the PWS method. As shown in Section 17.1, the latter uses the retarded Casimir–Polder potential and takes into account the nonzero skin depth at some average separation a by means of a multiplicative procedure. In contrast, the PFA when applied to surface roughness includes corrections due to the nonzero skin depth in a nonmultiplicative way. The disadvantage of the PFA in studying roughness is that it is applicable to a narrower region of separation distances, as compared with the roughness period or the correlation length. Below, we compare the two methods and discuss all the pros and cons in more detail.

17.2.1 Geometrical averaging for regular roughness

The PFA has some specific characteristic features when applied to regular and stochastic roughness. Here, we consider the application of this approximate method to surface roughness described by the regular functions (17.3) in the configuration of two plates shown in Fig. 17.1. The case of stochastic roughness is left for the next subsection.

If the functional form of $f_1(x, y)$ and $f_2(x, y)$ is fixed, this means that there is a correlation between the roughness profiles of the two surfaces. The separation distance d between any two points on the surfaces with the coordinates (x, y) is determined from eqn (17.3):

$$d = a + A_2 f_2(x, y) - A_1 f_1(x, y). \quad (17.85)$$

By replacing the surface elements in the vicinities of the points $z_1^{(s)}(x, y)$ and $z_2^{(s)}(x, y)$ with small plane plates perpendicular to the z-axis, we can approximately express the corresponding Casimir energy per unit area using eqn (17.8), which is exact for two infinite parallel plates:

$$\Delta E(d) = -\frac{\pi^2 \hbar c}{720 d^3} \eta_E(d), \qquad (17.86)$$

where the function $\eta_E(a)$ is defined in eqn (17.9). To obtain the Casimir energy between two rough plates, we now integrate eqn (17.86) over the plate area and arrive at

$$\begin{aligned} E_R(a) = -\frac{\pi^2 \hbar c}{720(2L)^2} \int_{-L}^{L} dx \int_{-L}^{L} dy \frac{1}{[a + A_2 f_2(x,y) - A_1 f_1(x,y)]^3} \\ \times \eta_E[a + A_2 f_2(x,y) - A_1 f_1(x,y)]. \end{aligned} \qquad (17.87)$$

A similar expression for the Casimir pressure in the framework of the PFA is given by

$$\begin{aligned} P_R(a) = -\frac{\pi^2 \hbar c}{240(2L)^2} \int_{-L}^{L} dx \int_{-L}^{L} dy \frac{1}{[a + A_2 f_2(x,y) - A_1 f_1(x,y)]^4} \\ \times \eta_P[a + A_2 f_2(x,y) - A_1 f_1(x,y)]. \end{aligned} \qquad (17.88)$$

An important feature of eqns (17.87) and (17.88), in comparison with eqns (17.40) and (17.41) obtained using the PWS method, is that the argument of the functions η_E and η_P depends on the separation. In eqns (17.40) and (17.41), the argument of η_E and η_P is calculated between the zero levels of the roughness (see Fig. 17.1). Because of this, the PFA provides a more exact, nonmultiplicative, method of taking account of the nonzero skin depth than does the PWS method, expecially at short separations, where the effect of the skin depth contributes up to several tens of percent to the energy and pressure. At relatively large separations, the functions η_E and η_P in eqns (17.87) and (17.88) are practically constant. In this case one can factor them out from the integrals. Expanding the remaining power-type functions under the integrals in powers of A_1/a and A_2/a, we return to the multiplicative eqns (17.40) and (17.41) for the Casimir energy and pressure obtained using the PWS method. Thus, at relatively large separations, the PFA when applied to surface roughness leads to the same results as does the PWS method applied under the condition $2\pi a/\Lambda \ll 1$ [in this case one can use the simplified eqn (17.33) for the expansion coefficients instead of the more exact values given by eqn (17.31)]. Note, however, that the PWS method has a wider application region when eqn (17.31) is used to calculate the expansion coefficients c_{kl}. In the case of an ideal metal, $\eta_E = \eta_P = 1$ and the PWS method leads to the same result as does the PFA at all separations satisfying the condition $2\pi a \ll \Lambda$.

We now consider the case where the roughness of the lower plate is described by a function $f_1(x,y)$ whereas the surface of the upper plate is flat, i.e. $f_2(x,y) = 0$. Let the functions η_E and η_P in eqns (17.87) and (17.88) be constant. This happens at sufficiently large separations a for plates made of real materials or at any separation for plates made of an ideal metal or materials with frequency-independent dielectric permittivities $\varepsilon_0^{(1)}$ and $\varepsilon_0^{(2)}$. Then, for the configuration with one plane plate, all of the interference-type coefficients given by eqns (17.31) and (17.33) are exactly equal to zero and the PFA leads to precisely the same results as does the PWS method [we note that in this case eqns (17.31) and (17.33) are equivalent]. In the general situation, where the functions η_E and η_P depend strongly on the separation, the PFA provides a more exact, nonmultiplicative, combined method of taking account of surface roughness corrections and realistic dielectric properties of the material bodies. This conclusion is important at short separations, where both the roughness corrections and the corrections due to the nonzero skin depth (for metals) are rather large.

As an example, we shall apply eqn (17.88) to calculate the Casimir pressure in the configuration of two plates with steps shown in Fig. 17.3(a). This is a typical situation where the deviations from flatness of the two plates are correlated. Substituting eqn (17.56) into eqn (17.88), we obtain

$$P_R(a) = -\frac{\pi^2 \hbar c}{480} \left[\frac{\eta_P(a-H)}{(a-H)^4} + \frac{\eta_P(a+H)}{(a+H)^4} \right]. \tag{17.89}$$

For small $H \ll a$, we assume $\eta_P(a-H) \approx \eta_P(a+H) \approx \eta_P(a)$. Substituting this into eqn (17.89) and expanding in powers of the small parameter H/a, we return to eqn (17.57). Thus, the results obtained using the PFA and PWS methods are in agreement. For the configuration in Fig. 17.3(b), both methods lead to the same result as for two parallel plates with flat surfaces at a separation distance a.

It is interesting to compare the two methods at shorter separations and larger H, where η_P is separation-dependent and simple analytic estimates, such as those above, are not applicable. We have performed computations using eqns (17.41) and (17.88) for the plates with steps shown in Fig. 17.3(a) made of Au (the dielectric permittivity along the imaginary frequency axis is given by the solid line in Fig. 13.2). It is convenient to present the computational results as a correction factor $X_{R,P}(a) = P_R(a)/P_{IM}(a)$, which takes into account the combined effect of the surface roughness and the nonzero skin depth of gold. In the case of the PWS method, this correction factor is equal to $\kappa_P(a)\eta_P(a)$. For the PFA, $X_{R,P}(a)$ does not have a simple representation in terms of $\kappa_P(a)$ and $\eta_P(a)$. At $a = 100$ nm for a step of height $H = 2$ nm, the correction factor $X_{R,P}(a) = 0.476$ and 0.474 if the computations are performed in the framework of the PWS multiplicative method and the PFA, respectively. At $a = 200$ nm, even closer results $[X_{R,P}(a) = 0.631$ and 0.630, respectively] are obtained. This means that both methods work well. The situation changes, however, if a step of height $H = 20$ nm is considered. In this case, at $a = 100$ nm, the PWS method results in $X_{R,P}(a) = 0.690$ and the PFA results in $X_{R,P}(a) = 0.513$, i.e. a 34.5% difference. At $a = 200$ nm, the PWS leads to $X_{R,P}(a) = 0.695$ and the PFA to $X_{R,P}(a) = 0.643$ (8.1% difference). Thus, the computational results differ significantly. The reason is that at short separation distances (about 10 to 20 times the amplitude of the step-like roughness), the multiplicative approach incorporated into the PWS method introduces large errors into the corrections to the Casimir pressure due to the nonzero skin depth. At short separations, a combined description of the corrections due to surface roughness and nonzero skin depth is provided by the PFA in accordance with eqns (17.87) and (17.88).

17.2.2 Geometrical averaging for stochastic roughness

The metallic films and semiconductor surfaces used in measurements of the Casimir force are covered with irregularly distributed peaks and valleys. These irregularities on the two plates are not correlated. Because of this, they cannot be described by eqn (17.3) by regular functions f_1 and f_2. The description by means of stochastic functions in Section 17.1.5 is based on the PWS method. It includes the dielectric properties by means of the multipliers $\eta_E(a)$ and $\eta_P(a)$ calculated at some mean separation distance. As shown in Section 17.2.1, this makes the method inaccurate at short separations, where the role of roughness corrections is most important. Here, we present another approach for stochastic roughness based on the PFA, which is applicable at short separation distances (Klimchitskaya et al. 1999).

As is customary in the PFA, we replace the elements of the two rough surfaces at a separation z with two small parallel plates. The fraction of the total plate area S spaced at separations from z to $z + \Delta z$ is $\Delta S(z)/S = \rho(z)\Delta z$. Here, $\rho(z)$ is the distribution function for the probability that the separation between points $z_1^{(s)}$ and $z_2^{(s)}$ with fixed (x, y) on the rough surfaces in Fig. 17.1 lies in the interval $[z, z+\Delta z]$. Using the exact expression (17.8) for the Casimir energy per unit area between two perfectly shaped parallel plates at a separation z, we can approximately represent the interaction energy per unit area of two rough surfaces in the form

$$E_R(a) = -\frac{\pi^2 \hbar c}{720} \int_{z_{\min}}^{z_{\max}} \frac{dz}{z^3} \rho(z) \eta_E(z), \qquad (17.90)$$

where $z_{\min}$ and $z_{\max}$ are the minimal and maximal separations between the roughness profiles of the two surfaces. The respective Casimir pressure is given by

$$P_R(a) = -\frac{\pi^2 \hbar c}{240} \int_{z_{\min}}^{z_{\max}} \frac{dz}{z^4} \rho(z) \eta_P(z). \qquad (17.91)$$

Equations (17.90) and (17.91) are applicable when the separation distance between the plates is at least several times smaller than the correlation length of the stochastic roughness of the two surfaces (see Section 17.4).

The distribution function $\rho(z)$ can be determined experimentally from atomic force microscope images of plate surfaces (see Chapters 19–21). From these images, the roughness of the plates (1) and (2) can be characterized by the fractions of plate area $v_i^{(1)}$ and $v_k^{(2)}$ with heights $h_i^{(1)}$ and $h_k^{(2)}$, respectivly ($i = 1, 2, \ldots, N_1$ for the first plate and $k = 1, 2, \ldots, N_2$ for the second plate). These heights are measured from the absolute minimum level on each plate $h_1^{(1)} = h_1^{(2)} = 0$. The data for the roughness heights and the respective fractions of the plate area allow one to determine the zero levels $H_0^{(1,2)}$, relative to which the mean values of the roughness profiles are zero. The zero levels of the roughness are found from the equalities

$$\sum_{i=1}^{N_1}\left[H_0^{(1)} - h_i^{(1)}\right] v_i^{(1)} = 0, \qquad \sum_{k=1}^{N_2}\left[H_0^{(2)} - h_k^{(2)}\right] v_k^{(2)} = 0. \qquad (17.92)$$

From the definition of the roughness amplitudes in eqn (17.3), one obtains

$$A_i = \max\left[h_{N_i}^{(i)} - H_0^{(i)}, H_0^{(i)}\right], \quad i = 1, 2. \qquad (17.93)$$

Taking into account the fact that there is no correlation between the roughness profiles of the two plates, the probability that there is a peak $h_i^{(1)}$ on the first plate and a peak $h_k^{(2)}$ at the same surface position on the second is given by $v_i^{(1)} v_k^{(2)}$. The discrete versions of eqns (17.90) and (17.91) are the following (Decca et al. 2005b):

$$E_R(a) = \sum_{i=1}^{N_1}\sum_{k=1}^{N_2} v_i^{(1)} v_k^{(2)} E[a + H_0^{(1)} + H_0^{(2)} - h_i^{(1)} - h_k^{(2)}],$$

$$P_R(a) = \sum_{i=1}^{N_1}\sum_{k=1}^{N_2} v_i^{(1)} v_k^{(2)} P[a + H_0^{(1)} + H_0^{(2)} - h_i^{(1)} - h_k^{(2)}]. \qquad (17.94)$$

Here, a is the separation between the zero levels of the roughness for the two plates, and $E(a)$ and $P(a)$ are defined in eqns (17.8) and (17.15). As is shown in Chapters 19–21, eqn (17.94) is best suited for taking account of corrections to the Casimir energy per unit area and the Casimir pressure at separations below 200 nm, where the multiplicative approach is not exact enough.

The variances describing the stochastic roughness are found from the formulas

$$\delta_1^2 = \sum_{i=1}^{N_1}\left[H_0^{(1)} - h_i^{(1)}\right]^2 v_i^{(1)}, \qquad \delta_2^2 = \sum_{k=1}^{N_2}\left[H_0^{(2)} - h_k^{(2)}\right]^2 v_k^{(2)}. \qquad (17.95)$$

Equation (17.94) can be generalized in a simple way to the case of nonzero temperature by replacing the energies with the free energies. At larger separations, $E(z)$ and $P(z)$ in eqn (17.94) are almost constant within some separation interval determined by the surface roughness. One can then put these quantities in front of the summation sign. In this case eqn (17.94) leads to the same computational results as does the PWS method.

17.3 Nonparallel plates as large-scale roughness

A configuration which provides an opportunity to compare the efficiency of different approximation methods and also allows exact calculation in the ideal-metal

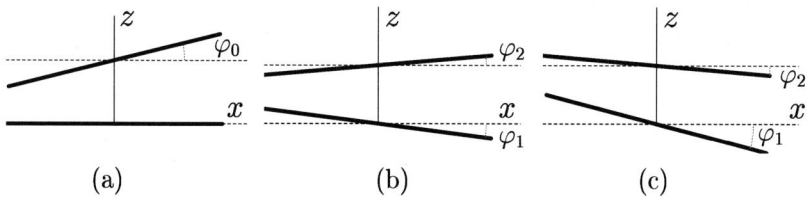

FIG. 17.4. A configuration of two plates inclined at a small angle φ_0 to one another. (a) One plate is parallel to the x-axis and the other is inclined. (b) One plate is inclined in the positive direction and the other in the negative direction. (c) Both plates are inclined in the negative direction.

case consists of two plates inclined at a small angle φ_0 to one another. The role of nonparallelity of the plates is important for measurements of the Casimir force (see Section 18.1.1). Small deviations from parallelity can be considered as a large-scale distortion with a characteristic length equal to the size of the plate. The same methods as were developed above to describe surface roughness can then be applied to calculate the corrections due to the nonparallelity of the plates. We start with an approach which uses the PWS method.

A configuration of two plates of area $2L \times 2L$ inclined at a small angle φ_0 to one another can be represented in three different ways, as shown in Fig. 17.4. In Fig. 17.4(a), the lower plate is not perturbed, and the upper plate is inclined at an angle φ_0 with respect to the lower one. In terms of eqn (17.3), the surfaces of the plates are given by the functions

$$f_1(x,y) = 0, \quad f_2(x,y) = \frac{x}{L}, \quad A_2 = \varphi_0 L. \tag{17.96}$$

Then, from eqns (17.36) and (17.41), we obtain

$$P_R(a) = -\frac{\pi^2 \hbar c}{240 a^4} \eta_P(a) \left(1 + 10\langle f_2^2 \rangle \frac{A_2^2}{a^2} - 20\langle f_2^3 \rangle \frac{A_2^3}{a^3} + 35\langle f_2^4 \rangle \frac{A_2^4}{a^4} \right). \tag{17.97}$$

Taking into account the fact that

$$\langle f_2^2 \rangle - \frac{1}{3}, \quad \langle f_2^3 \rangle - 0, \quad \langle f_2^4 \rangle - \frac{1}{5}, \tag{17.98}$$

the Casimir pressure between the nonparallel plates takes the form

$$P_R(a) = -\frac{\pi^2 \hbar c}{240 a^4} \eta_P(a) \left[1 + \frac{10}{3} \left(\frac{\varphi_0 L}{a} \right)^2 + 7 \left(\frac{\varphi_0 L}{a} \right)^4 \right]. \tag{17.99}$$

In this case, when only one of the two plates is perturbed, there are no interference-type contributions to the Casimir pressure.

The same configuration, however, can be represented in two other ways, as shown in Figs. 17.4(b) and 17.4(c). The analytic representation of the distortion functions related to Fig. 17.4(b) is given by

$$f_1(x,y) = -\frac{x}{L}, \quad f_2(x,y) = \frac{x}{L}, \quad A_1 = \varphi_1 L, \quad A_2 = \varphi_2 L, \tag{17.100}$$

where $\varphi_0 = \varphi_1 + \varphi_2$. In a similar way, Fig. 17.4(c) is described analytically by

$$f_1(x,y) = -\frac{x}{L}, \quad f_2(x,y) = -\frac{x}{L}, \quad A_1 = \varphi_1 L, \quad A_2 = \varphi_2 L, \tag{17.101}$$

where $\varphi_0 = \varphi_1 - \varphi_2$. It is evident that in both of the cases given by eqns (17.100) and (17.101), interference contributions to the Casimir pressure are present. To find the pressure, we can use eqn (17.37) [normalized by the factor (17.16)] because the equalities $f_2 = \mp f_1$ are satisfied. By calculating the mean values

$$\langle f_i^2 \rangle = \frac{1}{3}, \quad \langle f_i^3 \rangle = 0, \quad \langle f_i^4 \rangle = \frac{1}{5} \tag{17.102}$$

with the help of eqns (17.100) and (17.101) and substituting them into eqn (17.37), we return to eqn (17.99). Thus, in the configurations of Figs. 17.4(b) and 17.4(c), the interference terms contribute exactly the right amount to produce the result obtained when the interference is absent.

Now we consider the configuration of two plates inclined at a small angle φ_0 using the PFA. In the framework of this approximation, the Casimir pressure is given by eqn (17.88). Let the two plates be described by eqn (17.96) [see Fig. 17.4(a)]. Substituting eqn (17.96) into eqn (17.88) and making the change of variables $z = a + \varphi_0 x$, we obtain

$$P_R(a) = -\frac{\pi^2 \hbar c}{480 \varphi_0 L} \int_{a-\varphi_0 L}^{a+\varphi_0 L} \frac{dz}{z^4} \eta_P(z). \tag{17.103}$$

Now we represent the correction factor η_P due to the nonzero skin depth by using eqn (17.58), obtained in the framework of the plasma model. The integration with respect to z leads to

$$P_R(a) = -\frac{\pi^2 \hbar c}{1440 \varphi_0 L} \sum_{i=0}^{4} d_i \left[\frac{1}{(a - \varphi_0 L)^{3+i}} - \frac{1}{(a + \varphi_0 L)^{3+i}} \right]. \tag{17.104}$$

Expanding the right-hand side of eqn (17.104) in powers of the small parameter $\varphi_0 L/a$, we can represent the Casimir pressure in the form

$$P_R(a) = -\frac{\pi^2 \hbar c}{240 a^4} \left[1 + \frac{10}{3} \left(\frac{\varphi_0 L}{a} \right)^2 + 7 \left(\frac{\varphi_0 L}{a} \right)^4 + \frac{4 d_1}{3a} + \frac{20 d_1}{3a} \left(\frac{\varphi_0 L}{a} \right)^2 \right.$$
$$\left. + \frac{5 d_2}{3 a^2} + \frac{35 d_2}{3 a^2} \left(\frac{\varphi_0 L}{a} \right)^2 + 2 \frac{d_3}{a^3} + \frac{7 d_4}{3 a^4} \right]. \tag{17.105}$$

On comparing this with eqn (17.99), obtained using the PWS, it can be seen that for an ideal metal, where $\eta_P(a) = 1$ and $d_i = 0$ for $i \geq 1$, the two results coincide. Substituting eqn (17.58) for η_P into eqn (17.99), we find that it differs from eqn (17.105) by terms of the third order in the small parameters. Below, we estimate this difference numerically.

It is interesting to compare the approximate results (17.99) and (17.104) obtained using the PWS and PFA methods, respectively, with an exact result derived for ideal-metal plates inclined at an angle φ_0 using the configuration of a wedge. For this purpose, the exact energy density of the electromagnetic field $\varepsilon(\rho)$ inside the wedge must be integrated over the space between square plates of sides $2L = \rho_2 - \rho_1$. The energy obtained is given by

$$E^{\text{tot}}(a) = \int_0^{\varphi_0} d\varphi \int_0^{2L} dz \int_{\rho_1}^{\rho_2} \rho \, d\rho \, \varepsilon(\rho), \qquad (17.106)$$

where we have used the $\varepsilon(\rho)$ for a wedge defined in eqn (7.134) with an added factor $\hbar c$. Performing the integration in eqn (17.106) under the condition $\varphi_0 \ll 1$, we arrive at

$$E^{\text{tot}}(a) = -\frac{\pi^2 \hbar c L}{720 \varphi_0} \left[\frac{1}{(a - \varphi_0 L)^2} - \frac{1}{(a + \varphi_0 L)^2} \right]. \qquad (17.107)$$

Then the Casimir pressure is obtained from eqn (17.21):

$$P(a) = -\frac{\pi^2 \hbar c}{1440 \varphi_0 L} \left[\frac{1}{(a - \varphi_0 L)^3} - \frac{1}{(a + \varphi_0 L)^3} \right]. \qquad (17.108)$$

This result coincides with that in eqn (17.104) (PFA) when we put $d_0 = 1$ and $d_i = 0$ with $i \geq 1$, as is the case for an ideal metal. As can be seen from eqn (17.105), the result (17.108) also coincides with eqn (17.99) (PWS) after the expansion in the small parameter $\varphi_0 L/a$ is performed.

As an example, we have compared numerical results computed for Au plates inclined at an angle φ_0 such that $\varphi_0 L = 10$ nm by using eqn (17.99) (PWS) and eqn (17.103) (PFA). In both cases, the exact expression for the function $\eta_P(z)$ presented in eqn (17.16) was used in the computations. At a separation distance $a = 100$ nm, the correction factor $X_{\text{R},P}(a)$ to the ideal-metal pressure P_{IM} between parallel plates due to the nonzero skin depth and the tilt, computed using the PWS, was equal to 0.4899. The correction due to the tilt alone was equal to 1.034. If the PFA was used, $X_{\text{R},P}(a) = 0.4866$ and a value of 1.027 for the correction due to the tilt was obtained. The relative difference between the magnitudes of $X_{\text{R},P}(a)$ computed using the two methods was only 0.68%. At a separation $a = 200$ nm, we obtained $X_{\text{R},P}(a) = 0.6355$ (PWS) and $X_{\text{R},P}(a) = 0.6346$ (PFA), i.e. only a 0.14% difference. The correction due to the tilt varied from 1.008 (PWS) to 1.007 (PFA). Thus, for small tilt angles, the two methods lead to almost coincident results for real materials.

17.4 Various approaches for short-scale roughness

For roughness characterized by a small period or a small correlation length (not much less than the separation distance), the dependence of the roughness correction on the separation may become more involved owing to the effects of nonadditivity. The reflection of electromagnetic waves from short-scale distortions of the surface leads to diffraction-type and correlation effects which cannot be described adequately by the summation of interatomic potentials or by the replacement of the roughness profiles with a set of small, plane parallel plates. However, with decreasing roughness period or correlation length, the limits of applicability of the PWS method and the PFA are different. Below, we discuss how these limits can be estimated and what alternative methods can be applied for the description of short-scale roughness in the context of the Casimir effect.

In the case of an ideal metal, the path-integral approach allows a complete calculation of the roughness corrections to the second perturbation order in the roughness amplitudes (see Section 10.5). For uniaxial roughness on one plate (with the second plate perfectly smooth) described by the regular functions

$$f_1(x,y) = \cos\frac{2\pi x}{\Lambda_x}, \qquad f_2(x,y) = 0, \tag{17.109}$$

the roughness correction has been investigated analytically for any separation and roughness period (Emig et al. 2003). Thus, in the limiting case $a \ll \Lambda_x$, the Casimir energy per unit area is given by

$$E_R(a) = -\frac{\pi^2 \hbar c}{720 a^3}\left[1 + 3\frac{A_1^2}{a^2} + O\left(\frac{A_1^3}{a^3}\right)\right]. \tag{17.110}$$

This is in agreement with eqn (17.40), obtained using PWS, if one substitutes $A_2 = 0$ and $\langle f_1^2 \rangle = 1/2$ in eqn (17.39). However, in the opposite limiting case $a \gg \Lambda_x$, a rather different behavior of the Casimir energy follows from the path-integral approach (Emig et al. 2003):

$$E_R(a) = -\frac{\pi^2 \hbar c}{720 a^3}\left[1 + 2\pi\frac{A_1^2}{\Lambda_x a} + O\left(\frac{A_1^3}{a^3}\right)\right]. \tag{17.111}$$

This behavior cannot be obtained by the PWS method, which does not take into account the correlations between different points on the plates. The limiting case (17.111) was first derived by Karepanov et al. (1987). Novikov et al. (1990a, 1990b, 1992a, 1992b) developed a perturbation theory for the electromagnetic Green's function in terms of small roughness amplitudes, applicable to media with a frequency-dependent dielectric permittivity. The result (17.111) was rederived by Novikov et al. (1992b). Note that if both plates have roughness, eqn (17.40) is applicable under the more stringent condition $2\pi a \ll \Lambda_x$.

The validity limits of the PFA, as applied to surface roughness, have been investigated within the scattering approach for real metals described by the plasma model, using a second-order perturbation theory in the roughness amplitudes (Maia Neto et al. 2005). This approach describes roughness by stochastic functions $A_1 f_1(\boldsymbol{r})$ and $A_2 f_2(\boldsymbol{r})$ defined as in eqns (17.3) and (17.4), where $\boldsymbol{r} = (x, y)$, and assumes that the roughnesses of the two surfaces are statistically independent. It is also assumed that the roughness amplitudes are the smallest parameters, much less than the separation distance a, the plasma wavelength λ_p, and the correlation length Λ_c. The correlation length satisfies the condition $\Lambda_\text{c} \ll \sqrt{S}$, where S is the plate area. From the translational symmetry in the (x, y) plane, the correlation function of each plate depends only on $\boldsymbol{r} - \boldsymbol{r}'$:

$$C_i(\boldsymbol{r}, \boldsymbol{r}') = \text{corr}[A_i f_i(\boldsymbol{r}), A_i f_i(\boldsymbol{r}')] = C_i(\boldsymbol{r} - \boldsymbol{r}', 0), \quad i = 1, 2. \qquad (17.112)$$

From this, the correlation function for coincident arguments is expressed as a mean value,

$$C_i(\boldsymbol{r}, \boldsymbol{r}) = C_i(0, 0) = \left\langle [A_i f_i(\boldsymbol{r})]^2 \right\rangle. \qquad (17.113)$$

The roughness spectrum is defined as

$$\sigma_{ii}(\boldsymbol{k}) = \int d\boldsymbol{r}\, e^{-i\boldsymbol{k} \cdot \boldsymbol{r}} C_i(\boldsymbol{r}, 0), \qquad (17.114)$$

where $\boldsymbol{k} = (k_x, k_y)$ is the two-dimensional wave vector.

The Casimir energy per unit area for the rough plates can be represented in the form (Maia Neto et al. 2005)

$$E_\text{R}(a) = E(a) + \int \frac{d\boldsymbol{k}}{(2\pi)^2} G(\boldsymbol{k}) \sigma(\boldsymbol{k}), \qquad (17.115)$$

where $E(a)$ is the exact Casimir energy per unit area for flat plates defined in eqn (17.8), $\sigma(\boldsymbol{k}) \equiv \sigma_{11}(\boldsymbol{k}) + \sigma_{22}(\boldsymbol{k})$, and $G(\boldsymbol{k})$ is the roughness response function. Under the above conditions, this function can be expressed in a second-order perturbation theory in terms of the relative roughness amplitudes A_1/a and A_2/a through the specular and nonspecular reflection coefficients (Maia Neto et al. 2005).

Now we consider the connection between the Casimir energy per unit area in eqn (17.115) and in eqn (17.87), obtained with the PFA. For this purpose, we assume that the roughness profiles are sufficiently smooth that with the use of eqns (17.113) and (17.114), the roughness spectrum can be represented as

$$\sigma_{ii}(\boldsymbol{k}) \approx (2\pi)^2 \left\langle [A_i f_i(\boldsymbol{r})]^2 \right\rangle \delta^{(2)}(\boldsymbol{k}). \qquad (17.116)$$

Substituting this in eqn (17.115), we arrive at

$$E_\text{R}(a) \approx E(a) + G(0) \left\langle [A_1 f_1(\boldsymbol{r})]^2 + [A_2 f_2(\boldsymbol{r})]^2 \right\rangle. \qquad (17.117)$$

The limiting value of the response function $G(\boldsymbol{k})$ when $\boldsymbol{k} \to 0$ in the second-order perturbation theory satisfies the condition (Maia Neto et al. 2005)

$$G(0) = \frac{1}{2} \frac{\partial^2 E(a)}{\partial a^2}. \qquad (17.118)$$

Substituting this in eqn (17.117), we get

$$E_R(a) \approx E(a) + \frac{1}{2} \frac{\partial^2 E(a)}{\partial a^2} \left\langle [A_1 f_1(\mathbf{r})]^2 + [A_2 f_2(\mathbf{r})]^2 \right\rangle. \qquad (17.119)$$

On the other hand, the Casimir energy per unit area for rough surfaces in the PFA is given by eqn (17.87). This equation can be rearranged as

$$E_R(a) = \frac{1}{(2L)^2} \int_{-L}^{L} dx \int_{-L}^{L} dy\, E[a + A_2 f_2(x, y) - A_1 f_1(x, y)]. \qquad (17.120)$$

Here, the Casimir energy per unit area for the flat plates has the form

$$E(a) = \int_0^\infty k_\perp dk_\perp \int_0^\infty d\xi\, \Phi(k_\perp, \xi, e^{-2aq}), \qquad (17.121)$$

where the explicit expression for the function Φ is contained in eqn (12.30). Substituting eqn (17.121) into eqn (17.120) and expanding up to the fourth order in the small parameter $A_2 f_2 - A_1 f_1$, we get

$$E_R(a) = \frac{1}{(2L)^2} \int_{-L}^{L} dx \int_{-L}^{L} dy \int_0^\infty k_\perp dk_\perp \int_0^\infty d\xi\, \Phi\!\left[k_\perp, \xi, e^{-2(a + A_2 f_2 - A_1 f_1)q}\right]$$

$$\approx \frac{1}{(2L)^2} \int_{-L}^{L} dx \int_{-L}^{L} dy \int_0^\infty k_\perp dk_\perp \int_0^\infty d\xi \left[\sum_{n=0}^{4} \frac{\Phi^{(n)}(k_\perp, \xi, e^{-2aq})}{n!} (A_2 f_2 - A_1 f_1)^n \right], \qquad (17.122)$$

where (n) denotes the order of differentiation with respect to a. By changing the order of integrations and using the definitions of the mean values in eqn (17.4), eqn (17.122) can be rearranged to

$$\begin{aligned}
E_R(a) =\; & E(a) + \frac{1}{2} \frac{\partial^2 E(a)}{\partial a^2} \left(A_1^2 \langle f_1^2 \rangle - 2 A_1 A_2 \langle f_1 f_2 \rangle + A_2^2 \langle f_2^2 \rangle \right) \\
& - \frac{1}{6} \frac{\partial^3 E(a)}{\partial a^3} \left(A_1^3 \langle f_1^3 \rangle - 3 A_1^2 A_2 \langle f_1^2 f_2 \rangle + 3 A_1 A_2^2 \langle f_1 f_2^2 \rangle - A_2^3 \langle f_2^3 \rangle \right) \\
& + \frac{1}{24} \frac{\partial^4 E(a)}{\partial a^4} \left(A_1^4 \langle f_1^4 \rangle - 4 A_1^3 A_2 \langle f_1^3 f_2 \rangle + 6 A_1^2 A_2^2 \langle f_1^2 f_2^2 \rangle \right. \\
& \left. - 4 A_1 A_2^3 \langle f_1 f_2^3 \rangle + A_2^4 \langle f_2^4 \rangle \right). \qquad (17.123)
\end{aligned}$$

This representation of the energy per unit area for rough surfaces in the framework of the PFA, if extended to higher orders, is equivalent to the representation (17.87), but is not as convenient for applications. Note that in the case of ideal-metal plates, where $E(a) = E_{\mathrm{IM}}(a) = -C_E/a^3$, all derivatives are easily calculated and eqn (17.123) leads to the result in eqns (17.39) and (17.40) with $\eta_E(a) = 1$, obtained with the PWS method.

For stochastic roughness, $\langle f_1 f_2 \rangle = 0$. In this case eqn (17.119), obtained within the scattering approach in the second-order perturbation theory for smooth roughness profiles, coincides with eqn (17.123) if one neglects terms of the third and fourth orders. In this simplified form, eqn (17.123) was derived by Genet et al. (2003a). Then the deviation of the roughness response function $G(k)$ from $G(0)$ can be used as a measure of the disagreement between the second-order scattering approach and the second-order PFA (Maia Neto et al. 2005). Thus, it was found that for Au plates ($\lambda_p = 137$ nm) at a separation $a = 100$ nm, $G(k) \approx G(0)$ for $k \leq 0.02$ nm^{-1}. On the other hand, at a separation $a = 200$ nm and $k = 0.02$ nm^{-1}, $G(k)/G(0) \approx 1.6$. In the first case one would expect that for roughness with a sufficiently large correlation length, the second-order PFA would lead to the same results for the roughness correction as does the second-order scattering approach, whereas at larger distances, as in the second case, deviations between the two approaches are possible. This was confirmed by direct computations (Maia Neto et al. 2005). For a Gaussian spectrum, it was shown that the second-order PFA was in agreement with the second-order scattering approach if $a \leq 100$ nm and the correlation length $\Lambda_c > 150$ nm. Thus, the PFA appears to be applicable not under the general requirement $a \ll \Lambda_c$ but under the much less stringent requirement $a < 2\Lambda_c/3$. On the other hand, at a separation $a = 300$ nm (two times larger than the correlation length), the second-order scattering result for the roughness correction is approximately 50% larger than that given by the second-order PFA.

Thus, the scattering approach provides an opportunity to determine the accuracy of the second-order PFA. This approximation has been confirmed to be accurate at sufficiently small separations compared with the correlation length, which is the region where the effect of surface roughness is very important. However, in its present form the scattering approach is not preferred for comparison between experiment and theory. The reason is that at sufficiently large separations, where this approach provides a more accurate description of the effect of roughness than does the PFA, the total magnitude of the roughness correction is much less than a fraction of a percent. As a result, in this region the roughness correction amounts to very little and can be neglected. At relatively small separations of around 100 nm, where surface roughness plays an important role and must be taken into account, both the simple plasma model description of the dielectric properties and the second-order perturbation theory employed by the scattering approach are not accurate enough. At these separations, the complete (higher-order) PFA, taking into account the dielectric properties by means of tabulated data or the generalized plasma-like model, is both more accurate and convenient for practical computations (see Chapters 19–21).

17.5 Sinusoidally corrugated surfaces

The configuration of parallel plates with uniaxial sinusoidal corrugations is of special interest owing to the possibility of a nontrivial interplay between the geometry and the material properties. This configuration gives rise to both a normal and a lateral Casimir force. The latter is a function of the phase shift between the corrugations of the two plates. The first experimental demonstration of the lateral Casimir force (Chen et al. 2002a, 2002b) was performed using a configuration of a corrugated plate and a sphere and was related to the case of two corrugated plates by means of the PFA (see Section 21.2). The configuration of two plates made of real materials with periodically corrugated surfaces is an interesting subject from a theoretical point of view because the size of the corrugation amplitudes relative to the separation can be tailored. However, this creates the problem of how to simultaneously take into account both the dielectric properties of the plate materials and the deformed geometry with sufficient accuracy. As shown below, the PWS method is better suited for complicated geometrical shapes, whereas the PFA is more accurate when realistic dielectric properties are to be included, and the exact methods are presently soluble only in second-order perturbation theory in relation to the corrugation amplitudes. This problem has not been completely solved to date. In this section, we present the results obtained with an emphasis on those useful in the comparison between experiment and theory.

17.5.1 The Casimir energy and pressure

We consider the configuration of two parallel plates shown in Fig. 17.1. Now the surfaces of the two plates are covered with uniaxial sinusoidal corrugations of equal periods but different amplitudes, as shown in Fig. 17.5. These corrugations are described by eqn (17.3) with the following functions f_1 and f_2:

$$f_1(x,y) = \sin \alpha x, \qquad f_2(x,y) = \sin(\alpha x + \phi). \qquad (17.124)$$

Here, $\alpha = 2\pi/\Lambda$ and $\phi = 2\pi x_0/\Lambda$, where Λ is the corrugation period and x_0 is the shift of the corrugations on the upper plate in relation to those on the lower plate. We start with a consideration of separation distances satisfying the condition $2\pi a \ll \Lambda$. In this case the Casimir energy per unit area of the corrugated surfaces is given by eqns (17.39) and (17.40), derived within the PWS method. Below, we shall mostly discuss the energy rather than the normal force because the former is the quantity needed to calculate the lateral Casimir force. The nonzero mean values of the functions $f_{1,2}$, their products, and powers are

$$\langle f_i^2 \rangle = \frac{1}{2}, \qquad \langle f_1 f_2 \rangle = \frac{1}{2} \cos \phi, \qquad \langle f_i^4 \rangle = \frac{3}{8},$$

$$\langle f_i f_j^3 \rangle = \frac{3}{8} \cos \phi, \qquad \langle f_1^2 f_2^2 \rangle = \frac{1}{8}(2 + \cos 2\phi). \qquad (17.125)$$

Substituting eqn (17.125) into eqn (17.39), we obtain the correction factor to the Casimir energy due to the corrugations,

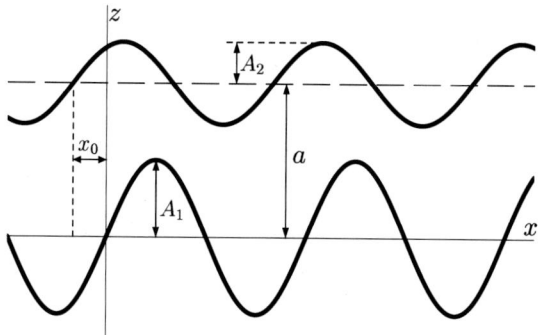

FIG. 17.5. Uniaxial sinusoidal corrugations of equal periods Λ but different amplitudes and with a phase shift $\phi = 2\pi x_0/\Lambda$ between the two corrugated plates.

$$\kappa_E(a) = 1 + 3\left[\frac{A_1^2 + A_2^2}{a^2} - 2\cos\phi\frac{A_1 A_2}{a^2}\right] \qquad (17.126)$$
$$+ \frac{45}{8}\left[\frac{A_1^4 + A_2^4}{a^4} - 4\cos\phi\frac{A_1 A_2^3 + A_1^3 A_2}{a^4} + 2(2+\cos 2\phi)\frac{A_1^2 A_2^2}{a^4}\right].$$

For ideal-metal plates, the correction factor due to the nonzero skin depth, η_E, is equal to unity, and exactly the same Casimir energy per unit area as in eqns (17.40) and (17.126) is obtained from eqn (17.87), derived using the PFA. Thus, in this case the two methods lead to the same results provided the separation distance satisfies the condition $2\pi a \ll \Lambda$.

It is interesting to determine the role of the fourth-order terms in eqn (17.126) for different corrugation amplitudes and phase shifts. For phase shifts $\phi = 0$ and π between the corrugations, eqn (17.126) results in

$$\kappa_E(a) = 1 + 3\frac{(A_1 \mp A_2)^2}{a^2} + \frac{45}{8}\frac{(A_1 \mp A_2)^4}{a^4}. \qquad (17.127)$$

Thus, in the case $\phi = 0$ and for equal amplitudes, i.e. $A_1 = A_2$, we get $\kappa_E^{(2)} = \kappa_E^{(4)} = 1$, where the superscript in brackets indicates in what perturbation order the calculation has been done. For $\phi = \pi$, $a = 200$ nm, and $A_1 = A_2 = 50$ nm, the magnitudes of κ_E computed up to the second and fourth perturbation orders are different: $\kappa_E^{(2)} = 1.750$ and $\kappa_E^{(4)} = 2.102$. For $\phi = \pi/2$, eqn (17.126) takes the form

$$\kappa_E(a) = 1 + 3\frac{A_1^2 + A_2^2}{a^2} + \frac{45}{8}\frac{(A_1^2 + A_2^2)^2}{a^4}. \qquad (17.128)$$

Then, with the same values of the separation and of the amplitudes, $\kappa_E^{(2)} = 1.375$ and $\kappa_E^{(4)} = 1.463$. It can be seen that for the equal amplitudes considered, the fourth perturbation order contributes up to 17% of the total result.

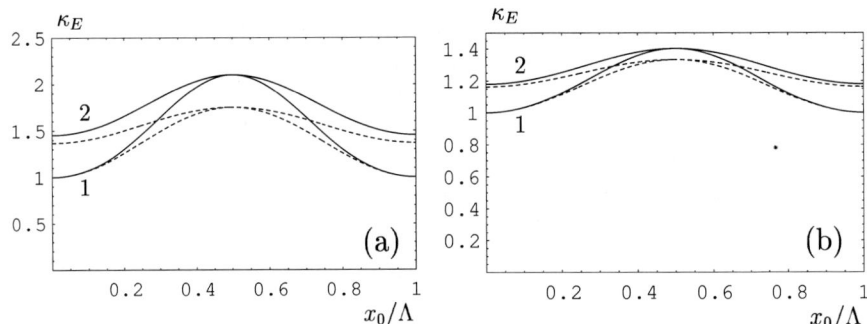

FIG. 17.6. The correction factor to the Casimir energy per unit area for plates due to sinusoidal corrugations with amplitudes $A_1 = A_2 = 50\,\text{nm}$ (solid and dashed lines labeled 1) and $A_1 = 85\,\text{nm}$, $A_2 = 15\,\text{nm}$ (solid and dashed lines labeled 2), versus x_0/Λ. The solid and dashed lines were computed up to the fourth- and second-order terms, respectively, in the corrugation amplitudes. The separation between the plates is (a) $a = 200\,\text{nm}$ and (b) $a = 300\,\text{nm}$.

Another example important for the consideration of experiments is the case of different amplitudes, for example $A_1 = 85\,\text{nm}$ and $A_2 = 15\,\text{nm}$, with the same sum as above, equal to $a/2 = 100\,\text{nm}$. In this case the following results are obtained: $\kappa_E^{(2)} = 1.368$, $\kappa_E^{(4)} = 1.452$ ($\phi = 0$), $\kappa_E^{(2)} = 1.559$, $\kappa_E^{(4)} = 1.754$ ($\phi = \pi/2$), and $\kappa_E^{(2)} = 1.750$, $\kappa_E^{(4)} = 2.102$ ($\phi = \pi$). The latter magnitudes are the same as those for equal amplitudes, in agreement with eqn (17.127). The largest contribution from the fourth perturbation order, equal to 16.7%, takes place at $\phi = \pi$. Computational results for arbitrary phase shifts obtained using eqn (17.126) are presented in Fig. 17.6(a), where the correction factor κ_E is plotted as a function of x_0/Λ at $a = 200\,\text{nm}$. The solid and dashed lines labeled 1 are related to the case of equal amplitudes $A_1 = A_2 = 50\,\text{nm}$. The solid and dashed lines labeled 2 are related to the case of different amplitudes $A_1 = 85\,\text{nm}$, $A_2 = 15\,\text{nm}$. In both cases the dashed lines were computed up to the second perturbation order and the solid lines up to the fourth perturbation order. As can be seen in Fig. 17.6(a), the relative contribution of the fourth order strongly depends on the phase shift and reaches its maximum value at $x_0/\Lambda = 1/2$ or, equivalently, $\phi = \pi$.

With increasing separation distance, the relative role of the fourth perturbation order decreases. As an example, the correction factor κ_E is plotted in Fig. 17.6(b) as a function of x_0/Λ for a separation $a = 300\,\text{nm}$ between the plates. The meanings of the solid and dashed lines are the same as those in Fig. 17.6(a). At $x_0/\Lambda = 1$ ($\phi = \pi$), where the relative contribution of the fourth perturbation order is the largest, $\kappa_E^{(2)} = 1.333$ and $\kappa_E^{(4)} = 1.403$. This leads to a 5% contribution from the fourth-order correction in the value of κ_E.

Now we compare computational results obtained with inclusion of the nonzero skin depth using the PFA (17.87) and the PWS method (17.40) (the latter up

TABLE 17.1. The correction factor due to the nonzero skin depth for corrugated surfaces with $A_1 = A_2 = 50\,\text{nm}$.

ϕ	$a = 200\,\text{nm}$			$a = 300\,\text{nm}$		
	PFA	PWS$_{(4)}$	PWS$_{(2)}$	PFA	PWS$_{(4)}$	PWS$_{(2)}$
0	0.6980	0.6980	0.6980	0.7747	0.7747	0.7747
$\pi/2$	0.9676	1.021	0.9597	0.8980	0.9173	0.9038
π	1.3925	1.4669	1.2215	1.0474	1.0867	1.0329

to the second- and fourth-order perturbative expansions in the corrugation amplitudes). The correction factor due to the nonzero skin depth η_E was computed by use of eqn (17.9) using the dielectric permittivity of Au along the imaginary frequency axis (the solid line in Fig. 13.2). Within the PFA, the factor η_E was integrated in agreement with eqn (17.87). When the PWS is used, one must multiply the result for κ_E by the value of $\eta_E(a)$ at the mean separation. The computational results for the correction factor $X_{R,E}(a) = E_R(a)/E_{IM}(a)$ in the case of corrugation amplitudes $A_1 = A_2 = 50\,\text{nm}$ are presented in Table 17.1 and those in the case of $A_1 = 85\,\text{nm}$, $A_2 = 15\,\text{nm}$ in Table 17.2, for various phase shifts ϕ (first column) and for separation distances $a = 200\,\text{nm}$ and $300\,\text{nm}$. The second and fifth columns of the two tables contain results computed using the PFA, the third and sixth columns contain results computed using the PWS method up to the fourth-order terms [PWS$_{(4)}$], and the fourth and seventh columns present similar results computed up to the second-order terms [PWS$_{(2)}$].

As can be seen in Tables 17.1 and 17.2, the second-order PWS method underestimates the influence of the geometrical factor on the Casimir energy. The PWS results for $X_{R,E}(a)$ obtained in the fourth perturbation order are typically larger than those computed using the PFA. This is because including the effect of the nonzero skin depth by means of multiplication for the mean separation [$\eta_E(a) = 0.6980$ at $a = 200\,\text{nm}$ and $\eta_E(a) = 0.7747$ at $a = 300\,\text{nm}$] underestimates the effect of the nonzero skin depth. The overall conclusion is that at separations $2\pi a \ll \Lambda$, where both the PFA and PWS methods are applicable, the former should be considered more accurate.

TABLE 17.2. The correction factor due to the nonzero skin depth for corrugated surfaces with $A_1 = 85\,\text{nm}$ and $A_2 = 15\,\text{nm}$.

ϕ	$a = 200\,\text{nm}$			$a = 300\,\text{nm}$		
	PFA	PWS$_{(4)}$	PWS$_{(2)}$	PFA	PWS$_{(4)}$	PWS$_{(2)}$
0	0.9586	1.0134	0.9544	0.8941	0.9141	0.9012
$\pi/2$	1.147	1.224	1.088	0.9661	0.9969	0.9671
π	1.388	1.4669	1.2215	1.0458	1.0867	1.0329

From here on, we disregard the constraint $2\pi a \ll \Lambda$. Now eqns (17.36) and (17.39) for the geometric correction factors $\kappa_P(a)$ and $\kappa_E(a)$ are not applicable. In this case we must use the more accurate version of the PWS method given by eqns (17.28) and (17.30)–(17.32) for the coefficients of the perturbative expansion,

$$\kappa_P(a) = \sum_{k=0}^{4}\sum_{l=0}^{4-k} c_{kl} \left(\frac{A_1}{a}\right)^k \left(\frac{A_2}{a}\right)^l. \tag{17.129}$$

Calculating the functions $G_{mn}^{(k,l)}$ in eqn (17.30) for the corrugation profiles (17.124), substituting the results obtained into eqn (17.31), and using eqn (17.28), we arrive at

$$c_{02} = c_{20} = 5, \qquad c_{04} = c_{40} = \frac{105}{8},$$

$$c_{11} = -10\cos\phi\, e^{-a_\Lambda}\left(1 + a_\Lambda + \frac{2}{5}a_\Lambda^2 + \frac{1}{15}a_\Lambda^3\right), \tag{17.130}$$

$$c_{13} = c_{31} = -\frac{105}{2}\cos\phi\, e^{-a_\Lambda}\left[1 + a_\Lambda \Pi(a_\Lambda)\right],$$

$$c_{22} = \frac{105}{2} + \frac{105}{4}\cos 2\phi\, e^{-2a_\Lambda}\left[1 + 2a_\Lambda \Pi(2a_\Lambda)\right],$$

where $a_\Lambda = 2\pi a/\Lambda$ and $\Pi(z)$ is defined in eqn (17.32). The respective coefficients for the correction factor to the energy,

$$\kappa_E(a) = \sum_{k=0}^{4}\sum_{l=0}^{4-k} \tilde{c}_{kl} \left(\frac{A_1}{a}\right)^k \left(\frac{A_2}{a}\right)^l, \tag{17.131}$$

are given by

$$\tilde{c}_{02} = \tilde{c}_{20} = 3, \quad \tilde{c}_{04} = \tilde{c}_{40} = \frac{45}{8}, \quad \tilde{c}_{11} = -6\cos\phi\, e^{-a_\Lambda}\left(1 + a_\Lambda + \frac{1}{3}a_\Lambda^2\right),$$

$$\tilde{c}_{13} = \tilde{c}_{31} = -\frac{45}{2}\cos\phi\, e^{-a_\Lambda}\left[1 + a_\Lambda Q(a_\Lambda)\right], \tag{17.132}$$

$$\tilde{c}_{22} = \frac{45}{2} + \frac{45}{4}\cos 2\phi\, e^{-2a_\Lambda}\left[1 + 2a_\Lambda Q(2a_\Lambda)\right].$$

Here,

$$Q(z) = 1 + \frac{17}{40}z + \frac{11}{120}z^2 + \frac{1}{120}z^3. \tag{17.133}$$

An important characteristic feature of the more exact coefficents (17.130) and (17.132) is that they depend on the corrugation period Λ. This dependence is neglected both in the PFA and in the less accurate version of the PWS method [eqns (17.36) and (17.39)] applicable under the condition $2\pi a \ll \Lambda$. Equation (17.131) will be used below for the calculation of the lateral Casimir force. In Section 17.5.3, we compare the result given in eqns (17.131) and (17.132) with exact computations performed for an ideal metal to estimate the area of its applicability.

At the end of this subsection, we now present a few computational results illustrating the dependence of the PWS correction factors $\kappa_E(a)$ and $\kappa_P(a)$ given in eqns (17.131) and (17.129) on the corrugation period. In Fig. 17.7, $\kappa_E(a)$ is plotted as a function of $\Lambda^{-1}a$ at a fixed separation $a = 200$ nm, for equal corrugation amplitudes $A_1 = A_2 = 50$ nm and for phase shifts $\phi = 0$ [Fig. 17.7(a)] and $\phi = \pi$ [Fig. 17.7(b)]. The solid and dashed curved lines show $\kappa_E^{(4)}(a)$ and $\kappa_E^{(2)}(a)$ computed using the more accurate PWS method (17.131) up to the fourth and second perturbation orders, respectively. For comparison purposes, the flat horizontal lines (solid and dashed) present $\kappa_E^{(4)}(a)$ and $\kappa_E^{(2)}(a)$ computed for an ideal metal up to the fourth and second perturbation orders, respectively, using the PFA (17.123). Note that in this case the second- and fourth-order PFA methods lead to the same result as the respective orders of a less accurate version of the PWS method (17.39). In the case $\phi = 0$ [Fig. 17.7(a)], the total result given by the PFA coincides with that computed up to the second or fourth order [see eqn (17.127)]. Because of this, the solid horizontal line in Fig. 17.7(a) coincides with the dashed line. As is seen in Fig. 17.7, the relative contribution of the fourth perturbation order depends on the phase shift. It changes the magnitude of $\kappa_E(a)$ significantly for $\phi = \pi$ [Fig. 17.7(b)]. The computational results obtained using the more accurate version of the PWS method are very close to the corresponding results obtained with the PFA for corrugation periods $\Lambda > 10a$.

The dependence of the correction factor to the Casimir pressure $\kappa_P(a)$ on the corrugation period is illustrated in Fig. 17.8. The corrugation amplitudes have

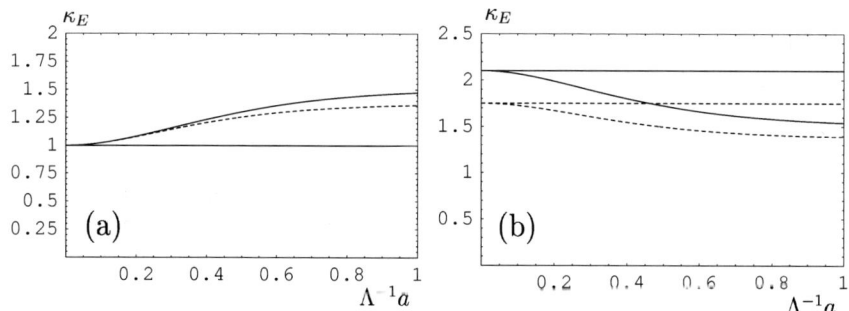

FIG. 17.7. The correction factor to the Casimir energy per unit area for plates due to uniaxial sinusoidal corrugations versus $\Lambda^{-1}a$ at a separation $a = 200$ nm, for equal corrugation amplitudes $A_1 = A_2 = 50$ nm and for phase shifts (a) $\phi = 0$, (b) $\phi = \pi$. The solid and dashed curved lines were computed using the PWS method up to the fourth- and second-order terms, respectively. The solid and dashed flat lines show the respective results computed in the PFA.

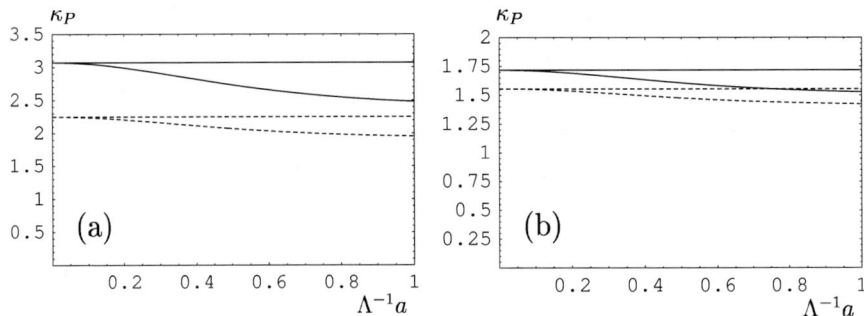

FIG. 17.8. The correction factor to the Casimir pressure due to sinusoidal corrugations versus $\Lambda^{-1}a$ for corrugation amplitudes $A_1 = 85$ nm and $A_2 = 15$ nm and a phase shift $\delta = \pi$, at separations (a) $a = 200$ nm, (b) $a = 300$ nm. The solid and dashed lines have the same meaning as in Fig. 17.7.

been chosen to be $A_1 = 85$ nm and $A_2 = 15$ nm, the phase shift is $\phi = \pi$, and the separation distance is $a = 200$ nm [Fig. 17.8(a)] and $a = 300$ nm [Fig. 17.8(b)]. As can be seen in both figures, there is a pronounced contribution from the fourth order terms in both the PWS and the PFA result (the differences between the solid and dashed curved and flat lines, respectively). The computational results using the PWS method and PFA are close for $\Lambda > 7a$.

As shown in Section 17.5.3, for $\Lambda \sim a$ or $\Lambda < a$, in the case of ideal-metal corrugated plates, the second-order PWS results deviate from those calculated in the second order of the exact path-integral theory (see Section 10.5). Because of this, the above computations within the more accurate version of the PWS method can be reliably used only under the condition $a \ll \Lambda$, which is, however, not as constrained as $2\pi a \ll \Lambda$, under which the PFA is applicable.

17.5.2 The lateral Casimir force

As noted in Section 17.1.3, the surface roughness described by eqn (17.43) leads to the existence of a lateral Casimir force (17.48). The configuration of parallel plates with uniaxial corrugations of equal period opens up an opportunity for the experimental observation of lateral displacements caused by zero-point oscillations. Because of this, we consider some important properties of the lateral force acting between two sinusoidally corrugated plates in more detail.

Using the geometric scale factor $\kappa_E(a)$ defined in eqn (17.131), the Casimir energy per unit area for a configuration of two corrugated plates made of real materials can be presented in the form

$$E_{\mathrm{R}}(a) = E_{\mathrm{R}}(a,\phi) = -\frac{\pi^2 \hbar c}{720 a^3} \eta_E(a) \sum_{k=0}^{4} \sum_{l=0}^{4-k} \tilde{c}_{kl} \left(\frac{A_1}{a}\right)^k \left(\frac{A_2}{a}\right)^l. \quad (17.134)$$

Substituting here the coefficients $\tilde{c}_{kl}$ from eqn (17.132), we obtain the lateral Casimir force in the framework of the PWS method,

$$F_R^{\text{lat}}(a,\phi) = -\frac{\partial E_R(a,\phi)S}{\partial x_0} = \frac{\pi^3 \hbar c S}{60 a^3 \Lambda}\eta_E(a)\frac{A_1 A_2}{a^2}\sin\phi \qquad (17.135)$$

$$\times \left\{ e^{-a_\Lambda}\left(1 + a_\Lambda + \frac{1}{3}a_\Lambda^2\right) + \frac{15}{4}e^{-a_\Lambda}[1 + a_\Lambda Q(a_\Lambda)]\frac{A_1^2 + A_2^2}{a^2} \right.$$

$$\left. - \frac{15}{2}\cos\phi\, e^{-2a_\Lambda}[1 + 2a_\Lambda Q(2a_\Lambda)]\frac{A_1 A_2}{a^2} \right\}.$$

An important condition for the existence of the lateral force (17.135) is that the periods of the corrugations on the two plates must be equal, as in eqn (17.124), or differ by a factor of an integer. Otherwise, on averaging over the plate area, all ϕ-dependent interference terms in the geometric factor $\kappa_E(a)$ would vanish. As a result, the lateral Casimir force would become zero.

The respective expression for the lateral Casimir force in the first four orders of the less accurate version of the PWS method is obtained from eqn (17.39) with the functions f_1 and f_2 given in eqn (17.124):

$$F_R^{\text{lat}}(a,\phi) = \frac{\pi^3 \hbar c S}{60 a^3 \Lambda}\eta_E(a)\frac{A_1 A_2}{a^2}\sin\phi\left[1 + \frac{15}{4}\frac{A_1^2 + A_2^2}{a^2} - \frac{15}{2}\cos\phi\frac{A_1 A_2}{a^2}\right]. \qquad (17.136)$$

It can be seen that eqn (17.136) follows from eqn (17.135) with $a_\Lambda = 0$. For an ideal metal, $\eta_E(a) = 1$ and eqn (17.136) coincides with the respective result obtained for corrugated ideal-metal plates using the PFA. Because of this, it is sometimes stated in the literature that the less accurate version of the PWS method is equivalent to the PFA. This is, however, not the case for real materials.

To obtain the lateral force in the PFA, one must substitute the mean values of the corrugation functions (17.125) into eqn (17.123), which leads to the Casimir energy per unit area calculated up to the fourth perturbation order of this approximation. Calculating the derivative of the resulting expression with respect to x_0, we get the lateral Casimir force,

$$F_R^{\text{lat}}(a,\phi) = -\frac{\pi A_1 A_2 S}{\Lambda}\sin\phi\left[\frac{\partial^2 E(a)}{\partial a^2} + \frac{1}{8}\frac{\partial^4 E(a)}{\partial a^4}(A_1^2 + A_2^2 - 2A_1 A_2 \cos\phi)\right]. \qquad (17.137)$$

This expression coincides with eqn (17.136) only if the multiplier $\eta_E(a)$ in eqn (17.8) does not depend on the separation. If η_E is separation-dependent, F_R^{lat} given by eqn (17.137) in the PFA leads to different results in comparison with eqn (17.136), derived using the less accurate version of the PWS. This can be easily verified analytically using the perturbative expansion for $\eta_E(a)$ (17.58) obtained in the framework of the plasma model, or numerically using the Lifshitz formula for $E(a)$. A comparison of the PWS and PFA methods with more exact approaches is performed in the next subsection.

Now we compare the calculation results obtained by using eqns (17.135) and (17.137) up to the fourth- and second-order terms. In Fig. 17.9(a), the ratio F_R^{lat}/N is plotted as a function of x_0/Λ for a separation between the plates $a = 200$ nm and a corrugation period $\Lambda = 1.2\,\mu\text{m}$, where the scaling factor is

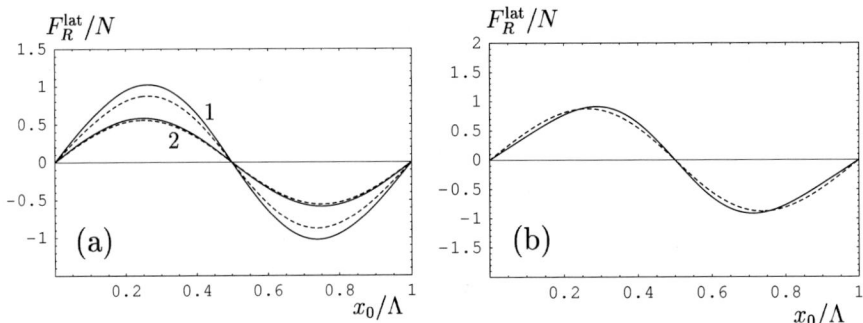

FIG. 17.9. The rescaled lateral Casimir force as a function of x_0/Λ at a separation $a = 200\,\text{nm}$ and with a corrugation period $\Lambda = 1.2\,\mu\text{m}$. (a) Corrugation amplitudes $A_1 = 85\,\text{nm}$ and $A_2 = 15\,\text{nm}$. The solid and dashed lines were computed using the PWS and PFA, respectively. The computational results up to the fourth- and second-order terms are labeled 1 and 2. (b) Corrugation amplitudes $A_1 = A_2 = 50\,\text{nm}$. The solid and dashed lines were obtained using the same methods up to the fourth perturbation order.

$$N = \frac{\pi^3 \hbar c S A_1 A_2}{60 a^5 \Lambda}. \tag{17.138}$$

In this figure, the corrugation amplitudes are different: $A_1 = 85\,\text{nm}$ and $A_2 = 15\,\text{nm}$. The solid and dashed lines labeled 1 show the results computed up to the fourth-order terms using the PWS method (17.135) and the PFA (17.137), respectively. The solid and dashed lines labeled 2 show similar results computed up to the second-order terms. The quantities $\eta_E(a)$ in eqn (17.135) and $\partial^2 E(a)/\partial a^2$, $\partial^4 E(a)/\partial a^4$ in eqn (17.137) were computed for Au by using the Lifshitz formula and the plasma model with $\omega_\text{p} = 9.0\,\text{eV}$. As can be seen in Fig. 17.9(a), for both methods the correction of the fourth order significantly increases the amplitude of the lateral force.

Equations (17.135)–(17.137) indicate that the dependence of the lateral Casimir force on the phase shift is not merely sinusoidal, because the phase-dependent $\cos\phi$ is present in the fourth-order term of the perturbative expansion in the corrugation amplitudes. The deviation of the lateral force from a sinusoidal shape is more pronounced for equal corrugation amplitudes. This is illustrated in Fig. 17.9(b), where the dependence of F_R^lat/N is shown as a function of x_0/Λ for $A_1 = A_2 = 50\,\text{nm}$, $\Lambda = 1.2\,\mu\text{m}$, and $a = 200\,\text{nm}$. The solid and dashed lines were computed in the fourth-order perturbation theory using the PWS method (17.135) and PFA (17.137), respectively.

17.5.3 Application regions of approximate methods

As repeatedly stated above, the PWS and PFA methods are approximate phenomenological approaches to the calculation of the Casimir force with uncontrolled accuracy. In applications to corrugated plates, they work well only at sufficiently small separation distances, much less than the corrugation period.

To fully determine the application regions of these approximate methods, one needs to compare the computational results obtained with exact results. Unfortunately, for the configuration of two parallel corrugated plates, exact results for the Casimir energy per unit area and the Casimir pressure are available only for ideal-metal plates and only in the lowest, second, order in the corrugation amplitudes. These results were obtained by Emig et al. (2003) and were presented in Section 10.5. Below, we compare them with corresponding results obtained in the second order using the PWS and PFA methods.

We start with the normal Casimir force in the configuration of parallel plates with sinusoidal corrugations (17.124) of equal amplitudes $A = A_1 = A_2$. The exact expression for the Casimir energy per unit area in the second-order perturbation theory can be presented in the form

$$E_R(a, \phi) = -\frac{\pi^2 \hbar c}{720 a^3} \left[1 + \tilde{c}_2 \left(\frac{a}{\Lambda}, \phi \right) \frac{A^2}{a^2} \right], \tag{17.139}$$

where, according to eqn (17.131), $\tilde{c}_2 = \tilde{c}_{20} + \tilde{c}_{02} + \tilde{c}_{11}$. Using the original notation (Emig et al. 2003),

$$\tilde{c}_2 = \frac{1440}{\pi^2} \left[G_{\text{TM}} \left(\frac{a}{\Lambda} \right) + G_{\text{TE}} \left(\frac{a}{\Lambda} \right) \right]$$
$$- \frac{720}{\pi^2} \cos \phi \left[J_{\text{TM}} \left(\frac{a}{\Lambda} \right) + J_{\text{TE}} \left(\frac{a}{\Lambda} \right) \right], \tag{17.140}$$

where the functions $G_{\text{TM,TE}}(z)$ and $J_{\text{TM,TE}}(z)$ are defined explicitly in eqns (10.164) and (10.171).

The coefficient $\tilde{c}_2$ in eqn (17.140) is plotted as a function of a/Λ in Fig. 17.10(a) as the solid lines 1 ($\phi = 0$), 2 ($\phi = \pi/2$), and 3 ($\phi = \pi$). In the same figure, the dashed lines 1 ($\phi = 0$), 2 ($\phi = \pi/2$), and 3 ($\phi = \pi$) show the same coefficient computed using eqn (17.132) within the framework of a more accurate version of the PWS method. As is seen in Fig. 17.10(a), the PWS method is in very good agreement with the exact results for $a \leq 0.2\Lambda$. At separations larger than 0.2Λ, the calculation results using the PWS method deviate from the exact results and this deviation increases with increasing separation.

In the simplest case $\phi = \pi/2$, the exact expression for the Casimir pressure is given by

$$P_R(a) = -\frac{\pi^2 \hbar c}{240 a^4} \left[1 + c_2 \left(\frac{a}{\Lambda}, \frac{\pi}{2} \right) \frac{A^2}{a^2} \right], \tag{17.141}$$

where, from eqn (17.129), $c_2 = c_{20} + c_{02}$ (recall that for $\phi = \pi/2$, $c_{11} = 0$). The coefficient c_2 can be found from eqns (17.139) and (17.140):

$$c_2 = \frac{2400}{\pi^2} \left[G_{\text{TM}} \left(\frac{a}{\Lambda} \right) + G_{\text{TE}} \left(\frac{a}{\Lambda} \right) \right]$$
$$- \frac{1440}{\pi^2} \frac{a}{\Lambda} \left[G'_{\text{TM}} \left(\frac{a}{\Lambda} \right) + G'_{\text{TE}} \left(\frac{a}{\Lambda} \right) \right], \tag{17.142}$$

where the prime means differentiation with respect to the argument. In Fig. 17.10(b), we plot the coefficient c_2 in eqn (17.142) versus a/Λ as the solid line.

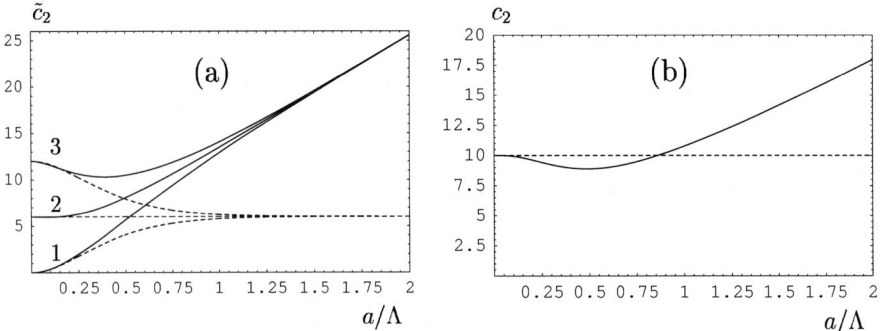

FIG. 17.10. (a) The coefficient $\tilde{c}_2$ in eqn (17.139) for the Casimir energy per unit area for corrugated plates as a function of a/Λ. The solid and dashed lines were computed using the exact theory and the PWS method, respectively, for $\phi = 0$ (lines labeled 1), $\phi = \pi/2$ (lines labeled 2), and $\phi = \pi$ (lines labeled 3). (b) The coefficient c_2 in eqn (17.141) for the Casimir pressure as a function of a/Λ. The solid and dashed lines were computed using the exact theory and the PWS method, respectively, for $\phi = \pi/2$.

In the same figure, the respective results obtained using the PWS method are shown by the dashed line. This line is flat because, for the phase shift $\phi = \pi/2$, the coefficients (17.130) do not depend on the separation or the corrugation period. As is seen in Fig. 17.10(b), the computational results obtained using the PWS method coincide with the exact results under the condition $a \leq 0.1\Lambda$. However, for the Casimir pressure, over a wide range of parameters $0 \leq a/\Lambda \leq 1$, the deviations between the exact results and those obtained using the PWS method do not exceed 12%.

Now we compare exact calculations of the lateral Casimir force with results obtained by using the PWS method and the PFA. The exact result for ideal-metal plates is obtained from eqns (17.139) and (17.140) by the negative differentiation of $E_R(a)S$ with respect to $x_0 = \phi\Lambda/(2\pi)$:

$$F_R^{\text{lat}}(a, \phi) = 2\pi \frac{\hbar c A^2 S}{a^6} \sin\phi \frac{a}{\Lambda} \left[J_{\text{TM}}\left(\frac{a}{\Lambda}\right) + J_{\text{TE}}\left(\frac{a}{\Lambda}\right) \right]. \qquad (17.143)$$

In Fig. 17.11(a), we plot the exact rescaled lateral Casimir force (17.143), $F_R^{\text{lat}}/\tilde{N}$, where the scaling factor is $\tilde{N} = \hbar c A^2 S/a^6$, as a function of a/Λ (solid line). In the same figure, the dotted line shows the respective results obtained up to the second-order terms using the PWS method. These results are given by eqn (17.135) with $A_1 = A_2 = A$ and $\eta_E = 1$, neglecting the contributions from the fourth-order terms. The dashed straight line in Fig. 17.11(a) represents the second-order results for the rescaled lateral Casimir force computed using the PFA. This is obtained from the first term on the right-hand side of eqn (17.137), where $E(a)$ is put equal to $E_{\text{IM}}(a)$. As is seen in Fig. 17.11(a), the results obtained using the PWS method practically coincide with the exact results within a wide

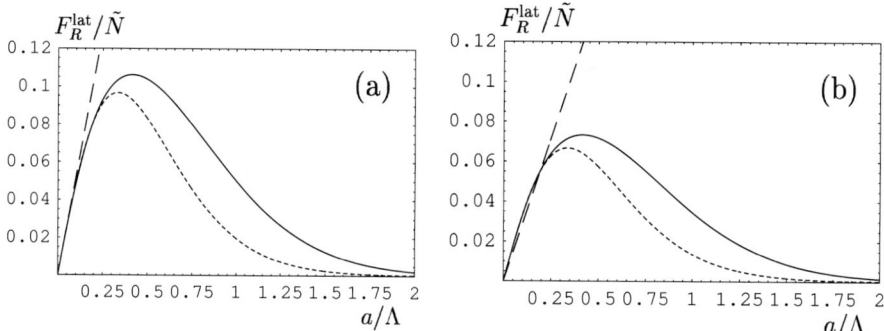

FIG. 17.11. The rescaled lateral Casimir force up to the second order in the corrugation amplitude using the exact theory (solid line), the PWS method (dotted line), and the PFA (dashed line), for (a) ideal-metal plates, (b) gold plates at $a = 200$ nm.

region $a \leq 0.25\Lambda$. The PFA reproduces the exact results in a narrower range of separations $a \leq 0.1\Lambda$.

The above results concerning the application region of the PWS method and the PFA should be considered as provisional because, as was shown in Sections 17.5.1 and 17.5.2, the higher-order terms in the corrugation amplitudes contribute significantly to both the vertical and the lateral force and cannot be neglected. These terms are easily calculated using the approximate phenomenological methods under consideration. However, their exact values remain unknown and, thus, a comparison between the exact and approximate values is not possible. The other obstacle is that the exact results used have been obtained for ideal-metal plates and their generalization to real materials is problematic. Because of this, there is nothing to compare with the computational results for real materials obtained using the phenomenological methods. Note that the comparison between the PFA and the more fundamental scattering approach performed by Maia Neto et al. (2005) and Rodrigues et al. (2007) using the plasma model was based on an intermediate spectral characteristic $G(k)$ in the lowest perturbation order (see Section 17.4), and does not provide the Casimir energy between rough or corrugated plates as a function of separation. We shall return to this comparison in Chapter 21 for the configuration of a corrugated sphere above a corrugated plate used in the measurement of the lateral Casimir force.

To illustrate the role of real material properties in the exact and PWS calculation methods, we have taken them into account in a multiplicative way. In so doing, the computational results in Fig. 17.11(a) shown by the solid and dotted lines (exact and PWS, respectively) were multiplied by the factor $\eta_E(a)$ for Au computed using the definition (17.9) with the dielectric permittivity of the plasma model for a fixed separation distance $a = 200$ nm between the zero levels of the corrugations on the two plates. Regarding the second-order PFA, real material properties were taken into account in a nonmultiplicative way, as given by the first term on the right-hand side of eqn (17.137). The computational results for the rescaled lateral Casimir force including the effect of real material properties are presented in Fig. 17.11(b) as a function of a/Λ (i.e. of the inverse corrugation period). The solid, dotted, and dashed lines indicate the results obtained using the exact path-integral method, the PWS method, and the PFA, respectively. As can be seen in the figure, all of these methods are in agreement for corrugation periods $\Lambda \geq 5a$. In general, for inclusion of material properties, the PFA should be considered as a more exact method than the PWS method because the latter incorporates this effect in a multiplicative way. As can be seen in Fig. 17.11(a), for ideal-metal plates the exact result for the lateral Casimir force obtained up to the second-order terms is constrained between the results derived using the approximate methods of the PWS method and the PFA. The exact result corrected for the nonzero skin depth almost coincides with both of the approximate results for $a/\Lambda < 0.2$ and is between them for $a/\Lambda \geq 0.2$ [see Fig. 17.11(b)].

The possibility of rigorous determination of the application regions of the two phenomenological methods depends on the availability of exact results which take into account both the nonplanar geometry and the material properties in higher perturbation orders (see Sections 21.3 and 21.5).

17.5.4 The role of roughness and corrugations in atom–plate interactions

Here, we consider corrections to the Casimir–Polder interaction between an atom and a plate with sides of length $2L$ due to imperfections in the plate geometry. This can be done using the approximate PWS method developed in Sections 17.1.2 and 17.1.4 for the cases of two parallel plates and a sphere above a plate, respectively. The application region of the PWS method (the separation a between the atom and the plate is much smaller than the roughness correlation length Λ_c or corrugation period Λ) can be estimated from a comparison with more fundamental calculations under conditions when those can be performed.

We start with the additive energy of the atom–plate interaction (17.19) calculated up to the fourth perturbation order, where the roughness profile of the plate is described by the function $f_1(x, y)$ introduced in eqn (17.3) and the height of an atom above the plate z_2 is replaced with a. Integration of eqn (17.19) with respect to x_1 and y_1 results in (Bezerra et al. 2000b)

$$E_{A,R}^{\text{add}}(x_2, y_2, a) = E_A^{\text{add}}(a) \sum_{i=0}^{4} h_i(x_2, y_2, a) \left(\frac{A_1}{a}\right)^i, \qquad (17.144)$$

where the additive energy of the interaction of an atom with a semispace is given by eqn (6.42), and $h_0(x_2, y_2, a) = 1$. The other expansion coefficients of the powers of $A_1 \ll a$ in eqn (17.144) are given by

$$h_1(x_2, y_2, a) = 4 \sum_{m,n=0}^{\infty} u_{mn}^{(1)}(x_2, y_2) e^{-z_{mn}} \left(1 + z_{mn} + \frac{z_{mn}^2}{3}\right), \qquad (17.145)$$

$$h_2(x_2, y_2, a) = 10 \left[g_{00}^{(2)} + \sum_{m,n=0}^{\infty} u_{mn}^{(2)}(x_2, y_2) e^{-z_{mn}} \left(1 + z_{mn} + \frac{2z_{mn}^2}{5} + \frac{z_{mn}^3}{15} \right) \right],$$

$$h_3(x_2, y_2, a) = 20 \left[g_{00}^{(3)} + \sum_{m,n=0}^{\infty} u_{mn}^{(3)}(x_2, y_2) e^{-z_{mn}} \right.$$

$$\left. \times \left(1 + z_{mn} + \frac{13 z_{mn}^2}{30} + \frac{z_{mn}^3}{10} + \frac{z_{mn}^4}{90} \right) \right],$$

$$h_4(x_2, y_2, a) = 35 \left[g_{00}^{(4)} + \sum_{m,n=0}^{\infty} u_{mn}^{(4)}(x_2, y_2) e^{-z_{mn}} \right.$$

$$\left. \times \left(1 + z_{mn} + \frac{19 z_{mn}^2}{42} + \frac{5 z_{mn}^3}{42} + \frac{2 z_{mn}^4}{105} + \frac{z_{mn}^5}{630} \right) \right],$$

where z_{nm} is defined in eqn (17.30). Here, we assume that $m + n \neq 0$; the same notation for the Fourier coefficients $[g_{00}^{(i)}, g_{l,mn}^{(i)}]$ of the function f_1 and its powers f_1^i as in Section 17.1.2 are used, and

$$u_{mn}^{(i)}(x_2, y_2) = g_{1,mn}^{(i)} \sin \frac{\pi m x_2}{L} \sin \frac{\pi n y_2}{L} + g_{2,mn}^{(i)} \sin \frac{\pi m x_2}{L} \cos \frac{\pi n y_2}{L}$$
$$+ g_{3,mn}^{(i)} \cos \frac{\pi m x_2}{L} \sin \frac{\pi n y_2}{L} + g_{4,mn}^{(i)} \cos \frac{\pi m x_2}{L} \cos \frac{\pi n y_2}{L}. \qquad (17.146)$$

In order to take approximate account of the effects of nonadditivity in eqn (17.144), we normalize it to the case of an atom interacting with a semispace with a flat boundary surface. In this case the exact interaction energy at zero temperature is given by eqn (16.21). This can be written in the form

$$E^{\mathrm{A}}(a) = -\frac{3\hbar c}{8\pi a^4} \alpha(0) \eta_E^{\mathrm{A}}(a), \qquad (17.147)$$

where

$$\eta_E^{\mathrm{A}}(a) = \frac{1}{12} \int_0^{\infty} e^{-y} dy \int_0^y d\zeta \frac{\alpha(i\omega_c \zeta)}{\alpha(0)} \left\{ 2y^2 r_{\mathrm{TM}}^{(1)}(i\zeta, y) \right. \qquad (17.148)$$
$$\left. -\zeta^2 \left[r_{\mathrm{TM}}^{(1)}(i\zeta, y) + r_{\mathrm{TE}}^{(1)}(i\zeta, y) \right] \right\}.$$

Then the normalization factor is

$$\tilde{K}_E^{\mathrm{A}}(a) = \frac{E_{\mathrm{A}}^{\mathrm{add}}(a)}{E_{\mathrm{A}}(a)}. \qquad (17.149)$$

Dividing both sides of eqn (17.144) into $\tilde{K}_E^{\mathrm{A}}(a)$, we obtain

$$E_{\mathrm{R}}^{\mathrm{A}}(x_2, y_2, a) = \frac{E_{\mathrm{A,R}}^{\mathrm{add}}(x_2, y_2, a)}{\tilde{K}_E^{\mathrm{A}}(a)} = -\frac{3\hbar c}{8\pi a^4} \alpha(0) \eta_E^{\mathrm{A}}(a) \kappa_E^{\mathrm{A}}(x_2, y_2, a), \qquad (17.150)$$

where the geometric factor is given by

$$\kappa_E^A(x_2, y_2, a) = \sum_{i=0}^{4} h_i(x_2, y_2, a) \left(\frac{A_1}{a}\right)^i. \qquad (17.151)$$

The energy E_R^A in eqn (17.150) includes the effect of the surface imperfections in a multiplicative way, similarly to eqns (17.40) and (17.41) for two parallel plates with roughness.

The component of the Casimir force acting on an atom normal to the plate has the form

$$F_R^A(x_2, y_2, a) = -\frac{3\hbar c}{2\pi a^5} \alpha(0) \eta_F^A(a) \kappa_F^A(x_2, y_2, a). \qquad (17.152)$$

This expression is obtained in the same way as for eqn (17.41) for the Casimir pressure between two rough plates (see the discussion concerning the alternative representation of the pressure within the PWS method in Section 17.1.3). The correction factor $\eta_F^A(a)$ is obtained from eqn (16.19) rewritten for $T = 0$,

$$\eta_F^A(a) = \frac{1}{48} \int_0^\infty e^{-y} y \, dy \int_0^y d\zeta \frac{\alpha(i\omega_c \zeta)}{\alpha(0)} \left\{ 2y^2 r_{TM}^{(1)}(i\zeta, y) \right. \qquad (17.153)$$
$$\left. - \zeta^2 \left[r_{TM}^{(1)}(i\zeta, y) + r_{TE}^{(1)}(i\zeta, y) \right] \right\}.$$

The geometric correction factor to the force, $\kappa_F^A(x_2, y_2, a)$, has the form of eqn (17.151), where the coefficients $h_i(x_2, y_2, a)$ are replaced with

$$\tilde{h}_i(x_2, y_2, a) = -\frac{a^{5+i}}{4} \frac{\partial}{\partial a} \left[\frac{1}{a^{4+i}} h_i(x_2, y_2, a) \right]. \qquad (17.154)$$

Equation (17.150) for the energy of an atom above a rough or corrugated surface shows that there is a lateral Casimir force acting on the atom. For example, by negative differentiation of both sides of eqn (17.150) with respect to x_2, we obtain

$$F_R^{A, \text{lat}}(x_2, y_2, a) = -\frac{3\hbar c}{8\pi a^4} \alpha(0) \eta_E^A(a) \sum_{i=1}^{4} h_{i, x_2}(x_2, y_2, a) \left(\frac{A_1}{a}\right)^i, \qquad (17.155)$$

where

$$h_{i, x_2}(x_2, y_2, a) = -\frac{\partial}{\partial x_2} h_i(x_2, y_2, a). \qquad (17.156)$$

Similarly to Sections 17.1.2 and 17.1.3, the above equations can be significantly simplified for large-scale roughness satisfying the condition $2\pi a \ll \Lambda$, where Λ is the corrugation period or the roughness correlation length. In this case the coefficients of the correction factor to the normal force, $\kappa_F^A(x_2, y_2, a)$, take the form (Bezerra et al. 2000b)

$$\tilde{h}_1(x_2, y_2) = 5 \sum_{m,n=0}^{\infty} u_{mn}^{(1)}(x_2, y_2), \quad \tilde{h}_2(x_2, y_2) = 15 \left[g_{00}^{(2)} + \sum_{m,n=0}^{\infty} u_{mn}^{(2)}(x_2, y_2) \right],$$

$$\tilde{h}_3(x_2, y_2) = 35 \left[g_{00}^{(3)} + \sum_{m,n=0}^{\infty} u_{mn}^{(3)}(x_2, y_2) \right], \tag{17.157}$$

$$\tilde{h}_4(x_2, y_2) = 70 \left[g_{00}^{(4)} + \sum_{m,n=0}^{\infty} u_{mn}^{(4)}(x_2, y_2) \right].$$

The expansion coefficients in eqn (17.155) are given by

$$h_{i,x_2}(x_2, y_2) = -c_i \sum_{m,n=0}^{\infty} \frac{\partial u_{mn}^{(i)}(x_2, y_2)}{\partial x_2}, \tag{17.158}$$

where $c_1 = 4$, $c_2 = 10$, $c_3 = 20$, and $c_4 = 35$.

As an example, we consider an atom above a plate with uniaxial sinusoidal corrugations described by the function f_1 in eqn (17.124). The coefficients $u_{mn}^{(i)}$ in eqn (17.157) can be easily calculated, leading to the following result for the Casimir force acting in the direction normal to the surface (Bezerra et al. 2000b):

$$F_R^A(x_2, a) = -\frac{3\hbar c}{2\pi a^5} \alpha(0) \eta_F^A(a) \left[1 + 5 \sin \frac{2\pi x_2}{\Lambda} \frac{A_1}{a} \right. \tag{17.159}$$

$$+ \frac{15}{2} \left(1 - \cos \frac{4\pi x_2}{\Lambda} \right) \frac{A_1^2}{a^2} + \frac{35}{4} \left(3 \sin \frac{2\pi x_2}{\Lambda} - \sin \frac{6\pi x_2}{\Lambda} \right) \frac{A_1^3}{a^3}$$

$$\left. + \frac{105}{4} \left(1 - \frac{4}{3} \cos \frac{4\pi x_2}{\Lambda} + \frac{1}{3} \cos \frac{8\pi x_2}{\Lambda} \right) \frac{A_1^4}{a^4} \right].$$

In the same way, from eqns (17.155) and (17.158) we obtain the lateral Casimir force acting between an atom and a corrugated plate,

$$F_R^{A,\mathrm{lat}}(x_2, a) = \frac{3\hbar c}{4a^4\Lambda} \alpha(0) \eta_E^A(a) \frac{A_1}{a} \left[4 \cos \frac{2\pi x_2}{\Lambda} + 10 \sin \frac{4\pi x_2}{\Lambda} \frac{A_1}{a} \right. \tag{17.160}$$

$$\left. + 15 \left(\cos \frac{2\pi x_2}{\Lambda} - \cos \frac{6\pi x_2}{\Lambda} \right) \frac{A_1^2}{a^2} + 35 \left(\sin \frac{4\pi x_2}{\Lambda} - \frac{1}{2} \sin \frac{8\pi x_2}{\Lambda} \right) \frac{A_1^3}{a^3} \right].$$

Note that the function F_R^A reaches a minimum at $x_2^{\min}/\Lambda = 0.25$ and a maximum at $x_2^{\max}/\Lambda = 0.75$ (i.e. the largest magnitude of the normal force is reached when the atom is above the position with the maximum value of the corrugation function). For a typical ratio $A_1/a = 0.1$, the perturbation orders from the first to the fourth contribute 50%, 15%, 3.5%, and less than 1%, respectively, of the magnitude of the normal force between an atom and a flat surface at the points of the maxima and minima. The lateral force $F_{\rm R}^{\rm A,lat}$ vanishes at the points of the maxima and minima of the function $F_{\rm R}^{\rm A}$. If the atom is in the vicinity of the point $x_2^{\rm min}$, it experiences a lateral Casimir force which attracts it to the point $x_2^{\rm min}$. After the atom crosses the point $x_2^{\rm min}$, the lateral force changes its sign. As a result, the atom oscillates around $x_2^{\rm min}$ with a decreasing amplitude. If the atom is in the vicinity of $x_2^{\rm max}$, the lateral Casimir force repels it from $x_2^{\rm max}$ in the direction of the nearest point $x_2^{\rm min}$.

The interaction of an atom with a corrugated plate has been considered within the scattering approach under the conditions $A_1 \ll a$, Λ, $\lambda_{\rm A}$, λ_0, where $\lambda_{\rm A}$ and λ_0 are the characteristic absorption wavelengths of the atom and of the material of the plate (Dalvit et al. 2008). Results obtained using the scattering approach in only the first perturbation order, A_1/a, were compared with corresponding results found using the PFA. A large deviation of about 30% was found for a corrugation period $\Lambda = 3.5\,\mu$m and an atom–plate separation of $a = 2\,\mu$m. This case is expected to be outside the application region of the PFA because $2\pi a > \Lambda$ (see the previous section). The computations performed in this section within the application regions of both the PFA and the PWS method demonstrate the important role of contributions to the Casimir force of higher order in powers of A_1/a. Thus, although the scattering approach in the lowest perturbation order can be used to indicate the applicability limits of the PFA, it is insufficient for configurations of experimental interest (see Chapter 21).

PART III

MEASUREMENTS OF THE CASIMIR FORCE AND THEIR APPLICATIONS IN BOTH FUNDAMENTAL PHYSICS AND NANOTECHNOLOGY

18

GENERAL REQUIREMENTS FOR CASIMIR FORCE MEASUREMENTS

The large body of material presented in the first two parts of the book shows that the measurement of the Casimir force is a complicated scientific and technological problem. Given the small value of the force for the experimentally accessible surface areas, the force sensitivity of the available measurement techniques has been a severe limitation. Another limitation is that the separation distances where the Casimir force becomes measurable are very small and their accurate determination has been difficult. Given that the force has a very strong dependence on the separation and on the geometrical and material properties of the boundary surfaces, the comparison between experiment and theory is a challenging task. In this chapter, we briefly consider older measurements of the Casimir force and formulate the general experimental requirements and best practices which follow from these measurements. Next we discuss rigorous procedures for comparison of experiment with theory in relation to force–distance measurements. Specifically, we elaborate on the presentation of experimental errors and precision and of the theoretical uncertainties for real materials. We also discuss the statistical framework for the comparison between experiment and theory. The concepts introduced in this chapter are used in all of the other chapters of Part III of this book, where the main experiments on the measurement of the Casimir force are considered.

18.1 Primary achievements of older measurements

In the half century after the theoretical prediction of the Casimir effect, there were only a few attempts to measure the Casimir force. Here, we briefly discuss the main experiments which were performed before the year 1997, when the modern stage in this field of research started. In all cases, special attention is drawn to one or a few of the necessary requirements for an ideal force–distance measurement that were met or not met in the experiment under consideration. This will allow us to summarize the experience gained from the older experiments in the next section.

18.1.1 *Experiment with parallel plates by Sparnaay*

Sparnaay (1958) made the first reported attempt to measure the Casimir force. He used a configuration of two flat metal plates. A force balance based on a spring balance was used in the final series of measurements. The sensitivity of the spring balance was between 10^{-4} and 10^{-3} dyn. The extension of the spring was measured through a measurement of the capacitance of the capacitor formed by the two flat plates. Calibration of this capacitance was done with the help of tungsten and platinum wires, though the uncertainties in this calibration were not reported. Care was taken with vibration isolation. It was noted that the knife-edges and the springs used led to a large hysteresis, which made determination of the separation distance of the surfaces difficult. This was reported to be the most severe drawback of the measurement technique. The plates were mounted such that they were electrically insulated from the rest of the apparatus. Sparnaay realized that even a small potential difference of 17 mV between the two parallel plates was sufficient to overwhelm the Casimir force.

To take care of any potential differences between the surfaces, the two plates were brought into contact at the start of the experiment. Three types of pairs of metal plates, namely two aluminum plates, two chromium plates, and a chromium plate and a steel plate, were used in the measurements. Even with a variety of electrical and mechanical cleaning procedures, dust particles larger than 2–3 μm were observed on the plates. The plates were aligned parallel by visual inspection with about 10% variation in the interplate distance from one of the plates to the other. Because of the presence of the dust particles it was estimated that even on contact, the plates were separated by 0.2 μm (the procedure used to determine this value was not provided). The pair consisting of chromium and steel plates and the pair consisting of two chromium plates both had attractive forces between them, whereas the pair of aluminum plates showed a repulsive force. The peculiar repulsive force noticed in the case of the aluminum plates was thought to be due to the presence of impurities on the aluminum surface. In the case of the attractive force for the chromium and chromium–steel pairs, given the uncertainties in the measurement of the interplate distance, only general agreement with the Casimir pressure formula (1.1) for perfectly reflecting boundaries could be achieved. If we ignore the repulsive forces measured with the aluminum plates, the following improvements other than an increase in the force sensitivity would have been desirable. The first was a more accurate measurement of the surface separation. The second desirable improvement was a more accurate measurement of the parallelism between the two surfaces. For plates of area 1 cm^2 ($L = 0.5$ cm) at a separation $a = 1\,\mu$m, the correction to the Casimir pressure due to the nonparallelity does not exceed 10% if an angle φ_0 less than 3.4×10^{-5} rad between the plates is guaranteed (see Section 17.3). As a third improvement, a measurement of any residual electrostatic potential difference present between the two surfaces was required, given the presence of the dust particles.

In conclusion, these measurements were the first indication of an attractive Casimir force between metallic surfaces, approximately in line with expectations. (Note that the aluminum plates showed a repulsive force and therefore the observation of an attractive force was not conclusive.) But, above all, the problems that needed to be overcome for a careful and conclusive measurement of the Casimir force were clearly elucidated.

18.1.2 Experiments by Derjaguin et al.

One of the major improvements that was pioneered by Derjaguin's group was the use of curved surfaces to avoid the need to maintain two flat plates perfectly parallel. This was accomplished by replacing one or both plates by a curved surface such as a lens, sphere, or cylinder. The first use of this technique was to measure the force between a silica lens and a plate (Derjaguin 1934, Derjaguin et al. 1956, Derjaguin and Abrikosova 1958). Sparnaay (1958) pointed out that this work did not take into account the presence of the "gel layer" which is usually present on such surfaces. Also, the possible substantial electrostatic forces due to the use of dielectric silica surfaces, which would result in systematic errors, were not reported in the experimental results. These experiments will not be discussed further here.

There was also related work with metallic surfaces by Derjaguin et al. (1956, 1987). In this case, the forces between platinum fibers and gold beads were measured. The force measurement was done by keeping one surface fixed and attaching the other surface to the coil of a galvanometer. The rotation of the galvanometer coil in response to the force led to the deflection of a light beam, which was reflected off mirrors attached to the galvanometer coil. The deflected light beam was detected through a resistance bridge, two of whose elements could be photoactivated. The measured data indicated a nonretarded van der Waals forces for distances below 50–80 nm and a retarded-force region for larger distances. However, more accurate modern theoretical results (Klimchitskaya et al. 2000) predict an unretarded force below distances of 2 nm in the case of gold. Derjaguin et al. (1987) reported a discrepancy in the force measurements of around 60%. Also, any possible electrostatic forces due to potential differences between the two surfaces appear to have been neglected. While the distance on contact of the two surfaces appears to have been taken as the zero distance (ignoring the role of surface roughness), it was mentioned that surface roughness might have affected the experimental measurements and have made the comparison with theory very difficult, particularly for distances less than 30 nm.

18.1.3 Experiments by Tabor, Winterton, and Israelachvili

In the intervening years between the experiments conducted by Sparnaay (1958) and by van Blockland and Overbeek (1978) using metallic surfaces, there were many force measurements on nonconductive surfaces. Of these, the experiments on muscovite mica will be discussed here (Tabor and Winterton 1968, Israelachvili and Tabor 1972, Israelachvili 1992, White et al. 1976). The major improvement in these experiments was the use of atomically smooth surfaces obtained from cleaved muscovite mica. This provides the possibility of very close approach of the two surfaces. As a result, it was possible to measure the transition region between the retarded and nonretarded van der Waals forces for this particular material. Cylindrical surfaces of radii between 0.4 and 2 cm, obtained by wrapping mica sheets around glass cylinders, were used to measure the force. The procedure used in the making of the mica cylinders led to uncertainties of 50%

in their radius. A spring-type balance based on the jump method was used (for large separations, modifications to this were made). Here, the force of attachment of one of the cylinders to an extended spring was overcome by the attractive force from the opposite cylindrical surface. By using springs of different extensions and different spring constants, a variety of distances could be measured. Multiple-beam interferometry was used for the measurement of the surface separation, with a reported resolution of 0.3 nm. In what appears to be the final work in this regard (Israelachvili and Tabor 1972), a sharp transition from the retarded to the nonretarded van der Waals force was reported at 12 nm [earlier work had measured a transition at larger surface separations (Tabor and Winterton 1968)]. Later reanalysis of the data with more precise spectral properties for the mica revealed that the data could be reconciled with the calculations only if errors of at least 30% in the radius of curvature of the mica cylinders were admitted (White *et al.* 1976). The possibility of changes in the spectral properties of the mica surface used to make the cylinders was also mentioned to explain the discrepancy. The separation on contact of the two surfaces was assumed to be zero, i.e. the surfaces were assumed to completely free of dust, impurities, and any atomic steps on the cleaved surface. Additionally, as mica is a nonconductor which can easily accumulate static charge, the role of electrostatic forces between the cylinders is hard to estimate, and was not considered in the experiments.

18.1.4   *Experiments by van Blockland and Overbeek*

The next major set of improved experiments with metallic surfaces was performed by van Blockland and Overbeek (1978). Here, many of the improvements achieved with the dielectric surfaces previously used were incorporated. Also, care was taken to address many of the concerns discussed above. [Earlier measurements by the same group (Rouweler and Overbeek 1971) with dielectric surfaces did not report on the effects of the chemical purity of the surfaces or the role of electrostatic forces between them.] The final improved version of the experiment using metallic surfaces was done by van Blockland and Overbeek (1978). The measurements were performed using a spring balance. The force was measured between a lens and a flat plate coated with either 100 ± 5 nm or 50 ± 5 nm of chromium. The chromium surface was expected to be covered with 1–2 nm of surface oxide. Water vapor was used to reduce the surface charges. This use of water vapor might have affected further the chemical purity of the metal surface. At the outset, the authors recognized several outstanding problems in Casimir force measurements. The first of them was that there was a potential difference between the two surfaces, leading to electrostatic forces which complicated the measurement. The second stubborn problem was that an exact determination of the separation distance between the two surfaces needed to be performed. The third problem, inherent in some measurement schemes, required an exact determination of the nonzero surface separation on contact of the two surfaces.

The above authors then tried to address these problems. The first problem was dealt with by two methods: by looking for a minimum in the Casimir force as a function of an applied voltage, and by measuring the potential difference from the intersection point of the electrostatic forces with application of positive and negative voltages. The two methods yielded approximately consistent values for the potential difference, of between 19 and 20 mV. This potential difference results in an electric force equal to the Casimir force at around 400 nm surface separation. Thus, to measure the Casimir force, the experiment had to be carried out with a compensating voltage present at all times.

The separation distance between the two surfaces was measured through a measurement of the lens–plate capacitance using a Schering bridge. This capacitance method was applicable only for a relative determination of distances, as the capacitances of cables and stray capacitances were of the same order as those between the spherical surface and the plate. Additional problems such as the tilt of the lens with respect to the plate were recognized by the authors. The distance was calibrated with the help of the electrostatic force at a few points. The force was measured for distances between 132 and 670 nm for the 100 nm thick metal coating. Only distances larger than 260 nm could be probed for the 50 nm metal coating.

The theoretical treatment of metallic chromium was noted to be problematic as it has two strong absorption bands around 600 nm. Given this, it was very hard to develop a complete description based on the Lifshitz theory, and some empirical treatment was necessary. The imaginary part of the dielectric permittivity corresponding to this absorption was modeled by a Lorentz atom (Krupp 1967). The two overlapping absorption bands were treated as a single absorption band. The strength of this absorption band could only be taken into account approximately in the theoretical modeling. However, this absorption band was found to cause about 40% of the total force. The long-wavelength response of chromium was modeled as that of a Drude metal with a plasma frequency based on an electron number density of 1.15×10^{22} cm^{-3}. With this theoretical treatment, the measured force was shown to be consistent with the theory.

The authors of the study estimated the effect of surface roughness (see Chapter 17), which was neglected in the theoretical treatment, to make a contribution of order 10%. The relative uncertainty in the measured force was reported to be around 25% near 150 nm separation but much larger around 500 nm separation. The authors reported that noise came from the force measurement apparatus. Given the above, we can estimate the experimental error to be of about 50%. But it is worth noting that this was the first experiment to grapple with all of the important systematics and other factors which are necessary to make a clear measurement of the Casimir force. This experiment can therefore be considered as the first unabiguous demonstration of the Casimir force between metallic surfaces. Thus, it was also the first measurement of surface forces, in general, where an independent estimate of the experimental precision could be attempted (though none was provided by the authors).

18.1.5 Dynamical measurements by Hunklinger and Arnold et al.

In theory, dynamical force measurements are more sensitive, as the signal (and the noise) in a narrow bandwidth is monitored. A dynamical force measurement technique was first used to measure the Casimir force between silica surfaces and silica surfaces coated with a thin layer of silicon (Hunklinger et al. 1972, Arnold et al. 1979). Here, a glass lens of radius 2.5 cm was attached to a metal-coated top membrane of a loudspeaker, while a flat glass plate was made the top surface of a microphone. In one case the glass surfaces were coated with silicon (Arnold et al. 1979). A sinusoidal voltage (at the microphone resonance frequency of 3 kHz) was applied to the loudspeaker such that the distance between the two surfaces also varied sinusoidally. The variation resulted in a sinusoidal oscillation of the flat plate on the microphone due to the Casimir force. This oscillation of the microphone was detected. The calibration was done by removing the plate and lens and applying an electrostatic voltage between the top of the loudspeaker and the microphone. A probable error of 20% (Hunklinger et al. 1972) or 50% (Arnold et al. 1979) in the force calibration was reported. The possible force sensitivity was about 10^{-7} dyn. The electrostatic force was minimized by use of water vapor and acetic acid vapor. No measurement of the residual electrostatic force was provided. The compensation of such forces was not possible, given the use of insulating surfaces such as silica. Even when Si surfaces were used, they were coated onto glass and not electrically connected. Given the glass manufacturers' roughness specifications of 50 nm for the surfaces, the surface separation on contact was estimated to be around 80 nm. Deviations from the expected behavior were found for separation distances below 300 nm and larger than 800 nm. Such deviations might be possible owing to the presence of the "gel layer" pointed out by Sparnaay (1958) and the role of electrostatic charges.

The change in the force on irradiation of Si-coated glass with white light was also reported. Surprisingly, illumination led to changes in the force only for large separation distances greater than 0.3 μm.

18.1.6 Measurements of the Casimir–Polder force by Sukenik and Hinds et al.

The magnitude and the distance dependence of the Casimir–Polder force acting between an atom and a cavity wall were measured by Sukenik et al. (1993) when they studied the deflection of ground-state Na atoms passing through a micron-sized gold cavity. The intensity of the atomic beam transmitted through the cavity was measured as a function of the plate separation. The measurement scheme was as follows. Na atoms at 180°C effused from a vertical oven slit into a vacuum of approximately 10^{-7} Torr. After traveling a distance of 18 cm, they entered a vertical gold cavity 3 cm high, 8 mm long, and adjustable in width from 0.5 to 8 μm (Sukenik et al. 1993). The cavity walls were made of thermally evaporated chromium (≈ 0.7 nm) followed by gold (42 ± 3 nm), on two flat silica plates. The plates were arranged so that they touched each other along one side to form a wedge. The distance between the plates on the opposite side was controlled by a nickel foil spacer. In the cavity, atoms were deflected by the atom–wall interaction. For ground-state atoms between parallel ideal-metal walls, a position-dependent interaction potential was found by Barton (1987a, 1987b). When the atom is not too close to one of the cavity walls, this potential depends on the distance between the walls as a^{-4}, i.e. in the same way as the Casimir–Polder potential (16.28). For a narrow cavity of plate separation a, the van der Waals dependence $\sim a^{-3}$ is reproduced, as given in eqn (16.24). In the experiment, the atom–wall interaction caused the Na atoms to stick to the walls of the cavity. Thus the fractional transmission of atoms through the cavity was a measure of the strength of this interaction. The atoms that exited from the cavity were resonantly excited and ionized by two laser beams so that they could be detected using an electron multiplier. The transmitted atom fraction (normalized to the transmission at 6 μm separation) was measured at different laser excitation positions along the wedge. The corresponding plate separation at those positions was determined using the interference pattern with Hg green light ($\lambda = 546$ nm) or Na yellow light ($\lambda = 589$ nm). The data were compared with a Monte Carlo calculation in which atoms having a Maxwell–Boltzmann velocity distribution propagated through a cavity under the influence of the Casimir–Polder or van der Waals potential. The data were shown to be in agreement with the Casimir–Polder potential over the measurement range from 0.75 to 7.5 μm. The same data within the range of separations from 0.75 to 1.15 μm were inconsistent with the hypothesis of a van der Waals interaction between an atom and the cavity wall. The magnitude of the Casimir–Polder interaction potential was confirmed with a relative error of 13%. The authors discussed a possible correction to the theoretical potentials used due to the nonzero skin depth of gold and found it negligible. The surface roughness correction and thermal effects at separations above 1.2 μm were not taken into account (the latter might be important, keeping in mind that the measurements were done to separation distances as large as 7.5 μm).

18.2 General requirements following from the older measurements

The first experiments dealing with the Casimir force clarified the problems to be solved so that precise and reproducible measurements could be performed. From the instrumental standpoint there are clear requirements, such as an extremely high force sensitivity and the capability to reproducibly measure the separation between the two surfaces. Other than these, there are material requirements necessary for a good measurement of the Casimir force. These fundamental requirements, as spelled out by Sparnaay (1958, 1989), are:

1. The plate surfaces must be completely free of chemical impurities and dust particles.
2. Precise, independent, and reproducible measurements of the separation between the two surfaces must be performed. In particular, the fact that the average distance on contact of the two surfaces is nonzero owing to the roughness of metal surfaces and the presence of dust must be taken into account.

3. Low electrostatic charges on the surfaces and a low potential difference between the surfaces are necessary. Note that there can exist a large potential difference between clean, grounded metallic surfaces owing to differences between the work functions of the materials used, and the cables used to ground the metal surfaces. Thus an independent measurement of the residual electrostatic force is absolutely necessary.

Each of the above instrumental and material requirements is difficult to comply with in practice and they are certainly very difficult to comply with together. They have bedeviled this field because at least one of the above was neglected in all of the early force measurements. Regarding the material requirements, as pointed out by Sparnaay (1958), requirement 1 was ignored in the experiments with glass and quartz surfaces (Derjaguin 1934, Derjaguin et al. 1956, Derjaguin and Abrikosova 1958), where surface reactions with moisture and silicone oil from the vacuum apparatus led to the formation of a "gel layer" on the surface. Sparnaay (1958) expected this gel layer to completely modify the forces for surface separation distances less than $1.5\,\mu$m. The last two requirements are particularly difficult to meet in the case of nonconductive surfaces or substrates such as glass and quartz (Derjaguin 1934, Derjaguin et al. 1956, Derjaguin et al. 1987, Hunklinger et al. 1972, Arnold et al. 1979), and mica (Tabor and Winterton 1968, Israelachvili and Tabor 1972, Israelachvili 1992, White et al. 1976). Yet all of these early measurements possibly neglected the systematic correction due to the electrostatic force in their experiments. Some other requirements, such as the necessity to determine the exact surface separation distance and to take account of surface roughness, were neglected in all but oldest experiments. Some experimenters have tried to use an ionized environment (Overbeek and Sparnaay 1954) to neutralize the static charges but have reported additional electrostatic effects. Also, all early measurements took the surface separation on contact to be zero. This can be a significant error for large flat surfaces or surfaces with large radius of curvature, as the inevitable presence of obstacles prevents achieving close contact between them. As stated by Sparnaay (1958), this is also true for some experiments with Pt metallic wires, where the point of contact was assumed to be zero separation distance (Derjaguin et al. 1987). Thus independent checks of the surface separation are necessary for correct analysis of the data.

Of the earlier experiments with metallic surfaces, only two meet at least some of the stringent criteria necessary for careful measurements of the Casimir force. The first one is that by Sparnaay (1958). The second is that by van Blockland and Overbeek (1978). It should be mentioned that both experiments were a culmination of many years of improvements, references for which are provided in the respective publications.

18.3 Rigorous procedures for comparison of experiment and theory

The comparison between experiment and theory for Casimir force measurements is a complicated problem which has been addressed in the literature in sufficient detail only recently (Decca et al. 2005b, Chen et al. 2006b, Klimchitskaya et al. 2006b). The difficulties which arise in such a comparison are connected with the impossibility of measuring the separation distance to the desired precision and with the fact that the Casimir force is a strongly nonlinear function of the separation. Insufficient information concerning the material properties of the test bodies also presents a challenge to the theorists. Below, we demonstrate that comparisons between measurement data and theoretical computations for Casimir force experiments can be performed in three independent steps. In the first step, the total experimental error in the measurement data must be ascertained, regardless of the theory to be used. Then the theoretical uncertainties must be analyzed and the total theoretical error calculated. Finally, we discuss various statistical approaches to comparing experiment with theory.

18.3.1 *Experimental errors and precision*

In the first step of the comparison between experiment and theory, we deal only with the experimental data for the measured quantity. Our aim is to characterize how precise these data are, regardless of the theory. To do this, the total experimental error of the measurement results must be calculated as a combination of systematic and random errors. Let the measured quantity be denoted by Π^{expt}. This may be, for instance, the separation distance between the two bodies, $\Pi^{\text{expt}} = a$, or the Casimir pressure (or force) as a function of the separation distance, $\Pi^{\text{expt}} = \Pi^{\text{expt}}(a) = P^{\text{expt}}(a)$ [or $F^{\text{expt}}(a)$, respectively]. In the two latter cases, the errors may also be separation-dependent.

We start with a discussion of *systematic errors*. We denote the absolute error of the physical quantity Π by $\Delta\Pi$. In so doing, systematic errors (and other types of errors) are denoted by an additional superscript, for example $\Delta^{\text{syst}}\Pi$. In each experiment, there are several (total number J) sources of systematic errors $\Delta_i^{\text{syst}}\Pi^{\text{expt}}$, which are usually called *absolute systematic errors*, where $1 \leq i \leq J$. The respective relative systematic errors are defined as

$$\delta_i^{\text{syst}}\Pi^{\text{expt}} = \frac{\Delta_i^{\text{syst}}\Pi^{\text{expt}}}{|\Pi^{\text{expt}}|}. \tag{18.1}$$

It is necessary to stress that both in metrology and in all natural sciences (physics, chemistry, biology, etc.) the term *systematic error* is used with two different meanings (Rabinovich 2000). According to the first meaning, a systematic error is some bias in a measurement which always makes the measured value higher or lower than the true value. Such systematic errors in the measurement results are usually removed using some known process, i.e. through a calibration. They can be also taken into account as corrections (see, for example, the description of the calibration procedure and an example of a correction in measurements of the Casimir force in Section 19.2.3). The systematic errors in this understanding are often called *systematic deviations*. Below, it is assumed that the experimental data under consideration are already free of such deviations.

Another meaning, which is used below in this book, defines the systematic errors as the errors of a calibrated measurement device. The errors of a theoretical formula used to convert a directly measured quantity into an indirectly measured one (see e.g. Section 19.3.2) are also considered as systematic. In accordance with common understanding, the error of a calibrated device is the smallest fractional division of the scale of the device. At the limits of this range, systematic errors are considered as random quantities characterized by a uniform distribution (equal probability). The total systematic error at a chosen confidence level is obtained from the following statistical rule (Rabinovich 2000):

$$\Delta^{\text{syst}}\Pi^{\text{expt}} = \min\left[\sum_{i=1}^{J}\Delta_i^{\text{syst}}\Pi^{\text{expt}}, k_\beta^{(J)}\sqrt{\sum_{i=1}^{J}\left(\Delta_i^{\text{syst}}\Pi^{\text{expt}}\right)^2}\right]. \quad (18.2)$$

Here, β is the confidence level, and $k_\beta^{(J)}$ is a tabulated coefficient depending on β and on the total number of systematic errors J. The same rule is also valid for the combination of the relative systematic errors (18.1) leading to the combined systematic error $\delta^{\text{syst}}\Pi^{\text{expt}}$. Note that in precise experiments, errors are usually determined at a confidence level of 95% or higher. In this case, for instance, $k_{0.95}^{(2)} = 1.1$ and $k_{0.95}^{(4)} = 1.12$. However, in Casimir force measurements much lower confidence levels are often used, and sometimes the experimental errors are not reported at all (see Chapter 19).

Next we discuss *random* experimental errors. Usually the measurement of the Casimir force or pressure at separations a_i ($1 \le i \le i_{\max}$) within a separation interval ($a_{\min}, a_{\max}$) is repeated several times, up to a total number n. This is done in order to decrease the random error and to narrow the confidence interval. All the measurement data from n sets of measurements can be represented as pairs $[a_{ij}, \Pi^{\text{expt}}(a_{ij})]$, where $1 \le i \le i_{\max}$, $1 \le j \le n$. If the separations with fixed i are approximately the same in all sets of measurements (i.e. $a_{ij} \approx a_i$), the mean and the variance of the mean at each separation a_i are obtained in the standard way (Rabinovich 2000):

$$\bar{\Pi}_i^{\text{expt}} = \frac{1}{n}\sum_{j=1}^{n}\Pi^{\text{expt}}(a_{ij}), \quad s_{\bar{\Pi}_i}^2 = \frac{1}{n(n-1)}\sum_{j=1}^{n}\left[\Pi^{\text{expt}}(a_{ij}) - \bar{\Pi}_i^{\text{expt}}\right]^2. \quad (18.3)$$

Direct calculations show that typical mean values $\bar{\Pi}_i^{\text{expt}}$ are uniform, i.e. they change smoothly with a change of i. The variances of the mean, $s_{\bar{\Pi}_i}$, needed for the determination of the random error, however, are not uniform. To smooth them, a special procedure is used in statistics (Brownlee 1965, Cochran 1954). This procedure is as follows. At each separation a_0, in order to find the uniform variance of a mean, we consider not only this given separation point, but also N neighboring points a_i on both sides of a_0. Then the smoothed variance of the mean at the point a_0 is given by (Brownlee 1965, Cochran 1954)

$$s_{\bar{\Pi}}^2(a_0) = \max\left[N\sum_{i=1}^{N}\lambda_i^2 s_{\bar{\Pi}_i}^2\right], \quad (18.4)$$

where the λ_i are statistical weights. The maximum in eqn (18.4) is chosen by comparing values from two sets of λ_i,

$$\lambda_i = \frac{1}{N}, \quad \lambda_i = \frac{1}{c_i \sum_{k=1}^{N} c_k^{-1}}, \tag{18.5}$$

where the constants c_k are determined from

$$s_{\bar{\Pi}_1}^2 : s_{\bar{\Pi}_2}^2 : \cdots s_{\bar{\Pi}_N}^2 = c_1 : c_2 : \cdots : c_N. \tag{18.6}$$

The number of points N is chosen such that any further increase of N does not influence the magnitude of $s_{\bar{\Pi}}(a_0)$ obtained. Note that the maximum in eqn (18.4) leads to the most conservative results, i.e. it overestimates the random error. Finally, the confidence interval for the quantity $\Pi^{\mathrm{expt}}(a_0)$, determined at a confidence level β, takes the form

$$\left[\bar{\Pi}^{\mathrm{expt}}(a_0) - \Delta^{\mathrm{rand}} \Pi^{\mathrm{expt}}(a_0), \bar{\Pi}^{\mathrm{expt}}(a_0) + \Delta^{\mathrm{rand}} \Pi^{\mathrm{expt}}(a_0) \right]. \tag{18.7}$$

Here, the random absolute error is given by

$$\Delta^{\mathrm{rand}} \Pi^{\mathrm{expt}}(a_0) = s_{\bar{\Pi}}(a_0) t_{(1+\beta)/2}(n-1), \tag{18.8}$$

where $t_p(f)$ can be found in tables for Student's t-distribution [see e.g. the textbook by Brandt (1976)].

Note that the experimental points used in the determination of the confidence interval should be checked for the presence of *outlying* results (Rabinovich 2000). For this purpose, at each point a_i it is necessary to consider the quantity

$$T_i = \frac{1}{\sqrt{n} s_{\bar{\Pi}_i}} \max |\Pi^{\mathrm{expt}}(a_{ij}) - \bar{\Pi}_i^{\mathrm{expt}}|, \tag{18.9}$$

where $\bar{\Pi}_i^{\mathrm{expt}}$ and $s_{\bar{\Pi}_i}$ are defined in eqn (18.3) and the evaluation of the maximum is done considering all measurement sets $1 \leq j \leq n$. Let us assume that the maximum is reached for the set where $j = j_0$. If the inequality $T_i > T_{n,1-\beta}$ is satisfied, where the $T_{n,1-\beta}$ are tabulated quantities, then the measurement result $\Pi^{\mathrm{expt}}(a_{ij_0})$ belonging to the set where $j = j_0$ is an outlying result with a confidence level β (Rabinovich 2000). If the measurement set $j = j_0$ contains outlying results at many different points a_i, it should be rejected at the confidence level β and not used in the data analysis.

Now we consider a more complicated experimental situation where the separations a_{ij} with fixed i but different j may be different. This happens when it is difficult to obtain approximately the same intermediate distances $a_1, a_2, \ldots$ in every set of measurements. In such a situation, the entire separation interval $(a_{\min}, a_{\max})$ is divided into partial subintervals of length $2\Delta a$, where Δa is the total absolute error in the measurements of the separations. Each subinterval, numbered k, contains a group of m_k points $a_{ij} \equiv a_{kl}$, where $1 \leq l \leq m_k$. Inside each subinterval, all points a_{kl} can be considered as equivalent, because within an interval of width $2\,\Delta a$ the random quantity representing the absolute separation is distributed uniformly. The mean and variance of the mean of the physical quantity $\Pi^{\rm expt}$ for the subinterval k are then defined as

$$\bar{\Pi}_k^{\rm expt} = \frac{1}{m_k}\sum_{l=1}^{m_k} \Pi^{\rm expt}(a_{kl}), \quad s_{\bar{\Pi}_k}^2 = \frac{1}{m_k(m_k-1)}\sum_{l=1}^{m_k}\left[\Pi^{\rm expt}(a_{kl}) - \bar{\Pi}_k^{\rm expt}\right]^2. \tag{18.10}$$

Similarly to the case where $a_{ij} \approx a_i$ considered above, the mean values $\bar{\Pi}_k^{\rm expt}$ are uniform, but the variances of the mean, $s_{\bar{\Pi}_k}$, are not uniform. To smooth them, the same statistical procedure as described above can be used. In this case one considers several subintervals (N altogether) situated to the left and to the right of the subinterval containing the point a_0. The smoothed variance of the mean at the point a_0, $s_{\bar{\Pi}}(a_0)$, is determined from eqn (18.4) with the statistical weights defined in eqns (18.5) and (18.6). Then, the confidence interval for the confidence level β is found from eqn (18.7) where, instead of eqn (18.8), the random error is

$$\Delta^{\rm rand}\Pi^{\rm expt}(a_0) = s_{\bar{\Pi}}(a_0)t_{(1+\beta)/2}(\min m_k - 1). \tag{18.11}$$

It is evident that eqn (18.11) leads to a larger random error than does eqn (18.8) because $\min m_k < n$, where n is the total number of measurement sets. In fact, if the intermediate separation distances are not approximately the same in each set of measurements, the averaging of the data can be performed only over a smaller number than the number of actual repetitions n.

In the case where the separations a_{ij} with fixed i but different j are different, one must also exclude outlying measurement results (if any) from the error analysis. For this purpose, instead of the quantity T_i obtained from eqn (18.9), one must use

$$T_k = \frac{1}{\sqrt{m_k}s_{\bar{\Pi}_k}}\max\left|\Pi^{\rm expt}(a_{kl}) - \bar{\Pi}_k^{\rm expt}\right|, \tag{18.12}$$

where the mean value and the variance are defined in eqn (18.10) and the maximum is taken over all m_k points belonging to subinterval number k.

Similarly to the relative systematic error, one can define the relative random error

$$\delta^{\rm rand}\Pi^{\rm expt} = \frac{\Delta^{\rm rand}\Pi^{\rm expt}}{|\Pi^{\rm expt}|}. \tag{18.13}$$

To find the total experimental error $\Delta^{\rm tot}\Pi^{\rm expt}(a)$ in the measurements of the quantity $\Pi^{\rm expt}(a)$, one must combine the random and systematic errors. The random error is described by the normal (or Student) distribution. The systematic error is described by a combination of uniform distributions. To be conservative (i.e. overestimating the errors), one can assume that the resulting systematic error is described by a uniform distribution as well (other assumptions would lead to a smaller total error). There are various methods in statistics to combine errors described by normal and uniform distributions (Rabinovich 2000). A widely used method is based on the value of the quantity

$$r(a) = \frac{\Delta^{\text{syst}} \Pi^{\text{expt}}(a)}{s_{\bar{\Pi}}(a)}. \qquad (18.14)$$

According to this method, at all separations where $r(a) < 0.8$, the contribution of the systematic error to the total experimental error is negligible. In this case one can put

$$\Delta^{\text{tot}} \Pi^{\text{expt}}(a) = \Delta^{\text{rand}} \Pi^{\text{expt}}(a) \qquad (18.15)$$

at a 95% confidence level, provided the random error is also determined with the same confidence using eqn (18.8) or (18.11). If the inequality $r(a) > 8$ is valid, the random error is negligible and the total experimental error determined at a 95% confidence level is

$$\Delta^{\text{tot}} \Pi^{\text{expt}}(a) = \Delta^{\text{syst}} \Pi^{\text{expt}}(a), \qquad (18.16)$$

where the systematic error is given by eqn (18.2) with $\beta = 0.95$. Note that eqn (18.16) is generally fulfilled for precise experiments in metrology, where the systematic error alone determines the total error and all necessary measures are undertaken to make the random error negligible. For the moment, there is only one experiment on the physics of the Casimir effect satisfying this condition (Decca et al. 2007a, 2007b), described in Section 19.3.4. In the range of separations where $0.8 \leq r(a) \leq 8$, the combination of the errors is performed using the rule

$$\Delta^{\text{tot}} \Pi^{\text{expt}}(a) = q_\beta(r) \left[\Delta^{\text{rand}} \Pi^{\text{expt}}(a) + \Delta^{\text{syst}} \Pi^{\text{expt}}(a) \right]. \qquad (18.17)$$

The coefficient $q_\beta(r)$ at a confidence level $\beta = 0.95$ varies between 0.71 and 0.81 depending on the value of $r(a)$. To be conservative, one can use $q_{0.95}(r) = 0.8$.

We emphasize that the total experimental error of the measurements of the Casimir force and pressure $\Delta^{\text{tot}} \Pi^{\text{expt}}(a)$ and the corresponding total relative error

$$\delta^{\text{tot}} \Pi^{\text{expt}}(a) = \frac{\Delta^{\text{tot}} \Pi^{\text{expt}}(a)}{|\Pi^{\text{expt}}(a)|} \qquad (18.18)$$

characterize the *precision* of an experiment on its own, without comparison with any theory.

18.3.2 Theoretical uncertainties for real materials

The theoretical values of the quantity $\Pi(a)$ (e.g. of the Casimir pressure between two parallel plates or the force between a sphere and a plate) are also burdened with some errors and uncertainties. The most important theoretical tool for calculating the Casimir force between real materials is the Lifshitz theory. For the simplest case of plane parallel plates it expresses the force in terms of the frequency-dependent dielectric permittivities of the plate materials. The dielectric permittivity of a material can be found using optical data for the complex index of refraction (see Sections 12.6.1 and 13.3). These data are determined with some errors and may also depend on the particular sample. As a result, the theoretical values $\Pi(a)$ are computed with some error $\Delta_1\Pi^{\text{theor}}(a)$. Note that there are different sets of optical data in the literature for films made of the same material but of different qualities and thicknesses which differ by far more than the error in the optical measurements. When such sets of the optical data are used in the Lifshitz theory, one can arrive at computational results for $\Pi^{\text{theor}}(a)$ differing by about 5% (Pirozhenko et al. 2006). But this, however, should not be confused with the relative theoretical error $\delta_1\Pi^{\text{theor}}(a) = \Delta_1\Pi^{\text{theor}}(a)/|\Pi^{\text{theor}}(a)|$ due to the errors in the optical data, because the former is usually an order of magnitude smaller (see Section 19.2.3). The point to note is that the choice of a specific set of optical data from those available in the literature is similar to a hypothesis, which, however, can be independently verified when the experiment is compared with theory (see the next subsection).

Another theoretical error, which we may denote by $\Delta_2\Pi^{\text{theor}}(a)$, is connected with the deviation of the boundary surfaces from the perfect geometrical shape (because of roughness, corrugations, etc.). The respective corrections to the Casimir pressure or force can be calculated only approximately (see Chapter 17), and this results in some error. There are other theoretical errors due to surface effects, for example uncertainties due to patch potentials (see Sections 19.2.3 and 19.3.3).

An important theoretical error arises when the proximity force approximation is used with the configuration of a sphere above a plate. In this case the proximity force approximation is part of the theoretical expression for the Casimir force acting between a sphere and a plate, and the inherent approximations result in some theoretical error $\Delta_3\Pi^{\text{theor}}(a)$. Note, however, that in some cases the error due to the use of the proximity force approximation should be included in the evaluation of the experimental systematic errors. This happens in the dymanic determination of the Casimir pressure between two parallel plates by means of the frequency shift of a sphere oscillating above a plate (see Section 19.3). Here, the exact Lifshitz formula for two parallel plates is used in the theoretical description. The proximity force approximation is applied only for the recalculation of one set of measurement data into another one in the process of making an indirect measurement.

Another example where the same error can be relevant to both experiment and theory is the error in the measurement of the separation distance Δa. Taken alone, this error is entirely experimental. However, in one of the approaches to the comparison between experiment and theory (see the next subsection), the theoretical values $\Pi^{\text{theor}}(a)$ are calculated not throughout the entire measurement range $(a_{\min}, a_{\max})$ but at the experimental separations a_i. In this case the values $\Pi^{\text{theor}}(a_i)$, in addition to the errors discussed above, are affected by one more error. This can be approximately estimated as (Iannuzzi et al. 2004a)

$$\delta_4 \Pi^{\text{theor}}(a_i) = \frac{\Delta_4 \Pi^{\text{theor}}(a_i)}{|\Pi^{\text{theor}}(a_i)|} = \alpha \frac{\Delta a}{a_i} \qquad (18.19)$$

if one accepts, for simplicity, the model of two parallel ideal-metal plates ($\alpha = 4$) or an ideal-metal sphere above an ideal-metal plate ($\alpha = 3$).

All of the theoretical errors described above must be combined together to obtain the total theoretical error representing the accuracy of the theory. For this purpose, we assume that all of them are described by a uniform distribution and in this sense are similar to systematic errors. Note that any other assumption would lead to a smaller total error. Then one can apply the statistical rule in eqn (18.2) with an appropriate value of J and $\beta = 0.95$ to obtain the total theoretical error $\Delta^{\text{tot}} \Pi^{\text{theor}}(a)$ at the 95% confidence level. Together with the total experimental error, this quantity can be used to find a rigorous measure of the agreement between experiment and theory.

18.3.3 Statistical framework for the comparison of theory with experiment

When experimental data from Casimir force measurements are compared with theory, the important question of how to quantitatively characterize the agreement between them has to be addressed. In many experiments (see Chapter 19), the agreement between data and theory was characterized by a global quantity, the root-mean-square deviation

$$\sigma_N = \left\{ \frac{1}{N} \sum_{k=1}^{N} \left[\Pi^{\text{theor}}(a_k) - \Pi^{\text{expt}}(a_k) \right]^2 \right\}^{1/2}, \qquad (18.20)$$

where $N = ni_{\max}$ is the total number of measurements in all measurement sets. It is known, however, that this method is not appropriate for strongly nonlinear quantities (Rabinovich 2000) such as the Casimir force, which changes rapidly with separation. It has been shown (Ederth 2000) that the calculation of the root-mean-square deviation between experiment and theory leads in this case to different results when applied in different separation intervals. Because of this, alternative, local methods have been suggested (Decca et al. 2005b, Chen et al. 2006b, Klimchitskaya et al. 2006b), using statistical methods in which the measures of agreement are separation-dependent.

In the first method, the experimental data are traditionally represented as crosses centered at points with coordinates $[a_i, \Pi^{\text{expt}}(a_i)]$ where the length of the horizontal arms is equal to $2\Delta a$ and the height of the vertical arms is equal to $2\Delta^{\text{tot}} \Pi^{\text{expt}}(a_i)$. Both of the absolute errors Δa and $\Delta^{\text{tot}} \Pi^{\text{expt}}(a_i)$ are meant to be determined at a common confidence level β (95%, for instance) as described above. The theoretical values $\Pi^{\text{theor}}(a)$ are computed over the entire measurement range $(a_{\min}, a_{\max})$ and presented in the form of a band as a function of separation. The width of this band is equal to $2\Delta^{\text{tot}} \Pi^{\text{theor}}(a)$, where the total theoretical error is determined according to the method in the previous subsection at the same confidence level β as for the experimental errors. If the theoretical band overlaps with the experimental data, including their errors, the data are *consistent* with the theory. If there is no overlap between the experimental crosses and the theoretical band over a wide separation range, one concludes that the theory is *excluded* by the data at a confidence level β within this range of separations.

Below, we consider situations where the measurement data are compared with the various theoretical approaches discussed in Chapters 12 and 14 (for instance, based on the Drude or the plasma model or with the inclusion or neglect of the small dc conductivity of a dielectric at nonzero temperature) and where the calculations are done with different sets of optical data available in the literature. Each theoretical approach using some definite set of optical data can be considered as a *hypothesis* and the process of its comparison with the data as a *verification of the hypothesis*. The procedures for such verification are well developed in statistics. To make a comparison of different hypotheses with experiment, one plots the theoretical bands and the experimental data as described above on one graph. A hypothesis whose corresponding band does not overlap with the data over a wide separation region must be rejected at a confidence level β. Importantly, the data and the theory are compared not at one point but over a wide separation region where the distance dependence is nonlinear. Thus, a set of optical data which is inconsistent with the experimental results when combined with any theoretical approach should be considered as irrelevant to the actual properties of the film used in the experiment. We note that in this first method of comparison between experiment and theory, the theoretical values $\Pi^{\text{theor}}(a)$ are computed over the entire measurement range $a_{\min} \leq a \leq a_{\max}$. They are not affected by the error Δa in the measurement of the separation distances.

A second, local method to compare theory with experiment is based on consideration of the confidence interval for the random quantity $\Pi^{\text{theor}}(a) - \Pi^{\text{expt}}(a)$ (Decca *et al.* 2005b, Chen *et al.* 2006b, Klimchitskaya *et al.* 2006b). Here, a is the separation at which the quantity Π^{expt} has been measured with an experimental error $\Delta^{\text{tot}}\Pi^{\text{expt}}(a)$. The theoretical value Π^{theor} must be computed at the same separation a, which is known with an error Δa. Because of this, in this second method of comparison between experiment and theory, $\Delta^{\text{tot}}\Pi^{\text{theor}}(a)$ includes the contribution (18.19), depending on the experimental error in the measurement of the separation. Note that if several measurements are performed at practically the same separation a, one should consider the random quantity $\Pi^{\text{theor}}(a) - \bar{\Pi}^{\text{expt}}(a)$ to characterize the deviations between theory and experiment.

The absolute error of the random quantity characterizing the deviations between theory and experiment, $\Xi_\Pi(a)$, at a confidence level β, can be obtained from the total theoretical and experimental errors, $\Delta^{\text{tot}}\Pi^{\text{theor}}(a)$ and $\Delta^{\text{tot}}\Pi^{\text{expt}}(a)$. To be conservative, we use for this purpose the composition rule (18.2) with $J = 2$, valid for two uniform distributions (otherwise the error would be smaller):

$$\Xi_\Pi(a) = \min \left\{ \Delta^{\text{tot}} \Pi^{\text{theor}}(a) + \Delta^{\text{tot}} \Pi^{\text{expt}}(a), \right. \quad (18.21)$$

$$\left. k_\beta^{(2)} \sqrt{[\Delta^{\text{tot}} \Pi^{\text{theor}}(a)]^2 + [\Delta^{\text{tot}} \Pi^{\text{expt}}(a)]^2} \right\}.$$

The confidence interval for the quantity $\Pi^{\text{theor}}(a) - \Pi^{\text{expt}}(a)$ at a 95% confidence level is given by $[-\Xi_\Pi(a), \Xi_\Pi(a)]$, where $\Xi_\Pi(a)$ is determined from eqn (18.21) with $k_{0.95}^{(2)} = 1.1$. When one compares a theoretical approach with experimental data, the differences between the theoretical and experimental values of the quantity Π may or may not belong to this interval at each fixed value of a. A theoretical approach for which no fewer than 95% of the differences $\Pi^{\text{theor}}(a) - \Pi^{\text{expt}}(a)$ belong to the interval $[-\Xi_\Pi(a), \Xi_\Pi(a)]$ within any separation subinterval $[a_1, a_2]$ of the entire measurement range is *consistent* with the experiment. In this case the measure of agreement between experiment and theory is given by $\Xi_\Pi(a)/|\Pi^{\text{expt}}(a)|$. In contrast, if for some theoretical approach a subinterval $[a_1, a_2]$ exists where almost all differences $\Pi^{\text{theor}}(a) - \bar{\Pi}^{\text{expt}}(a)$ are outside the confidence interval $[-\Xi_\Pi(a), \Xi_\Pi(a)]$, this approach is excluded by experiment at separations from a_1 to a_2 at a 95% confidence level. If the theoretical approach (hypothesis) is excluded by experiment at a 95% confidence level, the probability that it is true is at most 5%. It may happen that several theoretical approaches $i = 1, 2, \ldots$ are consistent with experiment, i.e. no fewer than 95% of the differences $\Pi_i^{\text{theor}}(a) - \bar{\Pi}^{\text{expt}}(a)$ $(i = 1, 2, \ldots)$ belong to the confidence interval $[-\Xi_\Pi(a), \Xi_\Pi(a)]$ (such situations are considered in Chapters 19 and 20). The statistical criteria used do not allow one to indicate the probability of the event that one of these approaches or all of them are false. The rejection of some of the experimentally consistent approaches can be done on a theoretical basis only. For example, if the measurement is performed at room temperature and theoretical computations done both at $T = 0$ K and at $T = 300$ K are consistent with the data, the computation at $T = 300$ K can be considered as preferable. In fact, to reliably discriminate between two experimentally consistent theoretical approaches, more exact measurements are desirable. These general criteria are illustrated below in Chapters 19 and 20 using the examples of various Casimir force measurements.

In this subsection, we have considered two local methods for the comparison of experiment with theory in Casimir force measurements. It should be noted that the conclusions concerning the consistency or rejection of hypotheses do not depend on the method of comparison used. In Chapters 19 and 20, both methods are applied repeatedly, and in all cases the results obtained are in agreement. This allows one to conclude that the local methods for the comparison of experiment with theory provide a satisfactory statistical framework for Casimir force experiments.

19

MEASUREMENTS OF THE CASIMIR FORCE BETWEEN METALS

Based on the older measurements of the van der Waals and Casimir forces considered in Section 18.1, it is clear that metallic test bodies have major advantages in comparison with dielectric ones as they enable one to ensure low electrostatic charges on the surfaces and a low potential difference between the surfaces. Thus, the first experiments in the modern phase of Casimir force measurements, performed by various research groups in several countries, exploited test bodies coated with metallic films. The thickness of these films was large enough to produce the same Casimir force as in the case of bulk metal bodies. Another idea taken from the older measurements is the use of a spherical lens or a sphere as one of the test bodies [only one modern experiment, by Bressi *et al.* (2002), considered in Section 19.4, uses the original Casimir configuration of two parallel metallic plates]. This has spared investigators the difficult necessity to preserve parallelity of the plates and allowed close approach of the surfaces. In the modern phase of Casimir force measurements started with the experiment performed by Lamoreaux (1997), the above experience was combined with advanced positioning and force measurement techniques utilizing recent technological achievements in microfabrication, micromechanics, and fiber optics.

In this chapter we present the main breakthroughs in the measurement of the Casimir force between metallic surfaces. We start with the first demonstration of the corrections to the Casimir force due to the nonzero skin depth and surface roughness, obtained by means of an atomic force microscope by Mohideen and Roy (1998), Klimchitskaya *et al.* (1999), Roy *et al.* (1999), and Harris *et al.* (2000). The second breakthrough was a series of precise indirect measurements of the Casimir pressure between two parallel plates by means of a micromechanical torsional oscillator by Decca *et al.* (2003a, 2003b, 2005b, 2007a, 2007b). These measurements allowed a definitive choice between different theoretical approaches to the thermal Casimir force with real metal surfaces. Many other experiments performed in the last few years using metallic test bodies are also presented. The chapter ends with a brief discussion of some proposed experiments using metallic surfaces.

19.1 Experiment with torsion pendulum

Chronologically the first experiment in the recent phase of Casimir force measurements was performed by Lamoreaux (1997). While this experiment rekindled interest in the investigation of the Casimir force and stimulated further development in the field, the results obtained contained several uncertainties. The

Casimir force between an Au-coated spherical lens and a flat plate was measured using a torsion pendulum. A lens with a radius of 11.3 ± 0.1 cm [later corrected to 12.5 ± 0.3 cm (Lamoreaux 1998)] was used. The surfaces of the lens and plate were first coated with $0.5\,\mu$m of Cu followed by a $0.5\,\mu$m coating of Au. Both coatings were deposited by thermal evaporation. The lens was mounted on a piezoelectric stack and the plate on one arm of the torsion balance in vacuum. The other arm of the torsion balance formed the center electrode of dual parallel-plate capacitors C_1 and C_2. Thus, the position of this arm and consequently the angle of the torsion pendulum could be controlled by application of voltages to the plates of the dual capacitor. The Casimir force between the plate and lens surfaces resulted in a torque, leading to a change in the angle of the torsion balance. This change in angle, in turn, resulted in changes in the capacitances C_1 and C_2, which were detected with a phase-sensitive circuit. Then, compensating voltages were applied to the capacitors C_1 and C_2 through a feedback circuit to counteract the change in the angle of the torsion balance. These compensating voltages were a measure of the Casimir force.

The calibration of the measurement system was done electrostatically. When the lens and plate surfaces were grounded, a "shockingly large" (Lamoreaux 1997) potential difference of 430 mV was measured between the two surfaces. This large electrostatic potential difference was partially compensated by application of a voltage to the lens. From the analysis, there appears to have been an electrostatic force even after this compensation. The latter was determined only by fitting to the total force, including the Casimir force, above $1\,\mu$m. "Typically, the Casimir force had a magnitude of at least 20% of the electrostatic force at the point of closest approach" (Lamoreaux 1997). The uncertainty in the measurement of the absolute separations Δa "was normally less than $0.1\,\mu$m" (Lamoreaux 1997).

The lens was moved towards the plate in 16 steps by application of a voltage to the piezoelectric stack on which it was mounted. At each step, the restoring force, given by the change in the voltage required to keep the pendulum angle fixed, was noted. The maximum separation between the two surfaces was $12.3\,\mu$m. The average displacement for a 5.75 V step was about $0.75\,\mu$m. A considerable amount of hysteresis was noted between the up and down cycles, i.e. the approach and retraction of the two surfaces. The displacement as a function of the 16 applied voltages was measured to $0.01\,\mu$m accuracy with a laser interferometer. The total force was measured for separations from $10\,\mu$m to contact of the two surfaces. The experiment was repeated, and 216 up/down sweeps were used in the final data set. The total measured force data was binned into 15 surface separation points. Two of the important experimental values needed, (a) the residual electrostatic force and (b) the surface separation on contact of the two surfaces, were obtained by curve fitting of the total measured force for separations greater than $2\,\mu$m to the sum of the Casimir force and the electrostatic force. Neither systematic nor random experimental errors in the measured forces were reported. The total experimental error in the force measurements can be estimated as $\Delta^{\text{tot}} F^{\text{expt}} \approx$

10^{-11} N on the basis of the graphical information (Lamoreaux 1997).

Data were compared with theory for an ideal-metal lens and plate using the least-squares method. This is a global method for the comparison of experiment and theory, as discussed in Section 18.3.3. In comparison with the use of the root-mean-square deviation, the least-squares method contains one or more fitting parameters, chosen so as to make the deviation between experiment and theory smaller. Thus, this method cannot provide an adequate characterization of the agreement between experiment and theory over a wide range of separations for a quantity which is a strongly nonlinear function of separation (see Section 18.3.3). Corrections to the Casimir force due to surface roughness or the skin depth were not taken into account.

Agreement between experiment and theory at the level of 5% was reported in the measurement range from 0.6 to 6 μm (Lamoreaux 1997). This conclusion is incompatible with the magnitude of the thermal correction to the Casimir force (Bordag et al. 1998b), because at separations of 4, 5, and 6 μm the thermal correction is 86%, 129%, and 174%, respectively, of the zero-temperature force (6.51). The data, however, were found to be "not of sufficient accuracy to demonstrate the finite temperature correction" (Lamoreaux 1997). The contradiction can be reconciled if one takes into account the fact that the actual relative error of the force measurements at, for example, $a = 6\,\mu$m was not 5% but 700% (Bordag et al. 1998c). From this it follows that agreement of the data with the theory at the level of 5%–10% could exist only at separations of about 1 μm. Here, the thermal correction is relatively small, and the larger corrections due to the nonzero skin depth and surface roughness have opposite signs and partially compensate each other. Keeping in mind that the theoretical forces calculated at the experimental separations were burdened with an additional error of about $3\Delta a/a \approx 30\%$ at $a = 1\,\mu$m (see Section 18.3.2), the errors in the differences $F^{\text{expt}}(a_i) - F^{\text{theor}}(a_i)$, as shown in Fig. 4, bottom, of Lamoreaux (1997), were significantly underestimated.

In conclusion, this experiment introduced the modern phase-sensitive detection of forces and thus brought possible increased sensitivity to the measurement of Casimir forces. Two of the three required criteria outlined by Sparnaay (the second and third, as discussed in Section 18.2) were only partially accomplished. By using piezoelectric translation of the lens towards the plate, reproducible measurements of the surface separation could be done. However, the value of the residual electrostatic force and the surface separation on contact could only be determined by curve fitting of part of the experimental data to the expected Casimir force. Such a procedure biases that part of the experiment with the input value of the Casimir force. Thus it cannot be considered as an independent measurement of the surface separation or the residual electrostatic force.

The results of this experiment at about 1 μm separation were used (Torgerson and Lamoreaux 2004, Lamoreaux 2005) to exclude a theoretical approach to the thermal Casimir force which uses the Drude model at low frequencies (see Section 14.3.1). The latter predicts a thermal correction of -18.9% at $a = 1\,\mu$m (to be compared with a 1.2% thermal correction in the case of an ideal metal), which was not experimentally observed. At a high confidence level, the conclusion that the Drude model approach is inconsistent with the data was obtained in the much more precise experiments of Decca *et al.* (2003b, 2004, 2005b, 2007a, 2007b) considered in Section 19.3.

19.2 Experiments with an atomic force microscope

The increased sensitivity of the atomic force microscope (AFM) was used by Mohideen and coworkers to demonstrate for the first time the influence of the nonzero skin depth and surface roughness on the Casimir force. In three successive measurements of the Casimir force between metallized surfaces of a sphere and a plate in vacuum, all of the requirements set forth by Sparnaay, i.e. the use of nonreactive and clean surfaces, the determination of the average separation on contact, and the minimization and independent measurement of electrostatic potential differences, were achieved independently of the Casimir force measurement. Below, we briefly consider the first two experiments in this series and then discuss in more detail the third one, which presents the most unambiguous results with respect to experimental precision and comparison with theory. A dynamic measurement using an AFM is also discussed.

19.2.1 *First AFM experiment with aluminum surfaces*

A schematic diagram of this experiment (Mohideen and Roy 1998) is shown in Fig. 19.1. A force between the sphere and plate causes the cantilever to flex. This flexing of the cantilever is detected by the deflection of the laser beam, leading to a difference signal between the photodiodes A and B. This difference signal from the photodiodes was calibrated by means of an electrostatic force.

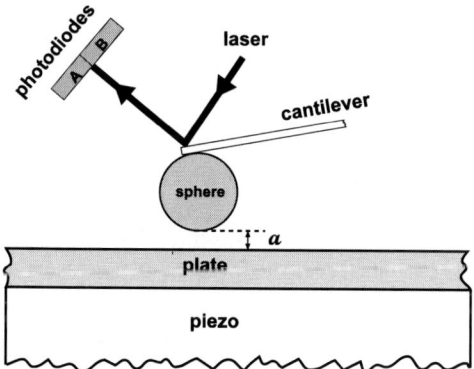

FIG. 19.1. Schematic diagram of the experimental setup using an atomic force microscope. Application of a voltage to the piezoelectric element results in movement of the plate towards the sphere (Mohideen and Roy 1998).

Polystyrene spheres of diameter $200 \pm 4\,\mu$m were mounted on the tip of metal-coated cantilevers with Ag epoxy. Such spheres are lightweight and can be easily attached to AFM cantilevers. They can be readily coated with smooth layers of a variety of materials, including metals such as Al, Cr, and Au, appropriate for Casimir force measurements. A 1 cm diameter optically polished sapphire disk was used as the plate. The cantilever (with sphere) and the plate were then coated by thermal evaporation with about 300 nm of aluminum. To prevent the rapid oxidation of the aluminum coating and the development of space charges, the aluminum was then sputter-coated with a 60%/40% Au/Pd coating of less than 20 nm thickness. In the first and second experiments, aluminum metal was used owing to its high reflectivity at short wavelengths (corresponding to small surface separations). Aluminum coatings are also easy to apply owing to the strong adhesion of the metal to a variety of surfaces and its low melting point.

To measure the Casimir force, the sphere and plate were grounded together with the AFM. The plate was then moved continuously towards the sphere and the corresponding photodiode difference signal was measured at time intervals corresponding to 3.6 nm. The signal obtained for a typical scan is shown in Fig. 19.2. Here "0" separation stands for contact between the sphere and plate surfaces. It does not take into account the absolute average separation between the Au/Pd layers due to the surface roughness, which was about 80 nm. If one also takes into account the fact that the Au/Pd cap layers can be considered transparent at small separations, the absolute average separation between the Al layers at contact was about 120 nm. Note that in the experiment, the separation distance on contact, $a_0 = 120 \pm 5$ nm, was found by fitting the experimental data for the measured force at relatively large separations above 250 nm with the sum of the theoretical electrostatic and Casimir forces. The electrostatic force arises because of the residual potential difference between the sphere and the plate. Thus, the second of the three requirements set forth by Sparnaay (see Section 18.2), concerning the independent measurement of the separation on contact, was only partially accomplished. At the same time, the first and third requirements, concerning the use of a clean surface and the independent measurement of the residual electrostatic force between the two surfaces, were satisfied (see below concerning the independent determination of the residual potential difference from the measurement of electric forces).

Region 1 in Fig. 19.2 shows that the force curve at large separations is dominated by a linear signal. This is due to increased coupling of scattered light into the photodiodes from the approaching flat surface. Embedded in the signal is a long-range attractive electrostatic force from the contact potential difference between the sphere and plate, and the Casimir force (small at such large distances). In region 2 (where the absolute separation varies from contact to 250 nm), the Casimir force is the dominant characteristic, far exceeding all the systematic errors. Region 3 corresponds to the flexing of the cantilever resulting from the continued extension of the piezoelectric actuator after contact of the two surfaces. Given the distance moved by the flat plate (the x-axis), the dif-

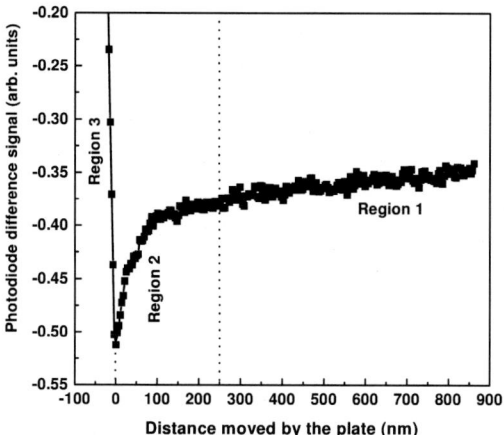

FIG. 19.2. Typical force curve as a function of the distance moved by the plate (Mohideen and Roy 1998).

ference signal of the photodiodes can be calibrated to a cantilever deflection in nanometers using the slope of the curve in region 3.

Next, the force constant of the cantilever was calibrated by an electrostatic measurement performed at separations greater than $3\,\mu\text{m}$, where the Casimir force is negligible. The sphere was grounded to the AFM, and different voltages V in the range $\pm 0.5\,\text{V}$ to $\pm 3\,\text{V}$ were applied to the plate. The exact expression for the electrostatic force between a charged sphere and a plate (in SI units) can be written as (Smythe 1950)

$$F_{\text{el}}(a) = 2\pi\epsilon_0 (V - V_0)^2 \sum_{n=1}^{\infty} \frac{\coth\alpha - n\coth n\alpha}{\sinh n\alpha} \equiv X(\alpha)(V - V_0)^2. \quad (19.1)$$

Here, V_0 represents the residual potential on the grounded sphere and $\cosh\alpha = 1 + a/R$. The function $X(\alpha)$ in the range of separations from $a = 100\,\text{nm}$ to $a = 6\,\mu\text{m}$ can be represented by the following polynomial with a relative error less than 10^{-4} (Chen et al. 2006b):

$$X(\alpha) = -2\pi\epsilon_0 \sum_{i=-1}^{6} c_i \left(\frac{a}{R}\right)^i, \quad (19.2)$$

where

$$c_{-1} = 0.5, \quad c_0 = -1.18260, \quad c_1 = 22.2375, \quad c_2 = -571.366, \quad (19.3)$$
$$c_3 = 9592.45, \quad c_4 = -90200.5, \quad c_5 = 383084., \quad c_6 = -300357.$$

From the difference in force for voltages $\pm V$ applied to the plate starting at separations sufficiently far from the sphere (to make the Casimir force negligible)

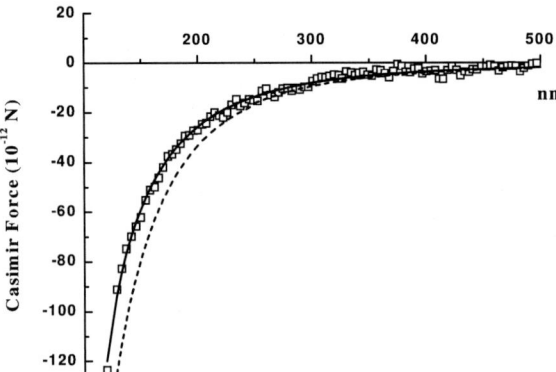

FIG. 19.3. Measured mean Casimir force as a function of plate–sphere separation, shown by open squares. The theoretical Casimir force with corrections due to surface roughness and nonzero skin depth is shown by the solid line, and for ideal-metal surfaces by the dashed line (Klimchitskaya et al. 1999).

to about 2 or 3 µm, the residual potential of the grounded sphere V_0 could be measured as 29 mV. This residual potential is a contact potential that arises from the different materials used to ground the sphere. The electrostatic-force measurement was repeated at five different separations and for eight different voltages V. The cantilever was electrostatically calibrated, and an equivalent force constant could be derived using Hooke's law and the force from eqn (19.1). The average value thus derived was k =0.0182 N/m.

In the original work (Mohideen and Roy 1998), the random and systematic experimental errors in the resulting data for the Casimir force were not specified and the force sensitivity of the AFM was about 10^{-12} N. The theoretical Casimir force, including the effect of the nonzero skin depth, was represented in a multiplicative way as in eqn (17.74). The factor $\eta_E(a)$ in this equation including the effect of the nonzero skin depth was found up to the second order in the small parameter δ_0/a [see eqn (13.8)]. The factor $\kappa_E(a)$ was defined according to eqn (17.39) up to the second order in the roughness amplitudes A_i/a with $\langle f_i^2 \rangle = 1/2$ and $\langle f_1 f_2 \rangle = 0$. Later, the theoretical Casimir force for this experiment was recalculated up to the fourth perturbation order in the parameter δ_0/a using the nonmultiplicative approach (17.94) to take account of surface roughness (Klimchitskaya et al. 1999). For this purpose the surface roughness, which was assumed to be the same on the sphere and the plate, was measured with the AFM. The scans obtained were used, as described in Section 17.2.2, to obtain the zero level of the roughness $H_0 = 12.6$ nm and the roughness amplitude $A = 27.4$ nm. At the shortest separation, the roughness correction contributed up to 17% of the magnitude of the Casimir force.

In Fig. 19.3, the mean measured Casimir force from the 26 measurement sets is shown by the open squares for the separation range from 120 to 500 nm.

The Casimir force data were obtained after subtraction from the total measured force of corrections due to two systematic deviations: the residual electrostatic force and the contribution to the force from the coupling of scattered light into the photodiodes. The solid line in Fig. 19.3 shows the theoretical Casimir force calculated as described above, including corrections due to the nonzero skin depth and the surface roughness (Klimchitskaya et al. 1999). The dashed line indicates the Casimir force (6.51) between an ideal-metal plate and an ideal-metal sphere of the radius $R = 100\,\mu$m used in the experiment. As is seen in Fig. 19.3, the solid line is in very good agreement with the data, thus demonstrating the role of the skin depth and roughness corrections. The data in Fig. 19.3 clearly deviate from the dashed line, which does not take these corrections into account.

The quantitative comparison between experiment and theory performed by Mohideen and Roy (1998) and Klimchitskaya et al. (1999) was based on the concept of the root-mean-square deviation. Later it was demonstrated that this global method may lead to inconsistent results when applied to strongly nonlinear functions of separation defined over a wide separation range (see Section 18.3.3). Nevertheless, it was shown (Klimchitskaya et al. 1999) that the root-mean-square deviation σ_N given in eqn (18.20) calculated using a more accurate theory, taking into account the skin depth corrections up to the fourth order and roughness corrections by the method of geometrical averaging, was a smoother function of the separation range (or the number of experimental points N) than the same σ_N calculated using the less accurate theory. The value of the root-mean-square deviation obtained, $\sigma_N \approx 1.5\,$pN (which is approximately 1% of the Casimir force at the closest separation), was considered as a measure of the agreement between experiment and theory at this separation. More rigorous methods of comparing experiment and theory at a given confidence level are considered below (see Section 19.2.3) in the case of the more conclusive, third experiment, on the measurement of the Casimir force between Au surfaces by means of an AFM.

19.2.2 Improved measurement with aluminum surfaces

An improved version of the measurement of the Casimir force between Al surfaces coated with Au/Pd layers was reported by Roy et al. (1999). The particular experimental improvements were (i) use of smoother metal coatings, which reduces the effect of surface roughness and allows closer separations between the two surfaces; (ii) implementation of vibration isolation, which reduces the total noise; (iii) independent electrostatic measurement of the surface separation; and (iv) reductions in the systematic errors due to the residual electrostatic force, scattered light, and instrumental drift. Also, the complete dielectric properties of Al were used in the theory. For this purpose, $\varepsilon_{\text{Al}}(i\xi)$ was determined from the tabulated optical data for the complex index of refraction of Al extending from 0.04 to 1000 eV (Palik 1985). This procedure was illustrated in Section 13.3 for the case of Au. Then the Casimir force between a perfectly shaped sphere and plate taking account of the nonzero skin depth was computed by using the proximity force approximation and the Lifshitz formula at zero temperature (12.30).

As in the previous experiment, the roughness of the metal surface was measured directly with the AFM. The metal surface was composed of separate crystals on a smooth background. The height of the highest distortions was 14 nm with intermediate distortions of height 7 nm, both on a stochastic background of height 2 nm, having fractional surface areas of 0.05, 0.11, and 0.84, respectively. The crystals were modeled as parallelepipeds. Then the roughness amplitude $A = 11.8$ nm was determined, as defined in Section 17.1.1. The role of the roughness was included by means of the multiplicative factor $\kappa_E(a)$ [see eqn (17.39)], calculated up to the fourth perturbation order in A/a. The roughness correction was found to be only about 1.3% of the measured force (i.e. there was more than an order of magnitude improvement over the previous measurement).

The preparation and measurement procedures of the improved experiment can be briefly described as follows. The same technique as before for the attachment of the sphere to the AFM cantilever was used. Then a 250 nm aluminum metal coating was evaporated onto the sphere and onto a 1 cm diameter sapphire plate. Next, both surfaces were sputter-coated with a 7.9 ± 0.1 nm layer of 60%Au/40%Pd. Thus, here, the Au/Pd coating was made much thinner, and also its thickness was precisely measured. The diameter of the sphere was measured using a scanning electron microscope to be $201.7 \pm 0.5\,\mu$m. The AFM was calibrated in the same manner as reported in the previous section. Next, the residual potential of the grounded sphere was measured as $V_0 = 7.9 \pm 0.8$ mV by the application of voltages to the plate at large separations from the sphere where the Casimir force could be neglected. Thus, the residual potential difference was a factor of 3.5 lower than in the previous experiment.

To measure the Casimir force between the sphere and the flat plate, they were both grounded together with the AFM. No voltage was applied to the plate. The data obtained for the photodiode difference signal were akin to those shown in Fig. 19.2. In this case the region 2 extended from contact to about 550 nm. Here, the Casimir force was the dominant characteristic, far exceeding all systematic deviations. These deviations were primarily due to the residual electrostatic force (less than 1.5% of the magnitude of the force at the closest separation) and a linear contribution from scattered light. The latter was observed and measured in region 1, which now started at separations $a > 550$ nm. The corrections arising from the systematic deviations were computed and subtracted from the force data.

In this experiment, a key improvement was that the electrostatic force between the sphere and the plate at large separation distances was used to arrive at an independent and consistent measurement of a_0, the average surface separation on contact of the two surfaces. This technique has now become the standard technique for precise determination of the separation distance between the zero roughness levels of two surfaces. The electrostatic-force measurement was done immediately following the Casimir force measurement, without breaking the vacuum and with no lateral movement of the surfaces. The flat plate was connected

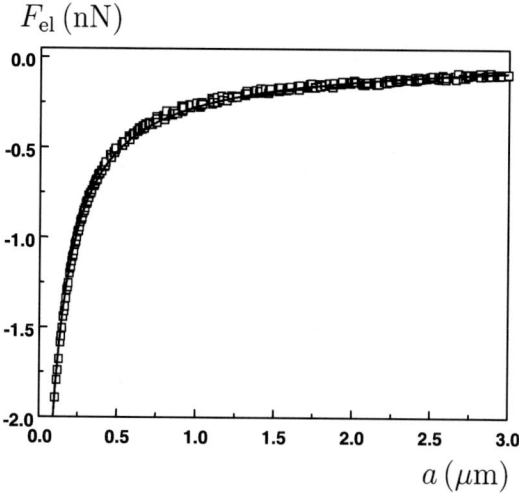

FIG. 19.4. The measured electrostatic force for an applied voltage of 0.31 V on the plate. The solid line shows a best χ^2 fit to the data (Roy et al. 1999).

to a dc voltage supply while the sphere remained grounded. The applied voltage V in eqn (19.1) was so chosen that at sufficiently large separations, the electrostatic force was much greater than the Casimir force. The open squares in Fig. 19.4 represent the measured force for an applied voltage of 0.31 V as a function of distance. This force can be considered as entirely electrostatic at $a \geq 300$ nm (at $a = 300$ nm, the contribution from the Casimir force is already less than 1% and it decreases rapidly with an increase of separation). The solid line, which is a best χ^2 fit for the data in Fig. 19.4, results in a value $a_0 = 47.5$ nm. This procedure was repeated for other voltages from 0.3 to 0.8 V, leading to an average value of $a_0 = 48.9 \pm 0.6$ nm. Given the 7.9 nm Au/Pd coating on each surface, this would correspond to an average surface separation 64.7 ± 0.6 nm in the Casimir force measurement. Thus, the experiment by Roy et al. (1999) was the first one satisfying all three requirements for a precise measurement of the Casimir force discussed in Section 18.2.

The measurement was repeated 27 times over the range of separations from 100 to 500 nm, and the mean Casimir force measured is shown by the open squares in Fig. 19.5 as a function of the surface separation. The solid line shows the theoretical Casimir force, including corrections due to surface roughness and nonzero skin depth computed as discussed above. Similarly to the first AFM measurement, the systematic errors of the force measurement were not reported here. However, the absolute random error from thermal noise was found to be $\Delta^{\mathrm{rand}} F^{\mathrm{expt}} = 1.3$ pN at a 67% confidence level, which leads to a relative random error $\delta^{\mathrm{rand}} F^{\mathrm{expt}}$ less than 1% at the shortest separation $a = 100$ nm. The experimental data were compared with the theory using the root-mean-squre deviation, which was found to be equal to 2.0 pN. This is also of the order of 1% of the force measured at the closest separation. The complete evaluation of the experimental

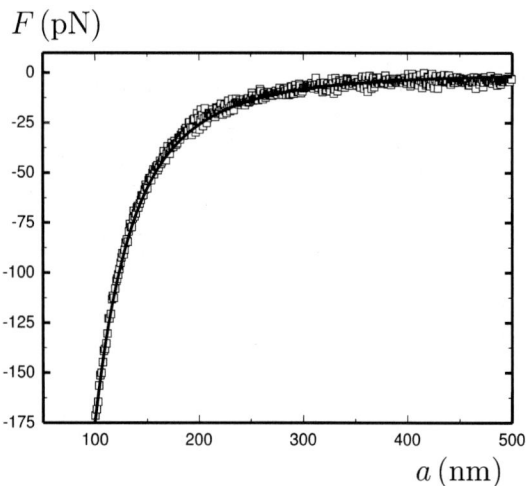

FIG. 19.5. The measured mean Casimir force, shown by squares. The solid line shows the theoretical Casimir force taking account of corrections due to surface roughness and nonzero skin depth (Roy et al. 1999).

and theoretical errors and a rigorous comparison between experiment and theory are considered in the next subsection, where we discuss the third experiment on the Casimir force, performed by means of an AFM using gold surfaces.

19.2.3 Precision measurement using gold surfaces

The primary difference in the third measurement of the Casimir force using an AFM (Harris et al. 2000) was the use of gold surfaces, which resulted in related experimental changes. The use of a thin Au/Pd coating on top of the aluminum surface to reduce the effects of oxidation in the previous two experiments prevented a complete theoretical treatment of the properties of the metal coating. For layers a few nanometers in thickness, the standard theoretical results derived for stratified media (see Section 12.2) are not applicable because of the effects of spatial dispersion (Klimchitskaya et al. 2000). Thus, it is important to use chemically inert materials such as gold for the measurement of the Casimir force.

The fabrication procedures had to be modified, given the different material properties of gold as compared with the aluminum coatings used by Mohideen and Roy (1998) and Roy et al. (1999). The $320\,\mu$m long AFM cantilevers were first coated with about 200 nm of aluminum to improve their thermal conductivity. This metal coating on the cantilever decreased the thermally induced noise when the AFM was operated in vacuum. Aluminum coatings were better, as applying thick gold coatings directly to these silicon nitride cantilevers led their curling owing to the mismatch in the thermal expansion coefficients. Next, a polystyrene sphere was mounted on the tip of a metal-coated cantilever with Ag epoxy. A 1 cm diameter optically polished sapphire disk is used as the plate. The

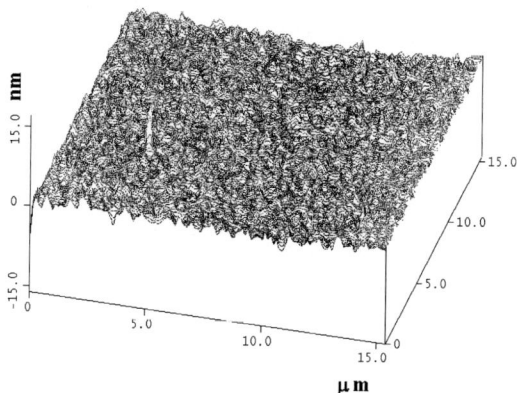

FIG. 19.6. Typical atomic force microscope image of an Au coating on a plate (Chen et al. 2004a).

cantilever (with sphere) and the plate were then coated with gold in an evaporator (uniformity was maintained by rotation). The sphere diameter after the metal coating has been deposited was measured using a scanning electron microscope to be $191.3 \pm 0.5\,\mu$m. The scanning electron microscope was calibrated using AFM calibration gratings, which were in turn calibrated independently. The roughness of the metal coating was investigated using an AFM. A typical scan of the surface is shown in Fig. 19.6. As is seen in this figure, the roughness is represented by stochastically distributed distortions and rare pointlike peaks. The amplitude of the random roughness $A_{\rm st}$ of the gold surface was determined to be 1.2 ± 0.1 nm. The thickness of the gold coating was measured using the AFM to be 86.6 ± 0.6 nm. Such a coating thickness is sufficient to reproduce the properties of an infinitely thick metal. To reduce the development of contact potential differences between the sphere and the plate, great care was taken to follow identical procedures in making the electrical contacts. This was necessary, given the large difference in the work functions of aluminum and gold. The residual potential difference between the grounded sphere and the plate was measured to be $V_0 = 3 \pm 3$ mV as described previously, with the application of voltages $\pm V$ to the plate. It was shown that the value of V_0 obtained did not depend on the separation at which it was determined. The residual potential led to forces which were less than 0.1% of the Casimir force at the closest separations.

Kim et al. (2008) have claimed that the residual potential V_0 obtained from an electrostatic calibration in the sphere–plate configuration is separation-dependent. This conclusion was obtained by using an Au-coated glass lens of 30.9 mm radius at separations of a few tens of nanometers above an Au-coated plate. The measurement data were fitted to the largest contribution to the electric force (19.1), given by the term of eqn (19.2) with $i = -1$. Decca et al. (2009a) demonstrated that for a centimeter-size spherical lens at such short separations from the plate, the electrostatic force law given by eqn (19.1) is not applicable, owing to the inevitable deviations from a perfect spherical shape of the mechanically polished and ground surface. Because of this, the anomalies in the electrostatic calibration observed by Kim *et al.* (2008) are irrelevant to the experimental results considered in this chapter. Small polystyrene spheres of about $100\,\mu$m radius made from the liquid phase are extremely smooth and almost perfectly spheroidal owing to surface tension (they have less than 10^{-3} asphericity). The investigation of the surface quality of such spheres in the scanning electron microscope did not reveal any scratches or bubbles, which are always present on the surface of centimeter-size glass lenses. The absence of the anomalous scaling in electrostatic calibrations for Casimir-force measurements was confirmed by de Man *et al.* (2009).

The measurement of the Casimir force was performed within the separation region from 62 to 350 nm in the same way as in the second experiment. The measurements were repeated 30 times, but three repetitions were later found (Chen *et al.* 2004a) to be outlying on the basis of the statistical criterion presented in Section 18.3.1. Because of this, the data analysis presented here used only 27 sets of measurement data. Note that the small correction due to the residual potential difference was the only correction which was subtracted from the force data. The small deviation due to the coupling of scattered light into the photodiodes discussed in the case of the first two experiments was eliminated here by using a very smooth gold coating on the cantilever surface and, thus, the respective correction was not needed.

In this experiment, special attention was paid to an independent and precise determination of the surface separation a, including the separation on contact of the two surfaces, a_0. The actual separation between the bottom of the gold coating of the sphere and top of the gold coating on the plate was given by

$$a = a_0 + a_\text{piezo} + S_\text{def} m. \tag{19.4}$$

Here, a_piezo is the distance moved by the plate owing to the voltage applied to the piezoelectric actuator, S_def is the photodiode difference signal, and m is the deflection coefficient. The last term on the right-hand side of eqn (19.4) represents a small additional correction to the separation, which results from the deflection of the cantilever in response to the attractive Casimir force. The latter leads to a decrease of the separation between the two surfaces ($S_\text{def} < 0$). The electrostatic force between the sphere and the flat plate was used to arrive at an independent measurement of the constant m in the deflection correction, of the force constant k, and of a_0, the average surface separation between the zero levels of the roughness on contact of the two surfaces. For this purpose, the flat plate was connected to a dc voltage supply while the sphere remained grounded. The applied voltage V in eqn (19.1) was so chosen that the electrostatic force was much greater than the Casimir force. The values $m = 8.9 \pm 0.3$ nm per unit photodiode difference signal, $km = 0.386 \pm 0.003$ nN per unit deflection signal, and $a_0 = 32.7 \pm 0.8$ nm were determined from many different applied voltages.

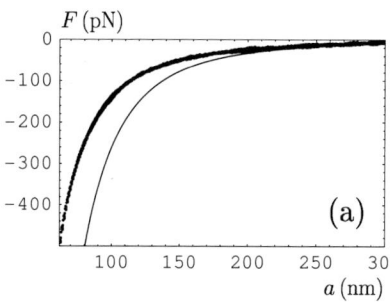 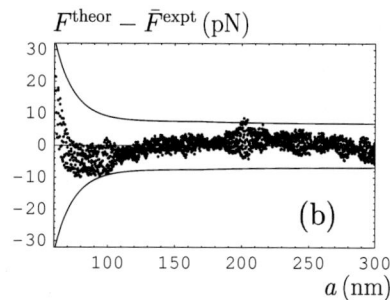

FIG. 19.7. (a) The mean measured Casimir force as a function of separation distance, shown by dots. The solid line shows the result for an ideal metal. (b) Differences between the theoretical and mean experimental Casimir forces, shown by dots. The solid lines indicate the borders of 95% confidence intervals (Klimchitskaya et al. 2007a).

It was confirmed that the value of a_0 obtained did not depend on the separation when it was determined within the range from 1 to 3 μm.

The mean Casimir force measured from the 27 scans is shown in Fig. 19.7(a) by dots. In the same figure, the solid line shows the Casimir force between an ideal-metal sphere and ideal-metal plate. It can be seen that the model of ideal-metal boundary surfaces is inconsistent with the data and a more sophisticated theory is required that takes real material properties into account. Before discussing such a theory, we consider the experimental errors and precision. In the original publication (Harris et al. 2000), the experimental random error was estimated based on all 30 measurement sets. Chen et al. (2004a) reanalyzed the data related to 27 sets of measurements (i.e. excluding the three sets of outlying measurements according to the statistical criterion of Section 18.3.1). A random error equal to $\Delta^{\mathrm{rand}} F^{\mathrm{expt}} = 5.8\,\mathrm{pN}$ at a 95% confidence level over the entire measurement range was obtained using eqn (18.8) (Chen et al. 2004a). The main contributions to the systematic error in this experiment were given by the error in the force calibration, $\Delta_1^{\mathrm{syst}} F^{\mathrm{expt}} \approx 1.7\,\mathrm{pN}$, by the noise when the calibration voltage was applied to the cantilever, $\Delta_2^{\mathrm{syst}} F^{\mathrm{expt}} \approx 0.55\,\mathrm{pN}$, by the instrumental sensitivity, $\Delta_3^{\mathrm{syst}} F^{\mathrm{expt}} \approx 0.31\,\mathrm{pN}$, and by the restrictions on the computer resolution of the data, $\Delta_4^{\mathrm{syst}} F^{\mathrm{expt}} \approx 0.12\,\mathrm{pN}$ (Chen et al. 2004a). By combining these systematic errors at a 95% confidence level using the standard rule (18.2) with $J = 4$, we obtain $\Delta^{\mathrm{syst}} F^{\mathrm{expt}} = 2.1\,\mathrm{pN}$ [note that the slightly larger value obtained by Chen et al. (2004a) can be explained by the use of a less rigorous combination rule that provides only an upper estimate of the magnitude of the resulting error]. This value of the combined systematic error is given by the second term on the right-hand side of eqn (18.2). The total experimental error at the 95% confidence level can now be calculated using eqn (18.17), with the result $\Delta^{\mathrm{tot}} F^{\mathrm{expt}} = 6.3\,\mathrm{pN}$. It does not depend on separation. The respective relative experimental error varies from 1.4% to 1.9% when the separation increases from 63 to 72 nm. It increases to 4.5% and 28% when the separation increases to 100 and 200 nm, respectively. We can conclude that the rigorous statistical approach leads to a 1.4% total experimental error in the measurement of the Casimir force using the AFM at the shortest separation. Regarding the error in the measurement of absolute separations, $\Delta a = 1$ nm, this is caused by the error in the determination of the separation on contact and by the error in the deflection coefficient m (the contribution to Δa owing to the error in a_{piezo} is negligible). We emphasize that the major contribution to Δa is given by the error in the surface separation on contact, $\Delta a_0 = 0.8$ nm. As a result, all experimental points relative to one another are uncertain not by 1 nm but by a negligible amount.

Now we compare the experimental results with theory. In the original publication (Harris et al. 2000), the Casimir force, including the effect of the nonzero skin depth, was calculated using the Lifshitz formula at zero temperature (12.30) and the proximity force approximation. The negligibly small corrections due to nonzero temperature and surface roughness (both much less than 1% of the calculated force) were taken into account in a multiplicative way. Chen et al. (2004a) also used the Lifshitz theory at $T = 0$ and the proximity force approximation, but presented a complete analysis of the surface roughness in the framework of the nonmultiplicative approach (17.94). For this purpose, the zero level of the roughness on the plate and on the sphere, $H_0 = 2.734$ nm, was found using eqn (17.92) with $N_1 = N_2 = 17$. If the roughness is defined by regular functions (17.3), this corresponds to a roughness amplitude $A = h_i^{\max} - H_0 = 13.266$ nm. If the roughness is described by stochastic functions (17.77), the respective variance is equal to $\delta = 0.837$ nm and, for the amplitude of a random process, one obtains $A_{\text{st}} = \sqrt{2}\delta = 1.18$ nm ≈ 1.2 nm, in agreement with what was determined from an AFM scan similar to that shown in Fig. 19.6. At the shortest separation $a = 62$ nm, where the surface roughness correction takes its maximum value, this correction contributes only 0.22% of the Casimir force between perfectly shaped bodies. This contribution decreases to 0.1% at $a = 90$ nm. At $a = 62$ nm, the relative difference between the combined effect of the surface roughness and of the nonzero skin depth corrections, calculated in a multiplicative and a nonmultiplicative way, is only 0.09% (Chen et al. 2004a). The role of diffraction-type and correlation effects in the effect of surface roughness was also investigated and found to be negligible. From AFM scans of the surface, the correlation length Λ_c was estimated to be ≈ 200 nm (see Section 17.4). The role of surface roughness, including correlation effects, was only 0.24% at $a = 62$ nm and 0.13% at $a = 90$ nm. Thus, although the role of the correlation effects in the effect of surface roughness increases with increasing separation, the contribution from the total roughness correction to the Casimir force decreases rapidly, thus making the role of the correlation effects negligible (Chen et al. 2004a).

Both the original paper (Harris et al. 2000) and the subsequent reanalysis (Chen et al. 2004a) used the global concept of the root-mean-square deviation between experiment and theory. A rigorous comparison of the experiment with theory using the local methods considered in Section 18.3.3 was performed by Klimchitskaya et al. (2007a). The Casimir free energy in the configuration of two plane plates (12.66) was computed at $T = 300$ K, i.e. at the laboratory temperature, using the dielectric permittivity $\varepsilon(i\xi)$ of the generalized plasma-like model (the solid line in Fig. 13.2). For gold, $\omega_p = 9.0$ eV was used. Note that the same value of ω_p was used by Harris et al. (2000) (the value indicated in the original publication is a typographical error). The results obtained were converted into a force between the sphere and the plate, $F^{sp}(a,T)$, by means of the proximity force approximation (6.71). Then, the final theoretical results taking roughness into account were obtained by means of geometrical averaging:

$$F^{\text{theor}}(a,T) = 2\pi R \sum_{i,k=1}^{17} v_i v_k \mathcal{F}(a + 2H_0 - h_i - h_k, T), \qquad (19.5)$$

where a table of the fractions of the surface area with heights h_i was presented in the paper by Chen et al. (2004a). Note that, here and below, we use the superscript "theor", to indicate that all corrections to the respective computed quantity, including that due to surface roughness, have been taken into account.

In Fig. 19.7(b), we plot as dots the differences between the theoretical Casimir force (19.5) and the mean experimental Casimir force versus separation. The solid lines indicate the borders of the 95% separation-dependent confidence intervals $[-\Xi_F(a), \Xi_F(a)]$ for the quantity $F^{\text{theor}}(a) - \bar{F}^{\text{expt}}(a)$, where $\Xi_F(a)$ is defined in eqn (18.21). To calculate $\Xi_F(a)$, one needs to know the total experimental error $\Delta^{\text{tot}} F^{\text{expt}}$ (already determined above) and the total theoretical error $\Delta^{\text{tot}} F^{\text{theor}}$. The latter has several parts, considered below as relative quantities (see Section 18.3.2). The first theoretical error, as discussed in Section 18.3.2, is due to the uncertainties in the optical data used to determine $\varepsilon(i\xi)$. Ignoring anomalous sets of data which lead to disagreement with all experiments performed on the Casimir force (see Section 19.3.4 for further discussion), the theoretical error introduced into the magnitude of the force by the errors in the optical data was estimated as $\delta_1 F^{\text{theor}} = 0.5\%$ (Chen et al. 2004a). The error introduced in the experiment under consideration due to the inaccurate description of the surface roughness was negligibly small, as demonstrated above.

As the second source of theoretical errors, we consider the uncertainty related to patch potentials. For the configuration of a sphere above a plate, the electric force due to random variation in the patch potentials is given by (Speake and Trenkel 2003)

$$F_{\text{patch}}(a) = -\frac{4\pi\epsilon_0 \sigma_v^2 R}{k_{\max}^2 - k_{\min}^2} \int_{k_{\min}}^{k_{\max}} dk \frac{e^{-ka} k^2}{\sinh ka}, \qquad (19.6)$$

where σ_v is the variance of the potential distribution, and $k_{\max}$ and $k_{\min}$ are the magnitudes of the extremal wave vectors corresponding to the minimum and maximum sizes of the grains. The values of the work function of gold are $V_1 = 5.47$ eV, $V_2 = 5.37$ eV, and $V_3 = 5.31$ eV for the crystallographic surface orientations (100), (110), and (111), respectively. Assuming equal areas of these crystallographic planes, one obtains

$$\sigma_v = \frac{1}{\sqrt{2}} \left[\sum_{i=1}^{3} (V_i - \bar{V})^2 \right]^{1/2} \approx 80.8 \, \text{mV}. \tag{19.7}$$

Using AFM images of the surface as in Fig. 19.6, the extremal sizes of the grains in the gold layers covering the test bodies were determined as $\lambda_{\min} \approx 68$ nm and $\lambda_{\max} \approx 121$ nm. This leads to $k_{\max} \approx 0.092 \, \text{nm}^{-1}$ and $k_{\min} \approx 0.052 \, \text{nm}^{-1}$. The grain sizes obtained are of the same order as the film thickness. Computations with eqn (19.7) using the above data lead to the *patch effect electric forces* $F_{\text{patch}}/R \approx -1.15 \times 10^{-8}$ N/m and -1.25×10^{-10} N/m at $a = 62$ nm and 100 nm, respectively. Comparing the results obtained with the values of the Casimir force at the same separations ($F^{\text{theor}}/R \approx -4.88 \times 10^{-6}$ N/m and -1.41×10^{-6} N/m, respectively), we conclude that the electric force due to the patch potentials contributes only 0.23% and 0.008% of the Casimir force at $a = 62$ nm and $a = 100$ nm, respectively (Chen et al. 2004a).

The third source of theoretical errors in this experiment is the proximity force approximation (6.71), which was used in the theoretical expression. The related error is estimated as $\delta_3 F^{\text{theor}} \approx a/R$.

The method used compares experiment with theory at the experimental separation distances. Because of this, the error $\delta_4 F^{\text{theor}} = 3 \Delta a/a$, as discussed in the context of eqn (18.19), with $\alpha = 3$, should be included in the calculation of the theoretical errors.

As one more theoretical error, the error arising from the uncertainty in the measurement of the sphere radius, $\delta_5 F^{\text{theor}} = \Delta R/R \approx 0.16\%$, should be considered. Finally, we estimate the theoretical error caused by the finiteness of the plate used in the experiment by Harris et al. (2000). The theoretical expressions (6.71) and (19.5) were derived for an infinite plate or a disk of infinite radius L. In the experiment under consideration $L = 5 \times 10^{-3}$ m. Using eqn (6.58), for the theoretical Casimir force taking into account the finiteness of the disk, we obtain

$$F_{\text{fin}}^{\text{theor}}(a, T) = \beta(a) F^{\text{theor}}(a, T), \tag{19.8}$$

where

$$\beta(a) = 1 - \frac{a^3}{R^3} \left(1 - \frac{R}{\sqrt{R^2 + L^2}} \right)^{-3}. \tag{19.9}$$

Here we have used the fact that for a sphere, in contrast to a spherical lens of some height H, one always has $Q = R/\sqrt{R^2 + L^2}$ in eqn (6.55). To make $\beta(a)$ have the maximum difference from unity, we put $a = 350$ nm and arrive at

$$\beta(a) = 1 - 8 \frac{a^3 R^3}{L^6} \approx 1 - 2.2 \times 10^{-17}. \tag{19.10}$$

Thus, the finiteness of the plate size leads to a negligible contribution to the Casimir force and does not play any role in this experiment.

By combining all the above-mentioned errors in accordance with the statistical rule (18.2) at a 95% confidence level, we arrive at the magnitudes of

$\delta^{\text{tot}} F^{\text{theor}}(a)$ and $\Delta^{\text{tot}} F^{\text{theor}}(a)$. As an example, at the separations 62, 100, and 200 nm the total theoretical error is equal to 5.5%, 3.4%, and 1.8%, respectively, of the theoretical magnitude of the force. At short separations the major contribution to the error is given by $\delta_4 F^{\text{theor}}$, which characterizes the comparison procedure, but not the theory itself. The total theoretical error $\Delta^{\text{tot}} F^{\text{theor}}(a)$, together with $\Delta^{\text{tot}} F^{\text{expt}}(a)$, results in the confidence intervals shown in Fig. 9.7(b) as the solid lines.

As is seen in Fig. 19.7(b), almost all of the dots are inside the error bars. Thus, the theoretical description by means of the generalized plasma model is consistent with the data. Note that a comparison of the data from this experiment with the zero-temperature theoretical force practically coincides with that shown in Fig. 19.7(b). This is because measurements of the Casimir force between metals by means of an AFM are still not of sufficient precision to measure the thermal effects predicted by theory at short separations.

Now we are in a position to quantify the agreement between experiment and theory. In accordance with Section 18.3.3, this agreement at a 95% confidence level can be characterized by the quantity $A = \Xi_F(a)/|F^{\text{expt}}(a)|$. This quantity depends strongly on separation. At $a = 61$ nm, it is equal to 6.2%, and it is approximately constant and equal to 5.4% within the interval 77.5 nm $\leq a \leq$ 85.5 nm. At separations of 100, 130, and 200 nm it is equal to 6.2%, 10%, and 30%, respectively. The relatively large magnitude of A at the shortest separations is explained by the contribution of the theoretical error, which dominates over the experimental error due to the error in the measurement of the separations. With increasing a, $\Delta^{\text{tot}} F^{\text{theor}}(a)$ decreases. At larger a, the value of $\Xi_F(a)$ is mostly determined by the total experimental error, which is separation-independent.

In Section 18.3.3, we also considered another local method for the comparison between experiment and theory, which we illustrate here in Fig. 19.8. In this method, the data are represented as crosses with horizontal and vertical arms equal to $2\Delta a$ and $2\Delta^{\text{tot}} F^{\text{expt}}(a_i)$, respectively, and the theoretical results are shown as a band with a width equal to $2\Delta^{\text{tot}} F^{\text{theor}}(a)$. The important point is that in this case the major error $\Delta_4 F^{\text{theor}}(a)$ due to the error in the measurement of the separations is not included in $\Delta^{\text{tot}} F^{\text{theor}}(a)$, because the theoretical forces $F^{\text{theor}}(a)$ are computed not at the experimental points but over the measurement interval. In Fig. 19.8, we present the experimental data as crosses and the theoretical results as a black band of the appropriate width over the range of separations from 62 to 100 nm. All errors are determined at a 95% confidence level. It can be seen that all crosses overlap with the theoretical band which means that the data are consistent with a theory based on the generalized plasma-like dielectric permittivity and the Lifshitz formula at nonzero T. These two methods of comparison between experiment and theory for the Casimir force measurements will be repeatedly used in the following sections and chapters.

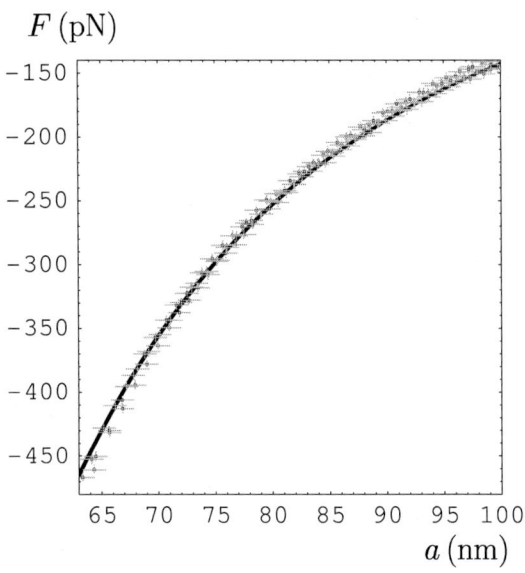

FIG. 19.8. The measured mean Casimir forces, together with the absolute errors in the separation and force, versus separation are shown as crosses. The theoretical Casimir force computed using the generalized plasma-like model is shown by the black band.

19.2.4 Dynamic measurement

A measurement of the Casimir force between a gold-coated sphere and a plate was performed by means of an AFM operated in the dynamic mode (Jourdan et al. 2009). Direct sphere–plate contact was avoided. The separation on contact was not needed for the determination of absolute separations as in the earlier experiments by Decca et al. (2005b, 2007a, 2007b) described in Sections 19.3.3 and 19.3.4. The AFM cantilever with the sphere glued to it was considered as a harmonic oscillator with a natural resonant frequency $\omega_0 = 2\pi \times 50182\,\text{rad/s}$ which was modified by the Casimir force. In the linear regime,

$$\frac{\partial F(a)}{\partial a} = k_0 - k_\text{eff}, \tag{19.11}$$

where $F(a)$ is the Casimir force acting between the sphere and the plate, $k_0 = m\omega_0^2$, $k_\text{eff} = m\omega_r^2$, m is the mass of the oscillator, and ω_r is its resonant frequency in the presence of the Casimir force. This experimental approach also resembles the earlier experiments of Decca et al. (2005b, 2007a, 2007b), performed by means of a micromechanical torsional oscillator to be discussed in Section 19.3. However, the uncertainty in the measurement of the absolute separation, $\Delta a = 2\,\text{nm}$, was more than twice as large as in the previous AFM experiments and three times larger than in the experiments using a micromachined oscillator. The calibration procedures and the determination of the absolute separation used electrostatic forces as in the experiments described previously. However, a much larger residual potential difference $V_0 = 75 \pm 3\,\text{mV}$ was reported. The calibration was done at separations $a < 500\,\text{nm}$ and the total force (electrostatic in addition to the Casimir force) was used in the fit. Thus, the second and third requirements for the precise measurement of the Casimir force proposed by Sparnaay (see Section 18.2) were only partially met.

The experimental data for the force gradient $\partial F/\partial a$ were compared with the Lifshitz theory at zero temperature (measurements were performed under a 10^{-6} Torr vacuum at $T = 300\,\text{K}$). The dielectric permittivity along the imaginary frequency axis was found using the complete optical data (Palik 1985), extrapolated to zero frequency by means of the Drude model with $\omega_p = 9.0\,\text{eV}$ and $\gamma = 0.035\,\text{eV}$ (Lambrecht and Reynaud 2000a). This permittivity is shown by the dashed line in Fig. 13.2. The variances of the surface roughness were measured using the AFM to be 3 and 2 nm on the sphere and on the plate, respectively. The effect of the roughness was not taken into account in the theoretical analysis. Both systematic errors and random errors at a 67% confidence level were discussed, but the total experimental error was not provided. As two major sources of theoretical errors, an uncertainty of the order of 5% due to the sample dependence of the optical data and an uncertainty of less than 1% due to the use of the proximity force approximation were considered. The differences between the force gradient measurements and the results of the theoretical computations described above were found to be within 3% of the theoretical force at separations between 100 to 200 nm. This 3% discrepancy, however, cannot be stated as a rigorous measure of agreement between the experiment and theory. This is because the authors of that study considered a 5% error in the calculated force gradient as part of the theoretical error (see Sections 18.3.2 and 18.3.3 for a discussion of this subject). In addition, in considering the differences between the force gradient measurements and theoretical computations at the experimental separations, one must include the contribution (18.19) with $\alpha = 4$ due to the error in the measurement of the separations in the theoretical Casimir force. At $a = 100\,\text{nm}$, this error is equal to 8% of the theoretical Casimir force. Thus, using eqn (18.2), the total theoretical error in this experiment reaches 10.4% at $a = 100\,\text{nm}$ and the agreement between experiment and theory needs to be characterized by a larger percentage.

The measured Casimir force gradient was also compared with the values computed using ideal-metal surfaces and a clear deviation was reported. On this basis it was concluded that the experimental data again demonstrated the effects of the skin depth on the Casimir force.

19.3 Experiments with a micromechanical torsional oscillator

Microelectromechanical systems are well adapted for the investigation of small forces acting between closely spaced surfaces. One such system, a *micromachined torsional oscillator*, was first used by Chan *et al.* (2001a, 2001b) to demonstrate the influence of the Casimir force on the static and dynamic properties of mi-

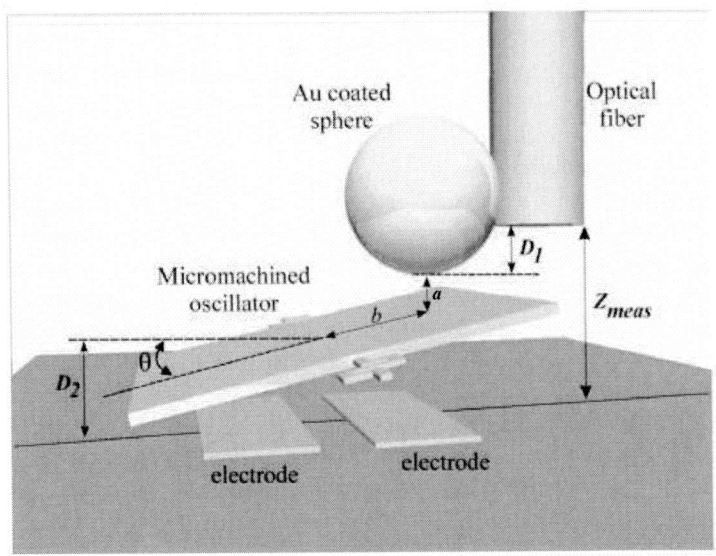

FIG. 19.9. Schematic diagram of the measurement of the Casimir force using a micromachined oscillator (Decca et al. 2005b).

cromechanical systems. These results are considered in Sections 23.2 and 23.3. Here, precision measurements of the Casimir force and pressure between metallic surfaces by means of a micromachined oscillator are described (Decca et al. 2003a, 2003b, 2004, 2005b, 2007a, 2007b). These measurements were accomplished in three successive experiments and resulted in the demonstration of the highest experimental precision ever achieved in Casimir force measurements. The data obtained have been successfully used to perform conclusive tests of various theoretical models of the thermal effects discussed in Chapter 14. We start with a presentation of the experimental setup and the measurement scheme used, with some modifications, in all three experiments. After a discussion of the first two experiments, we then concentrate on the third one, which incorporated improvements in both the experimental procedures and theoretical analysis. An experimental test of the proximity force approximation is also considered.

19.3.1 *Experimental setup and measurement scheme*

A schematic diagram of the experimental setup, including the micromechanical torsional oscillator, is shown in Fig. 19.9. The oscillator consisted of a 3.5 μm thick, $500 \times 500\,\mu\text{m}^2$ heavily doped polysilicon plate suspended at two opposite points on the midplane by serpentine springs. The springs were anchored to a siliconnitride-covered Si platform. When no torques were applied, the plate was separated from the platform by a gap of about $2\,\mu$m. Two independent electrically contacted polysilicon electrodes located under the plate were used to measure the capacitance between the electrodes and the plate. They could also be used to induce oscillation in the plate at the resonant frequency of the micromachined oscillator when the dynamic measurement regime was used. An 80 μm wide ribbon at the edge of the plate, below the sphere, was covered with metallic layers (different in different experiments). Above the oscillator, a large Al_2O_3 sphere of radius R was also covered with a metallic layer. The coated sphere was mounted with conductive epoxy on the side of an Au-coated optical fiber, establishing an electrical connection between them. The sphericity of the Al_2O_3 ball was measured with a scanning electron microscope. The metal-deposition-induced asymmetries were found to be smaller than 10 nm, the resolution of the scanning electron microscope. The entire setup (oscillator and fiber–sphere combination) was rigidly mounted into a can, where a pressure $< 10^{-5}$ Torr was maintained. The can had built-in magnetic-damping vibration isolation and was, in turn, mounted on an air table.

The separation distance between the metallic coatings on the plate and on the sphere was given by

$$a = z_{\text{meas}} - D_1 - D_2 - b\theta. \tag{19.12}$$

In this expression, z_{meas} is the separation between the end of the cleaved fiber and the platform (see Fig. 19.9), which was measured interferometrically with an absolute error $\Delta z_{\text{meas}} = 0.2$ nm. The quantities D_1, D_2, the rotation angle of the plate in response to the Casimir and electric force, and the lever arm b are also shown in Fig. 19.9. The value of b was determined optically. The value of θ was found by measuring the difference in capacitance $C_{\text{right}} - C_{\text{left}}$ between the plate and the right and left electrodes. In all reported cases, $\theta \leq 10^{-5}$ rad. Finally, the force between the two metallic surfaces separated by the distance a was $F(a) = k(C_{\text{right}} - C_{\text{left}})$, where k is a constant coefficient.

Before performing the Casimir force measurements, the surfaces of the plate and the sphere were characterized using an AFM probe with a radius of curvature of 5 nm in the tapping mode. Regions of the metal plate and of the sphere varying in size from $1 \times 1\,\mu m^2$ to $10 \times 10\,\mu m^2$ were scanned. The character of the roughness in the first experiment was very different from that in the second and third experiments. Because of this, it is discussed below with respect to each particular experiment.

An important part of the measurement scheme was system calibration. It was done electrostatically in all three experiments and used to find the proportionality constant k between the force and the difference in capacitance $C_{\text{right}} - C_{\text{left}}$, the residual potential difference V_0, the exact value of the sphere radius R, and the parameter $D = D_1 + D_2$. To perform the calibration, a known potential difference V was applied between the metal-coated sphere and the plate. This was done at separations $a > 3\,\mu m$, where the Casimir force was smaller than 0.1% of the total force. Thus, the total force could be approximated by only the electrostatic force $F_{\text{el}}(a)$ as given by eqn (19.1). The electric force, with different applied voltages, was measured and fitted to eqn (19.1). This allowed the values of all of the above parameters to be determined with high precision. The values obtained were different in different experiments. In each experiment, V_0 was observed to be constant for wide ranges of separations and it did not vary when measured over different locations of the sphere above the metal layer on the plate.

19.3.2 Static and dynamic measurements

Now we present the results of the first experiment on precision Casimir force measurement by means of a micromachined oscillator (Decca et al. 2003a, 2003b, 2004). In this experiment, the surface of the plate was coated with a 1 nm layer of Cr followed by 200 nm of Cu. The sphere, with a nominal radius of 300 μm, was coated with a 1 nm layer of Cr followed by a 203 nm layer of gold. Thus, in this experiment, the Casimir interaction between dissimilar metals was investigated. An analysis of AFM scans of the plate and sphere surfaces showed the presence of rather tall roughness peaks. Even peaks with $h \geq h_{59} = 98.5$ nm (but with $h \leq 100$ nm) were found, with a fractional area $v_{59} = 0.085$. The zero level of roughness was calculated to be $H_0 = 35.46$ nm using eqn (17.92) with $N_1 = N_2 = 59$, which leads to a roughness amplitude $A = 63.04$ nm, determined from eqn (17.93). This makes taking careful nonmultiplicative account of the roughness corrections necessary. A rather high residual potential difference $V_0 = 632.5 \pm 0.3$ mV was also found. This reflects the difference in the work functions for the dissimilar metals used (Au and Cu). The value of V_0 obtained was observed to be constant for a from 0.2 to 5 μm. Other calibration parameters were found from more than 100 force–distance relations with different applied voltages: $k = 50280 \pm 6$ N/F, $R = 294.3 \pm 0.1$ μm, and separation on contact $a_0 = 78.8 \pm 0.6$ nm [note that only in the first experiment with the micromachined oscillator were the absolute separations expressed not by using eqn (19.12) but by means of the separation on contact a_0].

The first experiment included two different measurements, a static one and a dynamic one. In the static regime, the sphere was brought into proximity with the plate and electrostatic measurements were performed. Without breaking the vacuum in the system, a measurement of the Casimir force between the sphere and the plate was then carried out. This was repeated 19 times within the range of separations from 190 nm to 1.15 μm. Each data point was obtained with an integration time of 10 s. The magnitudes of the Casimir force in one of 19 data sets are shown in Fig. 19.10(a). This set contains approximately 300 data points. The random error, estimated as $\Delta^{\text{rand}} F^{\text{expt}} \approx 0.3$ pN at a 95% confidence level, was much higher than the systematic errors and determined the value of the total experimental error. The relative total experimental error at the shortest separation was estimated to be 0.27% (Decca et al. 2003b).

The experimental results were compared with theoretical computations using the Lifshitz formula for the Casimir energy per unit area of two dissimilar plates (12.44) at zero temperature. The force F^{sp} was calculated by means of the PFA [eqn (6.71) at $T = 0$]. The dielectric permittivity of Au along the imaginary

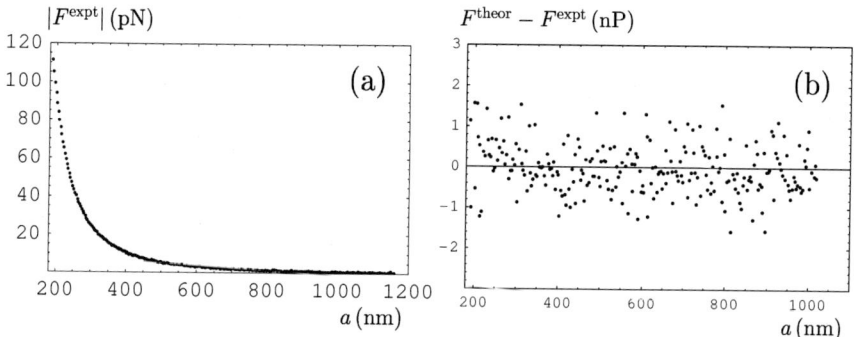

FIG. 19.10. (a) Absolute value of the measured Casimir force as a function of separation, obtained using the static regime. (b) Differences between the theoretical and experimental Casimir forces in a sphere–plate configuration versus separation (Decca et al. 2003b).

frequency axis as given by the dashed line in Fig. 13.2 (this line was obtained using the tabulated optical data extrapolated to low frequencies by means of the Drude model) was used. Almost coincident results follow from the solid line in Fig. 13.2 (the generalized plasma-like dielectric permittivity). The dielectric permittivity of Cu along the imaginary frequency axis was obtained in a similar way. The corrections due to the surface roughness were calculated using the nonmultiplicative approach (17.94). In Fig. 19.10(b), the differences between the theoretical and experimental force values $F^{\text{theor}}(a_i) - F^{\text{expt}}(a_i)$ as a function of surface separation a_i are shown as dots for one typical set of measurements. As can be seen from the figure, all of the data points are clustered around zero, demonstrating good agreement between experiment and theory. The quantitative measure of agreement between experiment and theory (Decca et al. 2003b) used the global concept of the root-mean-square deviation. A rigorous approach to the comparison of experiment with theory is discussed below for the second and third experiments using micromachined oscillators (see Sections 19.3.3 and 19.3.4).

The dynamic measurement also performed in the first experiment was an indirect measurement, resulting in the first precise determination of the Casimir pressure in the configuration of two parallel plates. In this measurement, the vertical separation between the sphere and the plate was varied harmonically,

$$a(t) = a + A_z \cos\omega_r t, \qquad (19.13)$$

where ω_r is the resonant angular frequency of the oscillator in the presence of the sphere. An amplitude A_z between 3 and 35 nm, depending on a, was chosen in such a way that the oscillator exhibited a linear response. In the presence of the Casimir force $F(a)$, the resonant frequency ω_r differed from the natural angular frequency of the oscillator, $\omega_0 = 2\pi \times 687.23\,\text{Hz}$. The solution for the oscillator motion yields (Chan et al, 2001a, 2001b)

$$\omega_{\rm r}^2 = \omega_0^2 \left[1 - \frac{b^2}{I\omega_0^2} \frac{\partial F^{\rm sp}(a)}{\partial a} \right], \qquad (19.14)$$

where I is the moment of inertia of the oscillator. The electrostatic measurements were used to find $b^2/I = 1.2978\,\mu{\rm g}^{-1}$.

The actual measured quantity in this dynamic measurement was the change of the resonant frequency of the oscillator, $\omega_{\rm r} - \omega_0$, under the influence of the Casimir force. Using eqn (19.14), the experimental data for $\omega_{\rm r} - \omega_0$ obtained at different separation distances can be transformed into $\partial F^{\rm sp}(a)/\partial a$. It is less useful, however, to recover the force $F^{\rm sp}(a)$ between the sphere and plate using the force gradient. A better approach is given by using the proximity force approximation (6.71). Differentiating it with respect to a and taking into account eqn (1.6), one arrives at an expression for the Casimir pressure in the configuration of two parallel plates,

$$P(a) = -\frac{1}{2\pi R} \frac{\partial F^{\rm sp}(a)}{\partial a}. \qquad (19.15)$$

From eqns (19.14) and (19.15), one can immediately convert the experimental data for the frequency shift into data for the Casimir pressure between two parallel plates. Thus, the dynamic measurement is in fact an *indirect measurement* of the pressure. Note that in metrological terms (Rabinovich 2000), the results of *direct measurements* are found only from experiment. The results of indirect measurements are obtained with the help of calculations using known equations [in our case eqns (19.14) and (19.15)] which relate the quantity under consideration (the pressure) to some other quantity measured directly (the frequency shift).

The Casimir pressure was dynamically measured five times within the range of separations from 260 nm to 1.15 μm. The data from one set of measurements are shown in Fig. 19.11(a) as a function of the separation. The random error

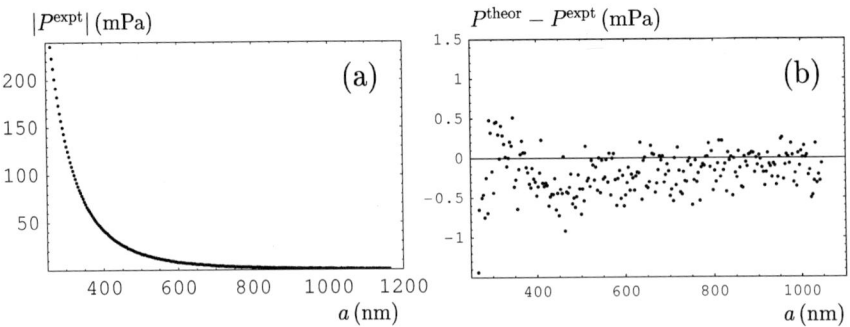

FIG. 19.11. (a) Magnitude of the measured Casimir pressure as a function of separation, obtained using the dynamic technique. (b) Differences between the theoretical and experimental Casimir pressures in a configuration of two plates versus separation (Decca *et al.* 2003b).

of the force measurements was estimated to be $\Delta^{\text{rand}} P^{\text{expt}} = 0.31\,\text{mPa}$. Here, the systematic errors should also be taken into account. The error in the measurement of the frequency was approximately $2\pi \times 10^{-2}\,\text{Hz}$, which, from eqn (19.14), leads to an absolute error in $\partial F/\partial a$ equal to $4.2 \times 10^{-7}\,\text{N/m}$. The latter, when combined with the error in the sphere radius (equal to $0.1\,\mu\text{m}$), leads via eqn (19.15) to a systematic error in the pressure $\Delta^{\text{syst}} P^{\text{expt}}$, which varies from $0.31\,\text{mPa}$ at $a = 260\,\text{nm}$ to $0.23\,\text{mPa}$ at all separations $a \geq 450\,\text{nm}$. As a result, the value of $\Delta^{\text{tot}} P^{\text{expt}}$ obtained by Decca et al. (2003b) at a 95% confidence level as a sum of random and systematic errors varies from $0.62\,\text{mPa}$ at $a = 260\,\text{nm}$ to $0.54\,\text{mPa}$ at $a \geq 450\,\text{nm}$. Hence, the relative error is $\delta^{\text{tot}} P^{\text{expt}} = 0.26\%$ at the shortest separation $a = 260\,\text{nm}$. A more accurate error analysis using the rigorous methods of Section 18.3.1 will be considered in application to the third experiment using the micromachined oscillator. This analysis includes the error of the proximity force approximation as a part of the measurement procedure for the indirect measurement of the Casimir pressure (see Section 19.3.4).

The theoretical Casimir pressure to be compared with the data from the dynamic measurement was calculated in the same way as for the Casimir force for the static measurement at $T = 0$ (Decca et al. 2003b). The only difference was that eqn (12.46) for the Casimir pressure was used instead of eqn (12.44) for the Casimir energy. As an example, the differences between the theoretical and experimental pressures, $P^{\text{theor}}(a_i) - P^{\text{expt}}(a_i)$, for one typical measurement set are presented in Fig. 19.11(b). This figure demonstrates that the mean difference pressure is equal to $-0.26\,\text{mPa}$, i.e. about half the absolute error of the pressure measurements. Thus, the theory is in a good agreement with the data.

The first dynamic measurement by means of the micromachined oscillator provided an experimental test of the large thermal corrections predicted in the various approaches to the thermal Casimir force (see Sections 14.3.1 and 14.3.5). Decca et al. (2003b) repeated the computations of the Casimir pressure using $\varepsilon(i\xi)$ obtained from the optical data extrapolated to low frequencies by use of the Drude model (the Drude model approach), but did so for $T = 300\,\text{K}$, the laboratory temperature. The differences $P_{\text{D}}^{\text{theor}}(a_i) - P^{\text{expt}}(a_i)$ obtained are plotted in Fig. 19.12(a) for the same measurement set as in Fig. 19.11. It is obvious that at separations $a < 700\,\text{nm}$ the quantity $P_{\text{D}}^{\text{theor}} - P^{\text{expt}}$ deviates significantly from zero. At the shortest separation $a = 260\,\text{nm}$, this deviation reaches $5.5\,\text{mPa}$, i.e. a factor of 9 larger than the total experimental error. Thus, the large thermal corrections to the Casimir pressure predicted by the Drude model approach are ruled out by the data. The more rigorous comparison between experiment and theory in the case of the second and third experiments (Sections 19.3.3 and 19.3.4) is in support of this conclusion.

The computations of the theoretical Casimir pressure were also repeated using the Lifshitz theory at the laboratory temperature and the same $\varepsilon(i\xi)$ as in the Drude model approach, but with the transverse electric reflection coefficient at zero frequency equal to unity instead of zero, as discussed at the beginning of Section 14.3.5. The differences $P_{\text{m}}^{\text{theor}} - P^{\text{expt}}$ obtained are shown in Fig. 19.12(b)

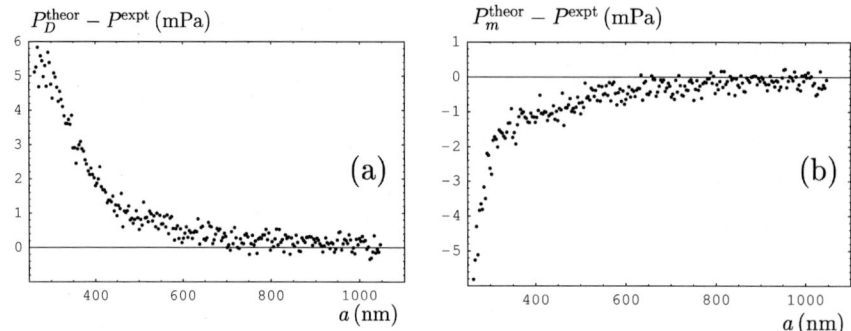

FIG. 19.12. Differences between the theoretical and experimental Casimir pressures in a configuration of two plates versus separation. The theoretical results were calculated using the Lifshitz theory at room temparature, (a) combined with the Drude model approach and (b) using a modified zero-frequency contribution from the transverse electric term (Decca et al. 2003b).

as dots. The same measurement set as in Fig. 19.11 was used. As can be seen in Fig. 19.12(b), at separations less than 600 nm the quantity $P_m^{\text{theor}} - P^{\text{expt}}$ deviates significantly from zero. It reaches 5 mPa at $a = 260$ nm. This is a factor of 8 larger than the total experimental error. Note that in the computations of P_m^{theor}, all contributions to the Lifshitz formula with the exception of the TE term at $\xi = 0$ were the same as in the Drude model approach. It follows that the large thermal correction arising from the modified TE reflection coefficient at zero frequency is also in contradiction with the first experiment using the micromachined oscillator. This conclusion is also strengthened by the results of subsequent experiments.

19.3.3 *Improved dynamic measurement*

The second experiment using a micromachined oscillator (Decca et al. 2005b, Klimchitskaya et al. 2005) was significantly improved in comparison with the first one. The results obtained were compared with theory using the rigorous methods presented in Section 18.3.3. One of the new features of the second experiment was that the Casimir attraction was measured between two layers of Au rather than between two dissimilar metals. The edge of the plate was coated with 10 nm of Pt followed by 150 nm of Au. A sphere with a nominal radius $R = 150\,\mu\text{m}$ was coated with a 10 nm layer of Ti followed by a 200 nm layer of Au. Characterization of the sample using an AFM was performed both before and after the Casimir force measurements, to check that the sample was not modified during the measurement. In Fig. 19.13(a,b), $10 \times 10\,\mu\text{m}^2$ AFM images of the film on the plate are shown before and after the experiment, respectively. Figure 19.13(c,d) contains similar $5 \times 5\,\mu\text{m}^2$ AFM images of the bottom of the sphere. Special measures were taken in this experiment before film deposition to obtain smoother surfaces. As a result, much lower roughness peaks on the Au coating than in the first experiment were obtained. From an analysis of the

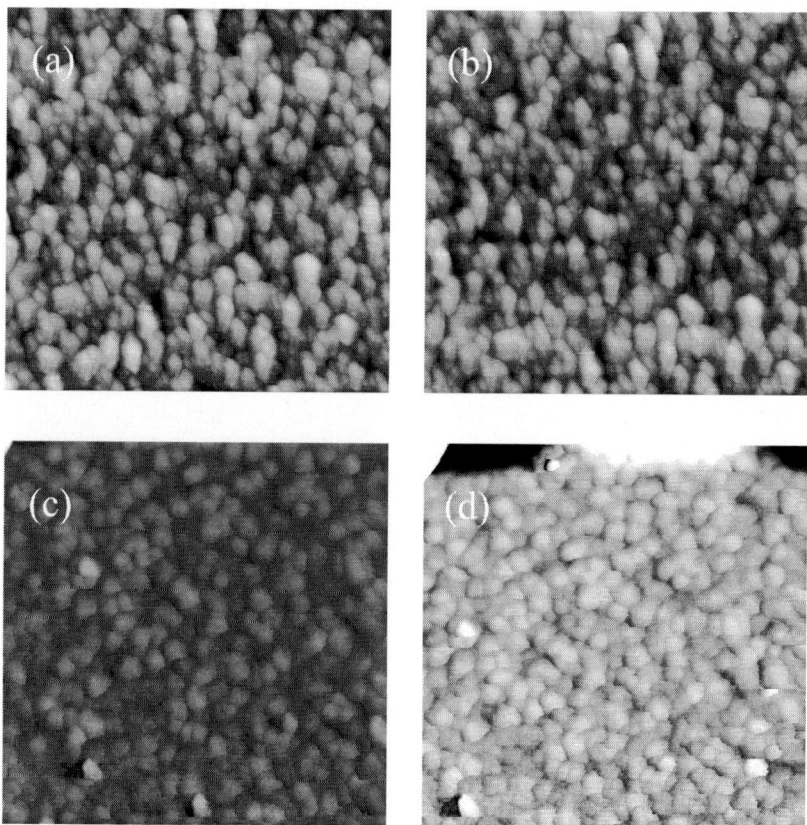

FIG. 19.13. Typical AFM images of the surface of the plate (a) before and (b) after the experiment, and of the sphere (c) before and (d) after the experiment (Decca et al. 2005b).

AFM images, numbers of bins $N_2 = 105$ and $N_1 = 112$ for the roughness heights were selected on the plate and on the sphere, respectively. The highest peaks had heights $h^{(2)}_{105} = 20.65$ nm on the plate and $h^{(1)}_{112} = 11.06$ nm on the sphere. Using eqn (17.92), zero levels of $H^{(2)}_0 = 9.72$ nm and $H^{(1)}_0 = 5.03$ nm on the plate and on the sphere, respectively, were obtained. From eqn (17.93), this leads to roughness amplitudes $A_2 = 10.93$ nm on the plate and $A_1 = 6.03$ nm on the sphere. The respective variances defined by eqn (17.95) are equal to $\delta_2 = 4.06$ nm and $\delta_1 = 1.91$ nm.

The calibration of the setup was performed by means of the electrostatic force as described in Section 19.3.1. For the residual potential difference, a much lower value, $V_0 = 17.5 \pm 0.1$ mV than in the first experiment was obtained. This value was observed to be constant for a in the range from 0.15 to $5\,\mu$m, and it did not vary for different locations of the sphere above the plate. Note that $F_{\rm el}$ was measured between the zero levels of the roughness on the plate and the sphere, relative to which the mean values of the roughness profiles were zero. As a result, the absolute separations obtained from eqn (19.12) were also determined between the zero levels of the roughness. For these reasons, there was no systematic error due to roughness in the measurements of the separations in addition to the uncertainties in the various parameters in eqn (19.12) discussed below. In particular, a set of 120 curves of $F_{\rm el}(a)$ was used to fit the parameter $D = D_1 + D_2$, with the result $D = 9349.7 \pm 0.5$ nm. Other parameters determined from the electrostatic-force measurements were $R = 148.7 \pm 0.2\,\mu$m and $k = 50455 \pm 7\,{\rm N/F}$. A measurement of θ through the difference in capacitances and the interferometric measurement of $z_{\rm meas}$ then yielded the absolute separation a. Unlike the first experiment, contact between the two surfaces was not made. In the second experiment, the separation on contact was not needed for determination of the absolute separations. In this experiment the lever arm was $b = 210 \pm 3\,\mu$m, $b^2/I = 1.2579 \pm 0.0006\,\mu{\rm g}^{-1}$, and $\omega_0 = 2\pi \times 702.92$ Hz. The absolute error in the measurement of $\omega_{\rm r}$ was $\Delta\omega_{\rm r} = 2\pi \times 6$ mHz, a factor of 1.7 smaller than in the first experiment.

The dynamic measurement of the Casimir pressure was performed as described in the previous subsection over a range of separations from 160 to 750 nm. The minimum value of the amplitude of harmonic oscillations given by eqn (19.13), $A_z = 1.2$ nm, was used at $a = 160$ nm. The possibility to perform measurements at lower separations than in the first experiment was connected with the smaller sphere radius used and the decreased surface roughness. The measurement was repeated 15 times at different positions on the sample (with 288–293 points in each set of measurements). One of the data sets is shown in Fig. 19.14(a). Each point of this figure was obtained with an integration time of 10 s. Decca et al. (2005b) performed rigorous error analysis of the experimental data obtained. In the measurement procedure employed, the separation step between two neighboring points was not uniform, even within one set of measurements, and additionally was quite different in different sets of data. This leads to the absence of even a few points (let alone 15) taken at the same separation which could be used for averaging during statistical analysis. Statistical methods adapted for such data are presented in Section 18.3.1.

The 15 available sets of measurements were analyzed for the presence of outlying results with the help of a statistical criterion (see Section 18.3.1) using the quantity given in eqn (18.12). One set of measurements was found to be outlying. As an example, at separations 170, 174, 180, and 250 nm the probabilities that the points of the rejected set were outlying were 80%, 98%, 95%, and 98%, respectively. For this reason, only 14 sets of measurements were used in the subsequent analysis. The random experimental error was found using eqns (18.10) and (18.11) (usually, it was necessary to consider four or five subintervals to the left and to the right of any point a_0 in order to find a uniform variance) over the entire measurement range at a 95% confidence level. The random er-

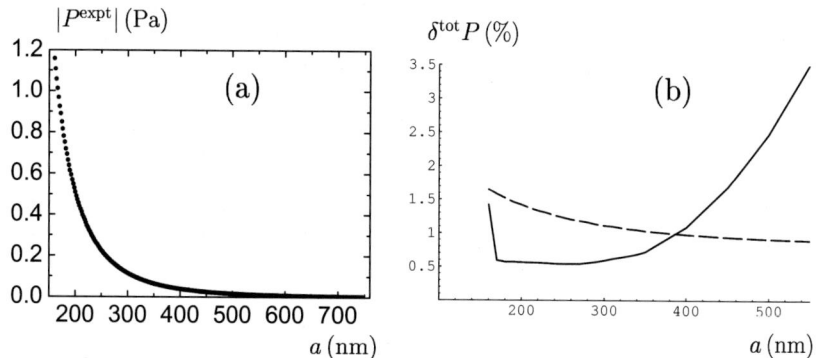

FIG. 19.14. (a) Magnitude of the measured Casimir pressure as a function of separation, obtained using the dynamic technique (Decca et al. 2005b). (b) The total experimental and theoretical relative errors of the Casimir pressure versus separation are shown as the solid and dashed lines, respectively (Klimchitskaya et al. 2005).

ror increased rapidly with decreasing separation. This is a consequence of eqns (18.10) and (18.11) and the decrease of m_k for subintervals k at the shortest separations. The main systematic errors were due to the error in the sphere radius $\Delta R = 0.2\,\mu\text{m}$ and the error in the angular frequency of the oscillator $\Delta \omega_r = 2\pi \times 6\,\text{mHz}$ [the quantities ω_0 and $b^2/(I\omega_0^2)$ were determined so precisely that their uncertainties did not contribute to the overall systematic error]. According to Section 18.3.2, the error introduced by the use of the proximity force approximation should also be considered as a systematic error in this indirect measurement of the pressure. However, in the original publication (Decca et al. 2005b) it was included as a part of the theoretical error. This does not change the resulting confidence interval used for the comparison of experiment and theory. The relative errors due to ΔR and $\Delta \omega_r$ were combined at a 95% confidence level using eqn (18.2) and the resulting systematic error was obtained. The statistical rule (18.17) was used to combine this result with the random error determined at the same confidence level. The resulting total experimental error defined at a 95% level is shown in Fig. 19.14(b) as the solid line. As can be seen in the figure, the total experimental error is almost constant (it varies between 0.55% and 0.6%) over a wide range of separations from 170 to 300 nm.

The experimental error Δa in the measurement of the surface separations should be considered as a systematic one. Using eqn (19.12), it can be found by combining the two errors $\Delta z_{\text{meas}} = 0.2\,\text{nm}$ and $\Delta D = 0.5\,\text{nm}$ [the third term on the right-hand side of eqn (19.12) does not contribute, because $\theta \leq 10^{-5}\,\text{rad}$]. The application of the statistical rule (18.2) leads to $\Delta a = 0.6\,\text{nm}$ at a 95% confidence level.

The theoretical Casimir pressure was computed by using the Lifshitz formula at nonzero temperature (12.70) with the Leontovich-impedance reflection coefficients (13.40). At all nonzero Matsubara frequencies $\omega_l = i\xi_l$ with $l \geq 1$, the Leontovich impedance was taken in the form (13.34), where $\varepsilon(i\xi)$ was given by the dashed line in Fig. 13.2 using the Drude parameters $\omega_p = 9.0\,\text{eV}$ and $\gamma = 0.035\,\text{eV}$. At $l = 0$, the impedance (13.33) for infrared optics was used. The surface roughness was incorporated using a nonmultiplicative approach [the second equality in eqn (17.94)], where the Casimir pressures $P(a)$ were replaced with the $P(a,T)$ defined in eqn (12.70). We emphasize that in this second experiment by means of a micromachined oscillator, the role of surface roughness was rather small. Thus, at the shortest separation $a = 160\,\text{nm}$, the roughness contributed only 0.65% and at $a = 200\,\text{nm}$ only 0.42% of the Casimir pressure. The contribution of correlation effects was negligible, as the correlation length was found to be rather large: $\Lambda_c \geq 600\,\text{nm}$ (Decca et al. 2005b). The contribution of patch effects was estimated similarly to Section 19.2.3, where, instead of eqn (19.6), the corresponding result for the pressure was used (Speake and Trenkel 2003),

$$P_{\text{patch}}(a) = -\frac{2\epsilon_0 \sigma_v^2}{k_{\max}^2 - k_{\min}^2} \int_{k_{\min}}^{k_{\max}} dk \, \frac{k^3}{\sinh^2 ka} \tag{19.16}$$

(see Section 19.2.3 for notation). Here, the extremal sizes of the grains in the Au layers covering the test bodies were $\lambda_{\min} \approx 25\,\text{nm}$ and $\lambda_{\max} \approx 300\,\text{nm}$, which lead to $k_{\max} = 0.251\,\text{nm}^{-1}$ and $k_{\min} = 0.0209\,\text{nm}^{-1}$. From eqn (19.16), this results in patch pressures $P_{\text{patch}} = 0.42$ and $0.25\,\text{mPa}$ at separations $a = 160$ and $170\,\text{nm}$, respectively. In comparison with the Casimir pressures at the same separations, the relative contributions of the patch effect are 0.037% and 0.027%, respectively, and decrease further with increasing a. As a result, the theoretical error in the second experiment can be estimated as a combination of three errors: $\delta_1 P^{\text{theor}} = a/R$ due to the use of the proximity force approximation, $\delta_2 P^{\text{theor}} = 0.5\%$ due to the uncertainties in the optical data, and $\delta_3 P^{\text{theor}} = 4\,\Delta a/a$ due to the calculation of the theoretical pressures at the experimental points. The total theoretical error obtained by the combination of these errors at a 95% confidence level using the statistical rule (18.2) is shown in Fig. 19.14(b) by the dashed line. It can be seen that at short separations the total theoretical error is larger than the experimental error (owing to the contribution of $\delta_3 P^{\text{theor}}$), but at $a > 380\,\text{nm}$ the experimental error becomes dominant.

In Fig. 19.15, the differences between the theoretical Casimir pressures computed as described above and the experimental values from 14 sets of measurements are shown by dots. The solid lines indicate the boundaries of the confidence intervals $[-\Xi_P(a), \Xi_P(a)]$ for the quantity $P^{\text{theor}}(a) - P^{\text{expt}}(a)$ determined at a 95% confidence level in accordance with eqn (18.21) using the total experimental and theoretical errors $\Delta^{\text{tot}} P^{\text{expt}}$ and $\Delta^{\text{tot}} P^{\text{theor}}$. As can be seen in the figure, fewer than 5% of all points fall outside the confidence interval. Thus, the theoretical approach using the Leontovich surface impedance at $T = 300\,\text{K}$ is consistent with the experimental data. The measure of agreement between experiment and theory is $A = \Xi_P(a)/|P^{\text{expt}}(a)|$. This quantity depends only slightly on a over

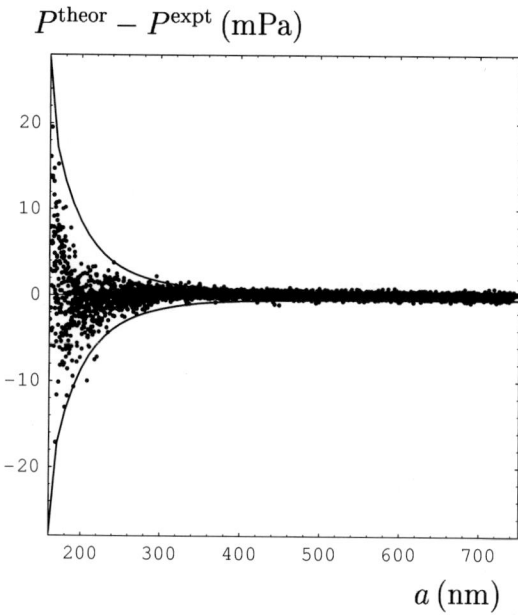

FIG. 19.15. Differences between the theoretical pressure computed using the impedance approach and the experimental Casimir pressure in a configuration of two plates versus separation, shown by dots. The solid lines indicate the borders of the 95% confidence interval (Klimchitskaya et al. 2005).

a wide separation range. Thus, it decreases from $A = 1.9\%$ at $a = 170$ nm to $A = 1.4\%$ in the interval $270\,\text{nm} \leq a \leq 370\,\text{nm}$, and then increases to $A = 1.8\%$ at $a = 420$ nm. The largest values of A are achieved at $a < 170$ nm ($A = 2.4\%$ at $a = 160$ nm) and $a > 420$ nm ($A = 13\%$ at $a = 750$ nm). The overall conclusion is that the impedance approach is in agreement with the data at the 1.5% level at a 95% confidence level for a wide range of separations.

The measurement data from the second experiment using the micromechanical oscillator were compared with two alternative theoretical approaches to the thermal Casimir force, based on the use of tabulated optical data extrapolated to low frequencies by means of the Drude model or by taking the TE reflection coefficient at zero frequency to be equal to unity (see Section 14.3.1 and the beginning of Section 14.3.5, respectively). In Fig. 19.16(a), the differences $P_D^{\text{theor}}(a) - P^{\text{expt}}(a)$ are shown by dots, using the experimental data from all 14 measurement sets. The solid lines indicate the confidence intervals determined at a 95% confidence level for the differences between theory and experiment (the confidence intervals are the same for both theoretical approaches). As can be seen in Fig. 19.16(a), in a wide range of separations from 230 to 500 nm all of the data fall outside the confidence interval. It follows that the Drude model approach to the thermal Casimir force is excluded by this experiment at a 95% confidence

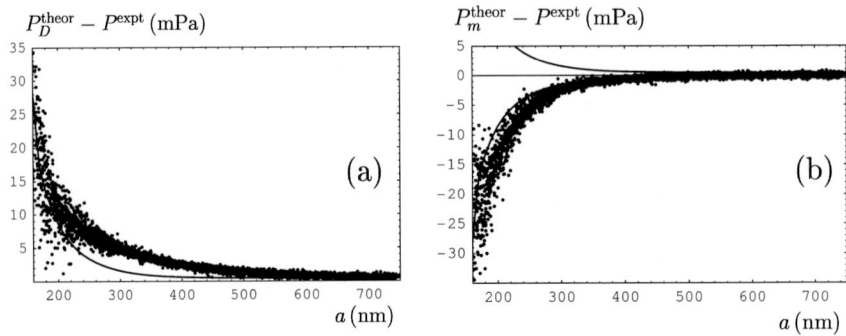

FIG. 19.16. Differences between the theoretical and experimental Casimir pressures in a configuration of two plates versus separation, shown by dots. The theoretical results were calculated using the Lifshitz thery at room temparature, (a) combined with the Drude model approach and (b) using the modified zero-frequency contribution from the transverse electric term. The solid lines indicate the borders of the 95% confidence intervals (Decca et al. 2005b).

level. In fact, this conclusion was obtained in an extremely conservative way. As can be seen in Fig. 19.16(a), even if the confidence intervals were widened to reach a 99% confidence level, the differences $P_D^{\text{theor}}(a) - P^{\text{expt}}(a)$ would still remain outside those intervals within some range of separations. To make this argument quantitative, we calculate the half-width of a new confidence interval from the following equality:

$$\frac{\Xi_P^{0.99}(a)}{\Xi_P^{0.95}(a)} = \frac{t_{(1+0.99)/2}(2)}{t_{(1+0.95)/2}(2)} \approx 2.31. \tag{19.17}$$

(To be conservative, we keep only two degrees of freedom, i.e. the minimum value obtained from the analysis of the random errors.) Using eqn (19.17), we find that $\Xi_P^{0.99} = 3.67$, 1.46, and 1.13 mPa at separations $a = 300$, 400, and 500 nm, respectively. From this it follows that within the range of separations from 300 to 500 nm, the Drude model approach is excluded experimentally at a 99% confidence level (Decca et al. 2005b, Klimchitskaya et al. 2005). In the next subsection it will be shown that the Drude model approach is excluded at even higher confidence level.

In Fig. 19.16(b), we present the differences $P_m^{\text{theor}}(a) - P^{\text{expt}}(a)$, where the Casimir pressure $P_m^{\text{theor}}(a)$ was computed by taking the TE reflection coefficient at zero frequency to be equal to unity (Svetovoy and Lokhanin 2001). The same experimental data and confidence intervals as in Fig. 19.16(a) have been used. As can be seen in Fig. 19.16(b), within the range of separations from 160 to 350 nm almost all dots fall outside the confidence interval. It follows that the theoretical approach based on the modified contribution from the TE term at zero frequency is excluded experimentally at a 95% confidence level.

19.3.4 More precise dynamic measurement, and conclusive test for some models of the thermal Casimir force

The third experiment in this series of dynamic determinations of the Casimir pressure by means of a micromachined oscillator (Decca et al. 2007a, 2007b) incorporated additional improvements which made it the most precise and reliable Casimir effect measurement performed to date. First, a new experimental procedure was implemented which permitted the measurements to be repeated over a wide range of separations, in such a way that data were acquired at practically the same points in each repetition. This made unnecessary the complicated statistical analysis employed in the second experiment. Second, the random experimental error was substantially reduced compared with the systematic error, as required for precision measurements. This was achieved by 33 repetitions of the measurement and a reduction in vibration noise by approximately 7%. Third, the plasma frequency of the Au films used in the experiment was determined using the measured temperature dependence of the resistivity of the films. The experimental data obtained were compared with various theoretical approaches to the thermal Casimir force using rigorous statistical methods.

A silicon plate of thickness 3.5 μm was first coated with a 10 nm thick layer of Cr and then with a 210 nm thick layer of Au. A sapphire sphere with a nominal radius of 150 μm was first coated with a 10 nm thick layer of Cr and then with a 180 nm thick layer of Au. Sample characterization was performed as described in the previous subsection. From AFM images of the surfaces, the fraction of surface area $v_i^{(1,2)}$ with height $h_i^{(1,2)}$ was determined for each surface. It was found that for the plate ($1 \leq i \leq N_2 = 85$), $h_i^{(2)}$ varied from 0 to 18.35 nm, and for the sphere ($1 \leq i \leq N_1 = 106$), $h_i^{(1)}$ varied from 0 to 10.94 nm. From eqn (17.92), this results in the zero levels of roughness $H_0^{(2)} = 9.66$ nm and $H_0^{(1)} = 5.01$ nm for the plate and the sphere, respectively. The respective roughness amplitudes were $A_2 = 8.69$ nm and $A_1 = 5.93$ nm. The roughness variances were $\delta_2 = 3.6$ nm for the plate and $\delta_1 = 1.9$ nm for the sphere. The Casimir pressure, including the effect of surface roughness, was calculated in a nonmultiplicative way using eqn (17.94). The contribution of the roughness correction to the Casimir pressure was found to be very small. For example, at the shortest separation $a = 162$ nm, the roughness correction contributed only 0.52% of the total pressure. At separations $a = 170, 200$, and 350 nm the roughness contributed only 0.48%, 0.35%, and 0.13%, respectively, of the Casimir pressure. The contributions from correlation effects and patch potentials to the Casimir pressure were negligibly small, as described in the previous subsection.

The calibration of the setup was performed as in the two earlier experiments and described by Decca and López (2009). The residual potential difference V_0 was found to be 15.29 mV, with a standard deviation equal to 0.13 mV. It was observed to be constant over the separation range from 160.4 to 5150.4 nm (Decca et al. 2009a). The resulting error in the measurement of the absolute separation was $\Delta a = 0.6$ nm at a 95% confidence level. The natural angular frequency

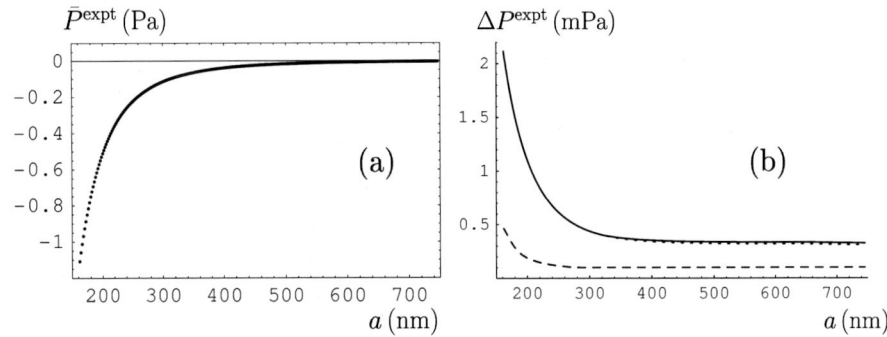

FIG. 19.17. (a) The mean measured Casimir pressure as a function of separation, obtained using the dynamic technique (Decca et al. 2007b). (b) The absolute total, systematic, and random experimental errors of the Casimir pressure are shown by the solid, short-dashed, and long-dashed lines, respectively (Decca et al. 2007a).

of the oscillator was equal to $\omega_0 = 2\pi \times (713.25 \pm 0.02)\,\text{Hz}$, the radius of the coated sphere R was $151.2 \pm 0.2\,\mu\text{m}$, and the parameter in eqn (19.14) b^2/I was $1.2432 \pm 0.0005\,\mu\text{g}^{-1}$.

Indirect measurements of the Casimir pressure were performed over the range of separations from $a_1 = 162.03\,\text{nm}$ to $a_{293} = 745.98\,\text{nm}$ and repeated 33 times. The values of the oscillator amplitude in eqn (19.13) were chosen in such a way that the oscillator exhibited a linear response. All data sets were analyzed for the presence of outlying results using the statistical criterion of Section 18.3.1 and no such results were found. Thus, all 33 measurement sets were used in the data analysis. Based on the advantage that in all measurement sets the data for the pressure were collected at approximately the same points a_i (up to the error Δa in the measurement of the absolute separations), the error analysis was simpler than in the second experiment. Here, eqns (18.3) and (18.8) were used for the determination of the mean experimental pressures $\bar{P}^{\text{expt}}(a)$ and the random error $\Delta^{\text{rand}} P^{\text{expt}}(a)$, at a 95% confidence level. The mean experimental pressures are plotted as dots in Fig. 19.17(a). The random error, as a function of separation, is shown by the long-dashed line in Fig. 19.17(b). It reaches a maximum value equal to $0.46\,\text{mPa}$ at $a = 162\,\text{nm}$, decreases to $0.11\,\text{mPa}$ at $a = 300\,\text{nm}$, and maintains this value up to $a = 746\,\text{nm}$.

The systematic error in this experiment is determined by the error in the measurement of the resonance frequency, $\Delta\omega_\text{r}$, the error in the radius of the sphere, ΔR, and also the error from using the proximity force approximation. In previous work (Decca et al. 2003b, 2004, 2005b), the latter was attributed to the theory, whereas the theoretical calculation of the Casimir pressure between two parallel plates is independent of the proximity force approximation. Also, the equivalent experimental Casimir pressure in eqn (19.15) requires the error of the proximity force approximation to be attributed to the experimental systematic errors. In Fig. 19.17(b), the systematic error $\Delta^{\text{syst}} P^{\text{expt}}(a)$, obtained by combining the above three errors with the help of the statistical rule (18.2), is shown by the short-dashed line. The total experimental error $\Delta^{\text{tot}} P^{\text{expt}}(a)$, obtained from eqn (18.17), is shown by the solid line. As is seen in Fig. 19.17(b), it is the systematic error which now dominates the magnitude of the total experimental error $\Delta^{\text{tot}} P^{\text{expt}}(a)$. This dominance of the systematic error over the random error had never before been achieved in Casimir force experiments. The total experimental relative error $\delta^{\text{tot}} P^{\text{expt}}$ defined in eqn (18.18) varies from 0.19% to 9.0% as the separation increases from 160 to 750 nm. Note that the contribution to the total experimental error from the use of the proximity force approximation varies from 0.04% to 0.5%, respectively. The precision achieved in the third experiment at short separations is a significant improvement over that of the second experiment.

The experimental data were compared with several theoretical approaches using the two different comparison methods described in Section 18.3.3 (Decca et al. 2007a, 2007b). We start with the approaches using the Leontovich surface impedance and the tabulated optical data extrapolated by use of the Drude model. These approaches have already been described in Section 19.3.3. Here, slightly different values of the Drude parameters were used in the extrapolation of the optical data, namely $\omega_p = 8.9$ eV and $\gamma = 0.0357$ eV, as determined from the measurements of the resistivity of the Au films (Decca et al. 2007b). The single theoretical error in the Casimir pressure of about 0.5% arises from the uncertainties in the tabulated optical data. When the first method of comparison between experiment and theory is used, the theoretical pressures are computed not at the experimental points but rather over the whole measurement range from 160 to 750 nm. Because of this, the error in the measurement of the separation Δa is irrelevant to the theory and should be included in the analysis of the experimental errors. The theoretical Casimir pressures, taking the surface roughness into account by means of the nonmultiplicative method of geometrical averaging, are shown in Fig. 19.18(a) by the light gray band (for the Leontovich impedance approach), and the dark gray band (for the Drude model approach) for separations from 500 to 600 nm. The width of the bands in the vertical direction is equal to twice the total theoretical error, $2\,\Delta^{\text{tot}} P^{\text{theor}}(a)$, determined at a 95% confidence level. The mean values of the Casimir pressure $\bar{P}^{\text{expt}}(a_i)$ averaged over all 33 sets of measurements are shown in Fig. 19.18(a) by crosses. The horizontal arms of the crosses are equal to twice the absolute errors, i.e. $2\,\Delta a = 1.2$ nm, determined at a 95% confidence level in the measurement of the separations between the zero levels of the roughness of the plate and the sphere. The vertical arms of the crosses are equal to twice the total experimental error, $2\,\Delta^{\text{tot}} P^{\text{expt}}(a_i)$, determined at a 95% confidence level in the measurement of the Casimir pressure. The latter is a function of separation and is shown by the solid line in Fig. 19.17(b). As seen in Fig. 19.18(a), the Leontovich impedance approach is consistent with the data over the entire separation range. In contrast, the Drude model approach, which leads to a relatively large thermal effect at

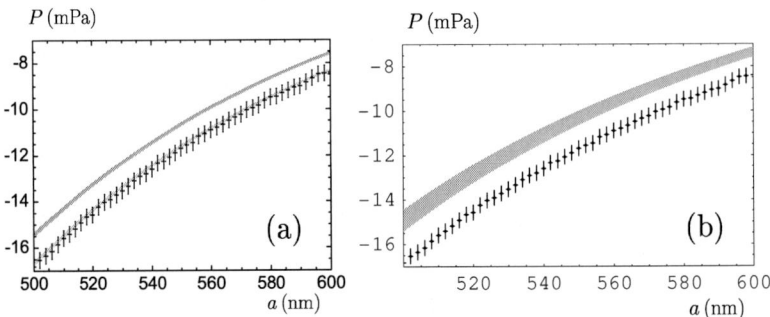

FIG. 19.18. The crosses show the measured mean Casimir pressures together with the absolute errors in the separation and pressure as a function of the separation. (a) The theoretical Casimir pressures computed using the generalized plasma-like model and the optical data extrapolated by use of the Drude model are shown by the light gray and dark gray bands, respectively (Decca et al. 2007a). (b) The theoretical Casimir pressures computed using various sets of optical data available in the literature, versus separation, are shown by the dark gray band (Klimchitskaya et al. 2009a).

$a < 1\,\mu$m, is excluded experimentally at a 95% confidence level over the entire separation range. The same conclusions are obtained for any range of separations from 160 to 750 nm, the range in which the measurements were performed [see Decca et al. (2007a), where appropriate figures are provided].

In Section 18.3.2, alternative sets of optical data were discussed which differ from the standard set (Palik 1985) by much more than the error in the optical measurements. It would be interesting to verify the hypothesis that the Drude model approach can be made consistent with the experimental data if some alternative set of optical data for Au films existing in the literature is used. To answer this question, the theoretical Casimir pressure was computed in the Drude model approach with ω_p varying from 6.85 to 9.0 eV, and the corresponding values of γ (Pirozhenko et al. 2006). The results obtained are plotted in Fig. 19.18(b) as the dark gray band. Note that the values of the relaxation parameter influence the Casimir pressure only slightly. The experimental data in Fig. 19.18(b) are the same as those in Fig. 19.18(a). As shown, the use of any alternative value of ω_p contained in the literature makes the disagreement even more acute between the Drude model approach and the measurement data. It is notable that in the framework of the Leontovich impedance approach, any value of $\omega_p < 8.8$ eV is excluded by the data.

We now apply another local method of comparison between experiment and theory, discussed in Section 18.3.3, which is based on consideration of the differences $P^{\text{theor}}(a_i) - \bar{P}^{\text{expt}}(a_i)$. Using this method, we first compare the data with the theoretical approaches based on the Leontovich surface impedance and on the generalized plasma-like dielectric permittivity. In Fig. 19.19(a), the differences between the theoretical Casimir pressure, computed as discussed above using

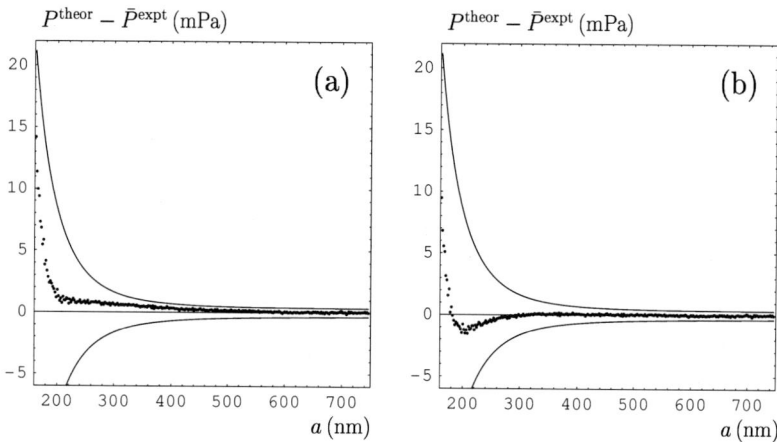

FIG. 19.19. The differences between the theoretical and mean experimental Casimir pressures versus separation are shown by dots, and the 95% confidence intervals are shown by the solid lines. The theoretical Casimir pressures were computed (a) using the Leontovich surface impedance approach and (b) using the generalized plasma-like model (Decca et al. 2007b).

the Leontovich impedance, and the mean experimental values are plotted. The solid lines represent the boundaries of the confidence intervals $[-\Xi_P(a_i), \Xi_P(a_i)]$ computed with eqn (18.21) at a 95% confidence level. Now the theoretical error $\Delta^{tot} P^{theor}(a_i)$ used for the determination of this interval includes the error (18.19) with $\alpha = 4$, due to the uncertainties in the measurement of the separations. As can be seen from the figure, the second method of comparison between experiment and theory demonstrates the consistency of the impedance approach with the data, similarly to the first method.

The dielectric permittivity of the generalized plasma-like model ε_{gp} along the imaginary frequency axis is given by eqn (13.46), where the oscillator parameters are presented in Table 13.3 and $\omega_p = 8.9\,\text{eV}$ (there is no need for a more accurate form of ε_{gp} at $a > 160\,\text{nm}$). Using eqns (12.70) and (17.94), one obtains the theoretical pressure $P^{theor}(a_i)$, including the effect of surface roughness, in the framework of the generalized plasma-like model. The differences between this pressure and the experimental data are plotted as dots in Fig. 19.19(b). As shown in the figure, all dots (and not only 95% of them, as required by the rules of mathematical statistics) are well inside the confidence interval at all separations considered. This means that the experimental data are consistent with the theory based on the generalized plasma-like dielectric permittivity, and that in our conservative error analysis the errors (especially at short separations) are overestimated. Thus, both of the theoretical approaches, using the Leontovich impedance and using the generalized plasma-like model, are consistent with the data. However, while in Fig. 19.19(b) (for the generalized plasma-like model) there are practically no deviations between experiment and

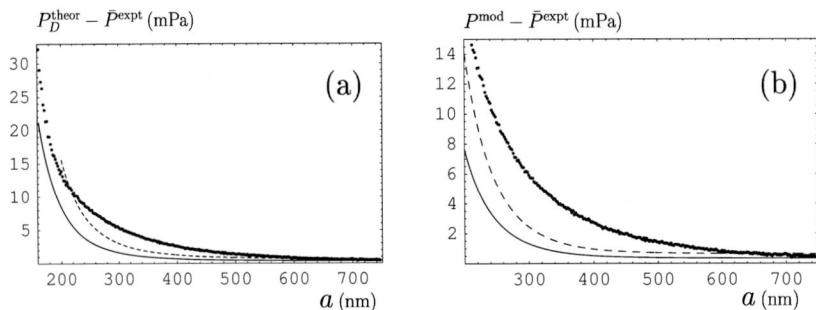

FIG. 19.20. Differences between the theoretical and mean experimental Casimir pressures in a configuration of two plates versus separation, shown by dots. The theoretical results were calculated using the Lifshitz theory at room temperature (a) combined with the Drude model approach (Decca et al. 2007b) or (b) using the modified transverse electric term at zero frequency. The solid and dashed lines indicate the boundaries of the 95% and 99.9% confidence intervals, respectively.

theory at $a > 350$ nm, in Fig. 19.19(a) (for the Leontovich impedance) deviations are noticeable up to $a = 450$ nm. Thus, the generalized plasma-like model is in somewhat better agreement with the data than is the surface impedance approach. As can be seen in Fig. 19.19, the largest deviations between the two theoretical approaches and the experimental data are at short separations from 160 to 200 nm. These deviations are not statistically meaningful, because they are well inside the confidence interval.

The method of comparison between experiment and theory using confidence intervals for the differences $P^{\text{theor}}(a_i) - \bar{P}^{\text{expt}}(a_i)$ is well adapted for tests of alternative approaches to the thermal Casimir force. Using the parameters of the Drude model determined by Decca et al. (2007b) from the resistivity measurements, the theoretical values of the Casimir pressure $P_{\text{D}}^{\text{theor}}(a_i)$ were computed in the framework of the Drude model approach. The calculation procedure, taking the surface roughness into account, was described in Section 19.3.3. The differences $P_{\text{D}}^{\text{theor}}(a_i) - \bar{P}^{\text{expt}}(a_i)$ at all experimental separations are shown as dots in Fig. 19.20(a). The confidence intervals $[-\Xi_P(a_i), \Xi_P(a_i)]$ at each a_i, determined at a 95% confidence level, are the same for all theoretical approaches. Once again, the limits of these confidence intervals are denoted by solid lines [only one such line, corresponding to the upper bound, is shown in Fig. 19.20(a) because almost all dots are above it]. As can be seen in Fig. 19.20(a), the Drude model theoretical approach is experimentally excluded at a 95% confidence level over the entire measurement range from 162 to 746 nm. This conclusion is in agreement with that obtained above using the first local method for comparison between experiment and theory.

The wide gaps between the solid line and the dots in Fig. 19.20(a) suggest that the Drude model approach is actually excluded experimentally at an even higher confidence level than 95%. To make this argument quantitative, we have calculated the half-width of the confidence interval at the 99.9% confidence level from

$$\frac{\Xi_P^{0.999}(a)}{\Xi_P^{0.95}(a)} = \frac{t_{(1+0.999)/2}(32)}{t_{(1+0.95)/2}(32)} \approx 1.85, \qquad (19.18)$$

where $t_p(f)$ is the Student coefficient. The boundaries of the 99.9% confidence intervals obtained from eqn (19.18) are shown in Fig. 19.20(a) by the dashed line. As shown in Fig. 19.20(a), the differences $P_D^{\text{theor}}(a_i) - \bar{P}^{\text{expt}}(a_i)$ are found to be completely outside the 99.9% confidence interval at separations from 210 to 620 nm. This conclusively demonstrates that the results of the third experiment by means of a micromachined oscillator are irreconcilable with the Drude model approach to the thermal Casimir force. This brings experimental confirmation to the phenomenological rule formulated in Section 14.6.3 that, in the application of the Lifshitz theory to real metals, the relaxation processes of conduction electrons should be disregarded.

It is instructive to apply this method of comparison between experiment and theory to the theoretical approach of Dalvit and Lamoreaux (2008) considered in Section 14.3.5. This approach attempts to include screening effects and diffusion processes in the Lifshitz theory. Computations of the theoretical Casimir pressures $P_m^{\text{theor}}(a_i)$ were performed using the Lifshitz formula (12.70) at $T = 300$ K with the TM reflection coefficient (14.53) and the standard TE reflection coefficient defined using the dielectric permittivity (14.54). The Thomas–Fermi screening length (14.61) was used, as this is appropriate for metals with a large carrier density n: for Au, $n \approx 5.9 \times 10^{22}$ cm^{-3}. In Fig. 19.20(b), the differences $P_m^{\text{theor}}(a_i) - \bar{P}^{\text{expt}}(a_i)$ are shown by dots (Mostepanenko 2009; Decca et al. 2008). In the same figure, the solid and dashed lines indicate the borders of the 95% and 99.9% confidence intervals, respectively. The theoretical approach taking screening effects and the diffusion current into account is experimentally excluded at a 95% confidence level over the entire measurement range from 160 to 750 nm and at a 99.9% confidence level at separations from 160 to 640 nm.

From the above discussion, it is clear that the three experiments performed with the micromechanical torsional oscillator satisfy all of the requirements for precision experiments of the Casimir force presented in Section 18.2. However, the role of the third experiment in this series is exceptional because it was the single one where the random error was made much smaller than the systematic error, and the comparison with theory here was performed in a most straightforward and transparent way. The results of this experiment are consistent with the Leontovich impedance approach and with the generalized plasma-like model. The computational results obtained using the Lifshitz formula at zero temperature are also consistent with the data. This means that the precision of Casimir force measurements is still insufficient to measure the small thermal effect at separations below 1 μm predicted by the traditional theoretical approaches, in qualitative agreement with the case of ideal metals. However, relatively large thermal effects, as predicted by the Drude model approach and by an approach based on the modification of the TE reflection coefficient at zero frequency, are excluded by all three successive experiments with increasing confidence. In addition to the figures, we present in Table 19.1 the magnitude of the mean experimental Casimir pressure at a few separations measured in the third experiment (column a). These data are compared with the theoretical Casimir pressure computed using the generalized plasma-like model (column b), the Leontovich surface impedance (column c), the Drude model approach (column d), and the half-width of the confidence interval for $P^{\text{theor}} - \bar{P}^{\text{expt}}$ determined at a 95% confidence level (column e). All of these quantities are given in mPa. The last

TABLE 19.1. Magnitude of the mean experimental Casimir pressure $\bar{P}^{\text{expt}}$ (column a) at different separations a compared with the magnitude of the theoretical pressure P^{theor} computed using the generalized plasma-like model (column b), the Leontovich surface impedance (column c), and the Drude model approach (column d). Column e contains the half-width Ξ of the 95% confidence interval for $P^{\text{theor}} - \bar{P}^{\text{expt}}$. All pressures are given in mPa. Column f contains the values (in %) of the quantity $\Xi/|\bar{P}^{\text{expt}}|$, which describes the agreement between experiment and experimentally consistent theories

a (nm)	a	b	c	d	e	f
162	1108.4	1098.4	1094.2	1076.2	21.2	1.9
166	1012.7	1007.1	1002.7	985.40	19.0	1.9
170	926.85	923.71	919.56	902.96	17.1	1.8
180	751.19	750.58	747.06	732.14	13.3	1.8
190	616.00	616.71	613.70	600.28	10.5	1.7
200	510.50	511.26	508.70	496.62	8.40	1.6
250	225.16	225.71	224.45	217.11	3.30	1.5
300	114.82	114.87	114.18	109.48	1.63	1.4
350	64.634	64.574	64.176	61.004	0.98	1.5
400	39.198	39.096	38.850	36.617	0.69	1.8
450	25.155	25.034	24.874	23.247	0.54	2.2
500	16.822	16.785	16.678	15.456	0.47	2.8
550	11.678	11.669	11.595	10.654	0.42	3.6
600	8.410	8.365	8.312	7.573	0.39	4.6
650	6.216	6.151	6.113	5.522	0.38	6.1
700	4.730	4.626	4.598	4.118	0.36	7.6
746	3.614	3.620	3.598	3.198	0.35	9.7

column (f) contains the quantity $\Xi_P/|\bar{P}^{\text{expt}}|$, determined at a 95% confidence level, which is the measure of agreement between experiment and the experimentally consistent theories. Comparing column (a) with columns (b) and (c), one can conclude that the generalized plasma-like model (column b) provides a more accurate description of the data than does the Leontovich impedance (column c). Bearing in mind that the Leontovich approach is not applicable to short-separation experiments performed using an atomic force microscope, we arrive at the conclusion that the generalized plasma-like dielectric permittivity provides the single approach that is consistent with the results of both short- and long-separation measurements of the Casimir force. The data in Table 19.1 confirm that the Drude model approach is excluded by experiment over the entire measurement range. Subtracting the magnitudes of the theoretical pressures $|P_D^{\text{theor}}|$ (column d) from the experimental values $|\bar{P}^{\text{expt}}|$ (column a), we obtain at all separations larger results than the half-width of the confidence interval $\Xi_P^{0.95}(a)$ (column e). The theoretical approach taking screening effects into account leads to approximately the same differences $|P_m^{\text{theor}}(a)| - |\bar{P}^{\text{expt}}(a)| > \Xi_P^{0.95}(a)$, as does the Drude model approach. Thus, both of these approaches are experimentally excluded.

Column (f) in Table 19.1 demonstrates that the agreement of the data with the theoretical approaches consistent with them varies with the separation. The best agreement, of about 1.5%, occurs at separations from 250 to 350 nm. At the shortest separations, an agreement of approximately 1.9% is achieved. This is much worse than the experimental precision and is mostly determined by the errors in the measurement of separations.

To conclude, the two local methods (see Section 18.3.3) for comparison between experiment and theory for Casimir force measurements lead to results in mutual agreement concerning both consistency with data and rejection of some theoretical approaches.

19.3.5 *Experimental test of proximity force approximation*

The micromechanical torsional oscillator described in Section 19.3.1 was used to perform an experimental test of corrections to the proximity force approximation (Krause et al. 2007). To date, exact results for the electromagnetic Casimir effect between a sphere and a plate are not available. Because of this, the accuracy of the proximity force approximation was estimated from the respective results for the scalar Casimir effect and the electromagnetic Casimir effect between a cylinder and a plate (see Section 10.3). Taking into account corrections to the proximity force approximation, the Casimir force between a sphere of radius R and a plate can be presented in the form (Scardicchio and Jaffe 2006)

$$F(a, R) = 2\pi R E(a) \left[1 + \beta \frac{a}{R} + O\left(\frac{a^2}{R^2}\right) \right]. \tag{19.19}$$

Here $E(a)$ is the Casimir energy per unit area of two parallel plates and β is a dimensionless parameter characterizing the lowest-order deviation from the proximity force approximation. Static measurements of the Casimir force between a sphere and a plate, as described in Section 19.3.2, can be used to obtain constraints on the parameter β in eqn (19.19).

Dynamic measurements of the Casimir pressure, considered in Sections 19.3.2, 19.3.3, and 19.3.4, are more precise than static ones. Substituting the Casimir force (19.19) into the right-hand side of eqn (19.15), one obtains the following expression for the effective Casimir pressure:

$$P^{\text{eff}}(a, R) = P(a) \left[1 + \tilde{\beta}(a)\frac{a}{R} + O\left(\frac{a^2}{R^2}\right)\right], \qquad (19.20)$$

where $P(a)$ is the Casimir pressure between two parallel plates and the dimensionless quantity $\tilde{\beta}(a)$ is given by

$$\tilde{\beta}(a) = \beta \left[1 - \frac{E(a)}{aP(a)}\right]. \qquad (19.21)$$

Note that for ideal-metal bodies, $\tilde{\beta}(a) = 2\beta/3 = \text{const}$.

To obtain constraints on β and $\tilde{\beta}$ in eqns (19.19) and (19.20), a series of experiments has been performed to measure and compare the Casimir force between five Au-coated spheres and Au-coated plates using a micromachined oscillator (Krause et al. 2007). A gold coating thickness of about 200 nm was used. An electrostatic calibration of the setup was performed separately for each sphere, as described in Sections 19.3.1–19.3.3. Spheres with radii $R = 10.5$, 31.4, 52.3, 102.8, and 148.2 µm were used. The surface roughness of both the spheres and the plate was measured using an AFM. In all cases the highest peaks were less than 21 nm in height and the respective variances were less than 5 nm.

A static measurement of the Casimir force between a sphere and a plate was performed at separations from 160 to 750 nm in 10 nm steps. The measurement was repeated 10 times. A dynamic determination of the effective Casimir pressure between two parallel plates was done at separations from 164 to 986 nm in 2 nm steps. The influence of the effects of the nonzero skin depth and of the surface roughness on the deviation from the proximity force approximation at $a = 160$ nm was estimated to be of order 10% and 1%, respectively, of the dominant a/R correction. A comparison between data and theory at separations $a < 300$ nm leads to the result $|\tilde{\beta}(a)| < 0.4$ at a 95% confidence level. In the same range of separations $|\beta| < 0.6$ was obtained (Krause et al. 2007).

These constraints are compatible with exact results (see Section 10.4.3) obtained for a sphere above a plate (for the scalar Casimir effect with Dirichlet and Neumann boundary conditions) and for a cylinder above a plate (for the electromagnetic Casimir effect). Note that in the error analysis in Sections 19.2.1–19.2.3 and 19.3.2–19.3.4 a less stringent estimate $\beta = 1$ was assumed, resulting in a relative error due to the use of the proximity force approximation equal to a/R. This confirms that the above analysis is indeed conservative.

19.4 Experiment using a configuration of two parallel plates

The only experiment in the recent series of Casimir force measurements which used the original configuration of two plane plates was performed by Bressi et al. (2002). A Si cantilever and a thick plate rigidly connected to a frame (the source), both covered with a Cr layer and with an adjustable separation distance between them, were used as the two plates. The coarse separation distance was adjusted with a dc motor and fine tuning was achieved using a linear piezo-electric transducer attached to the frame. Calibration was performed using the electrostatic force. As a result, the error in the absolute separation was found to be $\Delta a = 35$ nm. Small oscillations induced in the source by the application of a sinusoidal voltage to the piezoelectric transducer induced oscillations of the cantilever through the Casimir force. The motion of the resonator, placed in a vacuum, was detected by means of a fiber optic interferometer. After subtracting the electrostatic forces, the residual frequency shift was given by

$$\Delta\nu^2_{\text{expt}}(a) = \nu^2 - \nu_0^2 = -\alpha\frac{\partial P(a)}{\partial a}. \tag{19.22}$$

Here, $\nu_0 = 138.275$ Hz was the natural frequency in the absence of the Casimir pressure P, and $\alpha = S/(4\pi^2 m_{\text{eff}}) \approx 0.0479$ m^2/kg (where S is the area of the capacitor formed by the plates, and m_{eff} is the effective mass).

As discussed above, this measurement was a dynamic one and the directly measured quantity was the frequency shift (19.22) arising from the effect of the Casimir force. This frequency shift is related to the gradient of the Casimir pressure by eqn (19.22). Thus, although this experiment uses a configuration of two parallel plates, it is an indirect measurement of the Casimir pressure between the plates, similar to the experiments using a sphere–plate configuration in the dynamic regime for the same purpose (see Sections 19.3.2–19.3.4). Bressi et al. (2002) did not aim to recover the Casimir pressure from the pressure gradient. Instead, they fitted the experimental data for $\Delta\nu^2_{\text{expt}}(a)$ to $\Delta\nu^2_{\text{theor}}(a)$, computed from the theoretical dependence of the Casimir pressure between two ideal-metal plates, $-K_{\text{C}}/a^4$, with a free parameter K_{C}. The best fit resulted in $K_{\text{C}} = (1.22 \pm 0.18) \times 10^{-27}$ N m^2. This was compared with the exact Casimir coefficient for ideal-metal plates in eqn (1.1), $K_{\text{C}} = \pi^2 \hbar c/240 = 1.3 \times 10^{-27}$ N m^2. The conclusion was drawn that the related force coefficient had been determined at the 15% precision level (Bressi et al. 2002).

From the point of view of the general method for the comparison of experiment with theory (which was developed after this experiment was performed), it would be reasonable to compare $\Delta\nu^2_{\text{expt}}(a)$ with an exact expression for $\Delta\nu^2_{\text{theor}}(a)$ with no adjustable parameters such as K_{C}. In Fig. 19.21(a), the results of such a comparison are presented, where the experimental data are shown as crosses and the solid line shows $\Delta\nu^2_{\text{theor}}$ computed from the exact expression for the Casimir pressure $P_{\text{IM}}(a)$ between two ideal-metal plates, as given in eqn (1.1). As can be seen in the figure, at separations below 1 μm the experimental crosses only touch the solid line. This can be explained by the role of the nonzero skin depth.

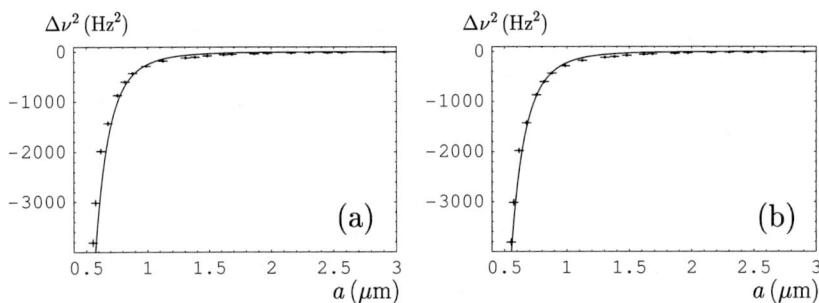

FIG. 19.21. The square of the measured frequency shift versus separation, together with the absolute errors in the separation and in $\Delta\nu^2$, is shown by the crosses. The solid lines indicate the theoretical frequency shift squared, calculated for (a) ideal-metal plates and (b) real metals with inclusion of the skin depth correction (Klimchitskaya et al. 2009a).

If, instead of $P(a) = P_{\text{IM}}(a)$, one uses the Casimir pressure with the first- and second-order corrections due to the nonzero skin depth from eqn (13.11),

$$P_{\text{p}}(a) = -\frac{\pi^2 \hbar c}{240 a^4} \left(1 - \frac{16}{3} \frac{c}{\omega_{\text{p}} a} + 24 \frac{c^2}{\omega_{\text{p}}^2 a^2}\right), \qquad (19.23)$$

there is improved agreement with the data. As an illustration, $\Delta\nu_{\text{theor}}^2$ has been recalculated using the theoretical Casimir pressure (19.23) with $\omega_{\text{p}} \approx 13\,\text{eV}$ for Cr. The results are shown by the solid line in Fig. 19.21(b). It is seen that the Casimir pressure taking the nonzero skin depth into account is in much better agreement with the data than is the ideal-metal Casimir pressure. Suggestions on how to improve the sensitivity of this experiment, as proposed in the literature, are discussed in Section 19.7.

19.5 Related experiments

Here, we briefly consider a few additional experiments on the measurement of the Casimir force between metallic bodies performed in the last few years. Some of them (Lisanti et al. 2005) confirmed earlier results on the role of the nonzero skin depth. The others (Ederth 2000, van Zwol et al. 2008a) were performed in an ambient environment and do not satisfy all of the requirements necessary for precise measurements of the Casimir force. The measurement of the Casimir force in liquids is also considered in connection with the experiment by Munday and Capasso (2007). This experiment has attracted new attention to the complicated and unresolved problem that arises in the case where there is a liquid between the interacting surfaces.

19.5.1 Thin metal layers

Lisanti et al. (2005) reported an observation of the effect of nonzero skin depth on the Casimir force between metallic surfaces. The Casimir force between a thick plate and a 100 μm radius polystyrene sphere coated with metallic films of different thicknesses was measured. The sphere was positioned above a micromachined torsional balance. The Casimir attraction between the sphere and the top plate of the balance induced a rotation angle, which was measured as a function of the separation between the surfaces. Without an indication of errors and confidence levels, it was reported that the Casimir attraction between the metallic plate and a sphere with a coating thinner than the skin depth was smaller than that for the same plate and a sphere with a thick metal coating. Physically, this is the same effect as that demonstrated in Figs. 19.3 and 19.7(a) in Sections 19.2.1 and 19.2.3, where the Casimir forces acting between ideal metals and between real metals were compared with corrections due to the nonzero skin depth (finite-conductivity corrections). Ideal metals are better reflectors, and the magnitude of the Casimir force between them is larger than between real metals. In a similar manner, *thick* real metal films are better reflectors than *thin* real metal films.

Lisanti *et al.* (2005) compared their experimental results with the Lifshitz theory at zero temperature, adapted for the description of layered structures as presented in Section 12.2. It was shown that the experimental forces obtained for films of thickness smaller than the skin depth had smaller magnitudes than those computed for such films using the Lifshitz theory. This result is not surprising, because, as it has been argued in the literature, for films of small thickness the effects of spatial dispersion, which are not included in the Lifshitz theory, should be taken into account (Klimchitskaya *et al.* 2000). The problem of spatial dispersion is discussed in more detail in Section 12.10.

19.5.2 *Ambient measurements*

In this recent period, two ambient (open to the air) experiments on measuring the Casimir force have been reported. Neither used a vacuum environment, and the presence of water layers on the interacting surfaces was reported in both cases. The first of these was an experiment by Ederth (2000), where the force was measured between two cylindrical template-stripped gold surfaces with 0.4 nm roughness, in a distance range from 20 to 100 nm. The 200 nm gold films were fixed to 10 mm radius silica cylinders using a "soft glue". In addition, a hydrocarbon layer of hexadecanethiol was applied to the interacting surfaces. It was noted that this top hydrocarbon layer was necessary to preserve the purity of the gold surface in the ambient environment. This hydrocarbon organic layer prevented a direct measurement of the electrostatic forces and an independent determination of the surface separation. Also, the presence of the hydrocarbon layer means that the experiment cannot be classified as a measurement of the force between two gold surfaces. The inability to measure the residual potential differences and the presence of water and hydrocarbon layers on the surface violate all of the requirements for precise Casimir force measurements discussed in Section 18.2. One of the cylinders was mounted on a piezoelectric actuator and the other on a bimorph deflection sensor. The charge produced by the bimorph in response to the deflection induced by a force on the interacting surfaces was measured by an electrometer amplifier. The soft glue led to a deformation which was estimated to be 18–20 nm for the glue thicknesses used. To compensate for this deformation, the two measured force–distance curves were shifted, one by 9 and the other by 12 nm, to overlap the calculations. The author reported that "it is not possible to establish with certainty" the validity of the displacement due to the deformation and that it "diminishes the strength of the measurement as a test of the Casimir force and also precludes a quantitative assessment of the agreement between theory and experiment."

A second experiment, done in an ambient environment using an AFM for separation distances between 12 and 200 nm, was reported recently (van Zwol et al. 2008a). Here, an Au-coated sphere of radius $R = 20\,\mu$m was fixed to a gold-coated AFM cantilever. The Au-coated plate was mounted on a piezoelectric actuator. Both the sphere and the plate were coated with 100 nm of Au. The optical properties of the Au-coated plate were measured with an ellipsometer in the wavelength region from 137 nm to 33 μm and fitted to obtain a plasma frequency of 7.9 ± 0.2 eV and a relaxation parameter of 0.048 ± 0.005 eV, since the corrections due to nonzero skin depth for the separation range considered are large. The roughness and the water layer reported to be present were not taken into account in the fit. The influence of stochastic roughness at short separations was explored by van Zwol et al. (2008b). The errors in the cantilever spring constant and the deflection coefficient were reported to be 4% and 3%, respectively, which together were reported to lead to errors from 4% to 10%. The calibration errors were reported to lead to an overall error in the range from 5% to 35%. The electrostatically measured contact potential, 10 ± 10 mV, was reported to lead to a 10% error. The authors reported that they were not able to independently determine the separation on contact of the two surfaces owing to the stiff cantilevers employed. Based on the roughness, they estimated a 1 nm error in the contact separation, "leading therefore to a 28% relative error at the smallest separations." However, a general 10% agreement with the theory was reported below 100 nm separation. Given the ambient nature of the experiment, water layers "typically a few nanometers" thick on both surfaces were present, but these were not treated in the theoretical comparison or systematic errors. The two factors here that do not meet the requirements for precise measurements, as discussed by Sparnaay (see Section 18.2), are the presence of the water layer and the lack of an independent measurement of the separation. Repeating the experiments in a vacuum environment should allow a more definitive comparison.

19.5.3 Measurements in liquids

The Casimir interaction between two thick plates (semispaces) with a liquid layer between them has long attracted attention because of the possibility of repulsive forces. In this case the interaction energy per unit area is given by eqn (12.47) with the reflection coefficients defined in eqn (12.53). In the nonrelativistic limit, where $R_{\text{TE}}^{(\pm)} \ll R_{\text{TM}}^{(\pm)}$ and $k^{(0)} \approx k^{(\pm)} \approx k_\perp$, eqns (12.47) and (12.50) lead to

$$E(a) = \frac{\hbar}{4\pi^2} \int_0^\infty k_\perp dk_\perp \int_0^\infty d\xi \qquad (19.24)$$
$$\times \ln\left[1 - \frac{\varepsilon^{(1)}(i\xi) - \varepsilon^{(0)}(i\xi)}{\varepsilon^{(1)}(i\xi) + \varepsilon^{(0)}(i\xi)} \frac{\varepsilon^{(-1)}(i\xi) - \varepsilon^{(0)}(i\xi)}{\varepsilon^{(-1)}(i\xi) + \varepsilon^{(0)}(i\xi)} e^{-2k_\perp a}\right].$$

It is evident that $E(a) > 0$ (i.e. the Casimir interaction is repulsive) if either of the inequalities

$$\varepsilon^{(-1)}(i\xi) < \varepsilon^{(0)}(i\xi) < \varepsilon^{(1)}(i\xi) \quad \text{or} \quad \varepsilon^{(1)}(i\xi) < \varepsilon^{(0)}(i\xi) < \varepsilon^{(-1)}(i\xi) \qquad (19.25)$$

is satisfied. For three-layer systems, the possibility of repulsive Casimir forces is also present in the relativistic limit when the interaction is described by the complete eqns (12.47) and (12.53).

In the region of nonretarded van der Waals forces, the observation of repulsion has been reported using an AFM (Milling et al. 1996, Meurk et al. 1997). Lee and Sigmund (2002) reported the observation of repulsive van der Waals forces at separations larger than 4 or 5 nm, where retardation effects contribute to the result obtained. No errors or experimental controls were reported in the above experiments. The Casimir repulsion in three-layer systems was measured at separations of about 30 nm by Munday et al. (2009). However, at larger separations, the Casimir repulsion in a three-layer system has not yet been investigated. The resolution of this problem is complicated by the presence of double layers of charges on the solid–liquid interfaces. Because of this, even the standard, attractive, Casimir forces in the presence of a liquid layer have not been investigated yet, and the Lifshitz theory for such systems has not been tested.

An interesting preliminary test of the Lifshitz theory for three-layer systems was the measurement of the attractive Casimir force between an Au-coated sphere and a plate immersed in ethanol using an AFM (Munday and Capasso 2007). The experimental data obtained were compared with the Lifshitz theory, taking into account the frequency dependence of the dielectric functions of Au and ethanol and the correction due to the surface roughness. Consistency of the data obtained with the Lifshitz theory was claimed, although at separations below 50 nm a disagreement was observed which increases with decreasing separation. However, as commented in the literature (Geyer et al. 2008b), theoretical computations of the Casimir force between smooth Au surfaces separated by ethanol done according to the method provided by Munday and Capasso (2007) [i.e. by the use of the Kramers-Kronig relations and tabulated optical data (Palik 1985)] lead to a discrepancy of up to 25% with respect to the reported theoretical results. The latter results can be reproduced if, at all imaginary frequencies, one uses the Drude dielectric function (13.14) for both the sphere and the plate material. A second drawback is that the effect of the residual potential difference between the sphere and the plate was calculated incorrectly and was significantly underestimated by a factor of 590. Finally, the possible interaction between the double layer formed in liquids owing to salt impurities, which would decrease the electrostatic force, was not taken into account, without any justification.

The resulting electrostatic force is of the same magnitude as the Casimir force to be measured. All this makes the interpretation of this experiment uncertain (Geyer et al. 2008b). In their reply, Munday and Capasso (2008) recognized that the original paper (Munday and Capasso 2007) did in fact use the Drude model. It was also recognized that the equation originally used to estimate the residual electrostatic force "is not strictly correct" and that salt contaminants exist even in the purest solutions, leading to the screening of electrostatic interactions. However, the reply claims that the experimental results are still consistent with the Lifshitz theory (Munday and Capasso 2008). Thus, this configuration remains interesting for future investigations. Later, Munday et al. (2008) investigated further the effect of electrostatic forces and Debye screening on the measurement of the Casimir force in fluids. The electrostatic force with account taken of Debye screening was calculated as outlined by Geyer et al. (2008b). The relative random experimental error of the Casimir force measurements in this experiment determined at a 67% confidence level was equal to approximately 7% at $a = 30$ nm and increased to 60% at $a = 80$ nm. Within this range of separations, the measurement data were found to be in qualitative agreement with the Lifshitz theory (for $a < 30$ nm, the authors admitted deviations between the theory and data).

19.5.4 Dynamic holography techniques

Another experiment used an adaptive holographic interferometer to measure periodic nonlinear deformations of a thin pellicle caused by an oscillating Casimir force due to a spherical lens (Petrov et al. 2006). Both test bodies were coated with a thin Al film and placed in a vacuum chamber. The lens was mounted on a vibrating piezodriver. As a result, the oscillations of the lens position led to a periodic modulation of the Casimir force. The experimental data were found to be in only qualitative agreement with a theory based on ideal-metal boundaries at separation distances of a few hundred nanometers. Corrections due to the finite skin depth and surface roughness were not provided. Also, the use of Al, whose surface undergoes rapid oxidation even in relatively high vacuum, adds uncertainty to the results of this experiment. However, the new measurement technique used may be promising for future measurements of the Casimir force. Later, an attempt was undertaken to calculate corrections to the Casimir force in this experimental configuration due to the nonzero skin depth (Bryksin and Petrov 2008). The dielectric permittivity of the plasma model (13.1) was used in the calculation. It was concluded, however, that the experimental results (Petrov et al. 2006) were consistent with theory only under the condition $\lambda_p > 1000$ nm.

19.6 Prospects for future measurements

As described in this chapter, there have been several important improvements in the measurements of the Casimir force between metallic bodies. However, the experiments performed to date have not been of sufficient precision to measure the magnitude of the thermal effect. The experiments using the micromachined oscillator (Section 19.3) possessed the highest experimental precision at separations below $1\,\mu$m. They have been used to exclude thermal effects predicted by approaches using the Drude model at low frequencies. However, the thermal effects predicted by the generalized plasma-like model at short separations remain below the experimental sensitivity. In this respect, large-separation measurements of the Casimir force would be of great interest. At separations of a few micrometers, the thermal regime is reached, where the Casimir free energy is entirely of thermal origin. Calculations using the plasma model or the generalized plasma-like model result in eqn (14.5), which is the same as for ideal metals. The Drude model approach to the thermal Casimir force leads to only one-half of the result for ideal-metal plates [see eqn (14.7)]. Thus, large-separation measurements of the Casimir force would bring direct information to bear on the magnitude of the thermal effect between macroscopic bodies (such a measurement in the atom–plate configuration has been performed already; see Section 22.1.1).

In Section 19.1 it was shown that the experiment using a torsion pendulum within the region from 0.6 to $6\,\mu$m was in fact uncertain at separations above $2\,\mu$m, where the thermal effects begin to make a substantial contribution. An analysis (Lamoreaux and Buttler 2005) shows that the torsion pendulum technique has the potential to measure the Casimir force between a plate and a spherical lens at $a = 4\,\mu$m with a relative error of 10%. Bearing in mind that at $a = 4\,\mu$m the thermal correction contributes as much as 86% of the zero-temperature Casimir force, such an experiment, if successfully performed, holds great promise.

There is a proposal aimed at measuring the Casimir force in the cylinder–plate configuration at separations around $3\,\mu$m (Brown-Hayes et al. 2005). This geometry can be considered as a compromise between the two-parallel-plate configuration (which is connected with serious experimental difficulties associated with the parallelity of the plates) and a sphere above a plate. In addition, as discussed in Section 10.3, an exact solution for the ideal-metal cylinder–plate configuration has been obtained recently. This gives us the possibility to determine the accuracy of the PFA and to apply it to real materials with high reliability. Finally, it was concluded that using a dynamic measuring scheme it is possible to measure the Casimir force between a plate and a cylinder at separations of about $3\,\mu$m with a precision of a few percent (Brown-Hayes et al. 2005).

Another proposal suggests the use of a highly sensitive torsion balance in the separation range from 1 to $10\,\mu$m to measure the Casimir force in the configuration of two parallel plates (Lambrecht et al. 2005). The construction of the balance is similar to that used in Eötvos-type experiments aimed at testing the equivalence principle. It is planned to measure the thermal Casimir force with an accuracy of a few percent and to discriminate between different theoretical approaches discussed in the literature.

One more experiment exploiting the two-parallel-plate configuration at separations larger than a few micrometers has been proposed (Antonini et al. 2006). The experimental scheme is based on the use of a Michelson-type interferometer and the dynamic technique, with one oscillating plate. Calibrations show that a force of 5×10^{-11} N can be measured in this setup with a relative error from about 10% to 20%. This would be sufficient to measure the thermal effect at a separation of $5\,\mu$m (Antonini et al. 2006).

The thermal effect on the Casimir force can be measured at short separations below $1\,\mu$m if the difference in the thermal forces $F(a, T_2) - F(a, T_1)$ at different temperatures rather than the absolute value of the thermal Casimir force is measured (Chen et al. 2003). For real metals, this difference in the thermal Casimir forces (in contrast to the relative thermal correction) does not increase but decreases with increasing separation distance. This allows the observation of the thermal effect on the Casimir force at small separations of about $0.5\,\mu$m, where the relative thermal correction, as predicted by the generalized plasma-like model, is rather small. Preliminary estimation shows that with a sphere of radius $R = 2$ mm attached to the cantilever of an AFM, measurable changes in the force amplitude of order 10^{-13} N are achievable from a 50 K change in the temperature. Such a temperature difference can be obtained by illumination of the sphere and plate surfaces with laser pulses of 10^{-2} s duration (Chen et al. 2004b). From the above it is clear that experimental investigations of thermal effects on the Casimir force appear feasible in the near future.

20

MEASUREMENTS OF THE CASIMIR FORCE WITH SEMICONDUCTORS

The experiments on measuring the Casimir force described in Chapter 19 dealt with metallic test bodies. However, the most important materials used in nanotechnology are semiconductors, with conductivity properties ranging from metallic to dielectric. The reflectivity of a semiconductor surface can be changed over a wide frequency range by changing the carrier density through variation of the temperature, using different kinds of doping, or via illumination of the surface with light. It should be noted that measuring the van der Waals and Casimir forces between dielectrics has always been a problem owing to the need to eliminate residual charges and contact potential differences. Semiconductors with a reasonably high charge carrier density have the advantage that, under appropriate conditions, they avoid accumulation of charge and screening effects but, at the same time, possess a dielectric-like dependence of the permittivity on the frequency over a wide frequency range. This makes it possible to examine the influence of material properties on the Casimir force and opens up new opportunities to modulate the magnitude and separation dependence of the force.

An early attempt to measure the van der Waals and Casimir forces on semiconductor surfaces and modify them by light was reported by Arnold *et al.* (1979). Attractive forces were measured between a glass lens and an Si plate and also between a glass lens coated with amorphous Si and an Si plate. The glass lens, however, was an insulator and therefore the electric forces, such as those due to work function differences, could not be controlled. This might explain why Arnold *et al.* (1979) found no force change occurred on illumination at separations below 350 nm, where it should have been most pronounced. One more attempt to modify the Casimir force was made by Iannuzzi *et al.* (2004b) when they measured the Casimir force acting between a plate and a sphere coated with a hydrogen-switchable mirror that became transparent upon hydrogenation. Despite expectations, no significant decrease of the Casimir force resulting from the increased transparency was observed. This negative result can be explained by the Lifshitz theory, which requires a change of the reflectivity properties over a wide range of frequencies in order to markedly affect the magnitude of the Casimir force. This requirement was not satisfied by hydrogenation.

In this chapter, we consider three experiments on measuring the Casimir force between an Au-coated sphere and an Si plate by means of an atomic force microscope. The first experiment (Chen *et al.* 2005a, 2006b) revealed that the measured Casimir force for a plate made of p-type Si was markedly different from the calculation results for high-resistivity dielectric Si. In the second experiment (Chen et al. 2006a), the difference between the Casimir forces for an Au-coated sphere and two plates made of n-type Si with different charge carriers densities was measured. Through this, a dependence of the Casimir force on the charge carrier density was demonstrated. The modification of the Casimir force through an optically induced change in the charge carrier density was first reliably demonstrated in the third experiment in this series, performed by Chen et al. (2007a, 2007b). This experiment was also used as a test of various theoretical approaches to the description of charge carriers in dielectrics and semiconductors (see Sections 12.5.2 and 16.4.3). Specifically, it was found that the inclusion of the static conductivity of a dielectric material in the model of the dielectric response was experimentally inconsistent. All of these results are presented below.

20.1 Experiment with gold-coated sphere and silicon plate

In this experiment, the Casimir force between an Au-coated sphere of diameter $2R = 202.6 \pm 0.3\,\mu$m and a single-crystal Si$\langle 100 \rangle$ plate was measured. The thickness of the Au coating was 105 nm. A schematic plan of the experimental setup was shown in Fig. 19.1 and described in Section 19.2. The Si plate (doped with B) had an area of $5 \times 10\,\text{mm}^2$ and a thickness of $350\,\mu$m. The resistivity of the plate, $\rho = 0.0035\,\Omega\,$cm, was measured using the four-probe technique. Note that the resistivities of metals are usually two or three orders of magnitude lower than this. Because of this, the Si plate used had a relatively large absorption, typical of semiconductors for all frequencies contributing to the Casimir force at the experimental separations.

The main improvements in the experimental setup in comparison with that described in Section 19.2.3 were the use of a much higher vacuum, and a reduction of the uncertainty in the determination of the absolute separations a. A much higher vacuum of 2×10^{-7} Torr was needed to maintain the chemical purity of the Si surface, which otherwise would oxidize rapidly to SiO$_2$. A high-vacuum system was also needed to prevent contamination. This vacuum system consisted of oil-free mechanical pumps, turbopumps, and ion pumps. To maintain the lowest pressure during data acquisition, only an ion pump was used. This helped to reduce the influence of mechanical noise. The absolute error in the determination of the absolute separations a was reduced to $\Delta a = 0.8$ nm in comparison with the value of $\Delta a = 1$ nm as described in Section 19.2.3. This was achieved by using a piezoelectric actuator capable of traveling a distance of $6\,\mu$m from the initial separation to contact of the test bodies (in the experiment described in Section 19.2.3, the movement of the piezoelectric actuator was used only for separations less than $2\,\mu$m, and the movement to larger separations of the plate from the sphere was done mechanically). Such large extensions of the actuator were also found necessary to allow time for the decay of noise associated with the separation of the gold sphere and plate after contact of the two surfaces. The complete movement of the piezoelectric actuator, a_{piezo}, was calibrated using a fiber optic interferometer (Chen and Mohideen 2001). To extend and contract the actuator, a continuous triangular voltage at 0.02 Hz was applied. Given that the experiment was done at room temperature, applying a static voltage would lead to creep of the actuator and loss of position sensitivity. The extension and contraction of the actuator were fitted to terms up to fourth order in the applied voltage. Because of this, the error in the calibration of the piezoelectric actuator did not contribute practically to Δa.

In contrast to Au, the Si surface is very reactive. Because of this, a special passivation procedure was needed to prepare the surface for force measurements. For this purpose, nanostrip (a combination of H_2O_2 and H_2SO_4) was used to clean the surface of organics and other contaminants. This cleaning, however, oxidizes the surface. Then a 49% HF solution was used to etch SiO_2 off the surface. This procedure also leads to hydrogen termination of the surface. The hydrogen termination prevents the reoxidation of the Si surface as long as it is kept in a high-vacuum environment. The termination is stable for more than 2 weeks under the vacuum conditions described above (Gräf et al. 1990, Arima et al. 2000). The effectiveness of the passivation technique in preventing the contamination of the Si surface was checked by the measurement of the distance dependence of the electrostatic force resulting from the residual potential difference between the interacting surfaces (see below).

To characterize the topographies of both surfaces, the Au coating of the sphere and the surface of the Si plate were investigated using an AFM. Images resulting from a surface scan of the Au coating demonstrated that the roughness was mostly represented by stochastically distributed distortions of heights $h_i^{(1)} \leq 25$ nm ($1 \leq i \leq N_1 = 26$). The fractions $v_i^{(1)}$ of the Au coating with heights $h_i^{(1)}$ were determined. The surface of the Si plate was much smoother, with heights $h_i^{(2)} \leq 1$ nm ($1 \leq i \leq N_2 = 11$); the respective fractions of the surface $v_i^{(2)}$ were again determined [the values of $v_i^{(1,2)}$ are presented by Chen et al. (2006b)]. From eqn (17.92), the zero levels of the roughness of the sphere and of the plate were $H_0^{(1)} = 15.352$ nm and $H_0^{(2)} = 0.545$ nm, respectively. The roughness variances determined from eqn (17.95) were $\delta_1 = 3.446$ nm for the sphere and $\delta_2 = 0.111$ nm for the plate.

Now we consider in sequence the calibration of the setup, together with related procedures such as the determination of the residual electrostatic force and the separation on contact; the measurement results and the error analysis; and the comparison between experiment and theory.

20.1.1 Calibration of the setup

All calibrations and determinations of the residual electrostatic force and of the separation on contact were done immediately before the Casimir force measurements in the same high-vacuum apparatus. As described in Section 19.2, the force was determined through the deflection of a cantilever. The calibration of the deflection signal S_{def}, which was negative for an attractive force and was measured by using two photodiodes either as a current or a voltage, was done by applying dc voltages to the Si plate. Care was taken to make ohmic electrical contacts to the silicon. Direct contact to the Si plate led to large residual potentials. Because of this, the electrical contact was made from a 100-nm-thick gold pad attached to the bottom of the plate. The electrical contact to the gold sphere was accomplished by applying a very thin gold coating to the cantilever. In addition, a small correction had to be applied to the separation distance between the gold sphere and the Si plate owing to the movement of the cantilever in accordance with eqn (19.4).

The measurement of the deflection coefficient m was performed by applying different dc voltages V between $+0.2$ to -0.4 V to the plate. To find the coefficient m, the cantilever deflection signal was measured as a function of the distance. A 0.02 Hz triangular wave was applied to the piezoelectric actuator to change the distance between the sphere and the plate. Larger applied voltages led to more cantilever deflection and therefore earlier contact of the two surfaces. The change in the contact position of the sphere and the plate as a function of the applied voltage could then be used to measure the deflection coefficient m (see Section 19.2.3). In order to determine the contact of the two surfaces precisely, 32 768 data points at equal time intervals were acquired for each force measurement. In distinction to the procedure previously used, in case where the contact was between two neighboring data points, a linear interpolation was used to identify the exact value. The deflection coefficient m was found to be 43.3 ± 0.3 nm per unit deflection signal. This value was used to correct the separation distance in all measurements.

The determination of the residual potential difference between the two surfaces, V_0, was performed by fitting electric-force measurements far away from contact (where the Casimir force is practically zero) to the exact force–distance relation. To measure the force, a calibration of the deflection signal was done. For this purpose Chen et al. (2005a, 2006b) used an improved method, rather than simple application of a dc voltage V to the plate. This method was employed to avoid systematic errors due to scattered laser light. In addition to the application of the dc voltage V to the Si plate, square voltage pulses with amplitudes in the range ± 0.4 V during the time interval corresponding to a separation distance between 1 and 5 μm were also applied to the plate. The dc voltage was close to the residual potential difference V_0 in order to decrease the systematic errors due to large deflections. Figure 20.1(a) shows the deflection signal of the cantilever in response to both the applied dc voltage and the square pulse as a function of the separation distance between the gold sphere and Si plate. By measuring only the difference in the signal during the pulse, it was possible to avoid the need for a background subtraction. Also, the large width of the pulse allowed checks of the distance dependence of the residual potential and any position dependence in the calibration.

The average values of the measured electric forces as a function of separation were used to fit the exact force–distance relation (19.1) or its approximate representation in eqns (19.2) and (19.3). Within the range of separations from

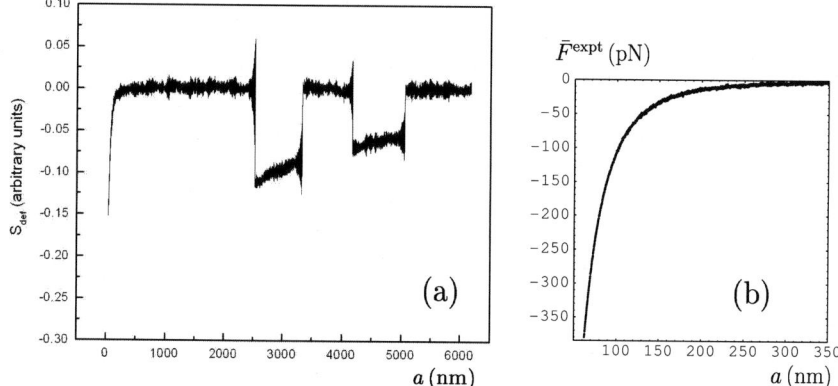

FIG. 20.1. (a) Deflection signal of the cantilever in response to a dc voltage and two square voltage pulses applied to the Si plate, as a function of separation. (b) Mean measured Casimir force between the Si plate and Au sphere, as a function of separation (Chen et al. 2006b).

1.8 to 5 μm, the relative error introduced by the use of $X(\alpha)$ obtained from eqn (19.2) instead of eqn (19.1) did not exceed 1.5×10^{-5}. Equation (19.2) at a fixed separation a was used to fit the difference signal, and the residual potential difference was determined to be $V_0 = -0.114 \pm 0.002\,\mathrm{V}$. The calibration of the deflection signal was also performed using the same procedure. The force calibration constant was determined to be $1.440 \pm 0.007\,\mathrm{nN}$ per unit cantilever deflection signal.

The value of V_0 was found to be independent of separation. This confirms the absence of localized charges and large screening effects, because these would lead to a dependence of F_{el} on a different from eqn (19.1), resulting in a residual potential difference varying with distance when we apply eqn (19.1) for the fit. As mentioned above, the relatively high conductivity of the Si plate used in this experiment was important in preventing the formation of localized charges and screening effects (see Section 20.2 for more details). The independence of V_0 of the separation also confirms the absence of any contamination of the Si or Au surface.

The average distance between the two zero levels of the roughness on contact of the two surfaces, a_0, needed to be independently determined for a comparison of the measured Casimir force with the theory. To achieve this goal, various dc voltages were applied to the Si plate (as in the measurement of m) and the electrostatic force was measured as a function of separation. This measurement at each voltage was repeated five times and the average signal curve was obtained. A compensation dc voltage equal to V_0 was applied to the plate and the resulting deflection signal was subtracted from the signal corresponding to the electrostatic-force curves at all other dc voltages. This procedure eliminated the need for subtraction of the background and Casimir forces from the electrostatic-force curves. In contrast to Harris et al. (2000), in the determination of a_0 a different procedure was used in order to reduce the role of uncertainties in V_0. This procedure also gives an additional way to determine V_0 and to check its distance independence. It was as follows. At a fixed separation a, different voltages V were applied to the plate and the electrostatic force was plotted as a function of V. The parabolic dependence of this force [see eqn (19.2)] was used to determine the values of V_0 and $X(\alpha)$ (van Blockland and Overbeek 1978). This was repeated for many different values of a. The value of V_0 was found to be the same as that determined earlier and to be independent of separation distance. Note that this determination was also independent of errors in the cantilever calibration. In order to determine a_0, $X(\alpha)$ was then plotted as a function of a and fitted to eqn (19.2). The value of a_0 so determined was 32.1 nm. The uncertainty in the quantity $a_0 + S_{\text{def}} m$ [see eqn (19.4)] due to both the uncertainty in m and the calibration was found to be 0.8 nm. As mentioned above, the error in the calibration of the piezoelectric actuator contributed negligibly to the error in the measurement of the absolute separations Δa. Because of this, using eqn (19.4), we arrive at $\Delta a = 0.8$ nm.

20.1.2 *Measurement results and experimental errors*

The Casimir force between the sphere and the plate was measured as a function of distance. In this process, the sphere was kept grounded while a compensating voltage V_0 was applied to the plate to cancel the residual electrostatic force. The distance was varied continuously from large to small separations by applying continuous triangular voltages at 0.02 Hz to the piezoelectric actuator. The actuator was extended to its maximum range of over 6 μm. The force data $F^{\text{expt}}(a_i)$ were collected at 32 768 equal time intervals as the distance between the sphere and plate was changed. This measurement was repeated for $n = 65$ times. A great advantage of the atomic force microscopy technique for the averaging was that the contact point between the two surfaces a_0 provided a starting point for alignment of all 65 measurements. Nevertheless, thermal noise in the cantilever deflection signal, S_{def}, leads to noise in the corresponding separations a. To account for this in the averaging, the separation distance was divided into a grid of 32 768 equidistant points separated by 0.17 nm. For each measured Casimir force–distance curve, the value of the force at each grid point was computed using linear interpolation of the neighboring two data points. Because the separation distance between neighboring points was as small as 0.17 nm, higher-order interpolation procedures were not required. Also, the noise spectrum and amplitude of the interpolated data were confirmed to be the same as those of the raw data. This allowed the averaging of the 65 Casimir force measurements, even including the effect of the change in the separation distance due to the thermal noise of the cantilever.

The measurement results were presented within the separation range from 62.33 nm to 349.97 nm. This distance range (containing 1693 points) was chosen because for larger separations the experimental relative error of the force mea-

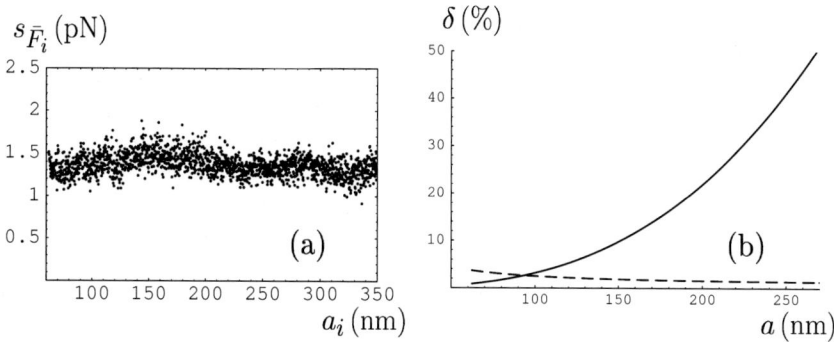

FIG. 20.2. (a) Variance of the mean measured Casimir force as a function of plate–sphere separation (Chen et al. 2006b). (b) Total relative experimental error $\delta^{\text{tot}} F^{\text{expt}}$ (solid line) and theoretical error $\delta^{\text{tot}} F^{\text{theor}}$ (dashed line) as a function of plate–sphere separation (Chen et al. 2005a).

surements caused by the noise exceeded 100% (see below), i.e. the data were not informative.

The mean values of the Casimir force computed using eqn (18.3) with $n = 65$ are plotted in Fig. 20.1(b) as a function of the separation. As seen from Fig. 20.1(b), at short separations the mean force $\bar{F}^{\text{expt}}(a)$ is uniform, i.e. it changes smoothly with changes of a. The measurement data in all 65 sets were checked for the presence of outlying results using the statistical criterion presented in Section 18.3.1. It was found that there were no outlying results among the measurement data and, thus, all of them could be used in the determination of experimental errors.

The variance of the mean was computed using eqn (18.3). The computational results are shown in Fig. 20.2(a). It is seen that $s_{\bar{F}_i}$ is not uniform, i.e. it changes stochastically with increasing separation. It was replaced with the smoothed variance $s_{\bar{F}} \approx 1.5\,\text{pN}$ defined in eqn (18.4). Then the random absolute error $\Delta^{\text{rand}} F^{\text{expt}} = 3.0\,\text{pN}$ at a 95% confidence level was calculated using eqn (18.6) with $t_{(1+0.95)/2}(64) = 2$. The respective relative random error $\delta^{\text{rand}} F^{\text{expt}}$ is given by eqn (18.13). It reaches its smallest value of 0.78% at the shortest separation $a = 62.33\,\text{nm}$ and increases with increasing separation.

There were the following four systematic errors in this experiment (Chen et al. 2005a, 2006b): $\Delta_1^{\text{syst}} F^{\text{expt}} \approx 0.82\,\text{pN}$ due to the error in the force calibration, $\Delta_2^{\text{syst}} F^{\text{expt}} \approx 0.55\,\text{pN}$ due to noise when the calibration voltage is applied to the cantilever, $\Delta_3^{\text{syst}} F^{\text{expt}} \approx 0.31\,\text{pN}$ due to the instrumental sensitivity, and $\Delta_4^{\text{syst}} F^{\text{expt}} \approx 0.12\,\text{pN}$ due to the restrictions on the computer resolution of the data. Combining these errors using the statistical rule (18.2) with $J = 4$ and $k_{0.95}^{(4)} = 1.12$, the value $\Delta^{\text{syst}} F^{\text{expt}} = 1.17\,\text{pN}$ was obtained at a 95% confidence level. The respective relative systematic error $\delta^{\text{syst}} F^{\text{expt}}$ takes its smallest value of 0.31% at the shortest separation. In fact, at all separations the magnitude of the relative systematic error is about 0.4 times the random error.

To find the total experimental error in the Casimir force measurements, it was necessary to combine the random and systematic errors. This was done using the statistical rule (18.17) with a conservative value of the coefficient $q_{0.95} = 0.8$. As a result, the total experimental error determined at a 95% confidence level is equal to $\Delta^{\text{tot}} F^{\text{expt}} = 3.33\,\text{pN}$. The total relative experimental error defined in eqn (18.18) is equal to only 0.87% at the shortest separation, increases to 5.3% at $a = 120\,\text{nm}$, and reaches 64% at a separation $a = 299.99\,\text{nm}$. At $a = 350\,\text{nm}$, it exceeds 100%. In Fig. 20.2(b), the total relative experimental error is shown as a function of separation (the solid line). At all separations, the major contribution to it is given by the random error.

20.1.3 Comparison between experiment and theory

As demonstrated in the preceding subsection, the lowest experimental errors are achieved at separations $a \leq 120\,\text{nm}$. At such short separations, thermal effects are not important. Because of this, to compare experiment and theory Chen et al. (2005a, 2006b) used the proximity force approximation (6.71) at $T = 0$ and the Lifshitz formula (12.44) with the reflection coefficients (12.45). The dielectric permittivity of Au along the imaginary frequency axis $\varepsilon^{(1)}(i\xi)$, as given by the dashed line in Fig. 13.2, was used. This was obtained from the tabulated optical data extrapolated to low frequencies by use of the Drude model. At the separations considered, practically the same results follow from using the solid line in that figure (the generalized plasma-like model).

The dielectric permittivity $\varepsilon^{(2)}(i\xi)$ of dielectric Si, with a resistivity $\rho_0 = 1000\,\Omega\,\text{cm}$, is presented in Fig. 12.2(a). These values of $\varepsilon^{(2)}(i\xi)$ can be used in precise computations of the Casimir and van der Waals interactions between test bodies made of high-resistivity (dielectric) Si. Note that the use of the analytical approximation for $\varepsilon^{(2)}(i\xi)$ suggested by Inui (2003) leads to an error of about 10% in the magnitude of the Casimir force and, thus, is not suitable for comparison with precise measurements. The same analytical approximation with different set of parameters (Inui 2006) leads to errors of less than 1% in the magnitude of the Casimir force. Another analytical expression for the dielectric permittivity of Si along the imaginary frequency axis was suggested by Lambrecht et al. (2007). This also results in errors of less than 1% if it is compared with computations using the tabulated optical data for Si. In the experiment under consideration, however, a Si plate of much lower resistivity, $\rho = 0.0035\,\Omega\,\text{cm}$, than ρ_0 was used. This resistivity corresponds to B-doped Si. The plasma frequency for such Si given by eqn (13.2) is equal to $\omega_p^{(2)} \approx 7 \times 10^{14}\,\text{rad/s}$. Here, the doping concentration leads to a carrier density $n \approx (2.9\text{--}3.2) \times 10^{19}\,\text{cm}^{-3}$. This value of n corresponding to a sample with a resistivity $\rho = 0.0035\,\Omega\,\text{cm}$ was obtained from Fig. 2.18 of the reference manual by Beadle et al. (1985). The optical effective mass for the B-doped Si used in this experiment is $m^* = m_p^* = 0.206 m_e$ (Hellwege 1982). The respective relaxation parameter of the Drude model $\gamma^{(2)}$ was determined from eqn (13.20):

$$\gamma^{(2)} = \frac{\omega_p^{(2)2}}{4\pi\sigma_0^{(2)}} = \frac{1}{4\pi}\rho\omega_p^{(2)2} \approx 1.5 \times 10^{14}\,\text{rad/s}. \qquad (20.1)$$

Since the optical properties of Si at the frequencies that make a nonnegligible contribution to the Casimir force depend on the concentration of charge carriers, the optical data of dielectric Si should be adapted for the case under consideration. This can be done (Palik 1985) by adding the imaginary part of the Drude dielectric function to the imaginary part of the dielectric permittivity $\varepsilon^{(2)}(\omega)$ (see Section 12.6.3). For the Si plate of lower resistivity ρ used in the experiment, this results in

$$\tilde{\varepsilon}^{(2)}(i\xi) = \varepsilon^{(2)}(i\xi) + \frac{\omega_p^{(2)2}}{\xi\left(\xi + \gamma^{(2)}\right)}. \qquad (20.2)$$

This permittivity is different from the solid line in Fig. 12.2 only at $\xi \leq 10^{15}\,\text{rad/s}$. Once the dielectric permittivities of Au and Si along the imaginary frequency axis have been computed, the Casimir force F between a smooth sphere and plate can be found by use of eqns (12.44) and (6.71). Note that experimentally indistinguishable results for F were obtained if the charge carriers were taken into account by means of the plasma model, i.e. by putting $\gamma^{(2)} = 0$ in eqn (20.2). The theoretical Casimir force F^{theor} taking account of surface roughness was computed by using the method of geometrical averaging presented in eqn (17.94), with the $H_0^{(1,2)}$ and $h_i^{(1,2)}$ discussed at the beginning of Section 20.1. The computational results show that in this experiment the influence of surface roughness is very small. For example, if the separation increases from 62.33 to 100.07 nm, the ratio F^{theor}/F decreases from 1.015 to 1.006. Thus, the contribution of surface roughness reaches its maximum value of 1.5% at the shortest separation and decreases to only 0.6% at $a = 100.07\,\text{nm}$. The contributions of diffraction-type and correlation effects to the roughness correction, which are not taken into account in the method of geometrical averaging, were found to be negligible at the separations considered (Chen et al. 2006b).

Now we discuss the theoretical errors which may occur in the computation of the Casimir force. These are the same errors as were considered in Section 19.2.3, devoted to the precision measurement of the Casimir force using gold surfaces. The first theoretical error, $\delta_1 F^{\text{theor}} = 0.5\%$, is due to uncertainties in the optical data used to determine $\varepsilon^{(1,2)}(i\xi)$. The anomalous sets of data which would lead to quite different force values were discussed in Section 19.3.4. This experiment does not support the hypothesis that the Au films used could be described by such data. The uncertainties due to patch potentials, $\delta_2 F^{\text{theor}}$, were shown to be equal to a fraction of a percent (see Section 19.2.3). The uncertainty due to the use of the proximity force approximation is $\delta_3 F^{\text{theor}} = a/R$. When theory is compared with experiment by considering the differences $F^{\text{theor}}(a_i) - F^{\text{expt}}(a_i)$, the error $\delta_4 F^{\text{theor}} = 3\,\Delta a/a$ should be taken into account (it should be, however, discarded when the theoretical forces are computed not at the experimental separations a_i but over the entire measurement range; see Section 18.3.3). The last theoretical

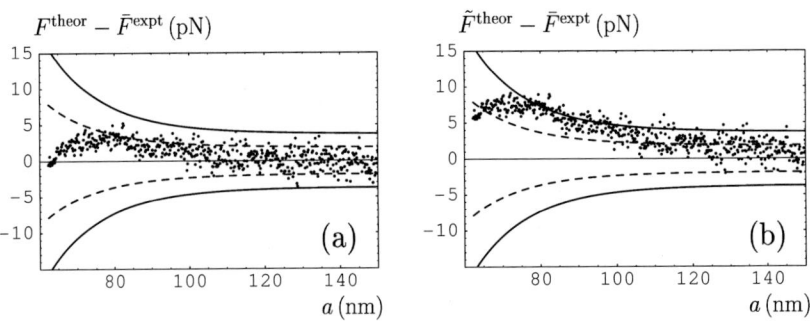

FIG. 20.3. Differences between the theoretical and mean experimental Casimir forces versus separation. The theoretical forces were computed for (a) the Si plate used in the experiment and (b) for dielectric Si. The solid and dashed lines indicate 95 and 70% confidence intervals, respectively (Chen et al. 2006b).

error is due to the uncertainties of the sphere radius $\delta_5 F^{\text{theor}} = \Delta R/R \approx 0.15\%$. When all theoretical errors are combined using the statistical rule (18.2) at a 95% confidence level, the total theoretical error obtained is that presented by the dashed line in Fig. 20.2(b). This error reaches its largest value of 3.8% at the shortest separation $a = 62.33\,\text{nm}$, decreases to 2% at $a = 119.96\,\text{nm}$, and decreases further to 1.2% at $a = 299.99\,\text{nm}$.

To compare the experimental data with the theoretical computations, the confidence intevals $[-\Xi_F^{0.95}(a), \Xi_F^{0.95}(a)]$ for the quantity $F^{\text{theor}}(a) - F^{\text{expt}}(a)$ were found at all experimental separations at a 95% confidence level using eqn (18.21). In the comparison of experiment with theory, the confidence intervals $[-\Xi_F^{0.7}, \Xi_F^{0.7}]$ obtained at a 70% confidence level were also needed. It is well known that for a normal distribution,

$$\frac{\Xi_F^{0.95}}{\Xi_F^{0.7}} = \frac{t_{0.975}(\infty)}{t_{0.85}(\infty)} = 2. \tag{20.3}$$

The distribution law of the quantity $F^{\text{theor}}(a_i) - F^{\text{expt}}(a_i)$ was investigated (Chen et al. 2006b) by using a method for testing hypotheses about the form of the distribution function of a random quantity (Rabinovich 2000). As a result, it was found that the hypothesis of a normal distribution was confirmed at all separations with probabilities larger than 70%. Note that for distributions, different from the normal distribution, $\Xi_F^{0.95}/\Xi_F^{0.7} > 2$ holds. Thus, putting $\Xi_F^{0.7}(a_i) = \Xi_F^{0.95}(a_i)/2$ in the error analysis is in fact conservative, as the confidence interval is wider than required.

Now the experimental data can be compared with theory. In Fig. 20.3(a), the differences $F^{\text{theor}}(a_i) - \bar{F}^{\text{expt}}(a_i)$ are plotted for all experimental points over the separation range from 62.33 to 150 nm, where the total experimental error is less than 10%. The theoretical forces in this figure were computed as described above for a Si sample of the conductivity ρ used in the experiment. The solid

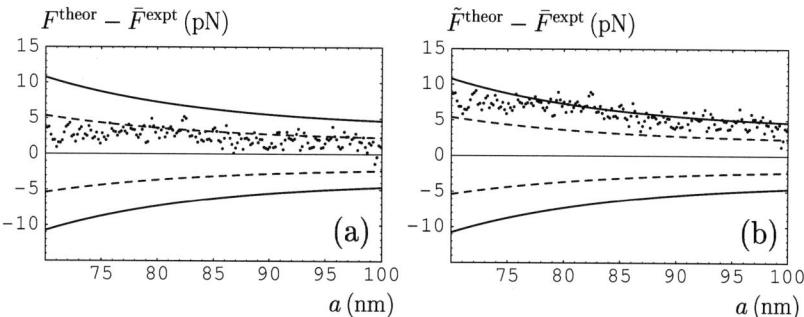

FIG. 20.4. Differences between the theoretical and mean experimental Casimir forces versus separation plotted on an enlarged scale. The theoretical forces were computed for (a) the Si plate used in the experiment and (b) for dielectric Si. The solid and dashed lines indicate 95 and 70% confidence intervals, respectively (Chen et al. 2006b).

lines indicate the confidence intervals $[-\Xi_F^{0.95}(a_i), \Xi_F^{0.95}(a_i)]$. The dashed lines show the confidence intervals $[-\Xi_F^{0.7}(a_i), \Xi_F^{0.7}(a_i)]$. As is seen from Fig. 20.3(a), experiment and theory are consistent, being inside the 95% confidence interval. In fact, not only 95% of individual points but all of them belong to the 95% confidence interval. What is more, not 30% (as is required at a 70% confidence level) but only 10% of all individual points are outside the 70% confidence interval. This is a clear manifestation of the fact that the theory is in excellent agreement with experiment and that the error analysis used is very conservative, overestimating the above-discussed errors and uncertainties. The main reason for the overestimation is that the exact magnitudes of the theoretical errors (such as those due to sample-to-sample variation of the optical data for Au and dielectric Si, the use of the proximity force theorem, and uncertainties in the surface separation) are not known, and they therefore were replaced with their upper limits.

In Fig. 20.3(b), the same information as in Fig. 20.3(a) is presented but the differences $\tilde{F}^{\text{theor}}(a_i) - \bar{F}^{\text{expt}}(a_i)$ were computed with the theoretical forces for dielectric Si. As is seen from Fig. 20.3(b), many points at all separations are outside the 70% confidence interval, and practically all of them are outside at $a < 100$ nm. In Fig. 20.4(a,b), the differences $F^{\text{theor}} - \bar{F}^{\text{expt}}$ and $\tilde{F}^{\text{theor}} - \bar{F}^{\text{expt}}$, respectively, are presented on an enlarged scale within the separation range from 60 to 100 nm. From Fig. 20.4(a) it can be seen that the theory for a sample of the conductivity ρ used in the experiment is consistent with the experimental data. There are no points outside the 95% confidence interval, and fewer than 3% of all points are outside the 70% confidence interval (once again, this is an indication that the errors have been overestimated). A completely different situation is observed in Fig. 20.4(b). Here almost all points representing the differences $\tilde{F}^{\text{theor}}(a_i) - \bar{F}^{\text{expt}}(a_i)$ computed for dielectric Si (all except two) are outside the 70% confidence interval. What this means is that the theory for

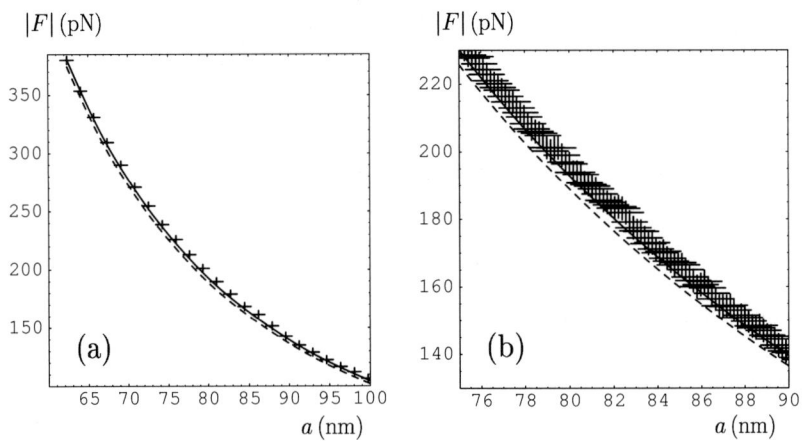

FIG. 20.5. Magnitudes of the experimental Casimir force with their error bars versus separation for (a) the points $1, 11, 21, \ldots$ and (b) all points on an enlarged scale. The solid lines show the theoretical dependence for the sample used in the experiment, and the dashed lines show the dependence for dielectric Si (Chen et al. 2006b).

dielectric Si is rejected by the experiment at 70% confidence within the separation range from 60 to 100 nm. The consistency of the experimental data with the theoretical forces $F^{\text{theor}}(a_i)$ and the rejection of the theory for dielectric Si at a 70% confidence level demonstrates how the density of charge carriers in a semiconductor influences the Casimir force between a metal and a semiconductor.

This conclusion was obtained using one of the two local methods for the comparison of experiment with theory presented in Section 18.3.3. The other method considered in Section 18.3.3 leads to similar results. In Fig. 20.5(a,b), the experimental points are plotted with their error bars ($\pm \Delta a, \pm \Delta^{\text{tot}} F^{\text{expt}}$) determined at a 95% confidence level, and the theoretical dependences for the conductive Si used and for dielectric Si are shown by the solid and dashed lines, respectively. It is not possible to plot all of the experimental points with error bars over the wide separation range from 60 to 100 nm. Because of this, in Fig. 20.5(a) the points with numbers $1, 11, 21, \ldots$ have been plotted. As can be seen from Fig. 20.5(a), the solid line is in very good agreement with experiment, whereas the dashed line deviates significantly from the experimental data. For a narrower range of separations from 75 to 90 nm, all experimental points have been plotted in Fig. 20.5(b) with their error bars, and, also, theoretical lines for conductive Si (solid line) and dielectric Si (dashed line). Once again, the solid line is consistent with experiment, whereas the dashed line is inconsistent (Chen et al. 2006b). Thus, as demonstrated in this first experiment on the Casimir effect using semiconductor surfaces, the measured force magnitude is consistent with the theory for a charge carrier density of about $3 \times 10^{19}\,\text{cm}^{-3}$ (as in the Si plate used) and inconsistent with the theory for a dielectric Si characterized by a zero density of charge carriers. This conclusion was directly confirmed in the second experiment, where the Casimir force between a gold-coated sphere and two Si samples with radically different densities of charge carriers was successfully measured.

20.2 Experiment on the difference Casimir force for samples with different charge carrier densities

The experiment by Chen et al. (2006a) pioneered the demonstration of the difference Casimir force between a gold-coated sphere and two Si samples which possess radically different charge carrier densities. In this experiment, a high-vacuum-based (2×10^{-7} Torr) atomic force microscope was used to measure the Casimir force between a gold-coated polystyrene sphere with a diameter $2R = 201.8 \pm 0.6\,\mu$m and two $4 \times 7\,\text{mm}^2$ size Si plates placed next to each other. The thickness of the gold coating on the sphere was measured to be 96 ± 2 nm. The details of the setup were as described in the previous section devoted to the experiment with one Si plate. For this second experiment two identically polished, single-crystal, $\langle 100 \rangle$ orientation Si samples were chosen, 500 μm thick and with a resisitivity 0.1–1 Ω cm. They were n-type and doped with P. The resistivity of the plates was measured using the four-probe technique to be $\rho_a \approx 0.43\,\Omega$ cm, leading to a concentration of charge carriers $n_a \approx 1.2 \times 10^{16}\,\text{cm}^{-3}$. One of these samples was used as the first Si plate in the experiment. The other one was subjected to thermal-diffusion doping to prepare the second, lower-resistivity, plate. A phosphorus-based spin-on-dopant (SOD) solution (P450, commercially available from Filmtronics Co.) was used. The wafer was spin-coated at a speed of $5000 \times 2\pi$ rad/min for 0.25 min, followed by prebaking at 200 °C for 15 min on a hotplate. The sample was then placed in a diffusion furnace. The diffusion was carried out at 1000–1050 °C for 100 hours in an $N_2(75\%)+O_2(25\%)$ atmosphere. A 49% HF solution was used to etch off the residual dopant after the diffusion process. The effectiveness of the above procedure was determined using both a four-probe resistivity measurement and a Hall measurement of a similarly doped 0.3 μm thick single-crystal $\langle 100 \rangle$ Si sample grown epitaxially on a sapphire wafer [in the original publication (Chen et al. 2006a), the substrate material was mistakenly indicated as Si]. This thin equivalent sample was homogeneously doped under the above conditions (Teh and Chuah 1989) and allowed a measurement of the carrier density. The resistivity and the carrier density were measured to be $\rho_b \approx 6.7 \times 10^{-4}\,\Omega$ cm and $n_b \approx 3.2 \times 10^{20}\,\text{cm}^{-3}$. Both plates, of the higher and the lower resistivity, were subjected to a special passivation procedure to prepare their surfaces for the force measurements. For this purpose, Nanostrip was used to clean the surface and a 49% HF solution to etch SiO_2 and to hydrogen-terminate the surface in the same way as described in the previous section. Finally, both plates were mounted adjacent to each other in the AFM.

The calibration of the spring constant k and the measurements of the residual electrostatic potential V_0, deflection coefficient m, and separation on contact a_0 were done using a significantly improved experimental technique as compared with previous experiments. All calibration and other measurements were done in the same high-vacuum apparatus as the Casimir force measurements. The actual separation distance a between the bottom of the gold sphere and the Si plates was given by eqn (19.4), where a_{piezo} is the distance moved by the piezoelectric actuator and S_{def} is the cantilever deflection signal from the photodiodes. First, the value of m was found for the higher-resistivity plate following the improved procedure. For this purpose, the sphere was grounded and 28 different voltages between -0.712 and -0.008 V were applied to the plate through a thick gold pad attached to the plate bottom. The change in the contact position between the sphere and the plate was used to find $m = 47.8 \pm 0.2$ nm per unit deflection signal.

Then the values of V_0, km (this product is needed for force measurements), and a_0 were found for the higher-resistivity plate by fitting the deflection signal S_{def} to the theoretical expression. From the definition of the deflection coefficient $a_d = mS_{\text{def}}$, it follows that

$$S_{\text{def}} = S_0 + \frac{F_{\text{el}}}{km}, \qquad (20.4)$$

where the electric force between the sphere and the plate, F_{el}, is given by eqn (19.1) and S_0 is the voltage-independent deflection caused by the contribution of the Casimir force to the signal. The value of S_0 was not used in the determination of V_0 and km. Because of this, it was not necessary to subtract the contribution of the Casimir force from the total deflection signal (uncertainties in the contact position due to drift of the piezoelectric actuator were found to lead to an error of 0.4% and corrected).

The parabolic dependence of the signal on the applied voltage, in accordance with eqns (19.1) and (20.4), was used to obtain the residual potential difference V_0 between the grounded sphere and the plate. For this purpose, at each separation distance a the deflection signal S_{def} was plotted as a function of the applied voltage V. In Fig. 20.6, two typical dependences of S_{def} on V are shown, one for the Si plate of higher resistivity ($\rho_a = 0.43\,\Omega\text{cm}$) and one for the plate of lower resistivity ($\rho_b = 6.7 \times 10^{-4}\,\Omega\text{cm}$) (Chiu et al. 2008). The applied voltage ranged between -0.712 and -0.008 V in Fig. 20.6(a) and between -0.611 and -0.008 V in Fig. 20.6(b). The least-squares method was then used to obtain the voltage value at the maximum of the parabola, which is V_0, and the value of $X(\alpha)$ in eqn (19.1). The values obtained for the particular sets of data shown in Fig. 20.6(a,b) are also shown in the figure. The plotting of the parabola was repeated for every a, and V_0 was measured as a function of a. The mean value of V_0 so determined was the residual potential difference. Note that at this point of the measurement, the exact value of a was uncertain as the mean separation between the zero levels of the roughness on contact, a_0, had not yet been determined. In Fig. 20.7(a,b), the residual potential differences V_0 obtained are plotted as a function of separation a for semiconductor plates of high and low resistivity, respectively (Chiu et al. 2008). As can be seen in the figure, the mean value of V_0 is relatively constant as a function of separation. The larger random error

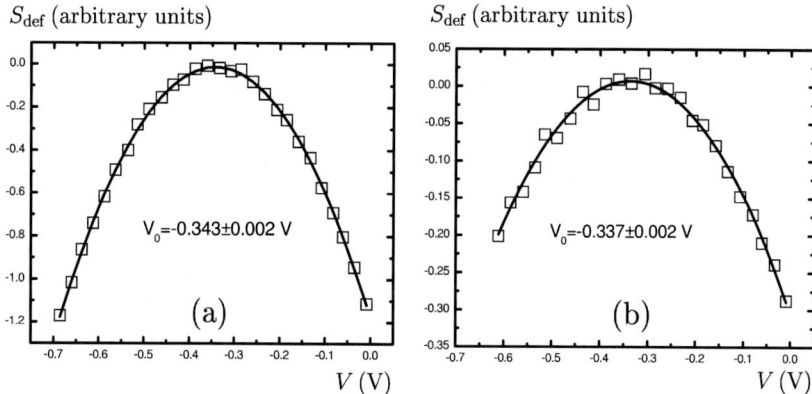

FIG. 20.6. Deflection signal plotted as a function of the voltage applied to the plate at a fixed separation distance for the silicon plate with (a) high and (b) low resistivity (Chiu et al. 2008).

with increasing separation is due to the decrease in the signal-to-noise ratio. For the sphere and the higher-resistivity plate, the mean residual potential difference was determined to be $V_0 = -0.341 \pm 0.002$ V.

The value of $X(\alpha)$ (we recall that α is a known function of a) obtained from fitting the parabolic curves in Fig. 20.6 was used to determine both a_0 and the cantilever spring constant multiplied by the calibration coefficient, km. This was done with the help of an iterative fitting procedure, requiring the output of the value of only one unknown parameter [details are presented by Chiu et al. (2008)]. For the higher-resistivity plate, the start point of the fit was from 300 to 400 nm, with the end point separation equal to 2.5 μm. The fit was repeated with different start points and the mean values of km and a_0 were determined. Typical values of a_0 obtained as a function of the start point of the fit are shown in Fig. 20.8(a,b) for the plates of higher and lower resistivity, respectively. The resulting mean values for the high-resistivity plate were $a_0 = 32.4 \pm 1.0$ nm

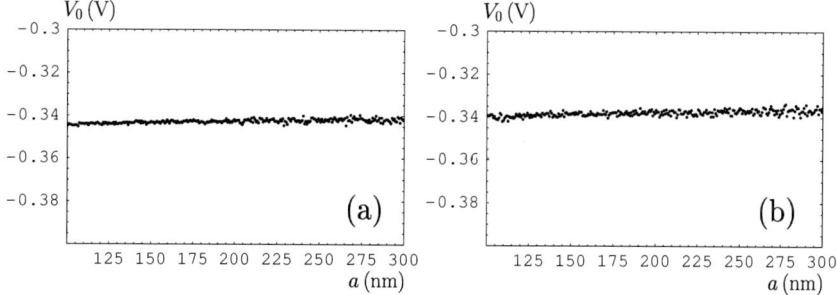

FIG. 20.7. Residual sphere–plate potential difference as a function of the separation where it was determined, for the silicon plate with (a) high and (b) low resistivity.

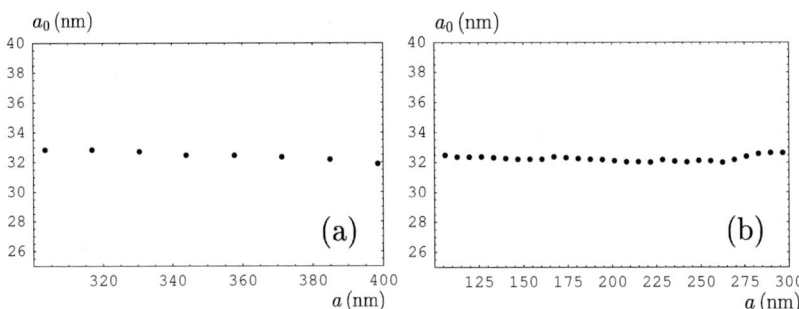

FIG. 20.8. Mean separation between the zero levels of the roughness on sphere–plate contact a_0, obtained from fitting of the electrostatic force, as a function of the start point of the fit for the silicon plate with (a) high and (b) low resistivity.

and $km = 1.646 \pm 0.004$ nN per unit deflection signal. The above measurements include corrections for thermal noise of the cantilever and mechanical drift of the plate and the piezoelectric actuator described in detail by Chiu et al. (2008).

An important point to note is that the expression for the electric force F_{el} used does not take into account the possible influence of the space-charge layer at the surface of high-resistivity Si. According to Bingqian et al. (1999), for n-type Si with a concentration of charge carriers of order 10^{16} cm^{-3}, the impact of this layer on the electrostatic force is negligible at separations larger than 300–400 nm. The use of the expression for the electric force between metallic surfaces may lead to nothing more than an increased error in the determination of a_0. The above fit was performed just within this range of separations, with the start point ranging within 300–400 nm to 2.5μm. However, as is seen in Fig. 20.8(a), the value of a_0 obtained does not depend practically on the start point of the fit. This demonstrates that eqn (19.1) is applicable. An independent experimental confirmation of the applicability of eqn (19.1) in the case of the higher-resistivity plate is given by Fig. 20.7(a). This demonstrates that the residual potential difference determined from eqn (19.1) at different separations is relatively constant at $a \geq 100$ nm. This, however, could not be the case if there were noticeable deviations from eqn (19.1) at the separations considered. In Figs. 20.7(b) and 20.8(b), it is clearly seen that for the lower-resistivity plate any possible deviations of the electric force from that given by eqn (19.1) are even more negligible. We emphasize that if V_0 were found to be separation-dependent, it would indicate the presence of electrostatic surface impurities, space charge effects such as screening, and/or electrostatic inhomogeneities (patch effects) on the sphere or plate surface.

After the calibration and related measurements for the higher-resistivity sample were done, the Casimir force between this sample and the sphere was measured from contact as a function of distance. Finally, the results at $a \geq 61.19$ nm were reported to avoid the influence of nonlinearities associated with the *jump*

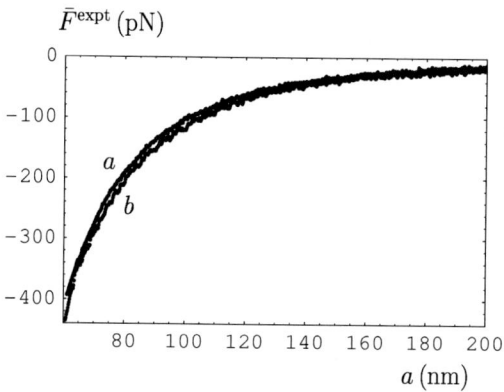

FIG. 20.9. The mean measured Casimir forces as a function of separation between the gold-coated sphere and the two Si plates of higher and lower resistivities are shown by the dots labeled a and b, respectively (Chen et al. 2006a).

to contact at shorter distances. For this purpose, the sphere was kept grounded while an appropriate compensating voltage was applied to the plate to cancel the residual electrostatic force. The distance between the sphere and the plate was varied continuously from large to small separations by applying triangular voltages at 0.02 Hz to the piezoelectric actuator. The force data $F_a^{\text{expt}}(a_i)$ were collected at equal time intervals corresponding to equidistant points separated by 0.17 nm. This measurement was repeated 40 times, and the forces obtained were averaged to reduce the influence of various random factors, including thermal noise. The mean values $\bar{F}_a^{\text{expt}}$ of the experimental Casimir force data as a function of separation are represented by the dots labeled a in Fig. 20.9. Note that the voltage-independent deflection S_0 in eqn (20.4) can be used for an independent determination of the Casimir force. The respective results were shown to be in agreement with those obtained from applying compensation voltages.

Next, all the above calibrations and measurements were repeated for the second, lower-resistivity, Si plate. In this case 25 different dc voltages were applied to the plate [see Fig. 20.6(b)]. The deflection coefficient m was found to be 47.9 ± 0.2 nm per unit deflection signal. Here again, the drift of the piezoelectric actuator was measured and corrected. After the same fitting procedure of the measured deflection signal, the following values of all related parameters were obtained: $V_0 = -0.337 \pm 0.002$ V, $km = 1.700 \pm 0.004$ nN per unit deflection signal, and $a_0 = 32.3 \pm 0.8$ nm. The fit was performed within the range of separations from 100–300 nm to 2.5 μm. Note that closer separations can be used here, as the effect of the space charge layer is negligible for the lower-resistivity sample. The values of km were slightly different in the two cases owing to changes in the level of the cantilever arm arising from minor deviations from the horizontal position in the mounting of the samples. Next, the Casimir force acting between the lower-resistivity sample and the sphere was measured from contact after application of the appropriate voltage to cancel the residual electrostatic force. The results in the linear regime at $a \geq 60.51$ nm were reported. This measurement was repeated 39 times. The resulting mean values $\bar{F}_b^{\rm expt}$ of the Casimir force data as a function of a are represented in Fig. 20.9 by the dots labeled b. As is seen from the figure, the dots labeled a and b are distinct from each other, demonstrating the effect of the different charge carrier densities in the two Si plates used in the experiment. The Casimir force measured by applying a compensating voltage was checked to be consistent with that obtained as the excess force leading to the deflection S_0 when different electric voltages were applied [see eqn (20.4)].

For the quantitative characterization of the deviation between the two measurements, the random errors were calculated using the procedure outlined in Section 18.3.1 based on Student's t-distribution. For the sample of higher resistivity (measurement a), the random error at a 95% confidence level is equal to 8 pN at $a = 61.19$ nm, decreases to 6 pN at $a = 70$ nm, and becomes equal to 4 pN at $a \geq 80$ nm. Measurement b, for the sample of lower resistivity, is slightly more noisy. Here the random error at a 95% confidence level changes from 11 pN at $a = 60.51$ nm and 7 pN at $a = 70$ nm to 5 pN at $a \geq 80$ nm. The systematic error determined at a 95% confidence level is equal to only 1.2 pN for both measurements (see Section 20.1.2 for details). Using the statistical criterion in Section 18.3.1, it was concluded that the total experimental errors $\Delta^{\rm tot} F_{a,b}^{\rm expt}$ determined at a 95% confidence level were equal to the random errors in each measurement. From Fig. 20.9, it is seen that the deviation between the two sets of data is several times larger than the total experimental error in the range of separations from 61.19 to 120 nm.

Then the force–distance relations measured for the two Si samples were compared with the theory. At separations below 150 nm, where the differences between the two measurements are most pronounced, the magnitudes of the predicted thermal corrections are negligible. At larger a, the relative contribution from thermal corrections is much less than the relative error of the force measurements. Because of this, the force between a smooth sphere and each of the plates was computed using the Lifshitz formula at zero temperature (12.44) and the proximity force approximation (6.71) (where T was also set equal to zero). The dielectric permittivity of gold $\varepsilon^{(1)}(i\xi)$ and that of dielectric Si $\varepsilon^{(2)}(i\xi)$ have already been considered in Section 20.1.3. For the higher-resistivity plate, the concentration of charge carriers n_a is below the critical value (see Section 12.5.2). Because of this, one should put $\varepsilon_a^{(2)}(i\xi) = \varepsilon^{(2)}(i\xi)$. For the plate with lower resistivity, its dielectric permittivity $\varepsilon_b^{(2)}(i\xi) = \tilde{\varepsilon}^{(2)}(i\xi)$ was obtained from eqn (20.2) with a plasma frequency $\omega_{p,b}^{(2)} \approx 2.0 \times 10^{15}$ rad/s and a respective relaxation parameter $\gamma_b^{(2)} \approx 2.4 \times 10^{14}$ rad/s [these values were calculated by use of eqns (13.2) and (20.1), with the effective mass of the electron $m^* = m_e^* = 0.26 m_e$]. Then the Casimir forces F_a and F_b were calculated over the experimental separation range.

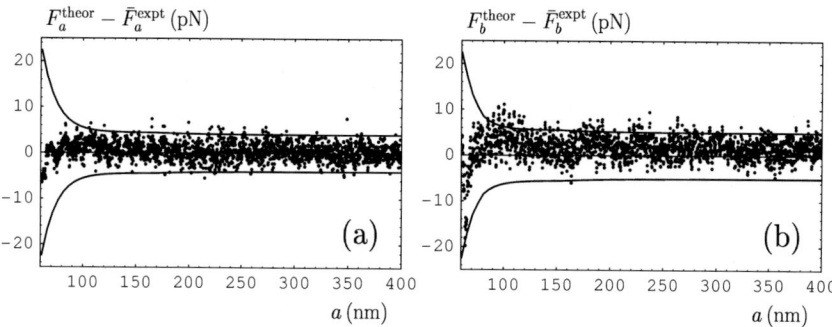

FIG. 20.10. Differences between the theoretical and mean experimental Casimir forces versus separation. The forces were computed and measured for (a) higher- and (b) lower-resistivity Si. The solid lines indicate 95% confidence intervals (Chen et al. 2006a).

The results obtained were corrected for the presence of surface roughness. To do this, the topographies of the sphere and of both Si samples were investigated with an AFM. Then the roughness data were used to compute the Casimir forces F_a^{theor} and F_b^{theor} starting from F_a and F_b, in accordance with the method of geometrical averaging as described in Section 20.1. The contributions from the roughness to the Casimir force for the two plates were equal. These contributions changed from 3.6% of the total force at $a \approx 60$ nm to 2.7%, 1.4%, and 0.65% at separations of 70, 100, and 150 nm, respectively. The surface distortion of the single-crystal Si was very low and did not contribute practically to the roughness correction, which was determined primarily by the roughness of the sphere.

The errors in the computation of the Casimir force between the gold-coated sphere and the Si plate were analyzed in Section 20.1.3. At the shortest a, they are mostly determined by the error $\Delta a = 1.0$ nm (for plate a) and $\Delta a = 0.8$ nm (for plate b) in the measurement of the separations a_i at which the theoretical values of the Casimir force were calculated for comparison with experiment. A 0.5% error due to the variation of the optical parameters was also included. At $a = 60$ nm, the total theoretical error determined at a 95% confidence level is equal to 19.6 pN (4.9% of the force) for plate a and 17.2 pN (4.0% of the force) for plate b. It decreases to 11 pN (4.2% of the force) for plate a and 9.6 pN (3.4% of the force) for plate b at $a = 70$ nm. The total theoretical error becomes less than the total experimental error at $a > 90$ and 85 nm for plates a and b, respectively.

The total theoretical error was combined with the total experimental error at a 95% confidence level using the statistical rule (18.21) to find the error $\Xi_F(a)$ in the difference between the theoretical and experimental forces. The confidence interval $[-\Xi_F(a), \Xi_F(a)]$ obtained, as a function of separation, is shown in Fig. 20.10 as solid lines. The differences $F_a^{\text{theor}} - \bar{F}_a^{\text{expt}}$ versus separation for the experiment with the higher-resistivity Si are plotted in Fig. 20.10(a) as dots. Similarly, the differences $F_b^{\text{theor}} - \bar{F}_b^{\text{expt}}$ for the lower-resistivity Si are shown as

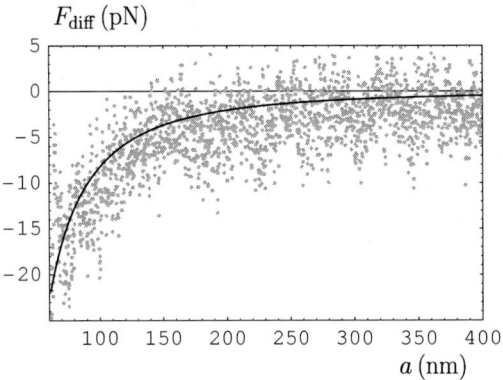

FIG. 20.11. The differences between the mean measured Casimir forces for the lower- and higher-resistivity Si samples versus separation are shown by dots. The corresponding theoretically calculated differences are shown by the solid line (Chen et al. 2006a).

dots in Fig. 20.10(b). As can be seen from Fig. 20.10, the two measurements are consistent with theories using the dielectric permittivity $\varepsilon_a^{(2)}(i\xi)$ [Fig. 20.10(a)] and $\varepsilon_b^{(2)}(i\xi)$ [Fig. 20.10(b)].

To illustrate the effect of the modification of the Casimir force by a change of carrier density, the differences between the measured mean Casimir forces for the plates of lower and higher resistivities, $F_{\text{diff}}^{\text{expt}} = \bar{F}_b^{\text{expt}} - \bar{F}_a^{\text{expt}}$, are plotted in Fig. 20.11 as dots, versus separation. In the same figure, the difference between the respective theoretically computed Casimir forces, $F_{\text{diff}}^{\text{theor}} = F_b^{\text{theor}} - F_a^{\text{theor}}$, versus separation is shown as a solid line. As can be seen from Fig. 20.11, the experimental and theoretical difference Casimir forces as a function of separation are in good agreement. It can be easily shown that the magnitude of the mean difference of the measured Casimir forces exceeds the experimental error in the force difference in the range of separations from 70 to 100 nm.

Thus, in the second experiment with semiconductors, the Casimir force between a gold-coated sphere and two Si plates with resistivities differing by several orders of magnitude has been measured. Each measurement was compared with theoretical results obtained using the Lifshitz theory with different dielectric permittivities and found to be consistent with it. The difference between the measured forces for the two resistivities is in good agreement with the corresponding difference between the theoretical results. The results of this experiment may find applications in the design, fabrication, and function of microelectromechanical and nanoelectromechanical devices (see Chapter 23).

20.3 Experiment on optically modulated Casimir forces

The most suitable method of changing the charge carrier density in a semiconductor is through the illumination of its surface with light (Opsal et al. 1987,

Vogel et al. 1992). The present section contains a description of an experiment on the modulation of the Casimir force by the irradiation of a Si membrane with laser pulses (Chen et al. 2007a, 2007b). In this experiment, the charge carrier density in the Si membrane was changed by the incident light, and the difference in the Casimir force acting between that membrane and a gold-coated sphere in the presence and in the absence of light was measured. An important feature of this experiment, in contrast with the experiment described in the previous section, is that the individual Casimir forces between a sphere and two plates with different charge carrier densities were not measured in sequence, but only their difference. This permits one to achieve much higher precision and to use the results obtained as a test of the various theoretical approaches to the thermal Casimir force between metallic and dielectric (semiconductor) surfaces.

20.3.1 *Experimental setup and sample preparation*

Here, the experimental setup used to demonstrate the modification of the Casimir force through a radiation-induced change in the carrier density is discussed. The general scheme of the setup is shown in Fig. 20.12. A high-vacuum-based AFM was employed to measure the change in the Casimir force between a gold-coated sphere of diameter $2R = 197.8 \pm 0.3\,\mu$m and a Si membrane (colored black) in the presence and in the absence of incident light. An oil-free vacuum chamber with a pressure of around 2×10^{-7} Torr was used. A polystyrene sphere coated with a gold layer of 82 ± 2 nm thickness was mounted at the tip of a $320\,\mu$m conductive cantilever. The Si membrane (see below for the process of preparation) was mounted on top of a piezoelectric actuator, which was used to change the separation distance a between the sphere and the membrane from contact to $6\,\mu$m. The excitation of the carriers in the Si membrane was done with 5 ms wide light pulses (50% duty cycle). These pulses were obtained from a cw Ar ion laser at 514 nm wavelength, modulated at a frequency of 100 Hz using an acousto-optic modulator (AOM). The AOM was triggered with a function generator. The laser pulses were focused on the bottom surface of the Si membrane. The Gaussian width of the focused beam on the membrane was measured to be 0.23 ± 0.01 mm.

The cantilever of the AFM flexed when the Casimir force between the sphere and the membrane changed depending on the presence or absence of incident light on the membrane. This cantilever deflection was monitored with a 640 nm beam from a second laser (see Fig. 20.12), reflected off the top of the cantilever tip. An optical filter was used to prevent interference of the 514 nm excitation light with the cantilever deflection signal. The transmission of this filter at 514 nm was 0.001%. Including the transmission of less than 1% through the Si membrane and the diode solid angle of 10^{-4}, the impact of leakage of the 514 nm light leads to a change in the force difference of less than 10^{-6} pN. These changes are negligibly small as compared with the measured cantilever deflection signal. The change in the Casimir force due to the incident light led to a difference signal between the two photodiodes. The resulting response to the carrier excitation was measured

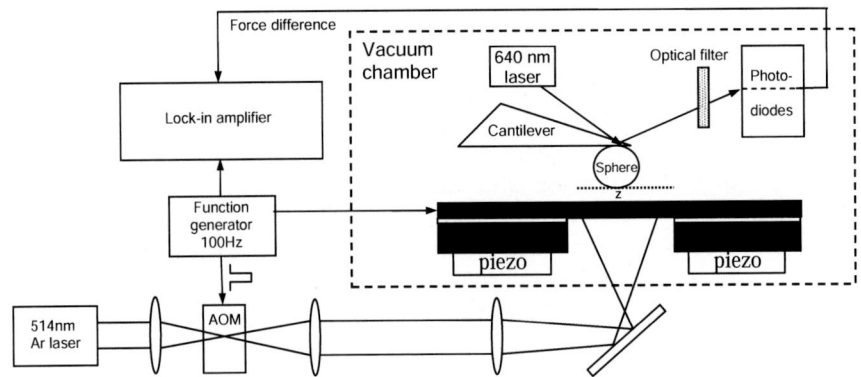

FIG. 20.12. Schematic of the experimental setup, showing its main components (Chen et al. 2007b).

with a lock-in amplifier. The same function generator signal used to generate the Ar laser pulses was also used as a reference for the lock-in amplifier.

The most important part of such a setup is the Si membrane. It should be sufficiently thin and of appropriate resistivity to ensure that the density of charge carriers increases by several orders of magnitude under the influence of the laser pulses. The Si membrane should be thick enough to make negligible the photon pressure of the transmitted light, as the illumination is incident on the bottom surface of the membrane. Therefore the thickness of the Si membrane has to be greater than $1\,\mu m$, i.e. greater than the optical absorption depth of Si at the wavelength of the laser pulses. The fabrication of an Si membrane a few micrometers thick with the necessary properties is described below.

A commercial wafer of Si grown on an insulator was used as the initial substrate. The insulator in this case was SiO_2, which is the native oxide of Si and thus leads to only small reductions of the excited-carrier lifetime in Si. The layout of the wafer is shown in Fig. 20.13. The wafer consisted of an Si substrate of thickness $600\,\mu m$ and a Si top layer of thickness $5\,\mu m$ (both were single crystals and had a $\langle 100 \rangle$ crystal orientation), with a buried intermediate SiO_2 layer of thickness $400\,nm$ [see Fig. 20.13(a)]. The Si was p-type doped with a relatively high nominal resistivity of about $10\,\Omega\,cm$. The corresponding carrier density was equal to $n \approx 5 \times 10^{14}\,cm^{-3}$ (Palik 1985).

The thickness of the Si substrate was reduced to about $200\,\mu m$ through mechanical polishing. Then, after RCA cleaning of the surface, the wafer was oxidized at $T = 1373\,K$ in a dry O_2 atmosphere for a duration of 72 hours. As a result, in addition to the buried SiO_2 layer, a thermal oxide layer with a thickness of about $1\,\mu m$ was formed on both (bottom and top) sides of the wafer [Fig. 20.13(b)]. This oxide layer served as a mask for subsequent tetramethylammonium hydroxide (TMAH) etching of the Si. First, a hole with a diameter

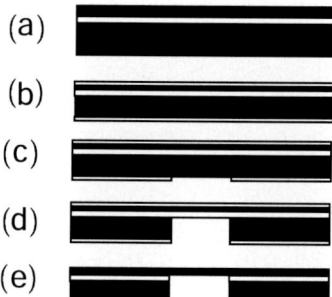

FIG. 20.13. Fabrication process of Si membrane. (a) The Si substrate (colored black) with a buried SiO$_2$ layer (white). (b) The substrate was mechanically polished and oxidized, and (c) a window in the bottom SiO$_2$ layer was etched with HF. (d) Next, TMAH was used to etch the Si. (e) Finally, the SiO$_2$ layer was etched away in HF solution to form a clean Si surface (Chen *et al.* 2007b).

of 0.85 mm was etched with HF in the center of the bottom oxide layer [Fig. 20.13(c)]. This exposed the Si substrate. Next, TMAH was used at 363 K to etch the Si substrate through the hole formed in the oxide mask [Fig. 20.13(d)]. Note that TMAH selectively etches Si, as its etching rate for Si is 1000 times greater than for SiO$_2$. TMAH etching led to the formation of a hole through the Si substrate. Given the selectivity of the etching, the buried 400 nm oxide acted as an etch stop layer. Finally, all of the thermal oxidation layers and the buried oxidation layer in the hole were etched away in HF solution to form a clean Si membrane over the hole as shown in Fig. 20.13(e). The thickness of this membrane was measured to be $4.0 \pm 0.3\,\mu$m using an optical microscope. In order for voltages to be applied to the Si membrane, an ohmic contact was formed by a thin film of Au deposited on the edge followed by annealing at 673 K for 10 min. The Si membrane was cleaned with Nanostrip and then passivated by dipping in 49% HF for 10 s. The passivated Si membrane was then mounted on top of the piezoelectric actuator as described above.

20.3.2 Calibration and excited-carrier lifetime measurement

All calibrations and other measurements were done during the same period of time as the measurement of the difference of the Casimir forces and in the same high-vacuum apparatus. The calibration of the deflection signal of the cantilever obtained from the photodiodes, $S_{\rm def}$, and the determination of the average separation on contact and residual potential difference between the gold-coated sphere and the Si membrane were done by measuring the distance dependence of an applied electrostatic force. For this purpose, the same function generator (see Fig. 20.12) was used for applying voltages to the membrane. For an attractive force, $S_{\rm def} < 0$ and this can be measured either as a current or as a voltage. In addition, a small correction had to be applied to the separation distance between the gold sphere and the Si membrane owing to the movement of the cantilever. The actual separation distance a between the bottom of the sphere and the membrane was given by eqn (19.4). It should be noted that the quantity a_0 in this equation is the absolute separation between the zero levels of the roughness at the position where physical contact between the two surfaces is achieved. The complete movement of the piezoelectric actuator was calibrated using a fiber optic interferometer. To extend and contract the actuator, continuous triangular voltages between 0.01 and 0.02 Hz were applied to it. Given that the experiment was done at room temperature, the application of static voltages would lead to creep of the actuator and loss of position sensitivity.

The gold sphere was kept grounded. Electrical contact to the sphere was accomplished by applying a very thin gold coating to the cantilever. The electrostatic force between the sphere and the membrane is given by eqn (19.1). First, 30 different dc voltages between 0.65 and −0.91 V were applied to the Si membrane. The cantilever deflection signal was measured as a function of the distance. A 0.02 Hz triangular wave was applied to the piezoelectric actuator to change the distance between the sphere and the membrane over a range of 6 μm. Larger applied voltages led to more cantilever deflection and, according to eqn (19.4), to contact of the two surfaces at a larger a_{piezo}. The dependence of a_{piezo} at contact of the sphere and the membrane on the applied voltage was then used to measure the deflection coefficient m. In order to determine the contact of the two surfaces precisely, 32 768 data points at equal time intervals were acquired for each force measurement (i.e. the interval between two points was about 0.18 nm). In cases where the contact point was between two neighboring data points, a linear interpolation was used to identify the exact value. The deflection coefficient m was found to be 137.2 ± 0.6 nm per unit deflection signal. The difference in this value of m from the measurements described in the previous sections is due to the use of the 514 nm filter, which reduced the cantilever deflection signal. The value of m obtained was used to correct the separation distance in all measurements as described in eqn (19.4). The electrostatic force resulting from the application of dc voltages was also used in the determination of the separation on contact of the two surfaces. The fit of the experimental force–distance relation to the theoretical eqn (19.1) was done as outlined in the previous sections. The separation distance on contact was determined to be $a_0 = 97$ nm. The uncertainty in the quantity $a_0 + mS_{\text{def}}$ obtained from eqn (19.4) was found to be 1 nm. This leads to the same error in the absolute separations $\Delta a = 1$ nm because the error in the calibration of the actuator was negligibly small.

For the calibration of the deflection signal and the determination of the residual potential difference between the two surfaces, an improved method, rather than simple application of dc voltages to the membrane, was used. This was done to avoid systematic errors due to scattered laser light. In addition to the application of a dc voltage to the membrane, described above, a square voltage pulse with an amplitude from 1.2 to −0.6 V and lasting for a time interval corre-

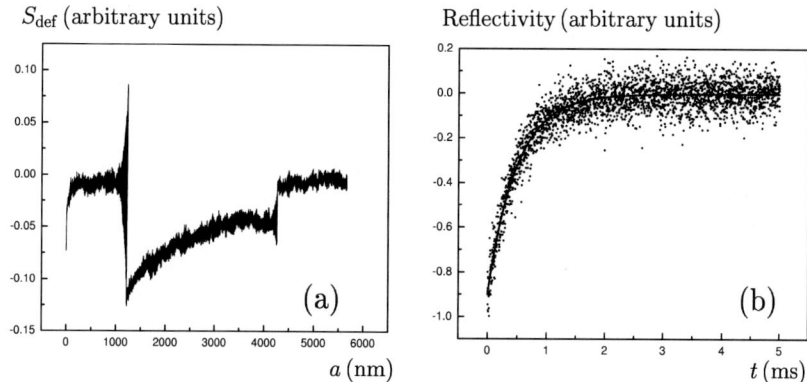

FIG. 20.14. (a) Deflection signal of the cantilever in response to a dc voltage and square voltage pulse applied to the Si membrane, as a function of separation. (b) Change in reflectivity after the termination of the laser pulse (Chen et al. 2007b).

sponding to a separation between 1 and 5 μm was also applied to the membrane. Figure 20.14(a) shows the deflection signal of the cantilever in response to both the applied dc voltage and the square pulse as a function of the separation between the gold sphere and the Si membrane. By measuring only the difference in the signal during the pulse, the investigators were able to avoid the need for a background subtraction. A fit of the difference signal to eqn (19.1) led to a value of the signal calibration constant $km = 6.16 \pm 0.04$ nN per unit deflection signal. The same fit was used to determine the residual potential difference between the sphere and the membrane, which was found to be $V_0 = -0.171 \pm 0.002$ V. The large width of the pulse applied in addition to the dc voltage allowed confirmation of the distance independence of the values of the calibration constant and the residual potential difference obtained.

An independent measurement of the lifetime of the carriers excited in the Si membrane by the pulses from the Ar laser was performed. For this purpose, a noninvasive optical pump–probe technique was used (Sabbah and Riffe 2002, Nagai and Kuwata-Gonokami 2002). The same Si membrane and Ar laser beam, modulated by the AOM at 100 Hz to produce 5 ms wide square light pulses, as used in the Casimir force measurement were employed as the sample and the pump, respectively. The diameter of the pump beam on the sample was measured to be 0.72 ± 0.02 mm. A cw beam with a power of 1 mW at a wavelength of 1300 nm was used as the probe. The probe beam photon energy was below the bandgap energy of Si and was thus not involved in carrier generation. This beam was focused to a Gaussian width $w_0 = 0.135 \pm 0.003$ mm. Thus the focal spot size of the probe beam was much smaller than the focal spot size of the pump light. As a result, it was possible to measure the lifetime in a homogeneous region of excited carriers. The change in the reflected intensity of the probe beam in the presence and in the absence of the Ar laser pulses was detected with an InGaAs photodiode. The change in the reflected power of the probe beam was monitored as a function of time and found to be consistent with the change of carrier density. Near-normal incidence for the pump and probe beams was used, with care taken to make sure that the InGaAs photodiode was isolated from the pump beam. The time decay of the reflected probe beam in response to the square Ar light pulses is shown in Fig. 20.14(b). The change in the reflectivity of the probe was fitted to an exponential of the form $-\exp(-t/\tau)$, where τ is the effective carrier lifetime. By fitting the whole 5 ms decay of the change in reflected power, the effective excited-carrier lifetime was measured to be $\tau = 0.47 \pm 0.01$ ms. Note that this time represents both surface and bulk recombination and is consistent with that expected for Si. Some dependence of the lifetime of the excited carriers on their concentration was observed. In the first 0.5 ms, while the charge carrier density was still high, the average value of the excited-carrier lifetime was measured to be $\tau = 0.38 \pm 0.03$ ms. The measured values of the carrier lifetime were used in theoretical computations of the change in Casimir force for many different incident laser powers.

20.3.3 *Experimental results and error analysis*

Here, the measured difference in the Casimir force resulting from the irradiation of the Si membrane with 514 nm laser pulses is presented. In fact, it was the difference in the total force (Casimir and electric) which was measured. As indicated above, even with no applied voltage there was some residual potential difference V_0 between the sphere and the membrane. A preliminary value of V_0 was determined during the calibration of the setup in the absence of laser pulses. In the presence of the pulses (even during the dark phase of the pulse train), the value of the residual potential difference could be different. These residual potential differences during the bright and dark phases of a laser pulse train (the latter was not exactly equal to the value determined during calibration) are denoted here by V_0^l and V_0, respectively. During the bright phases of the pulse train, a voltage V^l was applied to the Si membrane, and during the dark phases, a voltage V. Using eqn (19.1) for the electric force, we can represent the difference in the total force (electric and Casimir) for the states with and without carrier excitation in the following form:

$$F_{\text{diff}}^{\text{tot}}(a) = X(\alpha)\left[(V^l - V_0^l)^2 - (V - V_0)^2\right] + F_{\text{diff}}^{\text{expt}}(a). \qquad (20.5)$$

Here

$$F_{\text{diff}}^{\text{expt}}(a) = F^l(a) - F(a) \qquad (20.6)$$

is the difference between the Casimir forces F^l and F with and without light, respectively. The difference in the total force in eqn (20.5) was measured using a lock-in amplifier with an integration time constant of 100 ms, which corresponds to a bandwidth of 0.78 Hz. The measurement procedure is described below.

First, the voltage V was kept constant and V^l was changed. The parabolic dependence of $F_{\text{diff}}^{\text{tot}}$ on V^l in eqn (20.5) was measured at different separations

a. Care was taken to apply only small voltage amplitudes (up to a few tens of millivolts) so as to keep the space charge region negligible. At every measured separation distance, $F_{\text{diff}}^{\text{tot}}$ was plotted as a function of V^1. As can be seen from eqn (20.5), the value of V^1 where the parabola reaches a maximum is V_0^1 [recall that $X(\alpha) < 0$]. In this way, the value $V_0^1 = -0.303 \pm 0.002$ V was found and shown to be independent of the separation from 100 to 500 nm, where the difference in the Casimir force could be measured. This confirms that any possible deviations from eqn (19.1) due to screening effects were negligible and did not influence the results obtained. Next, V^1 was kept constant, whereas V was changed and the parabolic dependence of $F_{\text{diff}}^{\text{tot}}$ on V was measured at different separations. The value of V where the parabolas reached their minima was $V_0 = -0.225 \pm 0.002$ V. These values of the residual potential difference between the sphere and the membrane in the presence and in the absence of excitation light were substituted in eqn (20.5). The small change of around 78 mV in the residual potential difference between the sphere and the membrane in the presence and in the absence of excitation light is primarily due to the screening of surface states by a few of the optically excited electrons and holes. The above small value is equal to the change in band bending at the surface. It is consistent with the fact that almost flat bands are obtained at the surface with the surface passivation technique used here [see, e.g. Angermann (2002) and Kronik and Shapira (2001)].

Then, other voltages (V^1, V) were applied to the Si membrane and the difference in the total force $F_{\text{diff}}^{\text{tot}}$ was measured as a function of separation. Data were collected from contact, at equal time intervals corresponding to 3 points per 1 nm (i.e. at 1209 points within the separation interval from 100 to 500 nm). From these measurement results, the difference in the Casimir force $F_{\text{diff}}^{\text{expt}}(a)$ was determined from eqn (20.5). This procedure was repeated with some number J of pairs of different applied voltages (V^1, V), and at each separation the mean value $\bar{F}_{\text{diff}}^{\text{expt}}(a)$ was found. In Figs. 20.15(a), 20.16(a), and 20.17(a) the experimental data for $\bar{F}_{\text{diff}}^{\text{expt}}(a)$ as a function of separation are shown by dots for different absorbed laser powers: $P^{\text{eff}} = 9.3$ mW ($J = 31$), 8.5 mW ($J = 41$), and 4.7 mW ($J = 33$), respectively. The corresponding incident powers were 15.0, 13.7, and 7.6 mW, respectively. As expected, the magnitude of the Casimir force difference has the largest values at the shortest separations and decreases with increasing separation. It also decreases with decreasing absorbed laser power [the solid, short-dashed and long-dashed lines in Figs. 20.15(a), 20.16(a), and 20.17(a) are explained in Sections 20.3.4 and 20.3.5, devoted to the comparison with theory; Figs. 20.15(b), 20.16(b), and 20.17(b) are also discussed there]. We emphasize that although the difference Casimir force $\bar{F}_{\text{diff}}^{\text{expt}}$ obtained in this experiment is somewhat analogous to the $F_{\text{diff}}^{\text{expt}}$ considered in the experiment with two plates with different charge carrier densities (Section 20.2), the two quantities are in fact distinct. The point is that in the experiment on the optically modulated Casimir force, $\bar{F}_{\text{diff}}^{\text{expt}}$ is the mean value of the immediately measured quantity, the difference Casimir force. When this is measured, the individual Casimir forces between the sphere and the plate in the dark and bright phases remain undetermined.

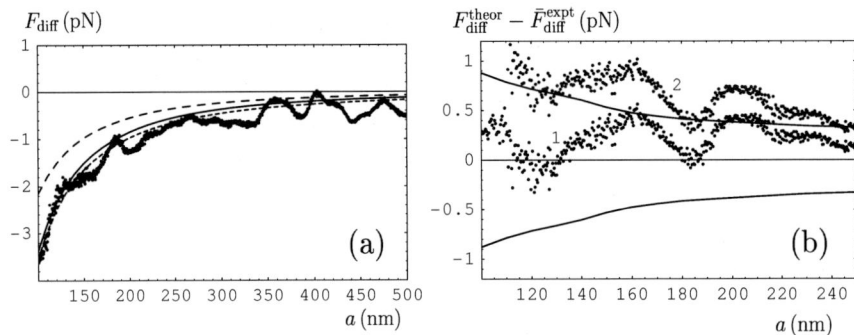

FIG. 20.15. (a) Differences between the Casimir forces in the presence and in the absence of light versus separation for an absorbed power of 9.3 mW. The measured differences $F_{\text{diff}}^{\text{expt}}$ are shown by dots; the differences calculated at $T = 300\,\text{K}$ and at $T = 0$ are shown by the solid and short-dashed lines, respectively; and the differences calculated including the dc conductivity of high-resistivity Si are shown by the long-dashed lines. (b) Theoretical minus experimental differences in the Casimir force versus separation. The results computed at $T = 300\,\text{K}$ using the model with a finite static permittivity for high-resistivity Si are labeled 1, and those including the dc conductivity are labeled 2. The solid lines show the 95% confidence intervals (Chen et al. 2007b).

For the experiment with two plates, the individual Casimir forces between the sphere and each of the plates were measured first, and $F_{\text{diff}}^{\text{expt}}$ was *calculated* as the difference between the mean measured forces afterwards. Additionally, the present experiment is effectively a dynamic measurement, where the change in the Casimir force only at the modulation frequency is measured. These characteristic features make the optical-modulation experiment much more sensitive than the experiments described in Sections 20.1 and 20.2.

Now, we proceed with the analysis of the experimental errors. The variance of the mean difference in the Casimir force is defined by eqn (18.3), where $\Pi^{\text{expt}} = F_{\text{diff}}^{\text{expt}}$; $n = J$; i is the number of points in one set of measurements, varying from 1 to 1209; and j is the number of pairs of applied voltages. Using Student's t-distribution with a number of degrees of freedom $f = 30$ (or 40 and 32 for the measurements with different absorbed powers) and choosing a confidence level $\beta = 0.95$, we obtain $p = (1 + \beta)/2 = 0.975$ and $t_p(f) = 2.00$. Then the absolute random error in the measurement of the difference Casimir force is given by eqn (18.8). This is presented in Fig. 20.18(a) as a function of separation for the three different measurements with different absorbed laser powers (the lines labeled a, b, and c correspond to decreasing power as indicated above). As can be seen from Fig. 20.18(a), the random error is rather different for different measurements. It is lowest for measurement b, which was done with 8.5 mW absorbed power. In this measurement, the random error decreases from 0.32 pN

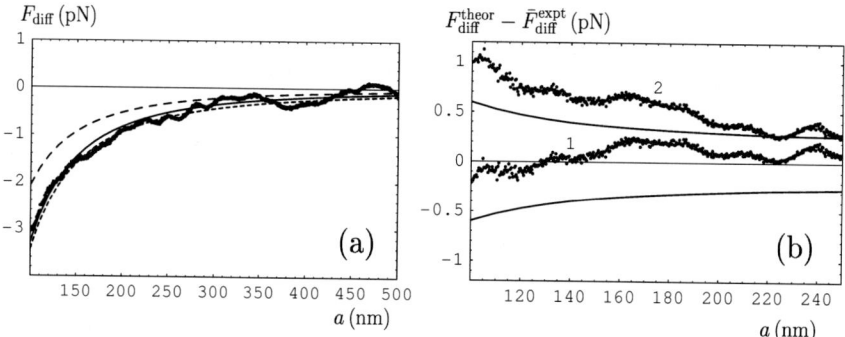

FIG. 20.16. (a) Differences between the Casimir forces in the presence and in the absence of light versus separation for an absorbed power of 8.5 mW. (b) Theoretical minus experimental differences in the Casimir force versus separation. All notation is explained in the caption to Fig. 20.15 (Chen et al. 2007b).

at $a = 100$ nm to 0.23 pN at $a = 250$ nm and preserves the latter value at larger separations. Figure 20.18(b) will be referred to in the discussion of the theoretical errors in the next subsection.

The main systematic error was due to the instrumental noise and was equal to $\Delta_1^{\text{syst}} F_{\text{diff}}^{\text{expt}} \approx 0.08$ pN, independent of separation. The systematic error determined from the resolution error in data acquisition, $\Delta_2^{\text{syst}} F_{\text{diff}}^{\text{expt}} \approx 0.02$ pN, also does not depend on separation. The calibration error, $\Delta_3^{\text{syst}} F_{\text{diff}}^{\text{expt}}$, depends on separation and was equal to 0.6% of the measured difference in the Casimir force. These systematic errors can be combined at a given confidence probability β using the statistical criterion (18.2). Choosing $\beta = 0.95$ (the 95% confidence

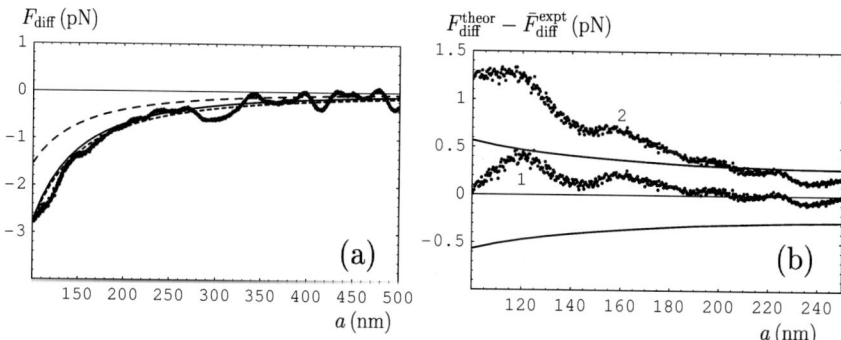

FIG. 20.17. (a) Differences between the Casimir forces in the presence and in the absence of light versus separation for an absorbed power of 4.7 mW. (b) Theoretical minus experimental differences in the Casimir force versus separation. All notation is explained in the caption to Fig. 20.15 (Chen et al. 2007b).

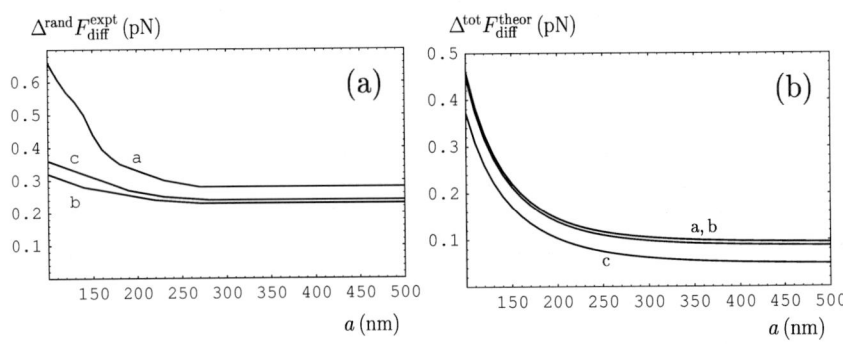

FIG. 20.18. (a) Random experimental errors (which are equal to the total error) and (b) total theoretical errors, versus separation. The cases of different absorbed powers 9.3, 8.5, and 4.7 mW are labeled a, b, and c, respectively (Chen et al. 2007b).

level), we arrive at a total systematic error for all three measurements varying from 0.092 to 0.095 pN.

The total experimental error in the force difference, $\Delta^{\rm tot} F_{\rm diff}^{\rm expt}(a)$, at a 95% confidence level can be found using the combination rule (18.15) because in this case the quantity $r(a)$ defined in eqn (18.14) is less than 0.8. Thus, the total experimental error in the values of $F_{\rm diff}^{\rm expt}(a)$ for all three measurements coincides with the random error, which is presented in Fig. 20.18(a). As a result, the relative experimental error varies from 10% to 20% at a separation $a = 100$ nm and from 25% to 33% at a separation $a = 180$ nm for the different absorbed laser powers. This allows us to conclude that modulation of the Casimir force with light has been demonstrated with high reliability and confidence. The observed effect could not be due to the thermally induced mechanical motion of the membrane. This is because membrane movement due to heating (in this case less than $1\,^\circ$C) would lead to a different force–distance relation for both the electrostatic force and the Casimir force, in disagreement with what was observed and with the distance-independence of V_0 and V_0^1. The temperature rise of less than $1\,^\circ$C was estimated based on the net thermal-energy increase in the Si membrane. The absorption of photons during the course of the optical pulse increases the thermal energy of the membrane, while the conductive and radiative heat outflow to the Si around the membrane and the surroundings leads to a decrease in its thermal energy. The net change is less than $1\,^\circ$C. The latter would lead to a negligible (less than 10^{-6}) relative expansion in the diameter of the membrane.

In order to account for roughness, the surface topography of the sphere and membrane was characterized using the AFM. Images obtained from a surface scan of the gold coating on the sphere demonstrated stochastically distributed roughness peaks with heights $h_i^{(1)} \leq 32$ nm ($1 \leq i \leq N_1 = 33$). A surface scan of the Si surface demonstrated much smoother relief, with heights $h_i^{(2)} \leq 1.68$ nm ($1 \leq i \leq N_2 = 17$). The fractions $v_i^{(1,2)}$ of the Au and Si surfaces with heights

$h_i^{(1,2)}$ were tabulated by Chen et al. (2007b). The roughness data were used in the theoretical computations of the differences in the Casimir forces.

20.3.4 Theoretical Casimir force differences and comparison with experiment

The Casimir force $F(a) = F(a,T)$ acting between a large gold sphere of radius R and a plane Si membrane at $T = 300$ K can be calculated by means of the Lifshitz formula (12.71), along with the use of the proximity force approximation (6.71). Keeping in mind that the difference force technique employed in this experiment was more sensitive, the thermal version of the Lifshitz theory at laboratory temperature was chosen for the comparison with data. The dielectric permittivities of gold, $\varepsilon^{(1)}(i\xi)$, and of high-resistivity Si, $\varepsilon^{(2)}(i\xi)$, in the absence of laser light have been repeatedly used above (see e.g. Section 20.1.3). On irradiation of the Si membrane by light, an equilibrium value of the carrier density is rapidly established during a period of time much shorter than the duration of the laser pulse. Therefore, one can assume that there is an equilibrium concentration of pairs (electrons and holes) when the light is incident. Thus, in the presence of laser radiation, the dielectric permittivity of Si along the imaginary frequency axis can be represented in the commonly used form (Vogel et al. 1992, Palik 1985)

$$\varepsilon_l^{(2)}(i\xi) = \varepsilon^{(2)}(i\xi) + \frac{\omega_{p(e)}^{(2)\,2}}{\xi\left[\xi + \gamma_{(e)}^{(2)}\right]} + \frac{\omega_{p(p)}^{(2)\,2}}{\xi\left[\xi + \gamma_{(p)}^{(2)}\right]}, \qquad (20.7)$$

where $\omega_{p(e,p)}^{(2)}$ and $\gamma_{(e,p)}^{(2)}$ are the plasma frequencies and the relaxation parameters for electrons and holes, respectively.

The values of the relaxation parameters are $\gamma_{(e)}^{(2)} \approx 1.8 \times 10^{13}$ rad/s and $\gamma_{(p)}^{(2)} \approx 5.0 \times 10^{12}$ rad/s (Vogel et al. 1992). The plasma frequencies were calculated using eqn (13.1), where the effective mass was replaced by $m_p^* = 0.2063 m_e$ (for holes) and $m_e^* = 0.2588 m_e$ (for electrons). The value of the concentration of charge carriers in the bright phase, which enters eqn (13.2), was calculated for the different absorbed powers in the following way. First, it was noted that for a membrane of thickness $d = 4\,\mu$m the concentration n does not depend on the depth. The reason is that a uniform concentration in this direction is established even more rapidly than the equilibrium discussed above (Vogel et al. 1992). In fact, the assumption of uniform charge carrier density in the Si membrane is justified by the long carrier diffusion lengths and the ability to obtain almost defect-free surfaces in silicon through hydrogen passivation (Yablonovitch et al. 1986). Next, the central part of the Gaussian beam, of diameter w, can be approximately modeled by a uniform cylindrical beam of the same diameter. The power contained in this cylindrical beam, P_w^{eff}, is equal to the power in the central part of a Gaussian beam with a diameter w. Elementary calculation using a Gaussian distribution leads to $P_w^{\text{eff}} = 0.393 P^{\text{eff}}$. The power P_w^{eff} is absorbed uniformly in the central part of the Si membrane of diameter w, which has a volume $V = \pi w^2 d/4$. Incidentally, the central region of the membrane with a diameter w contributes almost 100% [99.9999% using eqn (19.9)] of the total Casimir force acting between the membrane and sphere. At equilibrium, the number of created charge carrier pairs per unit time per unit volume $P_w^{\text{eff}}/(\hbar\omega V)$, where $\omega = 3.66 \times 10^{15}$ rad/s is the frequency of Ar laser light, is equal to the recombination rate of pairs per unit volume, n/τ. Thus, at equilibrium,

$$n = \frac{4 P_w^{\text{eff}} \tau}{\hbar \omega d \pi w^2}. \tag{20.8}$$

Equations (13.2) and (20.8) allow one to calculate the densities of charge carriers $n_a = (2.1 \pm 0.4) \times 10^{19}$ cm^{-3}, $n_b = (2.0 \pm 0.4) \times 10^{19}$ cm^{-3}, and $n_c = (1.4 \pm 0.3) \times 10^{19}$ cm^{-3} and the respective plasma frequencies

$$\omega_{\text{p(e)}}^{(2a)} = (5.1 \pm 0.5) \times 10^{14} \text{ rad/s}, \quad \omega_{\text{p(p)}}^{(2a)} = (5.7 \pm 0.6) \times 10^{14} \text{ rad/s},$$
$$\omega_{\text{p(e)}}^{(2b)} = (5.0 \pm 0.5) \times 10^{14} \text{ rad/s}, \quad \omega_{\text{p(p)}}^{(2b)} = (5.6 \pm 0.5) \times 10^{14} \text{ rad/s},$$
$$\omega_{\text{p(e)}}^{(2c)} = (4.1 \pm 0.4) \times 10^{14} \text{ rad/s}, \quad \omega_{\text{p(p)}}^{(2c)} = (4.6 \pm 0.4) \times 10^{14} \text{ rad/s},$$
$$\tag{20.9}$$

for all of the measurements a, b, and c with different powers of absorbed laser light. All of the above errors were found at a 95% confidence level. Note that in the original publication (Chen et al. 2007b,) the values of $\omega_{\text{p(e)}}^{(2c)}$ and $\omega_{\text{p(p)}}^{(2c)}$ contain misprints, corrected here. In the calculations of charge carrier densities using eqn (20.8), the values $\tau_a = \tau_b = 0.38 \pm 0.03$ ms and $\tau_c = 0.47 \pm 0.01$ ms were used, in accordance with the measurement results in Section 20.3.2, taking into account the fact that τ decreases when n increases. Recall that τ_a and τ_b were obtained from first 0.5 ms of the time decay. The value for τ_c, obtained using the whole 5 ms decay, may lead to a minor underestimation of the carrier density, a fact included in the resulting 21% error in the value of n_c. Note that the above values of the relaxation parameters $\gamma_{(e)}^{(2)}$ and $\gamma_{(p)}^{(2)}$ do not depend on the absorbed power (Vogel et al. 1992) and can be used in all measurements.

The values of $\varepsilon^{(1)}(i\xi)$ and $\varepsilon^{(2)}(i\xi)$ or $\varepsilon_l^{(2)}(i\xi)$ were substituted in the Lifshitz formula (12.71) combined with eqn (6.71) and the difference of the Casimir forces in the presence and in the absence of laser light, $F_{\text{diff}}(a)$, was computed at the laboratory temperature $T = 300$ K. The results obtained were corrected for the presence of surface roughness using the nonmultiplicative approach. The roughness data discussed at the end of Section 20.3.3, after substitution into eqn (17.92), led to $H_0^{(1)} = 20.0$ nm and $H_0^{(2)} = 1.1$ nm. Then the theoretical values of the difference Casimir force taking account of the surface roughness were calculated by geometric averaging:

$$F_{\text{diff}}^{\text{theor}}(a_i) = \sum_{k=1}^{33} \sum_{l=1}^{17} v_k^{(1)} v_l^{(2)} F_{\text{diff}}(a_i + H_0^{(1)} + H_0^{(2)} - h_k^{(1)} - h_l^{(2)}, T). \tag{20.10}$$

In the experiment under consideration, the contribution from the roughness correction was very small. Thus, at $a = 100$ nm, it contributed only 1.2% of the calculated $F_{\text{diff}}^{\text{theor}}(a)$. At $a = 150$ nm, the contribution from surface roughness decreased to only 0.5% of the calculated force difference. Similarly to Section 19.2.3, it could be easily seen that the contribution from nonadditive, diffraction-type effects to the roughness correction [which is not taken into account in eqn (20.10)] was negligibly small.

The results of the numerical computations of the difference Casimir force between rough surfaces, $F_{\text{diff}}^{\text{theor}}(a)$, are shown by solid lines in Figs. 20.15(a), 20.16(a), and 20.17(a) for the measurements with different powers of absorbed laser light. They are in very good agreement with the experimental data, shown by dots in the same figures (see below for the quantitative measure of agreement between experiment and theory).

For completeness, it is also reasonable to present the results of theoretical computations using the Lifshitz formula at zero temperature. These were obtained following the same procedure as at $T = 300$ K. These results are shown by the short-dashed lines in Figs. 20.15(a), 20.16(a), and 20.17(a). As can be seen from these figures, in all cases the short-dashed lines describe a slightly larger magnitude of the Casimir force difference than at $T = 300$ K and are in rather good agreement with the experimental data shown by the dots. Thus, this experiment is not sensitive enough to measure the small thermal effect which is predicted at the separations considered, where in the dark phase Si is characterized by zero conductivity [i.e. by the permittivity $\varepsilon^{(2)}(i\xi)$] and in the bright phase by the permittivity (20.7).

Importantly, the experimental data are also in agreement with theory if the dielectric permittivity of Si in the presence of light (the bright phase) is described by the generalized plasma-like model (see Sections 13.5 and 14.6). This permittivity can be obtained from eqn (20.7) by putting $\gamma_{(e)}^{(2)} = \gamma_{(p)}^{(2)} = 0$. Repeating the above computations of the thermal difference Casimir force $F_{\text{diff}}^{\text{theor}}$ under these conditions with $P_{\text{eff}} = 9.3$ mW, one obtains the results shown by the solid line in Fig. 20.19(a). In this figure, the same experimental data as in Fig. 20.15(a) are shown by dots. The dotted theoretical line reproduces the theoretical results computed using the dielectric permittivity (20.7) in the bright phase [in Fig. 20.15(a), this line is shown as a solid line]. The dashed lines in Figs. 20.19(a) and 20.19(b) are explained below. As can be seen in Fig. 20.19(a), both the solid and the dotted lines are in agreement with the experimental data (Mostepanenko and Geyer 2008). Thus, the experiment under consideration does not allow one to discriminate between the theoretical approaches using the dielectric permittivity (20.7) and the generalized plasma-like model. The same conclusions follow from a consideration of the measurement data obtained with the two other absorbed powers.

Before considering the quantitative comparison of data with various theoretical approaches, it is necessary to state the theoretical errors in the computation of the Casimir force acting between a sphere and a membrane. The major source

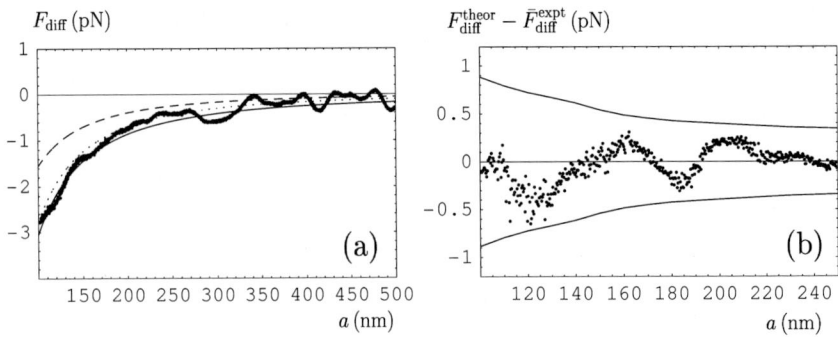

FIG. 20.19. (a) Differences between the Casimir forces in the presence and in the absence of light versus separation for an absorbed power of 9.3 mW. The measured differences $F_{\text{diff}}^{\text{expt}}$ are shown by dots; the differences calculated at $T = 300$ K using the generalized plasma-like permittivity and the permittivity (20.7) are shown by the solid and dotted lines, respectively; and those calculated including the dc conductivity of high-resistivity Si are shown by the long-dashed lines. (b) Theoretical minus experimental differences in the Casimir force versus separation. The results computed at $T = 300$ K using the model with a finite static permittivity in the dark phase and the generalized plasma-like permittivity in the bright phase are shown by dots. The solid lines indicate the 95% confidence intervals (Mostepanenko and Geyer 2008).

of the theoretical uncertainty in this experiment is the error in the concentration of charge carriers n when the light is on. From above, this error is about 20%. Calculations using the Lifshitz formula show that the resulting relative error in the difference Casimir force, $\delta_1 F_{\text{diff}}^{\text{theor}}$, is approximately equal to 0.12, i.e. 12%, and does not depend on separation. The error due to the uncertainty of the experimental separations a_i, at which the theoretical values $F_{\text{diff}}^{\text{theor}}$ should be computed, is equal to $3\,\Delta a/a$ and takes its maximum value of 3% of the Casimir force at the shortest separation of $a = 100$ nm (recall that $\Delta a = 1$ nm). This leads to only a 2% error in the difference of the Casimir force at $a = 100$ nm (so that $\delta_2 F_{\text{diff}}^{\text{theor}} \approx 0.02$) and to smaller errors at larger separations. The other theoretical errors discussed in Section 18.3.2, such as those due to sample-to-sample variation of the tabulated optical data for Au, the use of the proximity force theorem, patch potentials, and the finite thickness of the gold coating on the sphere, are negligible. Thus, for example, using the Lifshitz formula for a polystyrene sphere covered with a gold layer of 82 nm thickness instead of eqn (12.71) for a solid gold sphere, one would get only a 0.03% decrease in the Casimir force magnitude.

A specific uncertainty which is present in this experiment is connected with the pressure of the light transmitted through the membrane and incident on the bottom of the sphere. This effect is present only during the bright phase of the pulse train and can be easily estimated. The maximum intensity of laser light incident on a section of the sphere with radius $0 \leq r \leq R$ parallel to the membrane is

$$I(r) = \frac{2\alpha_0 P^{\text{eff}}}{\pi w^2} e^{-2r^2/w^2}, \qquad (20.11)$$

where α_0 is the fraction of the absorbed power transmitted through the membrane. The value of α_0 is given by

$$\alpha_0 = re^{-d/l_{\text{opt}}} \approx 0.00641, \qquad (20.12)$$

where $l_{\text{opt}} = 1\,\mu\text{m}$ (see Section 20.3.1) and the transmission coefficient r is approximately 0.35. The gold sphere is assumed to be perfectly reflecting.

The force due to the light pressure acting on the sphere takes the following form in spherical coordinates:

$$F_{\text{p}} = \frac{4\pi R^2}{c} \int_0^{\pi/2} d\vartheta\, I(R\sin\vartheta) \cos^2\vartheta \sin\vartheta. \qquad (20.13)$$

Substituting eqn (20.11) in eqn (20.13) and integrating, we obtain

$$F_{\text{p}} = \frac{2\alpha_0 P^{\text{eff}}}{c} \left[1 - e^{-2R^2/w^2}\, \frac{\sqrt{\pi} w\, \text{Erfi}(\sqrt{2}R/w)}{2\sqrt{2}R} \right], \qquad (20.14)$$

where $\text{Erfi}(z)$ is the imaginary error function.

For the absorbed powers used in the three experiments ($P^{\text{eff}} = 9.3$, 8.5, and 4.7 mW, respectively), eqn (20.14) leads to the following maximum forces which may act on the sphere owing to the light pressure: $F_{\text{p}} = 0.085$, 0.078, and 0.043 pN. The force due to light pressure can be taken into account as one more error in the theoretical evaluation of the difference Casimir force $F_{\text{diff}}^{\text{theor}}$. At a separation $a = 100$ nm, the respective relative error, $\delta_3 F_{\text{diff}}^{\text{theor}}$, is equal to 2.3%, 2.7%, and 1.5% for the three absorbed powers. At $a = 200$ nm, the relative theoretical error in $F_{\text{diff}}^{\text{theor}}$ due to light pressure increases to 8.9%, 8.7%, and 5.0%, respectively.

All three errors discussed above can be combined using the statistical rule (18.2). The resulting total absolute theoretical error, $\Delta^{\text{tot}} F_{\text{diff}}^{\text{theor}}$, is presented in Fig. 20.18(b) as a function of separation for the three experiments with decreasing power of absorbed laser light (lines a, b, and c, respectively). As can be seen from this figure, the total theoretical errors for the measurements a and b are almost equal, and for the measurement c, this error is slightly lower. The relative total theoretical error varies from 13.5% to 13.7% at $a = 100$ nm and from 13.7% to 14.4% at $a = 140$ nm for the three different absorbed powers. At $a = 200$ nm, the relative total theoretical error ranges from 14.9% to 17.2% for the different absorbed powers.

Now the quantitative comparison between experiment and theory can be done using the methods described in Section 18.3.3. For this purpose, the quantity

$F_{\text{diff}}^{\text{theor}} - F_{\text{diff}}^{\text{expt}}$ is considered. The absolute error of this quantity, $\Xi_{F_{\text{diff}}}^{0.95}(a)$, as a function of separation at a confidence level of 95%, is found by using eqn (18.21). The resulting confidence intervals $[-\Xi_{F_{\text{diff}}}^{0.95}(a), \Xi_{F_{\text{diff}}}^{0.95}(a)]$ are shown in Figs. 20.15(b), 20.16(b), and 20.17(b) by solid lines for the three measurements with the largest, intermediate, and smallest powers, respectively.

The differences between the theoretical values of $F_{\text{diff}}^{\text{theor}}$ computed at $T = 300\,\text{K}$ and the measured $\bar{F}_{\text{diff}}^{\text{expt}}$ are shown in Figs. 20.15(b), 20.16(b), and 20.17(b) by dots labeled 1 (once again, these dots are related to the three measurements with different power). As is seen in these figures, practically all of the dots labeled 1 are well inside the confidence intervals at all separation distances. This means that the Lifshitz theory at nonzero temperature, using the dielectric permittivity of high-resistivity Si $\varepsilon^{(2)}(i\xi)$ in the absence of laser light and the dielectric permittivity $\varepsilon_l^{(2)}(i\xi)$ given by eqn (20.7) in the presence of light, is consistent with the experiment. The consistency of the experiment with the theory is preserved when the theoretical values of $F_{\text{diff}}^{\text{theor}}$ are computed at zero temperature [see the short-dashed lines in Figs. 20.15(a), 20.16(a), and 20.17(a)]. The conclusion is that the thermal effect cannot be resolved, taking into consideration the experimental and theoretical errors reported above.

The same method of comparison between experiment and theory can be applied when the generalized plasma-like dielectric permittivity of Si is used in the presence of light instead of the permittivity (20.7). In this case the computational results for the difference $F_{\text{diff}}^{\text{theor}} - \bar{F}_{\text{diff}}^{\text{expt}}$ are shown by dots in Fig. 20.19(b) for an absorbed power $P_{\text{eff}} = 9.3\,\text{mW}$. It is seen that the dots are inside the 95% confidence intervals in the same way as the dots labeled 1 in Fig. 20.15(b), where $F_{\text{diff}}^{\text{theor}}$ was computed using the dielectric permittivity (20.7). The same occurs for the two other absorbed powers. To distinguish between the two models using the generalized plasma-like permittivity and the permittivity (20.7) of a metal-type semiconductor in the presence of light, more precise experiments are required (Mostepanenko and Geyer 2008).

For illustrative purposes, the agreement between experiment and theory is presented in Fig. 20.20 using another method from Section 18.3.3. Here, a narrower separation interval, from 150 to 200 nm, is considered and each third experimental point from measurement c is plotted together with its error bars $\left[\pm\Delta a, \pm\Delta^{\text{tot}} F_{\text{diff}}^{\text{expt}}\right]$, shown as crosses (there are too many points to present all of them in this form). The theoretical force difference $F_{\text{diff}}^{\text{theor}}$ computed with the Lifshitz formula at $T = 300\,\text{K}$ using the generalized plasma-like model is shown by the solid line, and that computed using the permittivity (20.7) by the dotted line. It can be seen that the experimental data are consistent with both theoretical approaches, in confirmation of the conclusion drawn above using Fig. 20.19.

20.3.5 Tests for the effect of charge carriers in dielectrics

In the calculations in the previous subsection, the dielectric response of high-resistivity Si in the absence of the incident laser light was described by the

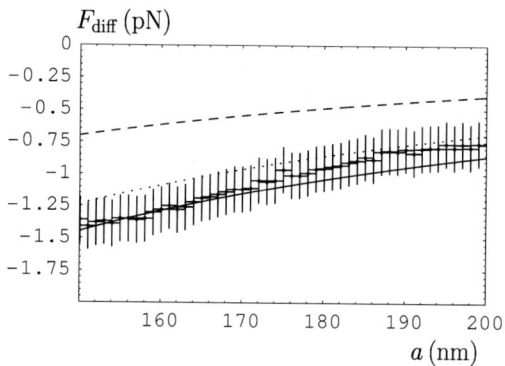

FIG. 20.20. The experimental differences in the Casimir force, with their experimental errors, are shown by crosses (the absorbed power is equal to 4.7 mW). The solid and dotted lines represent the theoretical differences computed at $T = 300$ K using a finite static permittivity for high-resistivity Si, but different models for Si in the presence of light (see text for further discussion). The dashed line represents the theoretical differences computed at the same temperature including the dc conductivity.

function $\varepsilon^{(2)}(i\xi)$, which has a finite static value $\varepsilon^{(2)}(0) \approx 11.66$. However, as discussed in Section 12.5.2, dielectrics have some nonzero dc conductivity σ_0 at any nonzero temperature. If this is included, the dielectric permittivity of the Si plate has the form (12.113) with the index $n = 2$. Using eqns (13.19) and (13.20), it can be represented in the form

$$\tilde{\varepsilon}^{(2)}(i\xi) = \varepsilon^{(2)}(i\xi) + \frac{\omega_{p(p)}^{(2)\,2}}{\xi\left[\xi + \gamma_{(p)}^{(2)}\right]}. \tag{20.15}$$

The value of the plasma frequency in eqn (20.15) can be found by substituting the carrier density $n \approx 5 \times 10^{14}$ cm^{-3} in eqn (13.2), with the result $\omega_{p(p)}^{(2)} \approx 2.8 \times 10^{12}$ rad/s. Note that for $n \leq 1.0 \times 10^{17}$ cm^{-3}, the value of the relaxation parameter has an insignificant effect on the magnitude of the Casimir force. Because of this, the same value of $\gamma_{(p)}^{(2)}$ as in eqn (20.7) was used in eqn (20.15).

The presence of a low dc conductivity in dielectric materials was used (Zurita-Sánches et al. 2004, Joulain et al. 2005) to obtain a large effect of van der Waals friction which could bring some observations (Stipe et al. 2001) into agreement with theory. In Section 12.5.2, for two dielectric plates, and in Section 15.4, for one metal and one dielectric plate, it was proved, however, that the inclusion of the dc conductivity of a dielectric in the Lifshitz theory leads to a violation of the third law of thermodynamics (the Nernst heat theorem). Thus, it is not acceptable from a theoretical point of view.

The experiment on the modification of the Casimir force with laser pulses clarified the problem of whether or not the dc conductivity of high-resistivity Si should be taken into account in the Lifshitz theory of the Casimir and van der Waals forces. For this purpose, the theoretical computations of the difference Casimir force presented in Section 20.3.4 were completely repeated, replacing the dielectric permittivity of Si $\varepsilon^{(2)}(i\xi)$ used above with $\tilde{\varepsilon}^{(2)}(i\xi)$ given in eqn (20.15). The theoretical results obtained for $\tilde{F}_{\text{diff}}^{\text{theor}}$ versus separation are shown by the long-dashed lines in Figs. 20.15(a), 20.16(a), and 20.17(a) for all of the three measurements with different powers of absorbed light. As can be seen in these figures, all of the long-dashed lines are far outside both the experimental data, shown by dots, and the solid lines calculated using the Lifshitz theory disregarding the dc conductivity of high-resistivity Si at laboratory temperature. Notice that the computational results at $T = 0$ [shown by the short-dashed lines in Figs. 20.15(a), 20.16(a), and 20.17(a)] do not depend on whether the dc conductivity is included in the dielectric permittivity used to describe the high-resistivity Si.

To draw a quantitative conclusion about the measure of agreement between the data and the two models with and without the inclusion of the dc conductivity of high-resistivity Si, the differences $\tilde{F}_{\text{diff}}^{\text{theor}} - F_{\text{diff}}^{\text{expt}}$, where $\tilde{F}_{\text{diff}}^{\text{theor}}$ was computed including the dc conductivity according to eqn (20.15), have been shown as dots, labeled 2, in Figs. 20.15(b), 20.16(b), and 20.17(b). As is seen in Figs. 20.15(b) and 20.16(b), the model including the dc conductivity of high-resistivity Si is excluded experimentally at a 95% confidence level within the region from 100 to 250 nm. From Fig. 20.17(b), it follows that this model is excluded at a 95% confidence level within the range of separations from 100 to 200 nm.

The conclusion that the model of high-resistivity Si which includes the dc conductivity is inconsistent with the experiment on the optically modulated Casimir force is confirmed also in Figs. 20.19(a) and 20.20, where the quantity $\tilde{F}_{\text{diff}}^{\text{theor}}$ versus separation is plotted as a dashed line. It can be clearly observed that the dashed line not only is far away from the solid line based on the theory neglecting the dc conductivity of Si in the absence of excitation light but is also distant from all of the error bars representing the experimental data.

In Section 12.5.2, a rule concerning the application of the Lifshitz theory to dielectric materials having zero conductivity at zero temperature was formulated. According to this rule, the dc conductivity of such materials that arises at nonzero temperature must be disregarded. Otherwise, owing to the violation of thermal equilibrium, the Lifshitz theory becomes inconsistent with thermodynamics. The experiment on the optically modulated Casimir force demonstrates that if this rule is not followed, the theoretical results obtained are excluded by the data. One can conclude that the experiment on the optically modulated force is complementary to the experiments by means of a micromechanical oscillator with metallic test bodies (see Section 19.3). In those experiments, the difference between the Casimir forces using the Drude and the plasma-like permittivities was resolved and the former was excluded. As mentioned above, the present experiment is not sensitive enough to discriminate between the Drude and plasma-like permittivities, but provides a radically new test of the inclusion of the dc conductivity of dielectrics in the Lifshitz theory.

The results of the experiment on the optically modulated Casimir force were also applied (Klimchitskaya et al. 2008a) to test the modification of the transverse magnetic reflection coefficient at zero frequency given by eqn (16.81) (Pitaevskii 2008a). As an example, in Fig. 20.21(a,b) the experimental data for the difference of the Casimir forces in the presence and in the absence of laser light are shown for the largest and smallest absorbed powers, 9.3 mW and 4.7 mW, respectively. Note that in contrast to Fig. 20.20, the absolute errors of the force measurements (the vertical arms of the crosses) are now indicated at a 70% confidence level. This was done to obtain more conclusive results. The solid lines in Fig. 20.21 were computed as explained above using the Lifshitz formula, with Si in the dark phase described as a dielectric with the permittivity $\varepsilon^{(2)}(i\xi)$. In the presence of light, the dielectric permittivity (20.7) was used, with $\gamma_{(e)}^{(2)} = \gamma_{(p)}^{(2)} = 0$ (the generalized plasma-like model). Almost the same results were obtained if eqn (20.7), with nonzero values of the relaxation frequencies, as indicated above, was used. The dashed lines were computed using eqn (16.81) for the modified transverse magnetic reflection coefficient at zero frequency, with corresponding concentrations of charge carriers n in the dark phase and $2n_a$ or $2n_c$ in the presence of light. At all nonzero Matsubara frequencies, the standard terms of the Lifshitz formula were used with the dielectric permittivity of Si as given by eqn (20.7) (note that in the dark phase, the Drude additions are negligible at all nonzero Matsubara frequencies).

As can be seen in Fig. 20.21, the experimental data are consistent with the theoretical results computed using the standard Lifshitz theory with the dc conductivity of dielectric Si neglected in the dark phase (the solid lines). The theoretical results computed using the modified TM reflection coefficient at zero frequency taking into account the effect of charge screening are excluded by the data at a 70% confidence level. The same conclusion follows from the third data set, obtained at 8.5 mW absorbed power.

According to Svetovoy (2008), the experimental data for the difference Casimir force are equally consistent with the nonlocal approach using the reflection coefficient (16.81) and the Lifshitz theory, with the dc conductivity neglected in the dark phase. To prove this, the experimental data of Fig. 20.21(a) at a 70% confidence level were used, but the dashed line was replaced with a theoretical band whose width was determined at a 95% confidence level using the corresponding uncertainty in the charge carrier density $\Delta n = 0.4 \times 10^{19}\,\mathrm{cm}^{-3}$. Such a mismatched comparison of experiment with theory is irregular. It can be easily seen that the theoretical bands related to the solid and dashed lines in Fig. 20.21(a) do not overlap if one uses in the computations the uncertainty in the charge carrier density $\Delta n = 0.3 \times 10^{19}\,\mathrm{cm}^{-3}$, determined at the same 70% confidence level as for the experimental errors (Klimchitskaya et al. 2008c, Mostepanenko et al. 2009).

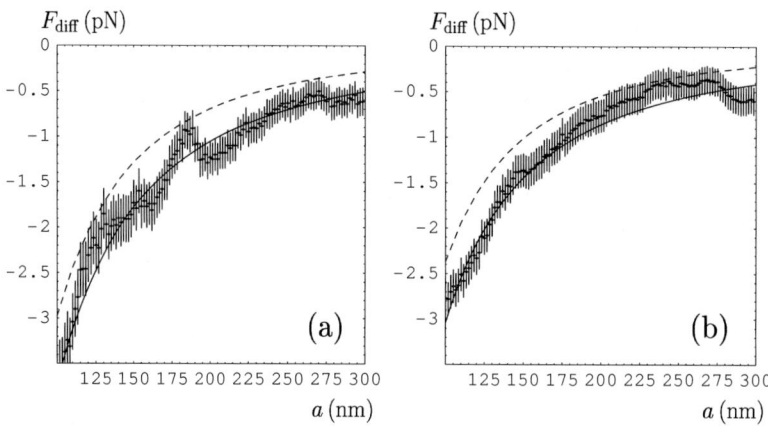

FIG. 20.21. Difference between the Casimir forces between an Au-coated sphere and an Si plate in the presence and in the absence of laser light on the plate, versus separation, for absorbed powers of (a) 9.3 mW and (b) 4.7 mW. The experimental data are shown by crosses. The solid and dashed lines were computed using the standard Lifshitz theory, with the dc conductivity of Si in the dark phase neglected and taking into account the effect of charge screening for the transverse magnetic mode, respectively (Klimchitskaya et al. 2008a).

The experiment on the optically modulated Casimir force also excludes the theoretical approach of Dalvit and Lamoreaux (2008) (see Section 14.3.5), as applied to high-resistivity semiconductor materials. The screening effects in these materials are characterized by the Debye–Hückel screening length as in the original formulation of the approach (Dalvit and Lamoreaux 2008). Note that the contribution of the zero-frequency TM reflection coefficient in this approach is the same as in the approach of Pitaevskii (2008a). The contributions of all other terms to the Lifshitz formula are almost equal in the two approaches. Because of this, the dashed lines in Fig. 20.21 reproduce the computational results for the difference Casimir force in the framework of the theoretical approach of Dalvit and Lamoreaux (2008). Thus, this approach that attempts to take the screening effects and the diffusion current into account in the Lifshitz theory is experimentally excluded not only for metals (see Section 19.3.4), but also for semiconductors.

20.4 Proposed experiments with semiconductor surfaces

The experiments presented in Sections 20.1–20.3 demonstrate the enormous potential of semiconductor materials for further investigation of the Casimir effect and the nontrivial interplay between zero-point and thermal fluctuations. Additional interest in this direction is being stimulated by the promise of nanotechnological applications. Because of this, several new experiments with semiconductor surfaces have been proposed. Some of them are considered below.

20.4.1 The dielectric–metal transition

An exciting possibility for modulation of the Casimir force due to a change in the charge carrier density is offered by semiconductor materials that undergo a dielectric–metal transition with an increase in temperature. From a fundamental point of view, the modulation of the Casimir force by phase transitions of various kinds offers one more precision test of the role of the conductivity and optical properties in the Lifshitz theory of the Casimir force.

An experiment has been proposed (Castillo-Garza et al. 2007) to measure the change of the Casimir force acting between an Au-coated sphere and a vanadium dioxide (VO_2) film deposited on a sapphire substrate, where the VO_2 undergoes a dielectric–metal transition with an increase in temperature. It has been known that VO_2 crystals and thin films undergo an abrupt transition from a semiconducting monoclinic phase at room temperature to a metallic tetragonal phase at 68°C (Zylbersztejn and Mott 1975, Soltani et al. 2004, Suh et al. 2004). The phase transition causes the resistivity of the sample to decrease by a factor of 10^4 from 10 to 10^{-3} Ω cm. In addition, the optical transmission for a wide region of wavelengths, extending from 1 μm to greater than 10 μm, decreases by more than a factor of 10, up to 100.

The increase in temperature necessary for the phase transition can be induced by laser light (Soltani et al. 2004, Suh et al. 2004). Thus, a setup similar to the one employed in the demonstration of optically modulated dispersion forces (see Section 20.3) can be used. In the initial stage of the experimental work, the procedures for film fabrication and the heating of the films were investigated (Castillo-Garza et al. 2007). Some preliminary theoretical results were also obtained (Castillo-Garza et al. 2007, Pirozhenko and Lambrecht 2008a) based on the Lifshitz theory and the optical data for VO_2 films (Verleur et al. 1968). Calculations of the difference Casimir force between an Au sphere and a VO_2 film on a sapphire substrate after and before the phase transition showed that the proposed experiment has much promise for the understanding of the role of conductivity in the Lifshitz theory of dispersion forces.

Interesting results can also be obtained when the change in the Casimir free energy associated with the phase transition of a metal to the superconducting state is investigated. The variation of the Casimir free energy during this transition is very small (Mostepanenko and Trunov 1997). Nevertheless, the magnitude of this variation can be comparable to the condensation energy of a semiconducting film and causes a measurable increase in the value of the critical magnetic field (Bimonte et al. 2005a, 2005b). Another proposed experiment is to measure the change of the Casimir force in a superconducting cavity due to a small change in temperature (Bimonte 2008).

20.4.2 Casimir force between a sphere and a patterned plate

Difference force measurements are very sensitive to relatively small variations of the Casimir force (see Section 20.3). Recently, an experimental scheme has been proposed (Castillo-Garza et al. 2007) which promises a record sensitivity

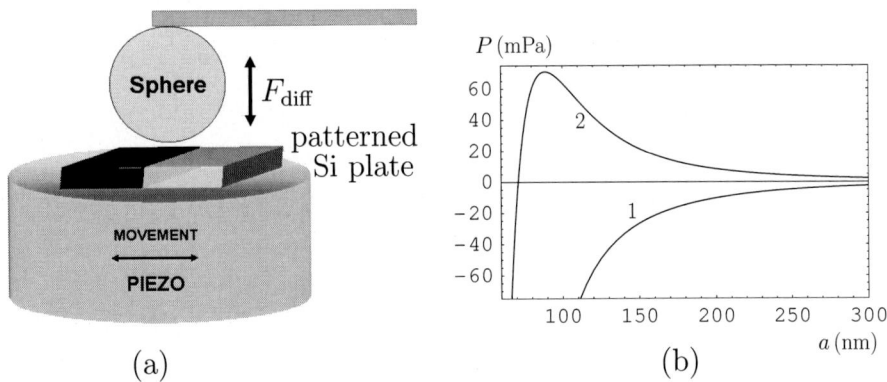

FIG. 20.22. (a) Schematic diagram of an experimental setup for the measurement of the difference Casimir force between an Au-coated sphere and a patterned Si plate (Klimchitskaya et al. 2009a). (b) Casimir pressure versus separation in a three-layer system α-Al_2O_3–ethanol–Si with no light on the Si plate (line 1) and with the Si plate illuminated (line 2). [See Klimchitskaya et al. (2007b).]

to a force difference, at the level of 1 fN. Here, a patterned Si plate with two sections of different doping concentrations [see Fig. 20.22(a)] is mounted on a piezoelectric actuator below an Au-coated sphere attached to the cantilever of an AFM. The actuator oscillates in the horizontal direction, causing flexing of the cantilever in response to the Casimir force above different regions of the plate. Thus, the sphere is subject to the difference Casimir force, which can be measured using the static and dynamic techniques. The patterned plate is composed of single-crystal Si specifically fabricated to have adjacent sections with two different charge carrier densities. A special procedure has been developed for the preparation of a Si sample with two sections having different conductivities (Castillo-Garza et al. 2007), where both p- and n-type dopants can be used (B and P, respectively). Sharp transition boundaries between the two sections of the Si plate, with a width less than 200 nm, can be achieved. Identically prepared but unpatterned samples can be used to measure the properties which are needed for the theoretical computations (with Hall probes for measuring the charge carrier concentration, and a four-probe technique for measuring the conductivity). The measurement of the difference Casimir force is planned as follows. The Si plate is positioned such that the boundary is below the vertical diameter of the sphere [see Fig. 20.22(a)]. The distance between the sphere and the Si plate, a, is kept fixed and the Si plate oscillates in the horizontal direction using the piezoelectric actuator such that the sphere crosses the boundary in the perpendicular direction during each oscillation [a similar approach has been exploited (Decca et al. 2005a) for constraining new forces using the oscillations of an Au-coated sphere above two dissimilar metals, Au and Ge, see Section 24.4.2]. The Casimir force on the sphere changes as the sphere crosses the boundary. This change corresponds to the differential force F_{diff}, equal to the difference between the Casimir forces due to the different charge carrier densities n_b and n_a. This causes a difference in the deflection of the cantilever. In order to reduce random noise by averaging, the periodic horizontal movement of the plate will be at an angular frequency $\Omega \sim 0.1$ Hz. The amplitude of the oscillations of the plate is limited by the characteristics of the actuator, but can be of order 100 μm, much larger than the typical width of the transition region, equal to 200 nm.

The proposed experiment holds promise for an investigation of the possible variation of the Casimir force associated with the dielectric–metal transition that occurs in semiconductors with an increase in doping concentration (Klimchitskaya and Geyer 2008). It has the potential to distinguish between the two models of the dielectric permittivity of a semiconductor with a concentration of charge carriers above the critical value [see eqn (20.7), with nonzero and zero relaxation frequencies].

20.4.3 Pulsating Casimir force

At present, a clear consensus has been reached that the applications of the Casimir force in the design, fabrication, and actuation of micromechanical and nanomechanical devices are ripe for exploitation. When the characteristic dimensions of a device shrink below a micrometer, the Casimir force becomes larger than the typical electric forces. Considerable opportunities for micromechanical design would be opened up by the use of pulsating Casimir plates, moving back and forth entirely because of the effect of the zero-point energy, without the action of mechanical springs. This can be achieved only through the use of both attractive and repulsive Casimir forces. In connection with this, it should be noted that while the existence of repulsive Casimir forces for a single cube or sphere is still debated, the Casimir repulsion between two parallel plates is well understood (see Section 19.5.3). Repulsion occurs when the inequalities in eqn (19.25) are satisfied.

Recently it was shown that the illumination of one (Si) plate in the three-layer systems Au–ethanol–Si, Si–ethanol–Si, and α-Al$_2$O$_3$–ethanol–Si with laser pulses can change the Casimir attraction to a Casimir repulsion and vice versa (Klimchitskaya et al. 2007b). The illumination can be performed in the same way as described in Section 20.3. Calculations show that in the system Au–ethanol–Si, the force is repulsive at separations $a > 160$ nm. The illumination of the Si plate changes this repulsion to attraction. In the system Si–ethanol–Si, the force between the Si plates is attractive; however, with one Si plate illuminated, the attraction is replaced with repulsion at separations $a > 175$ nm.

In the systems mentioned above, the magnitude of the repulsive force is several times less than the magnitude of the attractive force at the same separations. However, it is possible to design a system where the light-induced Casimir repulsion is of the same order of magnitude as the attraction. A good example is given by the three-layer system α-Al$_2$O$_3$–ethanol–Si, where the Si plate is illuminated with laser pulses. Computational results for the Casimir pressure versus separation in this system are presented in Fig. 20.22(b). The Casimir attraction (solid line 1) changes to repulsion at separations $a > 70$ nm when the Si plate is illuminated (solid line 2).

Note that the observation of a pulsating Casimir force requires that the plates be completely immersed in a liquid, far away from any air–liquid interfaces. This prevents the occurrence of capillary forces. Surface preparation of the plates is necessary to bring about intimate contact between the plates and the liquid. The only liquid-based force is the drag force due to the movement of the plates in response to the change of the force. For pressure values of around 10 mPa and typical spring constants of 0.02 N/m, the corresponding drag pressure from the plate movement would be six orders of magnitude less in value. Thus, in the near future one can expect an experimental confirmation of Casimir repulsion modulated by light.

21

MEASUREMENTS OF THE CASIMIR FORCE IN CONFIGURATIONS WITH CORRUGATED BOUNDARIES

As discussed in Section 17.5, configurations with sinusoidally corrugated boundaries present interesting opportunities for the observation of a new physical phenomenon, the lateral Casimir force. In this chapter, we consider the results of three experiments performed to date where the Casimir force due to corrugated boundaries has been measured. In the experiment of Roy and Mohideen (1999), the normal Casimir force between a sinusoidally corrugated plate and a smooth sphere was measured. Chen et al. (2002a, 2002b) first experimentally demonstrated the phenomenon of the lateral Casimir force in a sphere–plate configuration, where both bodies were covered with uniaxial sinusoidal corrugations. Chan et al. (2008) measured the normal Casimir force between a plate with rectangular corrugations and a smooth sphere. We also consider two theoretical approaches used for the interpretation of the experimental results, and the possibility to control the lateral Casimir force due to uniaxial corrugations of arbitrary shape.

21.1 Experiment with a sphere above a corrugated plate

The schematic diagram of the experimental setup of this experiment was the same as in Fig. 19.1, but the surface of the $7.5 \times 7.5\,\text{mm}^2$ plate was covered with uniaxial sinusoidal corrugations, described by

$$z = Af(x,y) = A\sin\frac{2\pi x}{\Lambda}. \tag{21.1}$$

Here, $A = 59.4 \pm 2.5\,\text{nm}$ was the amplitude of the corrugation and $\Lambda = 1.1\,\mu\text{m}$ was its period. The diameter of the sphere was $2R = 194.6 \pm 0.5\,\mu\text{m}$. Both the sphere and the plate were covered with 250 nm of Al and 8 nm of Au/Pd. The variances of the stochastic roughness of the corrugated plate and of the bottom of the sphere were measured to be 4.7 and 5 nm, respectively (Roy and Mohideen 1999).

The setup was calibrated electrostatically using the same procedures as described in Section 19.2.2. First, the residual potential of the grounded sphere was found. For this purpose, the electrostatic force between the sphere and the plate was measured for different applied voltages at five different separations $a \gg A$, leading to $V_0 = 14.9\,\text{mV}$. Next, by applying voltages V such that the electrostatic force was much larger than the Casimir force, the separation on contact $a_0 = 132 \pm 5\,\text{nm}$ was determined. The theoretical expression for the electrostatic force used in the fit was obtained from eqn (19.1) for a sphere above a

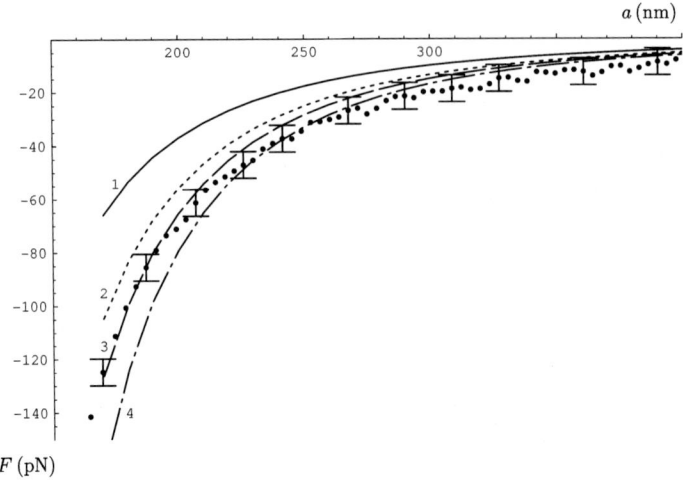

FIG. 21.1. The mean measured Casimir force between the sphere and the corrugated plate is shown by the circles. The theoretical lines numbered from 1 to 4 were computed under various alternative assumptions concerning the probability distribution of different sphere positions above a corrugated plate (Klimchitskaya et al. 2001). See text for further discussion.

smooth plate using the proximity force approximation. When the force was calibrated electrostatically, the average value of the force constant in Hooke's law was found to be $k = 0.021 \pm 0.001\,\text{N/m}$. The systematic corrections due to the residual potential difference and the coupling of scattered light from the plate into the photodiodes were subtracted from the measured total force. The experiment was repeated 15 times, and the mean measured Casimir force is presented by solid circles in Fig. 21.1. The errors in the force measurement $2\,\Delta F = 10\,\text{pN}$ are shown in the figure as error bars.

In cases where the surface of a plate is corrugated, the comparison between experiment and theory is not trivial. The reason is the uncertainty in the position of the sphere with respect to the corrugation; for example, the bottommost point of the sphere can be above a maximum or a minimum of the corrugation or in between. As a result, not only a vertical Casimir force but also a lateral force arises (see Sections 17.1.3 and 17.5.2). The lateral force may cause displacements of the sphere in the x-direction, so that the positions of the sphere above the points $0 < x < \Lambda$ within a corrugation period are not equally probable. Both the vertical and the lateral Casimir force acting on a sphere because of a corrugated plate were found by Klimchitskaya et al. (2001). It was shown that the lateral Casimir force has a zero value at the extremum points of the corrugation (21.1) and reaches a maximum at $x = 0$ and $\Lambda/2$, where the corrugation function is zero. If the center of the sphere is at a position $x < \Lambda/4$ (i.e. to the left of the corrugation maximum), the sphere experiences a positive lateral force $F_x > 0$.

If it is to the right of $x = \Lambda/4$, we have $F_x < 0$. In both cases the sphere tends to change its position in the direction of the corrugation maximum, which is a position of stable equilibrium (similarly to the case of an atom above a rough plate discussed in Section 17.5.4).

Using the proximity force approximation, the vertical Casimir force to be compared with the experimental data is given by

$$F^{\text{theor}}(a) = 2\pi R \int_0^\Lambda dx\, \rho(x)\, E_{\text{p}}[d(a,x)]. \qquad (21.2)$$

Here, $d(a, x)$ is the separation between the bottom of the sphere and the point x on the plate surface, including both the corrugations and the roughness; $\rho(x)$ is the probability distribution of the positions of the sphere above different points x within a corrugation period, and E_{p} is the Casimir energy per unit area of two parallel plates, calculated perturbatively in the framework of the plasma model using eqn (13.8). Computational results obtained using eqn (21.2) under different assumptions concerning the probability distribution $\rho(x)$ are presented in Fig. 21.1. The solid line 1 was computed with a uniform probability distribution, i.e. assuming that the positions of the sphere above all points on the plate are equally probable. This assumption was used in eqns (17.90) and (17.91), representing the application of the proximity force approximation to the problem of surface roughness. In the case of stochastic roughness, the average lateral force is equal to zero and the hypothesis of a uniform distribution is justified. The short-dashed line 2 was computed under the assumption that the sphere is located with equal probability above any point of the convex section of the corrugations. The long-dashed line 3 was computed with a distribution function which increases linearly when the sphere approaches a point of stable equilibrium, and the dashed-dotted line 4 is for a sphere situated above a point of stable equilibrium. As can be seen in Fig. 21.1, line 3 is consistent with the experimental data within the limits of the uncertainties $\Delta F = 5\,\text{pN}$ and $\Delta a = 5\,\text{nm}$. A complete comparison between experiment and theory would require simultaneous measurements of both the vertical and the lateral Casimir force.

21.2 Measurement of the lateral Casimir force

The first measurement of the lateral Casimir force between corrugated surfaces was performed by Chen et al. (2002a, 2002b) in the configuration of a sphere and a plate, both covered with uniaxial sinusoidal corrugations of equal periods. A schematic of the experiment is shown in Fig. 21.2. Measurements were performed at a pressure below $50\,\text{mTorr}$ at room temperature. The axes of the corrugations were held strictly parallel. A misalignment by $3°$ can lead to zero lateral force owing to the crossing of the corrugation axes. A plastic diffraction grating with uniaxial sinusoidal corrugations of period $\Lambda = 1.2\,\mu\text{m}$ and an amplitude $A = 90\,\text{nm}$ was used as the substrate for the corrugated plate. In order to obtain perfect orientation and a phase ϕ between the two corrugated surfaces,

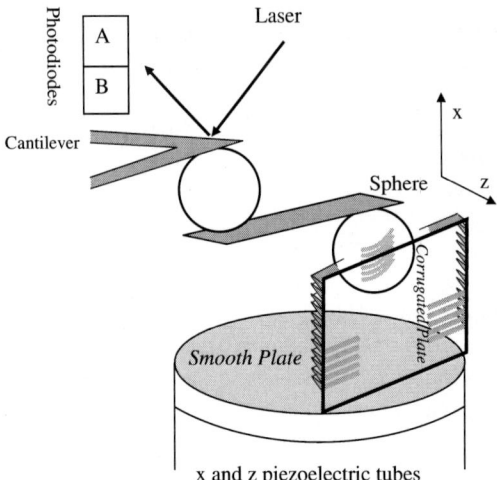

FIG. 21.2. Schematic diagram of the measurement of the lateral Casimir force between corrugated surfaces (Chen *et al.* 2002b).

a special *in situ* procedure was developed, where the corrugations of the plate were imprinted onto the gold-coated sphere by pressure.

A polystyrene sphere was attached to the tip of a 300 μm long cantilever with conductive silver epoxy. After this, a piece of freshly cleaved mica less than 10 μm thick, 100–200 μm wide, and 0.5 mm long was attached to the bottom of the sphere with silver epoxy. A second polystyrene sphere was then attached to the end of the mica plate. This second sphere was the one to be imprinted with corrugations and to interact with the corrugated plate. The cantilever (with mica plate and spheres), the corrugated plate, and a smooth flat plate (polished sapphire, see Fig. 21.2) were all coated with about 400 nm of gold in a thermal evaporator. A small region close to one edge of the corrugated plate was also coated with 100 nm of aluminum. As Al exhibits more hardness than gold, this region was used to imprint the corrugations from the plate onto the gold-coated sphere. The sphere and the plate were mounted as shown in Fig. 21.2. The addition of the first sphere and the mica plate was needed to isolate the laser reflection spot on the cantilever tip from the interaction region between the two corrugated surfaces.

To imprint the corrugations on the sphere, it was moved over to the region of the corrugated plate coated with Al. The other side of the sphere was mechanically supported and the corrugations were imprinted on the gold coating of the sphere by pressure using the piezoelectric tubes shown in Fig. 21.2. An AFM scan of the imprinted corrugations on the sphere, taken after the completion of the experiment, is shown in Fig. 21.3. The procedure used leads to uniaxial corrugations of equal period and to parallel allignment of the corrugations on the

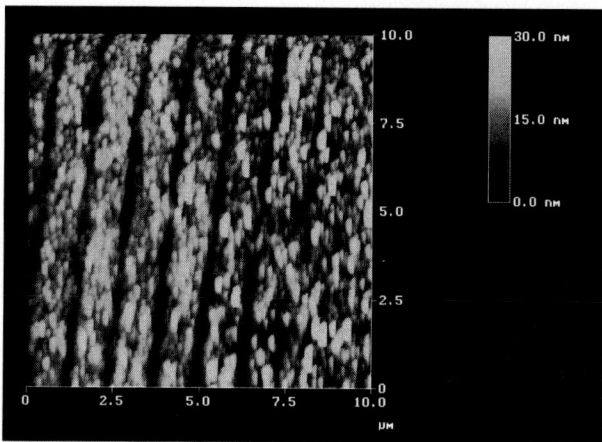

FIG. 21.3. AFM image of the imprinted corrugations on the sphere.

two bodies. If there is a nonzero angle θ between the axes of the corrugations, the phase shift along the x-axis becomes a periodic function of y with a period $\Lambda_y = \Lambda \cot\theta$. Within one period, $\phi(y)$ depends linearly on y, taking values from 0 to 2π. The resulting lateral force must be averaged over the period Λ_y. In the case of infinite bodies, this leads to a zero force at any $\theta \neq 0$. For real bodies of finite size with characteristic dimensions L and R, the lateral Casimir force is measurable only for small deviations of the corrugation axes from parallelity such that

$$\Lambda \cot\theta \gg \min(R, L). \tag{21.3}$$

The amplitude of the imprinted corrugations was measured from the AFM scan to be $A_2 = 8 \pm 1$ nm. The amplitude of the corrugations on the metallized plate, also measured using the AFM, was $A_1 = 59 \pm 7$ nm. After imprinting, the mechanical supports were removed and the sphere was translated over to the gold-coated area of the plate. Special care was taken to preserve the orientation of the corrugation axes during the translation.

The corrugated plate was mounted on two piezoelectric tubes that allowed independent movement of the plate in the vertical and horizontal directions; the two tubes used for this purpose will be referred to here as the "x-piezo" and "z-piezo", respectively. Movement in the x-direction with the x-piezo was necessary to achieve a lateral phase shift ϕ between the corrugations on the sphere and the plate. Independent movement in the z-direction was necessary for control of the surface separation between the sphere and the plate. The vertical position of the corrugated plate resulted in the usual bending of the cantilever in response to the lateral Casimir force, which acted tangentially to the plate, whereas a force acting normal to the sphere and the corrugated plate (the usual normal

Casimir force) would lead to a torsional deflection (rotation) of the cantilever. The torsional spring constant of this cantilever k_{tor} was much greater than the bending spring constant k_{ben}, making it much more sensitive to detecting the lateral Casimir force, while simultaneously suppressing the effect of the normal Casimir force.

The calibration of the cantilever (i.e. the determination of k_{tor} and k_{ben}) and the measurement of the residual potentials between the sphere and the plate were done by electrostatic means, as described in Sections 19.2.1–19.2.3. Here, in order to measure k_{tor}, the sphere was kept grounded and various voltages were applied to the corrugated plate. Using the proximity force approximation, the electrostatic force normal to the plate takes the form (Chen et al. 2002a, 2002b)

$$F_{el,z}(a,\phi) = -\pi R\epsilon_0 \frac{(V-V_0)^2}{a} \frac{1}{\sqrt{1-\beta^2}}, \qquad (21.4)$$

where V is the voltage applied to the corrugated plate, V_0 is the residual potential of the grounded sphere, and

$$\beta = \frac{b(\phi)}{a} = \frac{1}{a}(A_1^2 + A_2^2 - 2A_1 A_2 \cos\phi)^{1/2}. \qquad (21.5)$$

When V was applied to the corrugated plate, the electrostatic force on the sphere led to a torsional rotation of the cantilever. From the electrostatic force at different values of V, the torsional spring constant was measured to be $k_{tor} = 0.138 \pm 0.005$ N/m, and $V_0 = -0.135$ V was also measured. The measurement of k_{ben} was done after the measurement of the Casimir force but is discussed here to preserve continuity. For this purpose, the sphere was moved away from the vertical corrugated plate and brought closer to the smooth plate, which was positioned horizontally at the bottom as shown in Fig. 21.2. Again, various voltages V were applied to the bottom plate, and the electrostatic force led to the normal bending of the cantilever. By fitting of the measured force to eqn (21.4) with $A_1 = A_2 = \beta = 0$ owing to the smooth surface, the value $k_{ben} = 0.0052 \pm 0.0001$ N/m was obtained. Note that $k_{tor} \gg k_{ben}$ is required. The extension of the piezoelectric tube in the x-direction as a function of the applied voltage was calibrated by optical interferometry (Chen and Mohideen 2001). The horizontal extension of the tube in the z-direction was calibrated with AFM standards.

The separation on contact of the two surfaces was determined from the measurement of the lateral electrostatic force. This can be found using the proximity force approximation (Chen et al. 2002b),

$$F_{el,x}(a,\phi) = 2\pi^2 R\epsilon_0(V-V_0)^2 \frac{A_1 A_2}{\Lambda a^2} \frac{\sin\phi}{\sqrt{1-\beta^2}(1+\sqrt{1-\beta^2})}. \qquad (21.6)$$

The measurement of the lateral electrostatic force was repeated 60 times with two different applied voltages. The surface separation on contact of the two corrugated surfaces was measured to be $a_0 = 186 \pm 38$ nm.

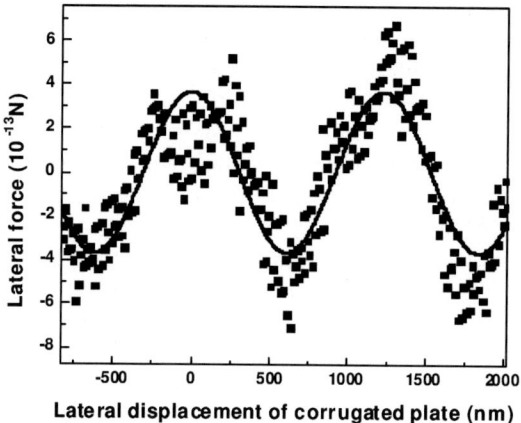

FIG. 21.4. The mean measured lateral Casimir force as a function of the lateral displacement is shown by solid squares. The solid line is the best-fit sine curve to the data, leading to a lateral-force amplitude of 0.32 pN (Chen et al. 2002b).

To measure the lateral Casimir force, the residual potential difference between the corrugated sphere and the plate was compensated by application of a voltage V_0 to the corrugated plate. The sphere was brought close to the corrugated plate and the separation distance was kept fixed. To measure the lateral force F^{lat} as a function of the phase shift ϕ, for a given sphere–plate separation, the corrugated plate was moved in the x-direction in average steps of 0.46 nm using the x-piezo and the lateral Casimir force was measured at each step. The measurement was repeated 60 times. The mean lateral Casimir force measured is shown by the solid squares in Fig. 21.4. The sinusoidal oscillations in the lateral Casimir force as a function of the phase difference ϕ expected from theoretical analysis (see the next section) are clearly observed. The periodicity of the lateral Casimir force oscillation is in agreement with the corrugation period of the plate. A sine curve fit to the observed data is shown by the solid line, with an amplitude $A_F = 3.2 \times 10^{-13}$ N. This corresponds to a separation distance between the two corrugated surfaces in the range from 218 to 221 nm (see the theoretical analysis in the next section).

The variance of the mean lateral-force amplitude is $s_{\bar{A}_F} = 0.22 \times 10^{-13}$ N. Using eqn (18.8), this leads to a random error determined at a 95% confidence level,

$$\Delta^{rand} A_F^{expt} = s_{\bar{A}_F} t_{(1+0.95)/2}(59) = 0.44 \times 10^{-13} \text{ N}. \quad (21.7)$$

The largest source of the systematic error was due to the resolution of the A/D board used in the data acquisition. This systematic error was $\Delta^{syst} A_F^{expt} = 0.33 \times 10^{-13}$ N. Combining the two errors in accordance with the statistical rule (18.17), one arrives at a total experimental error $\Delta^{tot} A_F^{expt} = 0.62 \times 10^{-13}$ N, determined at a 95% confidence probability. The resulting precision of the amplitude measurement at the closest point is around 19% [the larger value of 24% indicated in the original publication by Chen et al. (2002b) can be explained by the use of a nonrigorous combination rule for random and systematic errors).

The above lateral-force measurement was repeated for other surface separations. First, the separation between the sphere and the corrugated plate was increased by 12 nm with the z-piezo. The mean measured amplitude of the lateral force was 2.6×10^{-13} N. The separation distance was increased in 12 nm steps and the lateral Casimir force was measured for two more surface separations. The respective force amplitudes were found to be 2.1×10^{-13} N and 1.7×10^{-13} N. These data are consistent with an inverse fourth-power dependence of the lateral-force amplitude on a (Chen et al. 2002a, 2002b). Thus, the lateral Casimir force demonstrates a very different dependence on separation distance than the lateral electrostatic force (21.6), which has an inverse second-power dependence on a, which was checked independently.

21.3 Calculation of the lateral Casimir force in the configuration of a sphere above a plate

The starting point for the calculation of the lateral Casimir force between a corrugated sphere and a corrugated plate is the case of two parallel corrugated plates considered in Sections 17.5.1 and 17.5.2. In this case the surfaces of the corrugated plates are described by eqns (17.3) and (17.124). Using the proximity force approximation, the Casimir energy per unit area for the corrugated plates can be represented in the form

$$E_{\text{corr}}(a, \phi) = \frac{1}{\Lambda} \int_0^\Lambda E[z_2^{(\text{s})} - z_1^{(\text{s})}]\, dx. \tag{21.8}$$

Here, we integrate the Casimir energy per unit area for the configuration of two flat plates including the effect of the nonzero skin depth. We consider the latter effect in the framework of the plasma model. We can then use $E = E_{\text{p}}$ as given by eqn (13.8) because, at the separations in the experiment considered, the skin depth $\delta_0 \ll a$. The normal separation distance between opposite points of the corrugations on the two surfaces is

$$z_2^{(\text{s})} - z_1^{(\text{s})} = a + A_2 \sin\left(\frac{2\pi x}{\Lambda} + \phi\right) - A_1 \sin\frac{2\pi x}{\Lambda}. \tag{21.9}$$

This can be represented as

$$z_2^{(\text{s})} - z_1^{(\text{s})} = a + b(\phi) \cos\left(\frac{2\pi x}{\Lambda} - \alpha\right), \tag{21.10}$$

where the quantity $b(\phi)$ is defined in eqn (21.5) and

$$\tan \alpha = \frac{A_2 \cos\phi - A_1}{A_2 \sin\phi}. \tag{21.11}$$

Substituting eqns (13.8) and (21.10) into eqn (21.8) and integrating, we obtain

$$E_{\text{corr}}(a,\phi) = -\frac{\pi^2 \hbar c}{720 a^3} \sum_{n=0}^{4} c_n \left(\frac{\delta_0}{a}\right)^n X_n(\beta), \qquad (21.12)$$

where β is defined in eqn (21.5), and the following notation has been used:

$$X_0(\beta) = \frac{2+\beta^2}{2(1-\beta^2)^{5/2}}, \qquad X_1(\beta) = \frac{2+3\beta^2}{2(1-\beta^2)^{7/2}},$$

$$X_2(\beta) = \frac{8+24\beta^2+3\beta^4}{8(1-\beta^2)^{9/2}}, \qquad X_3(\beta) = \frac{8+40\beta^2+15\beta^4}{8(1-\beta^2)^{11/2}},$$

$$X_4(\beta) = \frac{16+120\beta^2+90\beta^4+5\beta^6}{16(1-\beta^2)^{13/2}}. \qquad (21.13)$$

Now the normal Casimir force between the corrugated plate and the corrugated sphere can be calculated approximately by use of the proximity force approximation, as

$$F_{\text{corr}}^{\text{norm}}(a,\phi) = 2\pi R E_{\text{corr}}(a,\phi), \qquad (21.14)$$

where $E_{\text{corr}}(a,\phi)$ is given in eqn (21.12). Note that eqn (21.14) is valid under the condition $a \ll R$, which is easily satisfied here. Regarding eqn (21.12), this equation was derived under the condition $2\pi a \ll \Lambda$ (see Section 17.5.1), which is not satisfied for the parameters used in this experiment. However, as shown below, in this particular case the proximity force approximation works well.

By integrating the normal force (21.14) with respect to the surface separation, the energy of a corrugated sphere above a corrugated plate is obtained. Then, by differentiating with respect to the phase shift, we get the lateral Casimir force acting between them,

$$F^{\text{lat}}(a,\phi) = -\frac{2\pi}{\Lambda}\frac{\partial}{\partial \phi}\int_a^{\infty} dz\, F_{\text{corr}}^{\text{norm}}(z,\phi). \qquad (21.15)$$

Substituting eqns (21.12)–(21.14) into eqn (21.15), after integration and differentiation, we finally obtain

$$F^{\text{lat}}(a,\phi) = \frac{\pi^4 R \hbar c}{120 a^4}\frac{A_1 A_2 \sin\phi}{\Lambda(1-\beta^2)^{5/2}}\left[1+\sum_{n=1}^{4} c_{n,x}\left(\frac{\delta_0}{a}\right)^n\right], \qquad (21.16)$$

where the expansion coefficients are given by

$$c_{1,x} = \frac{4+\beta^2}{3(1-\beta^2)}c_1, \qquad c_{2,x} = \frac{5(4+3\beta^2)}{12(1-\beta^2)^2}c_2, \qquad (21.17)$$

$$c_{3,x} = \frac{8+12\beta^2+\beta^4}{4(1-\beta^2)^3}c_3, \qquad c_{4,x} = \frac{7(8+20\beta^2+5\beta^4)}{24(1-\beta^2)^4}c_4.$$

It was shown (Chen et al. 2002b) that the corrections to eqn (21.16) due to surface roughness in the experiment under consideration were less than 1% and thus could be neglected. Note that eqn (21.16) takes into account all orders of perturbation theory with respect to the relative amplitudes A_i/a. However, this equation takes only approximate account of the material properties, up to the fourth-order term of the expansion series in the small parameter δ_0/a. This can lead to large errors at small separations. It is possible to take exact account of the material properties in accordance with the Lifshitz formula, but restrict oneself to the perturbative terms up to the fourth order in A_i/a. For this purpose, one must start with the representation of the Casimir energy per unit area (17.123) between two corrugated plates obtained in the framework of the proximity force approximation. Substituting this into eqn (21.14) instead of $E_{\text{corr}}(a, \phi)$ and performing the integration and differentiation in eqn (21.15), we get

$$F^{\text{lat}}(a, \phi) = 2\pi^2 R \frac{A_1 A_2}{\Lambda} \sin \phi \left[\frac{\partial E(a)}{\partial a} + \frac{1}{8} \frac{\partial^3 E(a)}{\partial a^3} b^2(\phi) \right]. \quad (21.18)$$

Before comparing eqn (21.16) for the lateral Casimir force with the experimental data, we shall derive an alternative theoretical expression using the method of pairwise summation. This method, as explained in Section 17.5.1, is applicable under the weaker condition $a \ll \Lambda$, which is met in the experiment. We start with eqn (17.40) as applied to ideal-metal corrugated plates, i.e. with $\eta_E(a) = 1$. Using the proximity force approximation, the normal force between the corrugated sphere and the corrugated plate is given by

$$F_{\text{corr}}^{\text{norm}} = -\frac{\pi^3 \hbar c R}{360 a^3} \kappa_E(a). \quad (21.19)$$

Note that the proximity force approximation is used only to relate the plate–plate to the sphere–plate configuration, whereas the corrugations are described only by using the pairwise summation method. Then, substituting eqn (21.19) into eqn (21.15), we find the lateral Casimir force betwen the corrugated sphere and corrugated plate,

$$F_{\text{IM}}^{\text{lat}}(a, \phi) = \frac{\pi^4 \hbar c R}{360 a^2 \Lambda} \frac{\partial}{\partial \phi} \left[\tilde{c}_{11}^{(s)} \frac{A_1 A_2}{a^2} + \tilde{c}_{13}^{(s)} \frac{A_1 A_2^3}{a^4} + \tilde{c}_{31}^{(s)} \frac{A_1^3 A_2}{a^4} + \tilde{c}_{22}^{(s)} \frac{A_1^2 A_2^2}{a^4} \right]. \quad (21.20)$$

The phase-dependent coefficients $\tilde{c}_{kl}^{(s)}$ are easily calculated by using eqns (17.131) and (17.132). They are given by

$$\tilde{c}_{11}^{(s)} = -3 \cos \phi \, S^{(1)}(a_\Lambda), \qquad \tilde{c}_{22}^{(s)} = \frac{15}{4}(2 + \cos 2\phi) S^{(2)}(2a_\Lambda),$$

$$\tilde{c}_{13}^{(s)} = \tilde{c}_{31}^{(s)} = -\frac{15}{2} \cos \phi \, S^{(2)}(a_\Lambda), \quad (21.21)$$

where

$$S^{(1)}(x) = e^{-x}\left(1 + x + \frac{1}{6}x^2 - \frac{1}{6}x^3\right) + \frac{1}{6}x^4 \Gamma(0, x),$$

$$S^{(2)}(x) = e^{-x}[1 + xQ_s(x)] - \frac{1}{480}x^6\Gamma(0,x), \qquad (21.22)$$

$$Q_s(x) = 1 + \frac{31}{80}x + \frac{13}{240}x^2 - \frac{1}{480}x^3 + \frac{1}{480}x^4.$$

Substituting eqns (21.21) and (21.22) into eqn (21.20), we arrive at

$$F_{\text{IM}}^{\text{lat}}(a,\phi) = \frac{\pi^4 \hbar c R}{120 a^4 \Lambda} A_1 A_2 \sin\phi \left[S^{(1)}(a_\Lambda) \right. \qquad (21.23)$$
$$\left. + \frac{5}{2}\frac{A_1^2 + A_2^2}{a^2} S^{(2)}(a_\Lambda) - 5\frac{A_1 A_2}{a^2}\cos\phi\, S^{(2)}(2a_\Lambda) \right].$$

Under the condition $a_\Lambda = 2\pi a/\Lambda \ll 1$, we have $S^{(1)} \approx S^{(2)} \approx 1$. In this case eqn (21.23) coincides with eqn (21.16) with $\delta_0 = 0$ (the ideal-metal case), expanded up to the terms of order β^2. To get the lateral Casimir force between a real material plate and sphere using the pairwise summation method, we must multiply eqn (21.23) by a correction factor defined by

$$-\int_a^\infty E(a')\,da' = \frac{\pi^2 \hbar c}{1440 a^2}\eta_E^{(\text{sp})}(a), \qquad (21.24)$$

where $E(a)$ is the Casimir energy per unit area of two parallel plates given by the Lifshitz formula (12.30). As a result, the lateral Casimir force between a corrugated sphere and plate made of real materials is given by

$$F^{\text{lat}}(a,\phi) = F_{\text{IM}}^{\text{lat}}(a,\phi)\eta_E^{(\text{sp})}(a). \qquad (21.25)$$

The interesting characteristic feature of eqns (21.16) and (21.25) is that the dependence of $F^{\text{lat}}(a,\phi)$ on ϕ is not exactly sinusoidal. This is because the quantity β in eqn (21.16) and the last term in the square brackets in eqn (21.23) also depend on ϕ which leads to some deviations from the exact sine function. The anharmonic dependence of the lateral Casimir force on the phase shift between the corrugations of both bodies was predicted by Chen et al. (2002b). This prediction awaits experimental confirmation.

Now we compare the computational results obtained by using eqns (21.16) and (21.25) with the experimental data. The measured amplitude of the lateral force at the closest sphere–plate separation a_1 is $A_F^{\text{expt}}(a_1) = 0.32\,\text{pN}$ (see Section 21.2). Using eqns (21.16) and (21.25), this corresponds to separation distances between the sphere and the plate equal to $a_1 = 221\,\text{nm}$ and $\tilde{a}_1 = 218\,\text{nm}$, respectively. Three other measurements of the lateral force were performed, each time with an increase in the separation of 12 nm. This was done with the z-piezo, so that the separation shifts of 12 nm are practically exact. This leads to the following values of the separations at which the measurements were performed: $a_2 = 233\,\text{nm}$, $a_3 = 245\,\text{nm}$, and $a_4 = 257\,\text{nm}$ [based on eqn (21.16)], or $\tilde{a}_2 = 230\,\text{nm}$, $\tilde{a}_3 = 242\,\text{nm}$, and $\tilde{a}_4 = 254\,\text{nm}$ [based on eqn (21.25)]. Computations using the proximity force approximation (21.16) result in $A_F^{\text{theor}}(a_2) =$

0.26 pN, $A_F^{\text{theor}}(a_3) = 0.21$ pN, and $A_F^{\text{theor}}(a_4) = 0.18$ pN. Alternatively, the pairwise summation method leads to $\tilde{A}_F^{\text{theor}}(\tilde{a}_2) = 0.25$ pN, $\tilde{A}_F^{\text{theor}}(\tilde{a}_3) = 0.20$ pN, and $\tilde{A}_F^{\text{theor}}(\tilde{a}_4) = 0.18$ pN. Both sets of theoretically computed lateral-force amplitudes are in very good agreement with the experimental data: $A_F^{\text{expt}} = 0.26$, 0.21, and 0.17 pN, respectively. We emphasize that two parameters of the experimental configuration are not small ($A_1/a_1 \approx 0.27$ and $A_1/\lambda_p \approx 0.43$). Specifically, the contribution of the fourth-order terms to the lateral-force amplitude is up to 18% of the second-order contribution. In the pairwise summation method, the contribution of the fourth-order terms is up to 22% of the second-order terms.

Recently a scattering approach for real metals described by the plasma model (see Section 17.4) was applied (Rodrigues et al. 2006b) to calculate the lateral Casimir force in the experimental configuration. Rodrigues et al. (2007) suggested that all of the experimental separations should be shifted by -20 nm in order to bring the measured force amplitudes into agreement with their theoretical approach based on the assumptions $A_{1,2}/a \ll 1$ and $A_{1,2}/\lambda_p \ll 1$, developed only up to order $A_1 A_2 / a^2$. This arbitrary shift of data is, however, without merit (Chen et al. 2007c) because, as demonstrated above, the higher-order terms in $A_1 A_2$ make a large contribution and cannot be neglected. In fact, as physical intuition suggests, the exact theoretical result taking proper account of both the nontrivial geometry and the material properties should lie between the predictions of the proximity force approximation and pairwise summation methods discussed above. More exact computations for sinusoidally corrugated surfaces within the scattering approach must be based on the general results of Chapter 10 without using the assumption that the corrugation amplitudes are small.

21.4 Control of the lateral Casimir force

Symmetric corrugations of sinusoidal shape were considered in both Section 17.5 and Section 21.3. Using the same methods, one can investigate the lateral Casimir force between two parallel plates covered with uniaxial grooves of arbitrary shape. This opens up opportunities to change the magnitude of the lateral Casimir force and obtain asymmetric lateral forces with a more complicated character of their equilibrium points. For the sake of simplicity, we consider the case of ideal-metal plates with the boundary functions f_1 and f_2 in eqn (17.3) depending on only one variable x. Let these functions be periodic with equal periods $\Lambda_1 = \Lambda_2 = \Lambda$, but have some phase shift x_0. In this case eqn (17.40) takes the form

$$E_R(a, x_0) = E_{\text{IM}}(a) \kappa_E(a, x_0), \qquad (21.26)$$

where κ_E is defined in eqn (17.39). The lateral Casimir force directed along the x-axis is

$$F^{\text{lat}}(a, x_0) = -\frac{\partial E_R(a, x_0)}{\partial x_0} S = P_{\text{IM}}(a) \frac{2 A_1 A_2}{a} S \qquad (21.27)$$

$$\times \left[2 \frac{\partial}{\partial x_0} \langle f_1 f_2 \rangle + 5 \left(\frac{A_1}{a} \frac{\partial}{\partial x_0} \langle f_1^2 f_2 \rangle - \frac{A_2}{a} \frac{\partial}{\partial x_0} \langle f_1 f_2^2 \rangle \right) \right.$$

$$+10\left(\frac{A_1^2}{a^2}\frac{\partial}{\partial x_0}\langle f_1^3 f_2\rangle - \frac{3}{2}\frac{A_1 A_2}{a^2}\frac{\partial}{\partial x_0}\langle f_1^2 f_2^2\rangle + \frac{A_2^2}{a^2}\frac{\partial}{\partial x_0}\langle f_1 f_2^3\rangle\right)\right].$$

This equation is applicable under the condition $2\pi a \ll \Lambda$.

Blagov et al. (2004) applied eqn (21.27) to the case of periodic sawtoothed corrugations with equal amplitudes [see Fig. 21.5(a)]. Within one period (from 0 to Λ for f_1 and from x_0 to $x_0 + \Lambda$ for f_2), the analytic representations of f_1 and f_2 are

$$f_1(x) = \frac{2x}{\Lambda} - 1, \qquad f_2(x) = 1 - \frac{2}{\Lambda}(x - x_0). \tag{21.28}$$

Calculating all of the mean values entering eqn (21.27), we arrive at

$$F^{\text{lat}}(a, x_0) = 8|P_{\text{IM}}(a)|\frac{A^2}{a\Lambda}\left(\frac{2x_0}{\Lambda} - 1\right)\left[1 + 10\frac{A^2}{a^2}\left(1 - 2\frac{x_0}{\Lambda} + 2\frac{x_0^2}{\Lambda^2}\right)\right]S. \tag{21.29}$$

As an illustration, Fig. 21.5(b) shows the dependence of $F^{\text{lat}}/(|P_{\text{IM}}|S)$ on x_0/Λ computed using eqn (21.29) with the values of parameters $A/a = 0.3$ and $a/\Lambda = 0.03$. Similarly to the case of sinusoidal corrugations, the lateral force in Fig. 19.26(b) is symmetric, and the points of unstable equilibrium $x_0/\Lambda = 0.5, 1.5, \ldots$ are midway between the points of stable equilibrium $x_0/\Lambda = 0, 1, 2, \ldots$. The lateral force is negative over one half of the period and positive over the other half. The magnitudes of the maximum and minimum values of the force are equal. At the same time, the case of a sawtoothed structure is different from that of sinusoidal corrugations because here the extreme values of the lateral force are obtained near the points of stable equilibrium, where the force is discontinuous. The points of stable equilibrium in this configuration are particularly stable. Even a small deviation from the stable equilibrium (where the value of the lateral force is taken as equal to zero, i.e. half the sum of the limiting values from the left and from the right) leads to a large lateral force, restoring the stable equilubrium. Blagov et al. (2004) also gave an example of a more asymmetric uniaxial corrugation that allowed one to obtain different magnitudes of the maximum and minimum values of the lateral Casimir force.

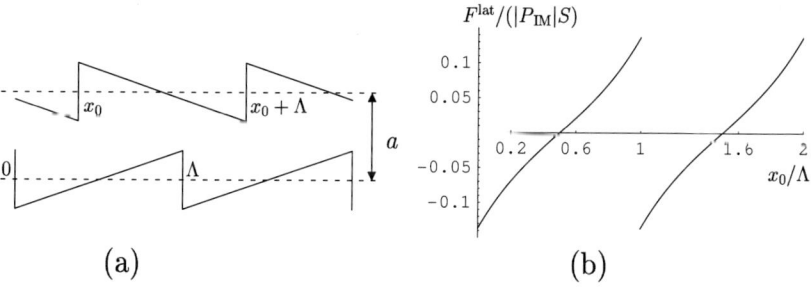

FIG. 21.5. (a) Periodic uniaxial sawtoothed corrugations. (b) Lateral Casimir force as a function of the phase shift (Blagov et al. 2004).

Periodic uniaxial corrugations of arbitrary shape on two parallel plates were considered by Emig (2007), and the lateral Casimir force was found for ideal-metal plates up to the second-order terms in the corrugation amplitudes. This force was shown to generate a ratchet effect, allowing periodic lateral motion of the plates. Ashourvan et al. (2007a, 2007b) suggested that the lateral Casimir force could be used in the configuration of a sinusoidally corrugated plate (rack) and a sinusoidally corrugated cylinder (pinion). The nonlinear dynamics of a pinion was studied for the case of a vibrating rack. Configurations with corrugated boundaries are attracting attention in connection with the possibility to experimentally investigate Casimir torques (see Section 12.8.2). One proposal is to observe the Casimir torque in a configuration of two corrugated Au plates with a small angle θ between the corrugation directions (Rodrigues et al. 2006a). From eqn (21.3), the lateral Casimir force and related torque are observable for an angle between the corrugation axes $\theta \ll \Lambda/L$, where L is the size of the plate along the corrugation axis. The optimum values of the corrugation period Λ and the plate separation were found in order to get larger values of the Casimir torque (Rodrigues et al. 2006a). Another experimental scheme, aimed at observing the Casimir torque using anisotropic test bodies, was proposed by Munday et al. (2005). The schematic of the setup includes an anisotropic disk placed above a barium titanate plate immersed in ethanol. The dielectric permittivities of the three materials are chosen in such a way that the Casimir force between the anisotropic bodies is repulsive (see Section 19.5.3). The disk would float parallel to the plate at a distance where its weight is counterbalanced by the Casimir repulsion, and would be free to rotate in response to the small torque. Detailed numerical calculations were performed demonstrating the feasibility of this experiment (Munday et al. 2005). Both of these experiments, if successfully performed, will provide important new information about the dispersion forces between real materials.

21.5 Experiment with a sphere above rectangular trenches

The two experiments considered in this chapter so far used sinusoidally corrugated boundaries. Chan et al. (2008) reported a measurement of the Casimir force between a gold-coated sphere of radius R and a silicon surface that had been structured with nanoscale rectangular corrugations (trenches). Measurements were performed in the dynamic regime using a micromechanical torsional oscillator (see Section 19.3). This means that the directly measured quantity was the change of the resonant frequency of the oscillator, which is proportional to the derivative of the Casimir force with respect to the separation distance. Using the proximity force approximation given by eqn (19.15), this derivative is equal to

$$F'(a) = -2\pi R P(a), \qquad (21.30)$$

where $P(a)$ is the Casimir pressure between one rectangularly corrugated plate and one plane plate (see Fig. 21.6).

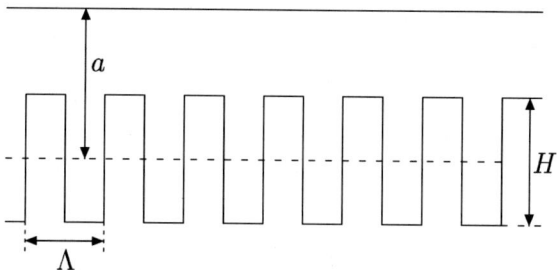

FIG. 21.6. The periodic uniaxial rectangular corrugations on one of the plates.

For the ideal-metal case, such a configuration was considered by Büscher and Emig (2004) using the exact methods presented in Chapter 10. It was shown that in the limiting case $a \ll \Lambda$ [recall that the separation is measured between the zero levels of the corrugations of the lower and upper plates defined in eqn (17.92)], the Casimir pressure when the width of a rectangular trench is equal to half the period is given by

$$P(a) = -\frac{\pi^2 \hbar c}{240} \frac{1}{2} \left\{ \frac{1}{[a - (H/2)]^4} + \frac{1}{[a + (H/2)]^4} \right\}. \qquad (21.31)$$

This is in fact the minimum value of the Casimir pressure between rectangular corrugated and plane plates. Under the condition $a \ll \Lambda$, for an ideal metal, the same result is obtained from the application of the proximity force approximation or the pairwise summation method (see Section 17.2).

In the opposite limit $a \gg \Lambda$, the exact result is given by (Büscher and Emig 2004)

$$P(a) = -\frac{\pi^2 \hbar c}{240} \frac{1}{[a - (H/2)]^4}, \qquad (21.32)$$

which is up to a factor of 2 larger in magnitude than the prediction of the proximity force approximation and the pairwise summation method in eqn (21.31). This is the maximum value of the Casimir pressure in the configuration of rectangular corrugated and plane plates made of an ideal metal. Note that the interpretation of the previous experiment on geometry dependence, considered in Section 21.1 (Roy and Mohideen 1999, Klimchitskaya et al. 2001), in the case of sinusoidal corrugations, was complicated by the presence of the lateral Casimir force. In the experiment by Chan et al. (2008), lateral movements between the surfaces were avoided by keeping the corrugation axis perpendicular to the torsional axis of the micromechanical torsional oscillator.

The trenches were fabricated in p-doped silicon (the density of charge carriers, 2×10^{18} cm^{-3}, was determined from the dc conductivity, equal to $0.028\,\Omega$ cm). Silicon oxide was used as the mask. Deep UV lithography followed by reactive ion etching was used to transfer the pattern. Trenches of depth $H \approx 1\,\mu$m were created. Three types of samples were made on the same wafer: sample A, with a period $\Lambda_A = 1\,\mu\text{m}$; sample B, with a period $\Lambda = 400\,\text{nm}$; and one sample with a flat surface. The sidewalls of the trenches were described as forming angles with 90.3° and 91.0° with the top surface in samples A and B, respectively, and the bottom corners were reported to have some degree of rounding in comparison with the top corners. Perfect corners were assumed in all of the analysis. Fractional areas of the top surfaces of the unetched regions were found to be $p_A = 0.478 \pm 0.002$ and $p_B = 0.510 \pm 0.001$ for the samples A and B using a scanning electron microscope. The residual hydrocarbons were removed by oxygen plasma etching and the oxide mask was etched with HF. Samples of size $0.7 \times 0.7\,\text{mm}^2$ were used for the force measurement. The corrugated silicon surface was passivated as in the papers by Chen et al. (2006a, 2006b) using hydrogen fluoride. However, after this the silicon chip was baked to 120 °C to remove the residual water from the bottom of the trenches, which might have also desorbed the passivation layer.

The torsional oscillator used in the measurements consisted of a $3.5\,\mu\text{m}$ thick, $500\,\mu\text{m}$ square silicon plate. Unlike in previous experiments with torsional oscillators, the spheres were attached to the plate. Two glass spheres of radius $R = 50\,\mu\text{m}$, sputter-coated with gold of about 400 nm thickness, were attached on top of each other to the torsional oscillator at a distance of $b = 210\,\mu\text{m}$ using conductive epoxy. The two spheres were used to provide a large distance between the corrugated surface and the top of the torsional oscillator. The resonant frequency ($\omega_0 = 2\pi \times 1783\,\text{Hz}$, quality factor $Q = 32\,000$) was excited by applying a voltage to one of the bottom electrodes. The oscillations were detected with additional voltages with an amplitude of 100 mV and a frequency of $2\pi \times 102\,\text{kHz}$, which were applied to measure the capacitance change between the top plate and the bottom electrodes. A phase-locked loop was used to detect the change in the resonant frequency as a function of the distance between the sphere and the corrugated plate, which was varied using a closed-loop piezoelectric actuator.

The measurements were done in a vacuum of 10^{-6} Torr using a dry roughing pump and a turbopump. The residual potential difference V_0 and the initial separation between the surfaces a_0 were determined, and a calibration was done using electrostatic forces. No value of a_0 or errors in its determination were provided. A residual potential difference $V_0 \sim -0.43\,\text{V}$ was found between the sphere and the flat Si plate and was noted to vary by 3 mV over the range of separations from 100 nm to $2\,\mu\text{m}$. Whether the same variation was found for the corrugated Si surfaces was not mentioned. Voltages between $V_0 + 245\,\text{mV}$ and $V_0 + 300\,\text{mV}$ were applied, and the calibration constant was found to be $628 \pm 5\,\text{m N}^{-1}\,\text{s}^{-1}$. Since no analytic expression for the electrostatic force is available for a trench geometry, a 2D numerical solution of the Poisson equation was used to calculate the electrostatic energy between a flat plate and the trench surface. This energy was then converted to a force between a sphere and the trench surface using the PFA. The electrostatic force was found to be insensitive to even 10% deviations in the trench depth.

The Casimir force gradients between the flat plate and samples A and B were measured after the application of compensating voltages to the plates. The main uncertainty in these measurements was reported to be that coming from thermomechanical noise, with a value of about 0.64 pN μm^{-1} at $a = 800$ nm. The phase-locked loop prevented operation for separation distances below 650 nm. The Casimir force gradient between the flat plate and the gold sphere F'_{flat} was first measured. Good agreement was found with calculation results using the Lifshitz theory. The tabulated data for gold and silicon (Palik 1985) were used, along with a modification corresponding to the carrier density of the silicon (Chen et al. 2006a). The roughness correction was taken into account using an rms roughness of 4 nm on the sphere and 0.6 nm on the silicon surface, measured using an AFM (Klimchitskaya et al. 1999). The larger roughness on the sphere was possibly due to the use of sputtering for the gold coating. Next, the force gradients $F'_{A,\text{expt}}$ and $F'_{B,\text{expt}}$ were measured on the corrugated surfaces using the same gold sphere.

As discussed in Section 17.2, for real material bodies the proximity force approximation and the pairwise summation method lead to different results even if only one body is covered with corrugations. For the configurations used by Chan et al. (2008), one can neglect the contribution of the remote bottom parts of the trenches. Then the proximity force approximation leads to the following force gradients for the samples A and B:

$$F'_{A,\text{PFA}}(a) = -2\pi R p_A P_R \left(a - \frac{H}{2}\right), \qquad F'_{B,\text{PFA}}(a) = -2\pi R p_B P_R \left(a - \frac{H}{2}\right), \tag{21.33}$$

where $P_R(a)$ is the Casimir pressure between two noncorrugated plates covered with a stochastic roughness, calculated using the Lifshitz formula.

To compare the experimental data with theory, Chan et al. (2008) considered the ratios

$$\rho_A = \frac{F'_{A,\text{expt}}}{F'_{A,\text{PFA}}}, \qquad \rho_B = \frac{F'_{B,\text{expt}}}{F'_{B,\text{PFA}}}. \tag{21.34}$$

It was shown that for sample A there were deviations of ρ_A from unity of up to 10% over the measurement range from $a = 650$ to 750 nm, exceeding the experimental errors. For sample B, there were deviations of ρ_B from unity of up to 20% over the same measurement range. This difference in the results of the comparison of the experimental data for the samples A and B with the PFA results is natural, as $a/\Lambda_A = 0.7$ and $a/\Lambda_B = 1.75$ at the typical separation considered, $a = 700$ nm. Thus, for sample B the applicability condition of the PFA, $a/\Lambda \ll 1$, is violated to a larger extent than for sample A.

The measurements were repeated three times for each sample, and consistent results were reported to have been observed. The data were also compared with values obtained from the path-integral approach for ideal-metal boundaries, which was converted to the case of a sphere and a trenched plate using the PFA.

However, the measured deviations from the PFA, as applied to the rectangular corrugations, were found to be 50% less than that expected for ideal-metal boundaries. This discrepancy was reported as being quite natural, owing to the interplay of the nonzero skin depth and geometrical effects.

In conclusion, the experiment by Chan et al. (2008) reported the measurement of a deviation resulting from the geometry for corrugated rectangular trenches of relatively small period. The depth of the trenches allowed good comparison with the results obtained using the proximity force approximation, taking into account only the fractional area of the top surface. Deviations of 10–20% from the PFA were reported. At present, no theoretical computations exist which would allow a comparison between experiment and theory for spherical and corrugated surfaces made of real metals at room temperature. For $T = 0$, such computations have been performed by Lambrecht and Marachevsky (2008) within the scattering approach. The metal was described using a simple plasma model. In future similar computations can be performed at nonzero temperature for metals described by the generalized plasma-like model.

22

MEASUREMENTS OF THE CASIMIR–POLDER FORCE

The influence of the Casimir–Polder force on the movement of atoms through a narrow cavity has been well investigated. Comparison of the measurement data with corresponding computational results allows one to verify the predictions of the Lifshitz theory concerning the character of the interaction potential (see Section 18.1.6, where one of the first of such experiments is discussed). Modern laboratory techniques make possible the investigation of the role of the Casimir–Polder force in experiments on Bose–Einstein condensation and quantum reflection. Thus, Antezza *et al.* (2004) demonstrated that the collective oscillations of a Bose–Einstein condensate of ultracold atoms provide a sensitive probe of Casimir–Polder forces. Later, the first measurement of the temperature dependence of the Casimir–Polder force was performed in this way (Obrecht *et al.* 2007). The Casimir–Polder interaction plays an important role in the scattering of atoms on various surfaces. Of special interest are situations where the wave nature of an atom becomes dominant in comparison with its classical behavior as a particle (this is referred to as *quantum reflection*). In this chapter, both types of experiments are considered in connection with the properties of the Casimir–Polder force.

22.1 Measurement of the thermal Casimir–Polder force

The present section contains the results of the experiment by Obrecht *et al.* (2007) on the measurement of the Casimir–Polder interaction between ^{87}Rb atoms and a SiO_2 plate, and a comparison of these results with various theoretical approaches in the framework of the Lifshitz theory. This experiment was the first one in Casimir physics where the thermal effect has really been measured. The experimental data were found to be in a very good agreement with theory if the dc conductivity of the SiO_2 plate was disregarded. However, if the dc conductivity of SiO_2 was included in the calculation, the theoretical results were shown to be inconsistent with the data (Klimchitskaya and Mostepanenko 2008b). Thus, this experiment provides additional confirmation of the rule on how to apply the Lifshitz theory to dielectric materials discussed in Sections 12.5.2 and 20.3.5.

22.1.1 *Measurement scheme and technique*

The experiment by Obrecht *et al.* (2007) was an indirect dynamic measurement of the Casimir–Polder force acting between approximately 2.5×10^5 ^{87}Rb ground state atoms, belonging to a Bose–Einstein condensate, and a dielectric substrate (a fused silica plate of size $2 \times 8 \times 5\,\text{mm}^3$ in the z, y, and x directions, respectively, where z is perpendicular to the plate). The substrate was placed on top of a monolithic Pyrex glass holder inside a Pyrex glass cell, which composed the vacuum chamber. The condensate was produced in a magnetic trap with frequencies of $\omega_0 = 2\pi\nu_0$, and $\nu_0 = 229\,\text{Hz}$ and $6.4\,\text{Hz}$, in the perpendicular and longitudinal directions, respectively. This resulted in respective Thomas–Fermi radii $R_z = 2.69\,\mu\text{m}$ and $R_l = 97.1\,\mu\text{m}$. The back face of the fused silica substrate (opposite to that interacting with the ^{87}Rb atoms) was painted with a $100\,\mu\text{m}$ thick opaque layer of graphite and treated in a high-temperature oven before placing in the vacuum chamber. By illuminating the graphite layer with laser light from an 860 nm laser, it was possible to vary the temperature of the substrate. The vacuum chamber used reached a residual gas pressure of about 3×10^{-11} torr in the region of the condensate. The temperature of the substrate was varied while the vacuum chamber walls were maintained at room temperature.

The cloud of ^{87}Rb atoms was shifted a distance a, in the range from 7 to $15\,\mu\text{m}$, from the substrate by the imposition of a vertical bias magnetic field. Then the condensate cloud was resonantly driven into a dipole oscillation by an oscillatory magnetic field (Obrecht et al. 2007). An oscillation amplitude $A_z = 2.50\,\mu\text{m}$ in the z direction was chosen, and was kept constant in all measurements. The unperturbed trap frequency ω_0 was measured at a separation $a = 15\,\mu\text{m}$ between the substrate and the center of mass of the condensate; this separation was sufficiently large to avoid the influence of the Casimir–Polder force. Then measurements of the perturbed oscillation frequencies ω_z were performed at separations a from 7 to $11\,\mu\text{m}$. The separation distances between the center of mass of the condensate and the plate were measured by means of an absorption imaging technique (Harber et al. 2005). For this purpose, ^{87}Rb atoms were illuminated with a light beam perpendicular to the long axis of the condensate. The fractional frequency difference

$$\gamma_z = \frac{|\omega_0 - \omega_z|}{\omega_0} \approx \frac{|\omega_0^2 - \omega_z^2|}{2\omega_0^2} \tag{22.1}$$

was measured as a function of a both in and out of thermal equilibrium.

In thermal equilibrium, the temperature of the fused silica plate T_P was equal to the temperature T_E of the environment: $T_P = T_E = 310\,\text{K}$. Out of thermal equilibrium, two sets of measurements were performed, for $T_P = 479\,\text{K}$ and $T_E = 310\,\text{K}$ and for $T_P = 605\,\text{K}$ and $T_E = 310\,\text{K}$.

Harber et al. (2005) carefully estimated the random, systematic, and total errors in the measured values of γ_z in similar experiments at a 66% confidence level. Obrecht et al. (2007) performed a corresponding analysis for each experimental point separately. In the next two subsections, where the experimental data obtained are compared with various theoretical approaches, the total absolute errors in the measurement of the separation distances a and of the relative frequency shifts γ_z are represented as crosses shown to true scale. The main sources of systematic errors discussed by Harber et al. (2005) were connected with the possible presence of spatially inhomogeneous electric or magnetic surface contamination or uniform magnetic and electric fields. Special investigations were done to obtain upper bounds on all systematic errors.

22.1.2 Comparison with theory in thermal equilibrium

The experimental data for γ_z obtained by Obrecht et al. (2007) in thermal equilibrium at separations below $10\,\mu$m are shown by crosses in Fig. 22.1 (the measurement results at $a = 11\,\mu$m are not statistically meaningful). To compare this data with theory, we need to calculate the change of the center-of-mass oscillation frequency of a Bose–Einstein condensate under the influence of the Casimir–Polder force $F^A(a,T)$ given in eqn (16.17) acting between the ^{87}Rb atoms and the fused silica wall. Using the description of a dilute gas trapped by means of a harmonic potential, we arrive at the mechanical problem solved by Antezza et al. (2004), with the result

$$\omega_0^2 - \omega_z^2 = -\frac{\omega_0}{\pi A_z m_a} \int_0^{2\pi/\omega_0} d\tau \cos(\omega_0 \tau) \qquad (22.2)$$
$$\times \int_{-R_z}^{R_z} dz\, n_z(z) F^A[a + z + A_z \cos(\omega_0 \tau), T].$$

Here, $m_a = 1.443 \times 10^{-25}$ kg is the mass of a rubidium atom, A_z is the amplitude of the center-of-mass oscillations in the z-direction, z is the vertical coordinate of each individual atom in the condensate cloud measured from its center of mass, and the distribution function of the gas density is

$$n_z(z) = \frac{15}{16 R_z}\left(1 - \frac{z^2}{R_z^2}\right)^2. \qquad (22.3)$$

The expansion of the function F^A in eqn (22.2) up to terms linear in A_z leads to the result (Antezza et al. 2004)

$$\omega_0^2 - \omega_z^2 \approx -\frac{1}{m_a} \int_{-R_z}^{R_z} n_z(z) \frac{\partial F^A(a+z,T)}{\partial a} dz, \qquad (22.4)$$

where terms proportional to A_z^2 and higher powers of A_z have been omitted on the right-hand side. In the form (22.4), the frequency shift of the oscillator under the influence of the Casimir–Polder force is analogous to the previously used eqns (19.11), (19.14), and (19.22) describing dynamic measurements of the Casimir force and Casimir pressure in sphere–plate and plate–plate configurations. However, in the case of the present experiment one must not use the linear approximation (22.4) but, instead, perform all calculations with the exact eqn (22.2), because the oscillation amplitude A_z is rather large.

Now we are in a position to discuss the computations of γ_z under various assumptions about the conductivity of fused silica. Following Antezza et al. (2004),

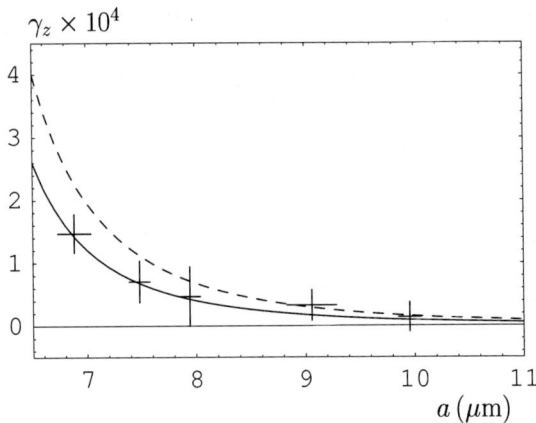

FIG. 22.1. Fractional change in the trap frequency versus separation in thermal equilibrium with $T_P = T_E = 310$ K, computed by neglecting (solid line) and including (dashed line) the conductivity of the dielectric substrate. The experimental data are shown by crosses (Klimchitskaya and Mostepanenko 2008b).

the static atomic polarizability of rubidium atoms $\alpha(0) = 4.73 \times 10^{-23}$ cm^{-3} was used in the computations. As discussed in Section 16.4.2, this allows one to obtain highly accurate results at the large separations under consideration. For fused silica, $\varepsilon(i\xi_l)$ as a function of ξ_l can be taken to be as in Fig. 12.2(b). This corresponds to the neglect of the dc conductivity and to the assumption that fused silica is a perfect insulator with a finite static permittivity $\varepsilon_0 = \varepsilon(0) = 3.81$.

Computations were performed by using eqns (22.1)–(22.3) and (16.17). In doing so, the averaging procedures could be performed analytically. By integrating with respect to z and τ, the quantity γ_z can be represented in the form (Klimchitskaya and Mostepanenko 2008b)

$$\gamma_z = \frac{1}{m_a A_z \omega_0^2} \left| \Phi_e^A(a, T) \right|, \qquad (22.5)$$

where

$$\Phi_e^A(a, T) = -2k_B T \alpha(0) \left[\frac{\varepsilon_0 - 1}{\varepsilon_0 + 1} \int_0^\infty k_\perp^3 dk_\perp e^{-2k_\perp a} I_1(2k_\perp A_z) g(2k_\perp R_z) \right.$$

$$\left. + \sum_{l=1}^\infty \int_0^\infty k_\perp dk_\perp h(\xi_l, k_\perp) e^{-2q_l a} I_1(2q_l A_z) g(2q_l R_z) \right], \qquad (22.6)$$

$$g(z) \equiv \frac{15}{z^5} \left[(3 + z^2) \sinh z - 3z \cosh z \right],$$

$$h(\xi_l, k_\perp) \equiv \left(2q_l^2 - \frac{\xi_l^2}{c^2} \right) r_{\text{TM}}^{(1)}(i\xi_l, k_\perp) - \frac{\xi_l^2}{c^2} r_{\text{TE}}^{(1)}(i\xi_l, k_\perp),$$

and $I_1(z)$ is a Bessel function. The computational results obtained in this way for γ_z are shown in Fig. 22.1 by the solid line (Obrecht et al. 2007, Klimchitskaya and Mostepanenko 2008b). Note that the results computed using $\varepsilon(i\xi_l) = \varepsilon_0$ (Obrecht et al. 2007) and with a frequency-dependent dielectric permittivity (Antezza et al. 2004, Klimchitskaya and Mostepanenko 2008b) are almost coincident at large separation distances (only small deviations are observed at $a < 8\,\mu$m). As is seen in Fig. 22.1, the theoretical computations are in excellent agreement with the data.

Although fused silica is a good insulator, at nonzero temperature it possesses a nonzero dc conductivity. The electrical conductivity of fused silica is ionic in nature and is determined by the concentration of impurities (alkali ions), which are always present as trace constituents. At $T_P = T_E = 310$ K, the conductivity varies within a wide region from $10^{-9}\,\mathrm{s}^{-1}$ to $10^2\,\mathrm{s}^{-1}$ (Bansal and Doremus 1986, Shackelford and Alexander 2001).

The inclusion of the dc conductivity in the model of the dielectric response, as in eqn (12.129), dramatically affects the calculational results. This changes the value of the reflection coefficient $r_{\mathrm{TM}}(0, k_\perp)$ and, consequently, the Casimir–Polder force (16.19), which leads to a corresponding change in the magnitude of γ_z computed using eqns (22.2) and (22.3). We emphasize that this change does not depend on the value of the conductivity σ_0 in eqn (12.129), but only on the fact that it is nonzero. The respective computational results for γ_z are shown in Fig. 22.1 by the dashed line. As is seen in this figure, the first two experimental points are in clear disagreement with the theory taking the conductivity of fused silica into account.

22.1.3 Comparison with theory out of thermal equilibrium

The experimental data for γ_z obtained by Obrecht et al. (2007) out of thermal equilibrium are shown by crosses in Fig. 22.2(a) ($T_P = 479$ K) and Fig. 22.2(b) ($T_P = 605$ K). In this case eqns (22.2)–(22.4) for the oscillator frequency shift remain valid, with the replacement of the Casimir–Polder force $F^A(a, T)$ with the force $F^A(a, T_P, T_E)$ obtained from eqn (16.94), acting between an atom and a plate in an out-of-thermal-equilibrium physical situation. Performing integration with respect to z and τ, we arrive at (Klimchitskaya and Mostepanenko 2008b)

$$\gamma_z = \frac{1}{m_a A_z \omega_0^2} \left| \Phi_e^A(a, T_E) + \Phi_n^A(a, T_P) - \Phi_n^A(a, T_E) \right|. \tag{22.7}$$

Here, the function $\Phi_e^A(a, T)$ is defined in eqn (22.6) and the function $\Phi_n^A(a, T)$ is expressed by eqn (16.95), where the function $f(\omega, x)$ is replaced with

$$\tilde{f}(\omega, x) = f(\omega, x) I_1\left(\frac{2A_z \omega x}{c}\right) g\left(\frac{2R_z \omega x}{c}\right). \tag{22.8}$$

The computational results for γ_z in the nonequilibrium situation, obtained by neglecting the dc conductivity of fused silica (Obrecht et al. 2007, Klimchitskaya and Mostepanenko 2008b) are presented in Fig. 22.2(a) ($T_P = 479$ K) and

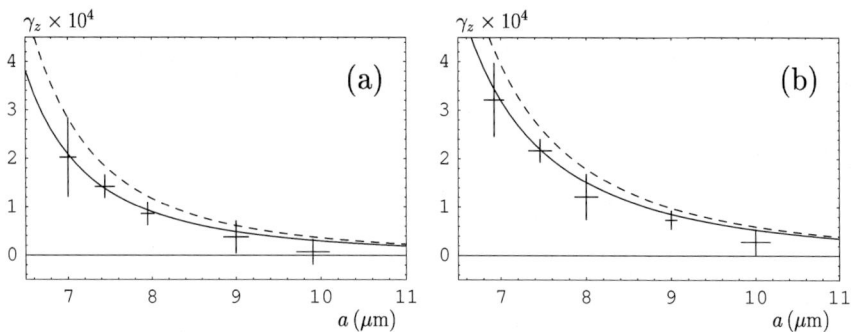

FIG. 22.2. Fractional change in the trap frequency versus separation for the out-of-thermal-equilibrium case (a) with $T_P = 479$ K and $T_E = 310$ K and (b) $T_P = 605$ K and $T_E = 310$ K. Computations were done by neglecting (solid line) and including (dashed line) the conductivity of the dielectric substrate. The experimental data are shown by crosses (Klimchitskaya and Mostepanenko 2008b).

Fig. 22.2(b) ($T_P = 605$ K) by solid lines. Note that the frequency dependence of $\varepsilon(\omega)$ does not affect the contribution to the frequency shift from the nonequilibrium terms in the total atom–wall force (16.94) (Antezza et al. 2005, Antezza 2006). As is seen in Fig. 22.2, the computations, performed with the neglect of the dc conductivity of fused silica, are in excellent agreement with the data.

The situation changes significantly when the dc conductivity of the substrate material is included. Direct computations using eqn (22.7) show that in the nonequilibrium case the disagreement between the experimental data and the theory widens further. The respective results are presented in Fig. 22.2 by the dashed lines (Klimchitskaya and Mostepanenko 2008b). As is seen in Fig. 22.2(a), the three experimental points for $T_P = 479$ K exclude the dashed line and the other two only touch it. The dashed line in Fig. 22.2(b) demonstrates that all data for $T_P = 605$ K exclude the theoretical prediction calculated with the inclusion of the dc conductivity of fused silica. Thus, the confidence with which the theoretical approach based on eqn (12.129) is excluded by the data increases with an increase of substrate temperature. A comparison of the complete set of data, as given by the crosses in Figs. 22.1 and 22.2, with the dashed lines shows that the inclusion of the static conductivity of fused silica in computations of the Casimir–Polder force is inconsistent with the measurement data of the experiment by Obrecht et al. (2007). This conclusion is consistent with a related but different experiment on the measurement of the Casimir force between a semiconductor plate and a gold sphere (see Section 20.3.5), where it was found that the dc conductivity of the semiconductor plate with a doping concentration below the critical value must be neglected.

We emphasize that the inclusion of the dc conductivity of SiO_2 in the model of the dielectric response does not affect the contributions to the frequency shift arising from the nonequilibrium terms $\Phi_n^A(a,T)$ in eqn (22.7) (Klimchitskaya and Mostepanenko 2008b). The magnitudes of $\Phi_n^A(a,T)$ computed with different values of σ_0 from 0 to $10^3\,\text{s}^{-1}$ coincide up to six significant figures. Thus, the conductivity influences the computational results only through the equilibrium term $\Phi_n^A(a,T)$.

It is pertinent to note that the modification of the transverse magnetic reflection coefficient considered in Section 16.4.3 (Pitaevskii 2008) was suggested just to describe the role of the free charge carriers in the experiment performed by Obrecht et al. (2007). However, as noticed in Section 22.1.2, the conductivity of fused silica and, consequently, charge carrier density at room temperature varies over a wide range. Because of this, from the experiment by Obrecht et al. (2007) it is not possible to discriminate within the limits of the experimental errors between the predictions of the standard Lifshitz theory with the dc conductivity neglected and a theoretical approach with a modified reflection coefficient $r_{\text{TM}}(0,k_\perp)$. As was discussed above, the latter approach is inconsistent with thermodynamics (Section 16.4.3) and is excluded by the experimental data from a more precise measurement of the difference Casimir force between an Au sphere and Si plate illuminated with laser pulses (Section 20.3.5).

22.2 Experiments on quantum reflection

Quantum reflection is a process in which a particle moving through a classically allowed region is reflected by a potential without reaching a classical turning point. In this section, we consider the quantum reflection of atoms interacting with a cavity wall. Usually, the interpretation of experimental data on quantum reflection is done with the help of a phenomenological potential which provides an interpolation between the van der Waals and Casimir–Polder interaction energies. Below, we discuss the exactness of the commonly used phenomenological potential by comparing it with more exact interaction energies obtained on the basis of the Lifshitz theory.

22.2.1 Main experimental results

The possibility of reflection of an ultracold atom under the influence of an *attractive* atom–wall interaction was predicted long ago on quantum-mechanical grounds [see e.g. Devonshire (1936)]. However, the experimental observation of this phenomenon has become possible only recently, owing to success in the production of ultracold atoms. First, it was investigated using liquid surfaces, by the reflection of He and H atoms from liquid He (Nayak et al. 1983, Berkhout et al. 1989) and on the basis of the sticking coefficient of H atoms on liquid He (Doyle et al. 1991, Yu et al. 1993). Later, the specular reflection of very slow metastable Ne atoms from Si and BK7 glass surfaces was studied (Shimizu 2001). The observed velocity dependence was explained by quantum reflection which was caused by the attractive Casimir–Polder interaction.

Quantum reflection becomes efficient when the motion of the particle can no longer be treated semiclassically (Friedrich and Trost 2004). The behavior of the particle is of quantum character when

$$\frac{\partial \lambda_B(z)}{\partial z} \geq 1, \qquad (22.9)$$

where $\lambda_B(z) = 2\pi\hbar/\sqrt{2m[E - V(z)]}$ is the local de Broglie wavelength for a particle of mass m and initial kinetic energy E moving in the potential $V(z)$. The same condition can be formulated as the condition that the variation of the local wave vector $k = 2\pi/\lambda_B(z)$ perpendicular to the surface, within a distance equal to the atomic de Broglie wavelength, is larger than k itself (Shimizu 2001),

$$\Phi = \frac{1}{k^2}\frac{dk}{dz} > 1. \qquad (22.10)$$

The reflection amplitude depends critically on the energy $V(z)$ of the atom–wall interaction. This has attracted considerable attention to the theoretical investigation of the dependence of the reflection amplitude on the atomic energy and the form of the interaction potential (Friedrich et al. 2002, Jurisch and Friedrich 2004, Madroñero and Friedrich 2007, Voronin and Froelich 2005, Voronin et al. 2005). The reflection probability tends to unity as the incident velocity tends to zero. Thus, a high probability of quantum reflection calls for small incident velocities, i.e. for cold atoms.

Advances in cooling techniques in the past decade have made it possible to perform experiments with cold atoms interacting with solid surfaces. Quantum reflection of ^{3}He atoms in scattering from an α-quartz crystal was observed at energies far from $E \to 0$ (Druzhinina and DeKieviet 2003). The observation of large reflection amplitudes for a dilute Bose–Einstein condensate of ^{23}Na atoms on an Si surface (Pasquini et al. 2004, 2006) allows the possibility of using quantum reflection as a trapping mechanism.

Practically all papers devoted to the investigation of quantum reflection [see e.g. Shimizu (2001), Friedrich et al. (2002), Druzhinina and DeKieviet (2003), Oberst et al. (2005)] use a phenomenological potential for the atom–wall interaction at zero temperature to calculate the reflection amplitude. In fact, the reflection amplitude depends critically on the form of this potential and on the atomic energy (Friedrich et al. 2002, Jurisch and Friedrich 2004, Madroñero and Friedrich 2007, Voronin and Froelich 2005, Voronin et al. 2005). The most often used phenomenological potential (interaction energy) has the form

$$E^A(a) = -\frac{C_4}{a^3(a+l)}, \qquad (22.11)$$

where l is a characteristic parameter with the dimensions of length that depends on the material. It is assumed that at short separations $a \ll l$ (typically at separations of the order of a few nanometers), $E^A(a)$ coincides with the van der Waals potential (16.24), so that $C_4 = lC_3$. Coincidence between $E^A(a)$ and the retarded Casimir–Polder potential (16.28) is achieved at separations of about

10 μm, where l is negligibly small in comparison with a. At such large separations, the correction factor to the Casimir–Polder energy due to the nonzero skin depth and the dynamic atomic polarizability is almost equal to unity. Comparison of computational results with measurement data for the reflection amplitudes allows one to estimate the parameters of the potential (22.11) (Druzhinina and DeKieviet 2003, Oberst et al. 2005). The increased precision of the measurements opens up new opportunities for a comparison with the more exact results for the energy of the atom–wall interaction given by the Lifshitz theory. Such a comparison could yield important new information on the role of the atomic and material properties in dispersion forces (Klimchitskaya et al. 2009a).

22.2.2 Accuracy of phenomenological potential

Here we compare the phenomenological potential (22.11) with more exact calculations of the atom–wall interaction energy on the basis of the Lifshitz theory (Bezerra et al. 2008). This allows us to determine the accuracy of this potential and its application region.

As an example, we consider an Au wall and an atom of metastable He*. Using the value of $\alpha(0)$ presented in Section 16.3.2, we obtain from eqn (16.28) the magnitude of the Casimir–Polder coefficient $C_4^{Au} \approx 1.1\,\mathrm{eV\,nm^4} \approx 1.8 \times 10^{-55}\,\mathrm{J\,m^4}$. The value of the van der Waals coefficient C_3 for Au can be computed from eqn (16.24). This has been done using the tabulated optical data for the complex index of refraction of Au (Palik 1985) in order to find the values of ε along the imaginary axis (Fig. 13.2), and the highly accurate data for the dynamic polarizability of He* atom presented in Fig. 16.3 [at short separations of a few nanometers, the plasma model (13.1) and the single-oscillator model (16.29) are not applicable in precise computations]. The computations (Caride et al. 2005) lead to $C_3^{Au} \approx 1.6\,\mathrm{a.u.} \approx 6.4 \times 10^{-3}\,\mathrm{eV\,nm^4} \approx 10.2 \times 10^{-49}\,\mathrm{J\,m^4}$. From this we obtain $l^{Au} = C_4^{Au}/C_3^{Au} \approx 172\,\mathrm{nm}$ for an Au wall and an He* atom.

In Fig. 22.3(a), the phenomenological interaction energy (22.11) multiplied by factor of a^4 is plotted as a function of separation for the case of an He* atom interacting with an Au wall (the dashed line). In the same figure, the solid line shows the computational results for the quantity $a^4 E^A(a)$, where the accurate interaction energy $E^A(a) = \mathcal{E}^A(a)$ is defined in Eq. (16.21) in accordance with the Lifshitz formula. As is seen in Fig. 22.3(a), at small and large separations the phenomenological potential (22.11) almost coincides with the accurate interaction energy as given by the Lifshitz formula. To give a better understanding of the correlation of the two potentials at separations below 1 μm, i.e. in the most important region for the experiments on quantum reflection, the two lines are shown in Fig. 22.3(b) on an enlarged scale. It can be seen that the solid and dashed lines coincide at $a \leq 50\,\mathrm{nm}$. The relative difference between the accurate and the phenomenological interaction energy,

$$\delta E^A(a) = \frac{E_{\mathrm{acc}}^A(a) - E_{\mathrm{ph}}^A(a)}{E_{\mathrm{acc}}^A(a)}, \qquad (22.12)$$

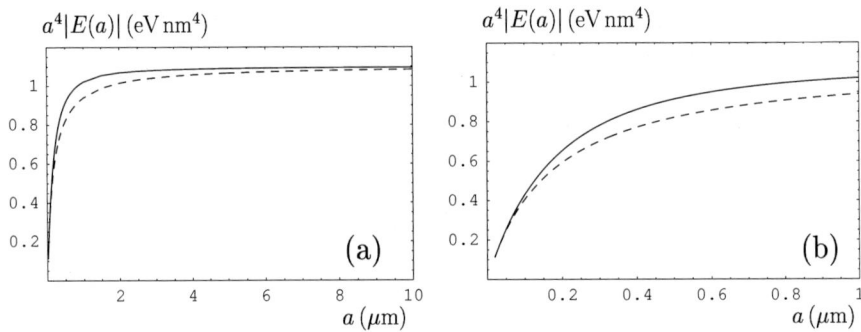

FIG. 22.3. Magnitude of the interaction energy between an atom of metastable He* and an Au wall multiplied by the fourth power of the separation, versus separation. Computations were performed using the Lifshitz formula at $T=0$ (solid lines) and the phenomenological potential (22.11) (dashed lines). (a) Separation varies from 20 nm to 10 μm. (b) Separation varies from 20 nm to 1 μm (Bezerra et al. 2008).

is a nonmonotonic function and varies from 5.7% at $a = 100$ nm to 7.9% at $a = 1$ μm. The largest values of δE^A happen at moderate separations, which are interesting from the experimental point of view: $\delta E^A = 10.2\%$, 10.4%, and 10.2% at separation distances $a = 300$, 400, and 500 nm, respectively.

Now we consider an He* atom near a high-resistivity Si wall (dielectric materials are often used in experiments on quantum reflection). In this case eqn (16.28) is not applicable. The value of the Casimir–Polder coefficient $C_4^{Si} \approx 0.75$ eV nm^4 was computed by Oberst et al. (2005) using the Lifshitz formula. The permittivity of dielectric Si along the imaginary frequency axis, with $\varepsilon^{Si}(0) = 11.66$, was computed from the tabulated optical data and the Kramers–Kronig relations [see Fig. 12.2(a)]. In a similar way, the value of the van der Waals coefficient $C_3^{Si} \approx 5.5 \times 10^{-3}$ eV nm^4 was obtained by Caride et al. (2005) and Oberst et al. (2005). This leads to $l^{Si} \approx 136$ nm for an He* atom near an Si wall.

For example, in Fig. 22.4(a), the phenomenological interaction energy (22.11) multiplied by a factor of a^4 is plotted as a function of separation for the case of He* atom interacting with an Si wall (the dashed line). The solid line presents the computational results for the quantity $a^4|E(a)|$ obtained using the Lifshitz formula as described above. In Fig. 22.4(b), the same lines are reproduced on an enlarged scale at separations below 1 μm. As can be seen in Figs. 22.4(a,b), at separations below 50 nm and at about 10 μm the limiting cases of the nonrelativistic and the relativistic potential V_3 and V_4, respectively, are achieved. The relative difference (22.12) between the accurate and phenomenological interaction energies varies from 9.4% at $a = 100$ nm to 8.6% at $a = 1$ μm. However, at intermediate separations δE^A reaches its largest values, which are equal to 12%, 12.5%, 12.2%, and 11.6% at separations $a = 200$, 300, 400, and 500 nm, respectively. Thus, for the Au wall the phenomenological interaction energy provides

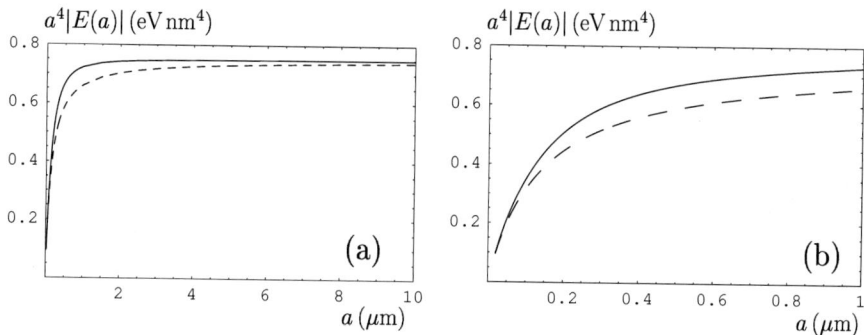

FIG. 22.4. Magnitude of the interaction energy between an atom of metastable He* and an Si wall multiplied by the fourth power of the separation, versus separation. Computations were performed using the Lifshitz formula at $T = 0$ (solid lines) and the phenomenological potential (22.11) (dashed lines). (a) Separation varies from 20 nm to 10 μm. (b) Separation varies from 20 nm to 1 μm (Bezerra et al. 2008).

a more accurate model of the atom–wall interaction than for the Si wall. This is connected with the fact that the strength of the atom–wall interaction for the Si wall is weaker than in the case of the Au wall.

The above computations using the Lifshitz formula were performed at zero temperature. It is instructive to compare the phenomenological potential (22.11) with the results of more accurate computations using the Lifshitz formula at the laboratory temperature $T = 300$ K. Computations were performed by substituting the above dielectric permittivity of Au and Si and the dynamic polarizability of the He* atom along the imaginary frequency axis into eqn (16.18). The computational results for the quantity $a^4|\mathcal{F}^A(a,T)|$ are shown by the solid lines in Fig. 22.5(a) for the Au wall and in Fig. 22.5(b) for the Si wall. In addition the same results as in Figs. 22.3(a) and 22.4(a) for the quantity $a^4|E^A(a)|$, where $E^A(a)$ is the phenomenological potential (22.11) for an Au or Si wall, are reproduced in this figure by dashed lines. From a comparison between Figs. 22.3(a) and 22.5(a), it can be seen that for an Au wall at separations $a \leq 2$ μm from the He* atom the relative differences between the accurate and phenomenological potentials are approximately the same in the cases where the accurate potential is computed at zero temperature and at $T = 300$ K. However, with an increase in separation the accurate potential, i.e. the free energy, computed at $T = 300$ K (the solid line) deviates significantly from the phenomenological potential in accordance with the classical limit in eqn (16.45). For Au [Fig. 22.5(a)], the largest deviation shown in the figure is equal to 31%, which is reached at $a = 5$ μm.

For an He* atom near an Si wall [Fig. 22.5(b)], the thermal effects play a more important role. A comparison between Figs. 22.4(a) and 22.5(b) demonstrates that here the differences between the accurate potential $\mathcal{F}^A(a,T)$ and the phenomenological potential $E^A(a)$ can be considered as temperature-independent

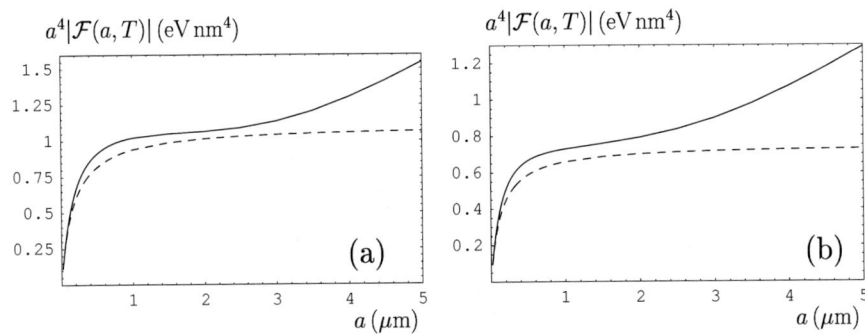

FIG. 22.5. Magnitude of the free energy between an atom of metastable He*
and (a) Au or (b) Si wall, multiplied by the fourth power of the separation,
as a function of separation (solid lines). The free energy was computed at
$T = 300\,\mathrm{K}$ using the Lifshitz theory. For comparison, the phenomenological
potential (22.11) for (a) Au and (b) Si walls is shown by the dashed lines
(Bezerra et al. 2008).

only below $0.5\,\mu\mathrm{m}$. Computations show that at $a = 1\,\mu\mathrm{m}$, the relative difference between them is equal to 9.6% (whereas, as indicated above, it is equal to only 8.6% when the zero-temperature Lifshitz formula is used). At a separation $a = 5\,\mu\mathrm{m}$ the relative difference between the accurate temperature-dependent potential and the phenomenological potential reaches 43.5%.

The larger deviations between the accurate temperature-dependent potential and the phenomenological potential for dielectrics than for metals are explained by the existence of temperature and separation regions where the Casimir–Polder entropy is negative (see Sections 16.3.1 and 16.3.3). The phenomenon of negative entropy occurs only for atoms near a metallic plate. As a result, within some range of temperature, the sign of the thermal correction to the Casimir–Polder energy is opposite to the sign of the energy, and the respective free energy becomes nonmonotonic. This makes the difference between the accurate free energy, as computed by use of the Lifshitz formula, and the phenomenological potential smaller. In contrast, for an atom near a dielectric wall, the Casimir–Polder entropy is always positive (see Section 16.4.1). This follows from the same property of the entropy in the configuration of two dielectric plates (Section 12.5.1). Then the thermal correction and the Casimir–Polder energy have the same sign, and the magnitude of the free energy is a monotonically increasing function of the temperature. Thus, with increasing temperature (or separation), the differences between the accurate free energy and the phenomenological potential can only increase. Thus, future experiments on quantum reflection need to use the accurate free energy of the atom–wall interaction, obtained on the basis of the Lifshitz theory, for the interpretation of measurement data.

23

APPLICATIONS OF THE CASIMIR FORCE IN NANOTECHNOLOGY

The immense technological promise and the possibility of novel physical phenomena offered by devices of small dimensions was anticipated by Feynman (1960) almost 50 years ago. The continual drive to increase functionality while minimizing energy consumption will inevitably lead to further shrinking of device sizes. Advances in integrated-circuit fabrication techniques based on photolithography and electron beam lithography and plasma and chemical etching have now allowed fabrication of mechanical and electromechanical devices with sizes ranging from microns to nanometers (Allen 2005). *Microelectromechanical systems* (MEMS) is a general term used in the present-day literature to describe nanofabricated devices. MEMS find applications in industries such as those associated with optical communication and cellular communication, and as variety of sensors. When MEMS dimensions shrink to submicron levels, these systems are usually called *nanoelectromechanical systems* (NEMS) (Ekinci and Roukes 2005).

The first papers, by Srivastava *et al.* (1985) and Srivastava and Widom (1987), anticipating the dominant role of Casimir forces in nanoscale devices appeared over 20 years ago but were largely ignored, as silicon chip fabrication dimensions were then on the order of many micrometers. Now, with device dimensions shrinking to nanometers, the important role of Casimir forces in nanoscale devices is well recognized. In this chapter, we discuss both the theoretical and the experimental aspects of the combined role of electrostatic and Casimir forces in MEMS. Then we consider the first MEMS actuated by the Casimir force, and the nonlinear micromechanical Casimir oscillator (Chan *et al.* 2001a, 2001b). Special attention is paid to the interaction of atoms with multiwalled and single-walled carbon nanotubes. At the end of the chapter, some prospective applications of the Casimir force in nanotechnology are discussed.

23.1 Combined role of electrostatic and Casimir forces in MEMS and NEMS

In present-day MEMS devices, the method of actuation is primarily electrostatic. Increasingly, as larger actuating forces and torques are demanded with the application of smaller voltages, the separations between the moving components and the fixed electrodes are shrinking. Thus the role of the Casimir effect has to be included for effective treatment of the device properties. Even without this consideration, we have to take account of the fact that the moving parts of MEMS, on close approach to the fixed electrodes, frequently jump into contact with the electrodes and adhere to them. This phenomenon, usually referred to as *pull-in* and *stiction*, leads to loss of functionality in devices. It has now become recognized that the Casimir force, owing to its strong distance dependence, is the primary cause of pull-in and stiction in devices. The opportunity to exploit anharmonicity and to understand and modify pull-in and stiction has motivated many theoretical and experimental investigations of the combined role of electrostatic and Casimir forces in MEMS, which are discussed below.

23.1.1 Modeling of the combined role of electrostatic and Casimir forces in MEMS and NEMS

Most nanomechanical devices are based on thin cantilever beams above a silicon substrate, fabricated by photolithography followed by dry and wet chemical etching (Allen 2005). Such cantilevers are usually suspended about 100 nm above the silicon substrate. The cantilevers move in response to the Casimir force and voltages applied to the substrate (Bishop *et al.* 2001) or in response to incoming radio-frequency signals (Lucyszyn *et al.* 2008). In the case of radio-frequency transmitters and receivers, a high quality factor Q is necessary for narrow-bandwidth operation of these devices. However, owing to the coupling to the substrate and neighboring cantilevers through the Casimir force, the vibration energy of the cantilever can be dissipated. This dissipation of mechanical energy leads to a decrease in Q and crosstalk with neighboring receivers, both leading to degradation in the signal. The problem will be exacerbated in the dense arrays of high-Q transmitters and receivers needed for future mobile communication. Thus effective incorporation of the Casimir force is necessary in the design of these devices to optimize their performance.

Serry *et al.* (1995) developed a simple model of the competitive interaction between the Casimir and elastic forces and showed that it might give rise to nonlinear behavior in MEMS devices. This model was called the *anharmonic Casimir oscillator*. The suggested model represents the MEMS device as a simple system of parallel plates at a separation distance a. One of the plates is movable. The elastic response of the movable plate is modeled as that of a linear spring with spring constant k, i.e. Hooke's-law-like behavior. A schematic of the model is shown in Fig. 23.1(a). For such a system, the total potential energy per unit area of the plates is given by the sum of the elastic and Casimir energies,

$$U(a) = \frac{k(b-a)^2}{2S} - \eta \frac{\pi^2 \hbar c}{720 a^3}. \tag{23.1}$$

Here, we have placed the stationary plate on the left in Fig. 23.1(a) at some distance b from an unstretched spring. The phenomenological prefactor η was introduced to account for the material properties. Serry *et al.* (1995) estimated the approximate value of η for two Si plates and considered it to be independent of separation (in reality, however, η depends on separation; see the computational results for Si in Section 12.6.2). In further calculations, ideal-metal plates with $\eta = 1$ were assumed.

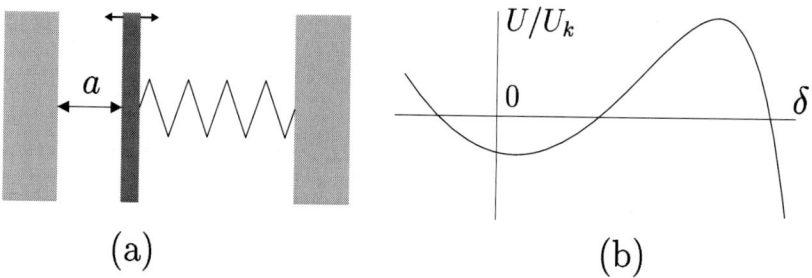

FIG. 23.1. An anharmonic Casimir oscillator: (a) schematic of the model and (b) a typical potential energy (see text for further details).

In terms of the dimensionless displacement $\delta = (b-a)/b$, the potential energy (23.1) can be rewritten as

$$U(\delta) = U_k \left[\frac{1}{2}\delta^2 - \frac{1}{3} \frac{C}{(1-\delta)^3} \right], \tag{23.2}$$

where $U_k \equiv kb^2/S$ is the unit of the elastic energy per unit area and $C \equiv \eta \pi^2 \hbar c S/(240 k b^5)$. The behavior of U as a function of δ is shown in Fig. 23.1(b) for sufficiently strong springs (large k), which means that C is less than some critical value $C_{\rm cr} \approx 0.0819$. As can be seen in Fig. 23.1(b), the potential energy has a metastable minimum followed by a barrier. For $\delta \to 1$, i.e. $a \to 0$, the potential energy goes to minus infinity because of the Casimir energy. Serry et al. (1995) pointed out that the depth of the potential well at $\delta = 1$ would be finite owing to the roughness of the plates, which prevents the two surfaces from having zero separation distance.

Based on numerical simulations, it was found that the metastable minimum was at the bottom of an asymmetric potential well and therefore would lead to an anharmonic oscillator and result in bistability. Serry et al. (1995) studied the height of this energy barrier by changing the spring constant. As the spring gets weaker (or the Casimir force is somehow increased), the energy barrier vanishes at the critical value $C = C_{\rm cr}$. This phenomenon could be used to open and close a Casimir effect switch. The switch would be in the closed state when the two parallel plates were in contact. It was proposed to accomplish this by effectively changing the ratio of the elastic force to the Casimir force. One proposal to modify the ratio of the forces and accomplish the closed phase was to decrease the separation a. The attainment of the open state was acknowledged to be a little more difficult but could be accomplished by moving the fixed end of the spring in Fig. 23.1(a). In order to reduce the height of the energy barrier but still retain the anharmonic potential well, a pneumatic approach was discussed, where a gas would be introduced between the two parallel plates. The mechanical compression of this gas during the period when the switch was in the closed position would decrease the depth of the potential well and facilitate the reopening of the switch. Another scheme, based on the application of dc voltages with insulating layers on the plate, was also proposed to facilitate the opening and closing of the switch.

Following the above study, Serry et al. (1998) performed a numerical estimation of the Casimir effect in the context of MEMS deflection and stiction. As the Casimir force between parallel surfaces has approximately a fourth-order dependence on the separation distance between the surfaces, it becomes comparable to other technologically important forces such as electrostatic forces when the separation falls below a micron. It was pointed out that the gravitational force and the Casimir force are equal for a $2\,\mu$m thick, highly doped silicon membrane suspended about $0.4\,\mu$m above an underlying rigid substrate. On the other hand, the Casimir force can exceed typical applied electrostatic forces in actuators for separation distances below 100 nm.

According to Serry et al. (1998), for typical MEMS device configurations the collapse of suspended membranes onto an underlying rigid substrate (stiction) due to the Casimir force can be avoided for some geometries. In MEMS fabrication, the problem of stiction is encountered during the wet etching of sacrificial layers between the membrane and the rigid substrate. Suspended membranes are used in many devices, such as accelerometers, micromirrors, and microdisplay actuators. The collapse of the membrane adversely affects the yield during fabrication. It also leads to device failure, thus severely limiting the lifetime of the device. For a model membrane strip of length L and thickness D ($L \gg D$), fixed at both ends, Serry et al. (1998) studied the static deflection at the center of the strip due to the Casimir force as a function of the physical and geometric parameters that are relevant to microfabrication of the device. As the membrane was curved between the supports owing to the combined action of the elastic and Casimir forces, the analysis was confined to a flat region (sufficiently small) near the center of the membrane where one could use the local-value approximation. The deflection of the membrane strip element from flatness under an external load was described by a differential equation of fourth order. The force balance was given by the flexural forces, the membrane-stretching forces, and the Casimir force between the suspended membrane and the rigid substrate in the parallel-plate approximation. The role of the flexural forces was neglected in comparison with the membrane-stretching forces.

It was shown that the stability of the membrane depends on the value of the quantity

$$K_c = \frac{\hbar c \pi^2 L^4}{240 E D b^7}, \qquad (23.3)$$

where E is the Young's modulus of the membrane and b is the initial separation distance between the membrane and the underlying rigid substrate. Thus, the membrane is unstable and will collapse if $K_c > 0.245$. This result can be somewhat modified in the presence of additional forces such as electrostatic or capillary forces in liquids during fabrication. For typical device parameters with gold-coated polymer membranes of $E = 10^9$ Pa , $L = 500\,\mu$m, and $D = 1\,\mu$m, collapse will result if b decreases below $0.8\,\mu$m.

Importantly, Serry *et al.* (1998) drew the attention of the MEMS community to the fact that the Casimir effect needs to be considered as a vital factor in the future design of MEMS. The results obtained served as a starting point for future developments of applications of the Casimir effect in MEMS systems, some of which are reviewed below. In particular, it was revealed that the Casimir effect might be the critical factor in the stiction failure of MEMS. New schemes for switches were also suggested that would take advantage of the same large Casimir force that leads to stiction. The theoretical modeling of pull-in and stiction phenomena and the role of the Casimir force in them is undergoing rapid development. Increasingly, the combined roles of material properties (Gusso and Delben 2006) and of roughness and electrostatic effects (Palasantzas 2007a, 2007b) in MEMS are being considered. Stiction and pull-in instabilities in micromembranes of various geometries under the influence of the Casimir force were investigated by Batra *et al.* (2007).

23.1.2 *Experimental investigation of the stability of MEMS*

Buks and Roukes (2001a) studied the combined role of the electrostatic and Casimir forces using an extremely thin nanofabricated cantilever. A gold cantilever fixed at both ends was fabricated on a silicon substrate using photolithography, electron beam lithography, and thermal evaporation. A cantilever of length $200\,\mu$m, width 240 nm, and thickness 250 nm fabricated between two gold electrodes was used. The thermal vibrations of the cantilever were studied in a scanning electron microscope. The scattered electron beam obtained by focusing the electrons on the side of the gold cantilever was collected by a photodiode and used to study the dynamical response. The resonant frequencies were found to be equally spaced at 176.5, 354.4, and 529.8 Hz and the quality factor was found to be 1800. Stiction (adhesion of the cantilever to the neighboring electrode) was introduced with a drop of deionized water. However, no estimate of the van der Waals interaction could be made experimentally. Only theoretical estimates were made. It was proposed that the methodology might allow calculation of the adhesion energies in the future.

In a follow-up to the above, using the same type of resonator, Buks and Roukes (2001b) investigated the metastability of cantilever resonators under the combined action of electrostatic and Casimir forces. The Au cantilever beam, separated from an adjacent counterelectrode by a $1\,\mu$m vacuum gap, jumped into contact at some critical distance and permanently adheres to the electrode. It was pointed out that, therefore, the free state of the cantilever was a metastable state and the state of contact had a lower energy owing to the large Casimir interaction. The height of the energy barrier separating these two states determined the lifetime of the oscillator in the metastable state. The mechanical properties of the beam were studied by applying dc voltages until contact occurred. By also applying small ac voltages, it was possible to study the nonlinear response as a function of the frequency and the applied dc voltage. An attempt was undertaken to model the potential-energy surface of the system and deduce the lifetime in the metastable state.

For this study a gold cantilever beam, fixed at both ends, 200 μm long, 0.28 μm wide, and 0.25 μm thick was used. The neighboring parallel counterelectrode had a 20 μm long rectangular protrusion, which was separated from the beam by a vacuum gap of 1.28 μm. All measurements were done at room temperature inside a scanning electron microscope. The first three mechanical resonances were excited by applying ac voltages and were noted to be 185.53, 372.4, and 563.8 kHz. This result was used to extract the elastic properties of the system. A dc voltage of 20 V was applied to the counterelectrode, and the nonlinear dynamic response for close proximity between the Au cantilever and the counterelectrode was studied by using ac frequencies. The oscillation amplitude was measured as a function of the frequency for a range of applied ac voltages. A nonlinear (skewed) response was observed for ac voltages of 25 and 50 mV. At larger ac voltages between 75 to 225 mV, hysteretic behavior was observed. Here, the oscillation amplitude dropped sharply at a critical frequency on the high-frequency side. On reversing the frequency scan, the amplitude remained low and then jumped at a frequency past the critical frequency. The mechanical nonlinearity parameter was extracted from this nonlinear response.

Using all of the above, the potential energy was represented as the sum of the elastic mechanical energy of the beam, the electrostatic energy (from the dc voltage), and the Casimir energy between the beam and the counterelectrode. The parallel-plate approximation was used for both the electrostatic and the Casimir energy. Then the system under consideration resembled that shown in Fig. 23.1(a), where, in addition, a voltage V was applied between the stationary plate on the left and the movable plate connected to the spring. In this situation, the potential energy per unit area (23.1) is replaced with

$$U(a) = \frac{k(b-a)^2}{2S} - \eta \frac{\pi^2 \hbar c}{720 a^3} - \frac{V^2}{8\pi a}. \qquad (23.4)$$

Introducing the dimensionless displacement $\delta = (b-a)/b$ once more, this can be rewritten in the form

$$U(\delta) = U_k \left[\frac{1}{2}\delta^2 - \frac{1}{3}\frac{C}{(1-\delta)^3} - \frac{C_1}{1-\delta} \right], \qquad (23.5)$$

where $C_1 \equiv V^2 S/(8\pi k b^3)$. It is easily seen that for relatively small values of C and C_1, the behavior of U as a function of δ is the same as that shown in Fig. 23.1(b), i.e. the system under consideration has a local minimum associated with a metastable state.

Buks and Roukes (2001b) calculated the position of the metastable minimum of the potential (23.5) in the approximation of small electrostatic and Casimir energies in comparison with the elastic energy. To overlap with the experimental results, the area had to be multiplied by a factor of 2.3. Based on the predicted potential-energy surface, the lifetime in the metastable state was calculated based on the Kramers model. The calculated results for the lifetime were found to be in "gross contradiction with experimental observations".

To conclude, Buks and Roukes (2001a, 2001b) studied the nonlinear mechanics of a MEMS device under the combined action of electrostatic, elastic, and Casimir forces. It was pointed out that the free state of the cantilever was indeed a metastable state, as was discussed in Section 23.1.1, and that the Casimir force would lead to a stable state of *stiction*, i.e. adhesion of free parts to neighboring substrates or electrodes. The nonlinearities resulting from this metastablility were measured and shown to lead to asymmetric resonances and hysteretic behavior. The size of the Casimir contribution in relation to the electrostatic one was not estimated. Based on the voltage of 20 V applied and the $20 \times 0.25\,\mu m^2$ area of the protrusion, the role of the Casimir effect in these experiments appears to be a factor of 1000 weaker than the electrostatic contribution.

23.2 Actuation of MEMS by the Casimir force

Chan et al. (2001a) achieved the first experimental demonstration of the actuation of a MEMS device by the Casimir force. The device was fabricated by standard nanofabrication techniques such as photolithography and chemical etching on a silicon substrate. This device consisted of a $3.5\,\mu m$ thick, $500\,\mu m^2$, heavily doped polysilicon plate freely suspended on its central axis by thin torsional rods. The ends of the torsional rods were anchored to the substrate by support posts. Two fixed polysilicon electrodes were located beneath the plate, symmetric with respect to the torsional axis (a similar device is shown in Fig. 19.9). An SiO_2 sacrificial layer was etched to create a $2\,\mu m$ gap between the plate on the top and the electrodes on the bottom. Thus, in response to an applied torque, the plate rotated freely around the torsional rod. The rotation of the plate was detected through measurement of the capacitance between the plate and the bottom electrodes. It was reported that the capacitances from the plate to the two bottom electrodes were almost equal. When the plate rotated in response to a torque, one of the capacitances increased and the other decreased. A small ac voltage of 100 mV was used on the bottom electrodes to measure the capacitances. A bridge circuit enabled the investigators to measure the change in capacitance to 1 part in 2×10^5, which was equivalent to a rotation angle of 8×10^{-8} rad, using an integration time of 1 s, when the device was in a vacuum of less than 1 mTorr.

In order to demonstrate actuation by the Casimir force, a polystyrene sphere of radius $R = 100\,\mu m$ was fixed by conductive epoxy to the end of a copper wire. A 200 nm thick gold film was evaporated on the sphere and on the top plate of the MEMS device. An additional 10 nm layer of gold was sputtered onto the sphere to provide electrical contact to the wire. The MEMS device was placed on a piezoelectric translation stage, with the sphere mounted closer to one side of the top plate. Extension of the piezoelectric stage moved the micromachined device toward the sphere. The measurements were performed at room temperature. The spring constant of the torsional rods was calibrated by measuring the electric force between the sphere and the top plate as a function of the sphere–plate distance. The top plate was grounded, while various voltages V were applied to the sphere. A theoretical expression for the electric force acting between the sphere and the plate was obtained using the PFA. It coincides with the first term ($i = -1$) in eqns (19.1) and (19.2),

$$F_{\text{el}}(a) = -\pi\epsilon_0 \frac{R(V - V_0)^2}{a}. \tag{23.6}$$

Here, the value of the separation a at the closest approach of the sphere and the plate is the separation on contact a_0, and V_0 is the residual potential difference. From a fit of the measured electric forces to eqn (23.6), the values $a_0 = 67.0$ nm and $V_0 = 30$ mV were found. From the electrostatic calibration, the proportionality coefficient α between the force applied to the top plate and its tilt angle, where $F_{\text{el}} = \alpha\theta$, was found to be 5.97×10^{-5} N rad^{-1}. No error bars in the above measurements were reported.

In the demonstration of the actuation of the top plate by the Casimir force, a compensating voltage corresponding to V_0 was applied to the sphere. The tilt of the top plate was measured as a function of the sphere–plate distance. The authors reported a tilting of the top plate in response to the Casimir force for separation distances less than 300 nm. The tilt angle reached a maximum of 6 μrad at the closest approach. The tilt angles were used to calculate the experimental values of the Casimir force. A theoretical expression for the Casimir force $F(a) = 2\pi R E(a)$, where $E(a)$ is the energy per unit area of two parallel plates defined in eqn (12.30), was obtained using the PFA. The values of $E(a)$ were calculated numerically using the tabulated optical data for the complex refractive index of gold. The surface roughness was considered as stochastic with a variance δ and an amplitude $A = \sqrt{2}\delta$. The roughness correction was taken into account in a multiplicative way. In accordance with eqn (17.84), this leads to

$$F_{\text{R}}(a) = 2\pi R E(a)\left(1 + 6\frac{A^2}{a^2} + 45\frac{A^4}{a^4}\right). \tag{23.7}$$

Note that the coefficient 15 given instead of 45 in the corresponding equation in the original publication, eqn (9) of Chan et al. (2001a), is in error. The roughness amplitude $A = 30$ nm was measured using an atomic force microscope.

The theoretical Casimir force (23.7) was fitted to the data, leaving the closest separation a_0 as a fitting parameter. From the fit, a_0 was reported to be 75.7 ± 1 nm. This was reported to be consistent with the 67.0 nm found from the electrostatic fit. It was stated that the difference was due to the large roughness. The authors reported that "relatively large surface roughness complicates an exact comparison between data and theory," as "the expansion parameter A/a is about 0.4 at the smallest separation, resulting in surface roughness corrections of more than 50% of the net force." However, eqn (23.7) with $A/a = 0.4$ leads to a roughness correction equal to 96%, if applied up to the terms of the second order, and to 211%, if applied up to the fourth-order term. Ideally, at such short separations and large roughness amplitudes one should use not eqn (23.7) but the method of geometrical averaging to take surface roughness into account (see Section 17.2.2).

In conclusion, Chan et al. (2001a) reported the first demonstration of the actuation of a MEMS device due to the action of the Casimir force. A clear difference in the response of the MEMS between that due to an electric force and the Casimir force was reported. The distance dependence of the force was fitted well by the Lifshitz theory. However, it was reported that the large roughness prevented a definitive calculation of the agreement between experiment and theory.

23.3 Nonlinear micromechanical Casimir oscillator

The nonlinearity of a MEMS due to the influence of the Casimir force was also investigated by Chan et al. (2001b). This effect is different from the one described in the previous section, where the static deflection of the plate was used to demonstrate the influence of the Casimir force. Here, the polysilicon plate of the MEMS was oscillated about its central axis with a frequency ω near the resonance frequency ω_0, and the shift of the resonance frequency under the influence of the Casimir force from the sphere was monitored. A simple model of this experiment is shown in Fig. 23.2. This was the first observation where the Casimir force was unambiguously shown to lead to frequency shifts and hysteretic behavior in a periodically driven MEMS. As predicted earlier by Serry et al. (1995) (see Section 23.1), MEMS display anharmonicity and bistability owing to the Casimir force. As shown in Section 23.1.2, nonlinearities in MEMS due to electrostatic effects have been previously demonstrated (Buks and Roukes 2001b, Krommer et al. 2000). Chan et al. (2001b) noted that while Buks and Roukes (2001b) considered the role of the Casimir force in the MEMS nonlinear response, its contribution was negligible. In the experiment reported, the movable plate of the oscillator was subject to a linear elastic restoring force and the nonlinear Casimir force. For separations larger than a critical value, the movable plate was bistable, i.e. the potential energy had a local minimum (primarily due to the elastic force) followed by a global minimum (adhesion due to the Casimir force), separated by a barrier whose width and height were determined by the elastic and Casimir force properties of the system [see Fig. 23.1(b)].

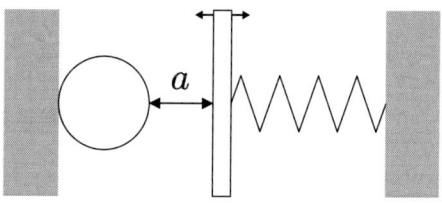

FIG. 23.2. Schematic diagram of the model of an oscillator actuated by the Casimir force.

The micromachined oscillator used was the same as that described in Section 23.2. The torsional oscillations were excited by supplying ac voltages to electrodes below the plate. In addition, a small dc bias was necessary to linearize the voltage dependence of the driving torque. The plate was grounded. Oscillations of the plate resulted in a time-varying capacitance between the plate and the electrodes. The change in capacitance was assumed to be linearly proportional to the amplitude of the oscillations. This change in capacitance at the excitation frequency was detected using a lock-in amplifier. The measurements were performed at room temperature and at pressures of less than 1 mTorr. The spring constant of the oscillator was reported as $k = 2.1 \times 10^{-8}$ N mrad^{-1}. The resonance frequency was found to be constant at 2753.47 Hz, independent of the amplitude of the excitation voltage, demonstrating that the oscillator had a linear behavior in the absence of other external forces.

In this experiment the Casimir force $F(a)$ arose owing to an Au-coated polystyrene sphere of diameter 200 μm, as in the experiment by Mohideen and Roy (1998) considered in Section 19.2.1. This sphere was mounted above one side of the top plate of the oscillator, at distance b from the torsional axis. The distance a between the sphere and the equilibrium position of the top plate was controlled by using a closed-loop piezoelectric stage. The oscillator equation describing the motion of the plate is (Chan et al. 2001b)

$$\ddot{\theta} + 2\gamma\dot{\theta} + \omega_r^2\theta = \frac{\tau}{I}\cos(\omega t) - \alpha\theta^2 - \beta\theta^3, \tag{23.8}$$

where $I = 7.1 \times 10^{-17}$ kg m^2 is the moment of inertia of the plate, γ is the damping parameter, τ is the amplitude of the external torque, $\alpha \equiv b^3 F''(a)/(2I)$, and $\beta \equiv -b^4 F'''(a)/(6I)$. The quantity ω_r defined by eqn (19.14) is connected with the resonance frequency $\omega_0 = \sqrt{k/I}$, where $k = 2.1 \times 10^{-8}$ N mrad^{-1} is the torsional spring constant. If the oscillations are small and one can neglect the terms $\alpha\theta^2$ and $\beta\theta^3$ in eqn (23.8), ω_r corresponds to the resonance frequency in the presence of the Casimir force. The same equation is valid in the absence of the Casimir force (for example, at sufficiently large separations) when an electrostatic force is applied between the sphere and the plate. In this case $F(a)$ in eqn (19.14) is replaced by $F_{el}(a)$ from eqn (23.6). From the electrostatic calibration, it was found that $V_0 = 75$ mV, $a_0 = 122.4$ nm, and $b = 131.0$ μm. Here, contact between the sphere and the plate was avoided, as the top plate of the oscillator adhered to the sphere when they were close. Thus, a_0 corresponds to some chosen closest approach of the sphere to the plate in the electrostatic measurements, but not the minimum achievable separation. The errors in the parameters a_0 and b were not reported.

A compensating voltage V_0 was then applied to the sphere, and the frequency shift due to the Casimir force was measured as a function of the separation distance between the sphere and the plate for the case of small oscillations. Using eqn (19.14), the results obtained were recalculated as the gradient of the Casimir force. The data for the force gradient were fitted to the theoretical expression obtained for ideal-metal surfaces from eqn (6.51),

$$\frac{\partial F(a)}{\partial a} = \frac{\pi^3 \hbar c R}{120 a^4}. \tag{23.9}$$

The value of the separation at the closest approach was determined from the fit to be $a_1 = 85.9$ nm (no errors were provided).

Next, to demonstrate the linear behavior of the oscillator when the Casimir force is negligible, the amplitude, as a function of the applied frequency, was measured for a sphere-plate separation of $3.3\,\mu$m. The amplitude profile was shown to be the same regardless of the direction of change of frequency, i.e. whether the frequency increased or decreased. To demonstrate the nonlinear behavior, the piezoelectric stage was extended to bring the sphere closer to the top plate (to a distance of 141 nm), while the excitation voltage was maintained constant at $55.5\,\mu$V. In this case a shift in the resonance to lower frequencies and an asymmetry in the resonance peak were observed. At even smaller sphere-plate separations of 116.5 nm, the asymmetry was shown to become stronger and led to hysteresis, where the amplitude response at the same driving voltage depended on the direction of change of the excitation frequency. This was shown to be most pronounced at the shortest separation distance considered, of 98 nm. It was reported that the response curve could be fitted for any particular a, using only the spatial derivatives of the ideal-metal Casimir force, without any adjustable parameters (Chan et al. 2001b). Some deviations from the fit were observed at 98 nm and were attributed to the truncation of the Taylor expansion of F to the third order in eqn (23.8). It was pointed out that this nonlinear response could be applied to demonstrate a spatial *memory effect*, which could be used for distance sensing. As the sphere-plate distance is changed, the resonance frequency shifts in response to the change in the derivatives of F. If the driving voltage and the excitation frequency are fixed and the sphere-plate distance is varied, the amplitude of the oscillator will trace an asymmetric curve which depends on whether the sphere-plate distance is being increased or decreased, owing to the nonlinearities. Thus the response retains a memory of the direction of the change of the sphere-plate distance.

In conclusion, Chan et al. (2001b) have reported the first clear observation of a nonlinear response of a MEMS due to the Casimir force. This observation bore out the expectations (see Section 23.1) that the nonlinearities introduced by the strong distance dependence of the Casimir force would lead to a metastable state and therefore might be responsible for the adhesion or stiction of the mobile parts of MEMS. These nonlinearites were shown to limit the operational capabilities of MEMS. It was also convincingly shown that the Casimir force has to be taken into account in the design and fabrication of MEMS and that the amplitude and operational frequency bandwidth of MEMS will be constrained by the effect of the Casimir force. It has since been shown that the material property dependence of the Casimir force can lead to even more interesting effects in MEMS, such as the pulsating Casimir force (Klimchitskaya et al. 2007b) considered in Section 20.4.3. It is certain that future work, taking into account the complete panoply of material dependences of the Casimir force, can exploit these nonlinearities further.

23.4 The Casimir–Polder interaction between atoms and carbon nanostructures

Carbon nanostructures (buckyballs, nanotubes, and nanowires) are attracting much attention in both fundamental science and nanotechnology owing to their unique electrical, optical, and mechanical properties [see e.g. the book by Harris (1999)]. One of the most attractive applications of carbon nanostructures is the proposed possibility to use them for the solution of the problem of hydrogen storage. According to the review by Nechaev (2006), there are conceptual possibilities to create carbon nanostructures capable of absorbing more than 10 mass percent of hydrogen. The solution of this fundamental problem requires a detailed investigation of the microscopic mechanisms of the interaction between hydrogen and graphite. In this section, we show how the Lifshitz formulas used in Section 16.7 for the investigation of the interaction of atoms with a graphite wall can be generalized to the case of atoms interacting with multiwalled carbon nanotubes (Blagov et al. 2005, Klimchitskaya et al. 2006a). Then we apply the normal modes and the reflection coefficients for the interaction of electromagnetic oscillations with a plasma sheet found by Barton (2004, 2005) to obtain Lifshitz-type formulas describing the interactions of atoms with graphene and single-walled carbon nanotubes (Bordag et al. 2006, Blagov et al. 2007, Klimchitskaya et al. 2008b). Some calculation results for the interaction of hydrogen atoms and molecules with carbon nanotubes are also presented.

23.4.1 Lifshitz-type formulas for the interaction of an atom with a multiwalled carbon nanotube

A multiwalled carbon nanotube can be modeled as a cylindrical shell of thickness d made of a uniaxial crystal. It is assumed that the crystal optic axis z is perpendicular to the cylindrical surface of the crystalline layers. The thickness d of the nanotube is assumed to be large enough that the nanotube contains sufficiently many layers. It is then possible to neglect the atomic structure of graphite and to describe it in terms of the dielectric permittivity. As is shown below, a nanotube with only three or four layers can be described in this approximation (the separation distance between two neighboring hexagonal layers in graphite is equal to 3.4 Å).

Now we shall obtain the Lifshitz-type formula describing the free energy of a microparticle located at a separation a from the external surface of a cylindrical shell of radius R and thickness d made of a uniaxial crystal. The crystalline material of the cylindrical shell is described by the dielectric permittivities $\varepsilon_x(\omega)$ and $\varepsilon_z(\omega)$ (see Sections 12.8.1 and 16.7.1). The derivation follows the same lines as in Section 16.2 for an atom above a plane plate. The cylindrical shape of the

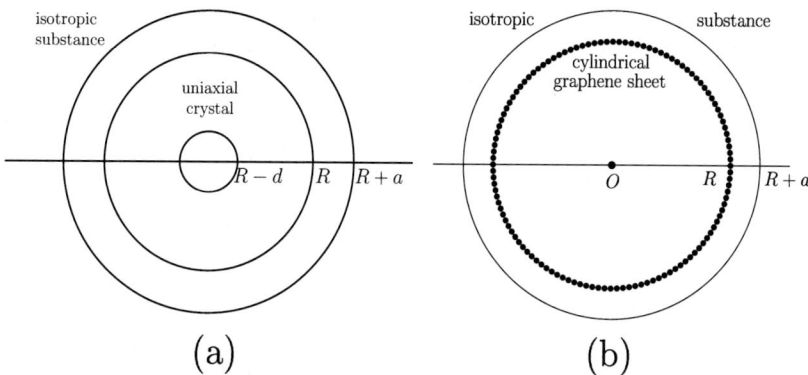

FIG. 23.3. Schematic illustration of (a) a cylindrical shell of radius R and thickness d made of a uniaxial crystal (Blagov et al. 2005) and (b) a cylindrical graphene sheet of radius R, which are concentrically placed into a cylindrical cavity of radius $R + a$ in an infinite space filled with an isotropic substance (Blagov et al. 2007).

nanotube is taken into account using the PFA. Let us consider an infinite space filled with an isotropic substance having a dielectric permittivity $\varepsilon(\omega)$, containing an empty cylindrical cavity of radius $R + a$. We introduce our cylindrical shell inside this cavity so that the cylinder axis coincides with the axis of the cavity [see Fig. 23.3(a)]. There is a gap of thickness a between the cylindrical shell and the boundary surface of the cylindrical cavity. Each element of the cylindrical shell experiences an attractive van der Waals (or Casimir) interaction from the boundary surface of the cylindrical cavity. With the help of the PFA, the free energy of this interaction can be represented in the form (Mazzitelli 2004)

$$\mathcal{F}^{c,c}(a,T) = 2\pi L \sqrt{R(R+a)} \mathcal{F}^{p,s}(a,T). \tag{23.10}$$

Here, $\mathcal{F}^{p,s}(a,T)$ is the free energy per unit area in the configuration of a flat plate of thickness d and a semispace separated by a gap of width a, where the plate is made of a uniaxial crystal and the semispace is made of a material with a dielectric permittivity $\varepsilon(\omega)$. In eqn (23.10), L is the length of the cylindrical shell, which is supposed to be much larger than its radius R.

As shown by Mazzitelli (2004) for the case of an ideal metal [see also similar results by Mazzitelli et al. (2003)], the accuracy of eqn (23.10) is rather high. For example, within the range of separations $0 < a < R/2$, the results calculated with eqn (23.10) coincide with the exact ones within 1% (for real materials, the accuracy may be only slightly different).

An explicit expression for the free energy $\mathcal{F}^{p,s}(a,T)$ is given by eqn (12.71), where the reflection coefficients $r^{(1)}_{\text{TM,TE}}$ are defined by eqn (12.137) and the reflection coefficients $r^{(2)}_{\text{TM,TE}}$ are defined by eqn (12.31). We now suppose that the isotropic substance is sufficiently rarefied, with N atoms or molecules per unit volume. Expanding the quantity $\mathcal{F}^{c,c}(a,T)$ on the left-hand side of eqn (23.10) in powers of N and using the additivity of the first-order term, we can write, in analogy with eqn (16.15),

$$\mathcal{F}^{c,c}(a,T) = N \int_a^\infty \mathcal{F}^{A,c}(z,T) 2\pi(R+z) L\, dz + O(N^2), \qquad (23.11)$$

where $\mathcal{F}^{A,c}(z,T)$ is the free energy of the dispersion interaction of a single atom of an isotropic substance with a cylindrical shell made of a uniaxial crystal (note that the separation z is measured perpendicular to the external surface of the cylindrical shell).

By differentiating both sides of eqn (23.11) with respect to a, we obtain

$$-\frac{\partial \mathcal{F}^{c,c}(a,T)}{\partial a} = 2\pi(R+a) L N \mathcal{F}^{A,c}(a,T) + O(N^2). \qquad (23.12)$$

The same derivative can be found by differentiating both sides of eqn (23.10)

$$-\frac{\partial \mathcal{F}^{c,c}(a,T)}{\partial a} = 2\pi L \sqrt{R(R+a)} \left[-\frac{\mathcal{F}^{p,s}(a,T)}{2(R+a)} + P^{p,s}(a,T) \right], \qquad (23.13)$$

where

$$P^{p,s}(a,T) = -\frac{\partial \mathcal{F}^{p,s}(a,T)}{\partial a} \qquad (23.14)$$

is the pressure between a flat plate made of a uniaxial **crystal** and a semispace with a dielectric permittivity ε. The expression for this pressure is given in eqn (12.71), with the reflection coefficients as specified above for the free energy $\mathcal{F}^{p,s}(a,T)$.

The dielectric permittivity of a rarefied isotropic substance can be expanded in a Taylor series in accordance with eqn (16.3) with $n=2$ (below, the index 2 on α and N is omitted). Then, for the reflection coefficients of an isotropic semispace, we obtain eqn (16.13). From this, the free energy $\mathcal{F}^{p,s}(a,T)$ is represented by eqn (16.14) with $r_{\text{TM,TE}}^{(1)} = r_{\text{TM,TE}}^{(u)}$, as defined in eqn (12.137). Then, using eqn (23.14), we obtain the following expression for the pressure:

$$P^{p,s}(a,T) = -k_B T N \sum_{l=0}^\infty{}' \alpha(i\xi) \int_0^\infty k_\perp dk_\perp q_l \qquad (23.15)$$

$$\times \left[\left(2 - \frac{\xi_l^2}{q_l^2 c^2}\right) r_{\text{TM}}^{(u)}(i\xi_l, k_\perp) - \frac{\xi_l^2}{q_l^2 c^2} r_{\text{TE}}^{(u)}(i\xi_l, k_\perp) \right] e^{-2aq_l} + O(N^2).$$

Substituting eqn (23.15) into eqn (23.13), we find

$$-\frac{\partial \mathcal{F}^{c,c}(a,T)}{\partial a} = -2\pi L N k_B T \sqrt{R(R+a)} \sum_{l=0}^\infty{}' \alpha(i\xi) \int_0^\infty k_\perp dk_\perp\, e^{-2aq_l} \qquad (23.16)$$

$$\times \left[q_l - \frac{1}{4(R+a)}\right] \left[\left(2 - \frac{\xi_l^2}{q_l^2 c^2}\right) r_{\text{TM}}^{(u)}(i\xi_l, k_\perp) - \frac{\xi_l^2}{q_l^2 c^2} r_{\text{TE}}^{(u)}(i\xi_l, k_\perp)\right] + O(N^2).$$

Finally, we substitute eqn (23.16) into the left-hand side of eqn (23.12), take the limit $N \to 0$, and arrive at the desired expression for the free energy of interaction of a microparticle and a cylindrical shell made of a uniaxial crystal (Blagov et al. 2005),

$$\mathcal{F}^{A,c}(a,T) = -k_B T \sqrt{\frac{R}{R+a}} \sum_{l=0}^{\infty}{}' \alpha(i\xi) \int_0^\infty k_\perp dk_\perp \, e^{-2aq_l} \qquad (23.17)$$

$$\times \left[q_l - \frac{1}{4(R+a)}\right] \left\{2r_{\text{TM}}^{(u)}(i\xi_l, k_\perp) - \frac{\xi_l^2}{q_l^2 c^2} \left[r_{\text{TM}}^{(u)}(i\xi_l, k_\perp) + r_{\text{TE}}^{(u)}(i\xi_l, k_\perp)\right]\right\}.$$

In the limiting case $R \to \infty$, this equation coincides with the known result (16.16) for the free energy of a microparticle near a plane surface. In Section 23.4.3, eqn (23.17) is applied to compute the free energy of the van der Waals and Casimir interaction of hydrogen atoms and molecules with multiwalled carbon nanotubes.

23.4.2 Lifshitz-type formulas for graphene and single-walled carbon nanotubes

The classical idealization of the dielectric permittivity used in the previous subsection to describe multiwalled nanotubes is not applicable for the description of single-walled nanostructures. This narrows the applicability of the standard Lifshitz theory and leads to application of phenomenological approaches such as density-functional theory (Dobson et al. 2006). There is, however, a possibility to obtain Lifshitz-type formulas in the case of single-walled nanostructures. The point is that some properties of a hexagonal monoatomic sheet of C atoms (graphene) admit a simplified model description in terms of a two-dimensional free-electron gas. In this description, the graphene sheet is characterized by some typical wave number K determined by the parameters of the hexagonal structure of graphite. Barton (2004, 2005) considered the interaction of electromagnetic oscillations with such a sheet and found the reflection coefficients. Bordag (2006b) obtained a Lifshitz-type formula for the van der Waals and Casimir interaction between two parallel plasma sheets. Using this model, the interaction between graphene and a material plate, between graphene and an atom or a molecule, and between a single-walled carbon nanotube and a material plate was also described by means of Lifshitz-type formulas (Bordag et al. 2006). Finally, Blagov et al. (2007) obtained a Lifshitz-type formula for the van der Waals and Casimir interaction between an atom or molecule and a single-walled carbon nanotube.

We begin by noting that the reflection coefficients for a sheet of graphene cannot be obtained from eqn (12.137) in the limit $d \to 0$ [in fact, the coefficients (12.137) go to zero when d vanishes]. The reason is that the case of a *thin* plate implies that d/a is sufficiently small, whereas d must be large enough for the validity of the macroscopic description in terms of ε.

Let us describe graphene as an infinitely thin plasma sheet, where the π-electrons are treated as a continuous, charged fluid moving in an immobile, overall neutralizing background of positive charge. Such plasma sheets have been considered by Fetter (1973) and, more recently, by Barton (2004, 2005) in connection with the Casimir effect for a fullerene and for a single basal plane of graphite. The plasma sheet model describes a charged planar fluid film by a two-dimensional displacement vector $\boldsymbol{R}(x,y)\exp(-i\omega t)$. The fluid has a charge ne and a mass nm per unit area of the film, where e and m are the electron charge and mass, respectively. For the hexagonal structure of a carbon layer, there is one π-electron per atom, resulting in two π-electrons per hexagonal cell. This leads to $n = 4/(3\sqrt{3}l^2)$, where $l = 1.421$ Å is the side length of the hexagon. The fluid provides a source for the Maxwell equations with the following surface charge and surface current densities:

$$\sigma = -ne\boldsymbol{\nabla}_t \cdot \boldsymbol{R}, \quad \boldsymbol{j} = -i\omega ne\boldsymbol{R}, \qquad (23.18)$$

where the operator $\boldsymbol{\nabla}_t$ acts in the tangential direction to the sheet [here and below, σ, $\boldsymbol{j}$, and the fields $\boldsymbol{E}$ and $\boldsymbol{B}$ depend on the coordinates; their dependence on time is obtained through multiplication by a common factor $\exp(-i\omega t)$]. The Maxwell equations are

$$\boldsymbol{\nabla} \cdot \boldsymbol{E} = 4\pi\sigma\delta(z), \quad \boldsymbol{\nabla} \times \boldsymbol{E} - \frac{i\omega}{c}\boldsymbol{B} = 0, \qquad (23.19)$$

$$\boldsymbol{\nabla} \cdot \boldsymbol{B} = 0, \quad \boldsymbol{\nabla} \times \boldsymbol{B} + \frac{i\omega}{c}\boldsymbol{E} = \frac{4\pi}{c}\boldsymbol{j}\delta(z).$$

By integration of these equations across the sheet, we obtain the matching conditions on the tangential and normal components of the fields (Barton 2005),

$$\boldsymbol{E}_{t,2} - \boldsymbol{E}_{t,1} = 0, \quad E_{z,2} - E_{z,1} = 2K\frac{c^2}{\omega^2}\boldsymbol{\nabla}_t \cdot \boldsymbol{E}_t, \qquad (23.20)$$

$$B_{z,2} - B_{z,1} = 0, \quad \boldsymbol{B}_{t,2} - \boldsymbol{B}_{t,1} = -2iK\frac{c}{\omega}\boldsymbol{j} \times \boldsymbol{E}_t.$$

Here $\boldsymbol{j} = (0, 0, 1)$ is the unit vector pointing in the z-direction, and the wave number of the sheet is

$$K = 2\pi\frac{ne^2}{mc^2} = 6.75 \times 10^5 \, \text{m}^{-1}. \qquad (23.21)$$

This is the main characteristic of graphene in the model under consideration. The value in eqn (23.21) corresponds to the frequency $\omega_K = cK = 2.02 \times 10^{14}$ rad/s. From eqn (23.19), outside the surface [i.e. with $\delta(z) = 0$], one obtains the usual Poisson equations (12.12) with $\varepsilon(\omega) = 1$ for all components of the fields. These equations together with the matching conditions (23.20) provide a complete description of the interaction of an electromagnetic field with the plasma sheet.

After the separation of variables in eqn (12.12) with $\varepsilon(\omega) = 1$ and (23.20), we arrive at a one-dimensional scattering problem. The solution of this problem leads to the following reflection coefficients for the graphene plasma sheet taken along the imaginary frequency axis (Barton 2005):

$$r_{\text{TM}}^{(1)}(i\xi_l, k_\perp) = \frac{c^2 q_l K}{c^2 q_l K + \xi_l^2}, \qquad r_{\text{TE}}^{(1)}(i\xi_l, k_\perp) = -\frac{K}{K + q_l}. \tag{23.22}$$

The reflection coefficients (23.22) can be substituted into eqn (12.71) for the Casimir free energy per unit area and the Casimir pressure. By choosing the reflection coefficients $r_{\text{TM,TE}}^{(2)}$ in these formulas, as defined in eqns (12.67) or (12.52), we obtain Lifshitz-type formulas for the free energy and the pressure in the configuration of a graphene sheet interacting with a semispace or with a plate of thickness d, respectively, made of an isotropic material. The substitution $r_{\text{TM,TE}}^{(2)} = r_{\text{TM,TE}}^{(1)}$ results in the free energy and pressure between two graphene sheets (Bordag 2006b). The substitution of $r_{\text{TM,TE}}^{(2)} = r_{\text{TM,TE}}^{(u)}$ obtained from eqn (12.135) or (12.137) allows us to calculate the Casimir free energy or pressure in the configuration of a graphene sheet interacting with a graphite semispace or a graphite sheet of finite thickness. A Lifshitz-type formula for an atom or a molecule interacting with graphene is obtained by the substitution of eqn (23.22) into eqn (16.16) or (16.17) for the free energy and force for the atom–wall interaction. Some of the computational results are presented in the next subsection.

Now we obtain the Lifshitz-type formula for the interaction of a single-walled nanotube of radius R with a thick material plate (Bordag et al. 2006). Let the nanotube lie along the y-axis at a separation a from the plate. For sufficiently small $a \ll R$, the desired result can be obtained using the PFA. We start with the Lifshitz-type formula (12.44) for the energy of the dispersion interaction at zero temperature between graphene and a material semispace. Here, the reflection coefficients $r_{\text{TM,TE}}^{(1)}$ are given by eqn (23.22) and the coefficients $r_{\text{TM,TE}}^{(2)}$ by eqn (12.45). Using the PFA, we replace the cylindrical surface of the nanotube by a set of long plane strips of width dx. The interaction of each strip, substituted for a part of the cylindrical surface, and the opposite strip belonging to the plate is calculated using eqn (12.44). The separation distance between the two opposite strips with coordinate x is

$$z = z(x) = a + R - \sqrt{R^2 - x^2}. \tag{23.23}$$

Expanding the logarithms in eqn (12.44) in a power series, the interaction energy per unit area between the strips can be presented in the form

$$E[z(x)] = -\frac{\hbar}{4\pi^2} \int_0^\infty k_\perp \, dk_\perp \int_0^\infty d\xi \sum_{n=1}^\infty \frac{1}{n} \tag{23.24}$$

$$\times \left[\left(r_{\rm TM}^{(1)} r_{\rm TM}^{(2)}\right)^n + \left(r_{\rm TE}^{(1)} r_{\rm TE}^{(2)}\right)^n\right] {\rm e}^{-2z(x)qn}.$$

To find the interaction energy $E_{\rm ns}$ per unit length between the semispace and the nanotube, we integrate eqn (23.24) from $x = -R$ to $x = R$ (the result of this is equal to twice the integral from zero to R). The integration variable $k_\perp$ is replaced with q, leading to

$$E_{\rm ns}(a) = -\frac{\hbar}{2\pi^2} \int_0^\infty q\,dq \int_0^{cq} d\xi \sum_{n=1}^\infty \frac{1}{n} {\rm e}^{-2aqn} \quad (23.25)$$

$$\times \left[\left(r_{\rm TM}^{(1)} r_{\rm TM}^{(2)}\right)^n + \left(r_{\rm TE}^{(1)} r_{\rm TE}^{(2)}\right)^n\right] \int_0^R dx\, {\rm e}^{-2qn(R-\sqrt{R^2-x^2})}.$$

By introducing a new variable $s = 1 - \sqrt{1 - x^2/R^2}$, the integral with respect to x in eqn (23.25) (which we denote by I) can be written in the form

$$I = R \int_0^1 \frac{(1-s)\,ds}{\sqrt{s(2-s)}} {\rm e}^{-2qnRs}. \quad (23.26)$$

The major contributions in eqns (23.25) and (23.26) come from $q \sim 1/a$. Taking into account the fact that the PFA works well for $R \gg a$, we conclude that the magnitude of the integral I is determined by the behavior of the integrand around the lower integration limit. Neglecting s in comparison with unity in eqn (23.26), we arrive at

$$I = \frac{R}{\sqrt{2}} \int_0^1 \frac{1}{\sqrt{s}} {\rm e}^{-2qnRs}\,ds = \frac{1}{2}\sqrt{\frac{\pi R}{qn}}\,{\rm erf}(\sqrt{2qnR}), \quad (23.27)$$

where ${\rm erf}(z)$ is the error function. Using once more the conditions $q \sim 1/a$ and $R \gg a$, we conclude that ${\rm erf}(\sqrt{2qnR}) \approx 1$ and obtain from eqn (23.25)

$$E_{\rm ns}(a) = -\frac{\hbar\sqrt{\pi R}}{4\pi^2} \int_0^\infty \sqrt{q}\,dq \int_0^{cq} d\xi \sum_{n=1}^\infty \frac{1}{n\sqrt{n}} {\rm e}^{-2aqn}$$

$$\times \left[\left(r_{\rm TM}^{(1)} r_{\rm TM}^{(2)}\right)^n + \left(r_{\rm TE}^{(1)} r_{\rm TE}^{(2)}\right)^n\right] \quad (23.28)$$

$$= -\frac{\hbar\sqrt{R}}{4\pi^{3/2}} \int_0^\infty \sqrt{q}\,dq \int_0^{cq} d\xi \left[{\rm Li}_{3/2}\left(r_{\rm TM}^{(1)} r_{\rm TM}^{(2)} {\rm e}^{-2aq}\right) + {\rm Li}_{3/2}\left(r_{\rm TE}^{(1)} r_{\rm TE}^{(2)} {\rm e}^{-2aq}\right)\right],$$

where ${\rm Li}_p(z)$ is the polylogarithm function.

In a similar way, for the force per unit length between a nanotube and a semispace, we get (Bordag et al. 2006)

$$F_{\rm ns}(a) = -\frac{\hbar\sqrt{R}}{2\pi^{3/2}} \int_0^\infty q^{3/2}\,dq \int_0^{cq} d\xi \left[{\rm Li}_{1/2}\left(r_{\rm TM}^{(1)} r_{\rm TM}^{(2)} {\rm e}^{-2aq}\right) + {\rm Li}_{1/2}\left(r_{\rm TE}^{(1)} r_{\rm TE}^{(2)} {\rm e}^{-2aq}\right)\right]. \quad (23.29)$$

It is apparent that the Lifshitz-type formulas (23.28) and (23.29) can be adapted to describe the interaction of a single-walled nanotube with an isotropic plate of finite thickness or with an anisotropic semispace or plate of finite thickness by choosing the reflection coefficients $r^{(2)}_{\text{TM,TE}}$ accordingly. By choosing $r^{(2)}_{\text{TM,TE}} = r^{(1)}_{\text{TM,TE}}$, one obtains from eqns (23.28) and (23.29) the interaction energy and force for a nanotube and a graphene sheet.

We conclude this subsection with the Lifshitz-type formula describing the interaction between a microparticle and a single-walled carbon nanotube. This case can be treated in perfect analogy to the case of a multiwalled nanotube (see Section 23.4.1). To repeat the derivation of Section 23.4.1, it is convenient to introduce a single-walled nanotube of radius R inside a cylindrical cavity of radius $R+a$ in an isotropic substance with a dielectric permittivity ε at temperature T [see Fig. 23.3(b)]. By rarefying the isotropic substance, as described in Section 23.4.1, we arrive once again at eqn (23.17) for the free energy of the interaction between an atom and a single-walled nanotube, $\mathcal{F}^{A,n}(a,T)$, where the reflection coefficients $r^{(u)}_{\text{TM,TE}}$ must be replaced with $r^{(1)}_{\text{TM,TE}}$ given by eqn (23.22).

The van der Waals and Casimir force acting between a microparticle and a single-walled carbon nanotube is obtained as the negative derivative of eqn (23.17) with respect to the separation distance,

$$F^{A,n}(a,T) = -k_B T \sqrt{\frac{R}{R+a}} \sum_{l=0}^{\infty}{}' \alpha(i\xi) \int_0^\infty k_\perp dk_\perp\, e^{-2aq_l} \qquad (23.30)$$

$$\times \left[q_l^2 - \frac{3}{8(R+a)^2} \right] \left\{ 2r^{(1)}_{\text{TM}}(i\xi_l, k_\perp) - \frac{\xi_l^2}{q_l^2 c^2} \left[r^{(1)}_{\text{TM}}(i\xi_l, k_\perp) + r^{(1)}_{\text{TE}}(i\xi_l, k_\perp) \right] \right\}.$$

This equation also gives the force $F^{A,c}(a,T)$ between an atom and a multiwalled nanotube if one replaces the reflection coefficients $r^{(1)}_{\text{TM,TE}}$ with $r^{(u)}_{\text{TM,TE}}$.

The above Lifshitz-type formulas describing the interaction of nanotubes with graphene and microparticles were obtained using the PFA, i.e. they are applicable only at sufficiently small separations. Bordag (2007) derived Lifshitz-type formulas describing the interaction of a cylindrical plasma sheet with a dielectric semispace and with graphene using the method of functional determinants (see Chapter 10). This allowed the first corrections beyond the PFA to be found for these configurations. A consideration of the vacuum energy of a spherical plasma shell, which is relevant to the description of buckyballs, was performed by Bordag and Khusnutdinov (2008).

It should be emphasized that the description of a single-walled nanotube in the approximation of a two-dimensional gas of free electrons used in this section is only a simplified model. It does not claim to be a complete description of all nanotube properties, for example the chirality of a nanotube. Specifically, it remains unclear whether it is possible to describe nanotubes with metallic or semiconductor surfaces by varying only one parameter K in the reflection coefficients (23.22).

23.4.3 Computational results for atom–nanotube interaction

We start with the calculation of the van der Waals and Casimir interaction between a hydrogen atom or molecule and a multiwalled carbon nanotube. In Section 16.7, the interaction of hydrogen with a graphite plate was considered. Here, we use the dielectric permittivities of graphite along the imaginary frequency axis $\varepsilon_x(i\xi)$ and $\varepsilon_z(i\xi)$ presented in Section 16.7.1 and the dynamic atomic polarizabilities of hydrogen atoms and molecules considered in Section 16.7.2. For convenience in numerical computations, we rewrite eqn (23.17) for the interaction of a cylindrical shell (which models a nanotube) with an atom in the form

$$\mathcal{F}^{A,c}(a,T) = -\frac{C_3^c(a,T)}{a^3}, \qquad (23.31)$$

where

$$C_3^c(a,T) = \frac{k_B T}{8}\sqrt{\frac{R}{R+a}}\left\{\frac{4R+3a}{2(R+a)}\alpha(0) \qquad (23.32)\right.$$
$$+ \sum_{l=1}^{\infty} \alpha(i\zeta_l \omega_c) \int_{\zeta_l}^{\infty} y\, dy\, e^{-y}\left[y - \frac{a}{2(R+a)}\right]$$
$$\left. \times \left[2r_{\mathrm{TM}}^{(\mathrm{u})}(i\zeta_l, y) - \frac{\zeta_l^2}{y^2}\left[r_{\mathrm{TM}}^{(\mathrm{u})}(i\zeta_l, y) + r_{\mathrm{TE}}^{(\mathrm{u})}(i\zeta_l, y)\right]\right]\right\}.$$

The dimensionless variables used here were introduced in eqn (12.89), and the reflection coefficients expressed in terms of these variables are given in eqn (16.106).

Computations of the van der Waals coefficient C_3^c using eqn (23.32) were performed for a hydrogen atom located at a separation $a = 5$ nm from the external surface of a nanotube of radius $R = 20$ nm at $T = 300$ K. The computational results are presented in Fig. 23.4 as a function of the nanotube thickness d. Note that $d = 20$ nm corresponds to the case of a solid cylinder. As can be seen in Fig. 23.4, at $d = 11$ nm the magnitude of C_3^c is only 1% lower than the value obtained for a solid cylinder of radius $R = 20$ nm. For thinner cylindrical shells, smaller values of the van der Waals coefficient are obtained. The same is also true for the hydrogen molecule. A comparison of the free energies of a hydrogen atom outside and inside a nanotube shows that a position inside a multiwalled nanotube is energetically preferable (Blagov et al. 2005, Klimchitskaya et al. 2006a).

Now we briefly consider the computational results for the interaction of hydrogen atoms and molecules with a graphene sheet (Bordag et al. 2006). In this case the free energy is given by eqn (16.16) with the reflection coefficients (23.22). In terms of the van der Waals coefficient, eqn (16.16) takes the form of eqns (16.66) and (16.67). The computational results for C_3 as a function of separation are presented in Fig. 23.5. The solid and dashed lines are related to the interaction of graphene with a hydrogen atom and a hydrogen molecule,

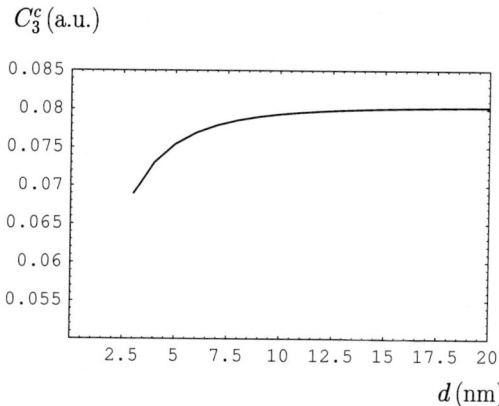

FIG. 23.4. Dependence of the van der Waals coefficient C_3^c on the thickness of a cylindrical shell with an external radius $R = 20\,\text{nm}$ for a hydrogen atom at a separation $a = 5\,\text{nm}$ from the shell (Blagov et al. 2005).

respectively. As can be seen in Fig. 23.5, the magnitude of the van der Waals coefficient for a hydrogen molecule interacting with graphene is larger than for an atom at all separations.

The next configuration considered in Section 23.4.2 is that of a single-walled nanotube near a semispace or a plate of finite thickness. In this case the interaction energy and force are given by eqns (23.28) and (23.29), respectively. We present the computational results for a nanotube of radius R in close proximity to an Au and an Si semispace. The reflection coefficients $r_{\text{TM,TE}}^{(1)}$ are given by eqn (23.22). The reflection coefficients $r_{\text{TM,TE}}^{(2)}$ are defined in eqn (12.31). The

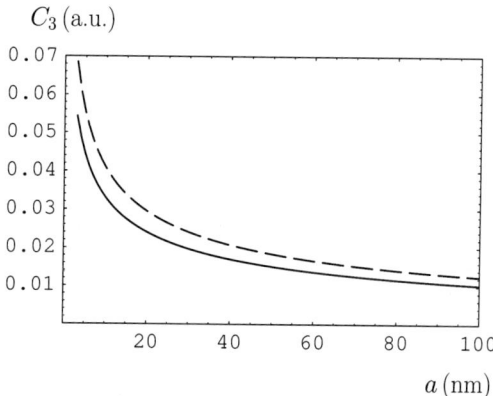

FIG. 23.5. Dependence of the van der Waals coefficient C_3 on separation for the interaction of a hydrogen atom (solid line) and molecule (dashed line) with graphene (Bordag et al. 2006).

TABLE 23.1. Ratios $E_{\text{ns}}(a)/E_{\text{IM}}^{\text{c}}(a)$ and $F_{\text{ns}}(a)/F_{\text{IM}}^{\text{c}}(a)$ for the van der Waals interaction of a carbon nanotube with Au and Si semispaces.

a (nm)	$E_{\text{ns}}(a)/E_{\text{IM}}^{\text{c}}(a)$ Au	Si	$F_{\text{ns}}(a)/F_{\text{IM}}^{\text{c}}(a)$ Au	Si
1	0.0151	0.0126	0.0114	0.00945
1.5	0.0193	0.0162	0.0147	0.0123
2	0.0230	0.0193	0.0175	0.0147
2.5	0.0262	0.0221	0.0201	0.0169
3	0.0291	0.0245	0.0224	0.0189

dielectric permittivities of Si and Au along the imaginary frequency axis are presented in Sections 12.6.1 and 13.3, respectively. A few computational results for the normalized interaction energy $E_{\text{ns}}/E_{\text{IM}}^{\text{c}}$ and the force $F_{\text{ns}}/F_{\text{IM}}^{\text{c}}$ per unit length, where E_{IM}^{c} and F_{IM}^{c} are the PFA results for the energy and force per unit length for an ideal-metal cylinder near an ideal-metal plane defined in eqn (6.54), are presented in Table 23.1. Column 1 shows the separation distance. Columns 2 and 3 contain the values of $E_{\text{ns}}(a)/E_{\text{IM}}^{\text{c}}(a)$ for Au and Si, respectively. Columns 4 and 5 contain the analogous values of $F_{\text{ns}}(a)/F_{\text{IM}}^{\text{c}}(a)$. As is seen in Table 23.1, the magnitudes of the normalized energies and forces for Si are smaller than those for Au and they are monotonically increasing functions with increasing separation. Note that the normalized magnitudes in Table 23.1 do not depend on the nanotube radius. However, keeping in mind that the largest diameter of a single-walled carbon nanotube is about 10 nm, the range of separations where the approximate equations (23.28) and (23.29) are applicable is very narrow.

The last point to be considered here is the calculation of the van der Waals coefficient $C_3^{\text{n}}(a,T)$ for the interaction of hydrogen atoms and molecules with single-walled carbon nanotubes. An expression for $C_3^{\text{n}}(a,T)$ is given by eqn (23.32), where the reflection coefficients $r_{\text{TM,TE}}^{(u)}$ are replaced with $r_{\text{TM,TE}}^{(1)}$ defined in eqn (23.22). The computations were performed for a nanotube with $R = 5$ nm at $T = 300$ K (Blagov et al. 2007). In Fig. 23.6(a), the computational results for C_3^{n} are plotted as a function of separation. The solid line 1 labels the case of an H atom and the solid line 2 the case of an H_2 molecule. For comparison, in the same figure, the previously computed van der Waals coefficients (see Fig. 23.5) for the interaction of an H atom (dashed line 1) and an H_2 molecule (dashed line 2) with a plane graphene sheet are included. As can be seen in Fig. 23.6(a) (solid lines 1 and 2), at all separations, the van der Waals coefficient for the molecule–nanotube interaction is larger than that for the atom–nanotube interaction. At the same time, the van der Waals coefficients for the interaction of an atom or a

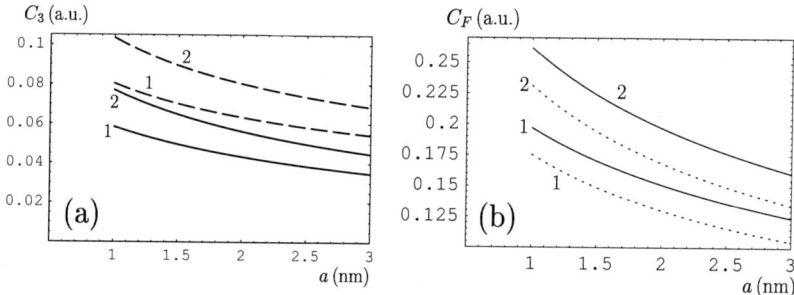

FIG. 23.6. The van der Waals coefficient (a) C_3 and (b) C_F as a function of separation for the interaction of a hydrogen atom (lines labeled 1) and a molecule (lines labeled 2) with a single-walled carbon nanotube of radius $R = 5$ nm (solid lines) and with a plane graphene sheet (dashed lines). [See Blagov et al. (2007).]

molecule with graphene (dashed lines 1 and 2, respectively) are larger than the respective coefficients for the interaction with a single-walled carbon nanotube. The increase is about 30%. Note that all results discussed in this subsection are almost independent of temperature in the temperature region from 0 to 300 K.

The force (23.30) acting between an atom or molecule and a single-walled carbon nanotube can also be presented in the form

$$F^{A,n}(a,T) = -\frac{C_F^n(a,T)}{a^4}, \qquad (23.33)$$

where

$$C_F^n(a,T) = \frac{k_B T}{8}\sqrt{\frac{R}{R+a}}\left\{\frac{3(2R+3a)(2R+a)}{2(R+a)^2}\alpha(0)\right.$$
$$+ \sum_{l=1}^{\infty} \alpha(i\zeta_l\omega_c)\int_{\zeta_l}^{\infty} dy\, y\, e^{-y}\left[y^2 - \frac{3a^2}{4(R+a)^2}\right]$$
$$\left.\times \left[2r_{\rm TM}^{(1)}(i\zeta_l,y) - \frac{\zeta_l^2}{y^2}\left[r_{\rm TM}^{(1)}(i\zeta_l,y) + r_{\rm TE}^{(1)}(i\zeta_l,y)\right]\right]\right\}. \qquad (23.34)$$

In Fig. 23.6(b), the coefficient C_F^n is plotted as a function of separation for an H atom (solid line 1) and an H_2 molecule (solid line 2) interacting with a nanotube of radius $R = 5$ nm. Similarly to the coefficients C_3, the values of C_F for a molecule are larger than for an atom at all the separations considered. The dotted lines in Fig. 23.6(b) (labeled 1 for an atom and 2 for a molecule) represent the calculation results obtained with the assumption that $C_F = 3C_3$, i.e. for the case where the van der Waals coefficient $C_3 = $ const and does not depend on separation distance. As is seen in Fig. 23.6(b), the differences between the solid and dotted lines are about 15%–20%. Thus, the dependence of the van der Waals coefficient on separation for a microparticle–nanotube interaction is important for obtaining precise computational results.

It is interesting to compare the van der Waals coefficients computed for the interaction of a hydrogen atom or molecule with single-walled and multiwalled carbon nanotubes. This permits us to determine how thick a multiwalled nanotube must be in order for the idealization of the dielectric permittivities of graphite to be applicable.

The van der Waals coefficient C_3^c for a multiwalled nanotube is given by eqn (23.32). The same equation gives the coefficient C_3^n for a single-walled nanotube if the reflection coefficients $r_{\text{TM,TE}}^{(u)}$ are replaced with $r_{\text{TM,TE}}^{(1)}$. The computational results for the coefficients C_3^c (for numbers of walls $k = 2, 3, 4, 5$) and C_3^n ($k = 1$) are presented in Fig. 23.7(a) for a hydrogen atom. The solid dots marked 1, 2, and 3 indicate the calculation results for C_3^n for a single-walled carbon nanotube of 5 nm radius at separations $a = 1, 2$, and 3 nm, respectively, from the atom. The solid dots connected by solid lines represent the magnitudes of C_3^c for multiwalled nanotubes of 5 nm external radius with $k = 2, 3, 4$, and 5 at the same separations from the atom. The dashed lines provide a smooth interpolation between the computational results for multiwalled and single-walled nanotubes. Note that the thickness of a multiwalled nanotube is related to the number of walls by $d = 3.4(k-1)$ Å. As can be seen in Fig. 23.7(a), the van der Waals coefficient C_3^n is different from C_3^c with $k = 1$. This is expected because the reflection coefficients in eqn (16.106) approach zero when the nanotube thickness d vanishes. At the same time, as Fig. 23.7(a) suggests, the coefficients C_3^c for a multiwalled nanotube with $k = 3$ walls at 1 nm from the atom and for a multiwalled nanotube with $k = 2$ walls at a separation of 2 or 3 nm from the atom are in proper proportion to the coefficients C_3^n computed for a single-walled carbon nanotube. This allows us to conclude that the macroscopic concept of the dielectric permittivity of graphite is applicable even for nanotubes containing only two or three walls, depending on the separation distance between the nanotube and the atom.

Figure 23.7(b) contains the same information as in Fig. 23.7(a) but for a hydrogen molecule. It can be seen that the approximation of the dielectric permittivities of graphite is a good approximation for a multiwalled nanotube with three walls at a separation of 1 nm from a molecule and for a two-walled nanotube at a separation of 2 or 3 nm from a molecule. The values of the van der Waals coefficient for a molecule are, however, larger than for an atom, as was discussed above.

The model of a continuous plasma sheet considered above is not applicable at separations below 1 nm between an atom and a nanotube, where other forces in addition to dispersion interactions should be taken into account. Using the method of phenomenological potentials and disregarding the role of chemical forces, Klimchitskaya *et al.* (2008b) have shown that below 1 nm, exchange repulsion gives rise to a lateral force that moves H atoms towards the cell centers. In the position above a cell center, the repulsive force cannot balance the van der Waals attraction. As a result, the atom penetrates inside the nanotube. This

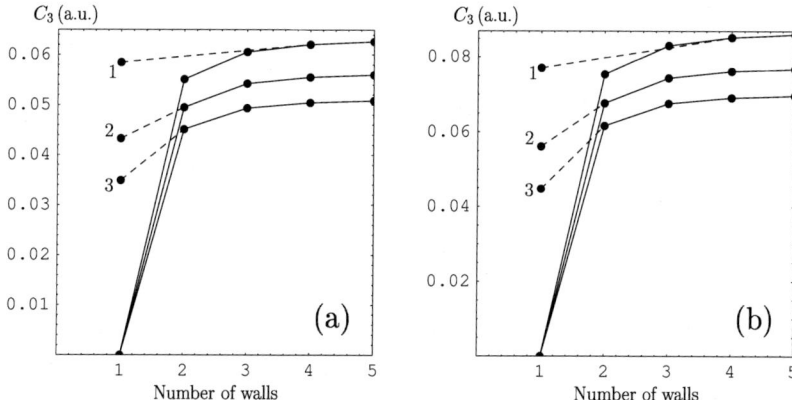

FIG. 23.7. The van der Waals coefficient C_3 as a function of the number of walls for the interaction of (a) a hydrogen atom and (b) a hydrogen molecule with multiwalled carbon nanotubes of different wall numbers and a single-walled carbon nanotube of the same radius (solid dots 1, 2, and 3) spaced at 1, 2, and 3 nm from the atom (or molecule), respectively (Blagov et al. 2007). See text for further discussion.

effect is analogous to the discontinuities in the constant-force surfaces that arise when a monoatomic tip of an atomic force microscope is scanned above a closely packed lattice in contact mode (Blagov et al. 1996, 1998, 1999).

23.5 Prospective applications

Many experimental and theoretical results on the Casimir effect considered in Parts II and III of this book may find applications in nanotechnology. In this chapter some of these applications, for instance MEMS actuators based on the Casimir force, have been discussed. In a reversal of roles, MEMS have already been exploited for precision measurements of the Casimir force (see Section 19.3). It should be noted that even the AFM cantilever used in the measurements of the Casimir force by means of an AFM (see Section 19.2) is one of the most important MEMS elements based on silicon chip nanofabrication technology. It has been shown (Chumak et al. 2004) that both electrostatic and Casimir forces have a strong effect on the vibrations of a cantilever, depending on the geometry of the tip. This demonstrates that the investigations in the fields of nanotechnology and Casimir effect are closely related.

In fact, some of the previously performed experiments on the measurement of the Casimir force discussed above may find nanotechnological applications. Thus, the effect of the optically modulated Casimir forces (Chen et al. 2007a, 2007b) considered in Section 20.3 could be used in the design and function of nanoscale actuators, micromirrors, and nanotweezers. The experimentally demonstrated phenomenon of the lateral Casimir force (Chen et al. 2002a, 2002b) gives the possibility to actuate not only normal but also lateral translations in nanodevices by means of electromagnetic zero-point oscillations (see Sections 21.2–21.4). The experiment by Chan et al. (2008) shows that by using a nanostructured silicon plate, it is possible to control the magnitude of the Casimir force (Section 21.5). The experiments by Chen et al. (2006a, 2006b) demonstrate that such control can also be achieved using different charge carrier densities in the semiconductor elements of microdevices (Sections 20.1 and 20.2). All these phenomena await future applications in nanotechnology.

In addition to the already demonstrated phenomena involving the Casimir force, there are some promising theoretical suggestions for nanotechnology. These are primarily about systems which suggest a possibility of a repulsive Casimir force, i.e. rectangular boxes (Chapter 8), spheres (Chapter 9), and three-layer systems (Sections 19.5.3 and 20.4.3). The realization of Casimir repulsion would help to resolve the problem of stiction, which is a challenge in the fabrication of ever more miniaturized microdevice elements. Specifically, there is always a liquid between the top and bottom Si microscopic elements during the manufacture of microdevices. If visible light of sufficient intensity was to shine on the device from the top, then nearly all of it would be absorbed by the top element. This could lead to a repulsive Casimir force, as described in Section 20.4.3. Thus, if the manufacture of MEMS is done under a source of bright light, stiction should be substantially reduced. However, many questions concerning the possibility of a Casimir repulsion between real materials still remain to be answered.

The introduction of micromechanical and nanomechanical devices has brought to light a host of new engineering problems not seen or anticipated before. In particular, owing to the large surface areas at the short separations involved, tribological effects (friction and adhesion) result in the wearing-out of device components. As traditional lubricants display large viscosities at the molecular scale, this problem needs an urgent solution. In this regard, proposals for nanodevices that transmit motion without contact are highly desirable. Two such proposals (Ashourvan et al. 2007a, 2007b, Emig 2007) for bringing about continuous linear motion of one corrugated surface by a periodic or linear motion of another were mentioned in Section 21.4.

Ashourvan et al. (2007a, 2007b) proposed the frictionless transduction of motion via the lateral Casimir force. They put forward a design of a nanoscale rack and pinion without intermeshing cogs. The suggested system consists of a corrugated plate (rack) and a corrugated cylinder (pinion) that are kept away from contact. Uniform sinusoidal corrugations with the same amplitude and period on both surfaces were considered. For uniform linear motion of the rack, it was found that the pinion velocity was locked to the rack velocity until a threshold was reached and the pinion underwent a skipping transition, after which it could no longer hold the corrugations in registry with those of the rack. In the skipping regime, it was found that the average pinion velocity could be positive or negative depending on the initial phase difference between the corrugations of the rack and pinion. The effect of an external load on the pinion and the effect of friction were also analyzed.

In the second such idea of a Casimir-force-driven ratchet, Emig (2007) suggested the use of asymmetric corrugations, which leads to the breaking of the reflection symmetry between the two surfaces, and therefore the surfaces could be set into relative lateral motion in the direction of the broken symmetry. The energy for this motion is provided by an external driver that periodically changes the surface separation between the corrugations in the normal direction. The transport velocity under the influence of the lateral Casimir force was shown to be stable across sizable intervals of the amplitude and frequency of modulation of the separation distance between the surfaces, even with the inclusion of damping. In these stable intervals the velocity of the lateral motion scales linearly with the frequency and is almost constant below some critical surface separation, beyond which it was shown to drop sharply.

If it is possible to generalize the above approaches to the case of real materials, they could be used for transferring motion between two MEMS without mechanical contact. This might be a solution to the tribological problems plaguing the MEMS industry.

24

CONSTRAINTS ON HYPOTHETICAL INTERACTIONS FROM THE CASIMIR EFFECT

Many extensions to the Standard Model of elementary particles predict the existence of long-range interactions between neutral macrobodies in addition to Newtonian gravity. The constraints on these interactions have traditionally been obtained from gravitational experiments (Fischbach and Talmadge 1999). Kuzmin *et al.* (1982) were the first to suggest that constraints on hypothetical Yukawa-type long-range interactions could be obtained from measurements of the van der Waals force. Mostepanenko and Sokolov (1987a, 1987b) have shown that measurements of the Casimir force lead to strong constraints on power-type long-range interactions. The availability of new precise measurements of the Casimir force, considered in Chapter 19, has provided further impetus for rapid progress in this direction. As a result, in the last few years, the previously known constraints on Yukawa interactions in the submicrometer range have been strengthened by up to ten thousand times. In this chapter we summarize the results obtained in comparison with parallel progress in gravitational measurements.

24.1 Long-range forces and constraints on them from gravitational experiments

Here, we briefly discuss two of the main theoretical schemes, which predict both Yukawa-type and power-type long-range interactions. These are the exchange of light and massless elementary particles between the atoms of two separate macrobodies and extra-dimensional unification theories with a low-energy compactification scale. We also list the strongest constraints on the parameters of the hypothetical long-range interactions obtained to date from Eötvos- and Cavendish-type gravitational experiments.

24.1.1 *Light particles and extra-dimensional physics*

It is common knowledge that the concept of a spontaneously broken symmetry leads to the prediction of massless bosons. If some symmetry is broken not only in the vacuum but in the Lagrangian as well, an initially massless particle acquires a nonzero mass. Because of this, there are many predictions of new massless and light bosons in various theoretical schemes, such as the arion (Anselm and Uraltsev 1982), scalar axion (Peccei and Quinn 1977), graviphoton (Ferrara *et al.* 1977), dilaton (Fujii 1991), goldstino (Deser and Zumino 1977), and moduli (Dimopoulos and Giudice 1996), among others. In fact, new light and massless elementary particles are predicted by almost every unification model (De Sabbata *et al.* 1992). They are electrically neutral and possess extremely small interaction constants. This makes it difficult to investigate these particles using the usual laboratory setups of elementary particle physics.

The exchange of light bosons of mass $m = \hbar/(\lambda c)$ between two atoms generates an effective Yukawa-type potential. This is a hypothetical long-range interaction with an interaction range λ that can vary from 1 Å (a separation distance much larger than the size of a nucleus) to hundreds of meters or even longer. Considering that the Yukawa-type hypothetical interaction coexists with gravitation, it is customary to describe it as an addition to the Newtonian gravitational potential. When this is done, the total interaction energy between the two neutral point masses m_1 and m_2 (atoms) at a separation r due to gravitation and the exchange of light bosons takes the form (Adelberger *et al.* 1991, 2003)

$$V(r) = -\frac{Gm_1m_2}{r}\left(1 + \alpha e^{-r/\lambda}\right). \tag{24.1}$$

Here, G is the Newtonian gravitational constant and α is a dimensionless constant characterizing the strength of the Yukawa interaction. Note that, in the literature, the hypothetical long-range interaction coexisting with gravity is often referred to as a *fifth force*.

The exchange of one massless particle between two atoms leads to an effective potential which is inversely proportional to the separation. This is just the usual Coulomb potential. Effective potentials inversely proportional to higher powers of the separation distance appear if the exchange of an even number of pseudoscalar particles is considered. Thus, it has been shown that the exchange of two arions leads to an interaction potential between atomic electrons falling as r^{-3} (Mostepanenko and Sokolov 1987a). Such power-type potentials with higher powers of the separation arise also in the exchange of two neutrinos, two goldstinos, or other massless fermions [reviews have been presented by Fischbach (1996) and by Mostepanenko and Sokolov (1993)]. Constraints on power-type hypothetical interactions were considered by Feinberg and Sucher (1979). As a correction to Newtonian gravity, power-type potentials between two point-like masses can be represented in the form

$$V_l(r) = -\frac{Gm_1m_2}{r}\left[1 + \Lambda_l\left(\frac{r_0}{r}\right)^{l-1}\right]. \tag{24.2}$$

Here, Λ_l is a dimensionless constant, l is a positive integer, and $r_0 = 10^{-15}$ m (the latter is introduced to preserve the correct dimension for the interaction energy).

The total force acting between two macrobodies due to the potential (24.1) or (24.2) can be obtained by integration over the volumes of the two bodies, and the subsequent negative differentiation with respect to separation distance.

Another prediction of Yukawa-type corrections to Newtonian gravity comes from those extensions of the Standard Model which exploit the Kaluza–Klein unification approach. According to this approach, the true dimensionality of space–time is $D = 4 + N$, where the additional N spatial dimensions are compactified at some small length scale (see Section 11.4). For a long time it was generally believed that the compactification scale should be on the order of the Planck length $l_{Pl} = (\hbar G/c^3)^{1/2} \sim 10^{-33}$ cm. The corresponding energy scale, $E_{Pl} = (\hbar c^5/G)^{1/2} \sim 10^{19}$ GeV, is so high that direct experimental observation of the effects of extra dimensions would seem impossible in the foreseeable future.

The situation changed dramatically with the proposal of models for which the compactification energy may be as low as the extra-dimensional Planck energy scale, $E_{Pl}^{(D)} = (\hbar^{1+N} c^{5+N}/G_D)^{1/(2+N)}$, which is assumed to be of the order of 1 TeV (Antoniadis et al. 1998, Arkani-Hamed et al. 1998, 1999). Here, G_D is the fundamental gravitational constant in the extended D-dimensional space–time. It is related to the Newtonian gravitational constant by the equality $G_D = G\Omega_N$, where $\Omega_N \sim R_*^N$, and R_* is the size of the compact manifold (see Section 11.4). Note that this proposal eliminates the hierarchy problem, since the characteristic energy scales of the gravitational and gauge interactions coincide. In order to be consistent with observations, the usual gauge fields of the Standard Model are presumed to exist on four-dimensional branes, whereas gravity alone propagates into the D-dimensional bulk. The characteristic size of the compact manifold is given by (Arkani-Hamed et al. 1999)

$$R_* \sim \frac{\hbar c}{E_{Pl}^{(D)}} \left(\frac{E_{Pl}}{E_{Pl}^{(D)}}\right)^{2/N} \sim 10^{(32-17N)/N} \text{ cm}. \qquad (24.3)$$

The usual Newton's law of gravitation is valid only in a four-dimensional space–time. If extra dimensions exist, it is modified. It has been shown (Floratos and Leontaris 1999, Kehagias and Sfetsos 2000) that at separations $r \gg R_*$ the resulting gravitational potential has the form of eqn (24.1). For extra-dimensional theories with $N \geq 1$ extra dimensions, $\alpha \sim 1$ and $\lambda \sim R_*$. Thus, for $N = 1$ (one extra dimension) one finds $R_* \sim 10^{15}$ cm from eqn (24.3). The existence of such a large extra dimension is excluded by solar-system tests of Newtonian gravity (Fischbach and Talmadge 1999). If, however, $N = 2$ or 3 the sizes of the extra dimensions are $R_* \sim 1$ mm and $R_* \sim 5$ nm, respectively. Note that at separations below $10\,\mu$m, corrections to Newton's law that are many orders of magnitude larger than the standard gravitational interaction are not excluded experimentally.

For extra-dimensional models with noncompact (but warped) extra dimensions (Randall and Sundrum 1999a, 1999b), the interaction energy takes the form of a Newtonian potential with a power-type correction,

$$V_3(r) = -\frac{Gm_1m_2}{r}\left(1 + \frac{2}{3k^2r^2}\right), \qquad (24.4)$$

where $r \gg 1/k$ and $1/k$ is the warping scale. This is a particular case of the potential (24.2) with $l = 3$.

24.1.2 Eötvos- and Cavendish-type experiments

Constraints on the parameters α, λ, and Λ_l of hypothetical interactions in eqns (24.1) and (24.2) can be obtained from gravitational experiments of the Eötvos and Cavendish types. In Eötvos-type experiments, the difference between the inertial and gravitational masses of a body is measured, i.e. the equivalence principle is tested. The influence of a hypothetical force that is not proportional to the masses of the interacting bodies would lead to an effective difference between the inertial and gravitational masses. Thus, if such a difference is not registered within the limits of experimental errors, this places some limits on the magnitude of the hypothetical force. Constraints on the parameters of hypothetical interactions following from Eötvos-type experiments have been reviewed by Adelberger et al. (1991, 2003), Mostepanenko and Sokolov (1993), and Mostepanenko (2002b). Here, we present only the strongest constraints, following from the experiments by Schlamminger et al. (2008) and Smith et al. (2000). The constraints on a Yukawa-type interaction based on the data by Schlamminger et al. (2008) are shown by line 1 in Fig. 24.1. In this figure, the permitted regions of the (λ, α)-plane lie beneath the lines, whereas the regions above the lines are excluded by the results of one experiment or another. In a similar way, the constraints following from the experiment by Smith et al. (2000) are indicated by line 2.

Strong constraints on hypothetical interactions can also be obtained from Cavendish-type experiments. These experiments aim to measure probable deviations of the gravitational force $F_{\rm gr}$ from Newton's law. In the case of two point-like bodies a distance r apart, such deviations can be characterized by a dimensionless parameter

$$\varepsilon = \frac{1}{rF_{\rm gr}} \frac{d}{dr}\left(r^2 F_{\rm gr}\right). \qquad (24.5)$$

Numerous Cavendish-type experiments have been performed to date. Here, we present only those which lead to the strongest constraints on hypothetical interactions within particular interaction ranges. It is notable that the theoretical predictions of the possibility of large extra dimensions have spurred the performance of new experiments at as short a separation as possible.

The best constraints at moderate separations of about 1 cm follow from the experiment by Hoskins et al. (1985). These are shown by line 3 in Fig. 24.1. Recent experiment by Kapner et al. (2007) placed the strongest constraints on the parameters of the Yukawa interaction in a wide interaction range from $\lambda = 9 \times 10^{-6}$ m to $\lambda = 4 \times 10^{-3}$ m (line 4 in Fig. 24.1). At smaller λ, from 4.7×10^{-6} m to 9×10^{-6} m, the best constraints, shown by line 5, were obtained by Smullin et al. (2005). At $\lambda < 4.7 \times 10^{-6}$ m, the strongest constraints on Yukawa interactions (line 6 in Fig. 24.1) follow from measurements of the Casimir force (see the next section, where, in Fig. 24.2, line 6 is shown over a wider interaction range). Note that gravitational experiments at short separation distances are making rapid progress. Thus, Weld et al. (2008) have proposed a new apparatus for detecting micron-scale deviations from Newtonian gravity. The first

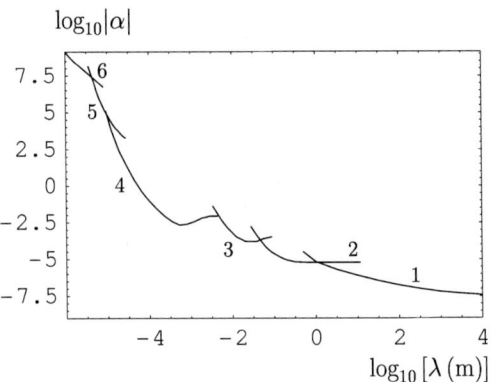

FIG. 24.1. The strongest constraints on Yukawa-type corrections to Newton's gravitational law following from various gravitational experiments (lines 1–5) and measurements of the Casimir force (line 6). The allowed values in the (λ, α)-plane lie beneath the lines (see text for further discussion).

constraints on the attractive hypothetical force obtained with the new apparatus at $\lambda = 5\,\mu$m are equal to the best existing constraints shown in Fig. 24.1. The constraints obtained on repulsive Yukawa-type forces are several times better. The prospective constraints that can be achieved by using the new apparatus are more than one order of magnitude stronger than the existing ones (Weld et al. 2008).

The constraints on Yukawa-type interactions shown in Fig. 24.1 place limits on the parameters of light hypothetical particles. Thus, for a dilaton in various models, $1 < |\alpha| < 2500$ (Kapner et al. 2007), and then line 4 restricts the possible range of the dilaton mass $m = \hbar/(\lambda c)$. In a similar way, various kinds of light hypothetical particles called *moduli*, used to determine the geometry of the extra dimensions, correspond to $\alpha < 10^5$ (Dimopoulos and Giudice 1996, Dimopoulos and Geraci 2003, Smullin et al. 2005, Kapner et al. 2007). Constraints on the ranges of masses of the fields of the moduli then follow from lines 1–4 in Fig. 24.1.

Figure 24.1 also constrains the possible number of extra dimensions. Thus, if N extra dimensions are compactified on a flat torus with the same radius R_* for each dimension, this leads to $\alpha = 8N/3$ and $\lambda = R_*$ (Adelberger et al. 2003). From Fig. 24.1, it can be seen that for $N = 1$ and 2 ($\alpha = 8/3$ and $16/3$) the permitted values of λ are less than 44 and $40\,\mu$m, respectively. These are much less than the respective values of R_* determined from eqn (24.3) (10^{15} cm and 1 mm). It then follows that the cases of $N = 1$ and 2 extra dimensions are experimentally excluded. However, for $N = 3$, we have $\alpha = 8$, and, from Fig. 24.1, $\lambda < 36\,\mu$m. This is consistent with the value $R_* \sim 5\,\mu$m obtained from eqn (24.3). Because of this, the possibility of three extra dimensions is consistent with the constraints presented in Fig. 24.1.

TABLE 24.1. Constraints on the constants of power-type potentials

l	$|\Lambda_l|_{\max}$	Source
1	1×10^{-9}	Gundlach et al. (1997)
2	4×10^8	Smith et al. (2000)
3	1.3×10^{20}	Adelberger et al. (2007)
4	4.9×10^{31}	Adelberger et al. (2007)
5	1.5×10^{43}	Adelberger et al. (2007)

We emphasize that the constraints listed above obtained by means of gravitational measurements are of high reliability. In particular, all possible sources of experimental errors were carefully analyzed and the final results were obtained at a 95% confidence level. No evidence for the presence of Yukawa-type corrections to Newtonian gravity in the millimeter range has been found. However, at scales $\lambda < 5 \times 10^{-5}$ m, gravity remains poorly tested. Thus, at $\lambda = 5 \times 10^{-5}$ m a Yukawa-type correction to Newton's law of order unity is not excluded experimentally, and at $\lambda = 10^{-5}$ m experiment allows a correction that is in excess of Newtonian gravity by a factor of 10^5. A general observation is that with decreasing λ, the strength of the constraints obtained by means of gravitational experiments rapidly decreases. As is discussed in the next sections, for an interaction range of order 1 μm or less the strongest constraints on Yukawa-type hypothetical interactions follow from measurements of the Casimir force, which is the dominant background at short separation distances.

Next, we briefly list in Table 24.1 the constraints on the power-type corrections to Newtonian gravitation presented in eqn (24.2). The strongest of them obtained to date follow from gravitational experiments. They are collected together in Table 24.1. For $l = 1$ and 2, the constraints presented in Table 24.1 were obtained from Eötvos-type experiments, and for $l = 3$, 4, and 5 from Cavendish-type experiments [note that Adelberger et al. (2007) used a different convention for the value r_0 in eqn (24.2)].

24.2 Constraints from older measurements of the Casimir force

Here, we present constraints on the parameters of Yukawa-type hypothetical interactions following from the older measurements of the Casimir force between dielectrics (Sections 18.1.2 and 18.1.3). In this section, we also consider constraints from the experiment performed by means of a torsion pendulum (Section 19.1) and the open-air experiment using hydrocarbon layers (Section 19.5.2). As explained in the respective sections, in all of these experiments the total experimental error at a given confidence level was not determined, and the comparison with theory was done in a qualitative manner. Because of this, the constraints obtained on the Yukawa-type hypothetical interactions are not of the same high reliability and confidence as those following from the Cavendish- and Eötvos-type experiments.

24.2.1 Constraints from measurements between dielectric test bodies

In the experiment by Derjaguin et al. (1956), which was briefly discussed in Section 18.1.2, the Casimir force acting between a spherical lens of thickness H and radius R and a plate of length L and thickness D made of quartz was measured. We consider one atom of the lens at a height l above the plate. The interaction potential of this atom with all atoms of the plate is obtained by the integration of the second term of eqn (24.1) over the plate volume:

$$v^{\text{hyp}}(l) = -2\pi G m^2 \alpha N \lambda^2 e^{-l/\lambda}\left(1 - e^{-D/\lambda}\right), \quad (24.6)$$

where $m_1 = m_2 = m$ and N is the number of atoms per unit volume of the plate and lens [below, we compare the Yukawa interaction between the lens and the plate with the Casimir interaction, which is much stronger than the gravitational interaction at separations less than 1 μm; because of this, the gravitational contribution in eqn (24.1) can be neglected].

The number of atoms in a horizontal section of the lens of thickness dl at a height $l \geq a$ above the plate is given by

$$d\sigma(l) = \pi N\left[2R(l-a) - (l-a)^2\right] dl. \quad (24.7)$$

Here, a is the separation distance between the plate and the point of the lens closest to it. By integrating the interaction potential (24.6) weighted with eqn (24.7) between the limits from a to $a + H$, we obtain the interaction potential between the lens and the plate. Then the hypothetical force is found as the negative derivative of this potential with respect to a,

$$F^{\text{hyp}}(a) = -4\pi^2 G \rho^2 \lambda^3 R \alpha \left(1 - e^{-D/\lambda}\right) e^{-a/\lambda} \quad (24.8)$$
$$\times \left[1 - \frac{\lambda}{R} + e^{-H/\lambda}\left(\frac{H}{R} - 1 + \frac{\lambda}{R} + \frac{H^2}{2R\lambda} - \frac{H}{\lambda}\right)\right],$$

where ρ is the density of the plate and sphere materials. In the above calculations, the inequalities $L, R, H, D \gg a$ have been used, which are satisfied in the experimental configuration. For values of λ belonging to the submillimeter range, $D, H, R \gg \lambda$ is also valid, so that eqn (24.9) is simplified to

$$F^{\text{hyp}}(a) = -4\pi^2 G \rho^2 \lambda^3 R \alpha e^{-a/\lambda}. \quad (24.9)$$

Using the interaction potential (24.6) it is easy to calculate the hypothetical energy per unit area in the configuration of two large parallel plates of thicknesses D_1 and D_2

$$E^{\text{hyp}}(a) = -2\pi G \rho^2 \lambda^3 \alpha \left(1 - e^{-D_1/\lambda}\right)\left(1 - e^{-D_2/\lambda}\right) e^{-a/\lambda}. \quad (24.10)$$

Under the condition that $D_1, D_2 \gg \lambda$, this leads to

$$E^{\mathrm{hyp}}(a) = -2\pi G \rho^2 \lambda^3 \alpha\, e^{-a/\lambda}. \tag{24.11}$$

As a result, the quantities (24.9) and (24.11) are connected by eqn (6.71). Thus, if the interaction range is much smaller than the sizes of the interacting bodies, the Yukawa-type force can be calculated (Decca et al. 2009b) using the proximity force approximation.

Based on the agreement between experiment and theory estimated to be within 10%, the constraints on the constants of the Yukawa-type interaction α and λ were obtained from the condition that the magnitude of the hypothetical force is less than 10% of the magnitude of the Casimir force (Mostepanenko and Sokolov 1989a, 1989b):

$$|F^{\mathrm{hyp}}(a)| < 0.1\, |E(a)|\, 2\pi R. \tag{24.12}$$

Here, the Casimir force between the sphere and the plate was calculated by using the PFA, where $E(a)$ is the Casimir energy per unit area in the configuration of two parallel plates at sufficiently large separation distances:

$$E(a) = -\frac{\pi^2}{720}\frac{\hbar c}{a^3}\Psi(\varepsilon_0), \tag{24.13}$$

with the correction factor $\Psi(\varepsilon_0)$ defined in eqn (12.43). The constraints obtained are shown by line 7 in Fig. 24.2. Note that the numbering of the lines in Fig. 24.2 continues from that in Fig. 24.1. Line 7 belongs to the region where the gravitational measurements do not place any constraints of comparable strength on the Yukawa-type hypothetical force.

24.2.2 Constraints from torsion pendulum experiment

As discussed in Section 19.1, the Casimir force between an Au-coated spherical lens and a flat plate was measured in an experiment with a torsion pendulum (Lamoreaux 1997). The experimental data were claimed to be consistent with the theoretical result $F_{\mathrm{IM}}^{\mathrm{sp}}$ in eqn (6.51) derived for an ideal metal within the limit of the absolute error of the force measurements $\Delta^{\mathrm{tot}} F^{\mathrm{expt}} = 10^{-11}$ N. However, the corrections to this result due to surface roughness, $\Delta_R F$, the nonzero skin depth, $\Delta_{\delta_0} F$, and nonzero temperature, $\Delta_T F$, which have different signs and may not lie within the limits of $\Delta^{\mathrm{tot}} F^{\mathrm{expt}}$, were not taken into account. For this reason, the constraints on the Yukawa-type interaction following from the torsion pendulum experiment were found from the inequality (Bordag et al. 1998b)

$$|F^{\mathrm{theor}}(a) - F_{\mathrm{IM}}^{\mathrm{sp}}(a)| \leq \Delta^{\mathrm{tot}} F^{\mathrm{expt}}. \tag{24.14}$$

Here, $F^{\mathrm{theor}}(a)$ is the theoretical value of the force including $F_{\mathrm{IM}}(a)$, all of the corrections to it mentioned above, and also the hypothetical Yukawa-type interaction:

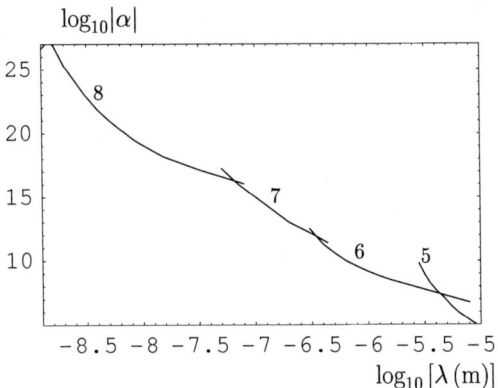

FIG. 24.2. The strongest constraints on Yukawa-type corrections to Newton's gravitational law following from the older measurements of the Casimir force between dielectric test bodies (line 7), from measurements by means of a torsion pendulum (line 6), and from the ambient measurement between two crossed cylinders (line 8). Line 5 was obtained from gravitational measurements (see Fig. 24.1). The allowed values in the (λ, α)-plane lie beneath the lines.

$$F^{\text{theor}}(a) = F_{\text{IM}}^{\text{sp}}(a) + \Delta_R F(a) + \Delta_{\delta_0} F(a) + \Delta_T F(a) + F^{\text{hyp}}(a). \qquad (24.15)$$

The hypothetical interaction of Yukawa type was calculated taking account of the layer structure of the plate and of the lens used in the experiment. The quartz lens and the plate, with respective densities $\rho' = 2.23 \times 10^3 \, \text{kg/m}^3$ and $\rho = 2.4 \times 10^3 \, \text{kg/m}^3$, were coated with Cu and Au layers of thickness $\Delta_1 = \Delta_2 = 0.5 \, \mu\text{m}$ and with densities $\rho_1 = 8.96 \times 10^3 \, \text{kg/m}^3$ and $\rho_2 = 19.32 \times 10^3 \, \text{kg/m}^3$, respectively. Calculation of the Casimir force can be performed under the same assumptions and following the same procedure as in Section 24.2.1. Under the condition that the sizes of the bodies are much larger than the interaction range λ, instead of eqn (24.9), we arrive at

$$F^{\text{hyp}}(a) = -4\pi^2 \alpha G \lambda^3 e^{-a/\lambda} \qquad (24.16)$$
$$\times \left[\rho_2 - (\rho_2 - \rho_1) e^{-\Delta_2/\lambda} - (\rho_1 - \rho) e^{-(\Delta_2 + \Delta_1)/\lambda} \right]$$
$$\times \left[R\rho_2 - (R - \Delta_2)(\rho_2 - \rho_1) e^{-\Delta_2/\lambda} - (R - \Delta_2 - \Delta_1)(\rho_1 - \rho') e^{-(\Delta_2 + \Delta_1)/\lambda} \right].$$

For thin layers with thicknesses $\Delta_1, \Delta_2 \ll R$ eqn (24.16) takes the form (Bordag et al. 1998b)

$$F^{\text{hyp}}(a) = -4\pi^2 \alpha G \lambda^3 e^{-a/\lambda} R \left[\rho_2 - (\rho_2 - \rho_1) e^{-\Delta_2/\lambda} - (\rho_1 - \rho) e^{-(\Delta_2 + \Delta_1)/\lambda} \right]$$
$$\times \left[\rho_2 - (\rho_2 - \rho_1) e^{-\Delta_2/\lambda} - (\rho_1 - \rho') e^{-(\Delta_2 + \Delta_1)/\lambda} \right]. \qquad (24.17)$$

In a similar way, for two large parallel plates with thicknesses $D_1, D_2 \gg \lambda$ each covered with thin layers with thicknesses Δ_1 and Δ_2, instead of eqn (24.11), we obtain

$$E^{\text{hyp}}(a) = -2\pi\alpha G\lambda^3 e^{-a/\lambda} \left[\rho_2 - (\rho_2 - \rho_1)e^{-\Delta_2/\lambda} - (\rho_1 - \rho)e^{-(\Delta_2+\Delta_1)/\lambda}\right]$$
$$\times \left[\rho_2 - (\rho_2 - \rho_1)e^{-\Delta_2/\lambda} - (\rho_1 - \rho')e^{-(\Delta_2+\Delta_1)/\lambda}\right]. \quad (24.18)$$

We see that the hypothetical force between a spherical lens and a plate from eqn (24.17) can be obtained as the energy per unit area from eqn (24.18) multiplied by $2\pi R$. This means that under the above conditions, the Yukawa-type force between layered bodies can also be found (Decca et al. 2009b) using the proximity force approximation (6.71).

The constraints obtained are shown by line 6 in Fig. 24.2. Typically, line 6 gives stronger constraints by a factor of 30 than the continuation of line 7, obtained from the old measurements of the Casimir force between dielectric surfaces. Lines 6 and 7 in Fig. 24.2 permit one to constrain the parameters of the gauge baryons predicted in many extra-dimensional models (Dimopoulos and Geraci 2003). The existence of these particles would result in a Yukawa-type interaction with $|\alpha|$ in the range from $\sim 10^{12}$ to $\sim 10^{15}$ and λ in the range from 10^{-8} m to 3×10^{-6} m (Decca et al. 2005a). Note that Long et al. (1999) made their own analysis of the constraints on the Yukawa-type interaction following from this torsion pendulum experiment and obtained a much weaker result than did Bordag et al. (1998b). The reason is that Long et al. (1999) did not take into account the various corrections to the ideal-metal result for the Casimir force between a lens and a plate, which significantly influence the computations in the case of real gold surfaces.

24.2.3 Constraints from ambient experiment with two crossed cylinders

This experiment was briefly considered in Section 19.5.2. The two cylinders, with $R = 1$ cm, were made of quartz with $\rho = \rho' = 2.23 \times 10^3$ kg/m^3. Each cylinder was coated first with a layer of Au with a density $\rho_1 = 18.88 \times 10^3$ kg/m^3 and thickness $\Delta_1 = 200$ nm, and then an outer layer of hydrocarbon with a density $\rho_2 = 0.85 \times 10^3$ kg/m^3 and thickness $\Delta_2 = 2.1$ nm. The absolute error of the force measurements was $\Delta^{\text{tot}} F^{\text{expt}} = 10$ nN (Ederth 2000). As shown by Mostepanenko and Novello (2001), under the assumption $\lambda, a \ll R$, where a is the closest separation between the cylindrical surfaces, the same eqn (24.17) which was derived for a sphere-plate configuration is also valid for the hypothetical interactions between crossed cylinders.

The Casimir force in the experimental configuration was computed taking account of the corrections due to the nonzero skin depth of Au and the stochastic roughness of the hydrocarbon layer [the temperature corrections were negligibly small at the separations from 20 to 100 nm used by Ederth (2000)]. This theoretical expression for the Casimir force between two crossed cylinders was found to be consistent with the measurement data within the limits of the experimental error $\Delta^{\text{tot}} F^{\text{expt}}$. Since no Yukawa-type interaction was observed within the limits of the experimental error, the constraints on the parameters of the hypothetical force were found from the inequality

$$|F^{\text{hyp}}(a)| \leq \Delta^{\text{tot}} F^{\text{expt}}. \tag{24.19}$$

The constraints obtained, calculated at $a = 20\,\text{nm}$, are shown by line 8 in Fig. 24.2. This experiment leads to the strongest constraints in the interaction range where $\lambda \sim 10^{-8}\,\text{m}$.

To conclude this section, we note that Casimir force measurements between metallic surfaces produce some constraints not only on Yukawa-type hypothetical interactions, but also on power-type interactions of the form (24.2) as well. However, the constraints on power-type interactions with $l \leq 5$ obtained from the Casimir effect (Mostepanenko and Sokolov 1987a, 1987b, Klimchitskaya et al. 1998, Mostepanenko and Novello 2001) are weaker than the respective constraints obtained from gravitational measurements of the Eötvos and Cavendish types (see Table 24.1). Because of this, we shall not consider this subject in more detail here. As was mentioned above, the constraints on hypothetical interactions from the older measurements of the Casimir force (including the first modern measurements considered in this section) rank below the constraints obtained from the gravitational measurements in reliability and confidence. In the next two sections, we consider the most precise measurements of the Casimir force, which result in constraints of the same reliability as those derived from the gravitational measurements.

24.3 Constraints from experiment with gold surfaces using an atomic force microscope

Constraints on the parameters of hypothetical long-range interactions of the Yukawa type were obtained from each of the three successive experiments using an atomic force microscope described in Sections 19.2.1–19.2.3 (Bordag et al. 1999b, 2000b, Fischbach et al. 2001). These constraints were reviewed by Bordag et al. (2001a), Mostepanenko (2002a, 2002b), Klimchitskaya and Mohideen (2002), and Chen et al. (2005b). Here, however, we consider in more detail only the strongest constraints on Yukawa-type interactions which follow from the most conclusive measurement of the Casimir force by means of an atomic force microscope between Au surfaces, performed by Harris et al. (2000) (see Section 19.2.3). The point is that this measurement was later reanalyzed using the rigorous methods of data processing and comparison between experiment and theory presented in Section 18.3 (Klimchitskaya et al. 2007a). As a result, the constraints on a Yukawa-type hypothetical interaction were determined at a 95% convidence level. These constraints are of the same reliability as those obtained from the gravitational experiments.

In this experiment, the Casimir force was measured between a sapphire disk and a polystyrene sphere (with a diameter $2R = 191.3\,\mu\text{m}$) coated with an

Au layer of thickness $\Delta_1 = 86.6$ nm and density $\rho_1 = 18.88 \times 10^3$ kg/m^3. The densities of the sapphire and polystyrene were $\rho = 4.0 \times 10^3$ kg/m^3 and $\rho' = 1.06 \times 10^3$ kg/m^3, respectively. Then, the Yukawa-type hypothetical force could be calculated using eqn (24.17) with $\Delta_2 = 0$ and $\rho_2 = 0$, because there was only one covering layer in this experiment.

Using rigorous procedures for the comparison of experimental data with theory (see Section 18.3), the constraints on the Yukawa-type interaction following from the experiment by Harris et al. (2000) were obtained by Decca et al. (2007b). For this purpose, the confidence intervals $[-\Xi_F(a), \Xi_F(a)]$ for the quantity $F^{\text{theor}}(a) - \bar{F}^{\text{expt}}(a)$ found by Klimchitskaya et al. (2007a) at a 95% confidence level were used (see Section 19.2.3). The theoretical Casimir force $F^{\text{theor}}(a)$ was computed with the help of the Lifshitz formula at the laboratory temperature $T = 300$ K, using the generalized plasma-like dielectric permittivity. Note that the total theoretical error $\Delta^{\text{tot}} F^{\text{theor}}$ includes all possible sources of errors, as discussed in Section 19.2.3. As a result, at each a the confidence interval $[-\Xi_F(a), \Xi_F(a)]$ lies between the solid lines shown in Fig. 19.7(b). For example, at $a = 61.08$ nm the half-width of the confidence interval is $\Xi_F = 31.6$ pN; with an increase of the separation to 100.15 and 200.46 nm, it decreases to 9.17 and 7.20 pN, respectively. The resulting constraints at a 95% confidence level were determined from the inequality

$$|F^{\text{hyp}}(a)| \leq \Xi_F(a). \qquad (24.20)$$

These constraints are represented by line 1 in Fig. 24.3. Note that the constraints given by line 1 in Fig. 24.3 are up to an order of magnitude weaker than those obtained by Fischbach et al. (2001) from the original publication by Harris et al. (2000). However, they benefit from high confidence, and they can be reliably compared with future work on the subject by using the same rigorous approach to the comparison of experiment with theory as that presented in Section 18.3, and also with constraints obtained from gravitational measurements. The first comparison with rigorous constraints obtained at a larger λ is contained in the next section.

24.4 Constraints from experiment using a micromachined oscillator

As discussed in Section 19.3, a set of experiments using a micromechanical torsional oscillator have achieved the highest precision in the measurement of the Casimir pressure. In the third experiment of this series, the random error was made much smaller than the systematic error, giving this measurement a metrological quality. Constraints on Yukawa-type hypothetical interactions were obtained from all measurements in this series (Decca et al. 2003b, 2004, 2005b, 2007a, 2007b, Klimchitskaya et al. 2005, Mostepanenko et al. 2008). Here, we present the strongest constraints on Yukawa-type interactions obtained by Decca et al. (2007a, 2007b) and describe the results of the so-called *Casimir-less experiment*, where the role of the Casimir force was largely excluded.

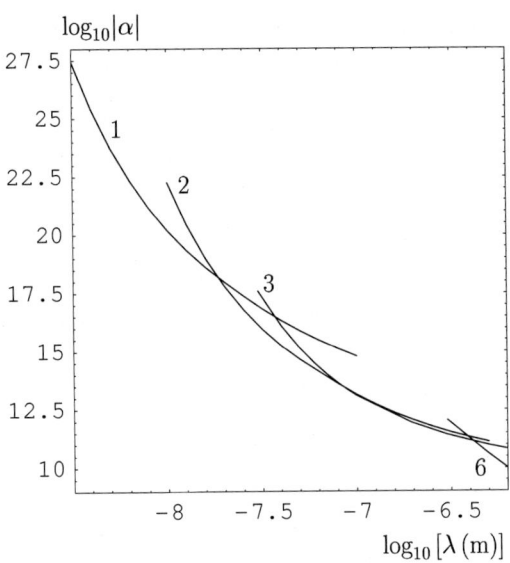

FIG. 24.3. The strongest constraints obtained on Yukawa-type corrections to Newton's gravitational law obtained at a 95% confidence level from the measurement of the Casimir force using an atomic force microscope (line 1), from the measurement of the Casimir pressure by means of a micromachined oscillator (line 2), and from the Casimir-less experiment (line 3). Line 6 indicates constraints obtained from the torsion pendulum experiment (see Fig. 24.2). The allowed values in the (λ, α)-plane lie beneath the lines.

24.4.1 Constraints from Casimir pressure measurement

The most precise indirect measurement of the Casimir pressure between gold-coated plates is described in Section 19.3.4. In this experiment, a sapphire sphere of density $\rho' = 4.1 \times 10^3 \, \text{kg/m}^3$ was first coated with a layer of Cr of density $\rho_1 = 7.14 \times 10^3 \, \text{kg/m}^3$ and thickness $\Delta_1 = 10 \, \text{nm}$, and then with an external layer of Au of density $\rho_2 = 19.28 \times 10^3 \, \text{kg/m}^3$ and thickness $\Delta_2 = 180 \, \text{nm}$. A Si plate of density $\rho = 2.33 \times 10^3 \, \text{kg/m}^3$ was also first coated with a layer of Cr of thickness $\tilde{\Delta}_1 = 10 \, \text{nm}$, and then with a layer of Au of thickness $\tilde{\Delta}_2 = 210 \, \text{nm}$. The equivalent Yukawa pressure between two parallel plates with the same layer structure as that of the above sphere and plate can be obtained in a similar way as in Section 24.2.2. The result is

$$P^{\text{hyp}}(a) = -2\pi G \alpha \lambda^2 e^{-a/\lambda} \left[\rho_2 - (\rho_2 - \rho_1) e^{-\Delta_2/\lambda} - (\rho_1 - \rho') e^{-(\Delta_2 + \Delta_1)/\lambda} \right] \\ \times \left[\rho_2 - (\rho_2 - \rho_1) e^{-\tilde{\Delta}_2/\lambda} - (\rho_1 - \rho) e^{-(\tilde{\Delta}_2 + \tilde{\Delta}_1)/\lambda} \right]. \quad (24.21)$$

The experimental data from this experiment have been analyzed and compared with various theoretical approaches using the rigorous statistical methods presented in Section 18.3. In this process a conservative analysis of the errors [and consequently of the width of the confidence interval $2\Xi_P(a)$] was used, which overestimated both the errors and the value of $2\Xi_P(a)$. The reason for this overestimation was that the error due to the uncertainty in the experimental separations Δa was included in the analyses of the theoretical errors when the second method for comparison between experiment and theory described in Section 18.2.3 was used. As a result, the theoretical pressures acquired an extra error of $\approx 4\Delta a/a$, which led to an enormous widening of the confidence interval at short separations (see Figs. 19.19 and 19.20). However, as is seen in Fig. 19.19, the actual width of the confidence interval is much less than that between the solid lines (recall that the actual confidence interval determined at a 95% confidence level is the narrowest interval which contains about 95% of the data dots). It is easily seen that if the theoretical error of $4\Delta a/a$ due to uncertainties in the experimental separations is disregarded, the resulting narrower confidence interval $[-\tilde{\Xi}_P(a), \tilde{\Xi}_P(a)]$ still contains all dots representing the differences $P^{\text{theor}}(a) - \bar{P}^{\text{expt}}(a)$ within the range of separations from 180 to 746 nm. At a separation $a = 180$ nm, the half-width of this interval is $\tilde{\Xi}_P = 4.80$ mPa. At the typical separations $a = 200, 250, 300, 350, 400$, and 450 nm, $\tilde{\Xi}_P$ is equal to 3.30, 1.52, 0.84, 0.57, 0.45, and 0.40 mPa, respectively. Thus, for 180 nm $\leq a \leq$ 746 nm, the magnitude of the hypothetical pressure should satisfy the inequality

$$|P^{\text{hyp}}(a)| \leq \tilde{\Xi}_P(a). \tag{24.22}$$

Bearing in mind that the half-width of the confidence interval $\tilde{\Xi}_P(a)$ is defined at a 95% confidence level, the same confidence also applies to the constraints following from the inequality (24.22).

Decca et al. (2007b) performed a numerical analysis of eqns (24.21) and (24.22) and determined the resulting region of the (λ, α)-plane where the inequality (24.22) was satisfied, such that the existence of a Yukawa-type interaction would be consistent with the level of agreement achieved between the experimental data and the relevant theory. The strongest constraints within the range of interaction distance 10 nm $\leq \lambda \leq$ 56 nm were obtained from a comparison of measurements with theory at a separation $a = 180$ nm. With an increase of λ, the strongest constraints on α were obtained from consideration of larger separations. Thus, constraints in the regions 56 nm $\leq \lambda \leq$ 71 nm, 71 nm $\leq \lambda \leq$ 89 nm, 89 nm $\leq \lambda \leq$ 140 nm, 140 nm $\leq \lambda \leq$ 220 nm, and 220 nm $\leq \lambda \leq$ 500 nm were obtained from the agreement between the Casimir pressure measurements and the theory at the separations $a = 200, 250, 300, 350$, and 400 nm, respectively.

The resulting constraints are plotted in Fig. 24.3 for different values of λ (line 2). It can be seen that the measurements of the Casimir pressure by means of a micromachined oscillator lead to the strongest constraints at larger values of λ than for the measurements of the Casimir force by means of an atomic force microscope. Line 2 in Fig. 24.3 places stronger limits on the parameters of gauge baryons and strange moduli (Decca et al. 2005a).

Recently, the measurements of the Casimir pressure by Decca et al. (2007a, 2007b) were used to constrain the predictions of chameleon field theories. Like many other extensions of the Standard Model, this theory introduces one or more scalar fields. A specific feature of these fields is that their masses depend on the local background matter density and that they can couple directly to matter with gravitational strength (Brax et al. 2007, Mota and Shaw 2007). The chameleon scalar field ϕ, if it really exists in nature, leads to an additional chameleon force acting between two nearby macrobodies. The form of this force as a function of separation distance is rather complicated and depends on the potential of the chameleon field. This potential can be chosen as (Brax et al. 2007)

$$V(\phi) = \Gamma_0^4 \left(1 + \frac{\Gamma^n}{\phi^n}\right), \qquad (24.23)$$

where n can be both positive and negative, and Γ_0 and Γ are some constants. To fit the data for the acceleration of the Universe, one requires $\Gamma_0 \approx 2.4 \times 10^{-3}\,\text{eV}$. The hypothetical pressure P_ϕ arising between two parallel plates in chameleon theories with a potential (24.23) and $\Gamma = \Gamma_0$ was calculated by Brax et al. (2007). It was shown that for $n > 0$, the experimental results of Decca et al. (2007a, 2007b) do not impose any constraints on the predictions of chameleon theories. The current limits in Fig. 24.3 must be strengthened by at least two orders of magnitude in order for constraints on chameleon theories with $n > 0$ to be obtained. At the same time, the experimental data of Decca et al. (2007a, 2007b) rule out chameleon theories with $n = -4$ and $n = -6$ (Brax et al. 2007).

24.4.2 Constraints from Casimir-less experiment

In the Casimir-less experiment (Decca et al. 2005a), the micromachined oscillator shown schematically in Fig. 19.9 was used to measure the difference between forces over two dissimilar materials, Au and Ge, which had been coated with a common layer of Au. This common Au top layer was of a sufficient thickness (more than the Au plasma wavelength $\lambda_p = 137\,\text{nm}$) to make the Casimir forces equal between an Au-coated sphere and the two halves of the patterned Au–Ge plate. However, Yukawa-type hypothetical forces between the sphere and the two halves of the plate should be different, owing to different densities of Au and Ge under the common Au top layer. Thus, when the micromechanical oscillator is moved back and forth parallel to the x-axis below the sphere, there should be some difference hypothetical force (if the latter exists in nature).

In the measurements, the static regime described in Section 19.3.2 was exploited, which meant that the vertical separation between the sphere and the plate was not varied harmonically. The difference hypothetical force between the sphere and the plate could be obtained in a way similar to that described in Section 24.1. The sapphire sphere of density ρ' was covered by a Cr layer of density ρ_1 with thickness $\Delta_1 = 1\,\text{nm}$, and by an outer Au layer of density ρ_2 and thickness $\Delta_2 = 200\,\text{nm}$. The silicon plate was covered by a Ti layer of thickness $1\,\text{nm}$. Then one half of it was covered by a layer of Ge with density ρ_{Ge} and thickness $\Delta_{\mathrm{Ge}} = 200\,\mathrm{nm}$, followed by a layer of Pt of thickness $\Delta_{\mathrm{Pt}} = 1\,\mathrm{nm}$ and an outer layer of Au with thickness $\tilde{\Delta}_2 = 150\,\mathrm{nm}$. The other half of the plate was covered by a layer of Au with thickness $\Delta_{\mathrm{Au}} = \Delta_{\mathrm{Ge}} = 200\,\mathrm{nm}$ followed by a common layer of Pt of thickness $\Delta_{\mathrm{Pt}} = 1\,\mathrm{nm}$ and a common outer layer of Au of thickness $\tilde{\Delta}_2 = 150\,\mathrm{nm}$. The resulting difference Casimir force is given by (Decca et al. 2005a)

$$\Delta F^{\mathrm{hyp}}(a) = -4\pi^2 \alpha G \lambda^3 e^{-a/\lambda} R \left[\rho_2 - (\rho_2 - \rho_1)e^{-\Delta_2/\lambda} - (\rho_1 - \rho')e^{-(\Delta_2+\Delta_1)/\lambda} \right]$$
$$\times \left[(\rho_2 - \rho_{\mathrm{Ge}})e^{-(\tilde{\Delta}_2+\Delta_{\mathrm{Pt}})/\lambda}\left(1 - e^{-\Delta_{\mathrm{Ge}}/\lambda}\right) \right]. \qquad (24.24)$$

Here, the radius R of the sphere was approximately $50\,\mu\mathrm{m}$, and the difference between the densities of Au and Ge is $\rho_2 - \rho_{\mathrm{Ge}} = 13.96 \times 10^3\,\mathrm{kg/m}^3$.

The measurements by Decca et al. (2005a) permitted limits on the quantity $|\Delta F^{\mathrm{hyp}}(a)|$ to be obtained, of the order of fN depending on the value of the separation distance in the region from 150 to 500 nm. This results in the constraints on the parameters (λ, α) of the hypothetical Yukawa interaction shown by line 3 in Fig. 24.3. These constraints were determined at a 95% confidence level. They are only slightly different from the constraints following from the absolute measurements of the Casimir pressure by means of a micromechanical torsional oscillator (see line 2). For comparison purposes, we also show in Fig. 24.3 a part of line 6 related to larger λ which was obtained from the torsion pendulum experiment (for this line, we have retained the numbering of Section 24.2.2). This line is presented in more detail in Fig. 24.2 (see also Fig. 24.1).

We can conclude that the increased precision of measurements of the Casimir force and the use of rigorous statistical procedures for data processing and for the comparison of experiment with theory permit the determination of new constraints on the parameters of Yukawa-type hypothetical interactions with the same reliability as for those obtained from gravitational measurements, but for shorter interaction ranges.

25
CONCLUSIONS AND OUTLOOK

This book was aimed at providing the reader with a comprehensive review of all of the important work that has been done in the wide area of Casimir physics. It has discussed material on the field-theoretical foundations of the Casimir effect, theoretical approaches to the description of the Casimir force between real material bodies, and experiments on measuring the Casimir force in different configurations. The above topics were presented sequentially in Parts I, II, and III of the book. As mentioned in Section 1.3 of the Introduction, the Casimir effect is a multidisciplinary subject which plays a major role in quantum field theory, condensed matter physics, atomic physics, astrophysics, gravitation and cosmology, and mathematical physics. It finds applications in nanotechnology and even for constraining the predictions of fundamental physical theories. Results from all these diverse areas have been presented in the book.

The main conclusion following from Part I of the book, devoted to the foundations of the Casimir effect for ideal boundaries, is that, 60 years after Casimir's discovery, we now have a general field-theoretical approach which allows investigation of the Casimir effect between two bodies of arbitrary shape with various boundary conditions on their surfaces. In the framework of this approach, all divergent contributions are separated, and finite expressions for the Casimir energy and force are obtained in terms of scattering matrices and functional determinants. In the case of single bodies (e.g. an ideal-metal rectangular box or a dielectric ball), no consensus has yet emerged. Specifically, the problem of divergences in the Casimir energy inside ideal-metal rectangular boxes is still being debated. According to one point of view, the divergences present in the calculations cannot be removed by a renormalization of the physical constants, as is done in quantum electrodynamics. As a result, the Casimir force depends on a cutoff function. From another point of view, the geometrical parameters of the configuration under consideration play the role of the constants to be renormalized. The latter point of view is based on renormalization procedures developed for quantum field theory in curved space-time (Birrell and Davies 1982, Grib *et al.* 1994).

The last ten years have witnessed rapid progress in the application of the Lifshitz theory to describe the van der Waals and Casimir forces acting between metallic, dielectric, and semiconductor test bodies. Some unexpected results obtained from this body of work are reflected in Part II. These have given rise to a lively debate in the literature. The many points of view have been presented in the book. An important conclusion obtained here is that real (drift or diffusion) currents of conduction electrons should not be included in the model of the dielectric response used in the Lifshitz theory. The presence of such a current violates the state of thermal equilibrium, which is the fundamental applicability condition of the Lifshitz theory. If a drift current of the Drude type (for metals) or the dc conductivity (for dielectrics) is included, this immediately results in contradictions between the Lifshitz theory and thermodynamics. However, if drift and diffusion currents are neglected and free carriers in metals are described by the plasma model, the Lifshitz theory is always found to be in perfect agreement with thermodynamics.

A good question to ask, then, is whether one should look for a new theory of the van der Waals and Casimir forces which incorporates the role of charge carriers in real materials on more fundamental grounds. As was demonstrated in Part II, there is a consistent way to deal with charge carriers in the standard Lifshitz theory. For any dielectric material (whose conductivity goes to zero when the temperature vanishes), the role of the free charge carriers that exist at nonzero temperature must be neglected. For metallic materials (whose conductivity does not go to zero when the temperature vanishes), the charge carriers must be included by means of the free-electron plasma model, which allows only displacement currents. The Lifshitz theory supplemented with this rule is in agreement with thermodynamics and universally applicable to any material in thermal equilibrium with its environment. By contrast, drift and diffusion currents violate thermal equilibrium owing to the existence of a unidirectional flux of heat from the system to a heat reservoir. Such currents can be created only by real electric fields. Fluctuating fields have a zero mean value. They lead to equal and mutual exchange of heat between a system and a heat reservoir, resulting in a zero mean heat flux. This suggests that fluctuating electromagnetic fields have *important differences* from real ones. It is common knowledge that when placed in an external electric field, a system goes out of thermal equilibrium and the fluctuation–dissipation theorem is violated. This should not and does not happen when the system is under the influence of fluctuating electromagnetic fields of any frequency, as are all physical systems at all times. At present there is no consensus in the published literature on the ideas that we have sketched here, but they find additional support from experiment.

Part II of the book contained many applications of the Casimir effect to real materials. First the Lifshitz theory was introduced. It was then applied to both wall–wall and atom–wall Casimir and Casimir–Polder interactions for various types of wall materials, such as metals, semiconductors, dielectrics (both polar and nonpolar), magnetic media, stratified media, and various atoms and molecules. All theoretical approaches proposed in the literature were reflected and the predictions made were carefully compared and analyzed. Special attention was paid to the role of surface roughness, and the methods used to incorporate it into the theoretical calculations were explained in detail. The inclusion of surface roughness allows a careful comparison of the experimental data with the theory.

Part III of the book presented all experiments to date on the measurement of the Casimir force. This section described general requirements for precise measurements of the Casimir force and the statistical framework for analyzing the data. A brief review of the older measurements was provided. The more modern measurements were discussed in detail. The methods for a rigorous comparison of the experiments with theory were presented. Observations of some new, interesting phenomena, such as the modulation of the Casimir force by light, the lateral Casimir force, the dependence of the Casimir force on geometry and the thermal Casimir–Polder force, were presented in detail. The main conclusion here is that the Lifshitz theory supplemented with the above rules is in very good agreement with all available experimental data. This has been demonstrated in three types of experiments performed by different experimental groups with metallic, semiconductor, and dielectric plates for both the Casimir and the Casimir–Polder interactions. By contrast, all attempts to include a drift or diffusion current in the model of the dielectric response have been found to be in contradiction with the experimental data.

An important conclusion following from the experiments presented in Part III is that the Casimir effect is central for nanotechnology. This has been confirmed by the actuation of micromechanical systems by the Casimir force and by the demonstration of a nonlinear micromechanical Casimir oscillator. Greater promise is offered by the possibility to change the magnitude of the Casimir force through the phenomenon of optical modulation. Additionally, the theoretical potential for using the lateral Casimir force in a rack and pinion arrangement has been put forward. This would be an important achievement if experimentally applied, as it would bring about a transfer of motion without contact, an important requirement in nanoscale devices, because lubrication cannot be used and direct contact will rapidly wear away small structures.

A no less important conclusion following from the experimental work on the Casimir effect is that it can be used as an effective test of theoretical predictions obtained from fundamental unification theories beyond the Standard Model. In the last few years the constraints on Yukawa-type corrections to Newtonian gravity, following from the exchange of light elementary particles and extra-dimensional physics, have been strengthened by up to a factor of 10^4 based on measurements of the Casimir force. In this respect, as the Casimir force dominates interactions at the nanometer scale, a complete understanding of it is required for effective measurements of deviations from Newtonian gravity at very short distances.

In this book, we have not discussed some Casimir-related phenomena which might have considerable promise for the future, but at present have not been completely explored theoretically and lack experimental confirmation. One example is the Casimir force between metamaterials, which has attracted much attention in electrodynamics and optics in the last few years (Vinogradov et al. 2008). Thus, Henkel and Joulain (2005) have determined the extent to which the Casimir interaction between two metamaterial plates can be manipulated by engineering their magnetodielectric response. Rosa et al. (2008) calculated the Casimir force between magnetodielectric and anisotropic metamaterials and found the possibility of repulsive forces in special cases. Leonhardt and Philbin (2007) and Yang *et al.* (2008) considered the Casimir effect between two idealized left-handed metamaterials. The possibility of obtaining Casimir repulsion using metamaterials based on metallic inclusions has been questioned by Pirozhenko and Lambrecht (2008b). Presently, extensive effort is being devoted to exploring the opportunities offered by realistic metamaterials for the Casimir effect.

The prospective applications of the Casimir effect in both fundamental physics and nanotechnology require the development of more powerful theoretical methods which would be applicable to real bodies of arbitrary geometrical shape and material properties. Specifically, a future theory of the thermal Casimir force should provide a more fundamental explanation for the rules presently used and incorporate a consistent description of spatial dispersion. This could be achieved on the basis of general representations of the Casimir energy and force as discussed in Chapter 10. The application of these representations to real materials should be accompanied, however, by the use of new physical ideas concerning the character of the reflection amplitudes of the fluctuating field from the interfaces of real bodies.

We believe that the infancy of this field of physics is over and that the Casimir effect is on the threshold of becoming a subject of major importance in both fundamental and applied science. Many new exciting developments are anticipated. We hope that this book will serve as a reference and resource for these future developments.

REFERENCES

Abramowitz, M. and Stegun, I. A. (1972). *Handbook of Mathematical Functions: With Formulas, Graphs, and Mathematical Tables.* Dover, New York.

Actor, A. A. (1995). Scalar quantum fields in rectangular boundaries. *Fortschr. Phys.* **43**, 141–205.

Actor, A. A. and Bender, I. (1995). Casimir effect for soft boundaries. *Phys. Rev. D* **52**, 3581–90.

Adelberger, E. G., Heckel, B. R., Stubbs, C. W., and Rogers, W. F. (1991). Searches for new macroscopic forces. *Ann. Rev. Nucl. Part. Sci.* **41**, 269–320.

Adelberger, E. G., Heckel, B. R., and Nelson, A. E. (2003). Tests of the gravitational inverse-square law. *Ann. Rev. Nucl. Part. Sci.* **53**, 77–121.

Adelberger, E. G., Heckel, B. R., Hoedl, S., Hoyle, C. D., Kapner, D. J., and Upadhye, A. (2007). Particle-physics implications of a recent test of the gravitational inverse-square law. *Phys. Rev. Lett.* **98**, 131104-1–4.

Aghababaie, Y. and Burgess, C. P. (2004). Effective actions, boundaries, and precision calculations of Casimir energies. *Phys. Rev. D* **70**, 085003-1–6.

Agnesi, A., Braggio, C., Bressi, G., Carugno, G., Galeazzi, G., Pirzio, F., Reali, G., Ruoso, G., and Zanello, D. (2008). MIR status report: An experiment for the measurement of the dynamical Casimir effect. *J. Phys. A: Math. Theor.* **41**, 164024-1–7.

Agranovich, V. M. and Ginzburg, V. L. (1984). *Crystal Optics with Spatial Dispersion, and Excitons.* Springer, Berlin.

Ahlfors, L. F. (1979). *Complex Analysis* (3rd edn). McGraw-Hill, New York.

Aliev, A. N. (1997). Casimir effect in the spacetime of multiple cosmic strings. *Phys. Rev. D* **55** 3903–6.

Aliev, A. N., Hortaçsu, M., and Özdemir, N. (1997). Vacuum fluctuations of a massless spin-2 field around multiple cosmic strings. *Class. Quantum Grav.* **14**, 3215–24.

Allen, J. J. (2005). *Micro Electro Mechanical System Design.* CRC Press, New York.

Alves, D. T., Farina, C., and Granhen, E. R. (2006). Dynamical Casimir effect in a resonant cavity with mixed boundary conditions. *Phys. Rev. A* **73**, 063818-1–8.

Ambjørn, J. and Wolfram, S. (1983). Properties of the vacuum. I. Mechanical and thermodynamic. *Ann. Phys. (N.Y.)* **147**, 1–32.

Angermann, H. (2002). Characterization of wet-chemically treated silicon interfaces by surface photovoltage measurements. *Analyt. Bioanalyt. Chem.* **374**, 676–80.

Anselm, A. A. and Uraltsev, N. G. (1982). A second massless axion? *Phys. Lett. B* **114**, 39–41.

Antezza, M. (2006). Surface–atom force out of thermal equilibrium and its effect on ultra-cold atoms. *J. Phys. A: Math. Gen.* **39**, 6117–26.

Antezza, M., Pitaevskii, L. P., and Stringari, S. (2004). Effect of the Casimir–Polder force on the collective oscillations of a trapped Bose–Einstein condensate. *Phys. Rev. A* **70**, 053619-1–10.

Antezza, M., Pitaevskii, L. P., and Stringari, S. (2005). New asymptotic behavior of the surface–atom force out of thermal equilibrium. *Phys. Rev. Lett.* **95**, 113202-1–4.

Antoniadis, I., Arkani-Hamed, N., Dimopoulos, S., and Dvali, G. (1998). New dimensions at a millimeter to a fermi and superstrings at a TeV. *Phys. Lett. B* **436**, 257–63.

Antonini, P., Bressi, G., Carugno, G., Galeazzi, G., Messineo, G., and Ruoso, G. (2006). Casimir effect: A novel experimental approach at large separation. *New J. Phys.* **8**, 239-1–13.

Arima, K., Endo, K., Kataoka, T., Oshikane, Y., Inoue, H., and Mori, Y. (2000). Atomically resolved scanning tunneling microscopy of hydrogen-terminated Si(001) surfaces after HF cleaning. *Appl. Phys. Lett.* **76**, 463–5.

Arkani-Hamed, N., Dimopoulos, S., and Dvali, G. (1998). The hierarchy problem and new dimensions at a millimeter. *Phys. Lett. B* **429**, 263–72.

Arkani-Hamed, N., Dimopoulos, S., and Dvali, G. (1999). Phenomenology, astrophysics, and cosmology of theories with submillimeter dimensions and TeV scale quantum gravity. *Phys. Rev. D* **59**, 086004-1–21.

Arnold, W., Hunklinger, S., and Dransfeld, K. (1979). Influence of optical absorption on the Van der Waals interaction between solids. *Phys. Rev. B* **19**, 6049–56.

Ashcroft, N. W. and Mermin, N. D. (1976). *Solid State Physics*. Saunders College, Philadelphia.

Ashourvan, A., Miri, M., and Golestanian, R. (2007a). Noncontact rack and pinion powered by the lateral Casimir force. *Phys. Rev. Lett.* **98**, 140801-1–4.

Ashourvan, A., Miri, M., and Golestanian, R. (2007b). Rectification of the lateral Casimir force in a vibrating noncontact rack and pinion. *Phys. Rev. E* **75**, 040103(R)-1–4.

Babb, J. F., Klimchitskaya, G. L., and Mostepanenko, V. M. (2004). Casimir–Polder interaction between an atom and a cavity wall under the influence of real conditions. *Phys. Rev. A* **70**, 042901-1–12.

Bachas, C. P. (2007). Comment on the sign of the Casimir force. *J. Phys. A: Math. Theor.* **40**, 9089–96.

Balian, R. and Bloch, C. (1970). Distribution of eigenfrequencies for wave equation in a finite domain. I. Three-dimensional problem with smooth boundary surface. *Ann. Phys. (N.Y.)* **60**, 401–47.

Balian, R. and Bloch, C. (1971). Distribution of eigenfrequencies for wave equation in a finite domain. II. Electromagnetic field. Riemannian spaces. *Ann. Phys. (N.Y.)* **64**, 271–307.

Balian, R. and Duplantier, B. (1977). Electromagnetic waves near perfect conductors. 1. Multiple-scattering expansions. Distribution of modes. *Ann. Phys. (N.Y.)* **104**, 300–35.

Balian, R. and Duplantier, B. (1978). Electromagnetic waves near perfect conductors. 2. Casimir effect. *Ann. Phys. (N.Y.)* **112**, 165–208.

Banach, R. and Dowker, J. S. (1979a). The vacuum stress tensor for automorphic fields on some flat space-times. *J. Phys. A: Math. Gen.* **12**, 2545–62.

Banach, R. and Dowker, J. S. (1979b). Automorphic field theory — some mathematical issues. *J. Phys. A: Math. Gen.* **12**, 2527–43.

Bansal, N. P. and Doremus, R. H. (1986). *Handbook of Glass Properties*. Academic Press, New York.

Barash, Yu. S. (1973). Van der Waals interaction between anisotropic bodies. *Izv. Vuzov., Radiofiz.* **16**, 1227–34 (*Radiophys. Quantum Electron.* **16**, 945–9).

Barash, Yu. S. (1988). *The Van der Waals Forces*. Nauka, Moscow (in Russian).

Barash, Yu. S. and Ginzburg, V. L. (1975). Electromagnetic fluctuations in a substance and molecular (van der Waals) interbody forces. *Usp. Fiz. Nauk* **116**, 5–40 (*Sov. Phys. Uspekhi* **18**, 305–23).

Barash, Yu. S. and Ginzburg, V. L. (1984). Some problems in the theory of van der Waals forces. *Usp. Fiz. Nauk* **143**, 345–89 (*Sov. Phys. Uspekhi* **27**, 467–91).

Barber, B. P., Hiller, R. A., Lofstedt, R., Putterman, S. J., and Weninger, K. R. (1997). Defining the unknowns of sonoluminescence. *Phys. Rep.* **281**, 65–143.

Barone, F. A., Cavalcanti, R. M., and Farina, C. (2004). Radiative corrections to the Casimir effect for the massive scalar field. *Nucl. Phys. B, Proc. Suppl.* **127**, 118–22.

Barton, G. (1987a). Quantum-electrodynamic level shifts between parallel mirrors: Analysis. *Proc. R. Soc. Lond. A* **410**, 141–74.

Barton, G. (1987b). Quantum-electrodynamic level shifts between parallel mirrors: Applications, mainly to Rydberg states. *Proc. R. Soc. Lond. A* **410**, 175–200.

Barton, G. (1999). Perturbative check on the Casimir energies of nondispersive dielectric spheres. *J. Phys. A: Math. Gen.* **32**, 525–35.

Barton, G. (2001). Perturbative Casimir energies of dispersive spheres, cubes and cylinders. *J. Phys. A: Math. Gen.* **34**, 4083–114.

Barton, G. (2004). Casimir energies of spherical plasma shells. *J. Phys. A: Math. Gen.* **37**, 1011–49.

Barton, G. (2005). Casimir effects for a flat plasma sheet: I. Energies. *J. Phys. A: Math. Gen.* **38**, 2997–3019.

Barton, G. (2006). Casimir piston and cylinder, perturbatively. *Phys. Rev. D* **73**, 065018-1–6.

Barton, G., Dodonov, V. V., and Man'ko, V. I. (2005). The nonstationary Casimir effect and quantum systems with moving boundaries. *J. Opt. B: Quantum Semiclass.* **7**, S1.

Batra, R. C., Porfiri, M., and Spinello, D. (2007). Effects of Casimir force on pull-in instability in micromembranes. *Europhys. Lett.* **77**, 20010-1–6.

Beadle, W. E., Tsai, J. C. C., and Plummer, R. D., eds. (1985). *Quick Reference Manual for Silicon Integrated Circuit Technology*. Wiley, New York.

Bender, C. M. and Milton, K. A. (1994). Scalar Casimir effect for a D-dimensional sphere. *Phys. Rev. D* **50**, 6547–55.

Bergström, L. (1997). Hamaker constants in inorganic materials. *Adv. Coll. Interface Sci.* **70**, 125–69.

Berkhout, J. J., Luiten, O. J., Setija, I. D., Hijmans, T. W., Mizusaki, T., and Walraven, J. T. M. (1989). Quantum reflection: Focusing of hydrogen atoms with a concave mirror. *Phys. Rev. Lett.* **63**, 1689–92.

Bernasconi, F., Graf, G. M., and Hasler, D. (2003). The heat kernel expansion for the electromagnetic field in a cavity. *Ann. Henri Poincaré* **4**, 1001–13.

Bezerra, V. B., Klimchitskaya, G. L., and Romero, C. (1997). Casimir force between a flat plate and a spherical lens: Application to the results of a new experiment. *Mod. Phys. Lett. A* **12**, 2613–22.

Bezerra, V. B., Klimchitskaya, G. L., and Mostepanenko, V. M. (2000a). Higher-order conductivity corrections to the Casimir force. *Phys. Rev. A* **62**, 014102-1–4.

Bezerra, V. B., Klimchitskaya, G. L., and Romero, C. (2000b). Surface roughness contribution to the Casimir interaction between an isolated atom and a cavity wall. *Phys. Rev. A* **61**, 022115-1–10.

Bezerra, V. B., Klimchitskaya, G. L., and Romero, C. (2001). Perturbation expansion of the conductivity corrections to the Casimir force. *Int. J. Mod. Phys. A* **16**, 3103–15.

Bezerra, V. B., Klimchitskaya, G. L., and Mostepanenko, V. M. (2002a). Thermodynamic aspects of the Casimir force between real metals at nonzero temperature. *Phys. Rev. A* **65**, 052113-1–7.

Bezerra, V. B., Klimchitskaya, G. L., and Mostepanenko, V. M. (2002b). Correlation of energy and free energy for the thermal Casimir force between real metals. *Phys. Rev. A* **66**, 062112-1–13.

Bezerra, V. B., Klimchitskaya, G. L., and Romero, C. (2002c). Surface impedance and the Casimir force. *Phys. Rev. A* **65**, 012111-1–9.

Bezerra, V. B., Klimchitskaya, G. L., Mostepanenko, V. M., and Romero, C. (2004). Violation of the Nernst heat theorem in the theory of thermal Casimir force between Drude metals. *Phys. Rev. A* **69**, 022119-1–9.

Bezerra, V. B., Decca, R. S., Fischbach, E., Geyer, B., Klimchitskaya, G. L., Krause, D. E., López, D., Mostepanenko, V. M., and Romero, C. (2006). Comment on "Temperature dependence of the Casimir effect". *Phys. Rev.*

E **73**, 028101-1-5.

Bezerra, V. B., Bimonte, G., Klimchitskaya, G. L., Mostepanenko, V. M., and Romero, C. (2007). Thermal correction to the Casimir force, radiative heat transfer, and an experiment. *Eur. Phys. J. C* **52**, 701-20.

Bezerra, V. B., Klimchitskaya, G. L., Mostepanenko, V. M., and Romero, C. (2008). Lifshitz theory of atom-wall interaction with applications to quantum reflection. *Phys. Rev. A* **78**, 042901-1-11.

Bezerra de Mello, E. R., Bezerra, V. B., and Khusnutdinov, N. R. (1999). Vacuum polarization of a massless spinor field in global monopole spacetime. *Phys. Rev. D* **60**, 063506-1-9.

Bezerra de Mello, E. R., Bezerra, V. B., and Saharian, A. A. (2007). Electromagnetic Casimir densities induced by a conducting cylindrical shell in the cosmic string spacetime. *Phys. Lett. B* **645**, 245-54.

Bimonte, G. (2006a). Comment on "Low-frequency character of the Casimir force between metallic films". *Phys. Rev. E* **73**, 048101-1-2.

Bimonte, G. (2006b). A theory of electromagnetic fluctuations for metallic surfaces and van der Waals interactions between metallic bodies. *Phys. Rev. Lett.* **96**, 160401-1-4.

Bimonte, G. (2007). Johnson noise and the thermal Casimir effect. *New J. Phys.* **9**, 281-1-9.

Bimonte, G. (2008). Casimir effect in a superconducting cavity and the thermal controversy. *Phys. Rev. A* **78**, 062101-1-9.

Bimonte, G., Calloni, E., Esposito, G., Milano, L., and Rosa, L. (2005a). Towards measuring variations of Casimir energy by a superconducting cavity. *Phys. Rev. Lett.* **94**, 180402-1-4.

Bimonte, G., Calloni, E., Esposito, G., and Rosa, L. (2005b). Variations of Casimir energy from a superconducting transition. *Nucl. Phys. B* **726**, 441-63.

Bimonte, G., Calloni, E., Esposito, G., and Rosa, L. (2007). Relativistic mechanics of Casimir apparatuses in a weak gravitational field. *Phys. Rev. D* **76**, 025008-1-10.

Bimonte, G., Klimchitskaya, G. L., and Mostepanenko, V. M. (2009). The impact of magnetic properties on atom-wall interaction. ArXiv:0904.0234; *Phys. Rev. A*, forthcoming.

Bingqian, L., Changchun, Z., and Junhua, L. (1999). Electrostatic force influenced by space charge in submicrometer or nanometer silicon microstructures. *J. Micromech. Microeng.* **9**, 319-23.

Birmingham, D., Kantowski, R., and Milton, K. A. (1988). Scalar and spinor Casimir energies in even-dimensional Kaluza-Klein spaces of the form $M^4 \times S^{N_1} \times S^{N_2} \times \cdots$. *Phys. Rev. D* **38**, 1809-22.

Birrell, N. D. and Davies, P. C. W. (1982). *Quantum Fields in Curved Space*. Cambridge University Press, Cambridge.

Bishop, D., Gammel, P., and Giles, C. R. (2001). The little machines that are making it big. *Phys. Today* **54**, 38-44.

Blagov, E. V., Klimchitskaya, G. L., and Mostepanenko, V. M. (1996). How to describe AFM constant force surfaces in repulsive mode? *Surf. Sci.* **349**, 196–206.

Blagov, E. V., Klimchitskaya, G. L., and Mostepanenko, V. M. (1998). Impact of atomic relaxation on the breaks of constant force surfaces in AFM. *Surf. Sci.* **410**, 158–69.

Blagov, E. V., Klimchitskaya, G. L., and Mostepanenko, V. M. (1999). Description of force surfaces in atomic force microscopy with allowance for the mobility of the lattice atoms. *Zhurn. Tekh. Fiz.* **69**, 111–17 (*Tech. Phys.* **44**, 970–6).

Blagov, E. V., Klimchitskaya, G. L., Mohideen, U., and Mostepanenko, V. M. (2004). Control of the lateral Casimir force between corrugated surfaces. *Phys. Rev. A* **69**, 044103-1–4.

Blagov, E. V., Klimchitskaya, G. L., and Mostepanenko, V. M. (2005). Van der Waals interaction between microparticle and uniaxial crystal with application to hydrogen atoms and multiwall carbon nanotubes. *Phys. Rev. B* **71**, 235401-1–12.

Blagov, E. V., Klimchitskaya, G. L., and Mostepanenko, V. M. (2007). Van der Waals interaction between a microparticle and a single-walled carbon nanotube. *Phys. Rev. B* **75**, 235413-1–8.

Blau, S. K., Guendelman, E. I., Taormina, A., and Wijewardhana, L. C. R. (1984). On the stability of toroidally compact Kaluza–Klein theories. *Phys. Lett. B* **144**, 30–6.

Blau, S. K., Visser, M., and Wipf, A. (1988). Zeta functions and the Casimir energy. *Nucl. Phys. B* **310**, 163–80.

Blocki, J., Randrup, J., Swiatecki, W. J., and Tsang, C. F. (1977). Proximity forces. *Ann. Phys. (N.Y.)* **105**, 427–62.

Bogoliubov, N. N. and Mitropolsky, Yu. A. (1985). *Asymptotic Methods in the Theory of Non-linear Oscillations*. Gordon and Breach, New York.

Bogoliubov, N. N. and Shirkov, D. V. (1982). *Quantum Fields*. Benjamin-Cummings, London.

Bohren, C. F. and Huffmann, D. R. (1998). *Absorption and Scattering of Light by Small Particles*. Wiley, New York.

Bordag, M. (1990). On the vacuum-interaction of 2 parallel cosmic strings. *Ann. Physik* **47**, 93–100.

Bordag, M. (1991). The vacuum interaction of magnetic strings. *Ann. Phys. (N.Y.)* **206**, 257–71.

Bordag, M. (2000). Ground state energy for massive fields and renormalization. *Commun. Atom. Nucl. Phys., Commun. Mod. Phys., Part D* **1**, 347–61.

Bordag, M. (2006a). Casimir effect for a sphere and a cylinder in front of a plane and corrections to the proximity force theorem. *Phys. Rev. D* **73**, 125018-1–14.

Bordag, M. (2006b). The Casimir effect for thin plasma sheets and the role of the surface plasmons. *J. Phys. A: Math. Gen.* **39**, 6173–85.

Bordag, M. (2007). Generalized Lifshitz formula for a cylindrical plasma sheet in front of a plane beyond proximity force approximation. *Phys. Rev. D* **75**, 065003-1–18.

Bordag, M. and Khusnutdinov, N. (2008). Vacuum energy of a spherical plasma shell. *Phys. Rev. D* **77**, 085026-1–12.

Bordag, M. and Kirsten, K. (2002). Heat kernel coefficients and divergencies of the Casimir energy for the dispersive sphere. *Int. J. Mod. Phys. A* **17**, 813–19.

Bordag, M. and Lindig, J. (1998). Radiative correction to the Casimir force on a sphere. *Phys. Rev. D* **58**, 045003-1–16.

Bordag, M. and Nikolaev, V. (2008). Casimir force for a sphere in front of a plane beyond proximity force approximation. *J. Phys. A: Math. Theor.* **41**, 164002-1–10.

Bordag, M. and Nikolaev, V. (2009). Beyond proximity force approximation in the Casimir effect. *Int. J. Mod. Phys. A* **24**, 1743–7.

Bordag, M. and Pirozhenko, I. G. (2001). Heat kernel coefficients for the dielectric cylinder. *Phys. Rev. D* **64**, 025019-1–7.

Bordag, M. and Scharnhorst, K. (1998). $O(\alpha)$ radiative correction to the Casimir energy for penetrable mirrors. *Phys. Rev. Lett.* **81**, 3815–8.

Bordag, M. and Vassilevich, V. D. (1999). Heat kernel expansion for semitransparent boundaries. *J. Phys. A: Math. Gen.* **32**, 8247–59.

Bordag, M. and Vassilevich, V. D. (2004). Nonsmooth backgrounds in quantum field theory. *Phys. Rev. D* **70**, 045003-1–7.

Bordag, M., Petrov, G., and Robaschik, D. (1984). Calculation of the Casimir effect for a scalar field with the simplest non-stationary boundary conditions. *Yadern. Fiz.* **39**, 1315–20 (*Sov. J. Nucl. Phys.* **39**, 828–31).

Bordag, M., Robaschik, D., and Wieczorek, E. (1985). Quantum field theoretic treatment of the Casimir effect. *Ann. Phys. (N.Y.)* **165**, 192–213.

Bordag, M., Dittes, F.-M., and Robaschik, D. (1986). The Casimir effect with uniformly moving mirrors. *Yadern. Fiz.* **43**, 1606–13 (*Sov. J. Nucl. Phys.* **43**, 1034–8).

Bordag, M., Klimchitskaya, G. L., and Mostepanenko, V. M. (1994). Casimir force between two parallel plates with small distortions of different types. *Mod. Phys. Lett. A* **9**, 2515–26.

Bordag, M., Klimchitskaya, G. L., and Mostepanenko, V. M. (1995a). The Casimir force between plates with small deviations from plane parallel geometry. *Int. J. Mod. Phys. A* **10**, 2661–81.

Bordag, M., Klimchitskaya, G. L., and Mostepanenko, V. M. (1995b). Corrections to the van der Waals forces in application to atomic force microscopy. *Surf. Sci.* **328**, 129–34.

Bordag, M., Klimchitskaya, G. L., and Mostepanenko, V. M. (1995c). Corrections to the Casimir force between plates with stochastic surfaces. *Phys. Lett. A* **200**, 95–102.

Bordag, M., Geyer, B., Kirsten, K., and Elizalde, E. (1996a). Zeta function determinant of the Laplace operator on the D-dimensional ball. *Commun. Math. Phys.* **179**, 215–34.

Bordag, M., Kirsten, K., and Dowker, J. S. (1996b). Heat-kernels and functional determinants on the generalized cone. *Commun. Math. Phys.* **182**, 371–93.

Bordag, M., Elizalde, E., Kirsten, K., and Leseduarte, S. (1997). Casimir energies for massive scalar fields in a spherical geometry. *Phys. Rev. D* **56**, 4896–904.

Bordag, M., Kirsten, K., and Vassilevich, D. V. (1998a). Path-integral quantization of electrodynamics in dielectric media. *J. Phys. A: Math. Gen.* **31**, 2381–90.

Bordag, M., Geyer, B., Klimchitskaya, G. L., and Mostepanenko, V. M. (1998b). Constraints for hypothetical interactions from a recent demonstration of the Casimir force and some possible improvements. *Phys. Rev. D* **58**, 075003-1–16.

Bordag, M., Gillies, T. G., and Mostepanenko, V. M. (1998c). Erratum: New constraints on the Yukawa-type hypothetical interaction from the recent Casimir force measurement [*Phys. Rev. D* **56**, R6 (1997)]. *Phys. Rev. D* **57**, 2024.

Bordag, M., Kirsten, K., and Vassilevich, D. V. (1999a). Ground state energy for a penetrable sphere and for a dielectric ball. *Phys. Rev. D* **59**, 085011-1–14.

Bordag, M., Geyer, B., Klimchitskaya, G. L., and Mostepanenko, V. M. (1999b). Stronger constraints for nanometer scale Yukawa-type hypothetical interactions from the new measurement of the Casimir force. *Phys. Rev. D* **60**, 055004-1–7.

Bordag, M., Geyer, B., Klimchitskaya, G. L., and Mostepanenko, V. M. (2000a). Casimir force at both non-zero temperature and finite conductivity. *Phys. Rev. Lett.* **85**, 503–6.

Bordag, M., Geyer, B., Klimchitskaya, G. L., and Mostepanenko, V. M. (2000b). New constraints for non-Newtonian gravity in nanometer range from the improved precision measurement of the Casimir force. *Phys. Rev. D* **62**, 011701(R)-1–5.

Bordag, M., Mohideen, U., and Mostepanenko, V. M. (2001a). New developments in the Casimir effect. *Phys. Rep.* **353**, 1–205.

Bordag, M., Vassilevich, D., Falomir, H., and Santangelo, E. M. (2001b). Multiple reflection expansion and heat kernel coefficients. *Phys. Rev. D* **64**, 045017-1–11.

Bordag, M., Nesterenko, V. V., and Pirozhenko, I. G. (2002). High temperature asymptotics of thermodynamic functions of an electromagnetic field subjected to boundary conditions on a sphere and cylinder. *Phys. Rev. D* **65**, 045011-1–16.

Bordag, M., Geyer, B., Klimchitskaya, G. L., and Mostepanenko, V. M. (2006). Lifshitz-type formulas for graphene and single-wall carbon nanotubes: van der Waals and Casimir interactions. *Phys. Rev. B* **74**, 205431-1–9.

Boström, M. and Sernelius, B. E. (2000a). Comment on "Calculation of the

Casimir force between imperfectly conducting plates". *Phys. Rev. A* **61**, 046101-1–3.

Boström, M. and Sernelius, B. E. (2000b). Thermal effects on the Casimir force in the 0.1-5 μm range. *Phys. Rev. Lett.* **84**, 4757–60.

Boström, M. and Sernelius, B. E. (2000c). Van der Waals energy of an atom in the proximity of thin metal films. *Phys. Rev. A* **61**, 052703-1–6.

Boström, M and Sernelius, B. E. (2004). Entropy of the Casimir effect between real metal plates. *Physica A* **339**, 53–9.

Boyer, T. H. (1968). Quantum electromagnetic zero-point energy of a conducting spherical shell and Casimir model for a charged particle. *Phys. Rev.* **174**, 1764–76.

Boyer, T. H. (1974). Van der Waals forces and zero-point energy for dielectric and permeable materials. *Phys. Rev. A* **9**, 2078–84.

Braggio, C., Bressi, G., Carugno, G., Lombardi, A., Palmieri, A., Ruoso, G., and Zanello, D. (2004). Semiconductor microwave mirror for a measurement of the dynamical Casimir effect. *Rev. Sci. Instrum.* **75**, 4967–70.

Brandt, S. (1976). *Statistical and Computational Methods in Data Analysis.* North-Holland, Amsterdam.

Brax, P., van de Bruck, C., Davis, A.-C., Mota, D. F., and Shaw, D. (2007). Detecting chameleons through Casimir force measurements. *Phys. Rev. D* **76**, 124034-1–31.

Bressi, G., Carugno, G., Onofrio, R., and Ruoso, G. (2002). Measurement of the Casimir force between parallel metallic surfaces. *Phys. Rev. Lett.* **88**, 041804-1–4.

Brevik, I. and Kolbenstvedt, H. (1982). Casimir stress in a solid ball with permittivity and permeability. *Phys. Rev. D* **25**, 1731–4.

Brevik, I. and Lygren, M. (1996). Casimir effect for a perfectly conducting wedge. *Ann. Phys. (N.Y.)* **251**, 157–79.

Brevik, I. and Nyland, G. H. (1994). Casimir force on a dielectric cylinder. *Ann. Phys. (N.Y.)* **230**, 321–42.

Brevik, I., Nesterenko, V. V., and Pirozhenko, I. G. (1998). Direct mode summation for the Casimir energy of a solid ball. *J. Phys. A: Math. Gen.* **31**, 8661–8.

Brevik, I., Marachevsky, V. N., and Milton, K. A. (1999). Identity of the van der Waals force and the Casimir effect and the irrelevance of these phenomena to sonoluminescence. *Phys. Rev. Lett.* **82**, 3948–51.

Brevik, I., Aarseth, J. B., and Høye, J. S. (2002). Casimir problem of spherical dielectrics: numerical evaluation for general permittivities. *Phys. Rev. E* **66**, 026119-1–9.

Brevik, I., Aarseth, J. B., Høye, J. S., and Milton, K. A. (2005). Temperature dependence of the Casimir effect. *Phys. Rev. E* **71**, 056101-1–8.

Brown, L. S. and Maclay, G. J. (1969). Vacuum stress between conducting plates—an image solution. *Phys. Rev.* **184**, 1272–9.

Brown, G., Celli, V., Haller, M., Maradudin, A. A., and Marvin, A. (1985).

Resonant light scattering from a randomly rough surface. *Phys. Rev. B* **31**, 4993–5005.

Brown-Hayes, M., Dalvit, D. A. R., Mazzitelli, F. D., Kim, W. J., and Onofrio, R. (2005). Towards a precision measurement of the Casimir force in a cylinder–plane geometry. *Phys. Rev. A* **72**, 052102-1–11.

Brownlee, K. A. (1965). *Statistical Theory and Methodology in Science and Engineering*. Wiley, New York.

Brühl, R., Fouquet, P., Grisenti, R. E., Toennies, J. P., Hegerfeldt, G. C., Köhler, T., Stoll, M., and Walter, C. (2002). The van der Waals potential between metastable atoms and solid surfaces: Novel diffraction experiments vs. theory. *Europhys. Lett.* **59**, 357–63.

Bryksin, V. V. and Petrov, M. P. (2008). Casimir force with the inclusion of a finite thickness of interacting plates. *Fiz. Tverdogo Tela* **50**, 222–6 (*Phys. Solid State* **50**, 229–34).

Buchbinder, I. L. and Odintsov, S. D. (1989). Effective action in multidimensional (super) gravities and spontaneous compactification (quantum aspects of Kaluza–Klein theories). *Fortschr. Phys.* **37**, 225–59.

Buenzli, P. R. and Martin, P. A. (2008). Microscopic theory of the Casimir force at thermal equilibrium: Large separation asymptotics. *Phys. Rev. A* **77**, 011114-1–15.

Buhmann, S. Y. and Scheel, S. (2008). Thermal Casimir vs Casimir–Polder forces: equilibrium and non-equilibrium forces. *Phys. Rev. Lett.* **100**, 253201-1–4.

Buhmann, S. Y. and Welsch, D.-G. (2007). Dispersion forces in macroscopic quantum electrodynamics. *Prog. Quantum Electron.* **31**, 51–130.

Buhmann, S. Y., Welsch, D.-G., and Kampf, T. (2005). Ground-state van der Waals forces in planar multilayer magnetodielectrics. *Phys. Rev. A* **72**, 032112-1–16.

Buks, E. and Roukes, M. L. (2001a). Stiction, adhesion, and the Casimir effect in micromechanical systems. *Phys. Rev. B* **63**, 033402-1–4.

Buks, E. and Roukes, M. L. (2001b). Metastability and the Casimir effect in micromechanical systems. *Europhys. Lett.* **54**, 220–6.

Bulgac, A., Magierski, P., and Wirzba, A. (2006). Scalar Casimir effect between Dirichlet spheres or a plate and a sphere. *Phys. Rev. D* **73**, 025007-1–14.

Büscher, R. and Emig, T. (2004). Nonperturbative approach to Casimir interactions in periodic geometries. *Phys. Rev. A* **69**, 062101-1–18.

Büscher, R. and Emig, T. (2005). Geometry and spectrum of Casimir forces. *Phys. Rev. Lett.* **94**, 133901-1–4.

Bytsenko, A. A., Cognola, G., Vanzo, L., and Zerbini, S. (1996). Quantum fields and extended objects in space-times with constant curvature spatial section. *Phys. Rep.* **266**, 1–126.

Callan, C. G., Coleman, S., and Jackiw, R. (1970). A new improved energy–momentum tensor. *Ann. Phys. (N.Y.)* **59**, 42–73.

Candelas, P. and Weinberg, S. (1984). Calculation of gauge couplings and compact circumferences from self-consistent dimensional reduction. *Nucl. Phys. B* **237**, 397–441.

Caride, A. O., Klimchitskaya, G. L., Mostepanenko, V. M., and Zanette, S. I. (2005). Dependences of the van der Waals atom-wall interaction on atomic and material properties. *Phys. Rev. A* **71** 042901-1–8.

Caruso, F., Neto, N. P., Svaiter, B. F., and Svaiter, N. F. (1991). Attractive or repulsive nature of Casimir force in D-dimensional Minkowski spacetime. *Phys. Rev. D* **43**, 1300–6.

Caruso, F., De Paola, R., and Svaiter, N. F. (1999). Zero-point energy of massless scalar fields in the presence of soft and semihard boundaries in D dimensions. *Int. J. Mod. Phys. A* **14**, 2077–89.

Casimir, H. B. G. (1948). On the attraction between two perfectly conducting plates. *Proc. K. Ned. Akad. Wet. B* **51**, 793–5.

Casimir, H. B. G. (1953). Introductory remarks on quantum electrodynamics. *Physica* **19**, 846–9.

Casimir, H. B. G. (1999). Some remarks on the history of the so called Casimir effect. In: M. Bordag (ed.), *The Casimir Effect 50 Years Later*. World Scientific, Singapore, pp.3–9.

Casimir, H. B. G. and Polder, D. (1948). The influence of retardation on the London–van der Waals forces. *Phys. Rev.* **73**, 360–72.

Casimir, H. B. G. and Ubbink, J. (1967). Skin effect. II. The skin effect at high frequencies. *Philips Tech. Rev.* **28**, 300–15.

Castillo-Garza, R., Chang, C.-C., Jimenez, D., Klimchitskaya, G. L., Mostepanenko, V. M., and Mohideen, U. (2007). Experimental approaches to the difference in the Casimir force due to modifications in the optical properties of the boundary surface. *Phys. Rev. A* **75**, 062114-1–13.

Cavalcanti, R. M. (2004). Casimir force on a piston. *Phys. Rev. D* **69**, 065015-1–5.

Cavero-Pelaez, I. and Milton, K. A. (2005). Casimir energy for a dielectric cylinder. *Ann. Phys. (N.Y.)* **320**, 108–34.

Cavero-Pelaez, I. and Milton, K. A. (2006). Green's dyadic approach of the self-stress on a dielectric–diamagnetic cylinder with non-uniform speed of light. *J. Phys. A: Math. Gen.* **39**, 6225–32.

Chan, H. B., Aksyuk, V. A., Kleiman, R. N., Bishop, D. J., and Capasso, F. (2001a). Quantum mechanical actuation of microelectromechanical system by the Casimir effect. *Science* **291**, 1941–4.

Chan, H. B., Aksyuk, V. A., Kleiman, R. N., Bishop, D. J., and Capasso, F. (2001b). Nonlinear micromechanical Casimir oscillator. *Phys. Rev. Lett.* **87**, 211801-1–4.

Chan, H. B., Bao, Y., Zou, J., Cirelli, R. A., Klemens, F., Mansfield, W. M., and Pai, C. S. (2008). Measurement of the Casimir force between a gold sphere and a silicon surface with nanoscale trench arrays. *Phys. Rev. Lett.* **101**, 030401-1–4.

Charlton, T. M. (1973). *Energy Principles in Theory of Structures.* Oxford University Press, Oxford.

Chazalviel, J.-N. (1999). *Coulomb Screening of Mobile Charges: Applications to Material Science, Chemistry and Biology.* Birkhauser, Boston.

Chen, F. and Mohideen, U. (2001). Fiber optic interferometry for precision measurement of the voltage and frequency dependence of the displacement of piezoelectric tubes. *Rev. Sci. Instrum.* **72**, 3100–2.

Chen, F., Mohideen, U., Klimchitskaya, G. L., and Mostepanenko, V. M. (2002a). Demonstration of the lateral Casimir force. *Phys. Rev. Lett.* **88**, 101801-1–4.

Chen, F., Mohideen, U., Klimchitskaya, G. L., and Mostepanenko, V. M. (2002b). Experimental and theoretical investigation of the lateral Casimir force between corrugated surfaces. *Phys. Rev. A* **66**, 032113-1–11.

Chen, F., Klimchitskaya, G. L., Mohideen, U., and Mostepanenko, V. M. (2003). New features of the thermal Casimir force at small separations. *Phys. Rev. Lett.* **90**, 160404-1–4.

Chen, F., Klimchitskaya, G. L., Mohideen, U., and Mostepanenko, V. M. (2004a). Theory confronts experiment in the Casimir force measurements: quantification of errors and precision. *Phys. Rev. A* **69**, 022117-1–11.

Chen, F., Mohideen, U., and Milonni, P. W. (2004b). Progress in the measurement of the difference thermal contribution to the Casimir force. In: K. A. Milton (ed.), *Quantum Field Theory under the Influence of External Conditions.* Rinton Press, Princeton, pp.5–10.

Chen, F., Mohideen, U., Klimchitskaya, G. L., and Mostepanenko, V. M. (2005a). Investigation of the Casimir force between metal and semiconductor test bodies. *Phys. Rev. A* **72**, 020101(R)-1–4.

Chen, F., Mohideen, U., and Milonni, P. W. (2005b). Limits on non-Newtonian gravity and hypothetical forces from measurements of the Casimir force. *Int. J. Mod. Phys. A* **20**, 2222–31.

Chen, F., Klimchitskaya, G. L., Mostepanenko, V. M., and Mohideen, U. (2006a). Demonstration of the difference in the Casimir force for samples with different charge-carrier densities. *Phys. Rev. Lett.* **97**, 170402-1–4.

Chen, F., Mohideen, U., Klimchitskaya, G. L., and Mostepanenko, V. M. (2006b). Experimental test for the conductivity properties from the Casimir force between metal and semiconductor. *Phys. Rev. A* **74**, 022103-1–14.

Chen, F., Klimchitskaya, G. L., Mostepanenko, V. M., and Mohideen, U. (2007a). Demonstration of optically modulated dispersion forces. *Opt. Express* **15**, 4823–9.

Chen, F., Klimchitskaya, G. L., Mostepanenko, V. M., and Mohideen, U. (2007b). Control of the Casimir force by the modification of dielectric properties with light. *Phys. Rev. B* **76**, 035338-1–15.

Chen, F., Mohideen, U., Klimchitskaya, G. L., and Mostepanenko, V. M. (2007c). Comment on "Lateral Casimir force beyond the proximity-force approximation". *Phys. Rev. Lett.* **98**, 068901-1.

Chernikov, N. A. and Tagirov, E. A. (1968). Quantum theory of scalar field in Sitter space-time. *Ann. Inst. H. Poincaré A* **9**, 109–41.

Chiu, H.-C., Chang, C.-C., Castillo-Garza, R., Chen. F., and Mohideen, U. (2008). Experimental procedures for precision measurements of the Casimir force with an atomic force microscope. *J. Phys. A: Math. Theor.* **41**, 164022-1–14.

Chodos, A. and Myers, E. (1984). Gravitational contribution to the Casimir energy in Kaluza-Klein theories. *Ann. Phys. (N.Y.)* **156**, 412–41.

Chodos, A., Jaffe, R. L., Johnson, K., Thorn, C. B., and Weisskop, V. F. (1974). New extended model of hadrons. *Phys. Rev. D* **9**, 3471–95.

Chumak, A. A., Milonni, P. W., and Berman, G. P. (2004). Effects of electrostatic fields and Casimir force on cantilever vibrations. *Phys. Rev. B* **70**, 085407-1–9.

Cochran, W. G. (1954). The combination of estimates from different experiments. *Biometrics* **10**, 101–29.

Cognola, G., Kirsten, K., and Vanzo, L. (1994). Free and self-interacting scalar fields in the presence of conical singularities. *Phys. Rev. D* **49**, 1029–38.

Contreras-Reyes, A. M. and Mochán, W. L. (2005). Surface screening in the Casimir force. *Phys. Rev. A* **72**, 034102-1–4.

Crocce, M., Dalvit, D. A. R., and Mazzitelli, F. D. (2002). Quantum electromagnetic field in a three-dimensional oscillating cavity. *Phys. Rev. A* **66**, 033811-1–9.

Dalvit, D. A. R. and Lamoreaux, S. K. (2008). Contribution of drifting carriers to the Casimir–Lifshitz and Casimir–Polder interactions with semiconductor materials. *Phys. Rev. Lett.* **101**, 163203-1–4.

Dalvit, D. A. R., Lombardo, F. C., Mazzitelli, F. D., and Onofrio, R. (2006). Exact Casimir interaction between eccentric cylinders. *Phys. Rev. A* **74**, 020101(R)-1–4.

Dalvit, D. A. R., Maia Neto, P. A., Lambrecht, A., and Reynaud, S. (2008). Probing quantum-vacuum geometrical effects with cold atoms. *Phys. Rev. Lett.* **100**, 040405-1–4.

Davies, B. (1972). Quantum electromagnetic zero-point energy of a conducting spherical shell. *J. Math. Phys.* **13**, 1324–9.

Davies, B. and Ninham, B. W. (1972). Van der Waals forces in electrolytes. *J. Chem. Phys.* **56**, 5797–801.

De Sabbata, V., Melnikov, V. N., and Pronin, P. I. (1992). Theoretical approach to treatment of non-Newtonian forces. *Prog. Theor. Phys.* **88**, 623–61.

Decca, R. S. and López, D. (2009). Measurement of the Casimir force using a micromechanical torsional oscillator: Electrostatic calibration. *Int. J. Mod. Phys. A* **24**, 1748–56.

Decca, R. S., López, D., Fischbach, E., and Krause, D. E. (2003a). Measurement of the Casimir force between dissimilar metals. *Phys. Rev. Lett.* **91**, 050402-1–4.

Decca, R. S., Fischbach, E., Klimchitskaya, G. L., Krause, D. E., López, D., and

Mostepanenko, V. M. (2003b). Improved tests of extra-dimensional physics and thermal quantum field theory from new Casimir force measurements. *Phys. Rev. D* **68**, 116003-1–15.

Decca, R. S., López, D., Fischbach, E., Klimchitskaya, G. L., Krause, D. E., and Mostepanenko, V. M. (2004). Precise measurements of the Casimir force and first realization of a "Casimir-less" experiment. *J. Low Temp. Phys.* **135**, 63–74.

Decca, R. S., López, D., Chan, H. B., Fischbach, E., Krause, D. E., and Jamell, C. R. (2005a). Constraining new forces in the Casimir regime using the isoelectronic technique. *Phys. Rev. Lett.* **94**, 240401-1–4.

Decca, R. S., López, D., Fischbach, E., Klimchitskaya, G. L., Krause, D. E., and Mostepanenko, V. M. (2005b). Precise comparison of theory and new experiment for the Casimir force leads to stronger constraints on thermal quantum effects and long-range interactions. *Ann. Phys. (N.Y.)* **318**, 37–80.

Decca, R. S., López, D., Fischbach, E., Klimchitskaya, G. L., Krause, D. E., and Mostepanenko, V. M. (2007a). Tests of new physics from precise measurements of the Casimir pressure between two gold-coated plates. *Phys. Rev. D* **75**, 077101-1–4.

Decca, R. S., López, D., Fischbach, E., Klimchitskaya, G. L., Krause, D. E., and Mostepanenko, V. M. (2007b). Novel constraints on light elementary particles and extra-dimensional physics from the Casimir effect. *Eur. Phys. J. C* **51**, 963–75.

Decca, R. S., Fischbach, E., Geyer, B., Klimchitskaya, G. L., Krause, D. E., López, D., Mohideen, U., and Mostepanenko, V. M. (2008). Comment on "Contribution of drifting carriers to the Casimir–Lifshitz and Casimir–Polder interactions with semiconductor materials". ArXiv:0810.3244; *Phys. Rev. Lett.*, forthcoming.

Decca, R. S., Fischbach, E., Klimchitskaya, G. L., Krause, D. E., López, D., Mohideen, U., and Mostepanenko, V. M. (2009a). Comment on "Anomalies in electrostatic calibration for the measurement of the Casimir force in a sphere–plane geometry". *Phys. Rev. A* **79**, 026101-1–4.

Decca, R. S., Fischbach, E., Klimchitskaya, G. L., Krause, D. E., López, D., and Mostepanenko, V. M. (2009b). Application of the proximity force approximation to gravitational and Yukawa-type forces. ArXiv:0903.1299.

de Man, S., Heeck, K., and Iannuzzi, D. (2009). No anomalous scaling in electrostatic calibrations for Casimir force measurements. *Phys. Rev. A* **79**, 024102-1–4.

DeRaad, L. L. and Milton, K. A. (1981). Casimir self-stress on a perfectly conducting cylindrical shell. *Ann. Phys. (N.Y.)* **136**, 229–42.

Derevianko, A., Johnson, W. R., Safronova, M. S., and Babb, J. F. (1999). High-precision calculations of dispersion coefficients, static dipole polarizabilities, and atom-wall interaction constants for alkali-metal atoms. *Phys. Rev. Lett.* **82**, 3589–92.

Derjaguin, B. V. (1934). Untersuchungen über die Reibung und Adhäsion, IV. Theorie des Anhaftens kleiner Teilchen. *Kolloid Z.* **69**, 155–64.

Derjaguin, B. V. and Abrikosova, I. I. (1958). Direct measurements of molecular attraction in solids. *J. Phys. Chem. Solids* **5**, 1–10.

Derjaguin, B. V., Abrikosova, I. I., and Lifshitz, E. M. (1956). Direct measurement of molecular attraction between solids separated by a narrow gap. *Q. Rev.* **10**, 295–329.

Derjaguin, B. V., Churaev, N. V., and Muller, V. M. (1987). *Surface Forces.* Plenum, New York.

Deser, S. and Zumino, B. (1977). Broken supersymmetry and supergravity. *Phys. Rev. Lett.* **38**, 1433–6.

Deutsch, D. and Candelas, P. (1979). Boundary effects in quantum field theory. *Phys. Rev. D* **20**, 3063–80.

Devonshire, A. F. (1936). The interaction of atoms and molecules with solid surfaces. V. The diffraction and reflection of molecular rays. *Proc. R. Soc. Lond. A* **156**, 37–44.

DeWitt, B. S., Hart, C. F., and Isham, C. J. (1979). Topology and quantum field theory. *Physica A* **96**, 197–211.

Dimopoulos, S. and Geraci, A. A. (2003). Probing submicron forces by interferometry of Bose–Einstein condensed atoms. *Phys. Rev. D* **68**, 124021-1–13.

Dimopoulos, S. and Giudice, G. F. (1996). Macroscopic forces from supersymmetry. *Phys. Lett. B* **379**, 105–14.

Dobson, J. F., White, A., and Rubio, A. (2006) Benchmarks for van der Waals energy functionals. *Phys. Rev. Lett.* **96**, 073201-1–4.

Dodonov, V. V. and Dodonov, A. V. (2006). The nonstationary Casimir effect in a cavity with periodical time-dependent conductivity of a semiconductor mirror. *J. Phys. A: Math. Gen.* **39**, 6271–81.

Dodonov, V. V. and Klimov, A. B. (1996). Generation and detection of photons in a cavity with a resonantly oscillating boundary. *Phys. Rev. A* **53**, 2664–82.

Dowker, J. S. (1984). Self-consistent Kaluza-Klein structures. *Phys. Rev. D* **29**, 2773–8.

Dowker, J. S. (1987). Casimir effect around a cone. *Phys. Rev. D* **36**, 3095–101.

Dowker, J. S. and Critchley, R. (1976). Covariant Casimir calculations. *J. Phys. A: Math. Gen.* **9**, 535–40.

Dowker, J. S. and Kennedy, G. (1978). Finite temperature and boundary effects in static space-times. *J. Phys. A: Math. Gen.* **11**, 895–920.

Doyle, J. M., Sandberg, J. C., Yu, I. A., Cesar, C. L., Kleppner, D., and Greytak, T. J. (1991). Hydrogen in the submillikelvin regime: Sticking probability on superfluid 4He. *Phys. Rev. Lett.* **67**, 603–6.

Druzhinina, V. and DeKieviet, M. (2003). Experimental observation of quantum reflection far from threshold. *Phys. Rev. Lett.* **91**, 193202-1–4.

Dutra, S. M. (2005). *Cavity Quantum Electrodynamics.* Wiley, New York.

Dzyaloshinskii, I. E., Lifshitz, E. M., and Pitaevskii, L. P. (1961). The general theory of van der Waals forces. *Usp. Fiz. Nauk* **73**, 381–422 (*Adv. Phys.* **10**,

165–209).
Eberlein, C. (1996). Theory of quantum radiation observed as sonoluminescence *Phys. Rev. A* **53**, 2772–87.
Ederth, T. (2000). Template-stripped gold surfaces with 0.4-nm rms roughness suitable for force measurements: Application to the Casimir force in the 20–100-nm range. *Phys. Rev. A* **62**, 062104-1–8.
Edery, A. (2006). Multidimensional cut-off technique, odd-dimensional Epstein zeta functions and Casimir energy for massless scalar fields. *J. Phys. A: Math. Gen.* **39**, 685–712.
Edery, A. (2007). Casimir piston for massless scalar fields in three dimensions. *Phys. Rev. D* **75**, 105012-1–9.
Efimov, N. V. (1980). *Higher Geometry*. Mir, Moscow.
Ekinci, K. L. and Roukes, M. L. (2005). Nanoelectromechanical systems. *Rev. Sci. Instrum.* **76**, 061101-1–12.
Elizalde, E. (1995). *Ten Physical Applications of Spectral Zeta Functions*. Springer, Berlin.
Elizalde, E. (2006). Uses of zeta regularization in QFT with boundary conditions: A cosmological Casimir effect. *J. Phys. A: Math. Gen.* **39**, 6299–307.
Elizalde, E., Odintsov, S. D., Romeo, A., Bytsenko, A. A., and Zerbini, S. (1994). *Zeta Regularization Techniques with Applications*. World Scientific, Singapore.
Elizalde, E., Bordag, M., and Kirsten, K. (1998). Casimir energy for a massive fermionic quantum field with a spherical boundary. *J. Phys. A: Math. Gen.* **31**, 1743–59.
Ellis, G. F. R. (1971). Topology and cosmology. *Gen. Rel. Grav.* **2**, 7–21.
Emig T. (2003). Casimir forces: An exact approach for periodically deformed objects. *Europhys. Lett.* **62**, 466–72.
Emig T. (2007). Casimir-force-driven ratchets. *Phys. Rev. Lett.* **98**, 160801-1–4.
Emig, T. (2008) Fluctuation-induced quantum interactions between compact objects and a plane mirror. *J. Stat. Mech.* P04007-1–33.
Emig, T. and Büscher, R. (2004). Towards a theory of molecular forces between deformed media. *Nucl. Phys. B* **696**, 468–91.
Emig, T., Hanke, A., Golestanian, R., and Kardar, M. (2003). Normal and lateral Casimir forces between deformed plates. *Phys. Rev. A* **67**, 022114-1–15.
Emig, T., Jaffe, R. L., Kardar, M., and Scardicchio, A. (2006). Casimir interaction between a plate and a cylinder. *Phys. Rev. Lett.* **96**, 080403-1–4.
Emig, T., Graham, N., Jaffe, R. L., and Kardar, M. (2007). Casimir forces between arbitrary compact objects. *Phys. Rev. Lett.* **99**, 170403-1–4.
Emig, T., Graham, N., Jaffe, R. L., and Kardar, M. (2008). Casimir forces between compact objects: The scalar case. *Phys. Rev. D* **77**, 025005-1–23.
Erdélyi, A., Magnus, W., Oberhettinger, F., and Tricomi, F. G. (1981). *Higher Transcendental Functions*, Vol. 1. Kriger, New York.

Esquivel, R. and Svetovoy, V. B. (2004). Corrections to the Casimir force due to the anomalous skin effect. *Phys. Rev. A* **69**, 062102-1–12.

Esquivel, R., Villarreal, C., and Mochán, W. L. (2003). Exact surface impedance formulation of the Casimir force: Application to spatially dispersive metals. *Phys. Rev. A* **68**, 052103-1–5.

Fabinger, M. and Hořava, P. (2000). Casimir effect between world-branes in heterotic M-theory. *Nucl. Phys. B* **580**, 243–63.

Fagundes, H. V. (1989). Quasar-galaxy associations with discordant redshifts as a topological effect. 2. A closed hyperbolic model. *Astron. J.* **338**, 618–29.

Faustov, R. N. (1970). Magnetic moment of the hydrogen atom. *Phys. Lett. B* **33**, 422–4.

Feinberg, G. and Sucher, J. (1979). Is there a strong van der Waals force between hadrons? *Phys. Rev. D* **20**, 1717–35.

Feinberg, J., Mann, A., and Revzen, M. (2001). Casimir effect: The classical limit. *Ann. Phys. (N.Y.)* **288**, 103–36.

Ferrara, S., Scherk, J., and Zumino, B. (1977). Algebraic properties of extended supergravity theories. *Nucl. Phys. B* **121**, 393–402.

Fetter, A. L. (1973). Electrodynamics of a layered electron gas. I. Single layer. *Ann. Phys. (N.Y.)* **81**, 367–93.

Feynman, R. P. (1948). Space-time approach to non-relativistic quantum mechanics. *Rev. Mod. Phys.* **20**, 367–87.

Feynman, R. P. (1960). There is plenty of room at the bottom. *Eng. Sci.* **23**, 22–36.

Fierz, M. (1960). Zur Anziehung leitender Ebenen im Vakuum. *Helv. Phys. Acta* **33**, 855–8.

Fischbach, E. (1996). Long-range forces and neutrino mass. *Ann. Phys. (N.Y.)* **247**, 213–91.

Fischbach, E. and Talmadge, C. L. (1999). *The Search for Non-Newtonian Gravity*. Springer, New York.

Fischbach, E, Krause, D. E., Mostepanenko, V. M., and Novello, M. (2001). New constraints on ultrashort-ranged Yukawa interactions from atomic force microscopy. *Phys. Rev. D* **64**, 075010-1–7.

Floratos, E. G. and Leontaris, G. K. (1999). Low scale unification, Newton's law and extra dimensions. *Phys. Lett. B* **465**, 95–100.

Foley, J. T. and Devaney, A. J. (1975). Electrodynamics of nonlocal media. *Phys. Rev. B* **12**, 3104–12.

Ford, L. H. (1975). Quantum vacuum energy in general relativity. *Phys. Rev. D* **11**, 3370–7.

Ford, L. H. (1976). Quantum vacuum energy in a closed universe. *Phys. Rev. D* **14**, 3304–13.

Ford, L. H. (1979). Casimir effect for a self-interacting scalar field. *Proc. R. Soc. Lond. A* **368**, 305–10.

Ford, L. H. (1980). Vacuum polarization in a non-simply connected spacetime. *Phys. Rev. D* **21**, 933–48.

Ford, G. W. and Weber, W. H. (1984). Electromagnetic interactions of molecules with metal surfaces. *Phys. Rep.* **113**, 195–287.

Friedrich, H. and Trost, J. (2004). Working with WKB waves far from the semiclassical limit. *Phys. Rep.* **397**, 359–449.

Friedrich, H., Jacoby, G., and Meister, C. G. (2002). Quantum reflection by Casimir–van der Waals potential tails. *Phys. Rev. A* **65**, 032902-1–13.

Frolov, V. P. and Serebriany, E. M. (1987). Vacuum polarization in the gravitational field of a cosmic string. *Phys. Rev. D* **35**, 3779–82.

Frolov, V. P., Pinzul, A., and Zelnikov, A. I. (1995). Vacuum polarization at finite temperature on a cone. *Phys. Rev. D* **51**, 2770–4.

Fujii, Y. (1991). The theoretical background of the fifth force. *Int. J. Mod. Phys. A* **6**, 3505–57.

Fulling, S. A. and Davies, P. C. W. (1976). Radiation from a moving mirror in 2 dimensional space–time—conformal anomaly. *Proc. R. Soc. Lond. A* **348**, 393–414.

Fulling, S. A., Kaplan, L., and Wilson, J. H. (2007a). Vacuum energy and repulsive Casimir forces in quantum star graphs. *Phys. Rev. A* **76**, 012118-1–7.

Fulling, S. A., Milton, K. A., Parashar, P., Romeo, A., Shajesh, K. V., and Wagner, J. (2007b). How does Casimir energy fall? *Phys. Rev. D* **76**, 025004-1–4.

Fursaev, D. V. (1994). The heat-kernel expansion on a cone and quantum fields near cosmic strings. *Class. Quantum Grav.* **11**, 1431–43.

Gal'tsov, D. V., Gratz, Y. V., and Lavrent'ev, A. B. (1995). Vacuum polarization and topological self-interaction of a charge in multiconic space. *Yadern. Fiz.* **58**, 570–6 (*Phys. Atom. Nucl.* **58**, 516–21).

Genet, C., Lambrecht, A., and Reynaud, S. (2000). Temperature dependence of the Casimir effect between metallic mirrors. *Phys. Rev. A* **62**, 012110-1–8.

Genet, C., Maia Neto, P. A., Lambrecht, A., and Reynaud, S. (2003a). The Casimir force between rough metallic plates. *Europhys. Lett.* **62**, 484–90.

Genet, C., Lambrecht, A., and Reynaud, S. (2003b). Casimir force and the quantum theory of lossy optical cavities. *Phys. Rev. A* **67**, 043811-1–18.

Geyer, B., Klimchitskaya, G. L., and Mostepanenko, V. M. (2001). Casimir force under the influence of real conditions. *Int. J. Mod. Phys. A* **16**, 3291–308.

Geyer, B., Klimchitskaya, G. L., and Mostepanenko, V. M. (2002). Perturbation approach to the Casimir force between two bodies made of different real metals. *Phys. Rev. A* **65**, 052113-1–7.

Geyer, B., Klimchitskaya, G. L., and Mostepanenko, V. M. (2003). Surface-impedance approach solves problems with the thermal Casimir force between real metals. *Phys. Rev. A* **67**, 062102-1–15.

Geyer, B., Klimchitskaya, G. L., and Mostepanenko, V. M. (2004). Reply to "Comment on 'Surface-impedance approach solves problems with the thermal Casimir force between real metals'". *Phys. Rev. A* **70**, 016102-1–8.

Geyer, B., Klimchitskaya, G. L., and Mostepanenko, V. M. (2005a). Thermal corrections in the Casimir interaction between a metal and dielectric. *Phys. Rev. A* **72**, 022111-1–9.

Geyer, B., Klimchitskaya, G. L., and Mostepanenko, V. M. (2005b). Thermal quantum field theory and the Casimir interaction between dielectrics. *Phys. Rev. D* **72**, 085009-1–20.

Geyer, B., Klimchitskaya, G. L., and Mostepanenko, V. M. (2006). Recent results on thermal Casimir force between dielectrics and related problems. *Int. J. Mod. Phys. A* **21**, 5007–42.

Geyer, B., Klimchitskaya, G. L., and Mostepanenko, V. M. (2007). Generalized plasma-like permittivity and thermal Casimir force between real metals. *J. Phys. A: Math. Theor.* **40**, 13485–99.

Geyer, B., Klimchitskaya, G. L., and Mostepanenko, V. M. (2008a). Analytic approach to the thermal Casimir force between metal and dielectric. *Ann. Phys. (N.Y.)* **323**, 291–316.

Geyer, B., Klimchitskaya, G. L., Mohideen, U., and Mostepanenko, V. M. (2008b). Comment on "Precision measurement of the Casimir–Lifshitz force in a fluid". *Phys. Rev. A* **77**, 036102-1–3.

Geyer, B., Klimchitskaya, G. L., and Mostepanenko, V. M. (2008c). Thermal Casimir effect in ideal metal rectangular boxes. *Eur. Phys. J. C* **57**, 823–34.

Geyer, B., Klimchitskaya, G. L., Mohideen, U., and Mostepanenko, V. M. (2008d). Comment on "Thermal Lifshitz force between an atom and a conductor with a small density of carriers". ArXiv:0810.3243; *Phys. Rev. Lett.*, forthcoming.

Geyer, B., Klimchitskaya, G. L., and Mostepanenko, V. M. (2008e). Comment on "Anomalous temperature dependence of the Casimir force for thin metal films". ArXiv:0810.3222.

Gies, H. and Klingmüller, K. (2006a). Casimir effect for curved geometries: Proximity-force-approximation validity limits. *Phys. Rev. Lett.* **96**, 220401-1–4.

Gies, H. and Klingmüller, K. (2006b). Worldline algorithms for Casimir configurations. *Phys. Rev. D* **74**, 045002-1–12.

Gies, H. and Klingmüller, K. (2006c). Casimir edge effects. *Phys. Rev. Lett.* **97**, 220405-1–4.

Gies, H., Langfeld, K., and Moyaerts, L. (2003). Casimir effect on the worldline. *J. High Energy Phys.* N06, 018-1–29.

Gilkey, P. B. (1995). *Invariance Theory, the Heat Equation and the Atiyah-Singer Index Theorem.* CRC Press, Boca Raton.

Ginzburg, V. L. (1985). *Physics and Astrophysics.* Pergamon, Oxford.

Ginzburg, V. L. (1989). *Applications of Electrodynamics in Theoretical Physics and Astrophysics.* Gordon and Breach, New York.

Golestanian, R. and Kardar, M. (1997). Mechanical response of vacuum. *Phys. Rev. Lett.* **78**, 3421–5.

Golestanian, R. and Kardar, M. (1998). Path-integral approach to the dymanic Casimir effect with fluctuating boundaries. *Phys. Rev. A* **58**, 1713–22.

Goncharov, Y. P. (1982). Casimir effect in a multiply connected space–time. Massless fields. *Izv. Vuzov, Fizika* **N9**, 30–2. (*Russ. Phys. J.* **25**, 791–4).
Gosdzinsky, P. and Romeo, A. (1998). Energy of the vacuum with a perfectly conducting and infinite cylindrical surface. *Phys. Lett. B* **441**, 265–74.
Gradshteyn, I. S. and Ryzhik, I. M. (1994). *Table of Integrals, Series, and Products*. Academic Press, New York.
Gräf, D., Grundner, M., Schulz, R., and Mühlhoff, L. (1990). Oxidation of HF-treated Si wafer surfaces in air. *J. Appl. Phys.* **68**, 5155–61.
Graham, N., Jaffe, R. L., Khemani, V., Quandt, M., Scandurra, M., and Weigel, H. (2002). Calculating vacuum energies in renormalizable quantum field theories: A new approach to the Casimir problem. *Nucl. Phys. B* **645**, 49–84.
Graham, N., Jaffe, R. L., Khemani, V., Quandt, M., Scandurra, M., and Weigel, H. (2003). Casimir energies in light of quantum field theory. *Phys. Lett. B* **572**, 196–201.
Graham, N., Jaffe, R. L., Khemani, V., Quandt, M., Schröder, O., and Weigel, H. (2004). The Dirichlet Casimir problem. *Nucl. Phys. B* **677**, 379–404.
Gray, A. (1997). *Modern Differential Geometry of Curves and Surfaces with Mathematica*. CRC Press, Boca Raton.
Greenaway, D. L., Harbeke, G., Bassani, F., and Tosatti, E. (1969). Anisotropy of the optical constants and the band structure of graphite. *Phys. Rev.* **178**, 1340–8.
Greiner, W., Müller, B., and Rafelski, J. (1985). *Quantum Electrodynamics of Strong Fields: With an Introduction into Modern Relativistic Quantum Mechanics* (2nd edn). Springer, Berlin.
Grib, A. A., Mamayev, S. G., and Mostepanenko, V. M. (1980). Vacuum stress-energy tensor and particle creation in isotropic cosmological models. *Fortschr. Phys.* **28**, 173–99.
Grib, A. A., Mamayev, S. G., and Mostepanenko, V. M. (1994). *Vacuum Quantum Effects in Strong Fields*. Friedmann Laboratory Publishing, St. Petersburg.
Gundlach, J. H., Smith, G. L., Adelberger, E. G., Heckel, B. R., and Swanson, H. E. (1997). Short-range test of the equivalence principle. *Phys. Rev. Lett.* **78**, 2523–6.
Gusso, A. and Delben, G. J. (2007). Influence of the Casimir force on the pull-in parameters of silicon based electrostatic torsional actuators. *Sens. Actuators A* **135**, 792–800.
Guth, A. H. (1981). Inflationary universe: A possible solution to the horizon and flatness problems. *Phys. Rev. D* **23**, 347–56.
Hacyan, S., Jáuregui, R., and Villarreal, C. (1993). Spectrum of quantum electromagnetic fluctuations in rectangular cavities. *Phys. Rev. A* **47**, 4204–11.
Hansson, T. H. and Jaffe, R. L. (1983a). The multiple reflection expansion for confined scalar, Dirac, and gauge-fields. *Ann. Phys. (N.Y.)* **151**, 204–26.

Hansson, T. H. and Jaffe, R. L. (1983b). Cavity quantum chromodynamics. *Phys. Rev. D* **28**, 882–907.

Harber, D. M., Obrecht, J. M., McGuirk, J. M., and Cornell, E. A. (2005). Measurement of the Casimir-Polder force through center-of-mass oscillations of a Bose-Einstein condensate. *Phys. Rev. A* **72**, 033610-1–6.

Hargreaves, C. M. (1965). Corrections to retarded dispersion force between metal bodies. *Proc. K. Nederl. Acad. Wet. B* **68**, 231–6.

Haro, J. and Elizalde, E. (2006). Hamiltonian approach to the dynamical Casimir effect. *Phys. Rev. Lett.* **97**, 130401-1–4.

Harris, P. J. F. (1999). *Carbon Nanotubes and Related Structures: New Materials for the Twenty-First Century*. Cambridge University Press, New York.

Harris, B. W., Chen, F., and Mohideen, U. (2000). Precision measurement of the Casimir force using gold surfaces. *Phys. Rev. A* **62**, 052109-1–5.

Heinrichs, J. (1973). Non-local effects in the macroscopic theory of Van der Waals and adhesive forces. *Sol. State Commun.* **13**, 1595–8.

Hellwege, K.-H., ed. (1982). *Semiconductors: Physics of Group IV Elements and III–V Compounds*. Springer, Berlin.

Helliwell, T. M. and Konkowski, D. A. (1986). Vacuum fluctuations outside cosmic strings. *Phys. Rev. D* **34**, 1918–20.

Henkel, C. and Joulain, K. (2005). Casimir force between designed materials: What is possible and what not. *Europhys. Lett.* **72**, 929–35.

Henkel, C., Joulain, K., Mulet, J.-P., and Greffet, J.-J. (2002). Radiation forces on small particles in thermal near fields. *J. Opt. A: Pure Appl. Opt.* **4**, S109–14.

Henkel, C., Joulain, K., Mulet, J.-P., and Greffet, J.-J. (2004). Coupled surface polaritons and the Casimir force. *Phys. Rev. A* **69**, 023808-1–7.

Hertzberg, M. P., Jaffe, R. L., Kardar, M., and Scardicchio, A. (2005). Attractive Casimir forces in a closed geometry. *Phys. Rev. Lett.* **95**, 250402-1–4.

Hertzberg, M. P., Jaffe, R. L., Kardar, M., and Scardicchio, A. (2007). Casimir forces in a piston geometry at zero and nonzero temperature. *Phys. Rev. D* **76**, 045016-1–13.

Hoskins, J. K., Newman, R. D., Spero, R., and Schultz, J. (1985). Experimental tests of the gravitational inverse-square law for mass separations from 2 to 105 cm. *Phys. Rev. D* **32**, 3084–95.

Hough, D. B. and White, L. R. (1980). The calculation of Hamaker constants from Lifshitz theory with applications to wetting phenomena. *Adv. Colloid Interface Sci.* **14**, 3–41.

Høye, J. S., Brevik, I., Aarseth, J. B., and Milton, K. A. (2003). Does the transverse electric zero mode contribute to the Casimir effect for a metal? *Phys. Rev. E* **67**, 056116-1–17.

Høye, J. S., Brevik, I., Aarseth, J. B., and Milton, K. A. (2006). What is the temperature dependence of the Casimir effect? *J. Phys. A: Math. Gen.* **39**, 6031–8.

Høye, J. S., Brevik, I., Ellingsen, S. A., and Aarseth, J. B. (2007). Analytical and numerical verification of the Nernst theorem for metals. *Phys. Rev. E* **75**, 051127-1–8.

Høye, J. S., Brevik, I., Ellingsen, S. A., and Aarseth, J. B. (2008). Reply to "Comment on 'Analytical and numerical verification of the Nernst theorem for metals'". *Phys. Rev. E* **77**, 023102-1–2.

Hunklinger, S., Geisselman, H., and Arnold, W. (1972). A dynamic method for measuring the van der Waals forces between macroscopic bodies. *Rev. Sci. Instrum.* **43**, 584–7.

Iannuzzi, D. and Capasso, F. (2003). Comment on "Repulsive Casimir forces". *Phys. Rev. Lett.* **91**, 029101-1.

Iannuzzi, D., Gelfand, I., Lisanti, M., and Capasso, F. (2004a). New challenges and directions in Casimir force experiments. In: K. A. Milton (ed.), *Quantum Field Theory under the Influence of External Conditions*. Rinton Press, Princeton, pp.11–6.

Iannuzzi, D., Lisanti, M., and Capasso, F. (2004b). Effect of hydrogen-switchable mirrors on the Casimir force. *Proc. Natl. Acad. Sci. USA* **101**, 4019–23.

Intravaia, F. and Henkel, C. (2008). Casimir energy and entropy between dissipative mirrors. *J. Phys. A: Math. Theor.* **41**, 164018-1–9.

Intravaia, F. and Lambrecht, A. (2005). Surface plasmon modes and the Casimir energy. *Phys. Rev. Lett.* **94**, 110404-1–4.

Intravaia, F., Henkel, C., and Lambrecht, A. (2007). Role of surface plasmons in the Casimir effect. *Phys. Rev. A* **76**, 033820-1–11.

Inui, N. (2003). Temperature dependence of the Casimir force between silicon slabs. *J. Phys. Soc. Jpn.* **72**, 2198–202.

Inui, N. (2006). Casimir force between a metallic sphere and a semiconductive plate illuminated with Gaussian beam. *J. Phys. Soc. Jpn.* **75**, 024004-1–6.

Isham, C. J. (1978a). Twisted quantum fields in a curved space-time. *Proc. R. Soc. Lond. A* **362**, 383–404.

Isham, C. J. (1978b). Spinor fields in four dimensional space-time. *Proc. R. Soc. Lond. A* **364**, 591–9.

Israelachvili, J. (1992). *Intermolecular and Surface Forces*. Academic Press, San Diego.

Israelachvili, J. N. and Tabor, D. (1972). The measurement of van der Waals dispersion forces in the range 1.5 to 130 nm. *Proc. R. Soc. Lond. A* **331**, 19–38.

Itzykson, C. and Zuber, J.-B. (2005). *Quantum Field Theory*. Dover, New York.

Jackson, J. D. (1999). *Classical Electrodynamics* (3rd edn). Wiley, New York.

Jaffe, R. L. (2005). Casimir effect and the quantum vacuum. *Phys. Rev. D* **72**, 021301(R)-1–5.

Jaffe, R. L. and Scardicchio, A. (2004). Casimir effect and geometric optics. *Phys. Rev. Lett.* **92**, 070402-1–4.

Jáuregui, R., Villarreal, C., and Hacyan, S. (2006). Finite temperature correction to the Casimir effect in rectangular cavities with perfectly conducting walls. *Ann. Phys. (N.Y.)* **321**, 2156–69.

Johnson K. (1975). The M.I.T. bag model. *Acta Phys. Polonica B* **6**, 865–92.

Johnson, L. G. and Dresselhaus, G. (1973). Optical properties of graphite. *Phys. Rev. B* **72**, 2275–85.

Johnson, R. E., Epstein, S. T., and Meath, W. J. (1967). Evaluation of long-range retarded interaction energies. *J. Chem. Phys.* **47**, 1271–4.

Joulain, K., Mulet, J.-P., Marquier, F., Carminati, R., and Greffet, J.-J. (2005). Surface electromagnetic waves thermally excited: Radiative heat transfer, coherence properties and Casimir forces revisited in the near field. *Surf. Sci. Rep.* **57**, 59–112.

Jourdan, G., Lambrecht, A., Comin, F., and Chevrier, J. (2009). Quantitative non-contact dynamic Casimir force measurements. *Europhys. Lett.* **85**, 31001-1–5.

Jurisch, A. and Friedrich, H. (2004). Quantum reflection times and space shifts for Casimir–van der Waals potential tails. *Phys. Rev. A* **70**, 032711-1–8.

Kaganova, I. M. and Kaganov, M. I. (2001). Effective surface impedance of polycrystals under anomalous skin effect conditions. *Phys. Rev. B* **63**, 054202-1–15.

Kapner, D. J., Cook, T. S., Adelberger, E. G., Gundlach, J. H., Heckel, B. R., Hoyle, C. D., and Swanson, H. E. (2007). Tests of the gravitational inverse-square law below the dark-energy length scale. *Phys. Rev. Lett.* **98**, 021101-1–4.

Kardar, M. and Golestanian, R. (1999). The "friction" of vacuum, and other fluctuation-induced forces. *Rev. Mod. Phys.* **71**, 1233–45.

Karepanov, S. K., Novikov, M. Yu., and Sorin, A. S. (1987). The influence of one-dimensional flat boundary distortions on the Casimir force. *Nuovo Cimento B* **100**, 411–5.

Kats, E. I. (1977). Influence of nonlocality effects on van der Waals interaction. *Zh. Eksp. Teor. Fiz.* **73**, 212–20 (*Sov. Phys. JETP* **46**, 109–13).

Kay, B. S. (1979). Casimir effect in quantum field theory. *Phys. Rev. D* **20**, 3052–62.

Kehagias, A. and Sfetsos, K. (2000). Deviations from the $1/r^2$ Newton law due to extra dimensions. *Phys. Lett. B* **472**, 39–44.

Kennedy, G. (1978). Boundary terms in the Schwinger–DeWitt expansion: Flat space results. *J. Phys. A: Math. Gen.* **11**, L173–8.

Kenneth, O. and Klich, I. (2006). Opposites attract: A theorem about the Casimir force. *Phys. Rev. Lett.* **97**, 160401-1–4.

Kenneth, O. and Klich, I. (2008). Casimir forces in a T-operator approach. *Phys. Rev. B* **78**, 014103-1–17.

Kenneth, O., Klich, I., Mann, A., and Revzen, M. (2002). Repulsive Casimir forces. *Phys. Rev. Lett.* **89**, 033001-1–4.

Kharchenko, P., Babb, J. F., and Dalgarno, A. (1997). Long-range interactions of sodium atoms. *Phys. Rev. A* **55**, 3566–72.

Khusnutdinov, N. R. and Bordag, M. (1999). Ground state energy of a massive scalar field in the background of a cosmic string of finite thickness. *Phys.*

Rev. D **55**, 3566–72.

Kim, W. J., Brown-Hayes, M., Dalvit, D. A. R., Brownell, J. H., and Onofrio, R. (2008). Anomalies in electrostatic calibrations for the measurement of the Casimir force in sphere–plane geometry. *Phys. Rev. A* **78**, 020101(R)-1–4.

Kirsten, K. (2000). *Spectral Functions in Mathematics and Physics*. Chapman and Hall/CRC, London.

Kittel, C. (1996). *Introduction to Solid State Physics*. Wiley, New York.

Kleinman, G. G. and Landman, U. (1974). Effect of spatial dispersion upon physisorption energies: He on metals. *Phys. Rev. Lett.* **33**, 524–7.

Klich, I. (2000). Casimir energy of a conducting sphere and of a dilute dielectric ball. *Phys. Rev. D* **61**, 025004-1–6.

Klich, I. and Romeo, A. (2000). Regularized Casimir energy for an infinite dielectric cylinder subject to light-velocity conservation *Phys. Lett. B* **476**, 369–78.

Kliewer, K. L. and Fuchs, R. (1968). Anomalous skin effect for specular scattering and optical experiments at non-normal angles of incidence. *Phys. Rev.* **172**, 607–24.

Klimchitskaya, G. L. (2009). Problems and paradoxes of the Lifshitz theory. *J. Phys.: Conf. Ser.* **161**, 012002-1–16.

Klimchitskaya, G. L. and Geyer, B. (2008). Problems in the theory of thermal Casimir force between dielectrics and semiconductors. *J. Phys. A: Math. Theor.* **41**, 164032-1–12.

Klimchitskaya, G. L. and Mohideen, U. (2002). Constraints on Yukawa-type hypothetical interactions from the recent Casimir force measurements. *Int. J. Mod. Phys. A* **17**, 4143–4152.

Klimchitskaya, G. L. and Mostepanenko, V. M. (2001). Investigation of temperature dependence of the Casimir force between real metals. *Phys. Rev. A* **63**, 062108-1–18.

Klimchitskaya, G. L. and Mostepanenko, V. M. (2007). Comment on "Effects of spatial dispersion on electromagnetic surface modes and modes associated with a gap between two half spaces". *Phys. Rev. B* **75**, 036101-1–4.

Klimchitskaya, G. L. and Mostepanenko, V. M. (2008a). Comment on "Analytical and numerical verification of the Nernst theorem for metals". *Phys. Rev. E* **77**, 023101-1–3.

Klimchitskaya, G. L. and Mostepanenko, V. M. (2008b). Conductivity of dielectric and thermal atom-wall interaction. *J. Phys. A: Math. Theor.* **41** 312002-1–8.

Klimchitskaya, G. L. and Pavlov, Yu. V. (1996). The corrections to the Casimir forces for configurations used in experiments: The spherical lens above the plane and two crossed cylinders. *Int. J. Mod. Phys. A* **11** 3723–42.

Klimchitskaya, G. L. and Shabaeva, M. B. (1996). Corrections to the Casimir force between plates with periodic roughness in the transition region. *Izv. Vuzov, Fizika* **N7**, 92–6 (*Russ. Phys. J.* **39**, 678–82).

Klimchitskaya, G. L., Bezerra de Mello, E. R., and Mostepanenko, V. M. (1998).

Constraints on the parameters of degree-type hypothetical forces following from the new Casimir force measurement. *Phys. Lett. A* **236**, 280–8.

Klimchitskaya, G. L., Roy, A., Mohideen, U., and Mostepanenko, V. M. (1999). Complete roughness and conductivity corrections for the recent Casimir force measurement. *Phys. Rev. A* **60**, 3487–97.

Klimchitskaya, G. L., Mohideen, U., and Mostepanenko, V. M. (2000). Casimir and van der Waals force between two plates or a sphere (lens) above a plate made of real metals. *Phys. Rev. A* **61**, 062107-1–12.

Klimchitskaya, G. L., Zanette, S. I., and Caride, A. O. (2001). Lateral projection as a possible explanation of the nontrivial boundary dependence of the Casimir force. *Phys. Rev A* **63**, 014101-1–4.

Klimchitskaya, G. L., Decca, R. S., Fischbach, E., Krause, D. E., López, D., and Mostepanenko, V. M. (2005). Casimir effect as a test for thermal corrections and hypothetical long-range interactions. *Int. J. Mod. Phys. A* **20**, 2205–21.

Klimchitskaya, G. L., Blagov, E. V., and Mostepanenko, V. M. (2006a). Casimir-Polder interaction between an atom and a cylinder with application to nanosystems. *J. Phys. A: Math. Gen.* **39**, 6481–4.

Klimchitskaya, G. L., Chen, F., Decca, R. S., Fischbach, E., Krause, D. E., López, D., Mohideen, U., and Mostepanenko, V. M. (2006b). Rigorous approach to the comparison between experiment and theory in Casimir force measurements. *J. Phys. A: Math. Gen.* **39**, 6485–93.

Klimchitskaya, G. L., Geyer, B., and Mostepanenko, V. M. (2006c). Universal behavior of dispersion forces between two dielectric plates at low-temperature limit. *J. Phys. A: Math. Gen.* **39**, 6495–9.

Klimchitskaya, G. L., Mohideen, U., and Mostepanenko, V. M. (2007a). Kramers–Kronig relations for plasma-like permittivities and the Casimir force. *J. Phys. A: Math. Theor.* **40**, F339–46.

Klimchitskaya, G. L., Mohideen, U., and Mostepanenko, V. M. (2007b). Pulsating Casimir force. *J. Phys. A: Math. Theor.* **40**, F841–7.

Klimchitskaya, G. L., Mohideen, U., and Mostepanenko, V. M. (2008a). Casimir-Polder force between an atom and a dielectric plate: Thermodynamics and experiment. *J. Phys. A: Math. Theor.* **41**, 432001-1–9.

Klimchitskaya, G. L., Blagov, E. V., and Mostepanenko, V. M. (2008b). Van der Waals and Casimir interactions between atoms and carbon nanotubes. *J. Phys. A: Math. Theor.* **41**, 164012-1–8.

Klimchitskaya, G. L., Mohideen, U., and Mostepanenko, V. M. (2008c). Comment on "Application of the Lifshitz theory to poor conductors". ArXiv: 0810.3247.

Klimchitskaya, G. L., Mohideen, U., and Mostepanenko, V. M. (2009a). The Casimir force between real materials: Experiment and theory. ArXiv: 0902.4022; *Rev. Mod. Phys.*, forthcoming.

Klimchitskaya, G. L., Blagov, E. V., and Mostepanenko, V. M. (2009b). Problems in the Lifshitz theory of atom–wall interaction. *Int. J. Mod. Phys. A* **24**, 1777–88.

Klucker, R., Skibowski, M., and Steinmann, W. (1974). Anisotropy in the optical transitions from the pi and sigma valence bands of graphite. *Phys. Stat. Sol. (b)* **65**, 703–10.

Kondepugi, D. and Prigogine, I. (1998). *Modern Thermodynamics*. Wiley, New York.

Kong, X. W. and Ravndal, F. (1997). Radiative corrections to the Casimir energy. *Phys. Rev. Lett.* **79**, 545–8.

Krause, D. E., Decca, R. S., López, D., and Fischbach, E. (2007). Experimental investigation of the Casimir force beyond the proximity-force approximation. *Phys. Rev. Lett.* **98**, 050403-1–4.

Krech, M. (1994). *The Casimir Effect in Critical Systems*. World Scientific, Singapore.

Krommer, H., Erbe, A., Tilke, A., Manus, S., and Blick, R. H. (2000). Nanomechanical resonators operating as charge detectors in the nonlinear regime. *Europhys. Lett.* **50**, 101–6.

Kronik, L. and Shapira, Y. (2001). Surface photovoltage spectroscopy of semiconductor structures: At the crossroads of physics, chemistry and electrical engineering. *Surf. Interface Anal.* **31**, 954–65.

Krupp, H. (1967). Particle adhesion: theory and experiment. *Adv. Colloid Interface Sci.* **1**, 111–239.

Kubo, R. (1968). *Thermodynamics*. North-Holland, Amsterdam.

Kuzmin, V. A., Tkachev, I. I., and Shaposhnikov, M. E. (1982). Restrictions imposed on light scalar particles by measurements of van der Waals forces. *Pis'ma v ZhÉTP* **36**, 49–52 (*JETP Lett.* **36**, 59–62).

Lachièze-Rey, M. and Luminet, J. P. (1995). Cosmic topology. *Phys. Rep.* **254**, 136–214.

Lambiase, G., Scarpetta, G., and Nesterenko, V. V. (2001). Zero-point energy of a dilute dielectric ball in the mode summation method. *Mod. Phys. Lett. A* **16**, 1983–95.

Lambrecht, A. and Marachevsky, V. N. (2008). Casimir interaction of dielectric gratings. *Phys. Rev. Lett.* **101**, 160403-1–4.

Lambrecht, A. and Reynaud, S. (2000a). Casimir force between metallic mirrors. *Eur. Phys. J. D* **8**, 309–18.

Lambrecht, A. and Reynaud, S. (2000b). Comment on "Demonstration of the Casimir force in the 0.6 to 6 μm range". *Phys. Rev. Lett.* **84**, 5672.

Lambrecht, A., Nesvizhevsky, V.V., Onofrio, R., and Reynaud, S. (2005). Development of a high-sensitivity torsional balance for the study of the Casimir force in the 1–10 micrometre. *Class. Quantum Grav.* **22**, 5397–406.

Lambrecht, A., Pirozhenko, I., Duraffourg, L., and Andreucci, P. (2007). The Casimir effect for silicon and gold slabs. *Europhys. Lett.* **77**, 44006-1–5.

Lamoreaux, S. K. (1997). Demonstration of the Casimir force in the 0.6 to 6 μm range. *Phys. Rev. Lett.* **78**, 5–8.

Lamoreaux, S. K. (1998). Erratum: Demonstration of the Casimir force in the 0.6 to 6 μm range. *Phys. Rev. Lett.* **81**, 5475–6.

Lamoreaux, S. K. (1999). Calculation of the Casimir force between imperfectly conducting plates. *Phys. Rev. A* **59**, R3149–53.

Lamoreaux, S. K. (2005). The Casimir force: Background, experiments, and applications. *Rep. Prog. Phys.* **68**, 201–36.

Lamoreaux, S. K. and Buttler, W. T. (2005). Thermal noise limitations to force measurements with torsion pendulums: Applications to the measurement of the Casimir force and its thermal correction. *Phys. Rev. E* **71**, 036109-1–5.

Landau, L. D. and Lifshitz, E. M. (1980). *Statistical Physics*, Part I. Pergamon, Oxford.

Landau, L. D., Lifshitz, E. M., and Pitaevskii, L. P. (1984). *Electrodynamics of Continuous Media*. Pergamon, Oxford.

Langbein, D. (1973). Macroscopic theory of van der Waals attraction. *Solid State Commun.* **12**, 853–5.

Langfeld, K., Schmuser, F., and Reinhardt, H. (1995). Casimir energy of strongly interacting scalar fields. *Phys. Rev. D* **51**, 765–73.

Law C. K. (1995). Interaction between a moving mirror and radiation pressure—a Hamiltonian formulation. *Phys. Rev. A* **51**, 2537–41.

Lee, S.-W. and Sigmund, W. M. (2002). AFM study of repulsive van der Waals forces between Teflon AF^{TM} thin film and silica or alumina. *Colloids Surf. A* **204**, 43–50.

Lennard-Jones, J. E. (1932). Processes of adsorption and diffusion on solid surfaces. *Trans. Faraday Soc.* **28**, 333–58.

Leonhardt, U. and Philbin, T. G. (2007). Quantum levitation by left-handed metamaterials. *New J. Phys.* **9**, 254-1–7.

Leskova, T. A., Maradudin, A. A., and Munõz-Lopez, J. (2005). Coherence of light scattered from a randomly rough surface. *Phys. Rev. E* **71**, 036606-1–10.

Letelier, P. S. (1987). Multiple cosmic strings. *Class. Quantum Grav.* **4**, L75–7.

Li, H. and Kardar, M. (1992). Fluctuation-induced forces between manifolds immersed in correlated fluids. *Phys. Rev. A* **46**, 6490–500.

Li, X.-Z., Cheng, H.-B., Li, J.-M., and Zhai, X.-H. (1997). Attractive and repulsive nature of the Casimir force in rectangular cavity. *Phys. Rev. D* **56**, 2155–62.

Liberati, S., Visser, M., Belgiorno, F., and Sciama, D. W. (2000). Sonoluminescence as a QED vacuum effect. I. The physical scenario. *Phys. Rev. D* **61**, 085023-1–18.

Lifshitz, E. M. (1956). The theory of molecular attractive forces between solids. *Zh. Eksp. Teor. Fiz.* **29**, 94–110 (*Sov. Phys. JETP* **2**, 73–83).

Lifshitz, E. M. and Pitaevskii, L. P. (1980). *Statistical Physics*, Part II. Pergamon, Oxford.

Lifshitz, E. M. and Pitaevskii, L. P. (1981). *Physical Kinetics*. Pergamon, Oxford.

Lim, S. C. and Teo, L. P. (2007). Finite temperature Casimir energy in closed rectangular cavities: A rigorous derivation based on a zeta function tech-

nique. *J. Phys. A: Math. Theor.* **40**, 11645–74.
Linde, A. D. (1990). *Particle Physics and Inflationary Cosmology*. Harwood Academic, Chur.
Linet, B. (1987). Quantum field theory in the space-time of a cosmic string. *Phys. Rev. D* **35**, 536–9.
Linet, B. (1992). The Euclidean thermal Green function in the spacetime of a cosmic string. *Class. Quantum Grav.* **9**, 2429–36.
Lisanti, M., Iannuzzi, D., and Capasso, F. (2005). Observation of the skin-depth effect on the Casimir force between metallic surfaces. *Proc. Natl. Acad. Sci. USA* **102**, 11989–92.
Lombardo, F. C., Mazzitelli, F. D., and Villar, P. I. (2008). Exploring the quantum vacuum with cylinders. *J. Phys. A: Math. Theor.* **41**, 164009-1–10.
London, F. (1930). Zur Theorie und Systematik der Molecularkräfte. *Z. Phys.* **63**, 245–279.
Long, J. C., Chan, H. W., and Price, J. C. (1999). Experimental status of gravitational-strength forces in the sub-centimeter regime. *Nucl. Phys. B* **539**, 23–34.
Loomis, J. J. and Maris, H. J. (1994). Theory of heat transfer by evanescent electromagnetic waves. *Phys. Rev. B* **50**, 18517–24.
Lucyszyn, S., Miyaguchi, K., Jiang, H. W., Robertson, I. D., Fisher, G., Lord, A., and Choi, J. Y. (2008). Micromachined RF-coupled cantilever inverted-micro strip millimeter-wave filters. *J. Microelectromech. Syst.* **17**, 767–76.
Lukosz, W. (1971). Electromagnetic zero-point energy and radiation pressure for a rectangular cavity. *Physica* **56**, 109–20.
Maclay, G. J. (2000). Analysis of zero-point electromagnetic energy and Casimir forces in conducting rectangular cavities. *Phys. Rev. A* **61**, 052110-1–18.
Madroñero, J. and Friedrich, H. (2007). Influence of realistic atom wall potentials in quantum reflection traps. *Phys. Rev. A* **75**, 022902-1–4.
Mahanty, J. and Ninham, B. W. (1976). *Dispersion Forces*. Academic Press, New York.
Maia Neto, P. A., Lambrecht, A., and Reynaud, S. (2005). Casimir effect with rough metallic mirrors. *Phys. Rev. A* **72**, 012115-1–14.
Mamayev, S. G. (1980). Vacuum expectation values of the energy–momentum tensor of quantized fields in a homogeneous isotropic space-time. *Teor. Matem. Fiz.* **42**, 350–61 (*Theor. Math. Phys.* **42**, 229–37).
Mamayev, S. G. and Mostepanenko, V. M. (1980). Isotropic cosmological models determined by vacuum quantum effects. *Zh. Eksp. Teor. Fiz.* **78**, 20–7 (*Sov. Phys. JETP* **51**, 9–13).
Mamayev, S. G. and Mostepanenko, V. M. (1985). Casimir effect in space-time with non-trivial topology. In: M. A. Markov, V. A. Berezin, and V. P. Frolov (eds.), *Proc. of the Third Seminar on Quantum Gravity*. World Scientific, Singapore, pp.462–78.
Mamayev, S. G. and Trunov, N. N. (1979a). Dependence of the vacuum expectation values of the energy–momentum tensor on the geometry and topology

of the manifold. *Teor. Matem. Fiz.* **38**, 345–54 (*Theor. Math. Phys.* **38**, 228–34).

Mamayev, S. G. and Trunov, N. N. (1979b). Vacuum averages of the energy–momentum tensor of quantized fields on manifolds of various topology and geometry. II. *Izv. Vuzov, Fizika* **N9**, 51–4 (*Russ. Phys. J.* **22**, 966–9).

Mamayev, S. G. and Trunov, N. N. (1979c). Vacuum averages of the energy–momentum tensor of quantized fields on manifolds of various topology and geometry. I. *Izv. Vuzov, Fizika* **N7**, 88–93 (*Russ. Phys. J.* **22**, 766-70).

Mamayev, S. G. and Trunov, N. N. (1980). Vacuum expectation values of the energy–momentum tensor of quantized fields on manifolds with different topologies and geometries. III. *Izv. Vuzov, Fizika* **N7**, 9–13 (*Russ. Phys. J.* **23**, 551–4).

Mamayev, S. G., Mostepanenko, V. M., and Starobinsky, A. A. (1976). Particle creation from vacuum near the homogeneous isotropic singularity. *Zh. Eksp. Teor. Fiz.* **70**, 1577–91 (*Sov. Phys. JETP* **43**, 823–30).

Marachevsky, V. N. (2007). Casimir interaction of two plates inside a cylinder. *Phys. Rev. D* **75**, 085019-1–6.

Maradudin, A. A. and Mazur, P. (1980). Effects of surface roughness on the van der Waals force between macroscopic bodies. *Phys. Rev. B* **22**, 1677–86.

Matsubara, T. (1955). A new approach to quantum statistical mechanics. *Prog. Theor. Phys.* **14**, 351–8.

Mazzitelli, F. D. (2004). Casimir interaction between cylinders. In: K. A. Milton (ed.), *Proc. of the 6th Workshop on Quantum Field Theory Under the Influence of External Conditions.* Rinton Press, Princeton, pp.126–32.

Mazzitelli, F. D., Sánchez, M. J., Scoccola, N. N., and von Stecher, J. (2003). Casimir interaction between two concentric cylinders: Exact versus semiclassical results. *Phys. Rev. A* **67**, 013807-1–11.

Mazzitelli, F. D., Dalvit, D. A. R., and Lombardo, F. C. (2006). Exact zero-point interaction energy between cylinders. *New J. Phys.* **8**, 240-1–21.

Mazur, P. and Maradudin, A. A. (1981). Effects of surface roughness on the van der Waals force between macroscopic bodies. II. Two rough surfaces. *Phys. Rev. B* **23**, 695–705.

Mehra, J. (1967). Temperature correction to Casimir effect. *Physica* **37**, 145–52.

Meurk, A., Luckham, P. F., and Bergström, L. (1997). Direct measurement of repulsive and attractive van der Waals forces between inorganic materials. *Langmuir* **13**, 3896–9.

Milling, A., Mulvaney, P., and Larson, I. (1996). Direct measurement of repulsive van der Waals interactions using an atomic force microscope. *J. Colloid Interface Sci.* **180**, 460–5.

Milonni, P. W. (1994). *The Quantum Vacuum. An Introduction to Quantum Electrodynamics.* Academic Press, San Diego.

Milton, K. A. (1980). Semi-classical electron models: Casimir self-stress in dielectric and conducting balls. *Ann. Phys. (N.Y.)* **127**, 49–61.

Milton, K. A. (1983). Fermionic Casimir stress on a spherical bag. *Ann. Phys. (N.Y.)* **150**, 432–8.

Milton, K. A. (2001). *The Casimir Effect: Physical Manifestations of Zero-Point Energy.* World Scientific, Singapore.

Milton, K. A. (2004). The Casimir effect: Recent controversies and progress. *J. Phys. A: Math. Gen.* **37**, R209–77.

Milton, K. A. and Ng, Y. J. (1998). Observability of the bulk Casimir effect: Can the dynamical Casimir effect be relevant to sonoluminescence? *Phys. Rev. E* **57**, 5504–10.

Milton K. A., Nesterenko, A. V., and Nesterenko, V. V. (1999). Mode-by-mode summation for the zero point electromagnetic energy of an infinite cylinder. *Phys. Rev. D* **59**, 105009-1–9.

Mitter, H. and Robaschik, D. (2000). Thermodynamics of the Casimir effect. *Eur. Phys. J. B* **13**, 335–40.

Mohideen, U. and Roy, A. (1998). Precision measurement of the Casimir force from 0.1 to 0.9 μm. *Phys. Rev. Lett.* **81**, 4549–52.

Moore, G. T. (1970). Quantum theory of electromagnetic field in a variable-length one-dimensional cavity. *J. Math. Phys.* **11**, 2679–91.

Moretti, V. (1999). Local zeta-function techniques vs. point-splitting procedure: A few rigorous results. *Commun. Math. Phys.* **201**, 327–63.

Morrish, A. H. (1965). *The Physical Principles of Magnetism.* Wiley, New York.

Mostepanenko, V. M. (2002a). Constraints on forces inspired by extra dimensional physics following from the Casimir effect. *Int. J. Mod. Phys. A* **17**, 722–31.

Mostepanenko, V. M. (2002b). Experimental status of corrections to Newtonian gravity inspired by extra dimensions. *Int. J. Mod. Phys. A* **17**, 4307–16.

Mostepanenko, V. M. (2003). Unexpected applications of Hill's differential equations in quantum field theory and cosmology. *Int. J. Mod. Phys. A* **18**, 2159–66.

Mostepanenko, V. M. (2008). The Casimir effect in relativistic quantum field theories. In: H. Kleinert, R. T. Jantzen, and R. Ruffini (eds.). *Proc. of the Eleventh Marcel Grossmann Meeting on General Relativity.* Part C. World Scientific, Singapore, pp.2707–26.

Mostepanenko, V. M. (2009). Experiment, theory and the Casimir effect. *J. Phys.: Conf. Ser.* **161**, 012003-1–18.

Mostepanenko, V. M. and Frolov, V. M. (1974). Production of particles from vacuum by a uniform electric field with periodic time dependence. *Yadern. Fiz.* **19**, 885–96 (*Sov. J. Nucl. Phys.* **19**, 451–6).

Mostepanenko, V. M. and Geyer, B. (2008). New approach to the thermal Casimir force between real metals. *J. Phys. A: Math. Theor.* **41**, 164014-1–12.

Mostepanenko, V. M. and Novello, M. (2001). Constraints on non-Newtonian gravity from the Casimir force measurements between two crossed cylinders. *Phys. Rev. D* **63**, 115003-1–5.

Mostepanenko, V. M. and Sokolov, I. Yu. (1987a). Restrictions on long-range forces following from the Casimir effect. *Yadern. Fiz.* **46**, 1174–80 (*Sov. J. Nucl. Phys.* **46**, 685–8).

Mostepanenko, V. M. and Sokolov, I. Yu. (1987b). The Casimir effect leads to new restrictions on long-range force constants. *Phys. Lett. A* **125**, 405–8.

Mostepanenko, V. M. and Sokolov, I. Yu. (1988). Casimir forces between complex shaped bodies. *Doklady Akad. Nauk SSSR* **298**, 1380–3 [*Sov. Phys. Dokl. (USA)* **33**, 140–1].

Mostepanenko, V. M. and Sokolov, I. Yu. (1989a). New restrictions on the spin-1 antigraviton following from the Casimir effect, Eötvos and Cavendish experiments. *Phys. Lett. A* **132**, 313–5.

Mostepanenko, V. M. and Sokolov, I. Yu. (1989b). Restrictions on the parameters of the spin-1 antigraviton and the dilaton resulting from the Casimir effect and from the Eötvos and Cavendish experiments. *Yadern. Fiz.* **49**, 1807–11 (*Sov. J. Nucl. Phys.* **49**, 1118–20).

Mostepanenko, V. M. and Sokolov, I. Yu. (1993). Hypothetical long-range interactions and restrictions on their parameters from force measurements. *Phys. Rev. D* **47**, 2882–91.

Mostepanenko, V. M. and Trunov, N. N. (1985). Quantum-field theory of the Casimir effect for real media. *Yadern. Fiz.* **42**, 1297–305 (*Sov. J. Nucl. Phys.* **42**, 818–22).

Mostepanenko, V. M. and Trunov, N. N. (1988). The Casimir effect and its applications. *Usp. Fiz. Nauk* **156**, 385–426 (*Sov. Phys. Uspekhi* **31**, 965–87).

Mostepanenko, V. M. and Trunov, N. N. (1990). *The Casimir Effect and Its Applications*. Energoatomizdat, Moscow (in Russian).

Mostepanenko, V. M. and Trunov, N. N. (1997). *The Casimir Effect and Its Applications*. Clarendon Press, Oxford.

Mostepanenko, V. M., Babb, J. F., Caride O. A., Klimchitskaya, G. L., and Zanette S. I. (2006a). Dependence of the Casimir-Polder interaction between an atom and a cavity wall on atomic and material properties. *J. Phys. A: Math. Gen.* **39**, 6583–7.

Mostepanenko, V. M., Bezerra, V. B., Decca, R. S., Fischbach, E., Geyer, B., Klimchitskaya, G. L., Krause, D. E., López, D., and Romero, C. (2006b). Present status of controversies regarding the thermal Casimir force. *J. Phys. A: Math. Gen.* **39**, 6589–600.

Mostepanenko, V. M., Decca, R. S., Fischbach, E., Klimchitskaya, G. L., Krause, D. E., and López, D. (2008). Stronger constraints on non-Newtonian gravity from the Casimir effect. *J. Phys. A: Math. Theor.* **41**, 164054-1–8.

Mostepanenko, V. M., Decca, R. S., Fischbach, E., Geyer, B., Klimchitskaya, G. L., Krause, D. E., López, D., and Mohideen, U. (2009). Why screening effects do not influence the Casimir force. *Int. J. Mod. Phys. A* **24**, 1721–42.

Mota, D. F. and Shaw, D. J. (2007). Evading equivalence principle violation, cosmological, and other experimental constraints in scalar field theories with a strong coupling to matter. *Phys. Rev. D* **75**, 063501-1–57.

Mota, B., Rebouças, M. J., and Tavakol, R. (2005). The local shape of the Universe in the inflationary limit. *Int. J. Mod. Phys. A* **20**, 2415–20.

Mott, N. F. (1990). *Metal–Insulator Transitions* (2nd edn). Taylor and Francis, London.

Müller, D., Fagundes, H. V., and Opher, R. (2002). Casimir energy in multiply connected static hyperbolic universes. *Phys. Rev. D* **66**, 083507-1–7.

Munday, J. N. and Capasso, F. (2007). Precision measurement of the Casimir–Lifshitz force in a fluid. *Phys. Rev. A* **75**, 060102(R)-1–4.

Munday, J. N. and Capasso, F. (2008). Reply to "Comment on 'Precision measurement of the Casimir–Lifshitz force in a fluid' ". *Phys. Rev. A* **77**, 036103-1–4.

Munday, J. N., Iannuzzi, D., Barash, Y., and Capasso, F. (2005). Torque on birefringent plates induced by quantum fluctuations. *Phys. Rev. A* **71**, 042102-1–9.

Munday, J. N., Capasso, F., Parsegian, V. A., and Bezrukov, S. M. (2008). Measurement of the Casimir–Lifshitz force in a fluid: The effect of electrostatic forces and Debye screening. *Phys. Rev. A* **78**, 032109-1–8.

Munday, J. N., Capasso, F., and Parsegian, V. A. (2009). Measured long-range repulsive Casimir–Lifshitz forces. *Nature* **457**, 170–3.

Nagai, M. and Kuwata-Gonokami, M. (2002). Time-resolved reflection spectroscopy of the spatiotemporal dynamics of photo-excited carriers in Si and GaAs. *J. Phys. Soc. Jpn.* **71**, 2276–9.

Narozhnyi, N. B. and Nikishov, A. I. (1973). Pair production by the periodic electric field. *Zh. Eksp. Teor. Fiz.* **65**, 862–74 (*Sov. Phys. JETP* **38**, 427–32).

Nayak, V. U., Edwards, D. O., and Masuhara, N. (1983). Scattering of 4He atoms grazing the liquid-4He surface. *Phys. Rev. Lett.* **50**, 990–2.

Nechaev, Yu. S. (2006). The nature, kinetics, and ultimate storage capacity of hydrogen sorption by carbon nanostructures. *Usp. Fiz. Nauk* **176**, 581–610 (*Phys. Uspekhi* **49**, 563–91).

Nesterenko, V. V. and Pirozhenko, I. G. (1999). Casimir energy of a compact cylinder under the condition $\varepsilon\mu = c^{-2}$. *Phys. Rev. D* **60**, 125007-1–6.

Nesterenko, V. V., Lambiase, G., and Scarpetta, G. (2002). Casimir effect for a perfectly conducting wedge in terms of local zeta function. *Ann. Phys. (N.Y.)* **298**, 403–20.

Nesterenko, V. V., Pirozhenko, I. G., and Dittrich, J. (2003). Non-smoothness of the boundary and the relevant heat kernel coefficients. *Class. Quantum Grav.* **20**, 431–56.

Newton, R. G. (1966). *Scattering Theory of Waves and Particles*. McGraw-Hill, New York.

Ninham, B. W., Parsegian, V. A., and Weiss, G. H. (1970). On the macroscopic theory of temperature-dependent van der Waals forces. *J. Stat. Phys.* **2**, 323–8.

Nojiri, S., Odintsov, S. D., and Zerbini, S. (2000). Quantum (in)stability of dilatonic AdS backgrounds and the holographic renormalization group with gravity. *Phys. Rev. D* **62**, 064006-1–8.

Novikov, M. Yu., Sorin, A. S., and Chernyak, V. Ya. (1990a). Fluctuation forces in a three-layer medium with rough boundaries. I. Principles of perturbation theory. *Teor. Matem. Fiz.* **82**, 178–87 (*Theor. Math. Phys.* **82**, 124–30).

Novikov, M. Yu., Sorin, A. S., and Chernyak, V. Ya. (1990b). Fluctuation forces in a three-layer medium with rough boundaries. II. Calculations in the second order of perturbation theory. *Teor. Matem. Fiz.* **82**, 360–5 (*Theor. Math. Phys.* **82**, 252–5).

Novikov, M. Yu., Sorin, A. S., and Chernyak, V. Ya. (1992a). Fluctuation forces in a three-layer medium with rough boundaries. III. Aspects of perturbation theory in the Casimir range. *Teor. Matem. Fiz.* **91**, 474–82 (*Theor. Math. Phys.* **91**, 658–63).

Novikov, M. Yu., Sorin, A. S., and Chernyak, V. Ya. (1992b). Fluctuation forces in a three-layer medium with rough boundaries. IV. Calculations in the second order of perturbation theory (Casimir range). *Teor. Matem. Fiz.* **92**, 113–8 (*Theor. Math. Phys.* **92**, 773–6).

Oberst, H., Tashiro, Y., Shimizu, K., and Shimizu, F. (2005). Quantum reflection of He* on silicon. *Phys. Rev. A* **71**, 052901-1–8.

Obrecht, J. M., Wild, R. J., Antezza, M., Pitaevskii, L. P., Stringari, S., and Cornell, E. A. (2007). Measurement of the temperature dependence of the Casimir-Polder force. *Phys. Rev. Lett.* **98**, 063201-1–4.

Opsal, J., Taylor, M. W., Smith, W. L., and Rosencwaig, A. (1987). Temporal behavior of modulated optical reflectance in silicon. *J. Appl. Phys.* **61**, 240–8.

Overbeek, J. T. G. and Sparnaay, M. J. (1954). Classical coagulation: London–van der Waals attraction between macroscopic objects. *Discuss. Faraday Soc.* **18**, 12–24.

Palasantzas, G. (2007a). Contact angle influence on the pull-in voltage of microswitches in the presence of capillary and quantum vacuum effects. *J. Appl. Phys.* **101**, 053512-1–5.

Palasantzas, G. (2007b). Pull-in voltage of microswitch rough plates in the presence of electromagnetic and acoustic Casimir forces. *J. Appl. Phys.* **101**, 063548-1–5.

Palik, E. D., ed. (1985). *Handbook of Optical Constants of Solids*, Vol. 1. Academic Press, New York.

Palik, E. D., ed. (1991). *Handbook of Optical Constants of Solids*, Vol. 2. Academic Press, New York.

Parsegian, V. A. (2005). *Van der Waals Forces: a Handbook for Biologists, Chemists, Engineers, and Physicists*. Cambridge University Press, Cambridge.

Parsegian, V. A. and Weiss, G. H. (1972). On van der Waals interactions between macroscopic bodies having inhomogeneous dielectric susceptibilities. *J. Colloid Interface Sci.* **40**, 35–41.

Parsegian, V. A. and Weiss, G. H. (1981). Spectroscopic parameters for computation of van der Waals forces. *J. Colloid Interface Sci.* **81**, 285–9.

Pasquini, T. A., Shin, Y., Sanner, C., Saba, M., Schirotzek, A., Pritchard, D. E., and Ketterle, W. (2004). Quantum reflection from a solid surface at normal incidence. *Phys. Rev. Lett.* **93**, 233201-1–4.

Pasquini, T. A., Saba, M., Jo, G.-B., Shin, Y., Ketterle, W., and Pritchard, D. E. (2006). Low velocity quantum reflection of Bose-Einstein condensates. *Phys. Rev. Lett.* **97**, 093201-1–4.

Peccei, R. D. and Quinn, H. R. (1977). CP conservation in the presence of pseudoparticles. *Phys. Rev. Lett.* **38**, 1440–3.

Peskin, M. E. and Schroeder, D. V. (1995). *An Introduction to Quantum Field Theory.* Addison-Wesley, Reading, MA.

Peterson, C., Hansson, T. H., and Johnson, K. (1982). Loop diagrams in boxes. *Phys. Rev. D* **26**, 415–28.

Petrov, V., Petrov, M., Bryksin, V., Petter, J., and Tschudi, T. (2006). Optical detection of the Casimir force between macroscopic objects. *Opt. Lett.* **31**, 3167–9.

Phipps, T. E. and Taylor, J. B. (1927). The magnetic moment of hydrogen atom. *Phys. Rev.* **29**, 309–20.

Pirozhenko, I. and Lambrecht, A. (2008a). Influence of slab thickness on the Casimir force. *Phys. Rev. A* **77**, 013811-1–8.

Pirozhenko, I. G. and Lambrecht, A. (2008b). Casimir repulsion and metamaterials. *J. Phys. A: Math. Theor.* **41**, 164015-1–8.

Pirozhenko, I., Lambrecht, A., and Svetovoy, V. B. (2006). Sample dependence of the Casimir force. *New J. Phys.* **8**, 238-1–16.

Pitaevskii, L. P. (2008a). Thermal Lifshitz force between an atom and a conductor with a small density of carriers. *Phys. Rev. Lett.* **101**, 163202-1–4.

Pitaevskii, L. P. (2008b). Reply on "Comment on 'Thermal Lifshitz force between an atom and a conductor with a small density of carriers'". ArXiv: 0811.3081v1; *Phys. Rev. Lett.*, forthcoming.

Planck, M. (1911). Eine neue Strahlungshypothese. *Verh. d. Deutsch. Phys. Ges.* **13**, 138–48.

Plunien, G., Müller, B., and Greiner, W. (1986). The Casimir effect. *Phys. Rep.* **134**, 87–193.

Polchinski, J. (1998). *String Theory*, Vols. 1 and 2. Cambridge University Press, Cambridge.

Polder, D. and Van Hove, M. (1971). Theory of radiative heat transfer between closely spaced bodies. *Phys. Rev. B* **4**, 3303–14.

Prudnikov, A. P., Brychkov, Yu. A., and Marichev, O. I. (1986). *Integrals and Series,* Vol. 2. Gordon and Breach, New York.

Raabe, C., Knöll, L., and Welsch, D.-G. (2003). Three-dimensional Casimir force between absorbing multilayer dielectrics. *Phys. Rev. A* **68**, 033810-1–19.

Rabinovich, S. G. (2000). *Measurement Errors and Uncertainties: Theory and Practice*. Springer, New York.

Rabinovich, Ya. I. and Churaev, N. V. (1989). Calculation of dispersion interaction forces between bodies with a rough-surface in a vacuum. *Kolloid. Zh.* **51**, 83–90 (*Colloid J. USSR* **51**, 65–71).

Rahi, S. J., Emig, T., Jaffe, R. L., and Kardar, M. (2008). Casimir forces between cylinders and plates. *Phys. Rev. A* **78**, 012104-1–11.

Randall, L. and Sundrum, R. (1999a). Large mass hierarchy from a small extra dimension. *Phys. Rev. Lett.* **83**, 3370–3.

Randall, L. and Sundrum, R. (1999b). An alternative to compactification. *Phys. Rev. Lett.* **83**, 4690–3.

Rauber, S., Klein, J. R., Cole, M. W., and Bruch, L. W. (1982). Substrate-mediated dispersion interaction between adsorbed atoms and molecules. *Surf. Sci.* **123**, 173–8.

Ravndal, F. and Thomassen, J. B. (2001). Radiative corrections to the Casimir energy and effective field theory. *Phys. Rev. D* **63**, 113007-1–7.

Razavy, M. and Terning, J. (1985). Quantum radiation in a one-dimensional cavity with moving boundaries. *Phys. Rev. D* **31**, 307–13.

Rechenberg, H. (1999). Historical remarks on zero-point energy and the Casimir effect (1911–1998). In: M. Bordag (ed.), *The Casimir Effect 50 Years Later*. World Scientific, Singapore, pp.10–9.

Renne, M. J. (1971). Microscopic theory of retarded van der Waals forces between macroscopic dielectric bodies. *Physica* **56**, 125–37.

Robaschik, D., Scharnhorst, K., and Wieczorek, E. (1987). Radiative corrections to the Casimir pressure under the influence of temperature and external fields. *Ann. Phys. (N.Y.)* **174**, 401–29.

Rodrigues, R. B., Maia Neto, P. A., Lambrecht, A., and Reynaud, S. (2006a). Vacuum-induced torque between corrugated metallic plates. *Europhys. Lett.* **76**, 822–8.

Rodrigues, R. B., Maia Neto, P. A., Lambrecht, A., and Reynaud, S. (2006b). Lateral Casimir force beyond the proximity-force approximation. *Phys. Rev. Lett.* **96**, 100402-1–4.

Rodrigues, R. B., Maia Neto, P. A., Lambrecht, A., and Reynaud, S. (2007). Lateral Casimir force beyond the proximity force approximation: A nontrivial interplay between geometry and quantum vacuum. *Phys. Rev. A* **75**, 062108-1–10.

Romeo, A. and Milton, K. A. (2005). Casimir energy for a purely dielectric cylinder by the mode summation method. *Phys. Lett. B* **621**, 309–17.

Romeo, A. and Milton, K. A. (2006). Note on a Casimir energy calculation for a purely dielectric cylinder by mode summation *J. Phys. A: Math. Gen.* **39**, 6703–10.

Romeo, A. and Saharian, A. A. (2002). Casimir effect for scalar fields under Robin boundary conditions on plates. *J. Phys. A: Math. Gen.* **35**, 1297–320.

Rosa, F. S. S., Dalvit, D. A. R., and Milonni, P. W. (2008). Casimir–Lifshitz theory and metamaterials. *Phys. Rev. Lett.* **100**, 183602-1–4.

Rouweler, G. C. and Overbeek, J. T. (1971). Dispersion forces between fused silica objects at distances between 25 and 350 nm. *Trans. Faraday Soc.* **67**, 2117–21.

Roy, A. and Mohideen, U. (1999). Demonstration of the nontrivial boundary dependence of the Casimir force. *Phys. Rev. Lett.* **82**, 4380–3.

Roy, A., Lin, C.-Y., and Mohideen, U. (1999). Improved precision measurement of the Casimir force. *Phys. Rev. D* **60**, 111101(R)-1–5.

Rumer Yu. B. and Ryvkin, M. S. (1980). *Thermodynamics, Statistical Physics, and Kinetics*. Mir, Moscow.

Rytov, S. M. (1959). *Theory of Electric Fluctuations and Thermal Radiation*. Air Force Cambridge Research Center, Bedford, MA.

Sabbah, A. J. and Riffe, D. M. (2002). Femtosecond pump–probe reflectivity study of silicon carrier dynamics. *Phys. Rev. B* **66**, 165217-1–11.

Safari, H., Welsch, D.-G., Buhmann, S. Y., and Scheel, S. (2008). Van der Waals potentials of paramagnetic atoms. *Phys. Rev. A* **78**, 062901-1–17.

Saharian, A. A. (2004). Surface Casimir densities and induced cosmological constant on parallel branes in AdS spacetime. *Phys. Rev. D* **70**, 064026-1–16.

Saharian, A. A. (2006a). Generalized Abel-Plana formula as a renormalization tool in quantum field theory with boundaries. In: A. A. Bytsenko, D. A. Sebastiao, J. A. Helayel-Neto and M. E. Guimaraes (eds.), *Proc. of the Fifth International Conference on Mathematical Methods in Physics*. Proccedings of Science, pp.019-1–15.

Saharian, A. A. (2006b). Bulk Casimir densities and vacuum interaction forces in higher dimensional brane models. *Phys. Rev. D* **73**, 064019-1–17.

Saharian, A. A. (2006c). Surface Casimir densities and induced cosmological constant in higher dimensional braneworlds. *Phys. Rev. D* **74**, 124009-1–19.

Santos, F. C. and Tort, A. C. (2000). Confined Maxwell field and temperature inversion symmetry. *Phys. Lett. B* **482**, 323–8.

Santos, F. C., Tenório, A., and Tort, A. C. (1999). Zeta function method and repulsive Casimir forces for an unusual pair of plates at finite temperature. *Phys. Rev. D* **60**, 105022-1–9.

Scardicchio, A. and Jaffe, R. L. (2005). Casimir effects: An optical approach I. Foundations and examples. *Nucl. Phys. B* **704**, 552–82.

Scardicchio, A. and Jaffe, R. L. (2006). Casimir effects: An optical approach II. Local observables and thermal corrections. *Nucl. Phys. B* **743**, 249–75.

Schaden, M. (2006). Sign and other aspects of semiclassical Casimir energies. *Phys. Rev. A* **73**, 042102-1–16.

Schaden, M. and Spruch, L. (1998). Infinity-free semiclassical evaluation of Casimir effects. *Phys. Rev. A* **58**, 935–53.

Schaller, G., Schützhold, R., Plunien, G., and Soff, G. (2002). Dynamical Casimir effect in a leaky cavity at finite temperature. *Phys. Rev. A* **66**, 023812-1–20.

Scharnhorst, K. (1998). The velocities of light in modified QED vacua. *Annalen d. Phys.* **7**, 700–9.

Schlamminger, S., Choi, K.-Y., Wagner, T. A., Gundlach, J. H., and Adelberger, E. G. (2008). Test of the equivalence principle using a rotating torsion balance. *Phys. Rev. Lett.* **100**, 041101-1–4.

Schram, K. (1973). Macroscopic theory of retarded van der Waals forces. *Phys. Lett. A* **43**, 282–4.

Schröder, O., Scardicchio, A., and Jaffe, R. L. (2005). Casimir energy for a hyperboloid facing a plate in the optical approximation. *Phys. Rev. A* **72**, 012105-1–9.

Schubert, C. (2001). Perturbative quantum field theory in the string-inspired formalism. *Phys. Rep.* **355**, 73–234.

Schütt, H. J. and Gerdes, E. (1992). Space-charge relaxation in ionicly conducting glasses. II. Free carrier concentration and mobility. *J. Non.-Cryst. Solids* **144**, 14–20.

Schwinger, J. (1992). Casimir energy for dielectrics: Spherical geometry. *Proc. Natl. Acad. Sci. USA* **89**, 11118–20.

Schwinger, J. (1993). Casimir light: Pieces of the action. *Proc. Natl. Acad. Sci. USA* **90**, 7285–7.

Schwinger, J., DeRaad, L. L., and Milton, K. A. (1978). Casimir effect in dielectrics. *Ann. Phys. (N.Y.)* **115**, 1–23.

Seeley, R. T. (1969a). The resolvent of an elliptic boundary value problem. *Am. J. Math.* **91**, 889–920.

Seeley, R. T. (1969b). Analytic extension of the trace associated with elliptic boundary problems. *Am. J. Math.* **91**, 963–83.

Sernelius, B. E. (2005). Effects of spatial dispersion on electromagnetic surface modes and on modes associated with a gap between two half spaces. *Phys. Rev. B* **71**, 235114-1–13.

Serry, F.M., Walliser, D., and Maclay, G. J. (1995). The anharmonic Casimir oscillator (ACO)—The Casimir effect in a model microelectromechanical system. *J. Microelectromech. Syst.* **4**, 193–205.

Serry, F.M., Walliser, D., and Maclay, G. J. (1998). The role of the Casimir effect in the static deflection and stiction of membrane strips in microelectromechanical systems (MEMS). *J. Appl. Phys.* **84**, 2501–6.

Shackelford, J. F. and Alexander, W., eds. (2001). *Material Science and Engineering Handbook.* CRC Press, Boca Raton.

Shimizu, F. (2001). Specular reflection of very slow metastable neon atoms from a solid surface. *Phys. Rev. Lett.* **86**, 987–90.

Shklovskii, B. I. and Efros, A. L. (1984). *Electronic Properties of Doped Semiconductors.* Solid State Series, Vol. 45. Springer, Berlin.

Smith, G. L., Hoyle, C. D., Gundlach, J. H., Adelberger, E. G., Heckel, B. R., and Swanson, H. E. (2000). Short-range tests of the equivalence principle. *Phys. Rev. D* **61**, 022001-1–20.

Smullin, S. J., Geraci, A. A., Weld, D. M., Chiaverini, J., Holmes, S., and Kapitulnik, A. (2005). Constraints on Yukawa-type deviations from Newtonian gravity at 20 microns. *Phys. Rev. D* **72**, 122001-1–20.

Smythe, W. R. (1950). *Electrostatics and Electrodynamics*. McGraw-Hill, New York.

Sokolov, D. D. and Shvartsman, V. F. (1974). Estimation of size of Universe from a topological viewpoint. *Zh. Eksp. Teor. Fiz.* **66**, 412–20 (*Sov. Phys. JETP* **39**, 196–200).

Soltani, M., Chaker, M., Haddad, E., Kruzelecky, R. V., and Nikanpour, D. (2004). Optical switching of vanadium dioxide thin films deposited by reactive pulsed laser deposition. *J. Vac. Sci. Technol. A* **22**, 859–64.

Sparnaay, M. J. (1958). Measurements of attractive forces between flat plates. *Physica* **24**, 751–64.

Sparnaay, M. J. (1989). The historical background in the Casimir effect. In: A. Sarlemijn and M. J. Sparnaay (eds.), *Physics in the Making: Essays on Developments in 20th Century Physics, in Honour of H. B. G. Casimir on the Occasion of His 80th Birthday*. North-Holland, Amsterdam, pp.235–46.

Speake, C. C. and Trenkel, C. (2003). Forces between conducting surfaces due to spatial variations of surface potential. *Phys. Rev. Lett.* **90**, 160403-1–4.

Srivastava, Y. and Widom, A. (1987). Quantum electrodynamic processes in electrical-engineering circuits. *Phys. Rep.* **148**, 1–65.

Srivastava, Y., Widom, A., and Friedman, M. H. (1985). Microchips as precision quantum-electrodynamic probes. *Phys. Rev. Lett.* **55**, 2246–8.

Starobinsky, A. A. (1980). A new type of isotropic cosmological models without singularity. *Phys. Lett. B* **91**, 99–102.

Stipe, B. C., Mamin, H. J., Stowe, T. D., Kenny, T. W., and Rugar, D. (2001). Noncontact friction and force fluctuations between closely spaced bodies. *Phys. Rev. Lett.* **87**, 096801-1–4.

Stratton, J. A. (1941). *Electromagnetic Theory*. McGraw-Hill, New York.

Suh, J. Y., Lopez, R., Feldman, L. C., and Haglund, R. F. (2004). Semiconductor to metal phase transition in the nucleation and growth of VO_2 nanoparticles and thin films. *J. Appl. Phys.* **96**, 1209–13.

Sukenik, C. I., Boshier, M. G., Cho, D., Sandoghdar, V., and Hinds, E. A. (1993). Measurement of the Casimir-Polder force. *Phys. Rev. Lett.* **70**, 560–3.

Svetovoy, V. B. (2004). Comment on "Surface-impedance approach solves problems with the thermal Casimir force between real metals". *Phys. Rev. A* **70**, 016101-1–4.

Svetovoy, V. B. (2008). Application of the Lifshitz theory to poor conductors. *Phys. Rev. Lett.* **101**, 163603-1–4.

Svetovoy, V. B. and Esquivel, R. (2005). Nonlocal impedances and the Casimir entropy at low temperatures. *Phys. Rev. E* **72**, 036113-1–8.

Svetovoy, V. B. and Lokhanin, M. V. (2001). Linear temperature correction to the Casimir force. *Phys. Lett. A* **280**, 177–81.

Svetovoy, V. B. and Lokhanin, M. V. (2003). Temperature correction to the Casimir force in cryogenic range and anomalous skin effect. *Phys. Rev. A* **67**, 022113-1-9.

Tabor, D. and Winterton, R. H. S. (1968). Surface forces: Direct measurement of normal and retarded van der Waals forces. *Nature* **219**, 1120-1.

Taylor, J. R. (1972). *Scattering Theory: The Quantum Theory of Nonrelativistic Collisions*. Wiley, New York.

Teh, S. T. and Chuah, D. G. S. (1989). Diffusion profile of spin-on dopant in silicon substrate. *Solar Energy Mater.* **19**, 237-47.

Titchmarsh, E. C. (1948). *Introduction to the Theory of Fourier Integrals*. Oxford University Press, Oxford.

Tomaš, M. S. (2002). Casimir force in absorbing multilayers. *Phys. Rev. A* **66**, 052103-1-7.

Tomaš, M. S. (2005). Casimir force between magnetodielectrics. *Phys. Lett. A* **342**, 381-8.

Toms, D. J. (1980a). Casimir effect and topological mass. *Phys. Rev. D* **21**, 928-32.

Toms, D. J. (1980b). Symmetry-breaking and mass generation by space-time topology. *Phys. Rev. D* **21**, 2805-17.

Torgerson, J. R. and Lamoreaux, S. K. (2004). Low-frequency character of the Casimir force between metallic films. *Phys. Rev. E* **70**, 047102-1-4.

Tutov, A. V., Maradudin, A. A., Leskova, T. A., Mayer, A. P., and Sánchez-Gil, J. A. (1999). Scattering of light from an amplifying medium bounded by a randomly rough surface. *Phys. Rev. B* **60**, 12692-704.

Uhlmann, M., Plunien, G., Schützhold, R., and Soff, G. (2004). Resonant cavity photon creation via the dynamical Casimir effect. *Phys. Rev. Lett.* **93**, 193601-1-4.

Valberg, P. A. and Ramsey, N. F. (1971). Hydrogen-maser measurements of atomic magnetic moments. *Phys. Rev. A* **3**, 554-65.

van Blockland, P. H. G. M. and Overbeek, J. T. G. (1978). Van der Waals forces between objects covered with a chromium layer. *J. Chem. Soc. Faraday Trans.* **74**, 2637-51.

van Bree, J. L. M., Poulis, J. A., Verhaar, B. J., and Schram K. (1974). Influence of surface irregularities upon van der Waals forces between macroscopic bodies. *Physica* **78**, 187-90.

van Kampen, N. G., Nijboer, B. R. A., and Schram, K. (1968). On macroscopic theory of van der Waals forces, *Phys. Lett. A* **26**, 307-8.

Van Vleck, J. H. (1932). *The Theory of Electric and Magnetic Susceptibilities*. Oxford University Press, Oxford.

van Zwol, P. J., Palasantzas, G., van de Schootbrugge, M., and De Hosson, J. T. M. (2008a). Measurement of dispersive forces between evaporated metal surfaces in the range below 100 nm. *Appl. Phys. Lett.* **92**, 054101-1-3.

van Zwol, P. J., Palasantzas, G., and De Hosson, J. T. M. (2008b). Influence of random roughness on the Casimir force at small separations. *Phys. Rev. B*

77, 075412-1-5.

Varshalovich, D. A., Moskalev, A. N., and Khersonskii, V. K. (1988). *Quantum Theory of Angular Momentum*. World Scientific, Singapore.

Vasiliev, A. N. (1998). *Functional Methods in Quantum Field Theory and Statistical Physics*. Gordon and Breach Science Publishers, Amsterdam.

Vassilevich, D. V. (2003). Heat kernel expansion: User's manual. *Phys. Rep.* **388**, 279–360.

Venghaus, H. (1975). Redetermination of the dielectric function of graphite. *Phys. Stat. Sol. (b)* **71**, 609–14.

Verleur, H. W., Barker, A. S., and Berglund, C. N. (1968). Optical properties of VO_2 between 0.25 and 5 eV. *Phys. Rev.* **172**, 788–98.

Vilenkin, A. and Shellard, E. P. S. (1994). *Cosmic Strings and Other Topological Defects*. Cambridge University Press, Cambridge.

Vinogradov, A. P., Dorofeenko, A. V., and Zouhdi, S. (2008) On the problem of the effective parameters of metamaterials. *Usp. Fiz. Nauk* **178**, 511–8 (*Phys. Uspekhi* **51**, 485–92).

Vogel, T., Dodel, G., Holzhauer, E., Salzmann, H., and Theurer, A. (1992). High-speed switching of far-infrared radiation by photoionization in a semiconductor. *Appl. Opt.* **31**, 329–37.

Volokitin, A. I. and Persson, B. N. J. (2001). Radiative heat transfer between nanostructures. *Phys. Rev. B* **63**, 205404-1–11.

Volokitin, A. I. and Persson, B. N. J. (2004). Resonant photon tunneling enhancement of the radiative heat transfer. *Phys. Rev. B* **69**, 045417-1-5.

Volokitin, A. I. and Persson, B. N. J. (2007). Near-field radiative heat transfer and noncontact friction. *Rev. Mod. Phys.* **79**, 1291–329.

Voronin, A. Yu. and Froelich, P. (2005). Quantum reflection of ultracold antihydrogen from a solid surface. *J. Phys. B: At. Mol. Opt. Phys.* **38**, L301–8.

Vonsovskii, S. V. (1974). *Magnetism*. Wiley, New York.

Voronin, A. Yu., Froelich, P., and Zygelman, B. (2005). Interaction of ultracold antihydrogen with a conducting wall. *Phys. Rev. A* **72**, 062903-1–11.

Weinberg, S. (1995). *The QuantumTtheory of Fields*, Vol. I. Cambridge University Press, Cambridge.

Weld, D. M., Xia, J., Cabrera, B., and Kapitulnik, A. (2005). New apparatus for detecting micron-scale deviations from Newtonian gravity. *Phys. Rev. D* **77**, 062006-1–20.

Wesson, P. S. (2006). *Five-Dimensional Physics: Classical and Quantum Consequences of Kaluza–Klein Cosmology*. World Scientific, Singapore.

Weyl, H. (1912). Das asymptotische Verteilungsgesetz der Eigenwerte linearer partieller Differentialgleichungen. *Math. Ann.* **71**, 441–79.

White, L. R., Israelachvili, J. N., and Ninham, B. W. (1976). Dispersion interaction of crossed mica cylinders: Reanalysis of Israelachvili–Tabor experiments. *J. Chem. Soc. Faraday Trans. I* **72**, 2526–36.

Wittmann, R. C. (1988). Spherical wave-operators and the translation formulas. *IEEE Trans. Antennas Propag.* **36**, 1078–87.

Wooten, F. (1972). *Optical Properties of Solids*. Academic Press, New York.

Yablonovitch, E. (1989). Accelerating reference frame for electromagnetic waves in a rapidly growing plasma: Unruh–Davies–Fulling–DeWitt radiation and the nonadiabatic Casimir effect. *Phys. Rev. Lett.* **62**, 1742–5.

Yablonovitch, E., Allara, D. L., Chang, C. C., Gmitter, T., and Bright, T. B. (1986). Unusually low surface-recombination velocity on silicon and germanium surfaces. *Phys. Rev. Lett.* **57**, 249–52.

Yampol'skii, V. A., Savel'ev, S., Mayselis, Z. A., Apostolov, S. S., and Nori, S. (2008). Anomalous temperature dependence of the Casimir force for thin metal films. *Phys. Rev. Lett.* **101**, 096803-1–4.

Yan, Z.-C. and Babb, J. F. (1998). Long-range interactions of metastable helium atoms. *Phys. Rev. A* **58**, 1247–52.

Yang, Y., Zeng, R., Xu, J., and Liu, S. (2008). Casimir force between left-handed-material slabs. *Phys. Rev. A* **77**, 015803-1–4.

Yu, I. A., Doyle, J. M., Sandberg, J. C., Cesar, C. L., Kleppner, D., and Greytak, T. J. (1993). Evidence for universal quantum reflection of hydrogen from liquid 4He. *Phys. Rev. Lett.* **71**, 1589–92.

Zel'dovich, Y. B. and Starobinsky, A. A. (1984). Quantum creation of a universe with nontrivial topology. *Pis'ma Astron. Zhurn.* **10**, 323–8 (*Sov. Astron. Lett.* **10**, 135–7).

Zhai, X.-H. and Li, X.-Z. (2007). Casimir piston with hybrid boundary conditions. *Phys. Rev. D* **76**, 047704-1–4.

Zhou, F. and Spruch, L. (1995). Van-der-Waals and retardation (Casimir) interactions of an electron or an atom with multilayered walls. *Phys. Rev. A* **52**, 297–310.

Zurita-Sánchez, J. R., Greffet, J.-J., and Novotny, L. (2004). Friction forces arising from fluctuating thermal fields. *Phys. Rev. A* **69**, 022902-1–14.

Zylbersztejn, A. and Mott, N. F. (1975). Metal-insulator transition in vanadium dioxide. *Phys. Rev. B* **11**, 4383–95.

INDEX

Abel–Plana formula, 22–26, 31, 32, 74, 76, 104, 106, 109, 127, 137, 138, 140–142, 146, 147, 149, 150, 153, 154, 156, 157, 263, 264, 267, 269, 299, 424
adhesion, 534, 659, 661, 663, 665, 680
analytic continuation, 57, 75, 82, 109, 139, 140, 146, 147, 150, 171, 172, 174, 180, 181, 183, 184, 204, 205
automorphic field, 265
auxiliary electrodynamic system, 288

black-body radiation, 78, 80, 163, 164, 212, 300
black-body radiation density, 78
Bogoliubov transformation, 134
Bohr magneton, 451
Boltzmann equation, 380
Bose–Einstein condensate, 643, 645, 650
boundary conditions
  antiperiodic, 26
  bag, 39, 115, 116, 207, 239
  continuity, 193, 222, 282, 285, 325, 339
  Dirichlet, 17, 38, 53, 70, 83, 85, 92, 103, 105, 107, 111, 136, 137, 143, 152, 153, 155, 156, 170, 172, 176, 185–187, 191, 192, 194, 212, 214, 230, 239, 240, 243–245, 253–255, 257, 258
  ideal-metal, 44, 232, 239, 328
  mixed, 26, 103, 106, 107, 111, 112, 135, 153–155
  Neumann, 38, 61, 70, 90, 106, 111, 136, 153, 155, 178, 179, 182, 183, 186, 191, 192, 228, 230, 236, 239, 244, 245, 248, 254–257
  periodic, 26, 116
  Robin, 38, 107, 178
boundary surface, 29, 38, 39, 41, 43, 44, 71, 72, 94, 99, 106, 139, 155, 165, 228–230, 326, 327, 375, 420, 467, 475, 481, 507, 513, 526, 543, 667
  curved, 29, 43
  nonideal, 40
braneworld cosmological scenarios, 278

calibration
  coefficient, 595
  constant, 585, 605, 640
  electrostatic, 541, 542, 572, 662, 664
  error, 576, 609, 622
  gratings, 541
  of the cantilever, 630
  of the deflection signal, 583–585, 603
  of the measurement system, 531
  of the piezoelectric actuator, 586
  of the setup, 543, 572, 583, 606
  of the spring constant, 593
  parameters, 552
  procedure, 521, 548
  voltage, 543, 587
  See also electrostatic measurement
Casimir
  energy, 7, 20–22, 25, 26, 31, 55, 79, 88, 90, 91, 95, 96, 98, 104–107, 109–113, 115, 116, 121, 127, 136, 139, 140–143, 145, 146, 149–151, 154, 158–162, 164, 167, 192, 193, 199, 200, 244, 247, 249, 255, 257–261, 286, 288, 291–294, 299–301, 316, 329, 331, 336, 342
    at nonzero temperature, 73, 300
    per unit length, 102, 240, 245, 249
  energy density, 6, 28, 104, 105, 129, 130, 263–265, 267, 269, 270, 273, 275–278
  entropy, 120, 121, 123, 301, 305, 307, 351, 360, 361, 370, 372, 373, 377, 379, 385, 387, 404, 410
  force, 20, 21, 23, 147, 162, 245, 249, 327, 328, 330, 332, 354, 355, 382, 387, 391, 392, 437, 533, 542, 553, 581, 589, 593, 599, 606, 623, 687
    at zero temperature, 343
    lateral, 323, 498, 500–502, 504, 505, 625–627, 631–636, 638, 679
    normal, 503, 625, 630, 633
    thermal, 125, 331, 343, 351, 355, 361, 365, 371, 373, 374, 381, 391–393, 451
  free energy, 11, 78–80, 103, 120, 122, 124, 156, 159, 162–164, 297,

302, 305, 307, 310, 312–315,
317, 318, 320, 351–354, 356,
360, 361, 365, 379, 381–388,
391, 393–395, 398, 399, 409,
416, 432, 579, 621
  at low temperature, 302
  pressure, 20, 107, 112, 120, 121, 123,
127, 131, 287, 289, 291–293,
297, 299–301, 304, 305, 307,
310, 313–318, 322, 330, 331,
334–336, 338, 342, 350, 353,
358, 363, 364, 380, 385, 388,
394, 395, 403, 413, 415, 470,
471, 474, 481, 489, 554, 558,
560, 564, 566, 568, 573, 574
    at low temperature, 305
    at nonzero temperature, 297, 364
    at zero temperature, 297, 315, 358
    repulsive, 107, 112
    thermal, 120, 121
  repulsion, 136, 146, 151, 577, 623, 624,
638, 680, 701
Casimir–Polder
  energy, 424, 426, 427, 429, 430, 433,
434, 452, 651, 654
  entropy, 425, 428, 429, 436, 440, 441,
446, 654
  force, 204, 419, 423, 424, 427, 431, 433,
436, 451, 518, 643–648, 654
  free energy, 423, 425, 429, 431, 433,
435, 437, 439–442, 444, 446
  potential, 204, 225, 463, 464, 466, 478,
482, 519, 650
characteristic frequency, 289, 329, 336,
381, 382, 411
classical limit, 11, 83, 88, 125, 160, 162,
215, 302, 307, 353, 354, 381,
382, 399, 429, 653
coefficients
  Bogoliubov, 133
  heat kernel, 60, 61, 63, 65, 66–72, 79,
81, 82, 84, 87, 167, 178, 179,
185, 197, 200, 201, 207, 210,
213–217, 224, 225, 236, 278
    See also expansion heat kernel
Minakshisundaram–Pleijel, 60
  reflection, 283, 287, 288, 291–294, 296,
298, 302, 303, 306, 308, 310,
318, 321, 324, 325, 329, 352,
354, 355, 366, 371, 378, 380,
383, 386, 402, 407, 411, 423,
457, 458, 668, 669, 671, 674
  Fresnel, 287, 325, 379
  Seeley–deWitt, 60
  van der Waals, 676, 678
compactification scale, 7, 14, 274, 684

conductivity
  dc, 12, 13, 309, 310, 316–318, 324, 352,
378, 380, 401, 416, 418, 419,
439–441, 453, 528, 618, 621,
648, 649, 699
  pure imaginary, 375, 376
  static, 307, 308, 318, 333, 390, 582, 648
confidence interval, 522–524
confidence level, 13, 363, 377, 380, 522,
524, 525
conformal anomaly, 272
conformal time, 268, 271
contact potential, 534, 536, 541, 576, 581
  See also residual potential
correction factor
  to the Casimir energy, 335, 337, 364,
393, 494, 499
  to the Casimir pressure, 334, 364, 394
  to the Casimir-Polder energy, 430, 651
correction factors, 334, 338, 350, 471, 498
cosmic strings, 262, 276–278
current
  diffusion, 377, 380, 569, 620, 700
  displacement, 375, 376, 400, 699
  drift, 375–377, 380, 392, 400, 418

dielectric
  ball, 71, 167, 193–196, 200–204, 206,
224, 225, 236
  cylinder, 178, 216, 221, 222, 224
  permittivity,
    generalized plasma-like, 343–345,
347, 349, 352, 376, 393, 399,
553, 566
    of the Drude model, 333
    of the normal skin effect, 332
  dilute approximation, 200, 201, 203, 207,
216, 225, 236
Dirac equation, 37, 125, 207, 266
direct measurement, 554, 575
Drude model, 331–338, 351, 361
dynamic magnetic susceptibility, 450, 451
dynamic polarizability, 423, 429

early Universe 8, 262, 276
effective action, 21, 49, 63, 67, 84, 91,
114, 116
effective temperature, 123, 124, 302
Einstein model, 268–270
Einstein relation, 378
Einstein's equations, 270, 272, 273, 276
  multidimensional, 274
electrostatic measurement, 537
elliptic boundary value problem, 39, 40

Index

energy-momentum tensor, 27, 34, 36, 37, 41, 45, 47, 106, 125, 130, 262–264, 266, 268, 270–275, 277, 283
error
  absolute, 521, 523, 527, 528, 548, 551, 555, 558, 565, 566, 574, 582, 587, 616, 644
  of the force measurement, 689, 691
  of the pressure measurement, 555
  random, 521, 522, 524, 525, 539, 543, 549, 552, 564, 565, 569, 587, 594, 598, 608, 610, 693
  systematic, 515, 521, 522, 524–527, 539, 549, 552, 555, 559, 564, 565, 587, 609, 645
  theoretical, 526
  total
    experimental, 521, 524, 525, 588, 610
    relative 525
    systematic, 522, 610
    theoretical, 527, 545, 547, 549, 560, 565, 590, 599, 615, 693
expansion
  heat kernel, 7–9, 55, 59, 61, 62, 69, 72, 81, 84, 189, 211
  high-frequency, 84
  high-temperature, 83, 214, 215
  large-mass, 69
  low-temperature, 212
  multiple-reflection, 84–88, 166
  semiclassical, 84, 88, 257
  Weyl, 63
extra dimensions, 265, 274, 684

Fermi–Dirac statistics, 378
Fermi energy, 281, 378
Fermi velocity, 339
fluctuation–dissipation theorem, 283, 323, 324, 699
force–distance
  curve, 576, 586
  measurement, 513
  relation, 552, 584, 598, 604, 610
frequency cutoff, 56
Friedmann model, 6, 262, 270, 272

gauge
  Coulomb, 39, 42, 44, 128, 148, 230
  Lorentz, 36, 39
gauge-fixing term, 36
gauge transformation, 35, 36
grand unification theories, 262

half-quanta, 1
Hamaker constant, 290, 463

heat kernel, 60, 70, 81, 89, 185
  local, 60
  See also expansion heat kernel
hypothetical interaction
  power-type, 683
  Yukawa-type, 683, 687, 688

identification conditions, 24, 262–264
impurities, 344, 372
indirect measurement, 526, 553–555, 559, 564, 573
inflation, 273
inflaton field, 273
infrared optics, 333, 340, 342

Johnson noise, 375
jump into contact, 656, 659

Kaluza–Klein theories, 7, 262, 274
Klein-Fock-Gordon equation, 17, 33, 40
Kramers–Kronig relations, 310, 331, 344
  generalized, 348

Lande factor, 451
Leontovich impedance, 339
  for the Drude model, 385
Lifshitz formula, 287, 299, 351
  at zero temperature, 287
  for an atom above a plate, 422
  for anisotropic plates, 321
  for stratified and magnetic media, 290
Lifshitz-type formula, 423, 425, 449, 450, 666, 669, 671–673
light pressure, 615

magnetic moment, 449, 451
magnetic permeability, 110, 293, 294, 449, 450
magnetodielectric, 294, 449, 450
mass-shell equation, 355
Matsubara frequencies, 75, 118, 295, 297
Maxwell equations, 35, 282, 325, 670
Maxwell–Boltzmann statistics, 378, 445
measure of agreement between experiment and theory, 529
Mobius strip, 20, 264
moduli, 682, 686, 695

Nernst heat theorem, 123, 305, 309, 355, 362, 365, 370, 373, 374, 377, 379, 380, 383, 399, 401, 404, 410, 416, 418, 419, 436, 444
Ninham–Parsegian representation, 312

748 Index

outlying results, 523, 558, 564, 587

paramagnetic atoms, 449, 450
parametric resonance, 134, 135
patch potentials, 526, 545, 563, 589, 614
penetration depth, 328, 340
    relative, 328
    See also skin depth
piezoelectric
    actuator, 534, 542, 575, 576, 582–584, 586, 594, 596, 597, 601, 603, 604, 622, 640
    element, 533
    stack, 531
    stage, 661, 664, 665
    transducer, 573
    translation, 532, 661
    tube, 628–630
plasma
    frequency, 329–332, 334, 348, 355, 356
    model, 329
    sheet, 666, 669, 670, 671, 673, 678
    wavelength, 329, 331, 334, 377
point splitting, 56
Poisson summation formula, 76, 120, 358
polar dielectrics, 283, 319
polarization
    electronic, 319, 320
    ionic, 319, 320
    orientation, 319, 320, 450
    state, 2, 42–44, 109, 117
    tensor, 115
    vector, 42, 43, 108, 117, 168, 169, 175, 176, 194
    See also separation of polarizations

quantum vacuum, 1, 7, 8
quasi-Euclidean models, 271

radiative heat transfer, 323
reflection amplitude, 650, 651
reflection property of the Riemann zeta function, 82
regularization, 17, 19, 20, 21, 23, 31, 55, 57, 62, 64, 65, 67–70, 72, 76, 77, 90, 136, 142, 147, 156, 168, 198, 200–204, 276
    cutoff, 55–57, 63, 67, 79, 189, 198, 203
    Epstein zeta function, 144
    point splitting, 56, 128
    zeta function, 55–57, 59, 62–66, 76, 92, 107, 109, 110, 128, 147, 156, 171, 178, 180, 185, 189, 197, 198, 201, 202, 217

regularization parameter, 20, 56, 109, 140, 180, 181, 226
relaxation parameter, 332–334
    See also residual relaxation
relaxation time, 451
renormalization, 17, 21, 32, 55, 56, 58, 60, 62, 64–66, 68–72, 110, 113, 142, 144, 147, 149, 156, 166–168, 178, 179, 185, 186, 188–190, 198, 207, 210, 211, 217
    finite, 68, 79, 158, 186, 189
    infinite, 68
    of the cosmological constant, 21
    of the geometrical object, 139, 147
    ultraviolet, 83
    zeta function, 65
renormalization condition, 292, 296
residual charge, 581
residual electrostatic force, 518, 520, 531, 532, 534, 537, 538, 578, 583, 586, 597, 598
residual gas pressure, 644
residual potential, 534–536, 538, 541, 542, 549, 551, 552, 557, 563, 575, 577, 583–585, 594–596, 603–607, 625, 626, 630, 631, 640, 662
residual relaxation, 372
residual resistivity, 372

Schwinger prescription, 352
screening length, 377, 378
    Debye–Hückel, 378
    Thomas–Fermi, 378
separation of polarizations, 168
    See also polarization state; polarization vector
skin depth, 328, 329, 340
    relative, 328, 329
skin effect
    anomalous, 326, 340–342, 382
    normal, 332, 339, 382
sonoluminescence, 167
spatial dispersion, 325–327
spectrum
    infrared, 319
    ultraviolet, 312, 319
static atomic polarizability, 425
stiction, 656, 659, 661, 665
surface impedance, 340
    exact, 343
    See also Leontovich impedance
systematic deviations 521, 537, 538

T-matrix approach, 228

Index

T-matrix operator, 233
T-matrix representation, 235
tabulated optical data, 310, 336
$TGTG$ representation, 235
   See also T-matrix representation
thermal correction, 77, 79, 80, 121, 161, 164, 212–214, 299–301, 310, 314–318, 328, 356–364, 368, 371, 374, 383, 385, 393, 410, 436, 440, 555, 556, 579, 598
   nonrenormalized, 156, 157, 161
   relative, 314–318, 320, 362–364, 383, 390, 394, 580
   to the Casimir energy, 164, 364, 382–384, 394
   to the Casimir force, 164, 354, 532
   to the Casimir–Polder energy, 434, 654
   to the Casimir–Polder force, 436
   to the Casimir pressure, 316, 317, 359, 360, 363, 382, 389, 391, 555
thermal equilibrium, 12, 73, 75, 117, 283, 286, 294, 297, 310, 316, 323, 324, 362, 376, 380, 392, 400, 418, 447, 449, 451, 452, 618, 644, 645, 646, 699
third law of thermodynamics, 123, 307
   See also Nernst heat theorem
three-layer system, 577, 622, 624, 680
time reversal symmetry, 376
topological defects, 262, 276–278
topology, 24, 33, 152, 153, 262, 264, 265, 270, 271, 273, 274
   Euclidean, 24
   non-Euclidean, 6, 7, 264–267, 273
   of a 2-torus, 152, 262
   of a 3-torus, 264, 265, 273
   of a circle, 24, 25, 262
   of a cylinder, 263
   of the Euclidean plane, 263
   of the Klein surface, 263
twisted field, 26, 265

ultraviolet divergences, 71, 76, 77, 84, 110, 113, 167, 185, 193, 197, 198, 200, 202, 207, 227, 232, 236
uniaxial crystals, 321
uniform distribution, 522, 524

vacuum
   energy density, 21, 27, 28, 41, 42, 44, 45, 47, 103, 106, 129, 131, 262, 263, 261, 265, 269, 277
   oscillations, 1, 3, 29
   polarization, 6, 7, 33, 270–272, 276
van der Waals force, 5, 6, 8, 167, 308, 338, 393, 437, 460, 515, 516, 577, 618, 682
Van Vleck determinant, 89
Van Vleck paramagnetism, 449
verification of the hypothesis, 528
virtual photons, 5
   See also zero-point oscillations

waves,
   cylindrical, 177, 241
   evanescent, 289, 299, 324, 382, 387–392
   plane, 78, 241, 251, 374–376
   propagating, 289, 299, 324, 387, 390
   spherical, 173, 251
   traveling, 11, 18, 351
wedge, 10, 103, 128, 129, 130, 277, 489, 518, 519
world line, 84, 92, 248, 257

Yukawa-type
   corrections, 683, 686, 687, 690, 700
   potential, 683

zero-point oscillations, 2–5, 7, 18, 33, 299, 500, 680
zeta function,
   Epstein, 8, 136, 139–141, 144–146, 149
   generalized, 57, 64, 192, 276
   Hurwitz, 111, 154, 184, 197, 201, 202
   Riemann, 57, 59, 82, 109, 140, 144, 145, 147, 213, 218, 225